Fatigue and Corrosion in Metals

Pietro Paolo Milella

Fatigue and Corrosion in Metals

Second Edition

Springer

Pietro Paolo Milella
Nuclear Engineering
University of Cassino
Rome, Italy

ISBN 978-3-031-51349-7 ISBN 978-3-031-51350-3 (eBook)
https://doi.org/10.1007/978-3-031-51350-3

This Springer imprint is published by the registered company Springer Nature Switzerland AG
The registered company address is: Gewerbestrasse 11, 6330 Cham, Switzerland

Paper in this product is recyclable.

To my wife Anna

Preface

This year I am celebrating 55 years of professional activity. There is a long story behind this book. A story that started with the continuous struggle to assure the structural integrity of nuclear power plants in any possible operating condition. «Safety first», we were used to say and that statement required an uncredible and continuous R&D effort in three main fields, in particular, that 50 years ago were not so advanced as they are today but were expanding fast under the pressure of events: Fatigue, Corrosion, and Fracture Mechanics. These were the subject of my work during the last 55 years, and these are the subjects of this book today. All the lessons and the experience learned are in this book. For 20 years I have been responsible for all nuclear safety research programs in the mechanical field in Italy. The arriving point was the Italian Design Criteria for the next generation of nuclear power plant in Italy, issued in mid-80s. However, the Chernobyl accident, which occurred in 1986, stopped the Italian Nuclear Power Program but not the criteria I wrote because they were judged enough advanced to be enforced by Westinghouse Nuclear and became part of the Westinghouse Nuclear Safety Criteria for the next generation of nuclear power plants and approved by the US NRC. In those years I became an adjoined professor of Machine Design at the University of Cassino and the research continued, also through a collaboration with the Italian Air Force, always in the field of Fatigue, Corrosion and Fracture Mechanics. Beginning of the New Century, I started a teaching activity for professional engineers in the field of Fatigue, Corrosion, and Fracture Mechanics which I still maintain. All those experiences are in this book. This is why this Second Edition of the book Fatigue and Corrosion in Metals has been issued. It represents my modest scientific legacy to the next generation of researchers, engineers and students. But there is also another reason for this publication. To the best of my knowledge, this is the only book that treats fatigue and corrosion in a rather comprehensive manner. Generally, in fatigue textbooks corrosion is given a modest attention limited to just few pages or more. There are four chapters in this book, namely Chaps. 17–20, dedicated to the most challenging mechanism of material failure that is known to men: corrosion and fatigue. Since corrosion, also known as static fatigue, is fundamentally an electrochemical process, I believed that it was not possible to fully understand it without a minimum knowledge of electrochemistry. Therefore, Chap. 17 is dedicated to the electrochemical treatment of corrosion and, in particular, to crevice corrosion and film/rupture-anodic dissolution process that is the base of any corrosion mechanism including hydrogen embrittlement. Those with an education in electrochemistry will probably find this treatment a little bit too simple, while those with an education in mechanics may find it a little bit difficult to digest. But the great successful accomplishment of the 70s was precisely to put together electrochemical and mechanical specialists trying to develop a common understanding and even a common language in a matter that, up to then, was almost precluded to mechanical engineers. The multidiscipline nature of stress corrosion had resulted, in fact, also in multilingual technical terminology used among experts having different education with the result that identical phenomena had been described differently and considered different while different phenomena could not be distinguished as such. Therefore, I tried to rebuild that effort to the benefit of students and young professional engineers or researchers hoping that they may be fascinated by this entangled subject matter as much as I was 45 years ago. Chapter 18 is then dedicated to the two most challenging stress corrosion processes that are the intergranular demolition of sensitized materials and hydrogen embrittlement. Hydrogen embrittlement is a mysterious universe; therefore, to try to approach it in just a limited number of pages is like to put the infinite space in a nutshell. "*O God, I could be bounded in a nutshell, and count myself king of the infinite space.*" But he was Hamlet and William Shakespeare, master of tragedy, did not know the tragedy of hydrogen embrittlement! Chapter 19 is the *revenge* of mechanical people because stress corrosion is finally treated with the tools of fracture mechanics. This was another great achievement of the 70s. Linear elastic fracture mechanics (LEFM) developed to shed some light over the brittle fracture of welded ships in World War II was found in the 60s to be applicable to fatigue crack growth and later, in the 70s, to corrosion predicting initiation and growth rate of stress corrosion cracking. Finally, Chap. 20 addresses the issue of corrosion-assisted fatigue trying to make some order and fix some key points in the confusing argument of whether corrosion is assisting fatigue or fatigue is providing the stress to corrosion and on the trickery crack growth rate effect associated with the lowering of the load frequency. But there are also other reasons that make me think this book can be useful. One of these is: process volume. Process volume is at the base of any design. It must be clearly understood that results obtained with small specimens cannot be directly applied to large structures. Chapter 6, dedicated to data scatter and statistical analysis, introduces also the issue of the process volume effect through the Weibull distribution and the weakest link criterion. I also tried to explain that process volume and probability of survival are precisely the same thing. Process volume effect is also used to explain why different types of loads result in different effects (Chap. 3) and why holes and notches can withstand fatigue loads more than expected (Chap. 11). Actually, the notch effect, so important in fatigue, cannot be fully understood without introducing the process volume effect. Another intriguing argument is multiaxial fatigue treated in Chap. 13. I tried to discuss as much as possible the non-proportional case of multiaxial fatigue and non-proportional hardening. Having a seismic education, I am familiar with Fourier transform. I still remember when in 1970 the first commercialization of the fast Fourier transform (FFT) shed an incredible enthusiasm among the scientific community. Computer made a dream that lasted two centuries come true. The passage from the space domain to the time and frequency domain was finally opened. In the last decade, in particular, I have been applying Fourier analysis to fatigue design with narrowband and wideband random processes. I consider this kind of analysis part of the future in fatigue design because variable load histories having complex spectra represent the most common form of fatigue cycling and I am a little bit surprised that practically no textbook is treating the subject. Therefore, I dedicated part of Chap. 12 on cumulative damage and life prediction to this kind of analysis that's time students and engineers become acquainted with. Also Miner rule or we should better say Palmgren–Miner rule is considered in Chap. 12 with its shortcomings because in real life damage progression and accumulation are never linear. Welds represent a very important sector of our technology that cannot be neglected in a fatigue textbook also because they still are the site of major cracks and defects formation. Therefore, Chap. 16 is dedicated to weld defects generation and classification and to fatigue weld design. But a major quality of the book, if I may say so, is its self-consistency in that each chapter follows in a logical fashion the previous one and introduces the next. The book starts with Chap. 1 describing the nature and phenomenology of fatigue and treating the four phases of fatigue. They go from the early phase in which the material changes its properties that mysteriously precedes and terminates right with the first damage of sub-microscopic size appearance in just one or few surface grains to the growth of this damaged embryo to become a microscopically small crack (MSC) and then a macroscopic crack no larger than 300–400 μm. With the macro-crack formation, conventional fatigue based on *S-N* curves terminates. The chapter also addresses the issues of non-propagating cracks and fatigue limits. Chapter 2 is dedicated to the morphology of the crack formation and propagation. Damage, whether of sub-microscopic, microscopic or macroscopic size, always leaves a permanent sign on the fatigue surface, like persistent slip bands or beach marks and striations. It is a kind of fingerprint that without ambiguity indicates who was the guilty and how he acted. There is a science to read these signs, fractography, that must be known because *post-mortem* analysis can reveal the causes and modes of fatigue failure. Chapter 3 tries to introduce the reader to this science at least indicating what he shall be looking for and why, in particular, in relation with what has been already discussed in Chap. 1. Chapter 4 introduces those factors that can affect and modify the design fatigue curve *S-N* obtained under standard experimental conditions, such as surface roughness and treatments and, in particular, nonmetallic inclusions which, in turn, introduces the arguments of Chap. 7 where the high cycle fatigue design approach is discussed illustrating the tremendous effect of the mean stress. The problem of the accuracy of the *S-N* curve is given full attention in Chap. 6 that is dedicated to data scatter and statistical analysis trying to make the reader understand that the classical Gauss approach can be used only when experimental data have a bell-shaped distribution, which is not always the case in fatigue, and the population of data is infinite. Also this concept of infinite population is given an operative definition. Other distributions are introduced and, in particular, the Student's t-distribution, the Weibull distribution and the extreme value Gumbel distribution, these two last of real powerful applications. The reader must also understand that the S-N curve obtained with few specimens can be considered at most as a mean design curve, but no designer wants to expose himself to the shame of a one-out-of-two workpieces failure! Today advanced design goals require at least 99.9% survival of the components. Chapter 8 is addressing low cycle fatigue design with the implications of a nonlinear analysis. The old Neuber approach is given a new life because it can provide sound results without running an elastic-plastic computer code calculation and in a very short time. This is connected with Chap. 11 dedicated to the notch effect that produces chilling shivers down the spine of designers. This effect is given a rather unusually extensive and comprehensive treatment. Chapter 9 is a rather new field of fatigue, that of very high cycle fatigue, that is receiving a particular attention with the introduction of new ultrasonic testing machines. The available test methods and how to construct a fatigue S-N curve with different degrees of accuracy even before having run any fatigue test are considered in Chap. 10. Conventional fatigue, based on S-N curves, terminates with the first appearance of the macro-crack. *Habemus fissuram* and from this moment on the residual life of a component can be assessed only through LEFM. Two chapters, namely Chaps. 14 and 15, are dedicated to this subject. 50 years already passed from the first formulation of the Paris' Law. Dough degraded to the role of simple postulate, the Paris–Erdogan equation has established a milestone in the study of fatigue. Yet many ambiguities still exist as to role of the R-ratio on the fatigue threshold and crack propagation, on the small cracks behavior and the effective ΔK opening a crack. Many questions as why a crack grows and how to consider the cumulative effect of variable amplitude loads are still to be fully answered and remain encrypted within the boundaries of the plastic zone set up ahead of the crack tip. The reading key is there. So Chap. 15 is dedicated to fatigue crack growth effects that depend on the plastic zone behavior. Finally, Appendix A provides a number of basic LEFM solutions that can be of some utility in fatigue crack growth rate calculations and tests preparation as well for fatigue and corrosion tests. Of the remaining chapters I have already said.

Before closing this preface allow me to indulge in a little, apparent digression. The history of man and his achievements is written in books. It is in books that we found the knowledge we need to continue that great history. The less we read, the less we know and ask, and the more we think to know. Today the introduction and development of computers and design codes are more and more sophisticated, feeds this great sensation of being able to design anything because we always obtain a precise answer by running a computer code. A precise answer, but not always the right answer. It shall, then, not come as somewhat of a surprise if, still today, structures fail as when there were no computers and we used rulers. Calculations are and will always be necessary, but if we don't know how materials and structures actually behave under the operating conditions or if we don't know what really means an operating condition, then any calculation will be just a mere theoretical exercise of approximation of reality. Nothing more than the term "know-how" convey this concept. And «I know-how» seems also to imply «You don't»! However, know-how is based on a continuous process of proving and doubting. Galileo Galilei, the founder of Mechanics, was used to say «Provando e Riprovando» (proving and disproving). The great Italian scientist read it in the Divine Comedy, 3rd Canto of the Paradise by Dante Alighieri:

> Quel sol che pria d'amor mi scaldò 'l petto,
> di bella verità m'avea scoverto,
> provando e riprovando, il dolce aspetto.

and he was fond of it. Proving and disproving became the base of the Scientific Method adopted by Galilei and by the Accademia del Cimento (Wrestle Academy), founded in Florence, Italy, in 1657 by Leopoldo de' Medici to promote science and knowledge also by making experiments, tests, and discussions. This is what we shall do: trying and trying again and doubting, always doubting to find the wright solution.

In the course of the years, I have been exposed to the privilege of some appreciations. I am pleased to acknowledge all those who did and, in particular, Prof. Jose Luis Arana who made the first check of the first edition of this book.

Rome, Italy Pietro Paolo Milella

Contents

Chapter 1
Nature and Phenomenology of Fatigue

1.1 History and Overview

For centuries man has been aware that the repeated application of loads would lead to the early failure of materials. It came as something of a surprise, however, when he also found, almost two centuries ago, that failure could occur under stresses of relatively low amplitude, lower than the yield strength σ_y of the material. The phenomenon, known as *fatigue*, has long been studied by researchers and given attention by engineers and designers since there are only few events other than fatigue that can cause so many failures, every year and in any sector of our technological society, sometime catastrophic also for the casualties involved. As matter of fact, among the five possible failure mechanisms, Fig. 1.1, that can jeopardize the strength of materials, which are:

(1) Cleavage or brittle fracture,
(2) Plastic flow or ductile fracture,
(3) Fatigue,
(4) Corrosion,
(5) Creep,

there is only one, namely *corrosion*, that can generate more damage than fatigue. Often, the two events combine giving rise to what can be considered the most devastating and unpredictable challenge of them all: *corrosion assisted fatigue* (CAF) also known as environmental fatigue. It has been estimated that 90% of all service failures of metal components are caused by fatigue and corrosion.

One hypothesis, certainly the first to come to mind to explain fatigue failure, is that each single cycle of load application introduces a sort of damage in the crystal lattice that, though negligible at the beginning, accumulates cycle by cycle leading eventually to the complete and unexpected fracture of the material. Whoever named it first, perhaps the Englishman Braithwaite [1] in 1854 or more probably the French Poncelet [2] who, already in 1839, described metals as being tired the term *fatigue*

P. P. Milella, *Fatigue and Corrosion in Metals*,
https://doi.org/10.1007/978-3-031-51350-3_1

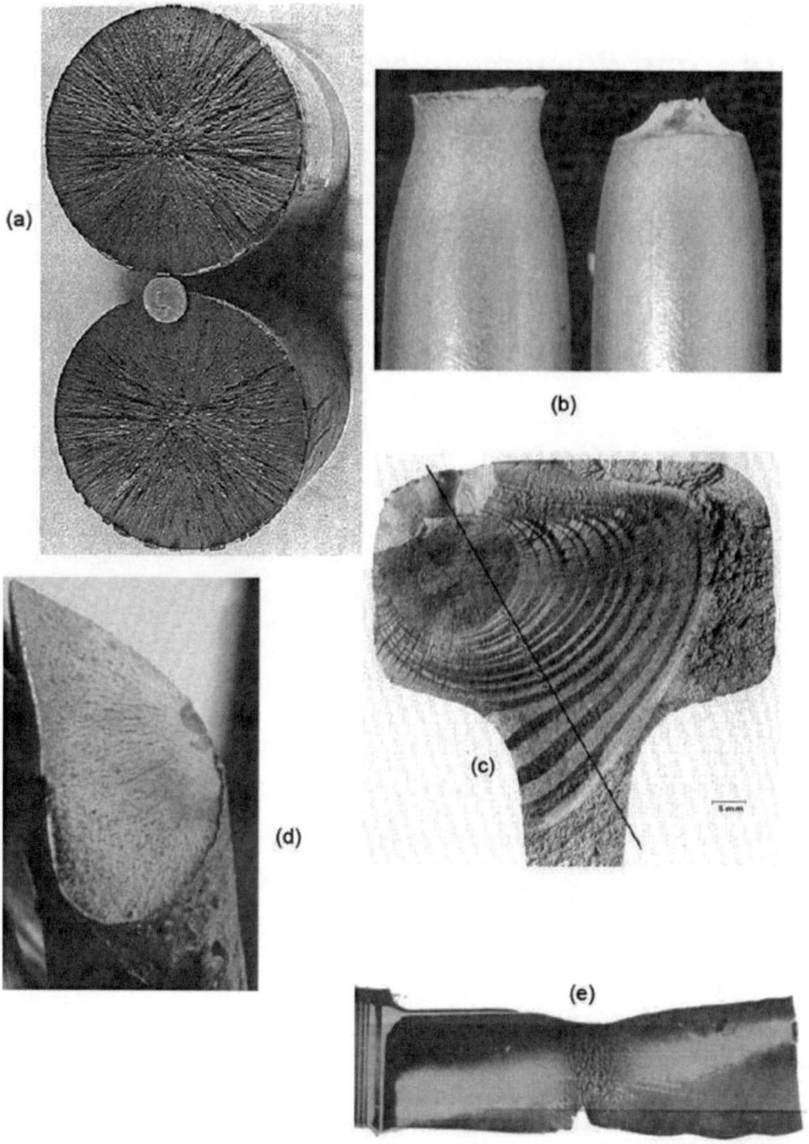

Fig. 1.1 The five basic mechanisms of failure: **a** Cleavage or brittle fracture, **b** Ductile fracture, **c** Fatigue failure, **d** Corrosion and **e** Creep fracture

is certainly the most appropriate to define precisely this continuous progression and accumulation of damage in the material that weakens and eventually fails. Unfortunately, this damage does not heal spontaneously as in humans after resting for a while. However, if a self-healing is not possible, an external intervention can activate the recovery so that the fatigue damage can be completely removed and the material restored. This can be done by just polishing the surface of the material, where the fatigue damage starts, at regular intervals of time during the cycling allowing the work piece to last endlessly. This surface polishing, just few tens of microns depth, is effective because it has been definitely recognized that the fatigue embryo is always an elementary plastic flow localized somewhere in some crystals on the surface of the material. The surface, in fact, works always in plain stress conditions that favour plastic flow to occur in some few crystals favourably oriented with respect to the external load. At variance, the flow of a single or few internal grains is always prevented by surrounding grains that don't want to yield leading to plain strain conditions.

This initial event or elementary surface damage has a submicroscopic size, much less than one tenth of a micron, and contained in a single slip line inside a single material grain. The damage evolves within the grain into a microscopic crack by the repeated application of loads and eventually becomes macroscopic breaking the grain border and joining other micro-cracks in the neighbouring crystals. The integrity of the component is indeed compromised when the first macroscopic crack, no larger than 300–400 micron, forms on the surface of the material. The lesson learned, therefore, is that designers shall be concerned with the surface of the component while they may forget the interior.

The written history of fatigue starts at the beginning of 1800 and follows the industrial development of that particular period, pushed by the two leading industries of the time, the mining and, in particular, the railway industry. Probably, the first article ever published on fatigue and the first fatigue experiment in the world was run in Germany by Wilhelm A.J. Albert [3] in 1837. It was concerned with the continuous failure of conveyor chain used in the Clausthal mines. Interesting enough, he tested the whole component, not just a specimen.

This approach that may seem pioneering can be simply explained by the fact that at the time there were no standards or rules or even indications on how to run fatigue tests on small specimens. Wöhler had still to come, therefore to check the entire chain behaviour was the most straightforward think to do!

But it was the railway sector that gave the greatest impulse to fatigue understanding and the most significant improvement to fatigue design. It was indeed a German railway engineer, A. Wöhler, who in 1858 initiated those studies that would bring to the discovery that fatigue damage depends on the amplitude of cyclic stress and to the formulation, in 1870, of the fundamental law named after him *Wöhler's law* [4]: *Materials can be induced to fail by many repetitions of stresses, all of which are lower than the static strength. The stress amplitude is decisive in the destruction of the cohesion of the material. The maximum stress is of influence only in that the higher it is the lower is the stress amplitude that leads to failure.* Wöhler, therefore, was the first to understand the role of stress amplitude and the importance of tensile mean stress on the fatigue resistance of materials. Wöhler never published his results in a graphical fashion, but in the form of data tables. It was just in 1874 that they will appear as graphics, nevertheless, we have to wait till 1910 when the American Basquin [5] first defined the shape of a typical S–N fatigue curve by using Wöhler test data and proposed the *log–log* relationship so familiar to today engineers.

It was not before 1924, 50 years after the end of Wöhler monumental work, that the first textbook on fatigue was published by the English Gough [6]. In the process of pushing forward, engineers frequently found themselves working at the very edge of what was technically possible. In many cases engineers didn't know where exactly the edge was until they passed it! There were significant gray areas and thus frequently engineers found themselves flying blind. One of those gray areas was precisely that of fatigue. In those early years of industrial development, many disastrous railroad accidents occurred. Probably, the most known fatigue disaster in the history of railroad was the so-called Versailles disaster that happened in France in 1842; a frightful event that took the lives of 40 to 80 people.

An unusually long train transporting between 1500 and 1800 passengers had an accident when the axle of the first engine, of the type shown in Fig. 1.2, broke. The second steam engine passed over it and the boiler burst. All the following carriages passed over the wreck and ignited. The many accidents occurred in England and the competitions between large monopolistic private companies that could have an impact on the safety of the public, suggested the Queen to establish in 1840 the HM Railway Inspectorate (Her Majesty's RI or HMRI) with the commitment to investigate accidents and report to Parliament. The HMRI collected tens of thousands of such episodes. Figure 1.3 presents one of the first drawings, by Joseph Glynn, showing a railway axle of a steam engine tender with a continuous circumferential crack all-round a journal, initiated at the corner of a keyway. Unfortunately, failures of axels and rail disasters did not finish with the pioneering years of railway transport and the number of casualties continued to grow.

As a tragic continuation of that list stands the accident occurred in Germany in June 3rd 1998 at Eschede, Fig. 1.4 [7]. This is certainly the worst high speed (200 km/h or 124 mph) train accident ever happened in the world. 101 people died and about

Fig. 1.2 Nineteen century steam-engine

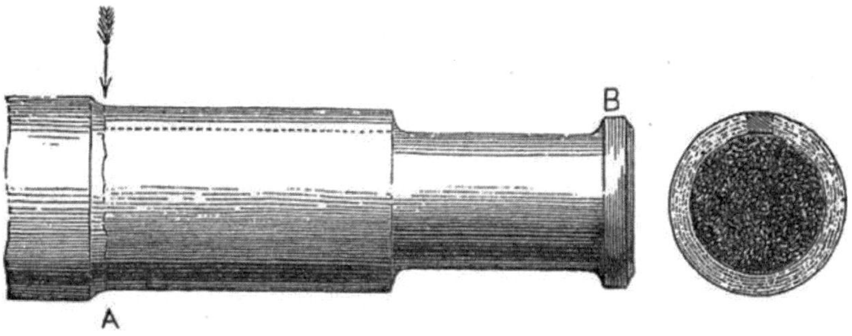

Fig. 1.3 1843 drawing by Joseph Glynn showing the failure of a railway axle of a steam-engine tender due to a crack initiated all-round a journal, indicated by the arrow as section **a**, at the corner of a keyway

100 were injured. The accident was the result of an improper design change of the steel wheels that was not tested in the field because at the time there were no test facilities in Germany. The original design used single-cast or *mono-block* wheels. However, this solution resulted in high vibrations at cruising speed affecting the comfort of passengers and causing metal fatigue. Therefore, designers changed the mono-block wheel design introducing a rubber damping ring 20 mm thick between the metal wheel rim and the wheel body, as schematized in Fig. 1.5. This new design weakened the wheel, causing cracks formation at the interface wheel-rubber ring, though reducing vibration and improving passengers comfort.

This new wheel-tire design actually was a heritage from streetcar application adapted for high speed rail use. But train accidents and axle failure continued to claim lives. A more recent accident occurred on May 2009 at the railway station

Fig. 1.4 Eschede (Germany), high speed train accident on June 1998 [7]

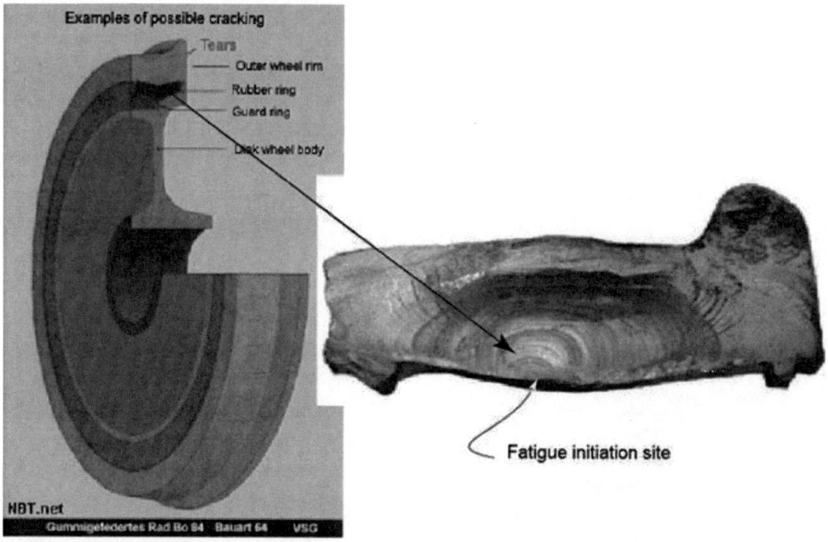

Fig. 1.5 Detail of fatigue crack on the outer wheel rim of the German high speed train

in Viareggio, Italy, where the failure of a spindle of an axle of the first carriage, transporting 80 ton of liquid gas, caused the derailing of the convoy, Fig. 1.6, with the explosion of the tank and the death of 33 people. The detail of Fig. 1.7 shows the typical rotating bending fatigue failure with beach marks and the terminal elliptical dark area of overload breakdown, see Fig. 2.13. In the early years of railroad constructions, Rankine [8] in England first outlined the importance of notches in the fatigue failure of materials. Treating the railway axel failures, he noted that fatigue fractures always occurred at sharp corners where cracks developed fairly soon (see Fig. 1.2) and recommended that journals be formed with a large curvature in the shoulders. But it took another German, H. Neuber [9], to fully treat the subject of discontinuities developing, before world war II, a method to calculate stresses and strains at sharp notches, method known as the *Neuber approach* or *hyperbola* and empirical tables to evaluate the impact of such notches.

The method was later followed by the American R. E. Peterson [10] who published a monumental handbook on stress concentration factors that can be still regarded as one of the best reference book. As early as 1903 the Englishmen Hewing and Humfrey [11] using an optical microscope observed that cyclic deformation leaded to the development of slip bands and fatigue cracks in crystals (see Fig. 1.40). This was probably the first metallurgical observation of fatigue damage in metals. Some years later, when small specimen testing was already established, the first full scale fatigue test was run in the U.K. by the Royal Aircraft Establishment on a large aircraft component [12].

The great acceleration in fatigue study appeared in the years between 1939 and 1960 when the attention of the scientific and technological world moved from the

Fig. 1.6 Viareggio (Italy), accident on May 2009. The failure of an axle of a tank carrying liquid gas (first carriage of the convoy) caused derailment followed by gas explosion

(a) (b)

Fig. 1.7 View, from both sides **a** and **b**, of the morphology of the spindle failure. The origin of the fatigue crack can be seen by following the beach marks opposite to the overload fracture area

railway sector to the air and space one, more advanced and strongly propelled first by war needs and later by military and commercial competition. Unfortunately, as it always happens in the history of mankind, the technological development is marked by unexpected disasters. Among these many, two became notorious in 1954, just two years after the introduction of the first commercial jet plane in the aviation history in 1952. Two de Havilland Comet passenger jet planes crashed within a few months of each other. In order to provide an economically payload and range at the high cruising speed which the turbo-jet engines offered, it was essential that the cruising height should be upwards of 10,500 m (35,000 ft), double than that of the then current airliners. This resulted in metal fatigue failure due to the high-pressure excursions. The fatigue crack originated at the corner of a far too sharp window, curiously not circular or elliptical, Fig. 1.8. The cracks once originated could not arrest since, even more curiously, the central section of the fuselage was made of one single piece of ultra-high strength aluminium alloy of the 7000 series, too brittle for that purpose.

It comes to mind the failures of the Liberty Ships and T2 tankers during World War II whose hull was monolithic due to introduction of welding in the ship construction technique and made of brittle steel. If a crack would develop at a square hatch corner on the welded deck it would not arrest.

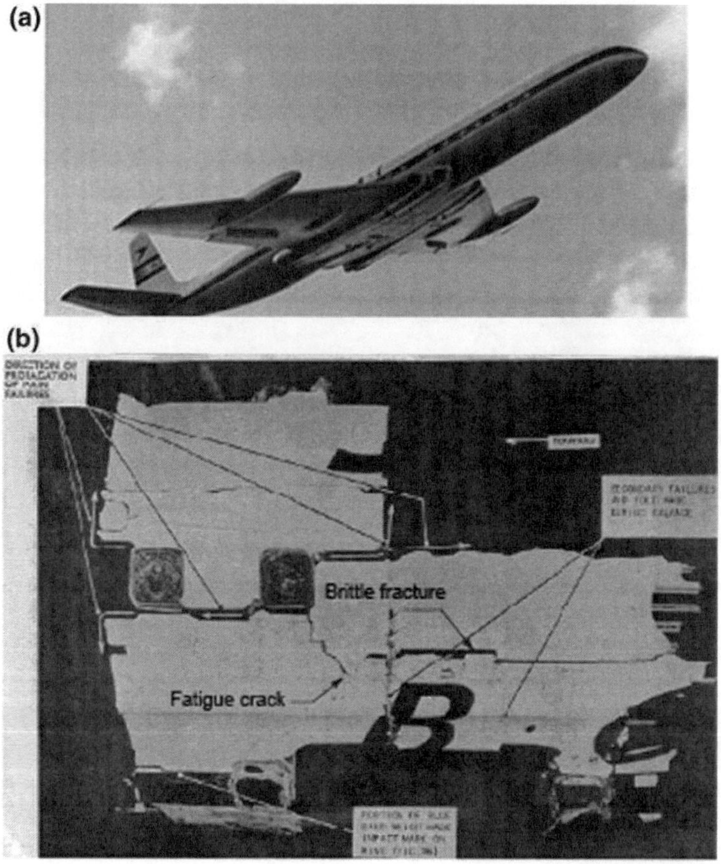

Fig. 1.8 **a** de Havilland Comet jet plane. **b** detail of a fatigue failure. The fatigue crack propagated to a point where it became unstable resulting in a brittle fracture

Many air accidents occurred in late years of last century because of fatigue initiated by a corrosion pit. Probably the most spectacular of them all was the Aloha Airlines accident occurred on 28 April 1988, shown in Fig. 1.9. The fuselage failure initiated in a lap joint, probably due to crevice corrosion. The failure was the result of a multiple site fatigue cracking. Fatigue initiated from the knife edge associated with the countersank lap joint rivet hole.

The study of fatigue continued after World War II under the impulse of new technological frontiers. It was noticed that fatigue cracks would propagate through the material at a rate that was somehow proportional to their dimensions and to the stress amplitude. This led to the formulation of several empirical formulas aiming at evaluating the residual life of the component. Often such formulas were not only inadequate, but also contradictoring since stresses alone were not suitable to describe the behaviour of cracks in metals. A fundamental impulse to fatigue study was then given by the introduction of fracture mechanics in the study of cracked bodies. With Fracture Mechanics the stress at a point was replaced by an entire stress field at the crack tip as reference value to assess the integrity of a structure, whose amplitude was given the name of stress intensity factor K by G. Irwin (see Chap. 5). Thanks to this new science, P. Paris and F. Erdogan in 1963 [13] postulated that the range of

Fig. 1.9 Aloha airline flight 243 accident

stress intensity factor ΔK might characterize sub-critical crack growth under fatigue loading through a *law* that takes their name. Nevertheless, even if its introduction raised great enthusiasm, the *Paris law* had an empirical derivation, almost as FAD (*fracture appearance diagram*) many years before in the study of brittle fracture. Even though declassed from a *law* to the rank of simple *postulate of Paris* we have to recognise that it opened great perspectives in the study of fatigue. It would explain the retardation of fatigue crack growth due to an overload and bring to the idea of crack closure introduced by Elber in 1968 [14], phenomenon this last that causes a decrease in the effective stress intensity excursion acting on the crack, leading to a reduction of the fatigue crack growth rate. But the potential of fracture mechanics did not end up with fatigue.

In 1968 Smith et al. [15] found that stress corrosion was a threshold phenomenon that started at a given value of the applied K_I, called K_{ISCC} (SCC → stress corrosion cracking), characteristic of a material, which we'll be discussing in Chap. 6, and later to the analytical treatment of the devastating coupling of fatigue with corrosion. At the time it appeared stupefying and in many respects it continues to be surprising even today that a phenomenon which has an electrochemical nature can be treated by a mechanical tool such as the stress intensity factor K_I.

Much has been done and even more needs to be done, yet the real merit of fracture mechanics has been to separate the fatigue crack initiation, still studied by conventional stress and strain, from propagation where it dominates. This first Chapter will address precisely the phenomenology and the theoretical fundaments of fatigue crack initiation, from the pioneering studies that opened the way based on typical concepts of metallurgy and engineering such as dislocations, plastic flow, stress–strain curve, cyclic hardening and softening, aging, striations etc., necessary to fully understand fatigue.

1.2 The S–N Curve

Conducting, over one hundred and fifty years ago, the first fatigue experiments on iron and steel specimens, A.Z. Wöhler [4] arrived to the conclusion that the fatigue resistance of a material depended on the applied cyclic stress amplitude σ_a, conventionally indicated also by the symbol S

$$S = \sigma_a = \frac{\sigma_{\max} - \sigma_{\min}}{2}. \tag{1.1}$$

The maximum stress σ_{max} has an effect on fatigue only in that the higher it is the lower the amplitude σ_a of the alternated stress that leads to failure. This last statement is addressing the effect of the mean stress σ_m on fatigue (see Sect. 7.3):

$$\sigma_m = \frac{\sigma_{\max} + \sigma_{\min}}{2}. \tag{1.2}$$

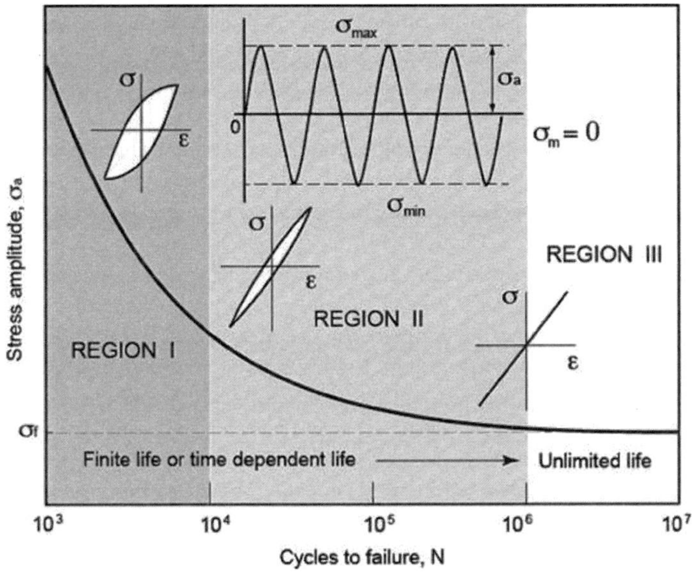

Fig. 1.10 S–N curve, cyclic stress σ_a versus number of cycles to failure N, also known as Wöhler diagram, showing the three regions that characterize fatigue behavior

Wöhler also noted that a lower limit σ_f existed for such amplitude below which the stress could be applied an unlimited number of times without causing fatigue failure of the material. Figure 1.10 is a schematic of the amplitude of the cyclic stress versus the cycles to failure *N*, in a log–log plot.

The plot, called *S–N* curve or *Wöhler diagram*, is referring to a symmetrical stress cycle, generally used as a reference cycle, in which the maximum stress is equal to the minimum one, except for the sign. The mean stress σ_m, in this case, is zero (see Eq. 1.2). The *S–N* curve is best fitting the experimental data not shown in Fig. 1.10. We can distinguish three characteristic regions on the *S–N* curve. In the first, extending from ¼ of a cycle to about 10^4 and sometime 10^5 cycles, indicated in Fig. 1.10 as REGION I, the material is loaded at a stress in the vicinity or over the yield strength σ_y of the material. The plastic strain dominates and controls fatigue life. The material undergoes a hysteresis loop whose area represents energy introduced in the crystal lattice that is converted, in most part, into heat and in part stored in the material as permanent deformation energy. Part if not all of this second component can be released and contribute to change the mechanical properties of the material through cyclic hardening or softening, as we'll discuss in Sect. 1.3. This first region is that of the so called *low-cycle fatigue*. In this region the use of a stress amplitude is rather an improper choice. First because stress is an ambiguous quantity in a domain where strain and stresses are not linearly related, secondly, as it will be discussed in Sect. 1.4, the material may change its characteristics under cyclic stress responding with strains different from those initially applied. Therefore, it is always convenient to run fatigue tests under strain controlled conditions and have an ε-*N* curve. The *S–N* curve shall be used only in the high-cycle region of fatigue where stress and strain are uniquely related and completely interchangeable. The term *S–N* curve or diagram should then be restricted to identify only that portion of curve. The second region, REGION II of Fig. 1.10, is generally more common to engineering applications. Here the material behaves elastically, at least on a macroscopic scale, and failure occurs under alternate stress whose amplitude σ_a is always lower than the yield strength σ_y.

This is the region of the so called *high-cycle fatigue* that extends up to 10^6 or 10^7 cycles and terminates, at least in BCC materials (ferrous alloys) and titanium, with a sharp *knee*. In this region the hysteresis loop is sharply reduced till it completely disappears in the vicinity of the fatigue limit, as schematized in Fig. 1.10, indicating the linear behaviour of the material. In nonferrous alloys the knee generally disappears and the *S–N* curve seems to continue even beyond 10^8 cycles, as schematized in Fig. 1.11. Beyond the knee of ferrous materials, REGION III begins. This is the region of unlimited life since the *S–N* curve flattens and fatigue life becomes independent of stress cycles. The corresponding stress is called the *fatigue limit* of the material, formerly referred to as *endurance limit*. Below this limit any load may be applied an infinite number of cycles without producing failure. Because of that, the first two regions where materials fail by fatigue cycling are also defined as time dependent life regions. In nonferrous materials or better in materials having a FCC lattice, such as copper, aluminium, gold, nickel or austenitic stainless steel and sometimes also in very low strength mild steel, this limit does not seem to exist and fatigue life continues to decrease indefinitely with continuing cycling, as schematically shown in Fig. 1.11. In these cases it is use to define a conventional fatigue limit at a given number *N* of cycles, usually 10^8 cycles.

However, under the continuous pressure of technological needs, the former value of 10^7–10^8 cycles that signed the limit of fatigue testing has been exceeded to enter the field of very-high cycle fatigue (VHCF). Nowadays an increasingly large number of applications in aircraft, automotive, railway and other industries requires the knowledge of fatigue behaviour beyond 10^8–10^9 cycles. This is the new technological frontier. Just think of gas turbine disks that can go as high as 10^{10} cycles or car cylinder heads and blocks (10^8 cycles), ball bearings, diesel engines of ships and high-speed trains (10^9 cycles) or wind-mill that converts wind power into rotational energy and then into electric energy.

What was practically impossible to achieve in a reasonably short time with conventional fatigue testing machines, has become possible today with the introduction of resonant-type fatigue testing machine that can reach 100–120 Hz. Lately, the introduction of the ultra-sonic systems operating at 20 kHz capable of producing 10^{10} cycles in less than one week (see Sect. 10.1 and Fig. 10.6) has sharply reduced the testing time allowing the collection of many more data that before, with conventional fatigue machine, were simply unthinkable. These new techniques are making possible the examination of the fatigue behaviour of materials in the VHCF domain. Their use has significantly modified the shape of the conventional Wöhler curve given in Fig. 1.11 for ferrous alloy that became what is known as duplex or multi-stage fatigue curve shown in Fig. 1.12. The unexpected failures of many structural BCC materials

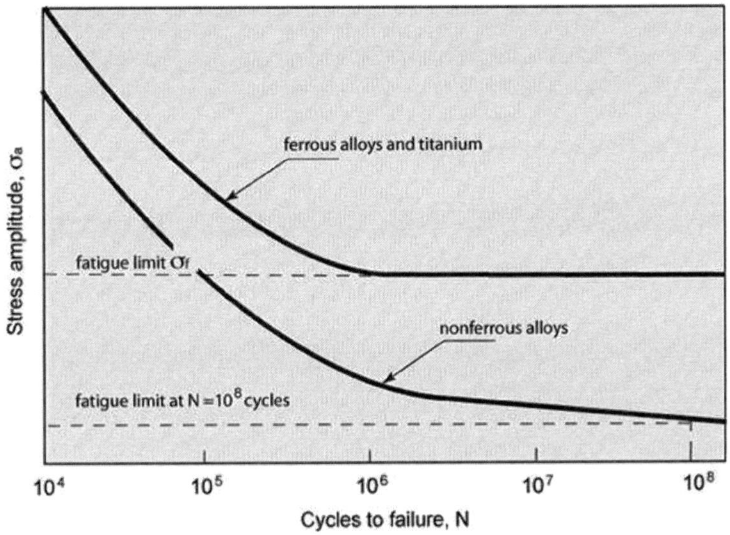

Fig. 1.11 S–N curve with and without a knee and a clear fatigue limit

in the VHCF regime have recently raised serious concerns. The very meaning of those curves is that also for BCC materials a fatigue limit may not necessarily exist, as for FCC materials, see Sect. 9.1.

Fatigue curves can be obtained under cyclic stress (stress controlled) or cyclic strain (strain controlled) tests depending on the region they belong. In REGION I

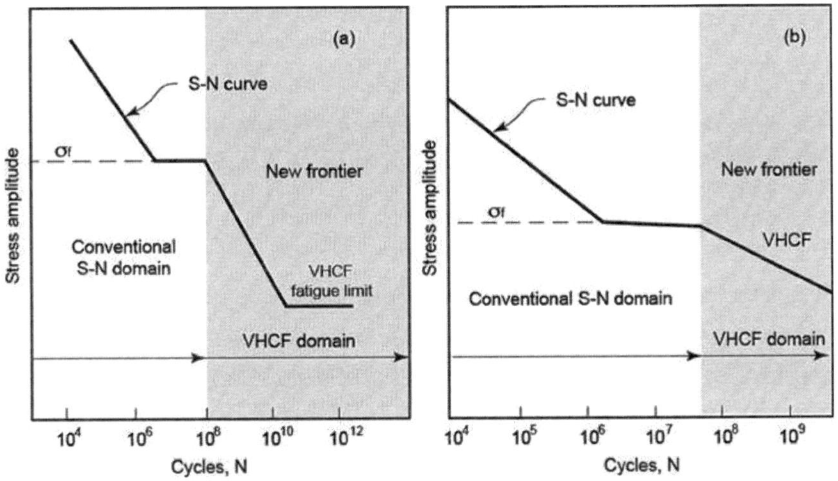

Fig. 1.12 Two possible types of VHCF response: **a** with apparent fatigue limit, **b** without fatigue limit

where fatigue is activated by stresses close or over the yield strength of the material, tests must be run under strain controlled condition obtaining an ε_a-N curve where:

$$\varepsilon_a = \frac{\varepsilon_{\max} - \varepsilon_{\min}}{2}. \tag{1.3}$$

In low-cycle fatigue the linear relationship between stress and strain given by Hook's low is lost. Strain is the controlling parameter while stress is depending on the prior history of deformation. In REGION II of high cycle fatigue, a one to one correspondence exists between stress and strain, therefore the condition under which tests are run is not important. Since the real extension of REGION I is not precisely known a priori, tests should be run under strain controlled conditions almost for the entire fatigue curve. Figure 1.13 is an example of ε_a–N_{25} curve obtained under stain controlled condition on type A 106 B and 333–6 carbon steel, used in piping construction, at room temperature [16, 17].

N_{25} is defined as the number of cycles at which the peak tensile stress drops by 25% from its initial value. For the specimens used, having 6 to 10 mm diameter, a 25% drop in peak tensile stress corresponds to a 3 mm-deep crack. This actually means that even dough the material was still resisting fatigue, a 3 mm-deep crack already formed in the specimens. The rationale behind this relatively new tendency to stop fatigue tests at some load drop is to distinguish between the two phases of crack formation and crack propagation, see Sect. 2.1 and Fig. 2.47. Crack formation is influenced only by material conditions in a small volume around the point of initiation, while propagation if affected by conditions throughout the cross section of the specimen or, in a real workpiece, is also a matter of geometry, shape and thickness. Therefore,

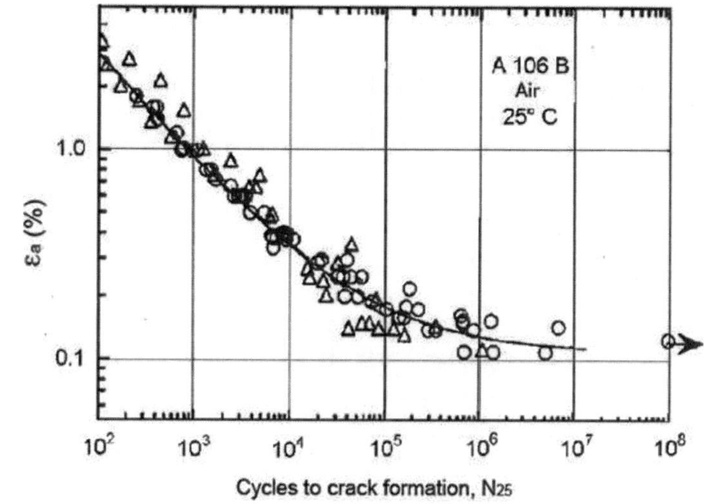

Fig. 1.13 ε_a–N fatigue curve of a carbon steel type A 106 B and A 333–6 obtained with cylindrical specimens under strain controlled alternate traction [16, 17]

what happens after macro-crack is formed shall not be considered by conventional fatigue analysis based on *S–N* or ε_a-*N* curves but must be studied only by Fracture Mechanics, see Chap. 14. In this respect, a 25% load reduction may be considered even too much and load drop be stopped before, let's say at a 5–10% already. ASTM Standard E 466 requires reporting the number of cycles N_{50} at which the initial load drops by 50%. As it can be seen in Fig. 1.13, REGION I of low-cycle fatigue for the steels considered extends until about 80,000–100,000 cycles where the applied cyclic deformation ε_a decreases to about the elastic strain limit.

The knee, typical of carbon steels, is present and the fatigue limit σ_f seems to be reached asymptotically at 0,1% cyclic strain that corresponds to about 200 MPa. Another example of $\varepsilon_a - N_{25}$ curve is shown in Fig. 1.14 for a low alloy carbon steel type A 533 B used in the construction of nuclear pressure vessels [17]. For this steel REGION II terminates at about 10^6 cycles where the curve shows the typical knee. Figure 1.15 shows several $\varepsilon_a - N$ fatigue curves for some aluminium alloys [18]: 7075-T6 (470 MPa yield strength and 530 MPa ultimate), 6061-T651 (266 MPa yield strength and 300 MPa ultimate), 5454–0 (120 MPa yield strength and 245 MPa ultimate) and 1100-H12 (105 MPa yield strength and 112 MPa ultimate).This time the fatigue curves do not show the knee typical of carbon steel, but continue to decrease without a definite asymptotic trend nor a knee. REGION III is almost non-existent.

As a general rule, the higher the strength of the material the higher its fatigue limit and, therefore, its behavior in REGION II of high-cycle fatigue and the lower that in REGION I of the *S–N* curve. In other words, materials that have higher fatigue limit and are more resistant to high cycle fatigue, behave worst in the low cycle regime, and vice versa. This is because the mechanism that triggers the fatigue process in

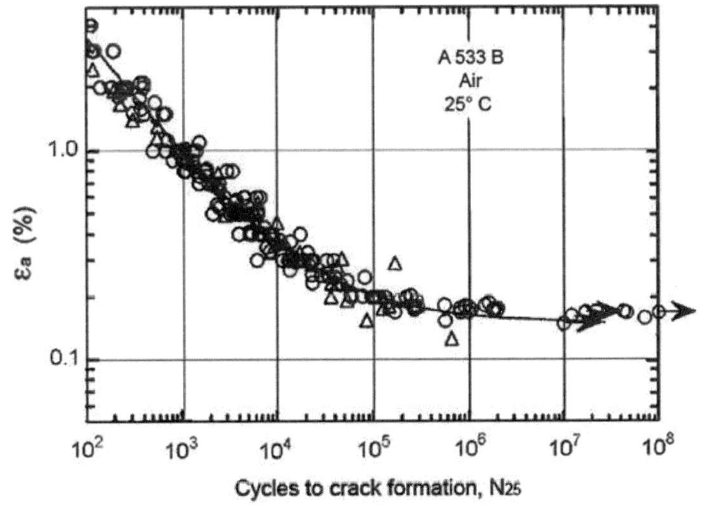

Fig. 1.14 $\varepsilon_a - N$ fatigue curve of a carbon steel type A 533 B obtained with cylindrical specimens under strain controlled alternate traction [17]

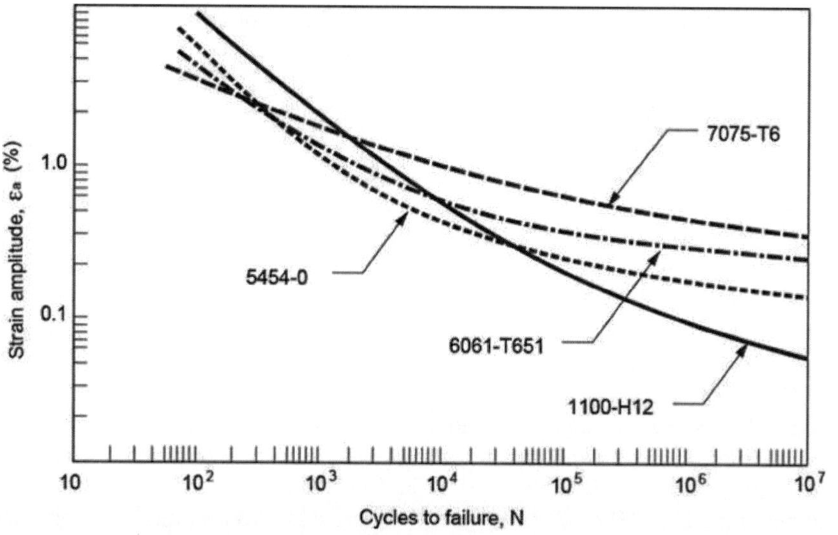

Fig. 1.15 ε_a–N fatigue curve of a series of aluminium alloys obtained with cylindrical specimens under strain controlled alternate traction [18]

metals is different in the two regions. In multi-crystals materials high-cycle fatigue always originates somewhere on the surface where just a single or few grains develop a severe plastic slip.

This is because the random orientation of surface grains makes some of them more prone to slip while all the others are still in the elastic state, as it will be described in Sect. 2.1. If slip is responsible for fatigue initiation, then materials having higher yield strength will behave better, since plastic slip is inhibited or retarded. On the contrary, in the low-cycle region the entire cross section of the specimen undergoes plastic slip and to resist fatigue requires differently oriented grains to adapt deformations. This request of uniform deformation is opposed by the grain boundary network that is generally much harder and brittle than individual grains. The result is that the grain boundary network will fail somewhere triggering the fatigue process. Therefore, softer materials that are more prone to general yielding will behave better. This characteristic change of behavior is schematized in Fig. 1.16 as a function of the corresponding hardness [19].

In the high-cycle fatigue domain harder steels present a better fatigue resistance than softer do. The opposite is true in the low-cycle fatigue region. Mild steels with a hardness of 200 HB that corresponds roughly to 600 MPa of ultimate strength, behave very poorly in the high cycle fatigue region, but they overwhelm all other steels below 10^3 cycles. In general, at room temperature in low-cycle fatigue cheap soft steels behave much better than expensive martensitic hard steels. When heat resistance is needed, as in jet turbines that work at about 1000 °C, carbon steels lose their fatigue properties and special nimonic alloys are required. All *S–N* curves cross at about 10^3 cycles at strain amplitude of about 0.01. This general trend is shown in

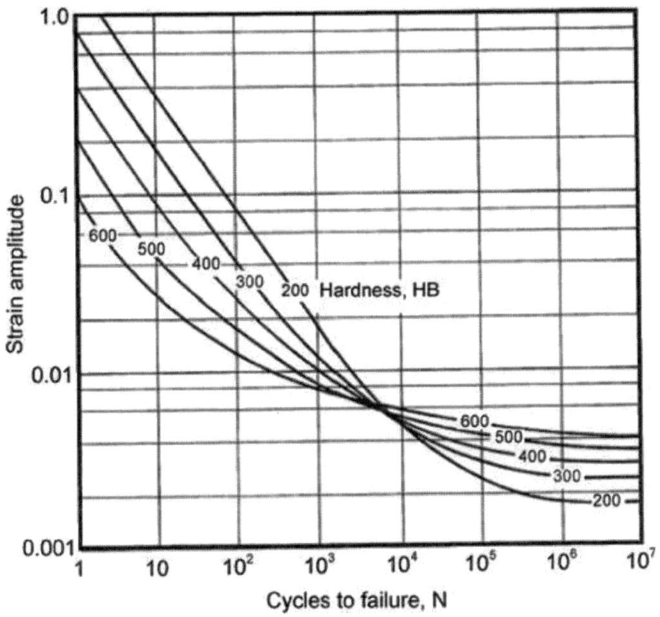

Fig. 1.16 Dependence of *S–N* curves on the hardness of the material [19]

Fig. 1.17a for steels classified as hard, tough and soft or ductile [20]. The difference between these three general categories of alloys appears in Fig. 1.17b in terms of stress–strain characteristic and full hysteresis loop [20]. The choice of the material becomes completely meaningless if it is to work around 10^3 cycles.

1.3 The Four Phases of Fatigue

The study of fatigue starts asking questions as why, how and where fatigue damage initiates, accumulates and, eventually, propagates. The answers require the knowledge of each single step through which the overall fatigue damage process proceeds. It is not just a heuristic issue but a question of fundamental importance to fatigue design and to avoid failure. You must have a full knowledge of your enemy if you want to be sure to beat him. Likewise, one cannot even tray to design against fatigue just by running calculation as long as all fatigue related mechanisms are not known; the risk would be designing something completely different from what you are actually manufacturing, since it is very possible that the real structure will not behave according to calculations. In this respect, paradoxically design by code rules may be even more accurate than by calculation since code rules tray to take into consideration, as much as possible, all factors affecting fatigue, almost relieving the designer from the burden of knowledge. But the price to pay is often too high because to comply with

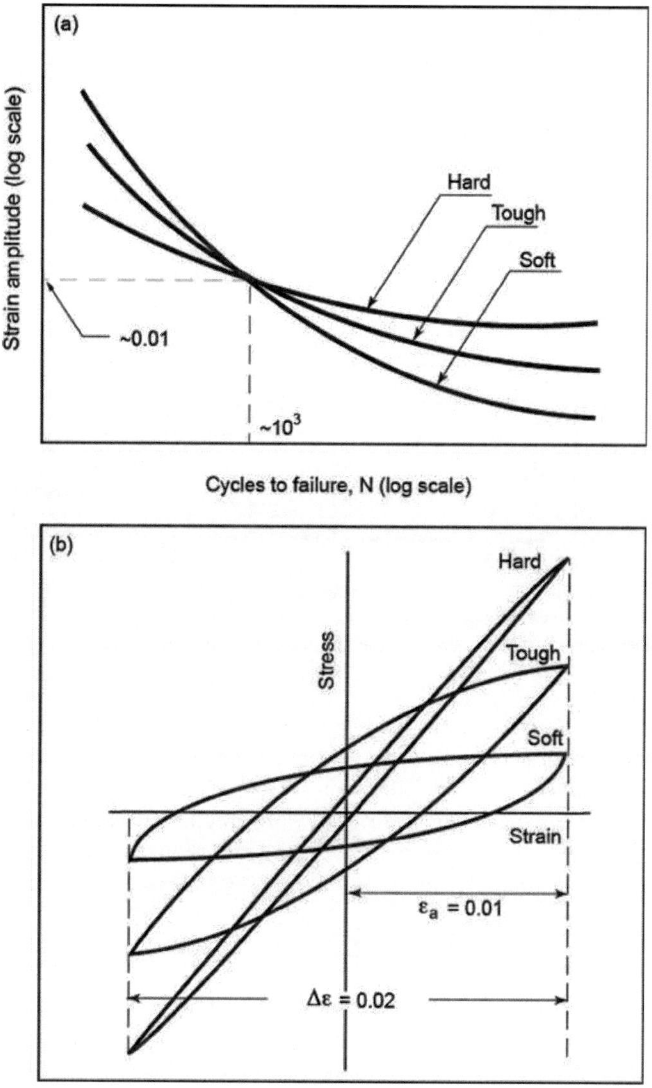

Fig. 1.17 **a** general trend of S–N curves for different alloys classified as hard, tough and soft, respectively, **b** this difference is presented in terms of the corresponding hysteresis loop [20]

all conservativisms built into rules may be unsustainable. To develop a real knowledge, it is fundamental to understand that the road to knowledge passes through a continuous downscaling from the macroscale of real structures to the nanoscale of what is inside a crystal, as schematized in Fig. 1.18. Small specimens, Fig. 1.18a, though necessary to know fundamental properties, such as static and cyclic strength, S–N curves, ductility etc., are not small enough to provide all information needed to design against fatigue. We must look down into the microscale of grains and phases, inclusions and slips, Fig. 1.18b and continue into that of dislocations and fractography, Fig. 1.18c to see characteristic features such as striations, corrosion facets, dimples.

Fig. 1.18 To reach a full knowledge of fatigue a multiscale analysis is necessary. The aim is at digging more and more to unveil the very beginning of fatigue damage and its progression to the macroscale of engineering applications

We could continue our journey in the extremely small and *see* crystals and eventually *atoms* with high resolution TEM or Field Ion Microscopy or Scanning Tunneling Microscope, Fig. 1.18d, but we will not go that far here. However, it is worth mentioning that the dots, black or white, we see in Fig. 1.18d are atoms of tungsten, silicon with their hexagonal structure and iron and these last are aligned to show their close-packed directions. It's a kind of Russian doll that we have to open, over and over again, discovering what is inside the next mini doll. In performing such a study, also the analysis of failures plays a fundamental role in understanding what really happened and why the metal failed. Again, it is necessary to acquire information at different scales from the macro down to the micro-scale and check for signs to reveal the actual causes of failure. Having this in mind, it can be very useful to proceed systematically identifying four fundamental phases that dominate and characterize fatigue:

- Phase 1—activation of a cyclic hardening process in the material or its opposite, the softening;
- Phase 2—activation of a system of elementary plastic flows localized somewhere on the surface of the material, at least in high cycle fatigue, where damage nucleates in some sort of submicroscopic appearance, less than 3 μm and eventually grows to become a microscopic crack, also called microstructurally small crack (3 μm < MSC < 300 μm);
- Phase 3—coalescence of MSC into a macroscopic crack, at least 300–400 μm long;
- Phase 4—fatigue growth of the macro-crack till failure occurs.

The time scale of the above mentioned four events or phases is schematically shown in Fig. 1.19 superposed to the Wöhler curve or *S*–*N* curve of the material. The time variable is represented by the number of applied cycles $N_{i,j}$ where the first subscript i refers to the phase ($i = 1, 2, 3, 4$) and second j to the stress amplitude S_j ($j = f, 1, 2, 3\ldots$).

The time duration (percentage of elapsed cycles) of each phase is varying depending on the stress amplitude S_j at which it occurs. In the high-cycle domain the first phase of cyclic hardening or softening may terminate after some thousands of cycles, as for $N_{1,2}$ in Fig. 1.19 at S_2, or hundreds of cycles near the low cycle fatigue region as for $N_{1,3}$ in Fig. 1.19 at S_3. In the very low-cycle region of fatigue this first phase may never terminate because fatigue failure comes first. This is the case where the so called saturation never occurs. In this case damage starts immediately after the first few application of load and phases 2 and 3 of damage nucleation and coalescence into a macro-crack in practice occur almost simultaneously ($N_{2,4} \sim N_{3,4}$). On the contrary, below the fatigue limit S_f, phase 1 doesn't start at all (hardening or softening doesn't occur) and phase 2, when and if it starts, takes an extremely large number of cycles to propagate, i.e., damage doesn't propagate but remains in one single surface grains. Between these two extremes all four phases develop completely, separately and in sequence.

In high-cycle fatigue, close to the fatigue limit, phase 3 may take even 90% or 95% of the whole life. The line joining the termination of phases 1 at various stress levels is called damage line since it represents the moment of persistent slip band

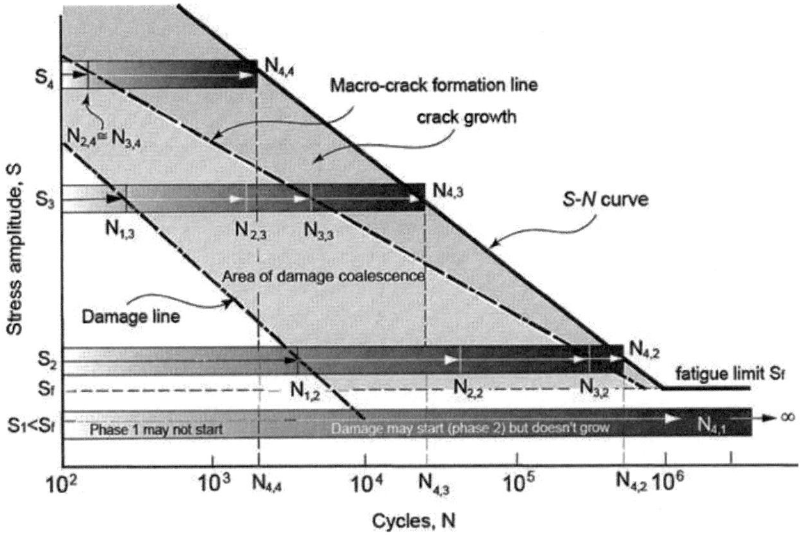

Fig. 1.19 Time distribution in terms of cycles $N_{i,j}$ ($i = 1, 2, 3, 4$) of the four fatigue phases: (1) cyclic hardening or softening, (2) slips with MSC formation, (3) macro crack formation and (4) growth of macro crack to failure, at four selected stress amplitude *Sj. Damage line represents the limit of the area where damage does not initiate while the macro-crack line limits the area where damage does not reach the macro-crack size. Below the fatigue limit Phase 1 may not start. Damage may start but does not accumulates*

formation, see Sect. 2.1, where damage initiates. It is rather interesting to note how the area of damage coalescence gets smaller and smaller as the stress amplitude S increases, indicating that higher stresses require less cycles to turn a submicroscopic damage into a macro-crack. The opposite happens as far as the propagation phase is concerned. At higher stress or strain amplitude that leads to low cycle fatigue, the distance between phase 3 and phase 4 is much higher than at lower stress amplitude. This indicates that at very high stress amplitude fatigue consists almost entirely of crack propagation. In the very low cycle fatigue it takes also 90% of the entire life to propagate a macro-crack and only a mere 10% to form it.

At variance with that, in high cycle fatigue and in particular close to the fatigue limit, it takes just 5 to 10% of the entire life to propagate a macro-crack that required about 90% or 95% of that life to form. For this reason and for what it has been said in Sect. 1.2, the fatigue curve which we should be really interested to is not the conventional Wöhler *S–N* curve but rather that of macro-crack formation, shown in Fig. 1.19. The N_{25} index curve of Figs. 1.13 and 1.14 is not precisely this latter curve but it is much closer to that than the Wöhler *S–N* curve. In other words, we should not be interested in the number of cycles N required to break a specimen but in that needed to break a real component. The fatigue strength of a real workpiece can be obtained only by studying the macro-crack formation separately from the crack propagation. Cycles needed to make a macro-crack become critical depends on the bulk characteristics of the workpiece and not on local material properties and stress state that affect the initiation phase. Bennett [21] clearly made it clear already in 1956 by saying: "Failure to distinguish between these two stages of the fatigue process leads to erroneous and sometimes dangerous results".

1.4 Cyclic Hardening or Softening

It is now more than 60 years that it has been understood that the very first phase of fatigue is represented by the cyclic hardening or softening of the material [22–29]. Almost all materials, carbon steel, austenitic steel, copper, aluminum, nickel, gold etc., undergo the phenomenon of cyclic hardening or softening, even though in different manner. Let's try to evidence the most important features. If a specimen of a given material is subjected to continuous cycling under strain controlled condition it may happen that its response, in terms of stress–strain characteristic, varies from cycle to cycle. This is shown in Fig. 1.20a for a specimen of annealed copper (soft copper) subjected at −73 °C to ten cycles of 0.04 total strain excursion (± 0.02 strain amplitude) [30]. The metal hardens since the stress by which it responds to the same imposed strain amplitude increases cycle by cycle.

A completely opposite effect is observed in copper initially hardened by cold-working that softens under the effect of cyclic loads. This is shown in Fig. 1.20b.

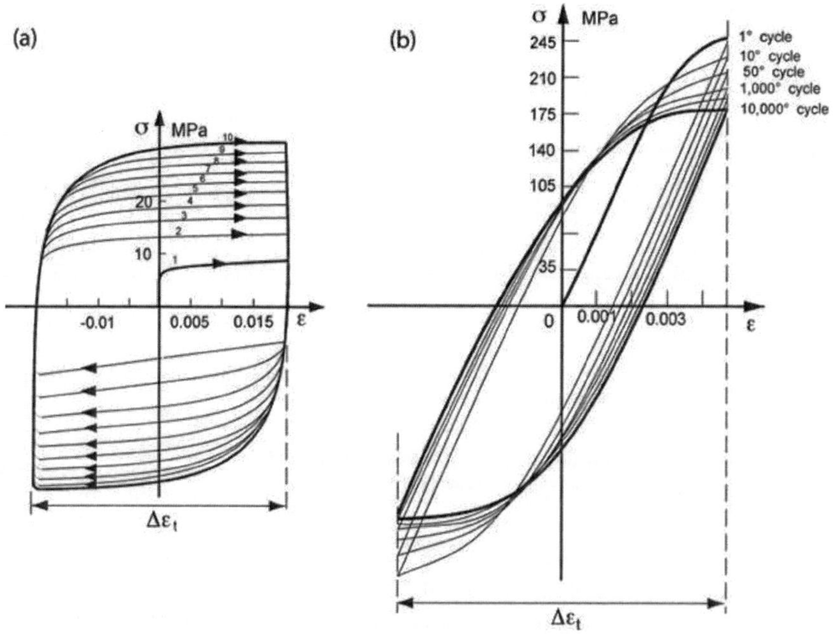

Fig. 1.20 **a** Cyclic hardening of a crystal of annealed copper subjected at −75 °C to ten strain controlled cycles; **b** cyclic softening of copper initially hardened by cold-working [30]

After a 1000 or more cycle the material is reacting to the same strain amplitude with a much lower stress. For a soft material, where initial dislocation density is low, the cyclic plastic straining sharply increases the dislocation density producing hardening. It is, then, evident that to generate hardening it is necessary to cyclically strain the material into the plastic domain. At variance, in hard materials strain cycling may result in softening even at stresses lower than the yield stress. Cycling, in fact, causes a rearrangement of barriers to dislocation motion so as to offer less resistance to deformation and the material softens. If the cycling is run under stress controlled conditions, softening is seen as a continuous increase of deformation accompanied by a widening of the hysteresis loop.

The opposite happens during cyclic hardening where the hysteresis loop shrinks and may even disappear making the metal behave elastically, see Fig. 1.39. Whether it hardens or softens, the material eventually arrives to what is called *saturation* and the process terminates. Even though saturation is an asymptotic process, from a practical point of view it occurs after a very small fraction of the entire fatigue life (see Fig. 1.19). As it can be seen in Fig. 1.20b, in the first 1000 cycles the material loses about 20% of its initial strength, but in the next 9000 cycles the further reduction is less than 5%. This initial variation of material characteristic followed by a rapid slowdown is particularly true in high-cycle fatigue cycling. Figure 1.21 summarizes the two fundamental material responses to cyclic straining and cycling stressing, respectively. Significant are the hardness measurements made by Kemsley as early as 1959 on copper hardened by cold-working or annealed [31]. He found that the surface hardness was effectively varying with the number of applied cycles reaching a final saturation value that was always the same, independently of the initial hardness or whether the virgin material was hard or soft. The results he obtained are shown in Fig. 1.22 for two different fully reversed applied stress amplitudes. The first stress amplitude in the elastic domain was equal to 112 MPa and led to fatigue fracture, at least in annealed copper, after about 10^6 cycles, the second in the plastic regime was equal to 170 MPa with a fatigue life reduced to about 10^4 cycles, always in annealed copper.

Two interesting features shall be noticed. First, in annealed specimens hardening takes place at a much higher initial rate than softening in cold-worked copper, then it levels off and saturates. Secondly, which is really stupefying, the hardness at saturation is always the same and equal to 75 or 80 HV whether starting from annealed copper with an initial hardness of 38 HV or cold-worked copper with initial hardness of 95 HV. Analogous results were obtained by Feltner and Laird [27] and Polakowski [24]. Note that this behavior is expected in materials having high stacking fault energy (SFE), such as aluminum and pure copper that promotes deformation by a wavy slip mechanism, as described in Sect. 1.4.1. This is schematically shown in Fig. 1.23a. At variance, materials having low SFE and are prone to planar slip, as austenitic stainless steels, show different cyclic saturation characteristics, depending on the initial conditions, whether cold-worked or annealed, as schematized in Fig. 1.23b. Generally, the number of cycles spent in hardening or softening is lower in the case of wavy slip materials, than in the case of planar slip materials. For example, the hardening of pure copper (wavy slip) cycled in the high cycle region of fatigue takes about 1–3% of the total number of cycles to failure, while the fatigue hardening of

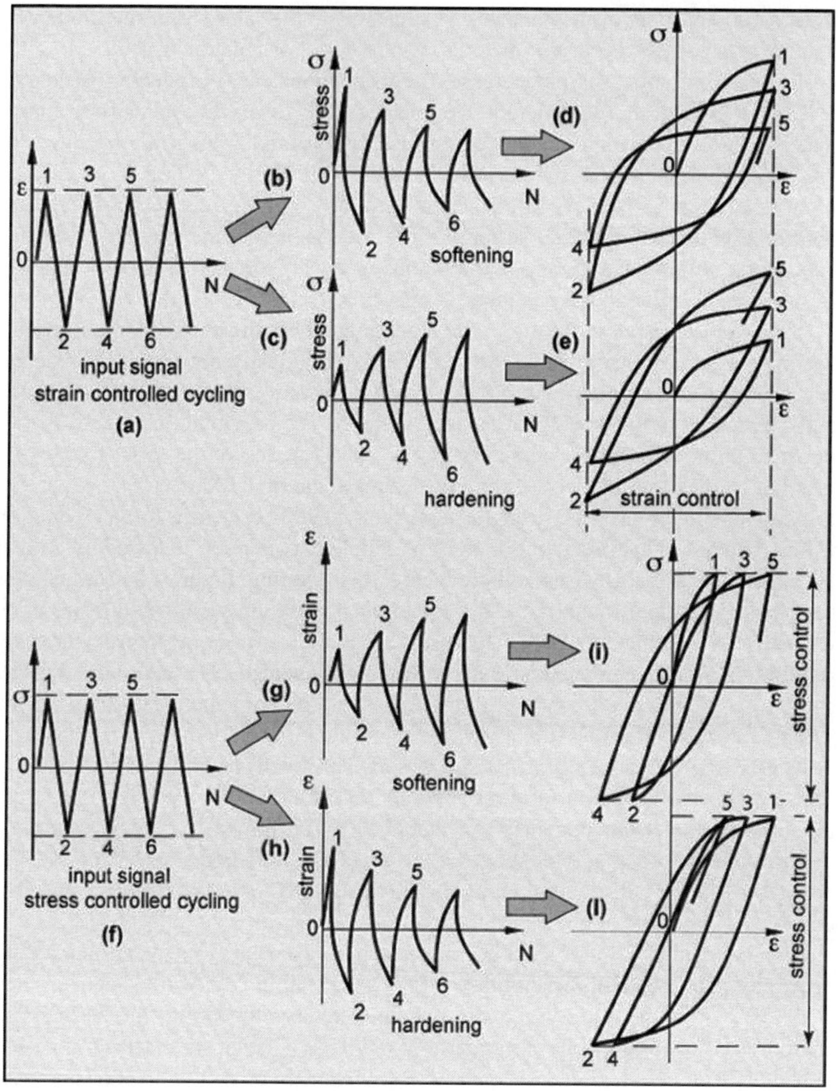

Fig. 1.21 **a** strain controlled cycling resulting in **b** softening or **c** hardening. **f** Stress controlled cycling resulting in **g** softening or **h** hardening. Label **d** through **l** represents same results in terms of stress-strain

Cu-30%Zn alloy (planar slip) under the same cycling conditions takes 30–40% of the total number of cycles. Note that the change of the material properties, whether softening or hardening, involves the entire volume of material. Therefore, it can be inferred through a tensile test on the same specimen after cycling. This procedure can be repeated with different specimens subjected to different cyclic strains, once the

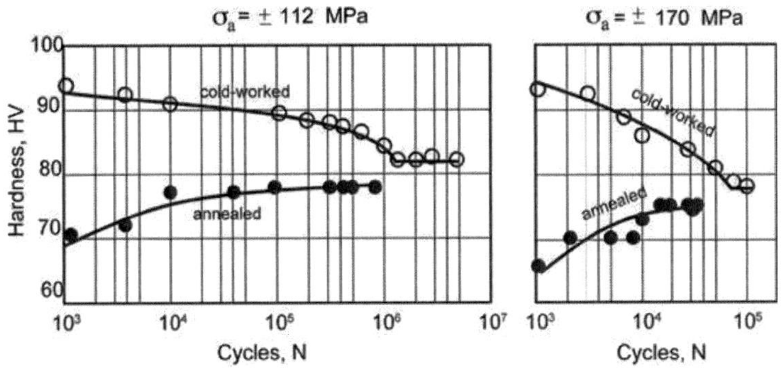

Fig. 1.22 Variation in the surface hardness measured by Kemsley on cold-worked and annealed copper at two different fully reversed stress amplitude [31]

saturation is estimated to have occurred almost entirely, to monitor the continuous change in material properties or it can be done by a single test procedure, as described in Sect. 1.4.2. Depending on whether softening or hardening takes place, the test will yield a different cyclic stress–strain curve. Also surface hardness can be used to infer the effect of cycling. If the material softens, its hardness decreases, while it increases in case of cyclic hardening. Laird [32] using this technique found the change of the material properties in nickel, both in the annealed and cold-worked conditions, as well. Figure 1.24 presents his findings on nickel initially hardened by cold-working. The two curves shown refer to two different initial hardness.

As it can be noticed, for the harder nickel the cyclic yield strength reduction reaches a maximum between $2{\cdot}10^4$ and 10^5 cycles (lower curve). In very low-cycles fatigue region the reduction of the yield strength is less severe because, as already said in Sect. 1.3, saturation may not be reached since fatigue failure occurs earlier. On the other side of the curve, after the maximum is reached at about 10^5 cycles, there is also a rapid decrease to zero of the softening, this time because as the material enters into the elastic regime the cycling loses more and more its softening effect. In the high-cycle fatigue region, in fact, the material, as previously said, cannot soften since the applied stress is below the yield strength of the material. Milella has found cyclic hardening in carbon steel and austenitic stainless steel by measuring surface hardness before and after fatiguing. This is shown in Fig. 1.25 for carbon steel specimens having 270 MPa yield strength and 500 MPa ultimate, after cycling at different strain amplitude. The increment, ΔHV, of surface hardness versus the applied strain amplitude appears to have the same trend as the cyclic stress–strain curve of the material, see Sect. 1.4.2.

Therefore once calibrated, the change of the mechanical characteristics of the steel due to fatigue cycling can be inferred from the parallel variation of its surface hardness. This technique of measuring the surface hardness before and after cyclic tests should be always used because it provides quick, cheap and very useful information. A large and reliable data base of hardness variations with cycling can be used as a reference standard to asses, for example, under what conditions a fatigue failure may have occurred.

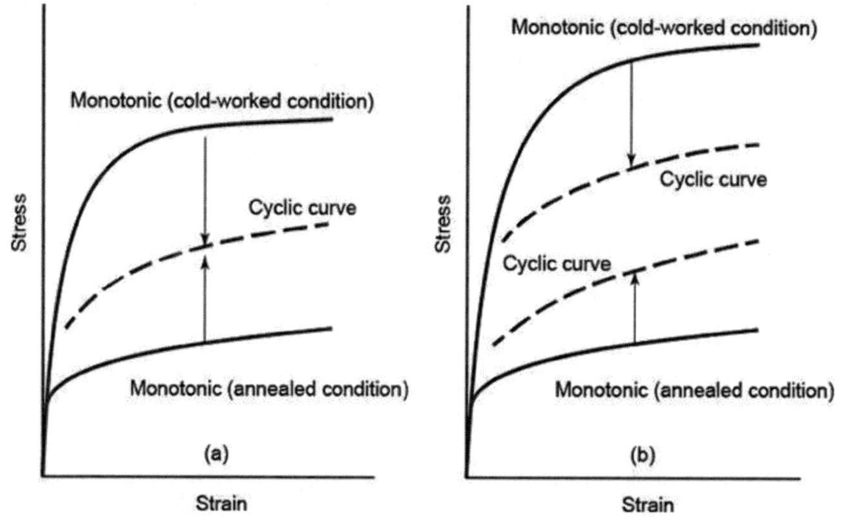

Fig. 1.23 **a** Materials that harden by a wavy slip mechanism develop the same cyclic stress strain curve independently of the initial condition, whether cold-worked or annealed, **b** materials that harden by a planar slip mechanism develop two different cyclic stress strain curves, depending on the initial condition

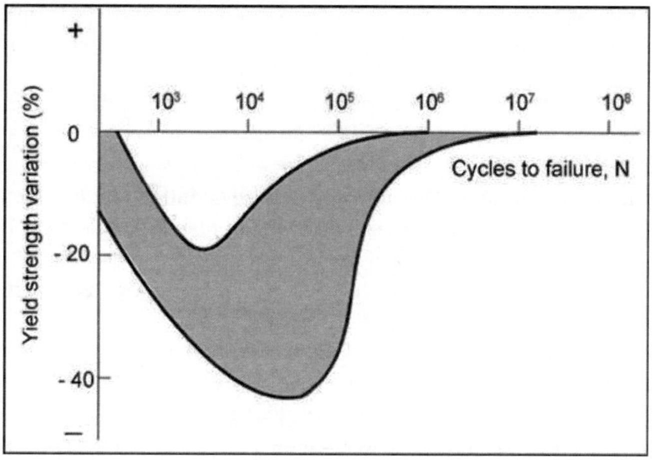

Fig. 1.24 Cyclic softening found by Laird conducting yield strength measurements on cold-worked nickel at different number of cycles [32]

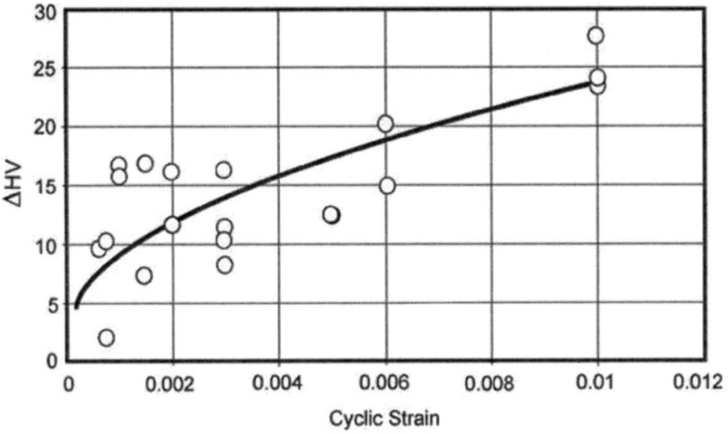

Fig. 1.25 Variation, ΔHV, of the surface hardness of specimens of low carbon steel found by Milella after fatigue cycling. The trend runs parallel to the elevation of the stress–strain curve by cyclic hardening

1.4.1 The Mechanism of Hardening and Softening

To understand how and why the repeated application of loads may trigger a mechanism of softening or hardening it is necessary to first understand how materials harden under monotonic loads. Hardening is a very complex mechanism due to many causes anyhow connected with dislocation motion. It starts with the plastic deformation of the material. This happens at the yield point when shear stresses begin to act prevailing over tension stresses. Shear stresses always exist in a material even when the only load applied is traction. Shear stresses activate plastic deformations through a process known as slip, as schematized in Fig. 1.26. At variance with elastic elongations, on a microscopic scale deformations by slip are not homogeneous, as shown schematically in Fig. 1.26. Slip is the microscopic evidence of a massive dislocation movement. The easier the slip process the lower the yield strength and the hardening of a material. However, dislocation movement can be hidden by obstacles producing hardening. Basically, there are two main types of hardening: strain or cold-work hardening and precipitation hardening. Strain hardening is a phenomenon that occurs in ductile metals as they are plastically deformed. It is also called cold work because the temperature at which plastic deformation takes place is much lower than the melting temperature, hereafter the term *cold*. Most ductile metals work hardens at room temperature. The most common work hardening process is cold rolling of carbon steel plates to make piping and pressure vessels. But also cold drawing produces the same effects as headed fasteners or extrusion products or any other cold forming technique, as well. The hardening is the result of a continuous displacement of dislocations that move until they are stopped by grain boundaries or large obstacles as impurities or foreign phases. Severe monotonic straining produce pile up of dislocations at these barriers. Figure 1.27 shows a train of dislocations piled up at a grain boundary in 304 austenitic stainless steel [33]. The pressure exerted over the boundary has activated new dislocations in the neighboring grain. Piled up dislocations interact to produce stress fields which resist the movement of additional dislocations toward the pile-up. As the stress fields intensify, the applied stress needed to move dislocations increases and the metal work hardens. Moreover, dislocations may also get pinned and locally stopped by other cross dislocations or by small foreign particles.

All these anchor points make dislocation lines to bulge on the slip plane and eventually break into new dislocation loops through a mechanism known, in material science, as Frank-Read source, as schematically sketched in Fig. 1.28 [34]. The Frank-Read source generates many dislocations in a single plane thus allowing the observation of dislocations that have sub-nanometer features. Figure 1.29 is a photomicrograph of massive plastic slips in polycrystal copper [35]. It appears that two slip systems operate for most of the grains, as evidenced by two sets of parallel yet intersecting sets of lines. Normal dislocation density in virgin annealed metals is about 10^4 to 10^6 mm^{-2}. In a cold-worked structure the dislocation density jumps to about 10^{11} mm^{-2}, i.e., 6 to 7 order of magnitude higher. As the deformation proceeds, this enormous and continuously growing dislocation density results in the formation of what is known as a dislocation *forest*. Forest is a very appropriate definition because it generates the idea of an intriguing matter difficult to unravel. As matter of fact, dislocations in the forest get continuously tangled as the loading phase proceeds, forming cellular arrangements whose boundaries have high dislocation density while the interior is almost free.

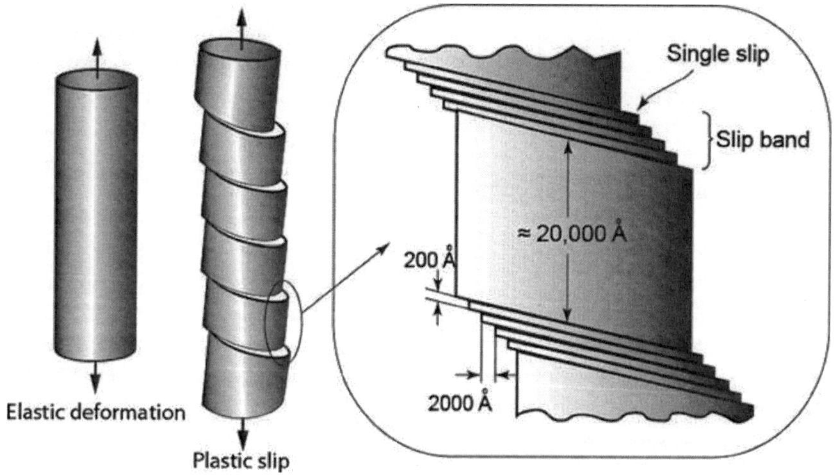

Fig. 1.26 Schematic of non-homogeneous plastic deformation by slip

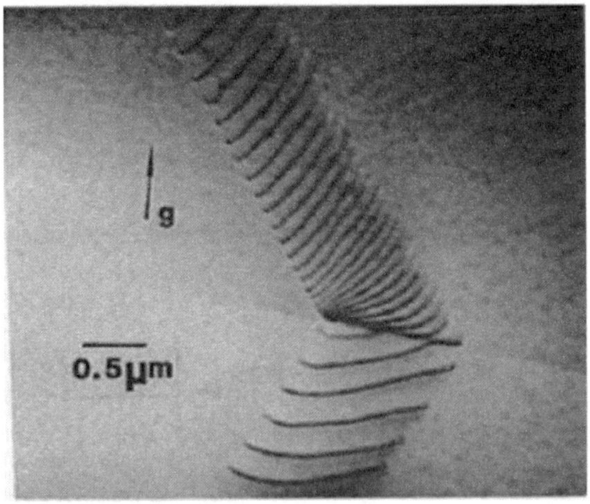

Fig. 1.27 TEM image of dislocations piling up at grain boundary in 304 austenitic stainless steel and emitted dislocations in the neighboring grain [33]

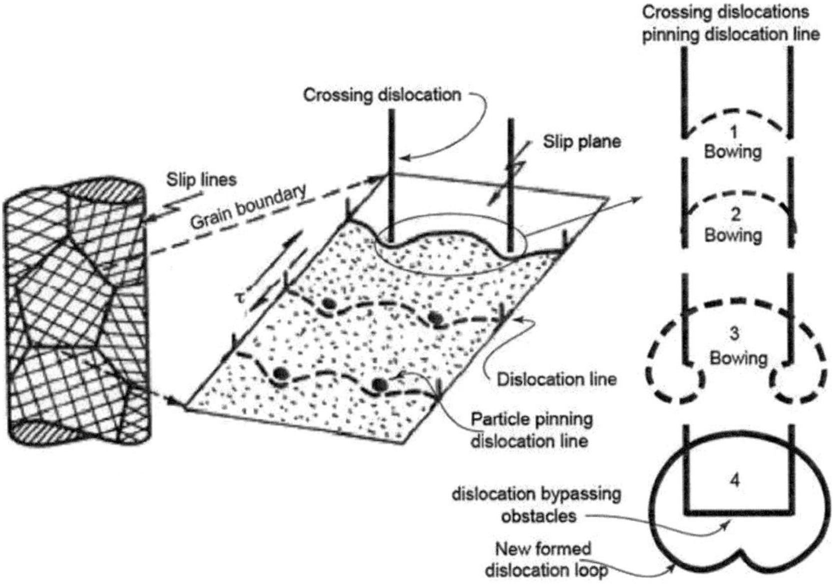

Fig. 1.28 Dislocation motion limited by bowing around discrete obstacles such as small particles or other dislocations and multiplication by the Frank-Read source mechanism (adapted from [34])

Fig. 1.29 Photomicrograph of slip lines developed in polycrystalline copper [35]

These cells that are generated inside grains are also called subgrains. An example of subgrains is shown in Fig. 1.31b. In many cases dislocations happen to be definitely blocked and will not move any longer. Slip process, then, becomes more and more difficult and takes higher and higher stresses to proceed. The overall result is hardening. However, cold working has two major drawbacks.

The first is that, for what it has been said, the hardening is obtained at the expenses of ductility which, as far as high cycles fatigue is concerned, is actually beneficial. Secondly, cold working is orienting grains, phases and inclusions along a preferential direction that is the working direction, see Fig. 4.11. Therefore, the material becomes strongly anisotropic. Another important mechanism by which dislocations can overcome barriers, in particular large obstacles, is cross slip. Cross slip is a mechanism by which a dislocation blocked on a slip plane can bypass the obstacle gliding on another plane where it continues to move, as schematized in Fig. 1.30.

In metals having high stacking fault energy, like aluminum and its alloys, iron and its alloys, nickel or copper, this mechanism of cross slip is very active. A stacking fault is an interruption of the normal stacking sequence of atomic planes, in particular in a face-centered-cubic crystal structure. These interruptions carry some energy called stacking fault energy (SFE). The movement of a dislocation in a face-centered-cubic (FCC) metal can result in a stacking fault. The dislocation line bifurcates into two partials whose distance or width depends on the SFE. High SFE materials have low width partial dislocations and vice-versa. The width or the SFE determines the ability of the two partials to recombine into the original extended dislocation. Cross-slip is possible only when the two partial dislocations recombine. High SFE means lower separation of the partials and greater attitude to cross-slip. This easiness to recombine and cross-slip results in higher hardening and lower hardening exponent. This process

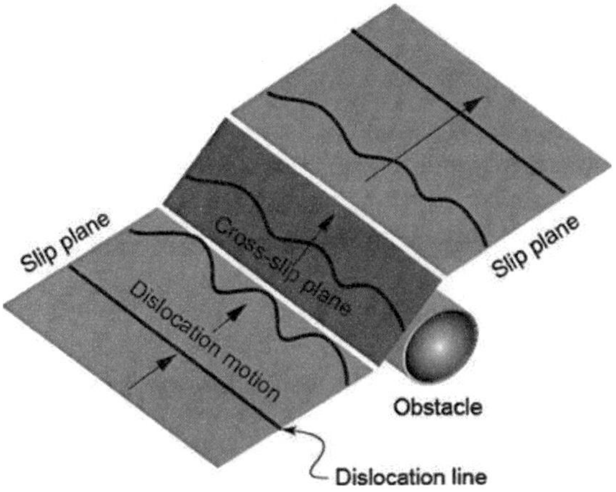

Fig. 1.30 Cross-slip mechanism by which dislocations can overcome an obstacle

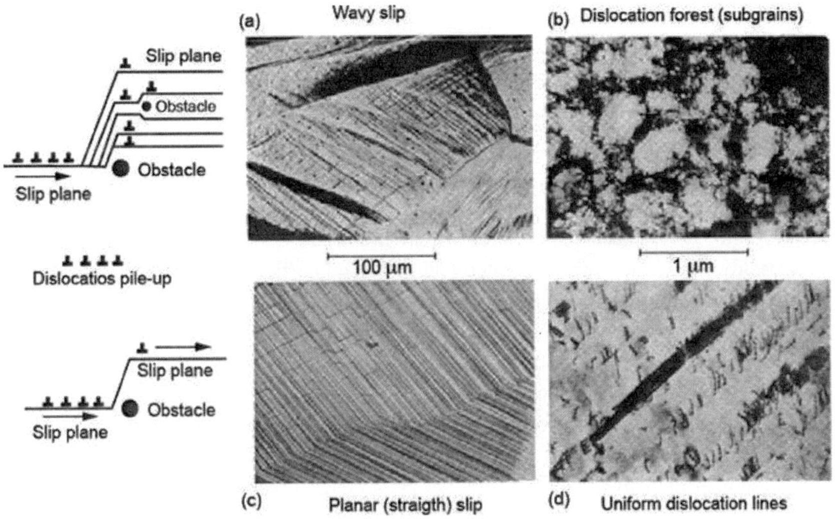

Fig. 1.31 Cross slip in **a** high stacking-fault energy materials (pure copper) and **c** low stacking-fault material (Cu-7.5wt% Al); **b** and **d** are high magnification views of **a**, **c** (modified from [36])

distributes dislocations uniformly in the matrix, which leads to fine wavy slip, see Fig. 1.31a and to the formation of cell structures and dislocation forests, Fig. 1.31b. Note the extremely small size of the subgrains which doesn't exceed 1 μm.

Aluminum, iron and nickel have the largest SFE in the range of 250–200 ergs/cm^2. At variance, low SFE means higher separation between partials. Cross-slip becomes difficult and planar slip occurs with straight slip lines, as shown in Fig. 1.31c where dislocations tend to stay in a planar array and their movement restricted, Fig. 1.31d.

Silver, austenitic stainless steels and α brass show the lowest SFE of the order of 25 ergs/cm^2 and even lower than 10 ergs/cm^2. Copper lies in between with about 90 ergs/cm^2. The second type of hardening is the so called precipitation hardening. Heat treatments can activate this mechanism in a very short time. Heat is an energy source of primary importance in any metallurgical transformation involving solid solutions.

The mechanism of hardening by heat treatments is quite complex and involve several stages of activation. The starting point is the presence of an element having very high solubility at high temperature and very low at ambient temperature, which is rather common. Solubility, in fact, depends on temperature and follows a rate of the Arrhenius relation type. What is an allowable concentration at high temperature becomes a supersaturated solid solution at low temperature. Cooling the metal causes the precipitation of solute atoms, in particular when this cooling is very fast as in quenching. Generally, solute atoms precipitate as second phases. Well known are the so called *ε phase* (carbides) and *α"phase* (nitrides) in carbon steels or the *θ phase* (Al_2Cu) in aluminum alloys. These second phases are much harder than the solvent matrix. They join in clusters that distort the matrix.

These areas are also known as GP zones from Guinier [37] and Preston [38] who, independently of each other, in 1938 interpreted features in diffuse X-ray scattering from aged aluminum alloys as evidence of clustering of atoms into very small zones. After almost 30 years the mystery of the hardening of aluminum alloys discovered by chance by Wilm was finally unveiled. The result is localized stress fields of adaptation. These precipitates are generally incoherent or semi-coherent and have limited effects on the metal hardening. If the alloy is tempered after quenching, then incoherent phases can turn into coherent phases. They are much more effective in hardening the material since their stress fields are non-symmetrical distortional stress fields. Dislocations are blocked and the metal hardens. Figure 1.32 [39] shows, schematically, the difference between a coherent and an incoherent precipitate. In a coherent phase there is a relationship between the rows of atoms in the matrix phase and the rows of atoms in the precipitate, as schematized in Fig. 1.32b. That is, you can follow a row of atoms in the matrix and you can more or less continue that course to follow the line of atoms precipitated in a coherent fashion. These two courses aren't straight, but bulge outwards.

This causes distortion between the two phases and this distortion introduces localized shear stresses that are the most effective in interacting and, therefore, blocking dislocations in their motion. Hence strong hardening effect. This occurs when the precipitate has the same crystal structure and lattice parameters as the matrix phase. At variance, an incoherent precipitate has no relationship with the surrounding matrix, see Fig. 1.32a. Crystal structure and parameters of the matrix are not similar to those of the precipitate. The corresponding strain field is more symmetrical or hydrostatic. An interesting image of coherent GP zone observed at the high resolution TEM in Al-0.59 Mg-0.0.71Ge (at%) aluminum alloy is offered in Fig. 1.33 [40]. Note the extremely small dimension of the precipitate that doesn't exceed 6 nm. Incoherent phases can always interact with dislocations, but not as much as coherent precipitates do. This mechanism of hardening induced by tempering the metal after quenching is

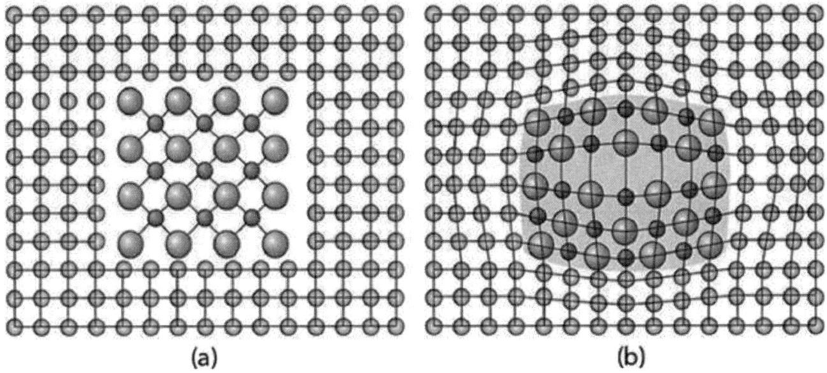

Fig. 1.32 **a** schematic of incoherent precipitate and **b** coherent precipitate [39]

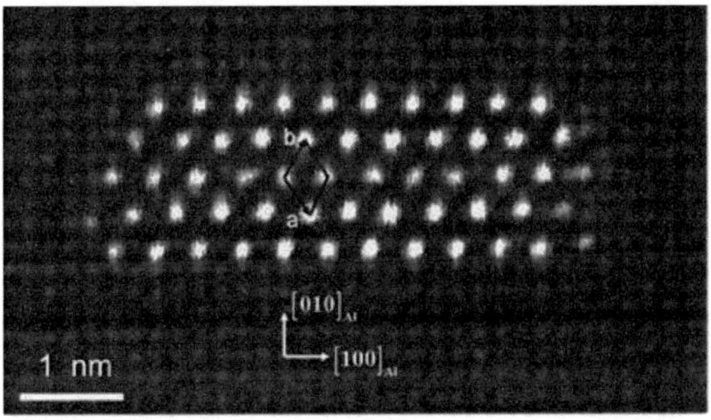

Fig. 1.33 HRTEM image of coherent precipitate in aluminum matrix [40]. The lattice directions in the Al matrix and the hexagonal sub cell of the Ge columns are shown

normally known as thermal ageing or temper ageing. The overall process is known as precipitation hardening.

Aluminum alloys own their strength to a precipitation hardening process. In these alloys strength and hardness are enhanced by the formation of extremely small uniformly dispersed second-phase particles within the aluminum matrix. In iron, the hardening effect may be the result of a stoichiometric compound formation, Fe_3C, called cementite, which is not a phase. Cementite in the α-iron matrix (ferrite) forms what is known as pearlite, Fig. 4.22. Cementite contains 6.67% C and is extremely hard and brittle. It is present in any commercial steel. By properly controlling the amount, size and shape of cementite, it is possible to control the degree of dispersion strengthening and the properties of steels. But carbon can also result in the formation a new phase called martensite, which is very hard and brittle, depending on the proper heat treatment given to the steel. Martensite can be regarded as a supersaturated solution of carbon in ferrite, and its hardness attributed to solid solution hardening. Once outlined, by general lines, what is hardening and how it is obtained, though through a relatively simple treatment, it is possible to explain how the repetition of loads operates making possible the softening or hardening of materials. In very hard materials, under cyclic loadings, dislocations move back and forth and can jump from one plane of atoms to another. These jumps from the one end can annihilate dislocations and from the other end enable some dislocations to bypass point defects that block dislocation motion on only one plane. As the number of piled-up dislocations decreases with each cycle, also the associated stress field decreases making the motion of dislocations in the following cycles much easier. But this jumping up and down can also result in a continuous rearrangement of the crystal lattice in a form more favorable to dislocation motion and, therefore, to plastic deformation.

Studies conducted on aluminum alloys [41–45] and in particular on 6061-T651, in which the monotonic hardening is obtained mainly through the precipitation of Mg_2Si in coherent form, have contributed to the understanding of this phenomenon of cyclic softening. Under cyclic loads the precipitates lose their coherence and dissolve into the lattice, thus reducing their capability to block dislocations. These start to move better cutting among precipitates now of smaller dimensions and distributed on more slip planes, changing completely the original hard structure. In materials with low SFE that develop fine wavy slip and cell structures, cyclic loads break down the cell structure to form persistent slip bands. Now on some crystallographic planes, slips are possible, which before were circumvented. It is there that plastic flows concentrate and become non-homogenous and highly localized inside single crystals. It generates a kind of composite material consisting of a hard matrix, which still behave elastically, and lamina of soft material where plastic flow develops without any obstacle resulting in a softening of the metal. On the contrary, cyclic hardening is the result of a continuous and repeated action of cold-working on a matrix full of foreign atoms such as nitrogen, oxygen and carbon in particular, which are elements notoriously responsible for dislocations blockage. No doubt remains on the role of these atoms and non-metallic phases.

It has been observed [29], for example, that a 0.1% C steel becomes softer and its fatigue life strongly reduces upon the removal of these elements. The fatigue limit and the *knee* of the S–N curve are no longer well defined and it has been supposed that it could eventually disappear upon the complete removal of carbon (decarburizing) and nitrogen. As matter of facts, it has been observed a remarkable reduction of fatigue limit in type 4340, 4140, 2340 and 5140 steel after decarburizing, as shown in Fig. 1.34 [46]. In particular, Fig. 1.34a refers to type 4340 steel hardened to 1600–1950 MPa in its initial state no decarburized and after decarburizing a surface layer from 0.07 to 0.7 mm thick. Decarburizing has resulted in the lowering of the fatigue life by a factor 1.4 analogous to a notch effect with a stress concentration factor k_t = 3 (see Sect. 3.4). In the other three alloys, Fig. 1.34b, initially hardened to 283 HRC, decarburizing had produced an even stronger fatigue life reduction. Always observations on aluminum alloys have evidenced the importance of non-metallic phases. A typical example is shown in Fig. 1.35 [47] for Al–Zn–Mg–Cu 0.5 alloy (type 7022).

The figure shows the stress amplitude versus the cumulative plastic strain for three initial conditions: free of precipitates, peak-aged and over-aged. The cumulative plastic strain is the total strain attained during the test and is equal to the plastic strain amplitude selected, indicated in Fig. 1.35 close to each curve, times the number

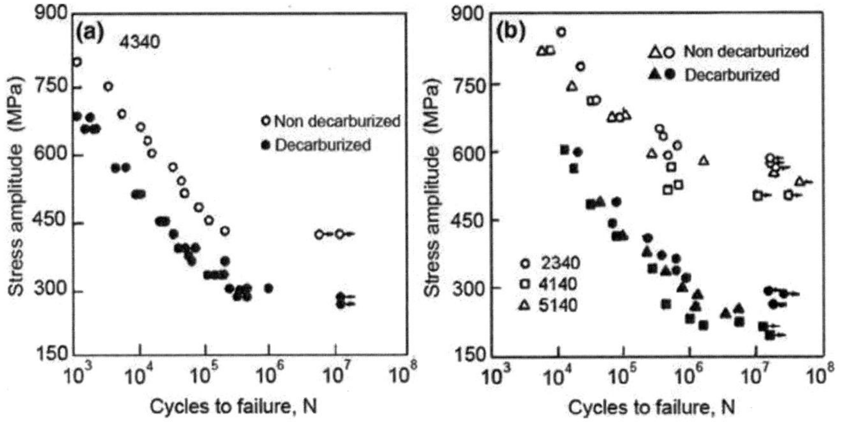

Fig. 1.34 **a** Effect of decarburization on fatigue life of **a** 4340 steel and **b** on three different steels hardened to 28 HRC [46]

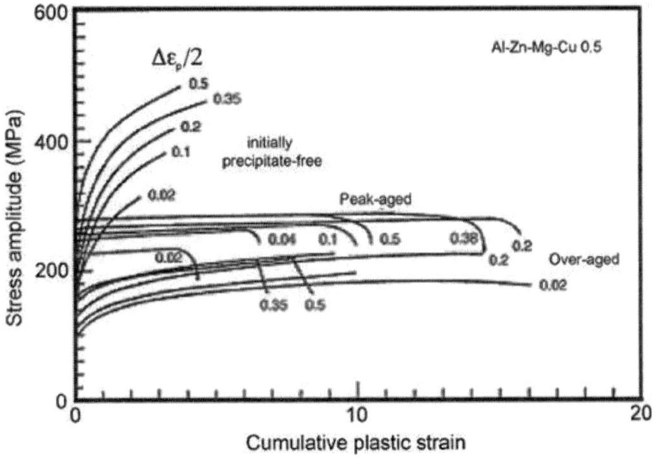

Fig. 1.35 Stress versus cumulative plastic strain obtained in a Al–Zn–Mg–Cu 0.5 alloy for different initial conditions

of applied cycles. The curves relative to the initial inclusion-free condition are of particular interest. They evidence a remarkable increase in hardening that doesn't take place in the other two conditions. The surface metallographic examination by transmission electron microscopy (TEM), Fig. 1.36 [48], reveals what actually happened. Figure 1.36a shows the situation before cycling. Just few inclusions are present that have not been dissolved by the solution-annealed process that brought the material in the initial inclusion-free situation. Second phases and precipitates are absent. Upon cycling a large number of fine dispersed second phases appear equally distributed in the matrix, Fig. 1.36b. The material hardening is due to these phases generated

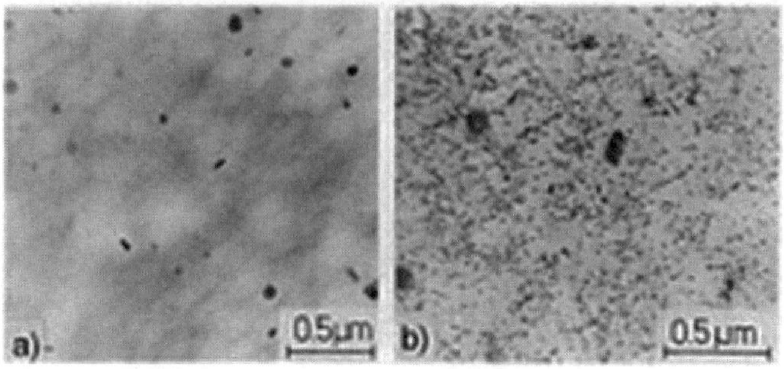

Fig. 1.36 Transmission electron microscopy examination of the free surface of Al–Zn–Mg–Cu 0.5 alloy. **a** before cycling and **b** after cycling [48]

by cycling even at the lower strain amplitude of 0.02. The effect, which is larger at higher strain amplitudes, is the opposite of the previously described dissolution of large coherent precipitates. Manson et al. [47, 48] have found that, in general, metals and alloys characterized by an ultimate-yield strength ratio equal or greater than 1.4 have tendency to harden whereas below 1.2 they soften.

Between the two values materials show a neutral behavior very close to the monotonic one

$$\begin{aligned} &\frac{\sigma_u}{\sigma_y} \geq 1.4 \rightarrow hardening \\ &1.2 \geq \frac{\sigma_u}{\sigma_y} \geq 1.4 \rightarrow neutral \\ &\frac{\sigma_u}{\sigma_y} \leq 1.2 \rightarrow softening\,. \end{aligned} \tag{1.6}$$

A similar check can be done on the monotonic strain hardening exponent n. In general, when $n > 0.20$ the material hardens, whereas for $n < 0.10$ it softens. As far as the hysteresis loop is concerned, Gough [49] observed that in soft materials like annealed copper subjected to reversed bending it could even disappear by cycling. This is shown in Fig. 1.37 where it can be seen that after 260,000 cycles the hysteresis loop is about to disappears, which eventually happens over 400,000 cycles when saturation occurs. Parallel to this, the total stress excursion $\Delta\sigma$ of the cycle that was initially equal to 72 MPa increases by cyclic strain hardening up to 84 MPa when the behavior becomes linear. The opposite is already known to occur in cyclic strain softening where the hysteresis loop continuously expands [50, 51], as shown in Fig. 1.21, and it is not observed any sudden increase of the specimen temperature [52] probably because the plastic work is spent in coherent precipitates and phase dissolution. What is really surprising is that during this first phase of fatigue where the material either softens or hardens there is no sign of fatigue damage at all [50–53], unless we enter the region of low cycle fatigue where, as said, damage is introduced

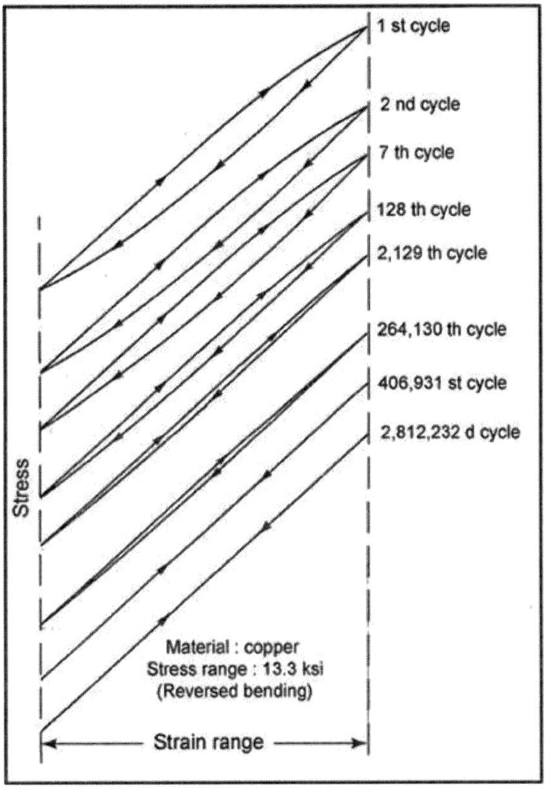

Fig. 1.37 Reduction of hysteresis loops in annealed copper after cycling in reversed bending [41]

in the material so early that phase 1 of cyclic hardening or softening has no time to develop and reach saturation.

As matter of facts, Alden and Backofen [54] found that the micro crack formation phase could be blocked and, therefore, the fatigue failure avoided simply maintaining the specimen under testing in a continuous state of predisposition to strain hardening. This can be done through the continuous re-annealing of the specimen that would remove the hardening introduced in the previous cycles bringing the material back to its initial state and retarding saturation of cyclic hardening that precedes the damage initiation phase. And in fact, they found that re-annealing was effective only if continuously practiced before saturation occurred. It should be recalled that saturation generally takes only a small percent (often less than 1%) of the fatigue life therefore in practice re-annealing may be difficult to perform in steels that have a fatigue life around 1×10^6–2×10^6 cycles. However, it can be done in light metals and alloys whose fatigue life goes beyond 10^8 cycles where even 1.0% means 1,000,000 cycles (see Fig. 1.37 where saturation occurs after about 400,000 cycles). Of course, this practice has only a demonstrative value in that it demonstrate that damage generally follows saturation, but has no practical application.

1.4.2 Cyclic Stress–Strain Curve Determination

We have seen in the previous paragraph, how fatigue cycling may change the mechanical behavior of materials. If this happens, the new stress–strain relationship, σ-ε, will differ from the monotonic curve of the virgin material. This is shown in Fig. 1.38 for high resistance 4340 steel [55]. There are several methods to derive the cyclic stress–strain curve. The most common is that making use of multiple specimens. A series of companion samples are cycled at various constant strain levels until the hysteresis loops stabilizes and saturation occurs. Each stable hysteresis loop at each given strain amplitude is then recorded and individual cycles superimposed to connect their tips, as shown in Fig. 1.30 (open circles and solid line). Figure 1.38 also shows the monotonic stress–strain curve for this material. As it can be seen, the steel softens. The method requires many specimens and is rather time consuming. An alternative method, which is quicker and produces good results, is the incremental step test [56]. Only one single specimen is required. A single specimen is subjected to repeated blocks of gradually increasing and decreasing strains, as shown in Fig. 1.39. Each half block contains some 20 cycles. After few blocks of these fully reversed cycles the material stabilizes. The hysteresis loops from half a stable block are then used to obtain the cyclic stress–strain curve. Normally, the strain is increased up to a maximum of $\pm$ 0.02 and few blocks are sufficient to reach stabilization. Fatigue failure generally occurs well before 20 blocks.

An example of stabilized cyclic stress–strain curve is shown in Fig. 1.40 for Cr–Mo-V steel of 660 MPa yielding strength and 820 MPa ultimate [57]. The method is described in details by Feltner and Mitchell [57]. Figure 1.41 presents some typical cases of cyclic hardening and softening [57, 58]. It is of great importance to remember that, at variance with cyclic hardening, cyclic softening takes place even before the monotonic yield strength of the material is reached. Because of that the material reduces its fatigue resistance.

Strain amplitude calculated to be elastic on the basis of the monotonic curve may be actually plastic, based on the cyclic one, which is the real one. If the effect is not taken into consideration, i.e. if the cyclic stress–strain curve is not found and used instead in calculation, the error may be tragic. This is particularly true for high strength materials. For example, the 4340 steel of Fig. 1.38 at a stress of 1000 MPa reacts elastically in monotonic condition, but under cyclic condition the corresponding strain is more than 2%. What may be thought to be a high-cycle fatigue behavior actually happens to be a low-cycle one with an early and unexpected failure because the material has become much softer. On the contrary, 2024-T4 aluminum alloy or waspaloy A, as indicated by Fig. 1.41, behave much better than expected. In this last alloy a stress of 570 MPa amplitude would cause a deformation over 2% if the monotonic curve were used, while the material would behave still elastic on the base of the cyclic curve. Figure 1.42 presents the stress strain characteristic of Fe510 carbon steel under monotonic and cyclic loads. The steel clearly hardens under cyclic conditions as expected according to Eq. (1.6), since the σ_u/σ_y ratio is about

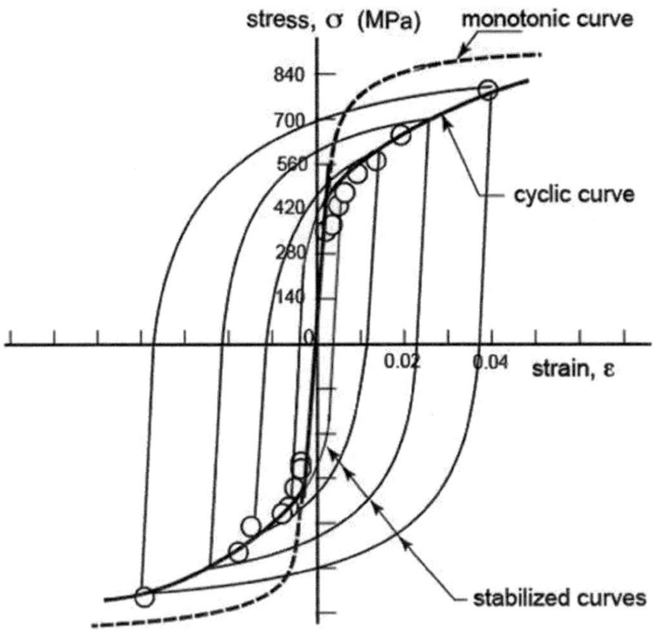

Fig. 1.38 Cyclic stress–strain curve (solid line) from stabilized hysteresis loops [55]. The monotonic stress–strain curve is also shown for comparison (dashed line)

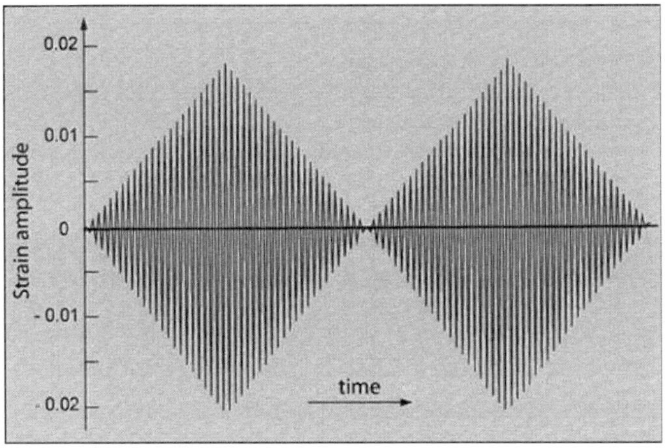

Fig. 1.39 Incremental step test. Blocks of increasing and decreasing strain amplitude in fully reversed cycles

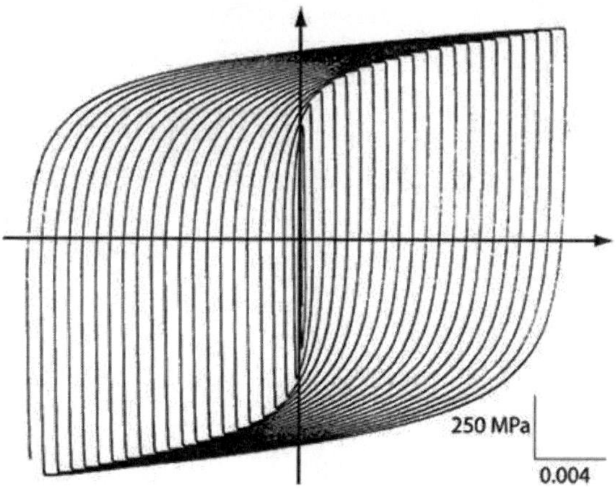

Fig. 1.40 Cyclic stress–strain curve obtained on Cr–Mo-V steel of 660 MPa yielding strength and 820 MPa ultimate with the incremental step test [57]

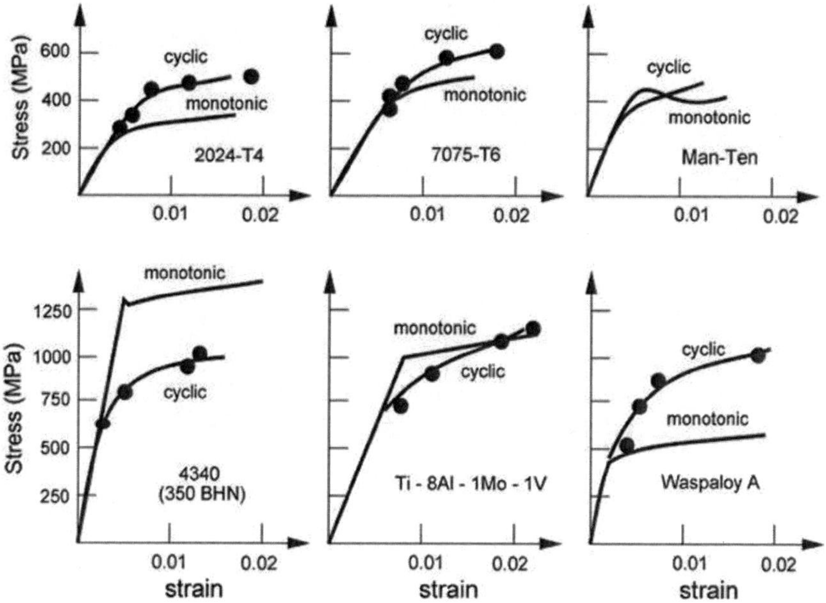

Fig. 1.41 Examples of cyclic hardening or softening in some alloys [57]

1.8. also shown is the true stress–strain characteristic of the steel, see Eq. (10.25). An interesting case is offered by ductile iron that can be thought as a composite material in that graphite microspheres are distributed in an α-iron matrix (ferrite). Ductile iron presents a different behavior whether tested in compression or in traction. The

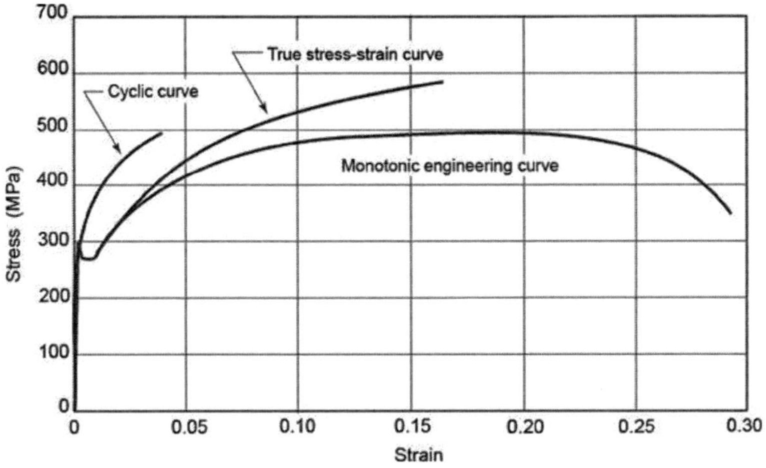

Fig. 1.42 Stress strain characteristic of Fe510 steel under monotonic and cyclic conditions

difference is due to residual stresses built up at the matrix-graphite interface during solidification.

These stresses are traction therefore they are harmful when the metal is put in traction, whereas they are beneficial in compression. However, when cyclic loads are applied, the material hardens and the cyclic characteristic is the same under either cyclic traction or compression, as shown in Fig. 1.43. Austenitic stainless steels are particularly prone to cyclic hardening. In these steels the addition of nickel over 7% in weight keeps the material fully austenitic on cooling at room temperature. Nevertheless, austenite at ambient temperature is metastable and can be transformed, in part, into hard and brittle martensite by any cold-working process. This strain-induced martensite transformation can be activated by cyclic plastic strains.

A particularly impressing example of such an attitude is given in Fig. 1.44 for austenitic stainless steels subjected to fatigue cycles under strain-controlled condition

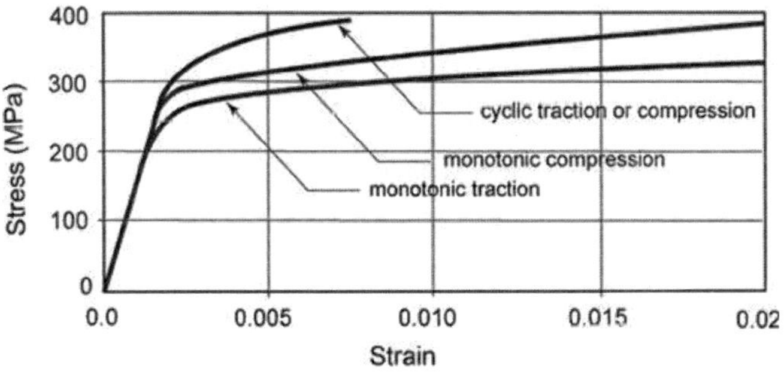

Fig. 1.43 Monotonic and cyclic characteristics of ductile cast iron obtained by Milella

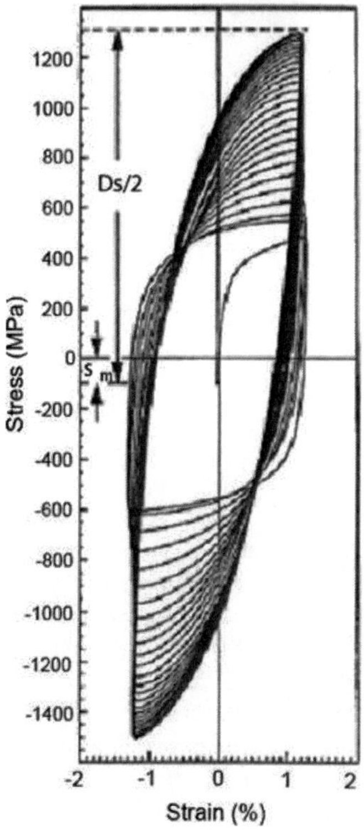

Fig. 1.44 Strain-induced hardening of austenitic stainless steel subjected to 30 cycles of ± 0.0126 total strain at −143 °C [59]

at $\varepsilon_a = \pm 0.0126$ [59]. The cycling has been run at −143 °C to facilitate the austenite–martensite transformation. Note how the metal response increases at each cycle. After 30 cycles it reaches 1300 MPa from an initial 430 MPa. Its strength has triplicated. The increase in strength is due to a parallel increase of volume fraction of martensite. The negative mean stress generated during the cycling is due to the volume expansion of martensitic phase.

References

1. Braithwaite, F.: In: On the Fatigue and Consequent Fracture of Metals, Institution of Civil Engineers, Minutes of Proceedings, vol. 13, pp. 463–474. London (1854)
2. Poncelet, J.V.: Introduction à la Mécanique Industrielle, Physique ou Expérimentale, Zweite Ausgabe. Paris, Imprimerie de Gauthier-Villars (1939)
3. Albert, W.A.J.: Über Treibseile am Harz, Archiv für Mineralogie. Georgnosie. Bergbau und Hüttenkunde **10**, 215–234 (1837)
4. Wöhler, A.: Über die Festigkeits-Versuche mit Eisen und Sthal, Zeitschrift für Bauwesen. vol. 20, pp. 73–106. (1870)
5. Basquin, O.H.: The exponential low of endurance tests. In: Proceedings Annual Meeting, American society for Testing and Materials. vol. 10, pp. 625–630. (1910)

6. Gough, H.J.: The Fatigue of Metals. Scott, Greenwood, London (1924)
7. Malcom, J.: Seconds from Disaster Derailment at Eschede, You Tube, 3 May (2013)
8. Rankine, W.J.M.: On the causes of unexpected breakage of the journals of the railway axles and on the means of preventing such accidents by observing the law of continuity in their construction. Institution of Civil Engineers, Minutes of Proceedings, vol. 2, pp. 105–108. London (1842)
9. Neuber, H.: In: Theory of Notch Stresses: Principle for Exact Stress Calculation. J.W. Edwards, Publishers, Incorporated, Ann Arbor, Michigan (1946)
10. Peterson, R.E.: Stress Concentration Factors. Wiley, New York (1973)
11. Ewing and Humfrey.: In: The Fracture of Metals Under Repeated Alterations of Stress. vol. 221, pp. 241–253. Philosophical Transactions of the Royal Society (1903)
12. Douglas, W.D.: Methods Employed at the Royal Aircraft Establishment for the Experimental Determination of the Ultimate Strength of aeroplane Structures. Advis. Comm. Aero. Rep. Memo. No. 476, June (1918)
13. Paris, P., Erdogan, F.: A critical analysis of crack propagation laws. J. Basic Eng. Trans. of ASME 528–534 (1963)
14. Elber, W.: Fatigue Crack Propagation: Some Effects of Crack Closure on the Mechanism of Fatigue Crack Propagation and Cyclic Tensile Loading. Ph.D. Thesis, University of New South Wales (1968)
15. Smith H.R., Piper, D.E., Downey, F.K.: A study of stress corrosion cracking by wedge force loading, Eng. Fract. Mech. I, 123–128 (1968)
16. Keisler, J., Chopra, O.K., Shack, W.J.: Statistical analysis of fatigue strain-life data for carbon and low-alloy steels. US-NRC, NUREG/CR-6237, Argonne Nat. Lab. (1994)
17. Environmentally Assisted Cracking in Light Water Reactors.: US NRC, NUREG/CR-4667, Vol. 22, Prepared by Chopra, O.K. et al. (ed) Semiannual Report, January 1996-June (1996)
18. Fatigue Design Handbook.: SAE, 2nd edn. pp. 41. (1988)
19. Fuchs, H.O., Stephens, R.I.: In: Metal Fatigue in Engineering. Wiley & Sons (1980)
20. Langraf, R.W.: The resistance of metals to cyclic loading. Achievement of High Fatigue Resist. Alloys, ASTM-STP 467, pp. 27. (1970)
21. Bennet, J.A.: The distinction between initiation and propagation of a fatigue crack. In: International Conference on Fatigue of Metals, London, The Institution of Mechanical Engineers, September (1956)
22. Pardue, T.E., Melcher J.L., Good W.B.: Proceeding of Society of Experimental Stress Analysis. vol. 1, pp. 27. (1950)
23. Yokobori, T.: J. Phys. Soc. Japan **6**, 81 (1951)
24. Polakowski, N.H.: In: Proceedings ASTM. vol. 52, pp. 1086. (1952)
25. Polakowski, N.H., Palchoudhuri, A.: Softening of certain cold worked metals under the action of fatigue loads. In: Proceedings of ASTM, vol. 54 (1954); Lipsitt, H.A., Horne, G.T.: Proceedings of ASTM, vol. 57, pp. 592. (1957)
26. Dugdale, D.S.: J. Mech. Phys. Solids **7**, 135 (1959)
27. Feltner, C.E., Laird, C.: Cyclic stress-strain I of FCC metals and alloys. Acta Metall. **15**, 1621–1653 (1967)
28. Burbach, J.: Zum Zyklischen Verformungsverhalten einiger Technischer Werkstoffe. Technischen Mittelungen Krupp Forschungsberichte, Bd. 28, H. 2, pp. 55–102. (1970)
29. Conway, J.B., Stentz, R.H.: Low-cycle and high-cycle fatigue characteristic of forged and cast 304 SS steel at room temperature and 427 °C, ASME MPC, Winner Annual Meeting, vol. 25. pp. 59–145 (1984)
30. Morrow, J.: Cyclic plastic strain energy and fatigue of metals. In: Internal Friction, Damping and Cyclic Plasticity, ASTM-STP 378, pp. 45. (1965)
31. Kemsley, D.S.: J. Institute of Metals **87**, 10–15 (1959)
32. Laird, C.: The influence of metallurgical structure on the mechanisms of fatigue crack propagation. In: 69th ASTM Annual Meeting, Atlantic City, N.J., Paper No. 32 (1966)
33. Hull, D., Bacon, D.J.: In: Introduction to Dislocations. 5th edn. Elsevier (2011)
34. Reed-Hill, R.E.: In: Physical Metallurgy Principles. D. Van Nostrand Co. (1973)

35. Callister, W.D.: In: Material Science and Engineering. Wiley and Sons
36. Guy, A.G.: In: Introduction to Materials Science. McGraw-Hill Book Co. (1971)
37. Guinier, A.: Nature **142**, 13 (1938)
38. Preston, G.D.: Proc. Roy. Soc. **A167**, 526 (1938)
39. Askeland, D.R.: In: The Science and Engineering of Materials; 3th SI edn. Chapman and Hall, London (1996)
40. Bjørge, R., Nakashima, P.N.H., Marioara, C.D., Andersen, S.J., Muddle, B.C., Etheridge, J., Holmestad, R.: Precipitates in an Al-Mn-Ge alloy studied by aberration-corrected scanning transmission electron microscopy. pp. 73. Acta Materialia Inc., Elsevier Ltd (2011)
41. Clark, J.B., McEvily, A.J.: Interaction of dislocation structures in cyclically strained aluminium alloys. Acta Metall. **12**, 1359 (1964)
42. Calabrese, C., Laird, C.: Mater. Sci. Eng. **13**, 141–150 (1974)
43. Calabrese, C., Laird, C.: Mater. Sci. Eng. **13**, 149–170 (1974)
44. Duva, J.M., Daeubler, M.A., Starke, E.A., Luetjering, G.: Acta Metallurgica **36**(3), 585 (1988)
45. Baxter, W.J., McKinney, T.R.: Metallurgical Trans. **19A**, 83 (1988)
46. Metals Handbook, Properties and Selection, ASM Vol. 1, 8th edn. pp. 223. (1975)
47. Smith, R.W., Hirschberg, M.H., Manson, S.S.: NASA TN D-1574 (1963)
48. Manson, S.S., Hirschberg, M.H.: In: Fatigue: an Interdisciplinary Approach. pp. 133. Syracuse University Press, Syracuse, N.Y., (1964)
49. Gough, H.J.: The Fatigue of Metals. Ernest Benn Ltd., London (1926)
50. Forrest, P.G.: In: In: International Conference on Fatigue, Institution of Mechanical Engineers, pp. 171. (1956)
51. Roberts, E., Honeycombe R.W.K.: J. Inst. Metals **91**, 134 (1962–63)
52. Haigh, B.P.: Trans. Farady Soc. **24**, 125 (1928)
53. Kocanda, S.: In: Fatigue Failure of Metals. Sijthoff & Noordhoff Int. Pubs (1978)
54. Alden, T.H., Backofen, W.A.: Acta Metall. **9**, 352 (1961)
55. Morrow, J.: Cyclic plastic strain energy and fatigue of metals. American Society for Testing and Materials, STP-378, pp. 45–87. (1965)
56. Landgraf, R.W., Morrow, J.D., Endo, T.: Determination of the cyclic stress-strain curve. J. Mater. JMLSA **4**(1), 176–188 (1969)
57. Feltner, C.E., Mitchell, M.R.: BASIC research on the cyclic deformation and fracture behavior of materials. In: Manual on Low-cycle Fatigue Testing, American Society for Testing and Materials, STP 465, pp. 27–66. (1969)
58. Landgraf, R.W.: Cyclic deformation and fracture of hardened steels. In: International Conference on Mechanical Behavior of Materials, Kyoto, Japan (1972)
59. Maier, H.J., Donth, B., Bayerlein, M., Mughrabi, H., Meier, B., Kesten, M., Metallkde, Z.: Low temperature fatigue induced martensitic transformation on the low cycle fatigue behaviour of stainless steel. **84**, 820–843 (1972)

Chapter 2
Damage Nucleation

2.1 High-Cycle Fatigue

We have seen, in Chap. 1 that the fatigue process develops through four fundamental steps that we called phases. The first phase is that of cyclic hardening or softening. Once phase 1 is concluded, at least in Region II of high-cycle fatigue, see Fig. 1.10, phase 2 starts with localized plastic slips (see Sect. 1.3 and Fig. 1.19). Again, it is necessary to distinguish between low-cycle and high-cycle fatigue because the damage initiation process may be completely different. Let's start with the latter. In high-cycle fatigue, where the applied stress amplitudeσ_a is lower than the material yield strength σ_y but higher than the fatigue limit σ_f, a series of highly localized processes start to take place, activated by shear stresses. It is important to realize that these processes do not involve the entire volume of material but just relatively few grains on the surface. This happens not only under bending conditions, where the maximum stress is effectively reached on the uppermost and lowermost fibers of the beam, precisely those on the surface, but also under cyclic traction, where the stress is acting uniformly over the entire cross section. The reason for that shall be found in two concomitant factors: the working conditions of the material and the polycrystalline nature of metals with a random orientation of crystal grains.

As to the former factor, it must be recalled that plastic flow occurs easier under plain stress conditions. Triaxiality, generated by plain strain conditions, can raise the yield strength by a factor of 2.5 or 3, depending on the value of the Poisson coefficient, making plastic slip harder. It is clear that the surface of a body, for example that of a specimen, is always working under plain stress conditions since there are no external forces applied. The more it goes into the thickness the more any single grain is subjected to the action of surrounding crystals. If any internal single grain wants to flow, the other grains would hinder that, generating triaxiality. Therefore, surface crystals are more prone to develop plastic slip.

Now it comes into action the second factor: the random orientation of crystals. Among all surface grains, millions or billions, some are more favorably oriented

P. P. Milella, *Fatigue and Corrosion in Metals*,
https://doi.org/10.1007/978-3-031-51350-3_2

relative to the applied external load. To better understand this it is worth reminding that plastic slip are not possible along any plane or direction inside a crystal grain but only on those few that have the highest atomic density. These close-packed plains and directions where metals can slip are called crystallographic planes and crystallographic directions, respectively. Among all surface crystals there are some where the crystallographic planes or slip planes and the slip directions are oriented as to maximize the effect of the so called *resolved shearing stress* that activates plastic flow. To this end, let's consider a specimen made of a single crystal having a cross-sectional area A subjected to an axial tensile force F, as shown in Fig. 2.1. Any active or crystallographic slip plane in the crystal is defined by its normal $\vec{n}$ that intersects the specimen axis of symmetry at an angle θ. The active slip directions on the slip plane are defined by the angle φ with the specimen axis of symmetry. With these definitions the area A_{sp} of the slip plane can be expressed as

$$A_{sp} = \frac{A}{\cos\theta} \tag{2.1}$$

The component F_{sd} of the tensile force F in the slip direction is

$$F_{sd}=F\cdot\cos\varphi \tag{2.2}$$

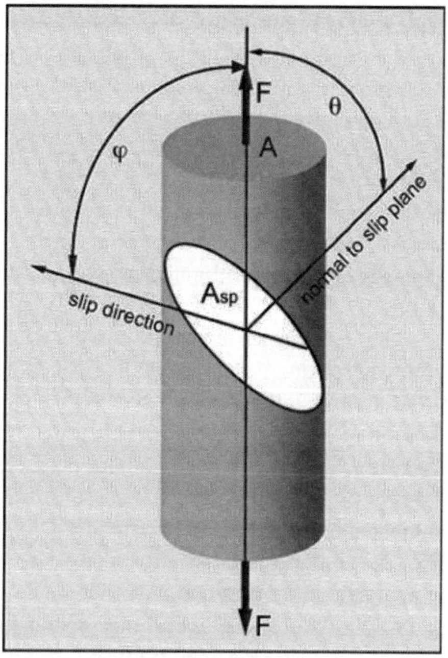

Fig. 2.1 Schematic of a slip plane and slip direction in a single crystal specimen subjected to an axial tensile force F

Therefore, the shear stress τ_r resolved on the slip plane and in the slip direction can be expressed as

$$\tau_r = \frac{F_{sd}}{A_{sp}} = \frac{F}{A} \cos \varphi \cos \theta \tag{2.3}$$

The term cos φ·cos θ is an orientation factor identified as Schmid's factor that ranges from zero to one. The rule expressed by (2.3) is known as Schmid's law. Schmid's law has been verified for a large number of metallic single crystals. The magnitude of the resolved shearing stress required to initiate slip in a crystal is known as *resolved critical shearing stress* $\tau_{r,cr}$ and represents the plastic flow stress of the crystal. According to Eq. (2.3) it can be said that plastic flow occurs preferably on planes and along directions where Schmid's factor is maximized. Therefore, if both the slip plane and the slip direction are normal to the external load F ($\varphi = \theta = 90°$) the resolved shearing stress τ_r is zero. If this is the case, plastic flow cannot take place whatever the value of the applied force may be.

On the contrary, when Schmid's factor reaches its maximum value, which is equal to one, slip in the crystal will take place under the minimum value of the external force F. This happens when slip planes and slip directions are at 45° to the external load F and the shear stress equals the resolved critical shearing stress. This actually means that the yield strength of a single crystal is not constant but depends on its orientation with respect to the direction of application of the external load. This is shown in Fig. 2.2 for a single crystals of zinc and aluminum, respectively [1, 2]. In practice, with very few exceptions like jet turbine blades, materials are polycrystalline in nature with grains randomly oriented. Figure 2.3 is a schematic of what happens on the surface grains of a polycrystalline material where grains are randomly oriented. As it can be noted, the random distribution results in an equally random distribution of the resolved shearing stress. It is just in those few grains, such as grain A of Fig. 2.3, in which the Schmid's factor is equal or very close to one that the resolved shearing stress may reach the critical value $\tau_{r,cr}$ and slip will occur, whereas the rest of material will behave elastically. In grain B, for example, the resolved shearing stress reaches a minimum value $\tau_{r,\min}$. Note that in Fig. 2.3 we have defined a width, called volume process, no larger than 500 μm, where slips can effectively occur. Beyond that limiting depth the plain stress conditions that favor flow to occur starts to be lost and plastic slip becomes more and more difficult. It doesn't matter whether these surface grains where slip occurs are just one or few or thousands because fatigue will find them and it will be there that damage nucleates, unless the external surface is already affected by defects (see Sect. 4.2). Figure 2.4 is a synoptic picture of slip systems in body-centered cubic metals (BCC), face-centered cubic (FCC) and Hexagonal (HE) crystal structure materials.

A slip system is the complex of possible slip planes and directions. Slip planes and directions are indicated by Miller's index in parenthesis {} and <>, respectively.

Apparently, the BCC crystal structure has many more slip systems (48) than FCC (12). However, only 12 of them are really active, the remaining requiring higher

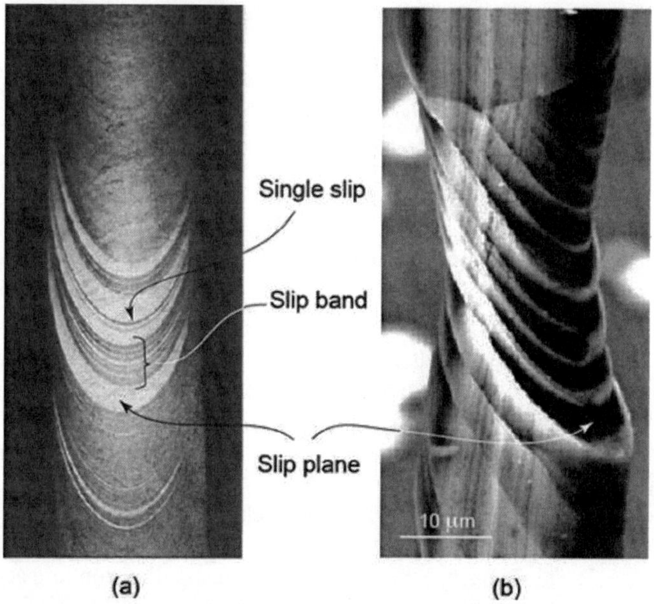

Fig. 2.2 Slips in single crystals of **a** zinc specimen along basal plane oriented at 45° to the traction axis [1] and **b** aluminum specimen showing rotation of slip planes due to constraint [2]

Fig. 2.3 Random distribution of surface grains in polycrystalline materials accompanied by an equally random distribution of resolved shearing stress

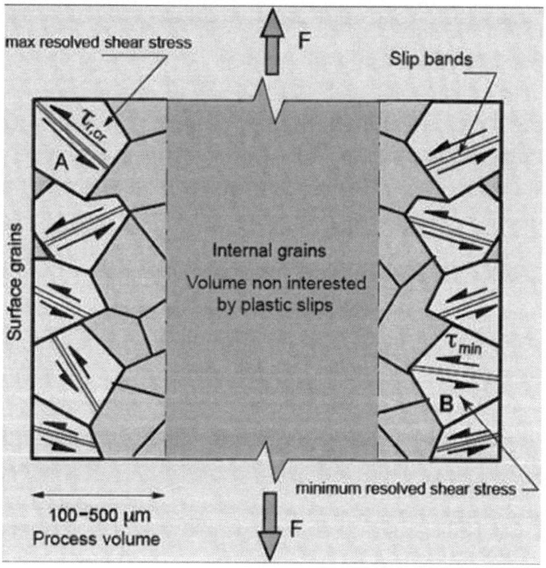

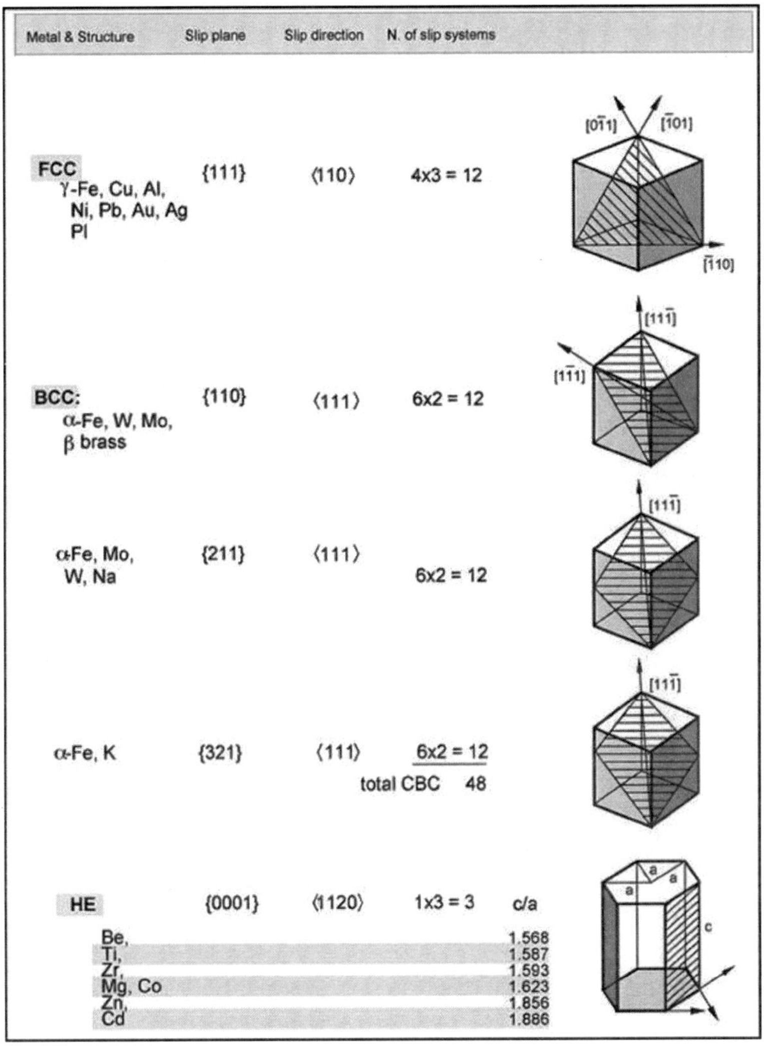

Fig. 2.4 Slip systems in metals. Shaded areas represent typical crystallographic planes

energy to be activated. Also note how the HE crystal structure is less rich in slip planes and directions than FCC and BCC structures.

Indeed, it has been well documented and verified that in high-cycle fatigue the first plastic slips occur wright on the surface and only in some crystals.

The most convincing evidence of that is that if a fatigue specimen is periodically removed from the load cell and its surface polished as to remove just a thin surface layer no more than 0.3–0.5 mm deep, its fatigue life apparently can be extended indefinitely. This is also a proof that the process volume, where fatigue first strikes the material, has a depth no greater than 0.5 mm from the surface.

This is schematically shown in Fig. 2.5 and represents a fundamental point to understand: all what is outside the process volume does not contribute to the fatigue life of the specimen or workpiece. It is only the outer skin that resists fatigue. It is there and only there that fatigue damage nucleates. The bulk material of the internal core plays no role. Any inclusion or defect in the process volume may sharply reduce the fatigue life. Any inclusion or defect in the internal core has no effect on the fatigue life of the specimen or workpiece. Initially, surface slips do not interest the entire length of the grain, from edge to edge, but only a part of it and are, therefore, called segmentary slips. Figure 2.6 is an example of segmentary slips observed in Armco iron 0.02% C having 220 MPa yield strength [3].

The discovery of slips in surface crystals dates back to the beginning of last century. Using an optical microscope Ewing and Humphrey [4], as early as 1903, observed for the first time slip lines on some surface crystals of fatigue specimens. Their findings

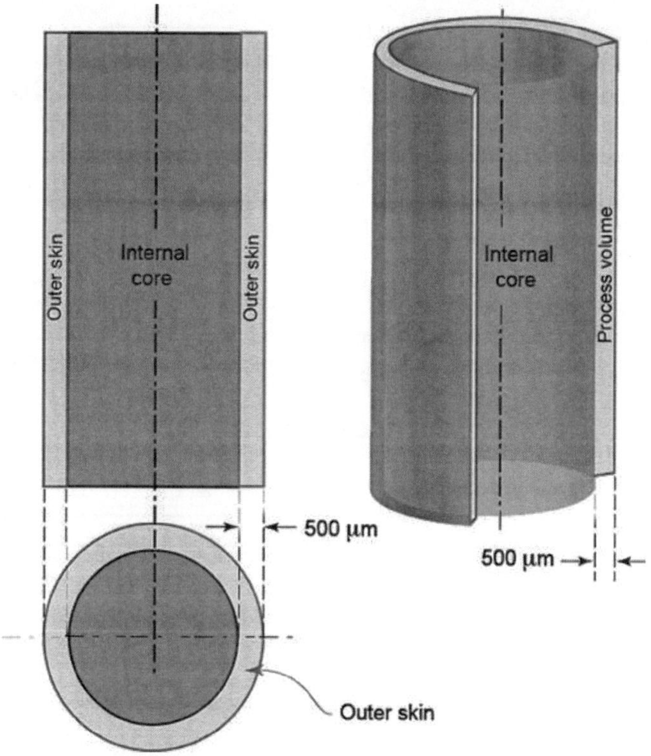

Fig. 2.5 The outer *skin* of a specimen, no more than 500 μm depth, represents the process volume where damage nucleates. The bulk material of the internal core plays no role in the fatigue initiation process

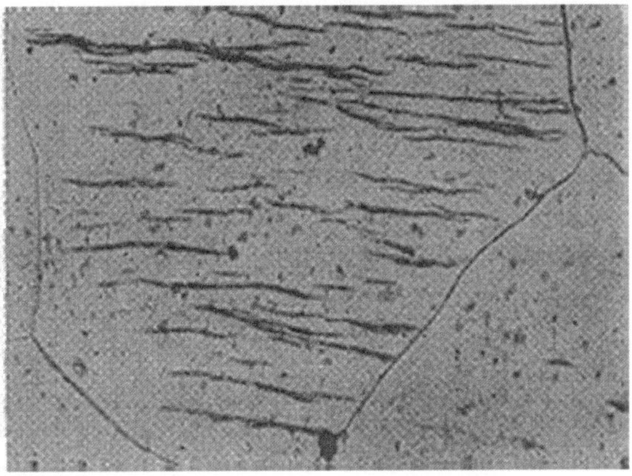

Fig. 2.6 Segmentary slip lines in an Armco iron 0.02% C [3]

are worth mentioning because they were exemplary. Some of the pictures they took, shown in Fig. 2.7, may not be so clear and well defined as a today optical image of the same subject, but they are the first ever and deserve to be reported.

The slip lines appeared after some 2000 cycles just after phase 1 of fatigue was concluded, see Sect. 1.3 and Fig. 1.19 and are of segmentary type After more reversals of stress, additional slip lines appear, still segmentary. Effectively, these plastic slips grow with the repetition of stress, but just in those crystals in which they initially appeared, where the Schmid's factor was the highest.

Slip lines are indicative of damage in that they contain the first embryo of damage. If the specimen were periodically polished, as to remove almost completely these superficial slip lines, and then cyclically reloaded, the same slip lines will reappear precisely on the same grains and direction as before. In the first phase of their appearance, these slips grow and arrest at the grain borders. This can be seen in Fig. 2.8 for mild steel [5]. Again, it shall be noted how slips occur only on some crystals where they accumulate. The surrounding grains are not affected at all by the slip process. This is a characteristic peculiar to high cycles fatigue process. Under monotonic traction, in fact, once plastic flow occurs, it takes place and develops through the entire cross section, in all crystals, as shown in Fig. 2.9a for Armco iron [3]. It affects the entire volume of material under the same stress.

At variance with that, in high-cycle fatigue, slips concentrate in separated bands, as shown in Fig. 2.9b for Armco iron and in Fig. 2.10 [6] for a copper crystal. Another difference relative to the monotonic case stems from the nature of load that in fatigue is alternated and often fully reversed.

The continuous repetition of load and, above all, its reversal gives the fatigue slip a particular appearance since slips that first occurred in one direction later are activated in the opposite direction, but never in a reversible manner. This leaves the surface where they formed in the state schematically shown in Fig. 2.11. The surface appears rough and made of ridges and grooves that act as localized stress risers.

We may argue that in perfectly homogeneous materials under elastic stress amplitude cycling the elastic deformation would completely reverse and thus the material

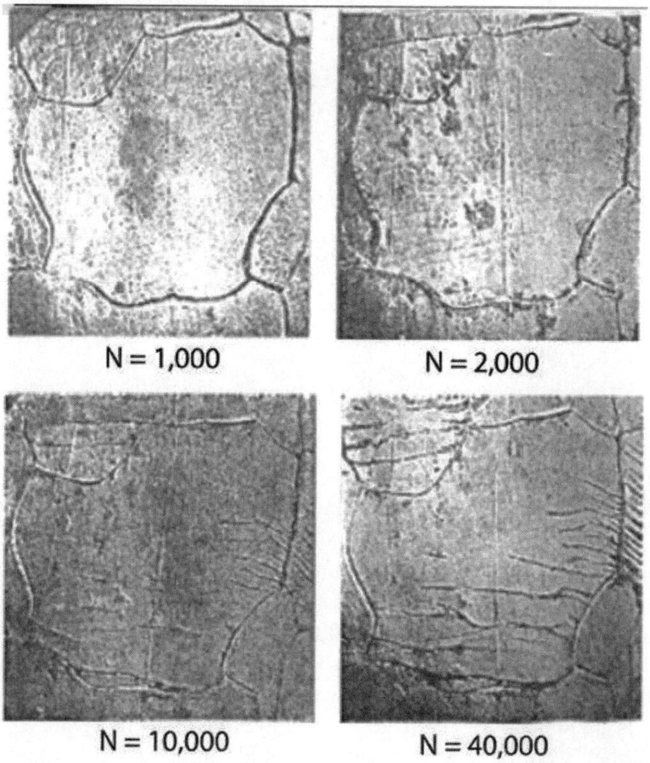

Fig. 2.7 Ewing and Humphrey examination in 1903 of the surface of a fatigue specimen. It revealed that after a few reversals of the stress slip lines appeared on some crystals. After more reversals of stress additional slip lines appeared [4]

would return to its original configuration, no matter how many times the stress is applied. However materials are really far from being perfectly homogeneous. Localized slips on the surface represent one of the most dangerous non-homogeneity we may have. It is this lack of homogeneity responsible for fatigue failure. Therefore, when loaded in the nominal elastic range, even below the fatigue limit, portions of dislocations will move and contribute to very small irreversible changes. Some extremely thin metal laminas may also emerge on the surface, called extrusions (see Sect. 3.2). Dimensions shown in Fig. 2.11 are extremely small, of the order of one tenth or hundredth of micron. Since the surface of these laminas or steps is exposed to the air, it oxidizes so that the process becomes irreversible.

We will see in Sect. 20.2, in fact, that fatigue in vacuum is much more forgiving since emergent faces do not oxidizes. In air, laminas or steps cannot recombine once the load is removed or reversed and the damage cannot heal. It is in this phase of fatigue that damage nucleates and, cycle by cycle, it accumulates and evolves till the formation of a micro crack, as it will be better described in Sect. 2.7.

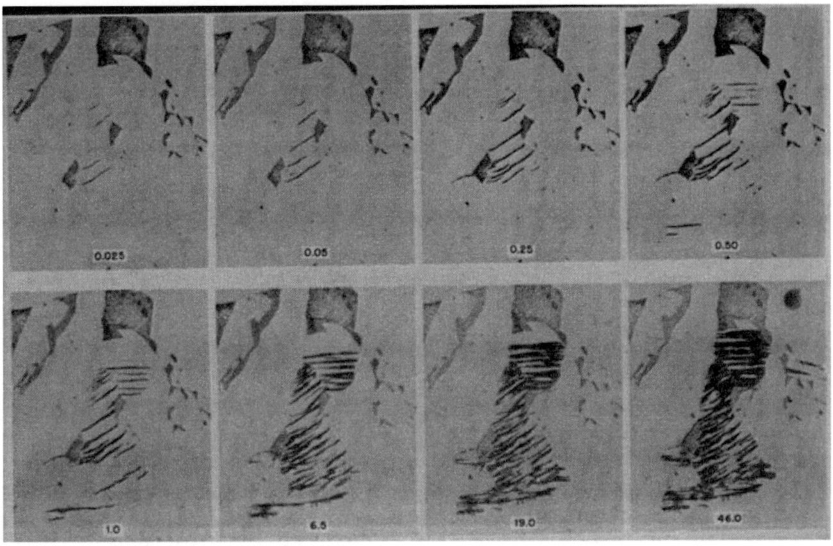

Fig. 2.8 Progressive activation of localized slips with increasing number of fatigue cycles on mild steel [5]. Slips develop only in some crystals and do not overcome the barriers represented by grain boundaries

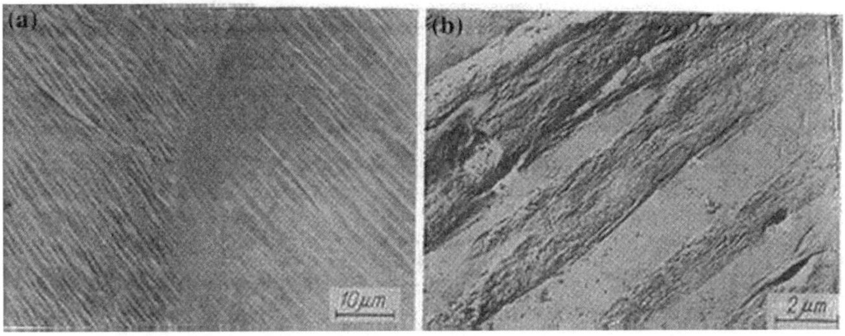

Fig. 2.9 Plastic flow in Armco iron: **a** under monotonic load slips interest the entire cross section, **b** under cyclic load they are grouped in separated bands [3]

Thompson, Wadsworth e Louat [7] testing mirror-polished specimens of polycrystalline high purity copper, found that if they would interrupt tests, from time to time, and remove a surface layer no more than 2 micron depth by electropolishing, most of slip lines and bands would disappear with the exception of some that, instead, would persist. This was the sign of a surface damage that, though sub-microscopic, was deeper than 2 μm. Since the slips and the bands remained on the surface even after a very light polishing, they called them *persistent slip bands*. Hempel [8, 9], testing mild steel with carbon content of the order of 0,09% in push–pull fatigue, had

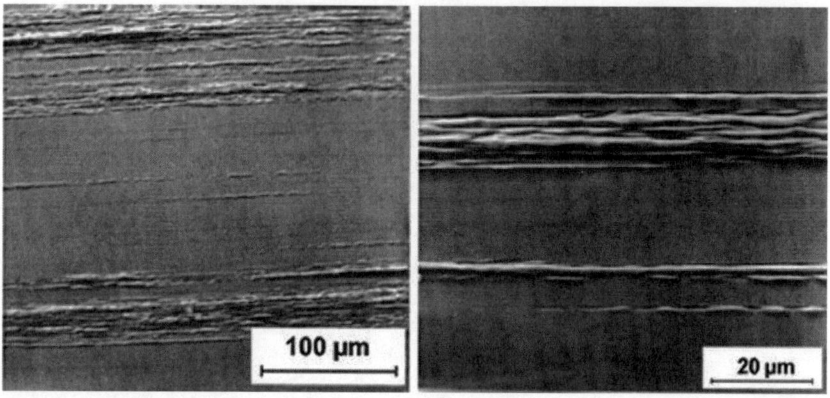

Fig. 2.10 Appearance of fatigue slip bands on the surface of a copper crystal [6]

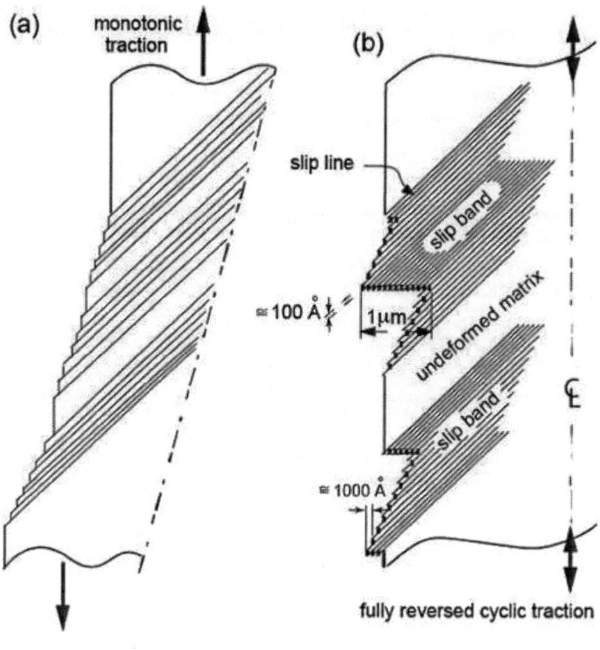

Fig. 2.11 Schematic of surface slips: **a** plastic flow in monotonic traction, **b** slips grouped in separated and superficial bands in fully reversed cyclic traction

analogous results; slip lines and slip bands would disappear by a light electropolishing of the specimen surface regularly during the test, but persistent slip lines and bands that contained dispersed micro cracks did not.

It is now clear that it is just on these slip bands that damage nucleates in the form of micro pores or cavities and progresses through a process of continuous coalescence [10–13]. It must be kept in mind that in these bands deformation is by far higher than in the rest of the surrounding matrix. For instance, it has been observed [14] that in a single crystal of copper the amount of deformation in persistent bands was about $7.5 \cdot 10^{-3}$ while in the surrounding matrix was of the order of $6 \cdot 10^{-5}$, i.e. 100 times lower. In such a process a sort of composite structure is created where the matrix is practically elastic and plastic deformation concentrates on slip bands. It is wright there, in the soft slip bands that cycle by cycle damage nucleates in the form of pores or cavities forming a void sheet. Figure 2.12 is very interesting a micrograph in that it shows a chain of submicroscopic pores found by Forsyth e Stubbington in a slip band of an aluminum alloy at room temperature and low amplitude stress [13]. Some beads already joined to form a tiny crack. However, though we know that fatigue damage doesn't initiate before phase 1 of hardening softening is completely finished, the moment or the number of cycles at which a persistent slip band starts nucleating a micro crack cannot be definitely established from metallographic observation. Such a crack can be better seen in the persistent slip bands of mild steel of Fig. 2.13 [15]. These bands appear corrugated and wavy in metals with high stacking-fault energy (SFE) such as iron, steels, copper, nickel and aluminum and are called wavy-slip lines, as described in Sect. 1.4.1 and Fig. 1.31a. In low stacking-fault energy materials, as Cu-Al and Cu–Zn alloys, Ag or austenitic stainless steel, planar cross slip prevails, evolving into many planes close to each other, as shown in Fig. 1.31c.

It has been noted that the formation of surface slip lines and bands under cyclic loads occurs only when plastic deformation exceeds a threshold value ranging from a minimum of 10^{-5} to 10^{-4}, depending on metals. This actually means that in metals that cyclically soften the persistent slip bands formation occurs under stress amplitudes σ_a well below the monotonic yield strength σ_y. However; persistent slip bands may also form under stress amplitudes below the fatigue limit. Therefore, the fatigue limit represents the stress amplitude below which any damage initiated in persistent slip

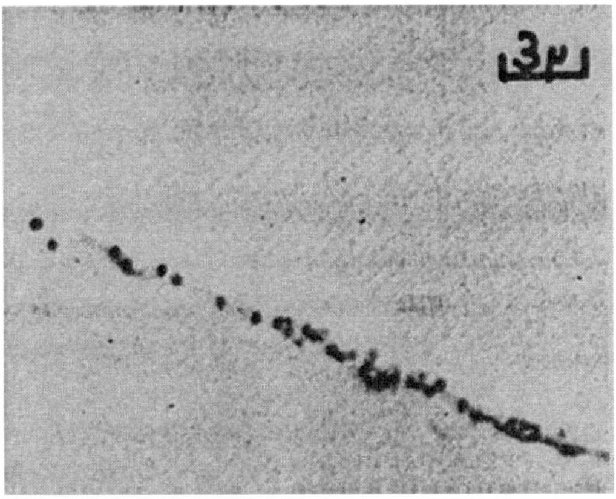

Fig. 2.12 Persistent slip line in aluminum alloy that reveals the nature of damage in the form of pore beads forming a void sheet [13]

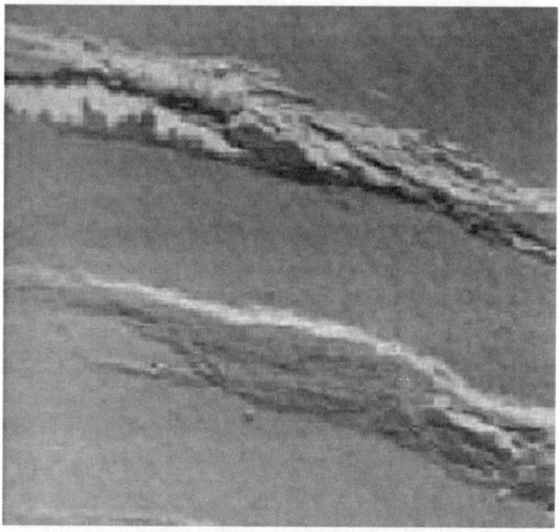

Fig. 2.13 Crack formation in wavy-slip bands of high stacking-fault material [15]

bands does not propagate and remains dormant and, therefore, it is the minimum stress amplitude that can still propagate a micro crack nucleated in a single grain to form a macro crack and produce failure. Alternatively, which is almost equivalent, the fatigue limit represents the highest stress amplitude that cannot propagate damage introduced in slip bands of single crystals.

2.2 Low-Cycle Fatigue

In low-cycle fatigue, strain and stress amplitude reach or exceed the elastic limit of the material. Therefore, the entire section of the specimen undergoes macroscopic plastic deformation and not just some grains on its surface, as in the case of high-cycle fatigue. Therefore, segmentary slip lines and bands now disappear, which had characterized high-cycle fatigue. Now the entire volume of the specimen develops plastic slips, as it can be seen in Fig. 2.14a for polycrystalline Cu30% Zn [16].

Now, the process volume is the entire volume of the specimen and not just the annular section, no thicker than 0.5 mm, schematically shown in Fig. 2.5. Under these conditions fatigue cracks develop along grain boundaries and not in grains, as shown in Figs. 2.14b or 2.15 for polycrystalline Cu30% Zn after 16,000 cycles at a plastic strain equal to $2.2\cdot10^{-3}$ [16]. Grain boundaries, in facts, are much harder than grains and, therefore, less prone to deform. They cannot accommodate the plastic deformation of crystals since they are rigid and brittle and the repeated request of deformation exerted by grains, cycle by cycle, results in micro cracks formation. Figure 2.16 shows schematically how micro cracks generated along grain boundaries grow to become a continuous intergranular macro crack. Once it becomes long

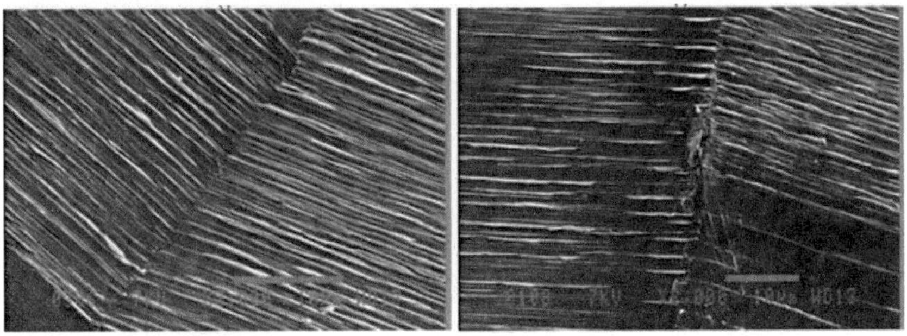

(a) (b)

Fig. 2.14 **a** slip lines in crystals of polycrystalline Cu30% Zn after 8000 cycles at a plastic strain equal to $3.4{\cdot}10^{-3}$ and **b** crack initiation along a grain boundary [16]

enough, about two or three grains or at least 300 μm, it starts to be directly driven by normal stresses and may growth in a plane normal to the applied load and continue to growth in an intergranular fashion if the stress amplitude is high enough.

Figure 2.17 schematizes this microstructural difference between low-cycle fatigue and high-cycle fatigue in that the latter is always transgranular while the former is intergranular and may switch into transgranular. Smith [11], working with aluminum at high temperature, found that in low-cycle fatigue persistent grain boundaries developed which were equivalent to persistent slip lines in high-cycle fatigue.

These boundaries were interspersed with micro pores that coalesced into a micro crack exactly as in high-cycle fatigue did slip lines and bands. Figure 2.18 is an

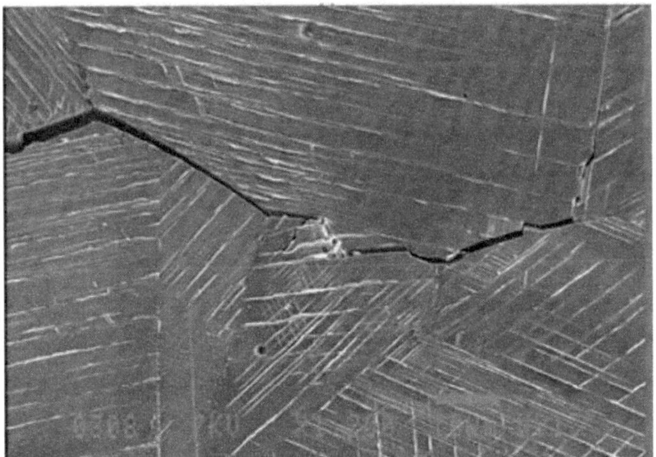

Fig. 2.15 Fatigue crack running along grain boundaries in polycrystalline Cu30% Zn after 16,000 cycles at a plastic strain equal to $2.2{\cdot}10^{-3}$ [16]

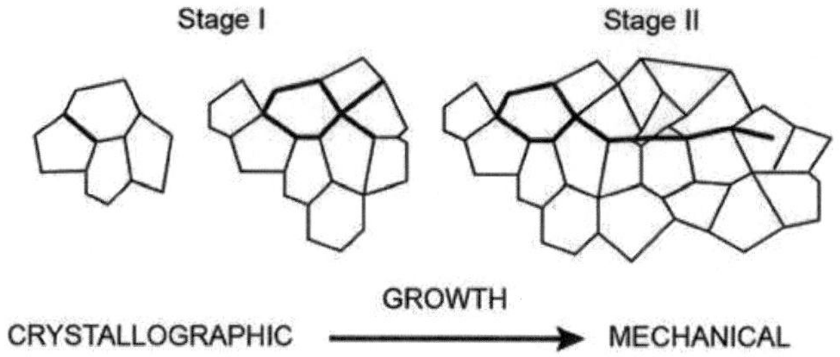

Fig. 2.16 First appearance and growth of an intergranular crack in low-cycle fatigue

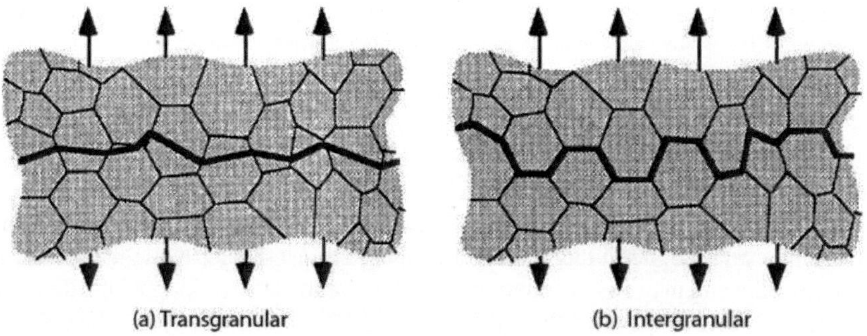

Fig. 2.17 Different microscopic appearance of a fatigue crack: **a** transgranular in high cycle fatigue, **b** intergranular in high strain low cycle fatigue

example of a persistent grain boundary developed in aluminum [11]. After cycling at high stress amplitude that developed heavily slips in crystals, Fig. 2.18a, the specimen was electropolished that left just those grain boundaries affected by deep damage produced by pores and tiny cracks, Fig. 2.18b. Electropolishing has canceled any trace of plastic flow in grains. The specimen was then recycled at high stress amplitude and a new system of slip lined developed, Fig. 2.18c, after which it was electropolished again canceling any trace of slip except the persistent grain boundary, Fig. 2.18d.

We have said that in low-cycle fatigue it is the entire volume of the specimen strained over the yield strength that represents the process volume and not just the outer skin as in high-cycle fatigue (see Fig. 2.5). This actually means that all defects like inclusions present in the entire volume of material, also those internal, may have an effect on fatigue strength. However, the easiness by which surface grains can flow with respect to internal grains makes the surface always more prone to fatigue damage, in particular when the strain amplitude is close to the yield strength. Figure 2.19a [29] shows an interesting example of an internal nonmetallic inclusion in high-strength steel that failed by low cycle fatigue at 292,000 cycles under a stress amplitude $\sigma_a = 1275$ MPa and a mean stress $\sigma_m = -784$ MPa.

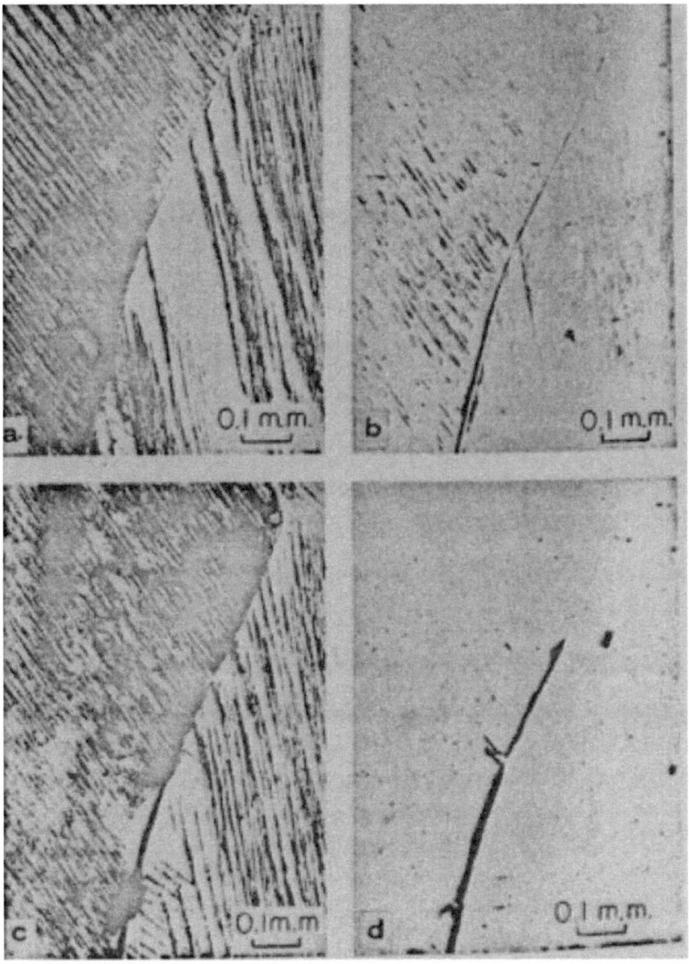

Fig. 2.18 Appearance of micro cracks along a persistent grain boundary in aluminum. **a** grains have heavily undergone plastic slips; **b** upon electropolishing the slip lines almost disappear, but grain border remains visible because affected by micro cracks; **c** and **d** after repeating cycling and electropolishing

It is worth noting that the inclusion, Fig. 2.19b, is more than 1 mm deep, i.e., it is not a surface or subsurface inclusion, yet it initiated a fatigue process that caused failure because of its extremely large size (> 150 micron). The signs of brittle fracture all around the inclusion may induce to believe that the initiation mechanism was not fatigue but hydrogen embrittlement. Than it grew to become a circular ODA (the dark area, see Sect. 9.3) that reached the surface causing the typical fish eye failure.

It is also worth reminding that, at variance with high-cycle fatigue, in the low-cycle fatigue region, damage may initiate well before the hardening or softening

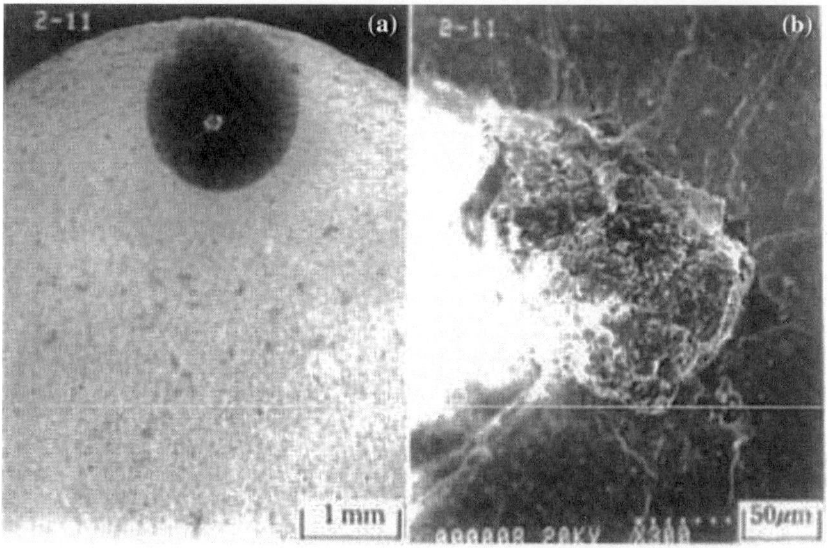

Fig. 2.19 **a** fish eye formation in low-cycle fatigue failure in high-strength steel and **b** close-up of the internal nonmetallic inclusion [29]

phase is ended. Saturation may not occur, as said in Sect. 1.3, since fatigue failure comes first.

2.3 Microscopically Small Cracks (MSC)

Fatigue damage that precedes macro crack formation originates in persistent slip bands well before the final failure occurs. It has been said in in Sect. 2.2 and shown in Fig. 2.5 that in high cycle fatigue, damage nucleates on the surface of the material as soon as the hardening/softening phase terminates and, in fact, it has been also said that if the cyclic hardening were periodically removed by annealing the specimen, the life of the specimen would increase enormously to almost become unlimited. Saturation takes a very small fraction of the entire life so that we shall see the first signs of material damage quite early. Figure 2.20 [17] shows how in a mild steel 0.09% C the first persistent slip lines that contain damage develop on the surface of the specimen at a small fraction of the expected fatigue life. The broken line of Fig. 2.20 represents the lower bound of damage nucleation and is, therefore, called *damage line* (see also Fig. 1.19). Apparently, below the damage line the material can be cycled indefinitely without producing any damage; any combination σ_a–N that falls below the damage line has no effect on the material.

For example, under a stress amplitude of about 185 MPa, which is close but a little bit higher than the fatigue limit of the material, the fatigue life is equal to about $6{\cdot}10^6$

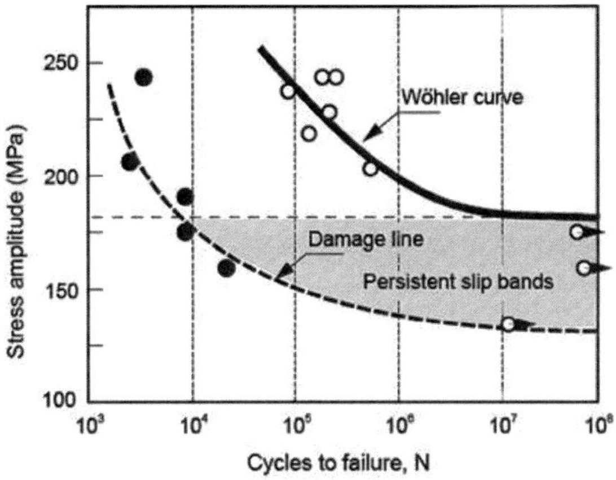

Fig. 2.20 Appearance of persistent slip bands (broken line) in a specimen of mild steel 0.09% C tested in reciprocating bending. Wöhler curve is also shown (full line) [17]

cycles (Wöhler curve). Under the same stress amplitude the first slip lines, where damage nucleates, appear already at about 7000 cycles (damage line), i.e. after a mere 0.1% of the entire life, but at 5000 cycles there is no evidence of slip lines at all. Interesting enough, plastic slips develop even below the fatigue limit σ_f of the metal, as already said in the previous section, in the region bounded by the fatigue limit and the damage line (gray area in Fig. 2.20). Their appearance below the fatigue limit σ_f can be considered as a general rule, something that pertains to all metals and alloys. Slip lines, initiated in some grain as segmentary, may grow the entire length of grain but they don't break the barrier to become a macrocrack. Slip lines are formed by moving dislocations. A long slip line, from border to border, means a long train of dislocations. Upon increasing the load, these moving dislocations pileup against the grain boundary that acts as a barrier. The pressure exerted by the leading dislocation on the barrier depends on the number of dislocations that pileup. This actually means that a long slip line formed in a large grain can generate a stress much higher and break the barrier much easier than short slip lines do in small grains. Large grains, then, offer a lower fatigue resistance.

Once the grain boundary has been broken, damage can proceed to the next grain. It is so that a micro-crack forms that we shall call microscopically small crack (MSC) since it has the morphology of a physical crack that breaks open or is about to break a grain but still represents a damage that can be seen only through the microscope. This process eventually gives rise to a macrocrack from an initial coalescence of pores. Damage nucleation and propagation is not really a new issue in fatigue, even though its study has been given a particular attention and an exceptional acceleration proportional to its importance only recently. Already in 1933, in fact, H. J. French introduced a method to derive what is called after him *French's curve* [18]. French's curve can be considered the stress dependent damage line that marks the points of initiation of MSC. The curve is obtained cycling a specimen at a given stress amplitude σ_a larger than the fatigue limit, σ_f, of the material for a given number of cycles, afterward the stress amplitude is reduced to σ_f and the cycling continued for at least 10^7 cycles. If the specimen fails it means that the preloading at the stress amplitude $\sigma_a > \sigma_f$ has introduced a micro-crack long enough to grew to failure during the following cycling at σ_f. At variance, if the preloading does not generate any such a micro-crack then the subsequent cycling at the fatigue level would not cause fatigue failure.

The procedure is repeated for different initial stress amplitudes of preload cycling and the French's curve drawn to separate couples of failure-non failure events, as shown in Fig. 2.21 for mild steel [18]. The non-failure events are marked with a downward arrow while those that caused failure have an upward arrow. The French's curve can be considered a curve obtained at constant crack length which is precisely the length that can be opened to failure by a stress equal to the fatigue limit. It is rather difficult to pick up the moment of coalescence of pores when they turn into MSC and that of passage from MSC to macro-crack without a precise definition of pore, MSC and macro-crack.

A compromise can be found between damage nucleation and micro crack formation in that we shall call MSC any crack whose length has reached at least 3 μm, but remains below about 300 μm. Below 3 μm damage will be called pore or micro cavity or micro void or, simply, submicroscopic defect. These definitions allow us to distinguish between damage initiation and MSC in persistent slip bands (PSB). Beyond 300 μm the damage will be called macrocrack. Based on such definitions, an interesting example of how damage evolves in Armco iron is shown in Fig. 2.22 [3]. The diagram is the end-result of a long lasting experimental program. A threshold stress, σ_{th}, of about 110 MPa can be marked below which no slip at all was found

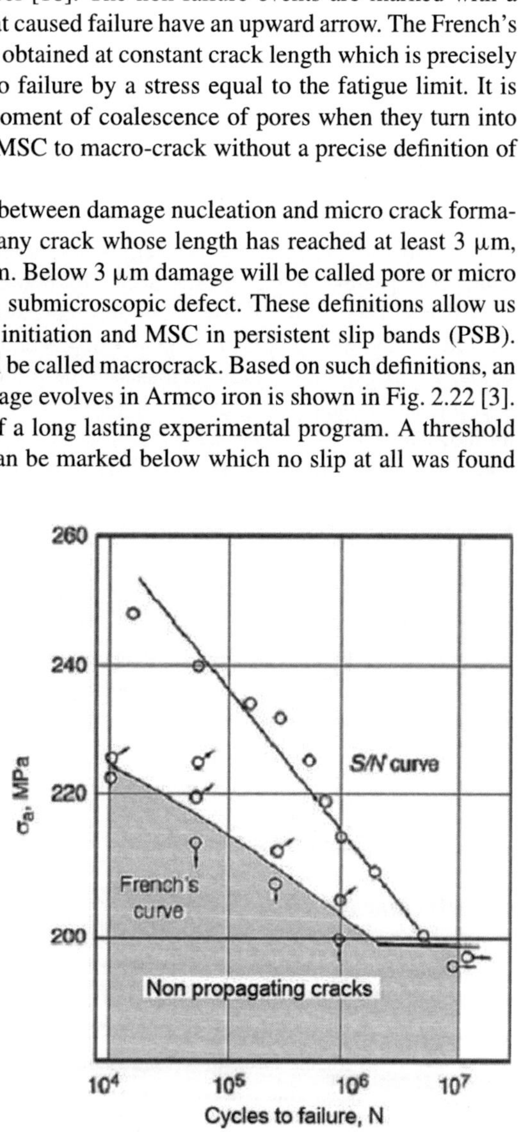

Fig. 2.21 *S–N* diagram and French's curve for mild steel [30]

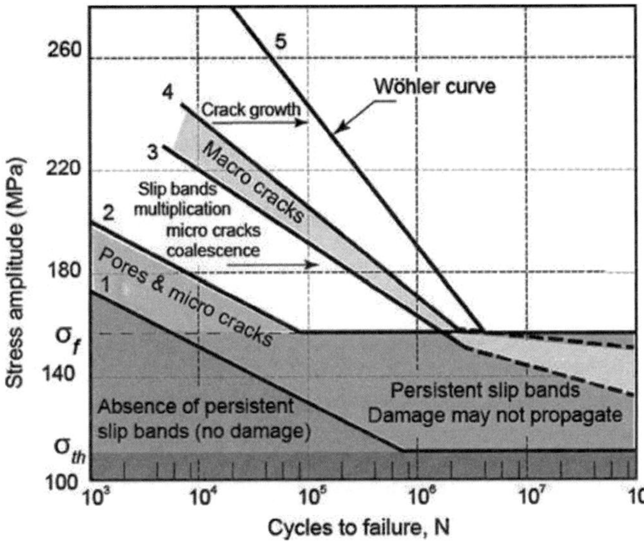

Fig. 2.22 Fatigue diagram for Armco iron evidencing zones of different damage evolution, from nucleation and micro crack formation in persistent slip bands (adapted from [3])

after 10^8 reversals. Between this limiting value and line 1, slip lines can be barely seen in just one or two grains, i.e., almost no damage occurs. It is above line 1 that PSB are generated, therefore line 1 is the submicroscopic damage line. Damage stats to occur in the form of pores no greater than 3 μm. Here we have to distinguish between the area above the fatigue limit $\sigma_f = 152$ MPa and that below it.

In the area aboveσ_f, between line 1 and line 2 sub-microscopic damages start to coalesce into tiny micro cracks that we called microstructurally small cracks. They are transgranular in nature and temporally blocked by grain boundaries as the slip bands in which they are formed. Therefore, line 2 can be considered equivalent to the French's line. In the area below the fatigue limit σ_f these micro-cracks or MSC do not grow to become a macro-crack and remain dormant, as long as the number of applied cycles N become very high ($N > > 10^7 - 10^8$ depending on the grain size; light grey area between lines 3–4 in the darker grey area of Fig. 2.22). From line 2 we assist to the multiplication of slip lines and growth of micro-cracks until at least one of them becomes a macrocrack. Macrocracks happen to develop between lines 3 and 4.

During all these stages it is the shear stress and, in particular, the resolved shearing stress that acts. Macroscopic cracks, that have a size equal or greater than 300 μm, are opened by normal stress and grow at each cycle along a plane normal to external loads till failure occurs at line 5 which is the Wöhler curve. Accordingly, linc 4 will be called initiation curve since it provides the number of cycles N_i necessary to initiate a macrocrack. It is precisely at this stage that fatigue testing should be terminated, as already said in Sect. 1.3, and the corresponding Wöhler curve drawn. What happens after initiation depends on the size, geometry and bulk material properties of the component. At this stage, it is the bulk material that determines the rate at which the macro crack will grow. Before, it was the outer skin with its local properties to be responsible for the MSC growth. A failure that in a small specimen usually happens with relatively few cycles after macrocrack initiation, in real components may actually take hundreds of thousands or even millions of cycles, depending on the size and geometry of the component. Figure 2.23 schematizes this sequence of events that leads to the macrocrack formation from an early slip in a single surface crystal that produces an extrusion or, anyhow, a groove or a ridge, Fig. 2.23a. It follows the MSC formation by pores coalescence, Fig. 2.23b that, once broken through the grain boundary, becomes a macrocrack whose length a exceeds 300 μm, Fig. 2.23c.

Also indicated in the figure are the pore/crack growth rates pertaining to the various stages of the growth process. Pores are likely to grow at an initial rate varying from 0.1 Å/cycle to 10 Å/cycle, which is very low requiring some tens of thousands of cycles to become an MSC. Figure 2.24 shows the variation of micro-cracks density in an α-iron (ferrite) for two different strain amplitudes versus the N/N_f ratio of applied cycles N to the expected fatigue life N_f [19]. In both cases the applied strain amplitude is above the fatigue limit so that the N/N_f ratio tends to one. The average micro crack length depends on the applied strain amplitude. For the lower strain, $\varepsilon_a = 1 \cdot 10^{-3}$, it is about 20 μm, for the larger strain amplitude, $\varepsilon_a = 6 \cdot 10^{-3}$, it reaches about 80 μm. As it can be seen in Fig. 2.23 for both strain amplitudes, the micro crack density rises as N increases up to a maximum, afterwards it decreases. It is interesting to discover that at a strain amplitude of $1 \cdot 10^{-3}$, which is already close to the elastic limit, 10 to 20 micro cracks per mm^2 can be counted at $N/N_f = 0.5$. Close to fatigue limit this density goes down to just one or two micro cracks per mm^2. Also for the higher strain amplitude the general trend remain the same. Actually the decrease in crack density after the maximum is reached is due to the fact that by increasing the density their interaction also increases. This interaction results in cracks coalescence producing a longer defect that ends up with the macro crack formation.

During this process new micro cracks are not formed. Ma and Laird [20] believe that stress redistribution following cracks generation arrests their multiplication allowing only some of them or even one single crack to grow further becoming a macrocrack. Note that if a micro crack is generated directly by the fracture of a surface inclusion, rather than through the slip bands process, the diagram of Fig. 2.23 presents saturation and does not show that typical peaked curve with increasing–decreasing trend. The very meaning of what has been said so far is that initiation must be avoided. Once an MSC is produced it is rather difficult to stop it growing and become a macrocrack. The barrier to this growth is provided entirely by the surface layer of Fig. 2.5. Any attempt to improve its resistance will retard or even avoid fatigue failure, as it will be shown in Sects. 2.4 and 2.5.

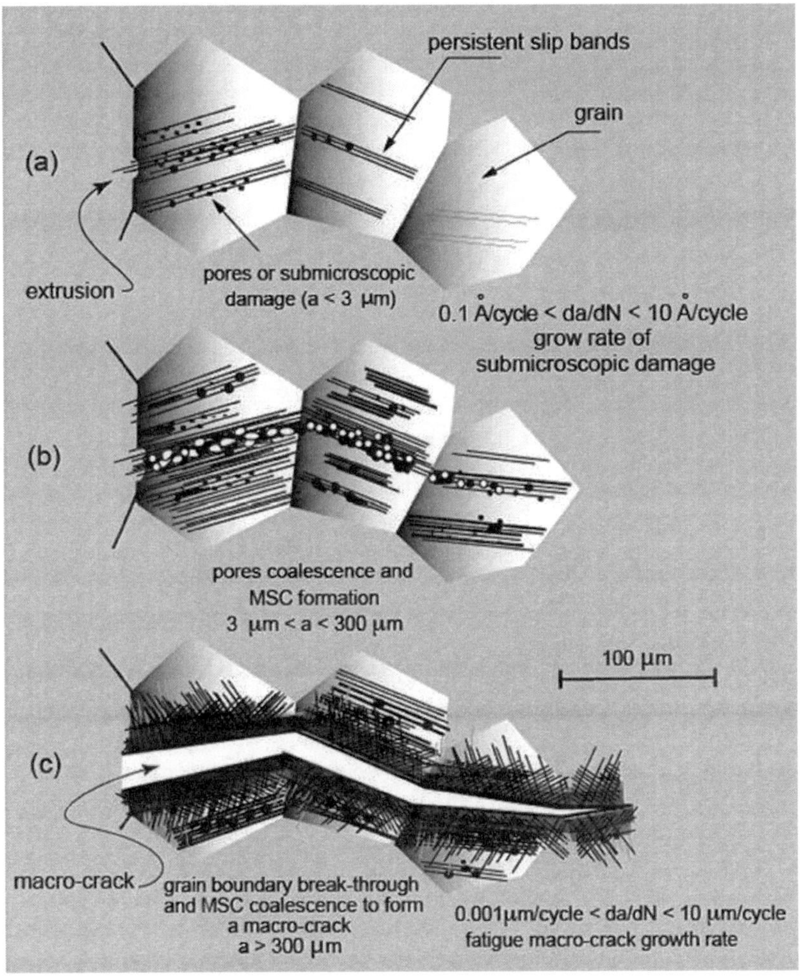

Fig. 2.23 Schematic of progression of events that lead to a macro-crack formation: **a** pore formation into persistent slip bands in some superficial grains, **b** growth and coalescence of submicroscopic damage with MSC formation and **c** grain boundary break-through by MSC and macrocrack formation by MSC coalescence

2.4 Very High Cycle Fatigue

The traditional subdivision of the Wöhler curve into the three regions of Fig. 1.10, at least for ferrous alloys or, better, for BCC crystal lattice materials has been recently extended to include a new region beyond some 10^7 or 10^8 cycles. It is the region of the so called very-high cycle fatigue or VHCF. The more demanding request for life from many technological applications has been the driving force to explore, also for

Fig. 2.24 Micro cracks density in α-iron (ferrite) at two strain amplitudes vs. relative fatigue life [19]

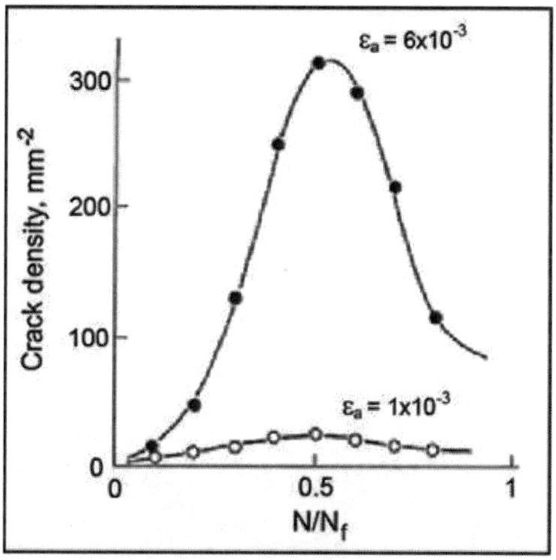

BCC materials, the field of 10^8 or even 10^9 cycles. The pressure to know how these materials perform in the VHCF region, if they really have a definite fatigue limit, has unveiled a new as well unexpected behavior that still remains argument of vigorous discussion among specialists.

However, it has also shed some light into the gray region of damage evolution. These new evidences appeared for the first time in 1984 and were reported by Japanese investigators [21, 22]. They would suggest that also for BCC materials a real fatigue limit may not exist, as for FCC materials, so that the characteristic S–N diagram of Fig. 1.10 now becomes one of the two shown in Fig. 1.12. The disappearance of the fatigue limit, long believed to exist for this class of materials, is today well documented, yet a question arises as to whether or not they still have a BCC crystal structure or, somehow, they turned into FCC materials. A rather comprehensive review of this issue can be found in Reference [23–31]. Below the conventional fatigue limit, persistent slip bands seem to be dormant and may apparently tolerate the action of an infinite number of cycles without degenerating. But an infinite number of cycles is merely theoretical matter without a real physical meaning.

It may seem that the applied stress amplitude is not large enough to propagate damage, but in reality the effects of cycling without interruption continue to be felt by the material that suffers, anyhow, and eventually fails. It's just a matter of time, which means cycles. The sum of extremely small damage per cycle, apparently undetectable, turns into a finite quantity when the number of cycles becomes extremely large. There seems to be also evidences of that lagging action. Consider the Armco steel of Fig. 2.25. We have seen that below 152 MPa, which is the fatigue limit, PSBs may form, but do not propagate. To this end, the sequence of Fig. 2.25 [3] results extremely interesting. At about 127 MPa, which is well below the fatigue limit, after $6{\cdot}10^4$ cycles there is almost no evidence at all of any damage, Fig. 2.25a. Just a single grain seems to have developed segmentary slips. At $N = 6{\cdot}10^5$ a first sign appears that those segmentary slips became wider, Fig. 2.25b. Continuing to

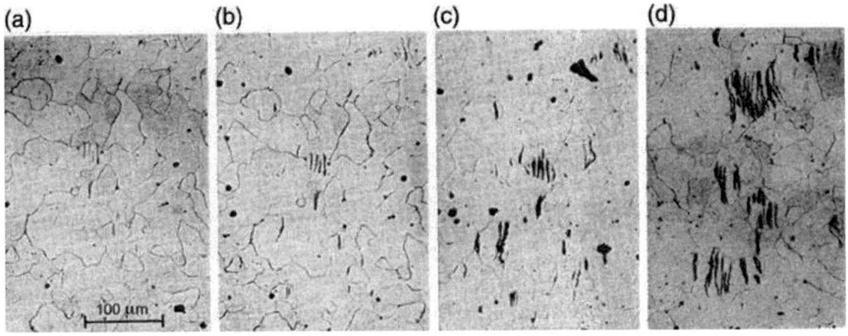

Fig. 2.25 Slip bands in Armco iron below the fatigue limit after: **a** $N = 6{\cdot}10^4$, **b** $N = 6{\cdot}10^5$, **c** $N = 6{\cdot}10^6$, **d** $N = 6{\cdot}10^7$ cycles (adapted from [3])

cycle, at $N = 6{\cdot}10^6$ the signs of slips are heavier. They seem to have broken the grain boundary and have also appeared in grains, Fig. 2.25c. Nobody would think of a possible degeneration of these innocuous slips. But just look what happens after ten more millions of cycles has been put into the material at the same stress amplitude, Fig. 2.25d.

Some PSBs joined and some other appeared in the neighboring grains. Note that the overall length is in the range of 300 μm: a macro-crack. Probably, at $N = 6{\cdot}10^8$ the specimen would fail, but, unfortunately, we haven't got any data to either support or reject that: the test was stopped at $N = 6{\cdot}10^7$ cycles. However, it is common opinion among researchers that the actual cause for failure at such high a number of cycles is the presence of subsurface inclusions. Based on large experimental evidence, these researchers believe that while in conventional high-cycle fatigue damage initiates on the surface of the specimen by a slip mechanism, in VHCF it starts below the surface at an inclusion site. A rather complete treatment of the VHCF subject is given in Chap. 9.

2.5 The Meaning of Fatigue Limit

All what has been said in the previous sections contributes to the definition of the fatigue limit in metals that actually have it, namely BCC crystal structure materials. Any limit different from infinity can be approached from two opposite sides. In our case, the sides are those above and below the fatigue limit, i.e., the area of finite life, Region I and II of Fig. 1.10, or that of unlimited life below the conventional fatigue limit. Moving from above, the fatigue limit σ_f can be defined as the lowest stress amplitude possible that can still produce fatigue failure, which actually means grow micro-cracks initiated in persistent slip bands (PSBs) into a macro-crack by breaking grain boundary barriers. Approaching from the other side, the fatigue limit is defined as the largest stress amplitude possible that will not drive a micro-crack

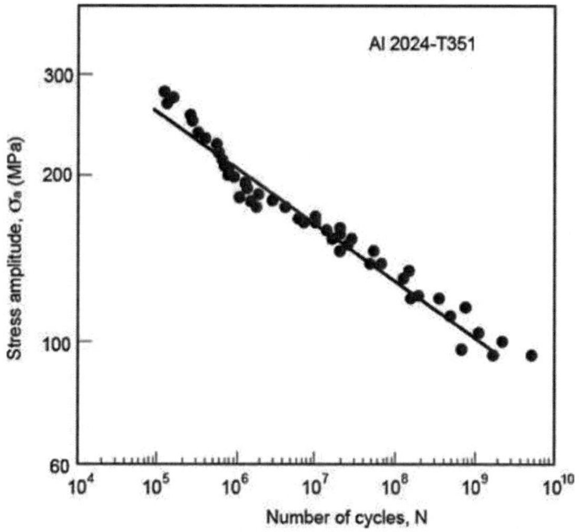

Fig. 2.26 Fatigue data for Al 2024-T351 tested in air at a frequency of 20 kHz (adapted from [33])

initiated into a PSB to become a macro-crack and produce failure. Two definitions of the same subject bearing opposite meanings. However, the second definition seems to be more appropriate since it actually defines a value that must be pursuit in order to avoid fatigue failure. In FCC crystal lattice materials such as copper, aluminum alloys or nickel alloys such a limit seems not to exist. This is shown in Fig. 2.26 for Al 2024-T351 [33] whose S–N diagram appears to proceed straight from 10^4 to 10^9 cycles without any lower limit. In these cases it is use to specify the stress amplitude σ_a at any given number of cycles N.

However, we have seen, in the previous section, how a real fatigue limit may not exist also for BCC materials. PSB may form below the conventional fatigue limit that can eventually turn into macro-cracks at very high number of cycles. But, in general, it is believed that in ferrous alloys below the fatigue limit, σ_f, micro-cracks, though nucleated, do not propagate (see Fig. 2.22). However, this is something that needs to be verified particularly for high strength steel having small size grains where micro-cracks can form remaining apparently dormant. Moreover, at VHCF, we have seen, hydrogen may play a fundamental role causing fatigue failure well below the conventional fatigue limit. Nevertheless, when this is the case, the danger should almost disappear by backing hydrogen out at a proper temperature. Forrest and Tate [34] found that in brass cycled below the fatigue limit no micro-crack developed when the grain size was large enough, but damage instead nucleated if the brass were fine-grained, between 10 μm and 40 μm, though it didn't propagate to contiguous grains.

Sinclair e Craig [35] found that in brass the fatigue limit increased as the grain size d decreased. It is well known that grain size has an effect on materials strength; therefore, it must have the same effect on fatigue. The famous Hall–Petch relation correlates the grain size of a material to its yield strength

$$\sigma_y = \sigma_o + kd^{-1/2} \tag{2.5}$$

where σ_o is a material property, namely the lattice friction stress, d the grain size and k a material characteristic. By increasing the grain size the yield strength of a material sharply decreases and so does the fatigue limit. This is because, we have explained in Sect. 2.3, in wider grains a large number of dislocations can form that pile-up against the grain boundary exerting an enormous pressure.

Therefore, slips that are the visible sign of dislocation trains can easily break the barrier or activate another source of dislocation in the neighboring grain, as discussed in Sect. 1.4.1 and shown in Fig. 1.27. Once formed, it is rather difficult to stop growing microcracks in large grains and this why PSB are not seen in large grain steels below the fatigue limit. At variance, small grains resist better the pressure of a lower number of dislocations formed and pileup within them and this is the reason why we can normally see PSB in high strength steel below the fatigue limit. In small grains PSB can be better tolerated since grain boundaries act as strong barriers.

Figure 2.27 shows the influence of grain size on the fatigue strength of AA 1050 aluminum in the condition of ultra-fine grain (UFG < 1 μm) and coarse-grain (20 μm < CG<170 μm) [35]. It is evident the quite flat appearance of the coarse-grain S–N curve compared to the UFG alloy that may still have same very tiny damage below the fatigue limit. This last aspect is well evidenced in Fig. 2.28 [36] that refers to the same alloys. In the CG aluminum alloy at a stress amplitude of 28 MPa, which is just below the fatigue limit, the surface damage by slip bands formation is evident already at $5\cdot10^6$ cycles, Fig. 2.28a and increases at larger number of cycles, namely $5\cdot10^7$ cycles, Fig. 2.28b and $5\cdot10^8$ cycles, Fig. 2.28c. Under such conditions it is difficult to believe in a levelled fatigue limit. Even at 13 MPa, Fig. 2.28d, the surface appears rushed. We shall expect that the S–N curve goes down almost endlessly without a fatigue limit. At variance, the UFG aluminum alloy shows very little effect of cycling below the fatigue limit at 13 and 28 MPa stress amplitude and $5\cdot10^8$ cycles, Fig. 2.28e, f.

2.6 Effect of Microcracks and Microstructurally Small Cracks on Fatigue Limit

The fatigue limit defined in the previous section is determined by tests performed using specimen as much as possible smooth to avoid the influence of possible preexisting surface cracks or defects of any type. When the fatigue specimen fails it is because a surface macrocrack developed during the cycling through the process, previously discussed, of persistent slip band formation and damage coalescence. Question arises as to what effect a preexisting crack may have on the fatigue limit of materials. Kunio, Shimuzu and Yamada [37] and Kunio e Yamada [38] working with ferritic-martensitic steels in which they introduced cracks whose length a was varying from 505 to 1010 micron, found that the threshold σ_{th} stress necessary to

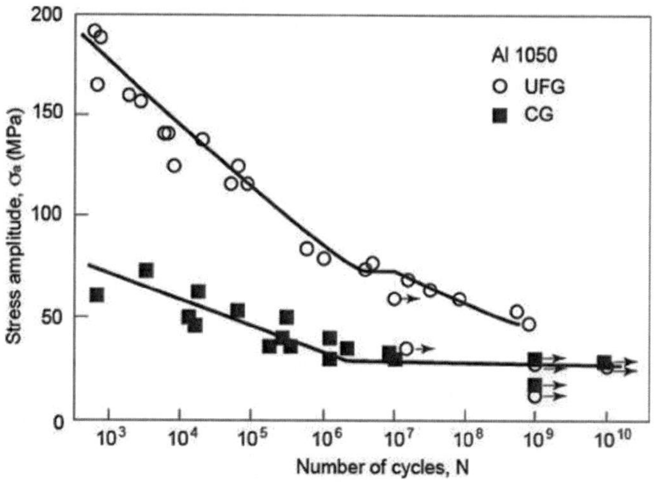

Fig. 2.27 Effect of grain size on fatigue S–N curve of AA 1050 aluminum alloy [36]

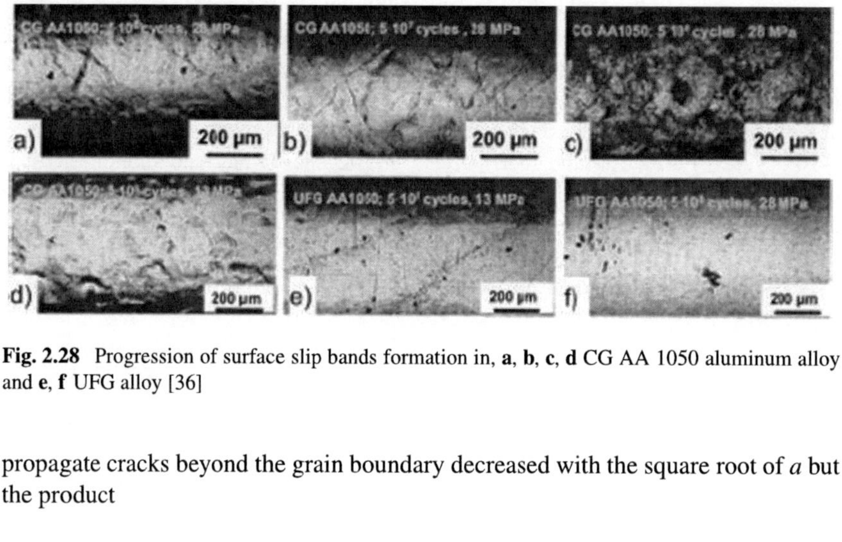

Fig. 2.28 Progression of surface slip bands formation in, **a**, **b**, **c**, **d** CG AA 1050 aluminum alloy and **e**, **f** UFG alloy [36]

propagate cracks beyond the grain boundary decreased with the square root of a but the product

$$K_c = \sigma_{\text{th}}\sqrt{\pi a} \tag{2.6}$$

remained constant and equal to 7.75 MPa$\sqrt{\text{m}}$. The subscript c in Eq. (2.6) indicates the critical value of K that is likely to represent the toughness of martensite plates that in martensitic steels represents the main barrier to dislocation motion, see Fig. 4.24. Equation (2.6) has the structure of the well-known fracture mechanics expression for the stress intensity factor K_I in an infinite plate containing a central crack of length $2a$ under remote biaxial load (see Appendix A). In our case the crack is not central but on the edge, the load is uniaxial and the specimen has a finite dimension so that border effects are generated and Eq. (2.6) needs to be changed into

$$K_c = \sigma_{\text{th}}\sqrt{\pi a} \cdot f(a/d) \tag{2.7}$$

where *f(a/d)* is a dimensionless factor already seen in Eq. (2.4). Linear elastic fracture mechanics (LEFM) predicts that for an edge crack the factor *f(a/d)* can be a little higher than 1, see Appendix A. Accordingly, the threshold stress σ_{th} may also decrease by a factor greater than 1. This let us understand how surface cracks are not only the most likely to occur for any reason, but also those that break the barriers easier. Solving Eq. (2.6) for σ_{th} and giving a the length d of the grain or the distance to the closest martensite platelet, we obtain that the fatigue limit takes the form

$$\sigma_f = C \cdot d^{-1/2} \tag{2.8}$$

where C is a constant representing a material property and, in the particular case studied by Kunio et al., is the martensite toughness. Equation (2.8) recalls Eq. (2.5) in that it suggests a dependency of the fatigue limit to the inverse square root of the grain dimension or something equivalent. Effectively, Klesnil et al. and Taira et al. [39, 40] measuring the fatigue limit of two ferritic steels having different grain size and carbon content, found that the fatigue limit was inversely proportional to the square root of d obeying an empirical relationship of the Hall–Petch type

$$\sigma_f = 11.0 \cdot d^{-1/2} + 110 \tag{2.9}$$

where d is expressed in mm. Equation (2.9) holds up to a maximum value of d equal to 76 μm, after which it would not decrease any longer remaining equal to about 150 MPa. This actually means that beyond a certain value of grain size the fatigue limit becomes independent of the grain size itself being determined again by the condition of crack non-initiation.

We realize, then, that there is a relation between macrocrack size and fatigue limit. This relation passes through the grain size. But grain size may not be the only characteristic dimension that has a role in the fatigue limit expression. Let's consider, for example, the tests made by Murakami and Matsuda [41]. They prepared cylindrical specimens containing micro–macro-cracks from 100 μm to 1 mm long. These cracks were generated from micro holes 40 μm in diameter and 40 μm depth, by cycling the specimens prior to run fatigue tests. A successive annealing at 600 °C ensured that any effect introduced by the previous cycling was removed. Each specimen was subjected to a single cycle at stress amplitude over the fatigue limit followed by 40 cycles below that limit; something like the French experience shown in Fig. 2.21. Cycles were symmetrical and given in a push–pull fashion with the stress ratio $R = -1$ ($R = \sigma_{\min}/\sigma_{\max}$). This basic package of cycles was repeated until failure occurred. Results are shown in Fig. 2.29 [41]. It can be seen how the fatigue limit depends on the initial crack size, as expected. The specimens with a crack 100 and 200 μm

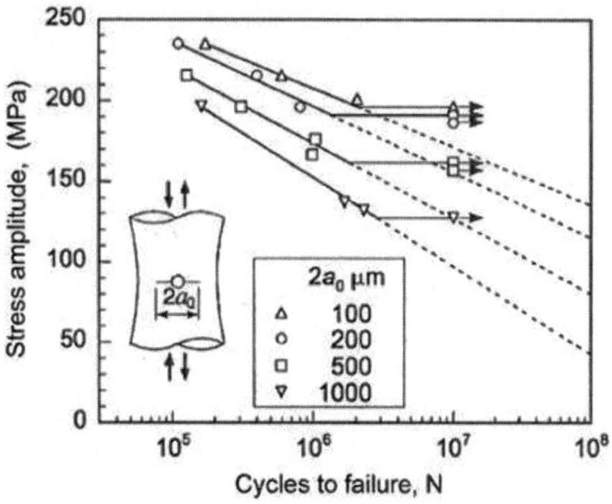

Fig. 2.29 Fatigue diagram obtained by Murakami and Matsuda with specimens containing cracks of different size [41]

long yielded values of the fatigue limit rather close to each other, yet lower but close to the real fatigue limit of the material, the one obtained with smooth specimens. This does not necessarily mean that cracks of about 100–200 μm can be driven to failure by cyclic stresses just below the fatigue limit, though very possible for such a rather long microcrack, since it is also reasonable that each first cycle of the package, the one above the fatigue limit, could have contributed to propagate the microcrack just a little bit so that at a certain moment, thanks to the initial overloads, the crack reached a size larger than 300 μm that is the threshold between a microcrack and a macrocrack.

At 10^7 cycles 250,000 cycles were overloads. With the 500 μm crack the fatigue limit dropped down to 160 MPa and it further decreases to about 125 MPa with the 1 mm long crack. In these two last cases, the cycle over the fatigue threshold was probably not necessary, since the macro-crack was already present. Besides the fact that the experiments outline the importance of the free surface of a material, already stressed in the previous sections that must be kept as much as possible smooth and without any macroscopic damage, they also indicate how the *apparent* fatigue limit decreases as result of the unwanted presence of a macro-crack of various sizes.

The cracks of Fig. 2.29 may well be inclusions that break or detach from the metal matrix in which they are embedded (see Fig. 4.20). Inclusions, in fact, can provide the initiation mechanism in hard or work-hardening metals, as it will be discussed in Sect. 4.4. In metals less prone to plastic slip, like high strength steels or precipitation-hardening aluminum alloys, slips in surface crystals are definitely much less an important crack initiation mechanism. Comparison between Figs. 2.30 and 2.31 is particularly exhaustive. Figure 2.30 [42] refers to a mild steel 0.13% C of 203 MPa yielding strength and 327 MPa ultimate under a cyclic stress of 196 MPa amplitude corresponding to a life of $1.5 \cdot 10^6$ cycles.

Its surface is deteriorating under the alternation of loads. After $5 \cdot 10^4$ cycles slip lines appear containing two small cracks that join after $2 \cdot 10^5$ cycles upon the failure

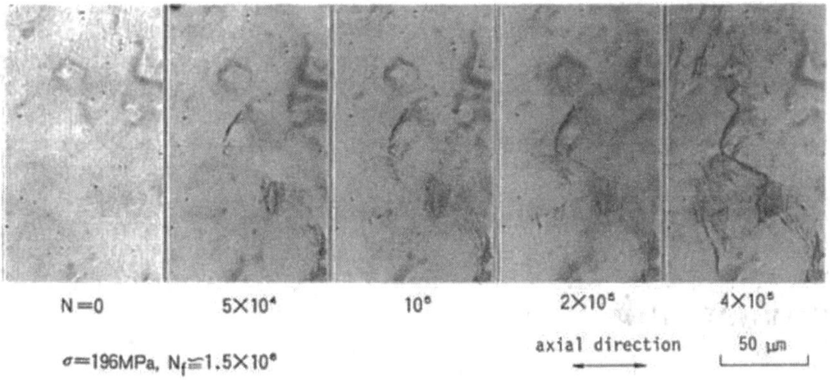

Fig. 2.30 Examination of the plain surface of a mild steel specimen 0.13% C [42]. Crack formation in slip lines after $5{\cdot}10^4$ cycles is visible. It can be also seen the ligament rupture between two small cracks leading to a longer one

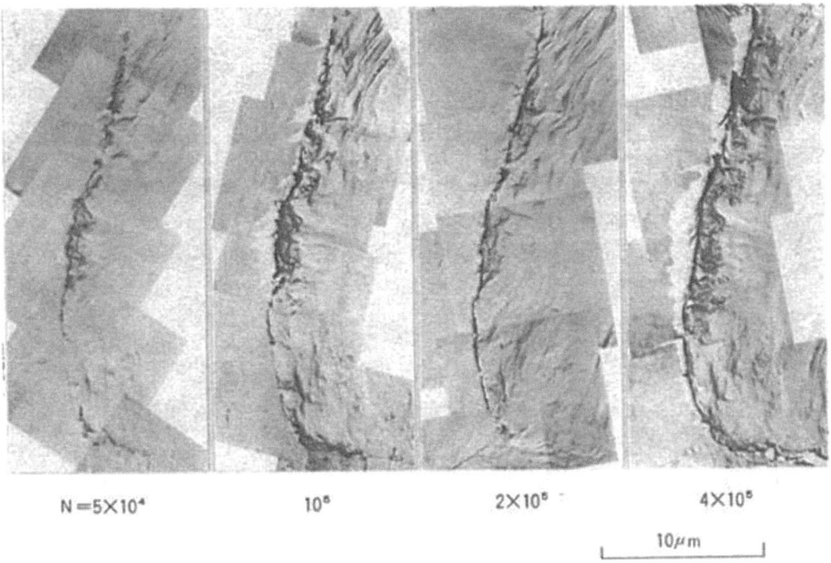

Fig. 2.31 Details of Fig. 2.35 obtained with SEM examination [42]

of the ligament. Figure 2.31 is a detail of Fig. 2.30 taken with SEM. At variance, Fig. 2.32 [42] shows the SEM examination of the surface of a precipitation-hardened or aged aluminum alloy under a cyclic stress producing failure after $4{\cdot}10^5$ cycles. After $2.6{\cdot}10^4$ cycles, corresponding to a mere 6.5% of the expected life, a crack is appeared on the surface whose length is about 3 μm. It is still a sub-microscopic defect. After $9{\cdot}10^4$ cycles, that is 22% of total life, the crack has reached a length

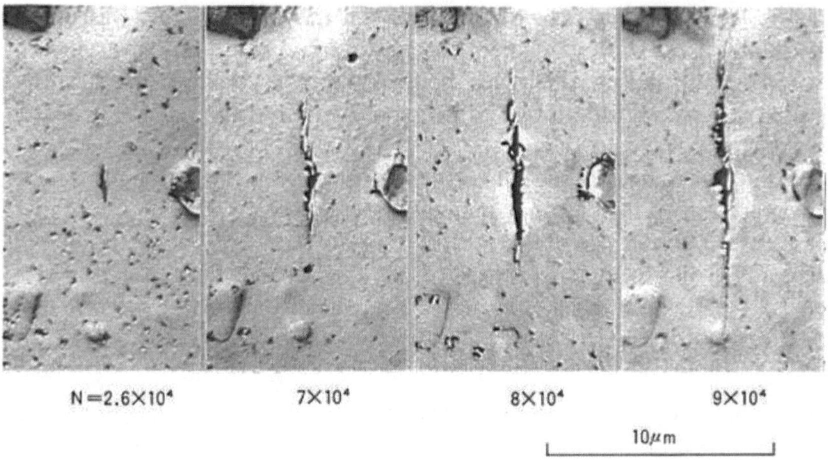

Fig. 2.32 SEM examination of the surface of a precipitation-hardened aluminum specimen that shows the formation of a sub-microscopic damage (< 3 μm) that grows under fatigue becoming a micro-crack still within the grain [42]. No sign of slip process is visible

of about 10 μm becoming a micro-crack or MSC smaller than the grain size that is about 100 μm.

What must be caught is the complete absence of slip lines which actually means that the crack is likely to be originated from a very small surface inclusion or void. Many small voids are actually seen in the background. The competition between slips and inclusions as initiation mechanism will be further developed in Sect. 4. 4. Results shown in Fig. 2.29 can be presented in a different fashion. Rather than plotting the various S–N curves, each for the corresponding crack size, we can plot the fatigue limit as a function of the crack size, *a*, yielding what is known as the Kitagawa-Takahashi diagram.

It roughly subdivides cracks into three categories according to their length: non-propagating microcracks, macroscopic transition cracks and large macrocracks. Figure 2.33 [43] presents the results obtained for mild steel by Kitagawa and Takahashi at two different stress ratio $R = -1$ (symmetrical cycle) and $R = 0$ (pulsating cycle), respectively. It is interesting to note how the fatigue limit of the material, σ_f, remains constant up to a micro-crack size, *a*, of about 0.1 mm, afterward it starts to switch into a downward curve whose slope increases with increasing crack length, *a*. This actually means that in this material a 100 μm crack cannot be grown to failure by a stress equal to the fatigue limit. After a transition behavior, the curve slope eventually becomes a straight line of slope ½. This straight line of slope ½ defines a relationship of the type expressed by Eq. (2.7) which is the basic expression of LEFM for long cracks. For shorter cracks, such as those we have called macrocracks, any relation based on LEFM cannot be applied. The ½ limiting slope marks the final passage from a mechanism of grain boundary brake through and coalescence of MSCs, fed by shear stresses, to that of small crack formation driven by normal stresses.

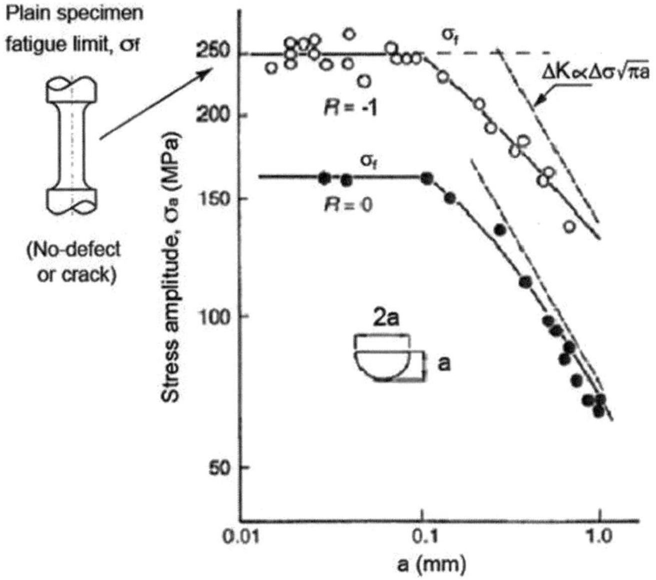

Fig. 2.33 Kitagawa-Takahashi diagram for surface micro cracks in a mild steel for two different stress ratio R (adapted from [43])

For the steel considered in Fig. 2.33, the initial crack size at which the fatigue limit starts to decrease is around 100 μm that in the particular case of mild steel corresponds to the retained austenite grain size. Beyond that value any crack already existing in the material cannot remain dormant but breaks metallurgical barriers and propagate at stress amplitude lower than the fatigue limit σ_f of the material. Also note the effect of the stress ratio on the fatigue limit of the material that sharply reduces. We'll discuss this issue of mean stress effect in Sect. 7.3. Analogous results were obtained by Lukáš et al. [44] on 2.25Cr/1Mo steel and electrochemical copper (OFHC). All these data suggest that a pre-cracked specimen possesses a definite fatigue limit which depends on the length of the crack and that below a certain value cracks may be non-propagating. The fatigue limit, in fact, is determined by the condition of non-propagation of a crack. For both alloys, the critical crack depth at which the fatigue limit started to decrease was again about 100 μm, but whether in copper the grain size was 50 μm in steel it was half that value and equal to 25 μm. The copper yield strength was equal to 40 MPa with an ultimate of 220 MPa, $\sigma_f =$ 73 MPa and a threshold $K_{\text{th}} = 2.5$ MPa$\sqrt{\text{m}}$. The steel had $\sigma_y = 420$ MPa, $\sigma_u =$ 580 MPa, $\sigma_f = 250$ MPa and $K_{\text{th}} = 5$ MPa$\sqrt{\text{m}}$.

This finding confirm the belief that a threshold crack exists below which a defect cannot propagate if the applied stress amplitude is equal or lower than the fatigue limit of a smooth specimen. In low strength steels such a dimension is about 100 μm and is equal to some grains or to the size of the prior austenite grain. These microcracks are non-propagating defects that do not lower the fatigue strength of the material.

The existence of such a critical crack size appears to be confirmed by a huge amount of experimental evidences [44–48].

Remarkable is the work done by Frost [50–52] and Frost and Dugdale [53] on non-propagating cracks that can be considered as pioneering. Frost was working with panels, 63 mm wide and between 5 and 12.7 mm thick, containing two sharp opposite-edge V-notches, 5 mm deep and 0.1 mm root radius, as shown in Fig. 2.34a.

The specimens were cycled to form and grow cracks from the sharp notch roots. They were then re-profiled so as to remove the notch and leave only two surface edge cracks of wanted depth, between 0.25 and 6 mm long. The specimens were given a suitable heat-treatment to minimize residual stresses induced by either the cracking or re-profiling processes and then subjected to a given fully reversal stress. Frost used also cylindrical specimens containing a circumferential V-notch that were subjected to the same procedure of cycling to form a root crack, re-profiling and heat-treating.

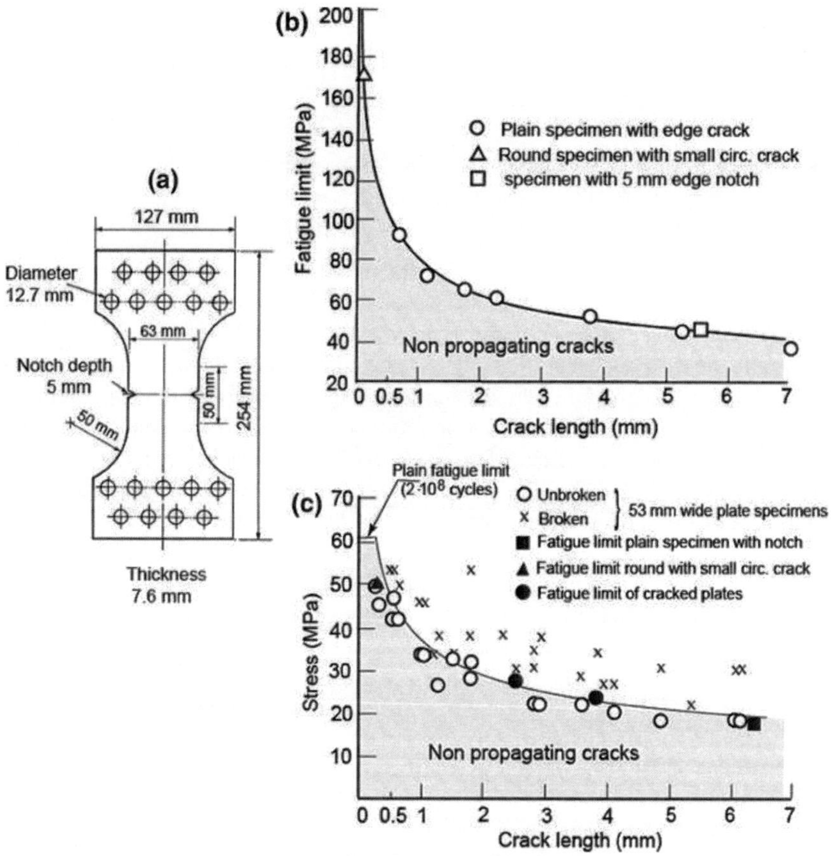

Fig. 2.34 **a** Wide plate specimen used by Frost and Dugdale, **b** fatigue limit of Ni–Cr steel specimens and **c** fatigue limit of copper specimens as a function of crack length [51, 52]

Tests were run on 12 different alloys, including nickel, iron, aluminum, copper and brass. Results are shown in Fig. 2.34b and c. What Frost actually noted was that cracks grew or remained dormant depending on the value of the product in the left side of the expression

$$\sigma_a^3 \cdot a = C. \tag{2.10}$$

in which σ_a was the applied stress amplitude and a the crack depth. C is a constant characteristic of the material. When the left side product in Eq. (2.10) exceeded the critical value C cracks would propagate. On the contrary, when it fell below that critical value cracks would lie dormant and would not propagate further. Table 2.1 provides a synoptic matrix of constant C values obtained by Frost et al. [54] with ΔK_{th} threshold values (see Sect. 5.7) for long cracks ($0.5 \leq l \leq 5$ mm) and short cracks ($0.025 \leq l \leq 0.25$ mm). Figure 2.34b presents the results obtained for Ni–Cr steel, 925 MPa ultimate and 500 MPa yield strength for which $C = 510$. Figure 2.34c shows the results obtained for copper, 225 MPa ultimate and 62 MPa fatigue limit, with $C = 56$, both at $5 \cdot 10^7$ cycles [52]. Note that the different slope of diagrams of Fig. 2.34 with respect to that of Fig. 2.33 is due to the different scale used that in Fig. 2.33 is bi-logarithmic.

Table 2.1 [54]

Material	Traction strength (MPa)	Fatigue limit (MPa)	C (l in m)	ΔK_{th} (Mpa$\sqrt{m}$) $0.5 \leq l \leq 5$ mm	ΔK_{th} (Mpa$\sqrt{m}$) $0.025 \leq l \leq 0.25$ mm
Inconel	655	220	750	6,4	–
Nichel	455	140	700	5,9	–
18/8 austenitic steel	685	360	540	6,0	–
Low-alloy steel	835	460	510	6,3	–
Mild steel	430	200	510	6,4	4,2
Ni–Cr alloy steel	925	500	510	6,4	3,3
Monel	525	240	360	5,6	–
Phosphor bronze	325	130	160	3,7	–
60/40 brass	330	105	94	3,1	–
Copper	225	62	56	2,7	1,6
4½ Cu-Al	450	140	19	2,1	1,2
Aluminum	77	27	4	1,02	–

Years later, in 1969 Kobayashi and Nakazawa [55] repeated Frost experiments reducing crack sizes that were ranging from 30 to 1.100 μm. They found a relation similar to that of Frost, but the exponent of the stress amplitude rose from 3 to 4.

$$\sigma_a^4 \cdot a = C. \tag{2.11}$$

Recently, also Murakami et al. [56–58] addressed the issue of non-propagating cracks or defect evidencing how it can be possible that micro cracks exist in materials under cyclic stress without propagating, when stress amplitude is at or below the corresponding fatigue limit. In a massive research program in which they would drill holes like those of Fig. 2.35 [58], having a diameter d equal to the depth h, Murakami and Endo derived the Kitagawa–Takahashi diagrams for a variety of alloys. For low strength metals to see any effect on fatigue strength, it was necessary to drill holes larger than 67 μm, very close to the 100 μm of the Kitagawa-Takahashi diagram of Fig. 2.23. Note the scale of Fig. 2.35 that is not logarithmical as in Fig. 2.23. As the hardness increased, as for the 0.46C steel, the critical size for non-propagation decreased to about 36 μm and reduced to 20 μm if the 0.46C steel was quenched and tempered and to just few microns if it was only quenched. Below the critical size, the fatigue limit was not altered at all by the presence of the hole even when cracks originated at its border, as shown in Fig. 2.36 [56].

They also derived an experimental relationship between the fatigue limit σ_f of a material and the dimension of an existing crack or defect that is going to change the approach to fatigue strength evaluation. The first major change was the introduction of the square root of the projected area of the defect on the growth plane, $\sqrt{area}$, as a better parameter characterizing a defect instead of its length a. Figure 2.37

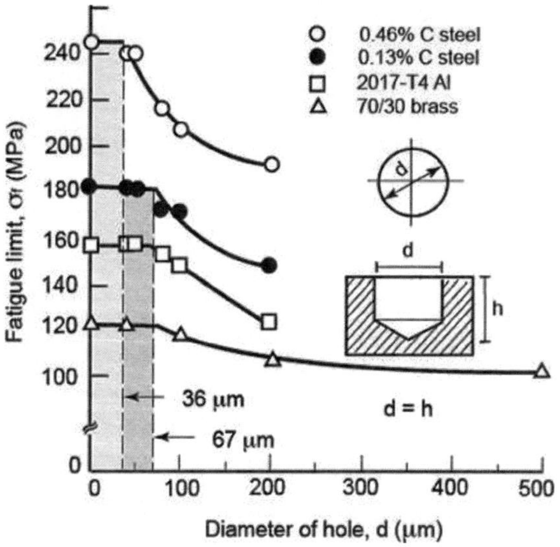

Fig. 2.35 Relationship between the size of an artificial hole and the rotating bending fatigue limit for some alloys (adapted from [58])

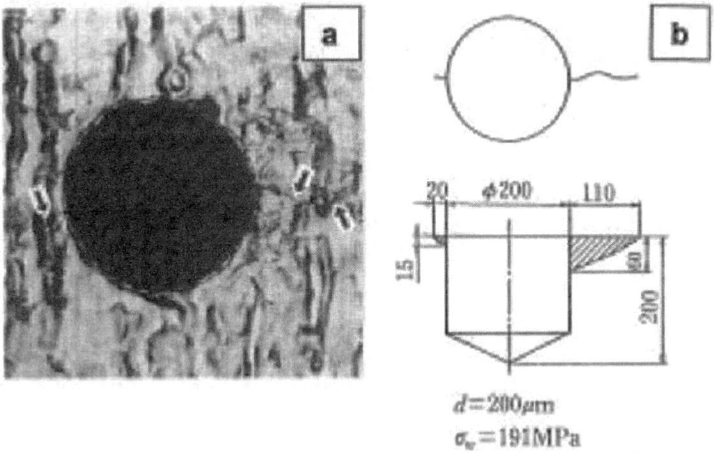

Fig. 2.36 Non propagating cracks observed at the borders of a hole drilled in 0.45% C steel at the fatigue limit [56]

presents the results obtained by Murakami et al. for the rotating bending fatigue limit of annealed 0.46% C medium carbon steel as a function of the projected area, $\sqrt{area}$, of the small defects introduced by drilling holes of various sizes [59]. The dependency of the stress amplitude, σ_f, on the $\sqrt{area}$ parameter is described by the relationship.

$$\sigma_a^6 = \sqrt{area} = C \tag{2.12}$$

The second important change introduced by Murakami and Endo was to relate the fatigue limit of a material to its hardness, instead of the grain size as done with

Fig. 2.37 Rotating bending fatigue limit of annealed medium carbon steel as a function of the projected area, $\sqrt{area}$, of defects introduced in the specimens (adapted from [59])

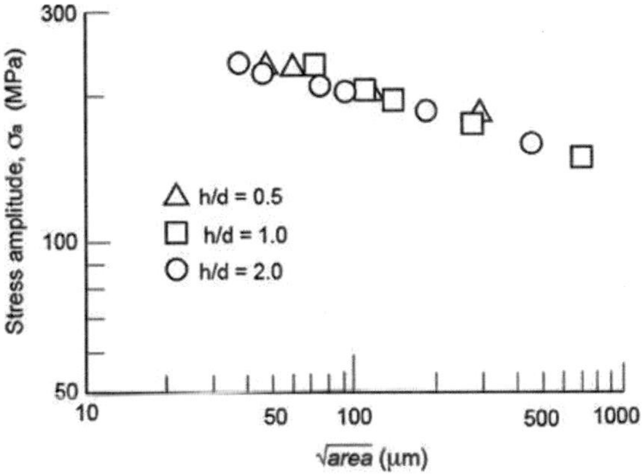

Eq. (2.8). Indeed, it is not a real change since the hardness is related to the grain size of the material and both are related to the microstructure.

However, hardness is more convenient a parameter than grain size for characterizing a material. Murakami and Endo chose the HV hardness but they could have chosen the HB or HC hardness very often used for correlation with the ultimate strength, σ_u (see Sect. 10.2). After another monumental work the authors came up with the empirical relationship

$$\sigma_{f,l} = C \cdot \frac{(HV + 120)}{\sqrt{area}^{1/6}} \tag{2.13}$$

where $\sigma_{f,l}$ is the fatigue limit associated with a defect of given size ($\sqrt{\text{area}}$), the crack $\sqrt{\text{area}}$ is expressed in micron and HV is the Vickers hardness of the material. The derivation of (2.13) will be given in Sect. 10.2. Here it is enough to note that the exponent of the stress amplitude has risen to 6 from that of Eqs. (2.11 and 2.10). The apparent incongruence among the stress amplitude exponents that is equal to 3 according to Frost, 4 for Kobayashi and Nakazawa and 6 for Murakami and Endo, must be found in the different range of micro–macro-crack dimensions they were introducing in the specimens. If we reduce all data to the same unit, namely the square root of the cross section area of the cracks analyzed, we realize that Murakami and Endo have investigated the range10 $\mu\text{m} \le \sqrt{\text{area}} \le 1000\ \mu\text{m}$, Kobayashi the other $500\ \mu\text{m} \le \sqrt{\text{area}} \le 2900\ \mu\text{m}$ and Frost the other one with $1400\ \mu\text{m} \le \sqrt{\text{area}} \le 7300\ \mu\text{m}$.

They represent complementary ranges with some light superposition. This difference in the size of the defects under study can be surely the key to solve the apparent puzzle of incongruence. Then, if we plot in a single diagram the predictions made for a high strength steel of 300 HV hardness and 950 MPa ultimate strength using Eqs. (2.2, 2.10, 2.11 and 2.13) for the respective field of application (namely: $\sqrt{\text{area}}$) we obtain the diagram of Fig. 2.38. As it can be seen, fatigue limit predictions versus damage dimensions according to the three formulas (2.2, 2.10, 2.11 and 2.12), respectively, merge into a continuous common curve that matches linear elastic fracture mechanics previsions of Eq. (2.7) for longer sizes when the micro crack has become a macro crack and LEFM can be effectively used.

The diagram of Fig. 2.38 is precisely the Kitagawa-Yakahashi diagram of Fig. 2.33. The Kitagawa-Yakahashi diagram and Eq. (2.13) have completely changed the approach to fatigue life. They provide, in fact, a practical and very useful tool to infer the fatigue limit of any material where the fatigue process is triggered by the presence of defects like inclusions and, as such, does not follow the conventional slip bands formation and micro-crack coalescence process described in the previous sections that leads to the conventional S–N curve. The action, then, becomes the prediction of the maximum defect size that can be expected to exist in a material with a given probability or, which is equivalent, in a given process volume. This procedure will be discussed in detail in Sect. 6.7. The real issue, then, has become the derivation of the analytical form of the nonlinear part of the diagram between

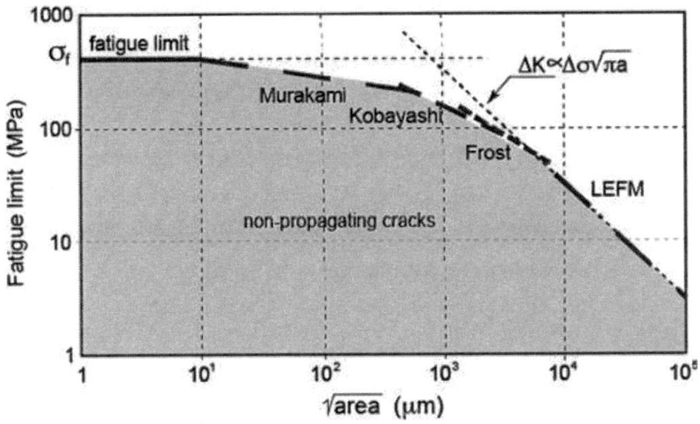

Fig. 2.38 Composite diagram of fatigue limit expectations versus defect size, in terms of $\sqrt{\text{area}}$, according to Murakami, Kobayashi and Frost, respectively, and asymptotic matching with LEFM prevision

the bilinear boundary given by the σ_f plateau and the LEFM line (dotted lines in Fig. 2.35). Note that the use of such a bilinear boundary would result too conservative in the range 50 μm $\leq \sqrt{\text{area}} \leq$ 2000 μm. A rather effective solution would be the use of the multilinear boundary represented by the initial horizontal line from σ_f up to $\sqrt{\text{area}} = 10$ μm, then the Murakami and Endo Eq. (2.13) till $\sqrt{\text{area}} = 1000$ μm and beyond that value the Kobayashi-Nakazawa expression (2.11) and so on.

El Haddad et al. [60] have proposed an empirical solution to model the boundary of the safe region of non-propagating micro–macro-cracks based on LEFM stress intensity factor $\Delta K'$ valid for any crack, micro or macro-crack, known as ETS model

$$\Delta K' = \Delta\sigma\sqrt{\pi(a + a_o)} \tag{2.14}$$

where a is the crack length and

$$a_o = \frac{1}{\pi}\left(\frac{\Delta K_{\text{th}}}{\sigma_{f,o}}\right)^2 \tag{2.15}$$

is a characteristic size which is estimated from the long crack fatigue threshold ΔK_{th} and the fatigue limit for smooth specimen under $R = 0$ conditions.

This kind of trick is needed to account for the fact that the fatigue crack propagation of a macrocrack cannot be modeled using LEFM and cannot have the same threshold as a crack obeying to LEFM rules otherwise a size $a \to 0$ would require $\Delta\sigma \to \infty$. But for $a \to 0$ there is the additional crack size to avoid the physical non-sense. El Haddad models the microcrack threshold though the explanation seems to be just a practical expedient than a convincing argument.

An example of the ETS model is presented in Fig. 2.39 for HT80 high strength steel [61] where $\Delta K_{\text{th}} = 11.2$ MPa$\sqrt{\text{m}}$ is the threshold for long cracks for the HT80

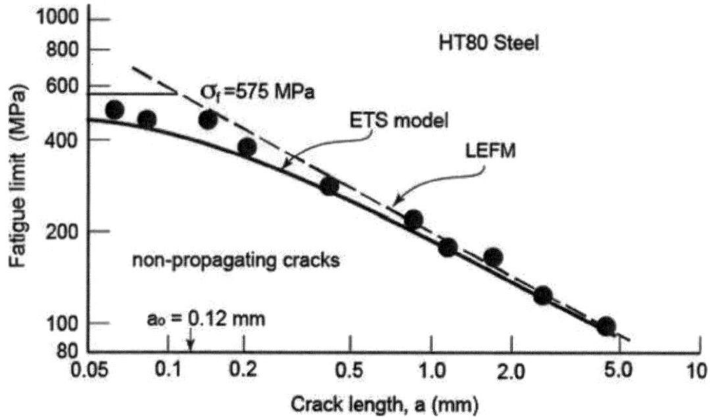

Fig. 2.39 Kitagawa–Takahashi diagram for HT80 steel using the ETS model (adapted from [61])

steel and $\sigma_{f,o}$ is the fatigue limit in cycles pulsating from zero, $R = 0$. The a_o value calculated with Eq. (2.15) seems to be the threshold size below which the fatigue limit is not influenced by the presence of a microcrack. But this should not be taken as a general case. The Kitagawa–Takahashi diagram can be conveniently used to identify the various crack or damage lengths we have been so far discussing. This is shown in Fig. 2.40. The initial plateau of constant stress amplitude corresponds to the presence of microscopically short cracks or MSC. The size of MSC that we have also called microcracks is of the same order of some characteristic microstructural features, such as grain size. This plateau where the stress amplitude is not affected by the presence of microcracks, terminates with cracks of from 100 μm to 300 μm. Beyond that size, it initiates the domain of the so called mechanically small cracks. This is the region where macrocracks start to form. Macrocracks affect the fatigue limit according to the previsions of Murakami, Kobayashi or Frost, depending on their respective length, as shown in Fig. 2.38.

2.7 The Three Stages of Fatigue

Damage evolution during the very first crystallographic phase of fatigue, see Fig. 2.23a, was defined by Forsyth [62] as Stage I of fatigue. The relative damage growth rate goes from some fraction of Angstrom to some tens of Angstrom per cycle, depending on the applied stress amplitude and micro pores density. Damage growth is very slow and difficult to observe. At very low stress amplitude, as in high cycle fatigue close to the fatigue limit, the break-through of two or more grains that leads to a macro crack formation of about 300 μm would then require some 10^5-10^6 cycles. This is why at stress amplitudes close to the fatigue limit even 90 or 95% of

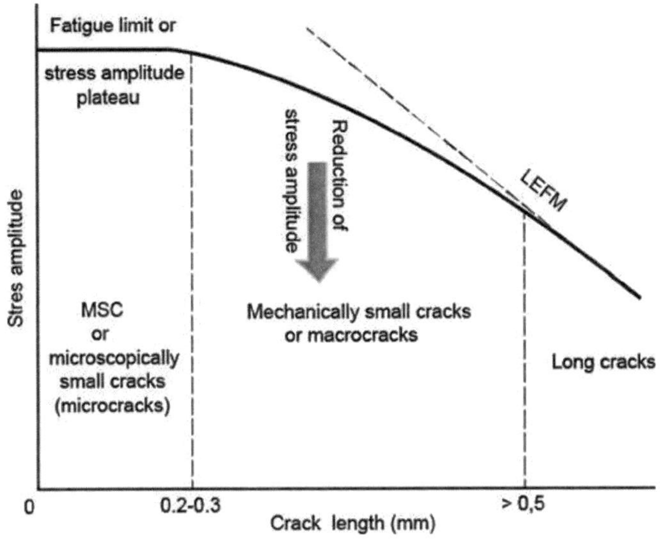

Fig. 2.40 Crack type or defect length and its effect on the fatigue limit

the total life may be spent in this first stage of fatigue just to generate a macro-crack, as already said.

The diagram of Fig. 2.41 [63] presents very clearly this concept of life fraction N_i/N spent just to *initiate* macro cracks (N_i = *cycles to initiation*) as function of stress amplitude σ_a. At variance, in low-cycle fatigue, in which the stress excursion $\Delta\sigma$ may be very high and failure may occur within few thousands or tens of thousands cycles, it takes a mere 20% of total life to initiate a macro crack. In Stage I of fatigue as we know, the generation and growth of micro cracks develop on crystallographic planes along those directions on which the resolved shear stress τ_{res} reaches a critical value (see Fig. 2.2). Due to their casual distribution, the planes of maximum shear stress don't have the same orientation in all surface grains, therefore the growth of a microcrack proceeds along a zigzag path, as schematized in Fig. 2.42 and shown in the detail of Fig. 2.42a.

Figure 2.43 shows the SEM observation of a microcrack growth in Stage I of fatigue in Al-Li 8090 alloy [64]. It can be noticed the zigzag transgranular propagation on crystallographic planes. In this stage of fatigue the growth is conditioned by metallurgical factors such as inclusions, second phases, metallurgical interphases, grain boundaries, etc. which may block or retard microcracks growth since they have a comparable size. Their growth will depend on local metallurgical factors till they cross lines 3 and 4 of Fig. 2.22 and become macrocracks reaching a size larger than those barriers. When a crack reaches a macroscopic size of at least 300–400 μm, i.e. two or three large grains in coarse grain alloys or less than 100 μm in high strength, small grain alloys, the control of its growth passes from local shearing stresses τ_{res} to normal stresses σ and the crack starts to growth on a plane normal to

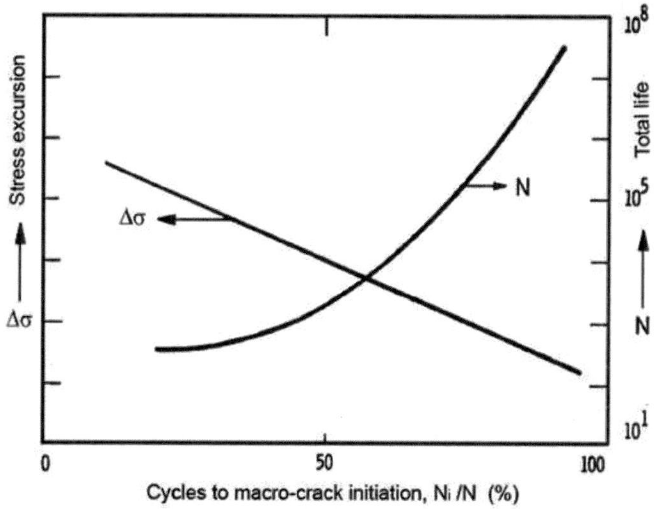

Fig. 2.41 Schematic of initiation life fraction *Ni/N* versus total life or stress excursion [63]

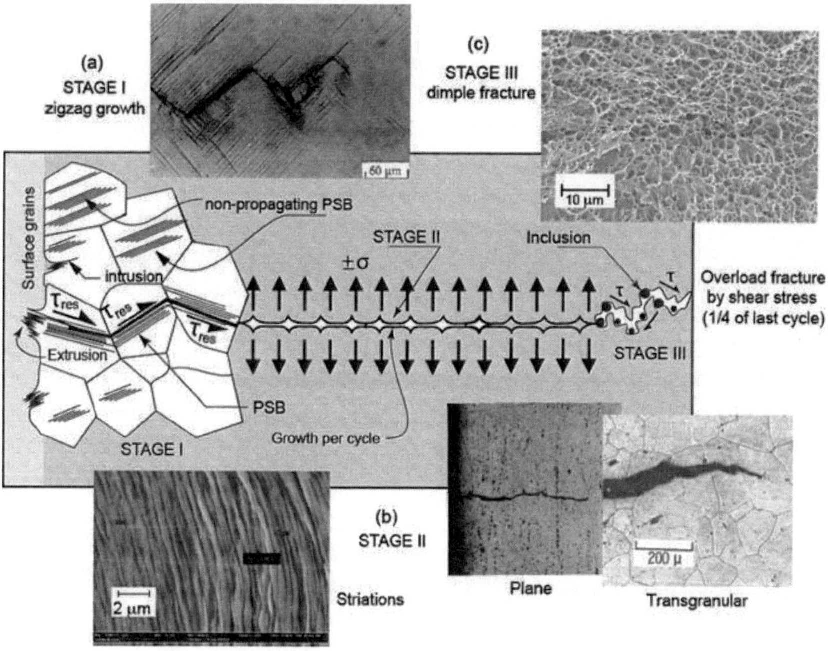

Fig. 2.42 Schematic view of Stage I, Stage II and Stage III of fatigue

Fig. 2.43 SEM view of fatigue crack in Al-Li 8090 alloy in its very first stage of growth on crystallographic planes having different orientation which leads to the characteristic zigzag morphology [64]

the direction of external load, Fig. 2.42. Now, the crack grows breaking grains in a transgranular fashion, as schematized in Fig. 2.42b giving life to Stage II of fatigue, as this stage was called by Thompson and Wadsworth [65]. In this stage the crack grows at each cycle leaving a characteristic mark on the surface, as fingerprints, known as striations, as shown in Fig. 2.42b. Striations can be seen by SEM analysis. An interesting example of the passage from Stage I to Stage II is offered in Fig. 2.44 [66]. Stage I of fatigue can be seen very clearly. The crack nucleated on the specimen surface by shearing stress and, therefore, is macroscopically oriented at 45° to the normal direction of external load. On a lower scale, the crack would show a zigzag profile, as already mentioned. Once grown to a macroscopic size, the crack turns and proceeds on a plane normal to external load driven by the principal normal stress (Stage II). Now it is the bulk properties of the material to determine the crack growth rate.

At the beginning of Stage II, the crack growth rate is of the same order of magnitude of the final grow rate of Stage I, which is about $10^{-7}-10^{-6}$ mm/cycle, but later with increasing crack length it may grow up to $10^{-1}-10^{-2}$ mm/cycle, depending on the applied ΔK (see Chap. 9). Some methods of analysis that make use of photo micrographs or surface replica through acetate films are capable to *see* details as small as one micron in size.

The replica method is two stage a process in which the external surface is first covered by liquid acetate that enters any tiny pore or scratch or roughness matching perfectly the surface profile. When the film dries and becomes solid it retains all details of the surface it matched. At this point, it is peeled off the surface and covered with a very thin layer of gold deposited by evaporation. A further layer of copper, 0.4 mm thick is then added by an electrolytic process. At this point, the acetate film

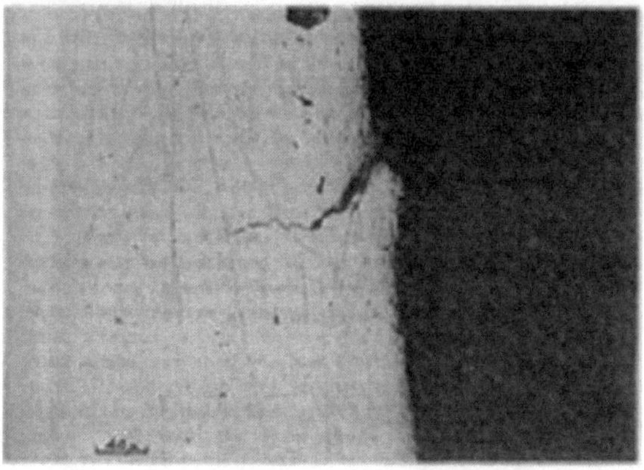

Fig. 2.44 Cross section of a crack nucleated on the surface of the specimen at about 45° under the control of shearing stress (Stage I of fatigue) and then grown on a horizontal plane normal to the maximum primary stress (Stage II of fatigue) [66]

is dissolved in acetone and the double layer copper–gold remains which contains, inverted, all details of the original surface. This replica can be easily analyzed by a scanning electron microscope. This technique allows the systematic observation and study of specific metallographic features such as persistent slip bands or MSC.

By growing in Stage II, the fatigue crack removes continuously resisting area to the remaining cross section that becomes more and more overloaded. Therefore, even though initiated as high-cycle fatigue, it will turn into low-cycle fatigue and grow very fast but this time the growth does not proceed on metallographic planes, as in Stage I, nor on normal plane but along persistent grain boundaries (see Fig. 2.14), as already mentioned in Sect. 2.2. Eventually, it enters Stage III of very rapid growth. Soon, the overload becomes so high to trigger a fast rupture that can be either by cleavage or by dimple process, depending on the material and temperature, as schematized in Fig. 2.42c for dimple fracture. When this is the case, fatigue comes back to be driven by shearing stresses that open, enlarge and join pores (dimples) around inclusions. Each pore, in fact, is the site of an inclusion that was debonded from the matrix leaving a hole or pore that grew under increasing load. Eventually, pores coalesce, and all the remaining resisting area is turned into a threadlike network of pores losing any strength, which leads to inevitable fracture.

Characteristic of this stage are the so called shear lips that are the visible sign of ductile failure, as schematized in Fig. 2.45 for a CT fatigue specimen. The initial plain strain conditions that characterized Stage II fatigue with its flat appearance change to plain stress conditions of STAGE III that makes crack propagate at 45° or in a slant fashion till final fracture by a dimple mechanism. Figure 2.51 presents a series of 1 T-CT specimens (see Appendix A) of A533B carbon steel, 1" thick, fatigue loaded at different increasing stress amplitudes [67]. Specimens 9 and 2 at extreme left have been tested under high-cycle fatigue conditions developing growth rates of the order of 2 μm/cycle. Those at extreme right underwent low-cycle fatigue at stress amplitudes higher than yield strength with growth rate of about 0.2 mm/cycle, hundred times higher. The specimens on the left have flat fracture surfaces, indicating plain strain conditions that characterize the planar fatigue crack growth perpendicular to the direction of external load. At variance, those on the right side tested in low-cycle fatigue have slant propagation surfaces with large shear lips typical of plain stress conditions (Fig. 2.46).

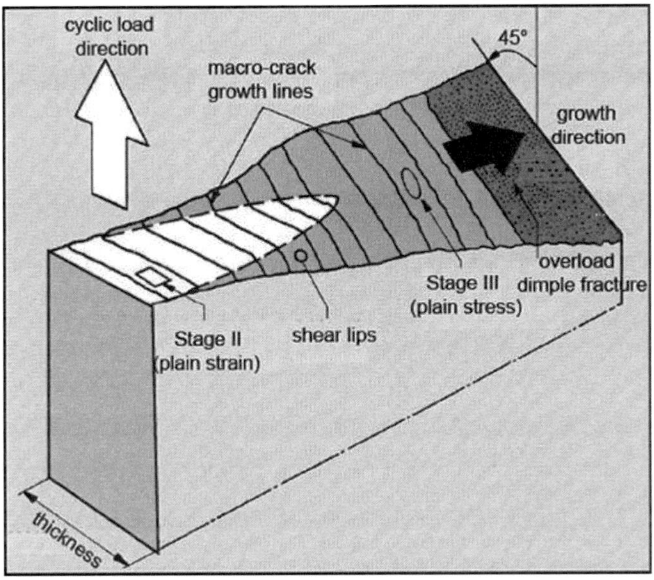

Fig. 2.45 Schematic view of Stage II and III crack front. Stage III introduces shear lips. Final failure happens under overload dimple fracture conditions

Fig. 2.46 1 T-CT fatigue specimens tested in linear elastic (high-cycle fatigue) and elastic–plastic regime (low-cycle fatigue) from left to wright. Shear lips are not present in specimens 9 and 2, but appear and develop in specimens 15, 16 and 18 [67]

2.8 Growth of MSC

Let's now follow the evolution of MSC relative to the life fraction expressed as the ratio between the number of cycles spent N and the total fatigue life N_f, N/N_f. This evolution is schematized in Fig. 2.52. The first two curves in the lower part of Fig. 2.47 start from a crack length between 10 nm and 3 μm, about, which is the limit of submicroscopic damage. As we have mentioned in the previous sections, damage initiates in localized slips in some surface crystals. Because the initiation and progression of surface damage is associated with localized surface regions of cyclic plastic strain, the fatigue strength of a metal or alloy will be increased by any metallurgical hardening process or decreased by the presence of any surface inclusion, scratch or irregularity larger than those 3 μm. As such, we will assume that the surface is clean and mirror polished. The initial part of the curves, from A to B, is characterized by an initial very high growth rate, though in the range of tens of Angstrom per cycle at most. However, this fast growth decelerates. If the stress amplitude is equal or below the fatigue limit σ_f, the growth saturates because the micro crack will never break through grain boundaries (see Fig. 2.56). Stage II of fatigue will never start and the crack behavior will be that described by curve ABD′ of Fig. 2.47. On the contrary, if the stress amplitude is above σ_f, crack growth saturation will not occur. In this case the initial submicroscopic crack after a period of apparent rest will accelerate and grow to become macroscopic, with a rate that varies from 10^{-6} mm per cycle to a tenth of mm per cycle.

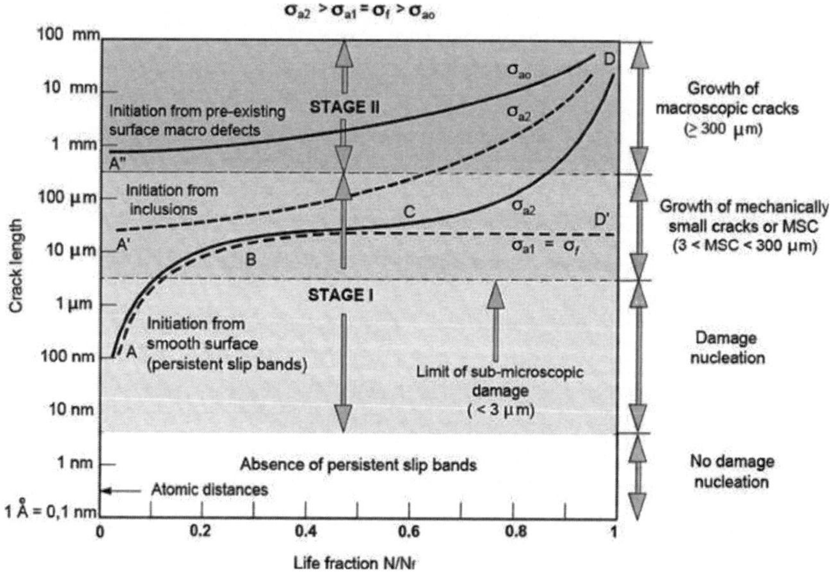

Fig. 2.47 Fatigue growth evolution of MSC and macrocrack relative to life fraction N/N_f

The new evolution curve is ABCD. The two phases AB and CD (initial and final growth) are joined by an intermediate phase BC in which fatigue cycles seem not be producing any apparent growth. We shall believe that during this phase nucleation episodes are happening in grains adjacent to the one in which the first micro crack appeared (phase between lines 2 and 3 of Figs. 2.22 or 2.23b. Growth restart will emerge with the break-through of grain boundaries by individual micro cracks leading to a single composite macro crack (see Fig. 2.23c). Stage II begins. However, a damage initiation mechanisms different from slips in single crystals may also take place. If an inclusion or a non-metallic phase exist on the surface or immediately below the surface, let's say within 30–50 μm depth, micro crack may originate directly from its rupture and, if this were the case, its initial length will be that of the inclusion. Inclusions in steel may reach a length equal to that of the grain in which are contained, even though with much lower probability (see Fig. 4.41). In this case, the first phase of crack nucleation will be skipped over and the evolution of the corresponding growth curve under the same stress amplitude σ_{a2} of the previous case will start directly from the inclusion size (point A′ in Fig. 2.47) and progress along the A′D path. Finally, if a large damage were present on the surface, such as a scratch or indentation, then the macro crack will be already there and both phases of nucleation and coalescence will not take place; it is not necessary any more. The new damage progression will be given by line A"D of Fig. 2.47. It must be observed that now the fatigue growth may happen under a stress amplitude σ_{ao} lower than the fatigue limit σ_f itself since there are no microcracks to form and overcome microstructural barriers. The new threshold stress excursion $\Delta\sigma_{\text{th}}$ will be given by the condition

$$\Delta K_{I,\text{th}} = \Delta\sigma_{\text{th}}\sqrt{\pi a} \cdot f(a) \tag{2.16}$$

where $\Delta K_{I,\text{th}}$ is the threshold value of the stress intensity factor excursion given by LEFM (see Chap. 9 and Appendix A), *a* is the length of the defect and *f(a)* is a geometry factor. Hobson [68], analyzing the results obtained by Lankford [69] on the propagation of MSC in grains of type 7075-T6 aluminum alloy under stress amplitudes larger than σ_f, found that the microcrack growth rate could be described by an expression of the type

$$\frac{da}{dN} = C \cdot (d - a)^{1-\alpha} a^{\alpha} \tag{2.17}$$

in which C and α are material characteristics, d is the grain dimension and a the microcrack length. Note that when $a = d$, that is to say when microcrack length reaches the grain size, the rate *da/dN* vanishes indicating that grain boundary has arrested the micro crack. For the alloy considered C was equal to $2.22 \cdot 10^4$, $\alpha = 0.4$ and grain size was about 16.5 μm. It is clear that if the stress amplitude is higher than σ_f the crack will not be arrested at grain boundary and propagation will continue as indicated by line BCD of Fig. 2.47. If this is the case, C is no longer constant, but becomes a function of the stress excursion $\Delta\sigma$

$$C = A \cdot (\Delta\sigma)^n \tag{2.18}$$

and Eq. (2.17) becomes

$$\frac{da}{dN} = A \cdot (\Delta\sigma)^n (d_1 - a)^{1-\alpha} a^\alpha \tag{2.19}$$

in which d_l is no longer the grain dimension, to avoid arrest, but a dimension characteristic of the material, equal to some grains, beyond which the microcrack becomes a macrocrack that enters Stage II of fatigue.

We may, then, assume that d_l be no larger than 300 μm. MSC and macrocrack growth measurements performed on smooth specimens of different types of carbon steel with yield strengths varying from 240 to 310 MPa, yielded the results presented in Fig. 2.48 [70–72]. The typical trend of line ABCD of Fig. 2.47 can be recognized. It can be seen how microcracks of only 10 μm are formed after a mere 10% of total life. The characteristic shelf BC of Fig. 2.47 can be recognized between 15 and 40% of life which hides the continuous effort of MSC to grow within the grains and join contiguous micro cracks. Grain size is about 40–50 micron. Accelerated growth will start with grain boundary break-through. The same diagram for a low-alloyed carbon steel type A 533 B is shown in Fig. 2.49 [73]. At variance with the previous case of Fig. 2.48, the growth starts from an MSC dimension between 20 μm and 30 μm, and its trend is analogous to that described by line A′D of Fig. 2.47. There is no apparent crack growth rest. This actually means that the MSC formed from a surface inclusion of the same size. Note how the applied deformation is practically non-influent on the fatigue growth rate, at least within the interval considered $0.005 \leq \Delta\varepsilon \leq 0.04$.

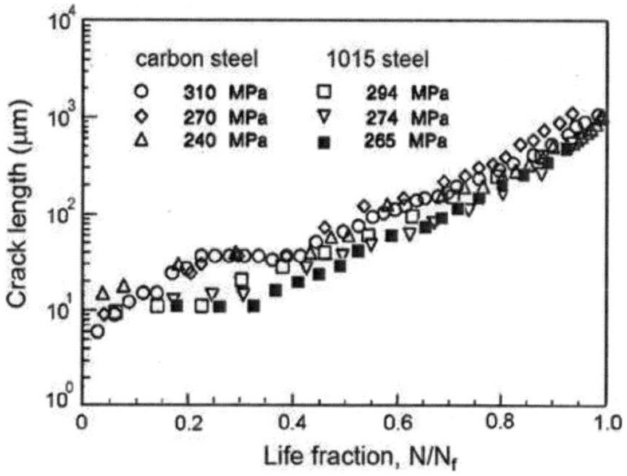

Fig. 2.48 MSC and macro cracks growth in carbon steels in air at ambient temperature versus life fraction N/N_i (adapted from [70–72])

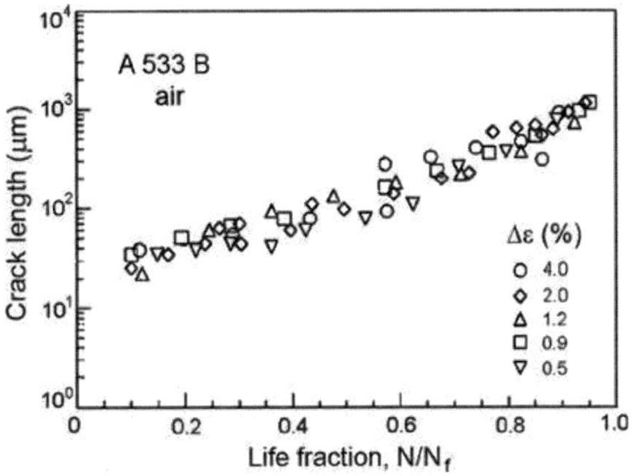

Fig. 2.49 MSC and macro cracks growth in a low alloyed carbon steel type A 533 B in air at ambient temperature vs. life fraction N/N_i [73]

A rather interesting example of MSC is given in Fig. 2.50 by De los Rios et al. [74] using the replica method previously described. Figure 2.50a is the optical microscopy of a steel sample 0.4% C before cycling. After about 21,000,000 cycles at a shear stress amplitude $\tau_a = 350$ MPa two MSC (Y-Y and X-X) are formed along a ferritic grain boundary, Fig. 2.50b. They are not joined. Upon increasing the shearing stress amplitude to 414 MPa, after 46,763 more cycles the two micro cracks joined, Fig. 2.50c and one of the two started to bifurcate at one boundary of the grain into two new microcracks whose tips are denoted as X_1 and X_2. After some additional 600,000 cycle at the same amplitude, also the second tip of the main crack Y-X bifurcates (tips are indicated as Y_1 and Y_2) and propagates, Fig. 2.50d. Hobson [75] found that for carbon steels the α exponent in (2.19) was equal to zero so that it can be written

$$\frac{da}{dN} = A(\Delta\sigma)^n (d_1 - a) \tag{2.20}$$

The result is that crack growth rate becomes a linear function of a. Table 2.2 lists the values of constants A and n obtained by Chopra et al. [76] at room temperature in air and in water at 288 °C. Once the crack has reached a macroscopic dimension Stage II of fatigue starts and the growth rate is governed, as we shall see in Chap. 14, by LEFM

$$\frac{da}{dN} = C \cdot (\Delta\sigma\sqrt{\pi a})^m f(a) \tag{2.21}$$

where C and n are material constants, $f(a)$ a non-dimensional function that depends on the geometry of the system and the product $\Delta\sigma\sqrt{\pi a}$ represents the Irwin's stress intensity factor K_I. At this point we can separate the MSC contribution to the total growth from that of macrocracks. This is schematized in Fig. 2.51. The broken line represents the behavior of a surface crack developed in a crystal of length d by slip.

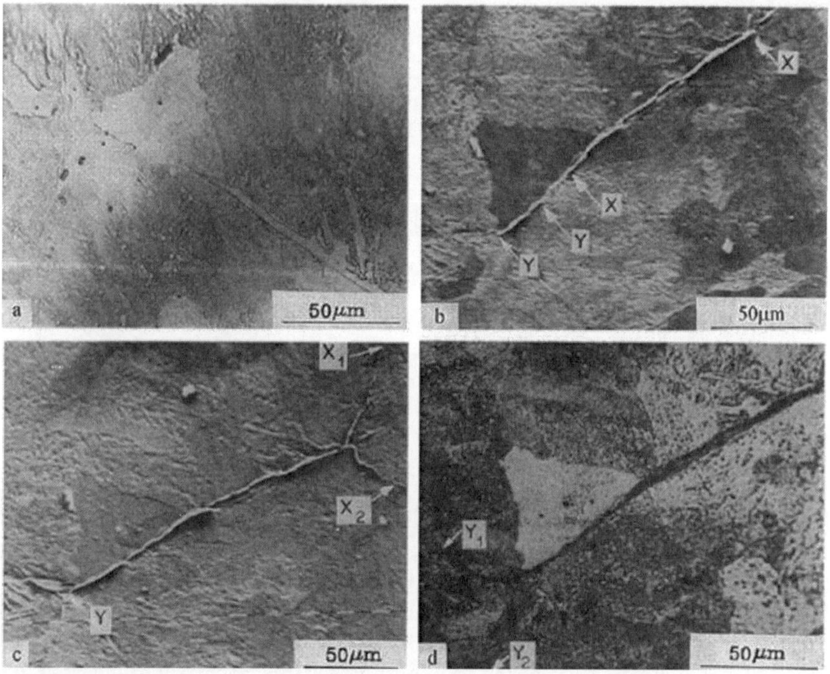

Fig. 2.50 MSC along a ferritic grain boundary in a 0.4% C steel: **a** before cycling, **b** two non-propagating cracks after about 21,000,000 low amplitude cycles, **c** increasing the stress level, after additional 46,763 cycles the two micro cracks join and propagate bifurcating (X_1 and X_2). Finally, **d** after 600,000 more cycles also the second crack tip bifurcates (Y_1 and Y_2) [74]

Under a stress range $\Delta\sigma$ lower than the fatigue life $\Delta\sigma_f$ the micro crack decelerates and stops at the grain boundary after traveling a distance *d*. have any effect. Stage II of fatigue starts at once.

At this point, if the stress range increases to $\Delta\sigma_1$ just a little bit over the fatigue limit, the grain boundary will be overcome and the same crack will start to grow again, but this time accelerating. At variance with that, the growth rate retardation previously observed decreases progressively and the grain boundary stop avoided as the initial applied stress increases: $\Delta\sigma_3 > \Delta\sigma_2 > \Delta\sigma_1$, with $\Delta\sigma_1 > \Delta\sigma_f$. For a sufficiently high $\Delta\sigma_4 > \Delta\sigma_3$ the deceleration will not occur at all. Finally, if a crack longer than 200–300 μm already exists on the surface, because a long surface inclusion broke or because a scratch has been inadvertently introduced or for whatever reason, both initiation phase and propagation to a macroscopic size will be skipped over, i.e. Stage I of fatigue, and even a stress range $\Delta\sigma_o$ lower than the fatigue limit $\Delta\sigma_f$ will be enough to drive the crack. For such a large crack microscopic barriers will not longer exist.

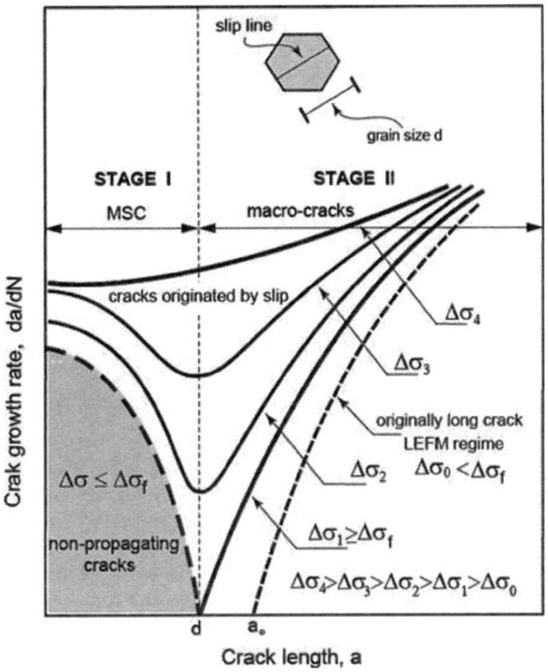

Fig. 2.51 Schematic of fatigue crack growth regimes relative to different combination of crack size and stress amplitude

Grain boundary, as said, may stop the growth but if the stress amplitude is above the fatigue limit σ_f and the cycling continues, the damage multiplication and accumulation will give rise to a stop and go situation in which the crack first arrests and then will jump again just to stop temporarily to the next barrier, and so on. This behavior can be seen in Fig. 2.52 from Akiniwa et al. [77] that describes this stop and go behavior in aluminum alloy 2024-T3. Interesting enough, the crack originated from a rather large Al_7Cu_2Fe inclusion, as already discussed for aluminum alloys. Other inclusions (points B, D and F), though much smaller, are also acting as barrier to the micro crack growth. The crack growth rate precipitates to zero when the barrier is reached and then restarts up to about 10^{-6} mm per cycle. It can be also seen how the micro crack once reached an overall macro size of about 300 μm doesn't stop any more. The method used to measure crack growth rates is always the replica procedure described in the previous section. The same behavior is seen in Fig. 2.53 for a surface crack growth in aluminum alloy 7475 [78]. Also in this case we can see how the micro crack stops at grain boundaries twice, but once it becomes a macro crack (~300 μm) its growth is no longer influenced by microstructural barriers and it doesn't arrest any more.

For this aluminum alloy 7074-T651, Lankford [79, 80] provides the diagram of Fig. 2.54 which is of the type shown in Fig. 2.51. In this case the threshold ΔK_{th}, Eq. (1.20), is about 4 MPa$\sqrt{m}$, very small. This value refers to a macro crack 160 μm long that had already broken three grains. In all cases shown, cracks originated from

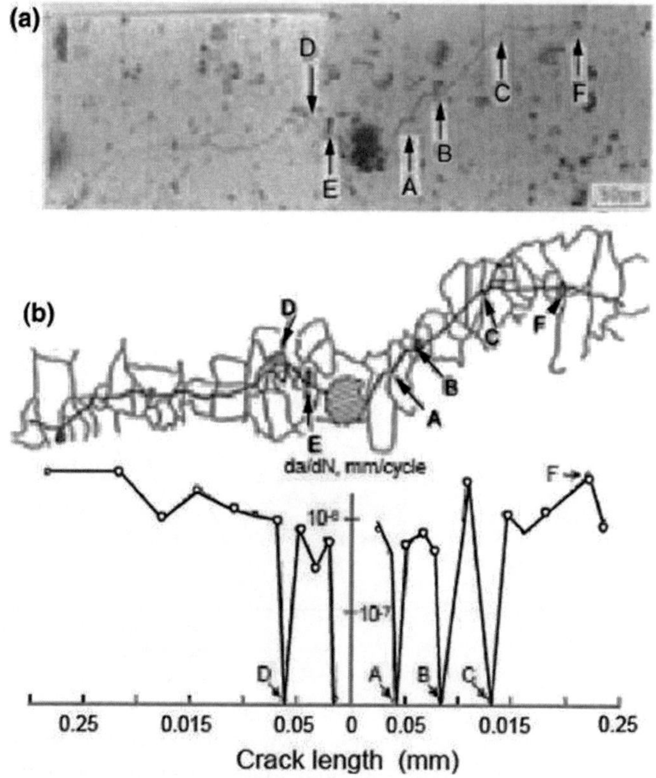

Fig. 2.52 Stop and go behavior of a micro crack originated from a rather large Al_7Cu_2Fe inclusion in 2024-T3 aluminum alloy [77]. Note that macro-crack formation (>300 μm) results in continuous propagation without any further arrest

Table 2.1 Values of constant *A* and *n* in (2.20)

Steel	T (°C)	A	n
Mild	25	$3.33 \cdot 10^{-41}$	13,13
	288	$9.54 \cdot 10^{-34}$	10,03
Low-alloyed	25	$1.45 \cdot 10^{-36}$	11,10
	288	$1.07 \cdot 10^{-43}$	

inclusions, which, again, is typical of aluminum alloys (see also Fig. 2.32). It can be seen how all curves merge into a single one relative to the long crack behavior. Interesting enough, the crack growth rates derived from striation spacing observed on the fracture surface by scanning electron microscope (SEM) do not agree with those calculated through the specimen compliance. This is because at those very low crack growth rates it takes more than a single cycle to leave a marking on the fracture

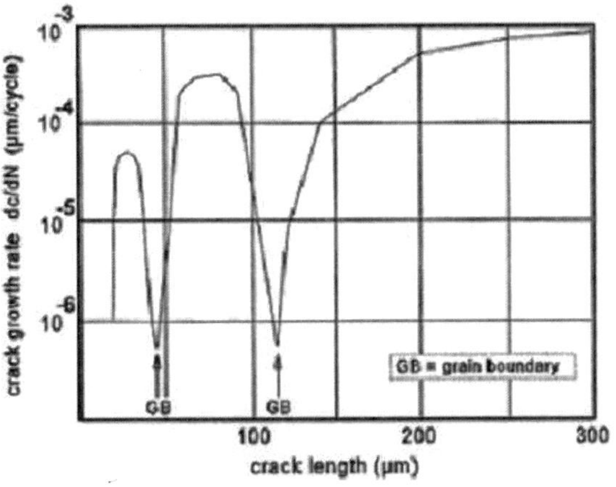

Fig. 2.53 Stop and go behavior of a micro crack in aluminum alloy 7075 [78]. Once it exceeds 300 m the crack doesn't stop any more

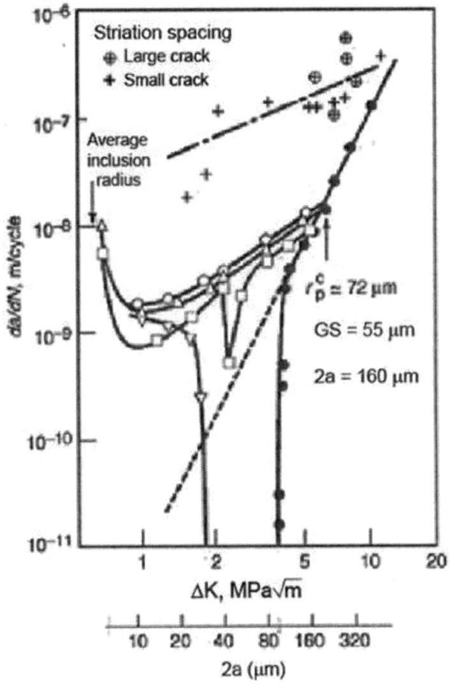

Fig. 2.54 Crack growth rates obtained by Lankford on aluminum alloy 7075-T651 [79, 80]

surface, as it will be discussed in Section. Finally, Fig. 2.55 summarizes the main features of this chapter.

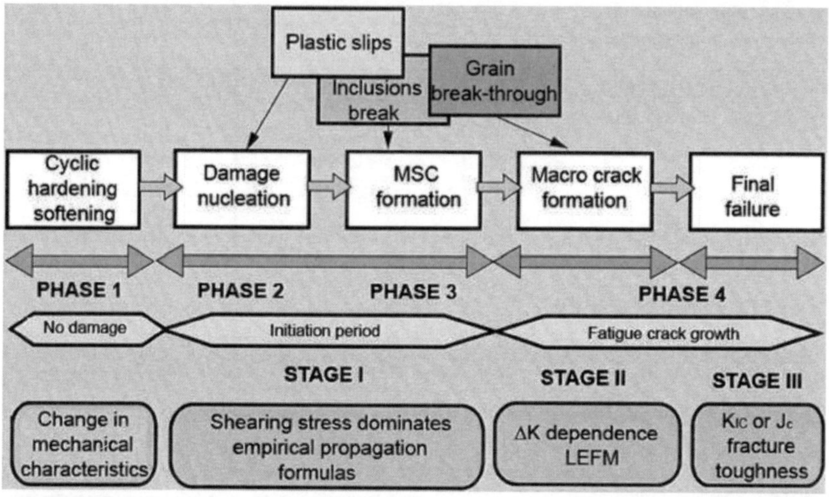

Fig. 2.55 Synthesis of the four phases and three stages of fatigue that characterize fatigue initiation and propagation

References

1. Parker, E., Washburn, J.: Deformation of single crystals, ASM Seminar on Modern Research Techniques in Physical metallurgy, pp. 186–204. (1952)
2. Emery, J.D., Shull, K.R., Rondinelli, J.M., Wolverton, C.M., Introduction to Material Science and Engineering, Noethwestern University
3. Ewing, J.A., Humphrey, J.C.: The fracture of metals under repeated alternation of stress. Phil. Trans. R. Soc. A **200**, 241–250 (1903)
4. Kocanda, S.: Fatigue Failure of Metals. Sijthoff & Noordhoff Int. Pubs. (1978)
5. Hempel, M.R.: Fracture, The Technology Press and Wiley & Sons, Inc., New York, pp. 376. (1959)
6. Polak, J.: Cyclic Plasticity and Low-cycle Fatigue Life of Metals. Elsevier (1991)
7. Thompson, N., Wadsworth, N.J., Louat, N.: Philosophical Magazine, **1**(113) (1956)
8. Hempel, M.R.: International Conference on Fatigue. Institution of Mechanical Engineers, pp. 543. (1956)
9. Hempel, M.R.: Fatigue in Aircraft Structure. Columbia University, Academic Press, New York, N.Y., pp. 83. (1956)
10. Forsyth, P.J.E.: In: Crack Propagation Symposium, Cranfield (1961)
11. Smith, G.C.: Proc. Royal Soc. **A**(242), 189 (1957)
12. Jaquet, P.A.: International Conference on Fatigue of Metals. Institute of Mechanical Engineers, ASME, London (1956)
13. Forsyth, P.J.E, Stubbington, C.A.: Nature, London, vol. 175, pp. 767. (1955)
14. Mughrabi, H., Ackermann, F., Herz, K.: American Society for Testing and Materials, STP 675, pp. 69 (1979)
15. Lawrence, F.V.: Mechanisms of fatigue crack initiation and growth. Metal Trans. **4**, 367–393 (1970)
16. Carstensen , J.V.: Structural evolution and mechanisms of fatigue in polycrystalline brass. Risø R-1500(EN), Risø National Laboratory, Roskilde, March (1998)
17. Fine, M.E., Kwon, I.B.: Fatigue crack initiation along slip bands. The Behaviour of Short Fatigue Cracks, EGF 1, Mechanical Engineering Publications, pp. 29–40 (1986)

18. French, H.J.: Trans. Amer. Chem. Soc. Steel Treatm. **21**, 899 (1933)
19. Vašek, A., Polák, J.: Kovové Materiály **29**, 113 (1991)
20. Ma, B.T., Laird, C.: Acta Metall. **37**, 337 (1989)
21. Ashami, K., Sugiyama, Y.: Fatigue strength of various surface hardened steels. Japanese Heat Treatm. Technol. Assoc. **25**(3), 147–150 (1985)
22. Kanazawa, K., Nishijima, S.: Fatigue of low alloy steel at ultra-high cycle regime under elevated temperature conditions. Japanese Soc. Mat. Sci. **46**(12), 1396–1401 (1997)
23. Kaziymyrovych, V.: Very high cycle fatigue of engineering materials—A literature review, Karlstad University Studies (2009)
24. Sakai, T., Sato, Y., Oguma, N.: Characteristic S-N properties of high-carbon-chromium-bearing steel under loadin in long life fatigue. **25**(8–9), 765–773 (2002)
25. Wang, Q.Y., Berard, J.Y., Rathery, S., Bathias, C.: High cycle fatigue crack initiation and propagation behaviour of high-strength spring steel wires. Fatigue Fract. Eng. Mater. Struct. **22**(8), 673–677 (1999)
26. Bathias, C., Drouillac, L., Le Francois, P.: How and why the fatigue S-N curves does not approach a horizontal asymptote. Int. Journal of Fatigue **2001**(23), 143–151 (2000)
27. Murakami, Y., Endo, M.: Effects of defects, inclusions and inhomogeneities on fatigue strength. Int. J. Fatigue **16**(3), 163–182 (1994)
28. Murakami, Y., Nomoto, T., Ueda, T.: Factors influencing the mechanism of superlong fatigue failure in steels. Fatigue Fract. Eng. Mater. Struct. **22**(7), 581–590 (1999)
29. Murakami, Y., Yokoyama, N.N.: Influence of hydrogen trapped by inclusions on fatigue strength of bearing steel. Bearing Steel Technology, American Society for Testing and Materials ASTM STP 1419, pp. 113–124. (2002)
30. Murakami, Y.: Metal Fatigue: Effects of Small Defects and Nonmetallic Inclusions. Elsevier Science Ltd, Kidlington, Oxford OX5 1GB, UK (2002)
31. Frost, N.E., Marsh, K.J., Pook, L.P.: Metal Fatigue. Clarendon Press, Oxford (1974)
32. Haibach, : A likely explanation of very high cycle fatigue in steel. In: 5th International Conference on VHCF, Berlin (2011)
33. Stanzl-Tschegg, S.E., Mayer, H.: Fatigue and fatigue crack growth of aluminium alloys at very high number of cycles. Int. J. Fatigue **23**, 231–237 (2001)
34. Forrest, P.G., Tate, A.E.L.: The Influence of grain size on the fatigue behavior of 70/30 brass. J. Instit. Metals **93**, 438 (1964–1965)
35. Sinclair, G.M., Craig, W.J.: Influence of grain size on work hardening and fatigue characteristics of alpha brass. ASM Trans. **44**, 929–948 (1952)
36. Höppel, H.W., Prell, M., May, L., Göken, M.: Influence of grain size and precipitates on the fatigue lives and deformation mechanisms in the VHCF-regime, Fatigue 2010, pp. 1025–1034. Elsevier Ltd (2010)
37. Kunio, T, Shimizu, M., Yamada, K.: Microstructural aspects of fatigue behavior of rapid-heat-treated steels. In: Proceedings of 2nd International Conference on Fracture, England, Chapman and Hall, pp. 630–642 (1969)
38. Kunio, T, Yamada, K.: Microstructural aspects of threshold condition for non-propagating fatigue cracks in martensitic-ferritic structures. Fatigue Mechanics, ASTM STP-675, pp. 342–370. (1979)
39. Klesnil, M., Holzmann, M., Lukáš, P., Ryš, P.: Some aspects of the fatigue behavior of rapid-heat-treated steel. J. Iron and Steel Instit. **203**, 47 (1965)
40. Taira, S., Tanaka, T., Hoshina, M.: Grain size effect on crack nucleation and growth in long life fatigue of low-carbon steel. Fatigue Mechan. Amer. Soc. Testing and Mater. ASTM STP **675**, 135–173 (1979)
41. Murakami, Y., Matsuda, K.: Proceedings of Fatigue 87. In: Ravichandram, K.S., Ritchie, R.O., Murakami, Y. (eds.) Small Fatigue Cracks, pp. 119–130. Elsevier, Oxford, UK (1999)
42. Nisitani, H.: Behavior of small cracks in fatigue and relating phenomena. The Society of Materials Science, Japan, Current Research in Fatigue Cracks, vol. **1**, pp. 1–26. Elsivier Applied Science (1987)

43. Kitagawa, H., Takahashi, S.: In: Proceedings Second International Conference on Mechanical Behavior of Materials, ASM, pp. 627. (1976)
44. Lukas, P., Kunz, L.: Mat. Sci. Eng. **47**, 93 (1981)
45. Kunio, T., Shimuzu, M., Yamada, K., Tamura, M.: In: Beevers C.J. (ed) Fatigue 84, Emas, Warley, pp. 817 (1984)
46. Miller, K.J.: Fatigue Fract. Eng. Mater. Struct. **10**, 93 (1987)
47. Perez Carbonell, E., Brown, M.W.: A study of short crack growth in torsional low cycle fatigue for a medium carbon steel. Eng. Mater. Struct. **9**, 15–33 (1986)
48. Lukáš, P.J., Kunz, L.: In: Miller K.J., De los Rios E.R. (eds) Short Fatigue Cracks, ESIS 13, Mechanical Engineering Publications, London, pp. 265. (1992)
49. Fine, M.E., Kwon, I.B.: Fatigue crack initiation along slip bands. The behaviour of short fatigue cracks, EGF 1, pp. 29–49. Mech. Eng. Publications (1986)
50. Frost, N.E.: Initiation stress and crack length in mild steel. In: Proceedings of Institution of Mechanical Engineers, vol. **173**, pp. 811. (1959)
51. Frost, N.E.: Stress analysis and growth of cracks. J. Mech. Eng. Sci. **2**, 109 (1960)
52. Frost, N.E.: Alternating stress required to propagate edge cracks in copper and nickel chromium alloy steel plates. J. Mech. Eng. Sci. **5**, 15 (1963)
53. Frost, N.E., Dugdale, D.S.: Fatigue tests on notched mild steel plates with measurements of fatigue cracks. J. Mech. Phys. Solids **5**, 182 (1957)
54. Frost, N.E., Grenan, A.F.: Cyclic stress required to propagate edge cracks in eight materials. J. Mech. Eng. Sci. **6**, 203–210 (1964)
55. Kobayashi, H., Nakazawa, H.: The effects of notch depth on the initiation, propagation and non-propagation of fatigue cracks. Trans. Japan Soc. Mech. Engrs. **35**, 1856–1863 (1969)
56. Murakami, Y., Endo, T.: Effect of small defects on the fatigue strength of metals Int. J. Fatigue **2**, 23–30 (1980)
57. Murakami, Y., Endo, M.: Quantitative evaluation of fatigue strength of metals containing various small defects or cracks. Eng. Fract. Mech. **17**(1), 1–15 (1983)
58. Murakami, Y., Endo, M.: Effects of defects, inclusion and inhomogenities on fatigue strength. Int. J. Fatigue **16**(3), 163–182 (1994)
59. Murakami, Y., Kodama, S., Konuma, S.: Quantitative evaluation of effects of non-metallic inclusions on fatigue strength of high strength steels. Trans. Jpn. Soc. Mech. Eng. A **54**(500), 688–696 (1988)
60. El Haddad, M.H., Topper, T.H., Smith, K.N.: Prediction of non-propagating cracks. Eng. Fract. Mech. **11**, 573–584 (1979)
61. Alarsón, M.V.G., Castro, J.T.P., Meggiolaro, M.A.: Fatigue specimens design to induce non-propagating short cracks. In: 67th ABM International Congress, July, 31st, Rio de Janeiro, Brazil (2012)
62. Forsyth, P.J.E.: Fatigue damage and crack growth in aluminium alloys. Acta Metallurgica 11, 703–715 (1963)
63. Clark, W.G.: How fatigue crack initiation and growth properties affect material selection and design criteria. Metals Eng. Quart. **16** (1974)
64. De los Rios, E.R., Sun, Z.Y., Miller, K.J.: The effect of hydrogen in short fatigue crack growth in an Al-Li alloy. Fatigue Fract. Eng. Mater. **16**(12), 1299–1308 (1993)
65. Thompson, N., Wadsworth, N.J.: Metal fatigue. Adv. Phys. **7**(25), 72 (1958)
66. Leis, B.J., Ahmad, J., Kanninen, M.F.: Effect of local stress state on the growth of short cracks. Multiaxial Fatigue, ASTM-STP 853, pp. 314–339 (1985)
67. Dowling, N.E., Begley, J.A. Fatigue crack growth during gross plasticity and the J-integral, ASTM-STP 590, American Society for Testing and Materials, pp. 99 (1976)
68. Hobson, P.D.: The formulation of a crack growth equation for short cracks. Fatigue of Eng. Mater. Struct. **5**, 323–327 (1982)
69. Lankford, J.: The growth of small fatigue cracks in 7075–T6 aluminium alloy. Fatigue of Eng. Mater. Struct. **5**, 233–248 (1982)
70. Brown, M.W.: Interference between short, long and non-propagating cracks. In: Miller J.M., de los Rios E.R. (eds) The Behaviour of Short Cracks, EGF Pub. 1, Mechanical Engineering Publication, London, pp. 423–439. (1986)

71. Tokaji, K., Ogawa, T., Osako, S.: The growth of microstructurally small fatigue cracks in a ferritic-pearlitic steel. Fatigue Fract. Eng. Mater. Struct. **11**, 331–342 (1988)
72. Kitagawa, H., Takahashi, S.: Applicability of fracture mechanics to very small cracks or the cracks in the early stages. In: Proceedings of 2nd International Conference Mechanical Behavior of Materials, Boston, pp. 627–631. (1976)
73. Dowling, N.E.: Crack growth during low-cycle fatigue of smooth axial specimens. ASTM STP **637**, 97–121 (1977)
74. De los Rios, E.R., Tang, Z., Miller, K.J.: Short crack fatigue behavior in a medium carbon steel. Fatigue Fract. Eng. Mater. Struct. **7**, 97–108 (1984)
75. Hobson, P.D.: The Growth of Short Fatigue Cracks in a Medium Carbon Steel, Ph.D. Thesis, University of Sheffield (1985)
76. Chopra, O.K.: Environmentally assisted cracking in light water reactors, US NRC NUREG/CR-4667, Vol. 30, Semiannual Report January 2000-June 2000, Prepared by Chopra, O.K. et al. (ed) June (2001)
77. Akiniwa, Y., Tanaka, K., Matsui, E.: Statistical characteristics of propagation of small fatigue cracks in smooth specimens of aluminum alloy 2024–T3. Mater. Sci. Eng. A **104**, 105–115 (1988)
78. Blom, A.E, Edlund, A., Zhao, W., Fathalla, A., Weiss, B., Stickler, R.: Short fatigue crack growth in Al 2024 and Al 747. In: Symposium on Behaviour of Short Fatigue Cracks, 37–76, EGF 1, Sheffield, September, (1985)
79. Lankford, J.: The growth of small fatigue cracks in 7075–T6 aluminum. Fatigue Fract. Engng. Mater. Struct. **5**, 233–248 (1982)
80. Lankford, J.: The influence of microstructure on the growth of small fatigue cracks. Fatigue Fract. Eng. Mater. Struct. **8**(2), 168 (1985)

Chapter 3
Morphological Aspects of Fatigue. Crack Formation and Growth

3.1 Introduction

The purpose of this chapter is to provide some basic information about the morphological aspects associated with the various fatigue processes that take place in materials during cycling. It represents an essential background for the study of fatigue and the comprehension of what may have happened in a work piece that failed by fatigue and why it failed. We have seen in the previous chapter that any single cycle of fatigue introduces in the material a damage that is initially confined to a single or, at most, some surface grain within a single persistent slip band in the form of cavities or pores (see Fig. 2.12). Though sub-microscopic in size, this damage may grow by the coalescence of pores to become a microcrack that, if the cyclic stress level is sufficiently high, will break through several grains producing a macrocrack through a continuing to-and-fro slip process. The macrocrack will eventually penetrate into the material. A macrocrack produced by persistent slip band formation is something not visible at naked eye since, at the beginning, it is just 300 μm long. When this macrocrack grows at each applied fatigue cycle it leaves on the fracture surface some characteristic features that can be detected with the aid of optical microscopy and electron microscopy. Therefore, a very important step in fatigue design, alas too often forgotten, is the *post mortem* examination of failed pieces.

On the fracture surface is written the complete story of its past. To know that story means to foresee the destiny of future tests or real workpieces. In the same way, these high resolution examination techniques used on test pieces are devoted to the experimental determination or verification of the lows and processes governing the entire process of fatigue failure in any operational condition and develop the necessary know how for fatigue fail-safe design. What we actually need is the key to decode and interpreter the characteristic features that can be seen at naked eye and, above all, those we cannot see at naked eye. Dough to day engineers know by general line what is a SEM or TEM and their applications, it may be convenient to spend few words. Electrons may go beyond what can be seen with visible light

P. P. Milella, *Fatigue and Corrosion in Metals*,
https://doi.org/10.1007/978-3-031-51350-3_3

(optical microscopy). Electrons can be used in transmission electron microscopy (TEM) used as X-ray in radiography. The difference is that they do not have such a high energy to go through a thick layer of steel or metal. High energy electrons, in fact, can cross just ultra-thin specimens no thicker than a thousand of angstroms (0.1 μm).

Therefore, fracture surface cannot be examined directly by a transmission electron microscope. Because of that, it is necessary to transfer the microscopic features we want to see to a thin, electron-transparent replica. This is done by covering the surface under investigation with a liquid plastic material as acetate. This liquid will enter any tiny detail of the wetted surface that will remain imprinted on the plastic when it hardens. After solidification, the plastic is stripped very carefully from the mating surface. The next step is to place the plastic replica in a vacuum chamber where a thin layer of carbon is deposited by evaporation of a pair of carbon electrode. Some heavy metal, such as gold or platinum, can be used instead, which will enhance contrast. At this point, the plastic is dissolved in acetone and the carbon replica remains, which has kept all details of the original mating surface. The replica is thin enough to be traversed by a high energy electron beam but it is also too thin to have any strength and be manipulated so that it is recovered on a copper grid of about 80 to 100 microns spacing that becomes its support any time we need to handle it. It is obvious that the original surface details will remain on the carbon replica upside down, so that original valleys will appear like dimples and holes like protruding. Magnifications approaching 1,000,000 × are possible with TEM. TEM is used to observe dislocations, as shown in Fig. 1.27.

However, there is another mode to use electrons as they were waves of very high frequency and, therefore, very low wavelength. Quantum mechanics has already come to age making clear the dual nature of electrons that can behave, at the same time, as particles or waves. As waves they are much more effective than light since their wavelength is some order of magnitude smaller enabling the vision of much smaller details that optical microscopy cannot *see*. In this relatively new use, electrons don't traverse the specimen but are reflected by its surface producing an image as done by light. With the introduction of the scanning electron microscope it has become possible to observe the fracture surface or the surface of the work piece directly. Actually, the electron beam impinging the surface to be examined is not precisely reflected. The high energy electron beam of small diameter scans across the surface and due to the excitation of these primary electrons, secondary electrons are emitted by the target surface as if they were reflected electrons. These secondary electron beam produce an image of the surface under investigation which can be collected and made visible.

The image that appears on the screen, which may be photographed, represents the surface features of the specimen. The surface must be electrically conductive. To this purpose, a very thin metallic coating must be applied to nonconductive materials. Magnifications ranging from 10 to in excess of 50,000 diameters and also very great depths of field are possible. Modern, advanced fatigue design cannot skip over the instrumental aspect and be confined to calculations, even dough to day designers may dispose of very advanced computer codes. Structures and components still fail, though computer designed, almost as they did when engineers used rulers. However, the today technological edge is rather much more advanced than yesterday ambitions. Too often engineers and technicians are running sophisticated calculations that do not match with what has been built. It is, then, fundamental to understand how and why they failed and in this effort optical and electron microscopy plays a fundamental role.

3.2 Extrusions and Intrusions

It has been said in Chap. 2 that in high cycle fatigue initial damage nucleates in some surface grains along persistent slip bands in the form of pores or submicroscopic cracks. In those bands plastic deformation concentrates and slip is an irreversible process due to damage. Slips are initiated by resolved shearing stress reaching the critical value $\tau_{r,cr}$ just in some grains on the surface (see Sect. 2.1 and Fig. 2.3) where plain stress conditions prevail. There are several models describing the micro crack formation in slip lines. Interesting is the model proposed by Neumann [1] and schematized in Fig. 3.1. According to this model, during the traction phase of the cyclic stress, Fig. 3.1a, a dislocation source is activated and slip occurs along a preferential direction 1 on a crystallographic plane. A submicroscopic discontinuity is then formed that influence another source of dislocation and a new slip takes place along line 2 perpendicular to the first (intersecting slip), Fig. 3.1b.

When load is inverted, both planes 1 and 2 are activated, but in the opposite direction living the surface in the condition depicted in Fig. 2.11. Apparently, when seen at naked eye, the surface has returned into the initial condition and the indentation disappeared, but the two faces that for a while have been exposed to air oxidizes, though very lightly, but this is enough to prevent surface re-cohesion and a damage remains indicated by arrow A in Fig. 2.1c. This let us believe that fatigue in vacuum may not exist or strike much less. This is partially true in that fatigue in vacuum is effectively much less damaging than in dry air (see Sect. 20.2), but not completely true since slips are the consequence of a plastic deformation that is not totally recoverable not just because of oxidation. Slips are, in fact, irreversible in nature and damage cannot completely heal. The repetition of loads keeps the basic process shown in Fig. 3.1a, b and c continue which produces further microscopic damage that accumulates and advances in a zigzag mode, as shown in Fig. 3.1j and eventually becomes a microcrack. But slip can also break free from the surface emerging and creating a microscopic defect called extrusion.

An extrusion is a metal lamina extremely thin that is extruded from the surface by the action of a fully reversed cycling. This is shown in Fig. 3.2 according to Forsyth [2]. Because extrusions are normally accompanied by cracks in the slip packet, they may be of significance in crack initiation. Intrusions are the inverse of extrusions and are narrow and shallow crevices. These surface discontinuities are approximately 10^{-4} to 10^{-5} cm in height and appear as early as some percent of the total life of a specimen.

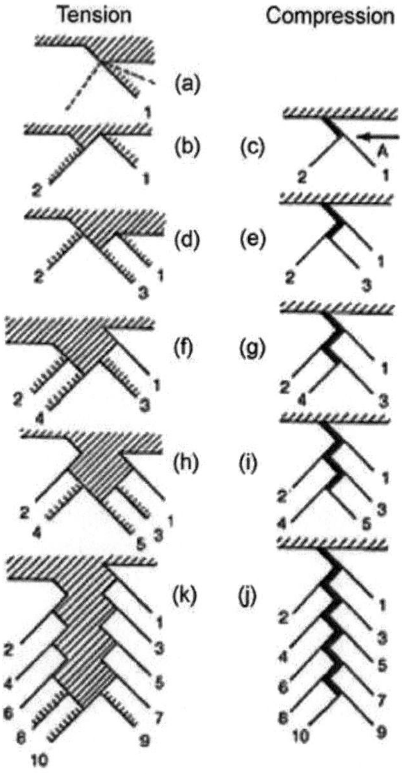

Fig. 3.1 Neumann model of micro crack formation during fatigue cycling [1]

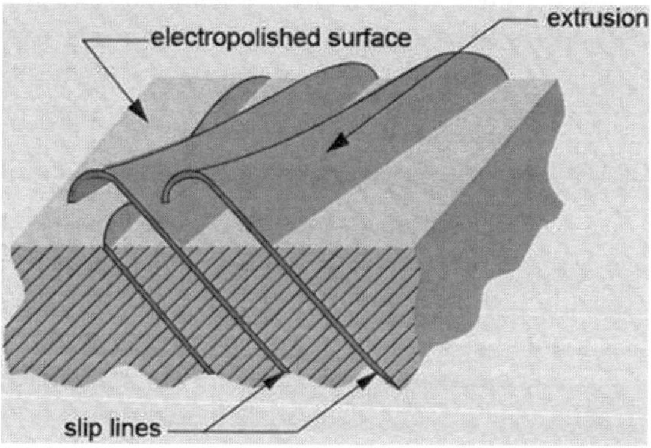

Fig. 3.2 Schematic of extrusions in copper according to Forsyth [2]

Forsyth first [2] and Forsyth and Stubbington later [3–6] report the extrusions formation on the surface of a 4½ Cu-aluminum specimen cold-working hardened in which slip lines were very fine. They were no thicker than 0.1 μm and about 10 μm long. Micro cracks were also present. At −196 °C they disappeared, but not the slip bands that became coarse. Also in 10% Zn-aluminum alloys they appeared, but this time also at −196 °C.

The presence of extrusion has been reported also in carbon steels [7] and hardened steels [8]. Cottrell and Hull [9] gave an interpretation of extrusions formation in FCC metals expanding the model of slips and slip bands already seen in Fig. 2.11. Their model is shown in Fig. 3.3. In the new scheme two slip lines perpendicular to each other are activated. First, an edge dislocation source S_1 becomes active and the crystal slips along the corresponding crystallographic plane and direction. This leaves the surface of the material in the form shown in Fig. 3.3a. Successively, but always in the same traction phase, a second source S_2 is activated having a slip plane at 45° with the external load, but perpendicular to the former plane (intersecting slips). The new appearance of the external surface is shown in Fig. 3.3b.

Note that in such a sliding the first slip plane moves upward relative to its original position. Therefore, in the successive phase of unloading when source S_1 on plane 1 is reactivated it creates a tooth A that in turn breaks the slip plane of source S_2, as it can be seen in Fig. 3.3c.When this second source is reactivated, during unloading, the surface has the appearance of Fig. 3.3d where the extrusion E and the intrusion I are formed. Without load inversion and consequent inversion of dislocation motion extrusions would not appear. A model of extrusion-intrusion formation activated by screw dislocations was also proposed by Mott [10]. An example of extrusion in copper and Fe-3 Si is offered in Fig. 3.4 [11]. Note that there is a difference in the appearance of slip bands in copper and Fe-3 Si. Copper exhibits wavy glide, as described in Sect. 1.4.1 and Fig. 1.31a, while Fe-3 Si exhibits more planar glide, as shown in Fig. 1.31c. Alloys of low stacking fault energy, such as Cu-7 Al, also develop slip bands that resemble those in Fe-3 Si rather than those in copper. It can be seen the loss of surface smoothness and the crack formation associated with extrusions. Another example of extrusion tongues a fracture surface of iron and copper under low cycle fatigue can be seen in Fig. 3.5a and b [12], respectively.

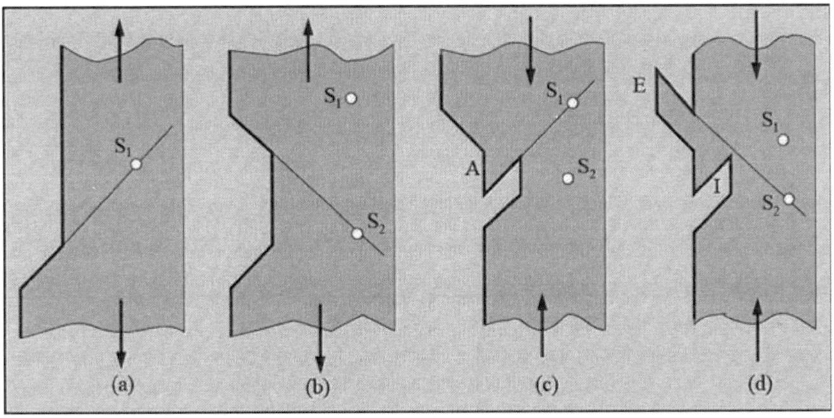

Fig. 3.3 Schematic of extrusion intrusion formation according to Cottrell and Hull

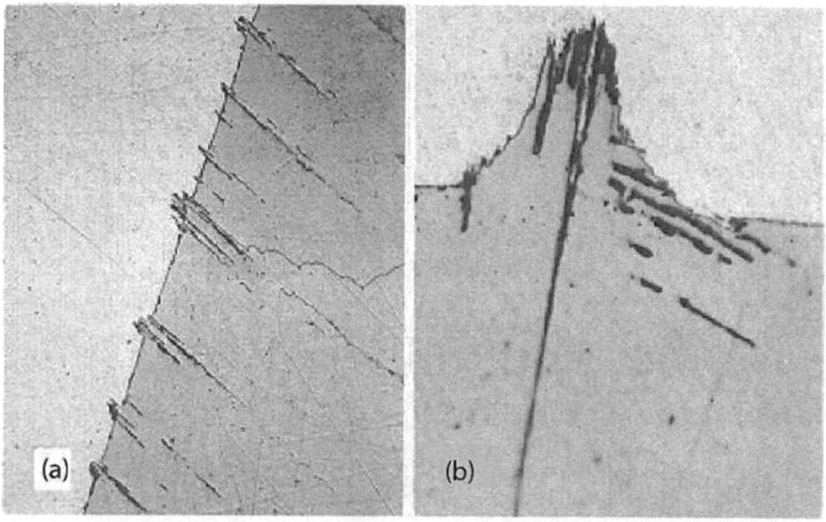

Fig. 3.4 Extrusions observed in a copper specimen **a** and in Fe-3 Si [11]

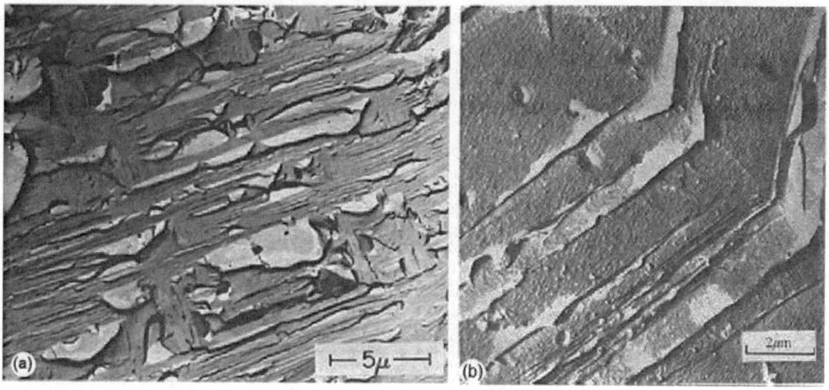

Fig. 3.5 TEM replica of extrusion tongues on a fracture surface of: **a** iron [12], **b** copper under low cycle fatigue [13]

3.3 Macroscopic Evidence of Fatigue Crack Propagation

Fatigue always leaves macroscopic signs on the fracture surface. These signs can be well distinguished by a naked eye observation. They appear as curved lines starting from the origin of fatigue somewhere on the external surface and moving towards the final fast fracture by overload. A rather comprehensive map of macroscopic appearance of possible fatigue failures in laboratory test specimens of various shape (cylindrical or prismatic) subjected to a wide range of different loading conditions, stress amplitude with or without stress concentration is offered in Fig. 3.6 [14]. The two characteristic features common to all cases listed in Fig. 3.6 are the white area within which less or more curved lines develops, also called beach marks or arrest lines, and the dark area that marks the final fast fracture by overload. The larger it is the lower the applied load. The density and width of beach marks is related with the applied stress level and number of cycles. Beach marks shall not be confused with striations.

Beach marks, also called clamshell marks, are macroscopic fatigue features that can be seen at naked eye while striations are microscopic in nature and can be seen only with the aid of electron microscopy. Their existence on the fracture surface may have different origins. As first, the beach marks are produced by changes in crack-growth rates when fatigue is applied in packages of consistent number of cycles. Each package produces a total crack extension that can be seen without any enlargement. But beach marks are also produced by other factors than simple changes in crack-growth rates during propagation. They are the external appearance of an entire block of applied cycles that has been marked by oxidation from air exposure of the free surface during the resting period. Oxidation marks the surface differently because of time of exposure and other environmental factors such as air temperature, humidity, pH, etc. and also loads spectra. In facts, many fatigue fractures produced under conditions of uninterrupted crack growth and without load variations do not exhibit beach marks. Therefore, each beach mark denotes a single package of cycles, not a single cycle, and signs the story of all rest periods and fatigue propagation periods.

At high magnification, thousands of fatigue striations (microscopic features) can be seen between two consecutive beach marks, as schematized in Fig. 3.7. Sometimes also the direction of stress cycling may have an additional effect on beach marks. This is the case shown in Fig. 3.8 [15] for a rail track that failed by fatigue. The rail shown was removed from service when a detail flaw was detected. The rail was subsequently installed in a test facility where it was tested using a train of 75 cars till fatigue failure occurred. The initial flaw is the dark circle from which the beach mark lines evolve as black and withe rings departing from the common origin. The reversal in the direction of the test convoy has produced the two different shadings of the beach marks.

Back to Fig. 3.6, we observe that beach marks always depart from a single or more initiation points. Generally, under moderate amplitude loads a single initiation site is observed, while high amplitude loads may generate two or more sites.

About the dark area that represents the final overload failure, in general when it is equal or even larger than the clear one interested by beach marks, as in the first column on the left of Fig. 3.6, it means that cyclic loads had high amplitude therefore after a relatively low number of fatigue cycles the remaining section failed by overload in the last 1/4 cycle (Stage III of fatigue).

The opposite happens when the load amplitude is low (columns on the wright of Fig. 3.6). We may say that low-cycle fatigue occurs when the ratio $A_{fatigue}/A_{collapse}$ between the surface interested by fatigue and that collapsed is less than 1. On the contrary, $A_{fatigue}/A_{collapse} > 1$ will be indicative of a long sufferance of the metal under high cycle fatigue that eventually fails. In unidirectional bending the initiation site is

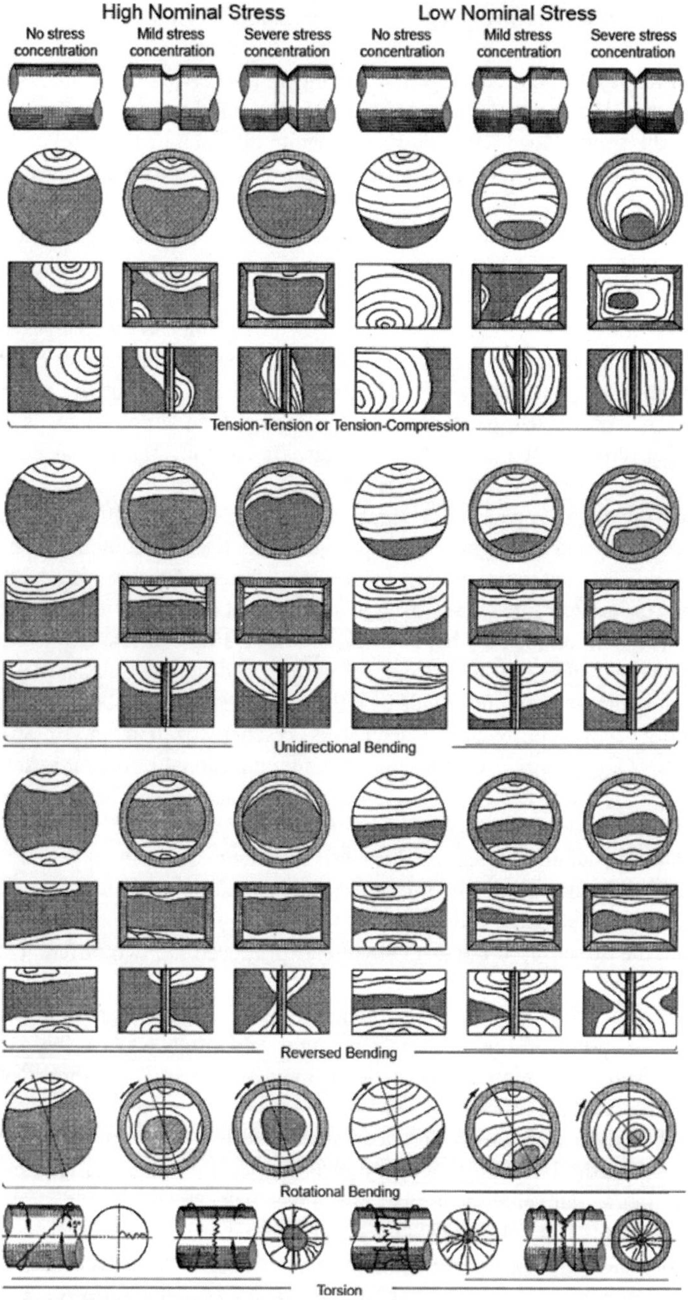

Fig. 3.6 Typical fracture surfaces for laboratory test specimens subjected to a range of different loading conditions [14]

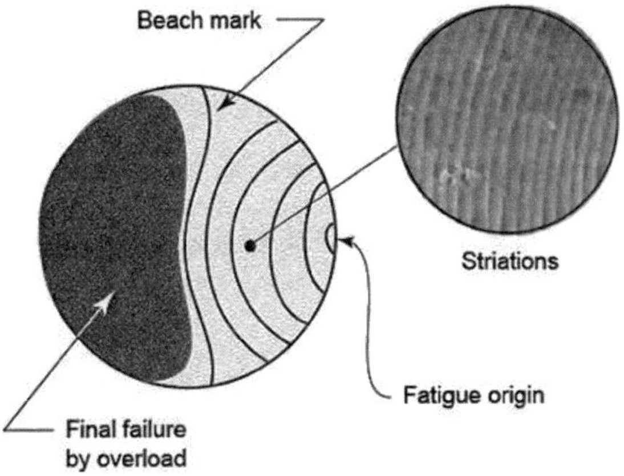

Fig. 3.7 Beach marks are packages of fatigue cycles visible at naked eye. By looking at SEM between two of them striations will appear that identify each individual cycle

Fig. 3.8 Beach marks in the form of circumferential rings visible at naked eye on the head of a rail failed by fatigue [15]. The initial crack is the dark circle at upper left, underneath the surface

located only on one side of the specimen where metal fibers are under traction, while in reverse bending the initiation site may appear on both sides. In rotational bending in which all fibers of the external circumference are equally stressed the initiation site is randomly located choosing the weakest point exactly as in traction.

Note how in rotating bending the collapse area has an elliptical shape with one of the two axes lightly inclined relative to fatigue direction. Experience teaches that an angle no lower than 15° is formed in the direction opposite to rotation.

The presence of concave beach marks denotes unidirectional bending therefore if they appear on the surface of a specimen tested in tension-tension or tension–compression we shall conclude that the traction was not perfectly axial and some bending occurred. Alike, if concave lines turn into convex beach marks we may conclude that unidirectional bending was not centered. Torsion is worth a discussion of its own. It produces a brittle fracture along a helical path, Fig. 3.9b, or a flat failure if the material behaves in a ductile fashion, as in Fig. 3.9c.The difference may be understood analyzing the stress state. Under pure torsional moment any element free body taken at an arbitrary point on the surface is subjected to pure shearing stress acting on longitudinal and transverse direction, whereas principal stresses acts at 45° being traction on one face of the element and compression on the other, Fig. 3.9a.

If principal stresses are high enough and the material is brittle, the traction stress will trigger a spiral fracture. Conversely, in case of ductile behavior it may take shearing stress to develop a fracture along a normal plane or axial, Fig. 3.9d. In this last case fissures will appear on the surface parallel to the axial direction. An example of helical fracture of a drive shaft of a scooter is shown in Fig. 3.10a [16]. The failure initiated at an inclusion indicated by the arrow. The second type of torsion failure (Fig. 3.9c or last row of Fig. 3.6, second case from the left) is shown in Fig. 3.11 [17]. It refers to an experimental 89 mm (3½ in.) diameter tractor axle of AISI 1041 steel that had been induction hardened. Note beach marks fanning out from the fatigue-crack origin (slightly to left of center, at top). Few beach marks are visible because cracking occurred soon after. It was of brittle type along the hardened perimeter, producing two sets of chevron marks pointing toward the crack origin, while the center (dark area), not hardened, failed in a ductile manner by overload.

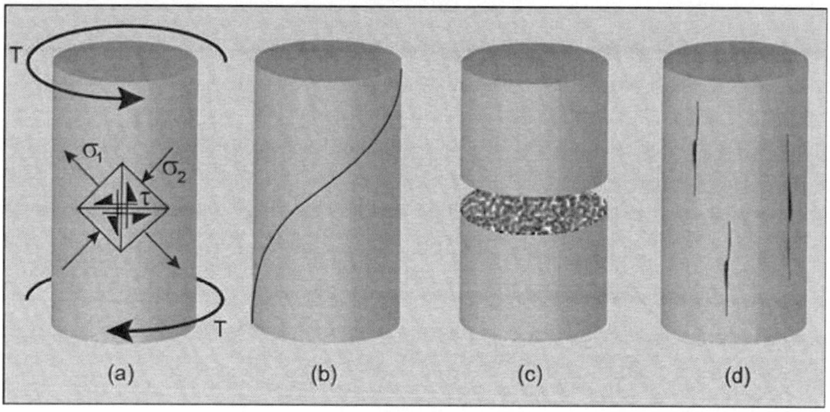

Fig. 3.9 Cylinder in pure torsion, **a** element free body stress state **b** brittle behavior with helicoidally failure, **c** and **d** ductile failure activated by either longitudinal or transverse shearing stress

Fig. 3.10 Helical fracture of the drive shaft of a scooter. The fracture was triggered by a surface inclusion indicated by arrow [16]; **a** mating halves, **b** separated halves

Fig. 3.11 Surface of a bending-plus-torsional-fatigue fracture in an experimental 89 mm (3½ in) diameter tractor axle of AISI 1041 steel that had been induction hardened [17]

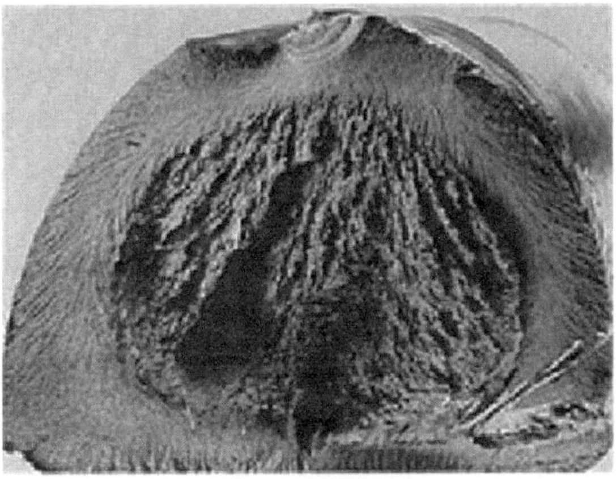

3.4 Microscopic Evidence of Fatigue Crack Propagation

In Sect. 2.7 and Fig. 2.47 we have clearly distinguished the three stages of fatigue. In the first stage that terminates with a macrocrack formation, the fatigue damage proceeds through a slip process on crystallographic planes and directions. It is fed by the resolved shear stress. Once a macrocrack is formed, its growth usually occurs through different mechanisms depending on the applied stress amplitude σ_a or, better, ΔK, which is the most proper parameter to use in the presence of a crack. This is shown in Fig. 3.12 for Ti-6Al-4 V alloy [18], but it can be considered as general.

At very low ΔK, close to the threshold ΔK_{th}, the crack grows through cleavage-like faceted and river pattern morphology, in particular in planar slip materials (see Sect. 1.4.1 and Fig. 1.31), classic evidence for crystallographic fracture. Some of these characteristic morphologies seen at SEM in ductile cast iron are shown in Fig. 3.13 and in Fig. 3.14 for alloy inconel 718 tested at room temperature [19] and 18–8 austenitic stainless steel in rotating beam specimen [20], respectively. Fatigue crack growth rate in inconel 718 occurred at $\Delta K = 14$ MPa$\sqrt{\text{m}}$ and $da/dN = 2{\cdot}10^{-3}$ μm/cycle. Metals as austenitic stainless steels and aluminum fracture in this way. The mode is transgranular. As the applied ΔK increases, Fig. 3.12, the

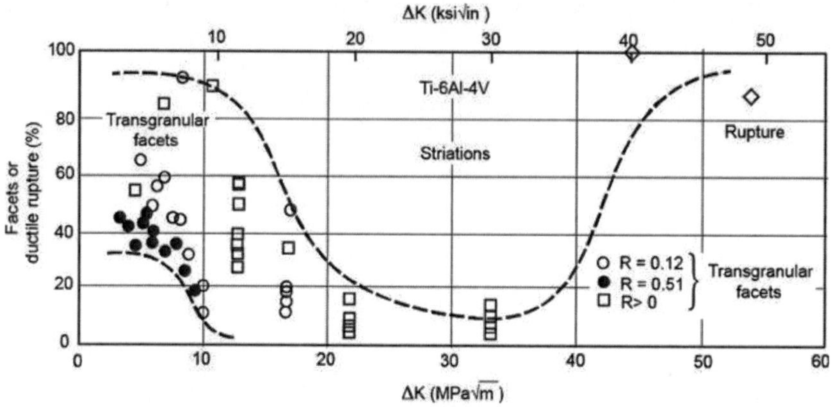

Fig. 3.12 Effect of ΔK on fatigue fracture mechanism in alpha–beta titanium alloy (adapted from [18])

morphology of crack propagation gradually changes into striations (see next section). Now the crack is already long enough (>300 μm) to escape shearing stress control and be driven by normal stress which produces a continuous growth, cycle by cycle, on a plane that is no longer a crystallographic plane (plane of higher close-packed atom stacking), but simply normal to the external applied loads.

At the beginning of striation appearance, between 2×10^{-2} and 10^{-1} μm/cycle, evidence of faceted growth still remains, the lower the stacking fault energy of the material. Then striations become the dominant morphological feature characterizing fatigue propagation (see Figs. 14.22 and 14.23). Increasing the applied ΔK the material approaches and eventually enters Stage III of fatigue, Fig. 2.47. Striations that dominated the fatigue propagation mechanism now tend to disappear. For Inconel 718 and 305 stainless steel this happens above 40 MPa$\sqrt{}$m, while for InconelX-750 or steels it starts at about 90 MPa$\sqrt{}$m. Stage III (see Fig. 2.47) that concludes fatigue

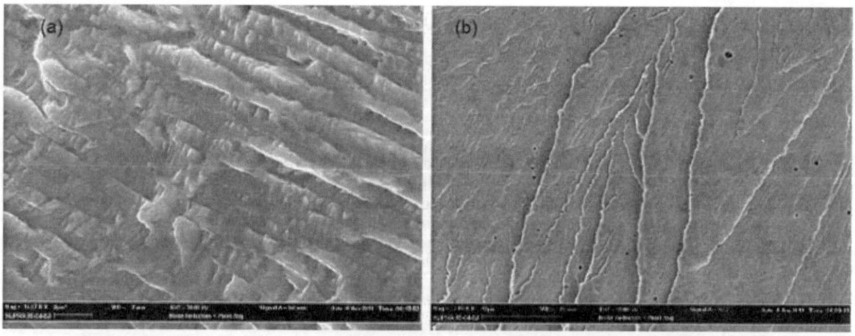

Fig. 3.13 SEM micrographs of the fracture surface of ductile cast iron at very low fatigue crack growth: **a** crystallographic striations, **b** river patterns fatigue (Milella)

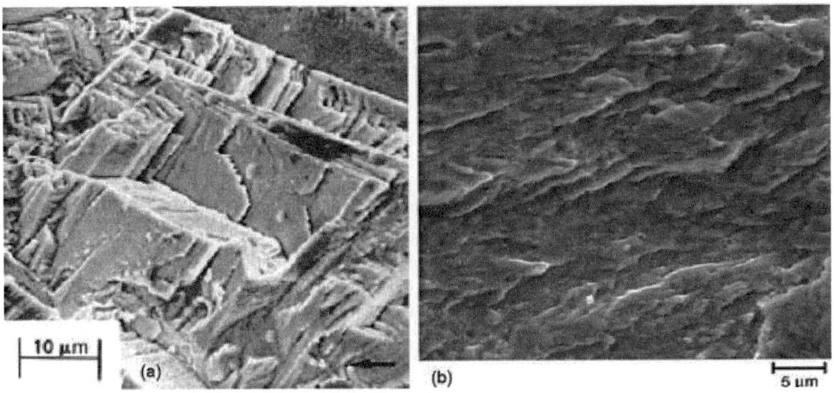

Fig. 3.14 SEM micrographs of fracture surface showing: **a** faceted stair-step appearance in Inconel 718 [19] and **b** crystallographic morphology in 18–8 austenitic stainless [20]

growth may happen in two steps. In the first, the crack continues to growth by fatigue, but the driving stress field increases continuously because the remaining ligament ahead of the crack decreases.

Therefore, fatigue, even when it starts as high-cycle fatigue, proceeds to become eventually a low-cycle high stress amplitude fatigue and again shearing stresses happen to be the driving stress, but this time crack growth doesn't proceed on crystallographic planes through a slip process as in Stage I, but simply in a direction at 45° relative to the maximum stress, as schematized in Fig. 2.47.

A schematic of this low-cycle crack propagation process was given by Forsyth for aluminum alloys [21] and is presented in Fig. 3.15. The crack leaves the crystallographic plane to be driven by shear stresses crossing other crystallographic planes. Now striations leave room to a form of microvoid growth and coalescence typical of ductile fracture, as shown in Fig. 3.16. Note that in the large foreground void of Fig. 3.16 it is still possible to see half of the inclusion detached from the metal matrix that originated the cavity. The second half is lost or remained in the other part of the specimen. The smaller cavity in the upper left still contains both halves of the broken inclusion. The inclusions were small MnS particles. In ferritic steels at temperatures close or below the nihil ductility transition temperature (NDT), final fracture may occur by cleavage, as shown in Fig. 3.17 for a medium strength steel. Cleavage is a net severance of two crystallographic planes without any apparent plastic deformation. From a morphological point of view the three stages of fatigue are schematized in Fig. 3.18. Stage I represents a system of surface slips activated by the resolved shear stress $\tau_{r,cr}$ on crystallographic planes (see Sect. 2.1). They are grouped in persistent slip bands in which voids nucleate and coalesce to form a microcrack, as schematized in Fig. 3.18a.

An extrusion tongue may appear on the surface. When a microcrack grows to become a macrocrack Stage II starts, Fig. 3.18b, in which the crack is opened by normal stresses. The growth now loses any crystallographic feature. The crack tip

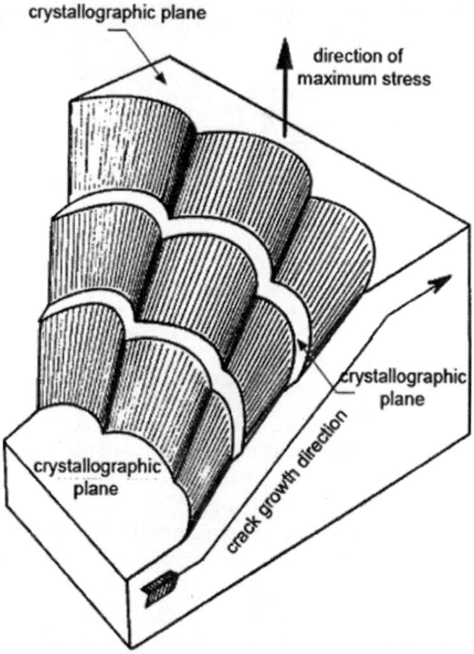

Fig. 3.15 Schematic of low cycle fatigue crack growth [21]. Crack propagation plane is not a crystallographic plane and is inclined relative to the maximum stress

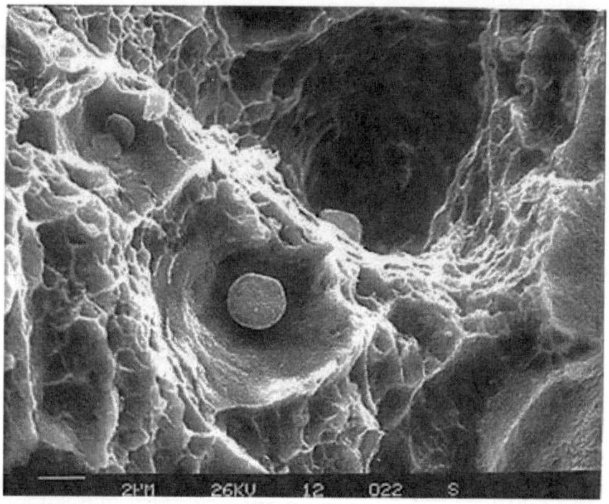

Fig. 3.16 SEM view of dimple fracture by decohesion of MnS inclusions from the metal matrix leading to voids formation, coalescence and fracture. Note the porous spongy texture (Milella)

blunts and two plastic lobes are formed at 60° to the direction of propagation. Even though these lobes are produced by shearing stresses they are orientated at about 60° with the crack plane, and not at 45°, as fracture mechanics claims for an edge crack. It is precisely this lobe formation that generates striations. Eventually, the high-cycle fatigue becomes a low-cycle fatigue due to the overload resulting from the continuous erosion of the resisting area. This process initiates Stage III of fatigue, as schematized

Fig. 3.17 SEM image of cleavage fracture in mild steel. Note the veined flat facets appearance (Milella)

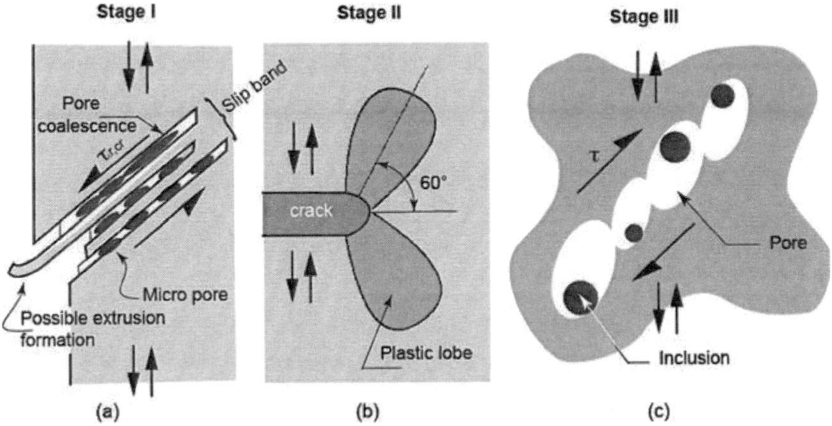

Fig. 3.18 Schematic of morphological features of: **a** Stage I, **b** Stage II and **c** Stage III of fatigue

in Fig. 3.18c that activates shear stresses large enough to tear the ligaments between voids grown around inclusions which coalesce. An interesting view of this kind of crack growth is given in Fig. 3.19 [22]. Due to stress concentration, two cavities formed ahead of the crack tip by inclusion decohesion have grown and coalesced, then the ligament between the tip of the main crack and these cavities failed by shear stress at about 45°. The ligament with the nearest cavity will fail next and the process continues.

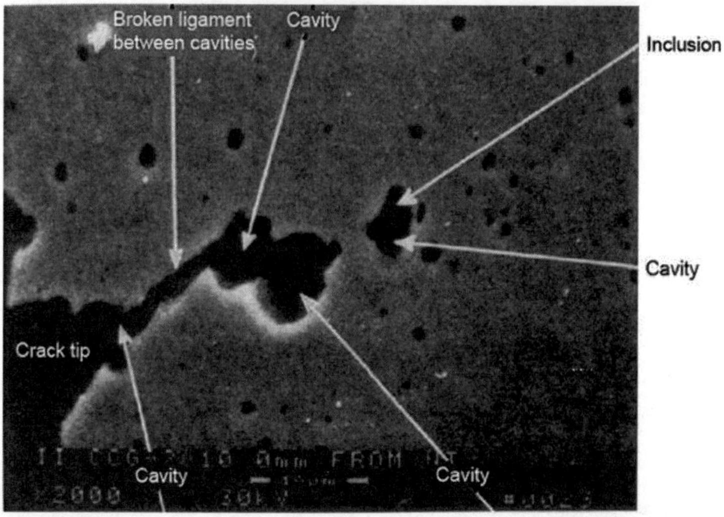

Fig. 3.19 Crack tip cavities formation and coalescence in a ductile fracture process (reproduced with permission of [22])

Now, the general picture is completed. Fatigue has claimed its victim. From a morphological point of view fatigue crack propagation in Stage II may be transgranular or transcrystalline in high-cycle fatigue and intergranular or intercrystalline in low-cycle fatigue as shown in Figs. 3.20 and 3.21 [23], respectively.

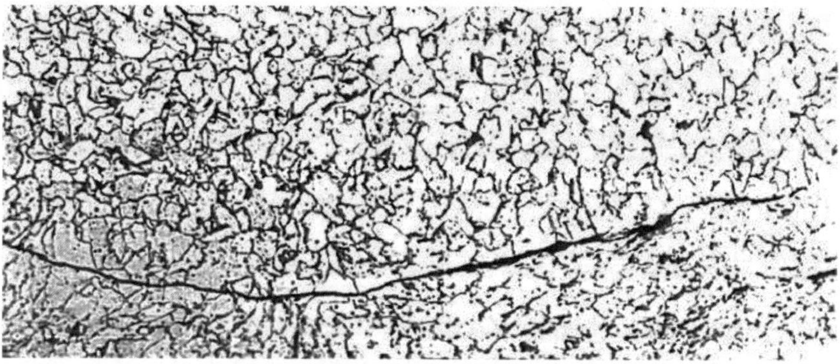

Fig. 3.20 Transgranular appearance of high cycle fatigue observed in low carbon steel (Milella)

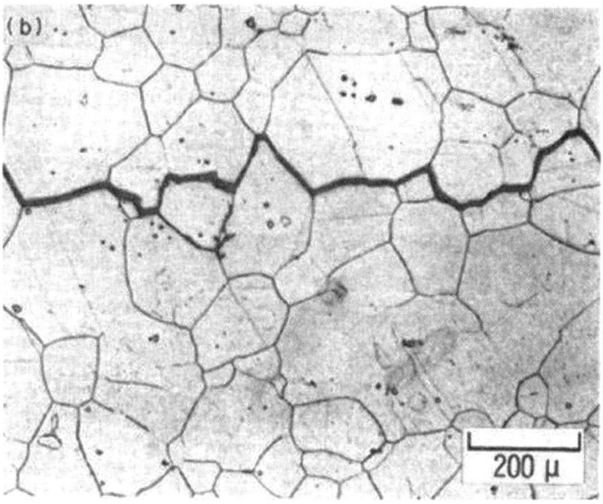

Fig. 3.21 Intergranular appearance of low-cycle fatigue in inconel X-750 at 650 °C [16]

3.5 Origin of Fatigue Striations

In the previous sections we have been discussing about striations that are the characteristic morphological feature of Stage II fatigue and beach marks being the visual evidence of packages of striations. At variance with beach marks that can be seen at naked eye, the trace of each single fatigue cycle can be caught only at high magnification, using an advanced optical or better SEM or TEM analysis. This trace was originally called slip band by Thompson e Wadsworth [24] and later striations by Nine e Kuhlmann-Wilsdorf [25].

Striations are the most striking fine scale feature left on the fatigue surface by the growth of a crack, a kind of microscopic fingerprint that identify fatigue crack propagation. The reference basic model in striations generation is that of crack tip plastic slip schematically shown in Fig. 3.22 [26]. During the loading phase the crack is opened by normal stress that activates plastic slips at the tip, see Fig. 3.18b. As said, fracture mechanics is predicting that this flow happens along two symmetrical directions at 60° with the crack propagation direction. In this phase crack tip blunts and grows by material decohesion associated to dislocations flowing into the tip or generated by stress concentration. This type of plastic deformation is not completely recoverable. Therefore, upon unloading the blunting is squeezed, but a new free surface remains head of the former crack with a new sharp tip, see Fig. 3.22. The irreversible process has made the crack grow by a quantity Δa. It is this step by step process of blunting and re-sharpening at each cycle that leaves on the crack path those footprints that we call striation, as schematized in Fig. 3.22. This process of crack opening, blunting, closing and re-sharpening can be fully appreciated in the sequence of Fig. 3.23 [27].

Laird has given a rather different interpretation of striation formation based on the so called plastic relaxation of crack tip that is schematized in the sequence of Fig. 3.24 [28]. The first picture, Fig. 3.24a, shows the initial state where crack is

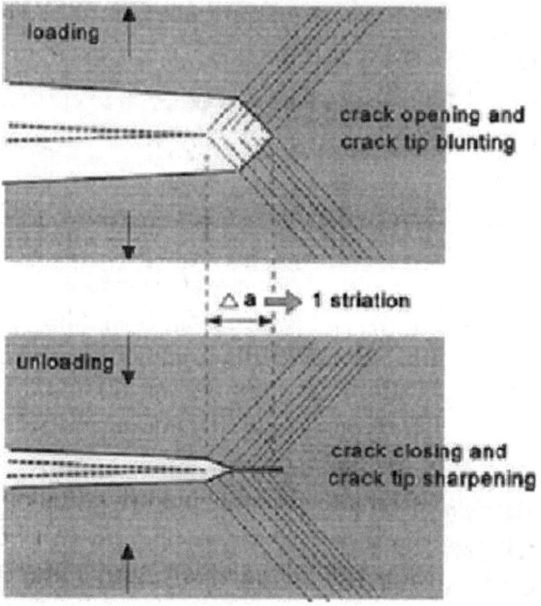

Fig. 3.22 Crack tip blunting and growth in during loading and closure upon unloading that leads to striation formation through the crack tip plastic slip mechanism [26]

resting. Then the load is applied, Fig. 3.24b and crack tip plastic slips occur on both sides. At maximum load, Fig. 3.24c, the tip is fully blunted. With load inversion, Fig. 3.24d, crack tip closure takes place by reverse plastic flow that leaves the crack in the same initial condition, but with a new formed elementary free surface that testifies the crack growth, Fig. 3.24e. Then, the process starts again, Fig. 3.24f, with a new loading phase.

Note that the process, that Laird called plastic relaxation, is based on the hypothesis of plastic collapse of crack tip during the unloading and closure phase that leaves the tip concave. Some characteristic profiles can be associated with crack propagation according to the Laird mechanism that are shown in Fig. 3.25 [28].

The first, Fig. 3.25a, indicates a fatigue propagation under relatively high stress amplitude, yet lower than yield strength. It consists of mating concave-convex parallel profiles on the two opposite fracture surfaces. Type (b) is characterized by the presence of small lateral cracks, indicated by arrows, and is characteristic of brittle striation, which will be discussed later. Types (c) and (d) are similar to type (a), but the smaller dimension of depressions indicates lower stress amplitudes. Types (a), (c) and (d) striations are called ductile striations since they are based on the mechanisms just seen of crack tip plastic slip.

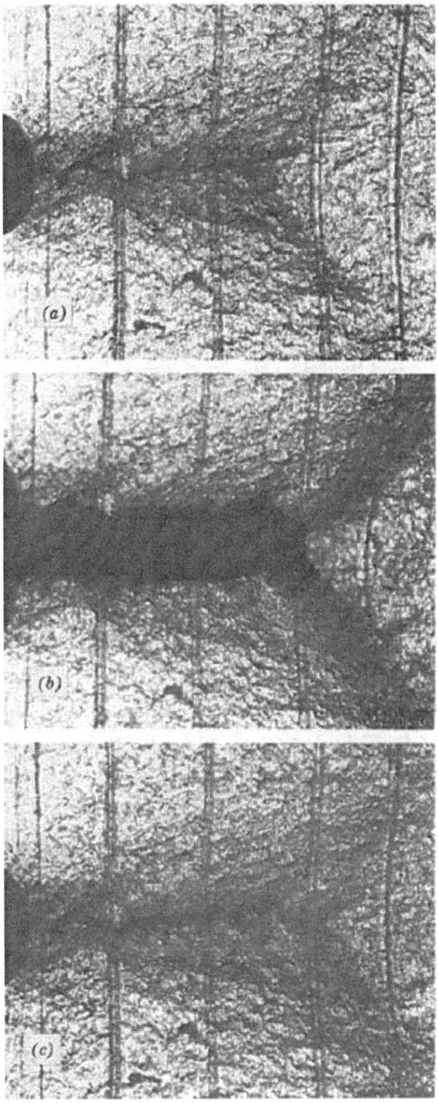

Fig. 3.23 Crack tip deformation (lobes) and growth during loading sequence [27]: **a** after compression, **b** after extension and **c** again after compression

3.6 Striation Observation

An example of ductile striations of the type presented in Fig. 3.25a, c and d is shown in Fig. 3.26a [29] for 7,5Zn-2,5 Mg aluminum alloy. Also brittle striations of the type schematized in Fig. 3.25b were observed by Forsyth on the same 7,5Zn-2,5 Mg aluminum alloy and are shown in Fig. 3.26b and c [29]. The micro cracks emanating from each propagation process can be clearly seen. Figure 3.26c shows the plastic

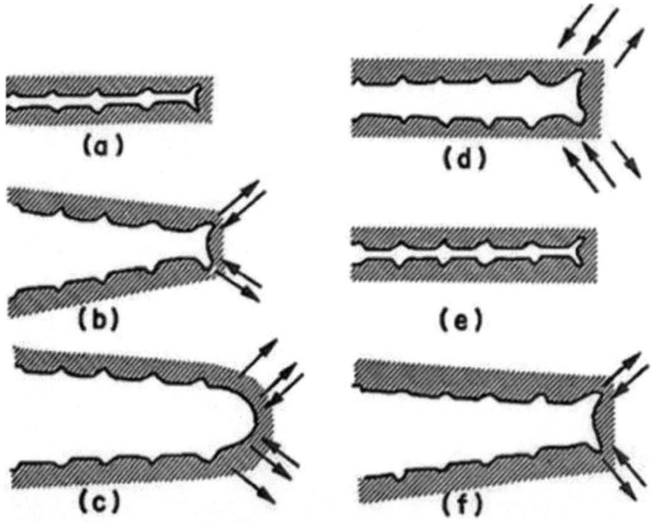

Fig. 3.24 Schematic of striations formation based on Laird crack tip plastic relaxation [28]

Fig. 3.25 Possible crack propagation profiles according to Laird's model of striations formation [28]

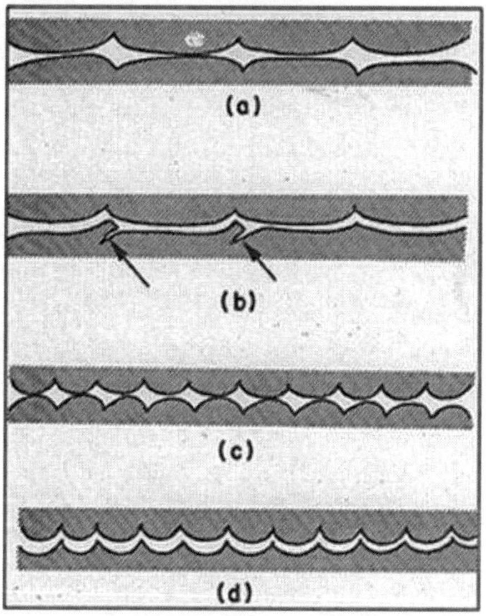

enclaves or lobes that appears on a free surface as tiny flakes, associated with the growth of brittle striations. The width so far measured of striations varies from a minimum of about 0.1 μm to a maximum of about 2.5 mm. It has been found [30] that in the transition between Stage I and Stage II of fatigue, when microcracks turn into macrocracks that can be opened by normal stresses (the so called fatigue threshold), striations width may be even some order of magnitude higher than the real crack propagation rate and keeps constant at about 0.1 μm (see Sect. 14.6). It is likely to happen that in this threshold region a single striation can be formed only after hundreds or thousands of cycles. Grinberg [30] measuring fatigue crack growth rate in annealed iron in humid air as a function of the excursion of the applied stress intensity factor ΔK_I and observing the striation spacing correspondent to each rate was the first to recognize that striation spacing was not following the rate of crack growth but remained constant in the order of some tenth of a micron, which was from one to three orders of magnitude higher than the corresponding growth per cycle. This is shown in Fig. 3.27 [30]. This apparent discrepancy has been found in many materials under various cycling conditions, including carbon steel, high strength steel, austenitic steel, iron, aluminum and titanium alloys, Cu-7.5%Al and Fe-3.7%Si.

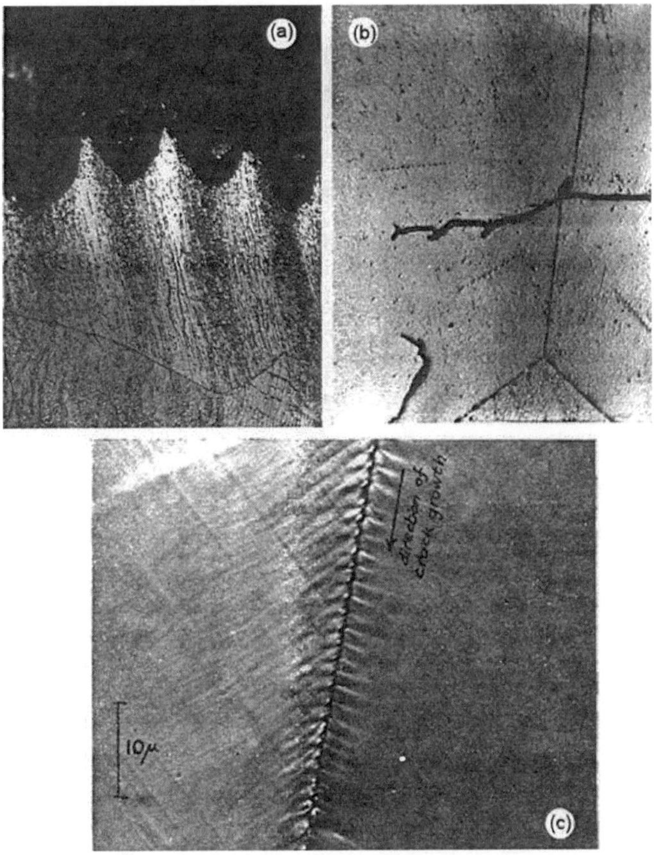

Fig. 3.26 **a** Ductile and **b** brittle striations observed by Forsyth on 7.5%Zn-2.5%Mg aluminum alloy; **c** Plastic flakes observed on a free surface associated with brittle striations [29]

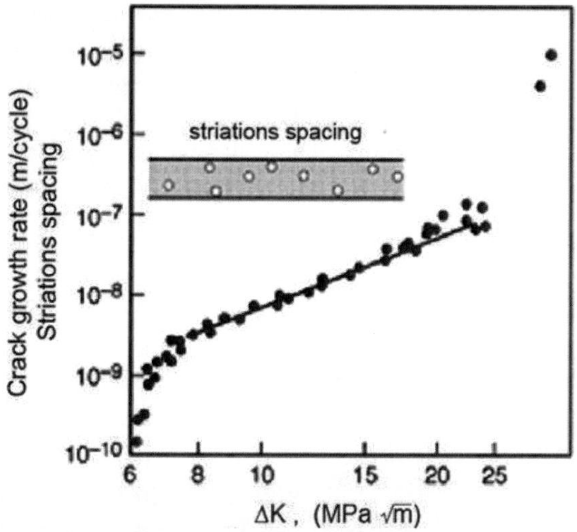

Fig. 3.27 Fatigue crack growth rate (circles and line) versus stress intensity factor ΔK in annealed iron in humid air. Also shown are the striations width measurements (open circles in gray band) [32]

This difference between the growth rate and the corresponding striation spacing appears for crack growth rates lower than some tenth of a micron. Figure 3.28 shows striations observed by Milella on carbon steel type Fe510 in the threshold zone. Striation width is about 0.1–0.2 μm, but the real growth rate should have been $10^{-4}-10^{-3}$ μm/cycle, i.e. from one thousand to one hundred folds lower. This actually means that for fatigue crack growth rates lower than about 0.1–0.2 μm it takes hundreds or even thousands of cycles to produce a single striation. Therefore, the one-to-one correspondence that exists between striations and cycles in Stage II fatigue well above the fatigue threshold appears to be lost at very low crack growth rate that characterize fatigue threshold. Sometimes striations having spacing below 0.1 μm are reported in technical papers.

Lund and Sheybany [20] warn that inexperienced or uninformed analysts too frequently interpret any set of roughly parallel markings observed in a SEM during examination of a fracture surface as fatigue striations. This must be considered any time total cycle counts for variable amplitude loading are accounted. Too often striationlike features have been mistaken for true striations. Features such as postfracture mechanical damage often exhibit fine, parallel marks or scratches that are not striations, as shown in Fig. 3.29 [20]. On a microscopic scale fatigue crack propagation evolves on different planes. This is due to metallurgical reasons and, in particular, to the fact that fatigue crack propagates along grains that are not necessarily coplanar, as schematized in Fig. 3.30 [31]. Striation are grouped in areas of homogenous propagation separated by tear ridges in which material fails by shearing resulting in the apparent continuity of the fatigue advancing front. The areas of homogenous propagation represent single grains.

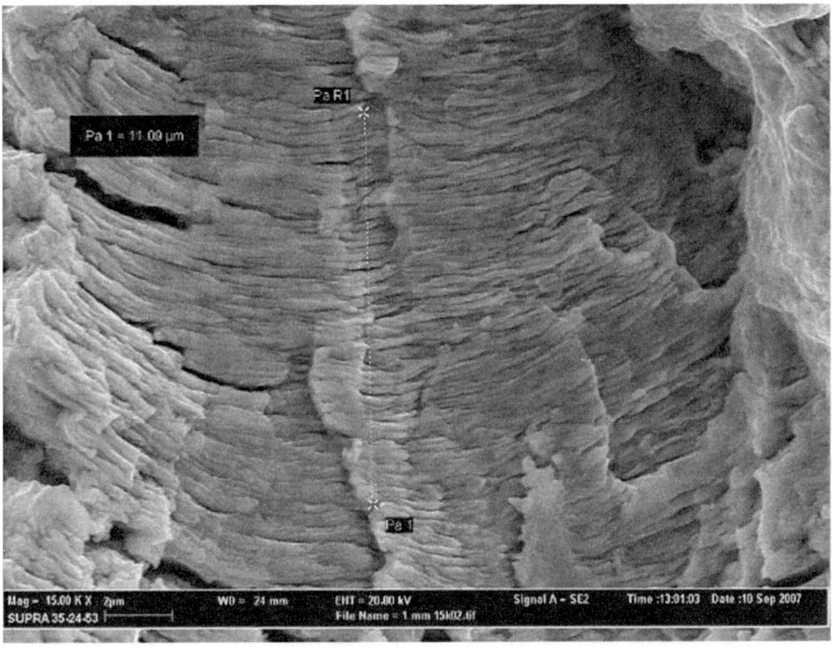

Fig. 3.28 Striations observed by Milella on Fe510 carbon steel in the fatigue threshold zone. Striations width is about 0.1–0.2 μm, but the real growth rate is significantly lower, about two orders of magnitude

Fig. 3.29 Post fracture mechanical damage can exhibit fine, parallel markings that are not striations, as illustrated in this SEM view of a steel sample [20]

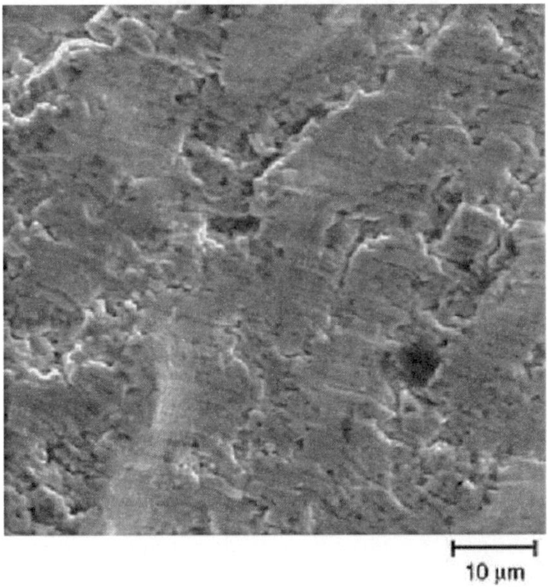

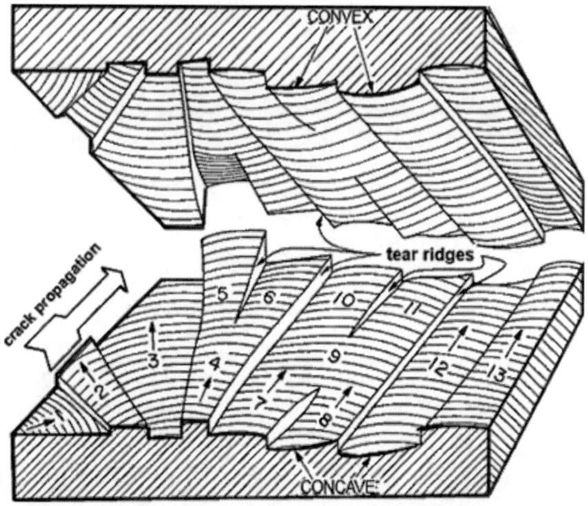

Fig. 3.30 Schematic of striation growth in grains and their merging or separation [31]

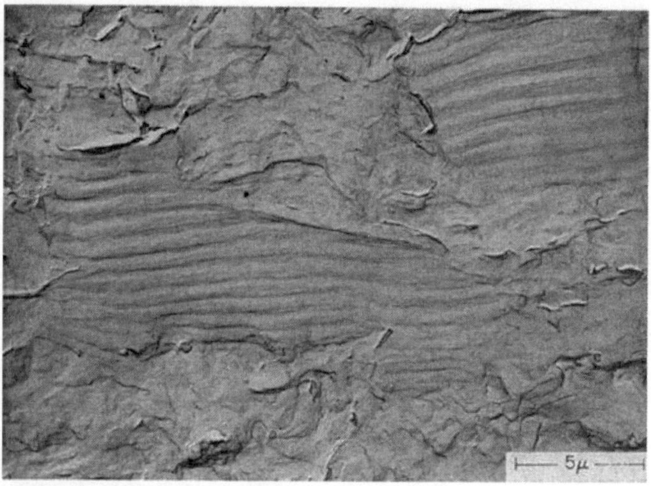

Fig. 3.31 Fatigue striations in 6Al-4 V titanium alloy [31]. Same area shows striations formation while some other shows signs of fracture

The various propagation areas may join as, for example, 7 and 8, in Fig. 3.30, that merge into 9 or separate as 4 that bifurcates into 5 and 6. In each of these individual areas the crack propagation direction, indicated by arrows, may not be exactly the same. Striations are always bowed out in the direction of crack propagation because their growth evolves more easily at center than on borders where they are blocked by grain boundaries and other barriers. Some areas of local propagation, as 7 and 13 in Fig. 3.30, are convex others, as 8 and 12, are concave on one side and the opposite on the other side of the specimen or workpiece, thus forming the concave-convex mating surfaces and profile as shown in Fig. 3.30.

It is rather common to observe areas of homogenous propagation separated by others that do not show any signs of fatigue, specifically striations. An example of that is shown in Fig. 3.31 for a 6Al-4 V titanium alloy [31]. In general, brittle fracture of intermetallic particles and ductile tear around large constituent particles are responsible for that appearance that results in a rapid and localized advance of the crack front. It must be said that striations, even dough they are the most characteristic morphological feature of fatigue, quite often are not so easy to be found. Many common steels exhibit ill-defined, unclear striations. Sometime they do not appear at all.

An example of non-resolved or ill-defined striations in quenched-and-tempered medium-carbon alloy steel tested in rotating bending is offered in Fig. 3.32 [20]. This mechanism, basically similar to that of Fig. 3.22, is based on a two stages process of blunting and closure re-sharpening of the crack tip. Compressive closure stress, Fig. 3.34b, may activate a series of slip lines ahead of the crack tip that leave a trace on the main striation. As shown in Fig. 3.34a, the crack opens on the rising-tension portion of the load cycle by slip on alternating slip planes.

Quite often it can happen to see two–three striations within a wider striation, as in Fig. 3.33. The origin of these closely-spaced striations may be explained with the mechanism proposed by Gross and schematized in Fig. 3.34 [32].

As slip proceeds, the crack tip blunts, but it is resharpened by partial slip reversal during the declining-load portion of the fatigue cycle. This process results in a compressive stress at the crack tip due to the relaxation of the residual elastic tensile stresses induced in the uncracked portion of the material during the rising load cycle (Fig. 3.34b). The closing crack does not reweld, because the new slip surfaces created

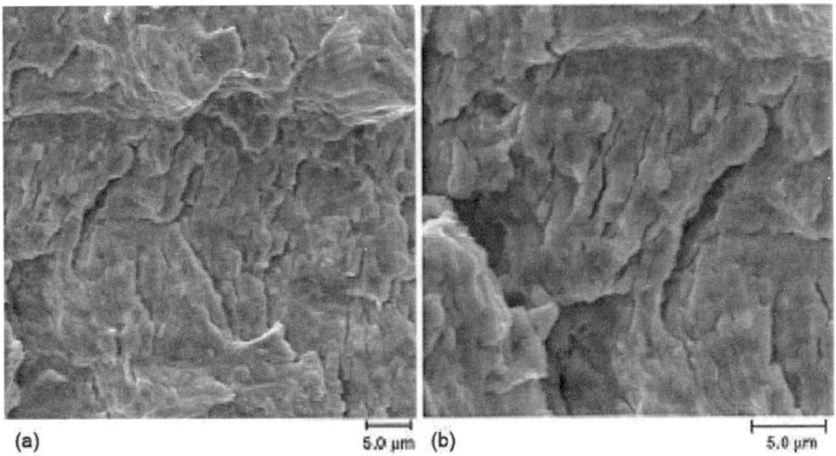

Fig. 3.32 Two SEM views at different magnifications showing fatigue appearance characteristic of many steels. Striations are not resolved or are ill-defined [20]

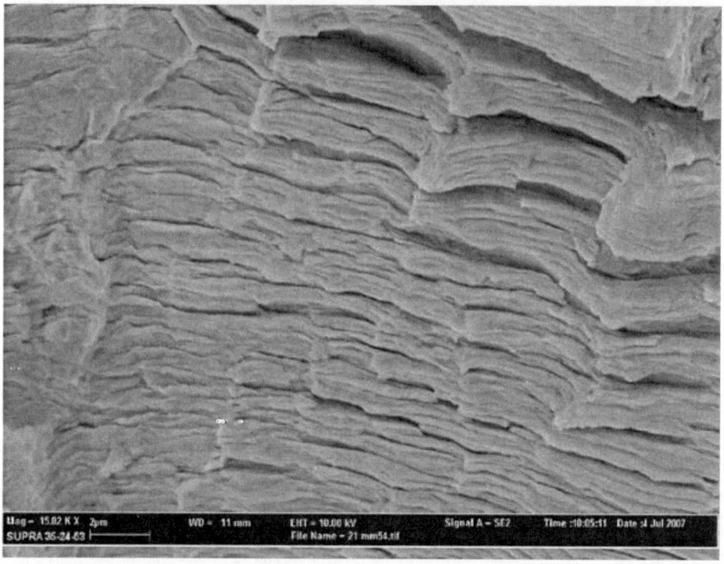

Fig. 3.33 Secondary cracking caused by high-cycle low amplitude fatigue observed by Milella. Cracks were produced after the crack had passed

during the crack-opening displacement are instantly oxidized, which makes complete slip reversal unlikely.

3.7 Ductile and Brittle Striations

Forsyth [29] recognized and described two general types of striations: ductile and brittle striations. They are schematized in Fig. 3.35 [33]. Both types of striations are transgranular. Ductile striations lay on different individual planes corresponding to single grains that macroscopically form, all together, the plane of propagation normal to the maximum tensile stress direction, as shown in Fig. 3.35a. They are called ductile because the material ahead of crack tip undergoes plastic deformations that produce the typical curved arrays by which they advance on the fracture surface. Brittle striations, instead, develop always on crystallographic planes, usually (100) planes [34], and appear as concentric circles departing from the initiation site or sites, quite often brittle inclusions, as shown in Fig. 3.35b. This gives brittle striations the typical flat appearance almost without any apparent (macroscopic) plastic deformation.

Some fracture surfaces containing widely spaced fatigue striations exhibit slip traces on the leading edges of the striation and relatively smooth trailing edges, as predicted by the model. Brittle striations are always associated with corrosion assisted fatigue and, in particular, with hydrogen embrittlement, as it will be discussed in Chap. 18. Examples of ductile and brittle striations are shown in Fig. 3.36 [33, 35]

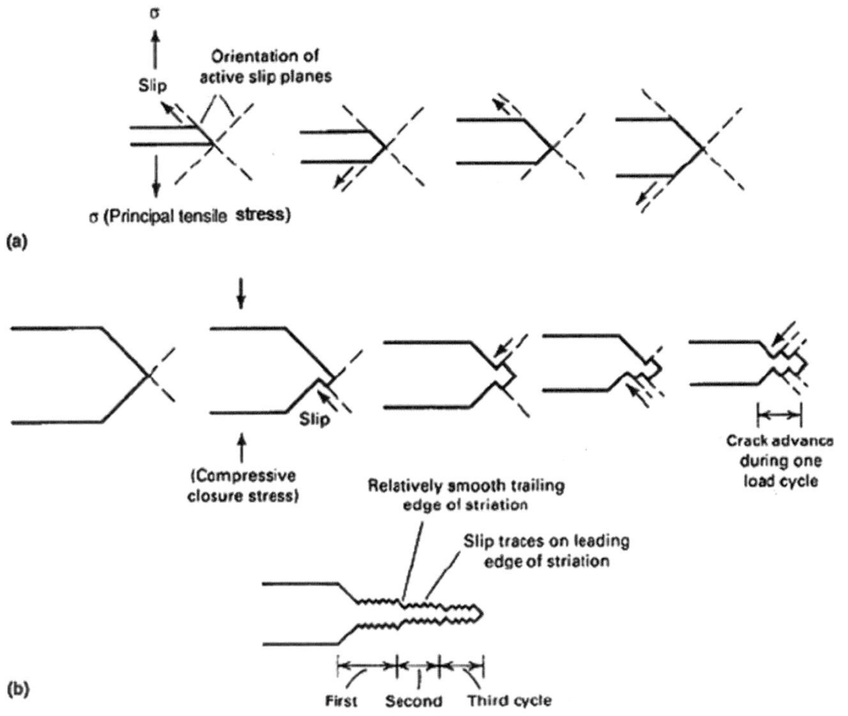

Fig. 3.34 Schematic mechanism of fatigue crack propagation by alternate slip at crack tip. **a** crack opening and crack-tip blunting by slip. **b** crack closure and crack-tip re-sharpening by partial slip reversal [32]

on 7.5Zn-2.5 Mg, 7178 and 2014 aluminum alloys. A characteristic feature of brittle striations is the uniform, flat and woody year-ring propagation surface that doesn't develop in single crystals but, as said, on crystallographic planes that are cleavage planes.

On these planes a continuous radial array of tiny cleavage steps appears that develops transvers to the circles that represent single propagation events which point back to fatigue fracture origin, as it appears in Fig. 3.36b and d. Originally observed by Forsyth on aluminum alloys, brittle striations have been found also on nickel, carbon and stainless steels. Figure 3.37 [36] is another example of brittle striations on admiralty alloy operating in a water solution of ammonia that provided the corrosion environment and, in particular, hydrogen. Four concentrically circles appear with the array of continuous cleavage steps. Under conditions of corrosion fatigue the crack tip doesn't open nor blunt as shown in Fig. 3.24c, but remains sharp. This allows the formation of an elastic stress field of high amplitude and triaxiality ahead of the crack tip that favors hydrogen adsorption, Fig. 3.38a. Hydrogen embrittles the metal through different mechanisms that will be discussed in Sect. 18. The high elastic

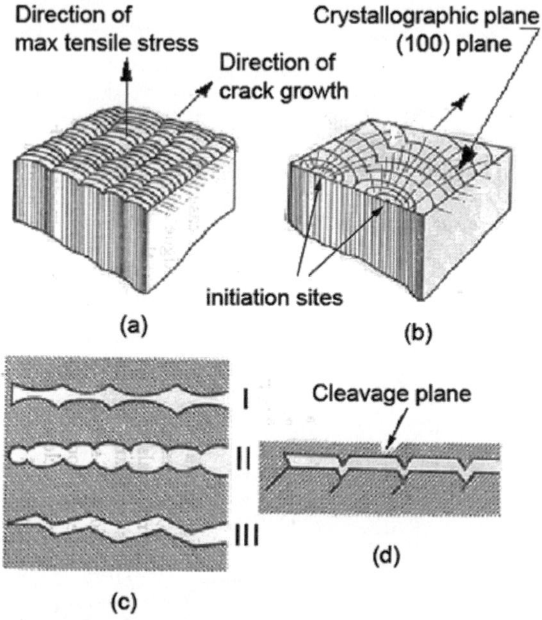

Fig. 3.35 Schematic difference between ductile and brittle striations; **a** ductile striations, **b** brittle striations, **c** ductile profiles, **d** brittle profiles ([29, 33])

stress field in a material embrittled by hydrogen breaks the metal ahead of the crack tip along a crystallographic plane or cleavage plane, Fig. 3.37b.

Note that brittle striations are difficult to detect just because the crack tip doesn't open, which is typical of brittle fracture, and no sign of plastic deformations (ductile striations) is left on the fatigue fracture surface. Striations of Fig. 3.37 are visible because every four to five cycles a light overload was applied just to open crack tip and mark the growth. The circles visible in Fig. 3.37 are precisely those overloads marks. An interesting example of how ductile fatigue striations can turn into a brittle propagation process under hydrogen embrittling conditions is offered in Fig. 3.39 [37]. The fatigue cycling in dry air has produced the characteristic ductile striations, Fig. 3.39a. Same steel, but in a hydrogen gas environment has completely changed its behavior developing brittle striations, Fig. 3.39b. Very often, hydrogen produces a characteristic grain boundary embrittlement, described in Sect. 18.4.1, which produces an intergranular separation of crystals.

This morphology of fatigue failure is so typical that any time an intergranular separation appears the first hypothesis made is always that of hydrogen embrittlement and in most cases it's the wright one. Figure 3.40 is an example of this kind of fatigue failure observed by Milella on a 1100 MPa NiCr high strength gear steel, hardened to HRC 55 that failed after 1,000,000 cycles with $R = 0$. The initiation site was a surface scratch produced during manufacturing.

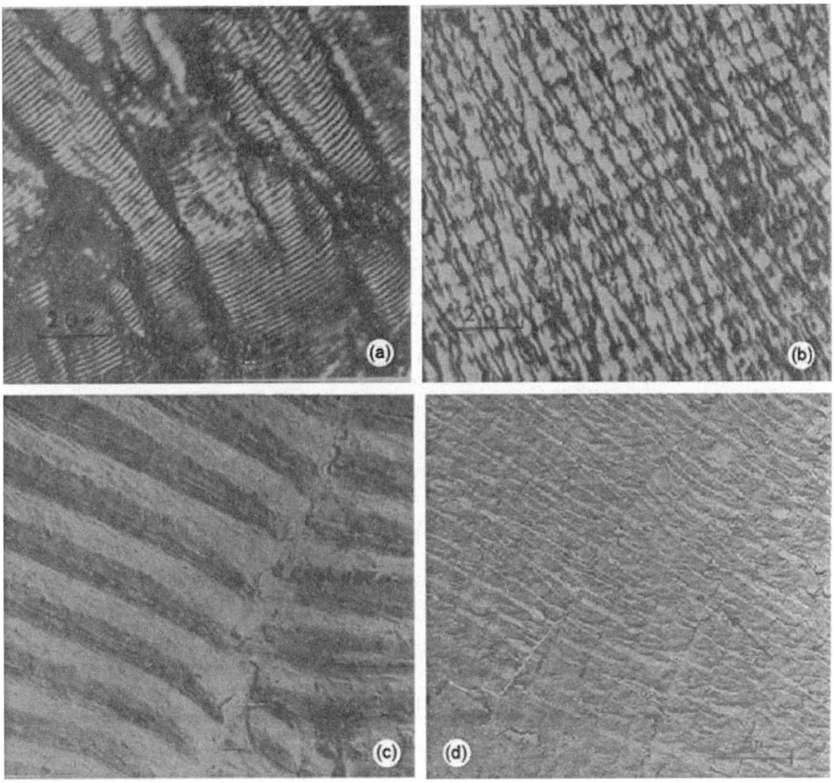

Fig. 3.36 **a** Ductile striations, **b** Brittle striations on 7.5Zn-2.5 Mg aluminum alloy [35], **c** Ductile striations on 7178 and **d** brittle striations on 2014 aluminum alloys [33]. a and b were observed with optical microscopy, (c) and (d) at TEM (two step replica). Note in b the array of groves transverse to striations circles better identify in (d) as cleavage steps

3.8 Striations and Fatigue Cycles

The question of whether or not a single striation corresponds to a single fatigue cycle, was first answered by Forsyth e Ryder [38] who used programmed constant amplitude fatigue cycles spaced by overload marking cycles. Fractographic observations of the fracture surface makes it possible to pick up these overload markers that separate the constant amplitude cycles and count these last. In this manner it was verified that indeed there was a correspondence between cycles and striations in that there were as many striations as the cycles imposed to the specimen and each striation had a width proportional to the corresponding cycle amplitude. An example of such a technique of programmed load and overloads used by Pelloux et al. on 2124-T351 aluminum specimens is shown in Fig. 3.41 [39].

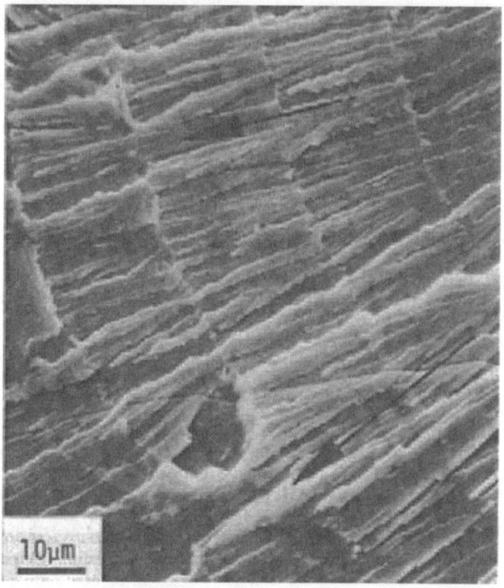

Fig. 3.37 Transgranular brittle striations by corrosion fatigue on admiralty alloy in water diluted ammonia [36]. Four concentrically circles appear intersecting the continuous array of cleavage steps. Black arrows indicate propagation direction

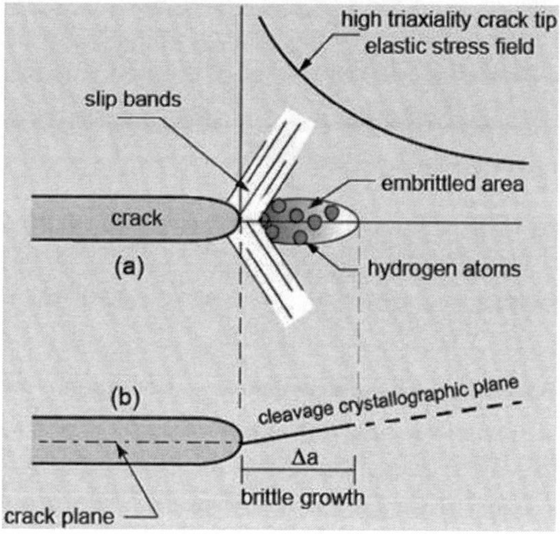

Fig. 3.38 Schematic of crack tip brittle growth by corrosion fatigue that leads to brittle striations formation; **a** hydrogen absorption driven by high triaxiality elastic stress field, **b** brittle growth

The load sequence A, B and C programmed in number and amplitude of cycles is repeated after an overload of two to four cycles. Each block of cycles is clearly detectable by SEM fractographic analysis. The two and four overload markers can be identified on the fracture surface. The number of applied cycles in each block is equal to the number of striations effectively counted on the fracture surface. The different amplitude of striations is also evident and is proportional to the stress amplitude. Moving from high-cycle fatigue to low-cycle fatigue regime, striations gradually

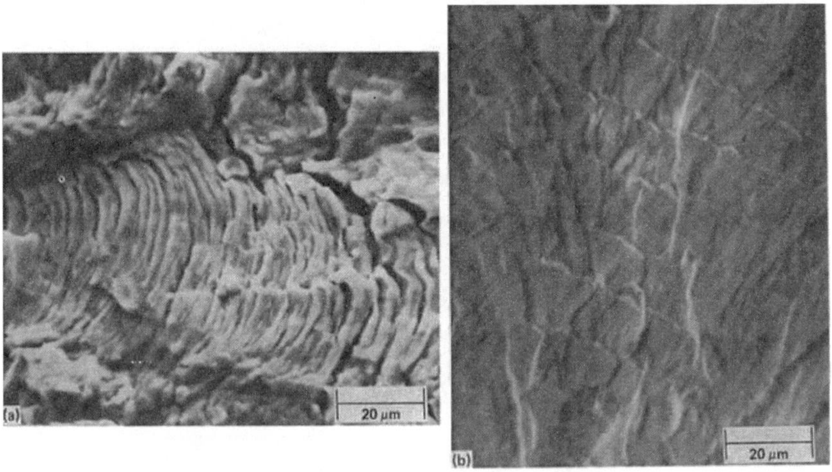

Fig. 3.39 **a** ductile striations and **b** brittle striations developed in ferritic low strength steel stressed **a** in dry air and **b** in hydrogen gas environment [37]

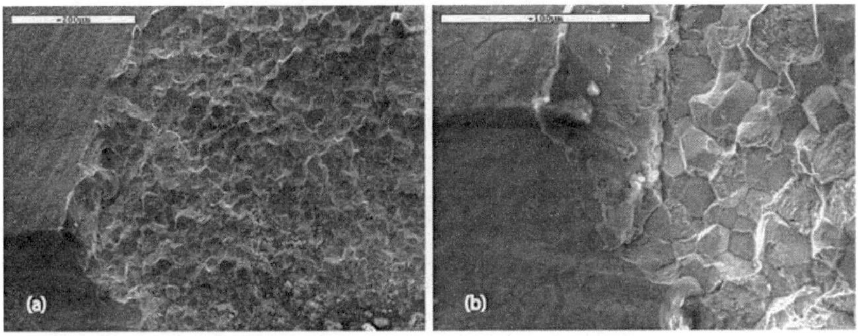

Fig. 3.40 **a** Morphology of intergranular fatigue failure of a gear tooth of NiCr steel hardened to HRC 55 observed at SEM by Milella. **b** Detail of the initiation site

disappear giving place to fractographic signs similar to those that characterize ductile fracture, according to what has been said about Stage III of fatigue in Sect. 3.4 (see Fig. 3.18).

A clear example is shown in Fig. 3.42 [40] for 6061-T651aluminum alloy. Figure 3.42a refers to Stage I damage nucleation phase almost featureless at low magnification. Figure 3.42b shows the region of crack propagation without striations, but with intergranular cracks and voids formed around inclusions. Intergranular separation is typical of low-cycle fatigue, as said in Sect. 2.2.

As already said fatigue striations are not present also in high-cycle fatigue during the early phase of macroscopic crack propagation at very low stress amplitudes close to the threshold stress intensity factor ΔK_{th} where crystallographic morphological

(a)

(b)

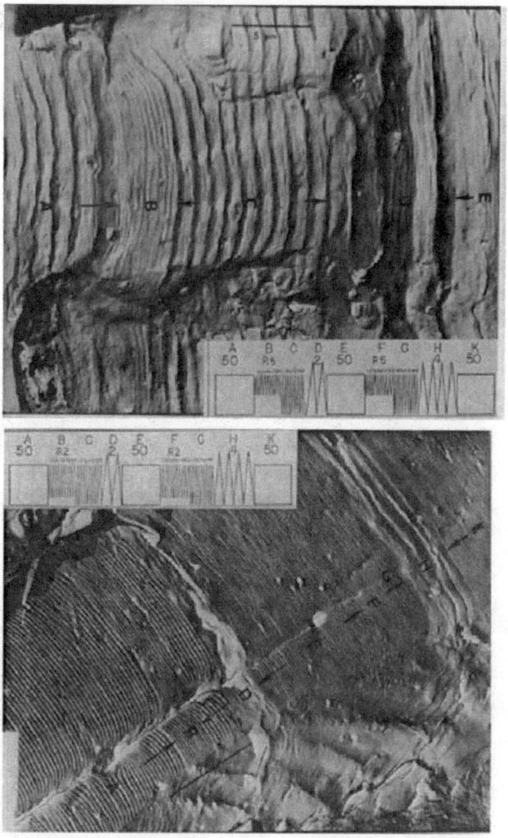

Fig. 3.41 Programmed loading of fatigue cycles in 2124-T351 aluminum alloy [44]. Loading sequence A, B and C is repeated after being marked by overload D. The number of cycles given in each block can be counted on the fracture surface by SEM analysis [44]

feature appears, as shown in Fig. 3.14 (see also Figs. 14.22 and 14.23). Even though striations are characteristic of high cycle fatigue their observation on the fatigue fracture surface may not be so simple.

This is the case of high strength materials having low deformation at fracture. These precipitation hardened metals are characterized by very small grains, 5 to 10 μm. They offer a thick grid of hard barriers to dislocations motion that are not allowed to extend and become so long to develop continuous arrays of ductile striations. Albeit fatigue striations are due to a plastic flow that opens and blunts crack tip, an important role in their visibility, at least with the aid of the electron microscope, is played by corrosion or, at least, oxidation. It has long been recognized that fatigue in vacuum leaves very light signs of striations when it doesn't leave any at all. Figure 3.43 is an example of fatigue striations observed at SEM by Milella on waspaloy at 500 °C in vacuum and air, respectively. Fatigue tests were initially run

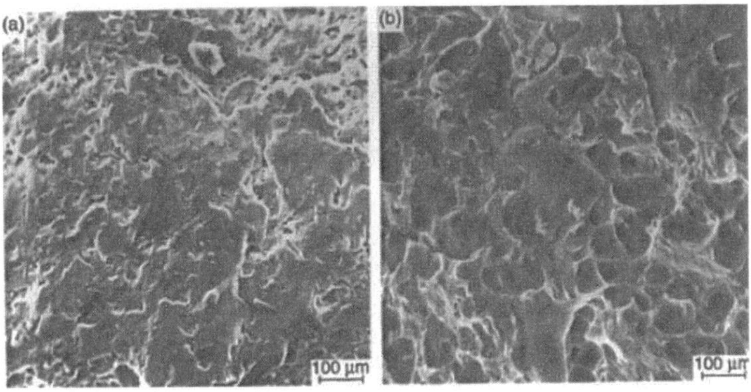

Fig. 3.42 SEM analysis of the fracture surface of 6061-T651 aluminum alloy failed by low cycle fatigue showing: **a** featureless crack initiation region, **b** Voids and intergranular cracks [40]

in vacuum where striations are barely visible inside grains, Fig. 3.43a. Tests were interrupted to check specimens and then started again under vacuum. During these resting periods air was allowed to enter into the environmental chamber. The lesson learned from experience is that in such cases tests at restart show all signs of fatigue in air, growth rate included, even though they are run in high vacuum. This can be seen in Fig. 3.43b where striations appear after air has been admitted in the test chamber and test restarted, though in high vacuum. Striations spacing in vacuum seems to be larger than in air when it is well known that fatigue in vacuum cannot be more dangerous than in air.

This apparent incongruence may be explained by saying that in vacuum each striation doesn't correspond to a single fatigue cycle since several cycles are needed to produce a single striation. Same result has been obtained on Ti-6Al-4 V by Pelloux [41, 42] and by Wadsworth and Hutchings on aluminum, copper and gold [43]. They found that the ratio between life in air and under high vacuum (10^{-5} mm Hg ~ $1.4 \cdot 10^{-4}$ atm) was about 1:20 for copper, 1:5 for aluminum and 1:1 for gold. This last result is particularly indicative since gold is very resistant not only to corrosion but also to simple oxidation. This actually confirms that in striations formation an environmental effect, namely surface oxidation, must take a role. It has been argued that in vacuum during the unloading phase and, in particular, during the load inversion a more or less complete re-fusion or healing of the sliding facets can take place because oxidation has not occurred. This delays or may even stop completely the fatigue process. This is particularly true under very low amplitude stress when plastic slips are very limited and irreversible damage is very low. McClintoc and Pelloux [44] proposed a striation formation mechanism based on a surface oxidation process. Finally, it is useful to remember that no striation will appear at all if the crack is opened by a cyclic torque. Shearing stresses do not produce striations. An example of macroscopic appearance of a carbon steel specimen broken under cyclic torque is presented in Fig. 3.44 [45]. The appearance of the fracture surface at naked eye

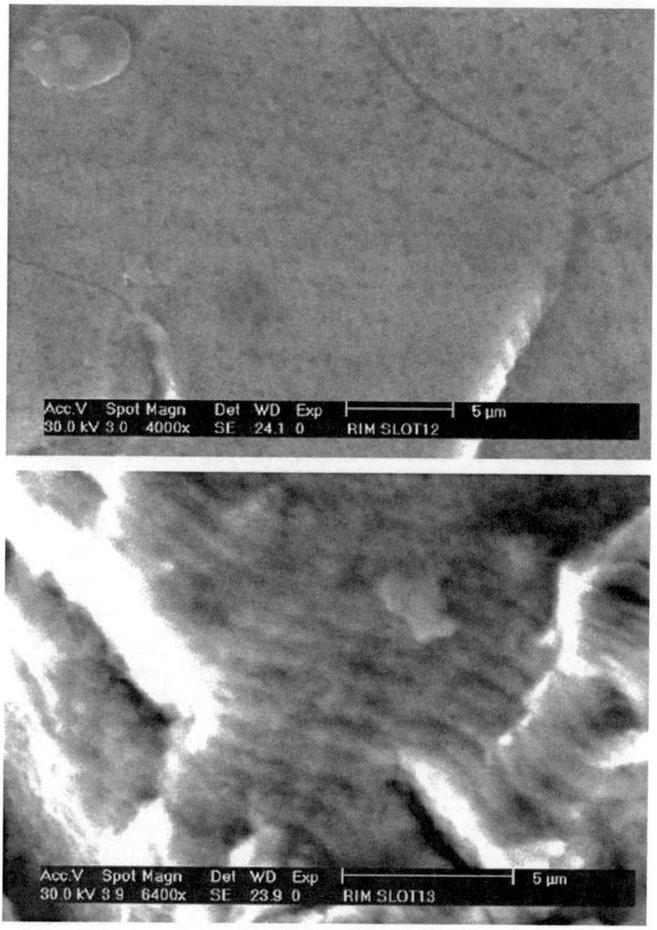

Fig. 3.43 SEM examination of the fracture surface of a waspaloy specimen fatigued at 500 °C **a** in vacuum and **b** after resting in air. Striations in vacuum are barely observed (Milella)

is called factory roof type. If one of those roofs were observed at SEM or TEM no sign at all of striations would appear at the analysis.

3.9 Real Cases

The following are just few yet significant examples of real cases fatigue fracture appearance that may be useful to designers:

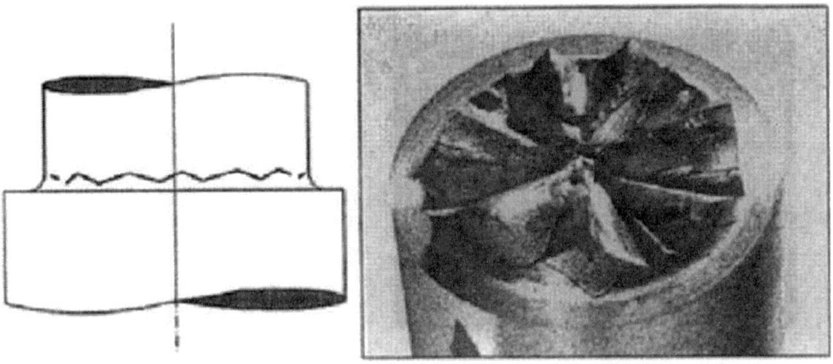

Fig. 3.44 Fatigue failure obtained under cyclic torsion in round carbon steel specimen (0.45%C) with a fillet notch [50]. Note the characteristic factory roof type fracture surface

1. Fatigue failure of a specimen of 7075-T6 aluminum alloy of rectangular section under unidirectional bending of random amplitude, Fig. 3.45 [46]. Fatigue originated on surface metal fibers in traction and propagated inside. The ratio $A_{fatigue}/A_{collapse}$ is definitely higher than 1 ($A_{collapse}$ covers about 40% of the total area) denouncing a high cycle fatigue with low stress amplitude. Beach marks are clear. Beach marks of different depths are due to crack growth rate variation following random amplitude cycling.
2. Service fracture of 4130 steel shaft with sharp circumferential notch subjected to unidirectional bending fatigue, Fig. 3.46a [47]. Fatigue initiated on the lower part where metal fibers were in traction. Beach marks, due to oxidation when material was idle, are initially concave and become wavy denoting final misalignment in load application. Fatigue has exceeded 50% of section indicating relatively low amplitude stress.

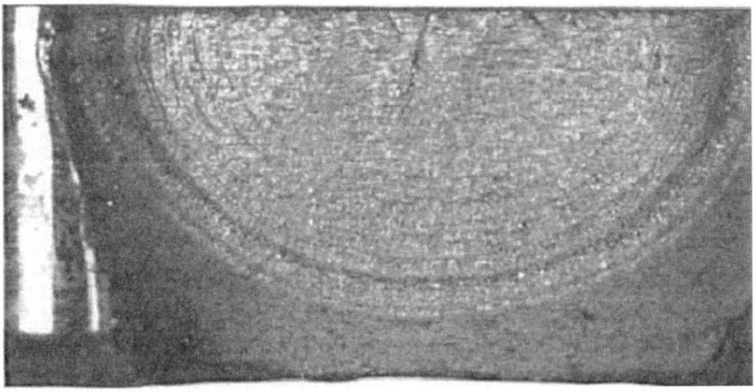

Fig. 3.45 Fatigue failure of a rectangular specimen of 7075-T6 aluminum alloy under unidirectional bending [46]

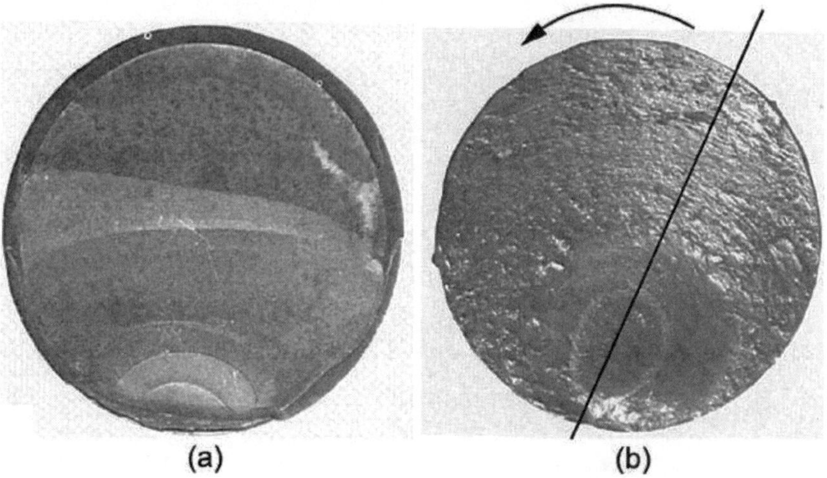

Fig. 3.46 **a** unidirectional bending fatigue failure of 4130 steel shaft with sharp circumferential notch [46], **b** fatigue failure of 4150 steel shaft subjected to rotating bending [47]

3. Fatigue failure of 4150 steel shaft subjected to rotating bending, Fig. 3.46b [47]. The section shows beach marks over a large area of the fracture surface. The oval dark area near the bottom marks the final fracture area. Its major axis is oriented at about 20° relative to the direction of beach marks evolution. This feature indicates that the shaft was rotating in counterclockwise direction. The final area is about 10 to 15% of the cross-sectional area which actually suggests low stress amplitude. Fatigue failure initiated at a point of stress concentration. Another example of rotating bending fatigue is that of Fig. 1.6 relative to a journal of an axle rail. Beach marks and the initiation site can be seen.
4. High temperature fatigue failure under axial loads of a valve stem of 21–2 valve steel (21% Cr, 2% Ni, 8% Mn, 0.5% C, 0.3N) solution-treated and aged and faced with stellite 12 alloy (30% Cr, 8% V, 1.35% C, rem Co), Fig. 3.47 [48]. Note the ratchet marks around the circumference that denote the presence of multiple initiation sites (indicated by arrows). The wavy shape of beach marks is indicative of off-axis load that has introduced a bending component.
5. Torsion fatigue failure of an axle of low carbon steel containing two holes, Fig. 3.48 [49]. Fatigue cracks initiated from holes that represent points of stress concentration. Cracks were fed by the traction principal stress acting on a plane at 45° to the axis (see Fig. 3.9a). The presence of two cracks at right angles to each other making an X suggests that the torque has been of a reversing character. Since cracks have approximately the same length the indications are that the torque reversals have been of equal magnitude. This applies, however, only as long as the cracks are in a comparatively early stage of development, since beyond this stage one crack usually takes the lead and such inferences are no longer justified.

Fig. 3.47 High temperature fatigue failure under axial loads of a valve stem of 21–2 valve steel [48]. Note the ratchet marks around the circumference that denote the presence of multiple initiation sites (indicated by arrows)

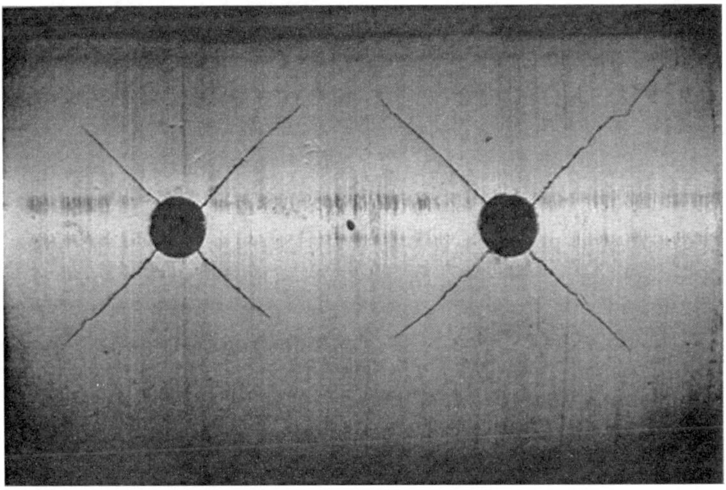

Fig. 3.48 Torsion fatigue failure of an axle of low carbon steel containing two holes [49]

6. Failure by torsional vibrations in oil engine crankshaft, Fig. 3.49 [50]. Initiation started at the journal nearest the flywheel from a longitudinal inclusion afterwards cracks propagated along two X directions as in the previous case.

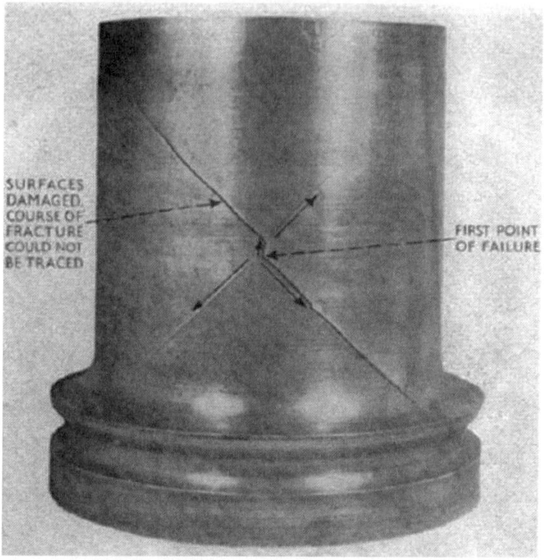

Fig. 3.49 Failure by torsional vibrations in an oil engine crankshaft [49]

7. Reversed bending failure in 1046 steel with a hardness of approximately HRC 30, Fig. 3.50 [51]. Rubbing has obliterated the early stages of fatigue cracking, but ratchet marks are present to indicate locations of crack initiation on both sides of the central region failed by overload that appear rougher than the fatigue areas.
8. Fracture surface of the piston rod of a pneumatic hammer, Fig. 3.51 [50]. It may seem a rotating bending failure because of the lateral center oval area denoting final fracture (see case 3), but it is not. The cracks were caused by tensile stresses developed due to non-axial loading, which were probably supplemented by compressive stress waves that became tensile when reflected. The small radial ridges near the periphery show that a series of cracks broke out there and joined together to form several crack fronts separated by the large ridges; ultimately, the crack fronts merged into a single annular one surrounding the zone of final failure.
9. View, Fig. 3.52, of a drive shaft (actual size shown) of AISI 1035 steel, with a hardness of 34 HRC, that fractured under torsional overload, twisting off in a fashion that produced a surface almost as flat as if it had been machined (see Fig. 3.9c) [51]. The threaded portion, at right, contains a fracture surface. A thin slice containing the mating fracture surface was cut from the portion at left; this slice is shown at center, with the fracture surface up. Shaft portion at left, including cut fracture surface, was hot etched in equal parts of hydrochloric acid and water to reveal deformation flow and forging flash lines.

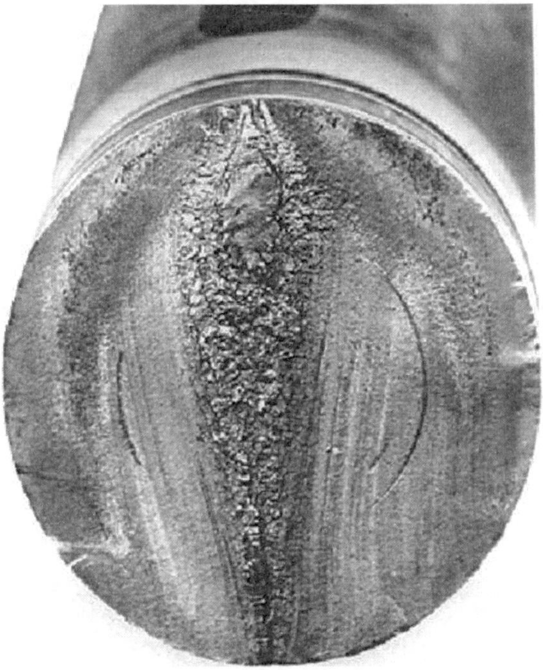

Fig. 3.50 Reversed bending failure in 1046 steel [50]. Rubbing has obliterated the early stages of fatigue cracking. Ratchet marks along circumference indicate locations of crack initiation

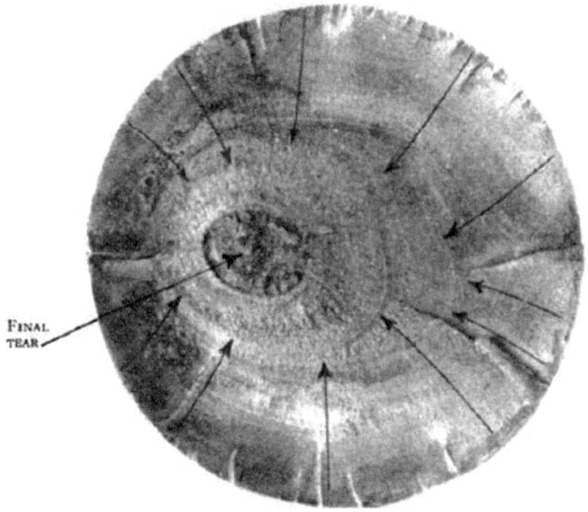

Fig. 3.51 Fracture surface of the piston rod of a pneumatic hammer failed under tension–compression loads [50]. The small radial ridges near the periphery show initiation sites that form several crack fronts that merged into a single annular one

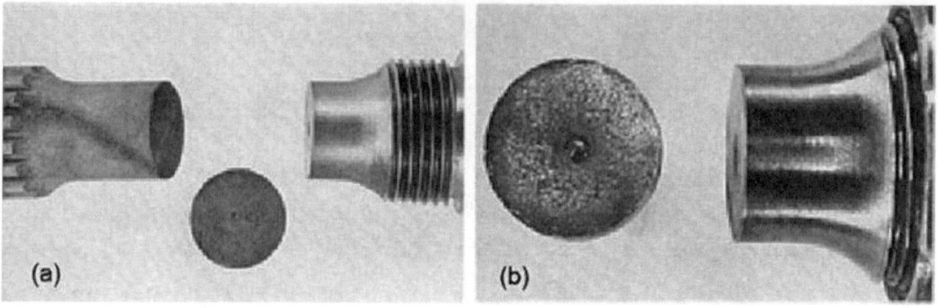

Fig. 3.52 View of a drive shaft (actual size shown) of AISI 1035 steel, with a hardness of 34 HRC, that fractured under torsional overload [51]

Fig. 3.53 Fatigue-fractured spring of 5-mm (0.200-in.) diameter of AISI 1060 steel wire (hardness, 43 to 48 HRC) [51]

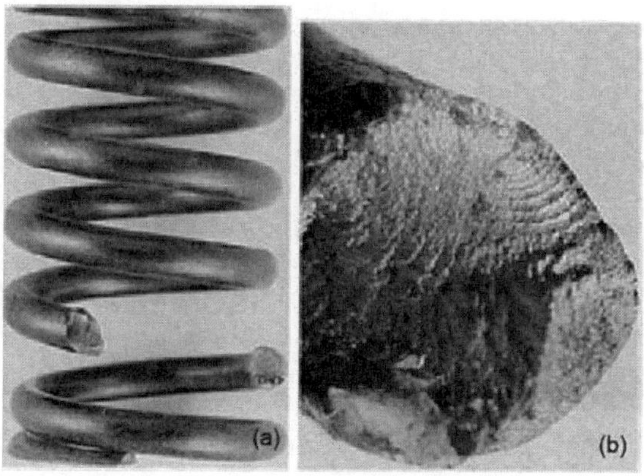

The plastic deformation is characteristic of torsional overload in ductile metals and is not typical of rotating fatigue. Figure 3.52b is a higher-magnification view (2×) of the threaded portion of the fractured drive shaft and the mating surface of the fracture [51].

10. Fatigue-fractured spring of 5-mm (0.200-in.) diameter of AISI 1060 steel wire (hardness, 43 to 48 HRC), Fig. 3.53 [51]. This fracture originated at the surface, 1.7× . The nucleus of the fatigue crack is clearly visible on the wire surface at top, with a succession of beach marks fanning out below it. The surface of the zone of final fast fracture appears to be quite woody, Fig. 3.52b. 15× (b) detail of the fracture site.

References

1. Neumann, P.: Bildung und Ausbreitung von Rissen bei der Wechselverformung. Zaitschrift f. Metallkunde H **11**, 780–789 (1967)
2. Forsyth, P.J.E.: In: International Conference on Fatigue. pp. 535. Institution of Mechanical Engineers (1956)

3. Forsyth, P.J.E., Stubbington, C.A.: The slip band extrusion effect observed in some aluminum alloys subjected to cyclic stress. Nature London **175**, 767 (1955)
4. Forsyth, P.J.E.: Some observation on the nature of fatigue damage, Phil. Mag. **2**, 437 (1957)
5. Forsyth, P.J.E.: Proceedings of Royal Society, **A242**, 198 (1957)
6. Stubbington, C.A., Forsyth, P.J.E.: Slip band extension effect observed in copper. J. Instit. Metals **86**, 90 (1957–58)
7. Klesnil, M., Lukáš, P.J.: Iron and Steel Institute **203**, 1043 (1965)
8. Cina, B.J.: Iron and Steel Institute **194**, 324 (1960)
9. Cottrell, A.H., Hull, D.: Extrusions and Intrusions by cyclic slip in copper. Proc. of Royal Soc. **A242**, 211–213 (1957)
10. Mott, N.T.: A theory of the origin of fatigue cracks. Acta Metall. **6**, 195–197 (1958)
11. Boettner, R.C., McEvily, A.J., Liu, Y.C.: Phil. Mag. **10**, 95 (1964)
12. Beachem, C.D.: Interpretation of electron fractographs, NLR Report 6330, Naval Research Laboratory, Washington D.C., pp. 49. (1966)
13. Yokobori, T., Kawasaki, T., Nakanishi, S., Kawaghishi, M.: Some experiments on heavy section specimen under low-cycle fatigue testing. Metal Sci. J. **5**(1), 25–33 (1969)
14. Metals Handbook.: In: Failure Analysis and Prevention. vol. 11, 8th edn, ASM, pp. 102. (1975)
15. Rice, R.C., Rungta, R.: Fatigue analysis of a rail subjected to controlled service conditions. Fatigue Fract. Eng. Mater. Struct. **10**(3), 213–221 (1987)
16. Schijve, J.: In: Fatigue of Structures and Materials. pp. 36. Kluwer Academic Publisher (2004)
17. Metals Handbook.: Fractography. vol. 12, 9th edn. ASM, pp. 483. (1987)
18. Gerberich, W.: Microstructure and fracture, mechanical testing. In: Metals Handbook, vol. 8, 9th edn, ASM, pp. 476–491. (1985)
19. Clavel, M., Pineau, A.: Frequency and waveform effects on the fatigue crack growth behaviour of alloy 718 at 298 K and 283 K. Metallurgical Trans. A, **9**, 471–480 (1978)
20. Lund, R.A, Sheybany, S.: Fatigue fracture appearances in ASM metals handbook. In: Failure Analysis and Prevention, vol. 11, 8th edn. ASM, 102 (1975)
21. Forsyth, P.J.E.: Fatigue damage and crack growth in aluminum alloys. Acta Metallurgica **11**, 713 (1963)
22. Liaw, P.K., Saxena, A., Schaffer, J.: Creep crack growth behavior of steam pipe steels: effects of inclusion content and primary creep. Eng. Fract. Mech. **57**(1) 112 (1997)
23. Mills, W.J., James, L.A.: Effect of temperature on the fatigue crack propagation behaviour of inconel X-750. Fatigue of Eng. Mater. Struct. **3**, 172 (1980)
24. Thompson, N., Wadsworth, N.J.: Metal fatigue. Adv. Phys. **7**(25), 72 (1958)
25. Nine, H.D., Kuhlmann-Wilsdorf, D.: Fatigue in copper single crystals in a new model of fatigue in face-centered-cubic metals. Canad. J. Phys. **45**(2), Part III, 865 (1967)
26. Schijve, J.: In: Fatigue of Structures and Materials. pp. 30. Kluwer Academic Publisher (2004)
27. McEvily, A.J., Johnston, T.L.: In: International Conference on Fracture, Sendai, Japan (1965)
28. Laird, C.: The influence of metallurgical structures on the mechanism of fatigue crack propagation. FORD Scientific Laboratory, Dearborn, Michigan, May 5 (1966)
29. Forsyth, P.J.E: Fatigue damage and crack growth in aluminum alloys. Acta Metallurgica **11**, 708 (1963)
30. Grinberg, N.M.: Int. J. Fract. **3**, 143 (1981) and **6**, 143–148 (1984)
31. Beachem, C.D.: Microscopic fracture processes. In: Liebowitz, H. (ed) Fracture an Advanced Treatise, **I**, vol. 311, (1968); Trans. ASM, **60**, 324 (1967)
32. Gross, T.S.: Micro mechanisms of monotonic and cyclic crack growth. In: Metals Handbook, vol. 19, Fatigue and Fracture, ASM (1998)
33. Beachem, C.D., Pelloux, M.N.: electron fractography—a tool for the study of micromechanisms of fracturing processes. In: 67th ASTM Symposium, STP-381, 236–237, June (1964)
34. Forsyth, P.J.E., Stubbington, G.A., Clark, D.: J. Instit. Metals **90** (1961)
35. Beachem.: Trans. AMS **60**, 325 (1967)
36. Becker, W.: Closed-form modelling of the unloaded mode I dugdale crack. Eng. Fract. Mech. **57**(4), 355–364 (1997)

37. Nelson, H.G.: Hydrogen embrittlement. In: Treatise on Materials Science and Technology, vol. 25, pp. 331. Academic Press (1983)
38. Forsyth, P.J.E., Ryder, D.A.: Some results of the examination of aluminum alloy specimen fracture surfaces. Metallurgia **63**, 117–124 (1961)
39. Pelloux, R.M.N., Faral, M., McGee, W.M.: Fractographic measurements of crack-tip closure. ASTM-STP **700**, 35–48 (1980)
40. Srivatsan, T.S., Shiram, S., Daniels, C.: Influence of temperature on cyclic stress response and fracture behavior of aluminum alloy 6061. Eng. Fract. Mech. **56**(4), 536 (1997)
41. Pelloux, R.M.N.: Corrosion fatigue crack propagation. In: II International Conference on Fracture, Brighton, Session V, Paper 64 (1969)
42. Pelloux, R.M.N.: Mechanisms of formation of striations. Trans. ASM **62**, 281–284 (1969)
43. Wadsworth, N.J., Hutchings, J.: Phil. Mag. **3**, 1154 (1958)
44. McClintoc, F.A., Pelloux, R.M.N.: Crack extension by alternating shear. Boeing Scientific Research Laboratories, D-1, July (1968)
45. Leger, J.: Fatigue life testing of crane drive shaft under crane-typical torsional and rotary bending loads. Schenck Hydropuls Magazine, Issue **1**(89), 8–11 (1989)
46. Metals Handbook.: Failure Analysis and Prevention, vol. 10, 8th edn. ASM, 97 (1975)
47. Metals Handbook.: Failure Analysis and Prevention, vol. 10, 8th edn. ASM, 100 (1975)
48. Metals Handbook.: Failure Analysis and Prevention. vol. 10, 8th edn. ASM, 275 (1975)
49. Frost, N.E., Marsh, K.J., Pook, L.P.: Metal Fatigue. Clarendon, Oxford (1974)
50. Hatchings, F.R., Unterweiser, P.M. (Ed.): Fatigue failure of diesel engine crankshaft, from Failure Analysis, the British Engine Technical Reports, ASM (1981)
51. Metals Handbook.: Fractography, Atlas of Fractographs, vol. 12, 9th edn. ASM, 97 (1987)

Chapter 4
Factors that Affect S–N Fatigue Curves

4.1 Introduction

In Chap. 1, basic design *S–N* diagrams have been introduced that can be referred to as standard *S–N* or Wöhler curves for plain specimens. The present chapter is concerned with modifying these *S–N* curves to account for all those factors that may have an effect on fatigue life. In doing this several important restrictions will be retained: first, the loading is completely reversed (no mean stress effect that will be fully treated in Sects. 7.3 and 7.4), second, the loading is simple and constant, for instance axial load or bending or torsion only (combined loading, load spectra and variable amplitude loads will be treated in Chap. 13), third, no point of stress concentration is present (stress concentration will be discussed in Chap. 11). Therefore, the factors that will be treated in this chapter are not stress related, but are metallurgical and physical factors. Fundamentally, there are just eight factors that affect fatigue resistance and *S–N* curves and four of them will be considered in this chapter:

- surface finish;
- inclusions;
- specimen or component size;
- texturing and hardening;
- load type;
- surface treatments (chemical, mechanical or thermal);
- temperature;
- environment (corrosion).

All of them have something to do, whether directly or indirectly, with the surface of the material. This should not come as something of a surprise since, as we already discussed in Chap. 2, see Figs. 2.3 and 2.5, in high cycle fatigue, in particular, the damage embryo that leads to macro crack formation and final failure always initiates on the surface within a localized plastic slip or at an extrusion-intrusion formation zone or, as we shall see, as a consequence of an inclusion that breaks but, with very

P. P. Milella, *Fatigue and Corrosion in Metals*,
https://doi.org/10.1007/978-3-031-51350-3_4

few exceptions, always on the surface. Therefore, any factor that affects that surface layer will certainly have a consequence on fatigue resistance. But a surface layer must have a depth within which initiation process will have a chance to take place and, therefore, a volume that we have called process volume.

Normally, this layer is no deeper than 0.5 mm, as shown schematically in Fig. 2.5, which actually means that the process volume depends only on the extension of the surface of the specimen or component. The larger the process volume, i.e., the wider the surface, the higher the probability that an initiation event be triggered by a stress amplitude lower than expected and, therefore, the shorter the fatigue life. The load type mentioned at point 5 of the above list will be considered just from this point of view since variations in fatigue strength arising by changing from a rotating bending load to simple reversed bending or to reversed axial load or torsional loading have to do with the process volume. This is not a new argument that this chapter will address, but a rather new approach to understand and consider load effects. The influence that each of the factors listed above may have over the *S–N* fatigue curve of the material is usually taken into consideration using corrective coefficients *C*: C_s, C_{in}, C_{sz}, C_{ld}, C_{st} and C_T where the various subscripts mean: S = surface finish, in = inclusions, sz = size, ld = load, st = surface treatment and T = temperature. Therefore, following the usual engineering approach to design, if σ_f or S_f is the fatigue limit measured with a standard plain specimen with a mirror-polished surface treatment, that will be the reference specimen, the effective fatigue limit σ'_f or S'_f resulting from the action of all modifying factors will be

$$\begin{aligned} \sigma'_f &= \sigma_f \cdot (C_S \cdot C_{sz} \cdot C_{ld} \cdot C_{st} \cdot C_T) \\ S'_f &= S_f \cdot (C_S \cdot C_{sz} \cdot C_{ld} \cdot C_{st} \cdot C_T). \end{aligned} \tag{4.1}$$

Note that in Eq. (4.1) it has not been included the corrective factor C_{in} due to inclusions since it is in competition with C_S due to surface finish, as it will be said in Sect. 4.4, therefore between the two only the one prevailing will be considered. Clear examples of such a case is offered by hard steels or aluminum alloys in which the presence of inclusions can nullify any benefit deriving from a mirror-polished surface or by gray iron and even ductile iron in which chunky graphite or graphite flakes and, in particular, fusion shrink voids are so detrimental and diffuse to relieve almost at all the importance of surface finish. As said, of the eight factors only four will be treated and given a numerical value in this chapter. Actually, it is not just the fatigue limit to be affected by those factors, but the entire *S–N* curve or Wöhler curve of the material, as we shall see in the following paragraphs.

4.2 Surface Finish

The surface finish of a specimen or component may well affect its fatigue life since it introduces some very localized stress concentration resulting from roughness. Roughness is a term that defines surface irregularities which result from the various machining processes. The actual roughness profile is that arising from the intersection of the surface workpiece with a plane normal to this surface, as schematized in Fig. 4.1. There are several parameters characterizing the surface roughness.

The most common surface roughness parameters, shown in Fig. 4.1, are: arithmetic average height parameter R_a or R_m, also known as the centerline average (CLA), root mean squared, R_q or RMS, maximum peak eight, R_p, maximum valley depth, R_v and total height of the profile, $R_t = R_{max}$, or peak-to-peak roughness. CLA is the average deviation from the mean line over a sampling length. R_a and R_q are defined as

$$R_a = R_m = \frac{1}{l}\int_0^l |y(x)|dx \quad or \quad R_a = \frac{1}{n}\sum_{i=1}^{n}|y_i|$$

$$R_q = \sqrt{\frac{1}{l}\int_0^l \{y(x)\}^2 dx} \quad or \quad R_q = \sqrt{\frac{1}{n}\sum_{i=1}^{n} y_i^2}. \tag{4.2}$$

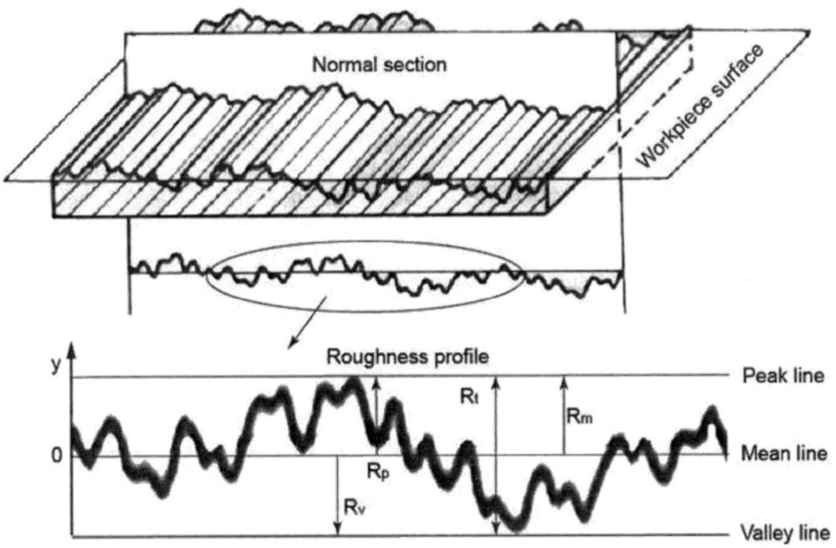

Fig. 4.1 Surface roughness and its profile

Nowadays, the importance of the surface finish is widely recognize and well documented [1–7]. Already in 1928 Houdremont and Mailänder [1] showed that the bending fatigue strength was dependent on the quality of the finishing. Siebel and Gaier [2] in 1957 testing various grades of steel obtained the diagram of Fig. 4.2, using a type of specimen shown in the lower right of the figure. There is a clear fatigue limit reduction with increasing surface roughness *R*. Particularly interesting is to note that in high strength quenched and tempered steels (see Sect. 5.3) the initial and apparently constant trend of the fatigue limit with roughness that we call plateau, terminates rather soon for $R_{max} \leq 2$ micron. In annealed steels this limit rises up to 6–10 μm. Also note the similarity of the Siebel and Gaier diagram with the Kitagawa-Takahashi diagram of Fig. 2.38. Actually, the Siebel and Gaier diagram appears to be precisely an *ante litteram* Kitagawa-Takahashi diagram. However, it is worth noting the incredible difference between the two in that the size of the surface irregularity R_{max} at which the fatigue limit starts to be *R* dependent is about one order of magnitude lower than the corresponding microcrack size *a* in the Kitagawa-Takahashi diagram. Superposed to the Siebel and Gaier diagram of Fig. 4.2 are data (closed symbols) obtained by Murakami [8] for carbon steel specimens in which they had introduced very small artificial defects.

Effectively there is a remarkable difference in the size of roughness or defect that terminates the *R* independent phase. This noticeable discrepancy may have a twofold

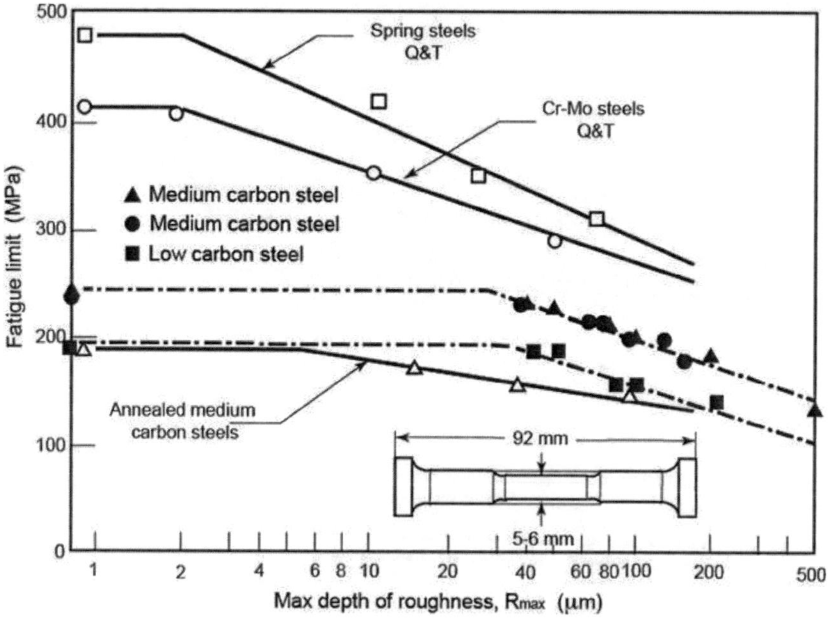

Fig. 4.2 Experimental data obtained by Siebel and Gaier on surface finish effect for Q&T and annealed steels (adapted from [2]). Close symbols are data obtained by Murakami for steels containing micro holes [8]

reason. First, finishing is affecting the entire surface of the specimen and not just a single point of that surface, as in the case of a crack deliberately introduced. Secondly, it may be the consequence of the fact that the growth of a microcrack depends on local microstructural properties of the material, particularly when the crack is still so small to be blocked by inclusions and grain boundaries that have a very strong retardant effect on MSC growth. At variance, when a microcrack is introduced in the specimen surface, it is very likely that the location chosen, absolutely irrespective of microstructural characteristics, is not the worst possible from a microstructural stand point. Conversely, a roughness that affects the entire surface of a test piece nucleates microcracks where and always where the slip process is easier, which actually means that natural microcracks always initiate where fatigue process finds the point where the material is metallurgically weaker. If I were allowed to use a quaint example, it is as if fatigue were scanning the entire surface of the test piece in search for the worst possible metallurgical conditions to initiate the damage. And it always finds them.

This may appear in contrast with what reported by Murakami and Endo [9] who found that specimens with a single notch experienced fatigue strength that was about 30% lower than that of specimens with multiple notches of the same size introduced by use of a lathe simulating surface roughness. This is because in their experiments the notches simulating roughness had all the same depth so that the interference among them resulted in the reduction or even absence of stress intensification, depending on the pitch among notches. In case of irregular roughness, like that of Fig. 4.1, this effect is by far reduced and individual peaks are effective in producing stress concentration. Therefore, roughness provides the most effective mean for initiating a microcrack through a natural slip process activated by the easiest fatigue mechanism. If this is the case, as it seems to be, then the importance of the surface finish become very apparent. Artificial mechanically small cracks may even lay dormant if the crack tip faces a very hard microstructural barrier. The same happens with inclusions as crack initiators since their distribution within the material is irrespective of the local metallurgical conditions, unless they are coupled with a surface roughness, see Sect. 4.4. Another aspect that should be caught by looking at diagrams like that of Fig. 4.2 is the slope of the curves of fatigue limit decay. For the spring steel it is about 7, while for the annealed medium carbon steel is about 100. These findings strengthen the need to have a mirror-polished surface with an average surface roughness $R_a \leq$ 1 μm to get the best fatigue strength.

A mirror-polished surface is required for specimen used to derive reference *S–N* curves. Intrusions and extrusions are surface defects of the order of some micron (see Sect. 1.4). If surface finish is rather rough leaving material plicas higher than these few microns then it will be not necessary to wait for extrusions formation or any surface slip process to get the initiation event. In such a condition there is already a surface damage and it is clear that fatigue life will be enormously reduced since most of cycles are spent just to initiate micro damage.

Figure 4.3 [4, 6] illustrates this circumstance for T − 1 high strength steel, 790 MPa yield and 850 ultimate strength, subjected to fatigue with *R* ratio ($R = \sigma_{min}/\sigma_{max}$) equal zero in two distinct surface conditions: fine ground that leaves a surface roughness R_m between 1.5 and 2 μm, and as received with a surface roughness $R_m \geq 20$ μm.

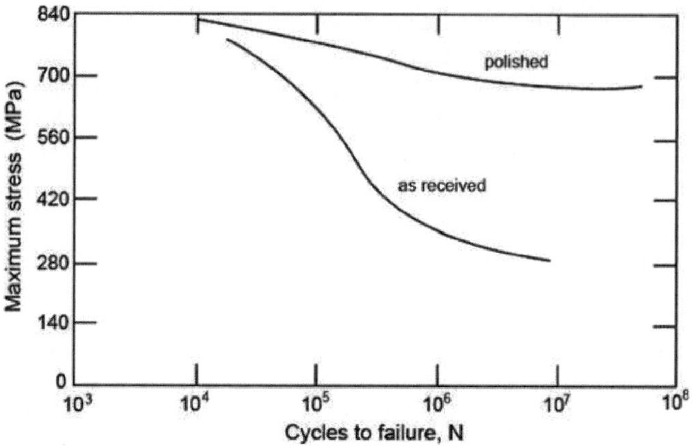

Fig. 4.3 *S–N* curves for T − 1 steel in polished [4] and as-received [6] conditions

In the high-cycle fatigue regime there is at least a factor of 2.5 affecting the fatigue limit of the two surface finish conditions for exactly the same material. Interesting enough, the allowable stress difference continuously reduces as the life *N* decreases till it disappears completely in the low-cycle fatigue regime at about 10^4 cycles. The reason is very clear: in low cycle fatigue stress amplitude is over the yield strength resulting in generalized plastic slips that destroy any given surface finish, typical is the so called orange-peel surface of stainless steel plastically deformed, therefore by increasing the amplitude of the applied stress the surface damage increases prevailing more and more over the initial finish effect. Another interesting example is shown in Fig. 4.4 [10]. This time fatigue specimens are made of Al–Mg-Si aluminum alloy, type 6082-T6 and are loaded in symmetrical four-point bending (pure bending).

Some specimens had a fine ground initial finish followed by electropolishing others were left as received. The first had a final surface mean roughness R_m no higher than 0.05 μm, the second had a residual mean roughness in the range of 1.5 μm, still very good but higher. Even though the difference in surface finish is extremely limited, there is a remarkable loss in fatigue limit ($N > 10^6$ cycles) that decreases from about 105 MPa to about 80 MPa. Also in this case, as in the previous one, the difference in the allowable stress amplitude reduces and eventually disappears as the material enters the low-cycle fatigue domain. Surface finish, then, reduces the fatigue limit, but it may also shift the knee of the *S–N* curve, as shown schematically in Fig. 4.5. The shift in life is due to the fact that the larger the surface roughness the lower the number of cycles to initiate the macro crack. As to the fatigue limit reduction it is rather difficult to understand since there should not be any difference whether a damage process initiates from a smooth surface or a rough one: they both lead to the same macrocrack. As we will see in Sect. 4.4, there is no decrease in fatigue limit when damage initiates from a small surface inclusion of a size comparable to roughness, but only a strong reduction of life (the knee of the S–N curve moves toward shorter lives), as shown in Figs. 4.30 and 4.33.

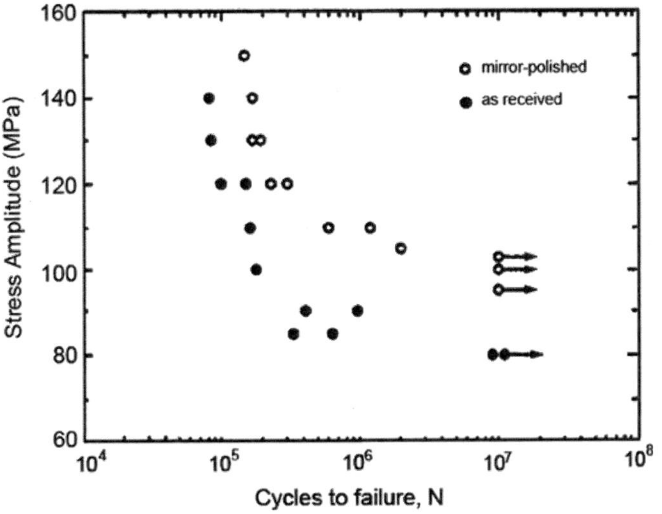

Fig. 4.4 *S–N* Curve for an aluminum alloy type 6082-T6, in two different surface conditions [10]

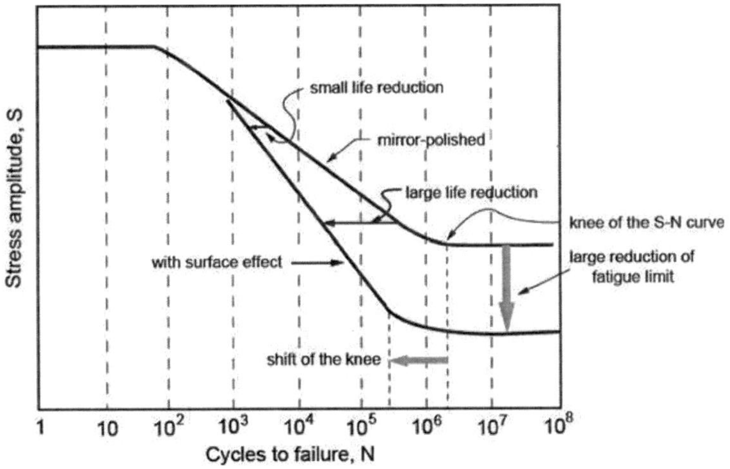

Fig. 4.5 Surface effects on *S–N* curve

Note that the opposite would hold if the surface damage were a macrocrack ($a >$ 300 μm) already existing in the material, as the Kitagawa–Takahashi diagram indicates, see Fig. 2.38. A macrocrack becomes dormant at stress amplitudes lower than the plain specimen fatigue limit. It is worth noting that for a long macrocrack, as already mentioned in Sect. 2.1, the growth resistance depends on the bulk properties of the material and not on local micro metallurgical conditions. Probably, we have to think that the fatigue limit reduction is due to the intrinsic characteristic of roughness that interests the entire surface and not just a point. This, as previously said, certainly increases the probability of finding multiple initiation sites having micro-metallurgically weak properties and resistance to crack growth. A combination this last that may results in the easier enucleation of larger macrocracks, hence the reduction of fatigue limit. Data like those presented in Fig. 4.2 indicate that the surface finish has more effects on high strength steels or alloys than in low strength ones.

Noll and Lipson [11] already in 1946, investigated the relationship between fatigue limit at 10^7 cycles and surface condition using ground, machined, hot rolled and forged specimens. Their tests, at Chrysler laboratories, provided what is probably one of the largest experimental campaign and data bank on the subject, in particular for the as forged condition and is still used in fatigue design and analysis.

The result of their tests in terms of fatigue limit versus tensile strength for the different surface conditions are reported in Fig. 4.6 [11]. While for a 400 MPa ultimate strength steel the refinement of the surface finish from the as-forged condition to the mirror-polished one increases the fatigue strength by a factor of about 2, in high strength steels of 1100 MPa ultimate strength the same refinement results in a much larger factor of about 3.5. However, beyond a certain limiting strength, which is about 1100 MPa, the same benefit is no longer achieved. The fatigue strength appears to decrease. The decrease is due to inclusions that start to play a role when the surface finish is no longer a problem (see Sect. 4.6) Therefore, for steels having 1100 MPa ultimate strength or higher it may be worthless to reduce the surface roughness to almost zero if a complete inclusions control is not achieved. A proper inclusion control may be pursued using, for example, an in vacuum re-melting technique (see Sect. 4.6) that strongly reduce the probability of leaving an inclusion right in the process volume where it really hurts.

Curious enough, the fatigue limit for each surface finish seems to depend on the tensile strength of the material rather than on the yield strength that activates plastic slips. The reason for that apparent incongruence is not completely clear. However, it can be said that though the yield strength represents the first barrier to slip, nevertheless once initiated, its continuation depends on the hardening properties of materials. Steels with very high hardening, which means very high ultimate strength, provide much more resistance to slip of a low hardening material with the same yield strength. Since tensile strength is related to hardness, then there must be a relationship between surface finish and hardness. In terms of surface finish factor C_S the results of numerous tests obtained by various investigators are summarized in Fig. 4.7 [12]. Finish factor $C_S \leq 1$ is defined as the ratio between the fatigue limit, σ'_f, obtained with specimens of given surface finish to that measured with mirror-polished specimens: $C_S = \sigma'_f/\sigma_f$ (see Eq. 4.1). The loss of fatigue strength as function of the surface roughness is particularly evident for high strength steels. In very hard steels, over 1200 MPa strength, even the micro roughness of a fine ground surface finish may jeopardize the fatigue life. This also anticipates the very high notch sensitivity of these steels, as it will be said in Sect. 11.2.

Another way to present surface factors is shown in Fig. 4.8 [13] in which surface factor is reported relative to the average surface roughness expressed in micron.

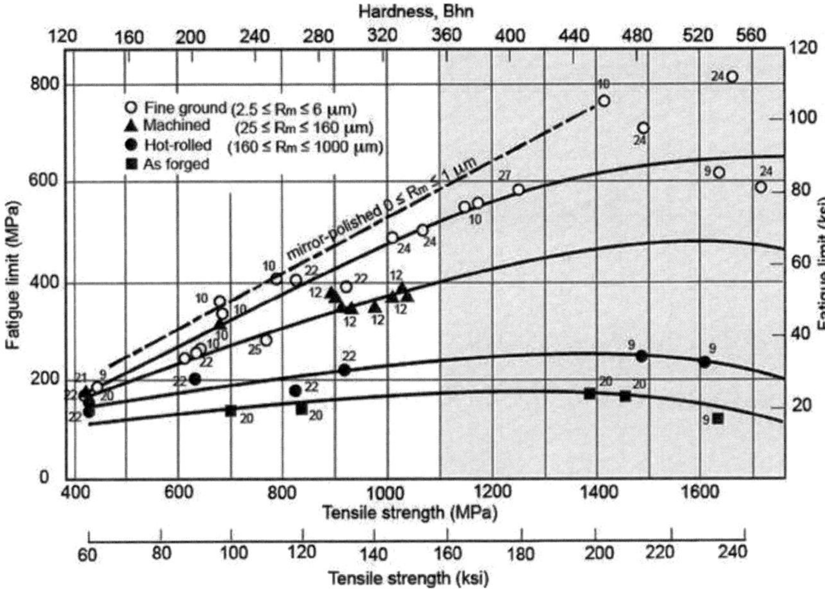

Fig. 4.6 Effect of surface finish on fatigue limit of steels (adapted from [11])

It is particularly significant observing that an average roughness of only 50 μm that would appear just insignificant and even nonvisible at naked eye, may reduce fatigue life of high strength steels to 50% of its original measured on mirror-polished specimens. For low strength steels this reduction still appears but is in the range of 25%. Indeed, it has been observed that the continuous refinishing of the surface of a fatigue specimen initially mirror-polished may increase fatigue life almost indefinitely [6, 14].

4.3 Decarburization

The reduction of the fatigue limit with surface roughness becomes a real breakdown passing from the fine ground or machined condition to the hot-rolled one and, in particular, to the forging condition, as clearly shown in Fig. 4.7. This is particularly true for high strength materials that are indeed non-forgiving materials. Surface roughness has less of an effect on low strength alloys. However, the fall of the fatigue limit cannot be justified by a simple matter of roughness; there must be something more than that. Note that for a steel of 1600 MPa ultimate strength, the fatigue limit of hot-rolled specimens drops to a mere 25% of the value obtained with reference mirror-polished specimens that further reduces to 15% for a forging condition. The reason of that tremendous loss of fatigue strength is twofold: decarburization and

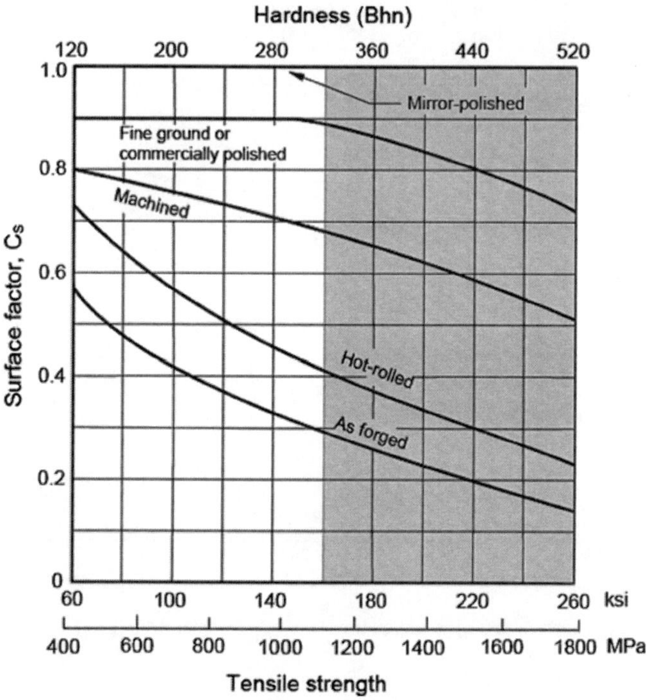

Fig. 4.7 Surface factor vs. tensile strength and hardness in steels for different surface finish (modified from [12])

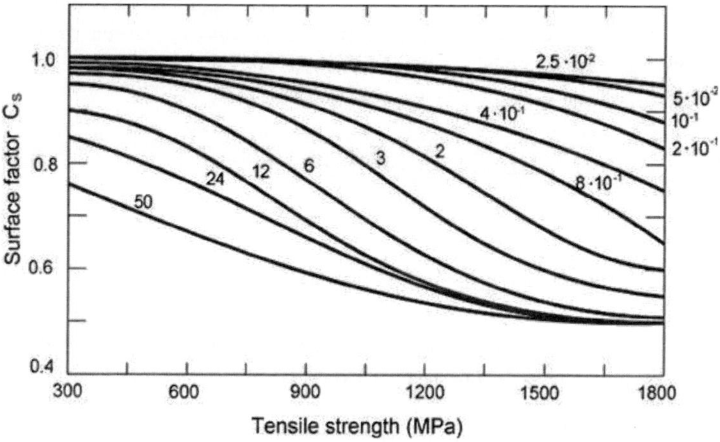

Fig. 4.8 Surface factor versus tensile strength in steels for different surface finish expressed in terms of mean roughness in micron (adapted from [13])

texturing. As to decarburization, there is a long history of study and research that dates back to the thirties of the last century.

Hankins and Becker [15], already in 1932, testing forged specimens of four types of steels having different initial hardness, under rotating cantilever bending, found that the surface was decarburized showing a hardness lower than that of the inner material. They also noted that there was less of a difference between surface and core hardness as specimen hardness decreased. Noll and Lipson [11] investigated, as mentioned in the previous section, the relationship between fatigue limit and surface condition using ground, machined, hot rolled and forged specimens. They also made metallurgical analysis of the as-forged specimens and found that the surface hardness was much lower than that of the inner material. This is shown in Fig. 4.9 [11]. The full transition from an average of 170 BH, measured on the surface, to about 350 BH occurs in some 0.75 mm. The carbon content in the surface decarburized zone was ranging from 0.1 to 0.2 wt%. Figure 4.10 is a magnified photograph of the specimen cross section showing the decarburized layer that appears lighter colored [11]. The macro-photograph corresponds to the hardness distribution of Fig. 4.9.

The reason of decarburization must be found in the hot-forging process itself. Decarburization is caused by high temperature surface oxidation. Forging, in fact, is a manufacturing process in which the material is shaped by hammering (automatically) the metal brought at a temperature above the crystallization temperature (hot forging), typically around 1200 °C for steel and 880–1230 °C for nickel alloys, or at temperature 30% below the crystallization temperature (warm forging). As an alternative to forge by delivering blows with hammer, close-die or open-die forging can be used in which dies move towards each other shaping the heated workpiece placed on the bottom die. The forging temperature is reached by means of induction heating or

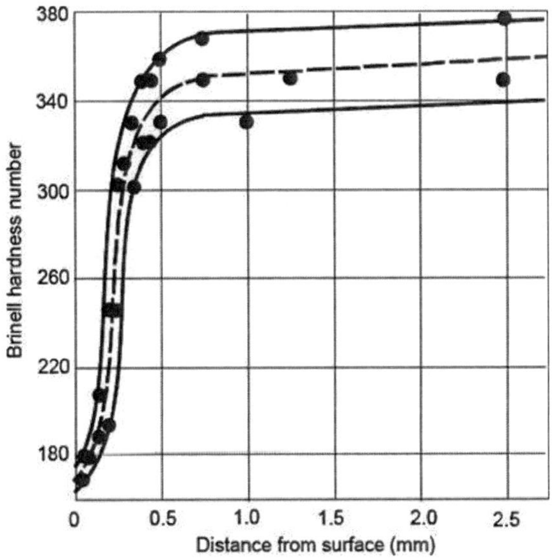

Fig. 4.9 Brinell hardness vs distance from surface for as-forged specimens with the average core hardness of 360 HB (adapted from [11])

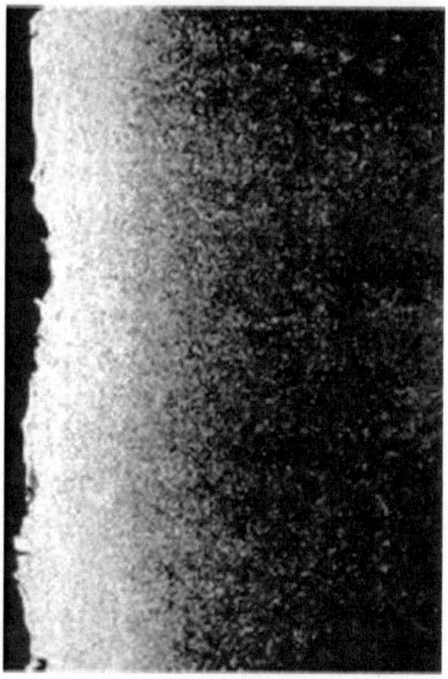

Fig. 4.10 Cross section of as-forged specimen showing surface decarburization [11]

gas furnace heating in which case more scale formation, oxidation, decarburization and grain coarsening is produced. Decarburization, in fact, occurs when carbon atoms at the steel surface interact with the furnace atmosphere and are removed from the steel as a gaseous phase. Oxygen penetrates the surface through pores and scratches that can be very common in forged workpieces. Oxygen has very high chemical affinity for carbon which combines immediately especially at higher temperatures forming carbon monoxide which is a gas that escapes the surface. Carbon from the interior diffuses towards the surface, moving from high to low concentration and continues to flow until the maximum depth of decarburization is attained. In the furnace hot-forging process the entire furnace atmosphere is brought to temperature, which accelerates the oxygen reaction with carbon having more of an effect on fatigue strength than induction heating hot-forging. Decarburization is a serious problem in fatigue because surface properties become inferior to core properties.

Decarburization results in lower hardness, as seen in Fig. 4.9 and this, in turn, results in lower strength and lower fatigue strength. The effects on fatigue strength depend on the depth of the decarburized zone. It is clear, then, that controlling the decarburization process will result in a better fatigue response. The target is to reduce the decarburization depth to less than 0.13 mm, compared to depth of 0.9 mm for the less controlled process. The second problem associated with hot-rolling or hot-forging is texturing. During hot-processes, inclusions reorient in a preferential direction, which is the direction of rolling or that normal to the hammering pressure or die casting pressure. Texturing that results in anisotropic fatigue properties, will be treated in the next section.

4.4 Texturing

Mechanical working operated at high temperature such as forging, rolling or extruding may result in a grain flow preferential direction. Also inclusions or second-phase particles become elongated giving rise to more or less material anisotropy depending, in particular, on the amount of mechanical working, on inclusions or non-metallic phase density and temperature. For instance, when a metal is fabricated by rolling at a red heat, grains and inclusions tend to be more malleable. They flow, deform and elongate in the rolling direction. This causes anisotropy. This can be clearly seen in Fig. 4.11 which refers to hot rolled A 106 carbon steel piping. The working process has plastically deformed and elongated iron carbides, Fe_3C, in bands along the working direction. This rather composite ferrite-carbide banded structure is strongly anisotropic. If a fatigue specimen is cut transverse to the rolling direction, inclusions lie along a plane perpendicular to cyclic stresses. It can be seen how inclusions can be 100 μm long or even 200 μm long and sharp so that, under the action of cyclic stresses, macrocracks develop rather soon, taking away most of the life of the material.

Moreover, these defects are all aligned almost continuously so that when just one of them nucleates a microcrack this jumps immediately to the next and the entire section may fall into a domino effect. The fatigue strength is severely jeopardized. On the contrary, specimens cut longitudinally, i.e., along the direction of grain and carbide flow, behave much better and fatigue strength is superior. Studies conducted on En25 steel [12] with tensile stress varying from 930 to 2000 MPa, depending on the heat treatment, have shown that in rotating bending the direction of specimen machining relative to grain flow was particularly important. In fact, the fatigue limit σ_f of specimens longitudinally oriented was growing from 510 to 770 MPa following tensile strength increase from 930 to 2000 MPa, as expected. On the contrary, in those specimens cut transverse to grain flow σ_f remained practically constant, independent of tensile strength and as low as 460 MPa, denouncing the heavy responsibility of elongated inclusions on the early fatigue death of metal. This anisotropy effect on fatigue response of metals had been already noted in the ’50s.

Love [13], for example, in his study performed on 40 different steels having fatigue limits ranging from 150 to 650 MPa, reported a systematic reduction of such a limit moving from specimen machined longitudinally to others in transverse direction. The amount of variation with specimen orientation ranged from a mere 5–25%. Kage and Nisitani [14] observed a remarkable reduction in fatigue strength in hot-rolled low carbon steel under rotating bending. The result of their experiments is shown in Fig. 4.12 [14]. Fatigue failure initiated from surface slip in longitudinal specimens while it initiated from elongated surface MnS or FeS inclusions in the transverse direction. At 45° specimens orientation with the loading direction the behavior was in between the two. Keeping the inclusions as spherical as possible reduces the anisotropy. The most common inclusions in steel are manganese-sulfite (MnS) or iron-sulfite (FeS). These inclusions are rather soft and follow grain flow during hot forming operations.

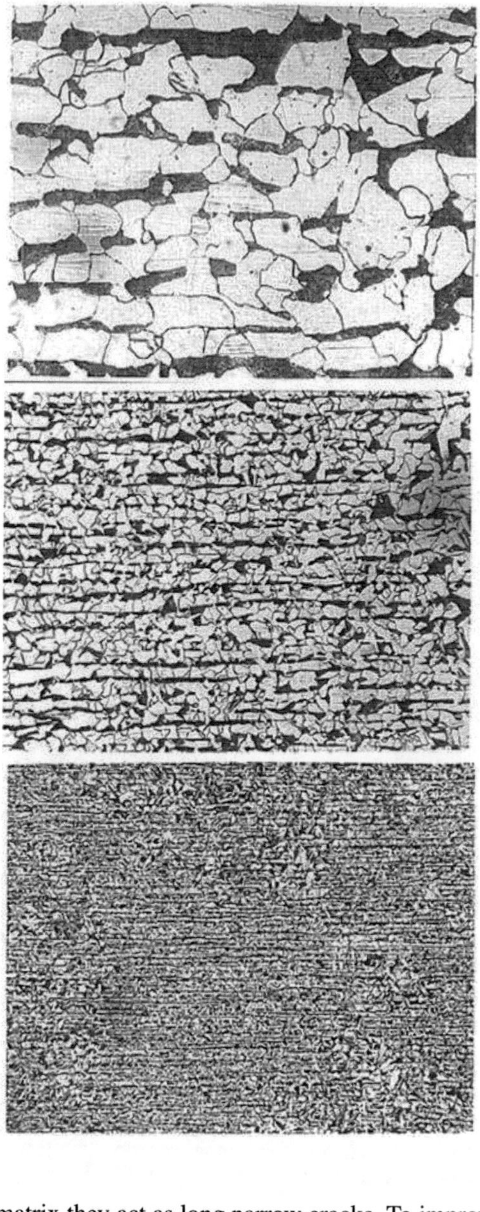

Fig. 4.11 Alternated bands of ferrite (white grains) and carbides (dark strings) stretched along the rolling direction, observed in A 106 steel piping

When debonded from the matrix they act as long narrow cracks. To improve the fatigue strength, then, it is necessary to perform a solid solution treatment adding, in particular, Ca which leads to more rounded MnS inclusions.

This is probably due to some chemical affinity of Ca for S that will prevent MnS formation. however, it is well known that calcium treatment may result in unwanted, large globular inclusions which results in in degraded fatigue strength and increase in

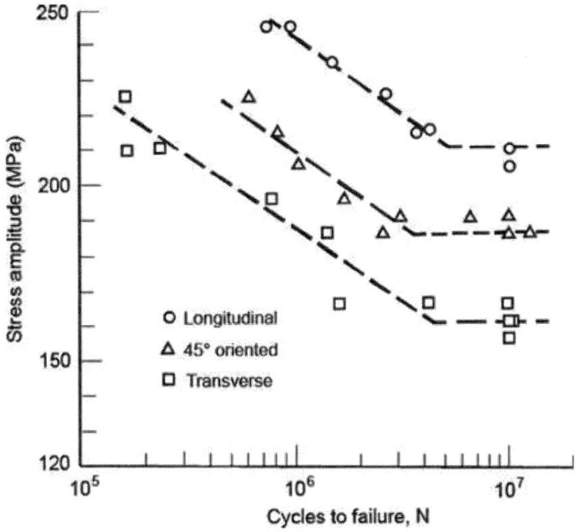

Fig. 4.12 Effect of anisotropy on the fatigue strength of rolled low carbon steel under rotating bending (adapted from [14])

data scatter. A solid solution strengthening treatment is also performed by addition of Cr, Ni or V and some rare earth elements that increase the hardness of MnS inclusions to about 400 HV or 40 HRC from about 170 HV, preventing or reducing their deformation during hot processes.The same anisotropy observed in hot-rolled steel can result from hot-forging operations, as shown in Fig. 4.13 [15]. Evans et al. [16] testing forged bars in rotating bending obtained results in some ways similar to those of Love. However, the fatigue limit of longitudinal specimens was not exceeding 10% of that measured on transverse specimens. Analogous results were obtained by Templin et al. [17] on sheets and bars of aluminum work-hardened and aged. The fatigue limit observed on the two perpendicular directions was quite similar. Probably it was due to the working process that produced spheroidal inclusions with consequent low anisotropy. Tests run by Ranson and Mehl [18] on 4340 forgings (0.4% C) evidenced a reduction of fatigue limit between 20 and 30% going from longitudinal to transverse direction.

Frost et al. [19] report that there is experimental evidence that even in those metals and alloys in which fatigue failure is initiated by cyclic slip in surface grains, the fatigue strength of specimens cut transverse to the working direction is inferior to that of longitudinal specimens.

It is a fact that the reduced fatigue strength observed in the transverse direction is due to inclusions that have been stretched by the hot working process. Also the surface appearance after fatigue failure occurs is different. In longitudinal specimens where fatigue initiated by a slip process it remains smooth. In transverse failures it is rough because fatigue originates from inclusions and there may well be many initiation sites that leave the surface rough, though only one of them will propagate to failure. Little difference has been reported between the torsional fatigue limit of specimens cut either longitudinally or transversely from bar aluminum alloys, 0.4 C steel and En25 steel on which rotating bending tests evidenced a dissimilar behavior [17, 20–22].

Fig. 4.13 Cross section of a forged steel bolt showing texturing by grain flow [15]

4.5 Residual Stresses

Actually, the issue of surface effect on fatigue life of materials appears to be more complex than expected from the considerations made in Sect. 4.2. To-day, in fact, there is the tendency to reconsider the effective consequences of the surface layer characterized just in terms of roughness. In the previous sections we have been dealing with decarburization and texturing which may influence fatigue strength more than expected by simple considerations on surface finish. Another effect must be considered that appears to be very important and this is the state of residual stresses introduced by machining or any other process that has interested the workpiece surface. The wrong choice of grindings parameters, for example, such as abrasive type (generally aluminum oxide or silicon carbide or diamond) and binder, grinding wheel speed, hardness, infeed and lubricant, or the use of procedures never validated through an experimental campaign just on the premises that it is common practice to do so, may result in a considerable risk of causing cracks in the workpiece or introducing high residual stresses. Incorrect grinding of a hardened tool can also result in such a high temperature at the ground surface to cause re-hardening of the material, as shown in Fig. 4.14. This produces a mixture of non-tempered martensite and tempered martensite with retained austenite.

The overall result is that very high residual stresses arise in the material because of the different specific volumes. Figure 4.15 [23] shows an example of residual stress state left on the surface of AISI 4340 quenched and tempered steel specimen hardened to 50 RC and finished by different grinding procedures labeled as gentle, conventional or abusive.

In the first case of gentle grinding, the peak stress is compression and is equal to about − 180 MPa, which is very favorable to fatigue strength. With conventional grinding it becomes traction rising at about 600 MPa and with abusive grinding it goes up to 700 MPa but, even more important, the tension stress state remains over 0.2 mm depth while in the previous case it terminated at about 0.1 mm, doubling the high residual stress state process volume. We know that the surface layer that

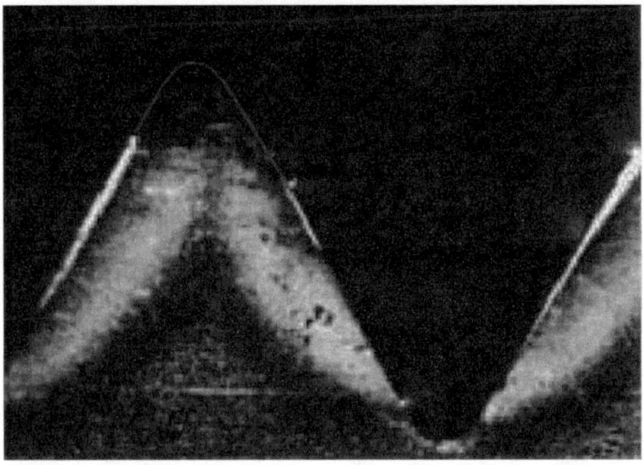

Fig. 4.14 Re-hardened surface layer due to incorrect ground tool (courtesy of Uddeholm)

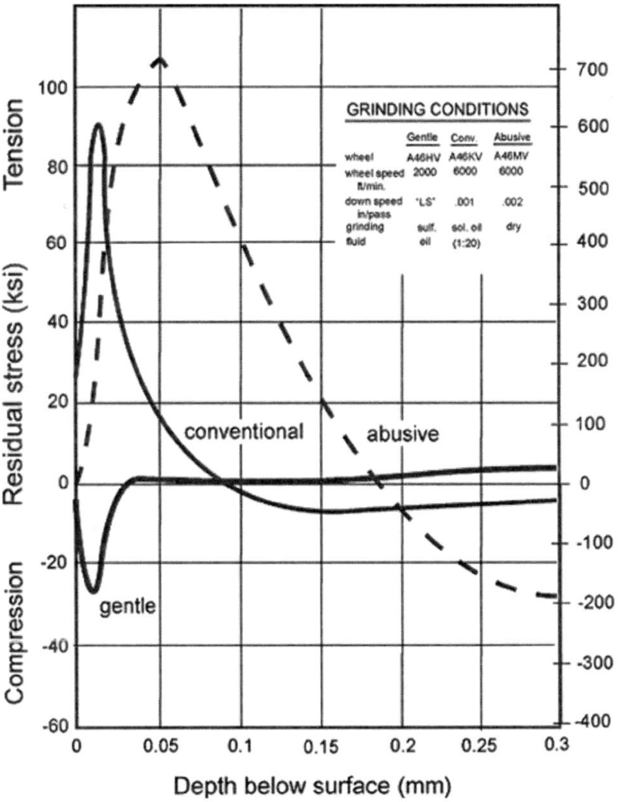

Fig. 4.15 Residual stresses in AISI 4340 finished by grinding (adapted from [23])

may have an effect on high cycle fatigue is no deeper than 0.5 mm (see Fig. 2.5) therefore the residual stress state will certainly have an effect on fatigue life. The three different grinding processes will result in three different fatigue lives, very high for the gentle type that leaves beneficial compression stresses on the surface and worst for the abusive grinding that results in very high traction stresses. This can be seen in Fig. 4.16 [23] for the same steel. When residual compressive stresses are generated on the surface the fatigue limit jumps to 900 MPa while with residual traction peak stress around 700 MPa (abusive grinding of Fig. 4.15) the fatigue limit reduces to 500 MPa, almost half of that. Note that for this steel hardened to 50 RC the fatigue limit should be about 766 MPa, based on Eq. (10.7). According to the diagram of Fig. 4.7, the surface factor C_S for the gentle grinding used and the hardness of the material should be equal to about 0.55 which yields a predicted fatigue limit of about $766 \cdot 0.55 = 421$ MPa whether the measured one is about 900 MPa. Another interesting diagram showing the combined effect of roughness and residual stress is offered in Fig. 4.17 [24].

Results were obtained for carbon steel, 0.29C, 0.21Si, 0.55Mn, 1.99Ni, 1.30Cr, 0.57Mo, quenched and tempered to obtain 844 MPa ultimate strength and 710 MPa yield strength. For specimen having the same accurate surface finishing the fatigue

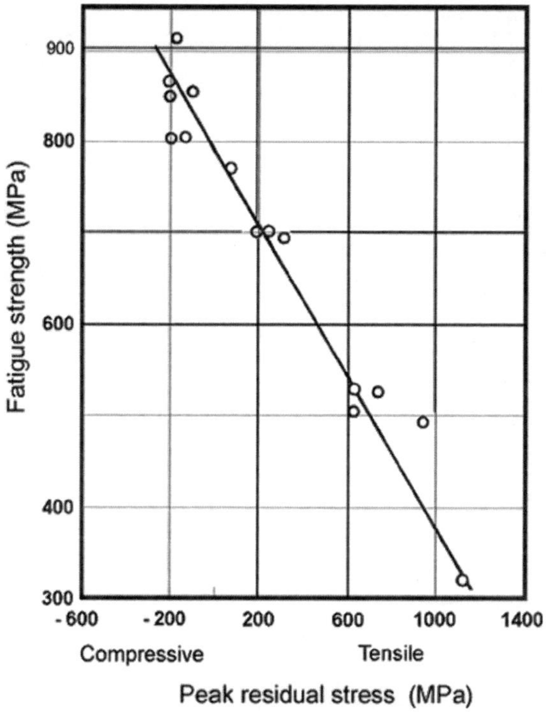

Fig. 4.16 Fatigue limit versus peak residual surface stress in ground AISI 4340 (modified from [23])

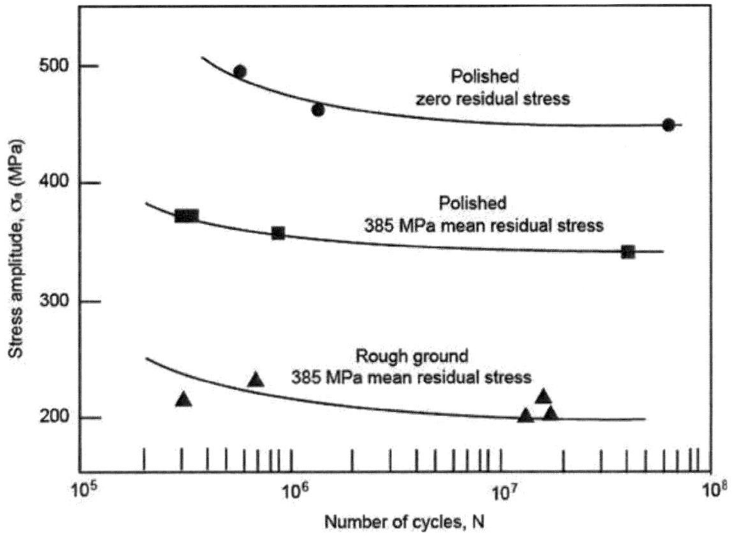

Fig. 4.17 S–N curves obtained for forging carbon steel as a function of surface finish and residual stresses (adapted from [24])

limit drops from about 450 MPa to about 340 MPa when a traction residual stress of 385 MPa is acting. With a rough ground and residual stresses the fatigue limit precipitates to about 200 MPa. This clearly indicates how the use of surface factors may under or over-estimate fatigue limit of materials if residual stresses are not known or not taken into consideration. Residual stresses can be considered to act as mean stresses and, as such, can be taken into consideration by calculation.

4.6 Effect of Inclusions

As already mentioned, damage usually develops in persistent slip bands activated in some surface grains by continuous fatigue cycling (see Sect. 1.5). A possible surface roughness can accelerate this process by providing a suitable local stress concentration that activates the slip process reducing the life of the specimen, as discussed in Sects. 4.2 and 4.3. Without any other competitor as damage initiator, the slip process activated by roughness starts at metallurgically weakest point on the surface. At variance, if inclusions or second-phase particles are present on the surface of the workpiece, then a kind of competition arises between these two factors, namely surface roughness and inclusions, as to which of them will initiate the first fatigue damage. Under monotonic traction, inclusions tend to reduce the allowable deformation. Figure 4.18 is an interesting example of the effect of shape and density of second-phase particles on steel strength [25]. It is worth noting how the complete absence of inclusions allows steels to reach 200% or higher true deformation at fracture, $\varepsilon_{t,f}$, or about 90% reduction of area. This should not come as something of a surprise. During deformation dislocations pile up against any obstacle that can be a grain boundary or an inclusion or second-phase particle inside the grain, exerting a local pressure. In order to move further, the slip process must overcome these obstacles. Depending on the relative hardness between inclusions and metal matrix, this pressure may create voids around inclusions by just deboning them from the metal matrix, in which case a void is formed or break them, in which case a crack is formed. Ductile failure is the consequence of this void formation followed by growth and coalescence, see Fig. 3.16. This type of ductile fracture is also called dimple fracture. When a crack is created, it may trigger a brittle fracture, depending on the material toughness at the working temperature. It is then clear that without any inclusion, fracture, whether ductile or brittle, is much retarded because the voids needed to trigger it are only those already existing in the lattice or created by dislocations movement. If the percent volume of inclusions or second-phase particles increases, cavities nucleation process proceeds faster and ductile fracture occurs under lower deformation, as Fig. 4.18 indicates. Therefore, oxides, sulfites, silicates, carbides Fe_3C and other second-phase particles such as Ni_3Al (phase γ'), Ni_3Ti (phase η), Co_3W (phase ε) etc., or Al_2O_3, Al_2CuMg and $Al7Cu_2Fe$ in aluminum alloys may strengthen the metal by blocking dislocation motion, but reduce deformation at fracture.

The form of inclusions or second-phase particles also has an effect on strength, as already mentioned in Sect. 4.3. Elongated and sharp inclusions, in facts, may compromise steel fracture strength. Figure 4.18 clearly indicates that spheroidal carbides cannot reduce fracture strength as much as pearlitic carbides or elongated sulfides, even increasing their concentration. Pearlite has a definite appearance under the microscope and can be clearly identified as a constituent and it is commonly

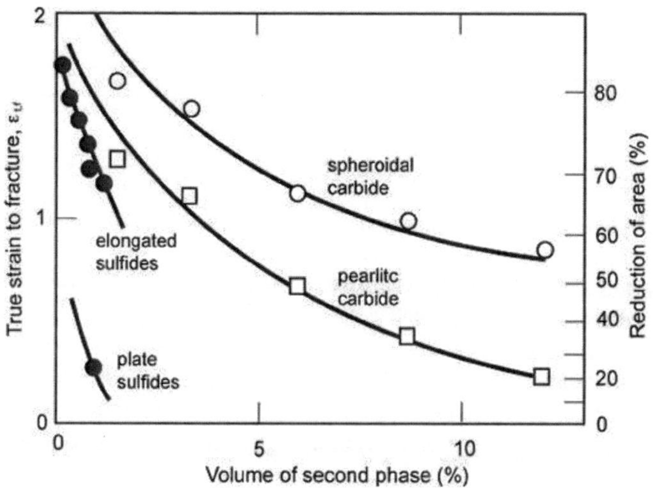

Fig. 4.18 Effect of the shape of second-phase particles on total ductility [25]

referred to as phase, though it is not. Pearlite, in fact, is rather a compound, a mixture of two phases: cementite (Fe_3C) and ferrite (α-iron) and consists of alternate plates (lamellae) of Fe_3C and ferrite, ferrite being the continuous phase. Figure 4.19 is an example of cementite-ferrite mixt structure that gives life to pearlite in a Fe 510 carbon steel. The cementite lamellae appear dark while the ferrite phase is white. Even though pearlite plates are elongated in shape, their borders and, in particular, their tips are not as sharp as those of graphite in gray iron or in ductile cast iron when it degenerates.

Figure 4.20 is an example of degenerated graphite in ductile cast iron. The morphology of graphite is no longer spheroidal as it should be, but assumes the pattern of a continuous network of needles that can jeopardize fatigue strength. Also pure martensite (not tempered), which is a super-saturated solution of carbon in α-iron, is characterized by sharp, needle-like or acicular cementite plates, as shown in Fig. 4.21a, but these plates cannot be considered second-phase particles, being an iron phase. The carbon trapped in the martensitic structure as fine needle-like plates can be released by heating the steel above 150° C for a convenient time, after quenching. The process, called tempering, allows the structure to deform plastically reliving some of its internal stresses. This reduces hardness and increases toughness, but it also reduces its tensile strength. The degree of tempering depends on temperature and time. Tempered martensite is shown in Fig. 4.21b, c.

The basic mechanism by which an inclusion acts on the strength of materials is, as previously mentioned that of generating a crack either by breaking or debonding. Inclusions work like barriers in the grain or along grain boundary. Slip bands can pile up against them, Fig. 4.22a, which may break or debond because of the pressure exerted over them, as shown schematically in Fig. 4.22. If the foreign phase is very

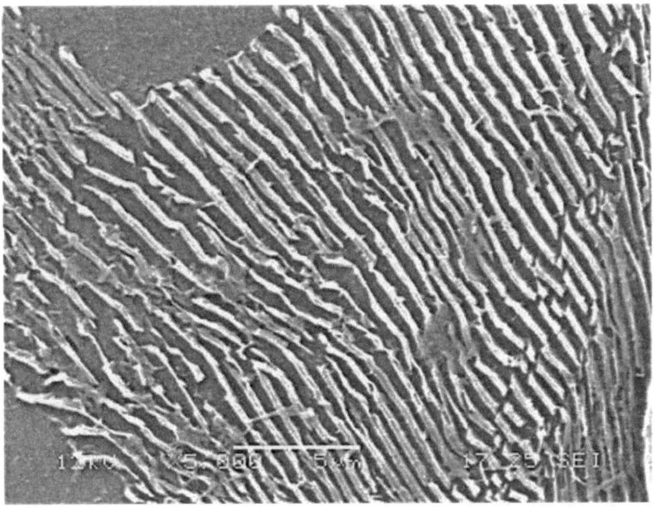

Fig. 4.19 Perlite lamellae (dark elongated laminas) in ferrite (white zones) in Fe 510 carbon steel

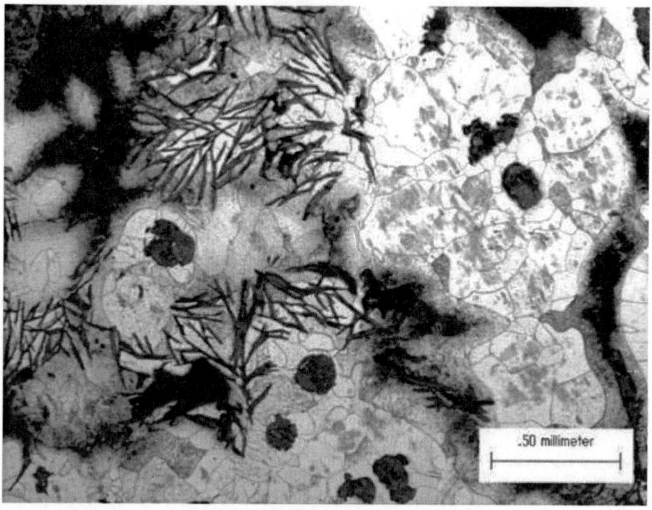

Fig. 4.20 Graphite needles in ductile cast iron (courtesy of M. Cova, SACMI, Imola, Italy)

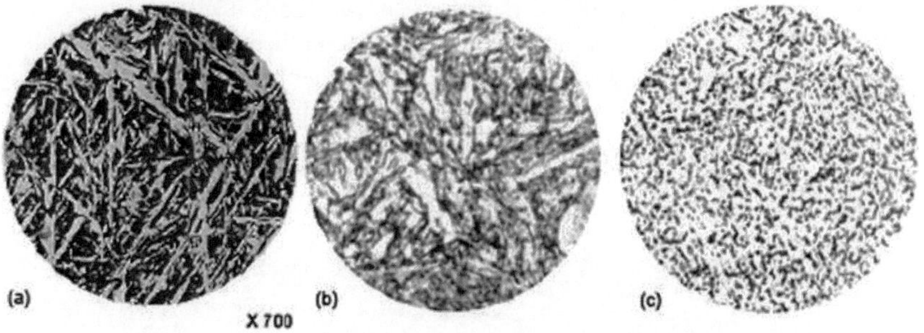

Fig. 4.21 **a** Acicular appearance of cementite plates in quenched martensite that **b** sharply reduces in tempered martensite and **c** disappears in heavy tempered martensite

brittle and the matrix hard, it is likely that it breaks, case (b) of Fig. 4.22, leading to brittle fracture or cleavage, see Fig. 3.12. Where the matrix is soft, debonding occurs, case (d) of Fig. 4.22, developing ductile or dimple fracture, see Fig. 3.16. Two interesting microscopic images of inclusion break and debonding are provided in Fig. 4.23 [26] for the case of static loads.

Note that once cracks or pores, perhaps with crack emanating as in Fig. 4.23b, are formed, the interference between the stress fields of two contiguous phases can lead to the failure of the bridging ligament. This can be seen in Fig. 4.24 [27] for two manganese-sulfide inclusions, MnS that are quite common in low-medium strength carbon steel. What the picture shows are two small voids of about 20 and 10 μm, respectively, at about 40 μm from each other that were the sites of two sulfide inclusions in a specimen of AISI 4340 steel linked by a narrow void sheet consisting

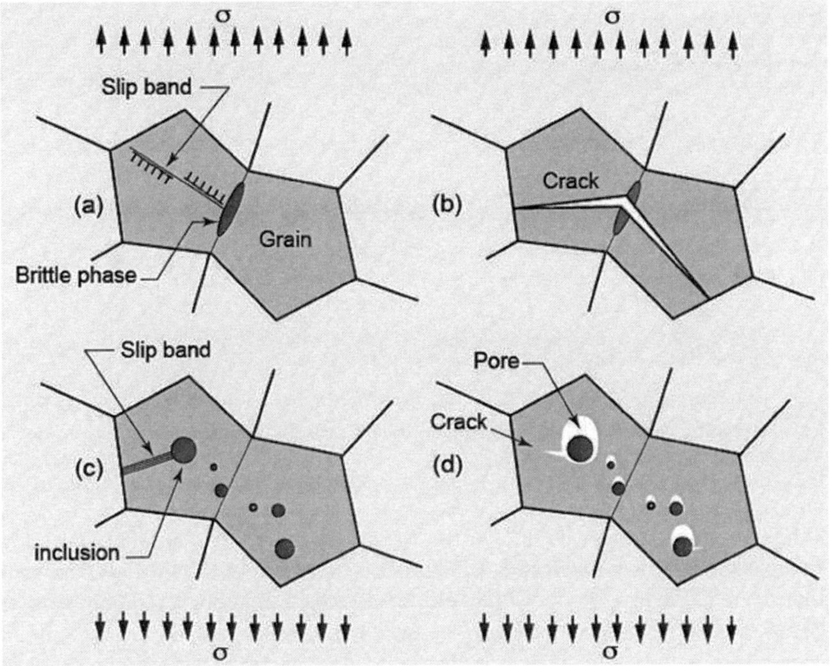

Fig. 4.22 **a** Dislocations piling up at a brittle phase; **b** the brittle phase breaks and cracks develop in grains. **c** Slip band pressing over inclusions **d** causing debonding, void formation and cracks

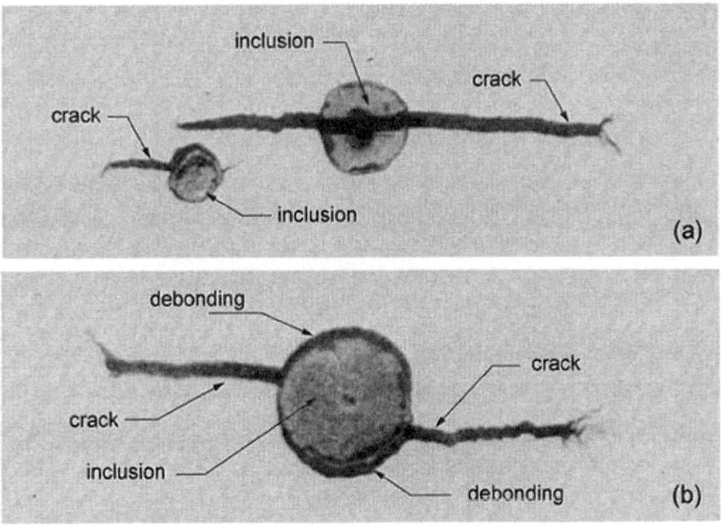

Fig. 4.23 **a** Spheroidal inclusion break and **b** debonding from iron matrix in high strength steel [26] creating a micro crack

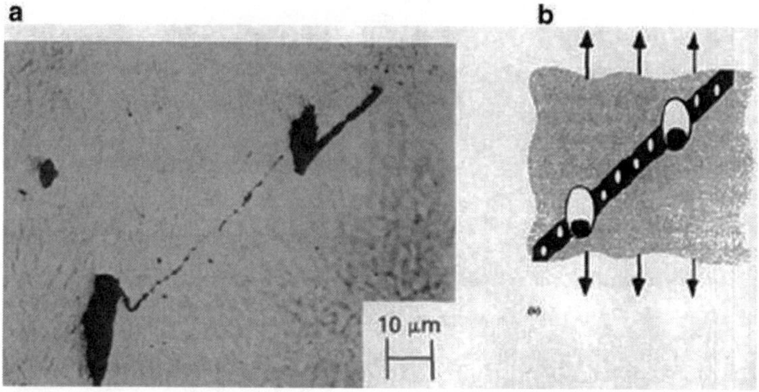

Fig. 4.24 **a** Section through the necked region of A 4340 steel specimen showing the formation of a void sheet between two large voids formed at inclusions. **b** Schematic of voids nucleation at smaller particles along the slip band [27]

of small submicroscopic cavities, like those shown in Fig. 2.12 for an aluminum alloy. On the right there is a schematic of the process occurred: two inclusions debonded from the metal matrix produced two elliptical voids. Stress concentration between the apexes of the two voids activated a slip band on which damage nucleated in the form of sub-microscopic voids.

It is rather evident that the same inclusions and second-phase particles that have a role in the fracture strength of a steel or metal alloy under static loads shall also affect their fatigue strength. This effect is not so easy to identify in its full extent. First of all, we shall consider that true deformation and true stress at fracture are related to fatigue strength through the Mason-Coffin and Basquin Eq. (8.10), respectively, as it will be discussed in Sect. 8.3. Therefore, it follows that the complete absence of inclusions and second-phases that would allow the steel to reach even 200% or higher true deformation $\varepsilon_{t,f}$ at fracture would also improve its fatigue strength by pushing up the S–N curve. If the process volume or grain dimensions are sharply reduced as to hinder or avoid any slip and reduce the probability to have any sort of inclusions, then fatigue life would be almost unlimited.

An irrefutable evidence of that can be found in whiskers or fibers that do have an almost infinite life because their infinitesimal volume does not allow slip to occur and reduces to practically zero the probability of getting an inclusion or defect, whatsoever. Precisely the same mechanism of slip bands pushing over inclusion that we have previously described as initiators of cracks under static loads and the interference of stress fields associated to voids that coalesce will work also under cyclic loads. The analogy with debonding and crack formation under static load presented in Figs. 4.22 and 4.23 can be seen for cyclic loads in Fig. 4.25 [28] that refers to 1190 MPa 4340 high strength steel having a smooth surface in which a crack nucleated at the matrix-inclusion interface.

At 2000 cycles, decohesion of calcium aluminate inclusion from the metal matrix is well visible, Fig. 4.27a. A tiny crack is already visible in the upper right side of the

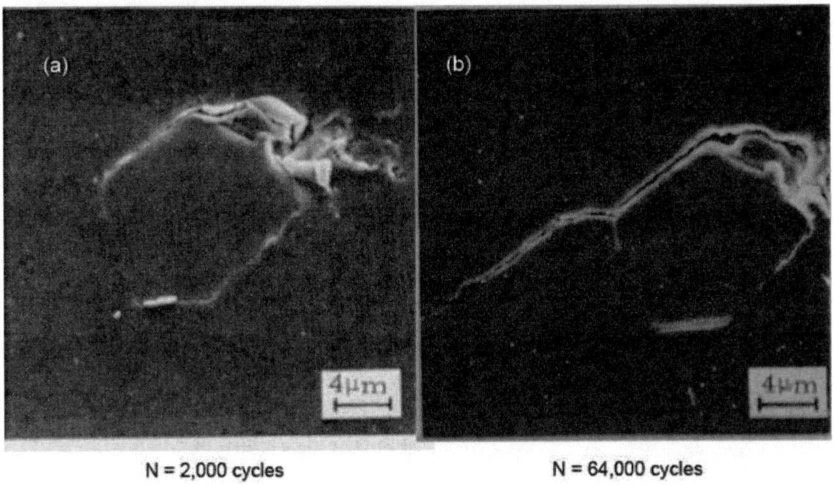

Fig. 4.25 **a** Debonding of a calcium aluminate inclusion in 1190 MPa strength 4340 steel after 2000 cycles and **b** crack formation at 64,000 cycles [28]

inclusion. At 64,000, Fig. 4.27b, this tiny crack has become a 30 μm microcrack penetrating the metal matrix. It is clear, however, that in the competition among surface roughness and inclusions or second-phase particles as initiation sites of microcracks, inclusions shall be comparable in size to the average roughness and lie on the surface. We have already mentioned this occurrence in Sect. 2.5 and in Figs. 2.35 and 2.37 about mild steel 0.13% C of 203 MPa yielding strength and 327 MPa ultimate and precipitation-hardened or aged aluminum alloy, respectively. In one case slips were responsible for crack formation, in the other case it took the inclusion to play this role.

Figure 4.26 [29] is an interesting example of what has been said about competition between inclusions and surface roughness. It refers to fatigue specimens of 13% Cr turbine blade steel, 580 MPa of yield and 850 MPa of tensile strength, under rotating bending. Figure 4.26a shows a micro crack originated by a small inclusion of 10 μm. In this case the specimen surface was mirror-polished with an average roughness of 1 μm. Such a small inclusion could then play a role in crack initiation. At variance with the polished specimens, in the rough specimen the fatigue crack also initiated from an inclusion, but this time its size was very large (> 40 μm), as it can be seen in Fig. 4.26b and was located in the high stressed region of the gage length. Any inclusion of smaller size or deeper located could not compete with surface roughness.

Eid and Thomason [30] observed the same behavior on carbon-molybdenum quenched and tempered steel containing Al_2O_3 inclusions of 25 μm diameter. Inclusion appeared either debonded from matrix or broken, but in either case they originated a fatigue microcrack. The authors underlined the importance of the Young's moduli ratio between phase and matrix. When this ratio is larger than 1, as in the case of alumina in iron, high stress concentration arises at the sharpest point of the

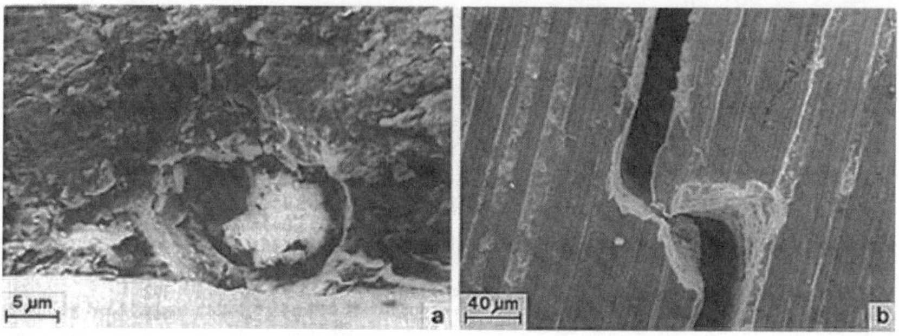

Fig. 4.26 Micrographs of crack initiation points: **a** small inclusion in polished specimen **b** large inclusion in abraded specimen [21]

Fig. 4.27 Dependence of stress concentration factor on size, shape and position of holes and inclusions in a plate under traction

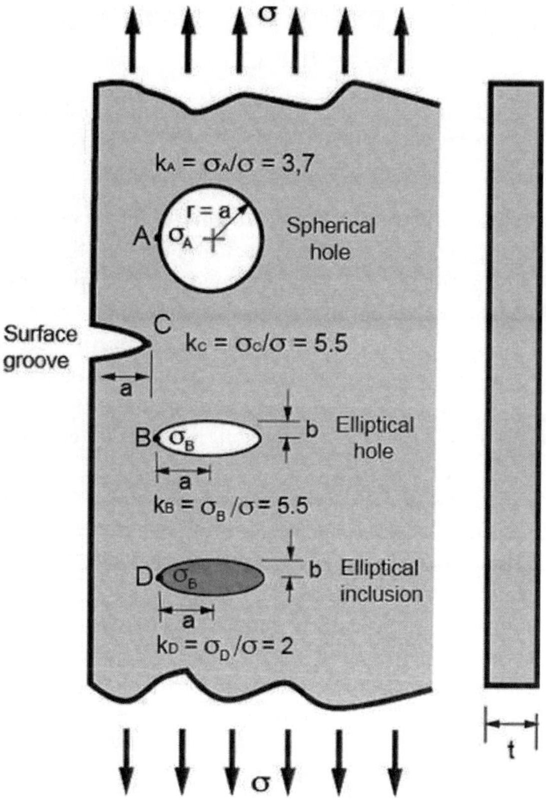

inclusion. Aluminous inclusions, Al_2O_3, for example, are rather dangerous because they are very hard and brittle and break rather soon. Titanium carbonitride that may be found in 4340 steel or 18 Ni maraging, are so brittle that break under very low deformation of the matrix crystals in which they are contained. Duckworth e Ineson [31] studying the fatigue behavior of 1846 MPa high strength steels type En24, in which they introduced surface and subsurface alumina inclusions of different sizes, found that inclusions had an impact on fatigue strength only when their size was equal or larger than 10 μm, dimension this last that competes with extrusion, intrusion or surface roughness.

Nevertheless, if those inclusions were just 100 μm below the surface their effect would diminish and to again compete with extrusion or surface roughness as initiation sites their size had to grow up to 30 μm. The reason for that has to be found in the stress concentration at inclusion apex that depends on inclusion size, shape and position. Figure 4.27 explains schematically this dependency. First consider a surface groove or notch 30 μm deep and 15 μm of surface opening simulating a surface defect, Fig. 4.27. the stress concentration factor at the deepest point C, k_C, is equal to 5.5. A hole of the same size (radius $r = 30$ μm) at the same distance from the surface of point C would have a stress concentration factor at point A, $k_A = 3.7$, much lower than k_C. To have the same effect of the surface notch the hole should have a radius of at least 42 μm. If the internal hole were of elliptical shape of the same length, $a = 30$ μm, then the stress concentration factor at B, Fig. 4.27, is $k_B = 5.5$, equating the surface notch. However, the elliptical hole cannot be considered to have the same concentration factor pertaining to an inclusion of the same length having the shape of an ellipsoid of revolution. A hole or cavity of any shape represents a more critical condition than a corresponding inclusion that represents a solid body adhering to the metal matrix. For a rigid inclusion, Donnell [32] obtained a stress concentration factor at point D, Fig. 4.27, $k_D = 2$. Actually, surface roughness cannot be represented by a single edge notch but, rather, by a row of notches of different depth and pitch, as schematized in Fig. 4.27b. It has long been recognized that a single edge notch generates a stress field by far higher than that existing ahead of series of closely spaced notches of the same kind, as schematized in Fig. 4.28. Passing from a single notch, Fig. 4.28a, to a continuous row of notches of the same kind, Fig. 4.28b, the stress concentration k decreases and, depending on the pitch, the reduction can be as high as 1.7 or 2. The actual stress concentration k_C would, then, decrease from 5.5 to 3.2 or 2.6, yet it will always be higher than that pertaining to an inclusion. The real situation is that sketched in Fig. 4.28c that represents a rough surface for which it may be difficult to define a single concentration factor unless a deep indentation or an extrusion appears. Anyhow, an inclusion cannot compete with a surface groove of the same size, unless it is very sharp and long and placed right on the surface. To reach the same concentration factor an inclusion should have a shape like that of a manganese sulfite (MnS) in steel of Fig. 4.29, but should be placed on the surface otherwise the same particle is no longer harmful.

To further evidence this point it may be useful the experimental program run by Murakami [33] on specimens with artificial surface roughness of mean value equal to 14.5, 37.5 and 53.3 μm, respectively and a maximum depth of 27.3, 66.4 and

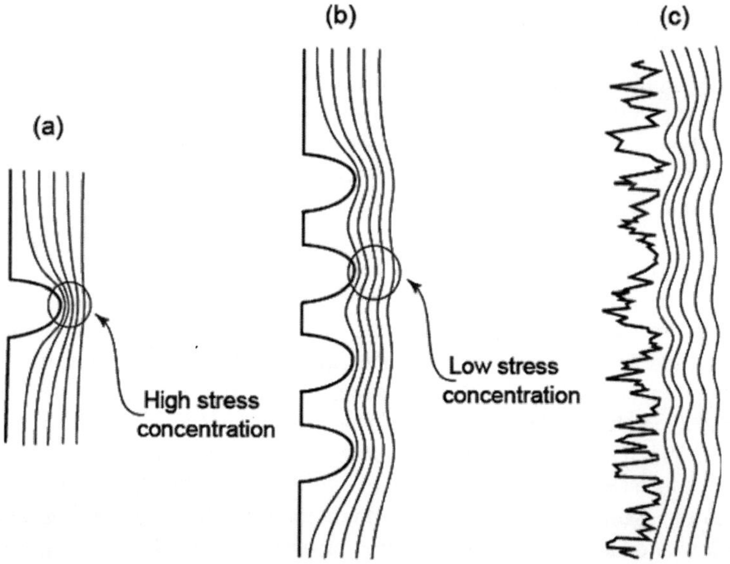

Fig. 4.28 Stress concentration ahead of: **a** single notch, **b** multiple notches of the same kind, **c** roughness

Fig. 4.29 Elongated manganese-sulfite (MnS) inclusions with a titanium nitride particle in carbon steel

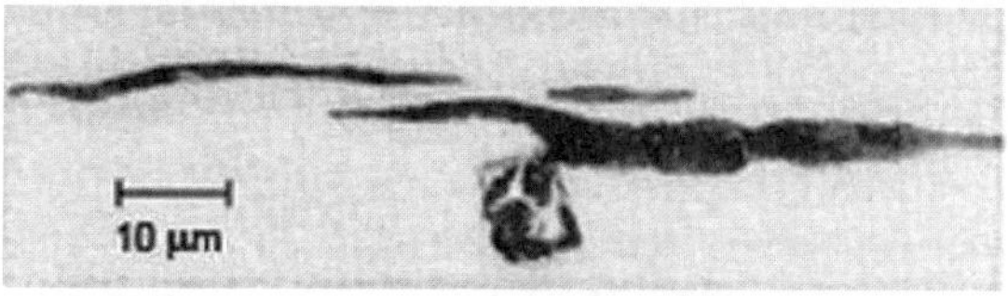

Fig. 4.30 *S–N* curves of specimens of 0.46% medium carbon steel (JIS S45C). The specimens were turned to shape after annealing for 1 h at a temperature of 844 °C [33]

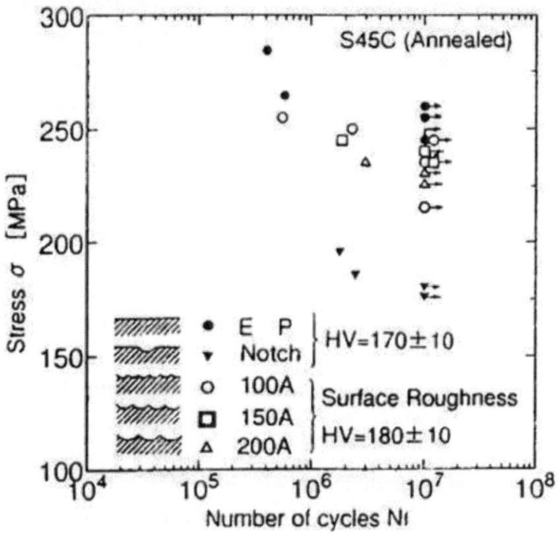

74.0 μm, respectively compared to specimens containing a single notch of 30 μm, Fig. 4.30. Although the mean roughness of specimen 150 A is 37.5 μm, which is of the same order as the depth of the single notch, 30 μm, the fatigue limit for specimen 150 A is 29.8% higher than that for the single notched specimen. The single notch, anyhow, is by far more dangerous than specimens 200 A containing a surface mean roughness of 53.3 μm. At 500 μm depth any inclusion of any size will not have any effect at all unless subjected to a high residual stress state. It is clear that if the specimen or workpiece bears a smooth surface of mirror-polished type any inclusion of any size will have a chance to become an initiation site, provided it is superficial. Vice versa, without any inclusion, initiation will always start by slip in grains having the worst orientation that is to say the higher Schmidt's factor, see Sect. 2.1. We can better understand now what has been said before that outside the process volume, see Fig. 2.5, local metallurgy doesn't play any role whatsoever on the initiation process. In terms of fatigue limit, we can then consider nonmetallic inclusions to be mechanically equivalent to microcracks. It is clear, in fact, that these microcracks originated by an inclusion will behave exactly as microcracks originated by a slip process in the absence of any inclusion. The only difference will be in the life expected that for specimens containing surface inclusions will be more or less reduced with respect to a smooth one, depending on the inclusion size and shape.

This actually means that the effect of a surface inclusion is that of reducing the life but not the fatigue limit of the specimen. This can be seen in Fig. 4.31 that presents some results obtained by Toryiama and Murakami [34] with high strength maraging steel (510 HV). They found that the fatigue limit was almost independent of initial acuity of defects they introduced in the specimens, were they holes or cracks generated from holes by fatiguing the specimens prior to run tests. The difference in the fatigue limit was almost negligible, at most some percent. Nevertheless, when defects were real cracks obtained by pre-cycling the specimens, fatigue limit was reached at much lower number of cycles, also two decades earlier. Specimens containing holes fail at about 10^7 cycles, whereas those with cracks do not reach 300,000–400,000 cycles. The difference is enormous. The rationale behind that is rather clear. The actual difference between a crack and a hole is just in the number of cycles needed to the latter to become a crack. It is also clear that these waiting cycles do not affect the fatigue limit of the specimens containing a hole with respect to that measured on specimens containing a crack, which is not that of a plain specimen, anyhow, but lower. Therefore, holes and cracks of the same size are equally affecting the fatigue strength of the material and differ only in their effects on the fatigue life.

Significant, in this respect, are the experiments by Natsume et al. [35] who prepared a series of specimens of alloy-tool steel containing carbide inclusions of the same shape. Depending on the preparation technique, some specimen had cracked carbides other had not. They found, Fig. 4.31, that there was no difference in the fatigue limit measured on all specimens, but on those with cracked carbides the fatigue limit was reached at lower lives, i.e., at a lower number of cycles N_f. It is clear that in the case of pre-cracked inclusions the crack formation phase was bypassed with a sharp reduction of life. In terms of cycles to failure N the impact of inclusions on the fatigue life is shown in the diagram of Fig. 4.32 that has been

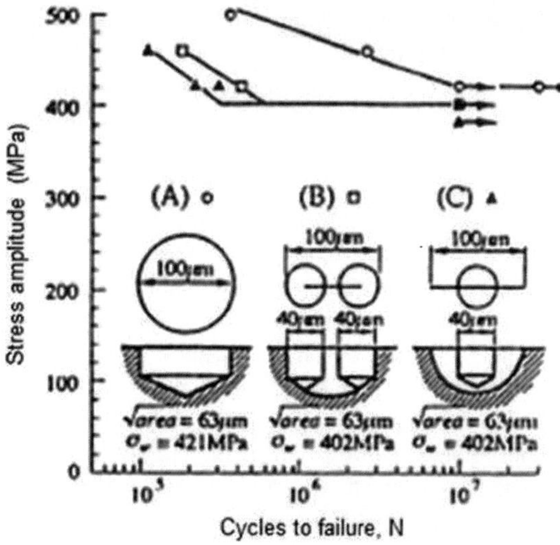

Fig. 4.31 S–N curves obtained in maraging steel specimens containing cracks or holes of the same size (area) [35]. Defect A is a hole, whereas B and C start from two or a single hole that originates a crack by pre-cycling the specimen

drown using experimental data obtained by Cummings et al. [36] with specimens of aeronautical grade type 4340 steel under stress amplitude equal to 600 MPa, in fatigue pulsing from zero. There is a clear dependence of the number of cycles to failure, N_f, on the dimension a of inclusions. If the inclusion size increases from few microns to 60 microns the number of cycles N_f decreases from 10^6 to 10^5, by a factor 10. This dependence for the case considered is given by

$$N_f \cdot \sqrt{a} = \text{costant} \approx 10^6 \tag{4.3}$$

with inclusion dimension given in microns.

Recalling Eqs. (2.10), (2.11) and (2.13) that link the inclusion dimension a to fatigue limit σ_f and indicating with n a generic exponent, it yields

$$\sigma_f^n = C' \cdot N_f^2 \tag{4.4}$$

with C constant characteristic of the material. This actually means that if the dimension a of the inclusion increases, there will be a parallel decrease of N_f, Eq. (4.3), then also σ_f shall decrease.

Assuming $n = 6$, as per Murakami, a threefold increase of length a results in a reduction of 1.7 of the cycles to failure, N_f and a reduction of 1.2 of the fatigue limit σ_f. This can be seen in Fig. 4.34 [37] that refers to two batches of specimens of the same steel 4340, 978 MPa strength, machined from two different bars. The specimens of the first batch were examined and no inclusions larger than about 20 μm was found to have triggered fatigue fracture. They were designated in Fig. 4.34 as *small inclusion.* The specimens of the second batch were found to have large spherical inclusions of about 127 μm that initiated failure. These are designated in

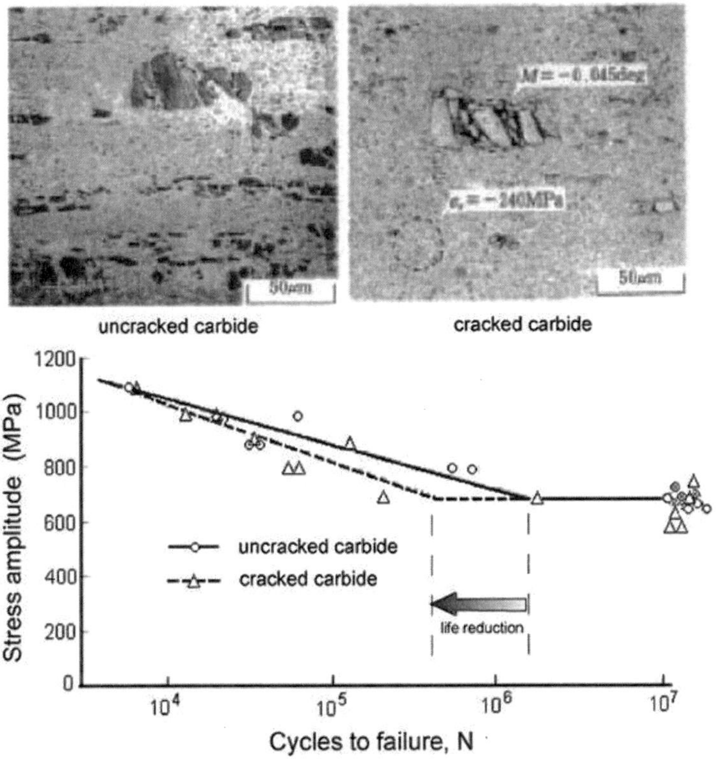

Fig. 4.32 Tests results obtained on specimens containing uncracked carbides and cracked carbides of the same size and shape (adapted from [36]). There is almost no difference in fatigue limit which is reached at shorter life in cracked specimens

Fig. 4.34 as *large inclusion.* Inclusions were identified as corundum and silicates. It appears that inclusions have the double effect of decreasing both the life N and the fatigue limit σ_f, as schematized in Fig. 4.35. This finding should not come as somewhat of a surprise at all.

It has been said that inclusions play the same role as roughness on the fatigue strength of materials and we have already discussed that roughness impinges on that strength reducing the fatigue limit, as shown in Fig. 4.5. However, we have to remember the meaning of the fatigue limit (see Sect. 2.4) that is defined as the highest possible stress amplitude that will not propagate an already existing crack. This actually means that not all inclusions have an effect on fatigue strength. We have already treated the issue of small cracks in Sect. 2.3 and seen the Kitagawa–Takahashi diagram for a mild steel, see Fig. 2.38, that was indicating how a crack of even 100 μm would be tolerated, remaining dormant. The limit size of a dormant crack, however, is not the same for any steel. It rather depends on its strength, as already shown in Fig. 4.3. This can be seen in Fig. 4.35 [24] that shows the Kitagawa–Takahashi

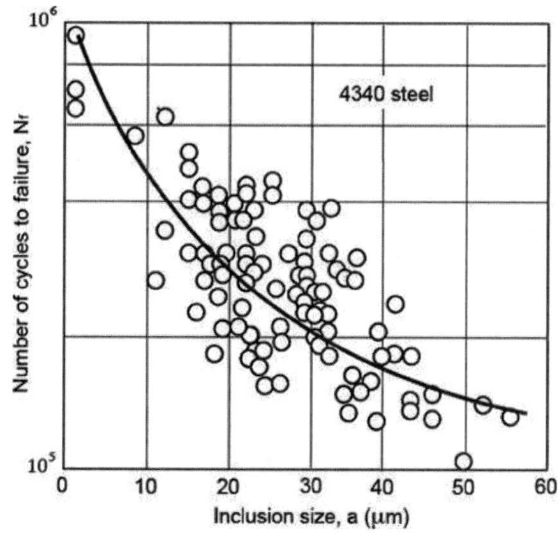

Fig. 4.33 Diagram made with data obtained by Cummings et al. showing the dependence of cycles to failure on inclusion size (data from [36])

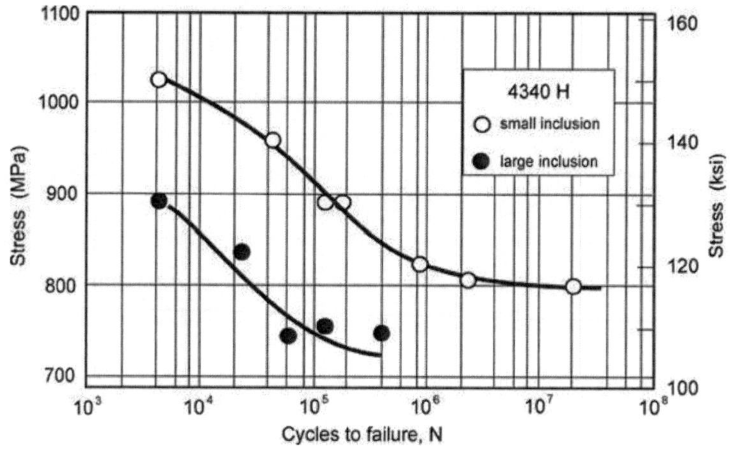

Fig. 4.34 Effect of inclusions on fatigue (modified from [37])

diagram this time for high strength carbon steel 0.29C, 1.99Ni, 1.30Cr quenched and tempered to 710 MPa yield and 844 MPa ultimate strength with hardness equal to 280 HV30, under cyclic loads applied with $R = -1$ and $R = 0$, respectively. It must be noted that the fatigue limit is already decreasing from its value obtained with plain specimens, for inclusion sizes just above a mere 2 μm. Such small a defect would have remained dormant without propagation in a mild steel specimen. Again, as for

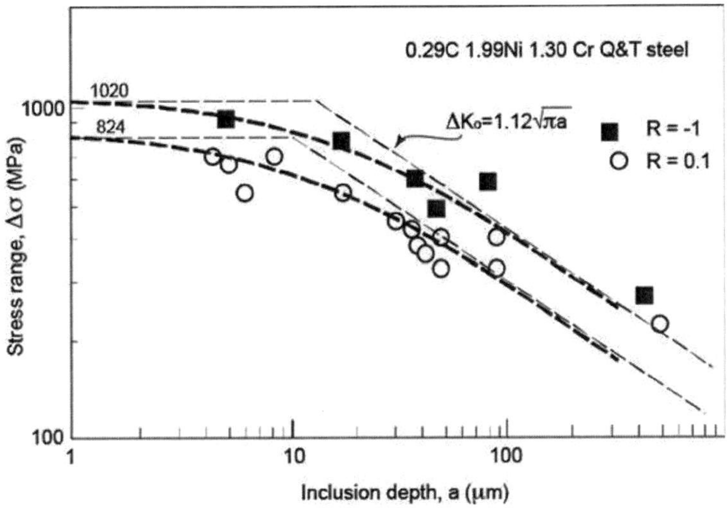

Fig. 4.35 Kitagawa–Takahashi diagram for high strength carbon steel (adapted from [24])

surface roughness, the dimension of inclusions has much more of an effect on high strength alloys than on low strength materials.

Murakami and Endo [9] have well documented this circumstance, as already shown in Fig. 2.40. It is quite clear that if fatigue failure is initiated by inclusions in surface grains any process capable of eliminating or at least reducing the presence of inclusions shall result in fatigue strength enhancement and reduction of anisotropy effect as well. This fact is illustrated in Fig. 4.36 [38], which provides three *S–N* curves for high-strength medium-carbon-alloy steel, type AISI 4340. Steel specimens heat-treated to the same 1586 MPa ultimate strength, were used to obtain each of the three curves. The upper and middle curves correspond to vacuum-melted steel, while the lower curve corresponds to steel that was air-melted. The inclusions in the vacuum-melted metal are smaller in size and less numerous than in air-melted steel. This also results in less anisotropy. This is because the vacuum-melting technique allows low melting temperature phases that during steel solidification originate inclusions, to leave the molten metal. On the contrary, in air-melted steel, gaseous element are absorbed and entrapped by the molten mass, creating inclusions. The lowest fatigue limit pertains to air-melted steel. The fatigue limit under vacuum-melted conditions is about 150 MPa higher than in air-melted conditions, indicating a much lower level of inclusions. However, though reduced, even under vacuum-melted conditions transverse specimens show an *S–N* curve below that of longitudinal specimens, indicating that anisotropy still persist. A second remelting would be necessary or a better control of foreign elements in the melting pool.

Consider that in the gear industry where very high strength steel are used and inclusion are not tolerated at all, often the ingot obtained under vacuum-melted conditions is remelted under vacuum. This technique was particularly used in the

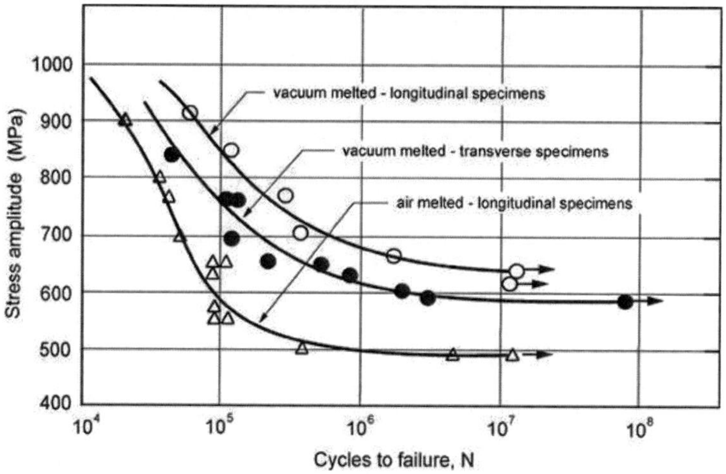

Fig. 4.36 Effect of nonmetallic inclusions and specimen orientation on the fatigue strength of 4340 steel (adapted from [38])

UK by the turn of 1800 while in Sweden it was preferred the open hearth furnace technique to produce high quality high-strength steel for bearing applications. Today the electric arc furnace can be conveniently used to obtain the desired quality.

Sulphur is one of the most devastating elements in fatigue strength as it forms MnS, as already mentioned. Moreover, sulfur contained in MnS inclusions may be dissolved in an aqueous environment or simply humid air creating an acid environment (see Section.), which stress corrosion is so fond of. Its elimination or, at least, significant reduction may completely change the fatigue response of metals. Figure 4.37 shows the S-N_{50} curves obtained in the transverse direction of three different batches of specimens of the same material SAE 4140 steel forging quenched and tempered to achieve a hardness of 52 HRC [39]. N_{50} denotes the number of cycles at 50% load drop (see Sect. 1.2). The only difference among the three batches was the sulfur content, namely 0.077% S, 0.012% S and 0.004% S, classified as high, low and ultra-low sulfur. The normal sulfur content for this steel is about 0.025%. Note the remarkable drop of the fatigue life and limit for the high sulfur steel.

To conclude this section about inclusions effects on fatigue strength it is worth showing the diagram of Fig. 4.38 [37] obtained testing 15 different steels quenched to 100% martensite and tempered. Specimens were standard 6.35 mm diameter mirror-polish finished under rotating bending fatigue. It can be seen how up to about 35 Rockwell C hardness the scatter is confined within 125 MPa. At higher hardness the scatter increases and becomes about 400 MPa at Rockwell C 55. The same relationship is obtained with specimens loaded axially or in bending, when the mean stress is zero. Note that Rockwell C 35 corresponds to a tensile strength of about 1100 MPa. In all cases data scatter is due to metallurgical variability, but whereas below

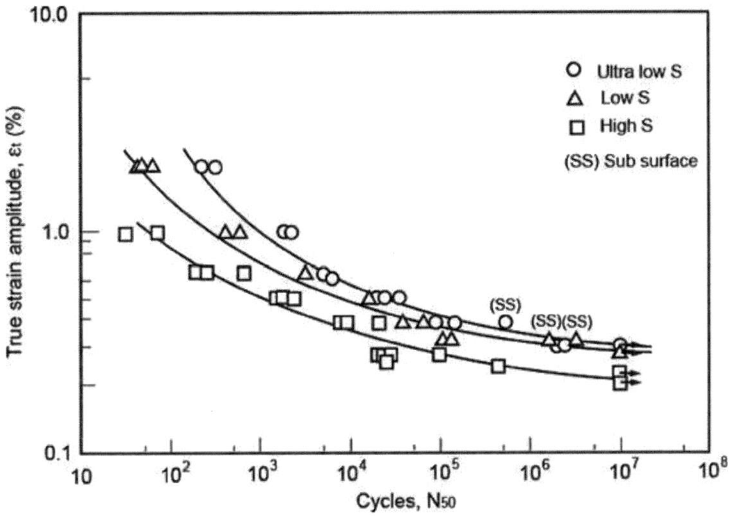

Fig. 4.37 S-N50 diagram in the transverse direction for 4140 forging hardened to 52 HRC for three different sulfur content (adapted from [39])

35 HRC it takes persistent slip bands to nucleate damage, above 40 HRC (~ 1280 MPa) it is the random distribution of surface inclusions to play a fundamental role.

Also note that these inclusions so detrimental to fatigue resistance of high strength steels have dimensions equal or smaller than 10 μm. Competition between persistent slip bands and inclusions as micro cracks initiators has a crossway at about 35 HRC. At Rockwell C 55 if the specimen fails by fatigue cycling at a stress amplitude of about 900 MPa it means that surface grains are practically inclusion free, but should an inclusion be present on a surface grain then the fatigue limit would drop down below 600 MPa. The control of inclusion is fundamental and absolutely necessary in high and very high strength alloys. The choice of high strength alloys to withstand fatigue loadings may become a real waste of many without any gain at all in fatigue strength if a total control of inclusions is not achieved. To better evidence this tremendous impact of inclusions on fatigue strength, it may be worth noting the test results obtained by Saito and Ito [40] for super clean spring steel tested under rotating bending fatigue, presented in Fig. 4.37 as black squares.

Their data fall along the upper bound of the data scatter obtained for conventional steels, at least up to 50 HRC. However, the inclusion control adopted by Sato and Ito, who were reducing the oxygen content to reduce inclusion size, appears not to be enough a good method when the hardness exceeds 50 HRC. The use of calcium in steel production is another method of deoxidizing and desulphurizing the alloy that dates back to the fifties of last century. Calcium can be used to for steels to lower the sulfur and oxygen content as well as to reduce the inclusion density and modify their morphology so to make inclusions become globular and not elongate, in particular,

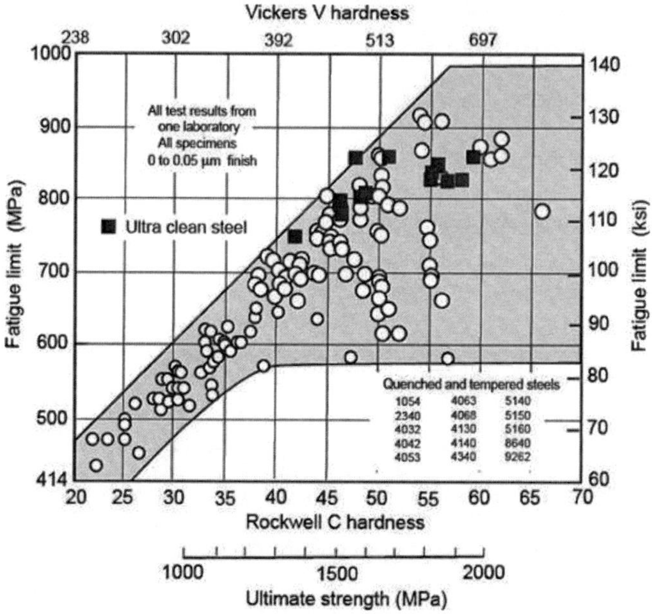

Fig. 4.38 Fatigue limit dependency on hardness and strength (adapted from [37]. Data for ultra clean steel are from [39])

during hot treatments of steel. Unlike manganese sulfides, calcium sulfides do not deform during hot rolling.

4.7 Effect of Hardness and Grain Size

It has been shown in the previous section how inclusions play a major role in fatigue life. Inclusions are normally contained in grains or at grain boundaries. Therefore, it is reasonable to believe that their size can equal that of the grain in which they are hosted. This is, actually, the maximum size that can be expected, to which it is associated the lowest probability of occurrence. We have already mentioned, in Sect. 2.4, that carbon steel yield strength σ_y is related to ferritic grain size d by the well-known Petch equation (see Eq. 2.5) [41]

$$\sigma_y = \sigma_o + k \cdot d^{-1/2} \tag{4.5}$$

where σ_o is the yield strength of a single crystal matrix and k a parameter that takes into consideration grain boundaries capability to block dislocation motion. A similar relationship has been found for titanium and its alloys [42]. In general, this is valid for planar slip materials (see Sect. 1.4.1) whereas wavy slip materials do not seem to show such a dependency. Formulas like that of Eq. (4.5) are mainly qualitative nevertheless they present the experimental evidence of the dependency of σ_y on d. Also tensile strength σ_u is inversely related to grain size. The dependency of the ultimate strength on grain size means that also the fatigue limit of a material should depend on grain size. This is definitively true, as Fig. 2.32 is indicating. In low strength alloys grain size is large and may be so large, even more than 80–100 μm, to allow internal persistent slip band formation of the same size or inclusions of the same size so to compete with slips as micro crack initiators. However, while it is always possible that persistent slip bands initiated by cyclic loads in a surface grain reach the grain size under any stress level larger than the fatigue limit, it is not possible, fortunately, that an inclusion of exactly that size always exists on a surface grain. The competition between slips and inclusions as fatigue damage initiators is unbalanced towards the former mechanism. Though with lower probability, such large an inclusion may be present if the process volume is large enough. It is worth remembering that the process volume in fatigue is just that of the external surface times the limiting thickness of 0.5 mm, see Fig. 2.5, times the number of workpieces or specimens that are supposed to be built. Figure 4.39 [42] shows two long micro cracks emanating after 3.75×10^5 cycles from a MnS inclusion in a low-alloyed carbon steel type 533 Gr. B Cl 1, 588 MPa yield strength and 703 MPa tensile strength, under a stress amplitude of 214 MPa and $R = 0.2$. The average grain size was about 100 μm.

The MnS inclusion has a length of about 150 μm and is 50 μm wide and the crack is already a macrocrack, being 400 μm long, from tip to tip. The surface of the specimen had been accurately finished by electropolishing, nevertheless the

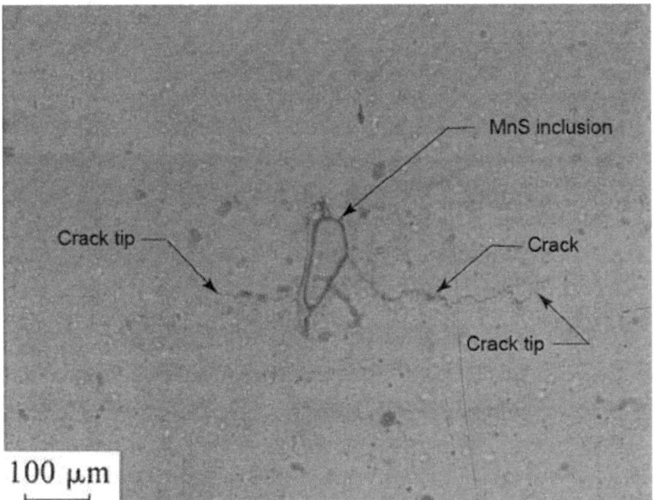

Fig. 4.39 Micrograph of the surface of 533B Gr. B Cl 1 specimen showing MnS inclusion emanating cracks after 3.75×10^5 cycles (replica) [43]

inclusion has such large a dimension that it would have prevailed over any other commercial finishing as damage initiator. However, inclusions of such large a size have an extremely low probability of existence, as it can be seen in Fig. 4.40, therefore it is more probable that in large grain steels macro crack forms through a persistent slip bands activation process. On the contrary, in high strength very hard steels grains are so small, even 5–10 μm, that slips are inhibited and, anyhow, they are of the same grain size. In this case, then, if the surface has a fine-grounded finishing or even better a mirror-polished one it takes an inclusion, even of small size, to behave as crack initiators, as we have already mentioned. As to the probability of existence of an inclusion of maximum length l, it depends on the inverse power of l according to an expression of the type

$$P = C \cdot \left(\frac{l}{l_o}\right)^{-m} \tag{4.6}$$

in which l_o is a normalizing factor and m an exponent that for a surface layer one grain deep, may assume a value between 1.5 and 3. Equation (4.6) is plotted in Fig. 4.40 (dashed curve) with $m = 2.5$ and is compared to experimental measurements of MnS inclusions density, represented by the number of inclusions of a given size per cm^2, on type A 533 Gr. B Cl 1 carbon steel with two different sulfur contents, very low 0.003% S and rather high 0.013% S. It can be seen how the probability to found a MnS inclusion of 60 μm is about 200 times lower than that associated to a 10 μm inclusion. Also the control of sulfur is fundamental since a 40 μm inclusion in high sulfur steels has the same probability of existence of a 20 μm inclusion in low sulfur one. Smaller inclusions are by large the most probable and below 6 μm they become almost absolute a fact. This is why increasing the tensile strength over 1250 MPa may not necessarily result in a parallel increase of fatigue limit, as shown in Fig. 4.38.

For example, Cummings et al. [44] reported that the origin of fatigue cracks in various alloy steels having tensile strength between 930 and 2100 MPa were always small silicate inclusions close to the surface, but while in the 930 MPa steel their size was larger than 60 μm, in those having 2100 MPa tensile strength inclusions down to a mean diameter of 6 μm were large enough to initiate fatigue cracks. Analogous results were found by other researches [45, 46]. Data scatter over 1200 MPa tensile strength is indeed due to the high probability of finding inclusions larger than 5 μm. Duckworth [45] collecting also data from other authors, noted that for all the high strength steels considered there was a relationship between fatigue limit at 10^8 cycles and the product of tensile strength and reduction of area, as shown in Fig. 4.41. Note how a fatigue limit σ_f of the order of 700 MPa is associated with a product σ_u·RA equal to about 70,000. With 1500 MPa ultimate strength steels such a value is possible only with RA = 40%. A 25% RA would yield 37,500 corresponding to about 400 MPa fatigue limit σ_f. But this limit would also be achieved with a 1000 MPa strength steel having 37.5% RA, which is much more convenient and economical.

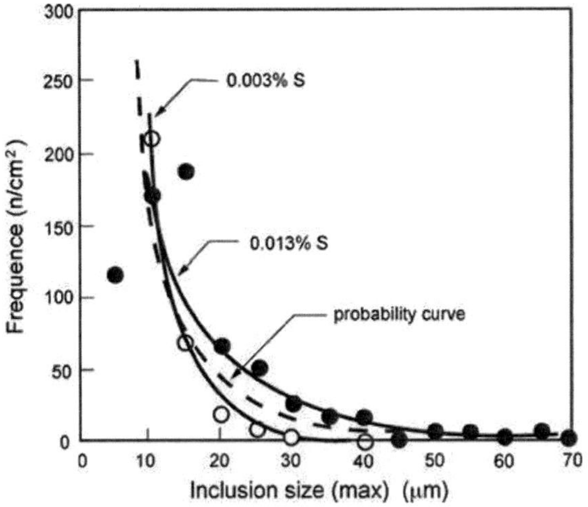

Fig. 4.40 Frequency distribution of MnS inclusions in carbon steels versus maximum size as function of sulfur content (in particles n per cm^2). Also show is the theoretical probability distribution

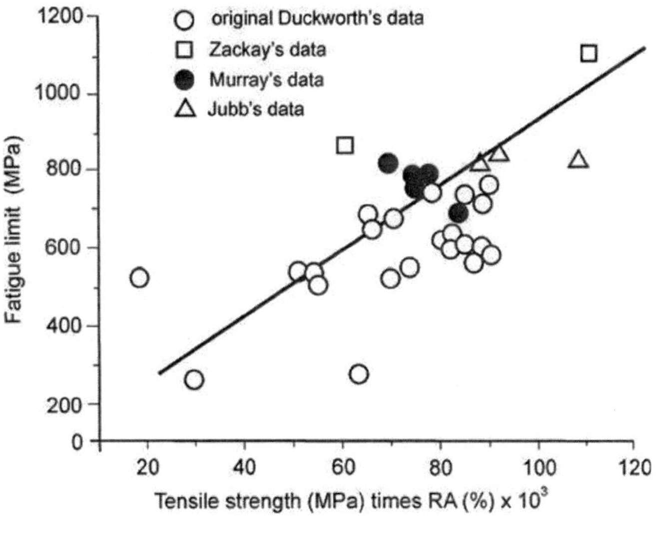

Fig. 4.41 Relation between fatigue limit at 10^8 cycles and the product of tensile strength times reduction of area (modified from [45])

Particularly interesting is Fig. 4.42 that shows the relationship between grain size, tensile strength of steel and critical inclusion size, i.e., the size of the inclusion that was found to be the initiation site of failure under monotonic tensile load (SEM analysis of fracture surface). Note how the average grain size decreases as tensile strength increases, as expected. For mild steels having a strength of 400–600 MPa,

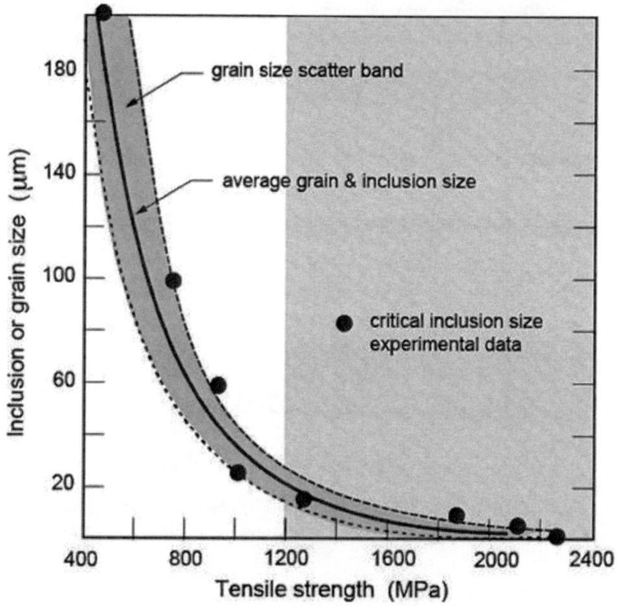

Fig. 4.42 Relationship between grain and critical inclusions size and tensile strength of steels

grains are very large with a size that ranges from 100 to 200 μm while in very high strength steels over 1600 MPa the grain size goes below 10 μm. Also inclusion size follows the same trend, being contained in the grain. The equation of the average grain size d or inclusion size l, solid curve of Fig. 3.36, is

$$l = 8 \cdot C \cdot \sigma_u^{-3} \tag{4.7}$$

with $C = 4000$. More recently, Murakami and Endo [9, 47, 48] have introduced a relationship that links the fatigue limit in rotating bending to the inclusion area intercepted by a plane normal to the stress direction

$$\sigma_{f,l} = C \cdot \frac{(HV + 120)}{\sqrt{area}^{1/6}} \tag{4.8}$$

in which HV is the hardness of the metal matrix, $\sqrt{area}$ is expressed in micron and C is a constant whose value is

$$C = \begin{cases} 1.43 \ \textit{for surface defects} \\ 1.56 \ \textit{for internal defects.} \end{cases} \tag{4.9}$$

Inclusions were assimilated to cracks, for the reasons we already know, see Figs. 4.25 and 4.39, simply because sooner or later they become cracks, it is just

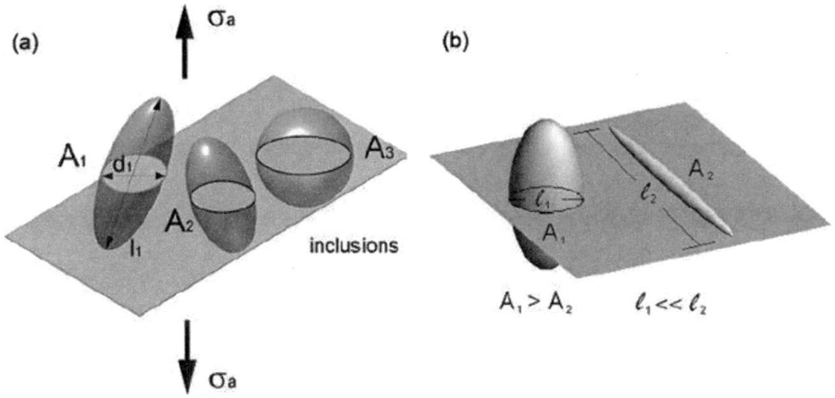

Fig. 4.43 **a** The largest area A cut by a plane normal to the applied stress direction determines the parameter $\sqrt{area}$ that enters into Murakami equation; **b** the inclusion that yields the largest area A_1 appears to be the most dangerous even when compared with a longer and sharper one, but having a smaller area A_2

a matter of cycles. In their experiments the defect sizes were ranging as 10 μm < $\sqrt{\text{area}}$ < 1.000 μm. The particular feature of Eq. (4.8) is exactly the net section area A of the inclusion cut on a plane normal to the loading direction, as shown in Fig. 4.43, at variance with Eqs. (2.10) and (2.11) where the fatigue limit is related to just its size.

The mathematical formalism of fracture mechanics (LEFM) (see Appendix A) behind that choice yields the corresponding stress intensity factor

$$\Delta K = 0.65 \cdot \Delta\sigma \cdot \sqrt{\pi \sqrt{area}}. \tag{4.10}$$

We can be somehow dubious about the use of linear elastic fracture mechanics with such small a defect. However, assuming that it can be used, it can also be said that crack propagation will occur when ΔK reaches a threshold value ΔK_{th}. Testing with 15 different materials, Murakami and Endo [48, 49] found that ΔK_{th} was proportional to

$$\Delta K_{th} \propto \sqrt{area}^{1/3}. \tag{4.11}$$

The exponent 1/3 comes from the slopes of lines interpolating the experimental results obtained in rotating bending and tension–compression tests on various materials, as shown in Fig. 4.44.

Therefore, Eq. (4.11) confirms that LEFM is not valid since the exponent in Eq. (4.11) should be 1/2 and not 1/3. However, the common trend shown by all experimental data allows the use of Eq. (4.11) in an empirical fashion. The choice of the $\sqrt{\text{area}}$ parameter resulted to be the most convenient to characterize the various notches, holes and cracks because it yielded a linear relationship with the other

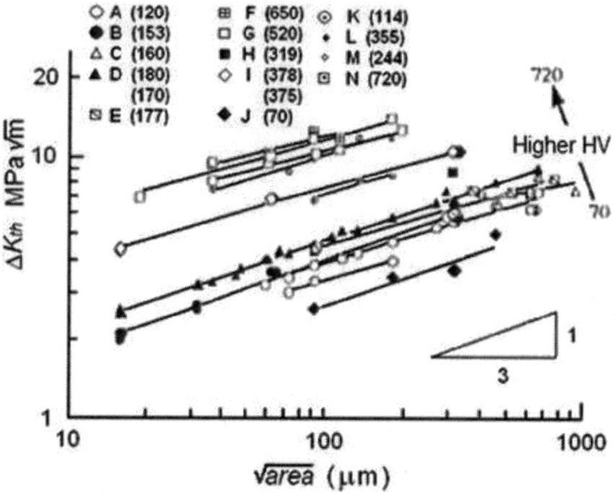

Fig. 4.44 Relationship between ΔK_{th} and $\sqrt{area}$ for various defects and cracks [49]. Numbers in parenthesis indicate hardness

physical parameter that characterized the material, namely the hardness. Murakami and Endo, in fact, trying to link the fatigue limit to the $\sqrt{\text{area}}$ parameter and to some physical material property, found that the most appropriate physical parameter was precisely the Vickers hardness HV. Murakami [9] expressed this occurrence by saying that "*It should be clear that the dependence of ΔK_{th} on material parameters can only be made clear after finding the most appropriate geometrical parameter. Although various material parameters such as yield strength (σ_y), ultimate tensile stress (σ_u), and hardness (Hv or H_B), may be correlated with ΔK_{th}, the Vickers hardness number, HV, is chosen after observing the trend of many data, and also for the sake of simplicity in measurement and available data*".

The relationship, always empirical, with the hardness HV derived from the analysis of a huge number of experimental results, shown in Fig. 4.44 [48, 49], was

$$\Delta K_{th} = 3.3 \cdot 10^{-3} \cdot (HV + 120) \cdot \sqrt{area}^{1/3}. \tag{4.12}$$

Equating Eqs. (4.12) to (4.10) yields Eq. (4.8). Note that the use of LEFM, though inappropriate, implies anyhow that the defect is already a crack. Equation (4.8) has been extended to a more general case in which is $R \neq -1$ ($R = \sigma_{min}/\sigma_{max}$) to become

$$\sigma_{f,l} = C \cdot \frac{(HV + 120)}{\sqrt{area}^{1/6}} \cdot (0,5 - R/2)^{0,226 + HV \cdot 10^{-4}}. \tag{4.13}$$

The artificial defects investigated to yield the Murakami-Endo model were very small drilled holes with diameters ranging from 40 to 500 microns and depth greater than 40 μm, very small and shallow notches with depth ranging from 5 to 300 μm, very shallow circumferential cracks with depth ranging from 30 to 260 μm, and Vickers hardness ranging from 70 to 720 Hv. The values of ΔK_{th} were obtained by substituting the stress range at the fatigue limit in Eq. (4.10). Equation (4.8) was found to be appropriate for most materials, except for two types of stainless steel, with an error less than 10%. Being an empirical relationship, Eq. (4.8) cannot be applied outside the limits in which it has been derived. The upper limit of defect sizes seems to be 1000 μm. The lower limit depends on materials properties and microstructure. It is rather interesting to apply the Murakami-Endo relationship (4.8) to interpreter the experimental data reported in Fig. 4.38. Entering Eq. (4.8) with a given hardness HV and the lower and upper fatigue limit $\sigma_{f,l}$ that in Fig. 4.38 marks the spread of experimental data corresponding to that given hardness, it is possible to derive the corresponding upper and lower inclusion sizes that would cause fracture if they were responsible for the fatigue initiation process. For example, at a hardness of 30 RC that corresponds to about 302 HV the spread of experimental data yields a lower fatigue limit of about 475 MPa and upper one of about 600 MPa. To these two extreme values of the experimental fatigue limit $\sigma_{f,l}$ it corresponds an upper value of inclusion size of 5 μm and a lower value of 2 μm, as shown by the gray arrow in Fig. 4.45. The grain size scale shown in Fig. 4.45 is indicative and is based on the approximate relationship (4.7) established with data of Fig. 4.42 and on the relationship

$$HV = 80.62 + 2001 \cdot d^{-1/2}. \tag{4.14}$$

Therefore, for each hardness, there is a grey band with two numbers indicating the possible grain size variation at that hardness. Also remember that the Murakami-Endo Eq. (4.8) has not been verified for hardness over 720 HV and, as such, cannot be used.

To remain in the corridor of experimental fatigue limits between 20 and 40 HRC the corresponding inclusion size variation, according to Eq. (4.8), is between 5 and 2 μm, as shown in Fig. 4.45. Since the surface finish of all the experimental data is reported to be varying from 0 to 2 μm, it means that fatigue strength in this hardness span depends on a slip process rather than on the presence of surface inclusions since, as already said in Sect. 4.4, inclusions cannot compete with roughness of the same size. The spread observed at 50 HRC, from 880 to 575 MPa, is obtained with inclusions between 2 and 20μm which, for what has been just said, means that the upper fatigue limit is due to surface slips and the lower to the presence of surface inclusions. Finally, at 60 HRC and above the fatigue limit spread corresponds to inclusion sizes between 10 and 2 μm. Note that at those levels of hardness no experimental data were reported below 770 MPa. If there were any confirming the same spread observed at 50 HRC, the inclusion size would go up to 60 μm.

However, at that level of hardness and strength the corresponding average grain size is of the order of 5 μm to a maximum of 10 μm, which actually means that it is very improbable to have an inclusion of 60 μm. Therefore, the grey area from 60 and 70 HRC is an exclusion area in which it is very unlikely that an inclusion larger than 10 μm can possibly exist and cause the corresponding loss of fatigue strength. In other words, the maximum loss of fatigue strength that can be expected due to a surface inclusion for steels over 2000 MPa ultimate strength is from about 1000 MPa to about 750 MPa. The fact that the upper limit of fatigue strength data must be due

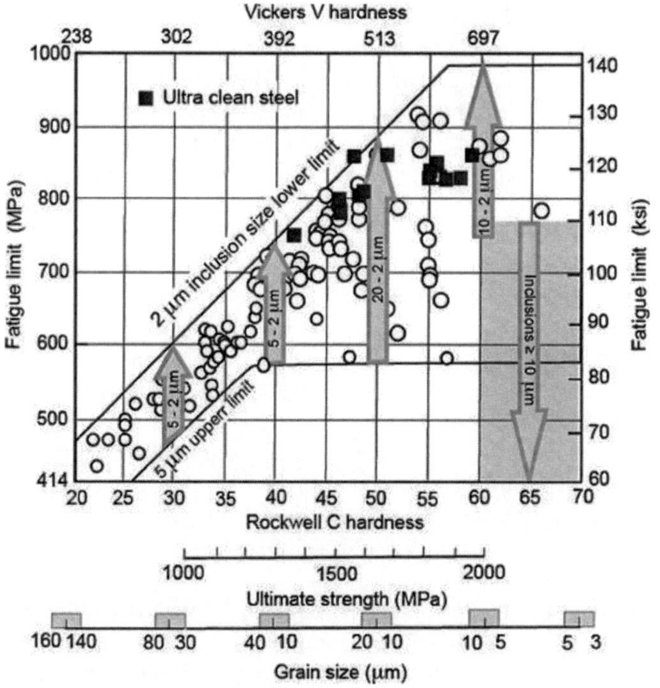

Fig. 4.45 Revisitation of Fig. 4.38 fatigue limit dependency on hardness and strength showing the grain sizes that would cause the experimental data scatter

to a surface slip process and not to surface inclusions of about 2 μm is confirmed by the data relative to ultra-clean steels that are just following that upper limit, at least up to 50 HRC.

4.8 Effect of Size

The dimension of the test piece may well affect the fatigue strength of materials. All previous sections have been dealing in one way or another with the metallurgy and condition of the surface layer recognized to play a fundamental role on fatigue strength. Metallurgical conditions, which actually means inclusions, foreign phases, defects such as voids or dislocations, phases, grains orientation and borders, etc., all of them have an influence on fatigue strength that depends on their random distribution in the metal surface. As the volume of the surface layer and, in particular, of the material under the highest stress state, that we have called process volume, increases also the probability of finding the worst metallurgical condition increases and, as such, the fatigue strength of the material decreases. Even dealing with non-metallurgical initiation mechanisms such as surface finish, the probability of having the possibly largest surface indentation increases as the surface of the test piece increases.

Fatigue strength, therefore, is controlled by the process volume or, better, by the process surface, since the depth of the process volume is constant and equal to about 0.5 mm, as shown in Fig. 2.5, unless we deal with a stress concentration problem in which the process volume is much smaller and related to the radius of the notch or indentation or, in general, to the geometry and stress conditions of the workpiece, as it will be discussed in Sect. 11.5. This is why, we have already mentioned, the fatigue strength of whiskers is almost unlimited; just because fibers are so thin, less than 10 μm in diameter, that their process volume is practically reduced to zero. It is the matrix into which carbon or glass fibers are bonded, usually a plastic resin that fails by fatigue not fibers. This is also why small ants or insects have the capability to sustain and carry loads well in excess of their weight. Imagine a man carrying with nonchalance some 500 kg or even one ton burden. It is now a proven fact that small specimens show strength and fatigue resistance better than larger ones.

For example, in cyclic bending or torsion, tests run by Faupel and Fisher [50] with steels having yield strengths varying from 350 to 1145 MPa, the fatigue limit S_f was depending on the specimen diameter almost in a step fashion. Below 10 mm diameter it was rather difficult to observe any size effect, but above 10 mm the fatigue limit S_f reduced all of a sudden by about 10%, to become practically constant up to a diameter of 5 cm. Probably one of the earliest test to infer the size effect on fatigue was run in 1945 by Moore [51] who observed a constant decrease of the fatigue limit with specimen diameter, as it can be seen in Fig. 4.46. Another example of size effect on fatigue can be observed in Fig. 4.47 for 1020 steel [51]. Specimens were cut from a 3½″ hot rolled bar.

The diameter d was varying from 0.125″ (3.17 mm) to 2″ (50.8 mm). It can be seen how the diameter is affecting the *S–N* curve of the material up to 25.4 mm (1″) after which the curve becomes less size dependent questioning whether the worst metallurgical condition has been eventually attained. This would suggest a

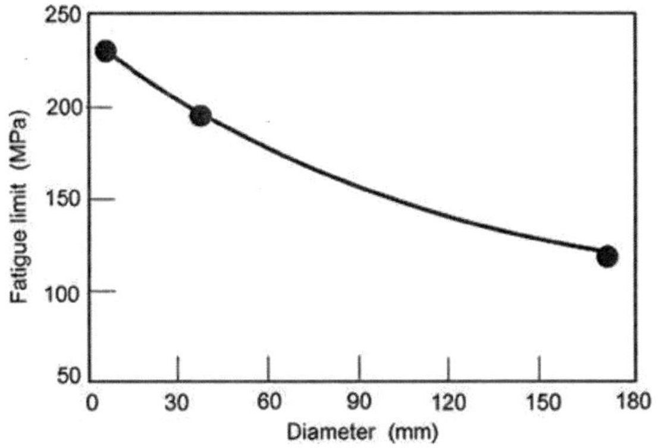

Fig. 4.46 Effect of specimen diameter on fatigue limit (modified from [51])

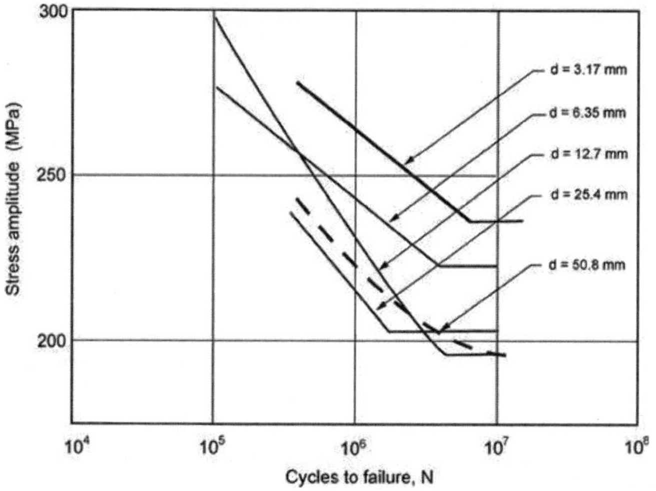

Fig. 4.47 Size effect on rotating-bending fatigue strength of SAE 1020 (modified from [51])

size saturation effect. Different empirical equations have been suggested to quantify size effects on fatigue by introducing a size factor C_{sz} that reduces the fatigue limit of material as the specimen diameter increases. One of the most common and rather conservative formula is that proposed by Shigley and Mitchell [52]

$$C_{sz} = \begin{cases} 1.0 \quad if\ d \le 8\ \text{mm} \\ 1.189 \cdot d^{-0.097}\ if\ 8\ \text{mm} \le d \le 250\ \text{mm} \end{cases} \tag{4.15}$$

where d is the diameter of the specimen or component. Being empirical in nature, Eq. (4.5) can be used only in the range of diameters considered, i.e., 8–250 mm and for the steels considered. When the section is not circular but rectangular of sides a and b Shigley and Mitchell suggest the use of a *diameter equivalent* d_{eq} to enter Eq. (4.5) in cyclic bending

$$0.0766 \cdot d_{eq}^2 = 0.05 \cdot a \cdot h. \tag{4.16}$$

Size effect may be quantified through rather simple, but refined considerations. Metallurgical variability that results in the dependence of fatigue limit σ_f or S_f on specimen size is also responsible for the dependence of tensile strength σ_u on specimen size. By using the Weibull theory of the weakest link (see Sect. 6.6.1) the ratio of the ultimate strength $\sigma_{u,2}$ of a specimens of volume V_2 to that $\sigma_{u,1}$ of a reference specimen of volume V_1 is inversely proportional to the power of the ratio of the respective volumes

$$\frac{\sigma_{u,2}}{\sigma_{u,1}} = \left(\frac{V_1}{V_2}\right)^{1/m} \tag{4.17}$$

where the exponent m, also called Weibull exponent, is a characteristic of the metal ranging from 9 to 50 and may be considered as an index of the metallurgical variability or quality of the metal. The higher the value of m the lower the metallurgical variability of the metal. Assuming (see Sect. 10.2) that the fatigue limit S_f is proportional to the ultimate strength of the metal, it is possible to use Eq. (4.17) also for S_f and write

$$\frac{S_{f,2}}{S_{f,1}} = \left(\frac{V_1}{V_2}\right)^{1/m}. \tag{4.18}$$

However, while in a specimen subjected to monotonic traction the entire volume, V, exerts a reaction, in high cycle fatigue only the surface is involved in the initiation process till a depth of about 500 μm, al already said many times, see Fig. 2.5. Below that depth, internal inclusions to be effective initiation sites must compensate for the reduced concentration factor (see Sect. 4.4 and Fig. 4.27) with an increase in size which reduces their probability of existence. Therefore in a cylindrical specimen of diameter d and length l the process volume to be considered from the fatigue point of view is

$$\begin{aligned} V &= \pi \cdot l \cdot \left(\frac{d^2}{4} - \frac{(d-1)^2}{4}\right) \\ &= \frac{\pi}{4} l \cdot (2d - 1) \end{aligned} \tag{4.19}$$

with d and l in mm. Equation (4.18) then yields

$$\frac{S_{f,2}}{S_{f,1}} = \left[\frac{l_1}{l_2} \cdot \frac{(2 \cdot d_1 - 1)}{(2 \cdot d_2 - 1)}\right]^{1/m}. \tag{4.20}$$

Equation (4.20) will be used to evaluate the size effect on fatigue limit S_f passing from a standard cylindrical specimen of 6 mm diameter to a larger one having a diameter of 20 mm. Assume that the l/d ratio remains constant and equal to 3 (fatigue specimens have $l/d \geq 3$) and $m = 20$ (poor quality material), Eq. (4.18) yields

$$\frac{S_{f,2}}{S_{f,1}} = \left[\frac{6}{20} \cdot \frac{11}{39}\right]^{1/20} = C_{sz} = 0.88. \tag{4.21}$$

Using the Shigley and Mitchell empirical formula (4.15)

$$C_{sz} = 1.189 \cdot 20^{-0.097} = 0.88 \tag{4.22}$$

which is precisely the value given by Eq. (4.21) based on the process volume and the Weibull analysis.

However, for a better quality steel having $m = 40$ Eq. (4.21) yields $C_{sz} = 0.94$ which is about 7% higher than 0.88. The comparison between size factors C_{sz} given by the empirical formula of Shigley and Mitchell and Eq. (4.20) versus specimen diameter is presented in Fig. 4.48 for different values of the Weibull exponent m and in Fig. 4.49 versus the process volume ratio where the reference volume V_o is that of a specimen of 8 mm diameter. As it can be seen the empirical formula of Shigley and Mitchell almost corresponds to the Weibull equation when $m = 20$ which has been said to refer to a quite poor material, but it has been also said that the Shigley and Mitchell formula is rather conservative.

The strong dependence of C_{sz} on the Weibull exponent, i.e., on the real metallurgical variability of the material under study, suggests the use of the Weibull Eq. (4.18) rather than the Shigley and Mitchell formula. It will be shown in Sect. 6.6, how to measure the Weibull exponent of a material. Figure 4.49a presents the size effect for two carbon steels and a CrNi steel [53] and Fig. 4.50b on 1.05 Cr steel [54]. In all of them the size effect seems to saturate over 50 mm. Also shown are the theoretical predictions according to the Shigley and Mitchell Eq. (4.15) and the Weibull analysis (4.18), respectively. For this latter prediction, the Weibull exponent m was assumed equal to 35 for the CrNi steel and equal to 30 for 0.35C steel. The Shigley and Mitchell formula follows rather well the experimental data for 0.18C steel, but not those relative to CrNi and 0.35C steels confirming what has been said about this formula being empirical and conservative.

It may be interesting to look at the size effect in nodular cast iron which is a typical metal of high metallurgical variability. This is shown in Fig. 4.51 where the line representing the Weibull analysis using $m = 11$ in Eq. (4.18) has been drawn on the experimental scatter-band relative to monotonic traction tests [54]. This dependence

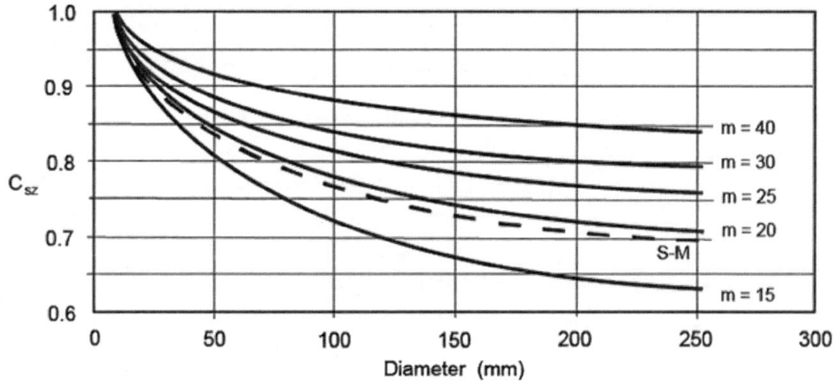

Fig. 4.48 Trend of size factor C_{sz} versus specimen diameter for several values of the Weibull exponent m. Also shown for comparison is the Shigley and Mitchell empirical formula (dashed curve named S-M)

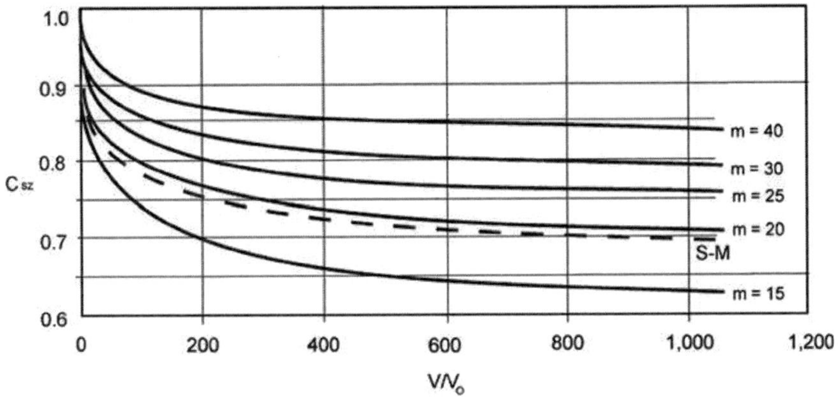

Fig. 4.49 Trend of size factor C_{sz} versus the volume ratio V/V_o for several values of the Weibull exponent m. Also shown for comparison is the Shigley and Mitchell empirical formula (dashed curve S-M)

of the size factor C_{sz} on the material has been recognized by the FKM-Guideline approach to stress based uniaxial fatigue analysis [56]. The following are the values of the factor C_{sz} suggested by FKM-Guideline for specific materials.

- wrought aluminum alloys

$$C_{sz} = 1.0 \tag{4.23}$$

- cast aluminum alloys

$$\begin{aligned} &C_{sz} = 1.0 \quad \text{for } d_{eff} \leq 12\,\text{mm} \\ &C_{sz} = 1.1 \cdot \left(d_{eff}/7.5\right)^{-0.2} \quad \text{for } 12\,\text{mm} < d_{eff} < 150\,\text{mm} \\ &C_{sz} = 0.6 \quad \text{for } d_{eff} \geq 150\,\text{mm} \end{aligned} \tag{4.24}$$

- gray cast iron

$$\begin{aligned} &C_{sz} = 1.207 \quad \text{for } d_{eff} \leq 7.5\,\text{mm} \\ &C_{sz} = 1.207 \cdot \left(d_{eff}/7.5\right)^{-0.1922} \quad \text{for } d_{eff} > 7.5\,\text{mm} \end{aligned} \tag{4.25}$$

- all steels, steel castings, ductile irons and malleable cast iron

$$\begin{aligned} &C_{sz} = 1.0 \quad \text{for } d_{eff} \leq d_{eff,min} \\ &C_{sz} = \frac{1 - 0.6786 \cdot a_d \cdot \log\left(d_{eff}/7.5\right)}{1 - 0.7686 \cdot a_d \cdot \log\left(d_{eff,\min}/7.5\right)} \quad \text{for } d_{eff} > d_{eff,min} \end{aligned} \tag{4.26}$$

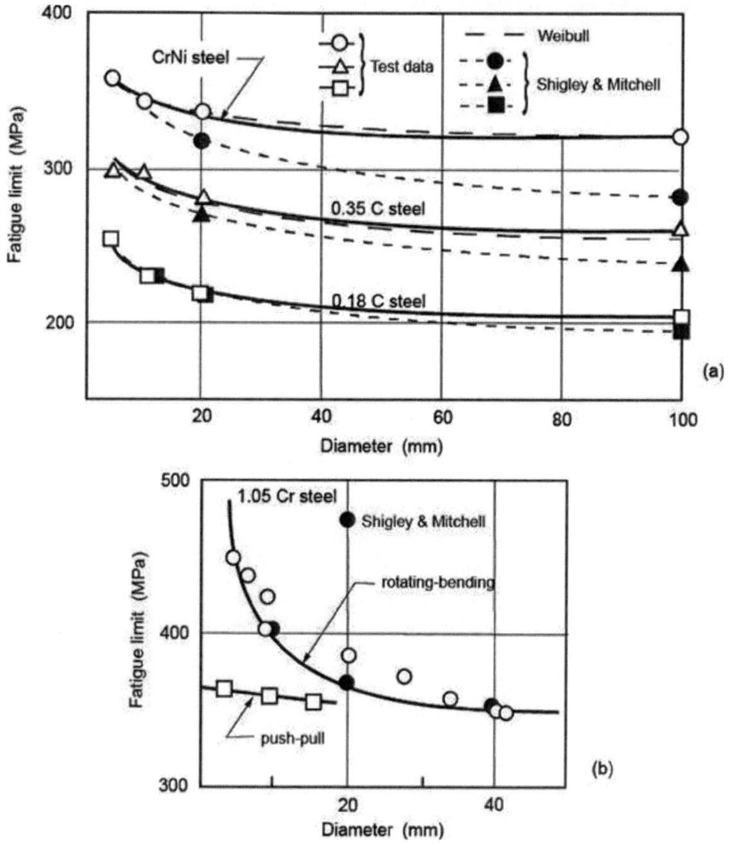

Fig. 4.50 Size effect on fatigue limit of several steels. Also shown are theoretical predictions according to Shigley & Mitchell and Weibull (adapted from [53, 54])

where d_{eff} is the effective diameter of a cross section, $d_{eff,min}$ and a_d are constant tabulated in Table 4.1

Two cases appear in Table 4.1 for d_{eff}. In Case 1, d_{eff} is defined as

$$d_{eff} = \frac{4V}{A} \tag{4.27}$$

where V and A are the volume and the surface of the section of the component of interest, respectively. in Case 2, d_{eff} is equal to the diameter or wall thickness of the component and applies to all components made of aluminum alloys. Example of calculation are shown in Table 4.2 [55].

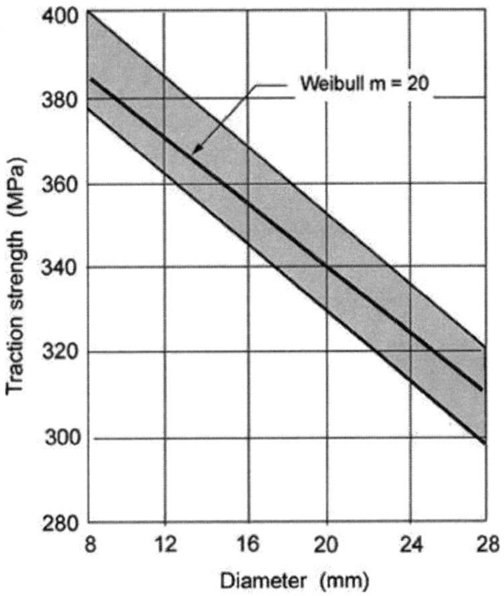

Fig. 4.51 Solid line shows the traction strength prediction using the Weibull approach with $m = 11$ traced on the experimental scatter band of traction tests on globular cast iron (experimental data scatter from [55])

Table 4.1 Constant used to estimate the size correction factor C_{sz}

Material	$d_{eff,min}$ (mm)	a_d	Case of d_{eff}
Plain carbon steel	40	0.15	2
Fine grained steel	70	0.2	2
Steel, quenched and tempered	16	0.3	2
Steel, normalized	16	0.1	2
Steel, case hardened	16	0.5	1
Steel, nitriding, quenched and tempered	40	0.25	1
Steel, forging, quenched and tempered	250	0.2	1
Steel, forging, normalized	250	0	1
Steel casting	100	0.15	2
Steel casting, quenched and tempered	200	0.15	1
Ductile iron	60	0.15	1
Malleable cast iron	15	0.15	1

4.9 Effect of Load Type

The use of the process volume concept in the previous section as the most significant parameter to evaluate size effects on fatigue can be usefully applied also to understand how different types of load can affect fatigue strength and quantify the corresponding load factor C_{ld}. Fatigue test are carried out with rotating bending, axial (push–pull) or torsional loading. The corresponding fatigue limits may bear significant differences.

Table 4.2 Calculation of the effective diameter d_{eff} (from FKM-Guidelines [55])

Cross section	d_{eff} Case 1	d_{eff} Case 2
Solid rod with round cross section 	d	d
Tubular rod with round cross section 	$2s$	s
Infinitely wide thin wall (tubes with large inner diameter) 	$2s$	s
Solid rod with rectangular cross section 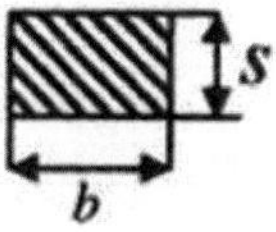	$\frac{2bs}{b+s}$	s
Solid rod with square cross section 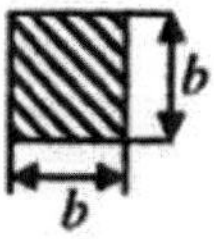	b	b

Figure 4.52 shows schematically the reason for such differences. Under cyclic traction (push–pull), case (a) of Fig. 4.52, the process volume, $V_{process}$, that accounts for fatigue damage initiation is represented by the annulus of constant thickness $t =$ 0.5 mm having the length l of the specimen: $V_{process} \approx \pi \cdot D \cdot t \cdot l$. This process volume is subjected to a constant stress amplitude, σ_a, and is represented in Fig. 4.52a by the cylinder filled in the outlines with the darker color. At variance with that, in cyclic plain reverse bending, case (b) of Fig. 4.52, the stress amplitude acting on the specimen section is not constant, but has a linear gradient, as shown in Fig. 4.52b, so that the maximum stress is acting only on the uppermost and lowermost metal fibers and reduces as the neutral axis is approached, both in the radial and tangential direction as well. As to this decrease, beyond an angle β the stress amplitude will no longer be high enough to jeopardize the fatigue strength.

This limiting stress amplitude may be assumed equal to 90% or 95% of the maximum bending stress σ_b. Therefore, the process volume reduces to the two upper and lower shells A and B dark tinted in Fig. 4.52b within the angle β. The process volume is, therefore, much smaller than in the cyclic traction case where the entire

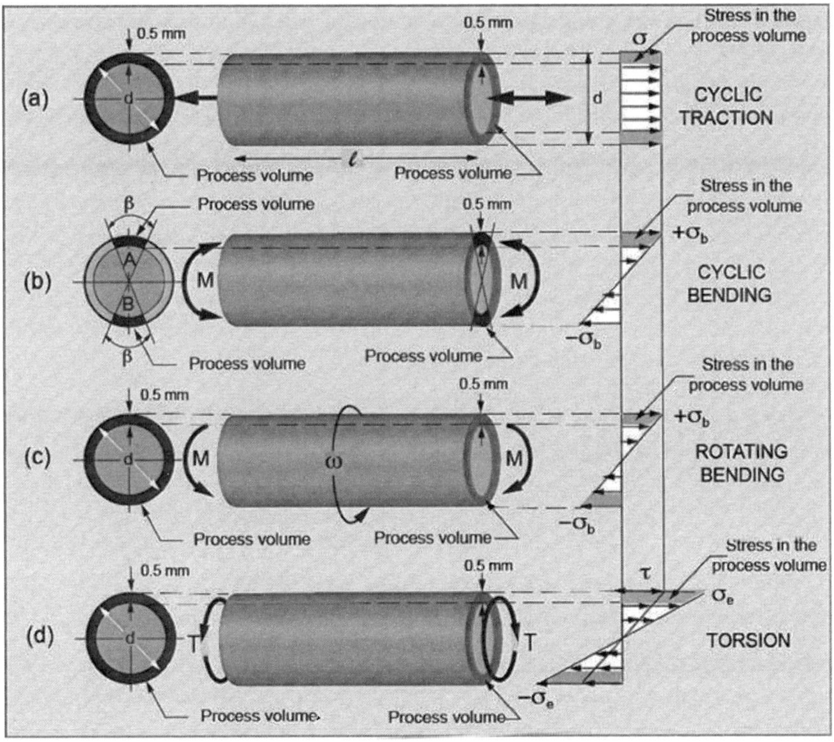

Fig. 4.52 Schematic of the various stress state and corresponding process volume (dark grey tinted) in a plain cylindrical specimen **a** under cyclic traction, **b** cyclic bending, **c** rotating-bending, **d** cyclic torsion

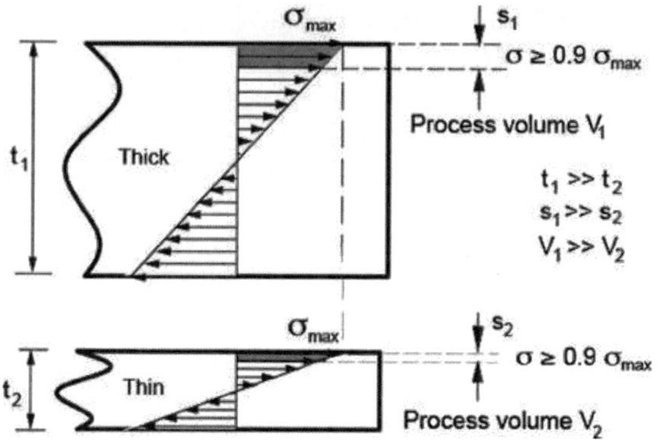

Fig. 4.53 Stress gradient in thick and thin specimens

annulus is involved. Consequently, also the probability of getting the worst condition as initiation site reduces which actually means that cyclic bending is somehow forgiving from the fatigue strength point of view, relative to cyclic traction. Therefore, the fatigue limit measured with this type of loading condition will be somewhat higher. It is clear that if the specimen is flat rather than cylindrical, the comparison between cyclic axial load and bending will yield almost the same process volume, the difference being always attributed to the gradient effect. Moving to rotating-bending condition, case (c) of Fig. 4.52, the process volume comes back to the cyclic traction case since rotation makes the entire annulus 0.5 mm thick be subjected to the same stress conditions, apart from the stress gradient that may introduce a beneficial effect. It has been said and will be farther discussed, that the stress gradient may be accounted for by considering only the thin layer of surface material subjected to 95% or 90% of the maximum stress, Fig. 4.53.

The precise percentage is not fully determined, but arguments that follow may help to decide which one to choose. Whatever percentage is considered, the resulting process volume will depend on the thickness of the test piece or real component. A thick component, in fact, has a less steep stress gradient and, hence, a larger volume of material subjected to high stress state. This is schematically shown in Fig. 4.53. Consequently, there will be a greater probability of initiating a fatigue crack in a larger component and, in this case, also the percentage of the maximum stress to consider may have an important role. It may happen, however, that, in thicker component in particular, the distance s at which the stress gradient reduces to 95% or 90% of the maximum stress be greater than those 0.5 mm that are recognized to be the depth of the surface layer where the fatigue damage initiates. In this case the gradient will be truncated at 0.5 mm. No farther volume must be considered. Finally, a particular attention must be dedicated to cyclic torsion, case (d) of Fig. 4.52. Apparently, from the process volume standpoint cyclic torsion should act as rotating bending and, besides the stress gradient effect, like traction since, once again, it is the entire annulus 0.5 mm thick to be subjected to almost the same stress state. However, a pure shearing-stress, τ, differ fundamentally from axial and bending stresses in that it generates a biaxial stress state. This is schematically shown in Fig. 4.54. Hence,

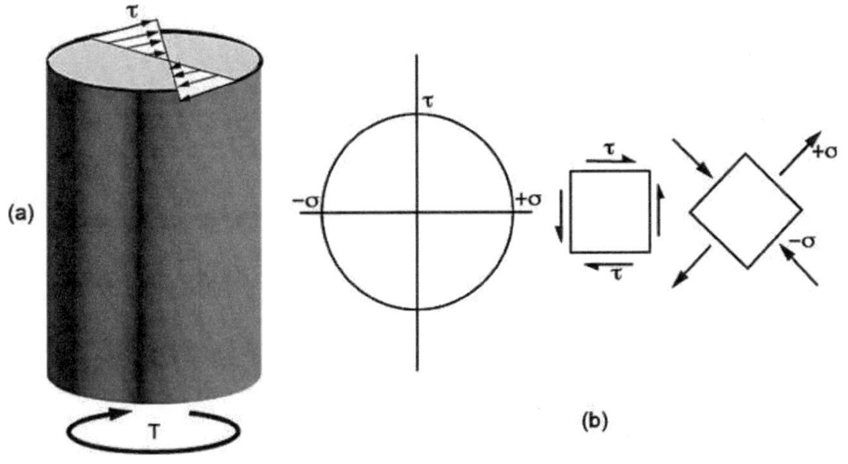

Fig. 4.54 **a** Stress distribution on a circular section subjected to a pure torsional moment T, **b** elemental free body taken at any point in the bar which shows the biaxial stress state generated by pure shearing-stress

to predict failure under torsional loads a failure theory is needed. It may be used, for example, the Tresca theory which yields an equivalent stress σ_e

$$\sigma_e = \sqrt{\left(\sigma_x - \sigma_y\right)^2 + 4\tau_{xy}^2} = 2\tau \tag{4.28}$$

or the von Mises theory

$$\sigma_e = \sqrt{\sigma_x^2 - \sigma_x\sigma_y + \sigma_y^2 + 3\tau_{xy}^2} = \sqrt{3}\tau. \tag{4.29}$$

Although some uncertainty still remains, these theories seem to hold reasonably well also for fatigue loading. In any case, the resulting stress state is higher, from 1.73 to 2 times, than that generated by a bending moment M of the same magnitude of the torsional moment T, as shown in Fig. 4.52d. Because of that torsional loads shall be regarded as more damaging than axial or bending loads. This finding raises the question of what value shall be given to the load factor C_{ld} to account for torsional loads when experimental fatigue data are obtained from specimens under cyclic traction or bending. For example, it has been found experimentally that the ratio of the axial load fatigue limit to the rotating bending one ranges from 0.6 to 0.9. These fatigue data actually may include a non-eliminable scatter and some error due to eccentricity in axial loading measurements that can be quite high using thin specimens.

A conservative estimate is

$$S_{f,axial} \approx 0.70 \cdot S_{f,bend}. \tag{4.30}$$

A more accurate estimate, avoiding alignment errors (eccentricity in axial loading), may be

$$S_{f,axial} \approx 0.90 \cdot S_{f,bend}. \tag{4.31}$$

Let's try to follow a different approach, analytical rather than empirical, based on the process volume and Weibull analysis. This analysis indicates, Eq. (6.59), that the ratio between the ultimate strength measured in traction and that under pure bending is given by

$$\frac{\sigma_{u,axial}}{\sigma_{u,bend}} = (2m + 2)^{-1/m} \tag{4.32}$$

where m is the usual Weibull exponent. Supposing, as stated in the previous section Eq. (4.18), that the fatigue limit is proportional to the ultimate strength of material, we derive the loading factor as

$$C_{ld} = \frac{S_{f,axial}}{S_{f,bend}} = (2m + 2)^{-1/m}. \tag{4.33}$$

Again, the ratio and therefore the load factor C_{ld} depend on the Weibull exponent m which actually means that C_{ld} is not a load constant but a constant characteristic of the material. Equation (4.33) yields

$$C_{ld} = \begin{cases} 0.73 \ for \ m = 10 \\ 0.85 \ for \ m = 25 \\ 0.89 \ for \ m = 40. \end{cases} \tag{4.34}$$

For a non-circular cross section Eq. (4.32) becomes

$$\frac{\sigma_{u,axial}}{\sigma_{u,bend}} = \left[2(m + 1)^2\right]^{-1/m} \tag{4.35}$$

so that Eq. (4.33) may be written as

$$C_{ld} = \frac{S_{f,axial}}{S_{f,bend}} = \left[2(m + 1)^2\right]^{-1/m} \tag{4.36}$$

from which

$$C_{ld} = \begin{cases} 0.58 \ for \ m = 10 \\ 0.75 \ for \ m = 25 \\ 0.82 \ for \ m = 40. \end{cases} \tag{4.37}$$

Table 4.3 Ratio between fatigue limits in torsion and traction [56]

Material	τ_f/σ_f
Steel	0.60
Aluminum alloys	0.55
Cu and its alloys	0.56
Mg alloys	0.54
Ti	0.48
Cast iron	0.90
Al and Mg castings	0.85

Equation (4.37) states that the passage from bending to axial load would penalize the fatigue limit by a factor 0.58 if $m = 10$ (cast iron) or by a factor ranging from 0.75 to 0.82 for medium to good materials ($25 \le m \le 40$). Note that the inverse passage from axial loads to bending would increase the fatigue limit of the material by inverse values of those given by Eq. (4.37), from 1.22 (1/0.82) to 1.72 (1/0.58), depending on the quality of steel, while Eq. (4.31) provides the value 0.9 for all materials. Moving to cyclic torsion, the Weibull analysis provides an expression that relates the ultimate strength to the volume of the specimen but not to the corresponding ratio

$$\frac{\sigma_{u,axial}}{\sigma_{u,tors}} \tag{4.38}$$

since torsion generates a biaxial stress state (see Fig. 4.53). Practice indicates that ultimate strength in torsion $\sigma_{u,tors}$ is roughly 0.8 that in traction for steel and cast iron and 0.7 for nonferrous metals. Table 4.3 provides a list of torsion fatigue limitτ_f to traction fatigue limit σ_f ratios for several metals [56]. While in most alloys this ratio τ_f/σ_f is kept well closed to the theoretical 0.58 forecasted by the von Mises criterion, for titanium it goes down to 0.48 and in the case of cast iron it goes up to 0.9.

Forrest [57] explanation for this last discrepancy is shown in Fig. 4.55. The random orientation of graphite needles in cast iron induces a similar behavior under traction and pure shear. There will always be a needle oriented at such an angle to produce the same peak stress at vertex in traction and pure shear. The same behavior is observed in large 7075-T6 aluminum castings, 570 MPa ultimate and 485 MPa yield strength, or Mg castings. In these alloys the role of graphite is taken by large elongated voids (lack of fusion) or second-phase particles formed during solidification. Table 4.4 provides a list of load factors C_{ld} of empirical nature (experimental) to be used in the absence of specific data to infer the fatigue limit of materials under a cyclic load type knowing that measured under a different loading condition.

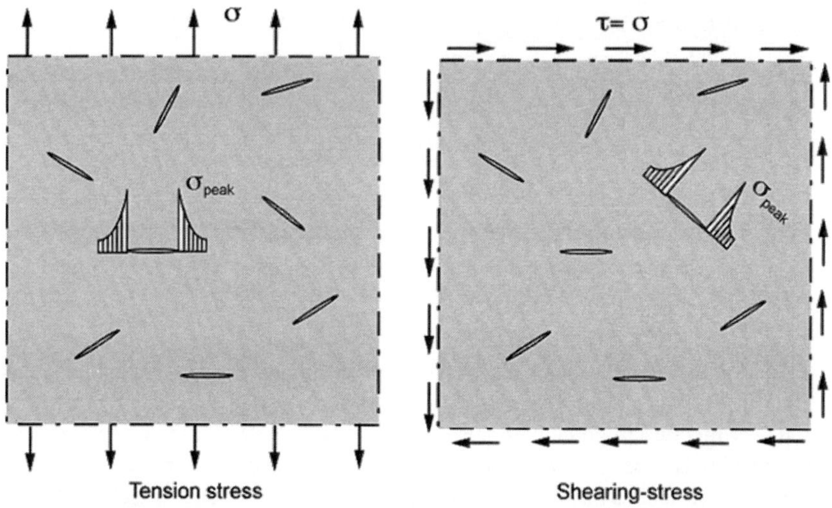

Fig. 4.55 Peak stress at the apexes needles [56]. The random orientation of graphite needles in cast iron induces a similar behavior under traction and pure shear

Table 4.4 Indicative load factors C_{ld} to be used in the passage from one type of load to another

From this load type (1)	To this load type (2)	Multiply by $C_{ld} = S_{f,2}/S_{f,1}$
Traction	Simple bending	1.2
Traction	Pure bending	1.2
Traction	Rotating bending	1.0–1.1[a]
Traction	Torsion	0.58
Rotating bending	Torsion	0.58[b] or 0.8[c]
Bending	Torsion	0.9
Torsion	Traction	1.73
Torsion	Rotating bending	1.73

[a]Thick or thin thickness, [b]Ductile metals, [c]Cast iron and brittle metals

References

1. Houdremont, E., Mailänder, R.: Bending fatigue tests on steels. Stähl and Eisen **49**, 833–892 (1929)
2. Siebel, E., Gaier, M.: The influence of surface roughness on the fatigue strength of steels and non-ferrous alloys. Eng. Digest **18**, 109–112 (1957) (Translation from VDI Zeitschrift, **98**, **30**, 1715–1723 (1956)
3. Thompson, N., Wadsworth, N.J., Br. J. Appl. Phys. **6**, 51 (1957)
4. Lutz, G.B., Wei, R.P.: US Steel Applied Research Laboratory TR Project No. 40112-011 (1) (1961)
5. Raymond, M.H., Coffin, L.F.: Trans. ASME J. Basic Eng. **85**, 548 (1963)
6. Munse, W.H., Stallmeyer, J.E., Rone, J.W.: University of Illinois Report (1965)

7. Neumann, P., Tonnessen, A.: Fatigue Crack Formation in Copper, pp. 41–47. The Technology Press and John Wiley & Sons Inc., New York (1986)
8. Murakami, Y.: Metal Fatigue: Effects of Small Defects and Nonmetallic Inclusions. Elsevier (2002)
9. Murakami, Y., Endo, M.: Effects of defects, inclusions and inhomogenities on fatigue strength. Int. J. Fract.Fract. **16**(3), 163–182 (1994)
10. Güngör, S., Edwards, L.: Effect of surface texture on fatigue life in a squeeze-cast 6082 aluminium alloy. Fatig. Fract. Eng. Mater. Struct. **16**(4), 391–403 (1993)
11. Noll, G., Lipson, C.: Allowable working stresses. Proc. Soc. Exp. Anal. **3**(2), 89–109 (1946)
12. Boyd, R.K.: Fatigue strength of an alloy steel: effect of tempering temperature and directional properties. Proc. Inst. Mechan. Eng. **179**, 733 (1965)
13. Love, R.J.: Motor Industry Research Association Report No. 1950/9 (1950)
14. Kage, M., Nisitami, H.: Fracture surface of anisotropic rolled steel in fatigue. J. Soc. Mater. Sci. Jpn/Zayrio **29**(321), 574–579 (1980)
15. Chastel, Y., Caillet, N., Bouchard, P.: Qualitative analysis of the impact of forging operations on fatigue properties of steel components. J. Mater. Process. Technol. **177**, 202–205 (2006)
16. Evans, E.B., Ebert, J., Briggs, C.W.: Fatigue properties of cast and comparable wrought steels. Proc. ASTM **56**, 979 (1956)
17. Templin, R.L., Howell, F.M., Hartman, E.C.: The effect of grain direction on the fatigue properties of aluminium alloys. Product Eng. **21**, 126 (1950)
18. Ranson, J.T., Mehl, R.F.: The statistical nature of the endurance limit. Proc. ASTM **52**, 779 (1952)
19. Frost, N.E., Marsh, K.J., Pook, L.P.: Metal Fatigue, vol. 81. Clarendon Press, Oxford (1974)
20. Findley, W.N., Mathur, P.N.: Modified theories of fatigue failure under combined stresses. ASTM **55**, 924 (1955)
21. Chodorowski, W.T.: International Conference on Fatigue. Institution of Mechanical Engineers, vol. 122 (1956)
22. Heywood, R.B.: Designing Against Fatigue. Chapman & Hall Ltd., London (1962)
23. Fatigue Design Handbook, SAE Fatigue Design and Evaluation Technical Committee, 2nd edn., vol. 90. SAE Inc. (1988)
24. Suhr, R.W.: The effect of surface finish on high cycle fatigue of a low alloy steel. In: The Behaviour of Short Fatigue Cracks, vol. 1, pp. 69–86. EGF Publisher (1986)
25. Gladman, T., Holmes, B., McIvor, I.D.: Effect of Second Phase Particles on the Mechanical Properties of Steels, vol. 68. Iron and Steel Institute, London (1971)
26. Jagannadham, K.: Debonding of circular second phase particles. Eng. Fract. Mech.Fract. Mech. **9**, 691 (1977)
27. Gross, T.S.: Micromechanisms of Monotonic and Cyclic Crack Growth, ASM Handbook, vol. 19, Fatigue and Fracture, Electronic Files (1998)
28. Lankford, J.: Initiation and early growth of fatigue cracks in high strength steel. Eng. Fract. Mech.Fract. Mech. **9**(4), 617–624 (1977)
29. Congleton, J., Wilks, T.P.: The air fatigue and corrosion fatigue of a 13% Cr turbine blade steel. Fatig. Eng. Mater. **11**(2), 139–148 (1988)
30. Eid, N.M.A., Thomason, P.F.: The nucleation of fatigue crack in low-alloy steel under high-cycle fatigue conditions and uniaxial loading. Acta Metall. **27**, 1239 (1979)
31. Duckworth, W.E., Ineson, E.: The effect of externally introduced alumina particles on the fatigue life of En24 steel. ISI Spec. Rep. **77**, 87–103 (1963)
32. Nisitani, H.: Method of approximate calculation for interference of notch effect and its application. Bull. Jpn. Soc. Mech. Eng. **11**, 725 (1968)
33. Murakami, Y.: Metal Fatigue: Effects of Small Defects and Nonmetallic Inclusions. Elsevier (2012)
34. Toryiama, T., Murakami, Y.: The area parameter model for evaluation of effects of various artificial defects and mutual interaction of small defects at the fatigue limit. J. Soc. Mater. Sci. Jpn. **42**, 1160–1166 (1993)

35. Natsume, Y., Muramatsu, T., Miyamoto, T.: Effect of carbide crack on fatigue strength of alloy-tool steel under cold working. Proc. JSME Meeting **900–86**, 323–325 (1990)
36. Cummings, H.N., Stulen, F.B., Schulte, W.C.: Relation of inclusions to the fatigue properties of SAE 4340 steels. Trans. ASTM **49**, 482 (1957)
37. ASM Metals Handbook: The selection of steel for fatigue resistance. In: Properties and Selection of Metals, 8th edn., vol. 1 (1975)
38. Aksoy, A.M.: Hot ductility of titanium alloy—a comparison with carbon steel. Trans. ASM **49**, 514 (1957)
39. Cyril, N.S.: Anisotropy an sulphide inclusion effects on tensile and fatigue behaviour of steels, MS Thesis in Mechanical Engineering, Advisor A. Fatemi, University of Toledo (2007)
40. Saito, M., Ito, Y.: Some properties of ultra clean spring steel. Trans. Jpn. Soc. Spring Res. **30**, 11–19 (1985)
41. Petch, N.J.: J. Iron Steel Inst. **174**, 25 (1953)
42. Jones, R.L., Conrad, H.: The minerals, metals & materials society. TMS-AIME **245**, 779 (1969)
43. Liaw, P.K., Yang, C.Y., Palusamy, S.S., Ren, W.: Fatigue crack initiation and propagation behavior of pressure vessel steels. Eng. Frac. Mechan. **57**(1), 90 (1997)
44. Cummings, H.N., Stulen, F.B., Schulte, W.C.: Proc. ASTM **58**, 505 (1958)
45. Duckworth, W.E.: Metallurgia. Br. J. Metals **69**, 53 (1964)
46. Atkinson, M.: The influence of non-metallic inclusions on the fatigue properties of ultra-high tensile steels. J. Iron Steel Inst. **195**, 64 (1960)
47. Murakami, Y., Endo, M.: Quantitative evaluation of fatigue strength of metals containing various small defects or cracks. Eng. Fract. Mech.Fract. Mech. **17**(1), 1–15 (1983)
48. Murakami, Y., Endo, M.: Effect of Hardness and Crack Geometries on ΔK_{th} of Small Cracks Emanating from Small Defects, The Behaviour of Short Fatigue Cracks, pp. 275–293. Mechanical Engineering Publications, London (1986)
49. Murakami, Y., Endo, M.: Effect of hardness and crack geometries on ΔK_{th} of small cracks emanating from small defects. Jpn. Soc. Mater. Sci. **35**(395), 911–917 (1986)
50. Faupel, J.H., Fisher, F.E.: Engineering Design. John Wiley and Sons, New York (1981)
51. Moore, H.F.: A study of size effect and notch sensitivity in fatigue test of steels. ASTM Proc. **45**, 507 (1945)
52. Shigley, J.E., Mitchell, L.D.: Mechanical Engineering Design, 4th edn. McGraw-Hill, New York (1983)
53. Buch, A.: Evaluation of size effects in fatigue tests on unnotched specimens and components (in German). Archivfür das Eisenhüttenwesen **43**, 885–900 (1972)
54. Kloos, K.H., Buch, A., Zankov, D.: Pure geometrical size effect in fatigue tests with constant stress amplitude and in program tests. Zeitshrift Werkstoftechniek **12**, 40–50 (1981)
55. Ankab, K.M., Shulte, O.E., Bidulia, P.N.: Isvestia Vishih Utchebnik Zavedenia-Tchornaia. Metallurghia **5**, 168 (1966)
56. Heibach, E.: FKM-Guidelines: analytical stress assessment of components in mechanical engineering, 5th Rev. edn., English version, Frankfurt (2003)
57. Forrest, P.G.: Fatigue of Metals. Pergamon Press, Oxford (1962)

Chapter 5
Surface Treatments and Temperature Effects

5.1 Introduction

The role that the external surface exerts on the fatigue strength of materials will never be stressed enough. In high cycle fatigue and to a less extent also in low-cycle fatigue, the first and most effective barrier to the demolishing action of cyclic loads is provided by the surface. A really well-treated surface will make a workpiece last fatigue as expected while an inaccurate and odd surface will lead to an early as well unexpected fatigue failure. Any process, like honing, polishing, grinding, milling, turning or plating, adopted to strengthen the free surface of the workpiece within a depth no larger than 500 μm, see Fig. 2.5, will sharply increase its fatigue life.

The first use of surface treatments runs parallel to the development of railroad technique and appeared already in nineteenth century. Around 1880, cold rolled shafts with improved surface finish and strength were successfully applied in railway industry. In Italy, at the beginning of last century Ansaldo discovered that after cold rolling plates to form piping the steel would harden if left resting for several months, even six to eight months, before being used and piping would better sustain the enormous pressure of more than one thousand meters of water in the hydroelectric plants. The hardening was due to an aging phenomenon that would have been found some years later by Alfred Wilm who, in 1906, made the accidental discovery of age hardening in the aluminum alloy he created, Al-3.5–5.5%Cu-Mg-Mn, known as duralumin.

However, we would wait some thirty years before scientist would understand the reason of that hardening. In the USA, cold rolling was already applied in the twenties of the last century as a surface treatment to strengthen axels of the world famous Ford T and in the thirties was applied to train axles. Today the most known application of cold-rolling is still in the automotive industry to improve crankshaft performance. However, the technical fields of cold-rolling applications have swelled to include almost any workpiece that has to withstand cyclic loads or even corrosion.

P. P. Milella, *Fatigue and Corrosion in Metals*,
https://doi.org/10.1007/978-3-031-51350-3_5

If the history of surface treatment goes back to nineteen century, we have to wait until 1928 when, as mentioned in Sect. 4.2, Mailänder [1] showed that differently polished surfaces led to differences in the fatigue behavior of steels. Section 4.2 has been devoted to surface finish. This chapter will be treating the subject of surface preparation, other than or in addition to polishing, aimed at protecting metals from fatigue damage. Basically, there are three different families of surface treatments and strengthening

- mechanical;
- thermal;
- thermochemical or metallurgical.

All of them will be fully treated in the following sections.

5.2 Prestressing

Prestressing may be regarded as the general basis for the great family of surface treatment other than polishing. Independently of the method of application, whether mechanical or thermal, its purpose is to generate a state of favorable residual stress state on the surface of the material to avoid or, at least, delay damage initiation. At variance with conventional surface finish, treated in Sect. 4.2, whose purpose is to leave the surface of the test piece or component in the smoothest condition possible to avoid stress risers, prestressing is actually generating a system of compressive residual stresses that can counteract the applied traction cyclic stresses. From a general point of view, prestressing may be divided into two distinct groups:

- mechanical,
- thermal.

5.2.1 Mechanical Prestressing

The first experiments that made it clear that cold rolling had a beneficial effect on fatigue strength of metallic materials were run in Germany by Föppl [2] in 1929. These experiments initiated a long series of testing that brought researchers to conclude that residual stresses were actually responsible for the beneficial effect [3–5]. Today no one would doubt about that conclusion, but in those days the question was not so evident. Consider a test piece that prior to fatigue cycling is subjected to an overload by bending that produces yielding in traction, but only on metal fibers farthest from the neutral axis, as schematized in Fig. 5.1a.

Imagine that the depth d of the metal layer exceeding yielding is negligible relative to the thickness t of the test piece: $d \ll t/2$, so that yielding, both in traction or compression, is localized just on the surface. Under these circumstances, upon unloading all surface fibers that yielded reaching any point between A and B (see

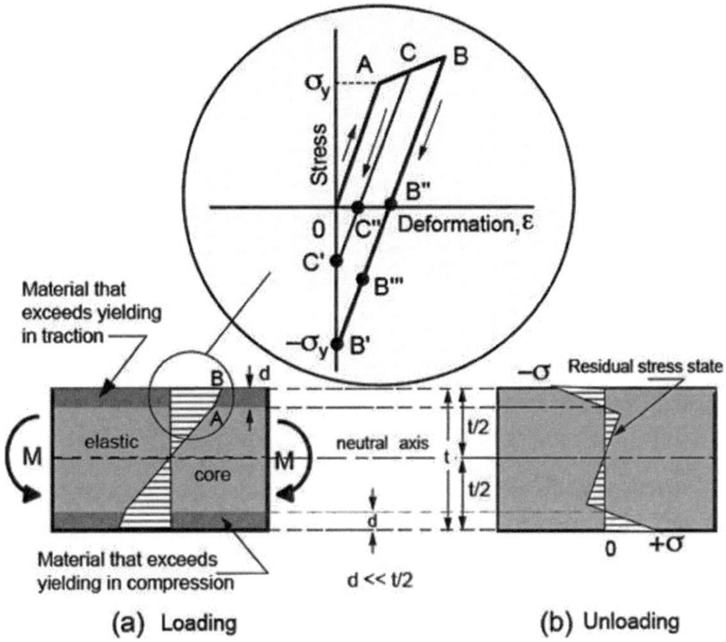

Fig. 5.1 Schematic of shake-down effect. Test piece subjected to bending: **a** outermost metal fibers exceed yield strength, **b** upon unloading these fibers remain in a compressive stress state

detail of Fig. 5.1a) will be forced by the elastic material (elastic core of Fig. 5.1a) to move along paths of the type BB′ or CC′, parallel to the elastic ramp, till they will arrive in B′ or C′, remaining in a state of permanent compression. This happens because, upon unloading, the elastic core returns to the initial state of zero deformation and zero stress. Its volume by far exceeds and, therefore, overwhelms that of the small plastic enclave so that it will force those fibers that yielded to go back to a zero deformation, because the congruence of deformations must be respected. Actually, any point like B will follow the unloading path BB′ arriving and resting somewhere between B and B′ (enlarged circle of Fig. 5.1a) depending on the $2d/t$ ratio.

The more this ratio is less than 1, i.e., the more yielding is localized, the more the generic element of material like B will be forced to approach point B′, passing through B″ and B′″ and the higher the compressive residual stress, $-\sigma$, will be. This spring back effect, called shake-down, is fully described in Sect. 5.3.2 and in Fig. 5.28. If just few outermost fibers yield, than point B will arrive exactly in B′. The same will happen to any other point element between A and B, such as C that will move to C′, remaining in a compression residual stress at load removal. At variance, when $2d/t \approx 1$ no shake-down effect will take place and by unloading all points that yielded will rest along the ε axis, like B″ or C″. The opposite will happen on the other surface where the entire layer of material plastically deformed in compression will reach a residual tension stress state $+\sigma$, as schematized in Fig. 5.1b. The overall residual stress state is necessarily self-equilibrating since it is composed of internal stresses, therefore

$$\int_{-t/2}^{t/2} \sigma_x dy = 0 \tag{5.1}$$

with t being the thickness of the test piece. Also moments must be zero

$$\int_{-t/2}^{t/2} y \cdot \sigma_x dy = 0. \tag{5.2}$$

Hence, the distribution of residual stresses σ_{res} following the initial prestressing will be that schematized in Fig. 5.1b. When the material is later cycled in the elastic regime (high-cycle fatigue) by a stress amplitude $\pm\,\sigma_a$ after the initial prestressing, residual stresses σ_{res} will sum with their sign to the stress σ_a generated by cyclic loads yielding

$$\sigma_{tot} = \sigma_a + \sigma_{res} \tag{5.3}$$

reducing the total stress state σ_{tot} where they are compressive and increasing it where they are of the same sign of cyclic stress σ_a. Localized yielding produced by overload bending may be also caused by any geometrical discontinuity and, in particular, by a sharp notch. This is shown in Fig. 5.2. In this case a cylindrical specimen carrying a circumferential notch is subjected to an overload F_I that causes a small volume of material at the notch tip to yield, as shown in Fig. 5.2b where the material has been assumed to be elastic-perfectly-plastic. If the volume of the plastic enclave so generated is very small, absolutely negligible with respect to the surrounding volume of elastic material, which actually means the more the notch is small or sharp, shake-down will take place leaving, upon unloading, the residual stress state illustrated in Fig. 5.2c with a compression peak stress at the notch tip $-\,\sigma_{nt}$ where before unloading the material yielded (point 2 of the notch tip stress history of Fig. 5.2). It is from this residual stress state $-\,\sigma_{nt}$ that the following cycling at a stress amplitude σ_a will operate. According to Eq. (5.3) the resulting cycles 2, 3, 4, 5, 6 … will produce a much lower peak stress σ_{max} and a negative mean stress σ_m, as shown by the notch tip stress history of Fig. 5.2.

The beneficial effect is shown schematically in Fig. 5.3 for 7075-T6 aluminum alloy [6]. It can be clearly seen that when the notched specimen is subjected to a monotonic axial prestressing the consequent fatigue S–N curve is improved with respect to that obtained with specimens that have not been prestressed.

The actual beneficial effect depends on the amount of prestressing that in Fig. 5.3 is expressed in percent of the ultimate strength of the notched specimen. At variance, for what has been said, too large a notch or hole may not develop a favorable residual

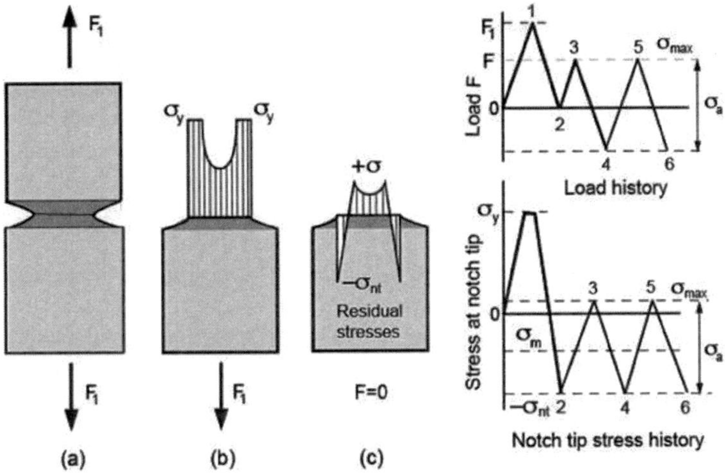

Fig. 5.2 Cylindrical notched specimen subjected **a** to a prestressing overload F_1 that **b** yields the material ahead of the notch and **c** leaves, upon unloading, a shallow surface layer in a compression stress state

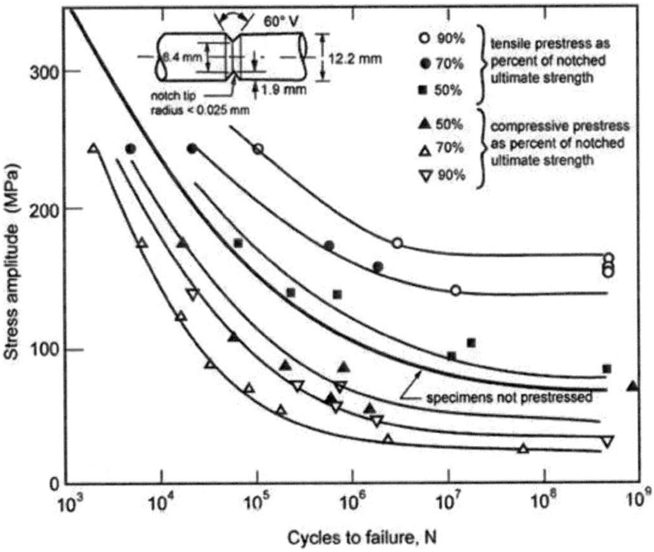

Fig. 5.3 Effect of axial monotonic prestressing on the S–N curve of 7075-T6 specimen tested in rotating bending (data from [6])

stress state because the associated plastic enclave can swell so much to become too large to allow a shake-down effect to occur. The fundamental prerequisite for prestressing to achieve the wonted target by shaking-down the plastically deformed zone is that this last be very small with respect to the volume of the elastic material and this can happen only when the stress gradient ahead of the discontinuity, hole or notch or whatever may be, is very steep so as to involve a very small volume of material. A large hole or notch and a small one may have the same concentration factor but the stress gradient is different: very smooth for the former and very steep for the latter. This is schematically shown in Fig. 5.4 where a small hole and a large hole are compared. A study conducted by Fuchs and Stephens [7] on plates with and without a central hole under axial loads has clearly shown that the fatigue limit S_f of the test pieces without hole did not change upon prestressing, as expected, while in those containing a hole it doubled arriving almost to match that of the plate without hole. Basically speaking, there are three different types of mechanical prestressing

- machining,
- rolling,
- shot peening.

Machining has already been discussed in Sect. 4.2. In the following sections it will be treated the issue of rolling and shot peening.

5.2.2 *Cold-Rolling*

An important mechanical process, probably the first, to improve the fatigue strength of materials is cold-rolling. The use of cold hammering metals has long been known. It was used as early as the Bronze Age. During the crusades (1100–1400 A.D.) it was widely used to strengthen armor and swords. The beneficial effects introduced by cold-rolling are threefold: first it leaves the surface of the test piece in a triaxial compressive stress state, secondly it hardens the surface by cold-working and last, but not least, it may leave the surface in a smooth condition that can approach the quality of a mirror-polishing treatment, especially in the case of deep rolling.

As already mentioned, the first experiments that demonstrated the usefulness of cold-working were performed by Föppl [2] in 1929 using the appliance shown Fig. 5.5a which is the original drawing appeared in a paper by Heydekampf [8]. The workpiece represented in the drawing is a plain specimen of 12 mm diameter steel, 1200 MPa ultimate strength that was subjected, after rolling, to reversed torsion showing an increase of fatigue strength of about 15%. The paper also presented some very interesting drawings anticipating what would have become evident decades later. It was observed, in fact, that in cold-rolled workpieces fatigue failure initiated from small slag inclusions beneath the surface, Fig. 5.5c, while in polished specimens that were not cold-rolled it always started from the surface, Fig. 5.5b. The reason for that, as we know, is in the residual stress state generated by cold-rolling. Figure 5.5d

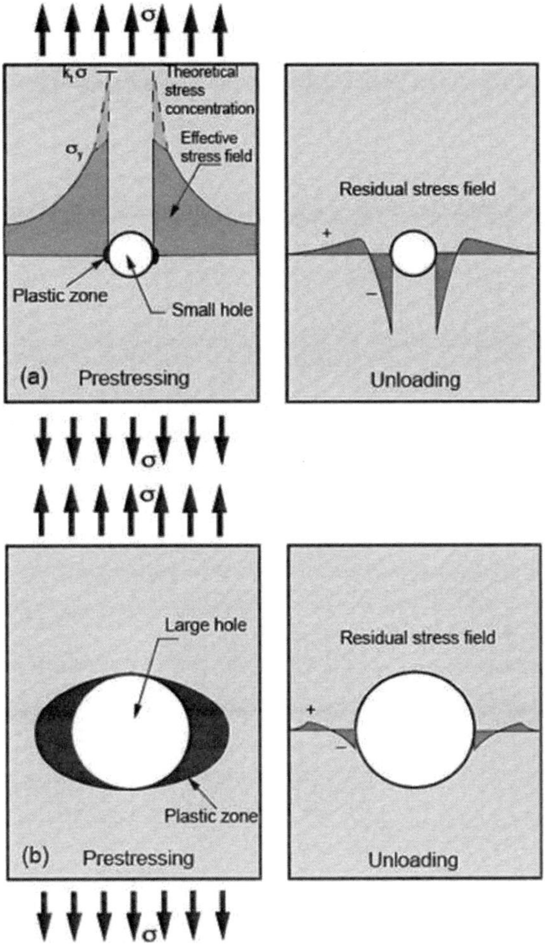

Fig. 5.4 Effect of prestressing on a plate: **a** small holes give raise to high stress gradient and small plastic zone, **b** large hole produce larger plastic zone. In the former case the shake-down effect fully develops, in the latter it doesn't

shows the measurements made by Fahrenhorst [9] of the triaxial stress state. In the left side of Fig. 4.5d it is presented the residual stress state after drawing, which may be particularly dangerous because it leaves traction stresses right on the surface where they affect fatigue the most. By applying the cold-rolled procedure, after drawing, the axial and tangential stress state is converted from traction into compression, becoming favorable to fatigue strength. Note the peak traction axial stress existing after cold-rolling which has its maximum at 3 mm below the surface of a 20 mm diameter shaft. It was precisely this peak tensile stress to cause fatigue failures when a large inclusion was present underneath the surface (Fig. 5.5c). Today, it is quite common practice to use large scale equipment like that shown schematically in Fig. 5.6 to cold-roll large component such as shaft over 170 mm in diameter. It is a three-roller device using rolls of very hard chromium steel, particularly suitable to withstand fretting. The rolls exert a pressure over the component that rotates along its longitudinal axis, as depicted in Fig. 5.6. Using this technique the external annulus of the component is put into compression and the resulting residual stress state is that of Fig. 5.5d.

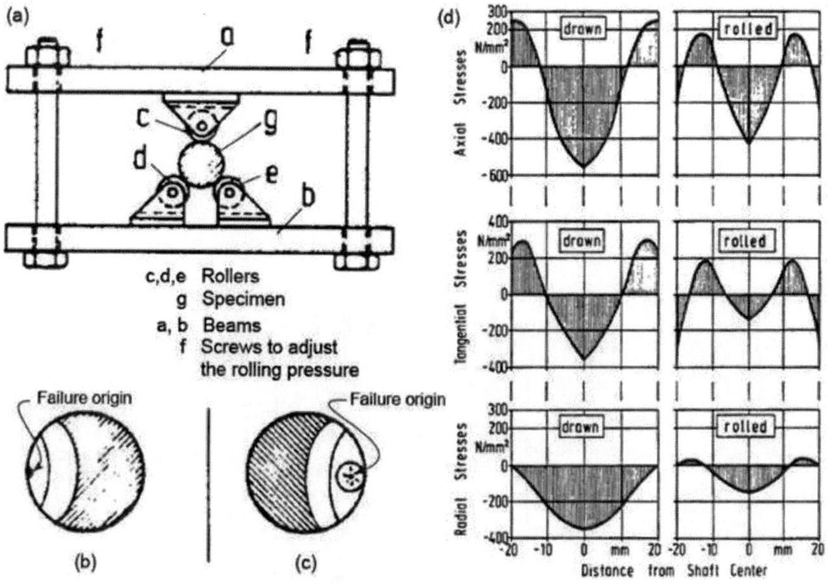

Fig. 5.5 Original drawings about **a** cold-rolling device used by Föppl [2], **b**, **c** indicates the origin of fatigue failure in a cold work piece and **d** residual stress state measured by Fahrenhorst [9]

Fig. 5.6 Schematic of cold-rolling apparatus

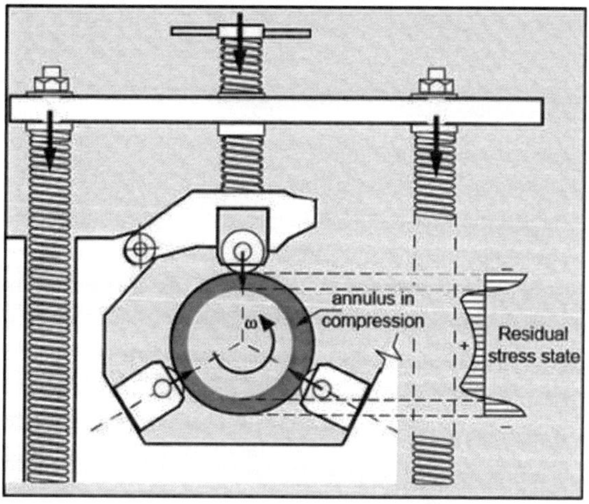

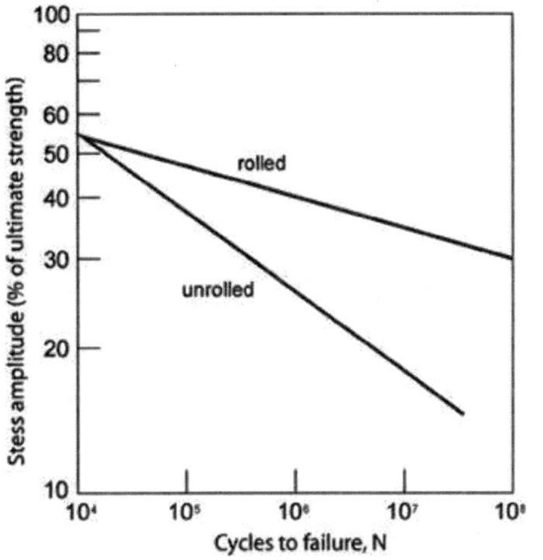

Fig. 5.7 *S–N* fatigue curves of a carbon steel in unrolled and cold-rolled conditions [10]

The method can introduce 40–80% cold-working with consequent hardening of the surface layer. Both residual compression and hardening improve, as well, the fatigue resistance of the material, as it can be seen in Fig. 5.7 [10] that indicates a two-fold increase of the fatigue limit at 10^8 cycles for carbon steel. Note that rolling with the purpose of improving fatigue strength cannot be applied in temperature. Hot-rolling, in fact, does not allow both surface hardening and compressive residual stresses. This is because high temperatures anneal cold-working effect and reduce the yield strength.

5.2.3 Deep Rolling and Burnishing

The apparatus shown in Fig. 5.6 requires the workpiece to have a perfect cylindrical shape and be capable of undergoing the treatment. When the workpiece is too wide and has a rather variable geometry, even though symmetrical, it cannot be cold-rolled with that kind of equipment. Still cold-rolling can be efficiently applied by using a technique called deep rolling. Deep rolling is a surface treatment that makes use of hard rolls or ball-point tools of any size to induce, by applying a proper pressure over the workpiece, deep plastic deformations and residual stresses in the surface layer. Its basic scheme is shown in Fig. 5.8. Figure 5.9 presents some typical applications of deep rolling on an axle, shoulder fillet or on any surface of various contours. It has all the benefits of cold-rolling and always leaves the surface in a state of finishing equivalent to that of a mirror-polishing treatment with roughness below 1 μm.

Figure 5.10 [11] shows the results obtained for Ni718 plate alloy compared to those obtained by a shot peening technique (see next section). It can be seen how deep rolling is by far more effective than shot-peening both for the residual compressive

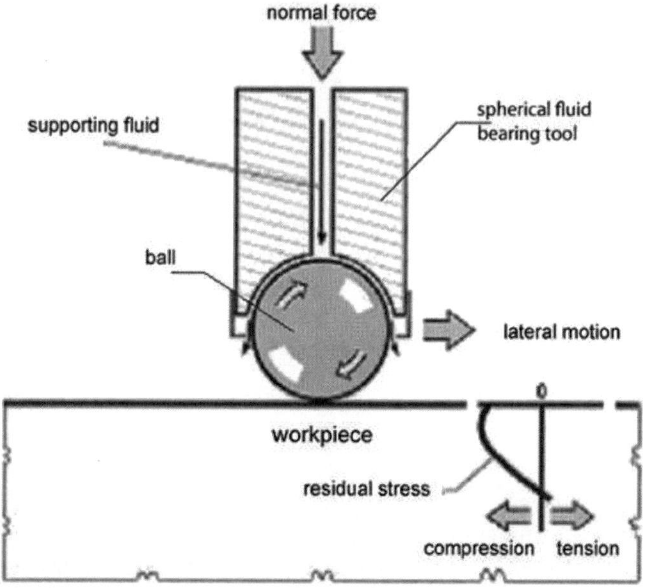

Fig. 5.8 Schematic of ball-point tool for deep rolling or low plasticity burnishing technique (reproduced with permission of Lamba Technology Group)

peak stress and the depth reached, as well. The effect is kept, dough reduced, even when temperature is maintained at 600 °C for 1000 h.

Moreover, as said, deep rolling leaves the surface in an excellent finish condition whilst shot peening does not, compare Fig. 5.9 with Fig. 5.13. Deep rolling is very effective also for forming and strengthening grooves, as shown in Fig. 5.11 [12] for a cylindrical bar of 37CrS4 steel (0.4 C, 1.1 Cr, 0.2 Ni, 0.1 Mo) 1150 MPa ultimate strength.

The difference between unrolled and deep rolled conditions is more than evident from the relative S–N curves. The fatigue limit is more than doubled from about 310 MPa in the unrolled condition to 700 MPa after deep rolling.

Indeed, the real great gain obtained using a deep-rolling technique is with notched members or workpiece having discontinuities like shoulder fillets or journey that can have a dramatic effect on fatigue strength. This can be seen in Fig. 5.12 [13] where the fatigue limit is plotted as a function of the yield strength for various steel in the as received and deep-rolled conditions for (a) unnotched members and (b) notched members with a stress concentration factor $k_t = 3$. In both cases there is a remarkable gain in fatigue strength, which increases as the material yield strength increases. As Fig. 5.12b clearly indicates, the method has much larger a benefit in notched members than in plain workpieces. Deep-rolling is also known as low plasticity burnishing (LPB). Nevertheless, a difference between the two processes exists. Generally speaking, the terminology deep-rolling refers to a surface rolling treatment having the purpose of generating deep plastic deformations, high strain

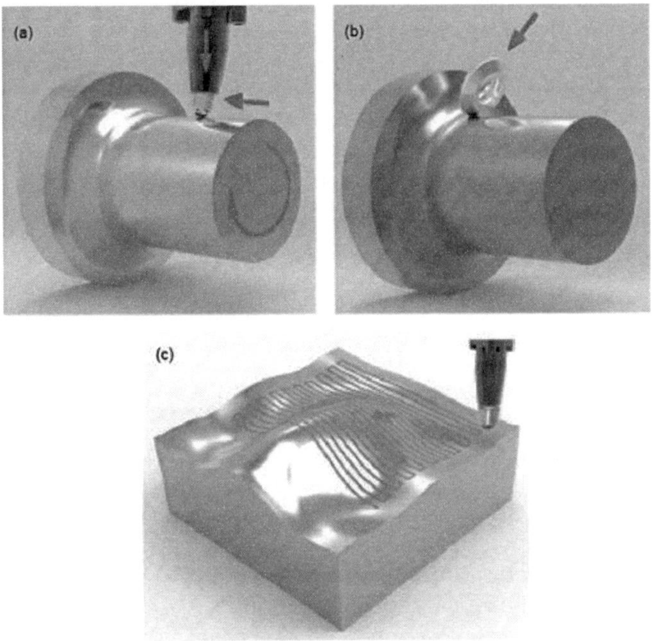

Fig. 5.9 Applications of deep rolling technique to **a** axles, **b** shoulder fillets or **c** any surface of variable contour

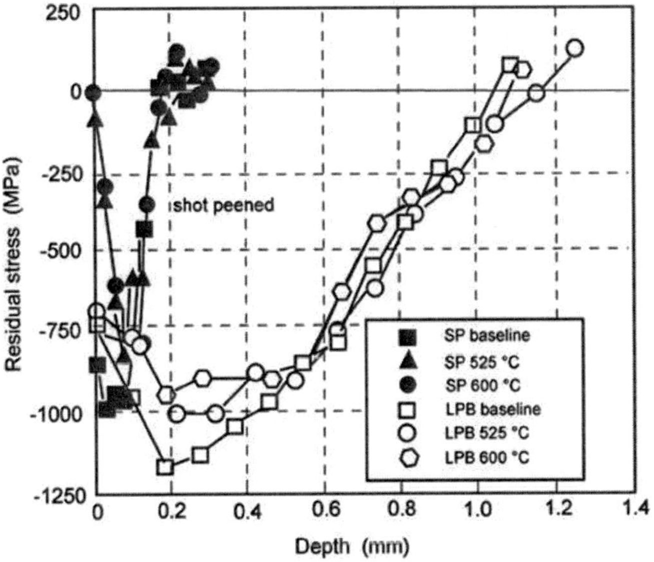

Fig. 5.10 Low plasticity burnishing residual stresses measured on a nickel alloy test piece compared to those relative to shot-peening [11]

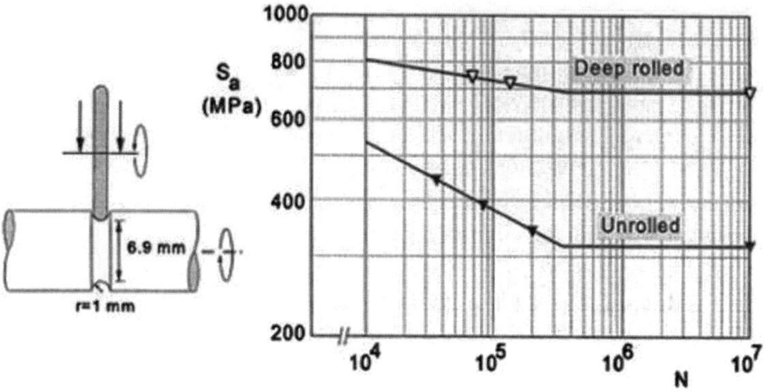

Fig. 5.11 Deep rolling used to form and strengthen a groove in a cylindrical bar and relative *S–N* curves [12]

hardening and residual stresses. At variance, LPB refers to a rolling technique usually applied with much lower pressure on the roller and mostly aimed at obtaining a quality smooth surface. In this respect, the data shown in Fig. 5.10 should be better referred to as typical of deep-rolling results rather than LPB's.

Depending on the material, deep-rolling can result in the formation of dislocation cell-structures, see Fig. 1.31b and may even actuate a martensitic transformation. The former is typical of aluminum alloys and carbon steel and, in general, of high SFE materials that deforms by wavy slip, see Sect. 1.4.1, while the latter can take place in austenitic stainless steel. This transformation, due to the fact that austenite is a metastable phase at normal operating temperatures, can have dramatic consequences

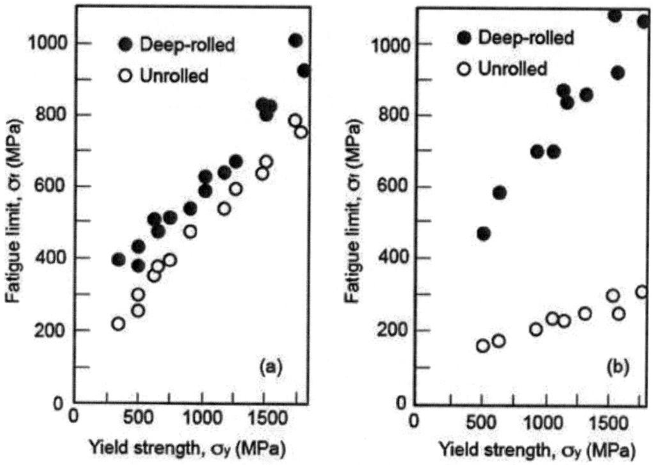

Fig. 5.12 Influence of yield strength on fatigue limit of steel before and after deep-rolling, **a** for plain specimens, **b** for notched specimens (adapted from [13])

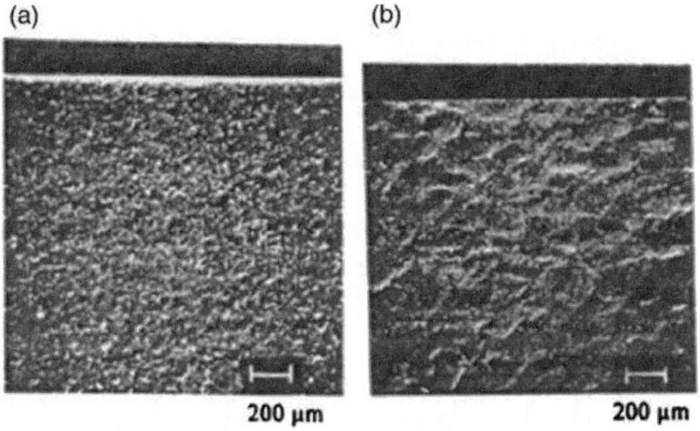

Fig. 5.13 Shot peening effect on the surface of a steel work piece. **a** small glass shot size, **b** large steel shot sizes [14]

because martensite is brittle and may initiate surface cracks particularly dangerous in corrosion assisted fatigue, as described in Sect. 18.2.2.

5.2.4 Shot Peening

Shot peening is not a new process. Gun barrels during the American civil war were subjected to peening to increase the hardness of the steel. In modern age it was used in Europe in the '20s of last century to improve the strength of crankshaft fillets using special made hammers and by General Motors, as well. Shot peening may be considered as the most versatile of all cold-working treatments, being applicable to virtually all metals and shapes. It consists on bombarding the surface of the work piece with high-velocity (about 30–40 m/s) cast iron or hard steel or even glass or ceramics shots discharged from a rotating wheel or nozzle. The diameter of the perfectly round balls ranges from 0.2 to 4 mm. This shooting produces a hammered or peened effect on the targeted surface, which wants to reduce its thickness and therefore to increase the area. Since this area increase is resisted by the elastic core whose volume is by far larger than that of the skin, it enters into a residual compressive stress state while the core metal remains in traction, in particular at the interface. The smaller shot sizes are capable to reach any small geometrical variation while larger sizes produce relatively deeper penetration and larger residual compressive stresses.

These latter can also harden the material to a larger extent. The compressive peak stress reaches its maximum just below the skin of the work piece, which is about half the yield strength of the material. Larger shot sizes may reach yielding. Shot peening is largely used in air and space industry, automotive industry, in the treatment of crank shafts, gears, leaf springs etc. ... However, as already mentioned, shot peening leaves a surface profile not perfectly smooth, as shown in Fig. 5.13 [14].

A double shot peening operation may then be required. In this case there will first be a heavy shot-peening that produces a considerable in-depth hardening followed by a micro-ball shot-peening operation to improve surface finish. Sometimes, the surface is also subjected to an electrochemical polishing, after shot-peening, to improve surface finish. Shot peening may be used on any metal to enhance fatigue strength, but its effective application is with very high strength steels over 35 HRC or 1200 MPa ultimate strength that have been said to have a fatigue limit jeopardized by inclusions with a remarkable loss of fatigue strength (see Sect. 4.6 and Fig. 4.37). With these steels shot peening may overcome the barrier to fatigue limit posed by inclusions or second-phase particles. Experimental results obtained by Nelson, Ricklefs and Evans [15] with SAE 86B4S steel, heat treated to different hardness, confirm this capability. This is shown in Fig. 5.14 where their experimental results have been superposed to the fatigue limit diagram of Fig. 4.37. It can be seen how in their experiments fatigue limit in the as received condition (without shot peening) is following the general trend of the diagram. At the hardness of about 40 HRC the fatigue limit is not continuing to grow with the same rate, as expected, but after reaching a maximum at about 50 HRC it goes down with increasing hardness. When shot-peened, the steel improves its fatigue response moving above the upper limit of the diagram, but still at about 35 HRC its growth slows down. But now, the use of conventional shot peening cannot suffice for industrial applications, a great deal of efforts has been made on the modification of shot peening treatment trying to optimize it. Optimized shot peening, in fact, avoids the downhill trend at higher hardness, reaching fatigue limits otherwise non achievable or achievable only with deep rolling (see previous section). This optimum use can be achieved by putting in traction the test piece or the actual work piece before shot peening.

After shot peening the tensile stress is released. With the elastic recovery, it results in a higher compressive residual stresses and better refined microstructure in the deformed layer, which strongly improves the potential of shot peening. Conversely, poor processing will do much more damage than benefits to harder materials. For instance, 302 stainless steel wire used for coil spring without a prestressing has a fatigue limit of about 110 MPa. With prestressing to block it moves up to 198 MPa. With warm prestressing the fatigue limit jumps to 247 MPa. The warm prestressing is generally performed at about 200 °C to reduce the yield strength of the material. If the coil is then glass peened the fatigue limit further improves exceeding 400 MPa. This is a well-known process in the coil industry. High strength steels used in manufacturing coil springs are scrapped at 200 °C to coil full length.

This plastic straining trigger the same process described in Fig. 5.4a. Helical springs, in fact, develop a very localized stress concentration along the inner radius which causes the development of residual stresses upon unloading. Dislocations to pile up at grain boundaries tangling and producing micro-residual block stresses. When the spring is settled, the residual stress state reduces the applied stresses as they act in the opposite direction. Another example of beneficial effect of shot peening if shown in Fig. 5.15 for carburized steel used in the gear industry [16]. Shot peening can be very useful in the spring industry. A systematic study performed on non-decarburized steel springs reveled, already in 1940 that the time at shot-blast was

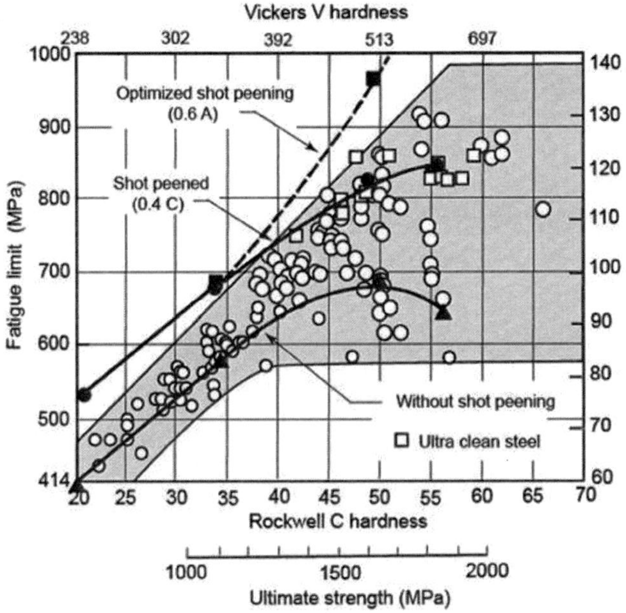

Fig. 5.14 Fatigue limit improvement by shot peening SAE 86B4S steel heat treated to different hardness (data from [15]). Also indicated is the Almen intensity

determining the effectiveness of the shot-peening process. Figure 5.16 [17] presents results obtained on springs of 0.65%C, 0.58%Mn heat treated carbon steel of 1380 MPa ultimate, in terms of shear stress fatigue limit, $\Delta\tau$, versus shot-peening duration.

The shear stress excursion increased from an initial value of about 500–800 MPa after 10 min, after which it remained constant. The springs of 4.1 mm wire diameter, were blasted with 1.19 mm diameter balls followed by 0.39 mm balls to reduce surface finish. Interesting enough is the diagram on Fig. 5.16b that reveals how annealing

Fig. 5.15 *S–N* diagram for shot peened and not shot peened carburized steel (adapted from Ref. [16])

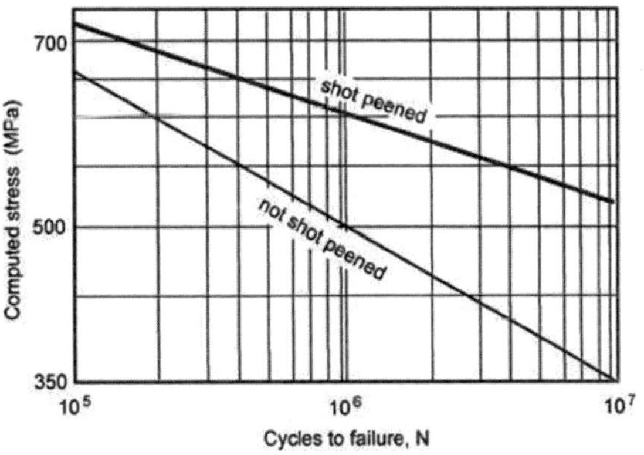

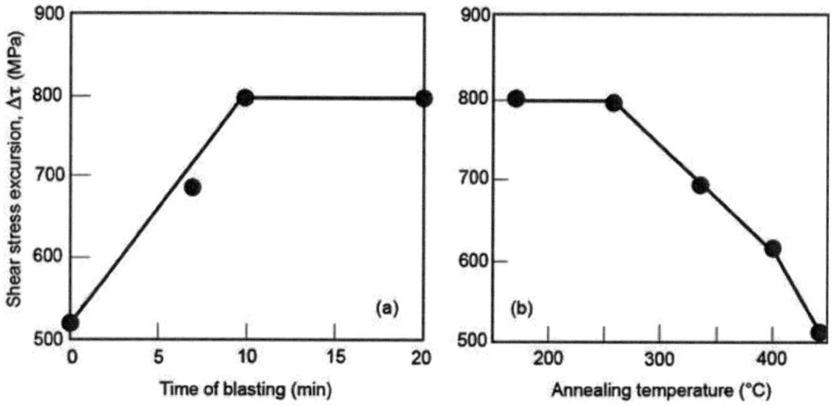

Fig. 5.16 **a** Fatigue limit of high carbon steel springs versus shot blasting time and **b** annealing temperature effect (adapted from [17])

can reduce or completely remove the beneficial effects of shot peening, depending on temperature. However, some problems may arise from residual stresses being self-equilibrating. Beyond a residual compressive stresses field there must necessarily exist a balancing tensile stresses field, as shown in Figs. 5.2 and 5.5d.

Considering that the compressive stress field develops within 200 μm or at most 500 μm from the surface, according to the Saint Venant's principle tensile stresses shall extinguish within 2–3 times this perturbation distance, i.e., within 600 μm to 1.5 mm with the peak acting at the interface with compressive stressesThese tensile stresses can trigger a fatigue failure if an inclusion happens to be present under the surface at some hundred microns, right inside the tensile stress field. Without residual tensile peak stresses such inclusions would not be important. An example of fatigue failure triggered by an under skin inclusion is offered in Fig. 5.17 [18]. The figure shows the failure origin at a nonmetallic inclusion located 0.2 mm (0.008 in.) below the wire surface (large withe arrow). This is well below the zone of peak compression produced by shot peening. Without shot peening the inclusion would not be harmful at all because too deep (see Sect. 4.2). However, it became harmful and triggered the fatigue failure because it happened to be located right in the residual traction stress field set up by shot peening. Consequently, the protective effect provided by superficial compression stresses faded leaving the traction stresses to operate their fatal action.

Note the beach marks forming nearly complete ellipses around the inclusion. A larger inclusion at the bottom of the fractography (small withe arrow marked A) took no part in the fatigue failure event because at about 700 μm; too deep and completely out of the tensile residual stress state. Also a small inclusion on the left (small withe arrow marked B), almost on the surface, took no part in the fatigue failure because it was right in the residual compression stress field and, therefore, was protected by the shot peening action. A method for measuring the intensity of shot peening is provided by the so called Almen test from whom who introduced it already in 1930

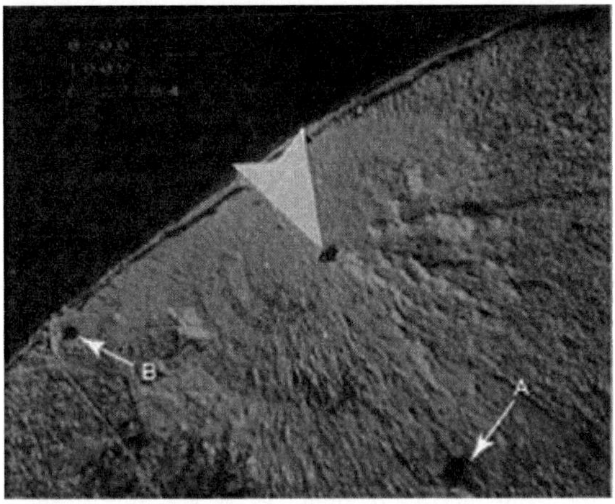

Fig. 5.17 Failure origin at an inclusion (large arrow) located 0.2 mm (0.008 in.) below the spring surface. Note beach marks surrounding inclusion. 110× [18]

[19]. The test makes use of a metal strip of standard size, 76 mm long and 19 mm wide, with three possible thickness indicated by letters N, A and C, as shown in Fig. 5.18.

The strip is firmly fixed by four screws on a holding block where a moving nozzle provides the shot stream, Fig. 5.18a. After shot peening, the Almen strip removed from the test block appears inflexed because residual stresses induce

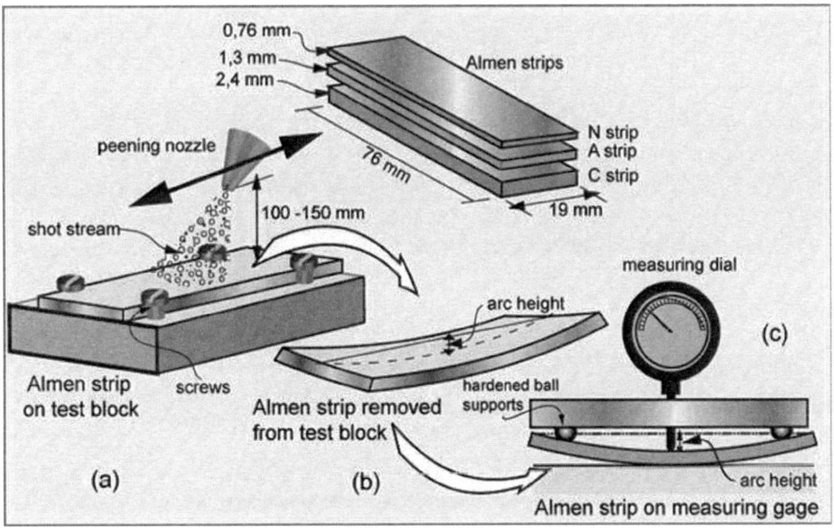

Fig. 5.18 Almen procedure to measure shot peening intensity

arching, Fig. 5.18b. The arch height is then measured by a dial, as shown if Fig. 5.18c, and taken as an indication of the shot peening intensity followed by the letter that specify the strip used:

Type A—medium intensity applications on casts and wires;

Type N—low intensity applications on glass or ceramics and non-ferrous metals;

Type C—high intensity applications (steels).

The depth of the compressed layer can be estimated from the corresponding Almen intensity reading, as shown in Fig. 5.19 for the three types of specimen A, C and N.

For example, a steel of 1500 MPa tensile strength having a measured Almen intensity of 0.4 mm C has a depth of the compressed layer between 0.6 and 1.8 mm, depending on its hardness. If it is very hard this depth is likely to be about 0.6 mm. An aluminum alloy with Almen intensity of 0.4 mm N has a depth of the compressed layer between 0.03 and 0.2 mm, depending on its hardness. The corresponding maximum compressive residual stress is given by the diagram of Fig. 5.20 [20].

Therefore, for the steel previously considered having a hardness of about 45 RC, the maximum compressive stress generated by shot peening would be about − 875 MPa. However, the very many parameters involved in the shot peening operation, among which there are:

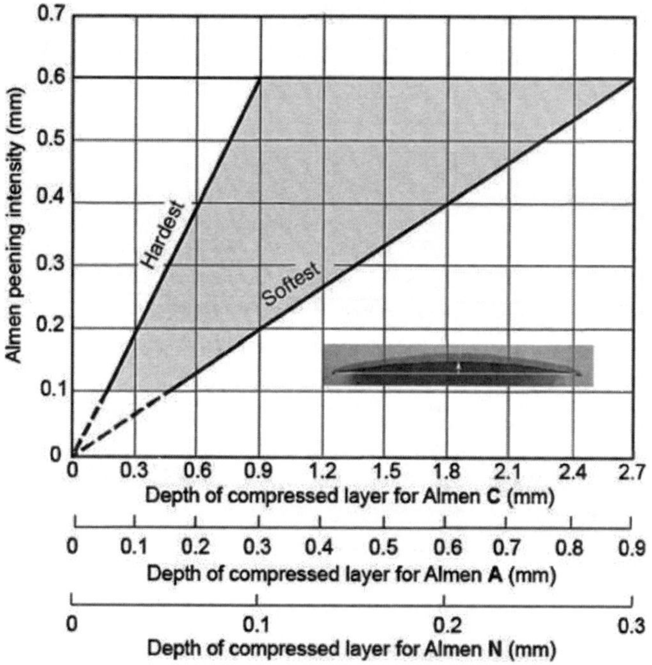

Fig. 5.19 Guideline diagram for the conversion of Almen intensity to depth of the compressed layer

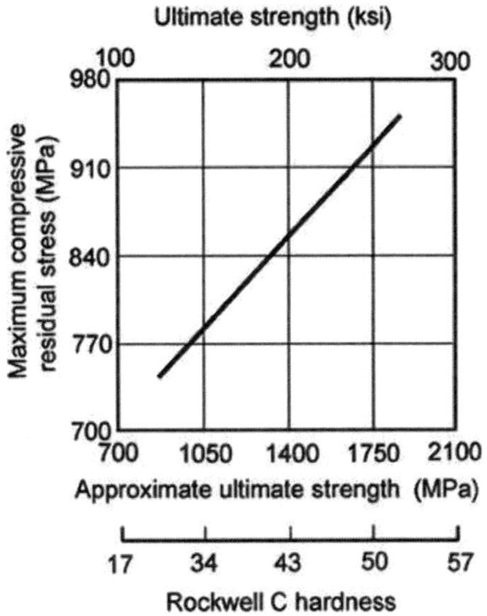

Fig. 5.20 Residual stress produced by shot peening versus ultimate strength of steel [20]

- shot speed,
- dimensions, shape, type and hardness of shot,
- angle of incidence,
- coverage,

make it be difficult to have a great accuracy on the real depth of the compressed surface layer. Tests performed at CETIM in France on Almen gauges have shown that, for identical Almen deflection measurements, it is possible to generate in the material quite different distribution of residual stresses [21]. This actually means that, besides granting the accuracy of the parameters chosen when performing the operation, a post operation stress measurement is always needed to check what has been really obtained, at least on a spot basis.

As to the question of surface roughness that has been said to represent the main drawback of shot peening technique, it can be possible to practice a fine ground operation after peening. However, the operator must be cautious in removing any surface layer because the residual stress field could also fade away. Normally, a 10% removal of the compressed layer depth is permitted as a reasonable maximum. The problem may be the correct evaluation of the thickness of the hardened layer. Table 5.1 lists the surface treatment factors C_{st} for various materials adopted by FKM-Guideline [22].

The decision of which value to choose for steel, whether the upper or the lower, is left up to the user. Therefore, one should be very cautious when choosing a value for the surface correction factor.

Table 5.1 Surface treatment factors for various materials [22]

Surface treatment	Diameter 8–15 (mm)		Diameter 30–40 (mm)	
	Cold rolling	Shot peening	Cold rolling	Shot peening
Steel	1.20–1.40	1.10–1.30	1.10–1.20	1.10–1.25
Cast iron	1.2	1.1	1.1	1.1
Aluminum	1.2	1.1	1.1	1.1

5.3 Thermal Treatments

Thermal treatments are suitable for increasing the fatigue strength of metals. At variance with the thermo-mechanical ones that create favorable residual stress states, thermal treatments are applied with the purpose of producing metallurgical transformations that can change, even completely, the mechanical characteristics of the alloy. It must be understood that, depending on the metal composition, heat treatments can be designed in almost infinite ways depending on what result the manufacturer is aiming at in terms of microstructure and material properties. As for steel, that is iron containing carbon in amount up to about 2.0% along with a number of impurities or alloying elements, five operations are the basis of heat treatment:

- full annealing,
- normalizing,
- hardening,
- tempering,
- stress relieving.

Full annealing is the process of softening steel by the application of heat, so that it may be bent or cut easier or the original properties restored that were altered by grinding process or turning and forming or welding. In annealing, steel is heated above the critical temperature afterwards it is cooled very slowly. The critical temperature of steel ranges from a minimum of about 723 °C to a maximum of 1150 °C, depending on the carbon content of the material. The critical temperature is that at which α-iron or ferrite (BCC crystal structure) transforms into γ-phase austenite (FCC structure).

The basic principle of annealing process of carbon steel is that the steel must be brought into the $\gamma + Fe_3C$ (carbide) region of the phase diagram and then controlled cooled, typically at 5–20 °C/h until all austenite has transformed into ferrite. The distinguish characteristic of full annealing is precisely the slow cooling that, usually, takes place in the furnace. Generally a temperature of 912 °C is chosen for heat treatments. A schematic *time–temperature* diagram for annealing heat treatment is shown in Fig. 5.21a. Adamaszek and Broz [23] studied the effects of decarburization due to the annealing process on hardness changes in carbon steels caused by high temperature surface oxidation. They stated that the decarburization causes the grain near the surface to grow. The authors explained that during the decarburization process oxygen penetrates the surface through cavities, pores and cracks. The

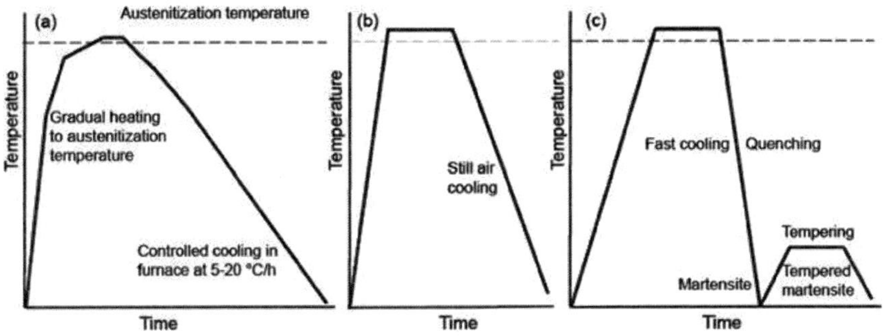

Fig. 5.21 Schematic of time–temperature heat treatments: **a** annealing, **b** normalizing and **c** quenching and tempering of steel

oxygen reacts with the different chemicals in the metal and, in particular, with carbon which it is so greedy for, causing the decarburization (see also Sect. 4.3). The authors also stated that there are hardness changes due to decarburization resulting in lower fatigue strength (see Sect. 1.4.1 and Fig. 1.34).

Normalizing is a heat treatment process for making materials softer. Normalizing is usually done to steel after hot forming as a grain refining operation. The steel is brought into the austenitic region of the Fe–C diagram by heating et al. least 55 °C above the critical temperature. After the steel has been completely austenitized, the steel is air cooled resulting in a refined, pearlitic structure (see Fig. 4.19). The treatment refines the grain size and improves the uniformity of microstructure and properties, in particular of hot rolled steel. Normalizing is used in the production of large forgings such as railroad wheels and axles and some bar products. The process is faster and less expensive than annealing. Normalizing also eliminates grain flow directions due to cold working and residual stresses as well. A schematic *time–temperature* diagram for normalizing heat treatment is shown in Fig. 5.21b.

Normalizing performed in improper manner, both from the starting temperature and time of cooling, may damage the fatigue response of a metal. This can be seen in Fig. 5.22 for mild steel ($\leq 0.1\%$ C) that has undergone several annealing operations including overheating at 1175 °C [24]. It can be seen the progressive deterioration of fatigue properties passing from as-drawn condition to overheating through reheating from about 950 °C followed by either rapid or slow cooling. Reheating at 950 °C not only has enlarged grain size, but has also removed residual compressive stresses, if they existed. Aluminum alloys can be annealed at about 315–430 °C for two or three hours, cool at a rate of about 30 °C per hour down to about 260 °C and then air cool to reduce strength and hardness level or eliminate residual stresses introduced by mechanical operations. In the temper designation of aluminum alloys, the letter O that may or may not be followed by one or more numbers, designates the annealing operation.

More complex is the hardening of alloys obtained by a quenching and tempering treatment that, at variance with normalizing and annealing, produces a significant change in the metallurgical structure of steel. These changes are improved by adding

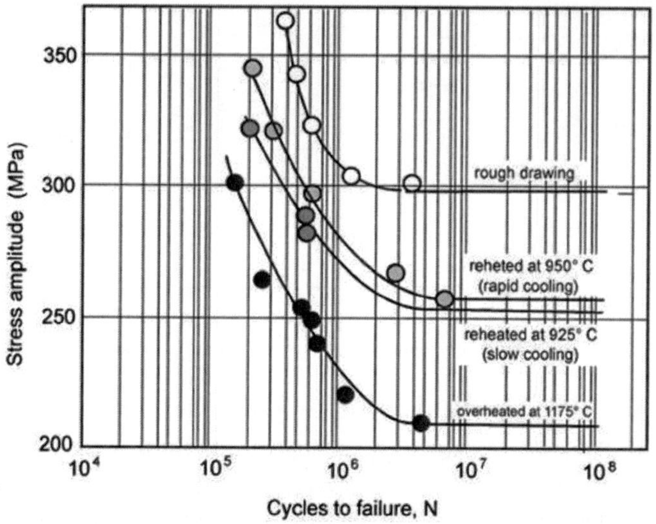

Fig. 5.22 Fatigue *S–N* curves of a rough-drawn mild steel. Influence of reheating and overheating [24]

foreign alloying elements. It can be said, in fact, that alloying elements are added right for the purpose of getting the wanted properties from the heat treatment. By quenching and tempering steel it may be obtained pearlite or martensite (see Fig. 4.21) with grain refinement and excellent hardness having high strength and, therefore, superior fatigue strength properties. Hardening is a twostep heat treatment in which the steel is first brought above the critical temperature (723–800 °C for steel 0.3–0.6% C) maintained for a convenient time at that temperature and then quenched in water or brine or salts for very rapid cooling below 200 °C just in one minute. Tempering from high temperature in water, particularly in still water, causes the formation of a vapor film that insulates the metal leading to the thermal crisis that prevents the rapid cool down.

To avoid the formation of such a film it can be used a weak organic acid diluted in the water of the tempering pool. It may appear as ridiculous a magic ritual the one followed by the world well famous sward makers in Toledo, Spain, during XIV and XV centuries. They had a secret practice that claimed that blond-haired children had to urinate in the water of the tempering pool early in the morning to obtain the best hardened steel.

As matter of fact, urine contains uric acid that is precisely a weak organic acid that avoid the formation of the vapor film. The children had to urinate early in the morning because when they got up had the bladder full. The reason for requiring blond-haired children still remains unclear to me. After steel is quenched, martensite forms that is usually very hard and brittle. Actually, this is the hardest condition for any given steel. The hardness of pure martensite depends uniquely on the carbon content and not on alloying elements. Therefore the maximum hardness, 69 HRC, is obtained in steel with only carbon, C = 0.9%. Bearing steel has 0.9–1.05% carbon. Since hardness is strictly related to strength, a steel composed of 100% martensite is at its strongest possible condition. Pure martensite is characterized by acicular plates of cementite, Fe_3C, as shown in Fig. 4.21a.

These plates offer a very effective barrier to dislocation motion increasing the yield strength of the material and reducing the deformation at fracture. Forasmuch the traction strength is increased by quenching, so too can the fatigue strength be improved.

However, the drawback of pure martensite is brittleness. The fracture of a single plate, quite easy under dynamic loads in particular, triggers the fracture of the entire workpiece. This is why a second step process is always required. To make the steel tough, it must be tempered. Tempering consists of reheating quenched steel to a suitable temperature, much below the critical temperature, for an appropriate time. This process reduces the strength but makes steel be tougher. A schematic *time–temperature* diagram for quenching and tempering heat treatment is shown in Fig. 5.21c. An example of this can be seen in Fig. 5.23 that shows the *S–N* curves obtained with cylindrical specimens of 0.45%C and 0.21% carbon steel, respectively, along with the microstructure in the longitudinal (L) and transverse direction (C) [25]. The 0.21% C steel had a normalizing treatment while the 0.45% C after normalizing was quenched and tempered.

The mechanical properties were 324 and 750 MPa lower yield strength, 498 and 833 MPa ultimate strength and 978 and 1510 MPa true fracture strength for the 0.21% C and 0.45% C steel, respectively. It can be seen the excellent grain refinement of the heat-treated 0.45% C steel that had a hardness of 270 HV and grain size much smaller

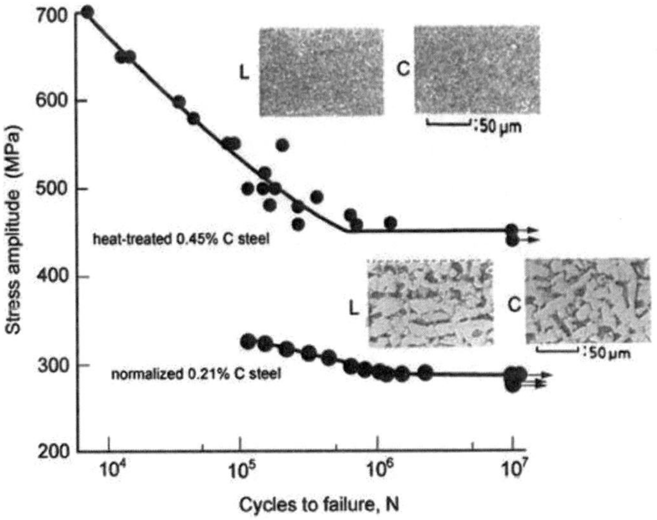

Fig. 5.23 Fatigue *S–N* curves and microstructures for normalized 0.21% C steel and heat-treated 0.45% C steel, respectively (composite drawing modified from [25])

compared to the normalized steel whose hardness was 145 HV. However, if the austenitization temperature in the hardening treatment is too high the resulting grain size may not be as mall as it should be. The smaller the grain size, the larger the fatigue strength, see Sect. 4.7. Figure 5.24a shows a typical brittle fracture of a coil spring made of high strength 55Cr3 quenched steel. The very hard and brittle non-tempered martensite has resulted in a multiple fracture that shed the metal into pieces. The same type of spring tempered after quenching broke with a single failure, Fig. 5.24b, revealing a helical appearance. During the quenching phase, when austenite turns into martensite, several processes occur that are responsible for generating residual stresses of different signs. First, rapid cooling of the surface causes a temperature gradient between the surface and the core. The surface that wants to cool faster is restrained by the core generating a state of residual traction stresses. But the rapid cooling also changes the surface material into martensite.

Martensite has a body-centered tetragonal structure (BCT) different from ferrite that has a body-centered cubic structure (BCC). BCT martensite has a coefficient of thermal expansion lower than that of the core material that still maintain a ferritic structure. Therefore, a system of compressive residual stresses develops wright on the surface counterbalancing the thermal gradient residual stresses. However, martensite has also a specific volume larger than ferrite so that the austenite–martensite transformation causes an expansion of the surface layer. But also in this case, the ferritic core will oppose that expansion generating a state of residual compression

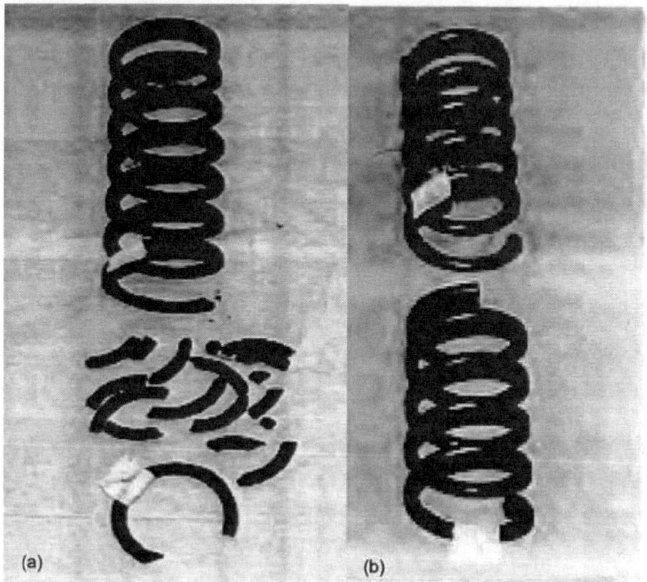

Fig. 5.24 Typical fatigue fracture in springs made of: **a** 55Cr3 quenched steel and **b** 55Cr3 quenched and tempered steel [25]

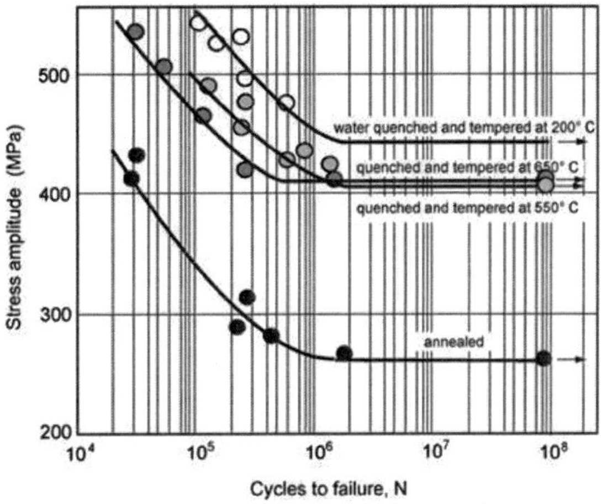

Fig. 5.25 *S–N* fatigue curves relative to a medium hard steel heat treated to four different conditions [26]

stresses. Finally, the last mechanism that contributes to residual stresses takes place during the final stage of cooling. While the surface layer has already reached its final martensitic structure, the core is still undergoing that transformation. Now, it is the core that, developing a martensitic structure, wants to expand releasing compression stresses previously generated. The overall process leaves the surface layer into a state of compression. But when pure martensite is tempered at high temperature, 550–700 °C, this compressive stress state is almost completely released.

This is why tempering a quenched steel at higher temperatures reduces the fatigue strength. This can be seen in Fig. 5.25 [26] for a medium-hard steel in four different heat-treated conditions, namely annealed and quenched and tempered at 200, 550 and 650 °C, respectively. The superior fatigue strength of low-temperature tempered martensite is evident. Thermal treatments such as quenching and tempering are normally used in the bearing industry, spring wires, beams, axles, gears etc. Figure 5.26 summarizes the hardness obtained by various heat treatments as a function of carbon content [27]. Not all aluminum alloys can be heat-treated. Non heat-treatable aluminum alloys are those with the following numbering designation: 2XXX, 6XXX and 7XXX.

5.4 Thermo-mechanical Treatments

Thermo-mechanical processes, in which the creation of a beneficial compressive residual stress state is produced by an imposed thermal gradient, are less known and practiced. Their use was introduced in the eighties of the last century in the nuclear industry to treat welds on large stainless steel pipes. The inner side of large pipes is left, after welding, in a high tension residual stress state, even approaching the yield strength of the material, and in a sensitize state, if it is made of austenitic stainless

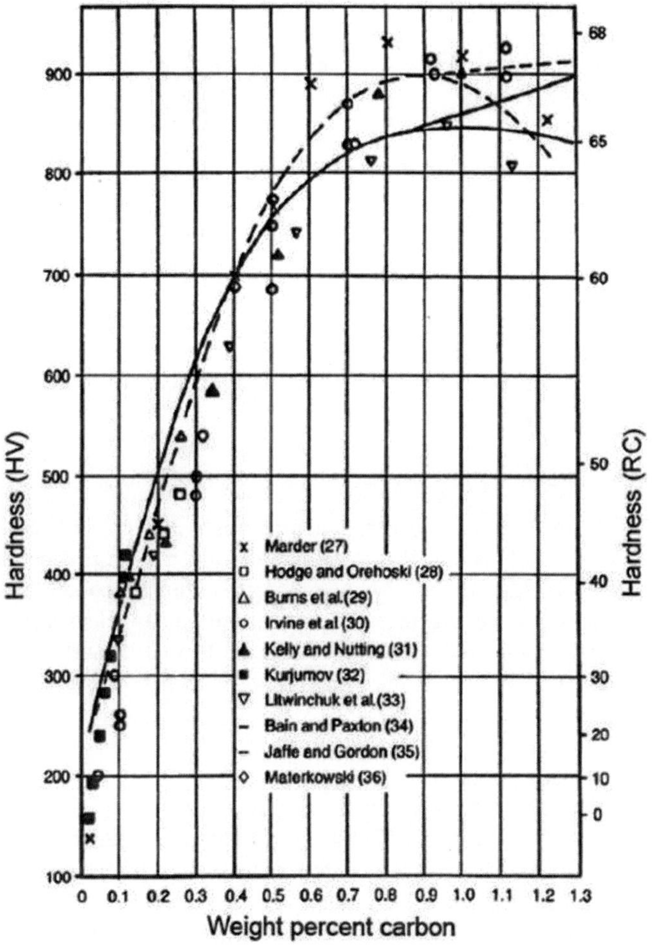

Fig. 5.26 Hardness of martensitic structures as function of steel carbon content [27]

steel. Both factors, high tensile stresses and sensitization, favor the occurrence of severe intergranular stress corrosion cracking (ISCC) when a fluid, even chemically pure water, is flowing inside, particularly at higher temperatures. Post-weld heat treatments cannot be performed on austenitic stainless steels after welding because the heat treatment would result in a massive sensitization of the all piping and not just of the heat-affected-zone (HAZ) (see Sect. 18.2). The rationale behind the process is shown schematically in Fig. 5.27. Assume that a steel wall, initially at uniform temperature T, has the upper surface heated at a temperature $T_e > T$ while the lower surface is kept at a temperature $T_i < T_e$ as to establish and maintain a temperature gradient $\Delta T = (T_e - T_i) > 0$. Because of the gradient, the beam will bent, Fig. 5.27a. If the bending is constrained, Fig. 5.27b, assume for example that the beam is the

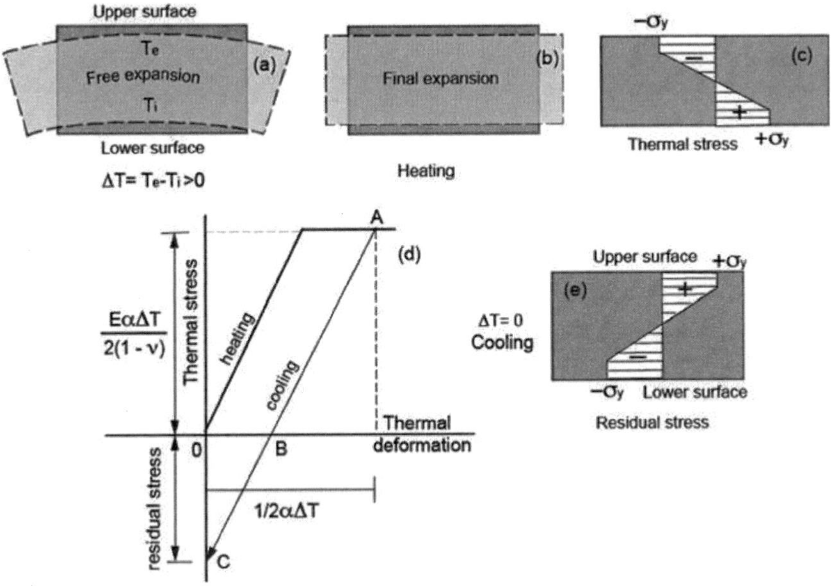

Fig. 5.27 Schematic of compressive residual stress generation by an imposed and restrained thermal gradient

wall of a pipe, then a stress gradient arises that will avoid the rotation of the section, Fig. 5.26c. Supposing that the thermal gradient is linear the stress is given by

$$\sigma = \pm\frac{E\alpha\Delta T}{2(1-\nu)} \tag{5.4}$$

where E is the Young's modulus of the metal, α the coefficient of thermal expansion of the material and ν its Poisson's coefficient. The stress is positive (traction) on the lower side and negative (compression) on the opposite side. Also assume that the material is elastic-perfectly-plastic, as shown in Fig. 5.27d and that the thermal gradient is so large as to start yielding just on the outermost fibers of the section, as shown in Fig. 5.27c. The elastic-perfectly-plastic assumption will not infringe the validity of the following considerations.

If thermal stresses were kept within the elastic field, upon removal of the thermal gradient they would return to zero, but if the temperature gradient ΔT is high enough to have produced a maximum stress, given by Eq. (5.4), beyond the elastic limit reaching point A in Fig. 5.27d and the entire plastically deformed zone is small compared to the volume of the core material that remains elastic, Fig. 5.27c, then the cool down at ambient temperature will force point A to move to C, by a shake-down effect, already discussed in Sect. 5.2.1 and shown in Fig. 5.1a and a residual compressive stress will remain on the external surface.

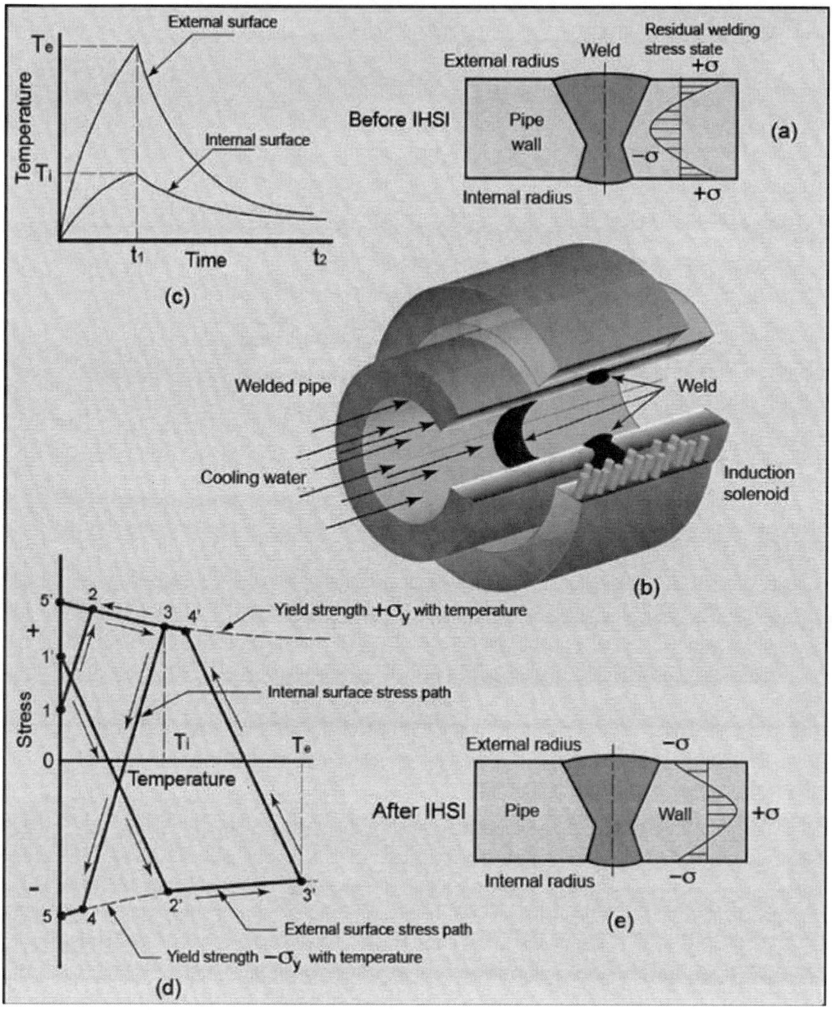

Fig. 5.28 Schematic of the IHSI treatment

Alike point A any other element of material gone beyond yield point will rest in a residual stress state, which will be compressive on one side and tensile on the other, as shown in Fig. 5.27e. This process was successfully applied for the first time in the eighties to treat large diameter austenitic stainless steel pipes of the primary cooling system of the BWR nuclear power plant of Caorso, in Italy. It is known as *induction heating stress improvement* (IHSI) [29–39]. The stress state introduced by butt-welding a large and thick pipe is shown in Fig. 5.28a. It is traction on both the external and the internal surface and compression at middle thickness. This tensile stress state on the internal surface is very dangerous since it may initiate a stress corrosion event, particularly in austenitic stainless steel sensitized by the welding process. In carbon steel it sustains fatigue damage. This residual stress state must be eliminated. IHSI consists of heating by an induction coil the external surface of a pipe, while the internal one, where compressive residual stress is desired, is cooled by a water flow and maintained at a temperature not higher than 95 °C, as shown in Fig. 5.28b. The temperature history is indicated in Fig. 5.28c. The thermal gradient so created puts under compression the warmer wall that wants to expand but is impeded by the internal cooler zones that enter into traction. This is schematized in Fig. 5.28d where the stress paths followed by a material point taken at an arbitrary point both on the external surface and the internal one, as well, is indicated. For the arbitrary point on the external skin the stress path starts from point $1'$ relative to the initial residual stress left by the weld process, Fig. 5.28d, and moves during heating towards $2'$ where yielding is reached.

Continuing the heating phase the maximum temperature T_e is reached and the stress of this hypothetic point moves along path $2'$–$3'$ following the decrease of yield strength with temperature, Fig. 5.28d. At this moment the induction heating terminates and the cooling of the system starts, Fig. 5.28c. The stress of the hypothetical point now inverts its motion following the path $3'$–$4'$ because the outer layer of material now wants to contract more than the core entering into a tensile stress state till yield is reached. From this point it follows the yield strength, which increases as temperature decreases till the final point $5'$. A final residual tensile stress is reached. The opposite happens for an arbitrary material point on the internal surface that will follow the stress path 1–2–3–4–5 remaining into a residual compressive stress state. The final situation is depicted in Fig. 5.28e. HISI may be applied any hollow geometry in which water can be flown to provide cooling while the induction coil is heating the external surface of the entire work piece.

5.5 Carburizing and Nitriding

Quench hardening of steel requires minimum carbon content, generally 0.15–0.2%. Below that level, steel cannot be quench hardened and requires carburization, see Fig. 5.26. Carburizing is the process by which carbon is introduced into a solid ferrous alloy in contact with a suitable carbonaceous substance that may be solid, liquid or gas. To be carburized the alloy is held for a convenient length of time at 850–950 °C in an environment containing carbon monoxide o directly carbon powder (generally 60% in weight charcoal and 40% barium carbonate) that penetrates by diffusion into surface layers. Figure 5.29a shows the carburizing effect on a gear tooth. The lighter layer represents the carburized zone or case. Its depth is about one millimeter, but fatigue, as we already know by now, is a process that initiates on the surface. The carburized alloy is usually quench hardened by single or double quenching: a first quenching is performed at 800 –900 °C in oil followed by a second at 750 –780 °C, generally followed by tempering at 150–80 °C. This treatment hardens the surface layer to about 700 HV leaving the core, non-hardenable because of low carbon

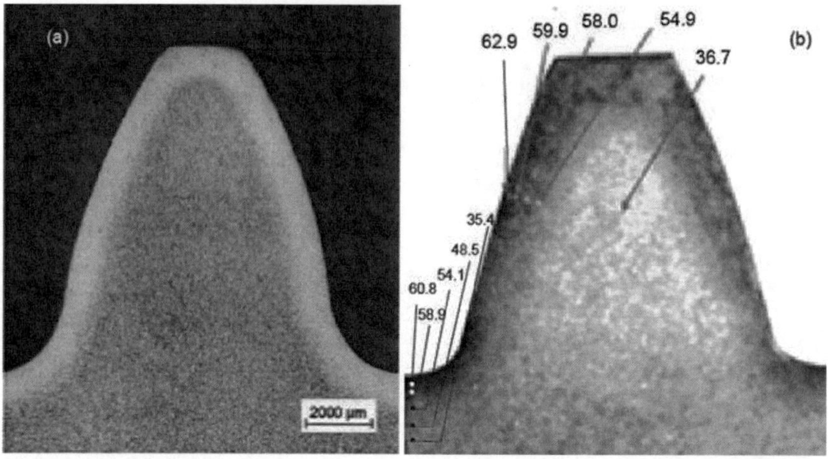

Fig. 5.29 **a** Carburized case of a gear that appears lighter, **b** nitriding of a gear showing hardness (HRC) at different points

content, still soft and ductile. Duration of caburizing determines the thickness of the hardenable layer, which may reach one millimeter. However, carburizing causes distortion when the rate of cooling is uneven, as the outermost surface cools down faster than the inner parts. In addition, the carburized part tends to be larger than before.

Therefore, carburized parts need to be machined. The second important thermochemical treatment is that of nitriding. Alike carburizing, nitriding is a process through which nitrogen is introduced into a solid ferrous alloy to create a case-hardened surface. The process was first developed in the early 1900s to avoid distortion caused by carburizing. The paternity of the process is still controversial. In the early years of the last century, the American metallurgist Adolf Machlet recognized that nitrogen was very soluble in iron where it produced relatively hard phases. He applied for a patent in 1908. In the same period, also Adolf Fry, in Germany, recognized that nitrogen was very soluble in iron at an elevated temperature. Fry first applied for his patent in 1921. However, Fry was more successful in Europe than Machlet in the States in convincing customers of the effectiveness of the nitriding process. The advantage that nitriding has over carburizing is that it does not requires a phase change from ferrite to austenite and eventually to martensite through a rapid cooling. In nitriding the metal is held at a proper temperature in contact with a suitable nitrogenous material, usually ammonia or molten cyanide. Generally, nitriding is performed after a quenching and tempering treatment.

The work piece is kept in a furnace at 500 °C for a period from 12 h to several days (50 h on the average) under an ammonia stream. Fundamentally, the dissolution of ammonia, NH_3, releases nascent nitrogen that, being a small atom like carbon, at high temperature diffuses into the surface layer of the steel for about 0.1 –0.2 mm.

Once in the alloy, nitrogen combines with iron, in particular, producing iron nitrides Fe_4N (γ' phase) and Fe_3N (ε phase) extremely hard. Steels containing aluminum are very prone to nitriding because of the affinity of aluminum for nitrogen forming very hard aluminum nitrides resulting in much a harder case. Nitriding, then, affects a less thick layer of metal than carburizing, but the hardness is by far higher, it can reach also 69 HRC.

From ammonia dissolution also hydrogen is released, which has a very severe corrosion potential. However, hydrogen is adsorbed by steel at much lower temperatures so that at such high temperatures needed for nitriding it can only leave the metal and not enter, without any negative effect. Nitriding, moreover, results in volume expansion that leaves the surface in a compressive stress state with further beneficial effects on fatigue strength. Figure 5.29b shows the ion-nitriding or plasma nitriding hardened layer in a carbon steel gear. Also indicated is the hardness in various points of the cast, in unit RC. Plasma or ion nitriding is a method of surface hardening that makes use of an electrical discharge to introduce elemental nitrogen in the metal. In a vacuum, high voltage electrical energy is used to form a plasma from a mixture of nitrogen and hydrogen. Plasma is a high temperature gas completely ionized, naming the technique. In the case of nitriding, the temperature range is 430–650 °C which results in very small dimensional deformation and distortion. An electric potential difference is maintained between the hardening workpiece and the plasma atmosphere favoring the nitrogen ions penetration into the metal. The process does not affect the surface finish.

With respect to normal nitriding, plasma nitriding creates a thicker case-hardened surface. An example of surface hardness increase and residual stress state induced in a CrMo steel, 1840 MPa ultimate strength, carbo-nitriding is offered in Fig. 5.30 [33]. Carbo-nitriding is more of a carburizing process than a nitriding one. small amounts of nitrogen are added to the gas carburizing atmosphere allowing both the nitrogen and carbon to diffuse into the component simultaneously. The process is done at temperatures around 870 °C which is slightly lower than the normal carburizing temperature. The added nitrogen increases the hardness of the component surface. the heating and cooling time of both carburizing and carbo-nitriding processes results in different degree of case depth and hardness. Figure 5.31 presents *S–N* fatigue curves for a CrMoV steel, 1100 MPa ultimate strength, used for nitriding. S–N curves were obtained using cylindrical specimens carrying a circumferential notch of 0.5 mm root radius under reversed bending. It is possible to recognize the significant fatigue limit increase due to nitriding and in particular to the beneficial effects of high compression residual stresses, which almost double. Furthermore, it is also interesting to note the trend inversion around 10^3 cycles where the much harder steel (skin) loses its fatigue resistance with respect to the softer one, as explained in Sect. 1.2 and shown in Fig. 1.16.

However, it must be remembered that the compressive residual stress state set up on the surface must be balanced by a tensile residual stress state. Once again, according to Saint Venant's principle this reaction should die within two or three perturbation distances. Since the perturbation is the surface layer under compression that is no larger than 0.2–0.4 mm, residual tensile stresses too will decease within

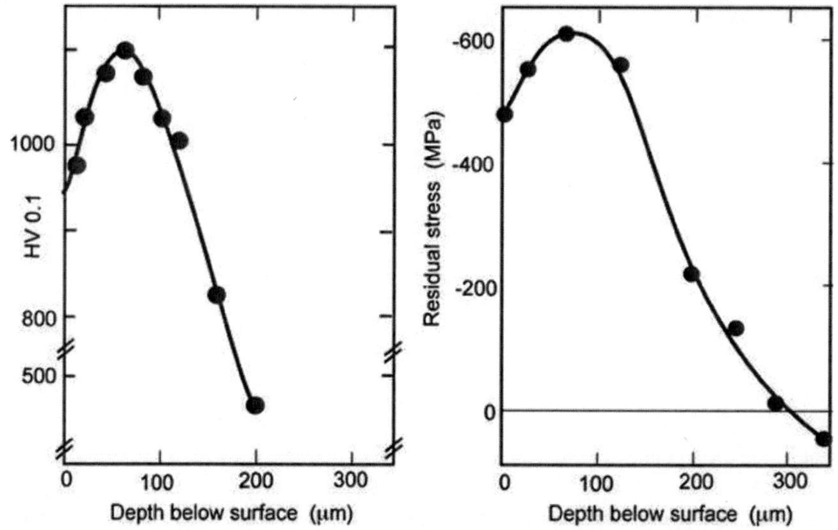

Fig. 5.30 Profile of Vickers Hardness and residual stress state in the surface layer of a nitricarbonised CrMo steel (adapted from [33])

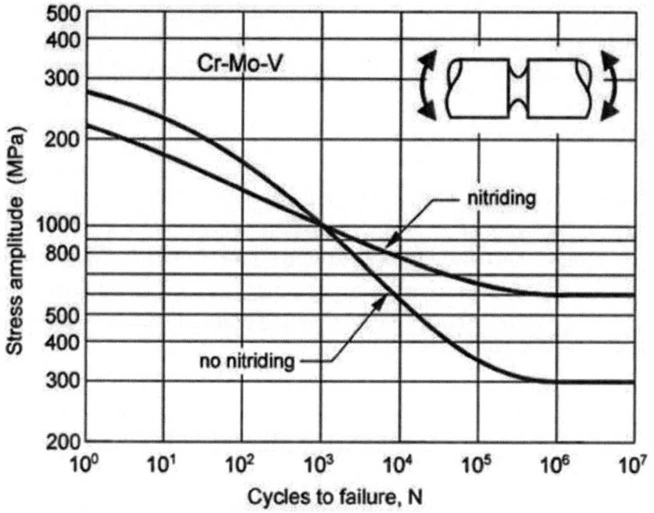

Fig. 5.31 *S–N* fatigue curves obtained on CrMoV steel specimens carrying a circumferential notch under reversed bending, with and without nitriding

a short range, independently of the core specimen or work piece dimension. This actually means that the tensile peak stress at the interface with compressive stresses may be quite high and extend for as much as a millimeter. By summing up to the applied stress, this may lead to the already cited episode of fatigue failure initiated by a subsurface inclusion, known as fish-eye (see also Figs. 2.19 and 2.29). Figure 5.32 [34] is a clear example of such an event. It shows a specimen of 34CrNiMo6 steel, 1100 MPa yield strength and 1250 MPa tensile strength that failed under push–pull fatigue from a subsurface inclusion located right in the residual traction stress zone. The small sub-surface inclusions that originated fatigue failure may be seen at the center of the dark round area and in the enlarged SEN image.

The specimens had been surface hardened by ion-nitriding treatment to about 900 $HV_{0.2}$ from an initial non-treated hardness of about 365 HV. The hardening effect vanished at about 400 μm depth. The treatment had increased the fatigue life from ~ 475 to ~ 600 MPa and the fretting fatigue limit, as well, from below 350 MPa to about 600 MPa, as shown in Fig. 5.33 [33]. Another interesting example of fish-eye fatigue failure is shown in Fig. 5.34 [34].

Here the material was plasma-nitrided high-strength low-alloy Cr-Al-Mo steel (EN 37CrAlMo6) annealed, quenched and tempered. The nitrided case was about

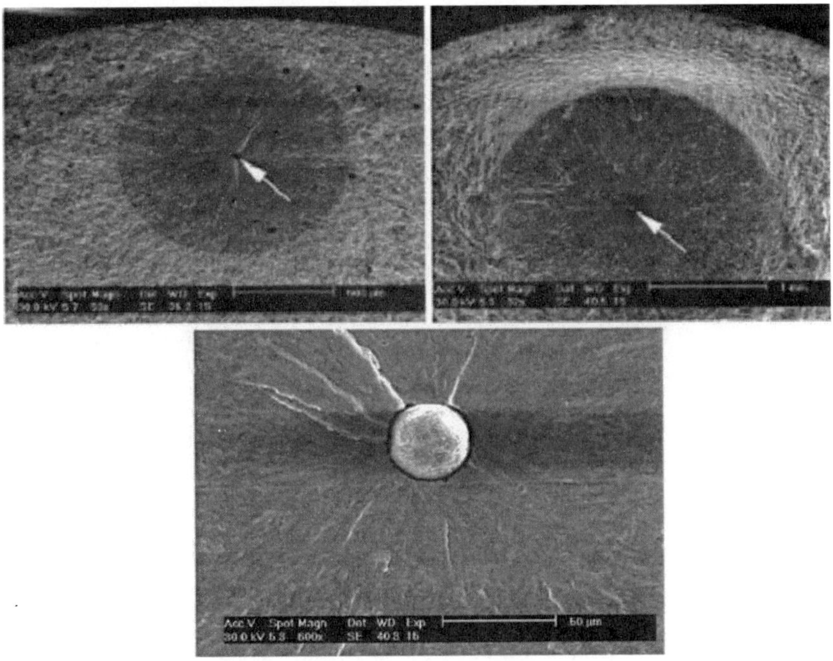

Fig. 5.32 SEM view at increasing enlargements of the subsurface crack nucleation from an inclusion in a fatigue specimen of 34CrNiMo6 steel, surface hardened by ion-nitriding treatment [33]

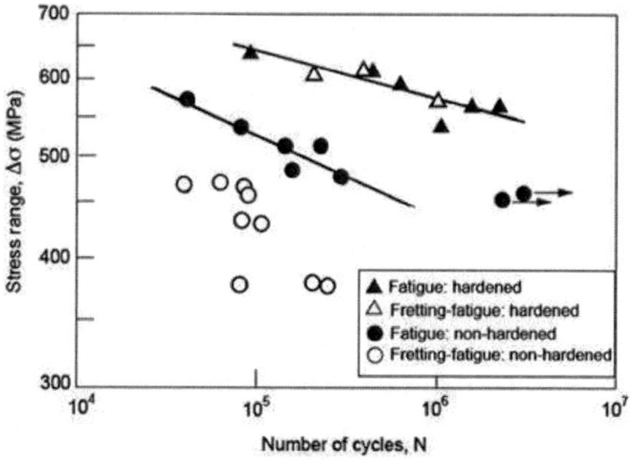

Fig. 5.33 Fatigue and fretting fatigue observed for 34CrNiMo6 steel specimens non-treated and after nitriding (adapted from [33])

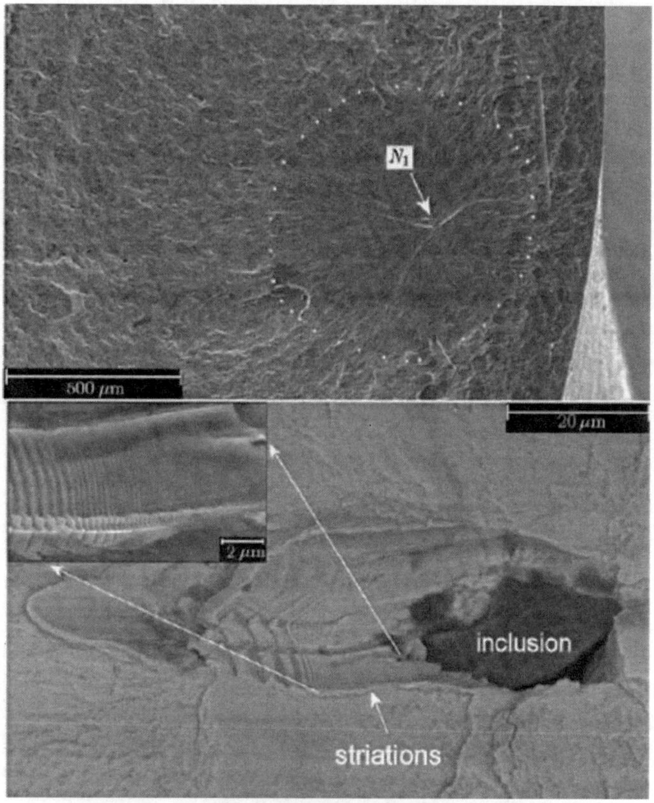

Fig. 5.34 SEM image of fish-eye fatigue failure in nitrided steel under combined bending-torsion loading. The enlarged view shows the inclusion with signs of striations beginning right at the inclusion border [34]

200 μm thick. The SEM examination revealed the presence of an inclusion located below the diffusion zone at about 470 μm. Measurements of residual stresses had shown a zero or very low traction stress existing at that depth. If there was any residual traction stress acting at the inclusion site, certainly the protective compression stress state exerted by nitriding was extinguished. Therefore the inclusion at the center of the fish eye zone, of about 30 μm shown in the detail of Fig. 5.34, could be effective in the fatigue process. What makes this case be interesting is the fact that, at variance with ODA formation examined in Sect. 2.4, Fig. 2.29, that we have described as due to a hydrogen embrittlement process and not to fatigue per se and to the case shown in Fig. 5.32, here there are clear signs of fatigue striations, as it can be seen in the enlarged SEM fractography of Fig. 5.34 [34]. Another advantage of nitriding with respect to carburizing is that the former lasts also at temperature up to 500 °C while the hardening of a carburized surface layer start to be lost at temperatures of about 200 °C. Even more, nitriding is the last treatment of an entire production cycle while the elevated temperature of carbonizing must be followed by quenching and tempering to get the wanted hardening.

It may also result in sensible deformations so it must be followed by grinding and subsequent surface finishing that will be performed on a very hard work piece. All what has been said in the previous Sections and this last about mechanical and thermal treatments and their effect on fatigue resistance of materials has been reassumed in Tables 5.2, 5.3 and 5.4, respectively.

Table 5.2 indicates that the fatigue limit σ_f in symmetric traction is about 0.5–0.5 the ultimate strength σ_u with a minimum of 0.32 for high strength steels and a

Table 5.2 σ_f/σ_u ratio for various steels relative to ultimate strength and type of applied cyclic stress

Steel	Ultimate strength σ_u (Mpa)	σ_f/σ_u ratio		
		Bending	Traction	Torsion
Mild steel (0.15% C)	400	0.55	0.5	0.34
Annealed steel with medium C	650	0.47	0.42	0.27
Quen. and temp. alloyed steel	1000	0.45	0.40	0.26
High strength steel	1600	0.35	0.32	0.19

Table 5.3 σ_f/σ_u ratio in steels under rotating bending relative to metallurgical structure

Structure	σ_f/σ_u
Perlite	0.40
Martensite	0.25–0.40
Kainite	0.48
Austenite (austenitic steels)	0.50
Tempered martensite	0.52
Quenced and tempered steels	0.55
Ferrite	0.60

Table 5.4 Fatigue strength increment obtainable with various process

Process	Application	Increment (%)
Nitriding	Smooth surfaces	15 –25
	Notched members	40 –80
Carbonizing and quenching	Gears	50
Quenching	Raw shafts	50–80
Rolling	Cylindrical work pieces	10–20
	Fillets and groves	30–50
	Smooth surfaces	10–20
Shot peening	Raw surfaces	50–80
	Notched members	30–40

maximum of 0.5 for mild steel. In cyclic bending this ratio is a little bit higher and maintains the same trend. In cyclic torsion it is much lower. Metallurgical structure, as said, plays a fundamental role, as indicated in Table 5.3. Apparently, the best structure from the point of view of fatigue strength is ferrite with a fatigue limit σ_f about 0.6 the ultimate strength σ_u. Unfortunately, ferrite has a rather low ultimate strength σ_u that sharply reduces this benefit limiting the final fatigue limit σ_f so that the best performing metallurgical structure is definitely that resulting from quenching and tempering alloyed steels. But it must be recalled (see Sect. 1.3) that mild steel generally harden under cyclic stress while hard steel softens so that with respect to cyclic properties the ratio σ_f/σ_u becomes almost the same and equal to about 0.5. Finally, Table 5.4 presents the effect various surface treatments on fatigue strength. It may appear singular the fact that best performances of nitriding and shot peening, up to 80% increase in fatigue limit, are obtained on notched members.

5.6 Plating and Coating

Plating is a process by which an adherent layer of metal is formed upon a work piece, specimen or component. Typically, metallic component are coated for a couple of reasons. First it may enhance the esthetic appearance of the component and second it provides protection against a hostile environment. The designer should be aware that both processes reduce more or less the fatigue resistance of components. Plating can be detrimental to fatigue strength because the plate material may be in a state of residual tension. Zinc and cadmium plating seem to have a relatively small effect in reducing fatigue strength whereas nickel and chromium plating have a substantial detrimental effect. In general, the thicker the layer of plating or coating, the more detrimental is the effect on fatigue properties. The damage is particularly relevant on parts made of high strength hard material. Figure 5.35 [36] is a clear example of such a damage operated by chromium plating on 4140 steel. Fatigue strength reduction is due to a residual tensile stress state following chromium plating. The same effect is seen with nickel plating. This is shown in Fig. 5.36 for a steel having 440 MPa ultimate strength [19]. Comparison is made with unplated steel and nickel. Since residual tensile stress is the reason of fatigue strength reduction, it is clear that any annealing or shot peening treatment that eliminate residual stress may have a beneficial effect.

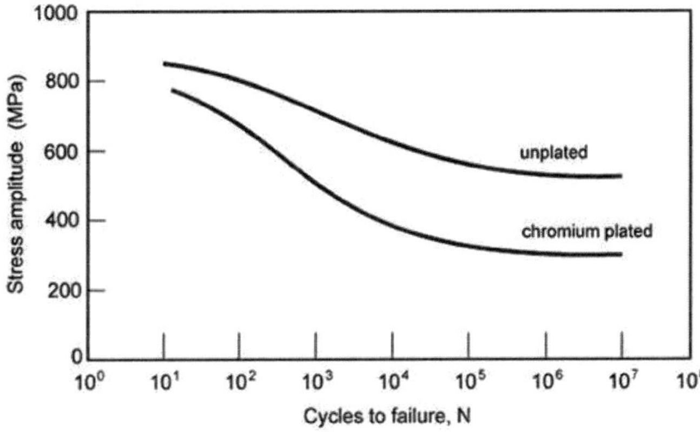

Fig. 5.35 Effect of chromium plating on fatigue strength of 4140 steel [36]

Fig. 5.36 Effect of nickel plating on fatigue strength of a steel having 440 MPa ultimate strength [19]

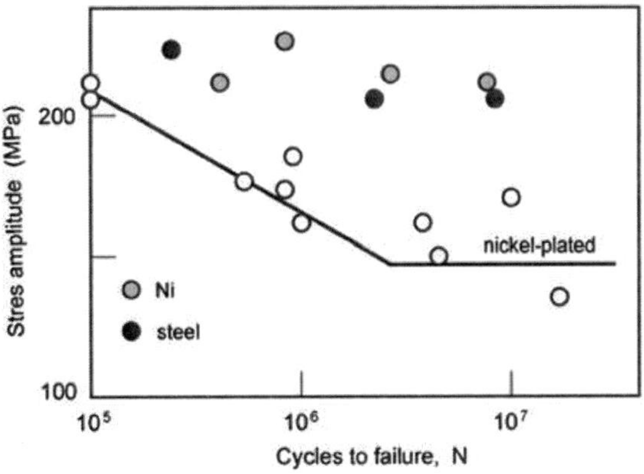

5.7 Effect of Temperature

Fatigue strength depends on the ultimate strength of materials (see Fig. 4.37). Any factor, then, having an effect on the ultimate strength shall modify the fatigue strength. This is the case of temperature. Generally, ultimate strength σ_u increases exponentially with decreasing temperature from about 0.3–0.4 times the melting temperature T_M of the metal. Above this last temperature range usually a sort of plateau is observed followed by a sudden drop of the mechanical characteristics of metals.

This happens for any metal whatever the structure may be: Body Centered Cubic (BCC), as in α-iron and ferritic steels, Face Centered Cubic (FCC), as in nickel, aluminum, copper, gold or silver or close-packed hexagonal, as in titanium and tin. A typical example of mechanical properties dependence on temperature is presented in Fig. 5.37 for a NiMoV steel forging. It can be noted how ultimate strength is varying from about 850 MPa at 200 °C to about 1150 MPa at – 180 °C. Accordingly, the fatigue strength, σ_f or S_f, should present a parallel increase and the fatigue limit shall be expected to grow from about 400 MPa to about 600 MPa, as for Table 5.2. Even more remarkable is the growth rate of the yield strength σ_y in the same temperature interval that, starting from a value much lower than ultimate strength, eventually reaches it at the very low temperature of about – 200 °C. The coincidence of yield and ultimate strength means the complete absence of ductility, as it can be seen in Fig. 5.37b at temperature T_1.

Ductility appears and develops in the temperature range T_1–T_2. Also shown is the true rupture stress $\sigma_{t,r}$ (see Eq. 10.25) that jumps and differentiate from the ultimate strength with the appearance of the first ductility. Another example of temperature dependence of mechanical properties is shown in Fig. 5.38 for cast molybdenum [37]. Again, it can be caught the significant increase of the ultimate strength σ_u with temperature decrease, which surges from 220 MPa at 600–700 MPa at – 100 °C. Together with ultimate strength also fatigue strength σ_f increases from 125 MPa at 600–300 MPa at 25 °C. Interesting enough, the ratio σ_f/σ_u remains constant over a large range of temperatures, dropping above 500 °C.

Figure 5.39 is a composite picture of fatigue strength versus temperature prepared by the author with data from open literature [27, 37–44] for a large variety of steels, ferritic, alloyed martensitic or austenitic, nimonic alloys, aluminum alloys and copper. Some characteristic feature may be caught. First, the incredible recovery of fatigue limit of ferritic steels both at very low and high temperature as well. The fatigue strength of these steels, in fact, is rather low at room temperature, but becomes so much high at very low temperature between – 100 and – 200 °C to match the fatigue limit of high strength steels. However, this low temperature fatigue strength shall not be misleading the designer because it is accompanied by a complete absence of ductility, as already seen in Fig. 5.37 and shown schematically in Fig. 5.40.

This actually means that at these low temperatures ferritic steels are so brittle that may fracture under static loads well below the elastic limit if just a tiny crack is present. This tiny crack may well be nucleated by fatigue cycling. This is why the use of ferritic steels is absolutely not recommended at low temperatures and, in

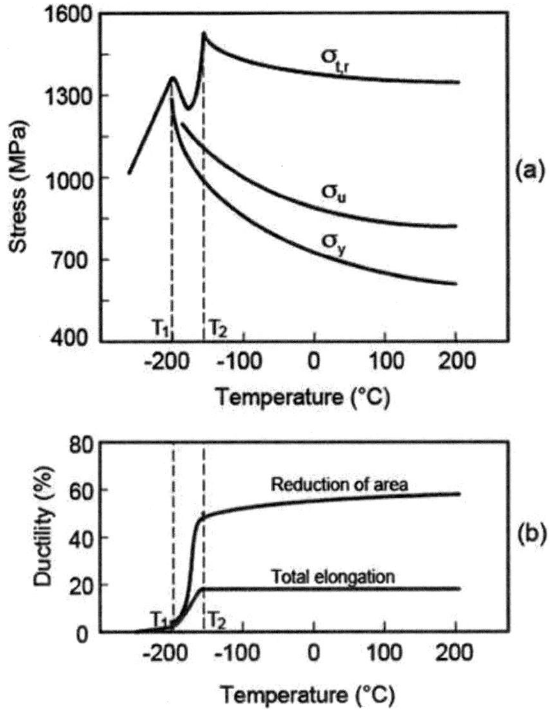

Fig. 5.37 **a** Yield strength σ_y, ultimate σ_u and true rupture stress versus temperature, **b** reduction of area (RA) and total deformation versus temperature for NiMoV forging

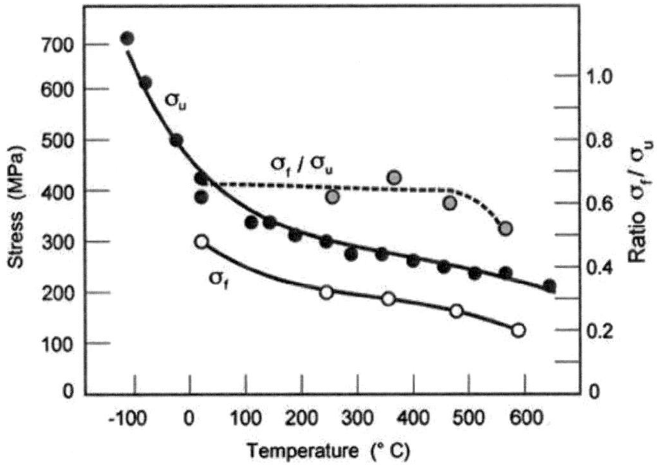

Fig. 5.38 Temperature variation of ultimate strength σ_u and fatigue strength σ_f in cast molybdenum [37]

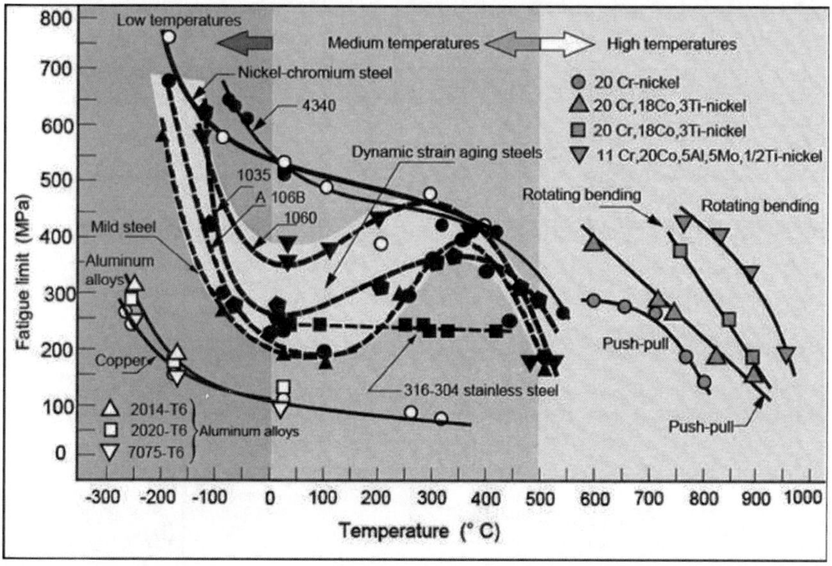

Fig. 5.39 Fatigue limit versus temperature of steels (ferritic, martensitic or austenitic), nimonic alloys, aluminum alloys and copper (data from [27, 37–44])

Fig. 5.40 Temperature effect on stress–strain curve of BCC materials

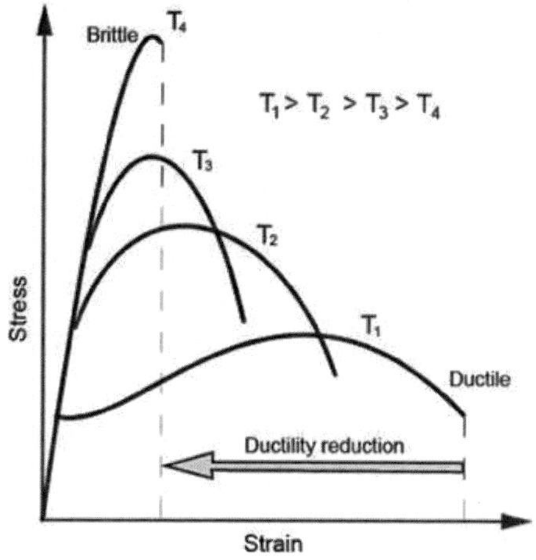

particular, for cryogenic applications. But even more surprising is the recovery that these steels show between 100 and 300 °C where they once again match the high strength steel fatigue performance. Afterwards, the fatigue limit slumps dramatically with temperature. This incredible strength recovery is due to dynamic strain aging. Dynamic strain aging (DSA) occurs in the range of temperatures between 100 and 300 °C owing to precipitation hardening of carbides as Fe_2C (carbide ε) or azides as and especially $Fe_{16}N_2$ (azide α″) and, as minor effect, of oxygen. This precipitation hardening occurs immediately, wright during the loading phase, that's why it is called *dynamic*. It may also be called blue brittleness because it happens in a temperature range where the metal covers by an oxide film whose color is blue. Figure 5.41 [45] presents a series of traction tests performed on a low carbon steel, 0.035%C, at different temperatures ranging from 50 to 350 °C. Already at 84 °C a first sign of a DSA effect can be seen which reaches its maximum at 185 °C. Serrations that characterize DSA, are evident on the stress strain diagram. Yielding has not been affected that much but ultimate strength, instead, has grown by some 50%. At 250 °C the DSA effect is still present, dough without serrations. At 346 °C the phenomenon is practically disappeared. Beyond that temperature restoration prevails over aging. The temperature dependence of fatigue limit of ferritic steels, having a BCC crystal structure, can be also understood from Fig. 5.40. These metals have a flow stress that consists of two basic component named: *athermal flow stress component* and *thermally activated component*, respectively

$$\sigma_y = \sigma_{th} + \sigma_E \tag{5.5}$$

were σ_{th} is the temperature dependent component of the yield strength, σ_y, and σ_E the athermal one that depends on dislocation structure existing in the metal.

The subscript *E* indicates that the only possible temperature dependence is through the Young's modulus *E*. While the yield strength decreases with temperature, ductility increases, as shown in Fig. 5.40. Opposite behavior may have FCC metals like austenitic stainless steels, aluminum and its alloys or copper, silver etc. in which flow stress has only the athermal component σ_E. The characteristic stress–strain relationship for this family of metals is shown in Fig. 5.42. Now, temperature has little or no influence at all on yield strength but ductility may even increase, which gives these metals a formidable capability to be used at cryogenic temperatures.

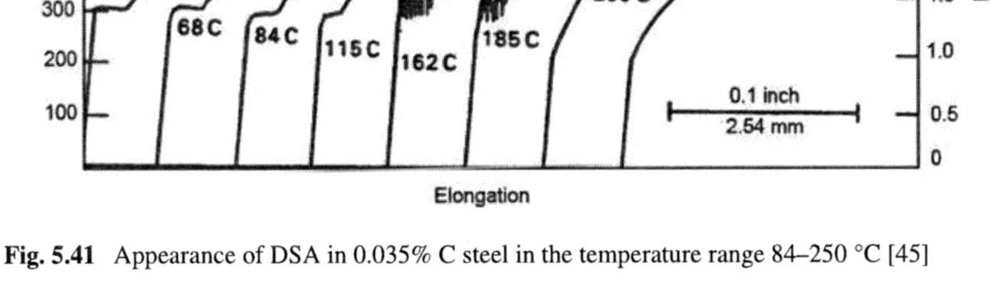

Fig. 5.41 Appearance of DSA in 0.035% C steel in the temperature range 84–250 °C [45]

Figure 5.43 is an example of true stress-true strain curves obtained with silver specimens of 17 μm grain size [45]. Note the enormous increase of ductility passing from 873 K (600 °C) to 20 K (– 253 °C). The appearance of dynamic strain aging in ferritic steels may alter the general trend indicated by Eq. (5.5). This is shown in Fig. 5.44 [46] for A 106B carbon steel that between 250 and 300 °C presents an unexpected upswing of ultimate strength, which becomes much higher than that at room temperature.

Also hardness increases with temperature instead of decreasing and this is another direct evidence of dynamic strain aging. If the tensile strength increases, then also the fatigue strength must increase. This can be seen in Fig. 5.45 [46] for a 0.17% C carbon steel. Monotonic ultimate strength σ_u is growing with temperature indicating that dynamic strain aging is taking place between 250 and 300 °C. But also the fatigue

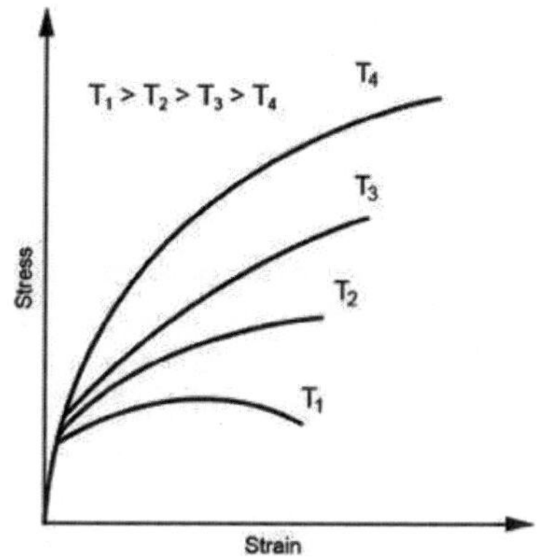

Fig. 5.42 Temperature effect on stress–strain curve of FCC materials

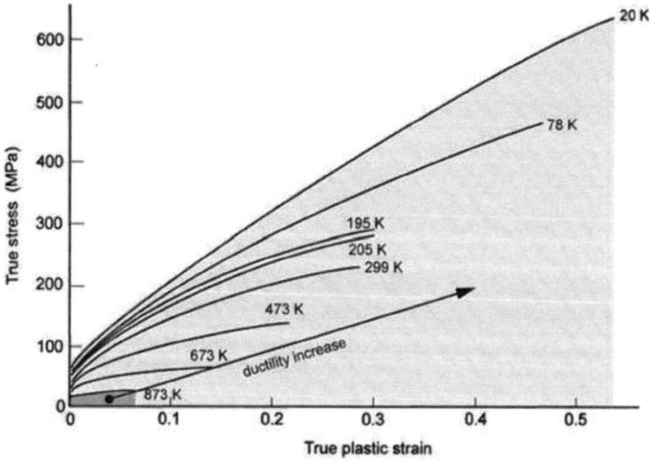

Fig. 5.43 True stress-true strain curves for silver specimens of 17 μm grain size [45]

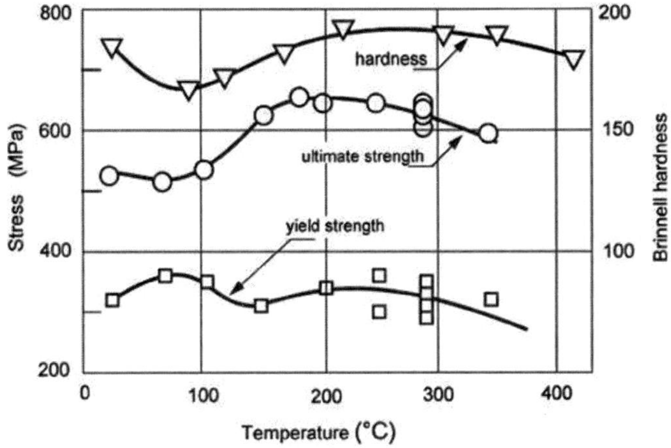

Fig. 5.44 Mechanical characteristics variation with temperature of A 106B 0.17% C carbon steel (adapted from [46])

strength σ_a or σ_f of the material has a parallel growth. This growth is actually more complex than expected because it introduces another factor of dependency: the strain rate $\dot{\varepsilon} = d\varepsilon/dt$. At 500,000 cycles and a cyclic frequency of 0.17 Hz, the cyclic stress amplitude σ_a is following almost the same trend of ultimate strength σ_u, but increasing the frequency to 33 Hz makes the maximum response shift of about 50 °C. This is because high frequency cyclic loads, reducing the time frame dt during which the maximum strain is reached, increase $\dot{\varepsilon}$. The same effect is seen on the fatigue limitσ_f of the material at 10^6 cycles. Another example of dynamic strain aging is offered in Fig. 5.46 [46] for a 0.58% C normalized steel.

Fig. 5.45 Variation of ultimate, yield and fatigue strength of 0.17% C steel with temperature and frequency of cyclic load application (adapted from [46])

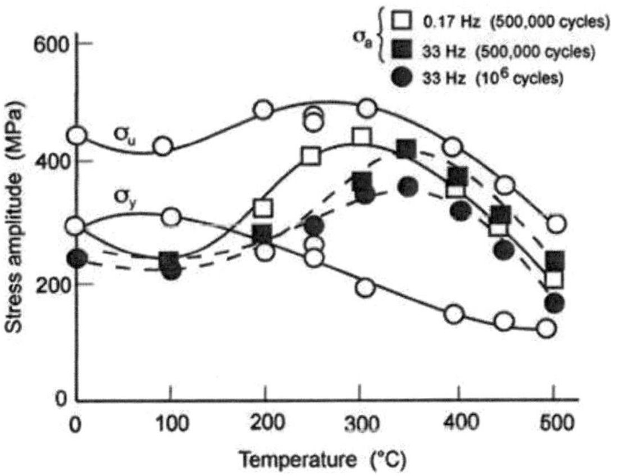

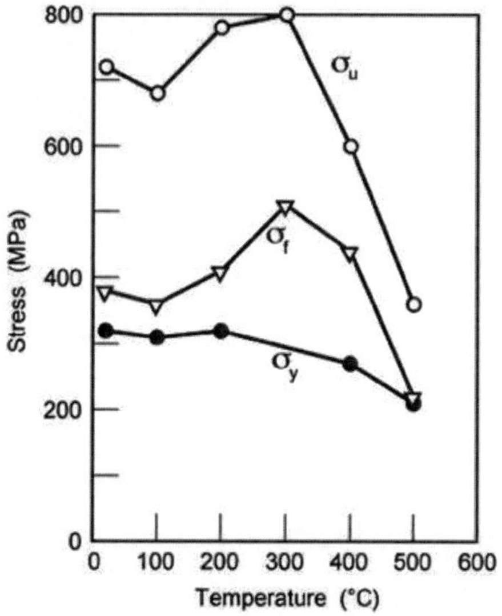

Fig. 5.46 Variation of yield, ultimate and fatigue strength for 0.58% C normalized steel as function of test temperature, according to Hempel and Tilmann [42]

Both ultimate strength σ_u and push–pull fatigue limit σ_f present a maximum at about 300 °C, afterward they sharply reduce because temperature solubilizes precipitated interstitials atoms hardening the metal. Interesting is also Fig. 5.47 [42] showing S–N fatigue curves of a 0.50% C normalized steel under traction fatigue pulsating from zero for various temperatures. The figure is relevant because it shows how the S–N curves go up and down in an apparently illogical fashion. Increasing the temperature from room temperature the fatigue strength of the material decreases at 100 °C, but then it returns to increase instead of decreasing as normally expected. But again, after a maximum performance at 300 °C it drops at 400 °C and continues to diminish at 500 °C. Dynamic strain aging is responsible for this up and down trend of the S–N curves. It can be possible to establish that the maximum hardening is reached by the steel at 300 °C.

This hardening may create some problems to the steel during cooling to room temperature when it is used at temperatures at which dynamic strain aging appears. In fact, the new high hardness that at 300 °C is not dangerous because the material is ductile, may instead mean low ductility and brittle behavior when the metal returns to room temperature during the cooling phase. Moreover, stress gradients accompanying the cooling phase put the surface layer in a high traction stress state while the toughness reduces with temperature.

The combined result may be brittle fracture especially if the fatigue cycling has provided the starter crack. This is why dynamic strain hardening steels are stabilized with the addition of alloying elements such as manganese and aluminum that combine with carbon, oxygen and, above all, nitrogen depleting the steel matrix and avoiding precipitation hardening. Also austenitic stainless steels are subjected to dynamic

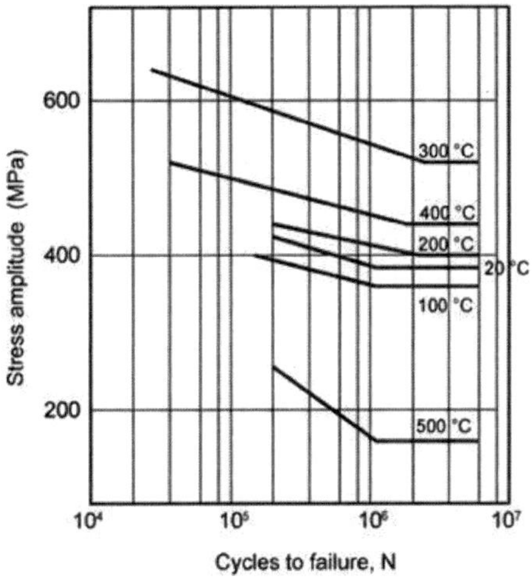

Fig. 5.47 *S–N* curves in cyclic traction pulsating from zero for 0.58%C normalized steel as function of temperature varying from 20 to 500 °C, according to Hempel and Tilmann [42]

strain aging that explains the flat trend of their fatigue limit with temperature between 25 and 430 °C that can be observed in Fig. 5.48 for types 304 and 316 [45].

These steels, in fact, in that particular temperature range undergo a rather heavy cyclic hardening from the partial transformation of soft metastable austenite into hard martensite. The consequence is that fatigue limit does not change with temperature. Actually, it is not only the fatigue limit that for these two steels remains constant, but the entire *S–N* curve, as Fig. 5.48 illustrates. The experimental data practically overlap at any temperature between 25 and 430 °C. The last feature of Fig. 5.39 to be discussed is the excellent high temperature fatigue behavior of nickel alloys or nimonic alloys. Above 400 °C, in fact, mechanical properties of steels, not only ferritic but also alloyed quenched and tempered steels, reduce progressively, becoming less and less capable to withstand cyclic loads and fatigue. Even nickel–chromium steels at 500 °C have a fatigue limit no higher than 250 MPa. Martensitic steel cannot do better, as it can be seen in Fig. 5.49 [41] for 4340 steel, 1120 MPa ultimate strength at room temperature. Nimonic alloys enter the space left open by these steels. It can be seen how nickel alloy type 11Cr,20Co,5Al,5Mo,½ Ti under rotating bending has a strength similar to that of 4340, but shifted by about 350 °C.

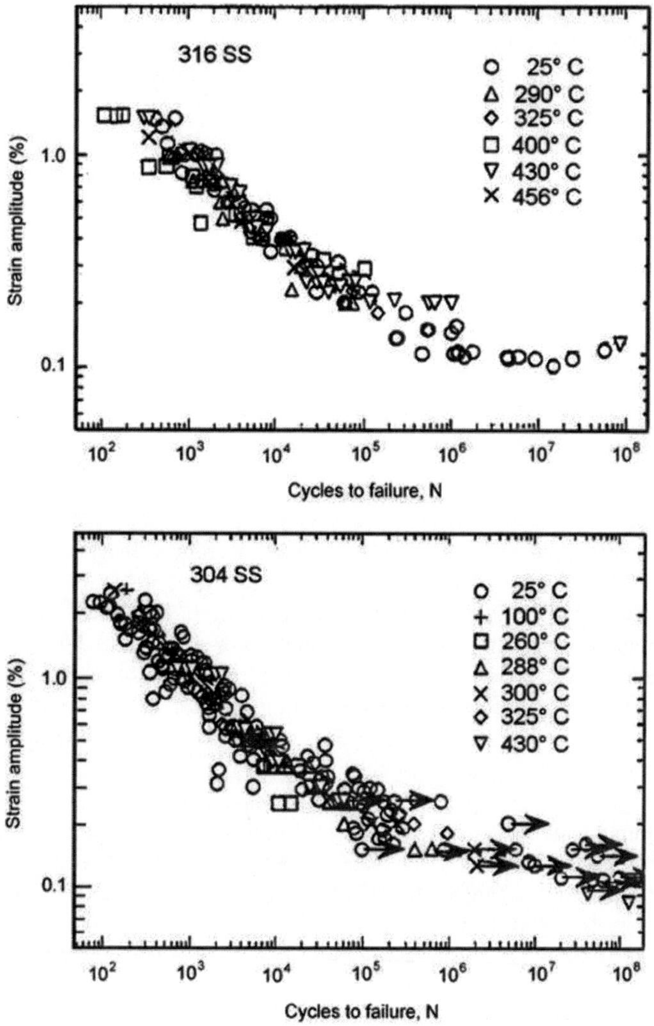

Fig. 5.48 *S–N* fatigue curves for 316 and 304 austenitic stainless steel in the temperature range 25–430 °C [44]

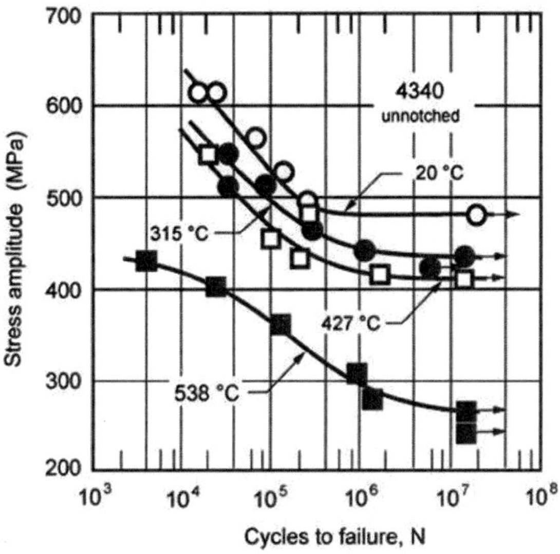

Fig. 5.49 Effect of temperature on the fatigue strength 4340 steel under symmetric traction-compression [41]

References

1. Houdremont, E., Mailänder, R.: Bending fatigue tests on steels. Stähl and Eisen **49**, 833–892 (1929)
2. Föppl, O.: Sthal u. Eisen **49**, 775 (1929)
3. Thum, A.: Die Entlwickung der Lehre von der Gestaltfestigkeit. Z. VDI **77**, 1337 (1933)
4. Thum, A., Ochs, O.: Z. VDI **76**, 951 (1932)
5. Horger, O.J.: Effect of surface rolling on the fatigue strength of steel. J. Appl. Mech. **57**, A-128 (1935)
6. Lyst, J.O.: The Effect of Residual Strain upon the Rotating Beam Fatigue Properties of Same Aluminum Alloys, Technical report No. 9-60-34, Alcoa, Pittsburgh (1960)
7. Fuchs, H.O., Stephens, R.I.: Metal Fatigue in Engineering. Wiley, New York (1980)
8. Heydekampf, G.S.V.: The Iron Age, vol. 775 (1930)
9. Fahrenhorst, W., Sachs, G.: Metallwirtsch **10**, 738–880 (1931)
10. Almen, J.O., Black, P.H.: Residual Stresses and Fatigue in Metals. McGraw-Hill, New York (1963)
11. Prevéy, P.S., Jayaraman, N.: Overview of Low Plasticity Burnishing for Mitigation of Fatigue Damage Mechanisms, Dr. John Cammett, Proceedings of ICSP 9 (Paper 260) Paris, Marne la Vallee, France, Sept. 6–9 (2005)
12. Kloos, K.H., Fuchsbauer, B., Adelmann, J.: Fatigue properties of specimens similar to components deep rolled under optimized conditions. Int. J. Fatig. **9**, 35–42 (1987)
13. Berstain, G., Fuchsbauer, B.: Z. Werkstofftech **14**, 103 (1982)
14. Franz, H.E.: X-ray measurements of residual stresses after surface machining of Ti6Al4V and Ti6Al6V2Sn (in German). VereinDeutscherIngenieure, Dusseldorf, VDI, Berichte **313**, 453–462 (1978)
15. Nelson, D.V., Ricklefs, R.V., Evans, W.P.: The role of residual stresses in increasing long life fatigue strength of notched machine members, achievements of high fatigue resistance in metals and alloys. ASTM STP **467**, 228–253 (1970)
16. Fatigue Design Handbook, AE-10, 2nd edn., vol. 78. Society of Automotive Engineers, Inc. (1988)

17. Zimmerly, F.P.: Mach. Des. **12**, 62 (1940)
18. Maker, J.H.: Failure Analysis and Prevention, ASM Handbook, Vol. 11. Fatigue of Springs (2002)
19. Almen, J.O., Black, P.H.: Residual Stresses and Fatigue in Metals. McGraw-Hill Book Co. (1963)
20. Fatigue Design Handbook, AE-10, 2nd edn., vol. 77. Society of Automotive Engineers, Inc. (1988)
21. Niku-Lari, A.: An Overview of Shot-Peening, International Conference on Shot Peening and Blast Cleaning
22. Heibach, E.: FKM-Guideline: Analytical Strength Assessment of Components in Mechanical Engineering, 5th Rev. edn., Frankfurt (2003)
23. Adamaszek, K., Broz, P.: Decarburization and hardness changes in carbon steels caused by high temperature surface oxidationin ambient air. Diffus. Defect Data: Defect Diffus. Forum **144**, 1701–1076 (2001)
24. Cazaux, R., Persoz, L.: La Fatigue de Métaux. Dunod, Paris (1937)
25. Hattingh, D.G.: The Fatigue Properties of Spring Steel, PHD Thesis, University of Plymouth, March (1998)
26. Goto, M.: Statistical investigation of the behaviour of microcracks in carbon steels. Fatigue Fract. Eng. Mater. Struct. **14**(8), 835 (1991)
27. Cazaux, R.: Fatigue of Metals. Chapman and Hall, London (1953)
28. Krauss, G.: Martensitic transformation, structure and properties in hardenable steels. In: Doane, D.V., Kirkaldy, J.S. (eds.) Hardenability Concepts with applications to Steels. AIME Warrendale, PA, 229–248 (1978)
29. Shimizu, T., Enomoto, K., Araki, S., Ikegami, T.: Induction heating stress improvement for welded pipes and its effectiveness. In: EPRI Seminar on Countermeasures for BWR Pipe Cracking, Vol. O, Session 3, January 22–24 (1980)
30. Tanaka, S., Umemoto, T.: Residual Stress Improvement by Means of Induction Heating, Ibidem, Vol. O, Session 3, January 22–24 (1980)
31. Futami, T., Matsumoto, T., Iwasaki, S., Umemoto, T.: IHSI Implementation to Actual Plants, Ibidem, Vol. O, Session 7, January 22–24 (1980)
32. Chrenko, R. M.: Residual Stress Measurements on Type 304 Stainless Steel Weld Pipes, Ibidem, Vol. O, Session 4, January 22–24 (1980); Van Wiggen, P.C., Rozendaal, H.C.F., Mittemeijer, E.J.: The nitriding behaviour of iron-chromium-carbon alloys. J. Mater. Sci. **20**, 4561–4582 (1985)
33. Costa, J.D., Ferreira, J.M., Ramalho, A.L.: Fatigue and fretting fatigue of ion-nitrided 34NiMo6 steel. Theor. Allied Fract. Mechan. **35**, 69–79 (2001)
34. Slámečka, K., Pokluda, J., Kianiková, M.: Fractographically-aided analysis of fish-eye crack growth in nitrided steel. J. Appl. Theor. Mechan. **51**(2), 439–446 (2013)
35. Osgood, C.C.: Fatigue Design, 2nd edn. Pergamon Press, Oxford (1982)
36. Brock, G.W., Sinclair, G.M.: Elevated temperature tensile and fatigue behaviour of unalloyed arc-cast molybdenum. Proc. ASTM **60** (1960)
37. McCammon, R.D., Rosemberg, H.M.: The fatigue and ultimate tensile strength of metals between 42 and 393 K. Proc. R. Soc. A **242**, 203 (1957)
38. Allen, N.P., Forrest, P.G.: International Conference on Fatigue. Institution of Mechanical Engineers, vol. 237 (1956)
39. Forrest, P.G.: Metal Fatigue. Chapman and Hall, London (1959)
40. Schwartzberg, F.R., Keys, R.D., Brown, M.J., Reightler, C.L., Martin-Marietta Corp., Rep. NASA-CR, 63-29 (1963)
41. Betteridge, W.: The Nimonic Alloys. Edward Arnold, London (1959)
42. Hempel, M., Tillmann, H.E.: Wechselzugversuche bei Holen Temperaturen. Max-Plank Inst., Eisenforschun, vol. 163 (1936)
43. Environmentally Assisted Cracking in Light-Water Reactors, NUREG/CR-4667, vol. 22, Prepared by O.K. Chopra et al., Semi-annual Report, January 1996–June 1996

44. Keh, A.S., Nakada, J., Leslie, W.C.: Dynamic Strain Aging in Iron and Steel, Dislocations Dynamics, pp. 381–408. A.R. Rosenfield and others, McGrow-Hill (1968)
45. Carreker, R.P., Jr.: Tensile deformation of silver as a function of temperature, strain rate, and grain size. Trans. AIME **209**, 112 (1957)
46. Loss, J.F. (ed.): Structural Integrity of Water Reactor Pressure Boundary Components, US NRC, NUREG/CR-3228, vol. 4, Annual Report (1985)

Chapter 6
Data Scatter and Statistical Considerations

6.1 Use of Statistics in Data Analysis

If five specimens of the same material, size and surface finish were subjected to the same fatigue test, it is likely that there would be five different results. If the test pieces were ten, then it is likely that there would be ten different results. Increasing the number of specimens will not change this general outcome and new values, higher and lower, will be added. Therefore, also the spread between the maximum and the minimum value will increase, albeit most values will appear closely-spaced. However, the search for the reasonably lower value when not for the lowest possible, which indeed may be the target of the designer, cannot be done by just increasing over and over the number of test specimens. Dough it is good practice to collect experimental data for better results, I know manufacturers who had collected more than 1,000,000 test data on a single steel, the problem of inferring what might be the target value with the higher confidence level possible is generally approached working on a limited availability of data.

Statistical methods based on probability theory, then, enter into action providing the most proper tools or, perhaps, the only tool to accomplish that goal. But there is also another reason why statistics are so important a tool in fatigue. The explanation of the size effect in fatigue, as already explained in Sect. 6.4. and the new approach to fatigue based on the forecast of what the largest inclusion size may be expected in a given volume of material, can be accomplished only with the aid of statistics. This chapter, dough in a limited number of pages where an entire book would be necessary, will address the issue of data analysis with a new ambitious target which is that of shifting the attention of the designer from a theoretical matter of probability of occurrence to a practical application of the concept of process volume. It will be shown that probability of occurrence and volume process are precisely the same thing, which actually means that when we test small specimens to search for the fatigue limit, for example, resulting in a probability of survival of 99.9% it is precisely the same object as to know the fatigue limit at which one specimen will fail in a batch

P. P. Milella, *Fatigue and Corrosion in Metals*,
https://doi.org/10.1007/978-3-031-51350-3_6

of 1000 specimens of the same type but it is not and this must be fully understood, the same thing as to know the fatigue limit at which one failure will occur in 1000 workpieces or components of the same material because the process volumes are not equal. In the following sections several among the most powerful statistical methods of analysis will be described that may give the designer the desired confidence on reliability of data he should be using in fatigue design starting from the limited number of tests initially performed. The adjective *initially* has been used on purpose since it must be understood once and for all that the lower the population of data available the larger the lack of confidence in the results obtained. Therefore, initial data paucity can be tolerated but a prolonged scarcity of data is unsustainable.

6.2 Samples, Population and Errors

Figure 6.1 [1] shows a typical set of fatigue data obtained with specimens of medium strength carbon steel, all of the same size, material and surface finish tested under the same cyclic loading conditions at various constant stress amplitudes S_i ($i = 1, 2, \cdots, 8$). Also shown are the mean trend (solid curve) and the upper and lower bound of experimental data (dashed lines). Note that they are not the true mean or absolute upper and lower bound *S–N* curves, but just those relative to the limited number of samples tested at each given stress amplitude. Increasing the number of specimens both curves might change. One of the purposes of this chapter is just to find them. It shall not come as a surprise the fact that the data scatter, in terms of cycles at any given stress amplitude, is rather high. This large variation of test results is due to metallurgical variability since each specimen, although similar to all the others, from the microscopic point of view is actually different and, in many respects, it may be even unique.

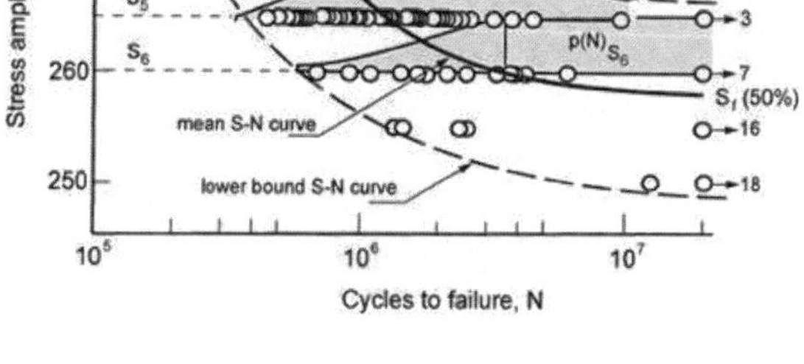

Fig. 6.1 Fatigue data and *S–N* mean curve obtained with specimens under the same test condition (modified from [1]). Also shown is the power density $p(N)S_i$ of the cycles to failure N obtained at constant stress amplitude S_i ($i = 1, 3, 5$ and 6)

Inclusions, second-phase particles, distribution of metallurgical phases, grain dimensions and boundaries, their distribution and orientation on the surface and, in particular, defects distribution are not exactly the same in all of them. They all are factors that contribute to the final result and, therefore, produce data scatter. It is then clear that the smaller the specimens the larger the data scatter since by reducing the process volume we also reduce its representativeness of all possible metallurgical variations. Generally speaking, fatigue specimens are rather small. Several general observations may be done looking at fatigue data of Fig. 6.1 and, in general, at any *S–N* diagram of any material. First, the dispersion of cycles to failure N_i is rather confined at very high stress amplitude *S* with respect to that pertaining to lower stress amplitudes, which eventually becomes almost unbounded in the vicinity of fatigue limit S_f. Arrows on some low stress amplitude data indicate, as usual, that specimens did not fail (so called run-out) and the test was suspended.

This dependency of data scatter on the stress amplitude *S* may be explained by saying that the higher the stress *S*, the lower the effect of metallurgical variations on test result. Stress overruns metallurgical variability. The same diagram is proposed in Fig. 6.2, but this time the scatter considered is that pertaining to the stress amplitude *S* at each constant life N_i selected. Now the opposite behavior is observed since tress amplitude scatter is larger at lower lives *N* and reduces at high cycles to failure *N*. This dispersion is shown Figs. 6.1 and 6.2 also in terms of *pdf* (probability density function or frequency). The $p(N)_S$ describes how the population of cycles to failure *N* at any constant stress amplitude *S* is distributed, while $p(S)_N$ refers to the stress amplitude distribution *S* at any given number of cycles *N*. the two distributions $p(N)_S$ and $p(S)_N$ become wider or shallower as the stress amplitude *S* decreases or the number of cycles *N* decreases, respectively.

Second, the $p(N)_S$ distribution looks like symmetrical at high stress amplitudes *S* (low cycle fatigue domain) but becomes strongly asymmetrical at low *S* values. As to the second distribution $p(S)_N$, it may also lose its symmetry. This last aspect of

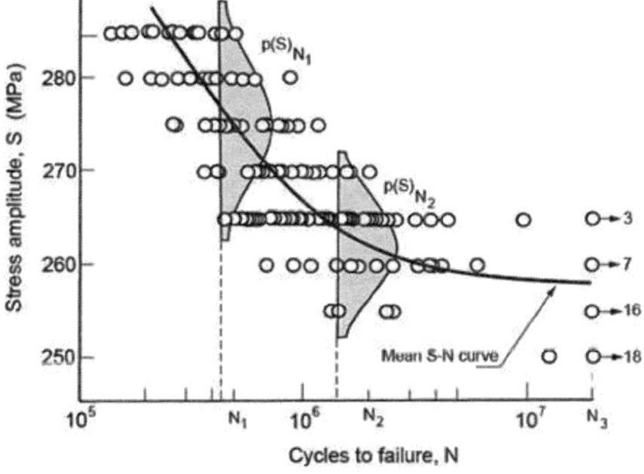

Fig. 6.2 Fatigue data and *S–N* mean curve of Fig. 4.1 (modified from [1]). Also shown is the power density $p(S)N_i$ of stress amplitudes S_i at constant number of cycles to failure *N*

pdf effective shape (more or less symmetrical) may have enormous consequences on statistical analysis since it determines the particular probabilistic technique that must be used. For instance, with a non-symmetrical distribution the well-known and potent Gaussian or normal distribution cannot be used. Very often or probably too often this is a fact forgotten by the designer. The starting point of a probability analysis is always the arithmetic mean. If a batch of n samples is extracted from a *total* population of measurable variables x_i, the arithmetic mean, $\overline{x}$, is given by

$$\overline{x} = \frac{x_1 + x_2 + x_3, \ldots, x_n}{n} = \frac{1}{n} \sum_{i=1}^{n} x_i. \tag{6.1}$$

This arithmetic mean $\overline{x}$ is not the *true* mean of the total population or, simply, *population* but just the *mean* of those n samples extracted from the population. Changing the samples drawn from the same population may yield a different mean. Also increasing the sample size may significantly change the mean value. Therefore, the question arises as to what is the population which we want to operate on. Actually, this is a question of fundamental importance to engineers, as it will be shown; therefore we'll be spending some words. The true mean value is that calculated over the population of a measurable variable x, which is given the special symbolμ. As n approaches the number of samples in the population, $\overline{x}$ becomes an increasingly precise measure of the true mean μ. The total population can be inferred from the process volume V_o under consideration. Assume, for instance, that n specimens having an individual volume V_{sp} are machined from a block of material of volume V_o and tested. Those n specimens represent $n \cdot V_{sp}/V$ of the entire population $n' = V/V_{sp}$ of specimens that can be machined out of the block of volume V_o for the purpose we want: traction, fatigue or whatever. If m blocks of the same size were used to build a component or workpiece, than the overall *process volume* becomes $m \cdot V_o$ and the population will be

$$N = \frac{m \cdot V_o}{V_{sp}}. \tag{6.2}$$

If we build p components of the same size, then the overall *process volume* will become $p \cdot m \cdot V$ and the population will surge to

$$N' = \frac{p \cdot m \cdot V_o}{V_{sp}}. \tag{6.3}$$

Consider that if V_{sp} is small, as it normally is, Eq. (6.3) yields incredibly high values for the entire or true population of samples. In the case of very large components or small components, but built in millions of pieces every year, the population may be 1,000,000 samples or even higher. Sometimes the process volume may be so large to require an almost unlimited number n of specimens so that the true mean will be given as a limit by

$$\mu = \lim_{n\to\infty} \frac{1}{n} \sum_{i=1}^{n} x_i. \tag{6.4}$$

What Eqs. (6.3) and (6.4) are actually saying is that the only way to scan for any possible metallurgical variation existing in a component of a given material is to test each possible piece of material. Of course, we cannot think to destroy the entire component or all the components we have built just to obtain specimens and get the total population of samples needed to infer the true mean value or what else nor we may think to run millions of tests. This is why we need statistics. The purpose of sections to come is explain how to manipulate a limited number of sample data to obtain what might be the mean value $\overline{x}$ closer to the true one μ given by Eq. (6.4) and derive what we shall reasonably expect to be the fatigue life of a large component with a wanted degree of confidence. The true mean, in fact, is not precisely the final target of a designer, because to design according to a reference mean value, insofar as close to the true one, means to accept the case where 50% of work pieces will fail, i.e., one out of two! So designer shall go deeper into the limited experimental data bunch to get, like a magician, what may be considered his final target: 99% survival, for example, or 99.99% survival, i.e., one failure out of 100 work pieces or out of 10,000 work pieces, respectively, with the required confidence. This is the goal of the following sections.

6.3 Normal Distribution

The result of the first test from a population of n specimens provides only an indication or *central tendency* of the values of the total population that can be the fatigue strength S_f or the life N at a given stress amplitude or whatever. The second test provides an indication of how uniform the results will be. Therefore, over the entire population of n specimens only n-1 contribute to the knowledge of dispersion and the bunch of data are said to have n-1 *degrees offreedom*. These concepts of central tendency, dispersion and degree of freedom find their representation in what is called probability density function $p(x)$ of the variable x. An example of probability density curve is shown in Fig. 6.3. The probability that a generic specimen yields exactly a given value x (plus or minus zero) is zero. The probability that its value lies within x and $x + dx$ is given by $p(x)dx$ (cross hatched area in Fig. 6.3).

Therefore, the integral of the probability density function, which is the area under the curve, is equal exactly to one, while the integral from $-\infty$ to a given value x, indicated as cumulative probability function $P(x)$, gives the probability $P(x)$ that a specimen yields a value lower than x. The curve shown in Fig. 6.3 is non-symmetrical and, as such, of general type. The three parameters indicative of the so called central tendency are shown on the curve. The first is the *mode* or modal value or peak value that corresponds to the peak of the probability density curve where the tangent is zero. The second is the mean value $\overline{x}$ or μ, Eqs. (6.1) and (6.4), that correspond

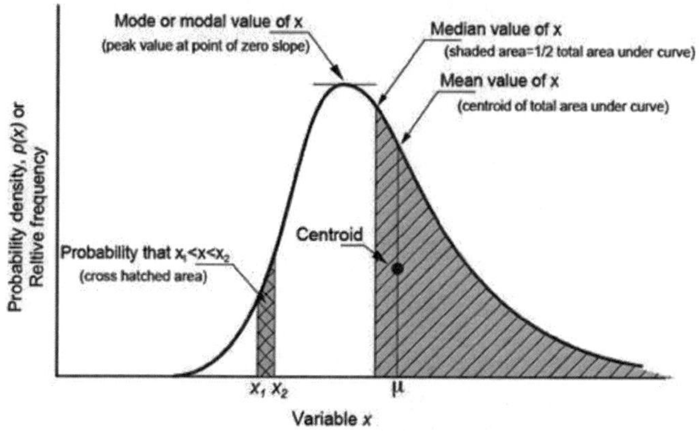

Fig. 6.3 Probability density or relative frequency curve showing three parameters indicative of central tendency

to the centroid of the probability curve and, third, the median that represents that value of x that is exceeded by half of the experimental results and, therefore, the value that divides the probability curve in two parts of equal area. If the curve were symmetrical, these three parameters would coincide and have the same value.

This is the case of the so called normal distribution or Gaussian distribution or Gauss' law of errors distribution. This distribution was actually proposed for the first time in 1733 by a great mathematician of the eighteen century: de Moivre and later extended by Laplace in 1812 so that to-day it is known as de Moivre-Laplace theorem. Carl Friedrich Gauss became associated with this distribution because he analyzed astronomical data using it and defined the equation of its probability density function. Its peculiarity is to be *bell shaped* that is to say perfectly symmetrical with the three parameters indicative of central tendency coincident. However, this is also its major drawback since not always experimental data are distributed exactly like a *bell* or in a perfectly symmetrical fashion. When this is not the case Gaussian distribution cannot be used, which actually means that the designer must absolutely check the real distribution of the data he is analyzing before using the Gaussian distribution. This is not an easy task since few specimens cannot provide such a piece of information. The equation of the normal distribution is

$$p(x) = \frac{1}{\Delta\sqrt{2\pi}} \exp\left[-\frac{(x-\mu)^2}{2\Delta^2}\right] \quad for \quad -\infty < x < \infty \tag{6.5}$$

in which μ is the true mean and Δ the standard deviation

$$\Delta = \sqrt{\frac{1}{n_t - 1}\sum_{i=1}^{n_t}(x_i - \mu)^2} \tag{6.6}$$

calculated over the entire population n_t given by Eq. (6.3), otherwise the standard deviation for a limited number of samples, also called standard error and indicated by the letter s, is

$$s = \sqrt{\frac{1}{n-1}\sum_{i=1}^{n}(x_i - \overline{x})^2} \tag{6.7}$$

where $\overline{x}$ is the mean value of the n samples given by Eq. (6.1). The variance Δ^2 is then

$$\Delta^2 = \frac{1}{n_t - 1}\sum_{i=1}^{n_t}(x_i - \mu)^2 \tag{6.8}$$

or, dealing with a limited number of data n drawn from the entire population

$$s^2 = \frac{1}{n-1}\sum_{i=1}^{n}(x_i - \overline{x})^2. \tag{6.9}$$

Figure 6.4 shows graphically the effect of the standard deviation Δ and of the mean μ on the probability density function $p(x)$ for a normal distribution.

As it can be seen, the standard deviation is a measure of data dispersion. Increasing Δ is representative of increasing data dispersion, while the peak value decreases. At variance, small Δ is associated with a narrow scatter and high peak value. But standard deviation has also a very precise meaning. In fact, on the distribution curve it identifies several particular points. This is graphically shown in Fig. 6.5 in which the entire *bell* shaped normal distribution curve is divided into sectors each of which has the width of a standard deviation. The characteristic feature is that:

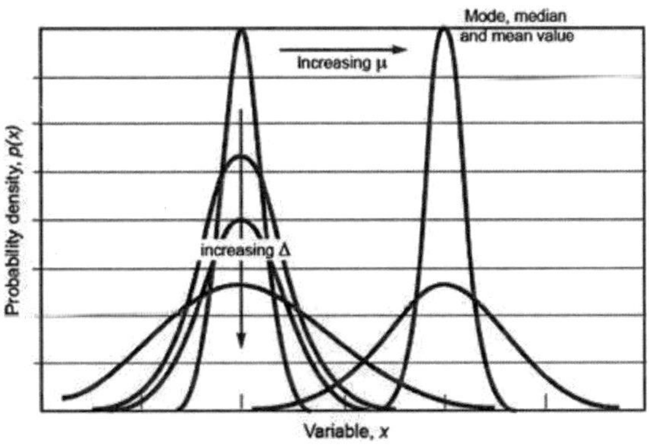

Fig. 6.4 Effect of the standard deviation Δ and of the mean μ on the probability density function for a normal distribution

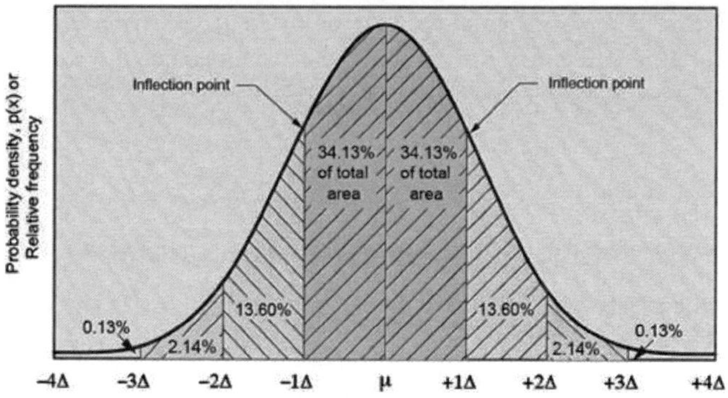

Fig. 6.5 Repartition of all experimental results within the *bell* or normal distribution

68.3% of all x values lie within the band $\mu \pm 1{\cdot}\Delta$.

95.4% of all x values lie within the band $\mu \pm 2{\cdot}\Delta$.

99.74% of all x values lie within the band $\mu \pm 3{\cdot}\Delta$.

99.99% of all x values lie within the band $\mu \pm 4{\cdot}\Delta$.

In the two extreme sectors beyond $\mu \pm 4{\cdot}\Delta$ (the tails of the distribution) it falls only 0.26% of the entire population of experimental data having a true mean given by μ. This actually means that just one over 10,000 samples will yield a value below $\mu - 4{\cdot}\Delta$ or above $\mu + 4{\cdot}\Delta$. The standard error of the mean (SEM) is the standard deviation of the sample mean estimate of a population mean.

SEM is usually estimated from the standard deviation divided by the square root of the sample size n

$$SEM(\overline{x}) = \frac{\Delta}{\sqrt{n}}. \tag{6.10}$$

The above formula assumes that the sample size n is much smaller than the total population size N, so that population can be considered infinite in size. The central-limit-theorem of statistics states that if $\overline{x}$ is the mean value of a random sample of n observations, $\overline{x}$ is normally distributed about the population mean with a standard deviation equal to SEM. With this as a basis, it is possible to determine the confidence limits within which the sample mean approaches the population mean to a certain level of certainty. When the sampling fraction is large (approximately 10% or more) the estimate of the error must be corrected by multiplying by a finite population correction (FPC) [2]

$$FPC = \sqrt{\frac{N-n}{N}} \tag{6.11}$$

to account for the added precision gained by sampling reaching a larger percentage of the total population. The effect of FPC is that the error becomes zero when the sample size n becomes equal to the total population size N. The integral of $p(x)$ from $-\infty$ to $+\infty$ yields the cumulative distribution function $P(x)$

$$P(x) = \frac{1}{\Delta\sqrt{2\pi}} \int_{-\infty}^{x} \exp\left[-\frac{(u-\mu)^2}{2\cdot\Delta^2}\right] du$$
$$= \frac{1}{2}\left[1 + erf\left(\frac{x-\mu}{\Delta\sqrt{2}}\right)\right] \tag{6.12}$$

$P(x)$ represents the probability that in a population of data there is one having a value lower or equal to a given x. In Eq. (6.12) *erf* is the so called error function defined as

$$erf(x) = \frac{2}{\sqrt{\pi}} \int_{0}^{x} e^{-t^2} dt. \tag{6.13}$$

Integral (6.13) cannot be expressed in closed form, but through a Taylor series expansion

$$erf(x) = \frac{2}{\sqrt{\pi}} \sum_{n=o}^{\infty} \frac{(-1)^n x^{2n+1}}{n!(2n+1)} = \frac{2}{\sqrt{\pi}}\left(x - \frac{x^3}{3} + \frac{x^5}{10} - \frac{x^7}{42} + \frac{x^9}{216} - \cdots\right). \tag{6.14}$$

The error function is plotted in Fig. 6.6 and tabulated in Table 6.1. Knowing $erf(x)$ the cumulative distribution function $P(x)$ given by Eq. (6.12) is immediately obtainable. An example of $P(x)$ is given in Fig. 6.7 for a population of yield stresses of a mild steel with a true mean $\mu = 276.5$ and three different values of the standard deviation $\varDelta$ equal to 10, 15 and 25, respectively. It can be seen how with a standard deviation equal to 25, 85% of the population will yield over 250 MPa that grows to 267 MPa when the standard deviation is only 10. It may be useful to plot the normal distribution function on a special coordinate paper which gives a straight-line relationship between life, relative to the mean value, and cumulative percent of failures or *probability of failure* or its inverse *probability of survival*. This is shown in Fig. 6.8. Note that *failure* or *survival* have the general meaning of falling outside an arbitrary boundary or within the boundary. The line yields how many standard deviations must be added or subtracted to the true mean μ to reach the target required. For example, if a 5% probability of failure $P(x)$ is required, which means 95% probability of survival $1 - P(x)$, then the number k of standard deviations $\varDelta$ to add to the true mean μ is $n = -1.65$.

Not always the data used by the designer are distributed according to the *bell* curve of Fig. 6.5. When this is the case, as for the fatigue life, it may be convenient

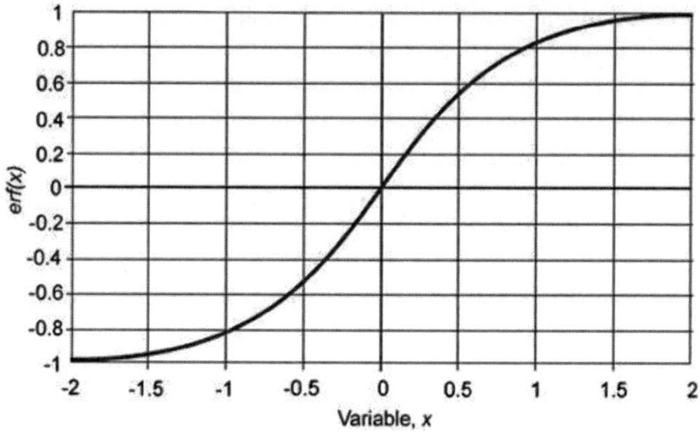

Fig. 6.6 Plot of the error function $erf(x)$

Table 6.1 Table of error function values

x	$erf(x)$	x	$erf(x)$	x	$erf(x)$	x	$erf(x)$
0.00	0.0000000	0.60	0.6038561	1.40	0.9522851	2.60	0.9997640
0.05	0.0563721	0.65	0.6420293	1.50	0.9661051	2.70	0.9998657
0.10	0.1124629	0.70	0.6778012	1.60	0.9763484	2.80	0.9999250
0.15	0.1679961	0.75	0.7111556	1.70	0.9837905	2.90	0.9999589
0.20	0.2227026	0.80	0.7421011	1.80	0.9890905	3.00	0.9999779
0.25	0.2763264	0.85	0.7706681	1.90	0.9927904	3.10	0.9999884
0.30	0.3286268	0.90	0.7969082	2.00	0.9953223	3.20	0.9999940
0.35	0.3793821	0.95	0.8208908	2.10	0.9970205	3.30	0.9999969
0.40	0.4283924	1.00	0.8427008	2.20	0.9981372	3.40	0.9999985
0.45	0.4754817	1.10	0.8802051	2.30	0.9988568	3.50	0.9999993
0.50	0.5204999	1.20	0.9103141	2.40	0.9993115		
0.55	0.5633234	1.30	0.9340079	2.50	0.9995931		

to approximate the normal distribution by taking the *log* of the quantity x rather than x itself. Normally, when the distribution function is not symmetrical it has a long *tail* on the right side. By taking the *log* of the variable x this tail is somehow compressed approximating the symmetrical distribution. A distribution in which the logarithms of a variable x follow the normal distribution is known as a *log-normal* distribution. However, also the log-normal distribution may not be perfectly symmetrical. In this case, the Gauss statistical analysis will be necessarily affected be an error and will not be correct. Unfortunately, the fact that generally in asymmetric distributions the left side is less expanded than the right one may results in a non-conservative error using the Gaussian distribution since it would assign a probability of survival, though small, to values that in reality have no chance to survive at all. Also the symmetrical

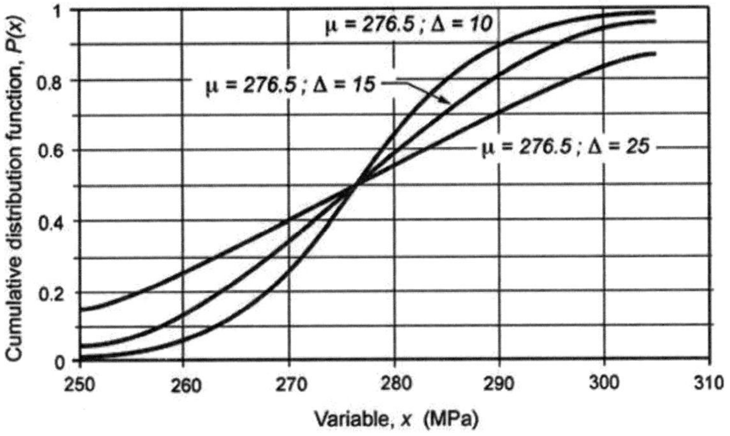

Fig. 6.7 Plot of cumulative distribution function $P(x)$ for a population with true mean $\mu =$ 276.5 MPa and three different standard deviations Δ equal to 10, 15 and 25, respectively

bell of the Gaussian distribution is not realistic since it goes from $-\infty$ to $+\infty$ having an infinitely long tale, which is not physically possible.

6.3.1 *Confidence Interval*

If the total population is known and, therefore, also the true mean μ and the standard deviation Δ, to infer the value that corresponds to a given percent of survival is rather an easy game. Assuming that a normal distribution holds, what shall be done is to evaluate the number k of standard deviation Δ to subtract to the true mean μ. But when the true mean is not known and the population of data is too large with respect to the sample size (see Eq. 6.3) and the mean $\overline{x}$ is just the sample mean, Eq. (6.1), the question arises as to how close or far we really are from the true one. There is no way to answer this question but in terms of confidence interval C, which actually means that we can only try to make an estimate of the true mean μ working with the available sample data. Its real value will never be determined exactly, but it may be included in a range of values whose size depends on the confidence we want to know it. This range of values is called a *confidence interval*. We associate a confidence interval with a *confidence level* that gives the probability that the interval selected includes the true value of the parameter in question. As the level of confidence decreases, also the size of the corresponding interval decreases. Common choices for the confidence level C are 0.90, 0.95 and 0.99. These levels correspond to percentages of the area under the normal density curve. For example, a 95% confidence interval covers 95% of the area under the normal curve. The probability $1 - C$ of observing a value outside this area is less than 0.05. Since the normal curve is symmetric, half of this

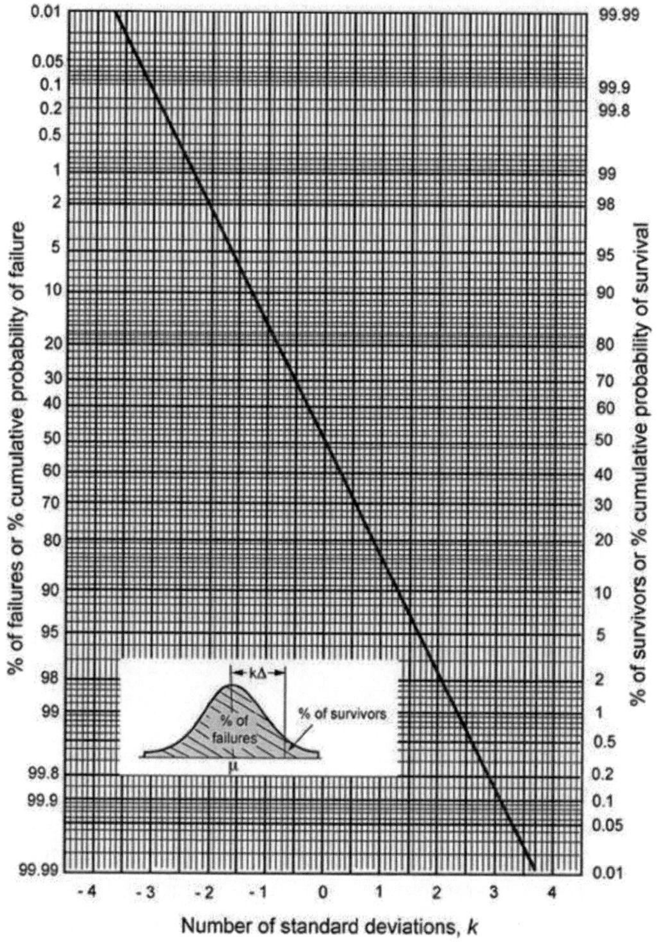

Fig. 6.8 Normal distribution curve plotted on special probability paper

area is in the left tail of the curve and the other half in the right one, as schematized in Fig. 6.8.

Therefore for a confidence interval of level C, the probability p is equal to $(1 - C)/2$. Points $- x*$ and x on the probability curve represent the limits that define the confidence interval, that is to say the points on the normal density curve such that the probability p of observing a value lower than $- x*$ or larger than x is $(1 - C)/2$: $p(x \leq -x^*) = (1 - C)/2$ and $p(x > x^*) = (1 - C)/2$. This limits $- x^*$ and x that define the confidence interval are called UCL or *upper confidence level* and LCL or *lower confidence limit*, respectively. Since they are on the probability density curve they represent a number k of standard deviations. For instance, if a 90% confidence interval is chosen, which means $p(x \leq LCL) = 0.05$ and $p(x > UCL) = 0.05$, their

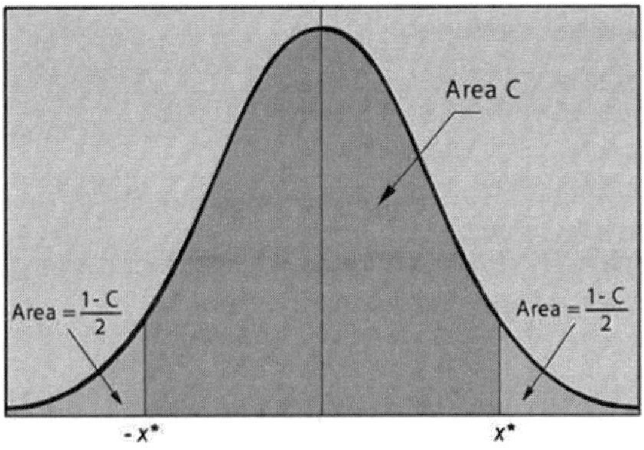

Fig. 6.9 Confidence interval of level C (area C) associated to a variable x

values can be inferred from the diagram of Fig. 6.9 entering with the ordinate 0.5 to which it corresponds ± 3.3. Therefore, for a sample of n data with a mean $\overline{x}$ the confidence interval of level $C = 90\%$ will be, recalling Eq. (6.10)

$$\overline{x} \pm k \cdot \frac{\Delta}{\sqrt{n}} = \overline{x} \pm 3.3 \frac{\Delta}{\sqrt{n}}. \tag{6.15}$$

Note that in writing Eq. (6.15), we have assumed to know the true standard deviation Δ. The ratio $UCL \cdot \Delta/\sqrt{n}$ in Eq. (6.15) is called margin of error m. Since the confidence level is given by the area under the normal distribution between the limits LCL and UCL or $-x^*$ and x^*, its value is given by

$$C = \frac{1}{\Delta\sqrt{2\pi}} \int_{-x*}^{+x*} \exp\left[-\frac{(x-\mu)^2}{2 \cdot \Delta^2}\right] dx \tag{6.16}$$

and recalling the definition of the cumulative distribution P, Eq. (6.12), it is

$$C = P(UCL) - P(LCL) = erf\left(\frac{k \cdot \Delta}{\sqrt{2}}\right). \tag{6.17}$$

The values of the error function $erf(k\Delta/\sqrt{2})$ are given in Table 6.2. The reverse relation of Δ multiples relative to same most used values of the area under the probability density curve are given in Table 6.3.

Table 6.2 Values of the error function for several $k\Delta$ values

$k\Delta$	$erf(k\Delta/\sqrt{2})$	$k\Lambda$	$erf(k\Delta/\sqrt{2})$	$k\Delta$	$erf(k\Delta/\sqrt{2})$
1	0.6826894	3	0.9973002	5	0.9999994
2	0.9544997	4	0.9999366	6	0.9999999

Table 6.3 Values of $k\Delta$ for several error function values

$erf(k\Delta/\sqrt{2})$	$k\Delta$	$erf(k\Delta/\sqrt{2})$	$k\Delta$	$erf(k\Delta/\sqrt{2})$	$k\Delta$
0.80	1.28155	0.98	2.32635	0.999	3.2905
0.90	1.64485	0.99	2.57583	0.9999	3.8906
0.95	1.95996	0.995	2.80703	0.99999	4.4172

6.4 Student's *t*-Distribution

In the previous section, the problem of estimating the population mean has been addressed when dealing with a sample mean $\overline{x}$, but knowing the population standard deviation Δ. Student's *t*-distribution or simply the *t*-distribution, is used when not only the population mean μ is unknown because the sample size is small, but also the population standard deviation Δ is unknown and has to be estimated from the data available that provide a sample standard deviation *s*. Its theoretical basis was published for the first time in 1908 by Gosset [3]. Gosset, however, was employed at a brewery that forbade the publication of researches and studies by its staff members. To circumvent this restriction, Gosset, that was a student, decided to issue his research under the pseudonymous of Student. Consequently, his distribution is known as the Student's *t*-distribution. He demonstrated that in a sample of *n* data x_i ($I = 1, 2, \ldots, n$) the pivotal quantity

$$T = \frac{\overline{x} - \mu}{s/\sqrt{n}} \tag{6.18}$$

has a normal distribution with a probability density function given by

$$p(t) = \frac{\Gamma\left(\frac{\nu+1}{2}\right)}{\sqrt{\nu\pi}\,\Gamma\left(\frac{\nu}{2}\right)}\left(1 + \frac{t^2}{\nu}\right)^{-\left(\frac{\nu+1}{2}\right)} \tag{6.19}$$

with $\nu = n - 1$ being the number of degrees of freedom. Γ is the Gamma function or Euler's function defined as

$$\Gamma(r) = \int_o^{\infty} e^{-x} x^{r-1} dx \tag{6.20}$$

or by the Stirling's series expansion

$$\begin{aligned}\Gamma(r) &= \left(\frac{r}{e}\right)^r \left[\sqrt{\frac{2\pi}{r}}\left(1 + \frac{1}{12r} + \frac{1}{288 \cdot r^2} - \frac{139}{51840 \cdot r^3} - \frac{571}{2488320 \cdot r^4} + \cdots\right)\right]. \\ &\approx \left(\frac{r}{e}\right)^r \sqrt{\frac{2\pi}{r}}\left(1 + \frac{1}{15r^2}\right)^{\frac{5}{4}r}\end{aligned} \tag{6.21}$$

When $r = \beta + 1$ with β integer the Γ function coincides with β factorial

$$\Gamma(\beta + 1) = \beta! \tag{6.22}$$

Values of the Γ function are tabulated in Table 6.4 [4]. The shape of the Student's t-distribution (6.19) resembles the bell shape of the Gaussian distribution with mean $\mu = 0$ and variance $\Delta^2 = 1$, except that it is a little lower and wider. As the number of degrees of freedom grows, the t-distribution approaches the normal distribution with mean $\mu = 0$ and variance $\Delta^2 = 1$. Figure 6.10 is a sketch of the Student's distribution. Its main feature is that it does not depend, any more, on μ or Δ or s, which gives the distribution an enormous importance both on theoretical and practical point of view, as well. Suppose, in fact, to select an interval of values of extremes $\pm$ A such that

$$C(-A < T < A) = 0.9. \tag{6.23}$$

This means that a confidence interval has been established for the parameter T with a level of confidence $C = 90\%$. In a sample of n data, A represents the 95 percentile, i.e., the value below which it falls 95% of all data

$$P(-\infty < T < A) = 0.95. \tag{6.24}$$

With a different symbols it may be written A $= t_{\alpha,\nu} = t_{0.05\nu}$, being $(1 - \alpha) = C$ and, as said, $\nu = n - 1$ the number of degrees of freedom. Likewise, it is also

$$P(-A < T < \infty) = 0.95 \tag{6.25}$$

so that the probability that T falls in the given interval is

$$C(-A < T < A) = 1 - 2a = 0.9. \tag{6.26}$$

From (6.18) and (6.26) it follows

$$C\left(-A < \frac{\overline{x} - \mu}{s/\sqrt{n}} < A\right) = 0.9 \tag{6.27}$$

that is to say

$$C\left(\overline{x} - A\frac{s}{\sqrt{n}} < \mu < \overline{x} + A\frac{s}{\sqrt{n}}\right) = 0.9. \tag{6.28}$$

As a consequence, the interval having $\overline{x} \pm As/\sqrt{n}$ as extremes represents the interval in which μ will fall with 90% of confidence. Let's consider the $(UCL)_{1-\alpha}$ of the mean given by

Table 6.4 Table of $\Gamma(r)$ function for argument values betweeen 1 and 2

r	$\Gamma(r)$	r	$\Gamma(r)$	r	$\Gamma(r)$	r	$\Gamma(r)$
1.00	1000	1.25	0.906	1.50	0.886	1.75	0.919
1.01	10.994	1.26	0.904	1.51	0.887	1.76	0.921
1.02	0.989	1.27	0.902	1.52	0.887	1.77	0.924
1.03	0.984	1.28	0.901	1.53	0.888	1.78	0.926
1.04	0.978	1.29	0.899	1.54	0.888	1.79	0.929
1.05	0.974	1.30	0.897	1.55	0.889	1.80	0.931
1.06	0.969	1.31	0.896	1.56	0.890	1.81	0.934
1.07	0.964	1,32	0.895	1.57	0.890	1.82	0.937
1.08	0.960	1.33	0.893	1.58	0.891	1.83	0.940
1.09	0.955	1.34	0.892	1.59	0.892	1.84	0.943
1.10	0.951	1.35	0.891	1.60	0.893	1.85	0.946
1.11	0.947	1.36	0.890	1.61	0.895	1.86	0.949
1.12	0.944	1.37	0.889	1.62	0.896	1.87	0.952
1.13	0.940	1.38	0.888	1.63	0.897	1.88	0.955
1.14	0.936	1.39	0.888	1.64	0.899	1.89	0.958
1.15	0.933	1.40	0.887	1.65	0.900	1.90	0.962
1.16	0.930	1.41	0.887	1.66	0.902	1.91	0.965
1.17	0.927	1.42	0.886	1.67	0.903	1.92	0.969
1.18	0.924	1.43	0.886	1.68	0.905	1.93	0.972
1.19	0.921	1.44	0.886	1.69	0.907	1.94	0.976
1.2	0.918	1.45	0.886	1.70	0.909	1.95	0.980
1.21	0.916	1.46	0.886	1.71	0.910	1.96	0.984
1.22	0.913	1.47	0.886	1.72	0.913	1.97	0.988
1.23	0.911	1.48	0.886	1.73	0.915	1.98	0.992
1.24	0.909	1.49	0.886	1.74	0.917	1.99	0.996
						2.00	1.000

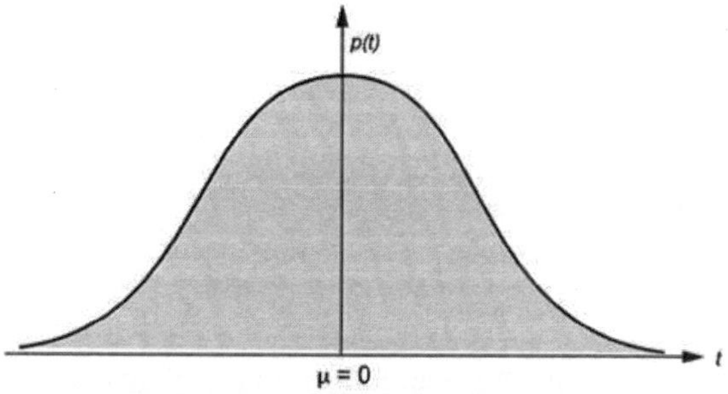

Fig. 6.10 Sketch of Student's t-distribution with mean $\mu = 0$ and variance $\Delta^2 = 1$

$$UCL_{1-\alpha} = \overline{x} + t_{\alpha,\nu}\frac{s}{\sqrt{n}}. \tag{6.29}$$

The value $t_{\alpha,\nu} = A$ that will be chosen depends on the confidence level $(1 - \alpha)$ relative to $UCL_{1\text{-}\alpha}$ (see Eq. 6.25) which, in turn, will depend on the confidence interval in which the true mean value μ will fall. Table 6.5 provides a set of $t_{\alpha,\nu}$ values for some selected values of the degrees of freedom $\nu = n - 1$. Percentages indicated in Table 6.5 refers to the quantity $(1 - \alpha)$ in %, that is to say to the probability that the true mean is lower than $UCL_{1-\alpha}$.

6.5 One-Sided Tolerance Limits for a Normal Distribution

It has been said that because of the limited number n of data in a sample that represent just a fraction of the entire population, the true mean μ and the standard deviation Δ cannot be assessed, but just *estimated.* It has been shown in Sect. 6.2 that when the true mean μ is not known, but the standard deviation Δ is available we can use an interval of the form given by Eq. (6.15) recurring to the concept of *confidence.* When both the true mean μ and the standard deviation Δ are not known, an interval of the form

$$\overline{x} \pm k \cdot s \tag{6.30}$$

can be used. Since both the sample mean $\overline{x}$ and standard deviation s vary depending on the sample size considered with respect to the entire population, it is impossible to determine k. It is, however, possible to partially overcome this limitation using again the concept of *confidence.* Most of the time engineers are not interested in the approximate interval of values in which a population parameter will fall, for instance the true mean, but just in the lower limit above which a certain percentage P will lie. In this case it is referred to as one-sided tolerance limit and this limit will be the lower one or LCL

$$LCL = \overline{x} - k \cdot s. \tag{6.31}$$

The values of k to use in Eq. (6.31) were found by Lieberman [5] and tabulated in Table 6.6. Later, also Williams et al. [6] developed the k factors to be used in Eq. (6.31), known as Owen factors. These last factors are tabulated in Table 6.7. From a comparative reading of Tables 6.6 and 6.7 two elements emerges:

1. increasing the confidence C for a lower limit associated to a given level of probability of survival is *paid* with an increasing value of the factor k;
2. k factors are not exact. It can be seen that with a sample of 15 data the k factor relative to $P99C90$ is 3.212 for Lieberman and 3.313 for Owen, with an increase of 3.4%. With $P99C95$ the k factor grows from 3.52 to 3.66 with an increase of 3.8%.

Table 6.5 Selected values for t-distribution with ν degrees of freedom for a range of one-sided tolerance limit

	Values of $(1-\alpha)$ in %									
ν	75	80	85	90	95	97.5	99	99.5	99.75	99.9	99.95
1	1.000	1.376	1.963	3.078	6.314	12.71	31.82	63.66	127.3	318.3	636.6
2	0.816	1.061	1.386	1.886	2.920	4.303	6.965	9.925	14.09	22.33	31.60
3	0.765	0.978	1.250	1.638	2.353	3.182	4.541	5.841	7.453	10.21	12.92
4	0.741	0.941	1.190	1.533	2.132	2.776	3.747	4.604	5.598	7.173	8.610
5	0.727	0.920	1.156	1.476	2.015	2.571	3.365	4.032	4.773	5.893	6.869
6	0.718	0.906	1.134	1.440	1.943	2.447	3.143	3.707	4.317	5.208	5.959
7	0.711	0.896	1.119	1.415	1.895	2.365	2.998	3.499	4.029	4.785	5.408
8	0.706	0.889	1.108	1.397	1.860	2.306	2.896	3.355	3.833	4.501	5.041
9	0.703	0.883	1.100	1.383	1.833	2.262	2.821	3.250	3.690	4.297	4.781
10	0.700	0.879	1.093	1.372	1.812	2.228	2.764	3.169	3.581	4.144	4.587
11	0.697	0.876	1.088	1.363	1.796	2.201	2.718	3.106	3.497	4.025	4.437
12	0.695	0.873	1.083	1.356	1.782	2.179	2.681	3.055	3.428	3.930	4.318
13	0.694	0.870	1.079	1.350	1.771	2.160	2.650	3.012	3.372	3.852	4.221
14	0.692	0.868	1.076	1.345	1.761	2.145	2.624	2.977	3.326	3.787	4.140
15	0.691	0.866	1.074	1.341	1.753	2.131	2.602	2.947	3.286	3.733	4.073
16	0.690	0.865	1.071	1.337	1.746	2.120	2.583	2.921	3.252	3.686	4.015
17	0.689	0.863	1.069	1.333	1.740	2.110	2.567	2.898	3.222	3.646	3.965
18	0.688	0.862	1.067	1.330	1.734	2.101	2.552	2.878	3.197	3.610	3.922
19	0.688	0.861	1.066	1.328	1.729	2.093	2.539	2.861	3.174	3.579	3.883
20	0.687	0.860	1.064	1.325	1.725	2.086	2.528	2.845	3.153	3.552	3.850
21	0.686	0.859	1.063	1.323	1.721	2.080	2.518	2.831	3.135	3.527	3.819
22	0.686	0.858	1.061	1.321	1.717	2.074	2.508	2.819	3.119	3.505	3.792
23	0.685	0.858	1.060	1.319	1.714	2.069	2.500	2.807	3.104	3.485	3.767
24	0.685	0.857	1.059	1.318	1.711	2.064	2.492	2.797	3.091	3.467	3.745
25	0.684	0.856	1.058	1.316	1.708	2.060	2.485	2.787	3.078	3.450	3.725
26	0.684	0.856	1.058	1.315	1.706	2.056	2.479	2.779	3.067	3.435	3.707
27	0.684	0.855	1.057	1.314	1.703	2.052	2.473	2.771	3.057	3.421	3.690
28	0.683	0.855	1.056	1.313	1.701	2.048	2.467	2.763	3.047	3.408	3.674
29	0.683	0.854	1.055	1.311	1.699	2.045	2.462	2.756	3.038	3.396	3.659
30	0.683	0.854	1.055	1.310	1.697	2.042	2.457	2.750	3.030	3.385	3.646
40	0.681	0.851	1.050	1.303	1.684	2.021	2.423	2.704	2.971	3.307	3.551
50	0.679	0.849	1.047	1.299	1.676	2.009	2.403	2.678	2.937	3.261	3.496
60	0.679	0.848	1.045	1.296	1.671	2.000	2.390	2.660	2.915	3.232	3.460
80	0.678	0.846	1.043	1.292	1.664	1.990	2.374	2.639	2.887	3.195	3.416
100	0.677	0.845	1.042	1.290	1.660	1.984	2.364	2.626	2.871	3.174	3.390

(continued)

Table 6.5 (continued)

	Values of $(1-\alpha)$ in %										
ν	75	80	85	90	95	97.5	99	99.5	99.75	99.9	99.95
120	0.677	0.845	1.041	1.289	1.658	1.980	2.358	2.617	2.860	3.160	3.373
∞	0.674	0.842	1.036	1.282	1.645	1.960	2.326	2.576	2.807	3.090	3.291

6.5.1 The Chi-Squared Function

It has been shown, in Sect. 6.2, how to estimate from a sample mean the limits within which the true mean of a population will fall with a desired level of confidence C. It may be desirable to estimate, now, the limits within which the population standard deviation will fall with a desired level of confidence C from the sample standard deviation s. The target is always the same: to estimate true values when the entire population of data is not available. To this purpose, the method of the so called χ^2 (chi-squared) function may be used. Again, as in all preceding sections, it must be assumed that the distribution is normal. The relation between s and Δ is given by

$$\frac{s^2}{\Delta^2} > \frac{\chi^2_{\alpha,n-1}}{n-1} \tag{6.32}$$

$$\Delta < \frac{s}{\sqrt{\chi^2_{\alpha,n-1}/(n-1)}}. \tag{6.33}$$

Values of χ^2 corresponding to a given probability α associated to (6.33) are shown in the diagram of Fig. 6.11. The confidence interval is that between α and $1-\alpha$.

6.6 Weibull Distribution

In all previous subsections the problem of determining the parameters characteristic of a population of data or of a sample of data has been addressed with the use of a *bell* shaped distribution or normal distribution. Not always experimental data are effectively distributed in a symmetrical fashion. This, for instance, is the case of fatigue life at stress amplitudes close to fatigue limit. A more general distribution function was introduced in 1939 by a Swedish engineer and researcher Weibull [7]. His probability density function in its simplest form (two parameters function) is defined as

$$p(x) = m\frac{x^{m-1}}{x_o^m}\exp\left[-\left(\frac{x}{x_o}\right)^m\right] \tag{6.34}$$

Table 6.6 Factors k such that at least a proportion P of the distribution has a confidence C greater than $\overline{x} - ks$

	C = 0.75					C = 0.90				
P n	0.75	0.90	0.95	0.99	0.999	0.75	0.90	0.95	0.99	0.999
3	1.464	2.501	3.152	4.396	5.805	2.602	4.258	5.310	7.340	9.651
4	1.256	2.134	2.680	3.726	4.910	1.972	3.187	3.957	5.437	7.128
5	1.152	1.961	2.463	3.421	4.507	1.698	2.742	3.400	4.666	6.112
6	1.087	1.860	2.336	3.243	4.273	1.540	2.494	3.091	4.242	5.556
7	1.043	1.791	2.250	3.126	4.118	1.435	2.333	2.894	3.972	5.201
8	1.030	1.740	2.190	3.042	4.008	1.360	2.219	2.755	3.783	4.955
9	0.984	1.702	2.141	2.977	3.924	1.302	2.133	2.649	3.641	4.772
10	0.964	1.671	2.103	2.927	3.858	1.257	2.065	2.568	3.532	4.629
11	0.947	1.646	2.073	2.885	3.804	1.219	2.012	2.503	3.444	4.515
12	0.933	1.624	2.048	2.851	3.760	1.188	1.966	2.448	3.371	4.420
13	0.919	1.606	2.026	2.822	3.722	1.162	1.928	2.403	3.310	4.341
14	0.909	1.591	2.007	2.796	3.690	1.139	1.895	2.363	3.257	4.274
15	0.899	1.577	1.991	2.776	3.661	1.119	1.866	2.329	3.212	4.215
16	0.891	1.566	1.977	2.756	3.637	1.101	1.842	2.299	3.172	4.164
17	0.883	1.554	1.964	2.739	3.615	1.085	1.820	2.272	3.136	4.118
18	0.876	1.544	1.951	2.723	3.595	1.071	1.800	2.249	3.106	4.078
19	0.870	1.536	1.942	2.710	3.577	1.058	1.781	2.228	3.078	4.041
20	0.865	1.528	1.933	2.697	3.561	1.046	1.765	2.208	3.052	4.009
21	0.859	1.520	1.923	2.686	3.545	1.035	1.750	2.190	3.028	3.979
22	0.854	1.514	1.916	2.675	3.532	1.025	1.736	2.174	3.007	3.952
23	0.849	1.508	1.907	2.665	3.520	1.016	1.724	2.159	2.987	3.927
24	0.845	1.502	1.901	2.656	3.509	1.007	1.712	2.145	2.969	3.904
25	0.842	1.496	1.895	2.647	3.497	0.999	1.702	2.132	2.952	3.882
30	0.825	1.475	1.869	2.613	3.454	0.966	1.657	2.080	2.884	3.794
35	0.812	1.458	1.849	2.588	3.421	0.942	1.623	2.041	2.833	3.730
40	0.803	1.445	1.834	2.568	3.395	0.923	1.598	2.010	2.793	3.679
45	0.795	1.435	1.821	2.552	3.375	0.908	1.577	1.986	2.762	3.638
50	0.788	1.426	1.811	2.538	3.358	0.984	1.560	1.965	2.735	3.604

	C = 0.95					C = 0.99				
P n	0.75	0.90	0.95	0.99	0.999	0.75	0.90	0.95	0.99	0.999
3	3.804	6.158	7.655	10.552	13.857	–	–	–	–	–
4	2.619	4.163	5.145	7.042	9.215	–	–	–	–	–
5	2.149	3.407	4.202	5.741	7.501	–	–	–	–	–
6	1.895	3.006	3.707	5.062	6.612	2.849	4.408	5.409	7.334	9.540

(continued)

Table 6.6 (continued)

	C = 0.95					C = 0.99				
P n	0.75	0.90	0.95	0.99	0.999	0.75	0.90	0.95	0.99	0.999
7	1.732	2.755	3.399	4.641	6.061	2.490	3.856	4.730	6.411	8.348
8	1.617	2.582	3.188	4.353	5.686	2.252	3.496	4.287	5.811	7.566
9	1.532	2.454	3.031	4.143	5.414	2.085	3.242	3.971	5.389	7.014
10	1.465	2.355	2.911	3.981	5.203	1.954	3.048	3.739	5.075	6.603
11	1.411	2.275	2.815	3.852	5.036	1.854	2.897	3.557	4.828	6.284
12	1.366	2.210	2.736	3.747	4.900	1.771	2.773	3.410	4.633	6.032
13	1.329	2.155	2.670	3.659	4.787	1.702	2.677	3.290	4.472	5.826
14	1.296	2.108	2.614	3.585	4.690	1.645	2.592	3.189	4.336	5.651
15	1.268	2.068	2.566	3.520	4.607	1.596	2.521	3.102	4.224	5.507
16	1.242	2.032	2.523	3.463	4.534	1.553	2.458	3.028	4.124	5.374
17	1.220	2.001	2.486	3.415	4.471	1.514	2.405	2.962	4.038	5.268
18	1.200	1.974	2.453	3.370	4.415	1.481	2.357	2.906	3.961	5.167
19	1.183	1.949	2.423	3.331	4.364	1.450	2.315	2.855	3.893	5.078
20	1.167	1.926	2.396	3.295	4.319	1.424	2.275	2.807	3.832	5.003
21	1.152	1.905	2.371	3.262	4.276	1.397	2.241	2.768	3.776	4.932
22	1.138	1.887	2.350	3.233	4.238	1.376	2.208	2.729	3.727	4.866
23	1.126	1.869	2.329	3.206	4.204	1.355	2.179	2.693	3.680	4.806
24	1.114	1.853	2.309	3.181	4.171	1.336	2.154	2.663	3.638	4.755
25	1.103	1.838	2.292	3.158	4.143	1.319	2.129	2.632	3.601	4.706
30	1.059	1.778	2.220	3.064	4.022	1.249	2.029	2.516	3.446	4.508
35	1.025	1.732	2.166	2.994	3.934	1.195	1.957	2.431	3.334	4.364
40	1.999	1.697	2.126	2.941	3.866	1.154	1.902	2.365	3.250	4.255
45	1.978	1.669	2.092	2.897	3.811	1.122	1.857	2.313	3.181	4.168
50	1.961	1.646	2.065	2.863	3.766	1.096	1.821	2.296	3.124	4.096

From: "Tables of one-sided statistical tolerance limits" by G. J. Lieberman, Industrial quality control, Vol. XIV, April 1958

in which m is a shape parameter, also known as Weibull exponent or slope, and x_o a scale factor to which, as it will be shown, it is associated a probability of failure of 63,2%. Equation (6.34) is actually defining a family of functions since its shape depends on the value of the Weibull exponent m. Figure 6.12 presents several Weibull probability density functions $p(x)$ related to different m values. Note how one of these approximates the Gaussian distribution for $m = 4$. To be more precise, the two distribution are equal for $m = 3.57$. Also for the Weibull distribution it is possible to define a mean $\overline{x}$, median $\breve{x}$ and mode $\tilde{x}$ value, which will coincide when $m = 3.57$, and a standard deviation s. The mean value is given by

Table 6.7 Factors k_{Owen} such that at least a proportion P of the distribution has a confidence C to be greater of $\overline{x} - ks$

	C = 0.90			C = 0.95		
P n	0.90	0.95	0.99	0.90	0.95	0.99
6	2.862	3.504	4.750	3.560	4.331	5.837
7	2.608	3.190	4.319	3.167	3.846	5.173
8	2.441	2.987	4.043	2.910	3.534	4.751
9	2.323	2.843	4.851	2.728	3.314	4.455
10	2.253	2.736	4.707	2.592	3.151	4.237
11	2.162	2.651	3.595	2.485	3.024	4.069
12	2.105	2.583	3.505	2.400	2.923	3.936
13	2.057	2.526	3.430	2.331	2.840	3.827
14	2.016	2.478	3.367	2.272	2.771	3.737
15	1.980	2.436	3.313	2.222	2.712	3.660
16	1.949	2.400	3.266	2.178	2.661	3.594
17	1.922	2.369	3.225	2.140	2.617	3.536
18	1.898	2.340	3.189	2.106	2.577	3.485
19	1.876	2.315	3.156	2.076	2.542	3.440
20	1.857	2.292	3.127	2.048	2.510	3.399
21	1.839	2.272	3.100	2.024	2.482	3.363
22	1.822	2.252	3.076	2.001	2.456	3.329
23	1.807	2.235	3.054	1.981	2.432	3.300
24	1.793	2.219	3.034	1.961	2.410	3.271
25	1.781	2.204	3.015	1.947	2.389	3.247
26	1.769	2.191	2.997	1.927	2.370	3.221
27	1.757	2.178	2.981	1.912	2.353	3.199
28	1.747	2.166	2.966	1.898	2.337	3.178
29	1.737	2.155	2.952	1.885	2.321	3.158
30	1.728	2.144	2.939	1.872	2.303	3.140

From: Williams et al. [6]

$$\overline{x} = x_o \cdot \Gamma\left(\frac{1}{m} + 1\right) \tag{6.35}$$

where Γ is the Gamma function already seen and given by Eq. (6.20) or (6.21) and tabulated in Table 6.4. The median value is given by

$$\breve{x} = x_o (\ln 2)^{1/m}. \tag{6.36}$$

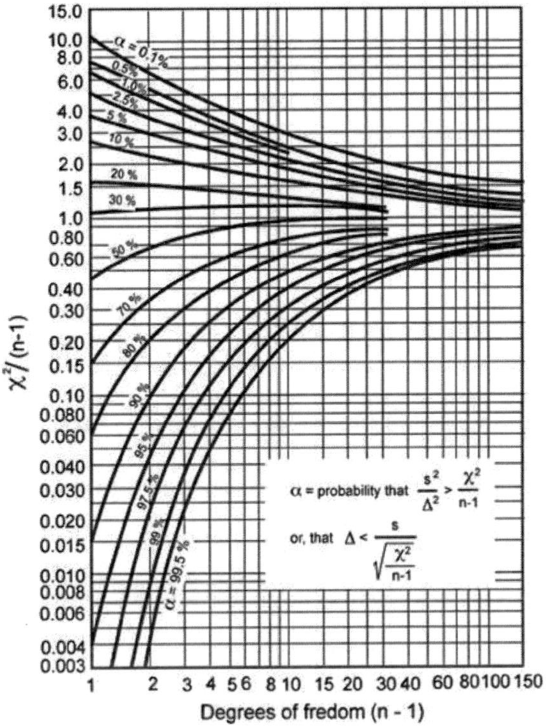

Fig. 6.11 Chi-squared functions

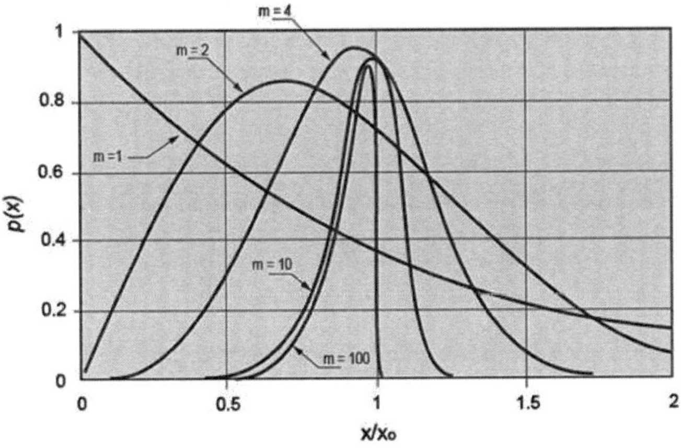

Fig. 6.12 Weibull probability density function $p(x)$ plotted for different m values

The mode is

$$\tilde{x} = x_o\left(1 - \frac{1}{m}\right)^{1/m} \tag{6.37}$$

and the standard deviation s or Δ

$$s = x_o\sqrt{\Gamma\left(\frac{2}{m} + 1\right) - \Gamma\left(\frac{1}{m} + 1\right)^2}. \tag{6.38}$$

From the probability density function it is derived the cumulative probability function $P(x)$ that is the probability that an element of the population assumes a value less or equal to a given x.

The cumulative probability function $P(x)$ is

$$\begin{aligned} P(x) &= \int_{-\infty}^{x} p(z)dz \\ &= 1 - \exp\left[-\left(\frac{x}{x_o}\right)^m\right] \end{aligned} \tag{6.39}$$

in which $p(x)$ is the probability density function (6.34). The cumulative probability function is plotted in Fig. 6.13 for some selected values of the Weibull exponent m equal to 1, 4, 10, 20 and 100, respectively. It shares some interesting features. First, all curves have the same inflection point of coordinates $x/x_o = 1$ and $P(x) = 0.632$ so that x_o represents, as said, the value below which 63.2% of the population falls. Secondly, it is interesting to note how data dispersion reduces progressively as the Weibull exponent m increases. For $0 < m \leq 1$ any value is possible from zero to infinity. For $m = 4$ the interval of possible values shrinks significantly, in practice it is given by $0.1 \leq x/x_o \leq 1.8$, while for $m = 100$ the cumulative probability function $P(x)$ approximate a step function that is to say a function with one single value: it is a constant. This actually means that all samples yield almost the same value. This, moreover, is confirmed by the same density probability function $p(x)$ that, as shown in Fig. 6.12, for m equal 1 or 2 indicates that any value is possible from 0 to infinity while for $m = 100$ the dispersion reduces to ½. Weibull exponent, therefore, may be thought as an index of the quality of the material or its homogeneity and metallurgical variability. The values of the Weibull exponent m and of the shape factor x_o are easily assessed. By taking twice the logarithm of Eq. (6.39) it yields

$$\begin{aligned} &\ln[1 - P(x)] = -\left(\frac{x}{x_o}\right)^m; \\ &\ln\ln\frac{1}{1 - P(x)} = m\ln(x) - m\ln(x_o). \end{aligned} \tag{6.40}$$

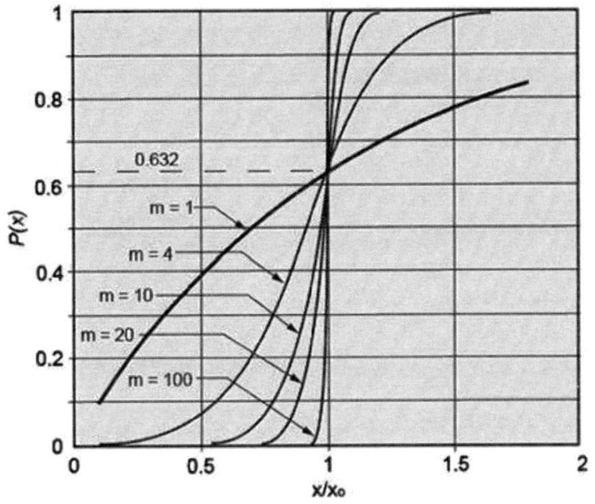

Fig. 6.13 Weibull cumulative probability function P for some values of the m exponent

In a bi-logarithmic scale Eq. (6.40) is represented by a line whose slope is m and the known term is $m\ln(x_o)$. In practice, the steps to determine the Weibull parameters m and x_o are the following: first, list the experimental results x_i ($i = 1, 2, 3, \ldots, n$) in ascending order; second, assign a rank to each results. The rank is indicating the failure probability in percent associated to each result, i.e., the frequency. Since the entire population is never available, we can only estimate the rank of the sample population. In this respect, several options are available. One of the most common estimate is the median rank, i.e., that rank that is as likely to err on the high side, with respect to the median value, as on the low side. The following equation gives a very good approximation of the estimated median ranks for any sample size n

$$\text{Estimated median rank} = \frac{0.7}{n+0.4}, \frac{1.7}{n+0.4}, \frac{2.7}{n+0.4}, \ldots, \frac{n-0.3}{n+0.4} \tag{6.41}$$

A valid alternative to the median rank is the mean rank, which is given by the equation

$$\text{Estimated mean rank} = \frac{1}{n+1}, \frac{2}{n+1}, \frac{3}{n+1}, \ldots, \frac{n}{n+1} \tag{6.42}$$

which may be written as

$$P(x) = \frac{N}{n+1} \tag{6.43}$$

where N is the serial number of the experimental result in the ascending order and n the sample size. The number n of experimental data plus 1 is used to avoid that the last value of the sample has a probability equal 1 ($n = N$), i.e., certainty. Another method to assign a probability of occurrence or rank to each result is

$$P(x) = \frac{N - 0.5}{n}. \tag{6.44}$$

or, alternatively

$$P(x) = \frac{N - 0.3}{n + 0.4}. \tag{6.45}$$

Whatever method is used, once a rank or probability of occurrence $P(x)$ is assigned to each datum of the sample, the experimental results x_i are plotted in a bi-logarithmic scale lnln[1/1−$P(x)$]−ln(x). The best-fitting line will have a slope equal to m. Once m is inferred also the scale factor x_o is obtained from the known term, NT, of the line equation

$$x_o = \exp\left[-\frac{NT}{m}\right]. \tag{6.46}$$

6.6.1 Process Volume Effect

The real power of the Weibull analysis is twofold: first it can follow the actual shape of the power density distribution of the sample data, while the Gauss distribution was asking the data to follow its symmetric shape and secondly, can treat the matter of the process volume with fascinating elegance. Any work piece or specimen has an individual process volume V that depends on the particular use or mechanical property that is to be measured. For example, if a cylindrical specimen of radius r and length l is used for measuring the yield or the traction strength of the material, the process volume V is the entire volume of the rod under traction: $V = \pi \cdot r^2 \cdot l$. If the specimen is tested in high cycle fatigue, then it is only the external surface no deeper than 0.5 mm to be interested by that loading, see Fig. 2.5. Therefore, the process volume of a specimen subjected to high cycle fatigue is much smaller than that involved in a traction test. This volume further reduces if the specimen is subjected to simple bending, as we already discussed in Section.

The concept of process volume is of extreme importance in engineering since it is strictly connected to metallurgical variability. The higher the process volume, the higher the probability to find a severe metallurgical condition or defect that will make the material fail at lower stress. The best evidence of the effects of this metallurgical variability through the process volume is provided by the so called whiskers that are fibers so thin, even less than 10 μm, to have a process volume negligible, almost zero. Fibers, then, have tensile strength so high that can reach the theoretical strength of the material that is almost one tenth of the Young's modulus simply because the probability to find a defect in a negligible volume is also negligible, practically zero.

Now, any process volume V may be thought to be formed by s statistically independent elemental parts of unit volume V_o, $V = s{\cdot}V_o$, as schematized in Fig. 6.14a, b. By applying the weakest link criterion, it is possible to assert that the entire system of volume V or V_1 or whatever volume fails if at least one of its elemental constituent parts of volume V_o fails. Alternatively, the probability of non-failure or reliability $R_s = 1 - P_s(x)$ of the system formed by s unit volumes V_o is given by the probability of non-failure of all s parts

$$[1 - P_s(x)] = [1 - P(x)]^s. \tag{6.47}$$

Using the Weibull cumulative probability function given by Eqs. (6.39) and (6.47), it yields

$$P_s(x) = 1 - \exp\left[-s\left(\frac{x}{x_o}\right)^m\right]. \tag{6.48}$$

What Eq. (6.48) is actually saying is that the probability of failure of a system formed by s pieces is higher than the probability of failure of each single piece. In other words, Eq. (6.48) states that *ceteris paribus* the larger the process volume the higher the probability of failure. This last result demonstrates what has been said, i.e., that the process volume and the effect of metallurgical variability are strictly related: increasing the volume also the probability of finding the worst metallurgical

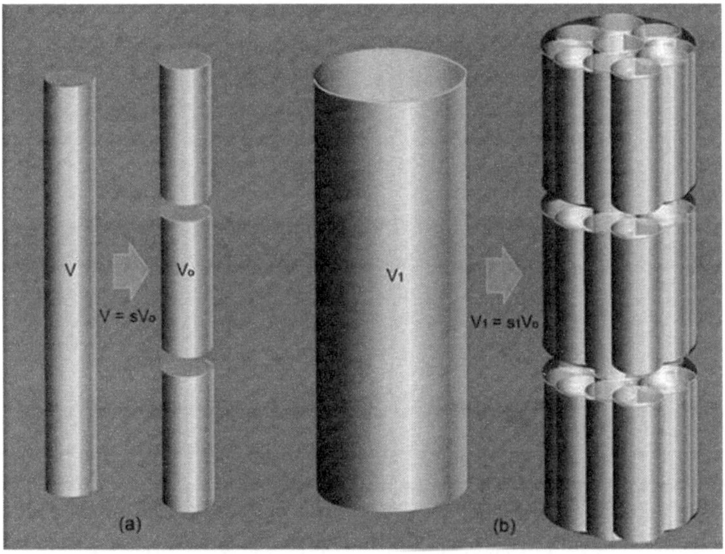

Fig. 6.14 **a** A generic test piece of volume V may be thought as formed by s elemental pieces of volume V_o and **b** any large volume V_1 may be thought as formed by s_1 independent elemental pieces of unit volume V_o

condition increases, hence we shall expect an early failure. In terms of probability of survival, Eq. (6.48) may be written as

$$R_s = \exp\left[-s\left(\frac{x}{x_o}\right)^m\right]. \tag{6.49}$$

Therefore, testing the same material first with n specimens of large volume V_1 that render a scale factor $x_{o,1}$, see Eq. (6.46) and then with n specimens of small volume V_o that yield a scale factor x_o, it will be found

$$\begin{aligned} \frac{x_{o,1}}{x_o} &= \left(\frac{V_o}{V_1}\right)^{1/m} \\ x_{o,1} &= x_o \cdot \left(\frac{V_o}{V_1}\right)^{1/m}. \end{aligned} \tag{6.50}$$

The scale factor $x_{o,1}$ obtained with a series of n large specimens is lower than that competing to the same number n of small specimens. It also indicates that the value corresponding to a probability of survival of 63.2% is lower for larger volumes. In general, it is possible to write:

$$\frac{x_2}{x_1} = \left(\frac{V_1}{V_2}\right)^{1/m}. \tag{6.51}$$

Equation (6.51) indicates that the ratio between two mechanical properties x_1 and x_2 coming from specimens of volumes V_1 and V_2, respectively, is inversely proportional to the power 1/*m* of their volumes. Figure 6.15 shows the variation of the generic quantity $x(s)$ measured with specimens of volume s-fold larger than that of reference versus s, for different values of the Weibull exponent m. For instance, testing two series of specimens of the same material having $m = 4$ but different volumes whose ratio is $s = V(s)/V = 600$, we shall expect that the mean value measured with the larger specimens will be a mere 5% only of that measured with the smaller ones. If m is equal to 40 that reduction will be much smaller, about 15%, $x(s) \sim 0.85x$. It must be remembered that the volume ratio may not be just that of two different test pieces, but, in practice, the other given by Eq. (6.2) or (6.3) between the total process volume and that of the specimens used.

Assuming that we are measuring the fatigue life of a single work piece at a stress amplitude close to the fatigue limit where m is seen to be very low, let's say about 5 and its process volume is 1000 times larger than that of the specimen, it is reasonable to expect that the life of the component be 35% of that measured with small specimens at the same stress amplitude. But if 1000 pieces of that component will be built, then the ratio of process volumes will be $1000 \times 1000 = 10^6$ so that the expected reduction will be about 95%! This actually means that one work piece out of 1000 will fail at a life N 95% lower than that obtained with small specimens. If the mean life measured with small specimens is about 10,000,000 cycles, then one work piece out of 1000

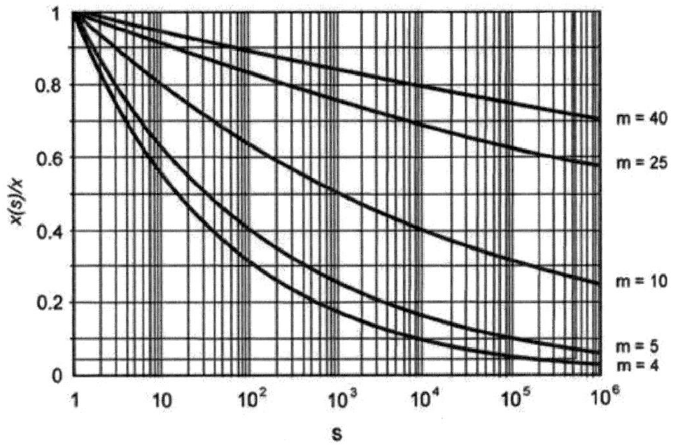

Fig. 6.15 Variation of the generic variable $x(s)$ measured with specimens of volume s-fold larger than that of reference versus s, for different values of the Weibull exponent m

will fail a 500,000 cycles only. However, Eqs. (6.48) and (6.51) are based upon the assumption that the exponent m is constant and don't depend on the volume itself. This is not precisely true. Simple reasoning can lead to the conclusion that m must be varying with the process volume: it is larger for larger process volumes and smaller for very small process volumes.

While almost any possible metallurgical variability, including the worst, may be found when the process volume is extremely large so that all specimens are practically alike, when the process volume reduces to almost zero we can imagine that each single specimen to be one of a kind. The resulting effect is that specimens having very large volume will yield results very close to each other with low scatter (standard deviation), independently of the quality and homogeneity of the material with high m values, while small specimens of the same material will yield very different results with very low m value and much larger scatter. Probably, fibers should have $m \sim 0$: almost any value is possible. This is shown in Fig. 6.16 that reports the probability density function, $pdf(\sigma_t)$, of the true failure strength obtained for DCI using specimens of different size, having diameters varying from 3 to 22 mm.

It is worth noting that the peak strength obtained with the smallest specimens, 560 MPa, is much larger than the peak strength of the largest specimen, 360 MPa, and the *tail* of the distribution of smaller specimens reaches the peak value of the distribution of the largest: what is a very low probability value for small volumes is the peak value of the largest. Therefore, the Weibull exponent m may be thought to be an index of the quality (metallurgical variability) of the material only comparing results obtained with specimens of the same size. This dependence of the Weibull exponent on the process volume was first evidenced by Milella et al. [8] considering process volumes of different sizes, from very small to that corresponding to a standard traction specimen. Specimens of standard size yielded a Weibull exponent ranging from about 45–50, but when the process volume was reduced using notched specimens of

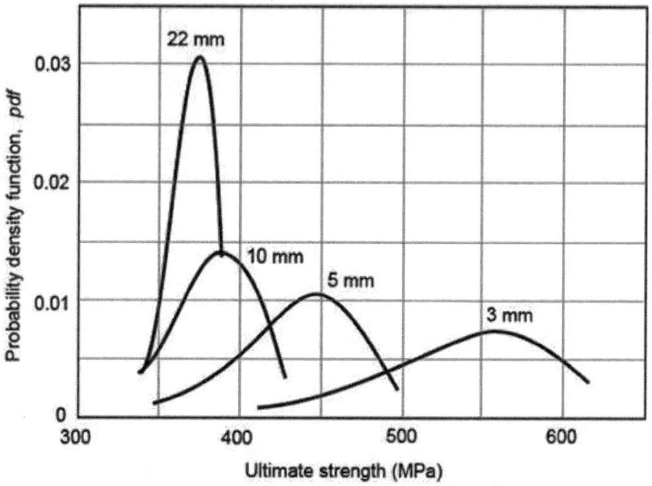

Fig. 6.16 Probability distribution, *pdf*, of the true failure strength, σ_t, of ductile cast iron (DCI) obtained using specimen of different diameters

different root radii the Weibull exponent for the same steel decreased to eventually reach the lower value of 4 connected to cracks. In a notched member the process volume is just the small plastic enclave at the notch tip where the stress state is higher because of stress concentration. It is there that fracture initiates. The sharper the notch, the lower the process volume and, therefore, the lower the Weibull exponent (see Sect. 11.5).

The dependence of the Weibull exponent on the root radius ρ was found by Milella for a medium strength steel is shown in Fig. 6.17 and expressed by the equation

$$m = 21 \cdot \rho^{0.25}. \tag{6.52}$$

The Weibull exponent seems to reach a saturation value. This tendency can be observed testing sample populations larger and larger or batches of specimens having increasingly larger process volumes. This has been done with ductile cast iron. A first testing was done using specimens having the same size, but their number increased from one testing campaign to the next. Batches of 30, 75, 150 and 300 specimens of the same size were tested with a total process volume varying from about 117,000 mm^3, for the 30 pieces set, to about 1,200,000 mm^3 for the 300 pieces set. Results were analyzed using the Weibull procedure. The m exponents so found are shown in Fig. 6.18 as open circles. A second experimental campaign was run that made use of specimens of different size (diameter and length) with a total process volume varying from about 4000 mm^3 to about 1,500,000 mm^3. Results were analyzed using the Weibull procedure (see Example 6.8.5) and the m exponent relative to each process volume inferred. Their values are reported in Fig. 6.18 as closed circles. The comparison of the two experimental campaigns is rather good. Therefore, the precise formulation of Eq. (6.48) is

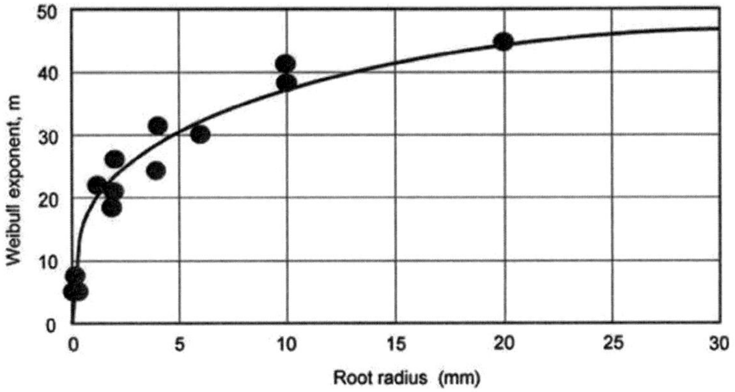

Fig. 6.17 Dependency of the Weibull exponent of a medium strength steel on notch root radius [8]

$$P_s(x) = 1 - \exp\left[-\int_V \frac{dV}{V_o}\left(\frac{x}{x_o}\right)^m\right] \tag{6.53}$$

that yields Eq. (6.48) when m is constant and does not depend on the volume. This dependency is rather beneficial since m growing with increasing volume leads to saturation earlier than expected, which results in lower decrease than predicted by Eq. (6.51). Since the Weibull exponent m increases as the process volume increases and the value of the shape factor x_o decreases, also the cumulative distribution curve $P(x)$ will get steeper and shift toward lower values of the ratio x/x_o as the process volume increases, as shown in Fig. 6.19. In practice if the material has rather high quality with m over 40 or 50 already with standard specimens, it can be assumed that m is constant and avoid integration of Eq. (6.53), but with poor materials like cast steel that may have m as low as 9 or 11, it cannot be avoided. However, it must be understood that the value of m depends also on the type of test and quantity measured. For instance, traction tests of high quality steel yield an m exponent over 40 or 50, while in fatigue tests the distribution of cycles to failure is characterized by an exponent m that depends on the stress amplitude and in vicinity of the fatigue limit can be as low as 2 or 3.

The Weibull distribution sheds more light on the issue of metallurgical variability and process volume showing that probability of failure and volumes ratio are exactly the same thing. Assume that we want to infer the value of a variable x with 99% probability of survival $R = 1 - P(x)$. By using Eq. (6.39) we may write

$$\ln[1 - P(x)] = -\left(\frac{x}{x_o}\right)^m;$$

$$x = x_o\{-\ln[1 - P(x)]\}^{1/m};$$

$$x(99\%) = x_o[-\ln(0.99)]^{1/m}$$

Fig. 6.18 Experimental values of the Weibull exponent obtained with specimens of DCI of the same size (same unit volume) and different size (different unit volume) as function of the total volume tested

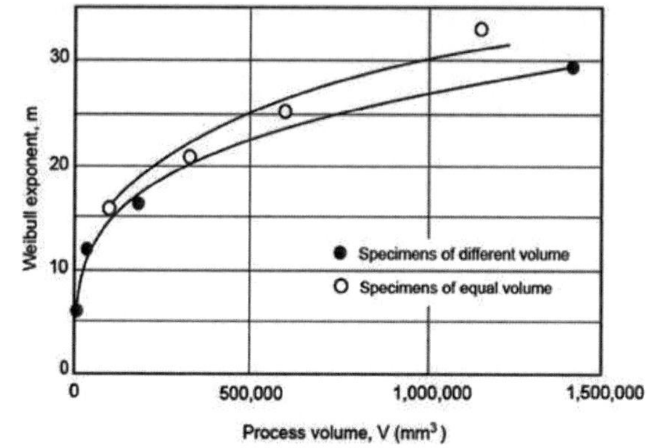

Fig. 6.19 Volume effect on the cumulative probability curve

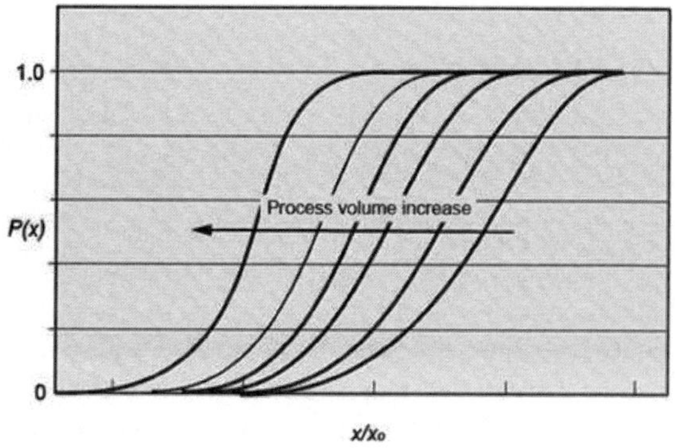

$$= x_o[-\ln(0.99)]^{1/m}$$
$$= x_o(0.01)^{1/m}. \qquad (6.54)$$

But a probability of 99% survival means that one piece out of 100 will fail at a value equal or lower than that. Let's try, then, to calculate what the value x_{100} of the variable x will be when the volume of the test piece is 100-times larger than the reference volume of the single specimen. We may use Eq. (6.50) and write

$$\frac{x_{o,100}}{x_o} = \left(\frac{1}{100}\right)^{1/m}$$
$$x_{o,100} = x_o(0.01)^{1/m} \qquad (6.55)$$

which is exactly the result given by Eq. (6.54).

This actually means that a wanted probability of survival $R = 1 - P$ represents the value at which a volume $1/(1 - P)$ larger than that of the specimen will fail. If, for example, the variable x is the fatigue limit measured with specimens of process volume V_o and we want to know what this limit will be with a probability of 99.9%, then this value is that at which a volume $V_{99.9}$ equal to $1/(1 - P)\cdot V_o = 1000\cdot V_o$ will fail.

This quantity is called *return period* T, see next section. It is also very important to realize when we calculate the probability of survival with the desired percentage from a sample data obtained with small specimens using any type of distribution, this calculation will yield the probability of survival of $1/[1 - P(\%)]$ pieces similar to the specimens used and not to the actual workpiece. We must be cautious not to mix up specimens with workpieces. The same is by drawing a lower bound curve from a bunch of experimental data as we did in Fig. 6.1. That lower bound curve refers to specimens and not to workpieces that have a much larger volume than specimens. A further passage is then needed to take into consideration the volumes ratio. Also the mean value of a batch of n specimens may be written considering the volume ratio as

$$\begin{aligned}\overline{x} &= \int_o^\infty x\partial[P(x)] \\ &= R\frac{I_m x_o}{V^{1/m}}\end{aligned} \tag{6.56}$$

x_o being, as usual, the scale factor, R a parameter and I_m the function

$$I_m = \Gamma\left(1 + \frac{1}{m}\right). \tag{6.57}$$

In writing Eq. (6.56) it has been assumed $V_o = 1$.

6.6.2 Loading Conditions

We know that process volumes depend also on the loading conditions or geometry of the component. This happens all the times high stress states are localized, as in notches or grooves, or non-uniform, as in torsion or bending (see Sects. 11.5 and 4.9). Bending loads or torsional loads, in fact, do not affect the same volume of material as traction loads. To take this effect into consideration, the cumulative probability function given by Eq. (6.53) can be written as

$$\begin{aligned}P_s(x) &= 1 - \exp\left[-\frac{V}{V_o}\left(\frac{x}{x_o}\right)^m\right] = 1 - \exp\left[-\left(\frac{1}{V^{-1/m}}\frac{x}{x_o}\right)^m\right] \\ &= 1 - \exp\left[-\left(\frac{1}{A}\frac{x}{x_o}\right)^m\right].\end{aligned} \tag{6.58}$$

Table 6.8 Values of parameters A ed R [9]

Load type	A	R
Tension of specimens of constant cross section	$V^{-1/m}$	1
Bending of specimens of rectangular cross section	$\left[\frac{V}{2(m+1)^2}\right]^{-1/m}$	$2^{1/m}(m+1)^{2/m}$
Pure bending of specimens of rectangular cross section	$\left[\frac{V}{2m+2}\right]^{-1/m}$	$(2m+2)^{1/m}$
Torsion of round bar	$\left[\frac{2V}{m+2}\right]^{-1/m}$	$\left(\frac{m+2}{2}\right)^{1/m}$

In deriving Eq. (6.58) it has been assumed $V_o = 1$. Referring to different loading conditions (tension, bending or torsion) the parameter A in Eq. (6.58) and R in Eq. (6.56) assumes different values, which are summarized in Table 6.8 [9]. If specimens of the same size are loaded under different conditions, the ratio of the respective strengths is inversely proportional to the power $1/m$ of the ratio of the correspondent parameters A. Therefore, according to Table 6.8, the ratio between bending strength and traction strength is given by

$$\frac{x_{bend}}{x_{trac}} = \left(\frac{A_{trac}}{A_{bend}}\right)^{1/m} = \left[2(m+1)^2\right]^{1/m}. \tag{6.59}$$

Alike, from pure bending to traction

$$\frac{x_{bend}}{x_{trac}} = \left(\frac{A_{trac}}{A_{bend}}\right)^{1/m} = (2m+2)^{1/m} \tag{6.60}$$

6.6.3 *Three Parameter Distribution*

According to the two parameters Weibull distribution (6.39) for low values of the m exponent the variable x in the population considered may be indefinitely low or even equal to zero, though with a very low probability of occurrence. In some cases this is not physically possible. Consider, as an example, the strength of a material or its fatigue limit. However small they may be, it is not possible that this quantities assume values indefinitely low. It is not credible that a material can break at a stress close to zero. For instance, consider a cast iron with $m = 9$, a maximum fatigue limit $S_f = 160$ MPa at 2,000,000 cycles and the scale factor $S_o = 140$ MPa. The probability that a small standard fatigue specimen fails at 2,000,000 by cycling at a stress amplitude of only 60 MPa is, according to Eq. (6.39), $P(60) = 0.0005$. This means that one specimens out of 2000 will fail at 60 MPa. The probability is small, but it is practically impossible to get so low a result. Therefore, the Weibull distribution may be corrected to account for the fact that physical variables may reach a minimum value, x_{min} also called location parameter, below which they cannot go. The corresponding Weibull distribution is known as three parameter distribution

$$P(x) = 1 - \exp\left[-\left(\frac{x - x_{\min}}{x_o}\right)^m\right]. \tag{6.61}$$

A possible variant of (6.61) is

$$P(x) = 1 - \exp\left[-\left(\frac{x - x_{\min}}{x_o - x_{\min}}\right)^m\right]. \tag{6.62}$$

However, the problem is only apparently solved because a third parameter x_{min} is introduced whose value is rather unknown. Note that the need to introduce a third parameter x_{min} may be mandatory if the Weibull exponent is low, as in the example before, or the quantity to be measured very low as in the case of toughness of brittle material that in ferritic steels may be as low as 50 MPa$\sqrt{}$m, below the transition temperature NTD. When m is high, things may change. For instance, going back to the previous example, if a good quality steel is considered with $m = 40$, a maximum fatigue limit of 400 MPa and $S_o = 350$ MPa, the probability to get a failure at let's say 250 MPa stress amplitude is $P(250) = 0.000001$ almost zero. In this last case to fix a minimum value at 250 would not have any practical effect.

6.7 Gumbel Distribution

In probability theory and statistics the Gumbel distribution [10] has become perhaps the most widely applied distribution in engineering. It is also known as the extreme value distribution since it is used to find the minimum or the maximum value of a number of samples having any possible distribution, normal or exponential. For example, it can be used to assess the minimum value of the fatigue limit having a sample of experimental values or the maximum defect or inclusion size that can be expected to exist in a workpiece or component. Named after the German mathematician Emil Julius Gumbel (1891–1966) the distribution is used in extreme value assessment and has two forms. One is aimed at finding the smallest extreme and the other the largest extreme. We call these extremes the minimum and the maximum case, respectively. The *pdf* relative to the minimum case is given by

$$p(x) = \frac{1}{\delta}\exp\left[\frac{x - \lambda}{\delta}\right]\exp\left\{-\exp\left[\frac{x - \lambda}{\delta}\right]\right\} \tag{6.63}$$

while the *pdf* for the maximum case is given by

$$p(x) = \frac{1}{\delta}\exp\left[\frac{-(x-\lambda)}{\delta}\right]\exp\left\{-\exp\left[\frac{-(x-\lambda)}{\delta}\right]\right\} \tag{6.64}$$

where λ is a location parameter and δ is the scale parameter. Figures 6.20a, b plot the Gumbel probability density function, $p(x)$, for the minimum case and the maximum case, respectively. The mean, median, mode and standard deviation of a population are given by

$$\mu = \lambda + \delta\gamma;\ median = \lambda - \delta\ln[\ln(2)];\ \text{mod}e = \lambda;\ \Delta = \frac{\pi\delta}{\sqrt{6}} \tag{6.65}$$

respectively, with γ being the Euler-Mascheroni constant = 0.5772. The cumulative distribution function $P(x)$ for the minimum value is

$$P(x) = 1 - \exp\left\{-\exp\left[\frac{x-\lambda}{\delta}\right]\right\} \tag{6.66}$$

and for the maximum value

$$P(x) = \exp\left\{-\exp\left[-\frac{(x-\lambda)}{\delta}\right]\right\}. \tag{6.67}$$

It is for this reason that the Gumbel distribution is also known as the log-Weibull distribution. Figure 6.20c plots the cumulative distribution function, $P(x)$, for the minimum case while Fig. 6.20d plots the Gumbel cumulative distribution function, $P(x)$, for the maximum case. To infer the value of both λ and δ from a batch of data we can proceed as for the assessment of the exponent m and the scale factor x_o of a Weibull distribution (see Eq. 6.40). Using the $P(x)$ expression (6.67) for the maximum case, it is

$$\begin{aligned} \ln P(x) &= -\exp\left[-\left(\frac{x-\lambda}{\delta}\right)\right] \\ -\ln[-\ln P(x)] &= \frac{x}{\delta} - \frac{\lambda}{\delta} \\ -\ln\left[\ln\frac{1}{P(x)}\right] &= \frac{x}{\delta} - \frac{\lambda}{\delta} \end{aligned} \tag{6.68}$$

while for the minimum case, using Eq. (6.66), it is

$$\begin{aligned} \ln[1 - P(x)] &= -\exp\left(\frac{x-\lambda}{\delta}\right) \\ \ln[-\ln(1-P)] &= \frac{x}{\delta} - \frac{\lambda}{\delta} \\ \ln\left[\ln\left(\frac{1}{1-P}\right)\right] &= \frac{x}{\delta} - \frac{\lambda}{\delta}. \end{aligned} \tag{6.69}$$

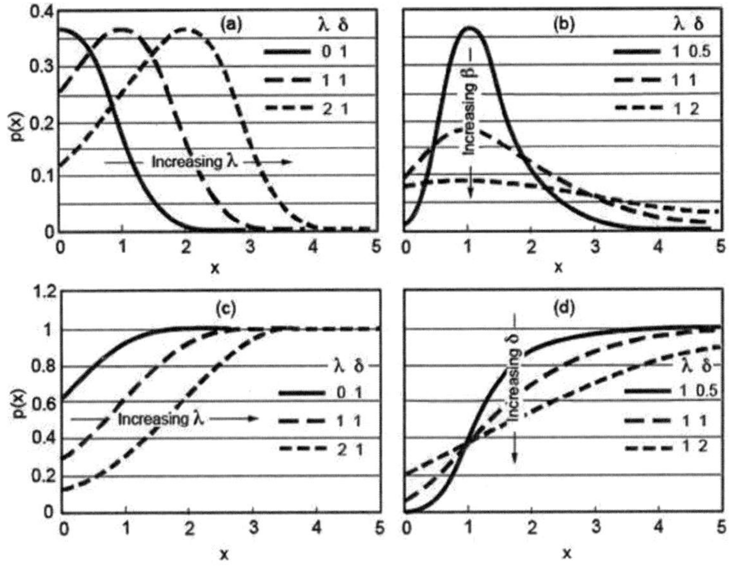

Fig. 6.20 Plots of **a** Gumbel $p(x)$for the minimum case and **b** maximum case, respectively, for varying λ and δ values. Plots of **c** Gumbel $P(x)$ for the minimum case and **d** maximum case, respectively, for varying λ and δ values

In a *lnln* versus x diagram Eqs. (6.68) or (6.69) represents a straight line whose slope is $1/\delta$ and the known term (NT) is $-\lambda/\delta$.

6.7.1 Extreme Value Statistics

Equation (6.68) can be solved for the variable x yielding

$$-\ln[-\ln P(x)] = \frac{x}{\delta} - \frac{\lambda}{\delta}$$
$$x = \lambda - \delta \ln[-\ln P(x)]. \tag{6.70}$$

that represents the maximum value, x_{max} that shall be expected with a given probability $P(x)$ once the cumulative probability function is known, i.e., the values of the parameters λ and δ. In Sect. 6.6.1 we have stressed the equivalence between probability and volume process, see Eqs. (6.54) and (6.55), by saying that to a certain probability $P(\%)$ it corresponds a volume process or a number of specimens equal to $1/(1 - P)$. Let's then introduce the parameter $T = 1/(1 - P)$. Equation (6.70) becomes together

$$x = \lambda - \delta \ln\left[-\ln\left(1 - \frac{1}{T}\right)\right]. \tag{6.71}$$

The parameter T is known as the *returning period* because the first application of the Gumbel extreme value statistic was to forecast when the maximum level of the waters of Rhone river would be reached that would cause flooding. As such, $1/T$ was not representing a process volume or a number of specimens, as in the case of a mechanical variable, but the years needed to have a flood; therefore it was called the returning period of a flooding. We are used to maintain that definition knowing, however, that in our case it represents the number of specimens or the number of elementary volumes. Murakami [12–15] has used the Gumbel distribution to address the issue of inclusion effects in metals using a method known as inclusions rating method by statistics of extreme (IRMSE). We have long been discussing this issue in Section. showing how inclusions are affecting the fatigue strength in competition with surface roughness. A perfectly clean surface is worthless if the material is not also perfectly clean, i.e., inclusion free, at least on the surface. The word *perfectly* or *perfection* actually attains the world of theory. In practice, by clean steel we mean a steel with a very low number of surface inclusions of any size. To have a clean alloy is particularly true for hard high strength streels and alloys of any kind whose fatigue strength is strongly jeopardized by the presence of surface inclusions. The Murakami approach is based on the assessment of the maximum size of an inclusion that may be reasonably expected in a given process volume on the base of the examination of a large number of small specimen surfaces used as sample population. Figure 6.21 explains the procedure that consists in:

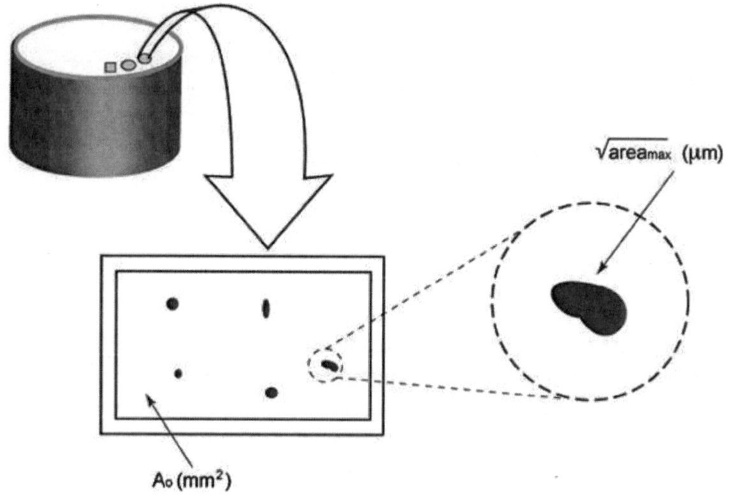

Fig. 6.21 IRMSE procedure for inclusion rating [14]

(1) cutting from a specimen a section perpendicular to the maximum principal stress. After polishing, the test surface is mirror-finished with buff;
(2) on this section a standard specimen area A_o is chosen as reference (in mm^2) and the inclusion of maximum size (area) selected;
(3) this operation is repeated n times on n equal areas of n different specimens (nA_o);
(4) the largest inclusions areas $area_{max}$ found on each of the selected areas $A_{o,i}$ ($i = 1, \ldots n$) are ranked from the smallest to the largest and a probability assigned using one of the ranking Eqs. (6.41) through (6.45);
(5) Eq. (6.69) is then used to plot the data on a probability paper and infer the λ and δ values.

A probability of occurrence, P, is then chosen or, better, a return period T to which it would correspond a given $\sqrt{area_{max}}$ according to Eq. (6.71)

$$area_{\max}(A) = \lambda - \delta \ln\left[-\ln\left(1 - \frac{1}{T}\right)\right] = \lambda - \delta \ln\left[-\ln\left(1 - \frac{A}{A_o}\right)\right] \quad (6.72)$$

where $T = A/A_o$ is the population of areas, i.e., the total target area divided by the reference area. To the expected maximum size a fatigue limit corresponds that can be assessed by the Murakami and Endo Eqs. (2.13) and (10.12) or using the ETS theory described in Fig. 2.44. Examples of the extreme value statistics are given in the next section.

6.8 Sample Problems

6.8.1 *Normal and Log-Normal Distribution-Population Mean and Sample Mean*

A low-alloyed carbon steel structure needs to be designed to withstand cyclic loads with a design life of at least 5,000,000 cycles with 99.9 and 99.99% probability of survival. Waiting for fatigue data obtained from fatigue tests on the actual material the designer decides to run a congruous number of fast monotonic traction tests on round specimens 10 mm diameter and 100 mm length to derive the fatigue limit S_f from a simple approximate equation of the type $S_f \approx 0.5\,\sigma_u$, see Eq. (10.1). A first series of 15 traction tests is run that yields the results listed in the following table. Since the designer knows that there are few chances to have results distributed according to a normal distribution, he decides to work on the logarithms for a log-normal distribution. The $log(\sigma_u)$ values for the first series of 15 specimens are listed in the second column of the same table. The sample mean for the normal and log-normal distribution is

$$\overline{\sigma}_u = 534.1\,\text{MPa}$$
$$\overline{\log}(\sigma_u) = 2.728$$

The sample standard deviation for the log-normal distribution is

$$s = 0.012$$

Using the probability diagram of Fig. 6.8 the number n of standard deviation to detract from the mean for 99.9 and 99.99% survival is $n = 3.1$ and $n = 3.7$, respectively. Therefore

$$R(99.9) = \overline{\log}(\sigma_u) - 3.1 \cdot 0.012 = 2.691$$
$$R(99.99) = \overline{\log}(\sigma_u) - 3.7 \cdot 0.0127 = 2.684$$

With these values the approximate fatigue limits are

$$S_{f,99.9} = 10^{2.691}/2 = 245.8\,\text{MPa}$$
$$S_{f,99.99} = 10^{2.684}/2 = 241.7\,\text{MPa}$$

First series of 15 specimens			Second series of 15 specimens			Third series of 15 specimens		
σ_u (MPa)	Log(σ_u)	S_f *(MPa)*	σ_u (MPa)	Log(σ_u)	S_f *(MPa)*	σ_u (MPa)	Log(σ_u)	S_f *(MPa)*
507.0	2705	253.50	501.2	2700	250.6	533.3	2727	266.7
515.2	2712	257.61	512.8	2710	256.4	533.9	2727	267.0
518.8	2715	259.40	514.2	2711	257.1	534.6	2728	267.3
524.8	2720	262.40	520.8	2717	260.4	537.0	2730	268.5
527.8	2723	263.92	522.6	2718	261.3	540.2	2733	270.1
527.7	2722	263.87	540.9	2733	270.5	540.9	2733	270.5
530.8	2725	265.41	530.9	2735	271.4	542.9	2735	271.4
533.9	2727	266.96	532.1	2739	274.4	543.3	2735	271.7
537.0	2730	268.52	533.3	2745	278.1	546.5	2738	273.3
540.2	2733	270.09	534.6	2746	278.9	548.8	2739	274.4
543.3	2735	271.67	509.6	2725	265.4	549.7	2740	274.9
546.5	2738	273.26	509.6	2726	266.1	552.9	2743	276.5
549.7	2740	274.86	509.6	2727	266.7	556.2	2745	278.1

The designer knows very well that the equations used for deriving *R*(*99.9*) and *R*(*99.99*) probability of survival are valid for a total population mean and standard deviation while the values he calculated on the sample of 15 data depend on the sample size. To check that he runs some 15 new traction tests on cylindrical specimens of the same size. The results are listed in the previous table. Using the new sample of 30 specimens he re-run all the statistics and gets

$$\overline{\sigma}_u = 532.5\,\text{MPa}$$
$$\overline{\log}(\sigma_u) = 2.726$$
$$s = 0.013$$

Effectively, as expected statistics are chanced, in particular the sample standard deviation s. With these new data the *R(99.9)* and *R(99.99)* probability of survival become

$$R(99.9) = \overline{\log}(\sigma_u) - 3.1 \cdot 0.013 = 2.687$$
$$R(99.99) = \overline{\log}(\sigma_u) - 3.7 \cdot 0.013 = 2.679$$

that yields

$$S_{f,99.9} = 10^{2.687}/2 = 242.3\,\text{MPa}$$
$$S_{f,99.99} = 10^{2.679}/2 = 237.7\,\text{MPa}$$

The new fatigue limits are 1.4% and 1.7% lower, respectively. The designer continues to run a third series of 15 additional traction tests, whose results are listed in the previous table. Also the bar diagram for the 45 data indicates a distribution close to a log-normal one, as shown in Fig. 6.22. This gives him the confidence that the experimental data are effectively following a log-normal distribution. More convincing than that is the plot of the logarithms of these 45 data on the normal probability paper. Effectively the data are almost aligned along a straight line as shown in Fig. 6.23. The designer re-run all the statistics on these 45 tests and gets

$$\overline{\sigma}_u = 537.8\,\text{MPa}$$
$$\overline{\log}(\sigma_u) = 2.723$$
$$s = 0.015$$

$$R(99.9) = \overline{\log}(\sigma_u) - 3.1 \cdot 0.015 = 2.675$$
$$R(99.99) = \overline{\log}(\sigma_u) - 3.7 \cdot 0.015 = 2.666$$

$$S_{f,99.9} = 10^{2.675}/2 = 236.9\,\text{MPa}$$
$$S_{f,99.99} = 10^{2.65}/2 = 231.9\,\text{MPa}.$$

The results are different from the 15 and 30 sample sizes analysis. Now the reduction of the expected fatigue limit is 3.6% for the 99.9% probability of survival and 4% for the 99.99% probability of survival that passes, this last, from 241.7 MPa to just 231.9 MPa. The designer continues his traction tests collecting a total of 140 data. The new distribution is shown in Fig. 6.24. Note the difference with that of Fig. 6.21. New entries have come to populate, in particular, the left tail of the

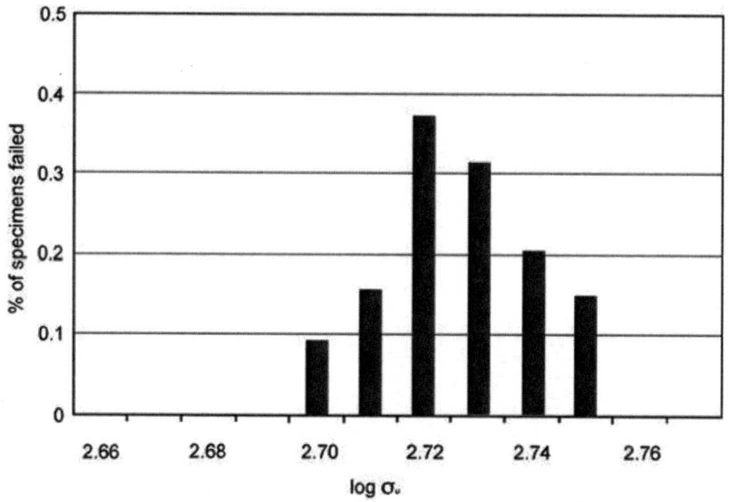

Fig. 6.22 Histogram showing the distribution of the 45 traction test results

distribution. The low probability data are extremely important to change the overall picture.

The new statistics yield

$$S_{f,99.9} = 223\,\text{MPa}$$
$$S_{f,99.99} = 216\,\text{MPa}$$

Now the reduction for the 99.99% probability of survival has grown to 10% from 241.7 MPa down to 216 MPa. As last, the designer wants to know how far or close these 140 samples are from the population size to use the finite population correction (FPC) given by Eq. (6.10). To that purpose he calculates the overall volume of all the work pieces that must be built with that steel to assess N using Eq. (6.2). The overall volume is equal to $2.2 \times 10^9\ \text{mm}^3$. Since the volume of the traction specimen used is equal to 7854 mm^3 the number N of specimens of the entire population is $N = 2.2 \times 10^9/7.854 = 280{,}112$. He concludes that 140 specimens represent only a mere 0.05% of the entire population size and he cannot apply any correction. This simple example evidences the difference that may arise when using a sample of limited size. Statistics results depend on the sample size. We may try, then, to use the Lieberman one-sided tolerance limit. The mean value of the logarithms of experimental data is $\overline{\log(\sigma_u)} = 2.717$ and $s = 0.02175$. From Table 6.6 we infer that the value of k for a sample of size 140 and a probability of survival of 99.9% with 0.90% confidence is: $k = 2.953$ therefore

$$\text{LCL}_{99.95} = 2.717 - 2.953 * 0.02175 = 2.653$$
$$S_{f99.95} = 10^{2.653}/2 = 224.7\,\text{MPa}.$$

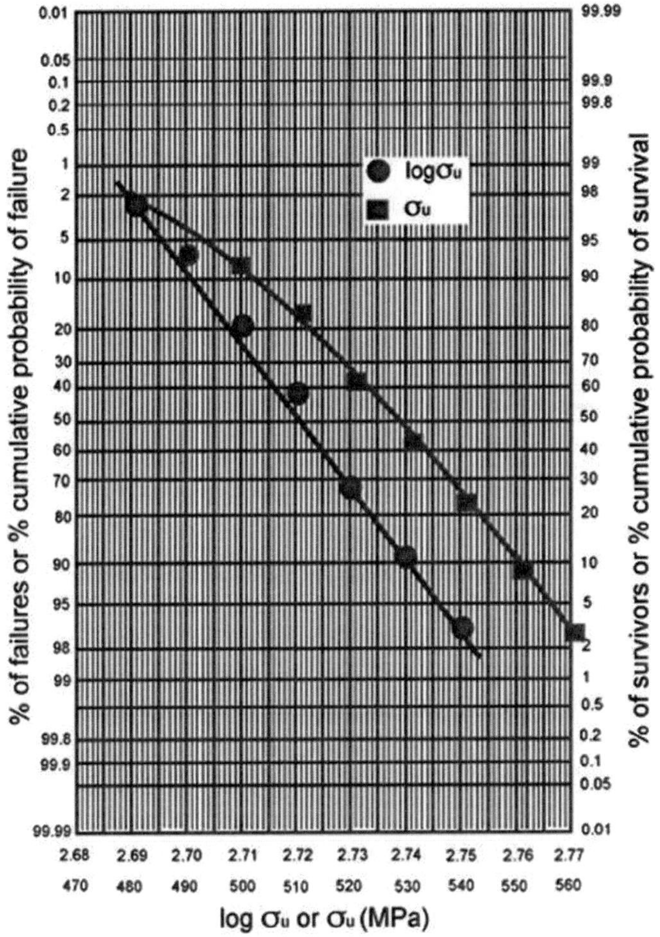

Fig. 6.23 Plot of the 45 traction test data revealing a log-normal distribution of the sample (closed circles). The traction test results (closed squares) do not follow a normal distribution

6.8.2 Student's Distribution-Yield Strength

The yield strength σ_y of carbon steel measured with 10 specimens of 5 mm radius and 100 mm long is shown in Fig. 6.25. Calculate the value of the yield strength corresponding to a probability of survival of 99%. The mean strength $\overline{\sigma}$ is 233 MPa. Assuming that the sample data are distributed according to a normal distribution, the sample standard deviation s calculated using Eq. (6.7) is equal to 5. Using the diagram of Fig. 6.8, the value of the yield strength with 99% probability of survival is

$$s_y = 233 - 2.9 \cdot 5 = 218.5\,\text{MPa}$$

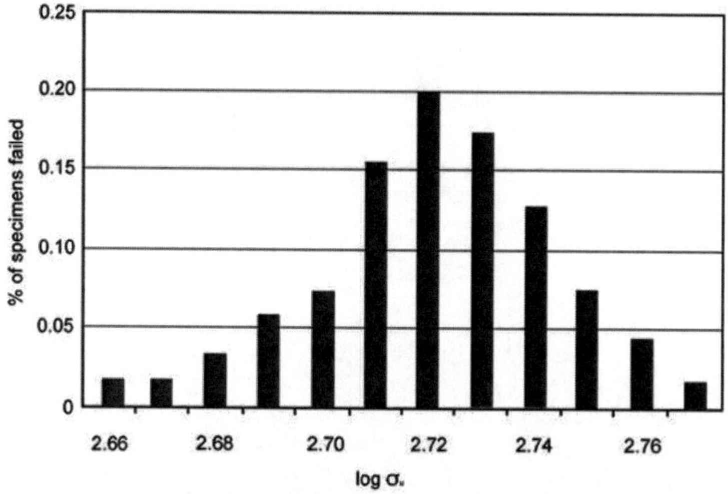

Fig. 6.24 Histogram showing the distribution of the sample of 140 experimental data

having subtracted to the mean value 2.9 standard deviations. But we know that 233 MPa is not the true mean of the population, but the sample mean (see previous problem). Also the standard deviation equal to 5 MPa is not that of the real population so that the final result 218.5 MPa depends on the size (10) of the sample considered. If more data were available increasing the sample size it is likely that the mean and the standard deviation would change, as seen in the previous problem, and, therefore, also the value of the yield strength with 99% probability of survival would change. We can try to assess what may be the interval in which the true mean will fall with 99% probability using the Student distribution. Table 6.5 indicates that with $n = 10$ the student parameter $t_{\alpha,\nu}$ is 2.821 so that

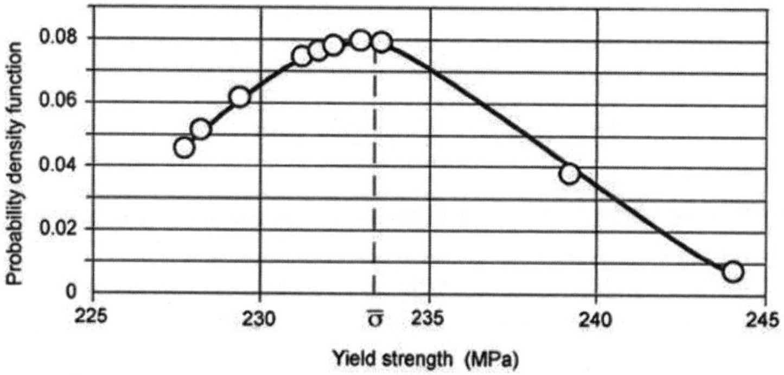

Fig. 6.25 Probability density curve of the ten traction tests

$$\bar{x} - 2.821\frac{5}{\sqrt{10}} \leq \mu \leq \bar{x} + 2.821\frac{5}{\sqrt{10}};$$
$$233 - 4.46 \leq \mu \leq 233 + 4.46;$$
$$228.54 \leq \mu \leq 237.46$$

Furthermore, as to the true standard deviation Δ by using the χ^2 function and the diagram of Fig. 4.11 with $n = 10$, with 99% it will be $\Delta < (5/\sqrt{0.24} = 10.2)$ while 1% of values will fall below $\Delta < (5/\sqrt{2.5} = 3.16)$. Therefore, the only thing we can say is that the true population mean will fall, with a probability of 99%, between 228.54 MPa and 237.46 MPa, with a 0.5% probability to fall below 228.54 MPa, and the true standard deviation will have the limits $3.16 < \Delta < 10.2$ with 98% confidence. An alternative is to use the Lieberman values k given in Table 4.6. According to the lower confidence limit procedure LCL the minimum yield strength of the material with 99% probability of survival and 99% confidence will be

$$LCL_{.99} = \bar{x} + k \cdot s = 233 - 5.075 \cdot 5 = 207.6\,\text{MPa}.$$

6.8.3 Log-Normal Distribution-Fatigue Life

The histogram of Fig. 6.26a represents the distribution of fatigue life N of 57 specimens of 75S-T aluminium alloy tested in reversed bending at 207 MPa (30 ksi) [10]. The dashed curve is a possible density probability distribution of the data $p(N)$. This distribution clearly does not correspond to a normal distribution. If the data are converted into log life and replotted as shown in Fig. 6.26b, the resulting distribution appears to be closer to a log-normal, which actually means that it can follow a Gauss distribution but it may not. The mean value is $\overline{\log N} = 7.212$ and the standard deviation is $s = 0.413$.

As in the previous problem both the mean and the standard deviation are sample values and not population values. Then, we may wish to assess the interval of 90% confidence level in which the true population mean will fall and the minimum value expected. Table 6.5 is showing that with $\nu = n$ -1 $= 56$ it is $t_{\alpha,\nu} = 2.395$, therefore the $\text{UCL}_{1-\alpha}$ and $\text{LCL}_{1-\alpha}$ are

$$UCL_{1-\alpha} = \bar{x} + t_{\alpha,\nu}\frac{s}{\sqrt{n}} = 7.21 + 2.395 \cdot \frac{0.413}{\sqrt{57}} = 7.341$$
$$LCL_{1-\alpha} = \bar{x} - t_{\alpha,\nu}\frac{s}{\sqrt{n}} = 7.21 - 2.395 \cdot \frac{0.413}{\sqrt{57}} = 7.079$$

Therefore

$$P(7079 < \mu < 7341) = 0.9$$

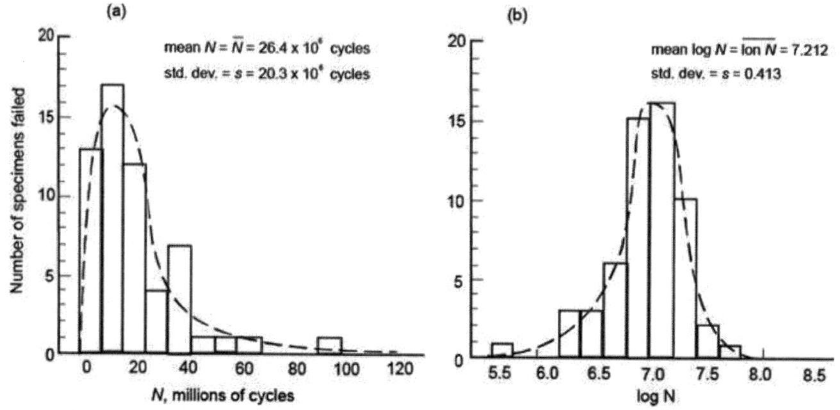

Fig. 6.26 Histogram showing: **a** fatigue life distribution of 57 specimens of 75S-T aluminum alloy tested in reversed bending at 206,7 MPa (30 ksi) and **b** log-normal distribution (adapted from [11])

and

$$P(7079 < \mu) = 0.95$$

This actually means that the true mean has 95% probability to be higher than 11,994,603 cycles against a sample mean of 16,292,960 cycles.

6.8.4 Weibull Distribution-Yield Strength

Let's consider sample problem 2 again. This time the ten experimental data will be analysed using the Weibull distribution. As first we list the ten results x_i ($i = 1, 2, 3, \ldots, 10$) in ascending order. Second, we assign a probability P to each result by considering its frequency of occurrence according to the mean rank Eqs. (4.40) or (4.41). Then we plot $\ln\{\ln[1/(1 - P)]\}$ versus the logarithm of the yield strength, as shown in Fig. 6.27a. It can be seen that the experimental data are aligned as expected. The slope of the line is the Weibull exponent, which is equal to 40.87 with the shape factor $\sigma_o = 236.4$ MPa. If we want to know the yield strength of the material with 99% probability of survival we may write

$$\sigma_y(99\%) = 236.4 \cdot [-\ln(1 - 0.01)]^{1/40.87} = 211.23\,\text{MPa}$$

instead of 218.5 MPa calculated with the normal distribution. We may also want to assess the volume effect and ask what the yield strength would be using specimens of 20 mm diameter and 200 mm length. The volume ratio is

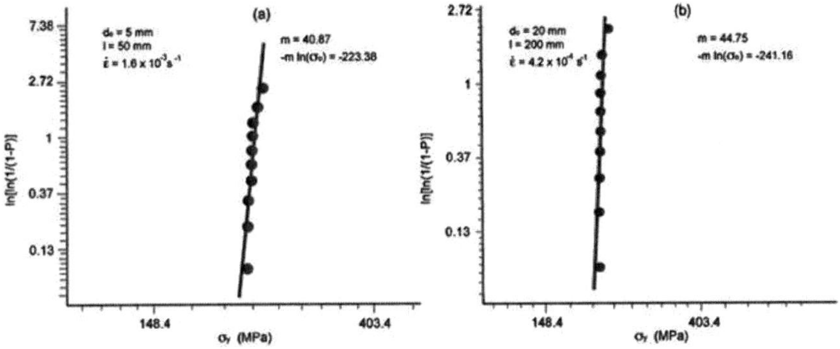

Fig. 6.27 Weibull plot of the ln{ln(1/(1 − P)} versus the yield strength **a** of the ten 5 mm diameter specimens and **b** of the 20 mm diameter specimens

$$s = \frac{V_{\varphi 20}}{V_{\varphi 5}} = \left(\frac{10}{2.5}\right)^2 \frac{200}{50} = 64.$$

We need to infer the scale factor $\sigma_{o,20\ \mathrm{mm}}$ for the 20 mm diameter samples. From Eq. (6.48)

$$\sigma_{o,20\ \mathrm{mm}} = \sigma_{o,5\ \mathrm{mm}} \left(\frac{1}{64}\right)^{1/40.87} = 213.5\ \mathrm{MPa}$$

therefore

$$\sigma_{y,20}(50\%) = 213.5 \cdot (-\ln 0.5)^{1/40.87} = 211.59\ \mathrm{MPa}.$$

To verify the expected value $\sigma_{y,20} = 211.59$ MPa of the yield strength for the 20 mm diameter samples, ten specimens of the same material have been prepared having a diameter of 200 mm and a gage length of 200 mm. The results are listed in the following table and plotted in Fig. 6.27b. The Weibull exponent is $m = 44.75$ and the scale factor is $\sigma_{\mathrm{o},20\ \mathrm{mm}} = 219$ MPa against the value of 213.5 MPa predicted by 5 mm diameter specimens. The predicted value is lower than the measured by 2.5% probably due to the fact in Eq. (6.48) it has been assumed m constant and not varying with the volume; therefore the calculated value is predicting a volume effect larger than the actual one. The experimental mean value is $\overline{\sigma}_{y,20} = 216.6$ MPa about 2.2% larger than predicted.

Yield strength of Φ 20 mm specimens		
213.8	215.1	218.1
214.0	216.1	226.5
214.0	216.6	
214.5	217.6	

6.8.5 Weibull Distribution-Volume Effect

Let's analyse the specimens of different size considered in Fig. 6.16. The Weibull probability plot of ln[ln(1/(1 − P))] as function of the true failure stress, σ_t, for the four sizes considered is shown in Fig. 6.28. As already presented in Fig. 6.18, the Weibull exponent m that indicates the slope of each strait line, increases as the size of the specimen increases, going from 3 to 22 mm. We can use Eq. (6.51) to pass from one specimen size distribution to the other. The ratio between the volume of the specimen of 5 mm diameter to that of 3 mm diameter is: $V_5/V_3 = 7.56$. This ratio must be elevated to $1/m$. but the Weibull exponent m is not constant but varies with volume. The most suitable m value is the mean between m_3 and m_5: $\overline{m} = 9.68$; therefore, if we move from the 3 mm specimen distribution that has a peak value of 557 MPa to that of the 5 mm, the predicted peak value is

$$\sigma_{5,peak} = \sigma_{3,peak}\left(\frac{V_3}{V_5}\right)^{1/\overline{m}} = 557 \cdot 0.13^{1/9.68} = 451\,\text{MPa}$$

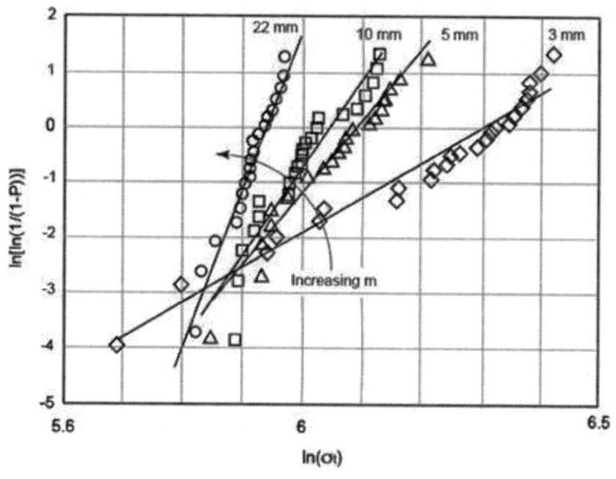

Fig. 6.28 Weibull probability plot for the four sizes tested

which is precisely the peak value of the 5 mm diameter distribution (see Fig. 6.16). Results for all passages from the smaller to larger specimens are listed in the table below

	V_3/V_i	$\overline{m}$	$(V_3/V_i)^{1/m}$	$\sigma_{i,peak}$	$\sigma_{o,i}$	σ_{peak}/σ_o
V_3/V_5	0.13	9.68	0.81	451	450	1.0
V_3/V_{10}	0.017	11.67	0.71	394	390	1.01
V_3/V_{22}	0.0023	16.12	0.69	383	374	1.02

where $\sigma_{o,i}$ represents the experimental peak stress of the i-distribution with $i = 5$, 10 and 22. The agreement between predicted values and experimental values is rather good.

6.8.6 *Weibull Distribution-Volume Effect*

In order to verify the volume effect it has been used a metal with a very low Weibull exponent like ferritic ductile cast iron (DCI). To that purpose, five sets of 15 specimens each have been prepared. Each set has specimens of the same volume that varies from set to set from a minimum of 500 mm^3 to a maximum of 400,000 mm^3. Each set of specimens has been subjected to a monotonic traction test and the yield strength analysed with the Weibull approach to infer the relative m exponent and the scale factor σ_o. Results are shown in Fig. 6.29 and listed in the following table. The minimum m value is 9, rather poor and typical of low quality DCI, the maximum is 40 typical of a very good quality material. The equation of the curve of Fig. 6.29 interpolating the five experimental results is

$$m = 2.2 \cdot V^{0.225}. \tag{6.73}$$

It is worth noting that Eq. (6.73) is well interpolating results obtained with the experimental campaign described in the former example. Since the m exponent is not constant, but varies with the volume of the specimens, to assess the volume effect Eq. (6.51) cannot be used because derived from Eq. (6.53) with m constant.

Because of the experimental evidence of a Weibull m exponent varying with the process volume according to Eq. (6.73), Eq. (6.53) must be integrated. The integral may be solved by finite differences method that yields the result shown in Fig. 6.30 by the solid line. The result that would be obtained considering $m = 9$ constant is also shown (dashed line). The difference is enormous. With a volume ratio of 1000, for example, the constant m analysis would indicate a reduction of the yield strength of about 45% while the variable m analysis is predicting a reduction of 25% only, almost a half of that. Furthermore, the variable m analysis is indicating that the saturation of the volume effect would take place at a process volume ratio of

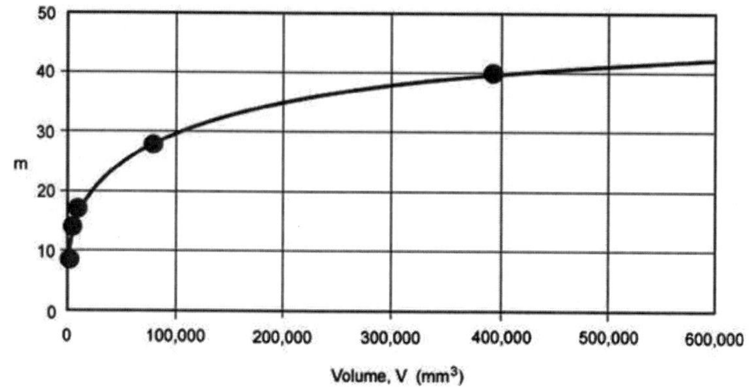

Fig. 6.29 Variation of the Weibull exponent m with specimen volume for a DCI

about 10,000 while the constant m analysis does not show any saturation at all. It must be pointed out the reason why in this example, as in the following one, it has been considered the yield strength of the material rather than the ultimate or the engineering fracture strength. This last strength, in fact, depends on the volume of the specimens, but also on the amount of necking at rupture, which is related to the area reduction of the specimen. Also necking depends on metallurgical variability of the material as the volume effect, but its contribution is actually influencing the true fracture stress. Necking has no effect on the yield strength. Therefore, when using the Weibull approach to infer the volume effect and derive what may be the strength of a given volume of material different from that of the small specimen it must be considered the true fracture stress and not the ultimate or the engineering fracture stress using the procedure outlined in Sect. 5.2.2.

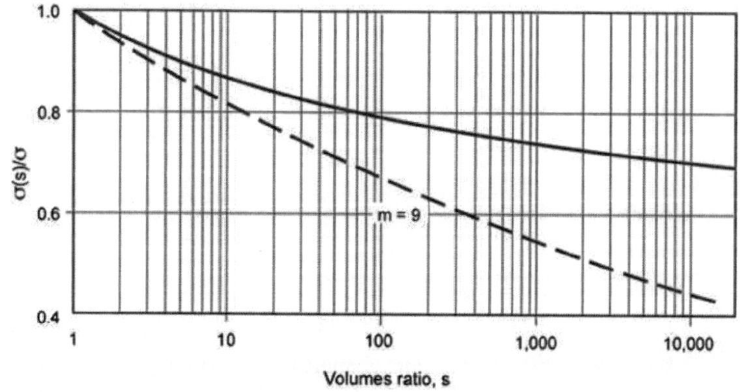

Fig. 6.30 Comparison between the volume effect evaluated with m constant equal 9 and that derived considering m variable according to Eq. (6.73)

Volume (mm^3)	m	σ_o (MPa)
500	9	540
4000	14	460
80,000	28	405
400,000	40	380

6.8.7 Fatigue S–N Curves—Weibull Distribution

Carbon steel specimens tested at four different stress amplitudes with maximum stress of 346, 324, 302 and 270 MPa, respectively, have yielded the fatigue lives listed in the following table.

270 MPa	302 MPa	324 MPa	346 MPa
4.57E+05	2.00E+05	8.00E+04	3.40E+04
7.20E+05	2.20E+05	8.90E+04	3.59E+04
8.27E+05	2.58E+05	1.08E+05	3.87E+04
1.13E+06	2.84E+05	1.18E+05	4.26E+04
1.26E+06	3.34E+05	1.26E+05	4.50E+04
2.40E+06	3.67E+05	1.35E+05	5.26E+04
3.07E+06	4.09E+05	1.50E+05	5.51E+04
> 7.20E+06	4.56E+05	1.58E+05	6.00E+04
> 7.20E+06	4.90E+05	1.81E+05	6.30E+04

Using the Weibull distribution derive the 90 and 99.9% probability of survival *S–N* curves and that at 90% probability of survival with 90% confidence P(90)C(90). The experimental data are reported in Fig. 6.31 with the average S–N curve marked as P(50). Analyzed with the Weibull procedure, already applied in the previous problem, the four different stress amplitudes yield the results shown in Fig. 6.29. Also shown are the best-fitting lines with the relative equations.

It is worth noting that the values of the m exponents are different from those found in the previous section that pertained to traction tests. It is also worth noting how the Weibull slope m decreases as the maximum stress decreases passing from $m = 5.16$ at 346 MPa to $m = 1.0$ at 270 MPa. This indicates how the dispersion of experimental data varies as inverse function of stress level, very large for low stress and lower for higher stress level. This circumstance is clearly indicated by the cumulative distribution of Fig. 6.32, see Eq. (6.39). The following table lists the scale factor, the Weibull exponent m, $n = 1/m + 1$, the gamma function $\Gamma(n)$, the mean $\overline{N}$, the median $\breve{N}$, the mode $\tilde{N}$ and the standard deviation s.

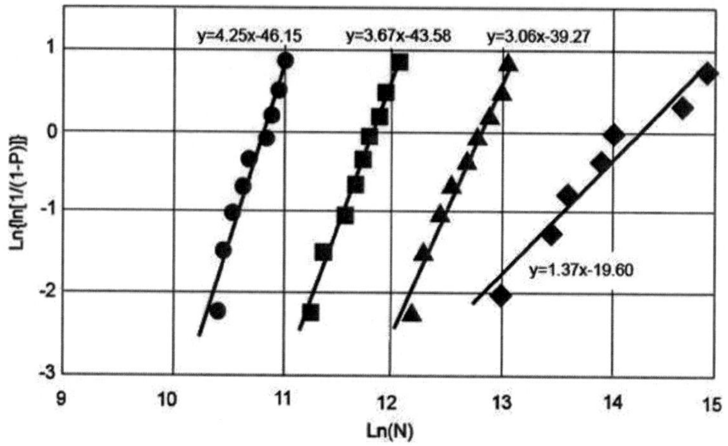

Fig. 6.31 Weibull ln{ln[1/(1 − P)]} versus ln(N) diagram for the four amplitude considered

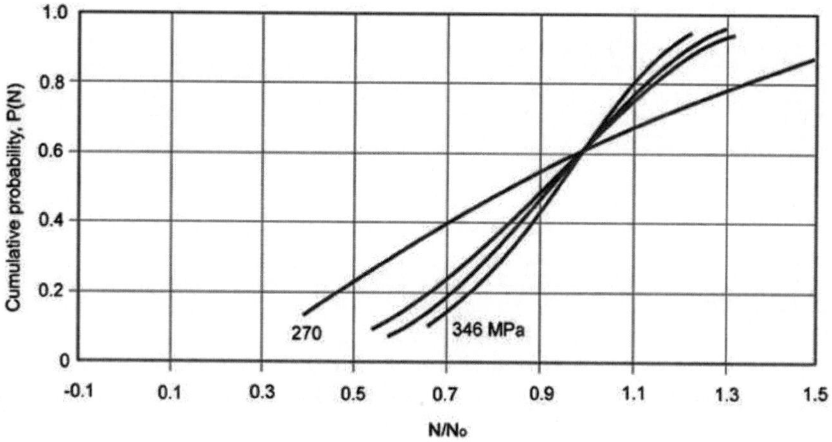

Fig. 6.32 Weibull probability distribution versus N/N_o with N_o being the scale factor

	270 MPa	302 MPa	324 MPa	346 MPa
N_o	1.653.625	377.775	141.308	52.136
m	1.369	3.058	3.673	4.249
$\Gamma(n)$	0.9144932	0.893392	0.901708	0.909137
$\overline{N}$	1,512,229	337,501	127,419	47,399
$\breve{N}$	1,265,221	335,105	127,888	47,827
$\tilde{N}$	634,644	331,887	129,595	48,945

(continued)

(continued)

	270 MPa	302 MPa	324 MPa	346 MPa
s	1,118,076	120,834	38,728	12,643

Interesting enough, the mean, median and mode becomes closer and closer as the stress level increases. At 346 MPa, in particular, the three parameters are practically equal to each other indicating that the experimental data (cycles) have a bell-shaped probability density $p(N)$, i.e., a normal distribution (see Fig. 6.3). This is not the case for the lowest stress level in the range of the fatigue limit where we know that cycles are not normally distributed. At 90% probability of survival it is

$$N_{270}(90\%) = 1{,}653{,}625 \times [-\ln(0.90)]^{1/1.369} = 319{,}552\,\text{cycles}$$
$$N_{302}(90\%) = 377{,}775 \times [-\ln(0.90)]^{1/3.058} = 201{,}483\,\text{cycles}$$
$$N_{324}(90\%) = 141{,}308 \times [-\ln(0.90)]^{1/3.673} = 76{,}574\,\text{cycles}$$
$$N_{346}(90\%) = 52{,}136 \times [-\ln(0.90)]^{1/4.249} = 30{,}699\,\text{cycles}$$

At 99.9% probability of survival it is

$$N_{270}(99.9\%) = 1{,}653{,}625 \times [-\ln(0.999)]^{1/1.369} = 10{,}647\,\text{cycles}$$
$$N_{302}(99.9\%) = 377{,}775 \times [-\ln(0.999)]^{1/3.058} = 39{,}470\,\text{cycles}$$
$$N_{324}(99.9\%) = 141{,}308 \times [-\ln 0.999)]^{1/3.673} = 21{,}550\,\text{cycles}$$
$$N_{346}(99.9\%) = 52{,}136 \times [-\ln(0.999)]^{1/4.249} = 10{,}260\ \text{cycles}$$

The relative *S–N* curves are shown in Fig. 6.33. It stems dramatically out the reduction of cycles to failure at 99.9% probability of survival. At a stress amplitude close to the fatigue limit one specimen out of 1000 is expected to fail at about 70,000–80,000 cycles or, at most, 100,000 cycles. Note that the 50% probability curve, solid line marked P(50), has been traced averaging experimental points. Also note that the P(90) point at 270 MPa is not on the trend line.

This is because the Weibull analysis at that stress amplitude was not considering the two specimens that did not fail at 7,000,000 cycles, when the tests were interrupted. Therefore, the analysis is unbalanced on the side of low cycles lacking the high cycle side. If the probability density $p(N)$ were normally distributed, to infer the 90% probability of survival with 90% confidence, $P(90)C(90)$, it may be used the LCL approach of Sect. 6.5 using the Lieberman's k values of Table 6.6

$$P(90)C(90) = \overline{N} - k \cdot s$$

With a sample size of 9 specimens it is $k = 2.133$ therefore for the three stress levels likely to have a probability density $P(N)$ that follows a normal distribution it is

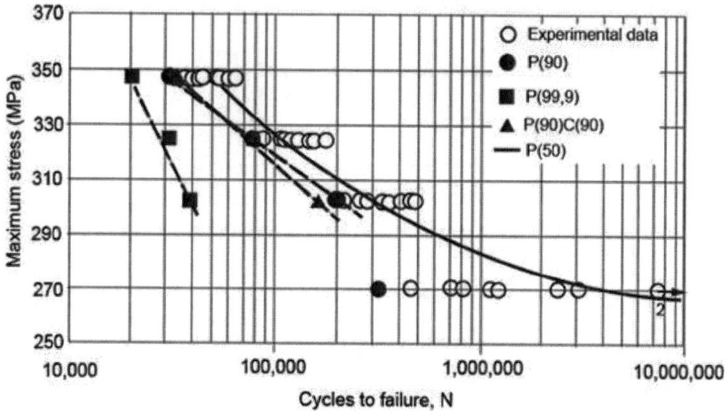

Fig. 6.33 S–N curves at 50% (mean curve), 90 and 99.9% probability of survival and P(90)C(90) S–N curve

$$P_{346}(90)C(90) = 48{,}173 - 2.133 \cdot 6383 = 34{,}557$$
$$P_{324}(90)C(90) = 129{,}263 - 2.133 \cdot 23{,}473 = 79{,}196$$
$$P_{302}(90)C(90) = 340{,}269 - 2.133 \cdot 84{,}340 = 160{,}371$$

These three points are shown in Fig. 6.33.

6.8.8 Weibull Distribution-Volume Effect

The same study presented in Sects. 6.8.5 and 6.8.6 on DCI has been conducted on a medium strength carbon steel using three batches of 20 specimens. Each batch had a different volume. The specimens were subjected to monotonic traction tests and the true strengths measured on specimens of the same size were reported in a Weibull plot $\ln\{\ln[1/(1-\mathrm{P})]\}$ versus $\ln\sigma_t$. Using this procedure, the m exponent for each batch has been inferred and, therefore, for each process volume. Results are listed in the following table and presented in Fig. 6.34.

Volume (mm^3)	m
8000	26
60,000	44
1,000,000	65

The interpolating curve has the equation

$$m = 5.5 \cdot V^{0.18}. \tag{6.74}$$

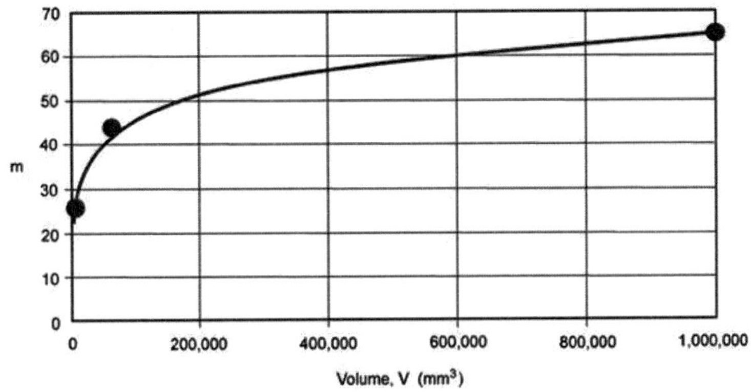

Fig. 6.34 Variation of the Weibull exponent m with specimen volume for a medium strength carbon steel

Equation (6.74) may be used to calculate the integral (6.53) to derive the volume effect. Again, the integral may be solved using the finite differences method. Results are shown in Fig. 6.35 by the solid line. Dashed line, instead, indicates the volume effect evaluated using $m = 40$ constant. This time, at variance with the DCI material, there is no significant difference between the two calculations. Again, as with the DCI material, considering a Weibull exponent varying with the volume leads to a saturation effect that is not seen with m constant.

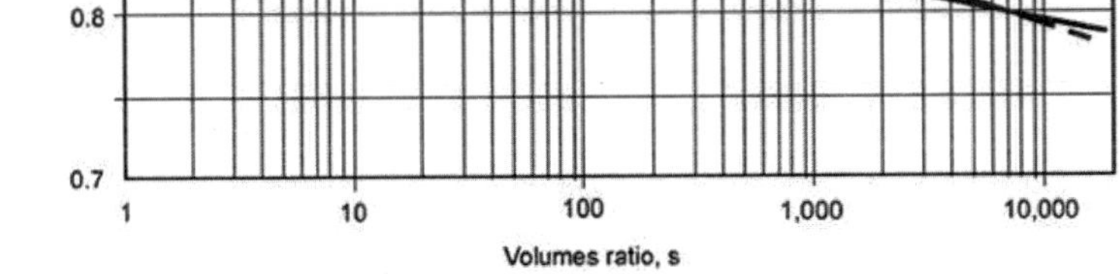

Fig. 6.35 Comparison between the volume effect evaluated with m constant equal 40 (dotted line) and that derived considering m variable according to Eq. (4.60)

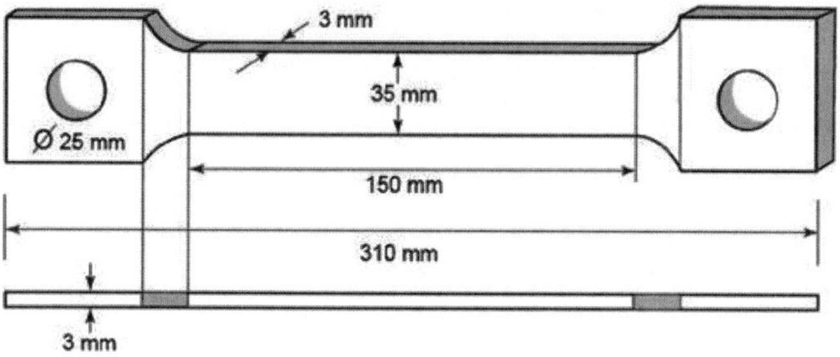

Fig. 6.36 Flat specimen used in the experimental campaign on GGG 40 ferritic cast steel

6.8.9 Gumbel Distribution-Maximum Expected Defect

A long lasting research program was run on ductile ferritic cast iron (DCI), type GC 40, of Italian, German and Chinese production. The aim was at assessing the fatigue limit lower bound using the statistic of extreme values (SEV) applied to a sample population of defects. To this purpose, it was decided to use of long, flat specimens of rather large process volume of the type shown in Fig. 6.36.

The flat surface would make possible an extensive metallurgical examination campaign while the 40-fold larger volume, with respect to smaller standard fatigue specimen, would increase the probability to sort out the real dangerous metallurgical damage initiators. This last choice was rewarding because it turned out that in all cases the initiation site was not chunky or degenerated, needle like graphite, as with smaller cylindrical specimens, but a much larger lack of fusion defect never seen before. Figure 6.37 is an example of such a defect.

As it can be seen, the lack of fusion void is about 2 mm long; it is like a cavern in comparison with the small graphite nodules. Some 25 specimen were cut from the same heat of ductile cast iron of the Italian production and tested in fatigue at $R = 0$. After the test, a section was removed from each specimen. The section was 30×35 mm wide and contained the fracture surface, as shown in the macrograph Fig. 6.38. Each section was examined at the optical microscope to check whether other lack of fusion signs could be found on the external surface. Apparently there was no sign at all of such a type of defect.

There were only clear signs of pearlite islands. However, a deeper examination revealed that each pearlite island contained the lack of fusion we were looking for. This is shown in the detail of Fig. 6.39. The specimens were sliced to check whether the entire volume of pearlite contained the lack of fusion. It turned out that the lack of fusion was an integral part of the pearlite phase. Pearlite was definitely containing the defect we were searching for. The next step was to use the Gumbel procedure to assess the maximum defect size that would possibly exist on a process volume equal to that of a single piece of machinery times the number of pieces to be built. In doing that a modified IRMSE procedure was followed (see Sect. 6.7). First of all, it was decided to measure the size of the pearlite island as being that of the defect we wanted to sample.

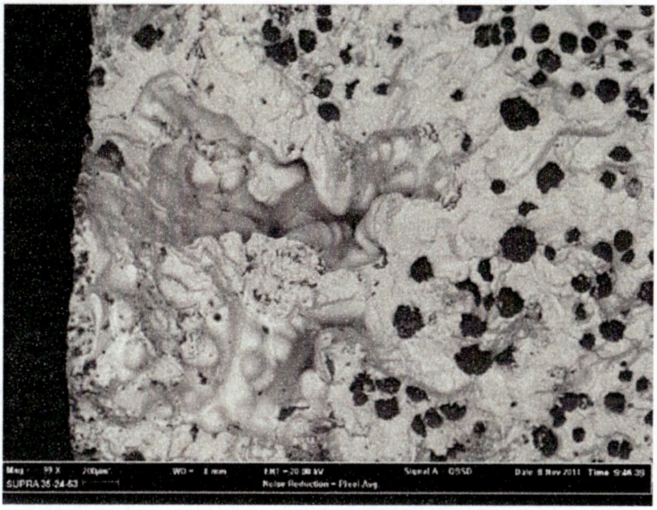

Fig. 6.37 SEM examination of the fracture surface of a flat specimen of ductile cast iron, 3 mm thick, of Italian production showing the origin of fatigue failure (Courtesy of M. Cova, SACMI Imola, Italy)

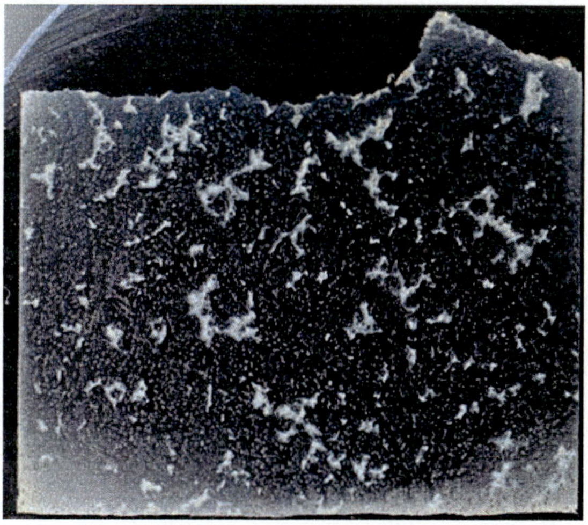

Fig. 6.38 Section 35 × 30 mm containing the fracture section etched to evidence pearlite islands. Graphite nodules appear as small withe points (Courtesy of M. Cova, SACMI Imola, Italy)

Probably, this size is somewhat larger than that of the internal void, but it was the simplest procedure to follow and was also leaning on the conservative side. Next step was to sample all the pearlite islands present on each specimen, like that of Fig. 6.38, having $\sqrt{area}$ larger than a threshold of 0.55: $\sqrt{area} \geq 0.55$ mm. Once the sampling of all pearlite islands was done, the pearlite areas were ranked in ascending order from the smallest to the largest and a probability assigned using one

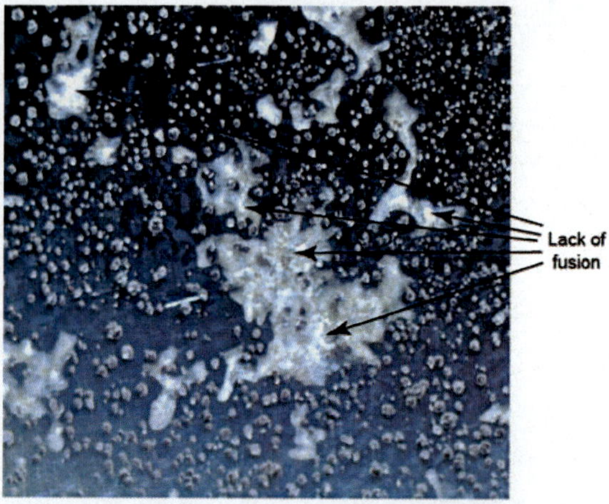

Fig. 6.39 Detail of pearlite islands in the ferrite matrix revealing the presence of lack of fusions (Courtesy of M. Cova, SACMI Imola, Italy)

of the ranking Eqs. (6.41) through (6.43). The Gumbel extreme value procedure was applied, Eqs. (6.68) and (6.69) and the values of δ and λ inferred. Results are shown in Fig. 6.42. The equation of the regression line is

$$-\ln\{\ln[1/P(x)]\} = 3.589\sqrt{\text{area}} - 2.769$$

with $\delta = 0.771$ and $\lambda = 0.278$.

The sampling surface, A_{spl}, is

$$A_{spl} = 35 \times 30\,\text{mm} = 1050\,\text{mm}^2$$

While the reference surface, A_{ref}, was assumed to be equal to 1 m^2 times 100 workpieces supposed to be built, so that

$$A_{ref} = 10^6 \times 100 = 10^8\,\text{mm}^2.$$

Therefore, the wanted return period T is

$$T = 10^8/1050 = 95{,}238$$

As such, the maximum expected $\sqrt{area}$ is 4.2 mm. Using the Murakami equation for assessing the fatigue limit Eq. (10.1) with R = 0 and $HV = 175$, it turns out

$$\sigma_f = 1.41 \cdot \frac{(175 + 120)}{(4,200)^{1/6}} \cdot \left(\frac{1}{2}\right)^{0.243} = 87\,\text{MPa}$$

which is lower than 114 MPa at 2×10^6 cycles and 98 MPa at 10^7 cycles measured with small cylindrical fatigue specimens containing chunky graphite. It must be said,

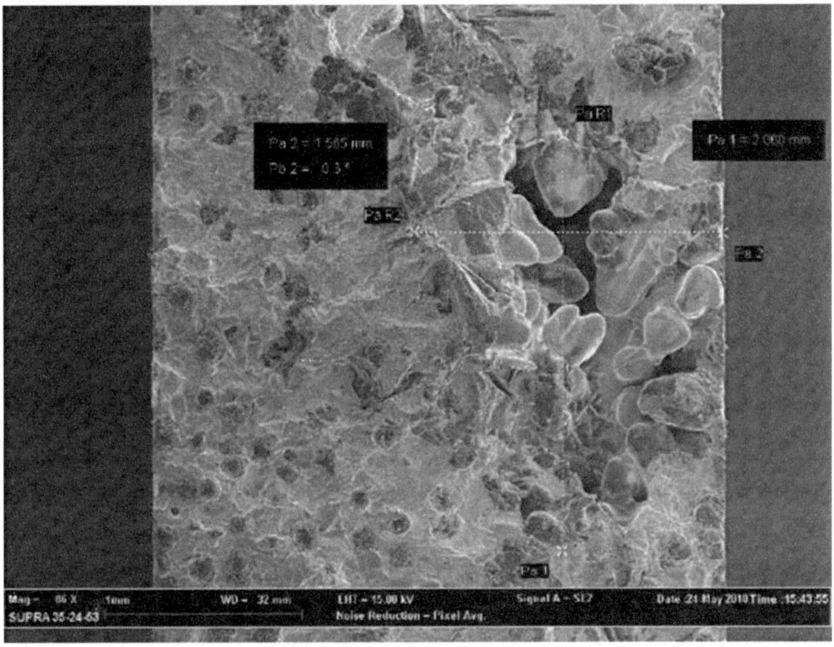

Fig. 6.40 SEM examination of the fracture surface of a flat specimen of ductile cast iron, 3 mm thick, of German production showing the origin of fatigue failure (Courtesy of M. Cova, SACMI Imola, Italy)

however, that the Murakami equation is valid for $\sqrt{area}$ no larger than 1000 μm while in our case the $\sqrt{area_{max}}$ was 4200 μm. Therefore, we may try using the ETS model and Eqs. (2.14) and (2.15). With a threshold $K_{th} = 12$ MPa$\sqrt{\text{m}}$ and $\sigma_f = 114$ MPa we have

$$a_o = 3.5\,\text{mm};$$
$$s'_f = 77\,\text{MPa}.$$

The same procedure was applied to ductile cast iron of German production. Again, as for the Italian production, the initiation site was an enormous lack of fusion, as shown in Fig. 6.40. The results of the SEV analysis are presented in Fig. 6.42 as curve 2. The data are practically overlapping the previous results obtained for the Italian production.

Finally, the same SEV procedure was applied to ductile cast iron of Chinese production. also in this case, as in the two previous cases, the initiation site was found to be an enormous lack of fusion, as shown in Fig. 6.41. The relative SEV analysis is shown in Fig. 6.42 as curve 3. This time the slope of the line is lower and the expected defect size for a return period T = 95,238 is $\sqrt{area_{max}} \approx 5300$ μm.

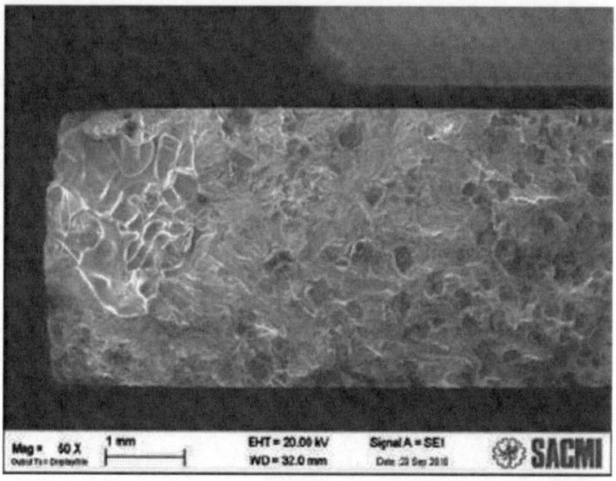

Fig. 6.41 SEM examination of the fracture surface of a flat specimen of ductile cast iron, 3 mm thick, of Chinese production showing the origin of fatigue failure (Courtesy of M. Cova, SACMI Imola, Italy)

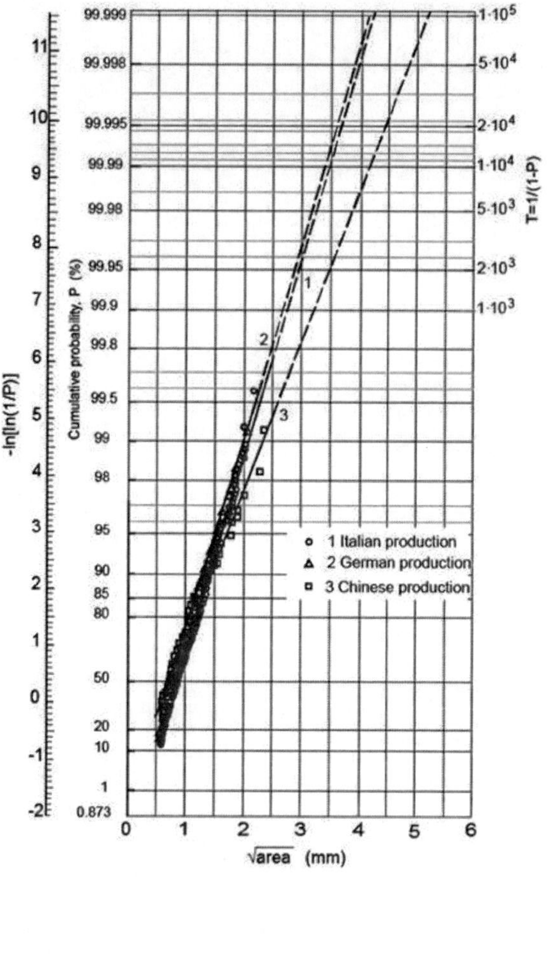

Fig. 6.42 SEV analysis of lack of fusion defects in ductile cast iron of (1) Italian, (2) German and (3) Chinese production

References

1. Härkegård, G.: Fatigue Assessment Based on Weakest-Link theory, Vortragim Rahmendes Kolloquiums Fuer Technische Wissenshaftenundess Seminars in Mechanik, 44, ETH Zürich, 18 Oktober (2007)
2. Hutchinson, T.P.: Essential of Statistical Methods, 41p. Rumbsby Scientific Publishing, Sidney (1993)
3. Student (Gosset, W. S.): The Probable Error of a Mean. Biometrica **6**, 1–25 (1908)
4. Beyer, W.H.: Handbook of Tables for Probability an Statistics, 2nd edn. CRC Press, Boca Raton (1966)
5. Lieberman, G.J.: Tables for one-sided statistical tolerance limits. Ind. Qual. Control **14**(10), 7–9 (1958)
6. Williams, C.R., Lee, Y., Rilly, J.T.: A practical method for statistical analysis of strain life fatigue data. Int. J. Fatigue **25**(5), 446–448 (2003)
7. Weibull, W.: Ingeniors Vetenskapf Akademien Handlingar. Stockolm **151**, 1–45 (1939)
8. Milella, P.P., Bonora, N.: On the dependence of the Weibull exponent on geometry and loading conditions and its implications on the fracture toughness probability curve using a local approach criterion. Int. J. Fract. **104**, 71–87 (2000)
9. Troshehenko, V.T.: Some comments to the construction of probabilistic models of structural integrity. In: NEA Workshop on Probabilistic Structure Integrity Analysis and Its Relationship to Deterministic Analysis, Report OECD/GD(96)124, 29–43, Stockholm, Sweden, 28 February–1 March (1996)
10. Gumbel, E.J.: The return period of flood flows. Ann Math Stat **12**(2), 163–190 (1941) also: Statistic of Extremes. Columbia University Press, New York (1957)
11. Sinclair, G.M., Dolan, T.J.: Effect of Stress Amplitude on Statistical Variability of 75S-T6 Aluminum Alloy. ASME Trans. **75** (1973)
12. Murakami, Y., Usuki, H.: Quantitative evaluation of effects of non-metallic inclusions on fatigue strength of high strength steel. Int. J. Fatigue **11**(5), 299–307 (1989)
13. Murakami, Y., Toriyama, T., Coudert, E.M.: Instruction for a new method of inclusion rating and correlation with the fatigue limit. J. Test. Eval. **22**(4), 318–326 (1994)
14. Murakami, Y.: Inclusion rating by statistic of extreme values and its application to fatigue strength prediction and quality control of materials. J. Res. Natl. Inst. Stand. Technol. **99**, 345–348 (1994)
15. Murakami, Y.: Metal Fatigue: Effects of Small Defects and Non-metallic Inclusions. Elsevier (2002)

Chapter 7
Stress-Based Fatigue Analysis—High Cycle Fatigue

7.1 Introduction

In Sect. 1.2 we have introduced the Wöhler curve and the *S–N* curve. Generally, they are considered to be the same thing, but they are not. The difference is in the life domain they cover. The Wöhler curve describes the fatigue behavior of a generic material from ¼ of a cycle to the fatigue limit, σ_f, or, in case of no precise fatigue limit as for non-iron alloys (see Fig. 1.11), up to 10^8–10^9 or more cycles. The Wöhler curve, therefore, spans the entire fatigue life from low-cycle fatigue to high-cycle fatigue and beyond (see Fig. 1.10) covering all the three regions of fatigue. The *S–N* curve, instead, describes the fatigue behavior of materials in the elastic domain or Region II, where the plastic contribution to stresses and strains is negligible. We may say, then, that the Wöhler curve is typically an ε-*N* curve obtained under strain controlled conditions while the *S–N* curve relates the stress amplitude σ_a given by Eq. (1.1), also indicated as *S*, to the number *N* of cycles to failure. When the stress amplitude is close to the fatigue limit, σ_f, or knee of the *S–N* curve the plastic component of deformation, ε_p, becomes negligible or even vanishes. Fatigue is driven by the elastic component of the strain amplitude that is proportional to the stress amplitude through the Young's modulus of the material. Therefore, as we approach the fatigue limit of the material, σ_f, we can abandon strains as the controlling parameter and relate the fatigue life directly to the stress amplitude. The hysteresis loop disappears, as shown schematically in Fig. 1.10, because the material behaves almost completely elastically. In this area of elastic or quasi-elastic behavior any analytical approach to fatigue is generally indicated as stress-life method. Historically, it has been the first approach to fatigue and has been the standard design method for almost 100 years. Stress-life methods are generally used when the designer pursues a target of infinite-life or safe-stress design. When the stress amplitude reaches the elastic limit and strains have a significant plastic component the *S–N* approach is no longer appropriate and a strain-based approach ε-*N* becomes necessary, as it will be described in the next section.

P. P. Milella, *Fatigue and Corrosion in Metals*,
https://doi.org/10.1007/978-3-031-51350-3_7

7.2 Basquin Line

In 1910 the American O. H. Basquin [1] observed that stress-life *S–N* data relative to high cycle fatigue domain (Region II of fatigue, see Fig. 1.10) had a linear trend when plotted in a log–log scale, which suggested a power law relationship for the elastic portion of the curve of the type.

$$\varepsilon_a E = \frac{\Delta\varepsilon_a E}{2} = \sigma_a = \sigma'_f (N)^b \tag{7.1}$$

where $\Delta\varepsilon_a$ is the total elastic strain excursion, $\varepsilon_a = \Delta\varepsilon_a/2$ is the elastic strain amplitude, E the Young's modulus, σ_a the elastic stress amplitude and N the life at σ_a. Equation (7.1) retains considerable value in fatigue design since it provides the linear portion of the Wöhler curve which actually means that if we work in the high cycle fatigue regime, over 200.000 ÷ 1.000.000 cycles, it is not necessary to have the entire *S–N* fatigue curve but just its linear potyion given by Eq. (7.1). Two new quantities enter Eq. (7.1): σ'_f, which is a fatigue strength coefficient and *b*, a fatigue strength exponent or Basquin exponent. Both σ'_f and b are material properties that can be obtained, as shown in Sect. 10.3 and Eq. (8.9). In particular the strength coefficient, σ'_f, is equal to the monotonic true failure stress, $\sigma_{t,f}$, since it is the intercept of the Basquin line with the *S* axis at one-quarter of a cycle, *i.e.*, $N = 1/4$, while the exponent *b*, Eq. (10.33), is given by

$$b = \frac{-n'}{1 + 5n'} \tag{7.1a}$$

where *n'* is the cyclic hardening exponent of the material. Figure 7.1 shows the Basquin line for high strength, quenched and tempered steel, for ductile cast iron (DCI) and 6066T-6 aluminum alloy, respectively.

The slope *b* of each line is given by the corresponding Basquin exponent in Eq. (7.1a). It is worth noting that all Basquin straight lines merge with the corresponding *S–N* curve, dotted line of Fig. 7.1, beyond some 10^5 cycles while they are underestimating the low cycle fatigue behavior which depends on the plastic strain amplitude. This actually means that beyond that value the fatigue behavior is almost elastic and we can replace the *S–N* curve of experimental derivation with the Basquin line without having run even one single fatigue test. As a first approximation, theσ'_f coefficient is equal to the monotonic true failure stress $\sigma_{t,f}$ while b varies from – 0.05 to – 0.12. It may appear surprising that a cyclic behavior can be assessed through the knowledge of a static property such as the true stress at failure $\sigma_{t,f}$. However, looking at Eq. (7.1), it turns out that σ'_f is the stress amplitude at $N = 1$. This actually means that the simplest fatigue failure is a monotonic failure that occurs during the rising ramp of the first cycle of load application. There will not be a second cycle. Therefore, the fatigue strength coefficient is the intersection of the Basquin line with the ordinate axis.

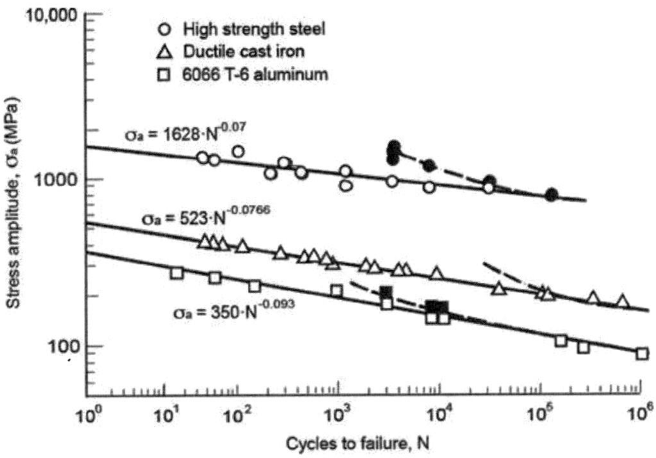

Fig. 7.1 Basquin lines for a high strength steel, DCI and 6066T-6 aluminum alloy, respectively. Closed symbols and dotted line refer to the Wöhler curve of each material

7.3 Mean Stress Effect

A standard Wöhler curve is obtained under fully reversed load conditions or $R = \sigma_{\min}/\sigma_{\max} = -1$, which is usually called *stress ratio*. The mean stress, σ_m, is zero. We have already mentioned (see Sect. 1.1) that already in 1858 Wöhler realized that the mean stress is of influence in that the higher it is, the lower is the stress amplitude that leads to failure. Today the primary role of the mean stress on fatigue life of materials has been definitely ascertained. Indeed, experience indicates that an alternating stress of amplitude σ_a may have completely different effects on fatigue life depending on whether or not a mean stress is acting. For example, observing the three cycles (a), (b) and (c) of Fig. 7.2 we note that they have the same amplitude $\sigma_a = \sigma_{\max} - \sigma_{\min}/2$ but not the same mean stress σ_m

$$\sigma_a = \frac{\sigma_{\max} - \sigma_{\min}}{2}, \quad \sigma_m = \frac{\sigma_{\max} - \sigma_{\min}}{2}$$

$$\sigma_a = \frac{\sigma_{\max}}{2}(1 - R), \quad \sigma_m = \frac{\sigma_{\max}}{2}(1 + R). \tag{7.2}$$

In the first case (a) of fully reversed cycle, the mean stress is zero. In the second case (b) where the stress is fluctuating from zero, it is $R = 0$ ($\sigma_{\min} = 0$) and third case (c) in which it is fluctuating from $\sigma_{\min} > 0$, it is $R > 0$. In the two last cases the cycle is non-symmetric with a positive, but different, mean stress, σ_m, that will affect the fatigue strength of the material. We will see how case (b) of Fig. 7.2 is worse than case (a) and both are less dangerous that case (c) which is the worse of them all.

An example of Wöhler's curves obtained on aluminum alloy type 2024-T351 for three different R ratios is shown in Fig. 7.3, based on data from [2].

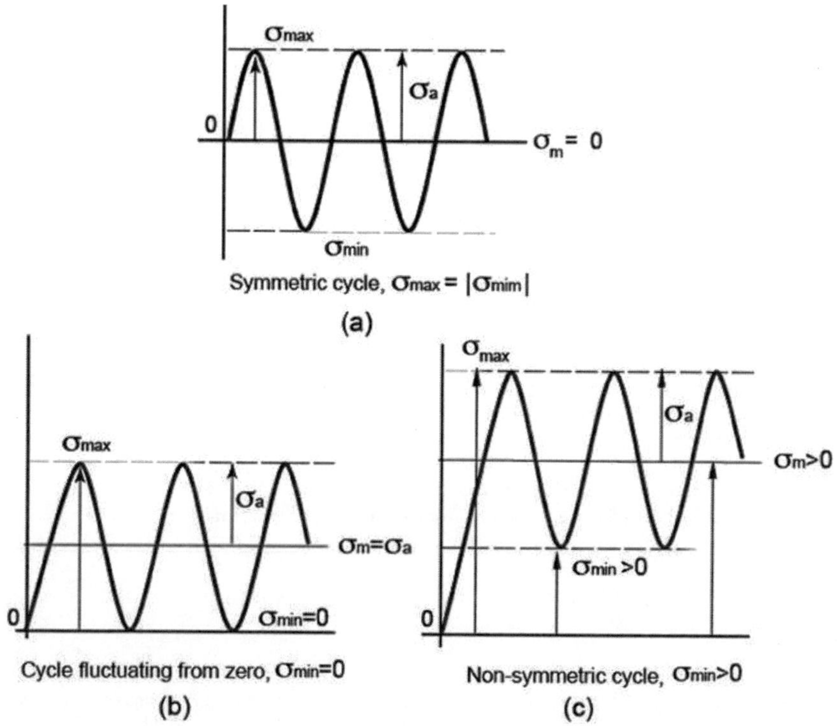

Fig. 7.2 **a** Fully reversed symmetric cycle with zero mean stress ($\sigma_m = 0$), **b** cycle fluctuating from the origin ($\sigma_m = \boldsymbol{\sigma_a} > 0$, $\sigma_{\min} = 0$) and **c** nonsymmetrical cycle ($\sigma_m > 0$ and minimum stress $\sigma_{\min} > 0$)

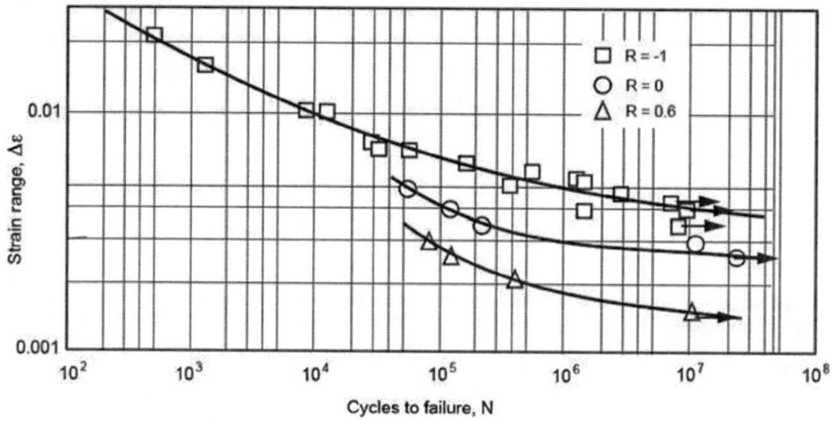

Fig. 7.3 Wöhler diagram for 2024-T351 aluminum alloy at various R ratios (data from [2])

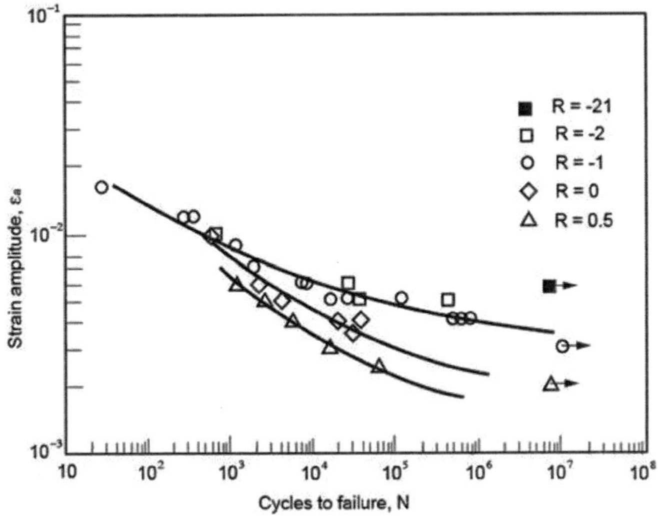

Fig. 7.4 Wöhler diagram for SAE 1045 hardened steel at various R ratios (data from [3])

The alloy has a tensile strength of 480 MPa and yield strength equal to 460 MPa. The first curve refers to a standard completely reversed traction fatigue test, $R = -1$, with a fatigue limit of 135 MPa at 10^7 cycles and $\sigma_f/\sigma_u = 0.28$ in compliance with Table 10.4 for that alloy. The fatigue limit decreases to an allowable strain range $\Delta\varepsilon$ of about 0.0026 ($\sigma_a \approx 92$ MPa) from about 0.004 ($\sigma_a \approx 135$ MPa) when the same specimens are subjected to fatigue tests with stresses fluctuating from zero, $R = 0$, and falls down below a $\Delta\varepsilon$ of 0.0016 ($\sigma_a \approx 53$ MPa) in non-symmetric fatigue traction tests with $R = 0.6$. The maximum stress is increasing from 135 MPa for $R = -1$ to 184 MPa for $R = 0$ to 250 for $R = 0.6$. The simple shifting from $R = 0$ to $R = 0.6$ has cut down the fatigue limit σ_f by a 50%. The same behavior can be seen in Fig. 7.4 for high strength SAE 1045 steel hardened to 55 HRC [3].

Two or three particular observations can be done just looking at both figures. The first is that for $R \geq 0$ tests have been run at constant R but not at constant mean σ_m stress. This is evident for the case $R = 0$ in which the stress amplitude σ_a and the mean stress are coincident so that σ_m decreases as $\Delta\varepsilon$ decreases. Just to give an example, on the $R = 0$ curve of Fig. 7.3 at a strain range $\Delta\varepsilon$ of about 0.004 ($\sigma_a \approx$ 132 MPa) that leads to fatigue failure after 100,000 cycles, the mean stress is equal to 132 MPa whereas at the fatigue limit of about 92 MPa also the applied mean stress is equal to 92 MPa. The second observation that anticipates what will be said in Section, is that, despite the apparent σ_m increase, all curves that are so far from each other at 10^6–10^7 cycles becomes closer and closer as the cycles to failure N reduce leaving credit to a possible merging at about 10^3 cycles. This would actually mean that in the low cycle fatigue domain mean stress may not have any effect on fatigue.

A third observation, also of particular relevance, is that a negative mean stress, R = − 2 and R = − 21 in Fig. 7.3, has either no effect or may improve the fatigue strength of the material. This feature, recognized to be true, will be discussed in Sect. 10.2. The dependence of *S–N* diagrams on mean stress would apparently result in the need to experimentally derive not just one, but an entire family of curves each corresponding to a particular value of the mean stress σ_m. Such an approach is very time consuming and expensive therefore not practical. Several proposals have been made that are more successful in accounting for mean stress effects. At the turn of nineteenth century, several models were developed to take care of the mean stress. The most known are those proposed in 1874 by the German researcher Gerber [4], in 1899 by the British Goodman [5], in 1917 by the American Haigh [6] and in 1930 by the American Soderberg [7]. The first two models, in particular, were interpreting with a certain accuracy the experimental evidence.

Gerber proposed a parabolic dependence of the fatigue limit σ_f on mean stress σ_m that could vary from zero to the ultimate strength σ_u, as shown in Fig. 7.5, known as Gerber parabola. As the mean strength σ_m increases the allowable fatigue limit $\sigma_{f.m}$ decreases from the initial value σ_f for $\sigma_m = 0$ to zero as σ_m approaches the ultimate strength σ_u. Gerber, to simplify testing procedures, measured the fatigue limit as the stress amplitude at 10^7 cycles. Any combination $\sigma_f - \sigma_m$ that would fall outside the domain enclosed by the Gerber parabola may give failure at 10^7 cycles. On the contrary, any combination $\sigma_f - \sigma_m$ that would fall within that domain may be considered safe. Goodman, instead, assumed a linear relationship between σ_f and σ_m, as shown in Fig. 7.5 by the line joining these two points, known as the Modified Goodman line. In both cases, the analytical relation that defines the allowable fatigue limit $\sigma_{f,m}$ when a mean stress σ_m is acting is

$$\sigma_{f,m} = \sigma_f \cdot \left[1 - \left(\frac{\sigma_m}{\sigma_u}\right)^n\right] \tag{7.3}$$

with the exponent n that is equal to 2 for the Gerber parabola and 1 for the modified Goodman line. For a given finite life N, point σ_f in Fig. 7.5 is replaced by the value of the stress amplitude σ_a corresponding to that given N on the *S–N* curve obtained under fully reversed loading ($R = -1$). In 1930 Soderberg [7] suggested that the mean stress would be limited by the yield strength, proposing a new safety domain delimited by his line shown in Fig. 7.5. Curious enough, the reason for the use of a parabola may stem, but it is not certain, from the fact that some very first experimental results obtained by Gerber with compressive mean stress would indicate a decrease of the fatigue limit. This might justify the use of the exponent $n = 2$ that would turn positive the term under parenthesis and reduce the value of the fatigue limit. Studies performed much later [8–10] have shown that the fatigue limit may not decrease at all under compressive mean stresses, at least till a certain limit value and it may even increase, as already said. This is shown in Fig. 7.6 [8–10] for steels and light alloys where σ_a is the allowable stress amplitude at constant number of cycles N. Therefore, the safety domain for negative mean stress or negative quadrant is larger than the positive one.

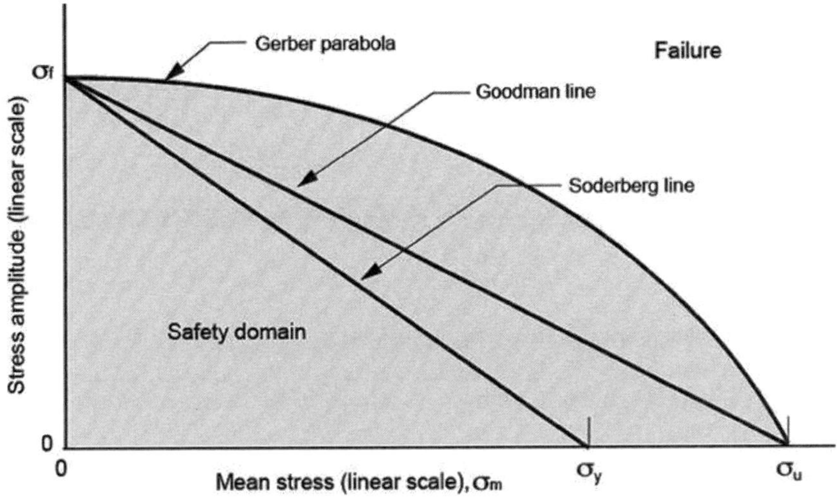

Fig. 7.5 Lines of equal fatigue life under uniaxial stress amplitude with tensile mean stress proposed by Gerber, Goodman and Soderberg, respectively

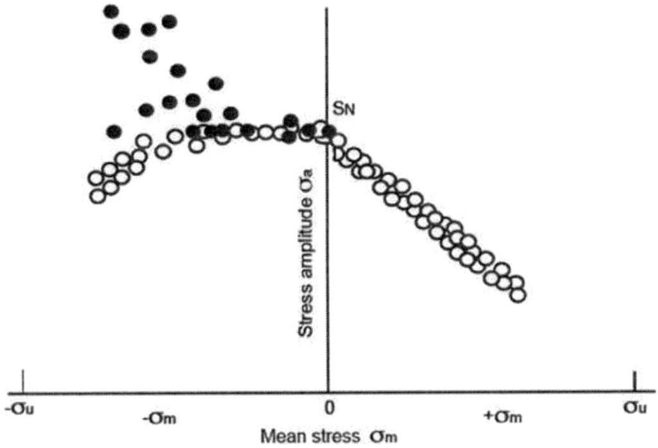

Fig. 7.6 Influence of negative mean stress on fatigue strength of steel and light alloys (data from [8–10])

Wilson and Haigh [11] introduced the line of constant yield as additional restrain for ductile materials by which they identified a safety domain in the negative quadrant of compression mean stress delimitated by the OCDB trapeze of Fig. 7.7. With that expedient it was excluded, among other things, the plastic buckling of the resisting section. Quite often some experimental data fall lightly outside the Gerber parabola indicating that even this criterion may be conservative.

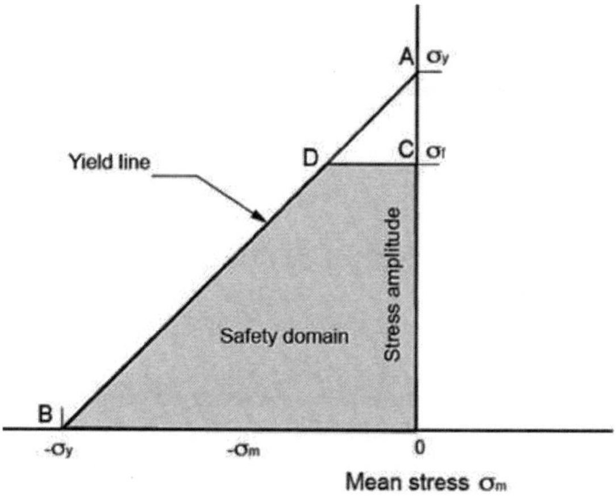

Fig. 7.7 Safety domain for compression mean stress

Figure 7.8a is an example of data obtained on 2024-T4 aluminum alloy that fall on the Gerber parabola, with some very light deviation [13]. In Fig. 7.8b, c and d other experimental data are reported, which refer to aluminum alloy [12], mild steel [13] and 4½%Cu aluminum alloy [14], respectively.

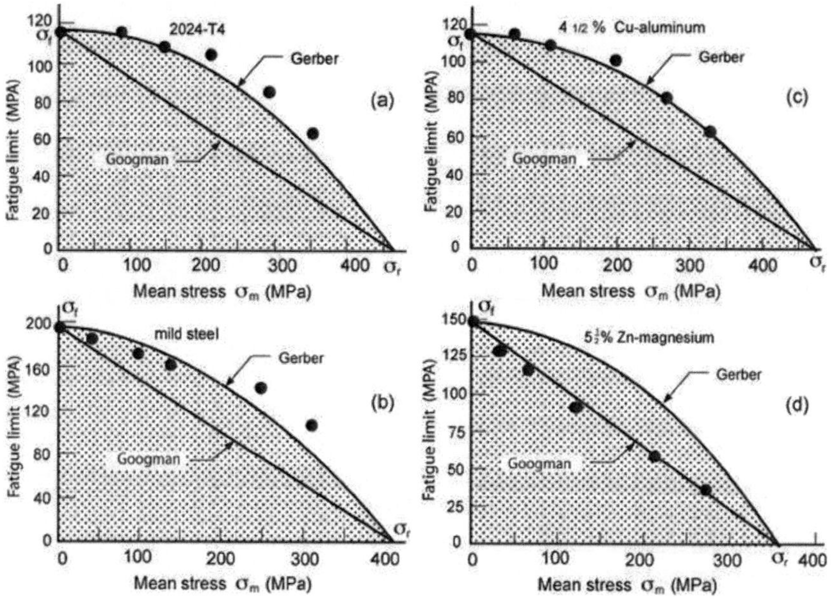

Fig. 7.8 Compressive mean stress effect. Experimental results obtained for **a** 2024-T4 aluminum alloy, **b** mild steel, **c** 4½% Cu aluminum alloy and **d** 5½% Zn Magnesium alloy (data from [12–14])

The use of the ratio of the alternating stress to the fully reversed fatigue strength, σ_a/σ_f, versus the ratio of the mean stress to the ultimate strength, σ_m/σ_u, as dimensionless parameters allows the plot of results obtained for different alloys on the same diagram. This is shown in Fig. 7.9 for steels and aluminum alloys at about 10^7 cycles [9]. It can be seen how even the Gerber parabola may be conservative, in particular for aluminum alloys. Generally, the Gerber parabola is better representing the effect of mean stress acting at stress amplitudes σ_a higher than the fatigue limit while the Goodman line is describing the same effect in the surrounding of the fatigue limit σ_f, as schematically shown in Fig. 7.10 [15]. However, the Gerber parabola has a major drawback in that it cannot predict a possible beneficial effect of the negative mean stress, as it will be discussed later.

Experience indicates [12] that about 90% of experimental data obtained on steel, aluminum, brass and other alloys when subjected to a compression mean stress is actually falling between the Gerber parabola and the Goodman line. Whatever the

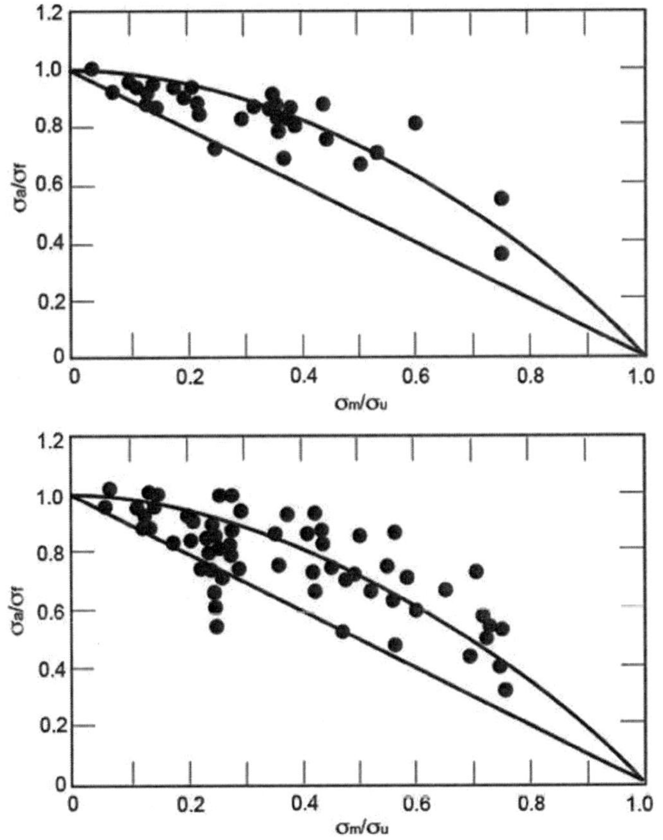

Fig. 7.9 Effect of mean stress on fatigue strength for: **a** steels and **b** aluminum alloys (adapted from [9])

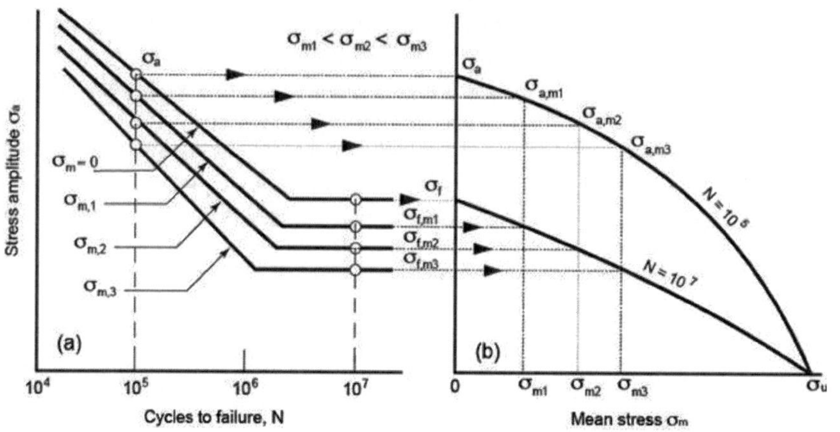

Fig. 7.10 **a** *S–N* diagrams obtained at different mean stress σ_m, **b** safety domain at constant *N* (modified from [15])

material response may be the Goodman line always represents the most conservative criterion. This is why the Goodman line is adopted by the ASME Code as a design criterion and in general by design codes worldwide.

But there is also another reason for using the Goodman line since it provides an increase of fatigue strength with a negative (compression) mean stress. Following the Goodman criterion, for any stress amplitude it can be written, as per Fig. 7.11.

$$\begin{aligned} \sigma_{f,m} &= \sigma_f\left(1 - \frac{\sigma_m}{\sigma_u}\right) \\ \sigma_{a,m} &= \sigma_a\left(1 - \frac{\sigma_m}{\sigma_u}\right). \end{aligned} \tag{7.4}$$

Fig. 7.11 Fatigue limit reduction from σ_f to $\sigma_{f,m}$ due to mean stress σ_m

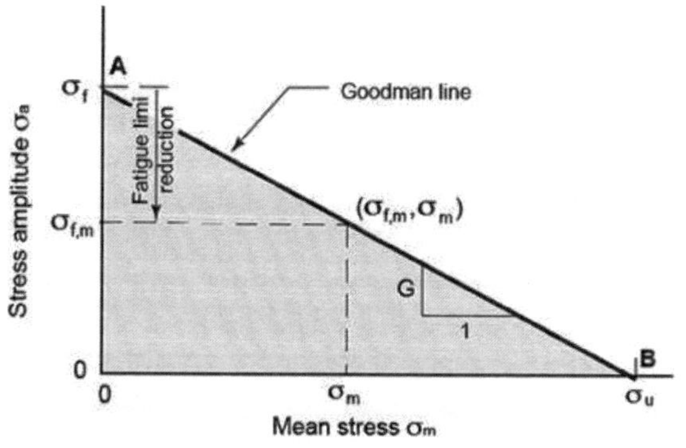

Alternatively, introducing the slope G of the Goodman line, Fig. 7.10, the reduced stress amplitude or fatigue limit can be expressed as

$$\sigma_{f,m} = \sigma_f - G \cdot \sigma_m. \tag{7.5}$$

For instance, in the case of the aluminum alloy considered in Fig. 7.3, for $R = 0.6$ and $N = 10^7$ cycles the mean stress is about 200 MPa so that the term between parenthesis in Eq. (7.4) is equal to 0.42. Since for $R = -1$ the fatigue limit, σ_f, is equal to 135 MPa, the new fatigue limit, $\sigma_{f,m}$, will be 55 MPa, very close to the one experimentally obtained of 53 MPa. Note that solved for σ_f, Eq. (7.4) also defines an equivalent fully reversed stress amplitude, $\sigma_{f,R\text{-}1}$, that is expected to cause the same fatigue life as the $\sigma_{a,m}$-σ_m combination

$$\sigma_{f,R-1} = \sigma_{a,m}(N) \cdot \left(1 - \frac{\sigma_m}{\sigma_u}\right). \tag{7.6}$$

Shear stress holds a treatment of its own. At variance with traction and bending stresses that have a sign, whether positive or negative, to which a physical state is associated traction or compression respectively, shear stress has no sign at all, but simply a direction conventionally assigned that has not a real physical meaning. Therefore, as far as shear stress is concerned there is no negative quadrant in the Goodman diagram. Moreover, shear stress amplitude τ_a or the same fatigue limit τ_f seem not to be influenced by a possible mean stress τ_m, provided the maximum stress is not causing yielding. Figure 7.12 shows that the allowable alternating stress for torsion members free from any stress riser is not influenced by mean stress, τ_m. When significant stress risers are present, as notches or shoulder fillets, the state of stress deviates from pure shear and experimental results indicate that a mean stress reduces the allowable alternating stress in about the same manner as for bending loads [16]. Thus, when a mean torsion stress is present the behavior of notched torsion members is similar to that pertaining to unnotched traction or bending fatigue members.

7.4 Shake-Down Effect on Fatigue

Going back to the Goodman line we may say that the safety domain of Fig. 7.11 is not totally accessible. Beyond a certain value the mean stress σ_m has no longer an effect on fatigue. Therefore, $\sigma_{f,m}$ does not fall down to zero as σ_m goes to σ_u, as Eq. (7.4) would predict. To understand it, suppose to be dealing with an elastic-ideally plastic material whose characteristic is that of Fig. 7.13. This apparent restriction is introduced just to simplify the considerations that follow, but it is not restrictive.

Also suppose that a flat thin specimen of that material is subjected to fatigue in bending fluctuating from zero ($R = 0$) with stress amplitude σ_a and mean stress σ_m high enough to cause just a thin surface layer to yield. Also this particular example is not restrictive. Fibers in this layer will undergo a cycle of plastic deformation.

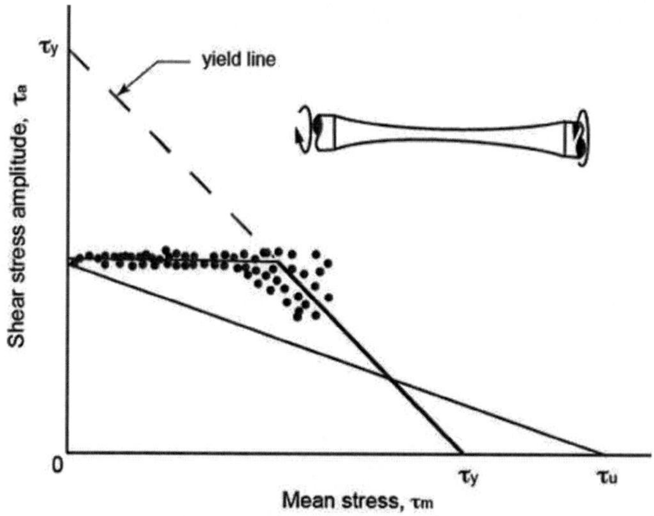

Fig. 7.12 τ_m–τ_a diagram for unnotched torsion members of ductile metal (from Juvinall [16])

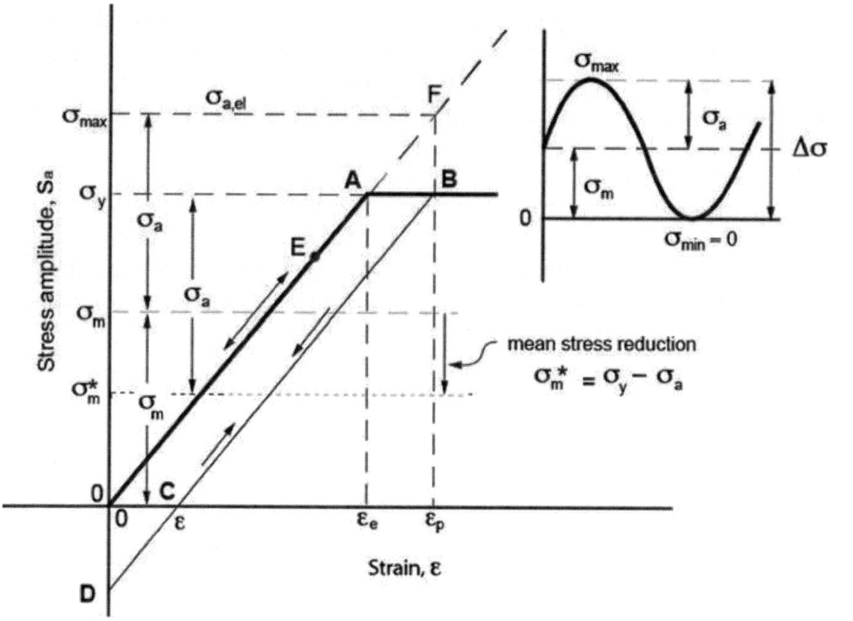

Fig. 7.13 Schematic of Shake-down effect in an elastic-ideally plastic material subjected to a cyclic stress amplitude σ_a and mean stressσ_m. Upon unloading the mean stress σ_m reduces to $\sigma_m{}^*$

Consider an arbitrary point P in one of these fibers farthest from the neutral axis of the specimen. Its state on the stress–strain curve of the material is shown by point B $\equiv (\varepsilon_p; \sigma_y)$ beyond the elastic strain ε_e, as shown in Fig. 7.13. The figure also shows the equivalent elastic stress $\sigma_{a,el}$ that would produce the same strain ε_p if the material were ideally elastic. In the case considered here the elastic equivalent stress $\sigma_{a,el}$ is equal to $\sigma_m + \sigma_a$, point F in Fig. 7.13. Upon first unloading, any material point from the neutral axis to the thin surface layer that yielded, like point E in Fig. 7.13 that did not reach the elastic limit ε_e or the yield strength σ_y, will return to the original state, i.e., to zero stress and zero strain, moving back on the elastic unloading line 0A.

As to point B, it will follow the unloading path BD parallel to the loading line 0A. In this backward path it is very likely that it will not stop in C. If it did, a residual positive non-zero strain ε would remain that would not be congruent with interface elastic material that has returned to zero stress and zero strain. Since the volume of the elastic material overwhelms that of the thin layer that yielded, it will dominate and control the unloading process causing the yielded fibers to return to zero strain. So that point B will not stop in C, but it will be pushed down to D where the strain is zero, as required by the principle of congruence of deformations. A compressive residual stress will, then, remain that is equal to –$(\sigma_{\max} - \sigma_y)$. This coercive action exerted by the elastic material on the plastic one is known as *shake-down effect.* Upon reloading, point B that moved to D will follow the path DB returning in B. This actually means that upon reloading the material that yielded in the first cycle will now respond elastically. In real life, it will take few cycles and not just one to respond elastically. Actually, it is as if the material that yielded in the first cycle increased its elastic limit by $(\sigma_{\max} - \sigma_y)$ and its new yield strength σ'_y is

$$\sigma_y' = \sigma_y + \left(\sigma_{\max} - \sigma_y\right) = \sigma_{\max} = \sigma_a + \sigma_m. \tag{7.7}$$

It is also evident that the shake-down effect has produced a decrease of the mean stress acting on this *new* material that will not be σ_m any more but $\sigma^*_m < \sigma_m$, median point of line DB of Fig. 7.13

$$\sigma_m^* = \sigma_y - \sigma_a. \tag{7.8}$$

It is indeed worth to recall that the shake-down effect is possible only when the plastic volume is very small with respect to the elastic one that has to control the process. This is the case of a thin beam in bending or a small hole or a sharp discontinuity whose associate plastic zone is embedded in the elastic surrounding and remains negligible. But when the plastic volume becomes too large and compares with the elastic volume the shake-down effect may not take place completely. Depending on the ratio of the plastic zone volume to elastic one, the unloading return point moves from D to C and the shake-down effect becomes less and less effective until it fails.

Furthermore, the effect persists as long as the equivalent elastic stress $\sigma_{a,el}$ does not go beyond $2\sigma_y$. It can be easily verified from Fig. 7.13 how beyond $2\sigma_{a,el}$ the material, upon unloading, undergoes an inverse plastic flow with consequent plastic loop formation that does not allow the material to follow a linear behavior any longer.

The consequence of the mean stress reduction by the shake-down effect is that not all points inside the safety domain below the Goodman line are effectively accessible. Any combination of stress amplitude σ_a and mean stress σ_m that falls inside the BCD triangle of Fig. 7.14, as for example point F, after few cycles will have the mean stress reduced according to Eq. (7.8) moving to G on the yield strength line. It is evident that when the stress amplitude σ_a becomes higher than the yield strength σ_y of the material the mean stress will go to zero. Reassuming, under conditions of validity of the shake-down effect the actual value σ^*_m of the mean stress after few cycles becomes

$$\begin{aligned} &1.\ \sigma_m^* = \sigma_m \quad \text{if} \quad \sigma_a + \sigma_m \le \sigma_y \\ &2.\ \sigma_m^* = \sigma_s - \sigma_a \quad \text{if} \quad \sigma_a + \sigma_m > \sigma_y \quad \text{and} \quad \sigma_a < \sigma_y \\ &3.\ \sigma_m^* = 0 \quad if \quad \sigma_a \ge \sigma_y. \end{aligned} \tag{7.9}$$

The reduction of the mean stress just described for a cycle fluctuating from zero ($R = 0$) remains valid also in the more general case of $\sigma_{\min} > 0$. It is worth noting that a very low stress amplitude superposed to a very high mean stress may not get any benefit from the shake-down effect (no mean stress reduction) if condition 2 in Eq. (7.9) $\sigma_a + \sigma_m > \sigma_y$ is not verified. If this is the case, all the *weight* of the mean stress will remain. Very often these low amplitude cycles are vibrations over posed to a main cycle and are applied at very high frequency. The combination of very high number of cycles N and very high mean stress may have a dramatic consequence on fatigue strength of the material if the mean stress is not reduced by shaking-down.

Going back to the Goodman line let's examine a generic allowable combination $\sigma_{a,1} - \sigma_m$ that lays in the BCD triangle of Fig. 7.14. Consider, for instance, the combination $\sigma_m - \sigma_{a,1}$ represented by point F. After few cycles the mean stress is reduced by the shake-down effect and point F will move to G. A new combination

Fig. 7.14 Shake-down effect on the mean stress and allowable stress amplitude σ_a at N cycles

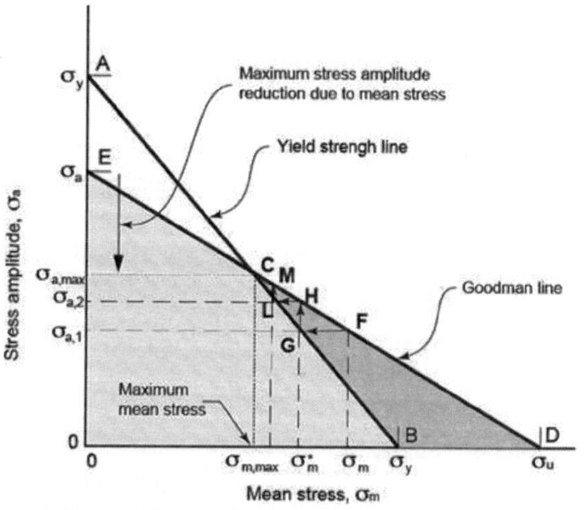

$\sigma_{a,1} - \sigma^*_m$ arises that lays inside the Goodman diagram on the yield strength line. The reduced mean stress will make it possible to increase the stress amplitude $\sigma_{a,1} - \sigma_{a,2}$, point H of Fig. 7.14 without jeopardizing the material integrity. But, again, the new combination $\sigma_{a,2} + \sigma_m{}^*$ is larger than σ_y and the effective mean stress will be further reduced moving point H to L and so on till it will reach C through the intermediate paths FG, GH, HL, LM arriving eventually in C where the process will stop since the mean stress cannot be further reduced. This value represents the maximum mean stress $\sigma_{m,\max}$ that may have any effect in fatigue cycling.

$$\sigma_{m,\max} = \sigma_y - \sigma_{a,\max}. \tag{7.10}$$

Therefore, the Goodman line turns into a two line diagram ECB called modified Goodman diagram. Looking at Fig. 7.14 and recalling Eq. (7.7) the new maximum possible stress amplitude $\sigma_{a,\max}$ that corresponds to the maximum allowable mean stress $\sigma_{m,\max}$ is given by

$$\sigma_{a,\max} = \sigma_a \cdot \left(1 - \frac{\sigma_{m,\max}}{\sigma_u}\right). \tag{7.11}$$

At the same time, the stress amplitude S or σ_a may be considered as the completely reversed stress amplitude $\sigma_{equ}(N)$ (R = -1 and $\sigma_m = 0$) equivalent to $\sigma_{a,\max}$. Substituting in (7.11) the value of the maximum stress given by (7.10) we get

$$\sigma_{a,\max} = \sigma_a \cdot \left(\frac{\sigma_u - \sigma_y}{\sigma_u - \sigma_a}\right) \tag{7.12}$$

when $\sigma_a < \sigma_y$.

7.5 Fatigue Strength Models

7.5.1 Morrow Mean Stress Sensitivity Factor

In 1968 Morrow [17], differing from Goodman and Gerber, underlined that the mean stress could go beyond the ultimate strength σ_u of the material reaching the value of the fatigue strength coefficient σ'_f introduced by Basquin, see Eq. (7.1). The fatigue strength coefficient σ'_f represents, as already said, the highest stress amplitude that can produce failure in ¼ of cycle, i.e., the true stress at fracture $\sigma_{t,f}$ of the material (see Sect. 7.2). The Resulting Morrow line is shown in Fig. 7.15 together with the Goodman line. Accordingly, Eq. (7.4) becomes

$$\sigma_{f,m} = \sigma_f \cdot \left(1 - \frac{\sigma_m}{\sigma'_f}\right). \tag{7.13}$$

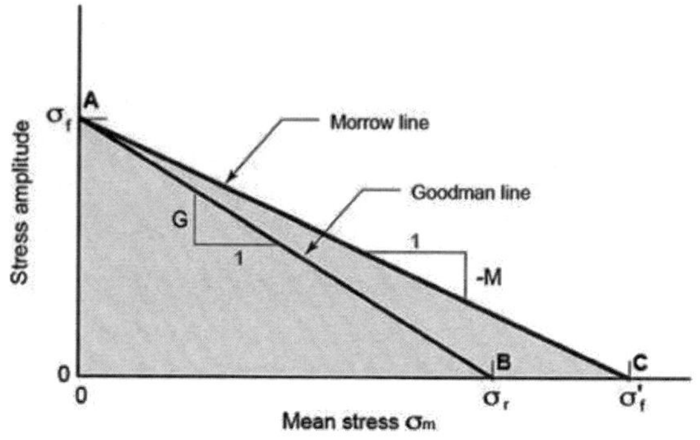

Fig. 7.15 Goodman and Morrow models to take into consideration the mean stress effect

Effectively, it can be observed experimentally that the Goodman line is not representing the actual behavior of the material because the mean stress is allowed to reach the true stress at fracture σ'_f, as shown in Fig. 7.21 for AISI 4340 steel.

However, for alloys that do not have a Wöhler curve that flattens down at about 10^6 or10^7 cycles, as aluminum alloys or more generally non-iron alloys, this extension of the mean stress does not seem to work, as shown in Fig. 7.16 for 2024-T3 aluminum having a fatigue strength coefficient or true fracture stress, σ'_f, equal to 1602 MPa [18]. As it can be seen, this aluminum alloy does not follow either the Goodman line or the Morrow line. Experimental data seem to be aligned along a line connecting the fatigue limit to the true maximum stress $\sigma_{u,t} = \sigma_u(1 + \varepsilon) = 610$ MPa (see section Eq. 10.25). Morrow also made the hypothesis that the mean stress would act only on the elastic component of the Wöhler curve so that Eq. (7.1) becomes

$$\sigma_{f,m} = \left(\sigma'_f - \sigma_m\right) \cdot N^b. \tag{7.14}$$

At the same time, referring to Fig. 7.14, Eq. (7.5) can be written as

$$\sigma_{f,m} = \sigma_f + M \cdot \sigma_m. \tag{7.15}$$

The Morrow M factor appears to be a complex function of the mean stress [19, 20] as schematized in Fig. 7.17, but also of the material and for that reason is also known as mean stress sensitivity factor. For low values of the mean stress ($-1 \leq R < 0$) the Morrow's factor is indicated by letter M and its value, referring to the diagram of Fig. 7.17 [20], is given by

$$M = \tan\phi = \frac{\sigma_{a,R=-1} - \sigma_{a,R=o}}{\sigma_{a,R=o}}. \tag{7.16}$$

Large M values indicate high σ_m sensitivity. In the negative mean stress quadrant ($-\infty \leq R < -1$) it is indicated by M_2 while for high mean stress values ($0 \leq R < 1$) it's referred to as M_3. M_2 varies from zero to M. Usually, M_3 is lower than M by a factor of 3 ($M_3 \approx 1/3\ M$). This is based on the empirical observation that loading

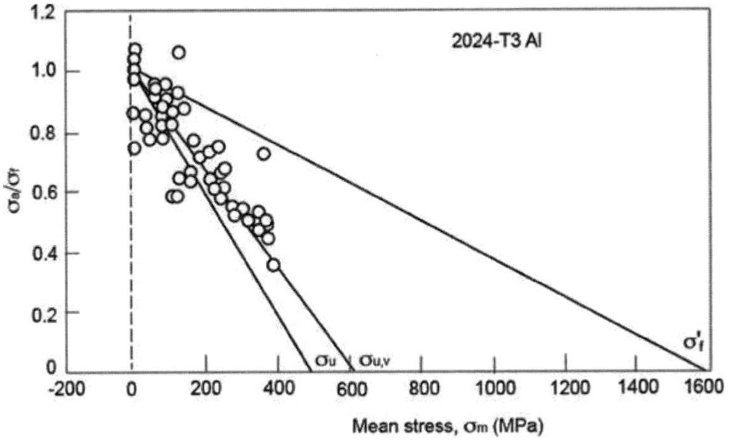

Fig. 7.16 Normalized stress amplitude versus mean stress for 2024-T3 aluminum (adapted from [18])

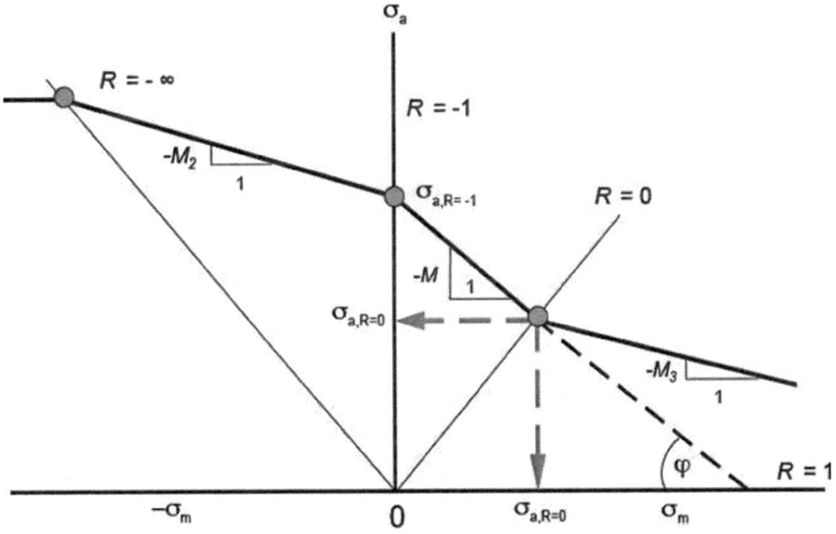

Fig. 7.17 Morrow diagram showing the mean stress sensitivity factor M and its dependence on mean stress [18]

with high mean stress and low stress amplitude has higher damaging effects than that predicted by M. Schütz [19] gathered numerous data for different materials and lives N, also in the range of fatigue limit, relating them to the corresponding monotonic traction properties σ_u. He observed a systematic dependence of M on the traction strength of the material. His results are shown in Fig. 7.18 [19].

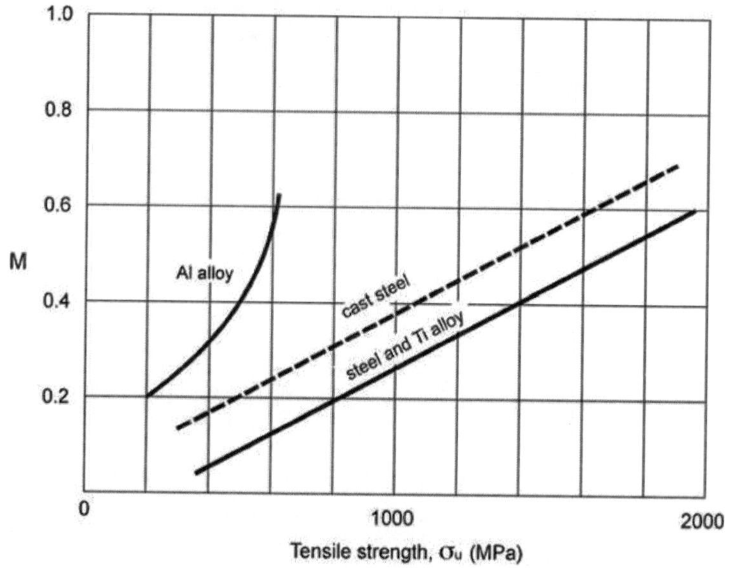

Fig. 7.18 Effect of traction strength on *M* factor [19]

7.5.2 *FKM-Guideline*

Fermér et al. [20–22] proposed a modification of the Morrow *M* factor diagram which later was developed as FKM mean stress correction method [20–22]. The method, shown in Fig. 7.19, uses 4 M factors which define the sensitivity to mean stress in 4 regimes as the sectors in which the diagram of Fig. 7.17 has been divided.

- Regime I is for stress ratio $R > 1$ where both the maximum and minimum stress are compression;
- Regime II defines the sector $-\infty \leq R \leq 0$ in which $R = -\infty$ corresponds to a maximum stress equal to 0. $R = -1$ defines the fully reversed cycling and $R = 0$ corresponds to a stress amplitude pulsating from zero, as usual;
- Regime III defines the sector $0 < R < 0.5$ in which both the maximum and minimum stress are tensile;
- Regime IV defines the sector $R \geq 0.5$ where high amplitude and mean tension stresses are acting. The corresponding fatigue strength can be written as follows:

Regimes I and IV

$$\sigma_{f,m} = \sigma_f \tag{7.17}$$

Regime II

$$\sigma_{f,m} = \sigma_f + M \cdot \sigma_m \tag{7.18}$$

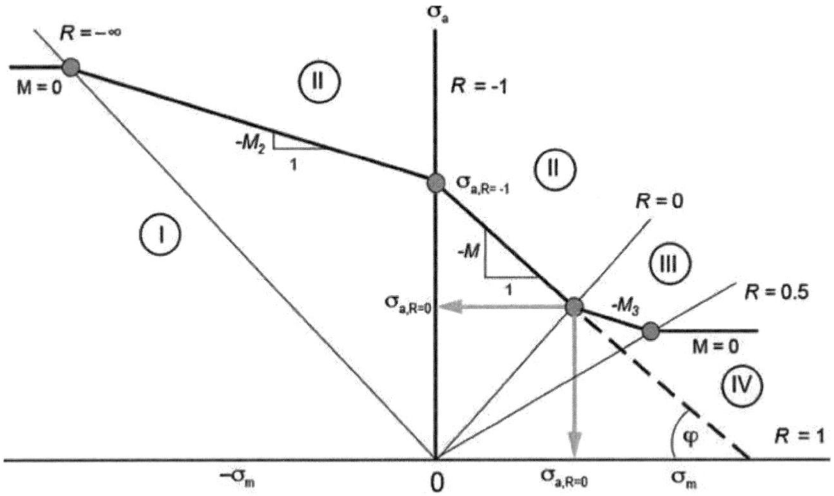

Fig. 7.19 FKM mean stress correction method [20–22]

Regime III

$$\sigma_{f,m} = (1 + M)\frac{\sigma_f + M/3 \cdot \sigma_m}{1 + M/3}. \tag{7.19}$$

The mean stress sensitivity factor that enters Eqs. (7.18) and (7.19) can be assessed from Fig. 7.19 or calculated as

$$M = a_M \sigma_u - b_M \tag{7.20}$$

with a_M and b_M material parameters given in Table 7.1. Note that Eq. (7.17) implies that in Region I and IV the mean stress has no effect, dough it is very high. Any combination $\sigma_a - \sigma_m$ falling inside the FKM diagram is safe. Figures 7.17 and 7.19 present a rather interesting an effect: a negative mean stress seems to be beneficial, as already shown in Fig. 7.6. Effectively, many evidences have been collected about the influence of the negative mean stress on the fatigue limit, σ_f, that support this argument. Figure 7.20 presents some results obtained by Sines on steels and aluminum alloys [23].

Very interesting are results obtained by Murakami [24] on a high speed tool steel SKH51, having a hardness Hv = 654, shown in Fig. 7.21. The introduction of a negative mean stress equal to –784 MPa is doubling the fatigue limit with respect to the R = –1 case (no mean stress, while a positive mean stress of the same value reduces the fatigue limit by half.

It is evident that the alloys considered seem to be gaining strength as the mean stress increases its compression action. Analogous results have been obtained by Dowling [18] with AISI 4340 steel, 1172 MPa ultimate strength, as shown in

Table 7.1 a_M and b_M for various materials (from FKM-Guideline [20])

Material	a_M	b_M
Steels	0.00035	−0.1
Steel casting	0.00035	0.05
Ductile irons	0.00035	0.08
Malleable cast irons	0.00035	0.13
Grey cast irons	0	0.5
Wrought aluminum alloys	0.001	−0.04
Cast aluminum alloys	0.001	0.2

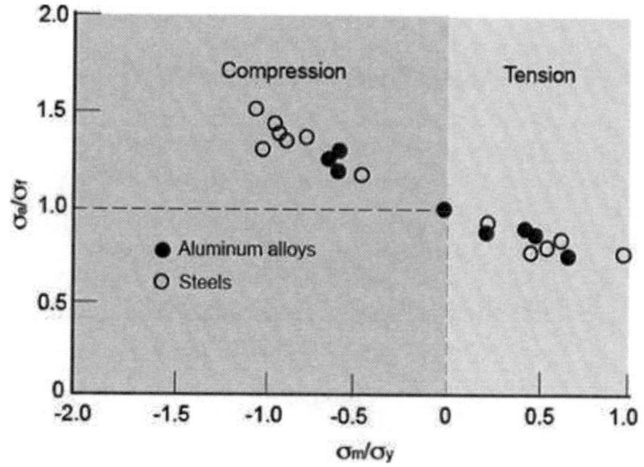

Fig. 7.20 Effect of mean stress in traction and compression on steels and aluminum alloys (adapted from [23])

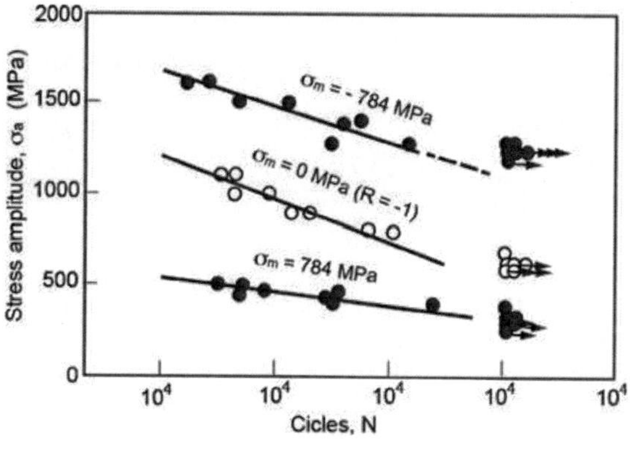

Fig. 7.21 Effect of mean stress on the fatigue behavior of high speed tool steel SKH51 [24]

Fig. 7.22. With a mean compression stress equal to − 200 MPa the fatigue limit grows by 15–20%.

It is worth noting how experimental points do not follow the Goodman line but they seem to be aligned along the Morrow line that reaches the true stress at fracture,

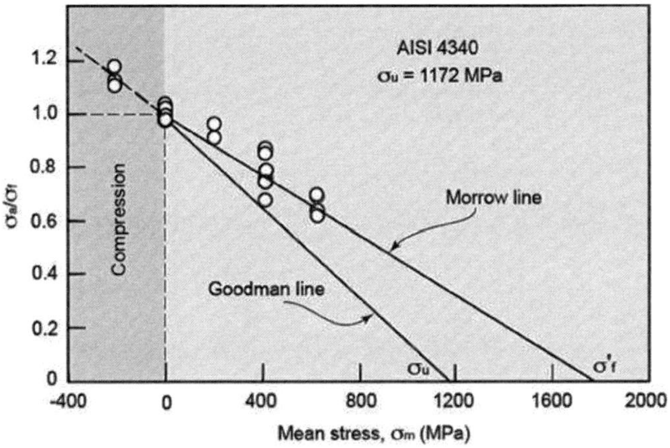

Fig. 7.22 Normalized stress amplitude-mean plot for AISI 4340 steel (adapted from [18])

σ'_f, as previously said about the Morrow hypothesis. For this material the true stress at fracture, σ'_f, is equal to 1758 MPa.

7.5.3 *Smith–Watson–Topper Approach*

Smith, Watson and Topper [25] proposed in 1970 a fatigue damage model under cyclic stress state with positive maximum stress, $\sigma_{\max} > 0$, also known as SWT model. The damage is expressed by the product $\sigma_{\max} \cdot \varepsilon_{a,m}$ in which $\varepsilon_{a,m}$ is the strain amplitude associated with the maximum stress $\sigma_{\max}$. They made the hypothesis that fatigue failure would happen when their model of damage, called SWT parameter, would be equal to the corresponding damage introduced in the material at failure under fully reversed fatigue

$$\sigma_{\max} \cdot \varepsilon_{a,m} = \sigma_{a,R-1}\varepsilon_{a,R-1} \quad \sigma_{\max} > 0 \tag{7.21}$$

in which $\sigma_{a,R\text{-}1}$ and $\varepsilon_{a,R\text{-}1}$ represent the stress amplitude and strain amplitude, respectively, of the completely reversed fatigue cycle having $R = \sigma_{\min}/\sigma_{\max} = -1$. For a stress-based fatigue analysis in which stress and strain are related by the Young's modulus, Eq. (7.21) can be expressed as

$$\sigma_{a,R-1} = \sqrt{\sigma_{\max}\sigma_{a,m}} \tag{7.22}$$

which, recalling Eq. (7.2) for σ_a as function of the R ratio, becomes

$$\sigma_{a,R-1} = \sigma_{\max}\sqrt{\frac{1-R}{2}} \quad \text{(a)}$$

$$\sigma_{a,R-1} = \sigma_a \sqrt{\frac{2}{1-R}} \quad \text{(b)} \tag{7.23}$$

Recalling Eq. (7.1) and Eq. (7.23), Eq. (7.21) becomes

$$\sigma_{\max} \cdot \sigma_{a,R-1} = \left(\sigma'_f N^b\right)^2$$

$$\sigma_{a,R-1} = \frac{\left(\sigma'_f N^b\right)^2}{\sigma_{\max}} \tag{7.24}$$

or

$$N = \left(\frac{\sigma_{\max}}{\sigma'_f}\sqrt{\frac{1-R}{2}}\right)^{1/b} \tag{7.25}$$

Note that for $R - 1$, $\sigma_{\max} = \sigma_{a,R\text{-}1}$ and Eq. (7.24) reduces to Eq. (7.1).

7.5.4 *Walker Approach*

In 1970 Walker proposed an approach to fatigue with non-zero mean stress that is similar to the SWT approach but provides a higher versatility due to the adjustable curve fitting parameter γ that is not always equal to ½ as in Eq. (7.22). His equation is

$$\sigma_{a,R-1} = \sigma_{\max}^{1-\gamma}\sigma_{a,m}^{\gamma} \tag{7.26}$$

which, recalling Eq. (7.2) for σ_a as function of the R ratio, becomes

$$\sigma_{a,R-1} = \sigma_{\max}\left(\frac{1-R}{2}\right)^{\gamma} \quad \text{(a)}$$

$$\sigma_{a,R-1} = \sigma_a\left(\frac{2}{1-R}\right)^{1-\gamma} \quad \text{(b)} \tag{7.27}$$

The parameter γ can be considered a sort of fitting constant characteristic of the material. Its value ranges from 0 to 1. When γ is equal to 0.5 Eqs. (7.26) and (7.27) reduce to the corresponding equation of the SWT approach. The drawback of the use of the fitting parameter γ is that it must be determined from experimental data. When this parameter is not known it is most commonly assumed as being equal to 0.5, which brings back to the SWT approach. Generally, for steels the γ parameter decreases as the material strength increases. To fit a set of experimental data to the Walker equation it can be used the following procedure that makes use of Eq. (7.1)

$$\sigma_{a,R-1} = \sigma'_f N^b = \sigma_{\max}\left(\frac{1-R}{2}\right)^{\gamma} \tag{7.28}$$

that solved for N yields

$$N = \left[\sigma_{\max}\left(\frac{1-R}{2}\right)^{\gamma}\frac{1}{\sigma'_f}\right]^{1/b}. \tag{7.29}$$

Next, take the logarithm to the base 10 of both sides

$$\log N = \frac{1}{b}\log\sigma_{\max} + \frac{\gamma}{b}\log\left(\frac{1-R}{2}\right) - \frac{1}{b}\log\sigma'_f \tag{7.30}$$

In a log–log diagram Eq. (7.30) is the equation of a straight line whose angular coefficient is given by $1/b$ and the known term, NT, provides the value of γ.

$$\gamma = \frac{b \cdot NT}{\log\left(\frac{1-R}{2}\right) - \frac{1}{b}\log\sigma'_f} \tag{7.31}$$

7.5.5 Correlation of Stress-Life Data

It is interesting to compare the various methods proposed to predict the fatigue response in the case of a mean stress acting on the fatigue cycle. In particular, we have been discussing in the previous sections, the modified Goodman line, the Morrow line, the FKM-guideline, the SWT approach and the Walker method. Dowling [18] has performed such a comparison on some steels and aluminum alloys, as well. Some of the results he obtained on AISI 4340 steel and 2024-T3 aluminum alloy are presented in Figs. 7.23 and 7.24, respectively. They are particularly instructive. For each material, the $\sigma_{a,R-1}$ values plotted in the figures are from Eqs. (7.4), (7.13), (7.24) and (7.27). In this plots the line fitted to Eq. (7.1) for the case $\sigma_m = 0$ is the Wöhler curve obtained in fully reversed loading conditions, $\sigma_{a,R-1}$.

The correlation is favorable to the extent that the data points fall very near the line. As for the Walker relation, the γ parameter for AISI 4340 steel and 2024-T3 aluminum alloy was equal to 0.650 and 0.460, respectively. as for 2024-T3 aluminum alloy, the traction strength used for the Morrow method was not the true fracture stress, σ'_f, but the true ultimate stress, $\sigma_{u,t} = \sigma_u(1+\varepsilon)$, as previously explained in Sect. 7.5.1, see Fig. 7.16. The conclusion drawn by Dowling [18] were:

- the Goodman mean stress Eq. (7.4) employing the ultimate tensile strength, σ_u, is inaccurate, being excessively conservative,

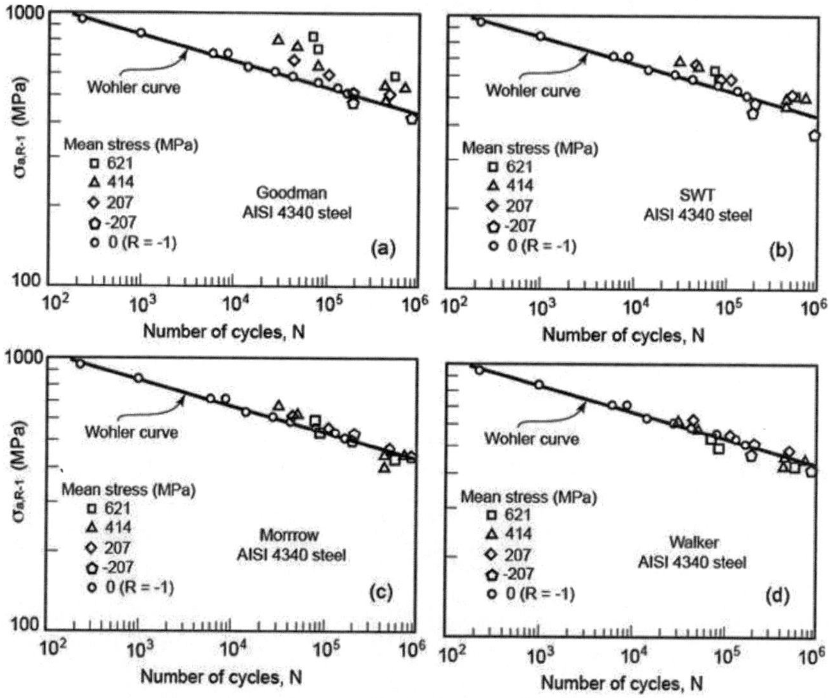

Fig. 7.23 Predictions for AISI 4340 steel using: **a** Goodman life correlation, **b** SWT life correlation, **c** Morrow life correlation and **d** Walker life correlation [18]

- the Morrow mean stress equation using the true fracture strength, σ'_f, works well for steels, but it is highly inaccurate and nonconservative for materials with log–log stress-life behavior that flattens at short lives. Thus it shroud not be used for aluminum alloys for which the true ultimate strength, $\sigma_{u,t}$, should be used,
- the SWT mean stress equation is a good choice for general use.

7.5.6 High and Smith Diagrams

Though with the restriction outlined at the end of the previous section stressing the large conservativism built-in the Goodman diagram, the Goodman diagram is largely used in a form that combines the diagrams of Figs. 7.7 and 7.14 providing the safety domain for compressive and tensile mean stresses, respectively. The result is the diagram of Fig. 7.25, also known as Haigh diagram. Any combination stress amplitude-mean stress that falls within the ABCDE area does not produce fatigue failure no matter what the number of cycles may be. Alike, if on the ordinate axis the allowable stress amplitude S or σ_a at N cycles is reported instead of the fatigue limit σ_f, the corresponding area delimitates the safety domain for those N cycles. m,

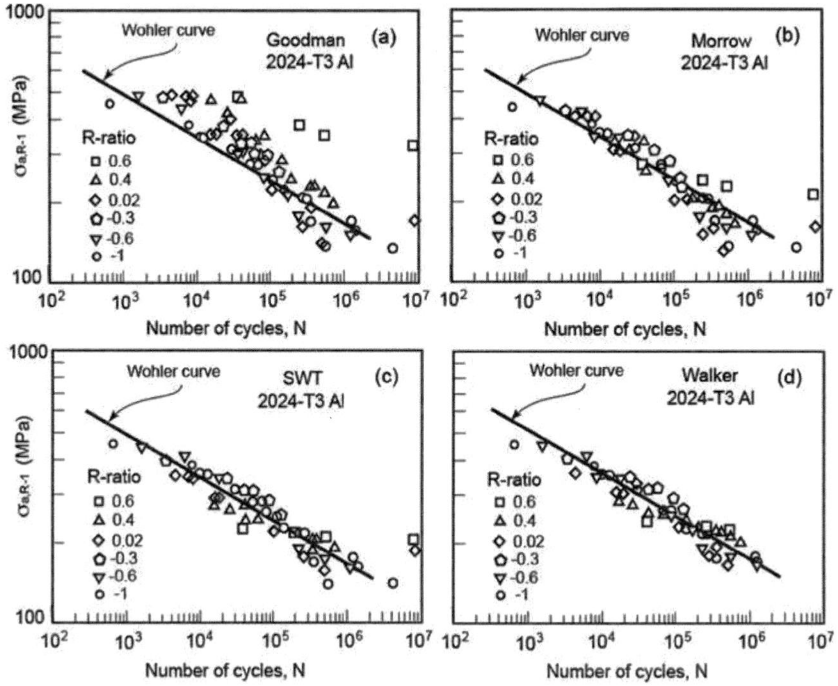

Fig. 7.24 Predictions for 2024-T3 Al using: **a** Goodman life correlation, **b** Morrow life correlation, **c** SWT life correlation and **d** Walker life correlation [18]

shown in Fig. 7.26. It provides the value of the total stress excursion $\Delta\sigma_a = \sigma_{max} - \sigma_{min}$ that may be safely applied when a given mean stress is acting. The abscissa axis is, as usual, that of the mean stress while the ordinate one is the axis of total stress excursion $\Delta\sigma = \sigma_{max} - \sigma_{min}$ so that stresses σ_{max} and σ_{min} are reported.

The diagram is built in the following way: start to trace the so called line of static load or mean stress which runs at 45° to the normal axes that have the same scale.

Fig. 7.25 Haigh diagram representing the safety domain for any combination $\sigma_m - \sigma_a$

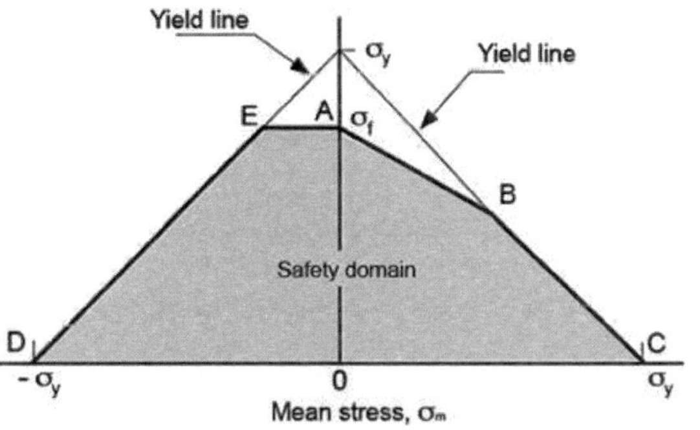

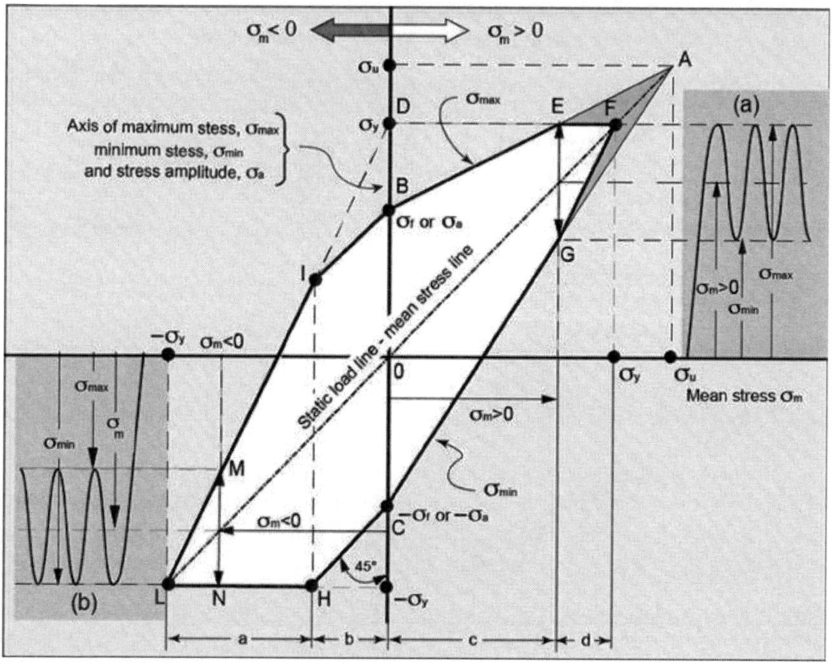

Fig. 7.26 Smith or Modified Goodman diagram for axial and bending fully reversed loads, ductile metals

On this line trace point A that has coordinates A≡(σ_u; σ_u)), σ_u being the tensile static strength of the material. Join point A with points B and C, respectively, on the ordinate that correspond to the fatigue limit $+\sigma_f$ and $-\sigma_f$, respectively, in complete reversed traction or bending ($\sigma_m = 0$).

Alternatively the allowable stress amplitude at a given N number of cycles $+\sigma_a$ and $-\sigma_a$ or $+S$ and $-S$ may be used. Note that line AB is the same AB line of the Haigh diagram of Fig. 7.25, in the new perspective. The Haigh diagram can be plotted in a different way known as Smith diagram or modified Goodman diagram. The ABC triangle is the locus of maximum allowable stress excursion, $\Delta\sigma = \sigma_{\max} - \sigma_{\min}$, at 10^6 cycles or 10^8 cycles for aluminum alloys or any other life N chosen to designate the fatigue limit. The value of the $\Delta\sigma$ excursion is given by the length of the vertical segment intercepted by the ABC triangle at the selected value of the mean stress σ_m. For instance, cycle (a) of Fig. 7.26 is represented by segment EG with positive mean stress, $\sigma_m > 0$, and cycle (b) by line MN, with negative mean stress, $\sigma_m < 0$.

The mean stress line is the locus of points having the same vertical distance from both sides of the triangle ABC. The abscissa of each point is the mean stress, the ordinate the maximum or minimum stress. Numerically they are the same because the scale, as said, is the same. The constraint on the yield strength introduced by the Soderberg line of Fig. 7.5 is represented by line EF of Fig. 7.26, while point G is the

Table 7.2 Regions, failure conditions and limiting values of σ_m

Region	Failure condition	Limiting value
a	$\sigma_{\max} - 2\sigma_m \geq \sigma_y$	$-\sigma_y \leq \sigma_m \leq (\sigma_a - \sigma_y)$
b	$\sigma_{\max} - \sigma_m \geq \sigma_a$	$(\sigma_a - \sigma_y) \leq \sigma_m \leq 0$
c	$\sigma_{\max} - (1 - r)\sigma_m \geq \sigma_a$	$0 \leq \sigma_m \leq \left(\frac{\sigma_y - \sigma_a}{1-r}\right)$
d	$\sigma_{\max} \geq \sigma_y$	$\left(\frac{\sigma_y - \sigma_a}{1-r}\right) \leq \sigma_m \leq \sigma_y$

intercept of E on line AC. The polygon AEFG represents the exclusion area where shake-down effect is active. As to the compression mean stress quadrant of Fig. 7.24 in the new representation of Fig. 7.26 it is represented by the course BIL above the mean stress line and by the course CHL on the lower side.

In practice, the modified Goodman diagram is obtained from the Haigh diagram of Fig. 7.25 by tracing the mean stress line at 45° to the coordinates axes and referring the diagram to this new axis so that lines AB, BC, AE and ED of Fig. 7.25 will keep the same angles to this new axis of Fig. 7.26 that they had before with respect to the abscissa axis. Figure 7.26 identifies four regions: a, b, c and d. Table 7.2 indicates the failure criterion and the limiting value of σ_m for each region. Fatigue diagram like that of Fig. 7.26 can be derived for any metal alloy from experimental data relative to fatigue limit and traction strength, as shown in Fig. 7.27. It is worth recalling that the construction of the Haigh diagram of Fig. 7.26 has made use of the ultimate strength, σ_u, while we have seen, in the previous section, that the mean stress can reach the true fracture strength, σ'_f, of the material or at least the true ultimate strength, $\sigma_{u,t}$. Also note that the Smith diagram of Fig. 7.26 has not considered the beneficial effect of mean stress discussed in Sects. 7.5.1 and 7.5.2.

7.5.7 *Master Curve*

The diagram of 17 and that of Fig. 7.19, as already said, can be traced for any allowable stress amplitude σ_a or S corresponding to any given number of cycles N. Let's rotate by 45° the mean stress and the stress amplitude axes. The result is shown in Fig. 7.28 for alloyed-steels with 862 to 1241 MPa ultimate strength (125 to 180 ksi), type 2330, 4130, 4340 or 8630 where coordinates giving minimum stress and maximum stress have been added. This is normally called Master curve or diagram. Thus the curves can be used as $\sigma_a - \sigma_m$ curves or $\sigma_{\min} - \sigma_{\max}$ curves, whichever is more convenient. For what has been said and shown in Fig. 7.9, the curves are seen to be intermediate between Goodman lines and Gerber parabolas, approaching the former as life increases and stress amplitude comes near to the fatigue limit. In particular, the diagram in Fig. 7.21 applies to axial cyclic loading. It must be understood that diagrams like that of Fig. 7.21 are of experimental derivation. Another example of master diagram is shown in Fig. 7.29 for 7075-T6 aluminum alloy with 565.5 MPa

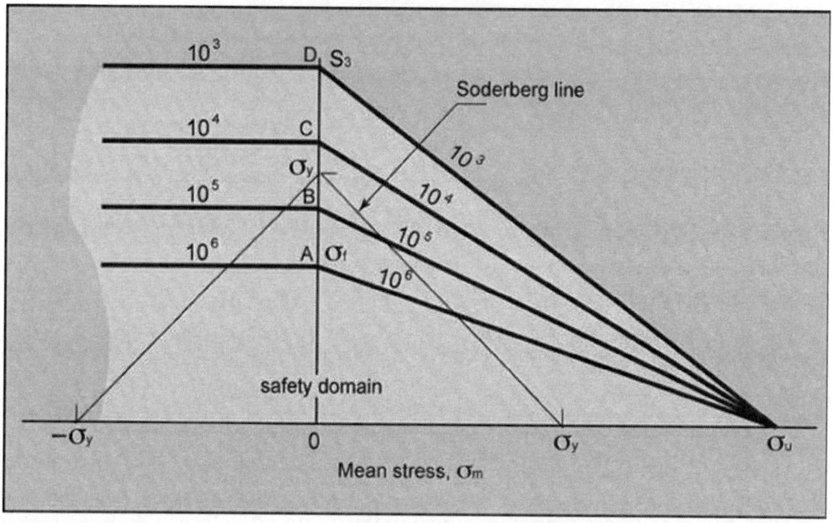

Fig. 7.27 Haigh diagram for allowable stress amplitudes corresponding to different fatigue lives *N*

(82 ksi) ultimate strength and 483 (70 ksi) yield strength. For example, referring to Fig. 7.28 with a mean stress equal to 20% of the ultimate strength S_u the allowable stress at 10^7 cycles is equal to 40% of the ultimate.

7.6 Sample Problems on Mean Stress Effect and Fatigue Strength Diagrams

7.6.1 *Shaft Under Torsion*

Consider the steel shaft or sample problem treated in Sect. 10.5.5 for which the *S–N* diagram of Fig. 10.34 has been derived under fully reversed torsional fatigue limit with τ_f and τ_3 at 1000 cycles, corrected to take into consideration factors affecting fatigue, equal to

$$\tau_f = C_{ld} \cdot C_{sz} \cdot C_s \cdot 640 = 0.58 \cdot 0.7 \cdot 0.9 \cdot 640 = 233.8\ MPa$$

$$\tau_{f,3} = C_{ld} \cdot 0.9 \cdot 0.8 \cdot 950 = 0.58 \cdot 0.9 \cdot 0.8 \cdot 950 = 396.7\ MPa$$

respectively. Assume that torsion is now applied to the shaft with a mean stresses of three different increasing values so that the R ratio from $R = -1$ becomes: (a) $R = 0$, (b) $R = 0.5$ and $R = 0.8$.

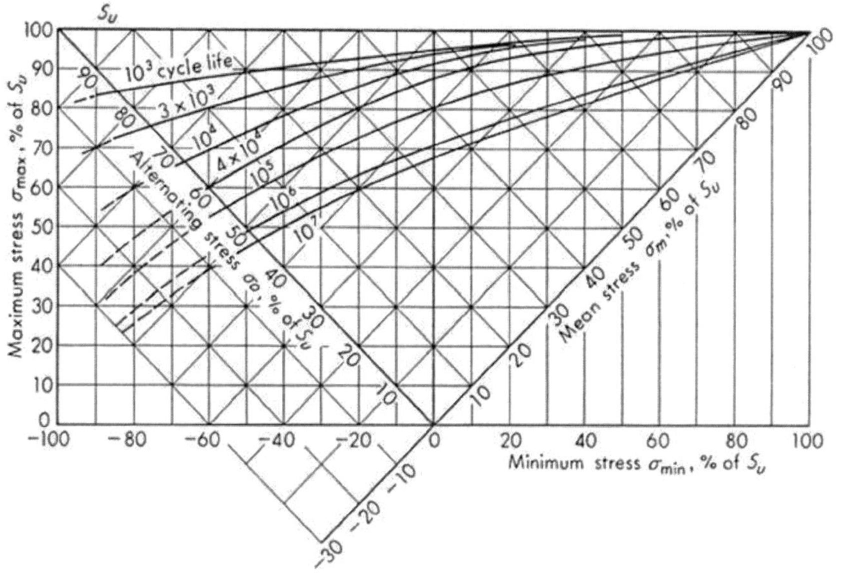

Fig. 7.28 Fatigue strength diagram for AISI 4340 steel (Courtesy of Grumman Aerospace Corporation)

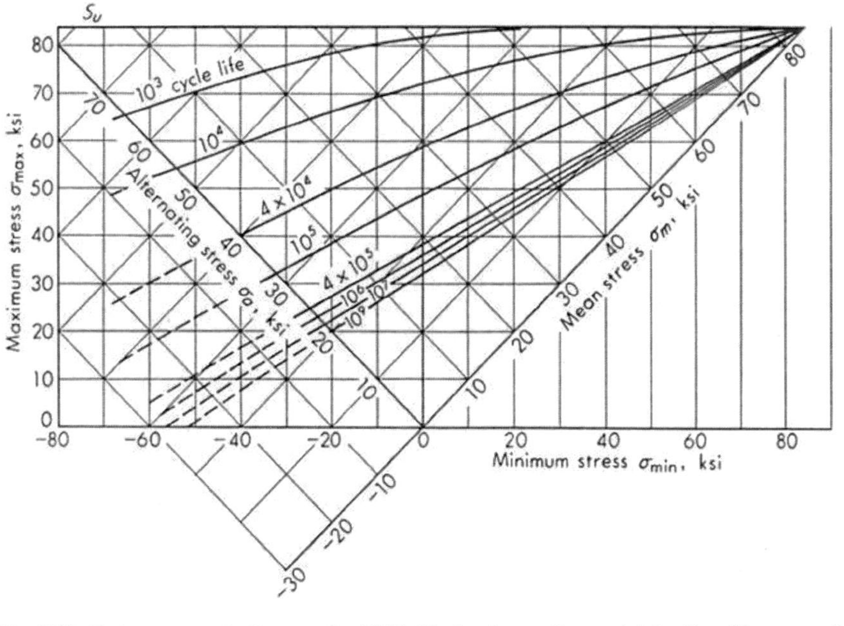

Fig. 7.29 Fatigue strength diagram for 7075-T6 aluminum alloy, axial loading (Courtesy of Grumman Aerospace Corporation)

Construct the relative *S–N* diagram using the Goodman approach. Let's start for case (a). In a cycle fluctuating from the origin it is $\tau_m = \tau_a = \tau'_f$ where τ'_f is the fatigue limit under the effect of mean stress. Therefore, the first of eq. (7.4) may be written

$$\tau'_f = \tau_f\left(1 - \frac{\tau_m}{\tau_u}\right) = \tau_f\left(1 - \frac{\tau'_f}{\tau_u}\right).$$

Resolved for τ'_f it is

$$\tau'_f = \tau_f\left(\frac{\tau_u}{\tau_u + \tau_f}\right) = 233.8 \cdot \frac{0.8 \cdot 950}{994} = 179\ MPa$$

$$\tau_m = \tau_a = \tau'_f = 179\ MPa.$$

The maximum stress $\tau_{\max}$ will be equal to $2 \cdot \tau_m = 2 \cdot \tau_a = 2\,\tau'_f = 358$ MPa. In case (b), $R = 0.5$, we shall assume that the fatigue limit is equal to the stress amplitude $\tau_a = \tau_{\max} \cdot (1 - R)/2$ with τ_m equal to

$$\tau_m = \tau_{\max}\frac{1+R}{2}.$$

Therefore

$$\tau'_f = \tau_f\left[1 - \frac{\tau_{\max}}{\tau_u}\frac{(1+R)}{2}\right] = \tau_a = \tau_{\max} \cdot \frac{(1-R)}{2}$$

$$\tau_{\max} = \tau_f \Big/ \left[\frac{1-R}{2} + \frac{\tau_f}{\tau_u}\frac{(1+R)}{2}\right].$$

Knowing $\tau_{\max}$ it is possible to infer τ'_f

$$\tau_{\max} = 233.8 \Big/ \left[\tfrac{1-0.5}{2} + \tfrac{233.8}{0.8 \cdot 950}\tfrac{(1+0.5)}{2}\right] = 486\ MPa$$

$$\tau'_f = 233.8 \cdot \left[1 - \tfrac{486}{0.8 \cdot 950}\tfrac{(1+0.5)}{2}\right] = 122\ MPa$$

$$\tau_m = 364.5\ MPa.$$

In the case I with $R = 0.8$ it is

$$\tau_{\max} = 233.8 \Big/ \left[\tfrac{1-0.8}{2} + \tfrac{233.8}{0.8 \cdot 950}\tfrac{(1+0.8)}{2}\right] = 620\ MPa$$

$$\tau'_f = 233.8 \cdot \left[1 - \tfrac{620}{0.8 \cdot 950}\tfrac{(1+0.8)}{2}\right] = 62\ MPa$$

$$\tau_m = 558\ MPa.$$

As to τ'_3 it remains exactly that found under fully reversed torsion since at 1000 cycles the mean stress has absolutely no effect at all

$$\tau_{f,3} = C_{ld} \cdot 0.9 \cdot 0.8 \cdot 950 = 0.58 \cdot 0.9 \cdot 0.8 \cdot 950 = 396.7\ MPa\,.$$

The *S–N* diagrams relative to the three new conditions are shown in Fig. 7.30. It is particularly interesting to observe the effect of the mean stress on the fatigue limit of the shaft, but it is above all important to note how the stress amplitude, dough reducing progressively from $R = -1$ to $R = 0$ and then to $R = 0.5$ and finally to $R = 0.8$, has an effect more and more devastating on the fatigue strength of the shaft because of the continuous increase of the mean stress that from the initial zero value has grown to 179 MPa to 364.5 MPa and eventually to 558 MPa.

The four cases examined are shown in Fig. 7.31. It must be also considered that fluctuating low amplitude stress are generally generated by medium–high frequency vibrations so that in few hours or days they may sum up to millions of cycles. Note how the diagram of Fig. 7.31 recalls the modified Goodman diagram of Fig. 7.26

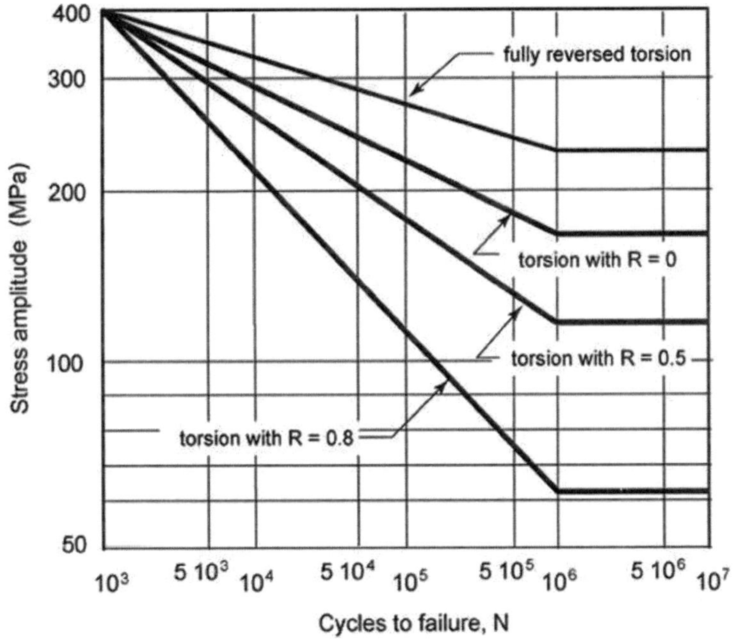

Fig. 7.30 *S–N* diagrams for high strength machined finish shaft under torsional fatigue with different *R* ratio equal to -1, 0, 0.5 and 0.8, respectively

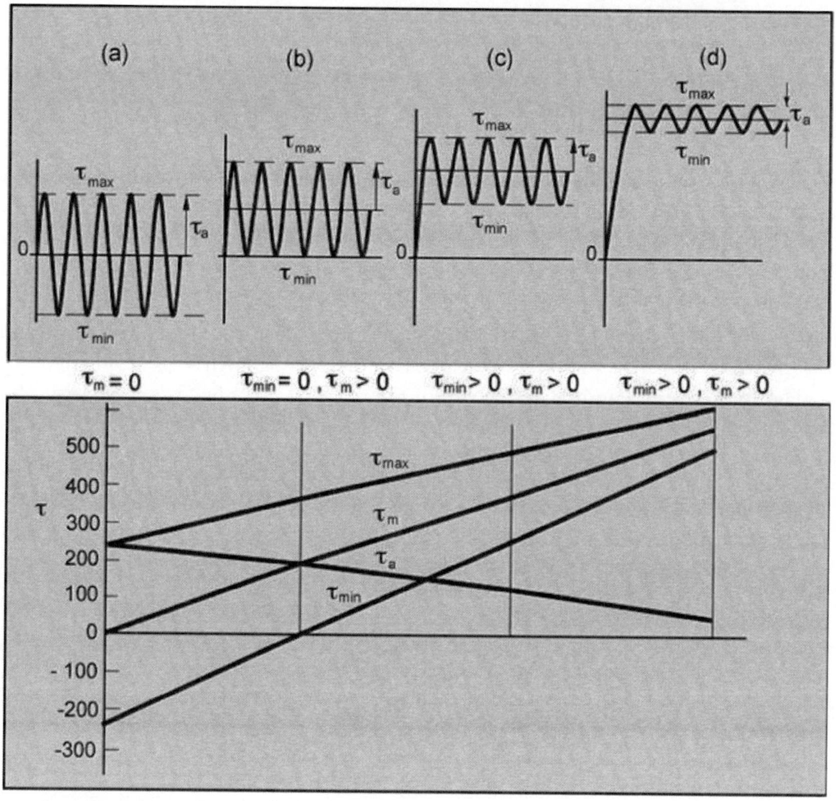

Fig. 7.31 Fatigue cycles relative to the four cases of Fig. 7.23

7.6.2 Mean Stress and Vibrations in Torsion-Bar

A splined torsion-shaft, Fig. 7.32, 7.2″ diameter is subjected to a constant torque $M_t =$ 180 KN. The material is high strength carbon steel heat treated to 1200 MPa ultimate strength and 313 MPa fatigue limit τ_f in completely reversed torsion. Service experience indicates continuous untimely failures along section A-A in correspondence of a shoulder fillet of radius $\rho = 2$ mm.

Therefore, it has been decided to monitor the shaft using strain gauges that have evidenced a torsional vibration variable from 420 to 740 MPa at 20 Hz. The shaft, then, is subjected to a shearing stress of amplitude $\tau_a = 160$ MPa and mean stress τ_m = 580 MPa and $R = 0.56$. Let's trace the Goodman line of Fig. 7.11 passing through points A≡(160;580) and B≡(τ_u;0), τ_u being the ultimate strength in torsion that may be appraised as $0.8{\cdot}\sigma_u = 960$ MPa. The intercept on the ordinate axis represents the equivalent torsion stress amplitude τ_{equ} at $R = -1$, i.e., the stress that causes the same effect as the actual stress with $R = 0.56$. It is $\tau_{equ} = 365$ MPa. Consider now the *S–N* diagram of Fig. 7.33b for the steel considered obtained with usual method of *S–N*

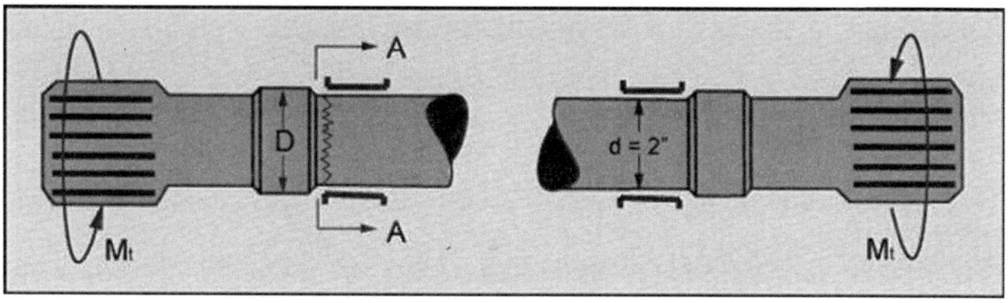

Fig. 7.32 Splined torsion shaft failed along section A-A at a shoulder fillet

construction. Entering the diagram with $\tau_{equ} = 360$ MPa it is found that life is $N \sim$ 200,000 cycles. Considered that the frequency of that vibration is 20 Hz it is deducted that the shaft life is about 5–6 h. In the previous consideration it has not been taken into account the notch effect that will be discussed in Sect. 11.6. Considering that effect the life would be further reduced.

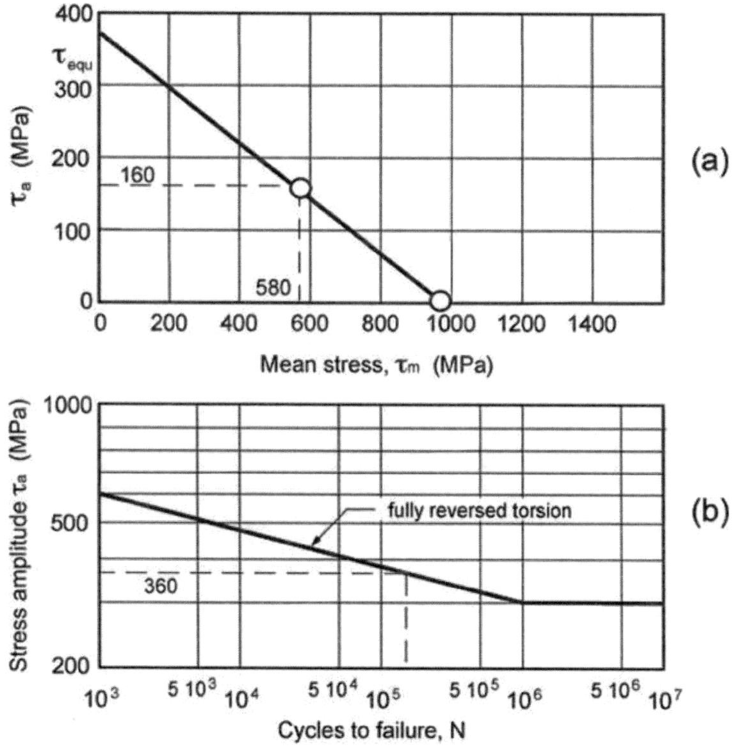

Fig. 7.33 **a** Goodman diagram and **b** *S–N* diagram in completely reversed torsion

7.6.3 Mean Stress and Vibrations in Torsion-Bar: Helical Spring

A helical spring of Fig. 7.34 with diameter $D = 50$ mm and circular section diameter $d = 16$ mm is made of high strength steel heat treated to 1400 MPa ultimate strength and fatigue limit $\sigma_f = 700$ MPa. It is subjected to a cyclic load F variable from 500 N to 6000 N. Verify whether it can sustain at least 10^6 cycles. The generic section is subjected to a torque M_t and a transverse component of force F the result in a maximum and minimum shear stress given by

$$\tau_{\max} = \frac{16}{\pi}\frac{F_{\max} \cdot R}{d^3} + \frac{16}{3\pi}\frac{F_{\max}}{d^2} = \frac{16}{\pi}\frac{6.000 \cdot 0.025}{0.016^3} + \frac{16}{3\pi}\frac{6.000}{0,016^2} \approx 227\ MPa$$
$$\tau_{\min} = \frac{16}{\pi}\frac{F_{\min} \cdot R}{d^3} + \frac{16}{3\pi}\frac{F_{\min}}{d^2} = \frac{16}{\pi}\frac{500 \cdot 0.025}{0.016^3} + \frac{16}{3\pi}\frac{500}{0.016^2} = 18.86\ MPa. \tag{7.32}$$

The American engineer A. M. Wahl found that the combined effect of curvature of the wire and transverse force resulted in a transverse shear stress that is not exactly given by the second term of the above equations. The complete equation for the maximum stress in a coil spring may be calculated as

$$\tau = \frac{8FD}{\pi d^3}k \tag{7.33}$$

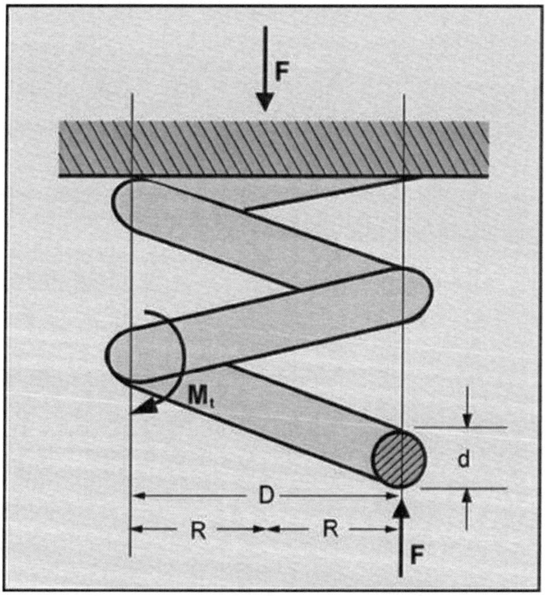

Fig. 7.34 Schematic of a helical spring

where k is expressed in terms of a *spring index* $C = D/d$

$$k = \frac{4C - 1}{4C - 4} + \frac{0.615}{C}. \tag{7.34}$$

The deflection of the helical spring is given by

$$\delta = \frac{8FD^3N}{d^4G} \tag{7.35}$$

with N being the number of coils. The shear stress amplitude τ_a and the mean τ_m value are

$$\tau_a = \frac{\tau_{\max} - \tau_{\min}}{2} = \frac{227 - 18.86}{2} \approx 104\ MPa$$
$$\tau_m = \frac{\tau_{\max} + \tau_{\min}}{2} = \frac{227 + 18.86}{2} \approx 123\ MPa.$$

The torsional fatigue limit $\tau_{f,m}$ with the mean stress τ_m is given by Eq. (7.3)

$$\tau_{f,m} = \tau_f\left(1 - \frac{\tau_m}{\tau_r}\right) = 0.58 \cdot \sigma_f \cdot \left(1 - \frac{\tau_m}{0.8 \cdot \sigma_u}\right) = 361\ \ MPa$$

where, as usual, the torsional fatigue limit is $\tau_f = 0{,}58{\cdot}\sigma_f$ and torsional strength is 80% of the tensile strength. The coil has a fine-ground finish with $C_s = 0.9$. The size effect results in $C_{sp} = 0{,}7$, therefore it is

$$\tau'_f = 0.9 \cdot 0.7 \cdot 0.58 \cdot \sigma_f = 256\ MPa.$$

Since the stress amplitude is $\tau_a = 104$ MPa the safety factor will be

$$v_f = \frac{256}{104} \approx 2.46$$

that may be acceptable.

7.6.4 *Mean Stress in a Cantilever Beam*

A high strength steel bar, 1400 MPa ultimate and 1200 MPa yield strength, operating in simple bending is subjected to vibrations that results in a stress excursion between a minimum of 420 MPa and a maximum of 600 MPa. The bar has a surface notch where local stress goes up to 840 MPa and 1200 MPa, respectively. Verify the fatigue behavior. The notch tip is the most stressed zone where fatigue failure is likely to occur. On the notch tip the stress amplitude an mean stress are

$$\sigma_a = \frac{\sigma_{\max} - \sigma_{\min}}{2} = \frac{1200 - 840}{2} = 180\ MPa$$

$$\sigma_m = \frac{\sigma_{\max} + \sigma_{\min}}{2} = \frac{1200 + 840}{2} = 1020\ MPa.$$

Construct the Goodman diagram for these working conditions. If we join points A≡(σ_m; σ_a) and B≡(σ_u;0), as shown in Fig. 7.35a, the intersection with the stress axis relative to $\sigma_m = 0$ will yield the equivalent stress limit $\sigma'_{f,equ}$ that in completely reversed bending ($R = -1$ and $\sigma_m = 0$) would produce the same effect of the actual stress conditions. This new limit is then equal to 700 MPa. If we enter the *S–N* diagram of the material, Fig. 7.35b, with this stress amplitude we obtain a life of about 300,000 cycles only. Let's try to evaluate whether or not a shake-down effect may take place. To this purpose consider the Soderberg line passing through points (0;1200) and (1200;0), as shown in Fig. 7.35a. The Goodman line intercepts the Soderberg line just in point A. Therefore the shake-down effect will not occur, first Eq. (5.57) and Fig. 7.14. The life of only 300,000 is realistic. A solution may be to change material reducing the yield strength to 950 MPa and maintaining the same ultimate strength of 1400 MPa. The new Soderberg line will cross the Goodman line in A″ so that the working point A will fall in the exclusion zone BA″C where the shake-down effect operates. After few cycles, then, point A will move back to A′ reducing the mean stress from 1200 to 780 MPa. This new working point will have coordinates A′≡($\sigma^*_{m,1}$; σ_a)≡(780;180). The new Goodman line passing through points A′ and B intercepts on the ordinate axis the new value $\sigma''_{f,equ} = 410$ MPa of the equivalent fatigue limit. This time is $\sigma''_{f,equ} < \sigma_f$ so that the fatigue life of the work piece will be non-limited.

7.6.5 *Fatigue Strength Diagrams*

Derive the modified Goodman diagram of a NiCrMo, type 40NiCrMo7, 850 MPa yield strength and 45% reduction of area RA, under.

a. Fully reversed traction (push-and-pull),
b. Rotating bending,
c. Torsion.

The mechanical properties of the steel are: 1100 MPa ultimate strength, 850 MPa yield strength, 45% RA. The steel is used to make shafts 85 mm diameter with fine-ground finish. Suppose that the fatigue limit has not been found jet. Lacking specific information about the fatigue limit, based on Table 5.2 of Sect 5.5 for quenched and tempered steels, it can be assumed σ_f as 0.40·σ_u = 440 MPa. Considering Fig. 4.37, to an ultimate strength of 1100 MPa it corresponds a mean fatigue life of 550 MPa. Referring to Fig. 4.40 since the product 1100·(RA) is equal to 49,500 the average value of fatigue limit is $\sigma_f \sim 500$ MPa. Conservatively we shall assume that $\sigma_f =$ 440 MPa, the lowest value.

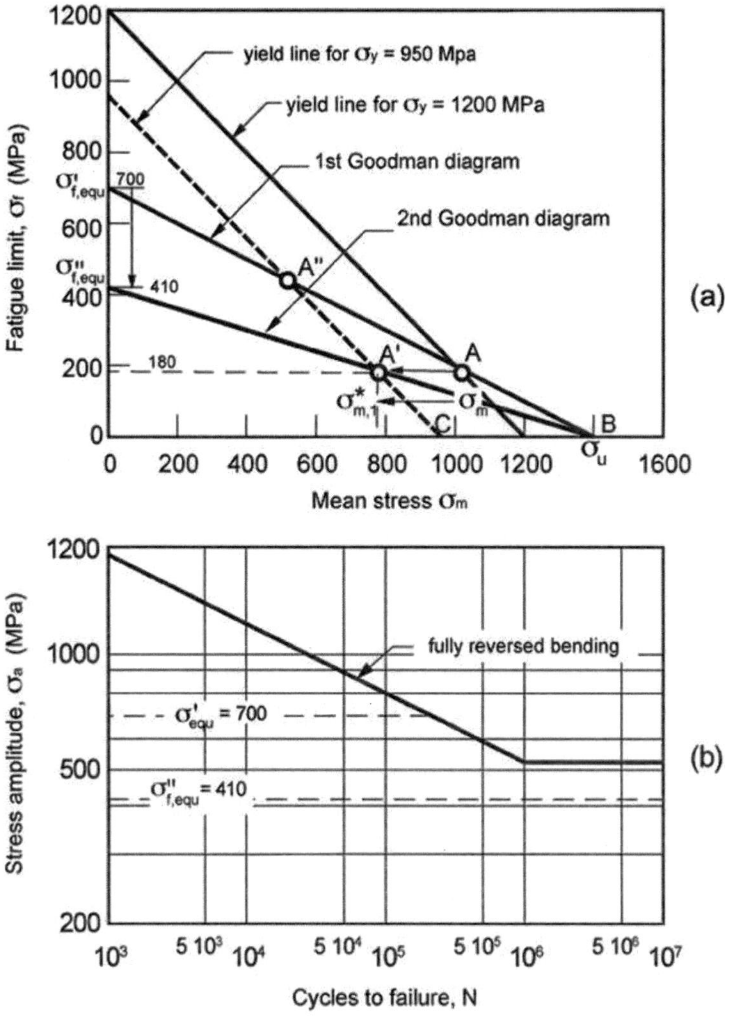

Fig. 7.35 **a** Goodman diagram and **b** *S–N* diagram for fully reversed bending

Nevertheless, it will be chosen $\sigma_f = 500$ MPa since this value is supported by two out of three estimates. Surface factor is $C_S = 0.9$ (see Fig. 3.11). Note that with $C_S = 0.9$ the fatigue limit becomes $\sigma_f \sim 450$ MPa conforming Fig. 4.7 prevision. The size factor may be inferred using Shingley and Mitchell Eq. (4.15), therefore

$$C_{sz} = 1189 \cdot (85)^{-0.097} = 0.77. \quad (7.36)$$

The final fatigue limit is

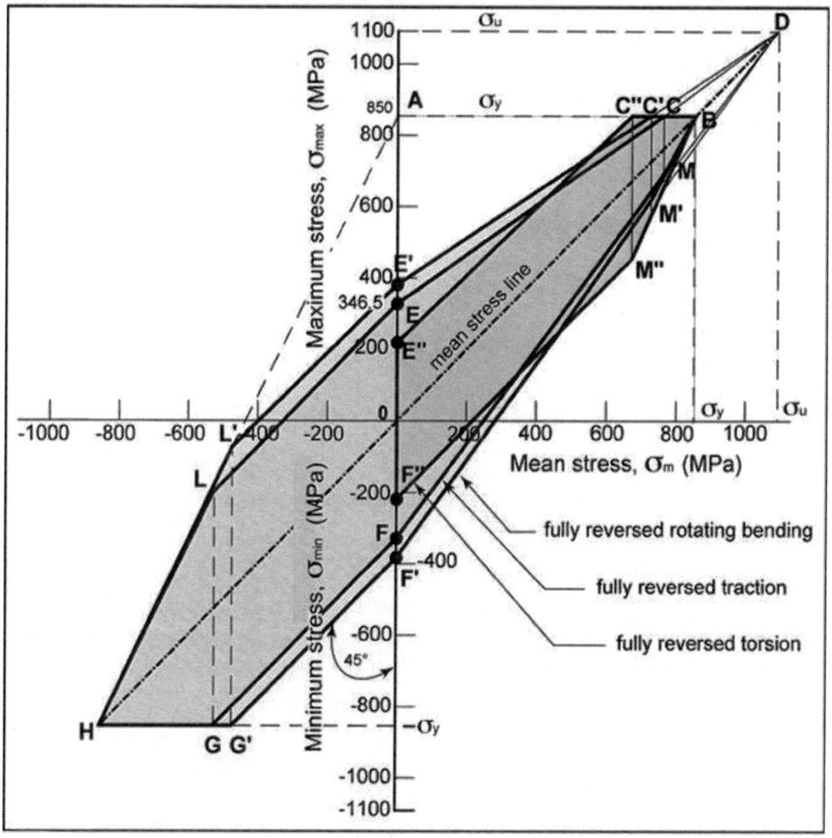

Fig. 7.36 Modified Goodman diagram for a 40NiCrMo7steel shaft, 85 mm diameter, subjected to fully reversed traction, rotating bending an rotating torsion, respectively

$$\sigma'_f = \sigma_f \cdot C_S \cdot C_{sz} = 346.5 \text{ MPa}. \tag{7.37}$$

The modified Goodman diagram is, therefore, the one BMFGHLECB shown in Fig. 7.36. Moving to rotating bending, Table 4.4 of Sect. 4.9 provides the load factor $C_{ld} = 1.1$ since the shaft is not thick, therefore

$$\sigma'_f = C_S \cdot C_{sz} \cdot C_{ld} \cdot \sigma_f = 381\ MPa.$$

The new modified Goodman diagram is BM′F′G′HL′E′C′B of Fig. 7.36. As last, considering fully reversed torsion the load factor is $C_{ld} = 0.58$ and

$$\sigma'_f = C_S \cdot C_{sz} \cdot C_{ld} \cdot \sigma_f = 201\ MPa$$

References

1. Basquin, O.H.: Proc. ASTM **10**(Part II), 625 (1910)
2. DuQuesnay, D.L., Topper, T.H., Yu, M.T., Pompetzki, M.A.: The effective stress range as a mean stress parameter. Int. J. Fatigue **14**, 45–50 (1992)
3. Wehner, T., Fatemi, A.: Effect of mean stress on fatigue behaviour of a hardened steel. Int. J. Fatigue 241–248 (1991)
4. Gerber, W.Z., Bayer. Archit. Ing., Vre. 6, 101 (1874)
5. Goodman, J.: Mechanics Applied to Engineering. Longman, Green and Co., London (1899)
6. Haigh, B.P.: Experiments on the fatigue of brasses. J. Inst. Met. **18**, 55–86 (1917)
7. Soderberg, C.R.: Fatigue of the safety and working stress. Trans. Am. Soc. Mech. Eng. **52**(Part APM-52-2), 13–28 (1939)
8. Sines, G., Weisman, J.L. (eds.): Metal Fatigue. McGraw-Hill, New York (1959)
9. Forrest, P.G.: Fatigue of Metals. Pergamon Press, Oxford (1962)
10. Dolan, T.J.: Stress range in Horger. In: Oscar J (ed.) ASME Handbook: Metals Engineering-Design, 2nd ed., part 2, sec 7.2. McGraw-Hill Book Company, New York (1965)
11. Wilson, J.S., Haigh, B.P.: Stresses in Bridges, Engineering. London, 446–448 (1923)
12. Howell, F.M., Miller, J.L.: Proc. Am. Soc. Test. Mater. **55**, 955 (1955)
13. Forrest, P.G.: International Conference on Fatigue, Institution of Mechanical Engineers, 171 (1956)
14. Lazan, B.J., Blatherwick, A.A.: Wright Air Development Center Technical Report No. 52-307 (1952); Rep. No. 53-181 (1953)
15. Schijve, J.: Fatigue of Structure and Materials, p. 120. Kluwer Academic Publishers (2004)
16. Juvinall, R.C.: Engineering Considerations of Stress, Strain and Strength, p. 275. Mc Graw-Hill, Inc. (1967)
17. Morrow, J.: Fatigue Design Handbook. Advances in Engineering, vol. 4, pp. 21–29. SAE, Warrendale, PA (1968)
18. Dowling, N.E.: Mean stress effect in stress-life and strain-life fatigue, Rep. F2004/51, Society of Automotive Engineers, Inc. (2004)
19. Schütz, W.: View Point of Material Selection for Fatigue Loaded Structures (in German), Laboratoriun für Betriebsfestingkeit LBF, Darmstadt, Bericht Nr. TB-80 (1968)
20. Lee, Y.L., Pan, J., Hathaway, R., Barkey, M.: Fatigue testing and analysis: theory and practice, Elsevier Butterworth-Heinemann (2005)
21. Lee, Y.L., Barkey, M, Kang, H.T.: Metal Fatigue Analysis Handbook: Practical Problem-Solving Techniques for Computer Aided Engineering, Elsevier (2012)
22. Haibach, E.: FKM-Guideline: Analytical Stress Assessment of Components in Mechanical Engineering, 5th Rev. Ed. VDMA (2003)
23. Sines, G.: Failure of Materials Under Combined Repeated Stress with Superposed Static Stresses. NACA TN 3945 (1955)
24. Murakami, Y.: Metal Fatigue: Effects of Small Defects and Nonmetallic Inclusions. Elsevier (2012)
25. Smith, K.N., Watson, P., Topper, T.H.: A stress-strain function for the fatigue of metals. J. Mater. **5**(4), 767–778 (1970)

Chapter 8
Strain-Based Fatigue Analysis—Low Cycle Fatigue

8.1 Introduction

Three regions were identified in Sect. 1.2 and shown in Fig. 1.10 as Region I, II and III concerning the fatigue strength of materials. In Sect. 2.5 Region III pertaining fatigue limit was analyzed. In that region the stress amplitude in not high enough to propagate a micro defect even though it may have been generated. In Chap. 5 we have been discussing of Region II, the region of high cycle fatigue where load levels are low and the main component of stress is elastic, when not totally elastic. In this chapter we will examine as last the first region or Region I of low cycle fatigue where strains and, in particular, plastic strains are the dominant factor. When load levels are low, stresses and strains are linearly related by Hooke's law. Under such conditions they are fully interchangeable. But when the plastic component cannot be neglected anymore and, indeed, prevails, stresses are no longer uniquely determined and must be put aside. Although most engineering structures and components are design such that nominal loads remain elastic, stress concentrations can easily introduce a non-negligible plastic strain component.

This is the case of discontinuities and, in particular, notches more or less sharp where local stresses can go well beyond the elastic limit of the material. Besides the case of stress concentrations, there are also cases where fatigue life is constrained in relatively few cycles, very far from the millions that characterize most structures designed to withstand fatigue. Think about turbine blades of jet engines subjected to elevated temperatures, in the range of 1000 °C, and very high centrifugal forces, whose life is some thousands of cycles only. The plastic domain is all another world. The material under cyclic stresses undergoes a hysteresis loop that may not only change its properties, as said in Sect. 1.4, but can initiate a micro crack almost immediately so that strain-life methods of analysis are also considered initiation life estimates. Fatigue is now controlled by strains so that also fatigue tests must be run under strain controlled conditions. The use of stress may lead to the paradox that in Region I of fatigue the stress amplitude σ_a appears to be independent of the number

P. P. Milella, *Fatigue and Corrosion in Metals*,
https://doi.org/10.1007/978-3-031-51350-3_8

of cycles N just because the material is undergoing a cyclic strain hardening that continuously changes its strength.

This very important domain of fatigue strength has reached a rather consolidated methodological approach both from design standpoint and experimental one, as well. A very important aspect that characterizes and distinguishes low cycle fatigue form high cycle fatigue is that initiation may occur anywhere in the work-piece or specimen and not just on the surface, as explained in Sect. 2.2 and shown in Fig. 2.19. Surface finish is no longer important, but what is really important is the fact that now the process volume is the entire volume of material affected by the *above yield* strain and not just its *skin*, as shown in Fig. 2.5. At the base of the strain-based approach stands the cyclic material response and the hysteresis loop that will be investigated in its manifolds implications in the next sections.

8.2 Stress-Plastic Strain Power Law Relation

It has been said in Sect. 1.3 that a metal subjected to load cycling close or above the elastic limit may experience a process of softening or hardening that changes, sometime significantly, its stress–strain characteristic with respect to that obtained under monotonic traction, as shown in Figs. 1.41 and 1.42. It has, also, been shown in Sect. 1.4.2, how this new characteristic cyclic response can be obtained. However, it must be held in mind that during cycling loading the material does not follow the cyclic strain–stress curve as it would do under monotonic traction with a static strain–stress curve. Rather, hysteresis loops are formed, such as that shown in Fig. 1.38. It is necessary, then, to characterize the shape and derive the equation of such hysteresis loops.

Consider, then, a stabilized stress–strain cycle as shown in Fig. 8.1 by loop OABCDA. The total deformation ε_a can be considered the sum of an elastic component ε_e and a plastic one ε_p

$$\begin{aligned}\varepsilon_a &= \varepsilon_e + \varepsilon_p \\ &= \frac{\sigma_a}{E} + \varepsilon_p\end{aligned} \tag{8.1}$$

where E is the Young' modulus of elasticity. The total excursion of plastic deformation $\Delta\varepsilon_p$ can be assumed, without any significant error, as that corresponding to the cycle amplitude on the strain axis. The elastic component can be numerically subtracted from the total deformation so to obtain the corresponding plastic cycle, as shown in Fig. 8.1b. This allows the measurement of plastic strains even when they are very small, provided that they are not lower than 0.0005 mm/mm. If the amplitude of plastic strain $\Delta\varepsilon_p/2$ is plotted in a log–log scale as function of the stress amplitude $\Delta\sigma_a/2$, usually a line is obtained. This is shown in Fig. 8.2 obtained using experimental data relative to hot rolled Man-Ten steel of 330 MPa yield and type RQC-100 steel quenched and tempered to 840 MPa yield strength from [1] and to

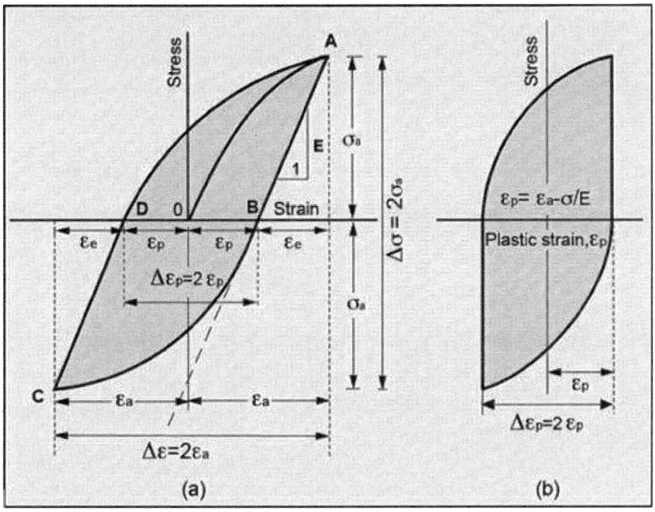

Fig. 8.1 **a** Hysteresis loop at saturation, stress–stain excursion, $\Delta\sigma - \Delta\varepsilon$, **b** plastic component of strain, $\Delta\sigma - \Delta\varepsilon_p$ loop

ductile cast iron (DCI) tested by Milella. The linear relationship that holds for almost all metals and alloys has induced Masing [2] to introduce a $\sigma_a - \varepsilon_p$ dependence of the type

$$\frac{\Delta\sigma_a}{2} = K'\left(\frac{\Delta\varepsilon_p}{2}\right)^{n'} \tag{8.2}$$

with K' being a sort of *plastic modulus* that is called cyclic strength modulus or coefficient and n' a cyclic strain hardening exponent.

Note that under the hypothesis made the value of the K' coefficient is obtained as intersection of the interpolating line with the vertical line passing through $\varepsilon_p = 1$ while the exponent n' represents its slope. For the two steels in Fig. 8.2 is $K' = 1120$ MPa for Man-ten, $K' = 1170$ MPa for RQC-100 and 715 MPa for DCI, respectively. Equation (8.1) can be written as

$$\begin{aligned} \varepsilon_a &= \varepsilon_e + \varepsilon_p \\ &= \frac{\sigma_a}{E} + \left(\frac{\sigma_a}{K'}\right)^{1/n'}. \end{aligned} \tag{8.3}$$

It is worth noting that the superscript $'$ on cyclic hardening exponent and plastic modulus reminds that they are cyclic quantities and not monotonic ones. Cyclic strength coefficient K' has, usually, values between 700 and 3000 MPa while n' is between 0.1 and 0.2. Table 8.1 offers a synoptic picture of n' and cyclic yield strength, $\sigma_{s'}$, compared to the correspondent monotonic values. Equation (8.3) represents the

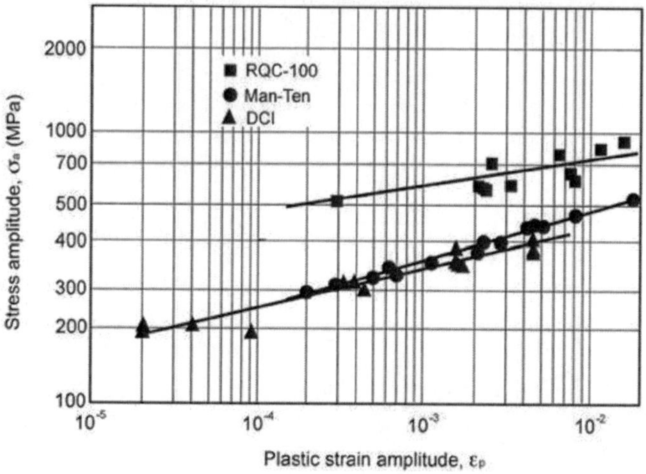

Fig. 8.2 Log–log plot of cyclic stress versus cyclic plastic strain (steel data from [1], DCI data from Milella)

equation of curve OA of the hysteresis loop of Fig. 8.1a. In 1926 Masing [2] observed that in a stabilized hysteresis cycle the unloading and load inversion curve, as ABC and CDA of Fig. 8.1a, were geometrically similar to the initial loading curve OA with a scale factor of two. Such an observation known as Masing's hypothesis has been verified by many investigators [5–8] who experimentally observed that both the increasing and decreasing strain portions of hysteresis loops fall near or often coincide with this locus along their entire length. It provides us with an empirical tool to write the equation of the hysteresis loop as if it were a regular stress–strain curve. To this purpose, consider stabilized cycles as AFA, BEB and CDC of Fig. 8.3, indicated as cycle 1, 2 and 3, respectively. Shift each cycle so that each point of load inversion, such as F, E and D (compressive tips) falls in the origin 0 of the coordinate axes. On the base of the Masing's hypothesis it can be said that also the tensile tips of loops A, B and C must lie on a cyclic strain–stress curve A′, B′ and C′. Thus, all stable hysteresis loop traces for a given metal can be thought to follow a unique curve which may be mathematically described as a factor two expansion of the cyclic stress–strain curve whose inversion points were shifted to the origin of axes. An analysis using as a strain–stress loading path the hysteresis loop curve cannot be distinguished from a static analysis using the cyclic stress strain curve except that in the latter case all quantities are smaller by a factor of two. With such a hypothesis, referring to symbolism of Fig. 8.3, the equation of the cyclic stress strain curve can be obtained using total excursions $\Delta\varepsilon$ and $\Delta\sigma$.

$$\frac{\Delta\varepsilon}{2} = \frac{\Delta\varepsilon_e}{2} + \frac{\Delta\varepsilon_p}{2}$$

$$= \frac{\Delta\sigma}{2E} + \left(\frac{\Delta\sigma}{2K'}\right)^{1/n'} \quad (8.4)$$

that reduces to

$$\Delta\varepsilon = \frac{\Delta\sigma}{E} + 2 \cdot \left(\frac{\Delta\sigma}{2K'}\right)^{1/n'}. \quad (8.5)$$

Therefore, the unloading portion CD of the cycle can be described by the equation

$$\varepsilon = \varepsilon_c - \frac{\sigma_c - \sigma}{E} - 2\left(\frac{\sigma_c - \sigma}{2K'}\right)^{1/n'} \quad (8.6)$$

where ε_c,σ_c are the coordinate of the generic point C of load reversal where the generic unloading path begins. Likewise, the generic reloading ramp DC has the equation

Table 8.1 Monotonic and cyclic behavior of some common alloys [3, 4]

Metal or alloy	Heat treatment	σ_y/σ_y' (MPa)(1)	n/n'(1)	Cyclic behavior
OFHC copper	Annealed	20/38	0.40/0,15	Hardening
	Partially annealed	255/200	0.13/0.16	Stable
	Cold work	345/235	0.10/0.12	Softening
Bras 70/30	Annealed	140/240	–/0.08	Hardening
	85% cold work	570/475	–/0.11	Softening
Aluminum				
2024	T4	305/450	0.20/0.11	Hardening
2024	T6	460/415	–/0.16	Softening
5456	H31, 95 BHN	235/360	–/0.16	Hardening
7075	T6	470/520	0.11/0.19	Hardening
Titanium				
8A1-1Mo-1 V	Annealed	1000/790	0.08/0.14	Mixt
Steel				
SAE 4340	QT-R* (1206 σ_u)	1170/790	0.07/0.14	Softening
SAE 1045	QT-R (2050 σ_u)	1860/1720	0.07/0.14	Stable
SAE 1045	QT-R (1725 σ_u)	1690/1275	0.05/0.12	Softening
SAE 1045	QT-R (1345 σ_u)	1275/758	0.04/0.17	Softening
SAE 4142	QT-R (1930 σ_u)	1690/1725	0.09/0.13	Stable
SAE 4142	QT-R (1310 σ_u)	1380/825	0.05/0.18	Softening
AISI 304	ELC 160 BHN	255/715	–/0.36	Hardening
AISI 305	0% cold work	250/405	–/0.05	Hardening

(1) First number refers to monotonic condition, second to cyclic one

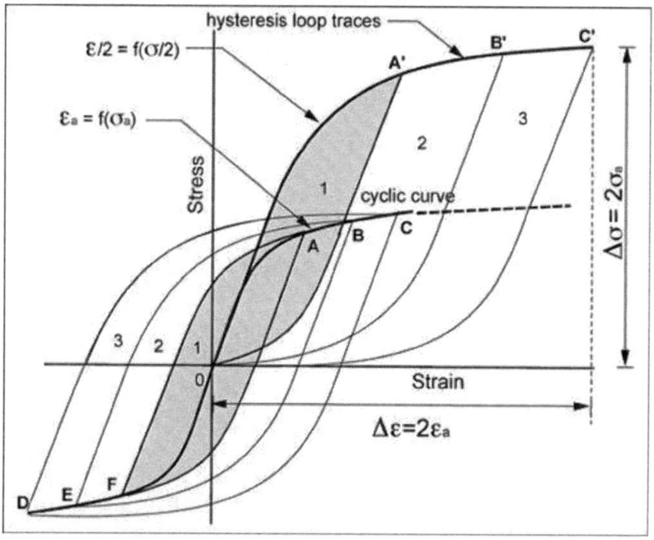

Fig. 8.3 According to Masing's hypothesis, all stabilized cycles, such as 1, 2 and 3, moved to have their points of load inversion F, E and D in the origin 0 describe a curve that can be mathematically expressed as the factor two expansion of the cyclic stress–strain curve

$$\varepsilon = -\varepsilon_D + \frac{\sigma_D - \sigma}{E} + 2\left(\frac{\sigma_D - \sigma}{2K'}\right)^{1/n'} \tag{8.7}$$

with $-\varepsilon_D$, $-\sigma_D$ being the coordinate of the generic point D.

8.3 Strain-Life Curve, ε–N

Once the analytical expression of the hysteresis loop has been derived, the following step is to infer the equation of the entire ε–N curve or Wöhler's curve. To this purpose, it is necessary to separate the two component of deformation, the elastic and the plastic one, as already done in Eq. (8.1). As to the elastic component ε_e, it has been said in Sect. 7.2 that in 1910 Basquin [9] found a power law relationship with the cycles to failure N, see Eq. (7.1), so that in a log–log scale the elastic component of the S–N curve is a line. Later in the'50, Manson and Coffin [10–13] independently found that also the plastic component of the cyclic strain was related to life cycles N through a power law function, known as the Manson-Coffin relationship

$$\varepsilon_p = \frac{\Delta\varepsilon_p}{2} = \varepsilon'_f (N)^c \tag{8.8}$$

where $\Delta\varepsilon_p$ is the plastic strain excursion, see Fig. 8.1, ε'_f a fatigue ductility coefficient and c a fatigue ductility exponent, both being material properties. In a log–log scale Eq. (8.8) is a line as it was le elastic curve or Basquin line. Alike the fatigue strength coefficient σ'_f, also the fatigue ductility coefficient ε'_f is one-reversal intercept of the plastic curve and, therefore, it is equal to the true strain at fracture $\varepsilon_{t,f}$, see Eq. (5.32), since it represents the strain at which failure occurs after just ½ cycle, i.e., under a monotonic load. The exponent c is the slope of the plastic curve that normally ranges from − 0.5 to − 0.8. This is shown in Fig. 8.4 that reports experimentally derived data [14, 15] of the ratio $\Delta\varepsilon_p/2\varepsilon'_f$ versus reversal N to failure (1 rev = ½ cycle). Note that at equal plastic strain moving from $c = -0.8$ to a larger value, $c = -0.5$, improves significantly the life of the material. Also note how at ½ cycle all data converge towards $\Delta\varepsilon_p/2\varepsilon'_f = 1$, i. e., $\varepsilon_p = \varepsilon'_f$. As for the Basquin exponent b, Eq. (), using energy-based arguments Morrow [12] found that the fatigue ductility exponent c was given by

$$c = \frac{-1}{1 + 5n'} \tag{8.9}$$

in which n' is the cyclic strain hardening exponent of the material. Equation (8.9) indicates that at variance with what happens in the elastic domain of the high cycle fatigue in the low cycle fatigue a higher strain hardening exponent increases the fatigue strength of the material. By using the Basquin elastic curve and the Manson-Coffin plastic one it is now possible to resolve the overall S–N, which now becomes

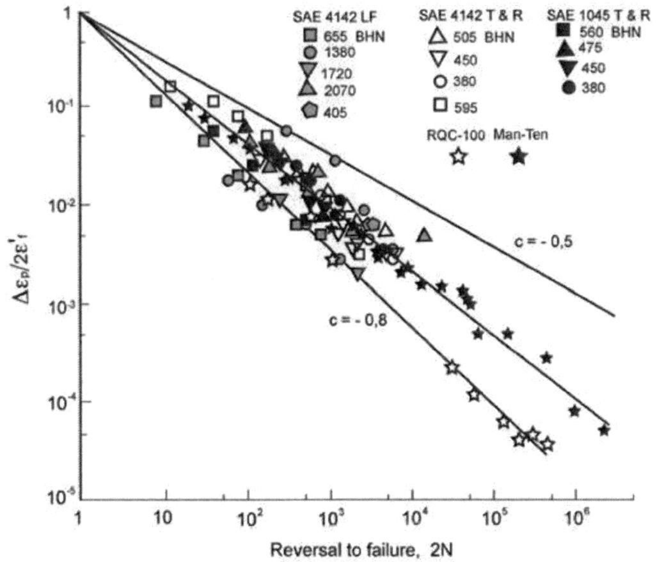

Fig. 8.4 Plastic strain amplitude versus reversal to failure (data from [14] and [15])

a ε_a–N curve into its elastic and plastic components given by Eq. (7.1) and Eq. (8.8), respectively.

$$\varepsilon_a = \frac{\Delta\varepsilon_t}{2} = \frac{\sigma'_f}{E}(N)^b + \varepsilon'_f(N)^c \tag{8.10}$$

where the fatigue strength exponent b is given by Eq. (7.1a). An example of S–N curve resolved into its plastic and elastic components is offered in Fig. 8.5 [15] for two steels Man-Ten and RQC-100, respectively and in Fig. 8.6 for AISI 316 austenitic stainless steel [16]. For the austenitic steel the total fatigue curve ε_a–N is given by the equation

$$\varepsilon_a = 5.5 \cdot 10^{-3} \cdot (N)^{-0.097} + 1.9 \cdot 10^{-1} \cdot (N)^{-0.325}. \tag{8.11}$$

Figures 8.5 and 8.6 show the so called transition point N_t where elastic and plastic components intersects. Beyond N_t it is the elastic component of strain that dominates and control the fatigue life of the material whereas below N_t the plastic strain prevails.

This actually means that beyond N_t high cycle fatigue becomes the dominant failure mode and the Basquin line may sufficiently well represent the S–N fatigue behavior of the material. Below N_t, instead, low cycle fatigue is the failure mode of the material and Mason-Coffin relationship based on strain amplitude is needed. In the surroundings of N_t it is necessary to consider both components. It is worth noting

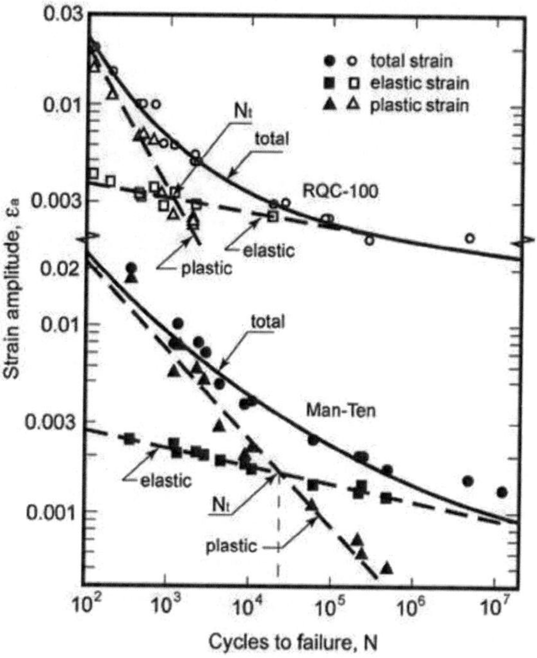

Fig. 8.5 Plastic, elastic and total strain versus life test data for Man-Ten and RQC-100 steels [15]

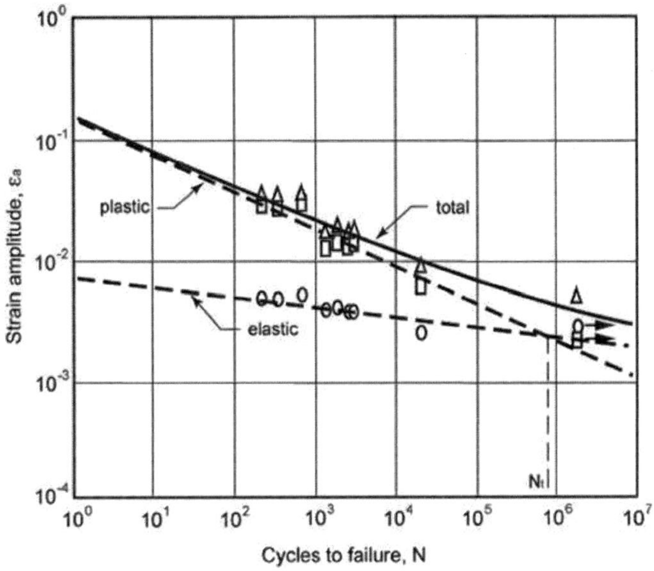

Fig. 8.6 Plastic, elastic and total strain versus life test data for A 6066-T6 aluminum alloy [16]

how for the softer steel Man-Ten the transition N_t from plastic to elastic behavior can be placed at about 2×10^4 cycles while for the harder steel RQC-100 it already happens at about 10^3 cycles. At 10^4 cycles the plastic component of strain is only a mere 1/30 of the elastic one. For the aluminum alloy this transition occurs even earlier at about 100 cycles. At about 10^3 cycles the total curve coincides with the elastic component. The coordinates of N_t can be found by putting $\varepsilon_p = \varepsilon_e$ and recalling Eq. (8.10) it yields

$$\frac{\sigma'_f}{E}(N_t)^b = \varepsilon'_f(N_t)^c$$

$$N_t = \left(\frac{\varepsilon'_f E}{\sigma'_f}\right)^{1/(b-c)} \tag{8.12}$$

The value of N_t depends, then, on the strength and, therefore, hardness of the material. Figure 8.7 [17] presents the dependency of N_t on hardness. It can be seen how mild steels and austenitic stainless steels with hardness lower than 200 HB may have the low cycle fatigue region (Region I) extended up to 10^6 cycles.

This gradual passage from plastic to elastic and the relative hysteresis loops are schematized in Fig. 8.8. The hysteresis loop corresponding to any number of cycles N shrinks as N decreases till it disappears completely in the fully elastic high cycle fatigue field. The more N_t is approached from both sides, the more the elastic and plastic component are needed to describe the total fatigue life behavior. Also shown

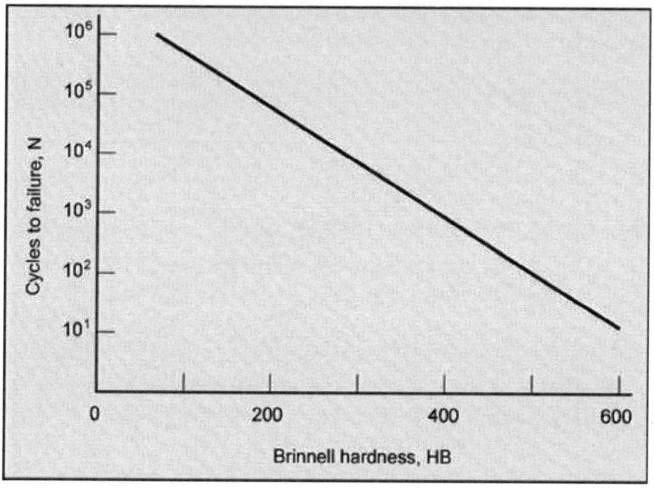

Fig. 8.7 Dependency of transition N_t on steel hardness [17]

are the one cycle intercepts of the plastic and elastic components, ε'_f and σ'_f/E, respectively. About this latter it has been said in Sect. 8.7.2 that σ'_f is given by the monotonic true stress at fracture $\sigma_{t,f}$. Indeed Fig. 8.9, taken from a publication of the American Society for Testing and Materials [14], shows how the fatigue strength coefficient σ'_f is varying with the monotonic true traction strength. It can be seen the linear relationship, almost 1:1, between σ'_f and $\sigma_{t,f}$. The line interpolating experimental data has the equation.

$$\sigma'_f = 18.4 + \sigma_{t,f} \tag{8.13}$$

From Eq. (7.1) and Eq. (8.8) it is found that

$$\frac{\Delta\sigma}{2} = \sigma'_f \left(\frac{\Delta\varepsilon_p}{2\varepsilon'_f}\right)^{b/c} \tag{8.14}$$

and recalling Eq. (8.1) it yields

$$K' = \frac{\sigma'_f}{\left(\varepsilon'_f\right)^{n'}} \; with \; n' = \frac{b}{c}. \tag{8.15}$$

It must be kept in mind that both Eq. (5.33) and Eq. (6.9) and therefore also their ratio

$$n' = \frac{b}{c} \tag{8.16}$$

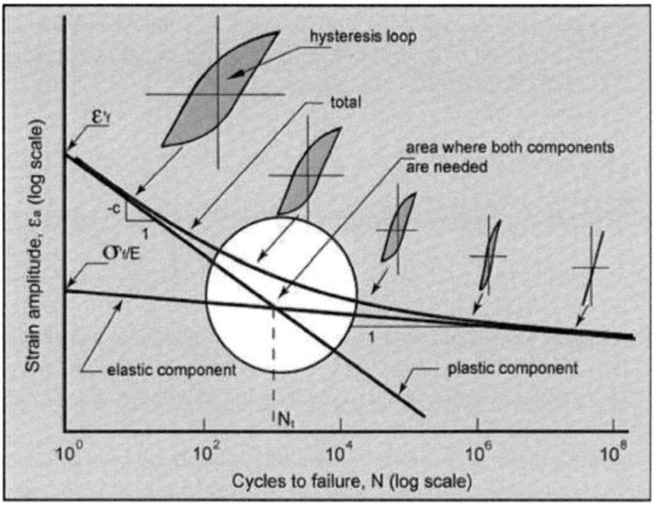

Fig. 8.8 Schematic of plastic, elastic and total strain versus cycles to failure and relative hysteresis loops

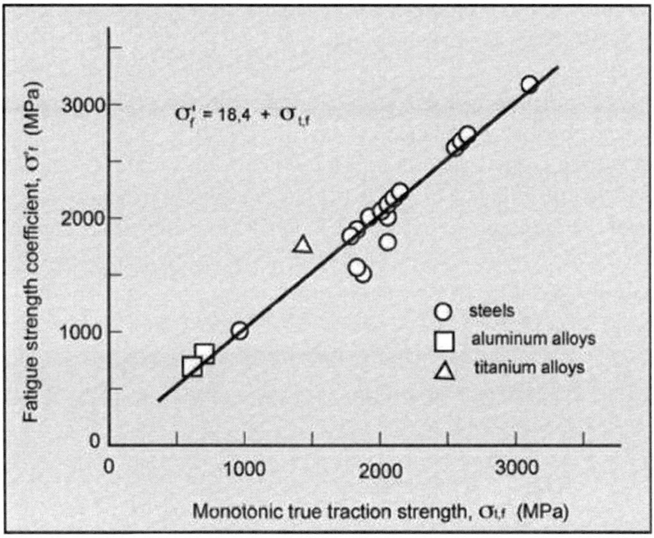

Fig. 8.9 Fatigue strength coefficient σ'_f versus true traction strength $\sigma_{t,f}$ of several metal alloys [14]

hold as long as is $R = -1$. For any other R, $R \neq -1$, the two relationships given by Eq. (7.1a) and Eq. (8.9) may not be valid and, therefore, also Eq. (8.16) may not be true.

8.4 Mean Stress in Low Cycle Fatigue

It has been shown in Sect. 7.3 the dangerous effect of the mean stress in high cycle fatigue. But it was also anticipated that this effect vanishes as the low cycle regime is approached, as shown in Fig. 7.3. As matter of fact, as the plastic strain component ε_p starts to play a role (for N lower than N_t) a stress relaxation phenomenon takes place that reduces mean stress and may even bring it to zero. This is schematized in Fig. 8.10 for a case of fatigue cycling under strain controlled condition. The phenomenon appears as a cyclic softening, but it is all another thing. It happens also in materials cyclically stable or subjected to cyclic hardening. An experimental verification of mean stress reduction to zero in low cycle fatigue is shown in Fig. 8.11 [18].

Figure is taken from experiments conducted by Terrell et al. on smooth specimens of A 106B carbon steel for piping at 288 °C. In the first case, Fig. 8.11a the specimen was tested under strain controlled condition at a total strain ε_a equal to 0.74% and failed after 605 cycles. It can be seen the progressive reduction of mean stress that after 605 cycles reduces from an initial 18% to something less than 5%. In the second case, Fig. 8.11b deformation was reduced to 0.19%. The specimen failed after 145,720 cycles. The strong reduction of the plastic component of strain is due to a cyclic strain hardening effect that could not take place within the previous 605 cycles. This has continuously reduced the hysteresis loop till its complete disappearance with linearization of the material response that, become elastic, has blocked the further reduction of the mean to a 16% from the initial 40%.

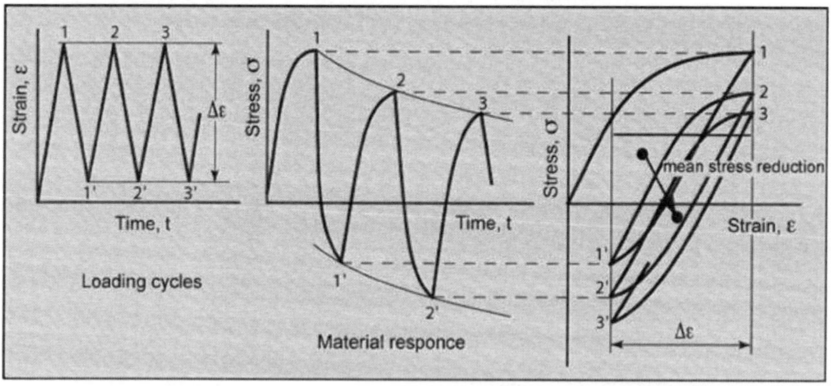

Fig. 8.10 Schematic of mean stress reduction resulting from plastic cycling

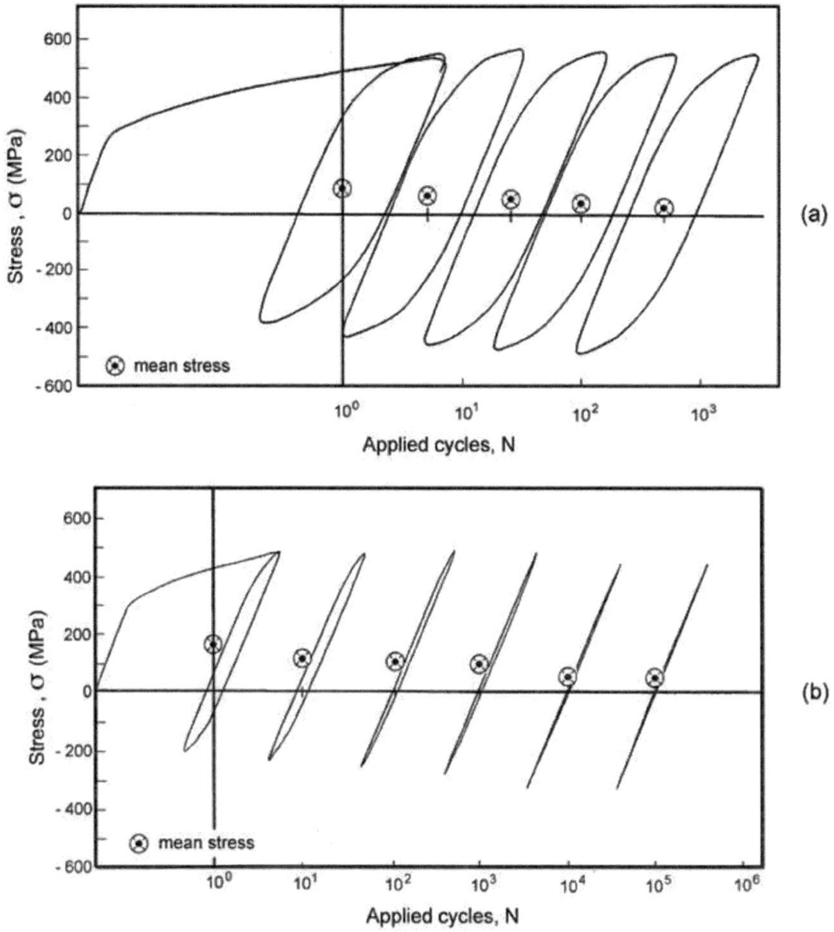

Fig. 8.11 Progressive reduction of mean stress and hysteresis loop in a smooth specimen of A106B carbon steel subjected to low cycle fatigue at 288 °C. **a** Strain amplitude $\varepsilon_a = 0.74\%$ and **b** $\varepsilon_a = 0.19\%$ [18]

8.4.1 Morrow Correction

To cope with mean stress effect and its continuous reduction as the material enters the low cycle fatigue region, Morrow [19, 20] has proposed the following correction that applies only to the elastic component of Eq. (8.10)

$$\varepsilon_a = \frac{\sigma'_f - \sigma_m}{E}(N)^b + \varepsilon'_f(N)^c. \tag{8.17}$$

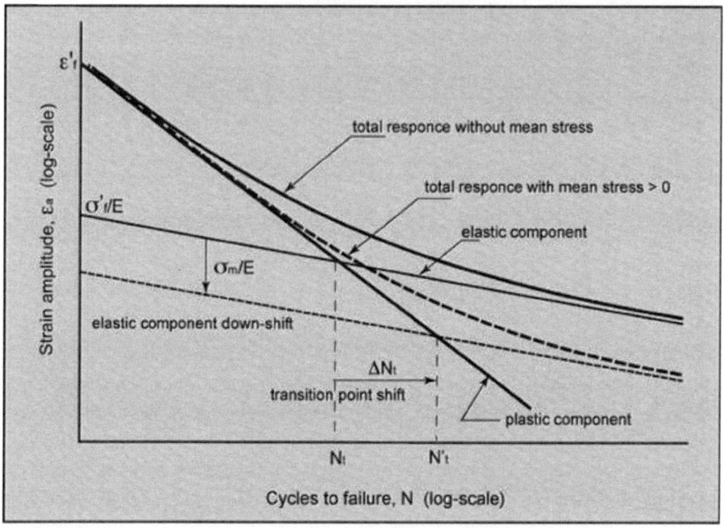

Fig. 8.12 Down-shift of fatigue strength curve according to Morrow correction that takes into consideration the mean stress only in the elastic component of total response

The graphic representation of (8.17) is offered in Fig. 8.12. The fatigue life reduction due to mean stress is actually affecting only the elastic component ε_e of the ε_a–N curve moving toward higher life the transition point N_t. However, the Morrow correction offers a drawback that clearly results from Fig. 8.13.

The ratio between the elastic component of strain and the plastic one, $\varepsilon_e/\varepsilon_p$, depends on the mean stress. It is clear that this cannot be necessarily true. Nothing prevents from having two hysteresis loops with the same ratio $\varepsilon_e/\varepsilon_p$, but different mean stress. For example, the two cycles 1 and 2 of Fig. 8.13 imposed inside the principal one have the same elastic and plastic strain excursions $\Delta\varepsilon_e$ and $\Delta\varepsilon_p$, respectively, but different mean stress; $\sigma_{m,1}$ in the first case and $-\sigma_{m,2}$ in the second. Because of that, Manson and Halford [21] proposed another correction that applies to both components of deformation

$$\varepsilon_a = \frac{\sigma'_f - \sigma_m}{E}(N)^b + \varepsilon'_f\left(\frac{\sigma'_f - \sigma_m}{\sigma'_f}\right)^{c/b}(N)^c \tag{8.18}$$

If the independence of the ratio $\varepsilon_e/\varepsilon_p$ of σ_m is restored it introduces an effect of mean stress in the plastic domain that experience does not confirm. As to the Morrow correction (8.17), Walcher, Gary and Manson [22] observed that in alloys like Ti-6Al-4 V the fatigue strength coefficient σ'_f can be so high as to mask the mean stress effect. Therefore, they proposed a correction to the fatigue strength coefficient of the type

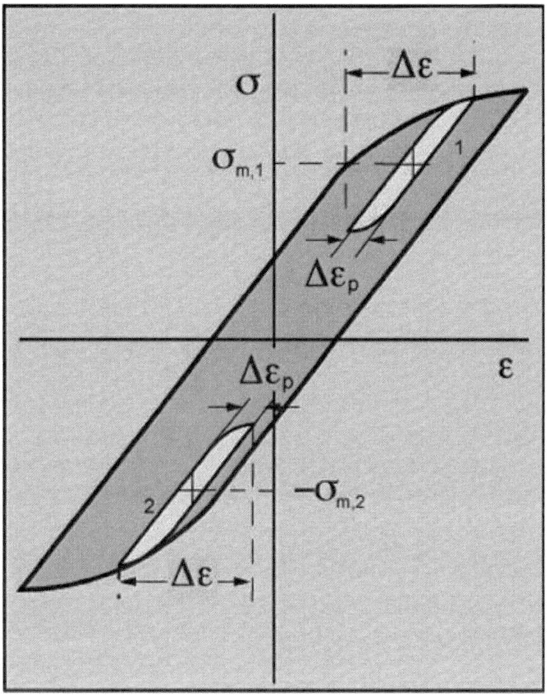

Fig. 8.13 Cycles 1 and 2 have the same elastic and plastic strain excursions $\Delta\varepsilon_e$ and $\Delta\varepsilon_p$ but different mean stress, σ_{m1} and $-\sigma_{m2}$, respectively

$$\varepsilon_a = \frac{k_m \sigma'_f - \sigma_m}{E}(N)^b + \varepsilon'_f (N)^c. \qquad (8.19)$$

The problem with Eq. (8.19) is that it needs the experimental derivation of k_m.

8.4.2 Smith, Watson and Topper (SWT) Model

Smith, Watson e Topper [23] in 1970 proposed a fatigue damage model under stress state with positive maximum stress, $\sigma_{\max} > 0$, also known as SWT model. The damage is expressed by the product $\sigma_{\max} \cdot \varepsilon_a$ in which ε_a is the strain amplitude associated with the maximum stress $\sigma_{\max}$. Moreover, they made the hypothesis that fatigue failure would happen when their model of damage, called SWT parameter, would be equal to the corresponding damage introduced in the material at failure under fully reversed fatigue

$$\sigma_{\max} \cdot \varepsilon_a = \sigma_{a,R-1}\varepsilon_{a,R-1} \quad \sigma_{\max} > 0 \qquad (8.20)$$

in which $\sigma_{a,R\text{-}1}$ and $\varepsilon_{a,R\text{-}1}$ represent the stress amplitude and strain amplitude, respectively, of the completely reversed fatigue cycle having $R = \sigma_{min}/\sigma_{max} = -1$. The value of $\varepsilon_{a,R\text{-}1}$ can be inferred from Eq. (6.10), where N is the number of cycles to failure in the fully reversed fatigue loading and $\sigma_{a,R\text{-}1}$ derived from the equation of the stabilized hysteresis loop (6.6). The SWT parameter does not foresee any damage if the maximum stress σ_{max} is equal to zero or becomes negative. Equation (6.20) can be solved by iteration. When Eq. (5.33) and (6.9) hold ($R = -1$) from Eq. (5.22) it is

$$\sigma_{a,R-1} = \sigma'_f \cdot (N)^b \tag{8.21}$$

therefore it is possible to write the SWT parameter as

$$\sigma_{\max} \cdot \varepsilon_a = \frac{\left(\sigma'_f\right)^2}{E}(N)^{2b} + \sigma'_f \varepsilon'_f (N)^{b+c} \tag{8.22}$$

in which N is the number of cycles to failure in fully reversed loading ($\sigma_m = 0$). The SWT parameter has been used successfully with cast iron (Fash and Socie [24]), with hardened carbon steels (Koh and Stephens [25] and Wehner e Fatemi [26]) and with alloyed-steels (Forsetti and Blasarin [27]). Figure 8.14, taken from [28], shows the results obtained on 4340 steel for different mean stress indexed to the SWF parameter. Results fall on the same curve that refers to the case $\sigma_m = 0$.

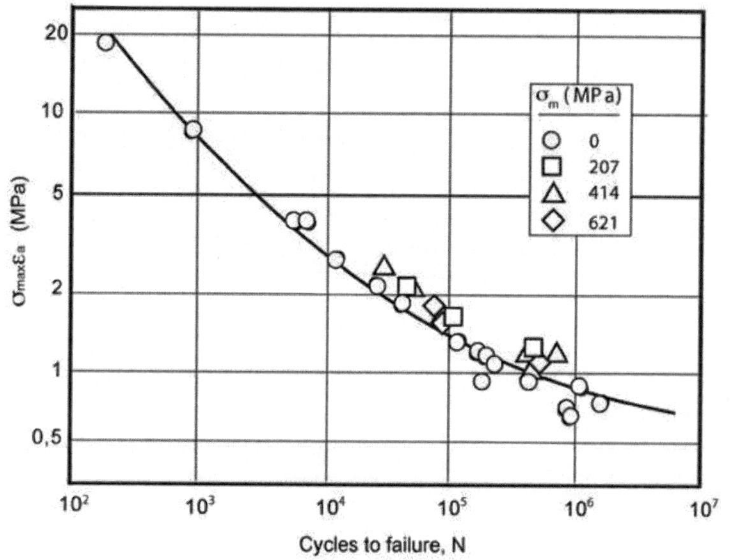

Fig. 8.14 SWT model used to analyze experimental results obtained at various mean stress on 4340 (from Mechanical Behavior of Materials [28])

8.5 Fatigue Life Prediction Based on Local Strain Approach

8.5.1 Equivalent Test Method

Machine and structures subjected to cyclic loadings quite often if not always present complex geometric shapes that favor the presence of localized plastic deformations. To evaluate the fatigue life of the structure one cannot avoid estimating strains and stresses that occur at the most critical location, which is usually a notch or other geometric discontinuity. The designer may also consider the possibility to verify the fatigue behavior of the real structure using an experimental approach. Suppose that the real structure is subjected to a time history of the type shown in Fig. 8.15a, which is common in the automotive industry being representative a of test treck [1]. He may use a SAE specimen, Fig. 8.15b, containing a notch of the type existing in the structure and subject the specimen to the same time history. Doing so, he will get the number of cycles to failure, *N*. The approach, dough sound, has a drawback since he will get just a single experimental point referring to the response of very small volume of material that may be even less than ½ cm^3, considering also the thickness of the specimen. It has been said over and over again about the fundamental role of the process volume on material response (see Sect. 4.8). Metallurgical variability is evidenced through the process volume. A too small process volume cannot be representative of material behavior.

This actually means that when using very small specimens or large specimens, but with a very small process volume as the SAE one, it is necessary to test a very

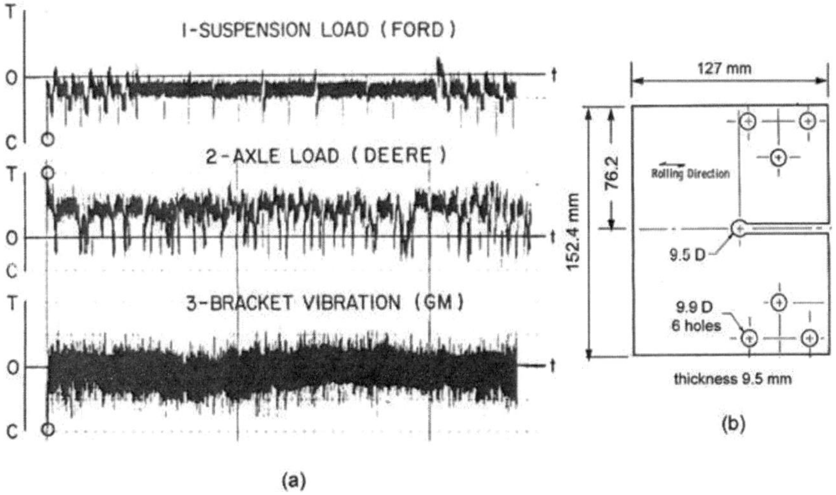

Fig. 8.15 **a** SAE test treck load history [1], **b** SAE specimen

large number of pieces to increase the process volume till it becomes representative of material behavior. To address this issue an equivalent specimen can be used, which may be considered a sort of test multiplier. Consider, for example a specimen containing a round notch like the SAE specimen presented in Fig. 8.15b subjected to load cycling having an irregular time history like that shown in Fig. 8.16a. If local plastic deformation occurs in A, the strain history will qualitatively resemble the load history, but the two will not be exactly proportional. The problem is that strains at the notch tip present such a high gradient over so short a distance that cannot be measured by any practical experimental mean. Suppose, then, to by-pass this drawback by using an analytical procedure, for instance, the Neuber hyperbola method described in Sect. 8.6.2. Then the strain fields ahead the notch will be calculated during the transient. This is shown in Fig. 8.16c.

The nonlinearity between loads and strains appears evident in Fig. 8.16d. Let's dwell for a moment on a particular feature of Fig. 8.16d represented by the so called *memory effect*. For instance, arrived in 2 from 1 and closed the hysteresis loop 2-3-2′ and returned in 2 the material point remembers that it was following path 1–4 before the load inversion never occurred. This is a general fact that can be expressed

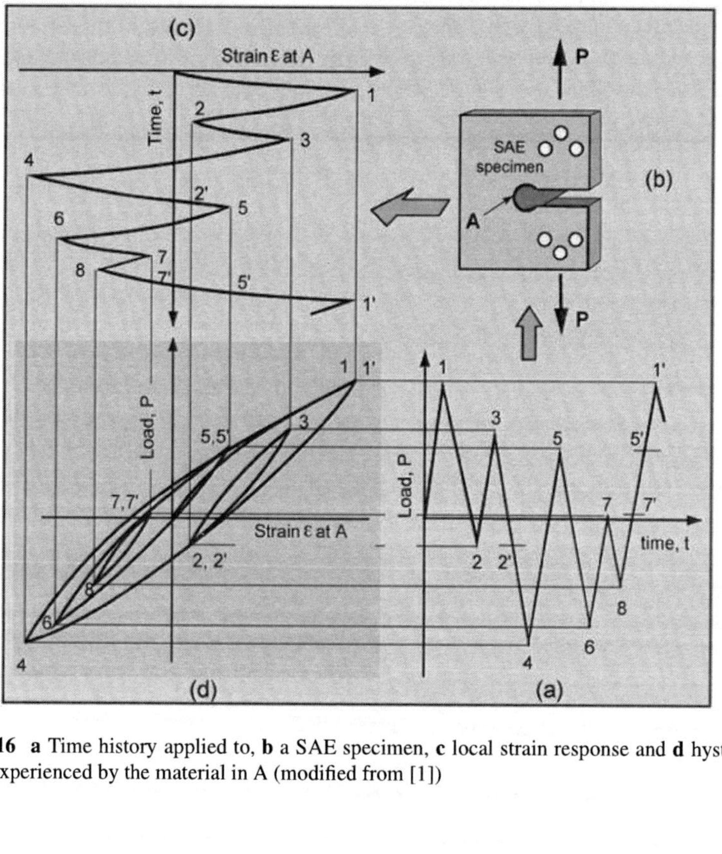

Fig. 8.16 **a** Time history applied to, **b** a SAE specimen, **c** local strain response and **d** hysteresis loops experienced by the material in A (modified from [1])

as follows: whenever the strain returns to a value previously reached in which the direction of load application was reversed (points 2′, 5′ and 7′ of Fig. 8.15c) also the stress returns to the same value it had before (corresponding to that load reversal 2-3-2′ and, therefore, it re-enters the original path as if strain) and the stress–strain path beyond this point continues without discontinuity as if the direction of straining had not been reversed. Note that the material subjected to the load history of Fig. 8.16a has experienced four hysteresis loops 1-4-1′, 2-3-2′, 5-6-5′ and 7-8-7′ that can be analyzed independently because they do not interfere among each other and it is on these cycles that the fatigue analysis will be run. Knowing the strain history at point A of stress concentration, Fig. 8.16c, it will be convenient to apply this strain history to a smooth axial specimen as schematically shown in Figure 8.17a and b. If the applied strain history is fully plastic (low cycle fatigue) then the entire volume of this new test piece will participate to the fatigue process and this equivalent smooth specimen will represent a sort of test multiplier because it will correspond to n SAE specimens where n is equal to the ratio of the smooth specimen volume to the small process volume associated to the SAE specimen. Note that if the applied strain history is elastic or mainly elastic the process volume of the smooth specimen will no longer be the entire volume, but only the external annulus not thicker than 0.5 mm (see Fig. 2.5) times the length of the specimen. High cycle fatigue, in fact, is affecting only a surface layer. The multiplier effect in high cycle fatigue is, therefore, much more contained. In low cycle fatigue, then, a result obtained with a single unnotched equivalent specimen is expected to represent a lower bound of values obtained with many notched SAE specimens. Figure 8.17c shows schematically the stress response time history and Fig. 8.17(d) the corresponding hysteresis loops. They are not exactly those of Fig. 8.16a and d, respectively, but the excursion $\Delta\varepsilon$–$\Delta\sigma$ are the same.

8.5.2 Fatigue Life Calculation

An important step in local strain fatigue analysis under irregular variation of load with time, as schematized in Fig. 8.16a, is the knowledge of the local notch strain–stress history, as shown in Fig. 8.16c. This, in turn, necessitates the knowledge of the cyclic stress–strain characteristic of the material (see Sect. 1.3.1). Advanced elastic–plastic analysis technique such as finite element computer code is needed or it can be used a simplified Neuber's rule as described in the next section.

Once the elastic–plastic response is known, Fig. 8.16c, using Eqs. (8.6) and (8.7) and the memory effect each hysteresis loop of Fig. 8.16d can be separately analyzed. In using Eqs. (8.6) and (8.7) it must be recalled what has been said in Sect. 8.2 about the Masing hypothesis and the factor of two that must be applied to handle the case as if the hysteresis loop traces were equivalent to a cyclic stress–strain curve. Therefore, Eqs. (8.6) and (8.7) becomes

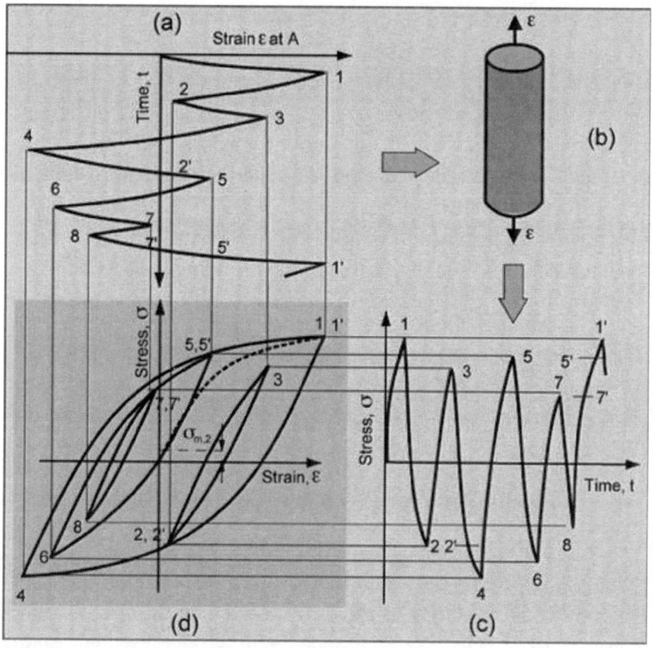

Fig. 8.17 **a** Deformation history applied to an unnotched specimen (**b**). **c** Stress response and **d** hysteresis loops generated in the material [1]

$$\frac{\varepsilon_a - \varepsilon}{2} = \frac{\sigma_a - \sigma}{2E} + \left(\frac{\sigma_a - \sigma}{2K'}\right)^{1/n'} \tag{8.23}$$

for the unloading portion of the cycle and

$$\frac{\varepsilon - \varepsilon_a}{2} = \frac{\sigma_a - \sigma}{2E} + \left(\frac{\sigma_a - \sigma}{2K'}\right)^{1/n'} \tag{8.24}$$

for the reloading ramp, where A≡(ε_a;σ_a) and B≡($-\varepsilon_a$;$-\sigma_a$) are the coordinate of tip traction and compression points, respectively. For instance, point A is point 1 or 3 or 5 or 7 in Fig. 8.16d and B coincides with point 2 or 4 or 6 or 8 of Fig. 8.16d. The next step will be to enter the strain-life curve of the material with the strain amplitude found and read the corresponding allowable cycles N_i for that hysteresis loop.

The global effect may be obtained using the Palmgren–Miner rule (see Section)

$$\sum_1^4 \frac{n_i}{N_i} \leq 1 \tag{8.25}$$

where N_i is the number of allowable cycles for loop cycle i and n_i the number of times the material has been subjected to loop i. For the example of Fig. 1.16 Eq. (8.25) becomes

$$\sum_{1}^{4} \frac{n_i}{N_i} = \frac{1}{N_{2-3}} + \frac{1}{N_{7-8}} + \frac{1}{N_{5-6}} + \frac{1}{N_{1-4}} \leq 1\,. \tag{8.26}$$

Furthermore, for repeating load time history it must be

$$m\left[\sum \frac{n_i}{N_i}\right] \leq 1 \tag{8.27}$$

in which m is the number of repetitions of the load time history or blocks applied to the test piece. If the summation in Eq. (8.26) exceeds one the component will fail. In the previous calculations hypothesis has been made about strains that were predominantly plastic. For this reason the effect of the mean stress on material response has not been taken into consideration. If this is not the case and the elastic component of strain is not negligible than a mean stress correction must be considered according to what has been said in Sect. 8.4.

8.5.3 Overstrain and Load History Effect

A particular attention in fatigue design must be given to overloads. Overloads, dough applied just one or few times at the beginning or during the life of a component, may have either devastating or beneficial effects, depending on the history of their application. The difference is found, in fact, in the mean stress the overload generates in the material that will affect the following cycles. An example of this effect is given in Fig. 8.18 [29]. It refers to type 2024-T4 aluminum panel having a central hole of about 5 mm diameter and 2.54 mm thick subjected to cyclic loads. Two similar, but not equal time histories have been applied. Both time histories have the same stress excursion $\Delta\sigma_2$ preceded by an overload $\Delta\sigma_I$ of the same amplitude, but applied so that in time history A the working cycles were initiated from the negative overload peak stress, while in history b they were following the positive overload peak stress. The overload resulted in quite a large hysteresis loop in the material around the hole. The following low amplitude cycles, instead, have been engaging the material predominantly in the elastic regime. In load history A, the stress range $\Delta\sigma_I$ had a traction-compression excursion, while in load history B it had a compression-traction one. As it can be seen the material response is completely different.

Under load history A material fails prematurely relative to load history B. The rational is to be found in the mean stress effect. In case A, in fact, the overload has generated a hysteresis loop of initial path 0-1-2 ending with stress reversal in 2 that brings the material back in traction (point 3). The following load sequence 3–7....of $\Delta\sigma_2/2$ amplitude is then applied in traction with positive mean stress σ_m. since the stress state is mainly elastic (the resulting hysteresis loop is very narrow indicating a prevalence of the elastic component) this mean stress is detrimental (see Sect. 5.4).

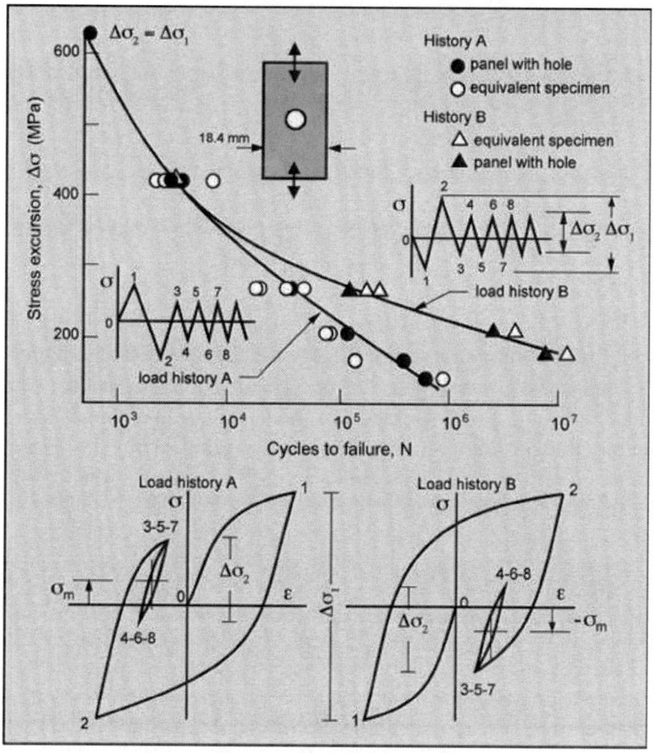

Fig. 8.18 Fatigue lives of specimens containing a hole resulting from two types of initial overload. Also shown are lives measured with equivalent unnotched specimens [29]

In time history B, instead, the large hysteresis loop caused by the overload starts in compression, reverses in traction, point 2, and then ends again in compression, point 3, where the following cycles engage the material (cycles 3–7…). The result is the same mean stress as before, but with negative sign, $-\sigma_m$. This time the mean stress effect is trivial or null.

This is clearly shown in Fig. 8.18 by the two material response stories. Interesting enough, the difference in material behavior between case A and B progressively reduces as the material around the hole approaches the plastic regime, low cycle fatigue Region I, where the mean stress has no effect at all and eventually disappears below $5 \cdot 10^3$ cycles (fully plastic regime).

Conversely, the gap between the two histories increases as the fatigue regime becomes more and more elastic, high cycle fatigue or Region II of fatigue. Also shown are the results obtained using equivalent unnotched specimens (see Sect. 8.5.1). There is quite an excellent agreement between the two methods (the panel carrying a hole and the equivalent specimen) confirming the validity of the method. What has been said in Sect. 8.5.1 about the effect of the process volume is now observed. Equivalent specimens yield lower bound curve with respect to the panel containing a hole, as

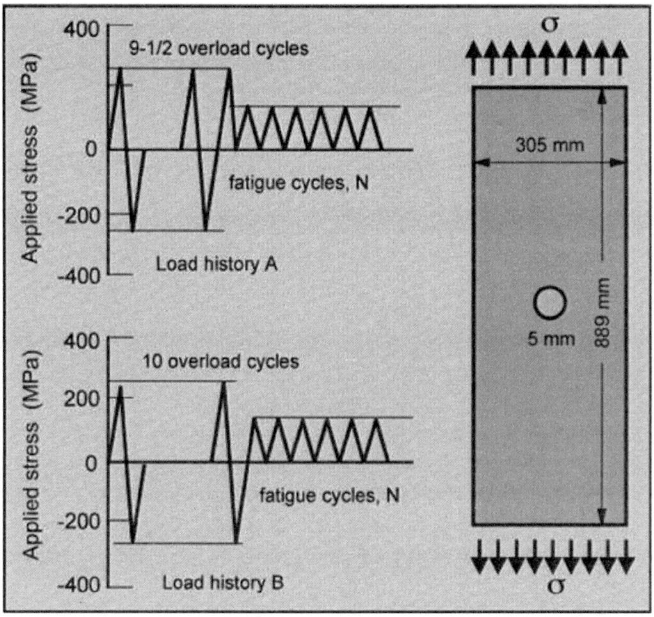

Fig. 8.19 Type 2024-T4 aluminum panel subjected to load histories having different final overload sign, which is traction in A and compression in B [30]

clearly shown in Fig. 8.18 for case history A. in case history B this seems not to be the case since the negative mean stress is playing a major role. Another significative example of overload effect on fatigue strength is given by Crews [30] and shown in Fig. 8.19. A plate of type 2024-T3 aluminum with a central hole is subjected to the two load time histories A and B, quite similar to those of Fig. 8.18.

Using the Palmgren–Miner rule no difference would be found between the two cases. Conversely, experimental results show a rather large difference because the plate would withstand 460,000 cycles under load history A and only 63,000 cycles under load history B. Interesting enough, without any overload application at all the pane would has a life of 115,000 cycles.

8.6 Neuber Approach to Strain-Life

We have been discussing in Sect. 8.5 about the need to estimate strains and stresses that occur at the most critical location of a structure which is usually a notch or other geometric structural discontinuity. The theory of elasticity treats the subject of notches or grooves introducing a theoretical stress concentration factor k_t as the ratio between the local or actual stress σ at the notch tip and the applied nominal stress σ_n far from the notch, $k_t = \sigma/\sigma_n$. Under conditions of validity of the theory of elasticity,

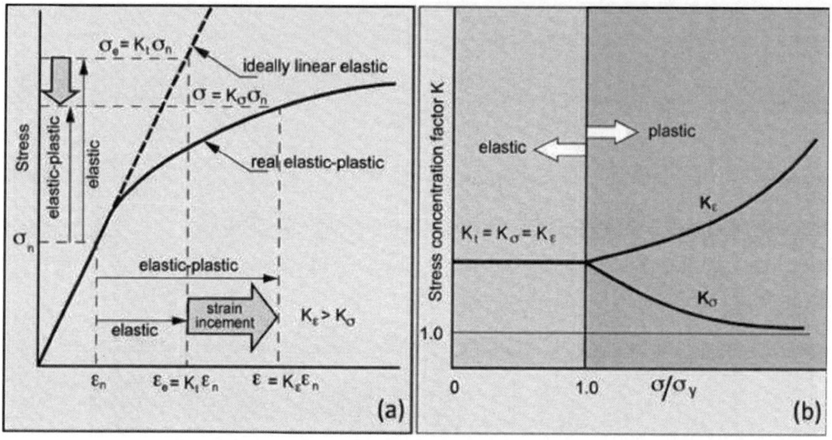

Fig. 8.20 **a** Elastic and elastic–plastic material response. **b** After yielding k_ε and k_σ diverge

k_t is the same for stress σ_n as for strain ε_n. This is because the theory considers both stress and strain elastic and linked by the Hooke's law with the result that a local concentration operated by a notch on stress must be equal to the concentration on strain, Fig. 8.20a. But beyond the elastic limit this no longer holds. The local stress amplification $k_\sigma = \sigma/\sigma_n$ of the nominal stress σ_n is no longer equal to the corresponding amplification $k_\varepsilon = \varepsilon/\varepsilon_n$ on nominal strain ε_n. The local strain ε grows much faster than stress σ with the result that is always $k_\varepsilon > k_\sigma$ and the two factors k_ε and k_σ diverge, as schematized in Fig. 8.20b. In practice, notch concentration effect lessens on stress and in ductile material goes asymptotically to unity, as shown in Fig. 8.20b, because the stress–strain curve flattens down. On the contrary, actual deformation starts to grow entering the plastic domain.

In the early 1937 Heinz Neuber [31, 32] made the hypothesis that, dough diverging, the product of the two concentration factors k_ε and k_σ would remain constant and be equal to the square of the theoretical stress concentration factor k_t

$$k_t^2 = k_\sigma \cdot k_\varepsilon \,. \tag{8.28}$$

In other words, Neuber assumed that the theoretical stress concentration factor k_t was the geometric mean of the actual k_ε and k_σ factors

$$k_t = \sqrt{k_\sigma k_\varepsilon}. \tag{8.29}$$

Equation (8.29) is known as Neuber's rule. Recalling the definition of k_t

$$k_\sigma = \frac{\sigma}{\sigma_n} = \frac{notch \cdot tip \cdot stress}{stress \cdot without \cdot notch}$$

$$k_\varepsilon = \frac{\varepsilon}{\varepsilon_n} = \frac{notch \cdot tip \cdot strain}{strain \cdot without \cdot notch} \tag{8.30}$$

and Hooke's law, Eq. (8.28) can be written as

$$\frac{(k_t \sigma_n)^2}{E} = \sigma \cdot \varepsilon. \tag{8.31}$$

Note that for any given value of stress level σ_n the first member of Eq. (8.31) is a constant, also known as Neuber's constant, so that Eq. (8.31) is the equation of a hyperbola. The intersection of the Neuber's hyperbola with the stress–strain curve of the material yields the solution of the elastic–plastic problem. This is schematically shown in Fig. 8.21 where the coordinates of point A≡(ε,σ) are the notch tip strain and stress, respectively.

Moving from monotonic loading to fatigue cycles Neuber's rule applies substituting the theoretical stress concentration factor k_t with the notch factor k_f (see Sect 7.1), as suggested by Wetzel [33] in 1968 and Topper et al. [34] in 1969, obtaining what is known as the Neuber modified equation

$$\frac{\left(k_f \sigma_n\right)^2}{E} = \sigma \cdot \varepsilon. \tag{8.32}$$

Equation (8.32) shall be used with the material cyclic stress–strain curve given by Eq. (8.3) instead of the monotonic one, so that for the initial loading ramp of the cycle Eq. (8.32) becomes.

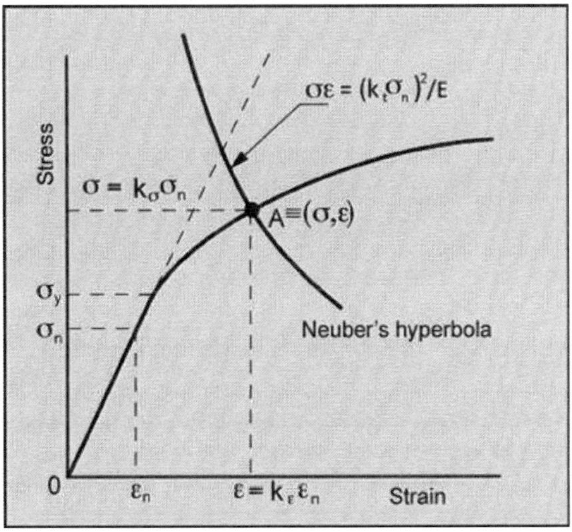

Fig. 8.21 Graphic representation of Neuber's rule. The working point A is the intersection of Neuber's hyperbola with the stress–strain curve

$$\frac{(k_f \sigma_n)^2}{E} = \frac{\sigma^2}{E} + \sigma \cdot \left(\frac{\sigma}{K'}\right)^{1/n'}. \tag{8.33}$$

Equation (8.33) may be solved using an iteration technique that yields the coordinates ε_A and σ_A of point A, which is the intersection between the Neuber's hyperbola and the cyclic curve as shown in Fig. 8.22. The nominal stress σ_{n1} is then reversed to a stress value σ_{n2}, as shown in Fig. 8.22. This results in a change of the notch root strain $\Delta\varepsilon$ and stress $\Delta\sigma$. The new intersection between the unloading ramp and the Neuber's hyperbola must be found, as shown in Fig. 8.22 by point B, using the hysteresis loop curve, instead of the cyclic stress–strain curve, as done before for the loading ramp. The origin of the axes is now at point A. The new equations to be used are Eq. (8.32) with $\Delta\sigma_n$, $\Delta\varepsilon$ and $\Delta\sigma$ instead of σ_n, ε and σ and Eq. (8.6) for the unloading ramp

$$\frac{(k_f \Delta\sigma_n)^2}{E} = \Delta\sigma \cdot \Delta\varepsilon \frac{n!}{r!(n-r)!}$$

$$\Delta\varepsilon = \frac{\Delta\sigma}{E} + 2 \cdot \left(\frac{\Delta\sigma}{2K'}\right)^{1/n'} \tag{8.34}$$

respectively, that combined yield

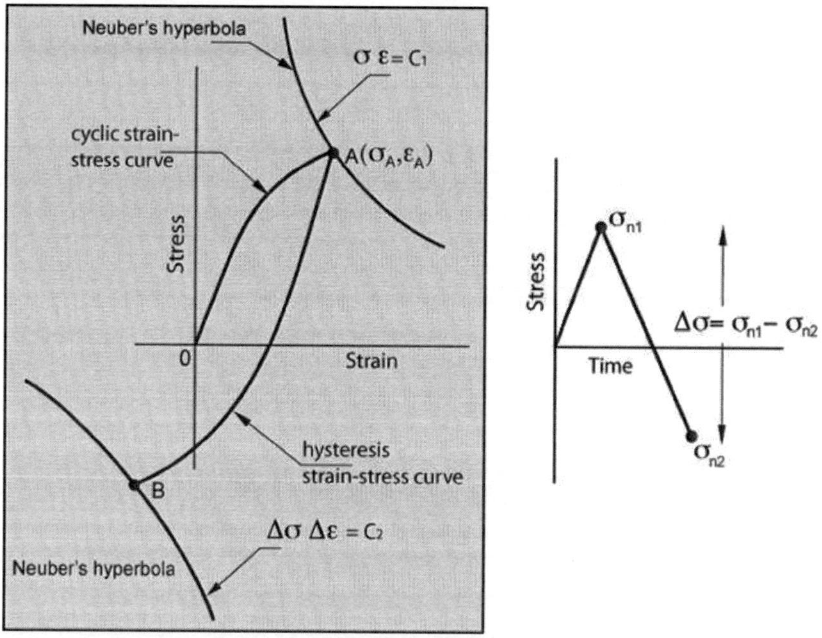

Fig. 8.22 Schematic view of the intersection of the hysteresis loop with the Neuber's hyperbola

$$\frac{(k_f \Delta\sigma_n)^2}{E} = \frac{\Delta\sigma^2}{E} + 2\Delta\sigma \cdot \left(\frac{\Delta\sigma}{2K'}\right)^{1/n'}. \tag{8.35}$$

Again, the values of $\Delta\varepsilon$ and $\Delta\sigma$ may be found using an iteration technique. Once determined, the values of ε and σ corresponding to the intersection at B can be inferred by subtracting $\Delta\varepsilon$ and $\Delta\sigma$ from the values ε_A and σ_A of point A, previously found.

8.6.1 Multiaxial Stress State and Neuber's Rule

It must be held in mind that notches as any discontinuity generate a localized multi-axial stress state the degree of which is as much intense as the notch is sharper. Under uniaxial stress state the nominal stress σ_1 is amplified at notch root because of stress concentration presenting a gradient that is responsible for the generation of the second stress component σ_2 in the direction of the gradient itself. The sharpness of the notch is actually influencing the stress gradient and, therefore, the value of the second component of stress σ_2. If the work piece is thick enough the contraction in the third direction z may be prevented ($\varepsilon_3 = 0$) and a third component of stress σ_3 is generated. This produces what is known as plain strain condition ($\varepsilon_1 \neq 0$, $\varepsilon_2 \neq 0$, $\varepsilon_3 = 0$) opposite to plain stress typical of thin and ductile components where the notch root contraction is allowed. Under plain strain condition eqs. (8.3) and (8.5) will be modified to become

$$\begin{aligned} \varepsilon_{1,a} &= \varepsilon_{1,e} + \varepsilon_{1,p} \\ &= \frac{\sigma_{1,a}}{E'} + \left(\frac{\sigma_{1,a}}{K'_{ps}}\right)^{1/n'} \end{aligned} \tag{8.36}$$

in which the subscript ps means plain strain, $\varepsilon_{1,a}$ and $\sigma_{1,a}$ represent the first component of strain and stress, respectively, equal to

$$\begin{aligned} \varepsilon_{1,a} &= \left(1 - \nu^2\right)\frac{\varepsilon_a}{\sqrt{1 - \nu + \nu^2}} \\ \sigma_{1,a} &= \frac{\sigma_a}{\sqrt{1 - \nu + \nu^2}} \end{aligned} \tag{8.37}$$

with ν being the Poisson's coefficient that in the plastic domain is equal to 0.5, and E' the Young's modulus in plain strain that is equal to

$$E' = \frac{E}{1-\nu^2}. \tag{8.38}$$

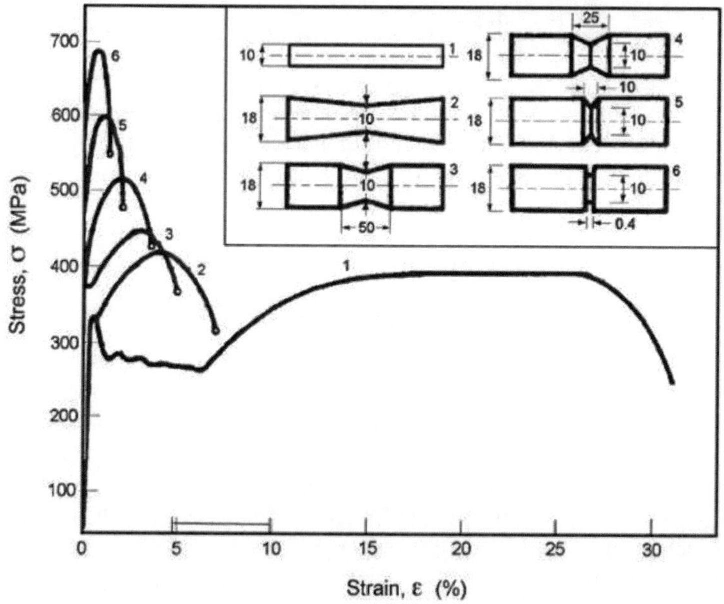

Fig. 8.23 Experimental results obtained by Lüdwick [35] using unnotched and notched specimens that evidences the effect of multi-axiality (dimensions in mm)

Triaxiality is actually changing the stress–strain curve of materials σ-ε and, therefore, $K'_{ps} \neq K'$. Figure 8.23 shows one of the first studies on material response to multiaxial stress state conducted by Lüdwick already in 1926 [35].

The material was a high ductility mild steel, as it can be deducted from the σ–ε characteristic obtained on the smooth unnotched specimen, curve 1. All notched specimens have the same minimum cross section equal that of the unnotched specimen. Notch acuity varies from specimen type 2 to type 6 becoming sharper and sharper. At the same time, also triaxiality generated by different notches changes becoming more and more intense influencing the stress–strain curve of the material. It can be noticed the progressive reduction of strain to failure coupled to a parallel yield strength increase that from about 270 MPa measured on the unnotched specimen grows up to 670 MPa in specimen 6 carrying the sharpest notch, with a 2.5 folds increase that represents the theoretical growth under complete plain strain condition. In fact, using for simplicity the Tresca criterion of the maximum shear stress it is

$$\tau_{\max} = \frac{\sigma_1 - \sigma_3}{2} = \frac{\sigma_y}{2} \tag{8.39}$$

with σ_y being the yield strength under uniaxial monotonic traction. Since under full plain strain condition it is $\sigma_1 = \sigma_2$ and $\sigma_3 = 2\nu\sigma_1$ (contraction in the z direction prevented, $\varepsilon_3 = 0$), it turns out that with $\nu = 0.3$ it is

$$\frac{\sigma_1 - \sigma_3}{2} = \frac{(1-2\nu)\cdot\sigma_1}{2} = 0.4\cdot\frac{\sigma_1}{2} = \frac{\sigma_y}{2} \tag{8.40}$$

and, therefore, $\sigma_1 = 2.5{\cdot}\sigma_y$. With $\nu = 0.33$ Eq. (6.40) would yield $\sigma_1 \sim 3{\cdot}\sigma_y$. This notch root hardening of the material has an effect on Neuber's rule results. Figure 8.24 schematizes this effect. It can be clearly seen the impact of a plain strain condition deriving from notch root triaxiality on the intersection of Neuber's hyperbola $\sigma{\cdot}\varepsilon$ = cost. In fact, the new intersection of hyperbola 1 has moved from A to A′ with a significant strain reduction and a large beneficial effect on fatigue strength. It is worth noting how in case of hyperbola 2 that with an uniaxial stress–strain curve would result in a plastic strain corresponding to point B, the hardening of the material following the notch root triaxiality would return the material response into the elastic regime, point B′. Triaxiality, therefore, is improving the fatigue strength of materials. Manjoine [36] using the triaxiality factor *TF* introduced by Davis and Connelly [37] as the ratio of hydrostatic stress to the von Mises equivalent stress

$$TF = \frac{\sqrt{2}}{3}\frac{\sigma_1 + \sigma_2 + \sigma_3}{\sqrt{(\sigma_1-\sigma_2)^2 + (\sigma_1-\sigma_3)^2 + (\sigma_2-\sigma_3)^2}} \tag{8.41}$$

found an empirical relationship between *TF* and strain at fracture. This is shown in Fig. 8.25 where the ductility is indexed to that measured on unnotched specimen under uniaxial monotonic traction, ε_{uni}. Figure 8.25 also shows some theoretical values found by Milella [38] that agree very well with Manjoine experimental results. The curve interpolating both the experimental and the theoretical results lies between the two curves having equation

$$\begin{aligned} \frac{\varepsilon}{\varepsilon_{uni}} &= (3\cdot TF)^{-1} \\ \frac{\varepsilon}{\varepsilon_{uni}} &= 2^{(1-3TF)} \end{aligned} \tag{8.42}$$

where ε is the actual strain and ε_{uni} the corresponding uniaxial strain that would be measured with an unnotched specimen. Note how a complete triaxiality with $\sigma_1 = \sigma_2$ and $\sigma_3 = 2\nu\sigma_1$, for which Eq. (8.41) yields $TF = 2.17$, is associated with an almost complete loss of ductility. The *TF* of an unnotched specimen under uniaxial monotonic traction, $\sigma_2 = \sigma_3 = 0$ and $\sigma_1 \neq 0$, is equal to 0.33 to which it corresponds a relative ductility obviously equal one. Values larger than one of the *TF* compete to multiaxial stress states in which at least one stress component is negative (compression).

Milella [39] has proposed an empirical relationship between the notch root radius ρ and the corresponding *TF*.

$$FT = 1.58 - 0.36\cdot\ln\rho\,. \tag{8.43}$$

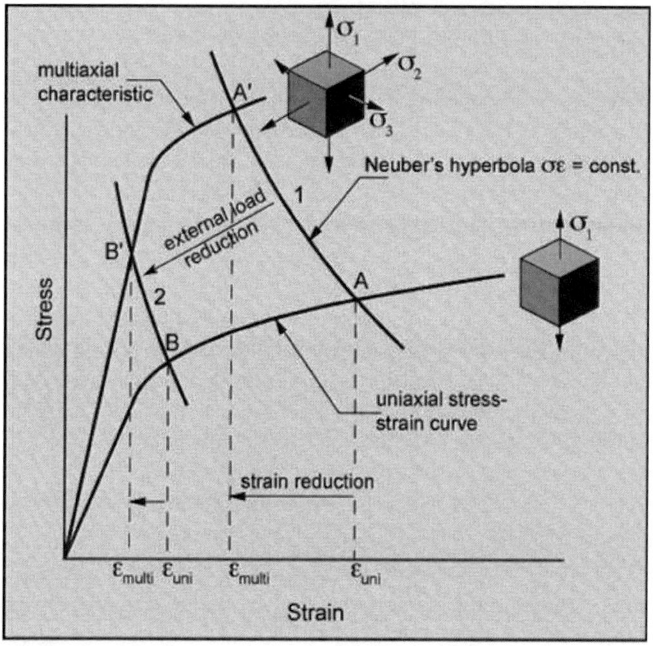

Fig. 8.24 Notch root strain reduction consequent the passage from a plain stress (uniaxial) to a plain strain condition (multiaxial)

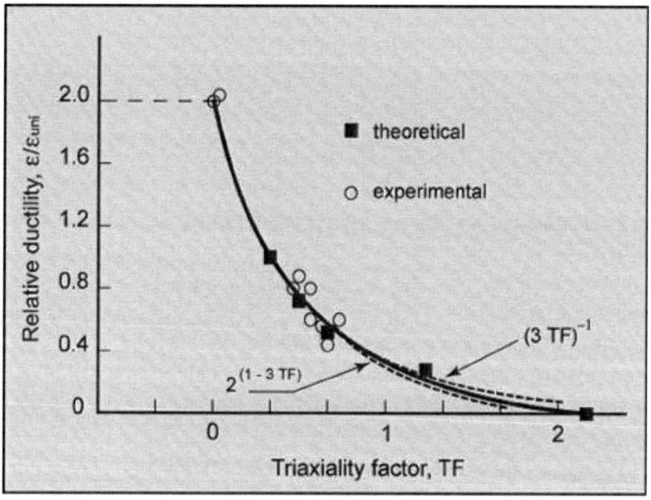

Fig. 8.25 Trend of relative ductility (ductility indexed to the uniaxial monotonic traction value) versus the triaxiality factor *TF*. Experimental data from Manjoine [36], theoretical data from Milella [38]

Equation (8.42) indicates, for example, that with a root radius of 0.5 mm the corresponding *TF* is 1.83. Entering the curve of Fig. 8.25 with this value it yields a ductility loss of about 90%. A rather big warning mark must be introduced at this time since this actually means that a material that under uniaxial traction fails at a deformation of 35% under plain strain condition will not reach more than 3.5% strain.

8.6.2 Neuber's Rule Applications

One of the first applications of Neuber's rule is shown in Fig. 8.26 [34] that refers to type 7075-T6 aluminum specimens, both unnotched and notched with notch factors k_f equal to 4.0, 3.0, 2.0 and 1.92, respectively. On the ordinate axis the equivalent elastic stress excursion, $\sigma_{a,el}$, times the Young's modulus, E, times the notch factor k_f. is reported

$$\sigma_{a,el} \cdot \varepsilon_a \cdot E \cdot k_t = \frac{1}{2}\Delta\varepsilon_t \cdot E \tag{8.44}$$

With this procedure all *S–N* fatigue curves relative to various notched specimens must overlap on a single curve. As matter of fact, all experimental results relative to notched specimens whose strain excursion $\Delta\varepsilon$ was evaluated using Neuber's rule fall on a single line that is the line of experimental results obtained with unnotched specimens.

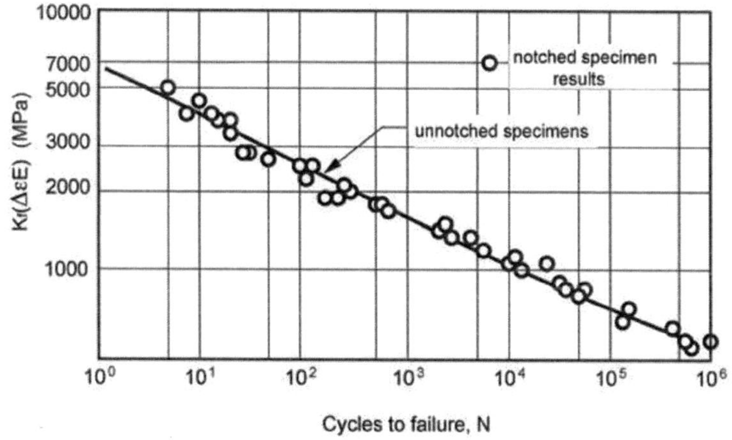

Fig. 8.26 Experimental results, reported in terms of equivalent elastic strain amplitude times the notch factor k_f obtained using Neuber's rule with notched 7075-T6 aluminum specimens compared to the *S–N* curve obtained with unnotched specimens of the same material under push-and-pull conditions [34]

However, the basic hypothesis of the constancy of the product $k_\sigma \cdot k_\varepsilon$ was verified by Neuber for just one particular notch geometry and, for such a reason, may not be considered universally valid. Finite elements calculations [40] have shown that Neuber's rule yields values of concentration factors k_σ and k_ε lightly higher than the real ones, or considered real, and can be conservative.

Figure 8.27 presents a comparison between FE calculations and Neuber's results [40]. Calculations refers to a fracture mechanics specimen type Compact, C(T) (see Appendix A), but without the sharp crack at the tip of the notch. Another example is that of a plate with a central hole examined by Huang [41] and reproduced in Fig. 8.28. It shows that Eq. (8.34) is systematically in error, depending on the material and nominal stress. Nevertheless, the most significant deviations pertain the field of nominal stresses higher than the yield strength and that of low values of the strain hardening exponent n. For nominal stress lower than yield strength, $\sigma_n/\sigma_y < 1$, discrepancies are rather contained. The issue of Neuber's rule application to fatigue has involved many researchers and study [33, 34, 40–45] with questions regarding whether it should be used k_t or k_f or the use of the most suitable stress–strain characteristic, whether the uniaxial cyclic one or the multiaxial cyclic curve. After more than thirty years of investigation it has come to the conclusion that the theoretical stress concentration factor k_t must be used when the notch root stress is still linear while k_f is applied when strains are well into the plastic domain.

Walker [44] using aluminum specimens with lateral groves to maintain triaxiality also on the external surface, found that when notch root stresses remain elastic an analysis based on the theoretical stress concentration factor k_t with a triaxial cyclic

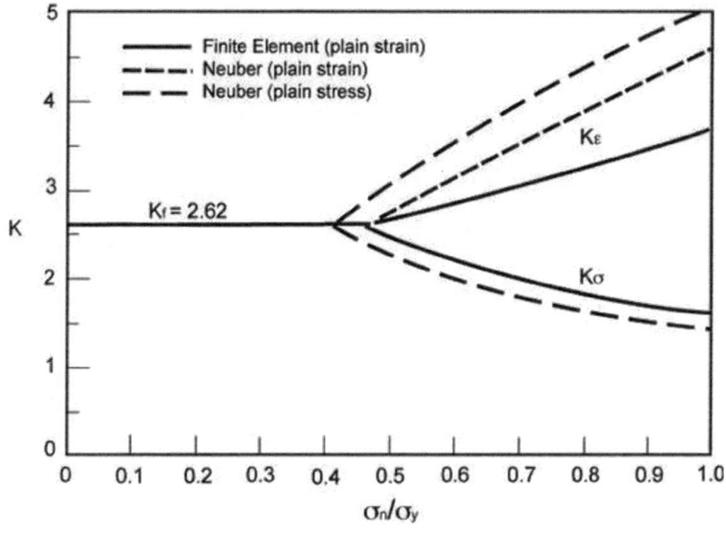

Fig. 8.27 Comparison between concentration factors k_σ and k_ε obtained with FE calculations and Neuber's rule, respectively, versus the non-dimensional ratio of nominal stress σ_n to yield strength σ_y [40]

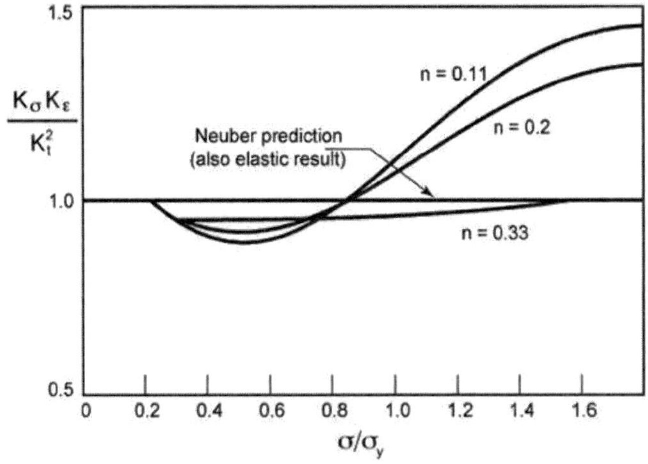

Fig. 8.28 Deviations from Neuber's rule found by Huang in a plate with a central hole for various applied nominal stresses and strain hardening exponents n (modified from [41])

stress–strain curve yields the same results as with the k_f notch factor and an uniaxial stress–strain cyclic curve, as schematized in Fig. 8.29.

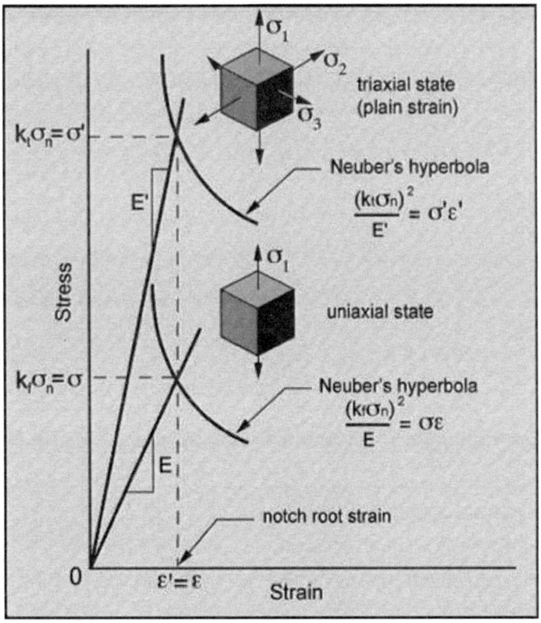

Fig. 8.29 Elastic strain calculated either with Neuber's rule using the theoretical stress concentration factor and a triaxial cyclic curve or with notch factor k_f and uniaxial cyclic characteristic of the material (modified from [44])

8.6.3 Residual Stresses with Neuber Approach

When a residual stress-strain state σ_r and ε_r exists ahead of the notch resulting, for instance, from welding it must be taken into consideration in Eq. (8.33) before any load reversal. To this end, three changes have been proposed

1. Lawrence et al. [46]

$$\frac{(k_f S_a+\sigma_r)^2}{E} = \frac{\sigma_a^2}{E} + \sigma_a \cdot \left(\frac{\sigma_a}{K'}\right)^{1/n'} \tag{8.45}$$

2. Reemsnyder [47]

$$\frac{(k_f S_a)^2}{E} = \left(1 - \frac{\sigma_r}{\sigma_a}\right)^2 \left(\frac{\sigma_a^2}{E} + \sigma_a \cdot \left(\frac{\sigma_a}{K'}\right)^{1/n'}\right) \tag{8.46}$$

3. Seeger [48]

$$\frac{(k_f S_a)^2}{E} = \sigma_a \cdot (\varepsilon_a - \varepsilon_r) \tag{8.47}$$

8.7 Sample Problems

8.7.1 Hysteresis Loop—Constant Amplitude Loads

A specimen of ductile cast iron (DCI) carrying a notch is subjected to a load cycling under strain controlled condition. On the notch root the strain excursion is equal to 20,000 με, as shown in Fig. 8.30a. The cyclic stress–strain characteristic of the material has been experimentally found to be

$$\frac{\Delta\sigma_a}{2} = 715\left(\frac{\Delta\varepsilon_p}{2}\right)^{0.118}. \tag{8.48}$$

It's Young's modulus is $E = 170{,}000$ MPa. Determine the hysteresis loop of the material. For the initial loading ramp Eq. (8.3) will be used

$$\varepsilon_a = \frac{\sigma_a}{E} + \left(\frac{\sigma_a}{K'}\right)^{1/n'} = 0.01 = \frac{\sigma_a}{170{,}000} + \left(\frac{\sigma_a}{715}\right)^{1/0.118}.$$

The entire loading curve can be found giving the strain ε growing values dividing the interval 0/0.01, for instance, in 5 steps of 0.002 each. Table 8.2 lists the results.

Therefore the final point A of the loading ramp has coordinates $\varepsilon_A = 0.01$ and $\sigma_A = 403$ MPa, Fig. 8.30b. The next step is to build the unloading ramp. To this purpose

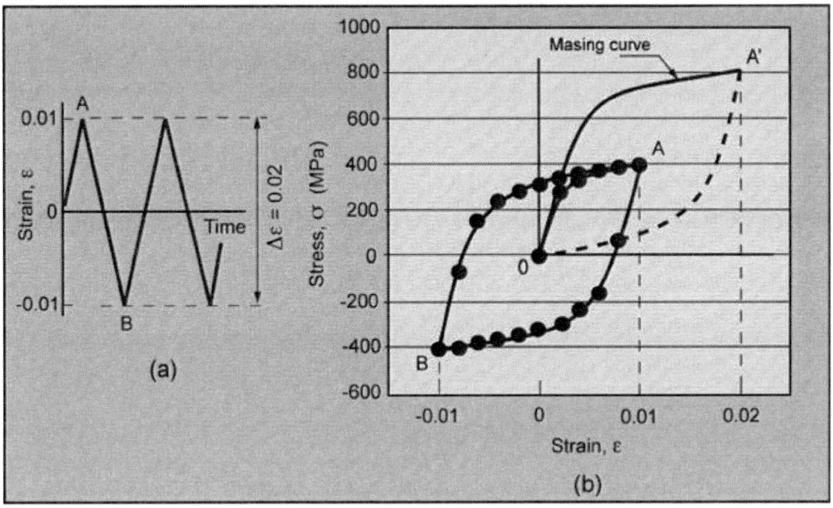

Fig. 8.30 **a** Load history under strain controlled condition, **b** Corresponding hysteresis loop and Masing curve

Table 8.2 Initial loading ramp

εa	σa (MPa)
0.002	280
0.004	344
0.006	371
0.008	389
0.01	403

Eq. (8.6) will be used with the starting point A≡(0.01;403). The equation will be solved by iteration with a step-by-step procedure using, as before, a strain unloading step $\Delta\varepsilon = 0.002$. For the first step it is.

$$0.002 = \frac{\Delta\sigma}{170{,}000} + 2\left(\frac{\Delta\sigma}{2 \cdot 715}\right)^{1/0.118}$$

which yields

$$\Delta\sigma = 340 \text{ MPa}.$$

The strain ε_1 and stress σ_1 values corresponding to the first unloading step can now be determined by subtracting the changes in strain and stress ($\Delta\varepsilon$, $\Delta\sigma$) from the values at point A (ε_A, σ_A).

$$\varepsilon_1 = \varepsilon_A - \Delta\varepsilon = 0.01 - 0.002 = 0.008 \text{ MPa}$$
$$\sigma_1 = \sigma_A - \Delta\sigma = 403 - 340 = 63 \text{ MPa}.$$

The results for each unloading step till point B reversal are listed in Table 8.3. Point B of load reversal has coordinates B≡(ε_B, σ_B) ≡(-0.01;-403). Point B will be the starting point for the reloading ramp that will be described by Eq. (8.7) with the same procedure used for the unloading ramp, but this time with a rising step $\Delta\varepsilon =$ 0.002. results are listed in Table 8.4. The closed hysteresis loop is that of Fig. 8.30b. Also shown is the Masing curve obtained moving point B reversal to the origin of coordinates.

Table 8.3 Unloading ramp

Δε	E	Δσ (MPa)	σ (MPa)
0.002	0.008	340	63
0.004	0.006	560	−157
0.006	0.004	642	−239
0.008	0.002	687	−284
0.010	0.000	718	−284
0.012	−0.002	742	−315
0.014	−0.004	761	−339
0.016	−0.006	778	−358
0.018	−0.008	792	−375
0.020	−0.01	806	−403

Table 8.4 Reloading from point B reversal

$\Delta\varepsilon$	ε	$\Delta\sigma$ (MPa)	σ (MPa)
0.002	−0.008	340	−63
0.004	−0.006	560	157
0.006	−0.004	642	239
0.008	−0.002	687	284
0.010	0.000	718	315
0.012	0.002	742	339
0.014	0.004	761	358
0.016	0.006	778	375
0.018	0.008	792	389
0.020	0.01	806	403

8.7.2 *Hysteresis Loop-Variable Amplitude Loads*

The same material of the previous sample problem is now subjected to a history of variable amplitude strains, as shown in Fig. 8.31a. Construct the corresponding hysteresis loops of the material. Cycle ABA is the same as before, but now the material ahead of the notch root once returned in A is partially unloaded to point C ≡ (0.002;−284) following the previously calculated path. Once in C the load is reversed so that Eq. (8.7) is used and solved by partial differences with the same strain step $\Delta\varepsilon = 0.002$. Results are listed in Table 8.5. The new reloading path is represented by curve CA of Fig. 8.31b and the corresponding hysteresis loop is ACA. Once back in A the system is unloaded to D following the usual unloading curve. From D the reloading Eq. (8.7) is used again with a strain step equal to 0.002. Results for this new reloading are shown in Table 8.6 and Fig. 8.31b.

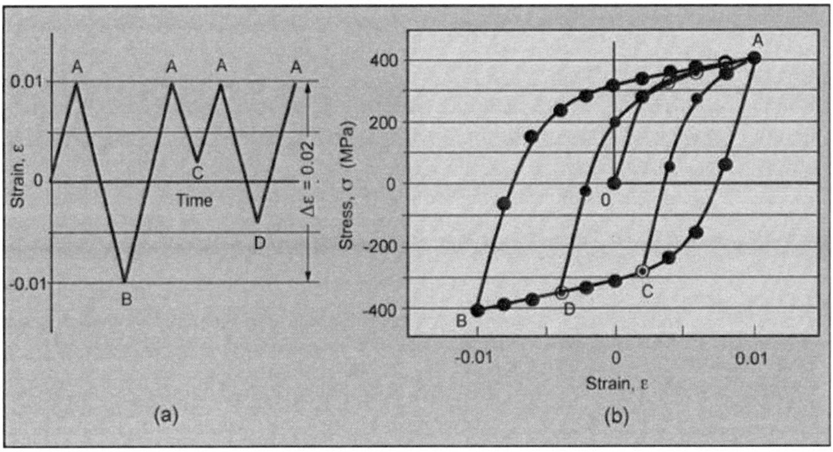

Fig. 8.31 **a** Variable strain amplitude time history and **b** Correspondent hysteresis loops

Table 8.5 Reloading from point C

$\Delta\varepsilon$	ε	$\Delta\sigma$ (MPa)	σ (MPa)
0.0	0.0	0	−284
0.002	0.004	340	56
0.004	0.006	560	276
0.006	0.008	642	358
0.008	0.010	687	403

Table 8.6 Reloading from point D

$\Delta\varepsilon$	ε	$\Delta\sigma$ (MPa)	σ (MPa)
	0.0	0	−358
0.002	−0.002	340	−18
0.004	0.000	560	202
0.006	0.002	642	284
0.008	0.004	687	329
0.01	0.006	718	360
0.012	0.008	742	384
0.014	0.01	761	403

8.7.3 Manson-Coffin and Basquin Curves

High strength steel, 1200 MPa ultimate, 1100 yield strength, Young's modulus $E =$ 205,000 MPa and Area Reduction $AR = 25\%$. Fatigue testing of unnotched smooth specimens yielded results listed in Table 8.7. Determine the constants K', n', σ'_f, b, ε'_f and c characteristic of the material. Trace the Manson-Coffin line in a log–log diagram together with the total strain amplitude ε. The difference between the total strain and the plastic strain ε_p will yield the Basquin component. This is shown in Fig. 8.32. The corresponding equation of the fatigue life (8.10) is

$$\varepsilon_a = \frac{\Delta\varepsilon_t}{2} = \frac{\sigma'_f}{E}(N)^b + \varepsilon'_f(N)^c$$
$$\varepsilon_a = 0.00794 \cdot (N)^{-0.07} + 0.46 \cdot (N)^{-0.617}. \tag{8.49}$$

Note that the exponents b and c are the slope of the two components, plastic and elastic, of the total life. Therefore the coefficients σ'_f, b, ε'_f and c under estimation are

$$\sigma'_f = E \cdot 0.00794 = 1628\ MPa$$
$$b = -0.07$$

$$\varepsilon'_f = 0.46$$
$$c = -0.617.$$

Knowing exponents b and c it will be possible to infer the cyclic strain hardening exponent n' from Eq. (8.9) that must yield the same value

$$n' = -\frac{b}{1+5b} = 0.107$$
$$n' = \frac{b}{c} = 0.113.$$

Table 8.7 Results of fatigue testing on unnotched smooth specimens

Total strain amplitude $\Delta\varepsilon/2$	Stress amplitude $\Delta\sigma/2$ (MPa)	Plastic strain amplitude $\Delta\varepsilon_p/2$	Cycles to failure N
0.040	1140	0.034	53
0.040	1134	0.034	65
0.031	1082	0.0245	119
0.020	1010	0.0152	247
0.021	1010	0.0153	340
0.0145	1000	0.0096	480
0.01	924	0.0059	1340
0.011	887	0.0061	1356
0.0072	850	0.003	4010
0.007	836	0.0028	3970
0.0053	800	0.0013	8890
0.0043	750	0.000	34,050
0.0035	595	0.000	141,500

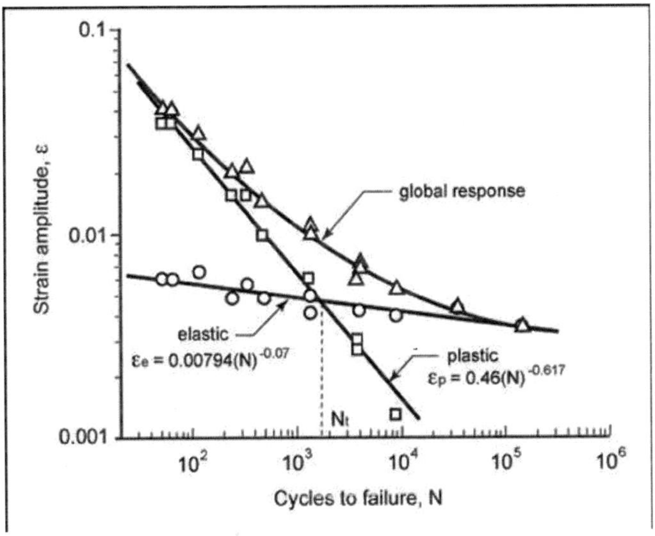

Fig. 8.32 Total strain-life curve and its components plastic and elastic

The small difference between the two values of n' obtained from b and c respectively shall not come as a surprise considering the approximate character of the linear interpolations. About the plastic modulus K', it can be inferred from Eq. (8.2) in different ways. A first evaluation is

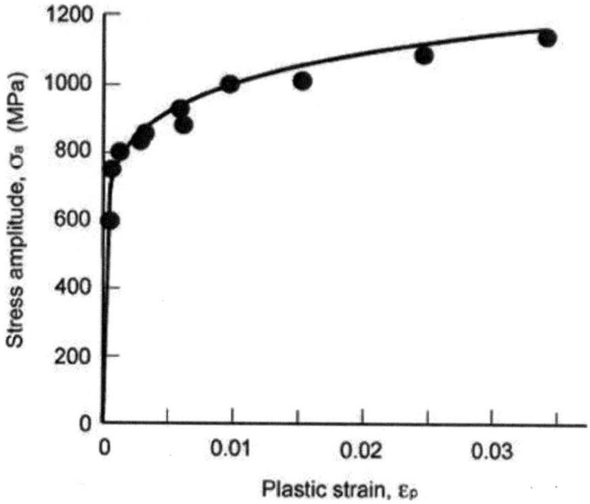

Fig. 8.33 Experimental cyclic stress–strain data and their interpolation

$$K' = \frac{\sigma'_f}{\left(\varepsilon'_f\right)^{n'}} = \frac{1628}{(0.46)^{0.107}} = 1769\ MPa.$$

The second evaluation is obtained from the interpolation of experimental cyclic stress–strain curve $\sigma_a - \varepsilon_p$ shown in Fig. 8.33.

The procedure yields

$$\sigma_a = 1750 \cdot \left(\varepsilon_p\right)^{0.121}$$

therefore it is

$$K' = 1750\ MPa$$
$$n' = 0.121.$$

Also in this case there is a slight difference between the two values of K', 1750 against 1769 previously obtained and the exponent n' is grown farther to 0.121. There is one more procedure to infer the K' value recalling that the cyclic modulus can be obtained by plotting the stress amplitude versus the plastic deformation, σ_a-ε_p, as the intersection of the interpolation line with the ordinate axis at $\varepsilon_p = 1$, as shown in Fig. 8.34 using the experimental data listed in Table 8.7. The intersection is actually yielding

$$K' = 1750\ MPa$$

confirming the previously found value.

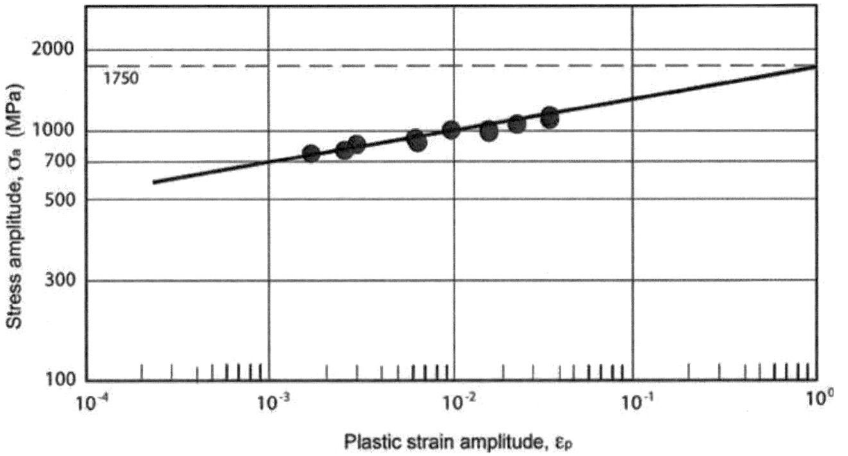

Fig. 8.34 The K' modulus can be inferred from the σ_a-ε_p diagram

8.7.4 Manson-Coffin and Basquin Curves

Derive the *S–N* curve for the steel considered in sample problem 1. of Sect. 5.3 using Basquin and Manson-Coffin equations. To build the *S–N* curve the cyclic strain hardening exponent of the material n' is needed together with the fatigue strength coefficient σ'_f and the fatigue ductility coefficient ε'_f. these last two coefficient have been already calculated in sample problem of Sect. 10.5.1 and are

$$\sigma'_f = 9.26\ MPa$$
$$\varepsilon_f = 1.10$$

About the strain hardening exponent of the material n' it is necessary to know the cyclic curve of the material. To this purpose, the multiple specimens procedure described in Sect. 1.3.1 (see Fig. 1.24) has been adopted using short test pieces of 8 mm diameter and 2 cm length. The cyclic curve is shown in Fig. 8.35 together with the monotonic one for comparison. It is possible to see how the steel underwent a strain hardening process with the complete loss of the upper and lower yield point. The cyclic curve equation is

$$\varepsilon_a = \frac{\sigma_a}{208{,}333} + \left(\frac{\sigma_a}{950}\right)^{1/0.165} \tag{8.50}$$

from which the cyclic strain hardening exponent can be inferred

$$n' = 0.165.$$

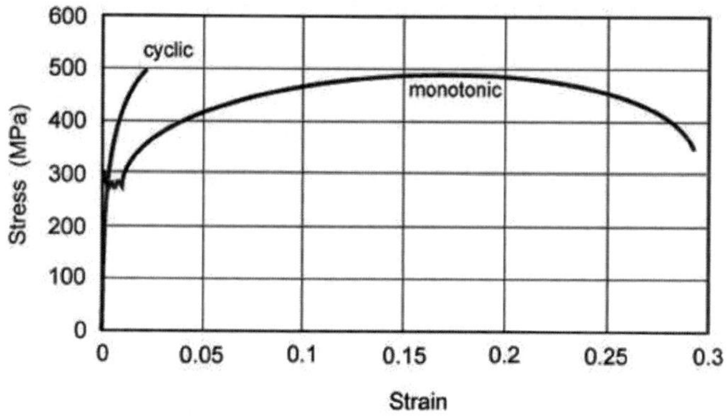

Fig. 8.35 Monotonic and cyclic stress–strain curves for the steel considered in the sample problem

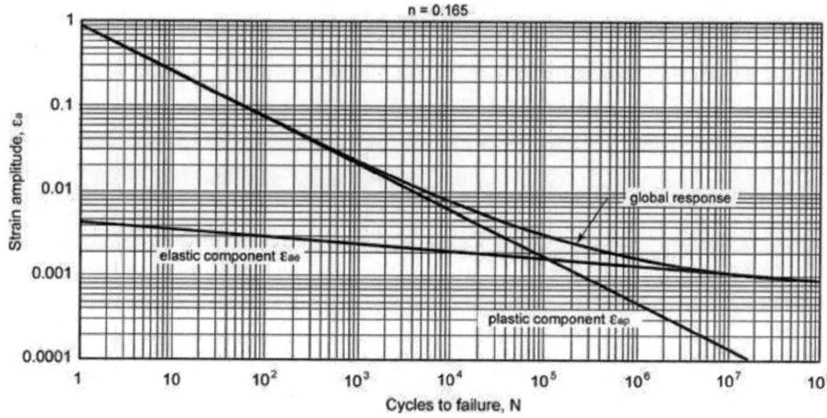

Fig. 8.36 *S–N* curve and its plastic and elastic components

Using Morrow Eqs. (5.33) and (6.9) the two exponent b and c are obtained

$$b = -0.09; c = -0.548$$

and both the Basquin and Manson-Coffin curves can be drown as shown in Fig. 8.36.

8.7.5 *Low Cycle to High Cycle Fatigue Transition*

A Medium–High Strength Steel Having $\sigma_U = 750$ MPa has a RA of 45% and a cyclic strain hardening exponent $n' = 0.2$. It is used in the construction of a component that will be subjected to fatigue cycles of 0.004 strain amplitude. It must be decided whether or not the component needs a shot peening treatment to withstand fatigue. The strain amplitude is anticipating a possible low cycle fatigue behavior of the material. If this is the case, the shot peening treatment will not be effective since low cycle fatigue is acting on the entire volume of the work piece and not only on the surface (see Sect. 1.4.2).

Therefore, any surface treatment will be irrelevant on the fatigue strength. Let's try to verify whether low cycle fatigue will affect the work piece. The number of cycles Nt at which the passage from low cycle fatigue to high cycle fatigue achieved is given by Eq. (6.12). To calculate N_t it is necessary to evaluate the four coefficients ε'_f, σ'_f, b and c knowing that Young's modulus E is equal to 210,000 MPa.

Equations (5.33) and (8.9) will be used, as usual, obtaining

$$b = -0.1$$
$$c = -0.5.$$

As far as ε'_f is concerned, it is given by

$$\varepsilon'_f = \ln\left(\frac{1}{1 - RA}\right) = 0.6$$

while σ'_f is given by Eq. (6.13) that yields

$$\sigma'_f \approx \sigma_r(1 + \varepsilon_r) + 18.4 = 750 \cdot 1.45 + 18.4 = 1100 \text{ MPa}.$$

Therefore, it will be

$$N_t = \left(\frac{\varepsilon'_f E}{\sigma'_f}\right)^{1/(b-c)} = \left(\frac{0.6 \cdot 210,000}{1100}\right)^{1/(-0.1+0,.)} = 140,425.$$

The plastic strain ε_p at 140,425 cycles is equal to the elastic one and is

$$\varepsilon_p = \varepsilon'_f (N_t)^c = 0.6 \cdot (140,425)^{-0.5} = 0.0016$$

so that the total strain $\varepsilon_p + \varepsilon_e$ is

$$\varepsilon_p + \varepsilon_e = 2 \cdot 0.0016 = 0.0032$$

which is lower that 0.004 of applied strain amplitude that will be acting on the component. Therefore, shot peening will not have any effect on the fatigue strength of the work piece, besides that the residual stress state will be cancelled by the plastic strain cycling.

8.7.6 Hysteresis Loops Resulting from Reloading—SWT and Morrow Model

Numerical results obtained on a notched component made of the same steel of the previous sample problem has given the time history of Fig. 8.37 on the notch root. Determine the.

- hysteresis loops,
- fatigue strength using the SWT model,
- fatigue strength using the Morrow correction.

The initial ramp of the hysteresis loop is given by Eq. (8.3). The values of the three exponents n', b and c are known

$$n' = 0.2$$
$$b = -0.1$$
$$c = -0.5.$$

The value of the plastic modulus K' is needed. The modulus can be obtained from the intersection of the interpolation line of σ_a–ε_p data with the vertical at $\varepsilon_p = 1$. Experimental results obtained using 14 hourglass specimens, Fig. 3.4a, having $\emptyset =$ 6 mm and $l = 18$ mm are shown in Fig. 8.38 with the relative interpolation line. These results arrive to 10% deformation to have a rather accurate extrapolation point at $\varepsilon_p = 1$ whose value yields $\sigma_a = 1210$ MPa. Now the loading ramp can be derived. It goes from 0 to 0.004 mm/mm of deformation amplitude, as shown in Fig. 8.37. Equation (8.3) can be solved by iteration. Table 8.8 lists the σ and ε values obtained at each iteration step. The initial loading ramp is shown in Fig. 8.39 as 0–1.

Consider now the first reversal. Equation (8.6) will be used. The new origin is point 1 of coordinates (0.004;359). The deformation at the end of the first unloading

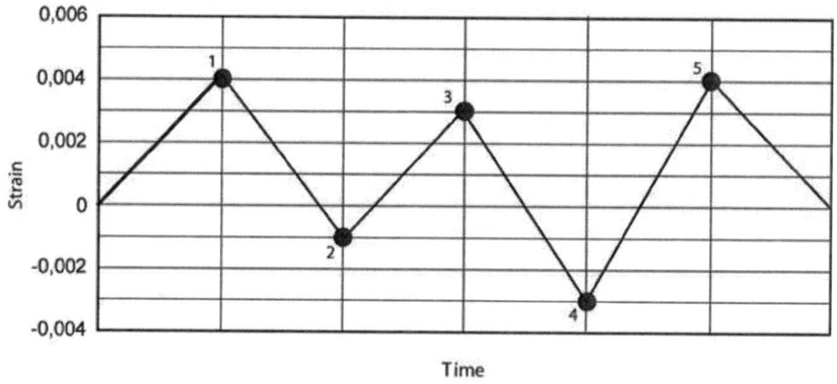

Fig. 8.37 Deformation time history at the notch root

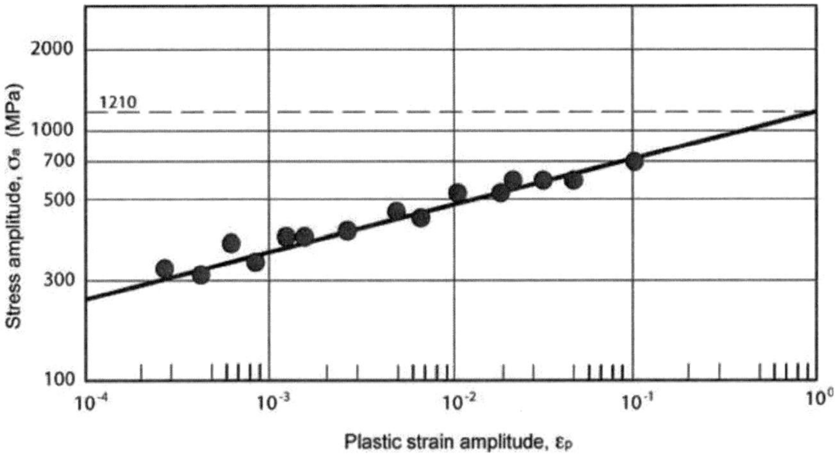

Fig. 8.38 Stress amplitude σ_a versus ε_p experimental data and interpolation line

Table 8.8 Initial loading rump

ε	σ (MPa)
0.00	0.00
0.0005	105
0.0007	143
0.0010	193
0.0015	245
0.0020	281
0.0040	359

is equal to -0.001 so that it is $\Delta\varepsilon = 0.004 + 0.001 = 0.005$. Therefore Eq. (8.6) can be written as

$$0.005 = \frac{\Delta\sigma}{210{,}000} + 2\left(\frac{\Delta\sigma}{2 \cdot 1210}\right)^{1/0.2}$$

that solved for $\Delta\sigma$ yields

$$\Delta\sigma = 613 \text{ MPa}.$$

Since Eq. (8.6) makes use of strain and stress variations $\Delta\varepsilon$ and $\Delta\sigma$ to $\Delta\sigma =$ 613 MPa it corresponds a stress σ_2 equal to

$$\sigma_2 = \sigma_1 - \Delta\sigma = 359 - 613 = -254 \text{ MPa}.$$

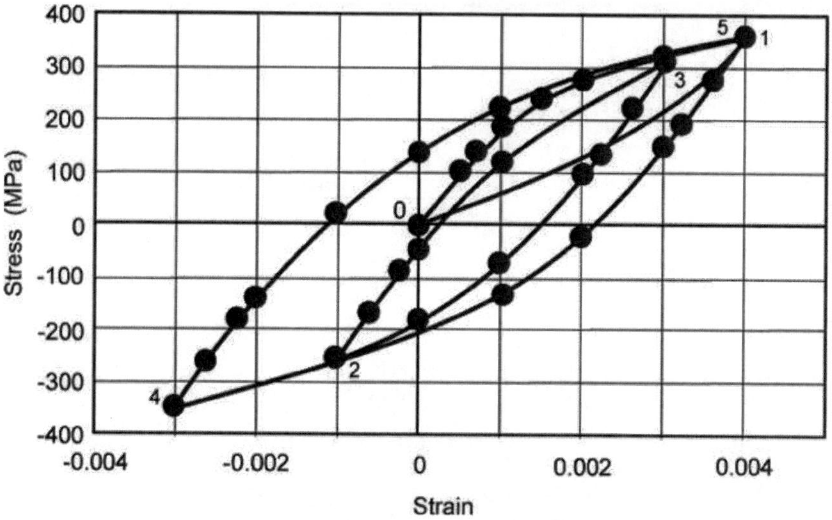

Fig. 8.39 Hysteresis loops produced by the time history of Fig. 8.37

This is shown in Fig. 8.39 by point 2. The corresponding unloading ramp from point 1 to point 2 is constructed by iteration of Eq. (8.6) using an initial $\Delta\varepsilon$ step of 0.0004 amplitude. The iterative process and its results are listed in Table 8.9.

The second cycle reversal starts from point 2≡(−0.001;−254) and goes up to point 3. The new strain excursion is $\Delta\varepsilon = 0.003 + 0.001 = 0.004$. For the reloading ramp it will be used Eq. (8.7) that yields

$$0.004 = \frac{\Delta\sigma}{210,000} + 2 \cdot \left(\frac{\Delta\sigma}{2 \cdot 1210}\right)^{1/0.2}$$

Solve for $\Delta\sigma$

$$\Delta\sigma = 560.5 \text{ MPa}.$$

Table 8.9 First cycle reversal

ε	Δε	Δσ (MPa)	σ (MPa)
0.0040	0.00	0	359
0.0036	0.0004	84	275
0.0032	0.0008	167.5	191.5
0.0030	0.0010	209	150
0.0020	0.0020	380	−21
0.0010	0.0030	489	−130
−0.0010	0.0050	613	−254

Table 8.10 Second reversal

ε	Δε	Δσ (MPa)	σ (MPa)
0.0040	0.00	0	359
0.0036	0.0004	84	275
0.0032	0.0008	167.5	191.5
0.0030	0.0010	209	150
0.0020	0.0020	380	−21
0.0010	0.0030	489	−130
−0.0010	0.0050	613	−254

The stress σ_3 corresponding to point 3 is.

$$\sigma 3 = \sigma 2 + \Delta\sigma = -254 + 560.5 = 306.5 \text{ MPa}.$$

The reloading ramp from point 2 to point 3 will be calculated by iteration using Eq. (6.7). Results are shown in Fig. 8.39 and listed in Table 8.10.

The third unloading will start from point 3≡(0.003; 306.5) to reach point 4. The strain variation is

$$\Delta\varepsilon = 0.003 + 0.003 = 0.006.$$

Following the same procedure for unloading it is

$$0.006 = \frac{\Delta\sigma}{210{,}000} + 2 \cdot \left(\frac{\Delta\sigma}{2 \cdot 1210}\right)^{1/0.2}$$

Solve for $\Delta\sigma$

$$\Delta\sigma = 654.4 \text{ MPa}.$$

Arriving at point 4 there will be a stressσ_4 equal to

$$\sigma_4 = \sigma_3 - \Delta\sigma = 306.5 - 654.4 = -347.9.$$

The curve relative to the third load reversal from point 3 to point 4 will be drown, as usual, solving by iteration Eq. (8.6). Results are listed in Table 8.11 and shown in Fig. 8.39. interesting enough, the new unloading from point 3 to point 4 crosses point 2 and continues exactly as if the unloading were fron point 11 to point 4 confirming the memory effect of the material. From point 4 a new reloading will reach point 5 coincident with point 1. Total strain is $\Delta\varepsilon = 0.004 + 0.003 = 0.007$. Solving Eq. (8.7) for $\Delta\sigma$ yields

$$\Delta\sigma = 717.6 \text{ MPa}.$$

Table 8.11 Third load reversal

ε	Δε	Δσ (MPa)	σ (MPa)
0,0030	0.00	0.00	306.5
0,0026	0.0004	84	222.5
0,0022	0.0008	167.5	139.0
0,0020	0.0010	209	97.50
0,0010	0.0020	380	−73.50
0,0000	0.0030	489	−182.5
-0,0010	0.0040	560.5	−254.0
-0,0030	0.0060	654.4	−347.9

Arriving at point 5 there will be a stress σ_5 equal to

$$\sigma_5 = \sigma_4 + \Delta\sigma = -347.9 + 717.6 = 369.7 \text{ MPa}$$

Finally, from point 1 there will be the last unloading to the origin. Let's count the applied cycles by using the rain flow rule (see Sect. 8.7.6). The procedure outlined in Fig. 8.13 is shown in Fig. 8.40. In the loading time history of Fig. 6.37 three independent cycles can be identified and separated, namely cycle (a), cycle (b) and cycle (c). These three cycles correspond to hysteresis loop 2-3-2, 1-4-5 and 0-1-0 of Fig. 8.39. The same cycles can be identified and separated using the rain flow procedure, as schematized in Fig. 8.41. Let's apply the SWT method to evaluate the allowable number N of cycles using Eq. (8.22) and the damage $D = 1/N$ introduced in the material by each of them. Table 8.12 reassumes the results obtained. The allowable number of cycles N are computed by iteration. The total damage D_{tot} is given by the sum of partial damages D_i

$$D_{tot} = \sum D_i.$$

The number of times that the time history of Fig. 8.37 can be applied is equal to $1/D_{tot}$

$$1/D_{tot} = 1/1.44^{-5} = 69{,}444.$$

Let's now use the Morrow correction (8.17) to take into consideration the mean stress. Table 8.13 lists the data relative to the three cycles to use in Eq. (8.17). Again, the number of cycles N has been calculated by iteration. Total damage D_{tot} is a little lower than that evaluated with the SWT method, therefore the allowable number of times that the time history of Fig. 8.37 can be applied is higher

$$1/D_{tot} = 1/1.30^{-5} = 77{,}118.$$

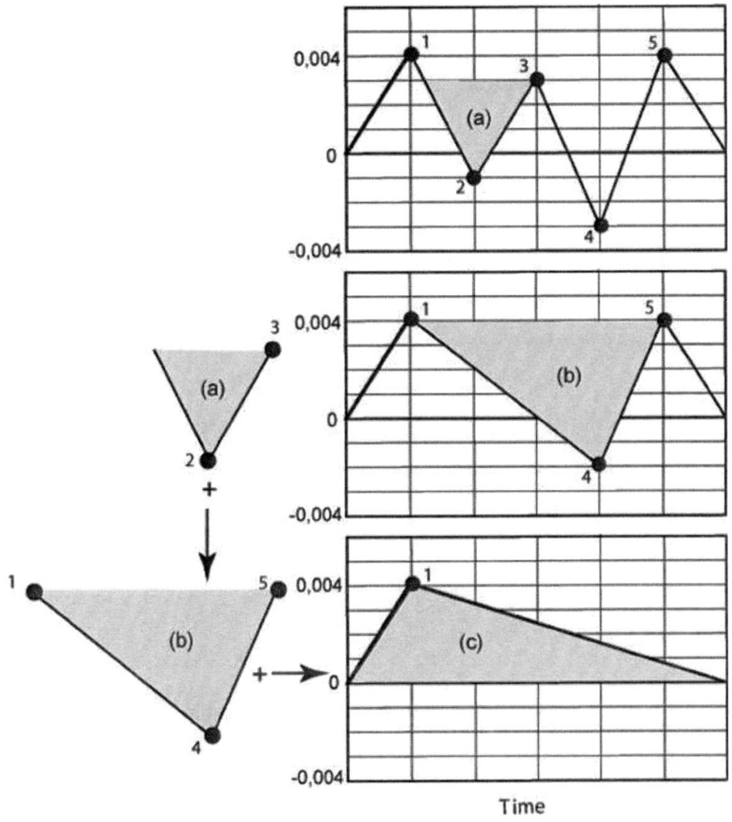

Fig. 8.40 Three independent cycles can be identified in the time history, namely cycle (**a**), cycle (**b**) and cycle (**c**)

Interesting enough, cycle 1-4-5 is the most damaging because of the higher strain excursion, but cycle 0-1-0 comes right after that because it has the highest mean stress of them all.

8.7.7 Manson-Coffin S–N Curve—Neuber Rule

A series of 17 smooth specimens of carbon steel type A 106 B, 6.35 mm (0.25″) diameter and 15.24 mm (0.6″) length, has been tested in push-and-pull traction (R = − 1) at 288 °C under strain controlled conditions. Results are shown in Fig. 8.42. A further testing with notched specimens having $k_t = 3$, root radius $\rho = 0.28$ mm and a diameter $D = 8$ mm with the reduced notch section having the same diameter

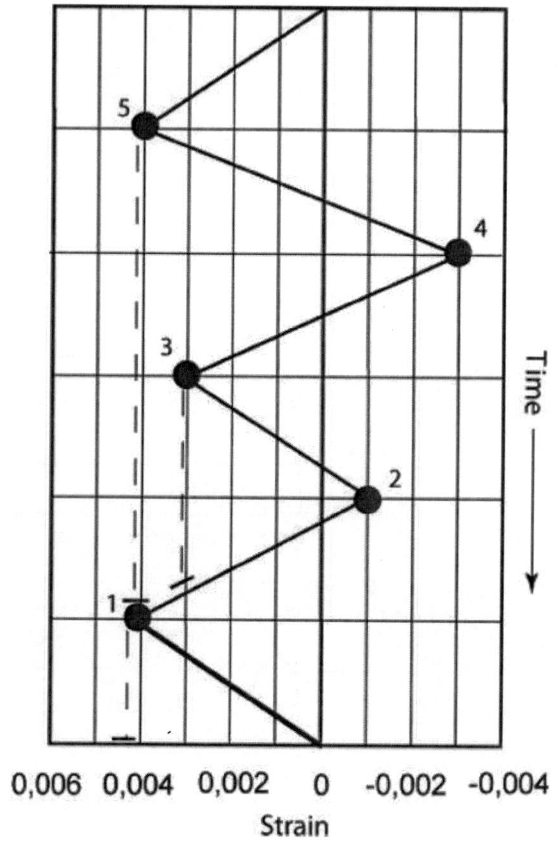

Fig. 8.41 The same cycles of Fig. 8.40 can be identified and separated using the rain flow procedure

Table 8.12 Allowable number of cycles N, partial and total damage associated with the three cycles

ε_{max}	σ_{max}	ε_a	$\sigma_{max}\,\varepsilon_a$	N	D	D_{tot}
0.003	306	0.0020	0.6120	645,000	1.55E-06	1.44E-05
0.004	359	0.0035	1.2560	96,200	1.04E-05	
0.004	359	0.0020	0.7180	410,000	1.19E-05	

Table 8.13 Data relative to the three cycles analyzed with the Morrow correction, Eq. (8.17)

ε_{max}	ε_{min}	σ_{max}	σ_{min}	ε_a	σ_m	N	D	D_{tot}
0.003	−0.001	306	−254	0.0020	26	700,000	1.43E-06	1.30E-05
0.004	−0.003	359	−347	0.0035	6	104,000	9.61E-06	
0.004	0.0	359	0	0.0020	179.5	520,000	1.92E-06	

$d = 6.35$ mm as smooth specimens, has yielded results listed in Table 8.14. Analyze the results using Manson-Coffin *S–N* curve and Neuber's rule.

Tests have been run to derive the cyclic stress–strain curve both at 25 °C and at the operating temperature of 288 °C. Results are shown in Fig. 8.43. It is possible to recognize that at 288 °C the material is undergoing a dynamic strain aging process that makes the steel behave much better than at room temperature. The Ramberg–Osgood Eq. (6.3) of the cyclic curve at 288 °C is

$$\varepsilon = \frac{\sigma}{E} + \left(\frac{\sigma}{747}\right)^{1/0.0866}. \tag{8.51}$$

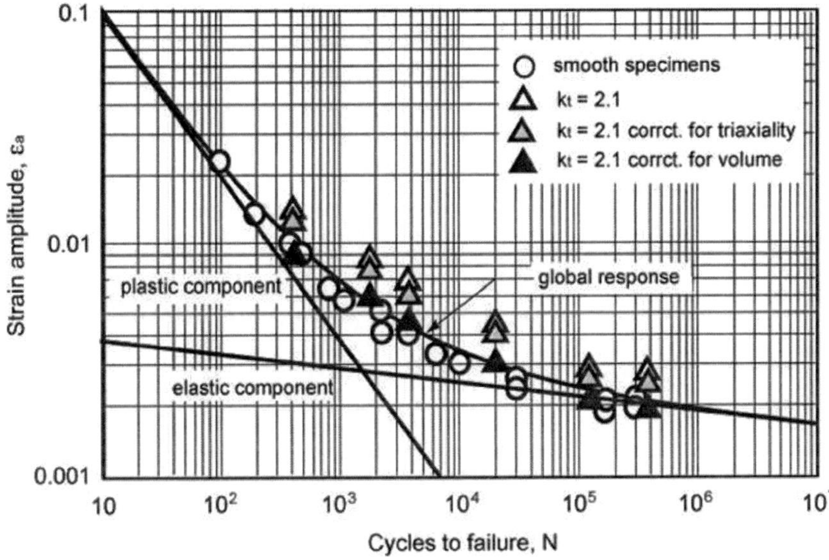

Fig. 8.42 Experimental fatigue life results obtained at 288 °C on unnotched and notched specimens of A106B carbon steel. Also shown is the *S–N* diagram (global response) found using Basquin and Manson-Coffin components

Table 8.14 Experimental and computed test matrix results

k_t	σ_a (MPa)	N	$\varepsilon_{a,notch}$	$\sigma_{a,notch}$ (MPa)	$\varepsilon'_{a,notch}$	$\varepsilon''_{a,notch}$
3	210	300,000	0.0026	395	0.01156	0.00868
3	220	115,000	0.0027	405	0.00770	0.00680
3	300	20,000	0.0047	440	0.00615	0.00544
3	380	3600	0.0071	470	0.00409	0.03625
3	430	1800	0.0089	480	0.00239	0.00211
3	540	400	0.0133	505	0.00223	0.00198

′corrected for triaxiality, ″corrected for process volume effect

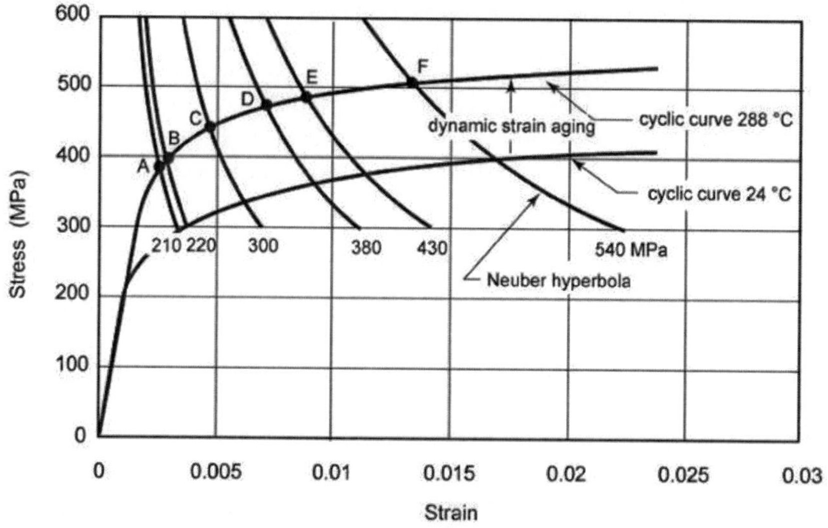

Fig. 8.43 Cyclic stress–strain curves obtained with A 106 B carbon steel specimens at room temperature and 288 °C. Also shown are the Neuber's hyperbolas for six different loading conditions

Also the true stress and true strain at failure, see Eqs. (5.27) and (5.32), have been calculated as $\sigma_{t,f} = 910$ MPa and $\varepsilon_{t,f} = 0.47$. Knowing from Eq. (8.50) the cyclic strain hardening exponent $n' = 0.0866$ it is possible to derive the b and c exponents of Eq. (8.10) that are

$$b = -\frac{n'}{(1 + 5n')} = -0.060; \; c = -\frac{1}{(1 + 5n')} = -0.6978. \quad (8.52)$$

Knowing $\sigma_{t,f}$, $\varepsilon_{t,f}$, n', b and c it is possible to use Eq. (6.10) and draw both the elastic and plastic component of the *S–N* curve of the material and the global *S–N* curve. This is shown in Fig. 8.42. It is interesting to note that experimental results obtained with smooth specimens (open circles) are perfectly aligned on the *S–N* curve. Let's consider now the notched specimens. Notch factors k_f can be inferred from Fig. 7.4 that yields a notch sensitity factor $q = 0.55$, with this last value Eq. (7. 5) yields $k_f = 2.10$ lower than $k_t = 3$, as expected.

Notch factor can be also inferred from Fig. 7.5 that yields a Neuber's constant $\sqrt{a_N} = 0.45$ that entering Eq. (7.12) yields $q = 0.54$. Using $k_f = 2.10$ the constant term of Neuber Eq. (6.31) gives

$$\frac{\left(k_f \sigma_a\right)^2}{E} = \frac{(2.10)^2 \sigma_a^2}{190,780} = 0.000023 \cdot \sigma_a^2 = \sigma_{a,notch} \varepsilon_{a,notch} \quad (8.53)$$

190,780 MPa being the Young's modulus E at 288 °C. Entering Eq. (8.52) with the six applied stress amplitudes $\boldsymbol{\sigma_a}$ listed in Table 8.14 it is possible to draw the six

Neuber's hyperbolas shown in Fig. 6.43. The intersections A, B, C, D, E and F with the cyclic stress–strain curve yields the values of the parameters $\sigma_{a,notch}$ and $\varepsilon_{a,notch}$ we wanted to know. Those values are listed in Table 6.14. The corresponding data points are shown in Fig. 8.42 (open triangles). The data falls over the *S–N* curve of the material for at least two reasons. First, Eq. (8.52) does not consider triaxiality that for the notch-specimen geometry is high and, according to Eq. (8.43), equal to two. Second, calculations do not consider the process volume effect that for the small stress concentration volume at notch root is much smaller than in the smooth specimen. As to the former, Eq. (8.37) can be used to correct the computed notch strain amplitude $\varepsilon_{a,notch}$. It must be recalled that in the plastic domain deformations (slips) occur without any volume change ΔV and is

$$\frac{\Delta V}{V_o} = \Delta\varepsilon_x + \Delta\varepsilon_y + \Delta\varepsilon_z = 0 \tag{8.54}$$

where V_o is the initial volume and $\Delta\varepsilon_x$, $\Delta\varepsilon_y$ and $\Delta\varepsilon_y$ are the strain variation following specimen deformation. To satisfy Eq. (8.53) the Poisson's coefficient ν must be necessarily equal to 0.5. The notch strain amplitude corrected for triaxiality are listed in Table 8.14 as $\varepsilon'_{a,notch}$ and shown in Fig. 8.42 with gray triangles. As to the much more important volume effect, the process volume associated with the notch can be evaluated considering a notch tip annulus of mean radius $R = 4.35$ mm and $r = \sqrt{a_N} = 0.45$. The corresponding process volume is

$$V_{notch} = 1.76 \text{ mm}^3$$

The process volume of the unnotched specimen used to derive the *S–N* curve of Fig. 8.42 is

$$V_{spcm} = 0.5 \cdot \pi \cdot 5.35/2 \cdot 15.24 = 128 \text{ mm}^3.$$

Therefore the notch strain amplitude $\varepsilon''_{a,notch}$ corrected for the process volume will be given by Eq. (4.48).

$$\varepsilon''_{a,notch} = \varepsilon_{a,notch}{}' \left(\frac{V_{notch}}{V_{spcm}}\right)^{1/m} \tag{8.55}$$

where m is the Weibull exponent of the material that is not constant, but depend on the process volume itself the in turn depends on the notch radius ρ. Using Eq. (4.49) with a notch radius $\rho = 0.28$ mm it is $m = 15.27$. Entering Eq. (8.54) with $m = 15.27$ it is possible to assess the values of $\varepsilon''_{a,notch}$ corrected for the process volume effect. These values are listed in Table 8.14 and shown in Fig. 8.42 with black triangles. It can be seen that these last values of the strain amplitude $\varepsilon''_{a,notch}$ are perfectly aligned on the theoretical *S–N* curve of the material and consistent with unnotched specimen data, as it should be.

8.7.8 Neuber's Rule—Cycle Reversal

A panel of carbon steel 10 mm thick, 15 mm wide carrying a center hole of 5 mm diameter is subjected to fatigue cycles from 0 to $F_{\max} = 28{,}000$ N. the steel is that already considered in sample problem 4. Assess the strain and stress amplitude of cycling. The cyclic stress–strain characteristic of the material is that of Fig. 6.35 having Eq. (8.49). The stress amplitude σ_a of the cycle is

$$\sigma_\alpha = F/A = 28,000/10 \cdot 15 = 186.66\ \text{MPa}.$$

Entering diagram of Fig. 7.4 with the steel property it is found that the notch sensitivity factor is $q = 0.75$ that renders a notch factor $k_f = 2.86$. With these data the Neuber equation is

$$\frac{\left(k_f\sigma_a\right)^2}{E} = \frac{(2.86 \cdot 186.66)^2}{208{,}333} = 1.368 = \sigma \cdot \varepsilon.$$

As suggested by Walker [44], it has been used the k_f and the cyclic curve of the material. The Neuber hyperbola for the loading phase is shown in Fig. 8.44. The intersection with the cyclic curve yields the working point A relative to the loading phase whose coordinates are

$$\varepsilon = 0.0039;\ \sigma = 350\ \text{MPa}$$

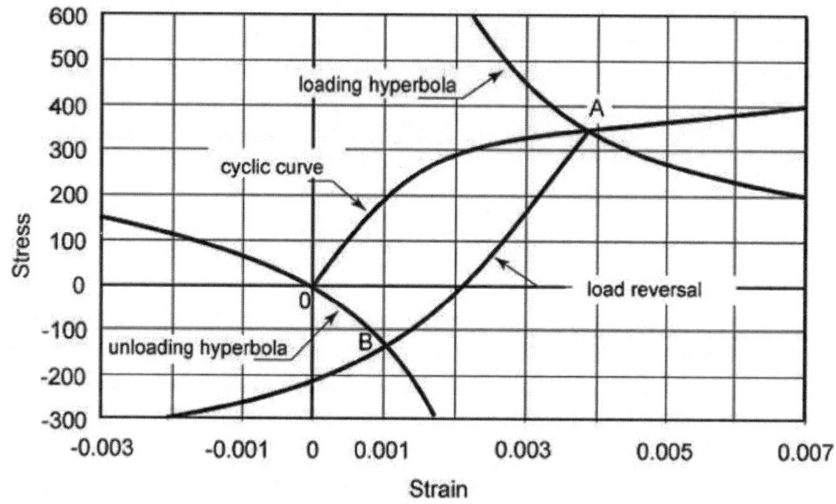

Fig. 8.44 Loading and unloading Neuber hyperbolas $\varepsilon = 0.0013;\ \sigma = -134$ MPa

Table 8.15 Load reversal. Stresses and strains corresponding to each couple of $\Delta\sigma - \Delta\varepsilon$

$\Delta\sigma$	$\Delta\varepsilon$	ε	σ
350	0.003909	−8.8E−06	0
370	0.003697	0.000203	−20
390	0.003508	0.000392	−40
410	0.003337	0.000563	−60
430	0.003182	0.000718	−80
450	0.00304	0.00086	−100
470	0.002911	0.000989	−120
490	0.002792	0.001108	−140
510	0.002682	0.001218	−160
530	0.002581	0.001319	−180
550	0.002487	0.001413	−200
570	0.0024	0.0015	−220
590	0.002319	0.001581	−240
610	0.002243	0.001657	−260

Note that the stress $\sigma = 350$ MPa can be also obtained solving Eq. (8.33). Moving to the unloading phase, Eq. (8.35) will be used. The value of the Neuber's constant does not change since $\Delta\sigma = \sigma_a = 186.66$ MPa. The new origin of coordinates is point A. the Neuber hyperbola for the load reversal is drawn using the first of Eq. (8.35). The hyperbola is shown in Fig. 8.44. For each $\Delta\sigma$ chosen the corresponding $\Delta\varepsilon$ is assessed using the first of Eq. (8.35). Results are listed in Table 8.15. From $\Delta\sigma$ and $\Delta\varepsilon$ the values of the corresponding σ and ε are computed subtracting $\Delta\varepsilon$ from the initial strain ε of point A. The intersection of the unloading hyperbola with the hysteresis loop yields the wanted working point B of coordinates.

Therefore, the stress excursion is

$$\Delta\sigma = 350 - (-134) = 484 \text{ MPa}$$

and the stress amplitude is

$$\sigma_a = 484/2 = 242 \text{ MPa}$$

References

1. Dowling, N.E., Brose, W.R., Wilson, W.K.: A discussion of local strain approach to notched member fatigue life prediction. Westinghouse Scientific Paper 76-1E7-PALFA-P1 (1976)
2. Tucker, L.E., Landgraf, R.W., Brose, W.R.: SAE Report 740279, Automotive Engineering Congress (1974)

3. Hickerson, J.P., Hertzberg, R.W.: Metall. Trans. **3**, 179 (1972)
4. Masing, G.: Eigenspannungen und Verfestigung Brim Messing. In: Proceedings of 2nd Congress on Applied Mechanics, Zurich, pp. 332–335 (1926)
5. Morrow, J.D.: Internal Friction, Damping and Cyclic Plasticity. American Society for Testing and Materials, ASTM STP-378, pp. 45–87 (1965)
6. Landgraf, R.W., Morrow, J.D., Endo, T.: Determination of the cyclic stress strain curve. J. Mater. **4**(1), 176–188 (1969)
7. Martin, J.F., Topper, T.H., Sinclair, G.M.: Computer based simulation of cyclic stress-strain behavior with applications to fatigue. Mater. Res. Standards **11**(2), 23–29 (1971)
8. Wetzel, R.M.: A Method of Fatigue Damping Analysis, Ph.D. Thesis, Dept. of civil Eng., University of Waterloo, Ontario, Canada, 1971 and Technical report No. SR 71–107, Scientific Research Staff, Ford Motor Co., Dearborn, Michigan (1971)
9. Basquin, O.H.: Proceedings ASTM, vol. 10, Part II, p. 625 (1910)
10. Coffin, L.F., Jr.: Transactions ASME, vol. 76, p. 931 (1954)
11. Tavernelli, J.F., Coffin, L.F., Jr.: Transactions ASM, vol. 51, p. 438 (1959)
12. Smith, R.W., Hirschberg, M.H., Manson, S.S.: NASA TN D-1574, NASA, April (1963)
13. Manson, S.S., Hirschberg, M.H.: Fatigue: An Interdisciplinary Approach, Syracuse University Press, Syracuse, New York, p. 133 (1964)
14. Landgraf, R.W., Morrow, J.D.: Achievement of High Fatigue Resistance in Metals and Alloys, ASTM STP-467, p. 3 (1970)
15. Tucker, L., Bussa, S.: The SAE Cumulative Fatigue Damage Test Program, Society of Automotive Engineers, Paper no. 750038, Automotive Engineering Congress exposition, Detroit, February (1965)
16. Milella, P.P., Pelilli, G. Traficante, M.: Comportamento a Fatica di un Acciaio AISI 316, (in Italian), ENEA, Rapporto QE00-1ET4B-84004-TBA, Roma, Aprile (1984)
17. Landgraf, R.W.: High Fatigue Resistance in Metals and Alloys, American Society for Testing and Materials, ASTM STP 776, pp. 33–43 (1982)
18. Terrell, J.B., Cullen, W.H.: Fatigue Life Response of ASME SA 106 B Steel in Pressurized Water Reactor Environments. In: Proceedings of the Third IAEA Specialist's Meeting on Subcritical Crack Growth, Moscow, USNRC NUREG/CP-0112 (1990)
19. Morrow, J.D.: Cyclic Plastic Strain Energy and Fatigue of Metals. International Friction, Damping and Cyclic Plasticity, ASTM, West Conshohocken, PA, pp. 45–86 (1965)
20. Morrow, J.D.: Fatigue Design Handbook, section 3.2, SAE Advances in Engineering, vol. 4, pp. 21–29 (1968)
21. Manson, S.S., Halford, G.R.: Practical Implementation of the Double Linear Damage Rule and Damage Curve Approach for Treating Cumulative Fatigue Damage. Int. J. Fracture **17**(2), 169–172, R35–R42 (1981)
22. Walcher, J., Gary, D., Manson, S.S.: Aspects of Cumulative Fatigue Damage Analysis of Cold End Rotating Structures. AIAA Paper 79–1190 (1979)
23. Smith, K.N., Watson, P., Topper, T.H., A Stress-Strain Function for the Fatigue of Metals. J. Mater. **5**(4), 767–778 (1970)
24. Fash, J.W., Socie, D.F.: Fatigue Behavior and Mean Effects in Gray Cast Iron. Int. J. Fatigue **4**(3), 137–142 (1982)
25. Koh, S.K., Stephens, R.I.: Mean stress Effects on Low Cycle Fatigue for High strength Steels. Fatigue Fract. Eng. Mater. Struct. **14**(4), 413–428 (1991)
26. Wener, T., Fatemi, A.: Effect of Mean Stress on Fatigue Behavior of Hardened Carbon Steel. Int. J. Fatigue **13**(3), 241–248 (1991)
27. Forsetti, P., and Blasarin, A.: Fatigue Behavior of Microalloyed Steels for Hot forged Mechanical Componentes. Int. J. Fatigue **10**(3), 153–161 (1988)
28. Mechanical Behavior of Materials, 2nd Edition by Dowling, Pearson Education, Inc. (1998)
29. Stadnick, S.J., Morrow, J.: Technique for Smooth specimen Simulation of the Fatigue Behavior of Notched Members. ASTM-STP 515, pp. 229–252 (1972)
30. Crews, J.H. Jr.: Crack Initiation at Stress Concentration as Influenced by Prior Local Plasticity. Achievement of High Fatigue Resistance in Metals and Alloys. ASTM-STP 467, p. 37 (1970)

31. Neuber, H.: Theory of Notch Stresses-Principles for Exact Stress Calculation. Julius Spring, Berlin, 1937. Translated and published by J.W. Edwards, Publishers, Incorporated, Ann Arbor, Mi (1946)
32. Neuber, H.: Theory of Notch Stresses: Principles for Exact Calculation with Reference to Structural Form and Strength. 2nd Edn. Springer-Verlag, Berlin, 1958. Translated and issued as AEC-TR-4547 by the US Office of Technical Information (1961)
33. Wetzel, R.M.: Smooth specimen Simulation of the Fatigue Behavior of Notches. J. Mater. **3**(3), 646–657 (1968)
34. Topper, T.H., Wetzel, R.M., Morrow, J.: Neuber Rule Applied to Fatigue of Notched Specimens. J. Mater. **4**(1), 200–209 (1969)
35. Lüdwick, P.Z., Ver. Dent. Ing (1926)
36. Manjoine, M.J.: Multiaxial Stress and Fracture. Fracture, vol. 3, Eng. Fund. and Environm. Effects, New York, Academic Press (1970)
37. Davis, E.A., Connelly, F.M.: Stress Distribution and Plastic Deformation in Rotating Cylinders of Strain-hardening materials. ASME, J. Appl. Mech. 25–30 (1959)
38. Milella, P.P.: Meccanica della Frattura Lineare Elastica ed Elastoplastica (in Italian). Ansaldo Nucleare Edithor, p. 71 (1999)
39. Milella, P.P., Bonora, N.: On the Dependence of the Weibull Exponent on Geometry and Loading Conditions and Its Implications on the Fracture Toughness Probability Curve Using a Local Approach Criterion. Int. J. Fracture, Kluwer Acad. Publishers, vol. 104, pp. 71–87 (2000)
40. Wilson, W.K.: Elastic-Plastic Analysis of Blunt Notched CT Specimens and Applications. J. Press Vessel Technol. **96**(4), 293–298 (1974)
41. Huang, W.C.: Theoretical Study of Stress Concentrations of Circular Holes and Inclusion in Strain Hardening Materials: Int. J. Solids Struct. **8**(2), 135–136, July (1972)
42. Saanouni, K., Bathias, C.: Study of Fatigue Crack Initiation in the Vicinity of Notches. Eng. Fract. Mech. **16**(5), 695–706 (1982)
43. Leis, B.N., Gowda, C.V.B. and Topper, T.H.: Cyclic Inelastic Deformation and the fatigue Notch Factor. Cyclic Stress-Strain Behavior—Analysis, Experimentation and Failure prediction. American Society for Testing and Materials, ASTM STP 519
44. Walker, E.K.: Multiaxial Stress-Strain Approximation for Notch Fatigue Behavior. J. Test. Eval. **5**(2), 106–113, March (1977)
45. Maiya, P.S.: Effects of Notches on Crack Initiation in Low Cycle Fatigue. Mater. Sci. Eng. **38**, 289–294 (1979)
46. Lowrence, F.V., Burk, J.V., Jung, J.Y.: Influence of Residual stress on the Predicted Fatigue Life of Weldments. ASTM STP 776 Residual Stress in Fatigue. West Conshohocken, PA, pp. 33–43 (1982)
47. Reemsnayder, H.S.: Evaluating the Effects of Residual Stresses of Unnotched Fatigue Resistance. Society of Environmental Engineers, Buntinford, U.K (1981)
48. Seeger, T., Heuler, P.: Generalized Application of Neber's Rule. J. Test. Eval. **8**(4), 199–204, July (1980)

Chapter 9
Very High Cycle Fatigue

9.1 Fatigue Limit in VHCF

The traditional subdivision of the Wöhler curve into the three regions of Fig. 1.10, at least for ferrous alloys or, better, for BCC crystal lattice materials has been recently extended to include a new region beyond some 10^7 or 10^8 cycles. It is the region of the so called Very-High-Cycle-Fatigue or VHCF. The more demanding request for life from many technological applications has been the driving force to explore, also for BCC materials, the field of 10^8 or even 10^9 cycles. The pressure to know how these materials perform in the VHCF region, whether or not they really have a definite fatigue limit, has unveiled a new as well unexpected behavior that still remains argument of vigorous discussion among researchers. It is a fact that vast majority of the data so far collected since Wöhler first tests with fatigue specimens, are confined within 10^7–10^8 cycles.

This is not only because, at least initially, most of the service requirements were not exceeding that limit, but also because of the technological boundary to the number of cycles, N, posed by the fatigue testing machines at the time available. Indeed, very many applications in the aircraft, automobile, railway, and other to-day industries exceeds 10^6 cycles where the conventional fatigue limit has been set. Typical are gas turbine disks, $N > 10^{10}$, car engine piston heads and blocks, $N > 10^8$, ball bearings, high frequency drilling machines, diesel engines of ships and high-speed trains, $N > 10^9$, etc. Even a bridge like the Morandi Bridge in Genova (Italy) exceeded 10^9 operational cycles in 50 years of service. Table 9.1 clearly indicates the unbearable waiting time needed to reach 10^9 cycles with conventional testing machines working at 10 Hz or even 100 Hz. Though it was already in 1950 that Mason introduced piezo-electric ultrasonic transducers that transformed 20 kHz electrical signals into mechanical vibrations of the same frequency, we had to wait more than 50 years to see ultrasonic resonant fatigue machines to be fully operative. With this new

P. P. Milella, *Fatigue and Corrosion in Metals*,
https://doi.org/10.1007/978-3-031-51350-3_9

type of fatigue testing machines working at 20 kHz, Fig. 10.6, the field of the so-called Very High Cycle Fatigue (VHCF) regime has become available to thoroughly investigation.

Almost all data so far available on VHCF seem to converge towards a common understanding that the fatigue limit conventionally posed at 10^6 cycles does not exists for any metal alloy, even for carbon steels and, in particular, for high strength steels. This new finding is particularly emotional for an old nuclear engineer like me who, some fifty years ago, used to work at Westinghouse Nuclear in Monroeville, Pittsburgh, USA, where senior engineers already came to the conclusion that the fatigue limit did not exist for any alloy. These new evidence appeared for the first time in 1984 and were reported by Japanese investigators [1, 2]. They would suggest that also for BCC materials a real fatigue limit may not exist, as for non-ferrous alloy or FCC materials, as shown in Fig. 1.11. However, as said, doubts still remains among researchers [3]. Figure 9.1 is an example of a possible double interpretation of facts. It shows the fatigue behavior of a low carbon ferritic steel tested at room temperature and R = 0.1. The trend is not really clear, and it can be argued that the apparent decrease in the fatigue life beyond 10^6 cycles is confined within a 5% to 10% that can be attributed to normal fatigue scatter results.

The disappearance of the fatigue limit, long believed to exist for high strength materials, is today well documented, yet a question arises as to whether or not they still have a BCC crystal structure or, somehow, they retained a somehow large percentage of FCC structure. A rather comprehensive review of this issue can be found in Ref. [4]. Figure 9.2 presents the fatigue results obtained by Sakai et al. [5] for high-carbon-chromium bearing steel tested under rotating bending using a 20 kHz ultra-sonic systems.

If the test program were run up to 10^7 cycles, a fatigue limit σ_f equal to 1265 MPa would be found, as expected. However, continuing to run tests much beyond 10^7 cycles a new descending trend was discovered after a plateau, which seems to propose a new fatigue limit, $\sigma_f(VHCF)$, equal to about 970 MPa at about 10^9 cycles. Whether or not this last is the final fatigue limit, there has already been a 25% reduction of the former limit. Wang et al. [6] investigated two types of ultra-high strength spring steel wires used in the automotive industry: a Cr-V steel of valve spring quality (ASTM A232) and a Cr-Si steel VSQ wire (ASTM A877) used for the same applications.

The tests were run using a 20 kHz ultra-sonic systems in traction with $R = -1$. The Cr-V spring steel effectively showed a fatigue limit equal to about 800 MPa at

Table 9.1 Waiting time needed to reach 10^9 cycles with conventional, resonance and ultrasonic testing machines

Number of cycles	Conventional tests (10 Hz)	Resonance tests (100 Hz)	Ultrasonic tests (20 kHz)
10^7	~10 days	~1 day	~10 min
10^8	~3 years	4 months	14 h
10^9	~30 years	3 years	6 days

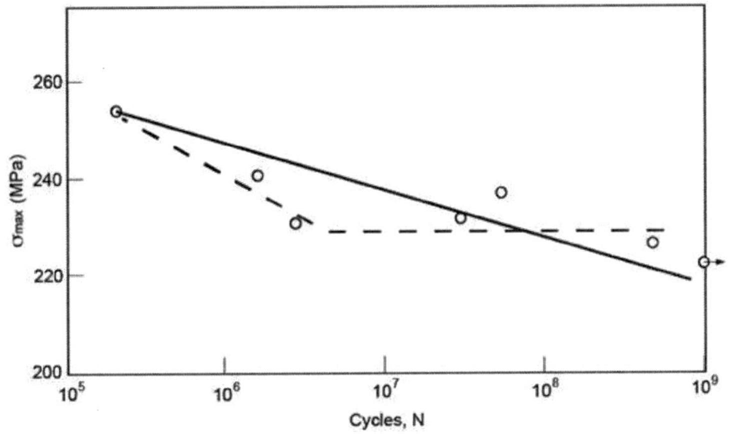

Fig. 9.1 VHCF behavior of low carbon sheet steel obtained R = 0.1 and 20 kHz. Question arises as to whether or not there is a fatigue limit (adapted from [1])

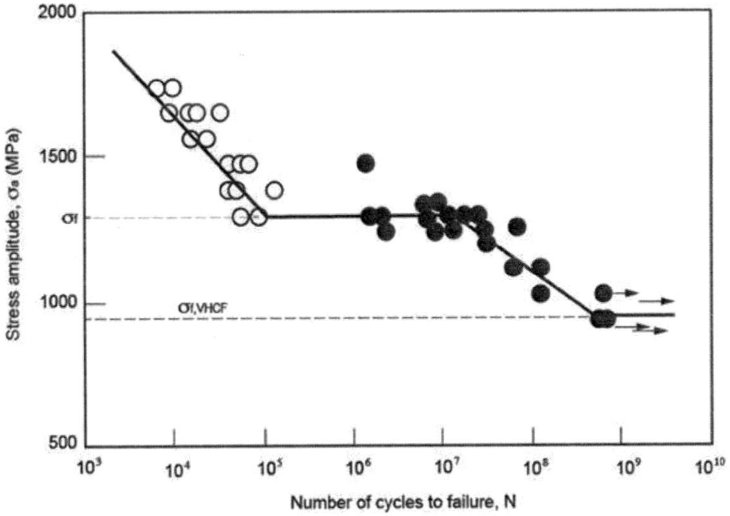

Fig. 9.2 VHCF behavior of high-carbon-chromium bearing steel tested under rotating bending at 20 kHz (adapted from [5])

about 10^6 cycles that did not change up to about 10^{10} cycles, as shown in Fig. 9.3. Conversely, the Cr-Si spring steel presented a continuously descending trend from 10^5 to 10^9 cycles, as shown in Fig. 9.3 [6]. Again, a reduction of about 27% was found in the fatigue limit at VHCF. Why this difference in material behavior exists between the two high strength steels? May the difference be attributed to a different crystal structure? Figure 9.4 is an example of VHCF behavior of a rail steel 700

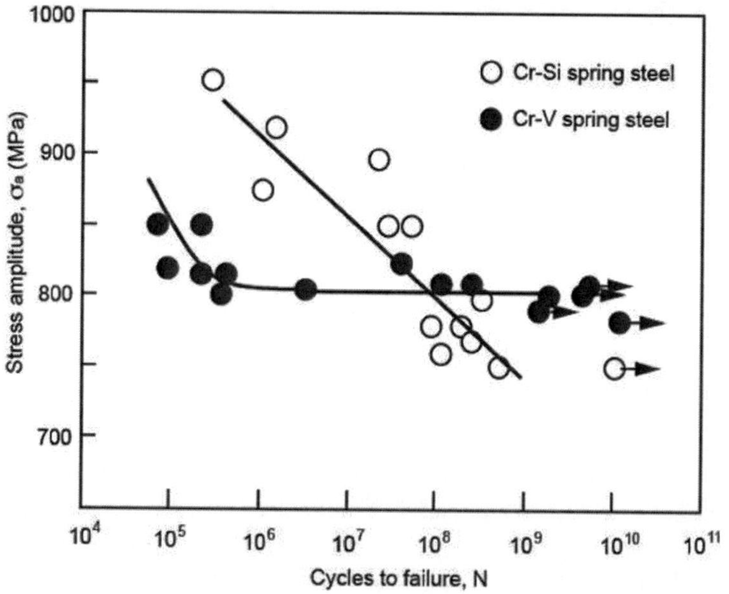

Fig. 9.3 VHCF behavior of two ultra-high strength spring steel wires tested at 20 kHz (adapted from [6])

obtained at 20 kHz, showing more than 25% decrease of the fatigue strength passing from 10^6 to 10^{10} cycles [7].

There is no apparent fatigue limit. However, Zettl, Mayer and Stanzl-Tschegg [8] reported that steels CK60, CK15 and 100CR6 did show a fatigue limit in the VHCF regime when tested in the normalized condition, Fig. 9.5. All these materials have a body-centered-cubic structure (BCC). High strength steel wires type Cr-Si VSQ (ASTM A877) used in the automotive industry, investigated by Wang et al. [6] yielded the continuous downward trend beyond 10^6 cycles, as shown in Fig. 9.6. This time, the loss of fatigue strength is of the order of 20%. The same 20% fatigue strength decrease is seen in a 17-4PH martensitic stainless steel at 20 kHz, RT and $R = -1$, as shown in Fig. 9.7 [3]. Haibach [9] tried to give an answer to the double behavior of high strength martensitic steels arguing that the decrease in the fatigue strength in the VHCF regime could be due to the low tempering temperature, below 200 °C, that resulted in a large amount of retained austenite.

A higher tempering temperature, over 200 °C, would cancel austenite and transform tetragonal martensite into BCC martensite and change the behavior of the steel reintroducing the fatigue limit at 10^6 cycles, typical of BCC steels. He made reference to a paper that was presented by Oguma et al. [10] relative to VHCF tests on bearing steel 100Cr6 (JIS SUJ2), quenched and tempered at three different temperatures: 180 °C, 220 °C and 280 °C, respectively.

The steel tempered at 180 °C is clearing presenting the downward trend of the fatigue strength in the VHCF regime, Fig. 9.8a [10] while the same steel tempered

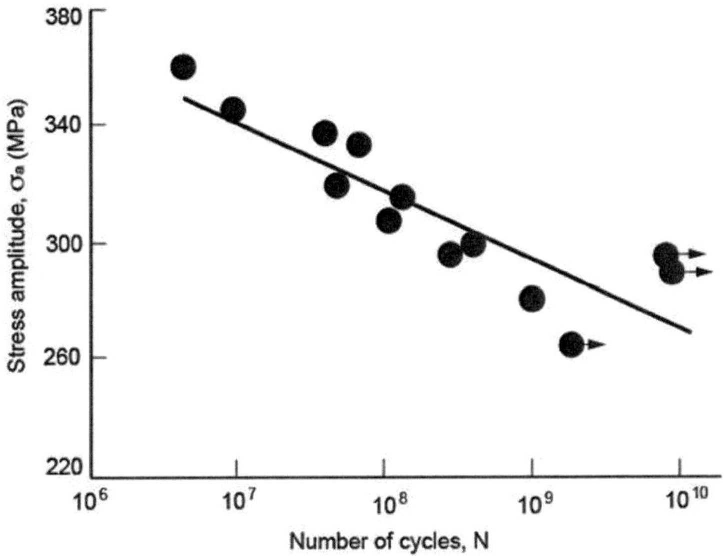

Fig. 9.4 VHCF behavior of rail steel 700 tested at 20 kHz (adapted from [7])

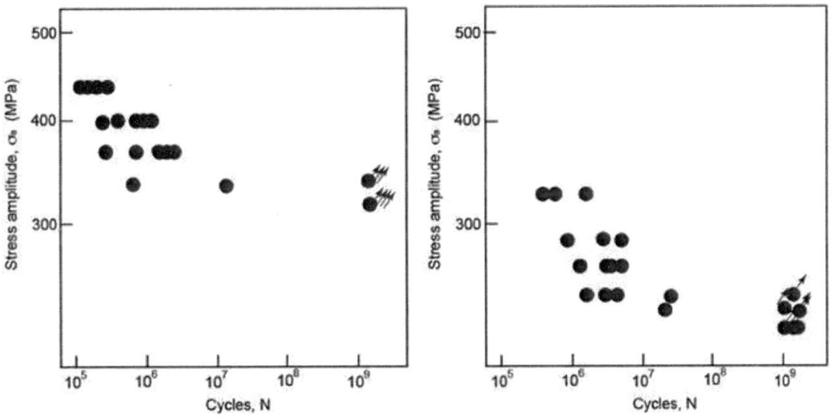

Fig. 9.5 VHCF behavior of carbon steel: **a** CK60 and **b** CK15 in the normalized condition (adapted from [8])

at 280 °C, which has a BCC martensite, shows the behavior typical of BCC carbon steels with a fatigue limit at 10^6 cycles, Fig. 9.8b [10].

According to these results, Haibach [9] argued, non-metallic inclusions cannot be the very reason of fatigue failures in the VHCF regime, because the inclusions must have been present in the same size and distribution in the specimen tempered at 180 °C or 260 °C. There is no doubt those inclusion could not be removed, nor their

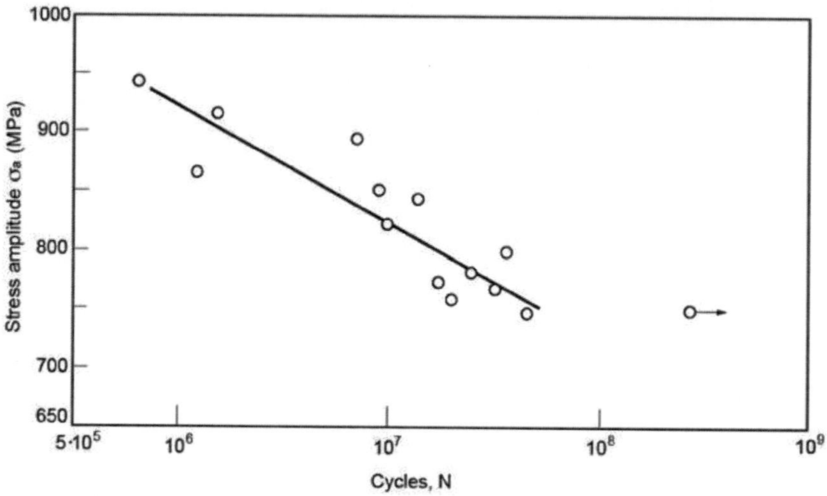

Fig. 9.6 VHCF behavior of Cr-Si spring steel at 20 kHz, RT and R = − 1 (adapted from [6])

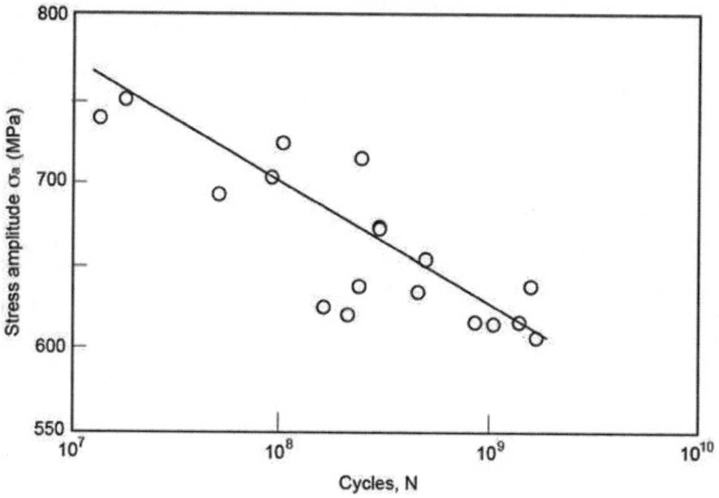

Fig. 9.7 VHCF behavior of 17-4PH martensitic stainless steel at 20 kHz, RT and R = − 1 (adapted from [3])

effect be eliminated by simply increasing the tempering temperature. As to titanium and titanium alloys that have a CPH structure below 800 °C (a-Titanium) and a BCC structure above 800 °C (b-Titanium), they do show a decisive downward trend in the fatigue limit beyond 10^6 cycles. This can be seen in Fig. 9.9 [7].

As to FCC materials, like austenitic stainless steel, aluminum alloys, nickel alloys, copper etc., it was already pointed out, in Sect. 1.2 and schematically shown in

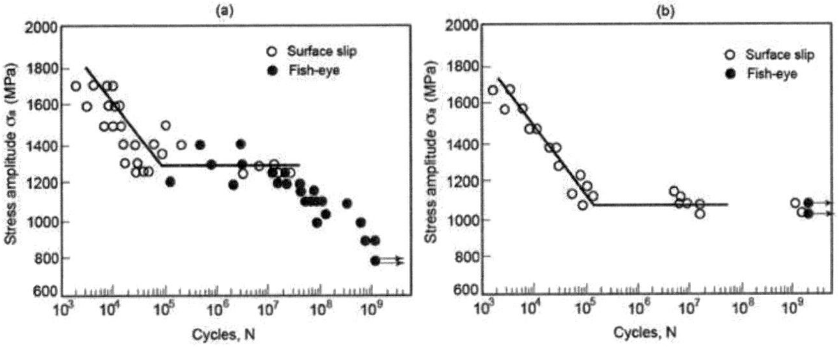

Fig. 9.8 VHCF behavior of 100CR6 steel. **a** quenched and tempered at 180 °C and **b** quenched and tempered at 260 °C (adapted from [10])

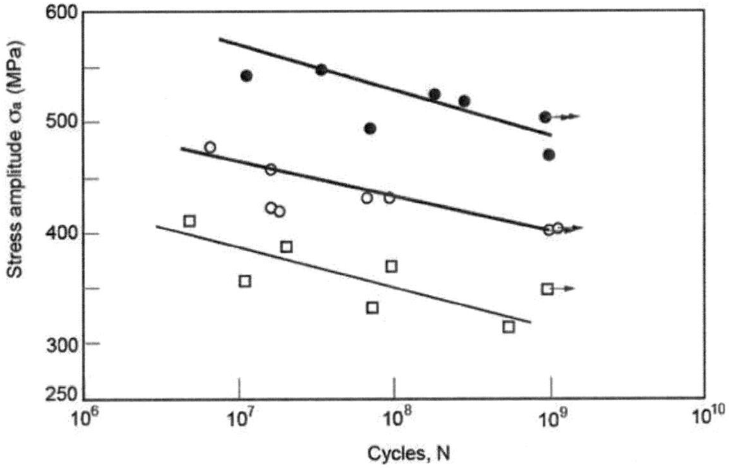

Fig. 9.9 VHCF behavior of Ti6246 alloy that received three different heat treatment, tested at 20 kHz, RT and R = − 1 (adapted from [7])

Fig. 1.11, that they do not present a fatigue limit at 10^6 cycles. For these materials a fatigue limit is assumed at 10^8 cycles as a conventional reference limit. An example of such a behavior is given in Fig. 9.10 [11] for 2024-T351 aluminum alloy tested at 20 kHz under fully reversed loading conditions, R = − 1, in ambient air at room temperature. It clearly indicates that no fatigue limit actually exists for this alloy. A possible explanation of that behavior could be, that in a material with a face centered cubic structure (FCC) normally multiple and much more slip lines can be activated, compared to those existing in materials with a body centered cubic lattice (BCC) and these slips will always activate and eventually give rise to a microscopic-turning-into-macroscopic damage, see Sects. 2.1 and 2.3.

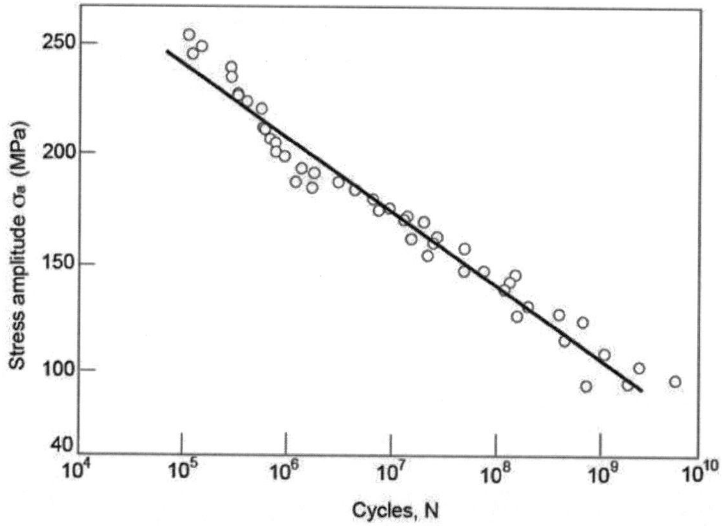

Fig. 9.10 VHCF behavior of 2024-T351 aluminum alloy, tested at 20 kHz, RT and R = − 1 (adapted from [11])

9.2 VHCF Stepwise S–N Curve. The Granular Bright Facet (GBF). Fish-Eye Formation

Sakai et al. [5] have studied the VHCF properties of high-carbon-chromium bearing steel JIS:SUJ2. The specimens were fatigue tested in push–pull axial loading, with load ratio R = − 1, at frequency of 50 Hz. Specimens were heated at 1108 K for 40 min and oil quenched then were tempered at 453 K (180 °C) for 120 min and air cooled. Results are shown in Fig. 9.11 [5]. The characteristic feature emerging from the VHCF testing of this high-carbon-chromium bearing steel is the so-called stepwise S–N curve or duplex S–N curve. The axial fatigue loading,

Figure 9.11a shows a light fatigue strength decrease already at 10^7 cycles that is more pronounced under rotating bending conditions, Fig. 9.11b, where a lower shelf appears from 10^5 to 10^7 cycles, followed by a further decrease in the fatigue strength.

This particular feature that is less pronounced in low strength carbon steel and more in high strength steel, has been referred to as *multi stage* fatigue behavior by Mughrabi [12] who has proposed the stepwise diagram of Fig. 9.12. The slight difference between the curves, Fig. 9.12a and b, reflects two viewpoints as to whether the fatigue limit in the VHCF range exists, Fig. 9.12a, or not, Fig. 9.12b. Broad experimental evidence suggests that in the absence of stress raises on the surface, like notch or rough finishing, with increasing number of cycles there is a gradual shift from fatigue failures initiating from the surface towards those starting at any internal material defects leading to the development of the so-called "fish-eye" on the fracture surface. In the time dependent part of the fatigue curve, indicated as line

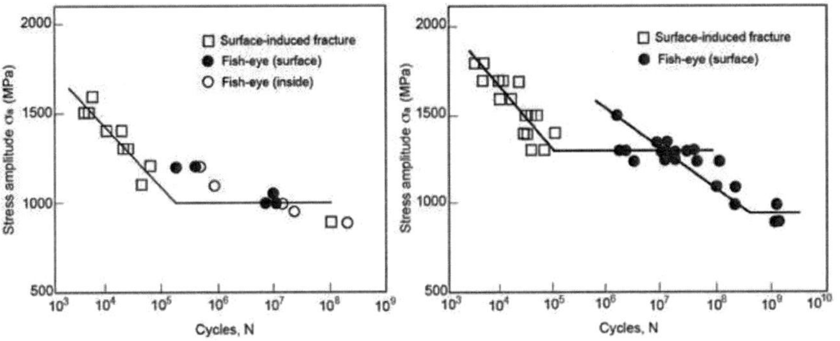

Fig. 9.11 VHCF behavior of JIS:SUJ2 bearing steel: **a** axial loading, **b** rotating bending (adapted from [5])

I in Fig. 9.12, that precedes the first lower shelf, curve II of Fig. 9.12, we know that fatigue failures always initiate on the surface by local plastic slips, as explained in Sect. 2.1. It is within this persistent slip bands or PSBs that the first microdamage occurs that develops into a macroscopic crack that leads to fatigue fracture.

Below the fatigue limit, line II of Fig. 9.12, these PSBs may exist but, apparently, they do not develop any macrocrack: the damage, dough existing does not progress to failure. However, it may be just a matter of time, which means cycles. It may seem that the applied stress amplitude is not large enough to propagate damage but the effects of cycling without interruption continue to be felt by the material that suffers, anyhow, and eventually fails. The sum of extremely small damage per cycle, apparently undetectable, turns into a finite quantity when the number of cycles becomes extremely large. There seems to be also evidences of this lagging action.

Consider the Armco steel of Fig. 2.22. We have seen that below 152 MPa, which is the fatigue limit, PSBs may form, but do not propagate. To this end, the sequence

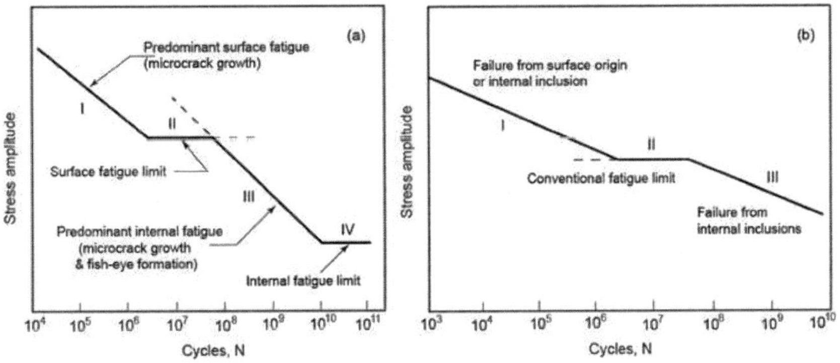

Fig. 9.12 Schematic illustration of stepwise S–N curve with (**a**) and without (**b**) fatigue limit in the VHCF range (from [12])

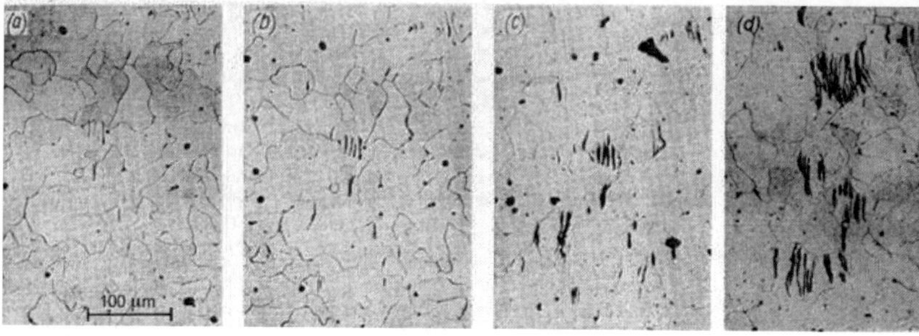

Fig. 9.13 Slip bands in Armco iron below the fatigue limit after: **a** $N = 6{\cdot}10^4$, **b** $N = 6{\cdot}10^5$, **c** $N = 6{\cdot}10^6$ and **d** $N = 6{\cdot}10^7$ cycles (adapted from [13])

of Fig. 9.13 [13] results extremely interesting. At about 127 MPa, which is well below the fatigue limit, after $6{\cdot}10^4$ cycles there is almost no evidence at all of any damage, Fig. 9.13a. Just a single grain seems to have developed segmentary slips. At $N = 6{\cdot}10^5$ a first sign appears that those segmentary slips became wider, Fig. 9.13b. Continuing to cycle, at $N = 6{\cdot}10^6$ the signs of slips are heavier. They seem to have broken the grain boundary and appeared in neighboring grains, Fig. 9.13c. Nobody would think of a possible degeneration of these innocuous slips. But just look what happens after ten more millions of cycles have been put into the material at the same stress amplitude, Fig. 9.13d. Some PSBs joined and some other appeared in new grains. Note that the overall length, if slips would join, is in the range of 300 μm: a macro-crack.

It is very possible that at $N = 6{\cdot}10^8$ the specimen would fail, but, unfortunately, we haven't got any data to either support or reject that hypothesis: the test was stopped.

In high and very high strength steels the resistance to plastic slip may be so high that it is easier for microcrack to develop from inclusions close to the surface or, as the number of cycles continues to grow, line III of Fig. 9.12, from more and more deep internal inclusions. Muzzoli [14], in Italy, was, probably, the first to report such type of failure mode and origin in very hard steels. The origin of fatigue failure appeared as a white circular area around a central dark spot, arriving to the surface and turning into a semi-circle where the fatigue crack growth rate sharply increased, Stage II of fatigue, see Sect. 2.7 and Fig. 2.47, with subsequent static fast fracture, Stage III of fatigue, as shown in Fig. 9.14 [14]. In 1955, Frifh [15] also found the same white circle, identifying an internal inclusion as the initiation site. Both Muzzoli and Frith used the term "white area" to identify the failure origin. This whitish circular area visible on the fracture surface represents the path of a crack growing radially outwards from an inclusion at its center. The white brilliant appearance is due to the fact that until the crack has grown to reach the surface of the specimen, air cannot enter and the oxidation of the faces is prevented. Once arrived at the surface, air is allowed to enter the flaw oxidizing the fatigue area, burnishing and producing the well-known gray appearance.

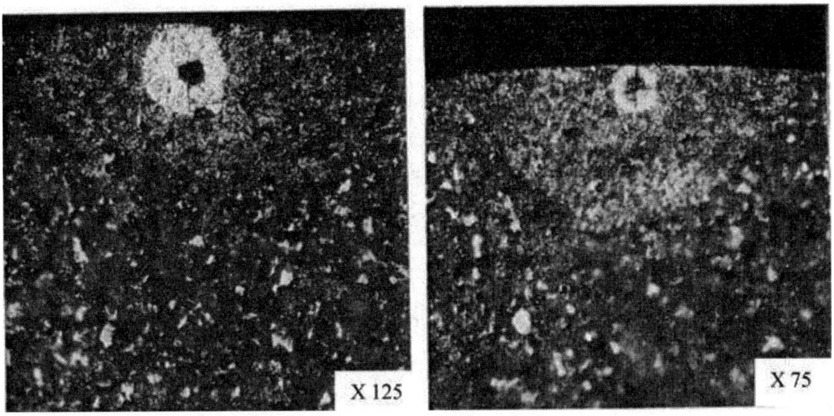

Fig. 9.14 Fatigue origin in very hard steel specimens as it appear at the optic microscope [14]

Recently, Shiozawa et al. [16] conducted a study of subsurface crack initiation and propagation mechanisms in specimens made from high carbon chromium bearing steel subjected to VHCF. They found, when observed by SEM, a white-bright facet area in the vicinity of the fracture origin, always an inclusion. It had a very rough and granular morphology. Because of its white appearance, the area was called "granular bright facet" (GBF). To-day, it is called "fish-eye" failure. It is during the formation of the GBF that the largest portion of fatigue life is consumed. Detailed analysis of the morphology of the GBF area using an electron probe microanalysis (EPMA), revealed the presence of a high concentration of carbon. In addition, the fracture surface roughness was measured and found that it was about 2.5 times greater than that of the outside area. At the very beginning of the fatigue process multiple micro-cracks are initiated by decohesion of spherical carbides from the matrix in the vicinity of the non-metallic inclusion, Step I of Fig. 9.15a [17].

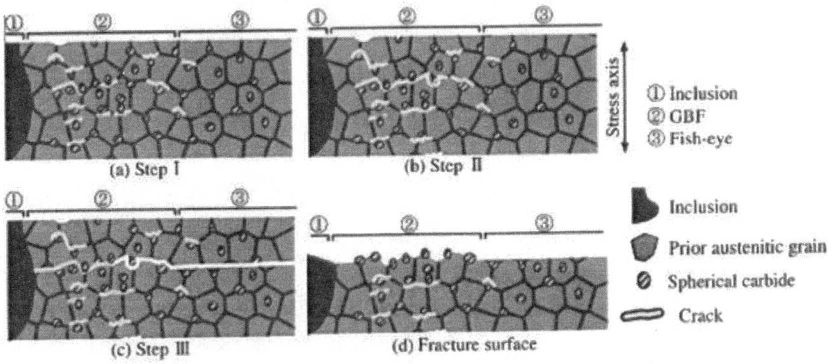

Fig. 9.15 Proposed model for the formation of GBF area around an inclusion during VHCF [15]

During fatigue process micro-cracks forms around the inclusion and eventually coalesce with each other forming the GBF area, Step II of Fig. 9.15b. The micro-cracks propagate along the boundaries between the matrix and the carbides. Therefore, the roughness generated at the fracture surface corresponds to the carbides size. At this stage the rate of micro-crack propagation is very low. However, having grown to a certain size, the crack marks the end of the GBF area and continues to propagate as an ordinary fatigue crack at each cycle, Step III of Fig. 9.15c. The roughness of the fracture surface in the GBF zone compared with the surrounding area is due to the left by carbides that have been peeled off the matrix. The roughness outside the GBF area is small because the fatigue crack propagates in a normal smooth fashion, as shown in Fig. 9.15d. Note that the GBF zone formed within a prior austenite grain.

The above model named "dispersive de-cohesion of spherical carbides" describes the development of the GBF area in the vicinity of non-metallic inclusion during VHCF, which subsequently leads to the macro-crack formation that results in fatigue fracture. The models validity was confirmed by authors [17] through an experimental study using high speed tool steel. GBF failures are usual of case-hardened steel specimens.

A typical GBF failure originated at the center of a circular white area is shown in Fig. 9.16 [18]. The crack grew from an inclusion situated underneath the hardened layer aided by a residual tensile stress state that was balancing the residual compressive stress state existing in the case. The carburized surface layer, about 400 μm thick, appears as a brighter surface ring.

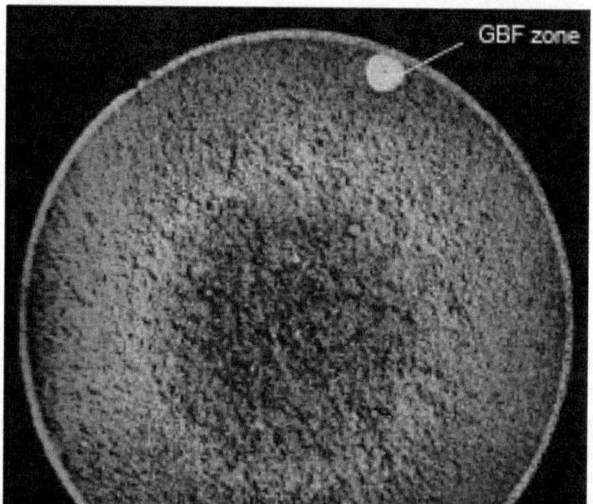

Fig. 9.16 Fatigue fracture appearance of carburized steel specimen, 25 mm diameter, showing the GBF zone with a central inclusion where the failure originated [18]

9.3 Hydrogen Assisted Fish-Eye Formation

Murakami [19] observed a different kind of fish-eye formation on a hard steels, Fig. 9.17. The specimen, rolled at Hv = 561, was loaded in rotating bending fatigue at a nominal stress amplitude of 843 MPa and failed after $1.03 \cdot 10^7$ cycles. The fracture origin was a nonmetallic inclusion of 16.7 μm size, located at 212 μm below the surface.

The bright white area is well visible around the internal inclusion, non-perfectly circular but lightly oval because of the bending conditions optically similar to the GBF fish-eye formation described in the previous section, except for the mechanism of crack initiation that, this time, is hydrogen embrittlement. This kind of fish-eye formation may also appear as a dark circle around a bright central inclusion, as shown in Fig. 9.18 [4]. Murakami called the dark area ODA (Optically Dark Area) [19]. The size of ODA increase with increase in fatigue life. It is interesting that ODAs are not found on the fracture surfaces of specimens which fail at a small number of cycles, Fig. 9.19 [19].

Scanning Electron Microscopy (SEM) observations reveled that ODAs had fracture surfaces typical of martensite lath structure. According to Murakami et al. the basic mechanism of ODA formation in the very-high-cycle regime, beyond $N = 10^7$ cycles is a hydrogen absorption embrittlement process [19]. This is schematically shown in Fig. 9.20. In the Murakami model, hydrogen absorbed by the hard steel

Fig. 9.17 Fish-eye formation observed in rotating bending fatigue on hard steel. The fatigued region within the fish-eye appears white under the optical microscopic, whereas the fatigue zone outside the fish eye appears grey [19]

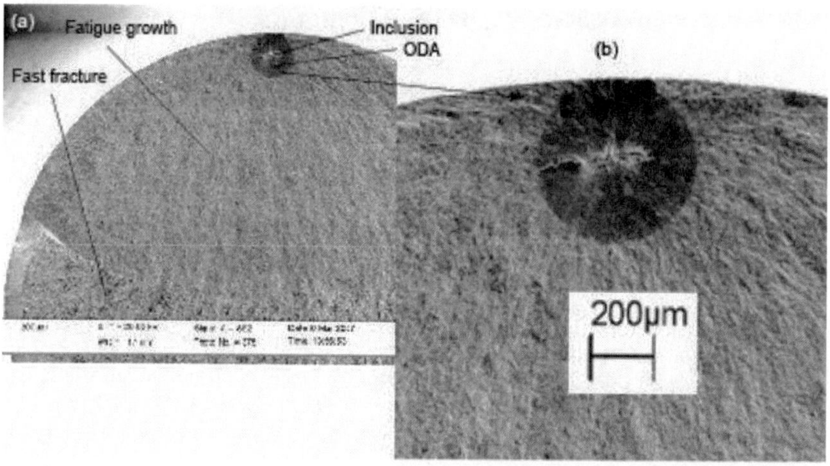

Fig. 9.18 **a** Typical VHCF fish-eye fracture surface appearance obtained in a specimen of AISI H13 steel showing the initiation site surrounded by ODA, **b** detail of (**a**) (adapted from [4])

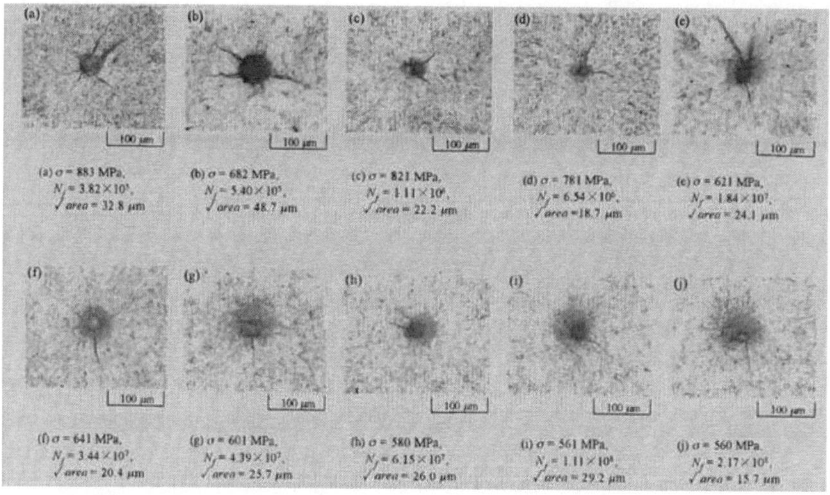

Fig. 9.19 Optical appearance of fracture origin depending on the fatigue life region in which it occurred, from 10^5 cycles to $2{\cdot}10^8$ cycles. The size of the Optically Dark Area (ODA) in the vicinity of an inclusion at a fracture origin increases with increasing fatigue life [19]

concentrates around the inclusion where it causes a corrosion fatigue crack growth up to the moment when the size of the ODA becomes large enough to continue to growth by normal fatigue in an inert environment (no-hydrogen). This happens when the stress intensity factor, K_I, at the ODA border equal the stress intensity factor threshold, $K_{I,th}$, for fatigue crack growth, see Sect. 15.2. This second stage of fatigue crack growth continues till the final failure of the specimen. Takai et al. [22, 23] directly verified the presence of hydrogen trapped in the ODA region, by Secondary Ion Mass Spectrometry (SIMS). Hydrogen concentration in the ODA has been found by Murakami to vary from 0.9 ppm to a mere 0.01 ppm, depending on the heat treatment applied to the steel. A hydrogen concentration of 0.9 ppm can be considered high enough to produce embrittlement and turn an inclusion into an ODA, while 0.01 ppm is not, as in cases (a), (b) and (c) of Fig. 9.19.

However, this mechanism of hydrogen assisted fatigue occurs because of the relatively high frequency fatigue testing that in the specific case was between 30 and 100 Hz. If the frequency of load application were reduced to just few cycle per minute, fatigue would disappear substituted by and intergranular stress corrosion cracking (IGSCC) process, see Sect. 20.3.1. In this last case, fatigue would just provide the load necessary to sustain the mechanism of stress corrosion cracking where the crack growth rate is independent of the applied stress, se Fig. 19.11. This actually means that, once the hydrogen assisted fatigue process starts, if the test frequency were reduced to some cycle per minute, the crack growth mechanism would change from corrosion fatigue to IGSCC. At this point, also the stress amplitude can be reduced

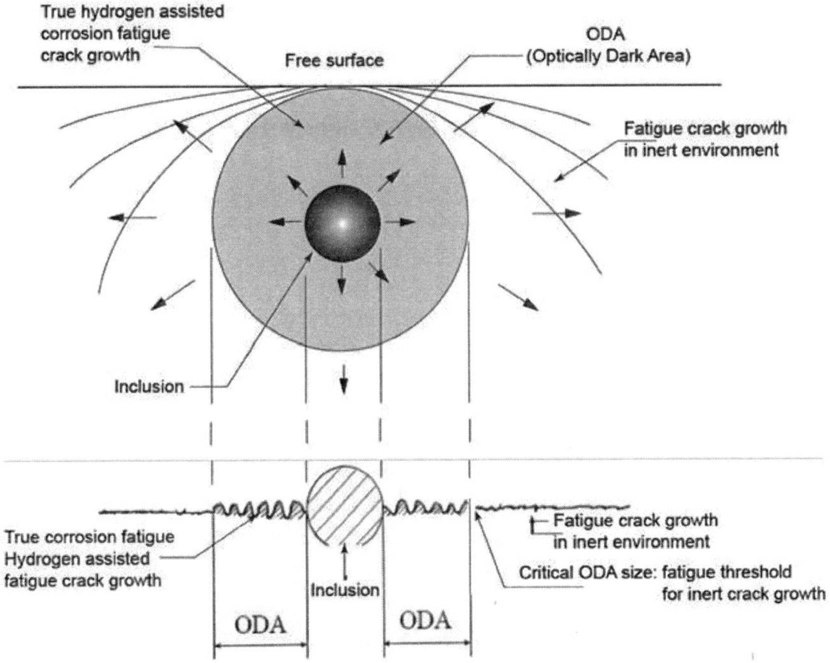

Fig. 9.20 Schematic illustration of hydrogen-assisted fatigue crack growth from non-metallic inclusion in the VHCF range (21) Scheme of fatigue fracture in ultra-long life regime (composite from [20, 21])

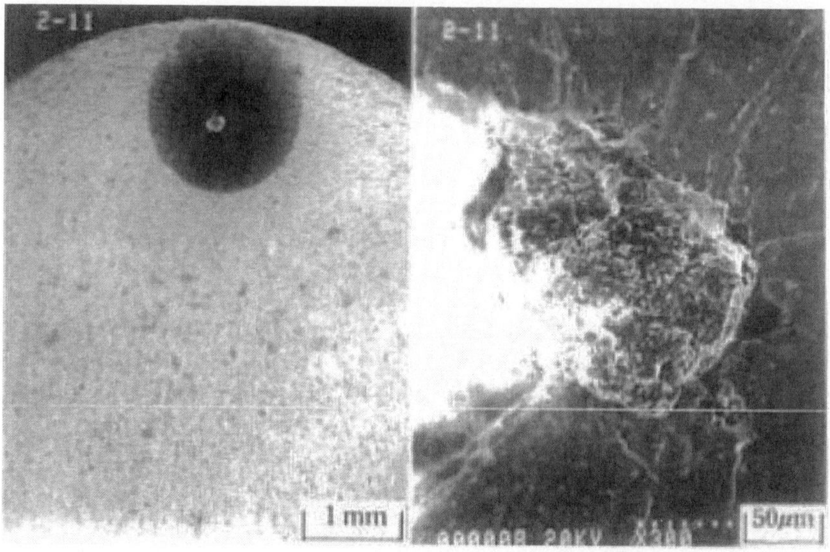

Fig. 9.21 **a** Fish-eye formation in high-strength steel of Hv = 654, failed at $N = 2.9 \cdot 10^5$ cycles under a stress amplitude $\sigma_a = 1275$ MPa and a mean stress $\sigma_{am} = -784$ MPa; **b** close-up of the internal nonmetallic inclusion [19]

and the growth of ODA would continue the same as long as it remains over the IGSCC threshold K_{IGSCC}.

Figure 9.21 is an interesting example of fatigue occurred at about $3 \cdot 10^5$ cycles in a specimen of high-speed chrome, molybdenum tool steel, SKH51, hardened at Hv = 654.

The fish-eye formation occurred from an inclusion at 1 mm below the surface and not on the surface or just below the surface because of the compression mean stress, $\sigma_{am} = -784$ MPa. Under this condition, fatigue could not start at the surface, strongly compressed, so it occurred underneath, finding a nonmetallic inclusion of unusually large size, i.e., about 200 μm.

Without the surface compression, that inclusion, though very large, would have not caused fatigue failure that would be occurred on the surface. Figure 9.22 is a schematic of the fatigue initiation site depending on the fatigue regime, whether low cycle fatigue (LCF) or high cycle fatigue (HCF) or VHCF. In the LCF regime, Fig. 9.22, fatigue damage always starts on the specimen surface as multiple initiation sites and is due to the hard, brittle nature of grain boundaries, Fig. 2.15. In the HCF regime, Fig. 9.22, fatigue damage always initiates on the surface but just in one single point, the weakest, that can be a surface slip, Fig. 9.22a, or a surface nonmetallic inclusion, Fig. 9.22b, or a fish-eye formation, Fig. 9.22c, this last occurs mostly in the range of 10^6–10^7 cycles, Fig. 9.19. In the VHCF regime, over 10^8–10^9 cycles, fatigue damage is always of the fish-eye type but initiates at any point inside the specimen and not necessarily on the surface or below the surface, as schematized in Fig. 9.22.

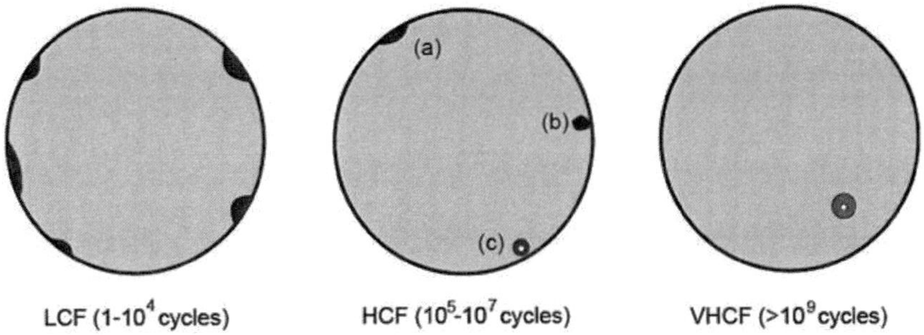

Fig. 9.22 Schematic representation of fatigue initiation sites, depending on the number of cycles to failure (modified from [24])

This is because with the low stress amplitude at which failure occurs, the dominating effect is the process volume. The nonmetallic inclusion needed the trigger fatigue failure must be particularly large. Therefore, the search for this weakest point, the large inclusion, is extended to the entire volume of the specimen.

References

1. Ashami, K., Sugiyama, Y.: Fatigue strength of various surface hardened steels. Jpn. Heat Treatment Technol. Assoc. **25**(3), 147–150 (1985)
2. Kanazawa, K., Nishijima, S.: Fatigue of low alloy steel at ultra-high cycle regime under elevated temperature conditions. Jpn. Soc. Mat. Sci. **46**(12), 1396–1401 (1997)
3. Marines, I., Bin, X., Bathias, C.: An understanding of very high cycle fatigue of metals. Int. J. Fatigue **25**(9), 1101–1107 (2003)
4. Kaziymyrovych, V.: Very High Cycle Fatigue of Engineering Materials—A Literature Review. Karlstad University Studies (2009)
5. Sakai, T., Sato, Y., Oguma, N.: Characteristic S-N properties of high-carbon-chromium-bearing steel under loading in long life fatigue. Fatigue Fract. Eng. Mater. Struct. **25**(8–9), 765–773 (2002)
6. Wang, Q.Y., Berard, J.Y., Rathery, S., Bathias, C.: High-cycle fatigue crack initiation and propagation behaviour of high-strength spring steel wires. Fatigue Fract Eng Mater Struct **22**(8), 673–677 (1999)
7. Bathias, C.: There is no infinite fatigue life in metallic materials. Fatigue Fract Eng Mater Struct **22**(7), 559–565 (1999)
8. Zettl, M., Stanzl-Tschegg, S.: Very high cycle fatigue of normalized steels. In: Proceedings of 3 International Conference on Very High Cycle Fatigue, pp. 398–405 (2004)
9. Haibach, E.: A likely explanation of very high cycle fatigue in steel. In: 5th International Conference on Very High Cycle Fatigue. Berlin (2011)
10. Oguma, N., Harada, Sakai, T.: Strength level dependence of long-life fatigue behavior for bearing steel in rotating bending. In: Proceedings of 3rd International Conference on Very High Cycle Fatigue, pp. 617–624 (2004)
11. Stanzl-Tschegg, S.E., Mayer, H.: Fatigue and fatigue crack growth of aluminium alloys at very high numbers of cycles. Int. J. Fatigue 231–237 (2001)
12. Mughrabi, H.: On 'multi-stage' fatigue life diagrams and the relevant life-controlling mechanisms in ultrahigh-cycle fatigue. Fatigue Fract. Eng. Mater. Struct. **25**(8), 755–764 (2002)

13. Ewing, J.A., Humphrey, J.C.: The fracture of metals under repeated alternation of stress. Philos. Trans. Roy. Soc. A **200**, 241–250 (1903)
14. Muzzoli, M.: La resistenza a fatica di acciai duri per cuscinetti a rotolamento e sua relazione con le caratteristiche strutturali. Atti del III Congresso Nazionale degli Ingegneri Italiani, Trieste (1935)
15. Frith, P.H.: Fatigue tests on rolled alloy made in electric and open-hearth furnaces. J. Iron Steel Inst. **180**, 26–33 (1955)
16. Shiozawa, K., Morii, Y., Nishino, S., Lu, L.: Subsurface crack initiation and propagation mechanism in high-strength steel in a very high cycle fatigue regime. Int. J. Fatigue **28**(11), 1521–1532 (2006)
17. Shiozawa, K., Morii, Y., Nishino, S.: Subsurface crack initiation and propagation mechanism under the super-long fatigue regime for high speed tool steel (JIS SKH51) by fracture surface topographic analysis. JSME Int. J. Ser. A: Solid Mech. Mater. Eng. **49**(1), 1–10 (2006)
18. Frost, N.E., Marsh, K.J., Pook, L.P.: Metal Fatigue. Clarendon Press, Oxford (1974)
19. Murakami, Y.: Metal Fatigue: Effects of Small Defects and Nonmetallic Inclusions. Elsevier Science Ltd, Kidlington, Oxford OX5 1GB, UK (2002)
20. Murakami, Y., Nomoto, T., Ueda, T.: On the mechanism of fatigue failure in the superlong life regime (N>107 Cycles). II. A fractographic investigation. Fatigue Fract. Eng. Mater. Struct. **23**(11), 903–910 (2000)
21. Murakami, Y., Nagata, J., Matsunaga, H.: Factors affecting ultralong fatigue life and design method for components. In: 9th International Fatigue Congress, Atlanta, USA (May 2006)
22. Takai, K., Honma, Y., Izutsu, K., Nagumo, M.: Identification of trapping sites in high-strength steels by secondary ion mass spectrometry for thermally desorbed hydrogen. J. Jpn. Inst. Met. **60**(12), 1155–1162 (1996)
23. Takai, K., Seki, J., Yamauchi, G., Honma, Y.: Observation of trapping sites of hydrogen and deuterium in high-strength steels with secondary ion mass spectrometry. J. Jpn. Inst. Met. **SS**(12), 1380–1385 (1994)
24. Bathias, C., Paris, P.C.: Gigacycle Fatigue in Mechanical Practice. Marcel Dekker Publisher Co., New York USA (2005)

Chapter 10
Fatigue Testing. Fatigue Curve Construction and Fatigue Limit Assessment

10.1 Fatigue Tests

At the base of fatigue design there is the Wöhler's curve *S–N* that is obtained through fatigue tests. Giving a stress amplitude *S* or strain amplitude ε_a to a series of specimens it is possible to derive the corresponding cycles to failure *N* or life. Considering the importance of such a curve and, therefore, of laboratory tests in general, it is convenient to spend some arguments on the matter just to evidence the most important features and the limits that may be associated, too often forgotten. Fatigue tests may be grouped in four fundamental types:

- simple or reversed bending;
- rotational bending;
- tension-tension or tension–compression (push–pull);
- alternate torsional loading.

In the first type of test it is recommended that specimen be flat and not cylindrical to maximize the process volume, i.e., the volume subjected to the maximum stress. The process volume plays a fundamental role in fatigue because both metallurgical variability and data scatter depend on its value (see Sect. 4.8 and Chap. 6). In bending the stress distribution over the cross section of a specimen has a triangular shape with the maximum value on the edge fibers, Fig. 4.52, remote from the bending neutral axis and a minimum equal zero on this axis. A cylindrical section concentrates most of volume of material wright on the neutral axis where it is useless while the edge fibers that should be subjected to the maximum stress have a near zero volume. The result is, ceteris paribus, that under bending loading a cylindrical specimen will yield a higher fatigue life with respect to a plain specimen, but for design purpose we need know the expected lower fatigue life and not the higher to be sure that we'll not run into unexpected fatigue failure. Figure 10.1 shows the test scheme of a flat specimen in fully reversed bending. As it can be seen, the specimen of the flat type is tapered so that the stress state is constant on any section independently of its distance from

P. P. Milella, *Fatigue and Corrosion in Metals*,
https://doi.org/10.1007/978-3-031-51350-3_10

the applied force. This allows using the entire length of the specimen increasing the process volume. The test rig can be used heather in completely reversed bending with no mean stress, $\sigma_m = 0$, or with a mean stress of given value, simply positioning at a due distance the eccentric crank drive with respect to the specimen clamping vise where displacement is actually applied. This test machine represents a typical constant deflection apparatus. Note that, at least in high cycle fatigue, it may be convenient to increase the specimen thickness depending on two conflicting effects: first, a larger thickness will require a higher force to produce the same stress state, secondly, increasing the thickness the stress gradient decreases so that the surface layer that will be working at the maximum or near the maximum stress increases and this maximize the process volume, Fig. 4.52. But to improve the process volume and, therefore, the probability of getting the worst metallurgical condition that will initiate the fatigue process, it is also convenient to use a rotating-bending testing machine with a cylindrical specimen because this time the whole circumference will work at the maximum stress, Fig. 4.51. The test apparatus may be of the type described in Fig. 10.2. The specimen is clamped by a couple of vices that sustain a load *F*. The system is loaded in four point bending (pure bending) with a constant moment *M* equal to $P{\cdot}L/2$, *L* being the distance between the support and the load bearing. When the system rotates along its longitudinal axis any point of the specimen surface on section A is subjected to the same stress, either in traction or compression. The hourglass shape of the specimen is required to provide a larger section in the clamping vices area where stress concentration increases the local stress state, avoiding that fatigue failure occurs there rather than in controlled section A. A cycle lasts $1/\omega$ minute, ω being the angular velocity in round per minute. The test rig can apply only symmetrical stresses and is not suitable for nonzero mean stress.

The advantage of bending fatigue testing machines of the type shown in Fig. 10.1 or 10.2 over conventional traction load cells is that they use an electrical engine that may reach angular velocities of the order of 4800 rpm with a frequency of 80 Hz which means that in a full day they may apply even seven million cycles. Therefore, only three days are needed to reach a fatigue life of 20,000,000 cycles whether with a conventional load cell it may require some 25 days. Over 100 Hz a cooling system may be necessary to avoid excessive specimen cycling overheating. High frequencies must be avoided in the study of corrosion fatigue to provide enough time to potential aggressive environment to exert its action. In such studies it may be necessary to use frequencies of the order of 0.1 Hz (see Sect. 6.11). An interesting device is the resonant-type direct-stress fatigue testing machine schematized in Fig. 3. 3. Two masses spring-coupled enter into resonance when one of them is excited at the proper frequency by a rotating system. The specimen is located between the two masses. Controlling the amplitude of oscillation, the specimen may be subjected to any desired stress level. The machine is best used in completely reversed stress conditions even though nonzero stress can be achieved by inserting an auxiliary spring in parallel with the specimen. Its main feature is the excitation frequency that may rise over 120 Hz giving tens of million cycles in a very short period of time (Fig. 10.3).

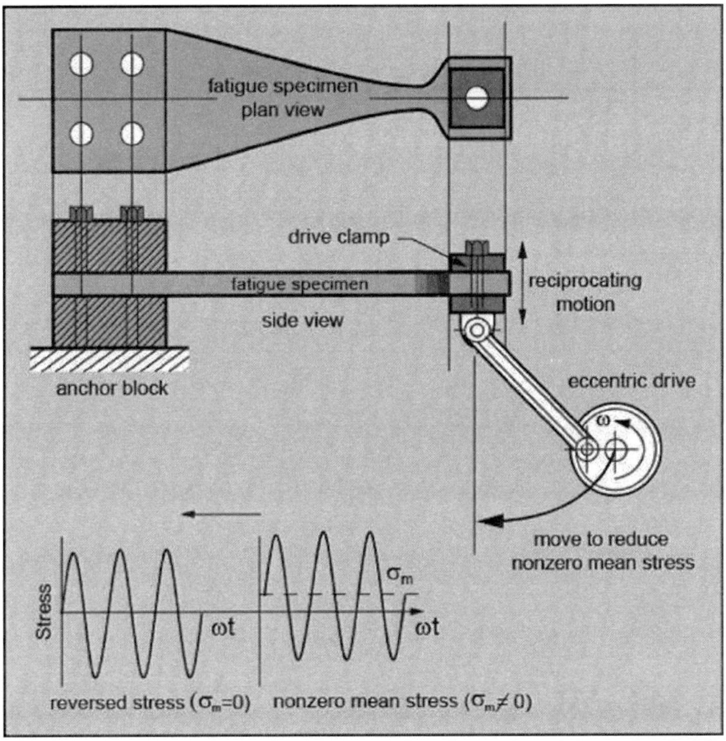

Fig. 10.1 Reciprocating-bending fatigue test machine

Typical fatigue specimens are shown in Fig. 10.4. Specimen (a) is the so-called hourglass specimen. Radius R is at least $8{\cdot}d$ where d is the section diameter. Its rather short and thick shape makes it suitable for reversed fatigue testing where the specimen undergoes high compression stress. A slant specimen, such as the one shown in Fig. 10.4, specimen (b), would go buckling. Therefore, it is used only in fatigue pulsing from zero (R = 0). Its length L is usually at least $3d$. Figure 10.4c shows a typical flat specimen used when volume effect is studied, and metallographic analyses are required. It must be recalled, in fact, that the process volume in high cycle fatigue is not the entire specimen volume, but only the surface times a depth no larger than 0.5 mm.

Therefore, to increase the specimen diameter is worthless: the core will not participate to the fatigue process. On the contrary, doubling the width W in a flat specimen would double the surface and therefore also the process volume and by reducing the thickness, t, the same net section area is maintained. Flat specimens are also used, as said, metallographic examination to check slip bands and damage formation during cycling (see Chap. 2), MSC formation (see Sect. 2.3) or to study inclusions, second phases or graphite nodules dimensions and morphology. These studies may be of great help in Very-High-Cycle-Fatigue analysis to relate a lower bound fatigue limit

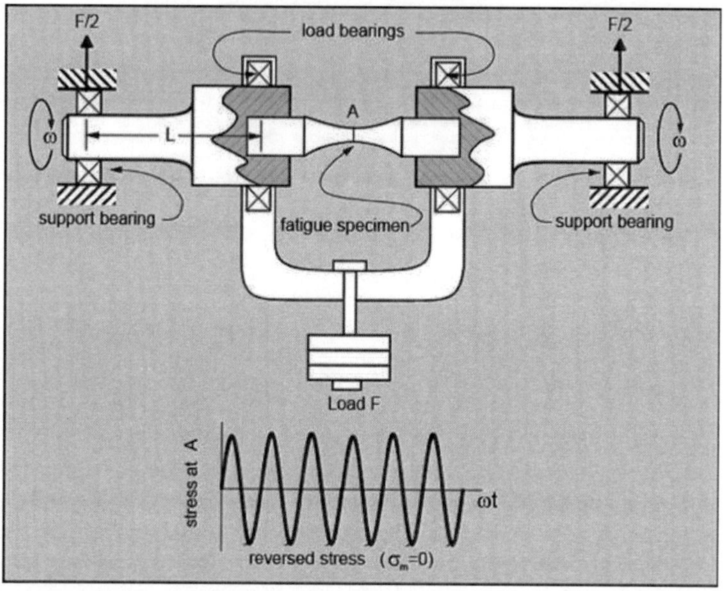

Fig. 10.2 Rotating-bending fatigue testing machine

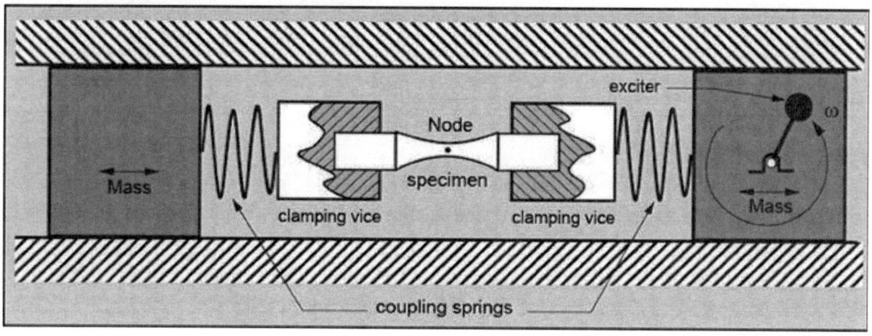

Fig. 10.3 Resonant-type direct-stress fatigue testing machine

with the size of a second phase or inclusion expected in the material with a certain probability (see Sect. 6.8.9). Flat specimens are also used when the material comes in sheets (less than 5–6 mm thick).

The four small open circles on the heads of the specimens of Fig. 10.4 represent positions where it may be convenient to perform hardness measurements before running fatigue tests. The resulting indentations will not affect fatigue test results. After test, hardness measurements may be carried out on any point on the surface that has been subjected to fatigue cycling. Comparing the two hardness values, before and after fatiguing, it may be possible to derive the hardening or softening effect operated

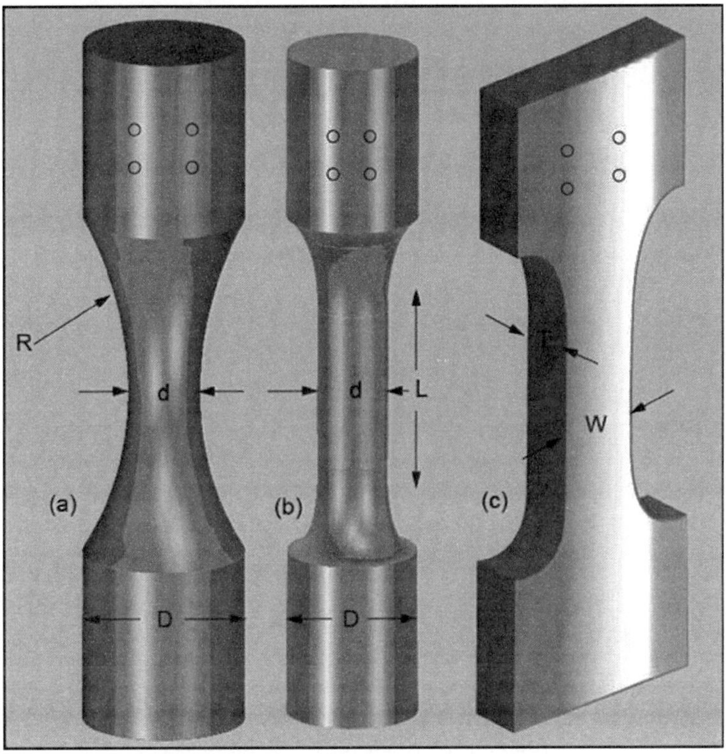

Fig. 10.4 Typical fatigue specimens. Open points on head of specimens indicate where hardness measurements should be made before running fatigue tests

by fatigue cycling (see Sect. 1.4 and Fig. 1.25). Including these measurements in all fatigue tests performed it is possible to accumulate precious data and knowledge on material behavior. The precise knowledge of material characteristic (hardness) variations with the applied cyclic stress amplitude may be used as diagnostic or checking tool on in service components made with same material to ascertain whether the operating stress amplitude was that calculated or not or to even infer it. Hardness measurements taken on cylindrical specimens cannot be directly compared with data coming from flat surfaces but must be converted into flat surface readings by use of conversion tables that consider the radius of the specimen.

Normally, fatigue specimens used to derive reference *S–N* curves have a mirror-polished finish to be sure that the fatigue data are not affected by surface finish. This kind of finish, that leaves an average roughness R_m less than a micron, is not required in low cycle fatigue where the stress amplitude exceeds the yield strength of the material. In these cases, in fact, the entire cross section and hence the entire surface undergoes plastic strains that may destroy almost immediately the original mirror-polished finish. If plastic strains are high enough an orange peel surface may develop

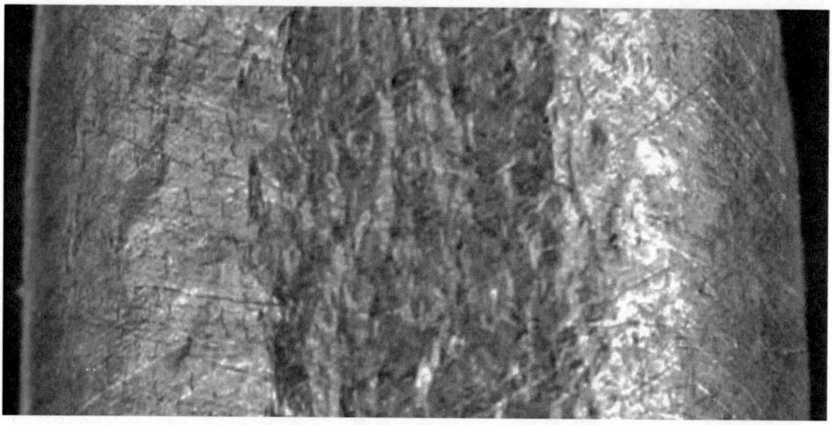

Fig. 10.5 Detail of an orange peel surface observed by Milella on a specimen of austenitic stainless steel subjected to a 15% total strain amplitude. Note slip lines intersecting at 120°

as observed on a specimen of austenitic stainless steel subjected to a 15% total strain amplitude and shown in Fig. 10.5. Slips lines can be clearly seen intersecting on the surface of the specimen at 120° which is typical of FCC lattice like austenite.

A new fatigue testing machine has been recently introduced to study the Very-High-Cycle-Fatigue behavior of materials (expected life over 10^9 cycles) in a reasonable short time of the order of just few hours. This is a resonant type, very high frequency ultrasonic testing machine capable of operating at 0.5–20 kHz.

In this type of device, a piezoelectric actuator converts a 50–60 Hz AC current into a ultrasonic pure sine wave that enters the resonant system schematically shown in Fig. 10.6. This is composed of an actuator, a booster, a horn, and the specimen itself.

The booster and the horn (typically, a conical sonotrode) are used to amplify the amplitude of the vibrations transmitted to the test specimen. Figure 10.6 shows a typical trend of the displacement and stress profile throughout the resonant system, with the stress having its maximum right at the center of the specimen. The specimen shape and dimensions are shown in Fig. 10.7. At the base of fatigue design there is the *S–N* diagram or Wöhler curve that relates the stress amplitude σ_a, to the number N of cycles to failure. In Chap. 4 it has been shown how this curve and the fatigue limit σ_f depends on several factors such as metallurgical structure, hardness, ultimate strength, surface conditions and process volume.

This actually suggests that the *S–N* diagram of a metal can be traced, in a first approximation needed as a starting point for initiating fatigue calculations, just knowing the corresponding relationships. This doesn't mean that fatigue testing is not necessary to build a Wöhler curve specific for our material and that we are released from running more and more tests to adjourn the *S–N* curves with the needed degree of reliability but it just means that in the absence of archive data and waiting for a reliable *S–N* curve of experimental derivation it is much better to construct our

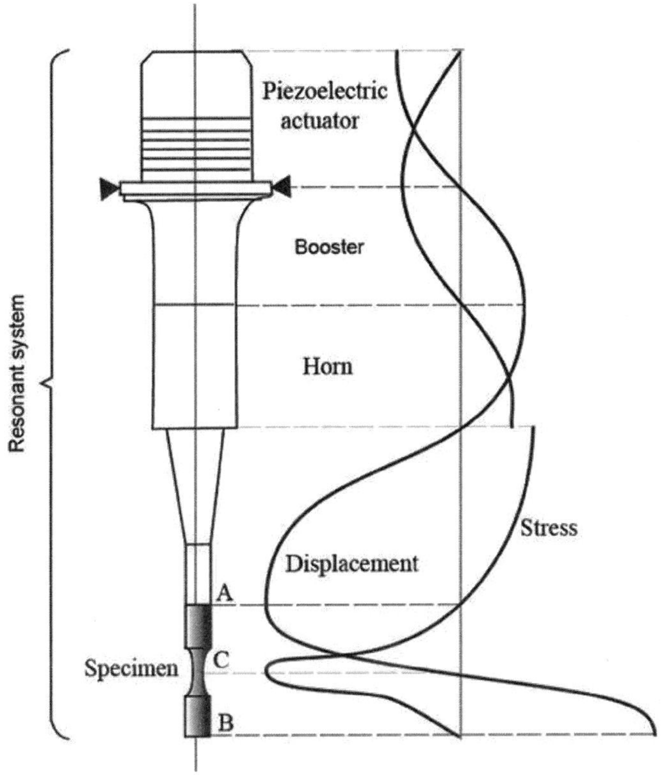

Fig. 10.6 Schematic of a resonant type ultrasound fatigue testing apparatus capable of working up to 20 kHz

own fatigue curve, rather than to use the so-called library reference fatigue curves of which we have no information at all about conditions and methods used to obtain them. With time and experience a first drawn *S–N* curve is more reliable than a library curve. The next sections deal with the issue of constructing simplified *S–N* diagrams and derive the fatigue limit of a material.

10.2 Empirical Equations

In the construction of *S–N* diagrams it is important to recall a characteristic shown in Figs. 1.16, 1.17 and 1.18a. All *S–N* curves relative to different metals or conditions intersect each other at about 10^3 cycles. Therefore, this common abscissa, $N = 10^3$ cycles, will be one of the starting points in the construction of *S–N* diagrams, in particular when we deal with high-cycle fatigue and are not interested in the field below 10^3 cycles, Region I of fatigue. The second and final point is the fatigue limit

Fig. 10.7 Typical VHCF specimen and dimension

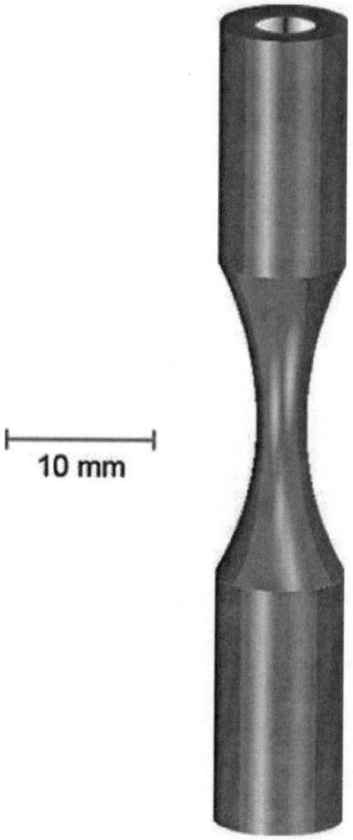

σ_f or S_f that will be conventionally posted at 10^6 cycles or, sometimes, at 2×10^6 cycles. We need know the ordinates S and σ_f of just these two points A $\equiv$ (1000; S_3) and B $\equiv$ (10^6;σ_f) and join them by a straight line in a log–log scale to build the preliminary fatigue *S–N* diagram. To this purpose, the composite diagram of Fig. 8.1 [1] provides important information. It shows some experimental results of fatigue allowable amplitude S normalized to the ultimate strength of the material, S/σ_u, obtained on standard wrought-steel with traction strength varying from 400 to 1400 MPa. Steels had received appropriate heat treatment producing a stress-relieved structure consisting essentially of tempered martensite. We can note that at 10^3 cycles the ratio S_3/σ_u falls between 0.9 and 1.0.

We shall remember that this ratio depends also on loading condition. Usually, the ratio S_3/σ_u is assumed to be equal to 0.9 for fully reversed bending and 0.75 for reversed traction.

In case of torsion load, lacking specific data, the ultimate strength in shear may be roughly approximated as $0.8{\cdot}\sigma_u$ for steel and cast iron and as 0.75 for nonferrous metals. Table 10.1 reports an estimate of S_3 for some metals and loading conditions. Information that stems out of Fig. 10.8 is that at 10^6 cycles the fatigue limit to the ultimate strength ratio σ_f/σ_u may be roughly approximated as 0.55.

As to the fatigue limit σ_f or S_f, experience shows that theσ_f/σ_u ratio varies from 0.4 to 0.6 depending on metallurgical structure with a minimum of 0.25 for pure

Table 10.1 S_3 estimate

Material	Load type	S_3
All materials	Bending	$0.9{\cdot}\sigma_u$
All materials	Axial	$0.75{\cdot}\sigma_u$
Steel	Torsion	$0.9{\cdot}\tau_u = 0.72{\cdot}\sigma_u$
Nonferrous materials	Torsion	$0.9{\cdot}\tau_u = 0.63{\cdot}\sigma_u$
Cast iron	Torsion	$0.9{\cdot}\tau_u = 0.72{\cdot}\sigma_u$

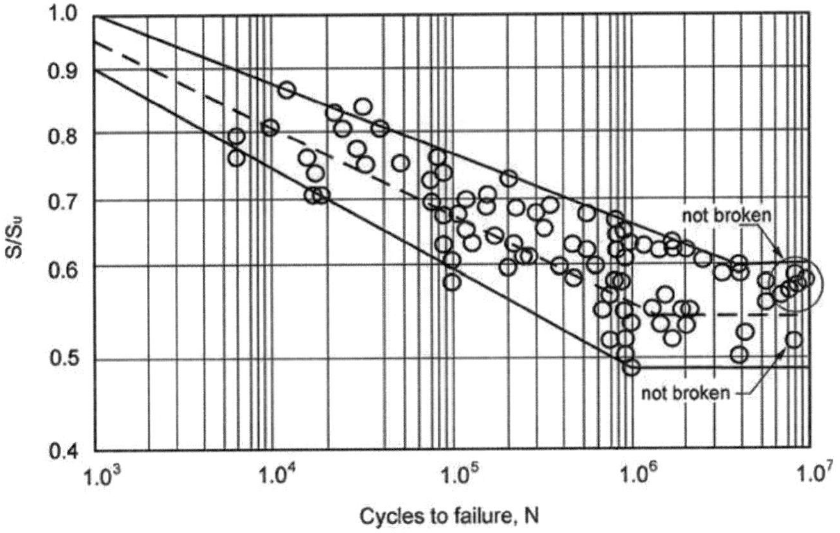

Fig. 10.7 Composite S–N curve for wrought steel [1]

martensite reaching values as high as 69 HRC that represents the hardest known structure. A collection of data obtained already in 1924 by Gough [2] on wrought steels with ultimate strength ranging from 310 MPa up to 2000 MPa indicated such a ratio as variable from 0.4 to 0.54, with an average of 0.46. Recent authors [3–5] report average values between 0.46 and 0.48. These values are aligned with Duckworth observation who found that up to 1200 MPa ultimate strength the average value of the fatigue limit σ_f was related to the traction strength σ_u as

$$\sigma_f = 0.5 \times \sigma_u. \tag{10.1}$$

The 1200 MPa ultimate strength limit considered is due to inclusions that in high strength steels play a fundamental role in microscopic damage nucleation, reducing the fatigue strength and life. Therefore, Eq. (10.1) can be assumed to be valid for strength not higher than 1200 MPa, afterwards the fatigue limit bends over, as shown in, Fig. 4.37. Besides Eq. (10.1), many empirical equations have been proposed that

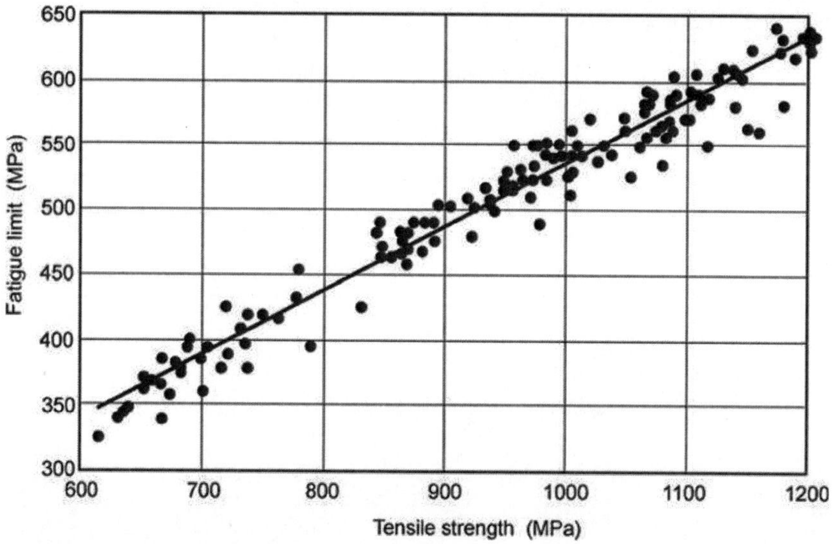

Fig. 10.9 Trend of fatigue limit at 10^6 cycles versus tensile strength observed by T. Svensson et al. (modified from [6])

relate fatigue limit to the mechanical properties of materials. Svensson et al. [6] working on a base of 166 data collected in Japan and relative to carbon steel, Cr–Mo steel and ferritic stainless steel proposed the diagram of Fig. 10.9. The regression line has equation

$$\sigma_f = 46.3 + 0.49 \times \sigma_u + \varepsilon_r \tag{10.2}$$

with ε_r being the residual with standard deviation estimated to be

$$\Delta(\varepsilon_r) = 16.1\,\text{MPa}. \tag{10.3}$$

Alternatively, knowing the steel hardness or yield strength σ_y, the Svensson et al. [6] equation is available

$$\begin{aligned} \sigma_f &= 32.0 + 1.58 \times HV + \varepsilon_{HV} \\ \sigma_f &= 168 + 0.41 \times \sigma_s + \varepsilon_s \end{aligned} \tag{10.4}$$

with standard deviation on residuals being

$$\begin{aligned} \Delta(\varepsilon_{HV}) &= 17.8\text{ MPa} \\ \Delta(\varepsilon_s) &= 16.3\text{ MPa}. \end{aligned} \tag{10.5}$$

The second of Eq. (10.4) is somewhat unusual since it introduces the yield strength, σ_s, in relation with the fatigue limit that, usually, is related to the ultimate strength, σ_u, of the material. Another empirical relationship that makes use of the Vickers hardness has been proposed by Murakami and Endo [7] for hardness less than 400 HV

$$\sigma_f = 1.6 \times HV \pm 0.1 \times \text{HV MPa} \tag{10.6}$$

or in terms of Rockwell C hardness (HRC)

$$\sigma_f = (9.16 \times \text{HRC} + 308) \tag{10.7}$$

Equation (10.7) is valid up to 40–50 HRC afterwards the fatigue limit stabilizes owing to inclusions always present in metals. Equations (10.6) and (10.7) are plotted in Fig. 10.10 on the Garwood et al. diagram [8]. As to the ultimate strength, σ_u, it may be useful the diagram of Fig. 9.4 [9] that provides a relationship between σ_u and the Brinell hardness, HB (Fig. 10.11)

$$\begin{aligned} \sigma_u &= 3.6 - 3.3(\text{HB}) \text{ for steel} \\ \sigma_u &= 3.48 - 3.21(\text{HB}) \text{ for brass} \\ \sigma_u &= 2.86 - 3.63(\text{HB})\text{for nodular iron} \end{aligned} \tag{10.8}$$

In the first of Eq. (10.8) the upper value, 3.6, is valid for mild steels and the lower, 3.3, for alloyed-steels.

Interesting enough, cast iron deviates rather considerably from steel. It may be adopted a correlation of the type

$$\sigma_u = 0.7 \times \text{HB} \tag{10.9}$$

while for ductile cast iron (DCI)

$$\sigma_u = 2.8 \times \text{HB} \tag{10.10}$$

The modest strength of cast iron is due to its particular composite structure in which carbon is present as pure graphite and not carbide, as in steel. Moreover, very high residual stress, built up during cooling, remains at the interface between graphite nodules and iron matrix. This stress state, being positive, reduces the ultimate strength of the material in traction while increases its response in compression. Often graphite nodules degenerate into vermicular or needle like shapes, as shown in Fig. 4.20, reducing also the fatigue life of the metal. When degenerated graphite reaches dimensions larger than 300 μm it may be considered as a real macro crack impinging also the fatigue limit. Macro-shrinkage cavities, shown in Figs. 6.37, 6.40 and 6.41 may have an even worst impact on fatigue since they can be even much longer than graphite needles. The duplex ferrite-perlite structure that DCI often presents may

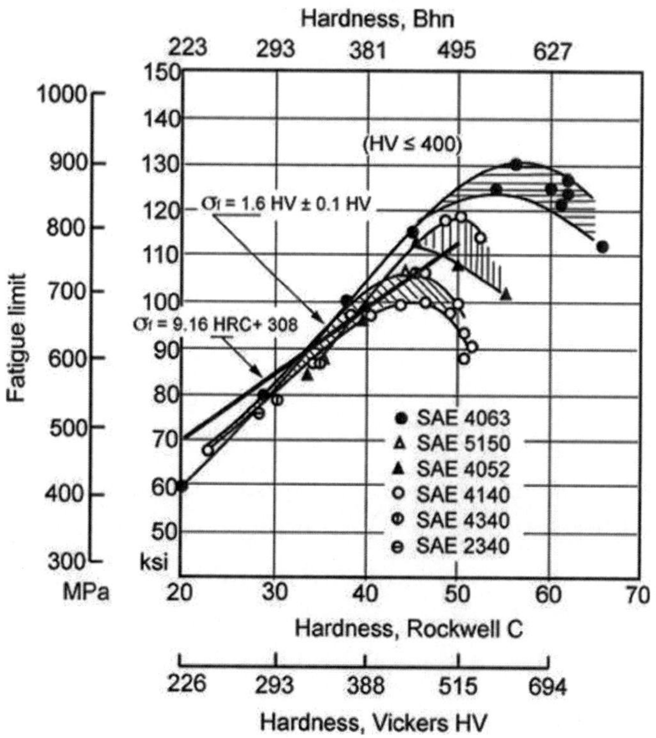

Fig. 10.10 Garwood et al. diagram [8] showing correlation between fatigue limit and hardness for several steels

contribute to ameliorate the ultimate strength, but at the expense of deformation at fracture that can precipitate below 1 or 2% becoming brittle. For austenitic steel with low yield strength and high hardening, the proportionality constant goes up to 5.2.

For reheated copper it becomes 5.5. Table 10.2 offers a synoptic picture useful for the passage from Vickers hardness (HV) to Rockwell C (HRC) or B (HRB) and Brinell (HB). It also provides an estimation of the corresponding traction strength of the material. Same relationships are presented in Figs. 10.12 and 10.13 as diagrams. Dowling diagram of Fig. 10.14 [10] widens the field of materials including also alloyed-steels, wrought steel, cast iron, DCI suggesting a relation always of the type

$$\sigma_f = C \times \sigma_u \tag{10.11}$$

but with the constant of proportionality varying from a maximum of 0.6 to a minimum of 0.3.

It has been already shown in Sect. 4.7, Eq. (4.12), the empirical equation of Murakami and Endo [11] that correlates the fatigue limit to the cross section area of the largest defect present in the work piece projected on the plane normal to the

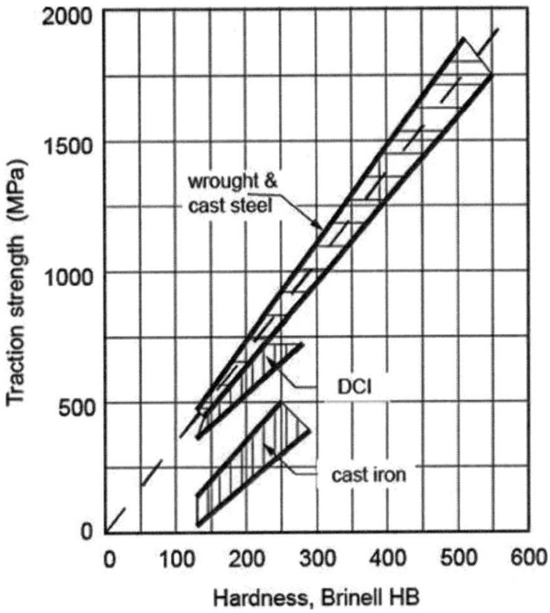

Fig. 10.11 Correlation between hardness and traction strength (modified from [9])

maximum tensile stress direction, $area_{\max}$, as shown in Fig. 4.42

$$\sigma_f = K \times (HV + 120) \times \left(\sqrt{area_{\max}}\right)^{-1/6}\left(\frac{1-R}{2}\right)^{\alpha} \quad (10.12)$$

with

$$K = \begin{cases} 1.43 & for\, surface\, defects \\ 1.43 & for\, sub\text{-}surface\, defects \\ 1.56 & for\, internal\, defects \end{cases}$$
$$\alpha = \left(HV \times 10^{-4} + 0.226\right). \quad (10.13)$$

Figure 10.15 shows Eq. (10.12) predictions for several defect areas and $R = -1$ on the Dowling diagram of Fig. 10.14. A certain consistency may be recognized.

The Dowling upper limit is reached with Murakami equation for $\sqrt{area_{\max}}$ very small, in practice with defect dimension of just 2 μm whereas the lower limit approximates the behavior of defects of about $\sqrt{area_{\max}} = 1.000$ μm. Figure 10.16 shows Eq. (10.12) predictions relative to different defects sizes and $R = -1$ compared with experimental data of Fig. 4.37. It is interesting to note how the 1/6 exponent of $\sqrt{area_{\max}}$ is matching the experimental trend. Recently, McGreevy [12] has formulated a model that extends the Murakami criterion of the square root of the cross section area to low strength steels. The model, called Hall–Petch-Murakami (HPM), starts from a surface semi-elliptical defect of major axis larger than $2a$ for which it

Table 10.2 Hardness conversion table for steels and corresponding ultimate strength

HV	HB[a]	HRC[b]	σ_u (MPa)	HV	HB[a]	HRC[b]	σ_u (MPa)
940		68		272	258	26	860
900		67		266	**253**	25	840
865		66		260	**247**	24	820
832		65		254	**243**	23	800
800		64		248	**237**	22	
772		63		243	**231**	21	770
746		62		238	**226**	20	760
720		61		234	228	**98**	
697		60		220	215	**96**	720
674		59		209	204	**94**	690
653	**615**	58	2180	198	194	**92**	650
633	**595**	57	2145	188	184	**90**	620
613	**577**	56	2130	179	176	**88**	600
595	**560**	55	2120	171	168	**86**	570
577	**543**	54	2050	164	161	**84**	550
560	**525**	53	1980	157	155	**82**	530
544	**512**	52	1920	151	149	**80**	510
528	**496**	51	1850	145	144	**78**	500
513	**481**	50	1800	140	139	**76**	
498	**469**	49	1730	135	134	**74**	460
484	**455**	48	1660	130	129	**72**	
471	**443**	47	1600	126	125	**70**	430
458	**432**	46	1550	122	121	**68**	
446	**421**	45	1500	119	118	**66**	
434	**409**	44	1450	115	114	**64**	400
423	**400**	43	1400	112	111	**62**	
412	**390**	42	1360	108	108	**60**	
402	**381**	41	1320	106	106	**58**	
392	**371**	40	1280	103	103	**56**	
382	**362**	39	1250	100	100	**54**	
372	**353**	38	1200	98	98	**52**	
363	**344**	37	1170	95	95	**50**	
354	**336**	36	1140	93	93	**48**	
345	**327**	35	1100	91	91	**46**	
336	**319**	34	1070	89	89	**44**	
327	**311**	33	1040	87	87	**42**	
318	**301**	32	1010	85	85	**40**	

(continued)

Table 10.2 (continued)

HV	HB[a]	HRC[b]	σ_u (MPa)	HV	HB[a]	HRC[b]	σ_u (MPa)
310	294	31	980	83	83	**38**	
302	**286**	30	950	81	81	**36**	
294	**279**	29	930	79	79	**34**	
285	270	28	900	78	78	**32**	
279	**264**	27	880	77	77	**30**	

[a] Values in bold refer to HB hardness obtained with carbide ball, the others with standard sphere
[b] Values in bold refer to HRB hardness

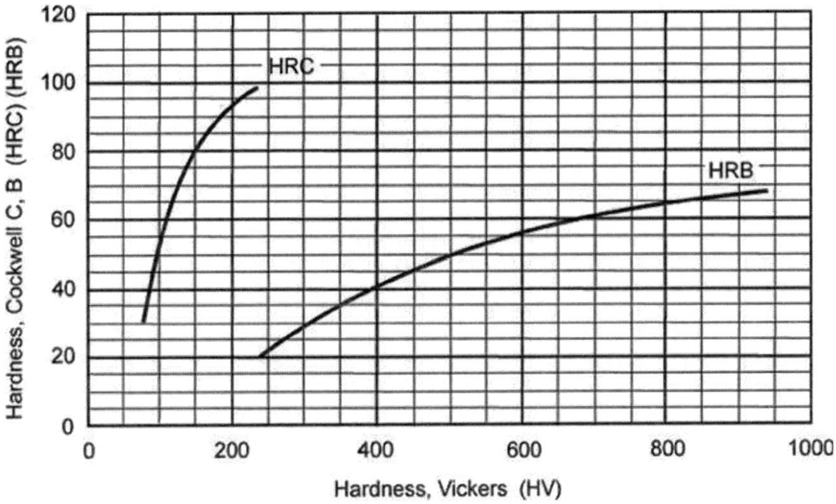

Fig. 10.12 Hardness correlation diagram

is

$$\sqrt{area} \cong a\sqrt{\frac{\pi}{2}}. \tag{10.14}$$

Introducing Eq. (10.14) in Eq. (10.12) yields for $R = -1$

$$\sigma_f = 1.43 \times (HV + 120) \times \left(a\sqrt{\pi/2}\right)^{-1/6}. \tag{10.15}$$

Recalling the well-known Hall-Petch relationship [13, 14] between yield strength σ_y and grain d dimension

$$\sigma_y = \sigma_o + k \times d^{-1/2} \tag{10.16}$$

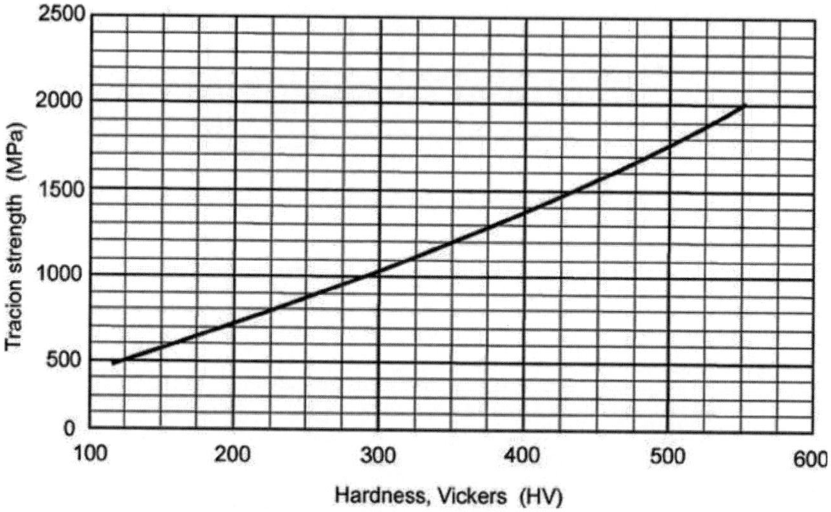

Fig. 10.13 Diagram showing the relation between hardness and traction strength

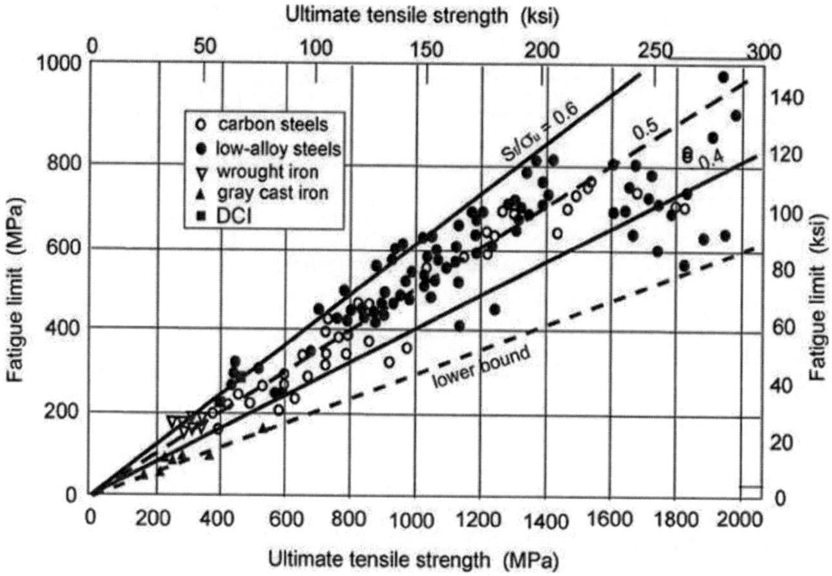

Fig. 10.14 Fatigue limit versus ultimate strength for various steels and cast iron [9]

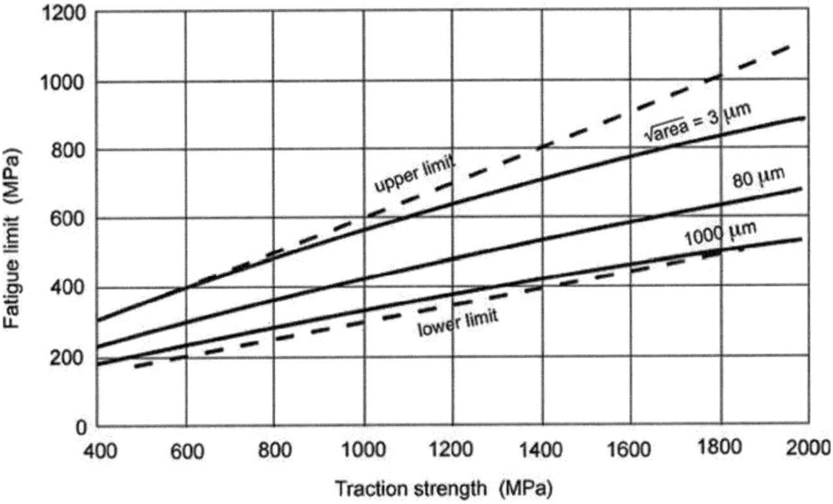

Fig. 10.15 Comparison between upper and lower bound curves of Fig. 10.14 and Murakami equation predictions versus defect area

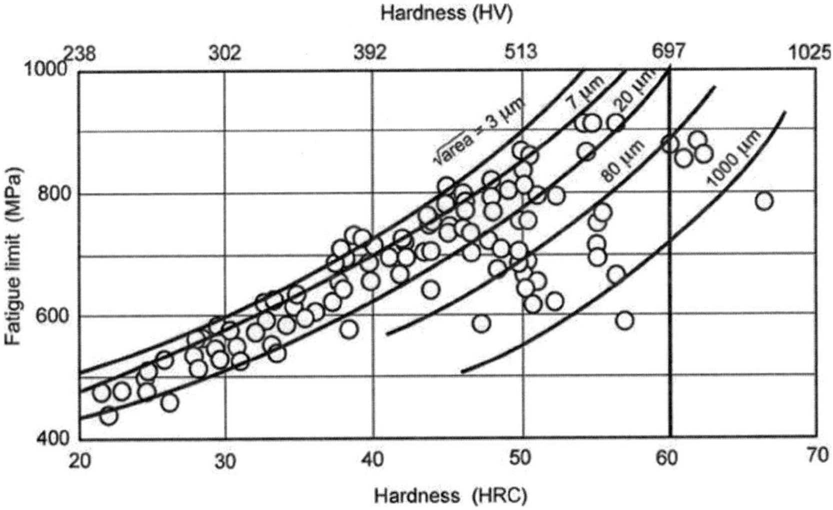

Fig. 10.16 Comparison between Murakami predictions for different defect area and experimental data of Fig. 4.37

in which σ_o is the yield strength of a single grain material and k a parameter that considers the grain boundary capability to block dislocations. Since for stain hardening materials HV hardness is linked to yield strength by

$$HV = \frac{3}{9.8}\sigma_y \tag{10.17}$$

Supposing that the $2a$ defect size is equal to the grain size d, we obtain that Eq. (3.10) for $R = -1$ may be written as

$$\sigma_f = 1.43 \times \left[\frac{3}{9.8}\left(\sigma_o + kd^{-1/2}\right) + 120\right]\left(\frac{d}{2}\sqrt{\frac{\pi}{2}}\right)^{-1/6}. \tag{10.18}$$

Equation (10.18) is rather difficult to apply since both parameters σ_o e k are not easy to derive. Therefore, it is more convenient to use the equation

$$\sigma_f = 1.43 \times \left(\frac{3}{9.8}\sigma_y + 120\right)\left(\frac{d}{2}\sqrt{\frac{\pi}{2}}\right)^{-1/6}. \tag{10.19}$$

As in the first of Eq. (10.4), also in Eq. (10.19) the yield strength of the material appears. Alternatively, considering the HV hardness Eq. (10.19) may be written as

$$\sigma_f = 1.43 \times (HV + 120)\left(\frac{d}{2}\sqrt{\frac{\pi}{2}}\right)^{-1/6}. \tag{10.20}$$

Table 10.3 provides a synoptic picture of the empirical equations presented for the assessment of the fatigue limit of steels. About light alloys we have been saying that it is sometimes difficult to find a fatigue limit. Experimental data show a continuous descent of the allowable stress amplitude σ_a even below 10^7–10^8 cycles. This is actually contributing to the apparent lack of a fatigue limit in light alloys. It is, in fact, rater time consuming to reach 10^8 cycles with a normal hydraulic or electromechanical load cell that cannot go beyond 10 cycles per second or 900,000 cycles per day. To reach 10^8 cycles and get one single result it takes at least four months. Today, with resonant-type testing machines shown in Fig. 3.3 that may go as high as 100–120 Hz this waiting time is sharply reduced to a week or ten days at most. However, it is not certain to reach fatigue failure also after 10^8 cycles so that usually tests are terminated at 10^8 cycles and the corresponding value of the stress amplitude is reported as if it were the fatigue limit σ_f with an indication of run out. Note that to go beyond 120 Hz may require a cooling system of the specimen under testing.

This is particularly true at the macro crack formation when energy is dissipated into heat at the crack tip plastic zone. The temperature may, therefore, significantly increase locally. As far as aluminum alloys is concerned, it must be said that they are usually divided into two groups, one of low-medium strength that depends mainly

Table 10.3 Synoptic table for the assessment of fatigue limit σ_f or S_f in steels

Reference	Formula	Note
Equation (10.1)	$\sigma_f = 0.5 \cdot r$	Steels
Equation (10.4)	$\sigma_f = 32.0 + 1.58 \cdot HV + \varepsilon_{HV}$	$\Delta(\varepsilon_{HV}) = 17.8$
Equation (10.6)	$\sigma_f = 1.6 \cdot HV \pm 0.1 \cdot HV$	
Equation (10.8)	$\sigma_u = 3.6\text{–}3.3 \cdot (HB)$	3.6 for mild steel; 3.3 for alloy steel
Equation (10.9)	$\sigma_u = 0.7 \cdot (HB)$	Gray cast iron
Equation (10.10)	$\sigma_u = 2.8 \cdot (HB)$	DCI
	$\sigma_u = 5.2 \cdot (HB)$	Austenitic stainless steels
	$\sigma_u = 5.5 \cdot (HB)$	Copper
Equation (10.7)	$\sigma_f = (9.16 \cdot HRC + 308)$	Hardness $\leq$ 45 HRC
Equation (10.4)	$\sigma_f = 168 + 0.41 \cdot \sigma_y + \varepsilon_y$	$\Delta(\varepsilon_s) = 16.3$
Equation (10.12)	$\sigma_f = K \times (HV + 120) \times \left(\sqrt{area_{\max}}\right)^{-1/6} \left(\frac{1-R}{2}\right)^{\alpha}$	$K = \begin{cases} 1.41 \ for \ surface \ defects \\ 1.43 \ for \ sub\text{–}surface \ defects \\ 1.56 \ for \ \text{internal} \ defects \end{cases}$ $\alpha = \left(HV \times 10^{-4} + 0.226\right)$
Equation (10.20)	$\sigma_f = 1.43 \times (HV + 120)\left(\frac{d}{2}\sqrt{\frac{\pi}{2}}\right)^{-1/6}$	d = grain size
Equation (10.19)	$\sigma_f = 1.43 \times \left(\frac{3}{9.8}\sigma_y + 120\right)\left(\frac{d}{2}\sqrt{\frac{\pi}{2}}\right)^{-1/6}$	d = grain size

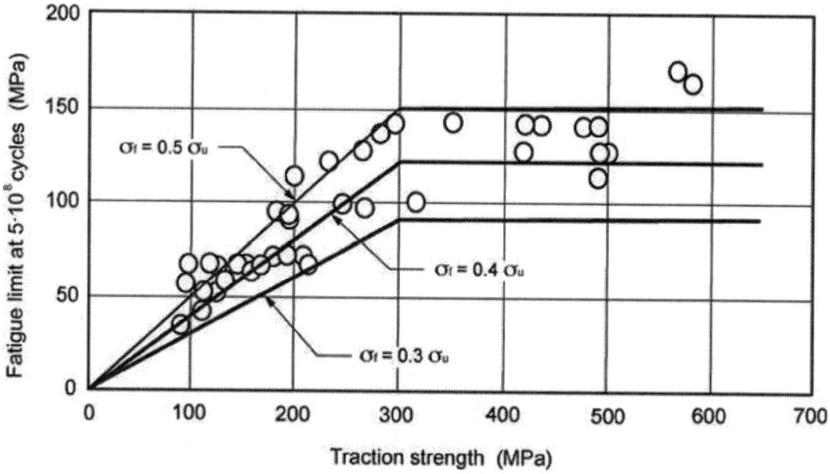

Fig. 10.17 Fatigue limit in aluminum alloys at 5×10^8 cycles [15]

on alloying elements and the other of high-very high strength obtained by precipitation hardening through opportune heat treatments. Alloys of the first group, which magnesium–aluminum alloys belong, normally show a precise fatigue limit almost as in steels and a better σ_f/σ_u ratio, usually equal to 0.4–0.5. Alloys of the second group generally withstand fatigue better, but with a lower σ_f/σ_u ratio and a fatigue limit that does not depend on the traction strength and stabilizes around 135 MPa. A compendium of fatigue limits versus tensile strength is offered in Fig. 10.17 for a series of aluminum alloys [15] among which there is 2014–0 T4 and T6, 2024-T3, 6061–0 T4 and T6, 6063–0 T42, T5 and T6 and 7075-T6. It is possible to catch what has been said: up to about 300 MPa ultimate strength the ratioσ_f/σ_u is about 0.4 in the average, afterwards the dependence of σ_f on σ_u flattens.

The reason for the lower bound ratio $\sigma_f/\sigma_u \cong 0.3$ observed on aluminum alloys having high strength obtained through appropriate heat treatments is to be ascribed to a cycling softening phenomenon (see Sect. 1.3) that actually dissolves coherent precipitates reducing the metal strength. Nevertheless, it is also possible that in these alloys dimensions, distribution and localization of intermetallic phases may act as inclusions do in high-strength steels that, as already said, cause the leveling off of the fatigue limit that beyond 1200 MPa ultimate strength does not continue to grow any more, but stabilizes on 700–800 MPa.

Typical is the extruded zinc-magnesium–aluminum alloy, type DTD 683, that presents a fatigue limit at 10^8 cycles exceptionally high: about 230 MPa [16]. It is not clear whether this is due to intermetallic particles extremely small or to the extrusion process that could introduce a residual compression stress state and, at the same time, squeeze the foreign particles farther reducing their dimensions. In general, we can say that metals having a BCC structure, like iron and its alloys, cobalt, niobium, tantalum, molybdenum and tungsten and its alloys, present *S/N*

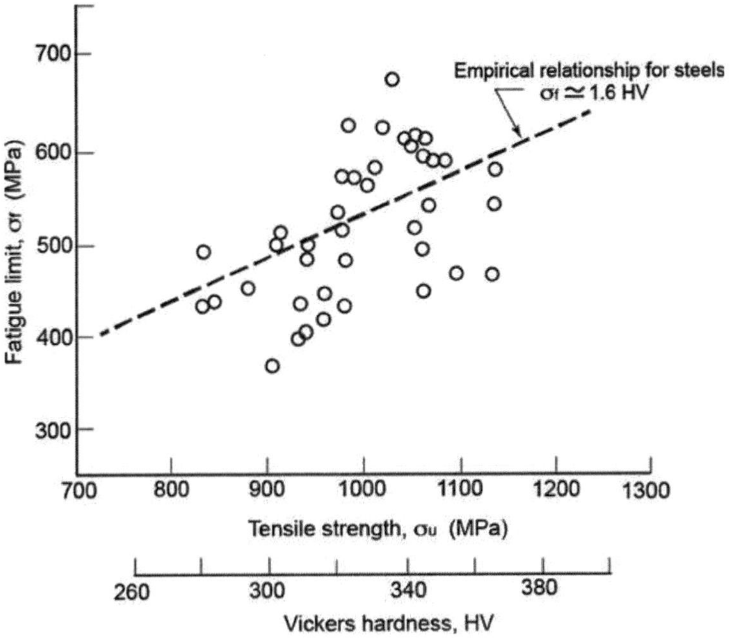

Fig. 10.18 Relationship between fatigue strength and ultimate tensile strength, or Vickers hardness, for Ti alloys [18]

curves having a well-defined *knee* before10^7 cycles and, at the same time, seem to have a ratio by far higher than those having FCC structure as nickel, aluminum, copper, silver and gold [17] (see Fig. 1.6).

Titanium, that has a close-packing hexagonal lattice (CPH), presents a behavior similar to steel, as it can be seen in Fig. 10.18 [18]. The large scatter of experimental data is due to the nonmetallic inclusions size variability. Table 10.4 is a compendium of results obtained by several researchers [19–26] on fatigue behavior of various metal alloys.

10.3 Empirical S–N Diagram-Wöhler Curve. Basquin Line

Once the values of σ_f and S_3 are assessed through the knowledge of the ultimate strength or the hardness of the material, using any of the empirical formulas previously shown and listed in Tables 10.3 and 10.4, it will be sufficient to draw a straight line connecting points (1.000; S_3) and (10^6 or 10^8;S_f) on an *S–N* diagram to have a first approximation fatigue curve for preliminary design purpose, as schematized in Fig. 10.19. It must be also remembered that σ_f is taken at 10^6 cycles for BCC metals and at 10^8 for FCC materials. The slope of this line is given by

Table 10.4 Compendium of ultimate strength and fatigue limit for steels and metal alloys [19–26]

Alloy	Ultimate strength σ_u (MPa)	Fatigue limit σ_f (MPa)	σ_f / σ_u
Mild steel	410	190	0.46
Nickel–chromium steel	865	480	0.55
SAE 4130 steel	800	338	0.42
0.35% C steel	620	210	0.34
Ferrite			0.58
Ferrite-perlite			0.38
Quenched martensite			0.26
Tempered martensite			0.55
Austenite			0.37
Cast iron			0.40
75S-T6 aluminum alloy	570	207	0.36
75S-T6 aluminum alloy	570	193	0.40
7075-T6 aluminum alloy	570	84	0.27
24S-T3 aluminum alloy	510	152	0.30
14S-T6 aluminum alloy	480	165	0.34
2014-T6 aluminum alloy	480	104	0.21
24S-T4 aluminum alloy	465	173	0.37
2024-T4 aluminum alloy	465	118	0.25
6061-T6 aluminum alloy	310	100	0.32[a]
5056–0 aluminum alloy	294	140	0.48[a]
5.6%Zn0.66%Zr magnesium alloy	355	151	0.42
2½%Zn0.65Zr magnesium alloy	260	90	0.35
Molybdenum (cast)	490	343	0.70

[a]Revers albending tests

$$b = \frac{\log S_3 - \log S_f}{\log 10^3 - \log 10^6} = -\frac{1}{3}\log\left(\frac{S_3}{S_f}\right). \tag{10.21}$$

In 1910 Basquin [27] observed that stress-life *S–N* data relative to high cycle fatigue domain (Region II of fatigue, see Fig. 1.10) could be plotted linearly in a log–log scale, which suggested a power law relationship for the elastic curve of the type (see Sect. 7.2)

$$\varepsilon_a E = \frac{\Delta\varepsilon_a E}{2} = \sigma_a = \sigma'_f (N)^b \tag{10.22}$$

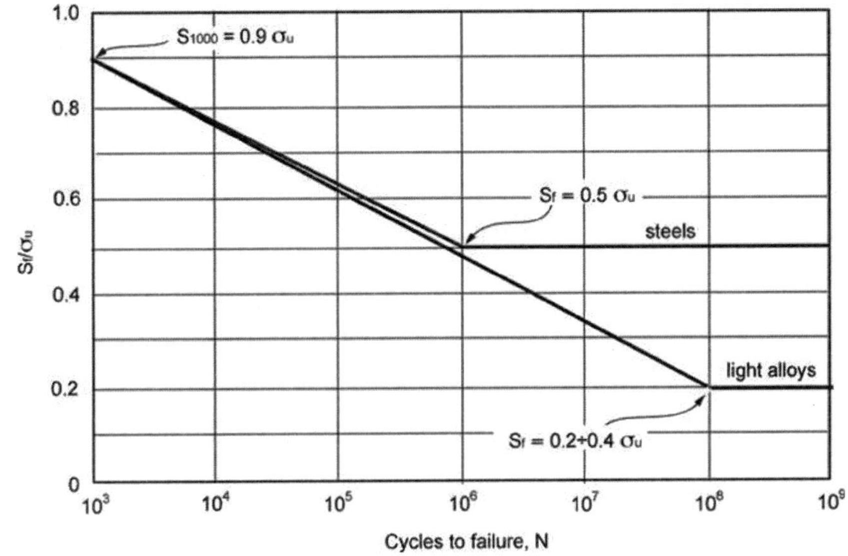

Fig. 10.19 Generalized *S–N* curve for steel and light alloys

where $\varepsilon_a = \Delta\varepsilon_a/2$ is the elastic strain amplitude, E the Young's modulus, σ_a the elastic stress amplitude, N the life at σ_a, σ'_f a fatigue strength coefficient and b a fatigue strength exponent or Basquin's exponent. Both σ'_f and b are fatigue properties of the material. Both σ'_f and b can be assessed from experimental *S–N* curve as one-reversal intercept of the elastic component. As a first approximation, the σ'_f coefficient is equal to the monotonic true failure stress $\sigma_{t,f}$ while b varies from –0.05 to –0.12. This actually means that metals with higher strength are the most suitable to withstand high cycle fatigue, as already said in Sect. 1.2 and shown in Figs. 1.15 and 1.16. It must be understood that the true failure stress$\sigma_{t,f}$ in a monotonic traction test is higher than the so called engineering tensile strength σ_u, while the true strain is always lower than the engineering strain ε based on the initial length l_o and final length l_f of the specimen

$$\varepsilon = \frac{l_f - l_o}{l_o} = \frac{\Delta l}{l_o}. \tag{10.23}$$

This is because the actual length l of the specimen changes during the loading phase and cannot be considered constant and equal to the initial one l_o. The change is particularly large as the elastic limit is exceeded and cannot be ignored. Therefore

$$\varepsilon_t = \int_{l_o}^{l} \frac{\partial l}{l} = \log\left(\frac{l}{l_o}\right) = \log\left(\frac{l_o + \Delta l}{l_o}\right) = \log(1 + \varepsilon). \tag{10.24}$$

Equation (10.24) assumes that from yielding to necking, point A of Fig. 10.16, the specimen experiences uniform deformations. Figure 10.20 is a schematic of the series of event that lead to final failure after necking. The elastic field terminates with yielding, specimen (A) of Fig. 10.20, afterwards the specimen undergoes a uniform dilatation process that does not change its volume, specimen (B) of Fig. 10.14. This phase terminates at point of maximum stress or ultimate strength, σ_u. It this just at this moment that necking starts to appear, specimen (C) of Fig. 10.20. Once necking occurs, deformations are no longer uniform, but concentrate in the necking region. Therefore, Eq. (10.24) is valid till necking occurs, afterwards the logarithm cannot be calculated. Also, the true stress, σ_t, is different and, this time, higher than the engineering stress, σ, because the specimen cross area A is continuously reducing, in particular during the phase of uniform deformations.

Beyond the elastic limit till the first appearance of necking, in practice till the maximum or ultimate stress σ_u, the passage from the engineering characteristics $\sigma - \varepsilon$ to the true quantities $\sigma_t - \varepsilon_t$ is, then, given by the relationships

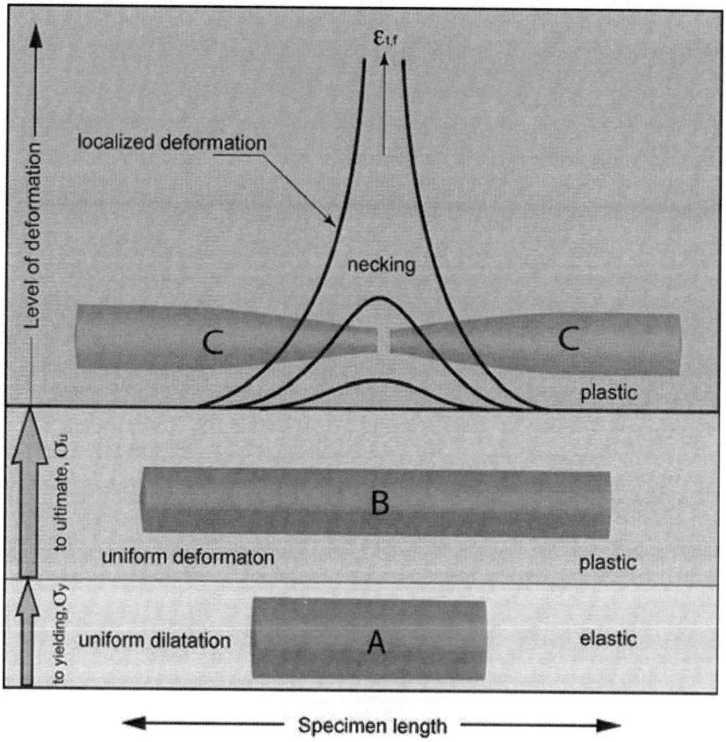

Fig. 10.20 Schematic of deformation stages that undergoes a specimen in monotonic traction from elastic dilatation to plastic deformation and eventually necking

$$\begin{aligned} \varepsilon_t &= \log(1+\varepsilon) \\ \sigma_t &= \sigma(1+\varepsilon)\,. \end{aligned} \tag{10.25}$$

As schematized in Fig. 10.22 where points D, C, B and A move at the same time back and up in D′, C′, B′ and A′. A simple way to measure the true stress at failure, $\sigma_{t,f}$, is to take the ratio between the load at failure P_f and the final area of the specimen section A_f that may be much smaller than the initial one A_o because of the occurrence of necking, depending on the ductility of the material

$$\sigma_{t,f} = \frac{P_f}{A_f}. \tag{10.26}$$

However, this is not exactly the true stress at failure, $\sigma_{t,f}$, because the occurrence of necking introduces a kind of discontinuity or local indentation that alter the local stress state that becomes triaxial. Another method may be used to infer the true stress at failure. The region where necking develops, in fact, may be thought as a small or reduced specimen within the main specimen, as schematized in Fig. 10.15b.

Its initial dimensions are: length l_{red} (*reduced*) and area A_{bn} (*before necking*). The final dimensions at failure are: $l_{red,f}$ and A_f. The cross section area A_{bn} of the specimen right before necking does not change till the final rupture because deformation and cross section reduction concentrate in the necking region. During the phase of uniform plastic elongation, the volume of the specimen does not change so that between two loading steps 1 and 2 it is always

$$\begin{aligned} l_1 A_1 &= l_2 A_2 \\ \frac{l_1}{l_2} &= \frac{A_2}{A_1}. \end{aligned} \tag{10.27}$$

Introducing the reduction of area RA

$$\begin{aligned} RA &= \frac{A_1 - A_2}{A_1} = 1 - \frac{l_1}{l_2} \\ (1+\varepsilon) &= 1 + \frac{l_2 - l_1}{l_1} = \left(\frac{1}{1-RA}\right) \\ &= \left(\frac{A_1}{A_2}\right). \end{aligned} \tag{10.28}$$

The same shall be supposed to hold also for the reduced specimen, Fig. 10.15c and, as first approximation, it can be written

$$l_{red} A_{bn} = l_{red,f} A_f. \tag{10.29}$$

Writing Eq. (10.25) for the reduced specimen and recalling Eq. (10.28) it is

$$\varepsilon_{t,red,f} = \ln(1 + \varepsilon_{red,f}) = \ln\left(\frac{1}{1 - RA}\right) = \ln\left(\frac{A_{bn}}{A_f}\right). \tag{10.30}$$

It will suffice, then, to measure the final area A_{bn} of the specimen far from necking and the area A_f at necking to infer the true deformation of the reduced specimen at failure, $\varepsilon_{t,f,red}$, see Fig. 10.21. As a first approximation, the total deformation $\varepsilon_{t,f}$ of the entire specimen will be that of the reduced specimen, $\varepsilon_{t,f,red}$, given by Eq. (10.30) plus the other one that has interested the specimen right before necking

$$\varepsilon_{t,f} = \ln\left(\frac{A_{bn}}{A_f}\right) + \ln\left(\frac{A_o}{A_{bn}}\right) = \ln\left(\frac{A_o}{A_f}\right). \tag{10.31}$$

From an operative point of view the procedure is the following: knowing the stress–strain curve of the material (the engineering curve) draw the true stress-true strain curve. Each point σ–ε of the engineering curve, like A, B, C and D will yield the corresponding A′, B′, C′ and D′ on the true stress-true strain σ_t–ε_t diagram using Eq. (10.25), as shown schematically in Fig. 10.22. This construction terminates at point A corresponding to the engineering ultimate stress or maximum stress $\sigma_u = \sigma_{max}$ where necking occurs. At necking, in fact, Eq. (10.25) lose their validity because

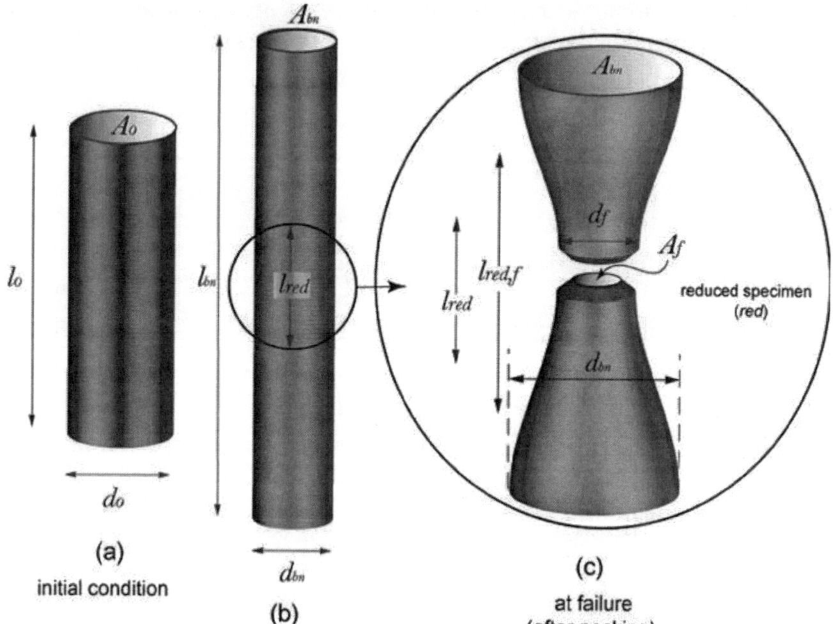

Fig. 10.21 **a** Specimen before testing, **b** specimen during testing right before necking, **c** enlargement and details of the area where necking occurred

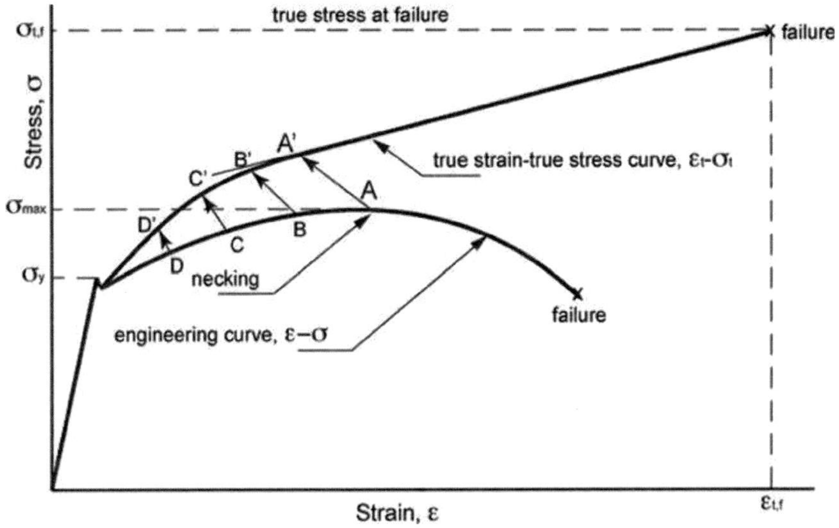

Fig. 10.22 Schematic showing engineering curve ε-σ and true strain-true stress curve ε_t-σ_t

Fig. 10.23 Schematic of traction specimen at necking. Symbols used in the Bridgman equation are shown

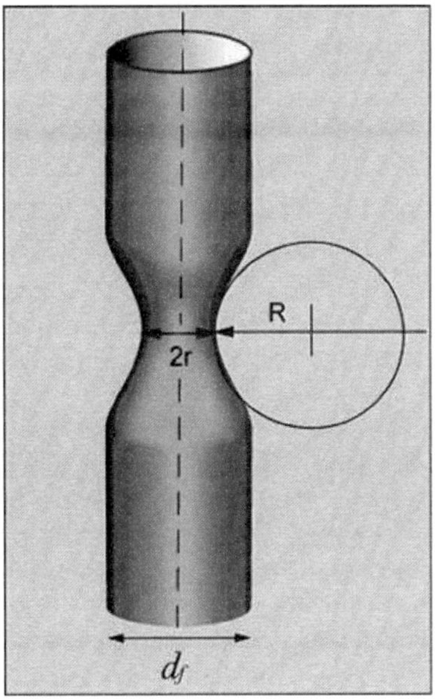

from this point on the specimen elongation is no longer uniform. From point A′ on the true curve, draw a line that terminates at the point of coordinates $\varepsilon_{v,f}$ given by (10.31) and $\sigma_{v,f}$ given by Eq. (10.26). There are arguments that would indicate this line as the tangent in A′ to the true strain-true stress curve. However, also the true stress at rupture assessed through Eq. (10.26) may not be the real one. Necking, in fact, introduces in the specimen a discontinuity that appears like a notch, more or less blunted. The specimen, in other words, initially cylindrical, at necking assumes the shape of a hourglass (Fig. 10.23).

The notch introduces a local stress concentration so that stress on the final cross section of the specimen is no longer uniform, as Eq. (10.26) is predicting. Moreover, the notch introduces a multiaxial stress state so much higher as necking is sharper. Triaxiality allows the metal section to react better because it reduces the equivalent stress σ_e. Bridgman [28, 29] was the first to address the issue in 1944. He proposed a solution for the equivalent stress σ_e on the reduced section at rupture that depended on the shape of necking

$$\frac{\sigma_{t,f}}{\sigma_e} = \left(1 + \frac{2R}{r}\right) \ln\left(1 + \frac{r}{2R}\right) = k \tag{10.32}$$

where R represents the radius of curvature of necking, r the specimen radius on the necked section, as schematized in Fig. 10.17, and $\sigma_{t,f}$ is the true stress at failure given by (10.26).

Equation (10.32) states that the Bridgman factor k is always larger than one, which means that the equivalent stress σ_e is always smaller than $\sigma_{t,f}$ obtained from Eq. (10.26) and tends asymptotically to one for $R \to \infty$, i.e., in complete absence of necking and, therefore, of triaxiality. The Bridgman solution is based upon a series of working hypothesis.

- axial deformation on the section of necking is uniform;
- the stress state on the section of necking is proportional;
- the material is ideal elastic–plastic (without strain hardening).

The first hypothesis is likely to be true. The second one is not completely true. It would be true if the radius of curvature were not varying during the loading phase, which is not the case since the specimen is continuously reducing its section as the load increases. The third hypothesis raises the question of whether or not strain hardening may have an effect on the effective response of the specimen to loading and how it may be taken into consideration. In highly strain hardening materials having also very high ductility, as austenitic stainless steel, Eq. (10.32) may underestimate the real equivalent stress at rupture. In high strength materials and, in particular, in metals showing a brittle behavior necking is generally little developed and its effect reduced. It shall not come as somewhat of a surprise the fact that a fatigue property such as the strength coefficient σ'_f is given by a monotonic traction test, being equal to $\sigma_{t,f}$. To understand this apparent discrepancy it is enough to recall that σ'_f is given by the intercept of the Basquin line with the ordinate axis in a *S–N* log–log diagram. In a log–log scale the ordinate axis is located at the abscissa equal one, $N = 1$. But

$N = 1$ actually means that fatigue failure occurs after just one single cycle or, better, at the end of the rising part of the cycle, which is precisely a monotonic loading. Basing on energy arguments, Morrow [30] found that the fatigue strength exponent b that appears in (10.22) could be expressed as

$$b = \frac{-n'}{1 + 5n'} \tag{10.33}$$

which actually means that in high cycle fatigue low values of the cyclic strain hardening exponent n' favor the fatigue strength of materials. From (10.22) it derives that the Basquin line may be expressed as

$$\begin{aligned} N &= N_f\left(\frac{\sigma_f}{\sigma_a}\right)^{-1/b} \\ &= N_f\left(\frac{\sigma_f}{\sigma_a}\right)^{k} \end{aligned} \tag{10.34}$$

with $k = -1/b$ and N_f being the life corresponding to fatigue limit σ_f. The Basquin line does not consider any knee or lower bound at 10^6 cycles for iron alloys or 10^8 cycles for light metals with a fatigue limit but run all the way beyond 10^6 cycles with the same slope given by Eq. (10.22). However, it is rather common opinion to-day that even stress amplitudes below the fatigue limit may be effective, see Chap. 9, in particular when preceded by cycles of higher amplitude (see Sect. 12.3.3). In the case of variable amplitude loading, that will be treated in Chap. 12, fatigue life does not terminate with a more or less sharp knee but continues beyond the fatigue limit although with a lower slope. To take into consideration this particular effect, Haibach [31] has suggested a correction to the Basquin line slope to apply beyond 10^6 cycles. The new slope reduces from $-1/b = k$ to $2\,k - 1$

$$N = N_f\left(\frac{\sigma_f}{\sigma_a}\right)^{(2k-1)}. \tag{10.35}$$

Figure 10.24 shows this bi-linear *S–N* diagram obtained with the Haibach correction. The Basquin line provides the simplest way to build the *S–N* diagram, provided that the slope b is known. However, it must be remembered that only the elastic component of the stress amplitude enters in the Basquin Eq. (10.22). Therefore, the Basquin line approximate the Wöhler curve only in the high cycle fatigue regime, as shown in Fig. 10.24. When the plastic component of the strain cannot be neglected and this happens in general below 10^4–10^5 cycles, the Basquin component must be integrated with the Mason-Coffin component, as it will be shown in the next chapter.

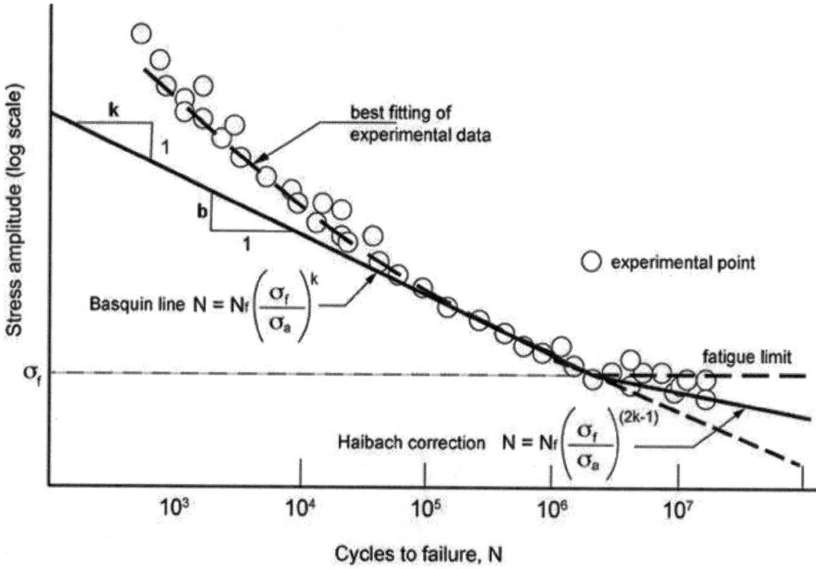

Fig. 10.24 Haibach correction to the Basquin *S–N* line

10.4 Staircase Test Method

As already said in Sect. 6.1, data scatter in fatigue becomes larger and larger as the stress amplitude decreases. Therefore, the determination of fatigue limit is characterized by the largest dispersion of experimental data. This, we know, is due to metallurgical variability and hence to specimen dimensions (see Sect. 6.6.1), to surface condition variability and other factors, included test conditions This rather high data scatter requires an equally high number of test specimens to operate any reliable statistics to infer the possible lower bound of fatigue limit with the required confidence. A testing method has been introduced, known as *stair case method* or *up-and-down method* that is intended to provide a reliable answer to the determination of fatigue limit with a limited number of specimens. The statistics make use of the Dixon and Mood analysis [32]. This actually means that its application is possible only if the experimental data have a normal or log-normal distribution (see Sect. 6.1.1). The method has also another important limitation in that it provides only the mean value of the fatigue limit, though with great confidence. To infer an extreme value of fatigue limit, for example 99% or 99.9% probability of survival, it may be used the Student distribution, but it is necessary to know the actual standard deviation. Developed by Nakazawa and Kodama [33] the method has been first adopted by the Japan Society of Mechanical Engineers in 1981, by ASTM in the States. Addressing the fatigue limit determination, the method does not consider at all the number of cycles at failure. It is an incremental method in the sense that any testing step is established on the base of the previous one.

In its standard format it utilizes a simple protocol based on 14 specimens, of which eight are used to determine the fatigue life and only six (the minimum) to infer the fatigue limit, according to the schematization of Fig. 10.25. In the first phase of fatigue life determination, it is recommended that at each stress level at least two specimens be tested. In this case it may be possible to express in terms of tests with replicable data. It may be also defined a percentage of replica PR based on the number of stress levels N_s and sample size N_{tot} used in the fatigue life determination

$$PR = 100 \times \left(1 - \frac{N_s}{N_{tot}}\right). \tag{10.36}$$

The PR must be equal to.

- 17–33 for preliminary tests
- 33–50 for research and development tests
- 50–75 for design tests
- 75–80 for reliable tests.

Once the fatigue life diagram has been determined, it is possible to make a first approximate estimate of the fatigue limit drawing an horizontal line that intersects the fatigue life diagram at 10^6 or 2×10^6 cycles, as shown in Fig. 10.25. It is up and down this line that the remaining six (minimum) specimens will be shot in sequence.

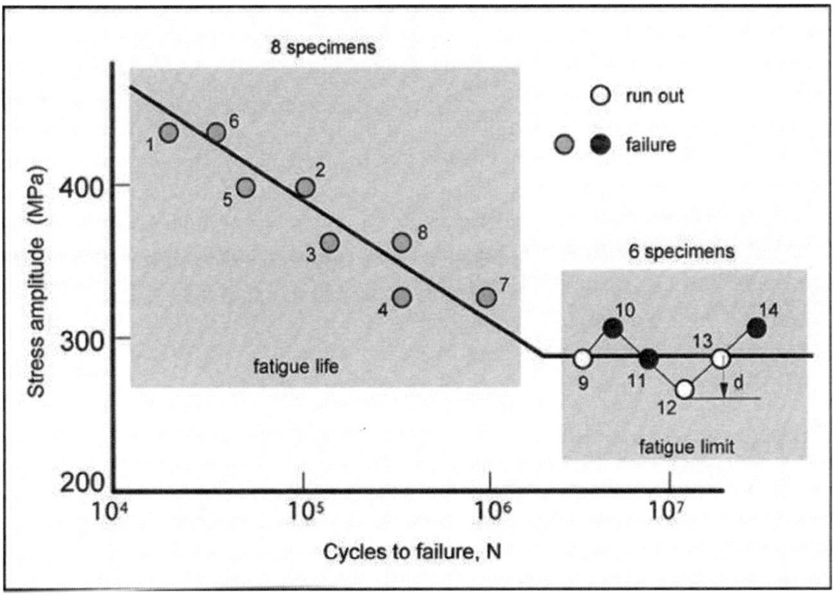

Fig. 10.25 Up-and-down or stair case fatigue test method used to determine mean fatigue limit at $2{\cdot}10^6$ cycles (the example refers to a carbon steel)

The first step is to fix the so called *run out*, i.e., the number of cycles at which it is intended to evaluate the fatigue limit, usually 2×10^6 or 10^7 for steels. The, n the initial stress amplitude is established that may be up or down the selected horizontal line, Fig. 10.25, as that amplitude at which it is thought that the specimen will or will not fail at the chosen number of cycles (failure or run out). The last preliminary choice is the interval or step size d by which the stress amplitude will be either incremented or reduced in each of the following tests. This interval d will be kept strictly constant throughout the test. It is fundamental that this interval d be maintained within a maximum of two standard deviations which means that the standard deviation must be known. If this is not the case it must be adopted an interval d small enough. If the first specimen does not break at the selected number of cycles (run out) the test is interrupted and a second specimen is then tested at stress amplitude incremented by a quantity d. If this second specimen fails within the given number of cycles, then the stress amplitude is reduced by d and the test continues with the remaining specimens either increasing or reducing the stress amplitude depending on the former result. Once all specimens are tested, the results are analyzed according to the Dixon e Mood analysis.

This analysis, it has been said, assumes that experimental data are distributed according to a Gaussian or, at least, a log-normal distribution. Often or may be too often this condition is forgotten. With only six specimens it is really arduous to check this compliance mainly because they are referring to even less stress levels and without knowing the number of cycles at failure, but just whether thy run out or not. We may try to give an answer to this important question by running enough traction tests checking whether the results are distributed or not according to a Gaussian. It has been shown in Sect. 10.2 how the fatigue limit is related to traction strength so that if the traction data have a Gaussian or log-normal distribution it is likely to assume that also fatigue limit data do. At this purpose, problem 1 of Sect. 6.8.1 Fig. 6.3 provides a suitable mean of identification. Under the restrictions cited above, the Dixon and Mood analysis provides the fatigue limit with 50% of probability of survival using the following procedure: the various stress amplitude used in the tests are numbered with $i = 0,1,2,..i_{\max}$ from the lowest to the highest. In the example of Fig. 10.19 it is $i = 0,1,2$. The less frequent event between the two possible run out or failure is selected and the sum A, B and N are calculated

$$N = \sum_{o}^{i_{\max}} n_i;\ A = \sum_{o}^{i_{\max}} i \times n_i;\ B = \sum_{o}^{i_{\max}} i^2 \times n_i \tag{10.37}$$

in which n_i is the number of occurrences of the selected event in the i-level. In the example of Fig. 10.19 the run out and the failure events have the same frequency: three out of six, therefore the choice is in influent. Let's choose the run out event. At level $i = 0$ is associated one single occurrence, therefore is $n_o = 1$. For the other two is $n_1 = 2$ and $n_2 = 0$. The 50% probability fatigue limit $\sigma_{f,50\%}$ is given by

$$\sigma_{f,50\%} = \sigma_{\min} + d \times \left(\frac{A}{N} \pm 0.5\right) \tag{10.38}$$

in which σ_{min} is the lowest stress amplitude used. In Eq. (10.38) the sign (+) is used if the event considered is run out or (−) if it is failure. The assessment of the standard deviation is given by:

$$s = 1.62 \times d \times \left(\frac{N \times B - A^2}{N^2} + 0.029\right) \tag{10.39}$$

when the ratio

$$\left(\frac{N \times B - A^2}{N^2}\right) \geq 0.3 \tag{10.40}$$

or equal to

$$s = 0.53 \times d \tag{10.41}$$

when it is

$$\left(\frac{N \times B - A^2}{N^2}\right) < 0.3. \tag{10.42}$$

In the case of Fig. 10.25 with $d = 20$ MPa and $s = 8$, the criterion $d \leq 2{\cdot}8 =$ 16 MPa is not respected so that we have to use Eq. (10.41) or increase the number of tests. We may also define an interval confidence within the true mean value of the fatigue limit $\overline{\sigma}_{f,50\%}$ will fall

$$\sigma_{f,50\%} - t_{\sigma,v}\frac{s}{N} < \overline{\sigma}_{f,50\%} < \sigma_{f,50\%} + t_{\sigma,v}\frac{s}{N}. \tag{10.43}$$

In Eq. (10.43) $t_{\alpha,\nu}$ is the student parameter relative to the chosen level of confidence $C = (1-\alpha)$, listed in Table 6.5. We may also evaluate the Lower Confidence level, LCL, relative to the wanted confidence

$$LCL = \sigma_{f,50\%} - k \times s \tag{10.44}$$

k being the Lieberman factor or the Owen factor listed in Tables 6.6 and 6.7, respectively.

10.5 Sample Problems on S–N Curves

The following examples are intended to describe how *S–N* diagrams of metals and steels, in particular, may be constructed making use of empirical equations presented in the previous sections. However, such diagrams must be considered as preliminary fatigue curves to be used in a starting activity of designing waiting for an extensive and comprehensive experimental program that shall be considered necessary and unavoidable. The procedures that will be used to build *S–N* diagrams follows the scheme of Fig. 10.26.

The starting point is the determination of points A≡(1000;S_3) e B≡(N_f;S_f) of Fig. 10.26. The resulting fatigue curve may be considered a mean fatigue curve obtained by smooth, mirror-polished specimens. This curve shall be modified to take into consideration all factors affecting fatigue strength such as load factor (see Sect. 4. 9), surface finish (see Sect. 4.2), size effect (see Sect. 4.8) and, eventually, also the requested probability of survival $(1-P)$.

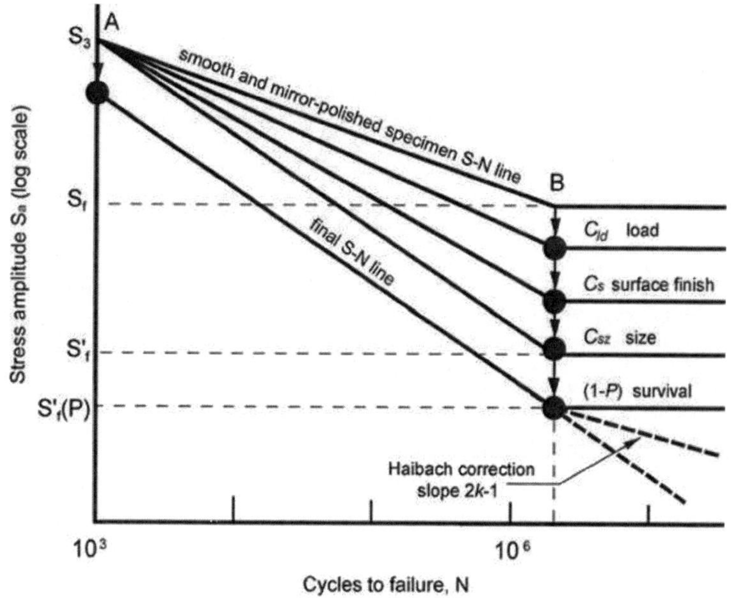

Fig. 10.26 Schematic construction of *S–N* diagram and its changes using modifying factors

10.5.1 Engineering Strain–Stress Curve and True Strain-True Stress Curve

The engineering strain–stress curve of a carbon steel specimen of medium strength and 10 mm diameter d_o and length $l_o = 50$ mm, is shown in Fig. 10.27. Construct the corresponding true strain-true stress curve. True strain ε_t and true stress σ_t are linked to the corresponding engineering quantities ε-σ by the relationships (5.25)

$$\varepsilon_v = \ln(1 + \varepsilon)$$
$$\sigma_v = \sigma(1 + \varepsilon)$$

Therefore, the true strain-true stress curve is that shown in Fig. 10.21 that run above the engineering one.

Note how the true curve terminates at 15.6% of true deformation. This is because beyond that value uniform deformation turns into necking and Eqs. (10.25) are no longer valid. Necking for this ductile steel is rather high as it can be seen in Fig. 10.28. True deformation at initiation of necking can be evaluated using Eq. (10.32) between A_o and A_{bn} with $A_{bn} = 66.47$ mm^2 measured at failure far from necking

$$\varepsilon_t = \ln(A_o/A_f) = \ln(78.54/66.47) = 0.166$$

while true stress at failure $\sigma_{t,f}$ is obtained as ratio between the load acting at failure, $P_f = 27{,}633$ N and the failure area P_f/A_f

$$\sigma_{t,f} = P_f/A_f = 27,633/26 = 1061 \text{ MPa}$$

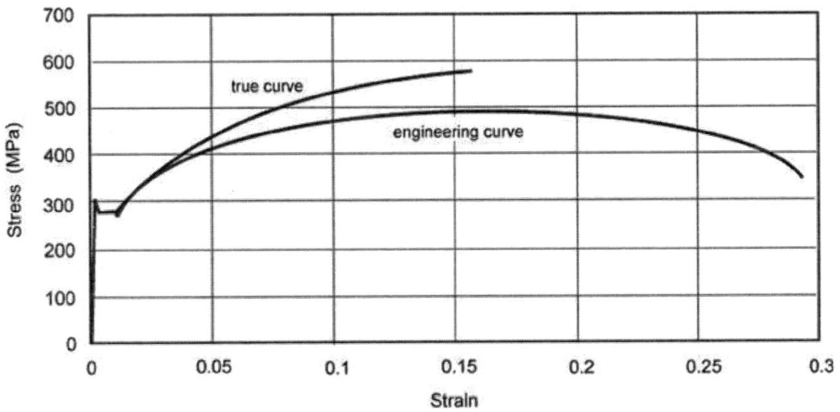

Fig. 10.27 Engineering strain–stress curve and true strain-true stress curve of a medium strength carbon steel specimen, type Fe510

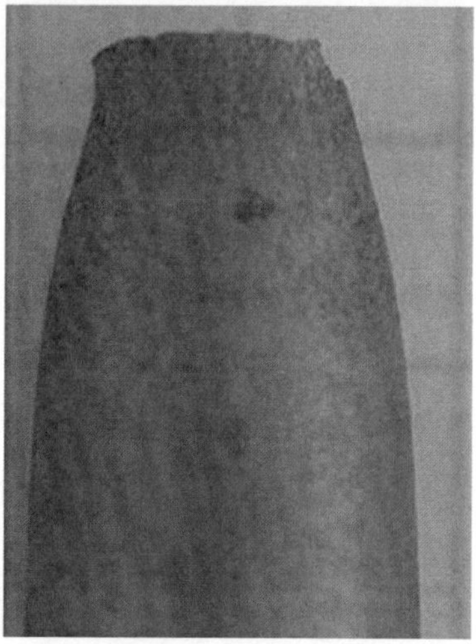

Fig. 10.28 Necking at failure in one of the carbon steel specimens of the sample problem

With symbols of Fig. 10.17, the necking shown in Fig. 10.22 gave the following data

$$R = 4.5 \text{ mm}; \; r = 2.88 \text{ mm}$$

then, the Bridgman correction (10.32) yields

$$\frac{\sigma_{t,f}}{\sigma_e} = \left(1 + \frac{2R}{r}\right) \ln\left(1 + \frac{r}{2R}\right) = k = 1.145.$$

Therefore, the equivalent stress at failure is

$$\sigma_e = \sigma_{t,f}/1.145 = 926 \text{ MPa}.$$

The corresponding deformation at necking is, from Eq. (10.31)

$$\varepsilon_{t,f} = \ln(A_o/A_f) = \ln(78.54/26) = 1.10.$$

Point of coordinates (1.1;926) is marked in Fig. 10.29 with letter A. Figure also shows point B of coordinates B $\equiv (\varepsilon_{t,f};\sigma_{t,f}) \equiv$ (1.18;1,061) obtained using $\sigma_{t,f} = P_f/A_f = 1061$ MPa. The two points have been joined to the true curve previously determined till necking. Note how the true stress given by the Bridgman correction is a little bit lower than that obtained considering a failure stress uniform on the entire cross section of the specimen and how true quantities are by far larger than the corresponding engineering values measured in a standard traction test. Deformation at failure that in a standard test happens at about 30% in real life exceeds 100% of the initial length of the specimen.

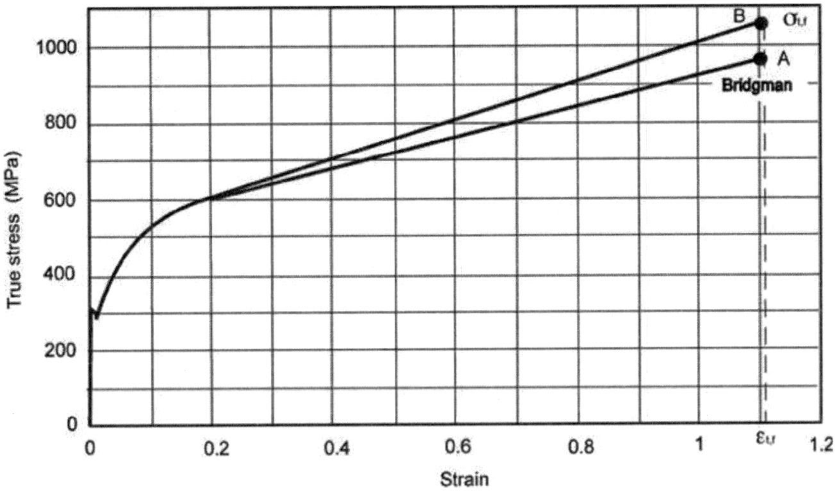

Fig. 10.29 Engineering and true stress–strain curves

10.5.2 S–N Curve from Hardening and Grain Size Data

With the steel of the previous problem 10.5.1, several cantilever beams 40 × 8 × 2 cm have been built that must withstand fully reversed bending loadings. Waiting for the necessary fatigue tests, for a first study determine the fatigue *S–N* curve knowing that the steel has a Vickers hardness of 157 HV and the mean ferritic grain size is about 40 μm and that the surface has a fine-ground commercially polishing. Moreover, cyclic tests have shown that the steel undergoes a cyclic hardening whose strain hardening exponent is $n' = 0.165$. The problem consists in the determination of S_3 and σ_f. Table 10.1 shows that S_3 is about $0.9{\cdot}\sigma_u$. Traction tests run on 10 specimens have confirmed the engineering curve of Fig. 10.27 yielding a mean ultimate strength $\sigma_u = 500$ MPa. Therefore

$$S_3 = 0.9 \times \sigma_r = 450 \text{ MPa at } 10^3 \text{cycles.}$$

The corresponding point is marked by letter A in Fig. 10.30. The second value to find is the fatigue limit σ_f. The simplest equation to use is (10.1) that yields

$$\sigma_f = 0.5 \times \sigma_u = 250 \text{ MPa.}$$

But we also know that the measured mean hardness is 157 HV so that Eq. (10.6) yields

$$\sigma_f = 1.6 \times 157 \pm 0.1 \times \text{HV} = 251 \pm 15.7 \text{ MPa.}$$

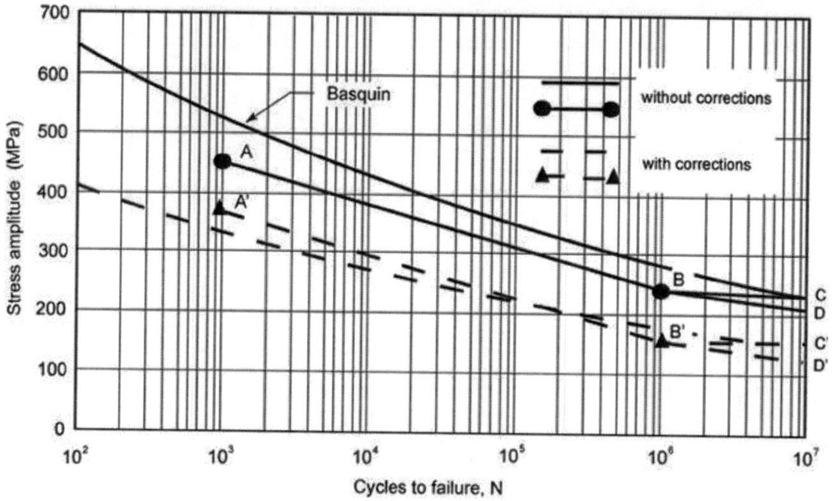

Fig. 10.30 *S–N* curves for the carbon steel considered in sample problem 2

Moreover, knowing that ferritic grain size is about 40 μm we may use the HPM Eq. (10.20) that yields

$$\sigma_f = 232 \text{ MPa}.$$

The four values obtained with different equations are close to each other, but not exactly equal. We may decide to take the average of the four obtaining

$$\sigma_f = 246 \text{ MPa}.$$

At this point we may trace also point B in Fig. 10.30 that will proceed horizontally in C from 10^6 cycles. However, the work pieces have a fine ground finish with a mean roughness of 10 μm. On the base of the abacus of Figs. 4.7 and 4.8 we may use a factor $C_s = 0.9$. Its volume is 640 cm^3, but the effective process volume it that limited to a depth of 0.5 mm at most since, as we know, high cycle fatigue is acting only on the surface. Therefore, the effective process volume will be $V_{eff} = 41.6 \text{ cm}^3$. The Shigley and Mitchell formula (4.15) with the equivalent diameter given by (4.16) yields a size factor C_{sz}

$$C_{sz} = 1.189 \times \left(\sqrt{\frac{0.05}{0.0766} 400 \times 80}\right)^{-0.097} = 0.73.$$

The Weibull exponent for this steel is $m = 29.64$. considered that the fatigue specimens have an effective reference process volume $V_o = 300 \text{ mm}^3$, it will be $V_{eff}/V_o = 139$. The abacus of Fig. 4.48 for $V_{eff}/V_o = 139$ and $m = 29.64$ yields C_{sz}

= 0.85. Considering that there will be built 100 pieces it may be convenient to use the Shigley and Mitchell value 0.73 as a saturation value. Therefore

$$S_3 = C_{sz} \times 450 = 0.73 \times 450 = 328.5\,\text{MPa}$$
$$\sigma'_f = C_s \times C_{sz} \times 246 = 0.9 \times 0.77 \times 246 = 170\,\text{MPa}.$$

Note that the surface finish factor has been considered only to correct the fatigue limit and not S_3 since we know that surface finish does not have any effect on low cycle fatigue. New points A′ and B′ are shown in Fig. 10.30. Let's evaluate the Haibach correction using Eq. (10.35). From Eq. (10.21) it can be inferred the slope b for line A′B′

$$b = -\frac{1}{3}\log\left(\frac{S_3}{S_f}\right) = -0.123.$$

Therefore

$$k = -1/b = 8.1$$

the final *S–N* curve is A′ B′ C′ shown in Fig. 10.30.

10.5.3 Basquin Line

An experimental campaign has been run on the carbon steel of the two previous sample problems to evaluate its cyclic characteristics. To this purpose 22 small hourglass specimens, so called button specimens, 8 mm minimum diameter and 2 cm long.

Specimens have been subjected to cyclic strain controlled loadings till stabilization of the hysteresis loop (see Sect. 1.3), as shown in Fig. 10.31 that refers to cycling with ± 0.3 total deformation. After 250 cycles the hysteresis loop stabilized. Stabilized cyclic strain–stress curve is shown in Fig. 10.32. Construct the *S–N* diagram using the Basquin procedure.

The experimental curve of Fig. 10.32 can be interpolated by the Ramberg–Osgood power low type curve of equation

$$\varepsilon_a = \frac{\sigma_a}{208,333} + \left(\frac{\sigma_a}{950}\right)^6. \tag{10.45}$$

Comparison with the monotonic curve of Fig. 10.27 indicates that the material undergoes cyclic hardening. The cyclic hardening exponent is $n' = 1/6 = 0{,}165$. Therefore, the b exponent of the Basquin line using the Morrow expression (10.33) is

$$b = \frac{-n'}{1 + 5 - n'} = -0.09.$$

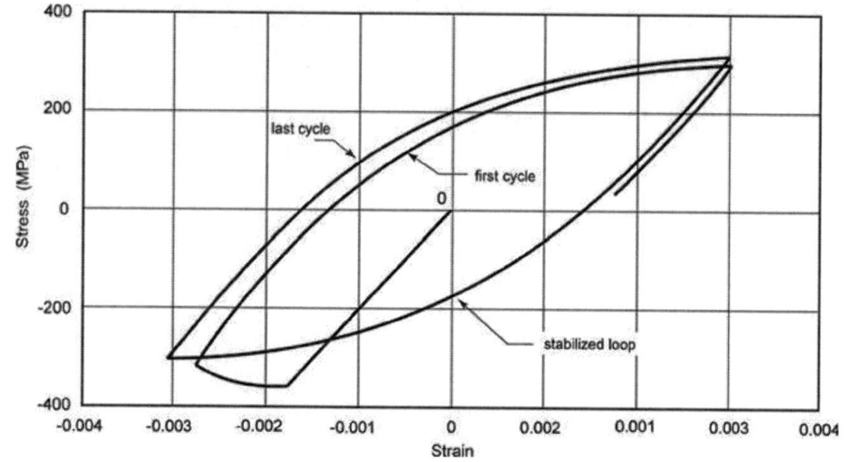

Fig. 10.31 Stabilized hysteresis loop under deformation controlled cycling from -0.3 to $+0.3$ mm/mm

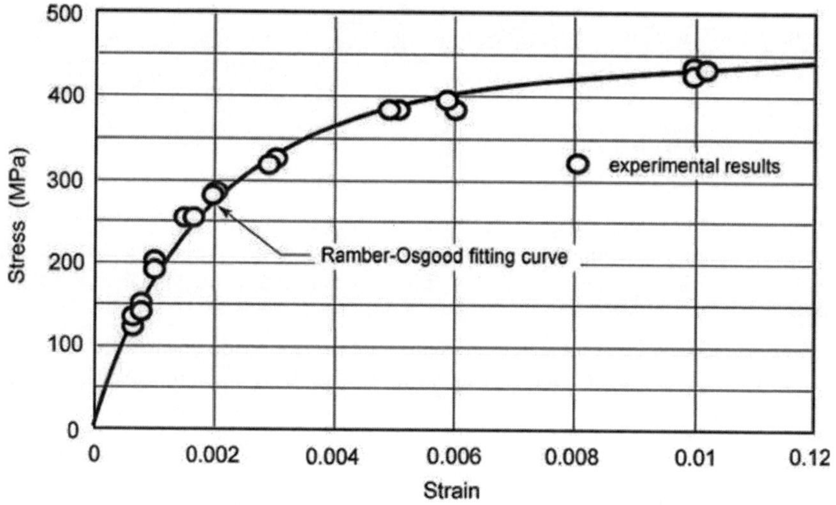

Fig. 10.32 Cyclic strain–stress curve

The Basquin line is shown in Fig. 10.30 by the full curve. Note that the Basquin curve is not a line because we have not been using a log–log scale, but a single log one. Also, in this case a correction must be applied to take into consideration the surface finish and the size effect multiplying the value of the fatigue limit σ_f times factors C_s and C_{sz} obtaining the dashed curve of Fig. 10.30.

10.5.4 Staircase Test Method

Always with the material of the previous sample problems, it has been run another experimental campaign to assess the fatigue limit of the carbon steel by the staircase test method. On the base of the *S–N* curves of Fig. 10.30 it has been chosen an initial stress amplitude of 225 MPa at a run out of 1,000,000 cycles and a step size $d =$ 10 MPa. Twenty-six button hourglass specimens of 8 mm minimum diameter have been used all having the same surface finish of the real work pieces to be built.

The results of the testing campaign are shown in Fig. 10.33. Fourteen specimens failed and 12 run out at 1,000,000 cycles. Therefore, the low frequency event of run out has been selected for calculations. The characteristic quantities are

$$N = 12,\ \mathrm{A}\ =\ 19\ \mathrm{and}\ \mathrm{B}\ =37(\mathrm{see\ Eq.10.37}).$$

Equation (10.38) yields

$$\sigma_{f,50\%} = \sigma_{\min} + d \times \left(\frac{A}{N} \pm 0.5\right) = 185 + 10 \times \left(\frac{19}{12} \pm 0.5\right) = 205.8\,\mathrm{MPa}.$$

The corresponding point is shown with a full square in Fig. 10.34 that is an enlarged detail of Fig. 10.30. Taking into consideration also the process volume effect $C_{sz} =$ 0.73 already found in sample problem N. 2 we obtain a point marked by the open square in Fig. 10.34.

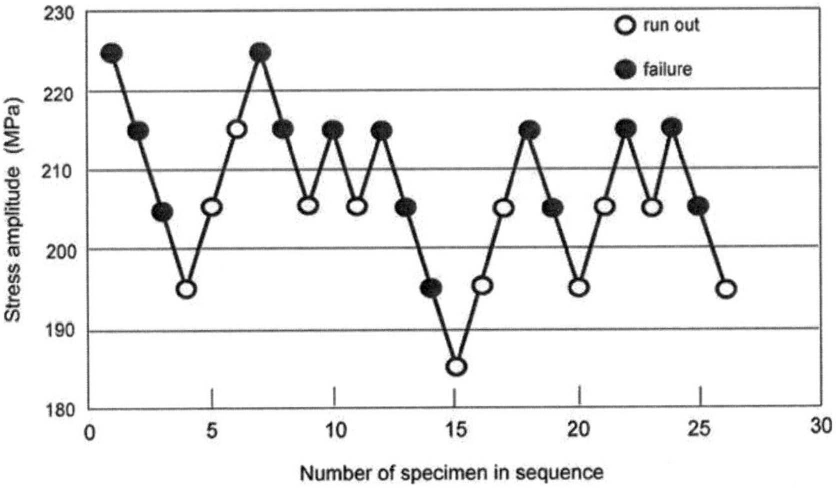

Fig. 10.33 Up-and-down fatigue test results for the carbon steel consider

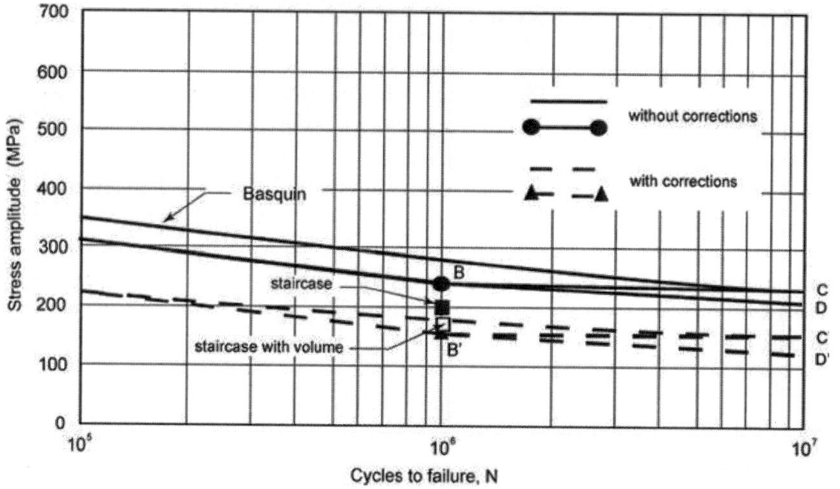

Fig. 10.34 Fatigue limit at 10^6 cycles inferred by staircase test method, shown on the previously determined *S–N* diagrams

10.5.5 Fatigue Life of Shafts Under Torsion

An alloyed-carbon steel shaft 850 MPa tensile strength having a Weibull exponent $m = 44$ is subjected to reversed torsional fatigue loading. The shaft of 15 mm diameter has been subjected to several heat treatments that produced a hardness of about 30 HRC followed by a shot peening process. Construct the relative *S–N* curve that in this specific case is a τ_a-*N* with a probability of survival $(1-P) = 99.9\%$.

The 30 HRC hardening has elevated the fatigue limit of the material to 583 MPa, as it can be deduced by Eq. (4.8) converting HRC hardness in HV and seen in Figs. 10.10 and 10.16. The following shot peening has farther brought that limit to 640 MPa. Note that without these treatments the fatigue limit of the steel would be much lower and assessed by considering the torsional strength τ_u equal to $0{,}8{\cdot}\sigma_u$, according to what has been said in Sect. 3.1, so that $\tau_f \sim ½\tau_r = ½{\cdot}0{,}8{\cdot}\sigma_r = 340$ MPa. Since the shaft withstand torsion its load factor is $C_{ld} = 0.58$, as it can be deducted from Table 4.4. Its diameter d of 15 mm introduces a size factor $C_{sz} = 1.189{\cdot}d^{-0.097} = 0.91$ (see Shingley and Mitchell formula 4.15). However, since many work pieces will be built the process volume will be conservatively assumed to be equal to 0.7 that corresponds to the upper limit of 250 mm of the Shingley and Mitchell formula. About the surface finish, since the shaft is fine-ground it will be assumed $C_s = 0.9$. Therefore, the torsional fatigue limit will be

$$\tau_f' = C_{ld} \times C_{sz} \times C_s \times 640 = 0.58 \times 0.7 \times 0.9 \times 640 = 233.8 \text{ MPa}$$

At 1000 cycles is $S_3 = 0.9{\cdot}\tau_u$ so that it would be

$$\tau'_{f,3} = C_{ld} \times C_{sz} \times 0.9 \times 0.8 \times 850 = 0.58 \times 0.70 \times 0.9 \times 0.8 \times 850 = 247.8\,\text{MPa}.$$

Actually, the true strength of shaft surface is no longer 850 MPa because of the heat treatments performed. Basing on the hardness, entering Table 10.2 with HRC = 30 we shall deduct that the surface strength is about 950 MPa, therefore

$$\tau'_{f,3} = C_{ld} \times C_{sz} \times 0.9 \times 0.8 \times 950 = 0.58 \times 0.70 \times 0.9 \times 0.8 \times 950 = 277.7\,\text{MPa}.$$

Since a 99.9% reliability is required, we may use Weibull Eq. (6.53). The Weibull exponent is known to be $m = 44$, but it is the exponent of the core metal not that of the surface. We don't know whether or not the surface material behaves better. To address the issue of reliability an experimental campaign has been run using 40 traction specimens of 5 mm diameter and 50 mm length. The specimens were given the same heat treatments of the shaft. Because of the reduced diameter of specimens, only 5 mm, it was assumed that the entire body had been affected by heat treatments and not just the surface, as in the shaft. It was checked that the hardness was effectively that of the shaft. Results indicated that the mean traction strength was $\sigma_u = 1010$ MPa. Using the log-normal distribution of the experimental data, it was found that the log-mean value was 2.97 with a standard deviation of 0.15. Entering the abacus of Fig. 6.8 we get that the 99.9% probability of survival is obtained subtracting 3.2 standard deviations from the mean, therefore

$$\log\sigma_{u,99.9} = 2.97 - 3.1 \times 0.15 = 2.51$$

$$\sigma_{u,99.9} = 10^{2.51} = 324 \text{ MPa}.$$

Than

$$\tau'_{f,99.9} = C_{ld} \times C_{sz} \times C_s \times 0.5 \times 324 = 0.58 \times 0.7 \times 0.9 \times 0.5 \times 324 = 60\,\text{MPa}$$

$$\tau'_{f,3,99.9} = 0.58 \times 0.7 \times 0.9 \times 0.8 \times 324 = 135\,\text{MPa}.$$

The resulting *S–N* curves are shown in Fig. 10.35. It must be remembered that the statistics performed cannot be considered reliable since the working data (40 specimens only) yield a sample mean and a standard deviation that are not the population mean (true mean) and the true standard deviation that stem from the entire population of data (see Sect. 6.1).

10.5.6 Bending Life of Multileaf Springs

A carbon steel is used in the construction of multileaf springs. Its traction strength is $\sigma_u = 450$ MPa and the AR = 35%. The material has a hardening exponent $n = 0.18$ with a tendency to harden under cyclic loads that results in a cyclic strain

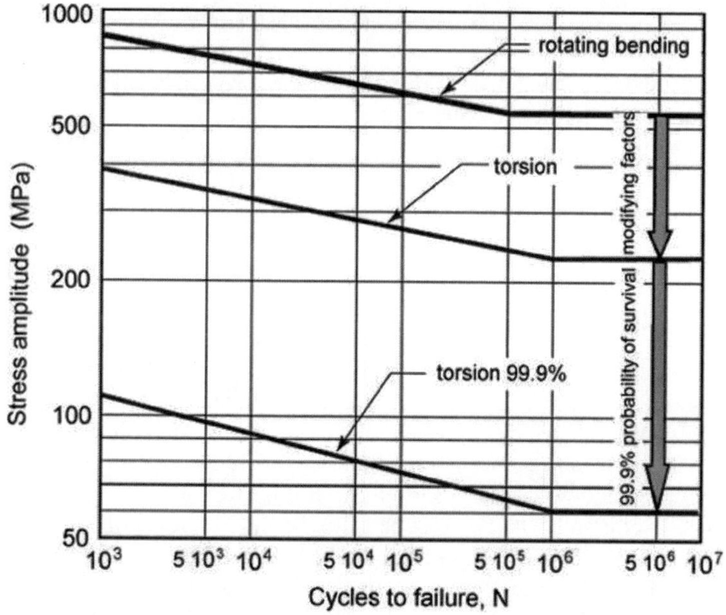

Fig. 10.35 *S–N* diagram for a carbon steel shaft in torsion. Also shown is the *S–N* curve in reversed bending

hardening exponent $n' = 0.14$. The elements of the spring are five leaves with $t =$ 5 mm thickness and $b = 60$ mm wide used as forged without any further machining. Construct the relative *S–N* curve using also the Basquin equation. The fatigue limit at 10^6 cycles may be evaluated as usual as

$$\sigma'_f = C_s \times C_{sz} \times C_{ld} \times (05 \times \sigma_u)$$

where C_s, C_{sz} and C_{ld} are the usual finish, size and load factors. As for diagram of Fig. 4.7, the first is equal to 0.55 (as forged). The second could be assessed on the base of its thickness using the Shingley and Mitchell formulas (4.15) and (4.16) that yield $C_{sz} = 0.92$. However, also in this case consideration about the number of work pieces holds, which actually means that the overall volume is increasing so that it is reasonable to use $C_{sz} = 0.7$. About the last factor since spring works in simple bending it will be $C_{ld} = 1.2$. With that it is

$$\sigma'_f = 0.55 \times 0.7 \times 1.2 \times (0.5 \times 450) = 95.29 \text{ MPa}.$$

At 1000 cycles

$$\sigma'_{f,3} = 0.7 \times 0.9 \times 450 = 315 \text{ MPa}.$$

The resulting *S–N* diagram is that og Fig. 10.30. Let's now use the Basquin Eq. (10.33). The engineering strength of the material is 450 MPa. According to the second of Eqs. (10.25) the true one is then

$$\sigma_{t,f} = 450 \times \left(\frac{1}{1 - RA}\right) = 692\,\text{MPa}.$$

From Eq. (10.33) it derives that $b = -0.08$ so that (10.22) yields for $N = 10^6$ cycles

$$\sigma_f = 0.55 \times 0.7 \times 1.2 \times 692 \times \left(10^6\right)^{-0.08} = 97\ \text{MPa}$$

rather close to the value 95.29 MPa found by other method. The relative Basquin line is shown in Fig. 10.36.

It can be seen how for this low strength steel the Basquin line is not well approximating the *S–N* diagram based on the S_3, S_f determination. This is because for this low strength material the contribution by plastic strain component to fatigue life is not negligible even at 10^5 cycles. The Basquin line is missing that contribution completely.

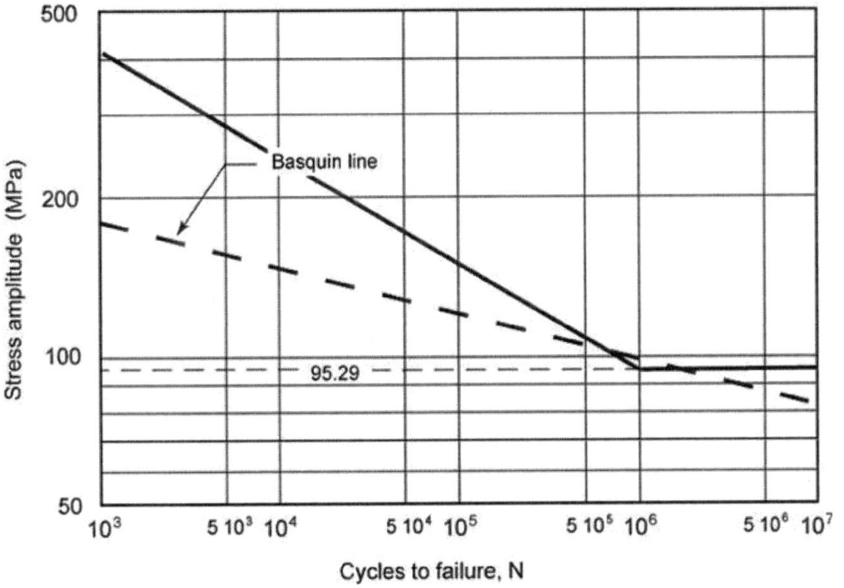

Fig. 10.36 *S–N* diagrams for the steel used in the multileaf spring construction. The first is based on the S_3, S_f determination. The second on the Basquin equation missing the plastic contribution to the strain

10.5.7 Application to Different Types of Load

Construct the *S–N* diagram for a high strength carbon steel bar 28 mm diameter and 1000 MPa tensile strength, heat treated to 400 HB, machine finish to be used in push-and-pull fatigue, rotating bending and torsional fatigue, respectively. From (10.6) and Table 10.2 it can be deducted that the fatigue limit in traction-compression and the traction strength are $\sigma_f = 677$ MPa and $\sigma_u = 1400$ MPa, respectively. From Fig. 4.7 is found that $C_s = 0.62$. as to the process volume, we assume that many pices will be built so that $C_{sz} = 0.7$. C_{ld} factors vary depending on load type and are

$$C_{ld}(rotating\,bending) = 1.1;\ C_{ld}(push - and - pull) = 1.0;$$
$$C_{ld}(reversed\,torsion) = 0.58.$$

At 10^3 cycles we have

$$S_3(tract.) = 0.75 \times \sigma_u;\ S_3(bend.) = 0.9 \times \sigma_u;\ S_3(tors.) = 0.9 \times \tau_r.$$

Therefore, the fatigue limits $\sigma'_f = C_{ld} \times C_{sz} \times C_s \times \sigma_f$ are

$$\sigma'_f(bend.) = 1.0 \times 0.7 \times 0.62 \times 677 = 293\,\text{MPa}$$
$$\sigma'_f(tract.) = 1.0 \times 0.7 \times 0.62 \times 677 = 293\,\text{MPa}$$
$$\tau'_f(tors.) = 0.58 \times 0.7 \times 0.62 \times 0.8 \times 0.5 \times 1400 = 141\,\text{MPa}$$

respectively. The stress $S_3 = C'_{ld} \times C'_{sz} \times \sigma_u$ at 1000 cycles are

$$S_3(bend.) = 0.90 \times 1400 = 1260\,\text{MPa}$$
$$S_3(tract.) = 0.75 \times 1400 = 1050\,\text{MPa}$$
$$S_3(tors.) = 0.9 \times 0.8 \times 1400 = 1008\,\text{MPa}$$

respectively. The three *S–N* diagrams are shown in Fig. 10.37.

10.5.8 Aluminum Alloys

Construct the *S–N* diagram of an aluminum alloy having 294 tensile strength and 69 HB hardness subjected to reversed bending. For this alloy the fatigue limit is 140 MPa, as it can be seen in Table 10.4. The S_3 value is $0.9 \cdot \sigma_u = 265$ MPa. The *S–N* diagram is then immediately traceable as shown in Fig. 10.38.

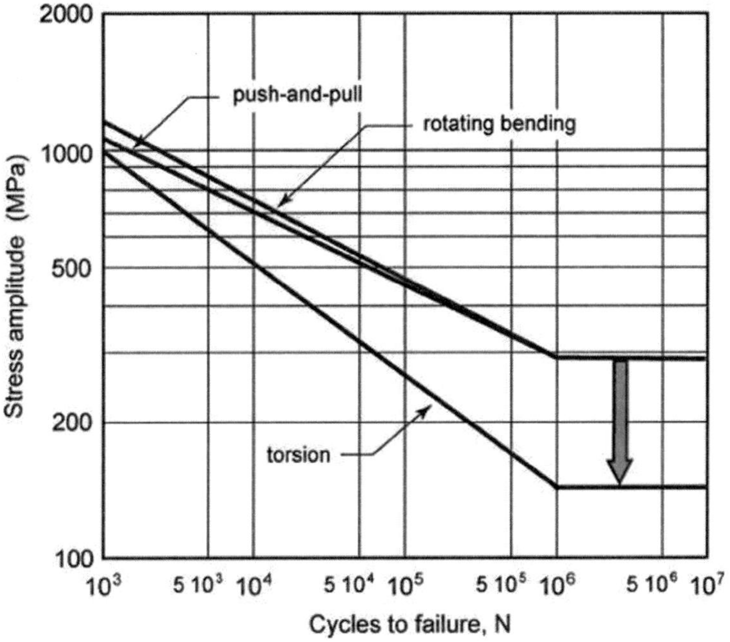

Fig. 10.37 *S–N* diagrams for a high strength, heat treated carbon steel rod used in rotating bending, push-and-pull and torsional fatigue, respectively

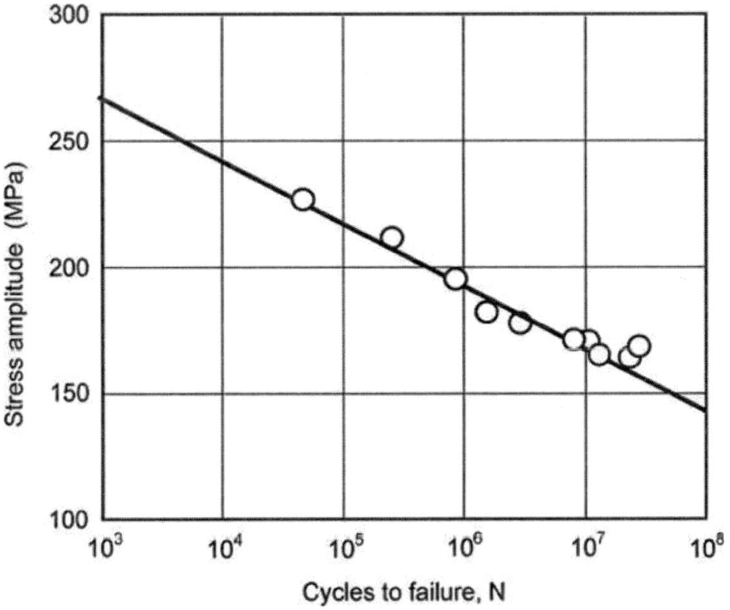

Fig. 10.38 *S–N* diagram for a 5056–0 aluminum alloy in reversed fatigue bending

References

1. Lipson, C., Juvinall, R.C.: Handbook of Stress and Strength. The Macmillan Company, New York (1963)
2. Gough, H.J.: The Fatigue of Metals. Scott, Greenwood and Son, London (1924)
3. Cazaud, R.: Fatigue of Metals. Chapman and Hall, London (1953)
4. Evans, E.B., Ebert, J., Briggs, C.W.: Proc. Am. Soc. Test. Mater. **56**, 979 (1956)
5. Heywood, R.B.: Designing Against Fatigue. Chapmann and Hall, London (1962)
6. Svensson, T., Maré, J., Wadman, B.: Determination of the Fatigue Limit, Methods and Problems.
7. Murakami, Y., Endo, T.: Effect of hardness on crack geometries on ΔK_{th} of small cracks emanating from small defects. In: Miller, K.J., Los Rios, E.R.D. (eds.) The Behaviour of Short Fatigue Cracks", EGF Publ. 1. Mechanical Engineering Publications, London, pp. 275–293 (1986)
8. Garwood, M.F., Zurburg, H.H., Erickson, M.A.: Correlation of laboratory tests and service performance, interpretation of tests and correlation with service.ASM, pp. 1–77 (1951)
9. Juvinall, R.C.: Engineering Considerations of Stress, Strain and Strength. Mc Graw-Hill, Inc., p. 214 (1967)
10. Dowling, N.E.: Mechanical Behavior of Materials: Engineering Methods for Deformation. Prentice Hall, Fracture and Fatigue (1993)
11. Murakami, Y., Endo, T.: Effect of defects, inclusions and in homogeneities on fatigue strength. Int. J. Fatigue 16:163–182 (1994)
12. McGreevy, T.E.: The Competing Role of Microstructure and Flaw Size on the Fatigue Limit of Metals. Mechanical Engineering, University of Illinois, Urbana-Champaign, Thesis (1998)
13. Sylvestrowicz, W., Hall, E.O.: Proc. Phys. Soc. London, B **64**, 495 (1951); Hall, E.O., Ibid. B **64**, 742–747 (1951).
14. Petch, N.J.: J. Iron and Steel Inst. **174**, 25; Progress in Metal Physics 5, Chap. 1, p. 1, Pergamon Press (1953).
15. Finney, J.M., Department of Supply, Australia, Aero. Research Lab. Struct. And Mats. Report 287 (1962)
16. Ferro, A., Mazzetti, P., Montalenti, G.: Phil. Mag. **12**, 867 (1965)
17. Gaugh, H.J., Wood, W.A.: Proc. Inst. Mech. Eng. **141**, 175 (1939)
18. Minakawa, K.: Fatigue of titanium alloys. Tetsu to Hagane **75**(7), 1104–1111 (1989)
19. Grover, H.J., Bishop, S.N., Jackson, L.R.: NACA Technical Note 2390 (1951)
20. Lazan, B.J., Blatherwick, A.A.: Wright Air Development Center, Technical Report No. 52–307 (1952)
21. Howell, F.M., Miller, J.L.: Proc. ASTM **55**, 955 (1955)
22. Low, A.C.: Royal Aeronaut. Soc. **59**, 629 (1955)
23. O'Conner, H.C., Morrison, J.L.: International conference on fatigue. Inst. Mech. Eng. 102 (1956)
24. Graham, J.A. (Ed.).: Fatigue Degign Handbook, Society of Automotive Engineers, Worrendale, Pa., (1968)
25. Brock, G.W., Sinclair, G.M.: Proc. ASTM, **60** (1960).
26. Pomp, A., Hempel, M.: Wechselfestingkeiten und Kerbwinkungszahlen von Inlegierten und Legierten Stählen bei +20° C und −78° C", Arch. F. Eisen-Hüttenw, pp. 191–195 (1950)
27. Basquin, O.H.: Proc. ASTM **10**(Part II), 625 (1910)
28. Bridgman, P.W.: The stress distribution at the neck of a tension specimen. Trans. Am. Soc. Metals **32**, 553–574 (1944)
29. Bridgman, P.W.: Studies in Large Plastic Flow and Fracture. McGraw-Hill Book Co., New York (1952)
30. Morrow, J.D., Internal Friction, Damping and Cyclic Plasticity. ASTM STP-378, p. 45 (1965)
31. Haibach, E.: Betriebsfestigkeit (Operational Integrity). VDI-Verlag Stahleisen GmbH, Düsseldorf (1986)
32. Dixon, W.J., Mood, A.M.: A method for obtaining and analyzing sensitivity data. J. Am. Stat. Assoc. **43**, 109–126 (1948)
33. Nakazawa, H., Kodama, S.: Statistical Research on Fatigue and Fracture, Elsevier, p. 56 (1987)

Chapter 11
Notch Effect

11.1 Stress Concentration

It has been shown, in Chap. 2 in particular, that nucleation damage in fatigue is always a localized sub-microscopic damage either on persistent slip lines in some surface grain or in persistent grain boundaries somewhere in the material, depending on whether fatigue is acting in the high cycle or low cycle regime. It is evident that any factor producing a localized stress amplification favors the premature outbreaks of such a damage and the subsequent crack formation. This local stress amplification is generally caused by a component notch or discontinuity. William Rankine of England was probably the first to understand this effect studying axels failures and stating in 1842 [1] "the fractures appear to have commenced with a smooth, regularly-formed, minute fissure, extending all-round the neck of the journal, and penetrating on an average to a depth of half an inch. … until the thickness of sound iron in the center became insufficient to support the shocks to which it was exposed". It is a fact that all fatigue failures initiate at a notch or discontinuity. Such a notch should not necessarily be an unwanted defect since it is normally introduced on purpose in the work piece. Just think about gear teeth, shaft keyways and shoulders, bolt threads, hub couplings, any hole in general and any geometrical variation or discontinuity more or less sharp that are generically called *structural notches*. Notch tip strain and stress amplification depend, obviously, on the shape and dimensions of the work piece, type and depth of the notch and, in particular, on its root radius that are generically referred to as *system geometry*. On system geometry depends the so called notch concentration factor k_t normally of theoretical derivation, hence the t subscript, based on linear elasticity. The stress concentration factor k_t determines the peak stress $\sigma_{\max}$ at the notch apex

$$k_t = \frac{\sigma_{\max}}{\sigma_n} \tag{11.1}$$

P. P. Milella, *Fatigue and Corrosion in Metals*,
https://doi.org/10.1007/978-3-031-51350-3_11

σ_n being the nominal stress far from the notch section and, therefore, not affected by the notch itself. Alternatively, it can be introduced a concentration factor k_t' defined as

$$k_t' = \frac{\sigma_{\max}}{\sigma_n} \cdot (l - a) = k_t \cdot (l - a). \tag{11.2}$$

Defined as the ratio between the peak stress $\sigma_{\max}$ and the stress that would act in the same point of the reduced section $(l - a)$ without considering the local stress amplification, with a being the notch length and l the width of the section containing the notch. The local stress intensification give rise to the suspicion that the fatigue limit σ_f or S_f must be consequently reduced by the same factor k_t to a value $\sigma_f{}'$ or $S_f{}'$

$$S_f' = \sigma_f' = \frac{\sigma_f}{k_t}. \tag{11.3}$$

Effectively, experience indicates a fatigue limit reduction following the presence of a notch, but not as much as Eq. (11.3) would indicate. Figure 11.1, obtained from experimental data made already in 1936 [2], is an example of a carbon steel shaft without or with shoulder fillets of different geometries subjected to rotating bending. It can be seen how *S–N* diagram and the fatigue limit S_f measured on the shaft of uniform section with $d = 12$ mm diameter, curve 1, are much higher than those measured on shafts of the same material, but with different diameters and shoulder fillets, curve 2, 3 and 4. The fatigue limit of the constant diameter shaft without any structural notch is $S_f = 364$ MPa. Moving to the first case of shaft with a shoulder of radius $\rho = 1$ mm the fatigue limit $S_f{}'$ goes down to 280 MPa, curve 2, with a reduction of 1.3 relative to the continuous shaft of 12 mm diameter. In case 3 the shoulder radius is $\rho = 1.8$ mm and the fatigue limit farther reduces to 252 MPa with a reduction of a factor 1.44, always with respect to the reference shaft without shoulder, curve 1. Finally, with a shaft of larger diameter $D = 81$ mm that reduces to $d = 54$ mm with a shoulder fillet of radius $\rho = 8$ mm the fatigue limit reaches its minimum of 240 MPa with a reduction of 1.5 times. The theoretical stress concentration factor k_t for the three geometries considered was 1.5 for case 2, 1.55 for case 3 and 1.5 for case 4, respectively. Therefore, the actual stress concentration factor that will be called shape or notch factor and indicated by k_f is equal or lightly lower than the theoretical k_t. Effectively, experimental studies [3–8] initiated in Germany in 1937 with the fundamental work of Heinz Neuber and translated into English in 1946 [3] confirmed this tendency. It has been observed that, in general, k_f approximates k_t the more this last is low and the material is hard. Interesting, at this purpose, are the diagrams of Fig. 11.2 elaborated by Milella on experimental data obtained by various researchers. Data of Fig. 7.2a were obtained by Phillips and Heywood [4] on mild steel and nickel–chromium specimens of 43 mm diameter, containing a circumferential V notch of 55°, 5 mm depth, with a root radius variable from 0.05 mm to 10 mm subjected to push-and-pull traction. Data of Fig. 11.2b were obtained by Templin [7], Mann [9] and Finney [10] with specimens of several aluminum alloys containing a V notch of various root radii, subjected to rotating bending. Also reported is the theoretical k_t vs. notch root radius. Two are the main features that must be caught. First, the notch factors k_f are generally lower than the corresponding theoretical concentration factors k_t. Second, the effective value of k_f depends on material. Interesting enough, the notch factor k_f deviates the more from the theoretical k_t the more this last is high. To a theoretical concentration factor $k_t = 18$ relative to $\rho = 0.005$, in practice a crack, it corresponds a $k_f \cong 2$–3. For $k_t = 14$ the factor decreases to 3 or 8 and 4 for steels of Fig. 11.2a. Eventually, the two factors almost merge when k_t becomes lower than 3 or 2, depending on the material. As to the k_f dependence on the material, at variance with k_t that is a pure geometrical factor based on linear elasticity, Fig. 11.3 obtained by Milella analyzing data by Marry and Dolan [6] shows how k_f varies with grain dimension in a family of brass 70/30. Brasses were given different heat treatments to change grain size and mechanical characteristics. Two geometries were considered having k_t equal to 2.9 and 1.8, respectively. It is worth noting how the k_f curve relative to $k_t = 2.9$ follows accurately that of the fatigue limit S_f obtained on unnotched specimens (dashed line). The k_f curve relative to $k_t = 1.8$ would also follow the S_f one if its origin (point B) would be moved in A by changing the k_t scale. That the grain size would influence the fatigue limit S_f was also found by Forrest and Tate [11] and does not come unexpected since, as already discussed in Sect. 4.7 and shown in Fig. 4.41 for steels, the strength of a material depends on grain size and, in turn, the fatigue limit depends on tensile strength, as said and shown in Figs. 4.37 and 4.40. The new fact is that material strength also influences the k_f factor. This is why Neuber [3] first in Germany and Peterson [12] later in the US thought it was convenient to introduce in fatigue a so called notch sensitivity index q or technical stress concentration factor, as defined by Neuber, to relate material property and notch geometry to notch factor k_f.

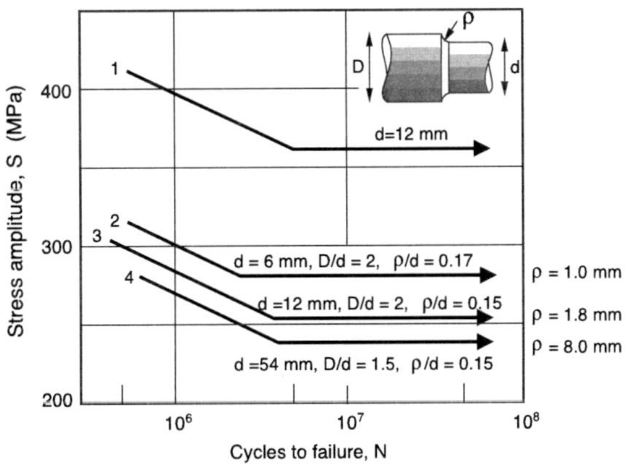

Fig. 11.1 Wöhler's curves for a carbon steel fillet shaft subjected to rotating bending (modified from tests done by Peterson et al. [2])

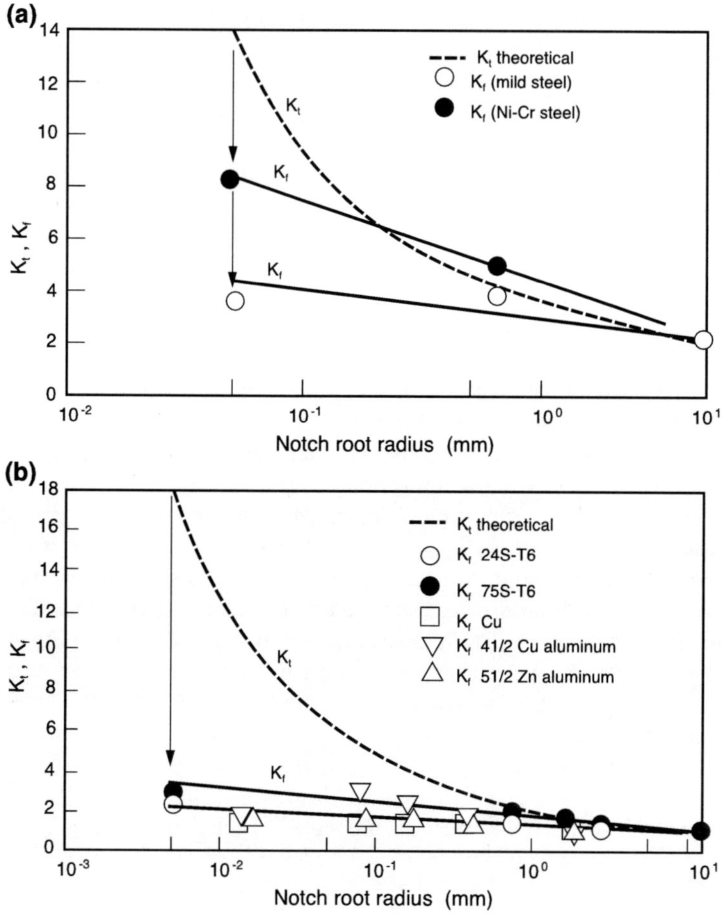

Fig. 11.2 Diagrams prepared by Milella with experimental data obtained by various researchers [3, 6, 8, 9]

11.2 Peterson Notch Sensitivity Index

Peterson [12–14] introduced a notch sensitivity index q, or simply q factor, defined as

$$q = \frac{k_f - 1}{k_t - 1} \tag{11.4}$$

from which it derives

$$k_f = 1 + q \cdot (k_t - 1). \tag{11.5}$$

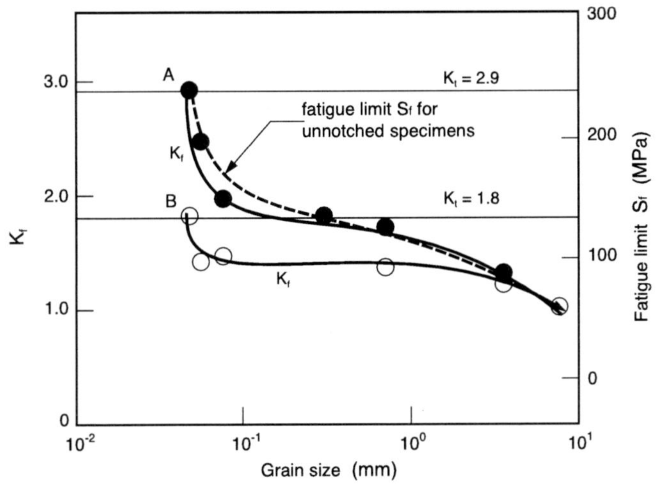

Fig. 11.3 Dependency of k_f factor on grain size in a 70/30 brass for to different k_t values. Also shown is the fatigue limit S_f of unnotched specimens (dashed curve). Diagram elaborated by Milella on Merry e Dolan experimental data [6]

The reason to subtract unity from both numerator and denominator in Eq. (11.4) is to provide a scale for q that ranges from zero for no notch effect to unity for full notch effect. In this latter case k_f is exactly equal to k_t. A deep analysis of fatigue limit experimental data brought Peterson to make the hypothesis that fatigue damage became critical in relation to the stress level σ_{peak} at a small distance δ ahead of notch root where the stress gradient is very steep

$$\sigma_{peak} \approx \delta\left(\frac{d\sigma_y}{dx}\right). \tag{11.6}$$

Assuming a linear decrease of the notch tip stress field, Peterson [13] formulated an empirical expression

$$q = \frac{1}{\left(1 + \sqrt{\frac{a_P}{\rho}}\right)} \tag{11.7}$$

in which, as usual, ρ is the notch root radius and a_P a critical distance that depends on material strength an ductility and, in the original Neuber formulation [3], also on grain dimension. Curious enough, Eq. (11.7) cannot be derived from Eq. (11.6), but it agreed quite well with experimental data available to Peterson on fatigue limit at $R = 0$ (fully reversed).

Introducing Eq. (11.7) in Eq. (11.5) it yields

$$k_f = 1 + \frac{k_t - 1}{1 + \sqrt{\frac{a_P}{\rho}}}. \tag{11.8}$$

For medium–high strength steels (σ_u > 560 MPa) the a_P values can be approximated by the following formulas

$$a_P = 0.0254 \cdot \left(\frac{2079}{\sigma_u}\right)^{1.8} \; traction\ and\ bendingloads \tag{11.9}$$

$$a_P = 0.01524 \cdot \left(\frac{2079}{\sigma_u}\right)^{1.8} \; torsional\ loads \tag{11.10}$$

where a_P is in mm and σ_u in MPa. Peterson q index are graphically shown in Fig. 11.4 [15]. Two new concepts emerge from the Peterson approach, as from that of Neuber that preceded it. The first is that it is the stress gradient and not just the maximum stress at notch root that influences fatigue strength. The second that also stems from Figs. 11.4 and 11.2, is that notch root radius influences fatigue strength and not just the k_t factor, which actually means that two similar discontinuities, for instance two holes having two different root radii, have also different effects on fatigue even though they have the same theoretical stress concentration factor k_t. In mild steel for a hole of radius $\rho = 0.2$ mm the q index is about 0.4 while with $\rho = 3$ mm it becomes equal to about 0.75. The apparent paradox that the larger radius hole has a more damaging effect than the hole with smaller radius is due to the corresponding process volumes, as will be discussed later in Sect. 11.4.

11.3 Neuber Approach

Peterson approach represented a simplification of the original Neuber one [3] of 1937. Neuber, studying V notches having flank angle φ and root radius ρ, proposed an effective notch factor k_f lower than the theoretical one k_t and linked to this last by the equation

$$k_f = 1 + \frac{k_t - 1}{1 + \frac{\pi}{\pi - \phi}\sqrt{a_N/\rho}} \tag{11.11}$$

in which a_N is a distance characteristic of the material related to the stress field acting ahead of the notch root in a very shallow surface layer where the average stress was assumed to be equal to the fatigue limit of the material. Actually, Neuber never explained how he derived Eq. (11.11). We owe it to Kuhn and Hardrath [16] the derivation of values of constant a_N characteristic of the material.

Its dependency on material strength is shown in Fig. 11.5 [17] for steel and aluminum alloys and both traction and bending loading. For $\varphi = 0$ Eq. (11.11)

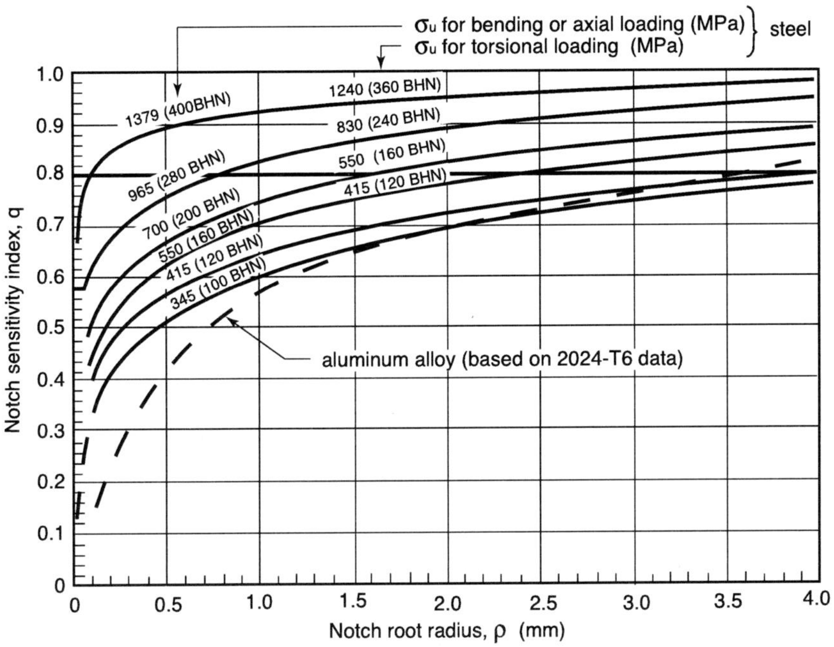

Fig. 11.4 Notch sensitivity index q for steels and aluminum alloys versus notch root radius (modified from [15])

becomes

$$q = \frac{k_f - 1}{k_t - 1} = \frac{1}{1 + \sqrt{\frac{a_N}{\rho}}}. \tag{11.12}$$

The value of Neuber's constant $\sqrt{a_N}$ is very close to that a_P of Peterson. Differences between the two Eqs. (11.12) and (11.7) appear to be contained within some percent for steels, with the Neuber equation underestimating the notch sensitivity index with respect to the Peterson expression, as presented in Table 11.1. In high strength steel this difference further reduces till becoming negligible. Both Eqs. (11.7) and (11.12) show that the notch sensitivity decreases as the strength or hardness or ductility of metals decrease, but increases as the root radius increases id est precisely with decreasing theoretical stress concentration factor k_t. It can be seen in Fig. 11.4 that moving from a high strength steel, 1200–1400 MPa ultimate, to a mild one, only 350 MPa ultimate, the sensitivity index for a sharp notch, for instance $\rho = 0.25$ mm, changes from about 0.85 to 0.45 with a factor two reduction, but when the notch root radius increases to 3 mm the difference in notch sensitivity between the two steels reduces to about 24% only. i.e., it cuts in half. At variance with high strength steels, aluminum alloys show the lowest sensitivity to notches, in particular for root radii

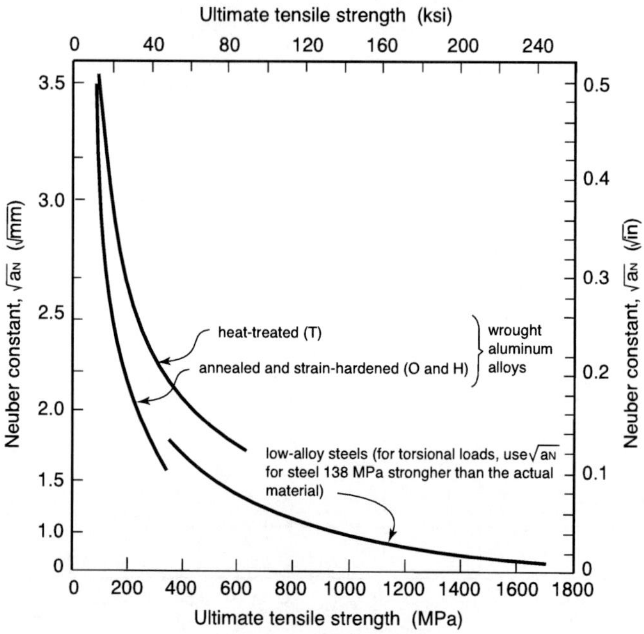

Fig. 11.5 Neuber constant $\sqrt{a_N}$ (in mm$^{1/2}$) for steels and aluminum alloys (modified from [17])

lower than 2 mm. Curious enough, gray iron seems not have any notch sensitivity since internal graphite needles have prevailed over external notches. Heywood [18] has proposed an empirical relation for cast iron

$$k_f = \frac{k_t}{1 + 2\sqrt{\frac{a'}{\rho}} \cdot \left(\frac{k_t - 1}{k_t}\right)} \tag{11.13}$$

in which a' represents the length of an equivalent crack. In cast iron with acicular graphite it is $\sqrt{a'} = 0{,}605$, in DCI it is $\sqrt{a'} = 173.6/\sigma_u$. for magnesium alloys it is $\sqrt{a'} = 0{,}0756$. In all cases a' and ρ are expressed in mm and σ_u in MPa.

Table 11.1 Comparison between Neuber and Peterson notch factors

σ_u (MPa)	400	800	1200	1600
$k_{f,\,Neuber}/k_{f,\,Peterson}$	0.91	0.95	0.97	0.99

11.4 Siebel and Stieler Method

Around mid '50 s, Siebel and Stieler [19] introduced a new parameter that they indicated with the symbol χ and called *relative stress gradient* (RSG) at notch root. The fundamental concept is always the same that is to say that fatigue is governed by the stress gradient rather than by the maximum stress at notch tip. Stress gradient means distance over which a stress level is maintained. The χ parameter takes the form

$$\chi = \frac{\left(\frac{d\sigma}{dx}\right)_{x=0}}{\sigma_{\max}}. \tag{11.14}$$

The stress gradient $d\sigma/dx$, schematized in Fig. 11.6, is in turn equitable through both the theoretical stress concentration factor k_t and the notch root radius ρ that, therefore, reenters into play

$$\left(\frac{d\sigma}{dx}\right)_{x=o} = \left(2 + \frac{1}{k_t}\right)\frac{\sigma_{\max}}{\rho} = \alpha\frac{\sigma_{\max}}{\rho}. \tag{11.15}$$

Finally

$$\chi = \frac{\alpha}{\rho}. \tag{11.16}$$

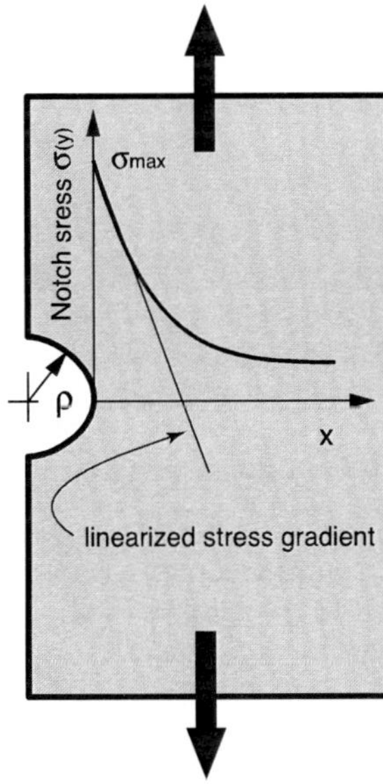

Fig. 11.6 Elastic stress gradient at the notch root

Variations of the α parameter are rather modest if we deal with structural notches having a theoretical stress concentration factor k_t between 2 and 5. Siebel and Stieler choose $\alpha = 2$ and considered the ratio k_t/k_f as being a function of χ

$$\frac{k_t}{k_f} = 1 + \sqrt{C \cdot \chi}. \tag{11.17}$$

Through a long lasting series of experimental tests run at 2×10^7 cycles on notched and unnotched specimens, Siebel and Stieler obtained a family of empirical curves that related the ratio k_t/k_f to RSG χ. Figure 11.7 [19] shows these curves. A comparison between Eq. (11.17) and the equivalent Eq. (11.7) by Peterson and Eq. (11.13) by Neuber reveals that, at variance with this last two, the Siebel and Stieler relation shows a ratio k_t/k_f that still depends on ρ, but not more on k_t. Still appears the concept of stress gradient that, as it will be shown, is fundamental in fatigue notch analysis. Several RSG values were calculated by Siebel and Stieler for various notched members on the base of theory of elasticity.

1. Plate of width w with a central notch of radius ρ
   (a) Axial loading

$$RGS = \frac{2}{\rho} \tag{11.18}$$

   (b) Bending loading

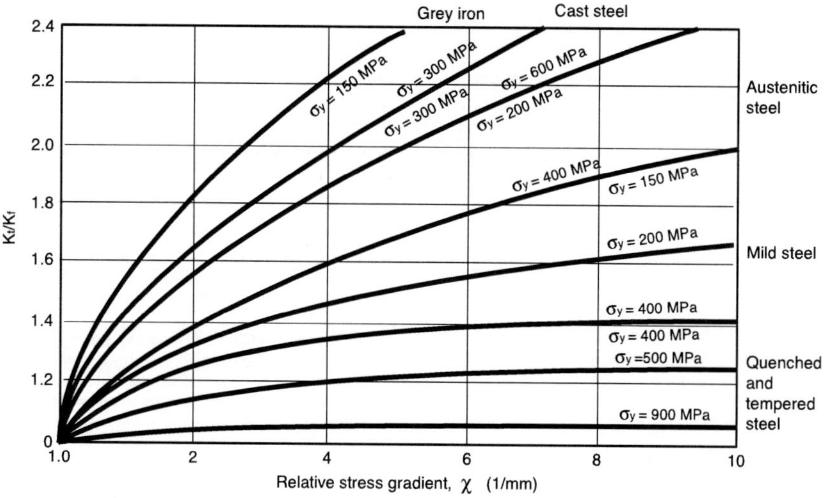

Fig. 11.7 Dependency of ratio k_t/k_f on RSG for different steels and cast iron (modified from [19])

$$RGS = \frac{2}{\rho} + \frac{2}{w} \tag{11.19}$$

2. Grooved shaft with notch radius ρ and minimum diameter d
   (a) Axial loading

$$RGS = \frac{2}{\rho} \tag{11.20}$$

   (b) Bending loading

$$RGS = \frac{2}{\rho} + \frac{2}{d} \tag{11.21}$$

   (c) Torsional loading

$$RGS = \frac{1}{\rho} + \frac{2}{D + d} \tag{11.22}$$

3. Shoulder shaft with notch radius ρ and minimum diameter d
   (a) Axial loading

$$RGS = \frac{2}{\rho} \tag{11.23}$$

   (b) Bending loading

$$RGS = \frac{2}{\rho} + \frac{4}{D + d} \tag{11.24}$$

   (c) Torsional loading

$$RGS = \frac{1}{\rho} + \frac{4}{D + d} \tag{11.25}$$

11.5 Process Volume and Notch Effect

From the reading of previous sections it is evident that notch sensitivity and its dependency on material characteristics are the result of two concurring factors completely different from each other. The first factor is related to material capability to relax higher peak stresses through a redistribution of root strains. The second, instead, is due to the stress or strain gradient on which, as it will be seen, metallurgical variability plays a fundamental role. Actually, it must be kept in mind that k_t is a

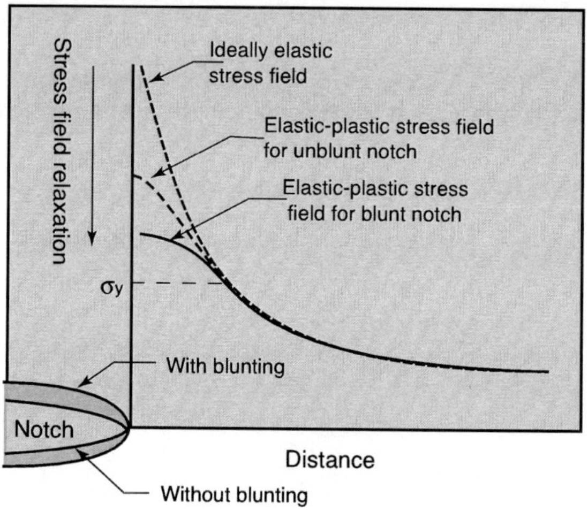

Fig. 11.8 Notch tip blunting in real elastic–plastic materials that results in stress relaxation with respect to ideally elastic materials

theoretical factor evaluated on the assumption of perfectly linear elastic behavior of the material that allows peak stress to go as high as to infinity, as in the case of a crack and does not allow any notch tip geometry change. In other words, the notch tip keeps its original root radius and does not blunt at all. At variance with the elastic case, in an ideal elastic–plastic material the peak stress cannot go higher than yield stress σ_y and notch tip blunts. In real elastic–plastic materials there will be an intermediate situation in that beyond yielding the notch root stress will continue to grow by strain hardening, though at lower rate. Now the notch tip blunts because of yielding reducing the sharpness and this geometry change contributes to lower the local stress, as schematized in Fig. 11.8.

The higher the metal ductility, the higher the geometry change and the consequent reduction of the local stress. The corresponding stress concentration factor k departs from the theoretical linear elastic k_t getting closer and closer to the notch factor k_f that appears in Eqs. (11.7) or (11.13) or (11.17). In practice, in metals having higher ductility already with a factor k_t higher than two or three there will not be any difference between sharp and blunted notches. The sharper notch will blunt and behave like the other. At variance, in very hard materials less prone to plastic slip a sharp notch will continue to keep part of its original sharpness, yet the stress concentration factor will not be as high as six or eight, at most. Nevertheless, the knowledge of the effective peak stress is not sufficient, by itself, to solve the problem of notch effect on fatigue.

It is important to realize that even if the actual peak stress is known, for example through a very refined finite element calculation, it cannot not be used without the knowledge of the entire stress field and, in particular, its gradient ahead of the notch tip. Two stress field having the same peak stress, but different gradient have certainly different effect on the fatigue strength of materials. This is shown in Fig. 11.9a relative to the stress fields in an infinite plate under remote traction ahead of four different holes. The elastic theory states that the stress concentration factor k_t is always three, independently of the hole diameter. But, though the maximum stress σ_{max} is equal on the root, the u four stress fields are much different from each other. The one associated with the smallest hole having a radius $r = 0.5$ mm is characterized by a sharper gradient and decays rapidly reaching the nominal stress σ_n already at a distance of 1 mm.

The stress field relative to the larger hole with $r = 10$ mm, instead, has the lower gradient and remains high on a longer distance leveling off to the nominal stress σ_n only after 20–25 mm. The other two stress fields relative to the holes of 2.5 mm and 5 mm radii show intermediate trends. These decays distance are consistent with the Saint Venant principle that states that any perturbation has a decay distance that is about two or three times the dimension of the cause of perturbation that in our case is

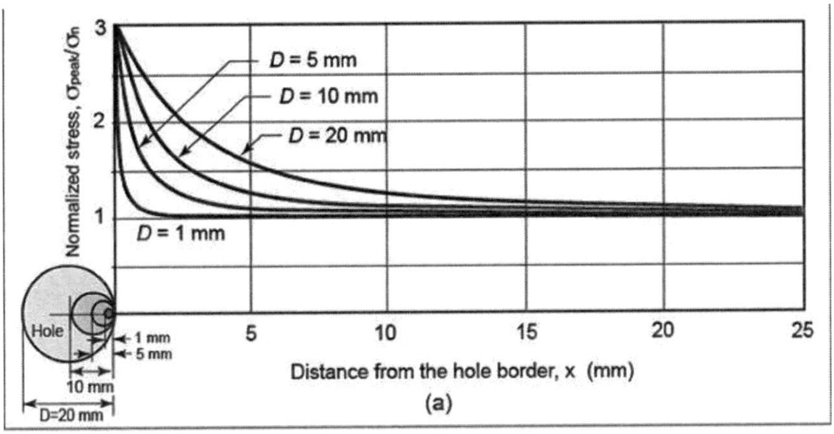

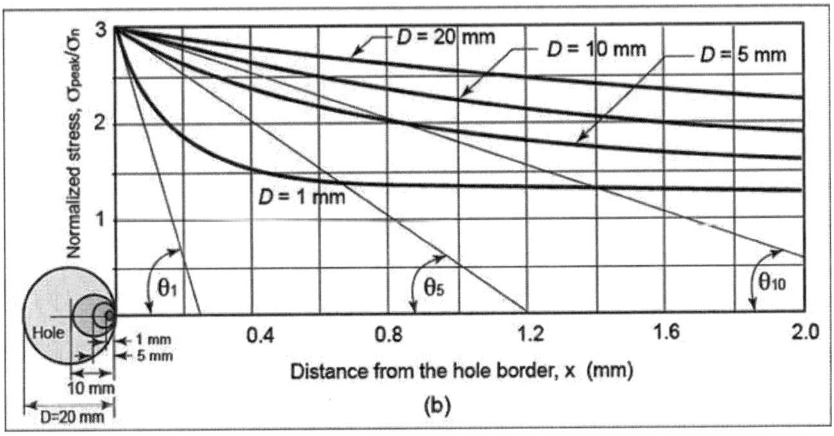

Fig. 11.9 **a** Stress field ahead of the root of four holes having different radii, 0.5 mm, 2.5 mm, 5 mm and 10 mm, respectively. **b** Detail of the first 2 mm from the hole surface showing the relative gradients

the radius of the hole. Figure 11.9b is a blow-up of the first 2 mm ahead of the holes. Stress field gradients are clearly visible. Both Peterson and Neuber were associating the stress field gradient to a critical quantity a_P/ρ and $\sqrt{a_N/\rho}$, respectively, (see Sects. 5.2.1 and 5.2.2). It is important to realize that this critical quantity is actually indicating a process volume. This is the true key to read the notch effect on fatigue.

The process volume, i.e., the volume of material where stresses are larger than a critical value, associated to a sharp notch is very small in comparison to that relative to a much larger notch root radius. Therefore, the probability to find a worst metallurgical condition ahead of the notch tip in the process volume is much higher for the blunted notch than for the sharper one. The consequence is that under the same peak stress fatigue life will be shorter for the blunted notch with respect to a sharp one. Consider, now, as reference process volume the volume $V_{1\text{-}x\%}$ associated to a distance from the surface of a notch root within which the peak stress σ_{peak} reduces by a small percentage X. It may be convenient to assume this small percentage as 10% only. Within the process volume $V_{90\%}$, then, the stress field amplitude will not be lower than 90% of the peak stress.

For what has been said, the larger this reference volume $V_{90\%}$ the higher the probability to have an earlier fatigue failure. Return, then, to the four holes considered in Fig. 11.9 and try to analyze this process or reference volume. This is done in Fig. 11.10. The process volume has been estimated on the base of the distance $x_{90\%}$ within which the local stress is higher or equal to 90% of the peak stress σ_{peak}. It comes out that the process volume $V_{90\%}^{(20)}$ associated with the hole of 20 mm is about 400 times larger than that $V_{90\%}^{(1)}$ ahead of the hole of 1 mm, sixteen times larger than $V_{90\%}^{(5)}$ relative to the hole of 5 mm and only four times larger than $V_{90\%}^{(10)}$. The distance $x_{90\%}$ can be assessed from the stress gradient considered, as a first approximation, as the tangent to the stress field in the origin (Fig. 11.9)

$$x_{90\%} = \frac{\sigma_{peak}}{\theta} \tag{11.26}$$

where θ is the angular coefficient or slope of the tangent. For instance, with reference to Fig. 11.9, the tangent to the stress field for the 1 mm hole has the angular coefficient equal to

$$\theta_1 = \frac{3}{0.24} = 12.5 \tag{11.27}$$

Therefore, the distance $X_{90\%}^{(1)}$ is equal to 0.024 mm, as shown in Fig. 11.10. Alike, for the 5 mm hole it is $X_{90\%}^{(5)} = 0.12$ and for the 10 mm hole it is $= X_{90\%}^{(10)}$ 0.24 mm.

Since the fatigue limit σ_f is proportional to the ultimate strength σ_u, see Figs. 10.8 and 10.13, and, in turn, the ultimate strength is inversely proportional to the process volume, knowing the process volumes $V_{90\%}$ relative to the various diameter holes and using Eq. (6.51) it is possible to infer how the fatigue limit σ_f of the material changes as the notch root radius varies

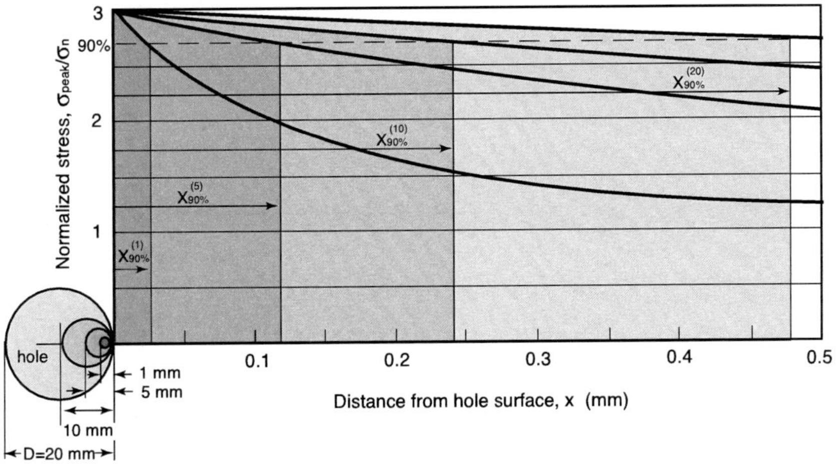

Fig. 11.10 Extent of the process volumes $V_{90\%}$ for the four holes examined in Fig. 7.9

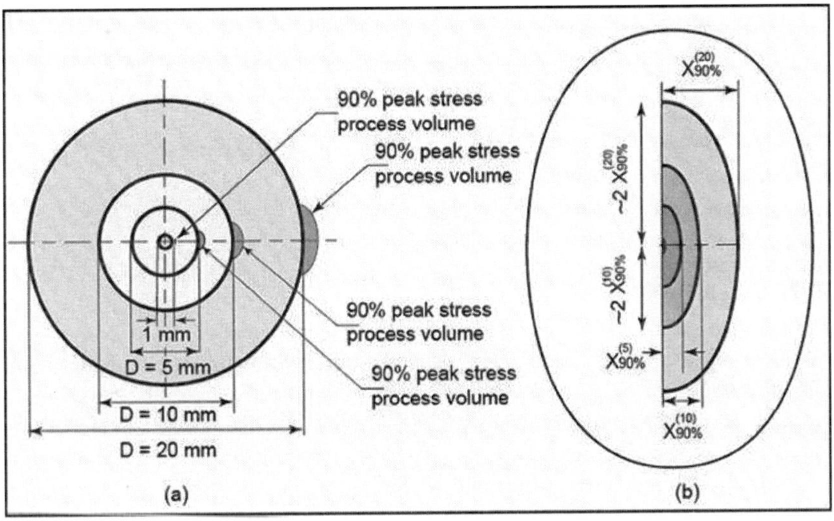

Fig. 11.11 **a** schematic of the areas where stresses are within the 90% of the peak stress for the two holes of 10 mm and 20 mm diameter, respectively, **b** the areas are approximately similar to a semi-ellipse

$$\frac{\sigma_{f,20}}{\sigma_{f,1}} = \left(\frac{\sigma_{u,20}}{\sigma_{u,1}}\right) = \left(\frac{V_{90\%}^{1}}{V_{90\%}^{20}}\right)^{1/m} \tag{11.28}$$

where m is the Weibull exponent. As schematized in Fig. 7.11, the shape of the 90% process volume $V_{90\%}$ ahead of the hole root surface is approximately equal to a semi-ellipse whose major axis is approximately equal to twice the minor one, according to the Saint Venant principle. Therefore, it is

$$\left(\frac{V_{90\%}^{1}}{V_{90\%}^{20}}\right)^{1/m} \approx \left[\left(\frac{X_{90\%}^{(1)}}{X_{90\%}^{(20)}}\right)^{2}\right]^{1/m} = \left(\frac{1}{400}\right)^{1/m}. \tag{11.29}$$

It has been shown in Sect. 6.6.1 Eq. (6.52) that the Weibull exponent depends on the notch root radius. For the 0.5 mm radius it is $m = 17.65$ and for the 10 mm one $m = 37.34$. Using an average value $m = 27.25$, Eqs. (11.28) and (11.29) yield

$$\frac{\sigma_{f,20}}{\sigma_{f,1}} = \left(\frac{1}{400}\right)^{1/27.25} = 0.80 \tag{11.30}$$

which actually means that a specimen containing a hole of 10 mm radius shall have a fatigue limit σ_f about 20% lower than that of the same material specimen carrying a hole of 0.5 mm radius. This result is exactly that obtained with the abacus of Fig. 11.4 for a medium strength steel of 560 MPa. In fact, for the 10 mm notch radius it is $q = 0.62$ while for the 10 mm one it is $q = 0.9$. Therefore

$$\frac{k_{f,1}}{k_{f,20}} = \frac{1 + 0.9(3 - 1)}{1 + 0.62(3 - 1)} = 0.8. \tag{11.31}$$

The choice of the 90% of the peak stress as a criterion to assess the process volume that may have an effect on fatigue is arbitrary, yet very reasonable. A higher value, such as 95% of the peak stress, could be too restrictive excluding from the fatigue process a still highly stressed region of material while a 85% choice would include material probably not enough stressed to contribute to fatigue crack initiation. The 90% criterion, in fact, has been successfully applied in the fatigue assessment of many structural elements such as vehicle wheels, pressure vessels, welded joints, rods, keyways, transmission boxes, etc.. However, it must be kept in mind that beyond 0.5 mm from the notch free surface the material does not contribute any more to the fatigue initiation process, if high cycle fatigue is expected on the work piece. This actually means that the distance that includes the 90% of the peak stress must be terminated any how at 0.5 mm, even dough calculation were predicting a longer distance.

11.6 Notch Effect on S–N Curve

Previous sections were dealing with fatigue limit reduction due to notches. A question arises as to whether or not the notch may have the same effect on the entire *S–N* curve. Actually, running fatigue tests on notched and unnotched specimens of the same material under stress controlled conditions, curves like those shown in Fig. 11.12 are normally obtained that indicate a notch effect not only on the fatigue limit, but on the entire *S–N* curve, dough with a different weight. In fact, the effect of a notch reduces as the low cycle (plastic) fatigue life region is approaches till dissolve around the S_3 point corresponding to 1000 cycles. A more specific example is shown in Fig. 11.13 [20] for a high strength steel type 4340.

This actually means that if a notch sensitivity factor is introduced also for the fatigue finite life that, usually indicated as k'_f , it will always be $1 \leq k'_f \leq k_f$ reaching eventually the unity at about 10^3 cycles.

Heywood [18] has shown that k'_f never becomes exactly equal to one, but it approaches the unity the more the alloy is soft and ductile. He, then, introduced a notch sensitivity factor q' at 1000 cycles analogous to the q index already seen in Sect. 7.2 for the fatigue limit, so that it is

$$k'_f = q' \cdot \left(k_f - 1\right). \tag{11.32}$$

The q' factor is of empirical derivation, like its homologues q. its values are shown in Fig. 11.14 for several families of alloys. It can be seen that for a steel of ultimate

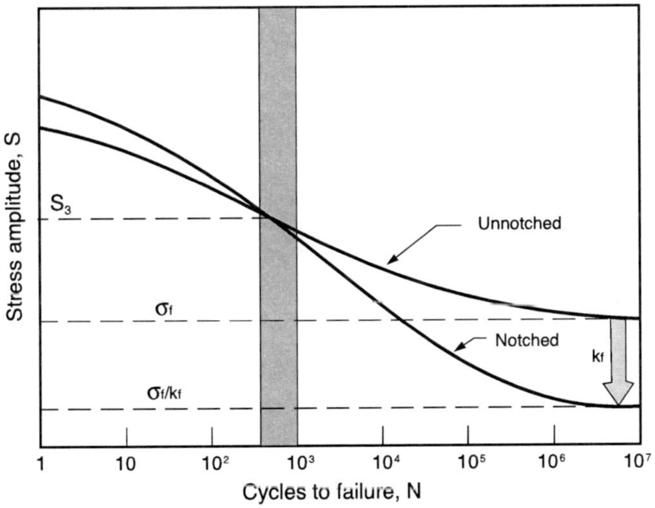

Fig. 11.12 Schematic of Wöhler's curve under stress controlled conditions for notched and unnotched specimens

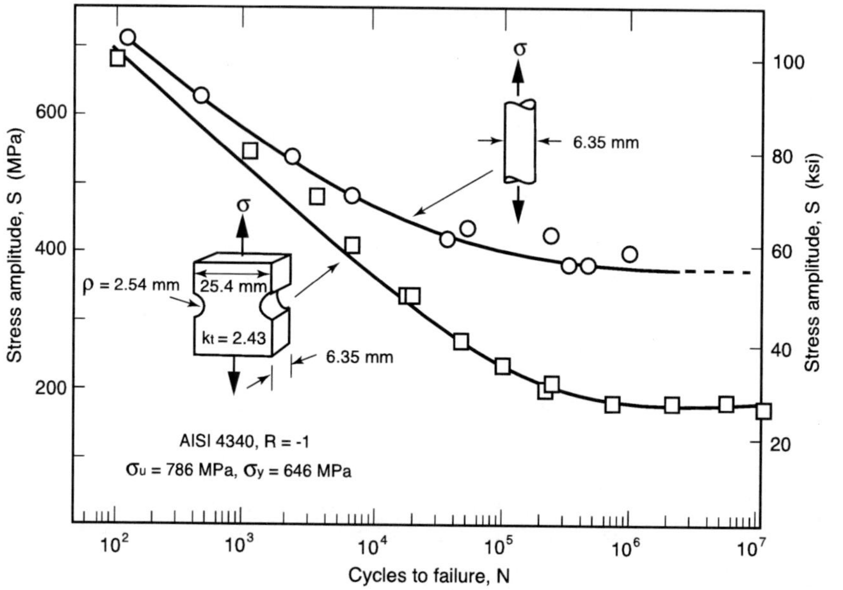

Fig. 11.13 Notch effect on the entire *S–N* curve for a high strength steel type 4340 (modified from [20])

strength $\sigma_u = 700$ MPa with a notch having a $k_t = 2.5$ and $k_f \sim 2.2$ it is $q = 0.2$. Accordingly, Eq. (11.32) yields $k'_f = 1.3$, close to unity. On the contrary, for a high strength steel with ultimate strength $\sigma_u = 2100$ MPa, the q' factor is about 0.7 and $k'_f \cong 2$. Even farer from unity k'_f would be if, for the 2100 MPa steel, the notch factor k_t were higher. However, it must be recalled that in low cycle fatigue, strain and not stress is the controlling factor.

Following the US ASME Code section III (Nuclear) and section VIII (Conventional) approach that make use of the equivalent elastic stress

$$S = \frac{1}{2}\Delta\varepsilon \cdot E \tag{11.33}$$

so that the the *S–N* curve appears as shown in Fig. 11.15, line A, where stress amplitudes can reach values of the order of 20,000 MPa that are not real. Imagine to run a series of fatigue tests using notched specimens with, for example, $k_f = 4$ under strain controlled conditions. The measured *S–N* curve converted into an elastic equivalent *S–N* curve will appear like curve B in Fig. 11.15. Curve B runs parallel to curve A. The gap between the two curves is precisely the notch factor k_f that does not vary any more with *N*. Therefore, knowing k_f it will be just sufficient to draw a curve parallel to the reference curve A shifted down by a factor k_f.

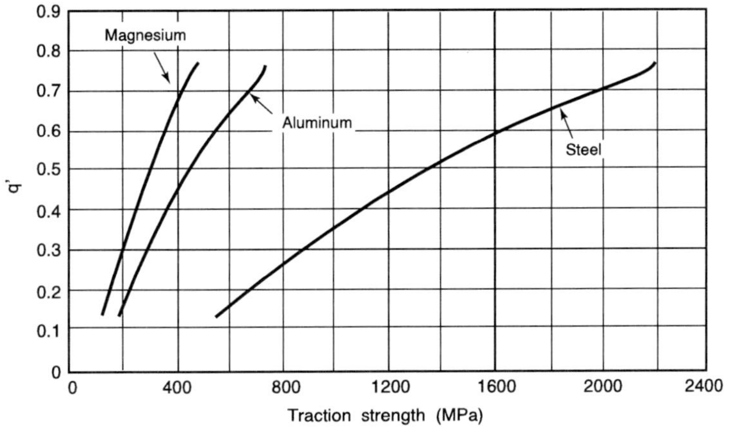

Fig. 11.14 Notch sensitivity factor q' at 10^3 cycles for various alloys of different strength (modified from [18])

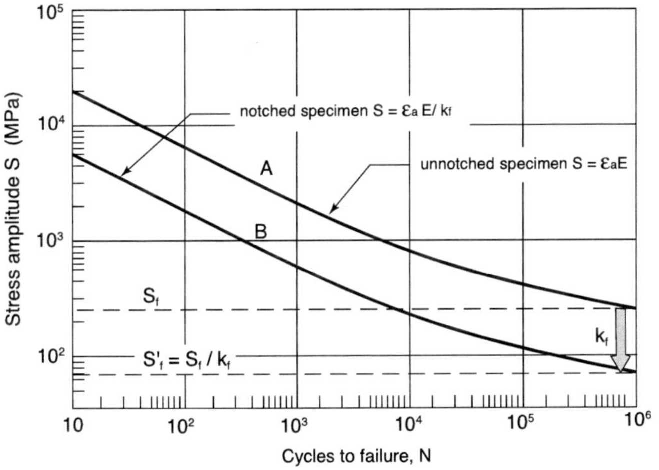

Fig. 11.15 Tipical elastic equivalent *S–N* curves for notched and unnotched specimens

11.7 Mean Stress Effect on Fatigue Limit of Notched Members

It has been shown, in the previous section, how the fatigue limit and the entire *S–N* curve are affected by the presence of a notch. We shall now discuss about the mean stress effect on a notched work piece, if any. A firs answer to the question may be given saying that whether the material behaves according to the Gerber parabola or the Goodman line it would be sufficient to scale down homotetically the relative

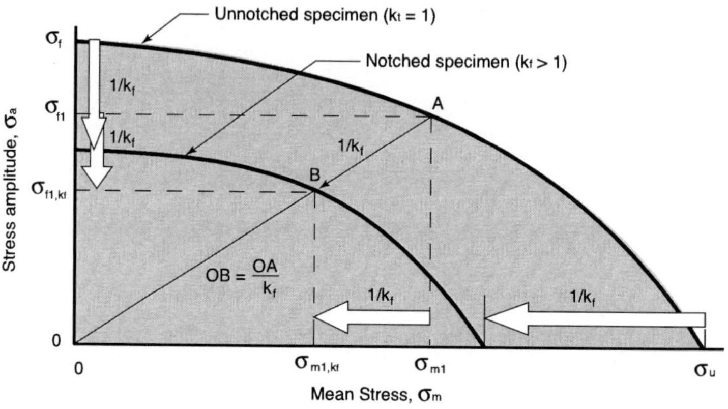

Fig. 11.16 Notch effect on notched and unnotched specimen applying a scaling factor k_f to both the fatigue limit and the ultimate strength

curve by a quantity equal to the notch factor k_f, as schematized in Fig. 11.16 for the Gerber parabola. The k_f factor has been applied to both the fatigue σ_f limit and the ultimate strength σ_u. A new safety domain is obtained so that the previous allowable combination σ_{fl} -σ_{ml} (point A) now becomes $\sigma_{fl,kf} = \sigma_{fl}/_{kf}$ and $\sigma_{ml,kf} = \sigma_{ml}/k_f$ (point B).

But such a simple procedure has its evident drawback for at least two reasons. The first reason is that when the mean stress σ_m is close to zero or anyhow very low, the peak stress on the notch root can be so low as to still remain in the elastic regime. If this is the case, there will not be any stress relaxation, the notch would not blunt keeping its original sharpness (see Sect. 11.5) and the notch factor k_f would not be different (lower) from the theoretical stress concentration factor k_t. therefore, for low values of the mean stress σ_m the theoretical notch concentration factor k_t shall be used rather than k_f.

The second reason is that it is known that a notched member may withstand loads even better than an unnotched one increasing its ultimate strength σ_u, in as long as brittle fracture can be excluded, as also shown in Fig. 8.23. This is due to triaxiality that can be generated at the notch root. Accordingly, there is no reason to reduce the ultimate strength σ_u by the k_f factor. These two arguments could induce to conceive a new safety domain, as schematized in Fig. 11.17.

It is also shown the 45° yield line of equation

$$k_t(\sigma_a + \sigma_m) = \sigma_y. \tag{11.34}$$

A generic point B will move to B*. As the mean stress reduces to zero, the k_f factor return to be k_t since it is likely that without a mean stress the elastic limit is not passed, as shown in Fig. 11.17. It is interesting to see how experimental data obtained on notched and unnotched specimens of aluminum alloy of 570 MPa ultimate strength and SAE 4130 steel of 807 MPa really behave. This is shown in

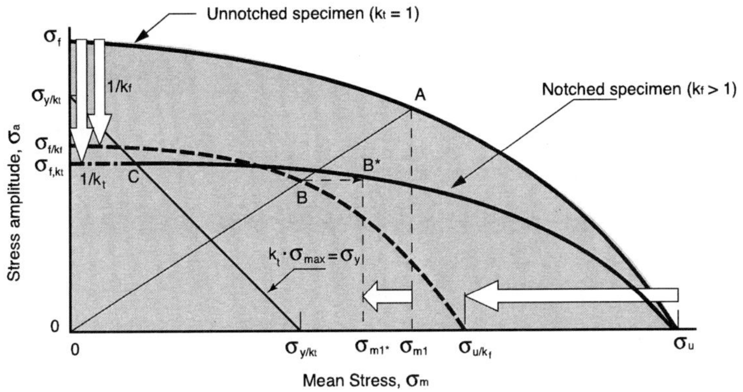

Fig. 11.17 Prediction of the safety domain of a notched member in which for low mean stresses the peak stress does not reach the elastic limit

Fig. 11.18 [21]. The behavior of the two alloys is completely different from each other and from that of Fig. 11.17. The unnotched specimens of aluminum alloy seem to follow the Goodman line while those of SAE 4130 steel the Gerber parabola as in Fig. 11.17. But differences do not finish with that. The notched specimens behavior of aluminum alloys seem to follow a curve scaled from the unnotched one by a factor equal to k_t. On the contrary, the behavior of notched specimens of SAE steel seems to follow exactly that predicted in Fig. 11.17. For mean stresses close to zero, $\sigma_m \sim 0$, the allowable stress reduces by a factor equal to the theoretical stress concentration factor k_t while the ultimate strength σ_u remains the same.

11.8 Saturation of Notch Effect

It has been explained in Sect. 11.6 why the notch factor k_f is always lower than the theoretical stress concentration factor k_t. This is, it has been said, because the synergic effect of notch root stress relaxation due to plasticity and reduced process volume of material stressed in the range of peak stress (for example 90% of peak stress). Figure 11.2 clearly shows how the two factors k_t and k_f become equal only when the theoretical factor k_t goes down to three for aluminum alloys and four for steels. This saturation of the stress concentration effects was very well documented by Frost [24, 25, 26] and Frost & Dugdale [25]. Figure 11.19 shows the results obtained by the two authors using cylindrical and plane specimens of the type shown in Fig. 2.39, carrying V notches of 5 mm and 1.3 mm depth, respectively.

It is interesting to note how the notched specimen behavior follows exactly the fatigue limit of the material σ_f divided by a factor k_t till $k_t \sim 4$. More precisely, for the specimens containing the 5 mm notch, Fig. 11.19a, saturation occurred in the proximity of $k_t = k_f \sim 4$ where the fatigue limit leveled to about 54 MPa, while for the

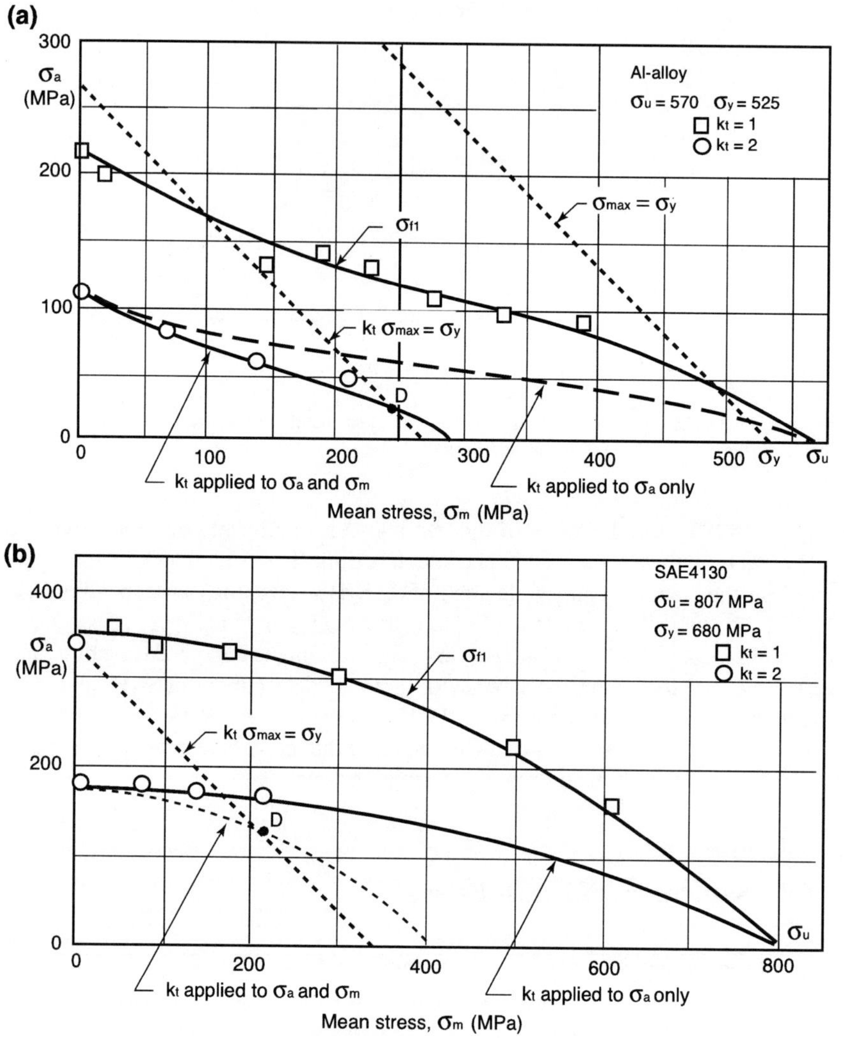

Fig. 11.18 Fatigue safety domain for notched and unnotched specimens of **a** aluminum alloy and **b** SAE 4130 steel [21] (data from [22, 23])

1.3 mm notch, Fig. 11.19b, saturation occurred in the proximity of $k_t = k_f \sim 3$ with $\sigma_f \sim 90$ MPa. Analogous curves were obtained by Frost [25] with similar specimens of 4½% Cu aluminum alloy in which the fatigue limit was following a σ_f/k_t trend up to a certain value beyond which σ_f/k_f leveled off to a constant value. Also in this case the fatigue limit depended on the notch depth being equal to 19 MPa for the 5 mm depth notch and 39 MPa for the 1.3 mm one. What Frost actually found was that between the two curves of the leveled and constant σ_f and that continuing to follow the $\sigma_f/$

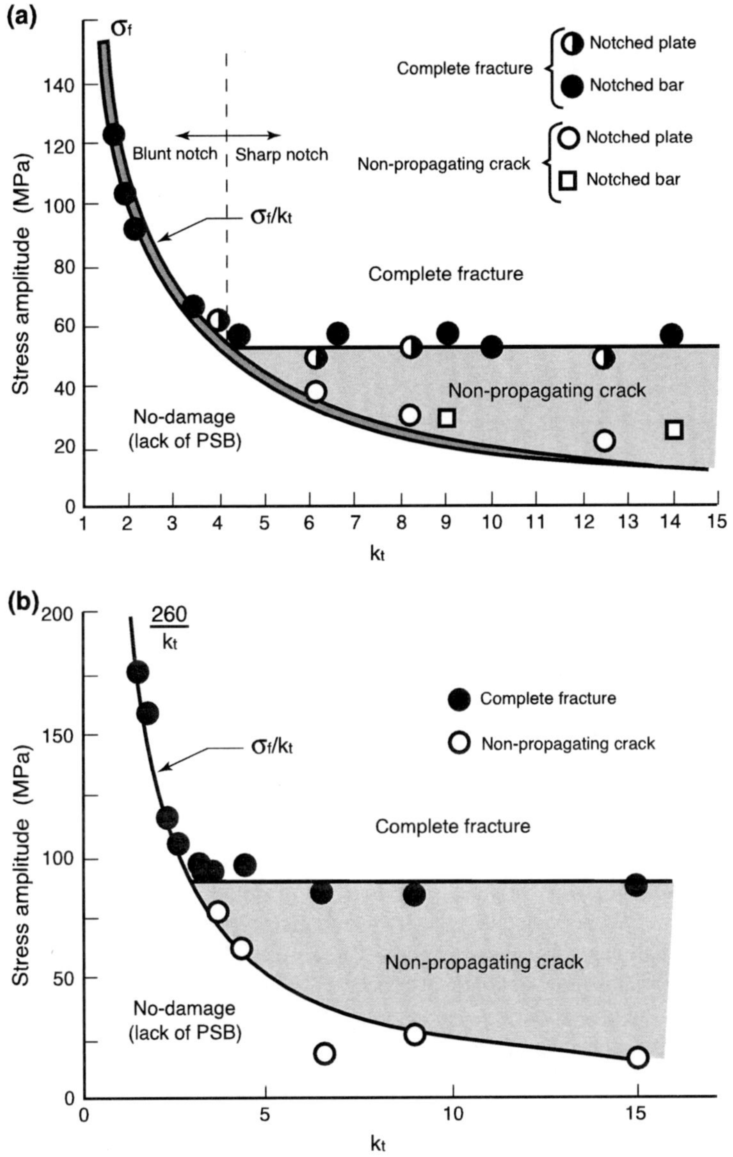

Fig. 11.19 Trend of the fatigue limit σ_f vs. the theoretical stress concentration factor k_t in mild steel specimens containing a V shaped notch **a** 5 mm depth and **b** 1.3 mm depth, respectively [26]

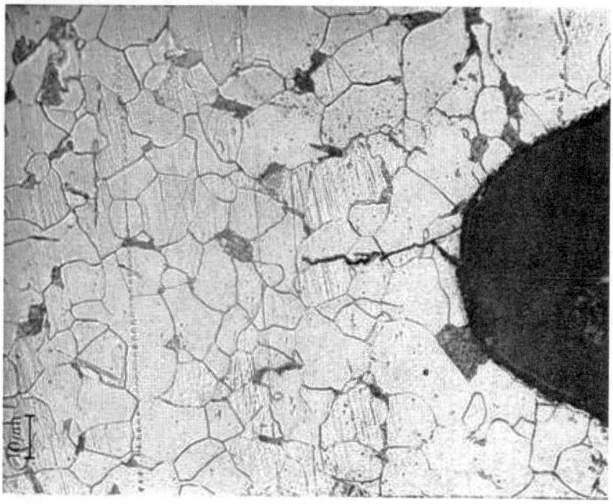

Fig. 11.20 Dormant crack at the notch root 1.3 mm depth in a mild steel after 24×10^6 cycles [27]

k_t trend, damage would nucleates anyhow on the notch root in persistent slip bands (see Sect. 2.1), but they would not propagate remaining *dormant*. This is perfectly consistent with the definition of fatigue limit as the higher possible stress amplitude that cannot propagate damage even dough it can still generate damage. Figure 11.20 [27] is a clear example of dormant crack in mild steel at the root of a V shaped notch, 1.3 mm depth and root radius equal to 0.035 mm, after 24×10^6 cycles at a stress amplitude below the fatigue limit.

The micro-crack is transgranular. It broke through four small grains to a distance of about 40 μm (MSC crack, see Sect. 1.5) and did not propagate any further because the stress amplitude was below the fatigue limit of the material. First evidence of these non-propagating cracks date back to 1949 [28] when they were found in in mild steel specimens containing a V shaped notch of 1.3 mm depth subjected to 8×10^7 fatigue cycles at 45% of the fatigue limit. Referring to Fig. 11.19, it is interesting to note that below the σ_f/k_t curve no crack or damage has nucleated. In other words, no persistent slip band has been activated. The dependency of the fatigue limit σ_f on the sharpness of the notch and, therefore, on k_t shown in Fig. 11.19, cannot be fully described without considering also the notch depth.

This second dependency is a known fact to-day [29–33] and is pointed out in Fig. 11.21 [33]. Here, the ratio between the fatigue limit of the notched specimen S'_f to the fatigue limit of the unnotched specimen S_f, is plotted versus the theoretical stress concentration factor k_t. For very shallow cracks, Fig. 11.21a, in practice the entire k_t field above the S_f/k_t curve is a non-propagating domain. It is an two-fold situation in that a defect either exists, but does not propagate or does not nucleate at all below the fatigue limit σ_f that does not decrease at all with the apparent k_t. The dependency of σ_f on k_t starts to appear beyond a certain initial notch depth, Fig. 11.21b and follows the trend already seen in Fig. 11.19. By further increasing the notch depth beyond a critical value, the non-propagating domain disappears completely. The reason for the three different behaviors is explained by saying that with shallow cracks the k_t based on elasticity actually loses completely its meaning.

Fracture mechanics (see Appendix A) predicts, in fact, that the notch is not capable to generate any triaxiality ahead of the notch tip and the theoretical stress concentration factor k_t is based also on full plain strain conditions (full triaxiality). A plastic enclave may develop all around the notch root, as schematically shown in Fig. 11.22a. A crack, whether it nucleates, would be eventually embedded into this plastic shelter and would remain closed without propagating and appear dormant, or propagating very slowly almost appearing dormant.

What the notch would possibly change if eventually would exit the plastic zone is only the fatigue life N of the specimen, but not the fatigue limit. Under this situation a k_f based dependency, if still possible, would appear as described in Fig. 11.21. The last case shown in Fig. 11.21c refers to a situation where the notch depth is

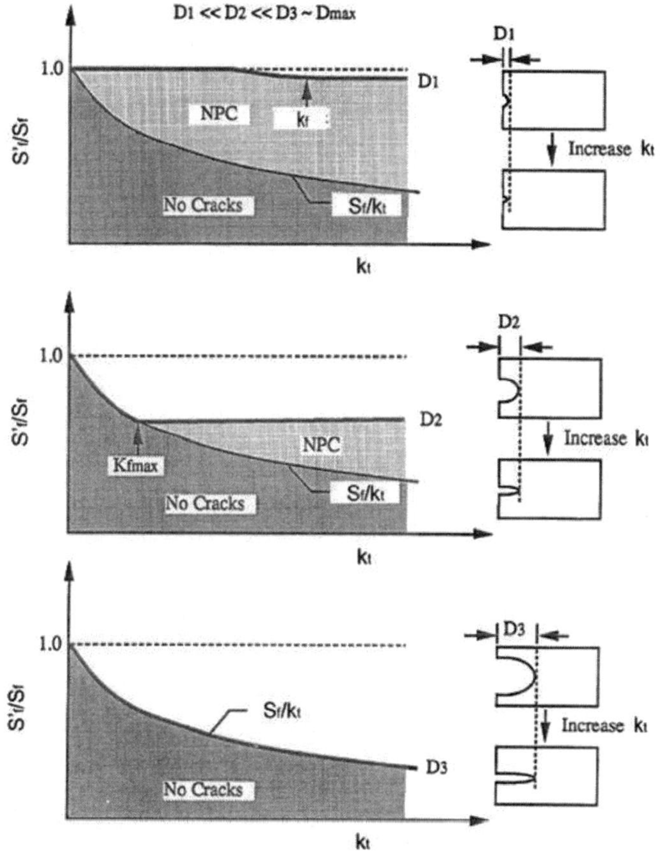

Fig. 11.21 Combined effect of notch depth and stress concentration factor k_t on the Frost σ_f -- k_t curve [33]

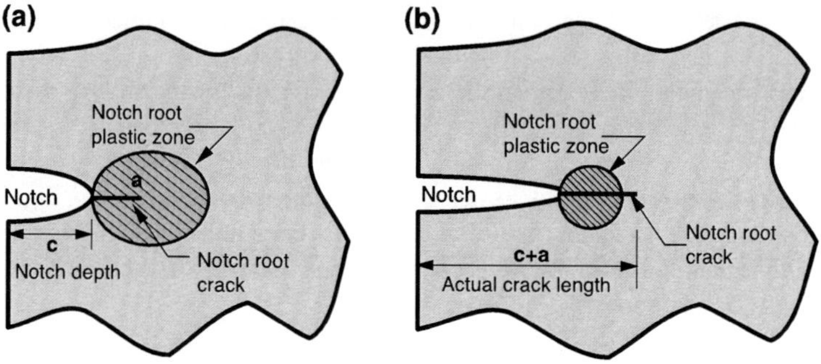

Fig. 11.22 Phenomenological difference between **a** shallow crack and **b** deep crack

some millimeters long. Now the plain strain conditions may be maintained so that the notch root plastic zone is small, as schematized in Fig. 11.22b. As the first micro-crack nucleates at the root it overcomes the plastic zone and becomes immediately a macro-crack being opened at each load cycle. Now the dependency is that predicted by fracture mechanics. In this situation notch sharpness will have an effect on both fatigue life and fatigue limit.

11.9 Notch Strain Hardening Effect

It has been said Sect. 1.4 that cyclic softening that normally takes place in hard metals and alloys may happen also before the elastic limit is reached. On the contrary, cyclic strain hardening operated on soft metals can occur only above the elastic limit. It may happen, then, that the stress concentration produced by a sharp notch, in particular, is so high to force the material to enter the plastic domain triggering the strain hardening process locally, right on the notch root. This process can produce the paradox of a notched member behaving as or even better than an unnotched one. This is the case of austenitic stainless steel that, as shown in Fig. 1.43, is particularly prone to cyclic strain hardening.

Tests run on notched and unnotched cylindrical specimens of SA 316 subjected to rotating bending have actually shown this effect, as it can be seen in Fig. 11.23. Notched specimens used in the experimental campaign had a diameter of 6.5 mm in the reduced section, the same as for the unnotched specimens, and were carrying a circumferential V shaped notch of 0.2 mm root radius, 1 mm depth with theoretical stress concentration factor k_t equal to 2.7. The steel had been used in to two different initial conditions. The first was consequent to a cold-worked strain hardening that resulted in a yield strength of 500 MPa and an ultimate strength of 640 MPa, respectively. As it can be seen in Fig. 11.23a, the presence of a notch actually reduces the fatigue limit of the material that goes down from 238 to 133 MPa at 10^8 cycles with a reduction of 1.8, lower than the theoretical k_t equal to 2.7. It is worth noting that the abacus of Fig. 11.4 is actually yielding for a 640 MPa ultimate strength steel a sensitivity index $q = 0.56$ to which it corresponds, see Eq. (11.5), a notch factor $k_f = 1.9$ against the found value of 1.8. In the second case, Fig. 11.23b, the stainless steel had been given an annealing heat treatment that resulted in a yield strength of 238 MPa and an ultimate strength of 574 MPa, respectively. This heat treatment had biased the steel to cyclic strain hardening. And in fact, the strain hardening process had been triggered by the strain concentration at the notch root, forcing the material to cycle beyond the elastic limit. The apparent paradox is that notched specimens are actually behaving like their unnotched companion.

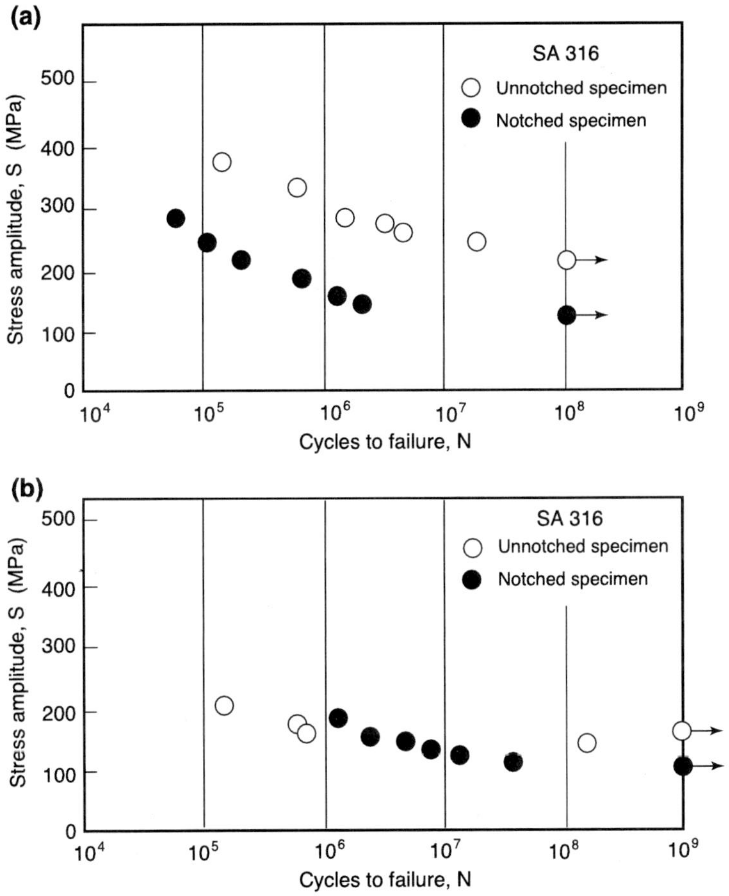

Fig. 11.23 Rotating bending fatigue on notched and unnotched specimens of austenitic stainless steel type A 316 **a** cold-worked conditions **b** annealed conditions

11.10 Sample Problems

11.10.1 Railway Axle in Rotating Bending-Shoulder Fillets

The railway axle shown in Fig. 11.24 is made with UNI Aq50 (0.35% C) steel of 550 MPa ultimate strength and 25% RA. It sustains a static weight of 25 t. Verify its fatigue integrity (Fig. 11.25).

The most stressed regions are certainly the shoulders (section I, II and III of detail A of Fig. 11.24) because they represent areas of discontinuity and act as stress raisers. The axle has a machine finish whose surface factor is $C_S = 0.78$ (see Fig. 4.7).

The static load acting on the axle journal is 25/2 = 12.5 t. A lateral component L exists that is equal to 5 t. Let's assess the bending moments M_1, M_2, M_3 and M_4 acting on sections I, II, III e IV:

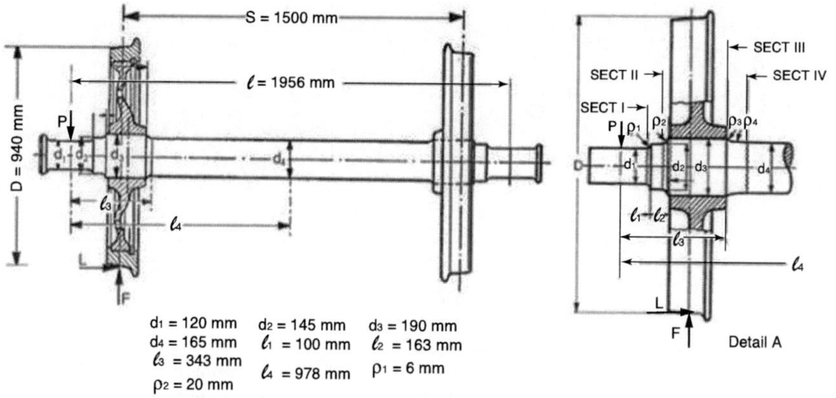

Fig. 11.24 Sketch of the railway axle with relative dimensions

Fig. 11.25 Diagram of bending moments acting on the axle

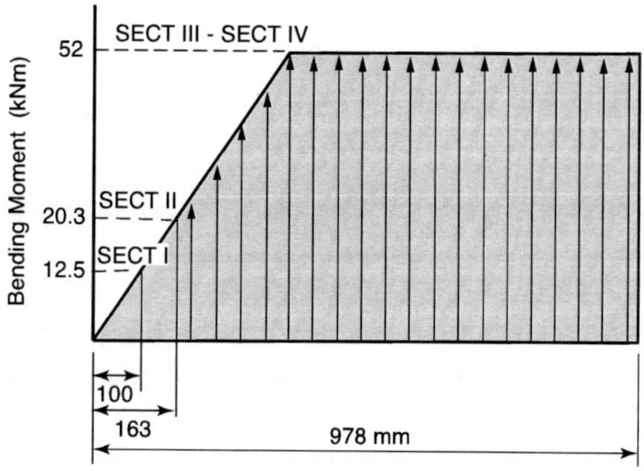

$$M_1 = P \cdot l_1 = 125 \times 0.1 = 12.5\,\text{KNm}$$
$$M_2 = P \cdot l_2 = 125 \times 0.163 = 20.37\text{KNm}$$
$$M_3 = P \cdot l_3 + L \cdot \frac{D}{2} - F \cdot \left(l_3 - \frac{l-S}{2}\right)$$
$$= 125 \times 0.37 + 50 \times 0.47 - 125 \times 0.142 = 52\,\text{KNm}$$
$$M_4 = P \cdot l_4 + L\frac{D}{2} - F\left(l_4 - \frac{l-S}{2}\right)$$
$$= 125 \times 0.978 + 50 \times 0.47 - 125 \times 0.75 = 52\,\text{KNm}$$

The diagram of bending moments is shown in Fig. 7.25.
Evaluates the stress state.

Section I nominal stress

$$\sigma_I = \frac{M_1}{W_1} = \frac{32 \times 12,5}{\pi \cdot 0.12^3} = 73.68\text{ MPa}$$

Fatigue limit S_f for the steel is about 200 MPa, as it can be deducted from Fig. 4. 6 and Fig. 4.40. The reduced fatigue limit S'_f that takes into consideration all the corrective factors is

$$S'_f = \frac{C_S \cdot C_{ld} \cdot C_{sz}}{k_f} S_f = \frac{0.75 \times 1.0 \times 0.75}{1.8} \times 200 = 62.5\text{ MPa}$$

The shoulder fillet near bearing, where the axle diameter change from $d_I =$ 120 mm to $d_2 = 145$ mm has a radius $\rho = 6$ mm. The theoretical stress concentration factor is $k_t = 1.95$, as it can be inferred from Fig. 11.26 [12]. The corresponding notch factor k_f cannot be assess using Fig. 11.4 because the radius $\rho = 6$ mm is out of the range of the figure. We can use Eq. (11.9) to assess a_P and Eq. (11.8) to calculate k_f

$$a_P = 0.276$$
$$k_f = 1.8$$

As for the other sections

Section II:

$$\sigma_{II} = \frac{M_2}{W_2} = \frac{32 \times 20.37}{\pi \times 0.145^3} = 68\text{ MPa}$$
$$S'_f = \frac{C_S \cdot C_{ld} \cdot C_{sz}}{k_f} S_f = \frac{0.75 \times 1.0 \times 0.75}{1.18} \times 200 = 95.34\text{ MPa}$$

Section III:

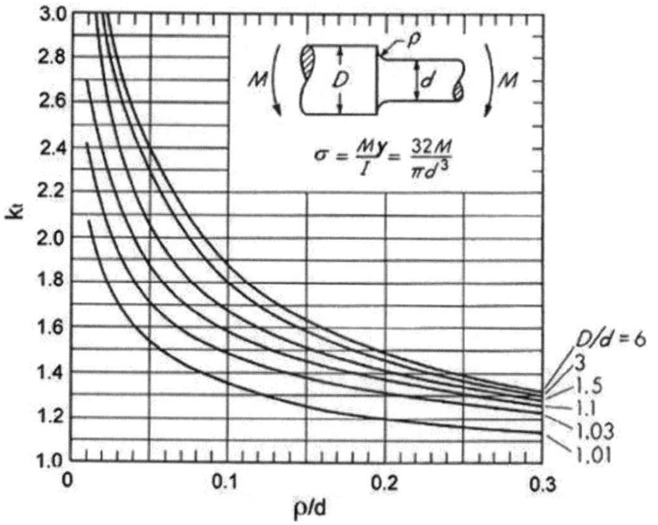

Fig. 11.26 Theoretical stress concentration factor k_t in the shoulder fillet of a shaft in bending (Modified from [12])

$$\sigma_{III} = \frac{M_3}{W_3} = \frac{32 \times 52}{\pi \times 0.19^3} = 77\ \text{MPa}$$

$$S'_f = \frac{C_S \cdot C_{ld} \cdot C_{sz}}{k_f} S_f = \frac{0.75 \times 1.0 \times 0.75}{1.17} \times 200 = 96.15\ \text{MPa}$$

Section IV:

$$\sigma_{IV} = \frac{M_4}{W_4} = \frac{32 \times 52}{\pi \times 0.165^3} = 117\ \text{MPa}$$

$$S'_f = \frac{C_S \cdot C_{ld} \cdot C_{sz}}{k_f} S_f = \frac{0.75 \times 1.0 \times 0.75}{1.0} \times 200 = 112.5\ \text{MPa}$$

Final result is that the axle is not adequately designed because it is stressed beyond the fatigue limit of the material. The axle is likely to fail also because there is a forcing stress between the axle and the bearing that has not been taken into consideration. The resulting fracture will be of the type shown in Fig. 3.47b in the absence of a notch (*Section IV*) and Fig. 11.27 [23] in the area of the shoulder fillet (*Section I*). Figure 11.27b schematizes the characteristic features visible on the failed surface such as: radial ridges that denote multiple initiation site characteristic of the presence of a notch (shoulder fillet), beach marks that indicate the growth between two consecutive metal idling (see Sect. 3.3 and Fig. 3.6), the direction of rotation opposite to 20° axis of the final dark area of overload fracture. The relative small size of this area indicates that loads were not high. Both failures were already analyzed in

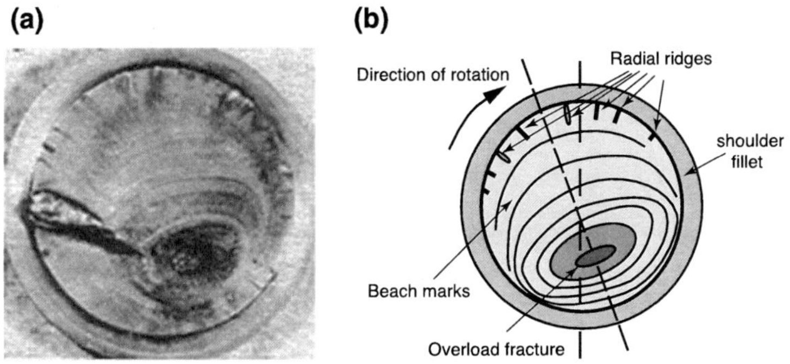

Fig. 11.27 **a** Fatigue failure of a railway axle in the area of a shoulder fillet [23]. **b** Sketch of the fracture surface

Sect. 3.3. The axle must be redesigned. It can be started from the material and surface finish using an Aq50 or Aq60 steel with ultimate strength equal to 700 MPa and RA = 20% and a fine ground with a coefficient C_s from 0.75 to 0.9, respectively. The new fatigue limit will be $S_f = 300$ MPa and the reduced one corrected for all factors

$$S'_f = \frac{C_S \cdot C_{ld} \cdot C_{sz}}{k_f} S_f = \frac{0.9 \times 1.0 \times 0.75}{1.85} \times 300 = 109 \text{ MPa.}$$

The safety coefficient will be for *Section I*

$$\nu_{sf,I} = \frac{109}{73.68} = 1.5$$

considered just acceptable, while for *Section IV* it is

$$\nu_{f,IV} = \frac{168.5}{117} = 1.44$$

11.10.2 Fatigue in a Freight Elevator Axle-Shoulder Fillets

The axle of a freight elevator, shown in Fig. 11.28, that must sustain a load $P = 43$ kN is made of Aq50 steel and fine machined in the areas of shoulder fillets. Verify its fatigue strength.

The most critical section is the A–A. The maximum bending moment is

$$M_{\max} = Pl_2 = 43 \times 0.21 = 9 \text{ KNm}$$

While on section A-A it is

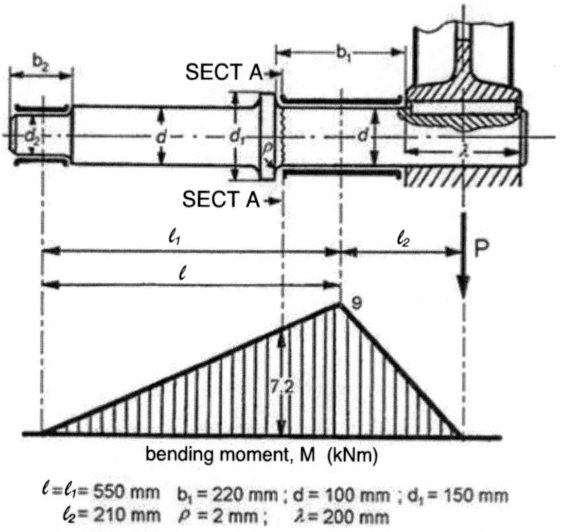

Fig. 11.28 Schematic of a freight elevator axle af a poligonal chain wheel and moment

$$M_A = 7.2\ \text{kNm}.$$

and the corresponding stress is

$$\sigma_A = \frac{M_A}{W_A} = \frac{32 \times 7.2}{\pi \cdot 0.1^3} = 73.33\ \text{MPa}$$

The fatigue limit S_f of Aq50 steel is (see sample problem 1) 200 MPa, the reduced one S'_f must consider the corrective factors C_i and the shoulder fillet. The stress concentration factor for the geometry under investigation is $k_t = 2.7$, as it can be inferred entering the abacus of Fig. 11.26 with $\rho/d = 0.02$ e $d_1/d = 1.5$. the corresponding notch factor can be assessed from the abacus of Fig. 11.4 that yields a notch index $q = 0.78$. Entering Eq. (11.5) with the q value it is $k_f = 2.3$. Almost the same value would be obtained from Eq. (11.8) with a_P given by Eq. (11.9)

$$k_f = 1 + \frac{k_t - 1}{1 + \sqrt{\frac{a_P}{\rho}}} = 1 + \frac{1.7}{1 + \sqrt{\frac{0.278}{2}}} = 2.24$$

or using the Neuber $\sqrt{a_N} = 0{,}45$ from Fig. 11.5 and Eq. (11.12)

$$q = \frac{k_f - 1}{k_t - 1} = \frac{1}{1 + \sqrt{\frac{a_N}{\rho}}} = 0.67$$

therefore it is

$$S_f' = \frac{C_S \cdot C_{ld} \cdot C_{sz}}{k_f} S_f = \frac{0.9 \times 1.0 \times 0.75}{2.24} \times 200 = 59.2 \text{ MPa}.$$

It can be concluded that the geometry considered is not capable to withstand fatigue for more than 10^6 cycles. It will be necessary, at least, to increase the shoulder fillet radius that is too small, resulting in a notch factor too high.

11.10.3 Plates Containing Bore Holes-Process Volume Effect

Three plates of Fe510D contain three bore holes of different radius and equal to 20 mm, 10 mm and 5 mm, respectively. Their geometries, shown in Fig. 11.29, were chosen to generate the same theoretical stress concentration factor $k_t = 3.48$, as it can be inferred from the abacus of Fig. 11.30 [12] for $a/w = 0.33$. The fact that it can be $k_t > 3$ is due a border effect. In fact, the well-known theoretical value $k_t = 3$ is valid for a hole in an infinite plate. Assess the fatigue behavior of the three holes when the plate is subjected to cycling with local peak stress amplitude equal to 282 MPa.

The three hole will show three different fatigue limit because, as discussed in Sect. 11.5, their process volumes $V_{90\%}$ where the local stress remains within the 90% of the peak stress are different.To assess these process volumes it is necessary to appraise the stress fields corresponding to the three different geometries. A finite

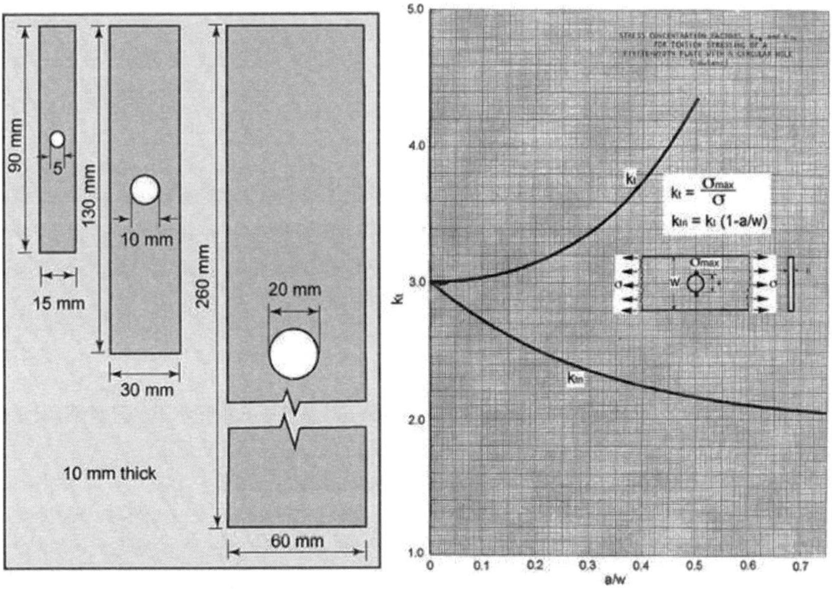

Fig. 11.29 Plates carrying bore holes of different diameter and relative stress concentration chart [12]

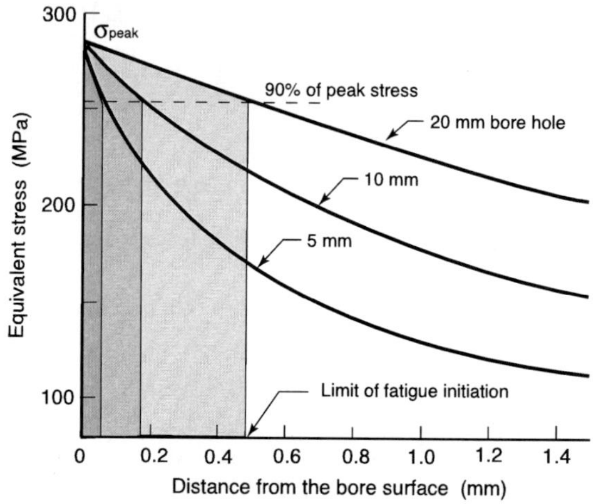

Fig. 11.30 Stress fields ahead of the bore surface for the three bored plates and relative distances for assessing the corresponding process volumes $V_{90\%}$

element calculation yields the results shown in Fig. 11.31 where the von Mises equivalent stress is shown. All three fields start from the common peak value of 282 MPa that corresponds to a peak strain $\varepsilon = 0.0013$. However, dough starting with the same value, the three stress fields have different trends.

The 5 mm hole stress field extinguishes very rapidly and reach the 90% peak stress at a distance equal to 0.06 mm, while the 10 mm hole stress field decays less sharply and reach the 90% peak stress at a distance equal to 0.16 mm. Finally, the 20 mm hole stress field maintains over the other two for a longer distance reaching the 90% peak stress at 0.5 mm from the hole surface. This last value, by the way,

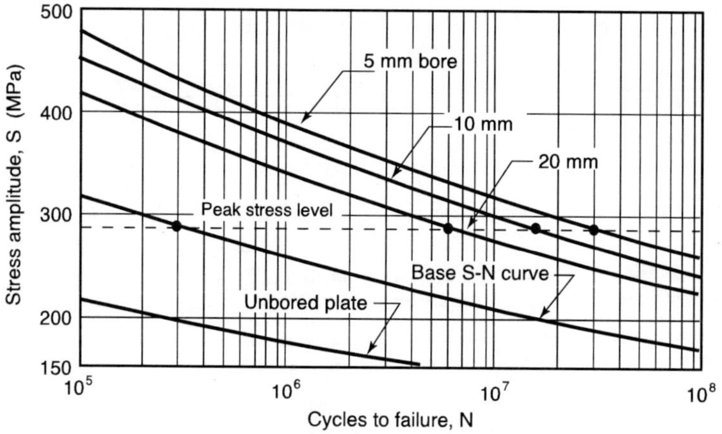

Fig. 11.31 Base *S–N* fatigue curve and its variation depending on the process volumes of the four different geometries considered (three bored plates and one unbored plate)

is just the limit of the surface layer that participates to the fatigue initiation process. Beyond that limit the material does not contribute to high cycle fatigue nucleation damage any more. The three different decay distance found by FE analysis are then

$$d_5 = 0.06 \text{ mm} \; ; d_{10} = 0.16 \text{ mm} \; ; d_{20} = 0.5 \text{ mm}.$$

Assuming the corresponding process volumes as ellipses having, according to the Saint Venant principle, semi-axis $a = d$ and $b = 2{\cdot}d$ it yields

$$V_{90\%}^{D=20} = 1.57 \text{ mm}^3; V_{90\%}^{D=10} = 0.16 \text{ mm}^3; \; V_{90\%}^{D=5} = 0.0226 \text{ mm}^3.$$

Using the Weibull analysis and Eq. (7.28) it is possible to assess the ratio between the fatigue strength of the plates containing the 10 mm and 20 mm holes

$$\frac{\sigma_{f,10}}{\sigma_{f,20}} = \left(\frac{V_{90\%}^{20}}{V_{90\%}^{10}}\right)^{1/m} = (9.76)^{1/30} = 1.08$$

where $m = 30$ is the Weibull exponent of the steel measured with standard traction specimens. The fatigue limit of the plate containing the 10 mm hole will be about 8% higher than that of the 20 mm bore hole plate. Alike, the ratio between the 5 mm hole plate and the 20 mm hole plate is

$$\frac{\sigma_{f,5}}{\sigma_{f,20}} = \left(\frac{V_{90\%}^{20}}{V_{90\%}^{5}}\right)^{1/m} = (69.4)^{1/30} = 1.15.$$

This time the variation is about 15%. The question arises now as to the knowledge of the fatigue limit $\sigma_{f,20}$ of the plate containing 20 mm hole with respect to that σ_f of plate without hole. The Wöhler's curve of the material is that already assessed in sample problem 4 of Sect. 6.3. It can be seen how the transition point N_t between low cycle fatigue and high cycle fatigue, i.e., between plastic and elastic behavior, is at about $\varepsilon = 0.0015$ so that we can assume that in our case the behavior is mainly elastic and neglect the plastic component and draw the *S–N* curve in terms of stress amplitude. The Basquin line contain a parameter, the fatigue strength coefficient σ'_f , that is the true stress at failure (see Sect. 7.2). Its value is normally measured with standard specimens of 10 mm diameter and 10 cm lenth. Therefore, its process volume, that we will assume as reference volume, is $V_{ref} = 7853 \text{ mm}^3$. Therefore

$$\frac{\sigma_{f,20}}{\sigma_f} = \left(\frac{V_{eff}}{V^{20}}\right)^{1/m} = \left(\frac{7854}{1.57}\right)^{1/30} = 1.32$$

which actually means that a plate with a 20 mm hole will have a fatigue limit 32% higher than that measured with small specimens at the same stress level of the peak stress. The plate with the 10 mm hole will have an additional 8% fatigue limit increase reaching a total of 43% increase and that with the 5 mm hole a total gain of 52%.

The final step will be to assess the fatigue limit of a plate of the same size as that of the bored one carrying a hole of 20 mm, but without any hole. The process volume of this plate is

$$V^{20}_{unbored} = 10 \times 60 \times 260 = 156,000\ \text{mm}^3.$$

Therefore its fatigue strength is likely to be

$$\frac{\sigma_f}{\sigma_{f,20}} = \left(\frac{V^{20}}{V^{20}_{unbored}}\right)^{1/m} = \left(\frac{1.57}{156,000}\right)^{1/30} = 0.68.$$

Figure 11.31 presents the overall result obtained for the three bored plates with respect to the reference *S–N* curve. Also shown is the fatigue life prediction for the unbored plate. Results indicate that the plate carrying the 20 mm bore would fail at after $6 \cdot 10^6$ cycles, while the 10 mm and 5 mm bore hole plates would fail at 1.6×10^7 cycles and 3×10^7 cycles, respectively. Interesting enough is that specimens fatigued at the same peak stress of 282 MPa would fail after only 300,000 cycles and the unbored plate well below 100,000 cycles. However, it must be understood that comparison among fatigue lives has been made on the assumption that both fatigue specimens and the unbored plate are cycled at a nominal stress amplitude equal to local peak stress of 282 MPa amplitude existing at the hole root.

11.10.4 Leaf Spring with Hole-Process Volume

A multileaf spring of the type schematized in Fig. 11.32 is subjected to load cycles variable from 700 to 160 kg. Leaves have a width $l = 80$ mm and thickness $s = 6$ mm and are made of quenched and tempered steel that has been subjected to shot peening that has introduced a residual compression stress equal to about – 400 MPa and surface hardness of about 260 HB with a mean surface roughness of 2.0 μm. Verify leaf spring to fatigue. The most stressed section of each leaf is the midsection where a hole of 4 mm radius has been drilled to insert a bolt to keep leaves together (Fig. 11.33).

The maximum force acting is $F_{\max} = 700/2 = 350$ kg and the minimum one is $F_{min} = 160/2 = 80$ kg. The amplitude and the mean force are

$$F_a = (350 - 80)/2 = 135\ \text{kg}$$
$$F_m = (350 + 80)/2 = 215\ \text{kg}.$$

If friction between leaves and shear stress due to force F are neglected, considering the moment M alone acting on the leaf midsection, then the stress amplitude σ_a acting on each leaf is

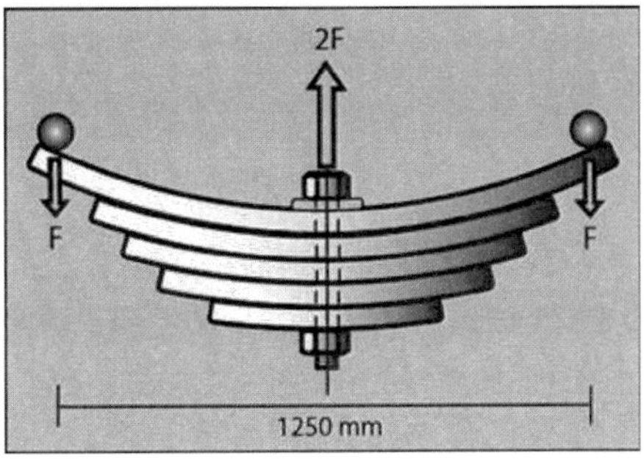

Fig. 11.32 Schematic of multileaf spring

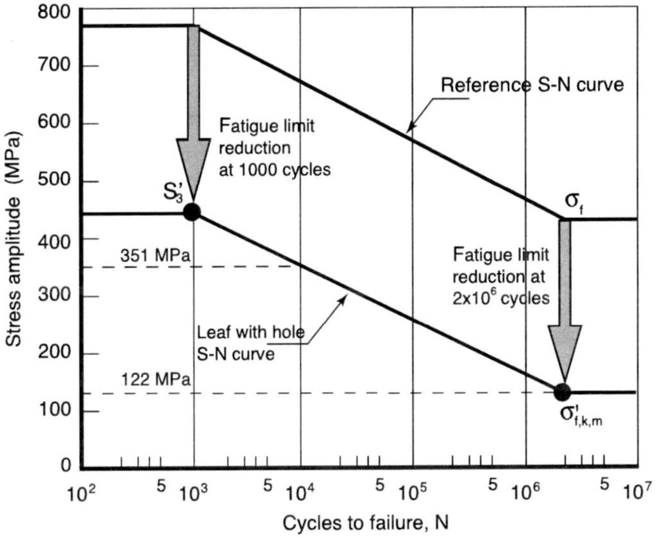

Fig. 11.33 S–N curve for the multileaf spring

$$\sigma_a = \frac{1}{5}\frac{Ms}{2 \cdot I} = \frac{1}{5}\frac{12 \cdot F_a Ls}{2 \times 2 \cdot ls^3} = \frac{3}{5}\frac{F_a L}{ls^2}$$
$$= \frac{3}{5}\frac{1350 \times (1250)}{80 \times (6)^2} = 351 \text{ MPa.}$$

The mean stress σ_m is

$$\sigma_m = \frac{1}{5}\frac{Ms}{2 \cdot I} = \frac{1}{5}\frac{12 \cdot F_m Ls}{2 \times 2 \cdot ls^3} = \frac{3}{5}\frac{F_m L}{ls^2}$$
$$= \frac{3}{5}\frac{2150(1250)}{80 \times (6)^2} = 559 \text{ MPa.}$$

The surface residual stress introduced by the shot peening process will not have any effect on the stress amplitude, but only on the mean stress that becomes.

$$\sigma_{m,tot} = 559-400 = 159\text{MPa}.$$

Let's find the fatigue *S–N* curve of the material. On the base of Eq. (5.8) the traction strength σ_u of the shot peened material is

$$\sigma_u = 3.3 \times \text{HB} = 3.3 \times 260 = 858 \text{ MPa}$$

while using Eq. (5.1) the fatigue limit σ_f is

$$\sigma_f = 0.5 \times 858 = 429 \text{ MPa}.$$

Note that by using Eq. (5.7) and the convertion Table 5.1 or Fig. 5.5 the fatigue limit would be

$$\sigma_f = (9.16 \times 2.5 + 308) = 555 \text{ MPa}$$

higher than 429 MPa found before. We will choose the lower value 429 to be conservative. Since a bore hole is present it must be considered the stress concentration factor that for the geometry of the leaf is $k_t = 3.3$ (see chart of Fig. 11.29). With a hardness of 260 HB and a hole of 4 mm radius the sensitivity index id $q = 0.95$ (see Fig. 11.4). Therefore it is

$$k_f = 1 + q \cdot (k_t - 1) = 3.18.$$

the new fatigue limit $\sigma_{f,k}$ is.

$$\sigma_{f,k} = \sigma_f / k_f = 429/3.18 = 135\text{MPa}.$$

The next step is to consider the effect of mean stress σ_m. Using the Gerber criterion, Eq. (5.47) wit $n = 2$, the fatigue limit will be

$$\sigma_{f,k,m} = \sigma_f/\sigma_{f,k} \times [1 - (\sigma_m/\sigma_u)^2] = 135[1 - (159/858)^2] = 130.4 \text{ MPa}.$$

However, this is not the last fatigue limit since there are still the corrective factor C_S, C_{sz} and C_{ld} to apply. As to the surface finish C_S, Fig. 4.8 indicates that with a mean roughness of 6 μm and a tensile strength of 858 MPa it is $C_S = 0.8$. As to the size effect, since many springs will be considered it can be conservatively assumed $C_{sz} = 0.75$. as to load factor, since the leaves are working in simple bending it is $C_{ld} = 1.2$ (see Table 4.4). The last factor to consider is the process volume. Using as reference sample problem 3, for the 8 mm radius hole the process volume effect is judge to be about 1.30.

Therefore, the final fatigue limit will be

$$\sigma'_{f,k,m} = \sigma_{f,k,m} \cdot C_S \cdot C_{sz} \times C_{ld} \cdot 1.3 = 130.4 \times 0.8 \times 0.75 \times 1.2.1.3 = 122 \text{ MPa}.$$

At 1,000 cycles, the S_3 limit can be derived from the sensitivity index q' of Fig. 7.14 whose value is

$$q' = 0.27$$

and so

$$k'_f = 1 + q' \times (k_t - 1) = 1.62$$

and then

$$\begin{aligned} S'_3 &= 0.9 \times \sigma_u \times C_S \times C_{sz} \times C_{ld}/1.62 \\ &= 0.9 \times 858 \times 1.0 \times 0.75 \times 1.2/1.62 = 429 \text{ MPa}. \end{aligned}$$

Note that the process volume effect has not been considered since at 1.000 cycles the material is working in full plastic regime with low cycle fatigue. The process volume, then, is the entire volume ahead of the hole, from both sides, which is practically equal to that of the fatigue specimen. The final fatigue *S–N* curve is, then, that of Fig. 7.33. As it can be seen, at the working stress amplitude leaves cannot resist longer than 10^4 cycles. Redesign the spring to withstand infinite life.

References

1. Rankine, W.J.M.: On the causes of unexpected breakage of the journals of the railway axles and on the means of preventing such accidents by observing the law of continuity in their construction. In: Institution of Civil Engineers, Minutes of Proceedings, vol. 2, pp. 105–108, London (1842)
2. Peterson, R.E., Wahl, A.M.: Two and three dimensional cases of stress concentration and comparison with fatigue tests. Trans. Am. Soc. Mech. Eng. 15–22 (1936)
3. Neuber, H.: Theory of Notch Stresses: Principle for Exact Stress Calculation. J.W. Edwards, Publishers, Incorporated, Ann Arbor, Michigan (1946)
4. Phillips, C.E., Heywood, R.B.: Proceedings Instit. Mech. Eng. **165**, 113 (1951)
5. Lazan, B.J., Blatherwick, A.A., Wright Air Development Centre, Technical Report No. 52-307 (1952)
6. Marry, R.W., Dolan, T.J.: Proc. ASTM **53**, 789 (1953)
7. Templin, R.L.: Proc. ASTM **54**, 641 (1954)
8. Grover, H.J., Hyler, W.S., Jackson, L.R.: NASA Tech. Note D **111** (1959)
9. Mann, J.Y.: Proc. ASTM **60**, 602 (1956)
10. Finney, J.M.: Australian defence scientific service, department of supply. Aero. Res. Lab., ARL/SM Report No. 287 (1962)
11. Forrest, P.G., Tate, A.E.L.: The Influence of Grain Size on the Fatigue Behavior of 70/30 Brass. J. Inst. Metals **93**(438):1964–1965
12. Peterson, R.E.: Stress Concentration Factors. John Wiley & Sons, New York (1973)
13. Peterson, R.E.: Metal Fatigue. Chap. 13. In: Sines, G., Waisman, J.L. (eds.) Notch Sensitivity, McGraw-Hill Book Co. New York (1959)

14. Peterson, R.E.: Analytical approach to stress concentration effects in aircraft materials. Technical Report 59–507, U.S. Air Force-WADC Symposiumon Fatigue in Metals, Dayton, Ohio (1959)
15. Sines, G., Waisman, J.L.: Metal fatigue Chap. 13. In: Peterson R.E. (eds.) Notch Sensitivity. Mcgraw-Hill Book Company, New York (1959)
16. Kuhn, P., Hardrath, H.F.: An Engineering Method for Estimating Notch-Size Effect in Fatigue Tests of Steel. Report NACA TN 2805 (1952)
17. Kuhn, P.: The Prediction Of Notch And Crack Strength Under Static Or Fatigue Loading. SAE-ASME paper 843C, April (1964)
18. Heywood, R.B.: Designing Against Fatigue. Chapman and Hall Ltd., London (1962)
19. Siebel, E., Stieler, M.: Significance of Dissimilar Stress Distributions for Cyclic Loading" (in German), Z.V.D.I., vol. 97, pp. 146–148 (1955)
20. Dowling, N.E.: Mechanical Behavior of Materials: Engineering Methods for Deformation, Fracture, and Fatigue, 2nd edn. Prentice Hall, New York (1998)
21. Schijve, J.: Fatigue of Structures and Materials. Kluwer Academic Publisher, p. 170 (2004)
22. Grover, H.J., Bishop, S.M., Jackson, L.R.: Fatigue Strengths of Aircraft Materials. AxialLoad Fatigue Tests on Notched Sheet Specimens of 24S-T3 and 75S-T6 Aluminum Alloys and ofSAE 4130 Steel. Report NACA TN 2324 (1951)
23. Grover, H.J., Bishop, S.M., Jackson, L.R.:Fatigue strengths of aircraft materials. axialload fatigue tests on notched sheet specimens of 24S-T3 and 75S-T6 aluminum alloys and ofSAE 4130 steelwith stress-concentration Factors of 2.0 and 4.0", Report NACA TN 2389,1951
24. Frost, N.E.: Proc. Inst. Mech. Eng. **173**, 811 (1959)
25. Frost, N.E., Dugdale, D.S.J.: Mech. Phy. Solids **5**, 182 (1957)
26. Frost, N.E., Marsh, K.J., Pook, L.P.: Metal Fatigue. Clarendon Press, Oxford (1974)
27. Jacques, H.E.: Bureau of Ships, Navy Dep. Report N. SSC-31 (1949)
28. Lukas, P., Kunz, L., Weiss, B., Stickler, R.: Non damaging notches in fatigue. Fatigue Fract. Eng. Mater. Struct. **9**, 195–204 (1986)
29. Tanaka, K., Akiniwa, Y.: Notch geometry effect on propagation threshold of short fatigue cracks in notched components. Fatigue '87, Vol. II, Ritchie, R.O., Starke E.A. (eds.) Third International Conference on Fatigue and Fatigue Threshold, pp. 739–748 (1987)
30. Topper, T.H., El Haddad, M.H.: "Fatigue Strength Prediction of Notches Based on Fracture Mechanics", Fatigue Thresholds: Fundamentals and Engineering Applications, vol. II, pp. 777–798. Camelom Press Ltd., London (1981)
31. DuQuesnay, D.L., Topper, T.H., Yu, M.T.: The effect of notch radius on the fatigue notch factor and the propagation of short cracks. EGF Publ **1**, 323–335 (1986)
32. Ting, J.C., Lawrence, F.V.J.: A crack closure model for predicting threshold stresses of notches. Fatigue Fract. Eng. Mater. Struct. **16**(1), 93–114 (1993)
33. Lipson, C.: Why Machine Parts Fail. Penton Publishing Co., Cleveland (1951)

Chapter 12
Cumulative Damage: Cycle Counting and Life Prediction

12.1 Load Spectra

Loads so far considered had constant amplitude (CA), σ_a or S. This constitutes a particularity that rarely happens in real life. Usually, cyclic loads have variable amplitudes (VA) that are referred to as the load spectrum or time history. Even apparently simple operating conditions actually result in a more or less frequent variability of load amplitudes with time. Think about a pressure vessel that is heated up and cooled down every working day. During normal functioning at the nominal pressure the vessel can experience some pressure variations that depend on numerous factors, often aleatory. These pressure variations are generally small amplitude fluctuations, but they are applied to the vessel shell with high mean stress provided by the heat up and cool down cycle. In addition, from time to time there might be large temperature and pressure transients that results in major stress excursions that do not occur too often, but in a 20 or 30 years of expected life they can sum up to a consistent number. Also think about the wing of an airplane subjected to continuous flexural vibrations of variable amplitude during take-off, flying cruise and landing, as schematized in Fig. 12.1.

Wing lift is a force that sustains the plane and is distributed over the entire wing surface. It generates a variable bending moment with the maximum at the wing-fuselage joint that turns out to be the most critical point where the lower wing skin is under tension fluctuations. Atmospheric turbulences at lower altitude during take-off and landing induce low frequency vibrations with relatively high amplitude while during cruise at high altitude there is a more stable situation with high frequency low amplitude vibrations. The point is that these vibrations are applied to the wing-fuselage connection with a rather high mean stress that is provided by half the airplane weight, at least during the flight. Another example is the load spectrum on the leaf springs or shock absorbers of a car running on a bumpy road or on the fork of a motorbike or brake disks as shown in Fig. 12.2. These types of loads have unpredictable random spectra or time histories. They are generally inferred by instrumenting the

P. P. Milella, *Fatigue and Corrosion in Metals*,
https://doi.org/10.1007/978-3-031-51350-3_12

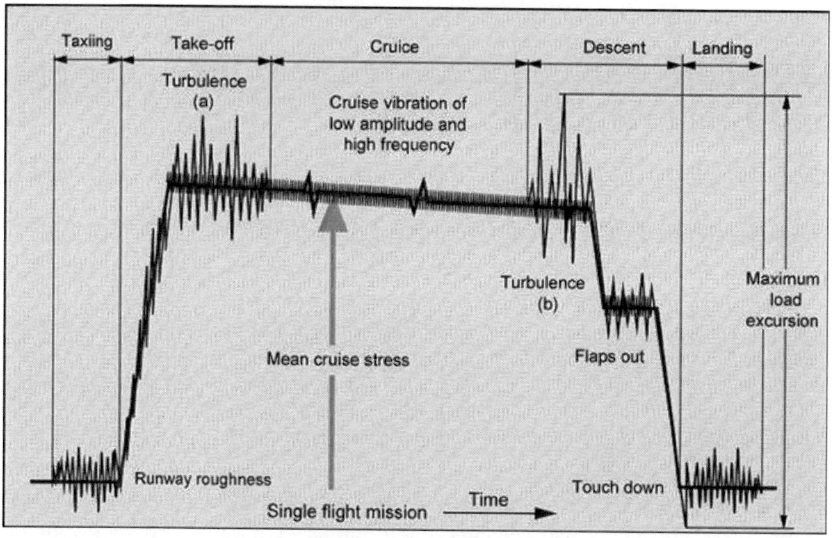

Fig. 12.1 Schematic of flexural loads induced in an airplane wing during a flight

work piece and can be analyzed and compared in terms of *pdf* (probability density function, see Chap. 4).

Sometimes they are replaced by the so called umbrella loads that are deterministic loads foreseen to be more restrictive than the real ones expected to act on the component and are thought to cover all possible situations that may happen in real life. Their advantage is that despite the inherent conservatism they have simple spectra and can

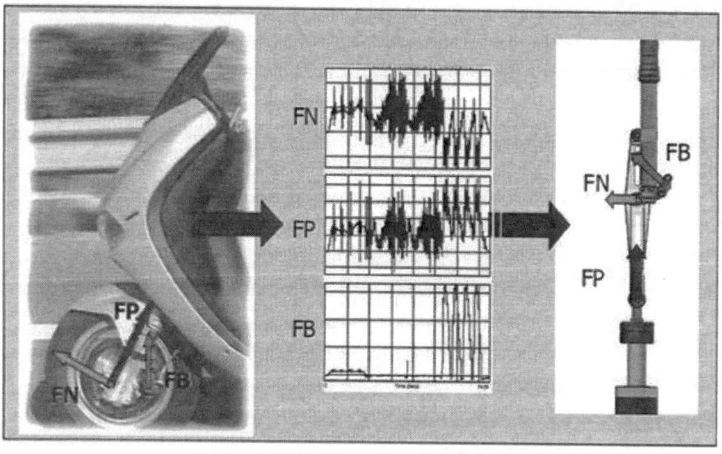

Fig. 12.2 Response spectra in a motorcycle: FP is the force parallel to the steering axle, FN the normal component and FB the force on disks [1]

be easily analyzed. Their drawback is that the conservatism necessary to guaranty the complete coverage of all possible situations may be too much to comply with. Dealing with variable amplitude loads a main question arises as to whether cycles must be treated individually or can be analyzed as single loading sequence that is repeated many times and how to combine their effects, which deals with the problem of damage progression within each single sequence, for instance the airplane *mission* or a daily pressure vessel operation or a standard one or two kilometer car trail and accumulation among different sequences. The following sections will be addressing precisely these questions of cycle counting, damage progression and accumulation. It will be shown that, despite the enormous knowledge so far acquired and the continuous efforts to understand and improve our capability to design, these questions have no single or simple answer.

12.2 Load Spectra Representation and Counting

12.2.1 Level Crossing Cycle Counting

Variable amplitude loadings are generally divided in two types:

- stationary spectra;
- non stationary spectra.

Spectra whose *pdf* has a mean value, median value and a standard deviation independent of time variable are stationary. On the contrary, the spectra whose statistical properties vary with time are said to be non-stationary.

A *spectrum* or load history can be considered a sequence of peaks and troughs of variable amplitude. The peaks can be symmetrical with respect to a mean value or not. An example of symmetrical spectrum is shown in Fig. 12.3. The corresponding *pdf* is relatively large and indicates a prevalence of peaks above the mean level. The frequency relative to each load level is that of the histogram of Fig. 12.4. If peaks below the mean level prevail, as schematized in Fig. 12.5, the relative *pdf* appears close and high, as in the histogram of Fig. 12.6. The number of peaks n_i found in the interval between two consecutive levels is given by the number $n_{i,cros}$ of peaks the are crossing level i minus that of level $i + 1$, $n_{i+1,cros}$

$$n_i = n_{i,cros} - n_{1+1,cros} \tag{12.1}$$

For a constant amplitude spectrum the corresponding *pdf* reduces to a single peak without tails with the standard deviation to mean value ratio equal zero, $s/\overline{x} = 0$. If the load history is non-symmetric also the relative peaks distribution appears non-symmetric, as shown in Fig. 12.7. A regular load time history as the one shown in Fig. 12.3 is rather particular. A measure of the regularity of a spectrum is given by

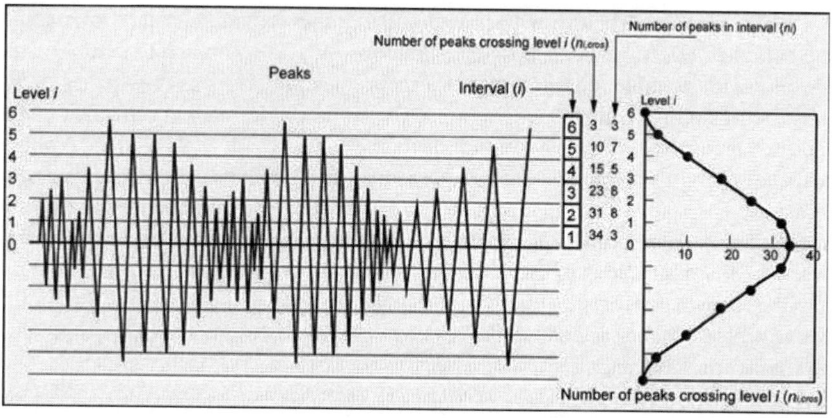

Fig. 12.3 Symmetric load spectrum and its representation in terms of peaks crossing the generic i-level

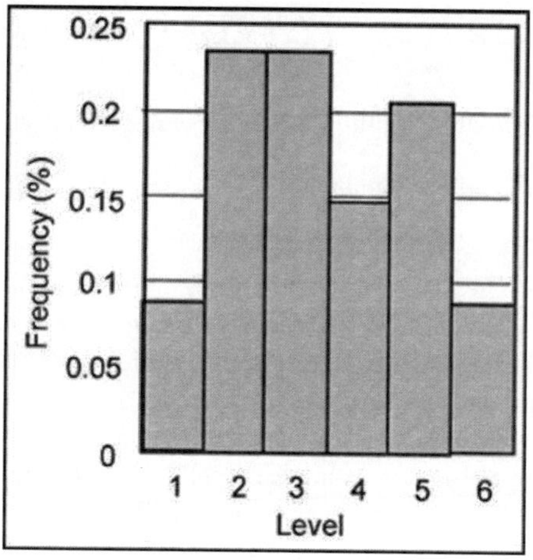

Fig. 12.4 Histogram of frequency distribution in each level

the ratio of the zero level upward crossing E[0] to the peaks in the sample. This ratio is known as the irregularity factor γ

$$\gamma = \frac{E[0]}{E[P]} \tag{12.2}$$

Figure 12.8 shows a simple example of how to calculate the irregularity factor. The number of the zero upward crossing is E[0] = 4 and the number of peaks is E[P] = 7, therefore it is $\gamma = 4/7 = 0.57$. Irregularity factor is found in the range of 0 to 1. Note that when $\gamma = 0$ there is an infinite number of peaks for every zero up crossing. This is characteristic of a wide-band random process. The value of $\gamma = 1$,

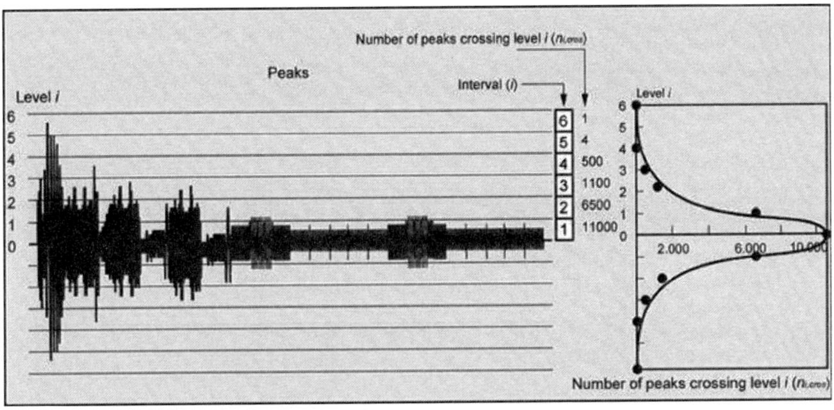

Fig. 12.5 Symmetric load spectrum with low level peaks prevalence and its representation in terms of peaks crossing the generic i-level

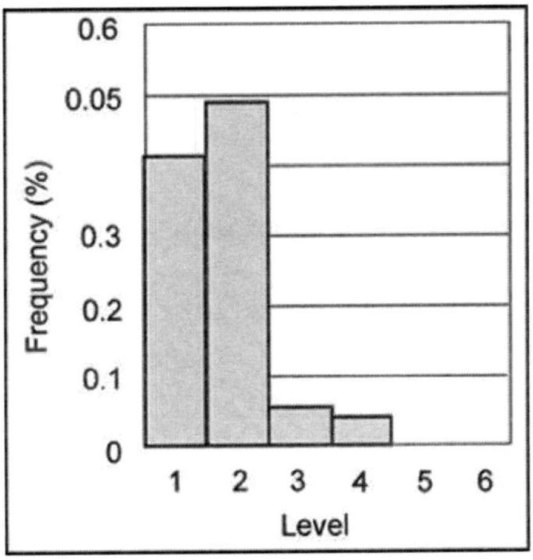

Fig. 12.6 Frequency distribution of peaks of Fig. 12.5

instead, corresponds to one peak per one zero up crossing and represents a narrow-band random process. Alternatively, a narrow-or-wide-band process can be judged by the width of its spectrum. To this purpose, the width parameter λ is introduced as

$$\lambda = \sqrt{1 - \gamma^2} \tag{12.2a}$$

Note that $\lambda \to 0$ represents a narrow-band random process. A narrow-band process is smooth and harmonic. For every peak there is a corresponding zero up crossing, meaning that E[0] is equal to E[P]. In these cases it is important to determine the relative *pdf* or the spectrum shape expressed as percentile of the maximum load in a given time length. For example, referring to the load spectrum of Fig. 12.3 for which

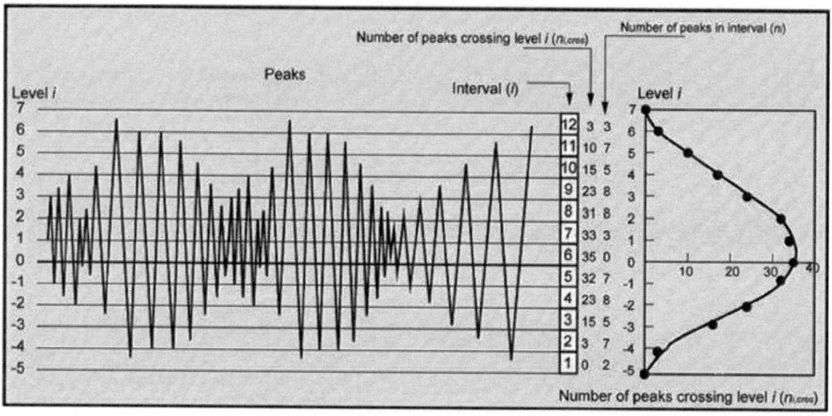

Fig. 12.7 Non-symmetric load spectrum and its representation in terms of peaks crossing the generic *i*-level

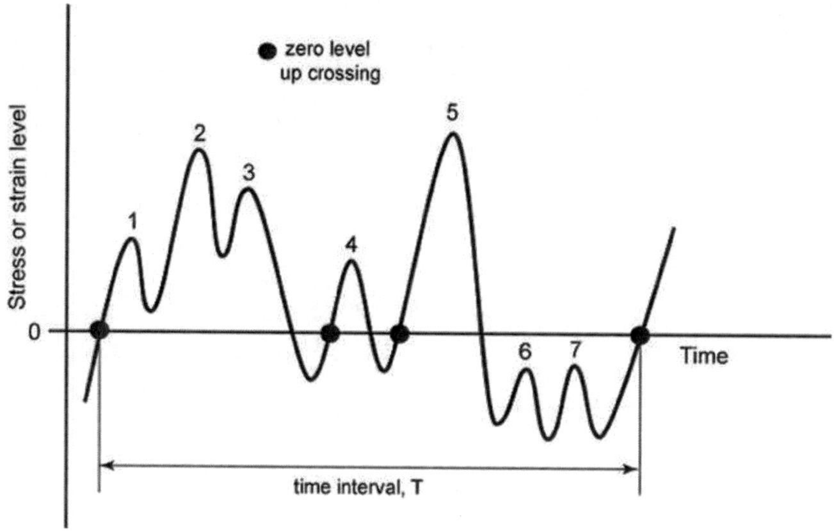

Fig. 12.8 Irregular load spectrum

it is $\gamma = 1$, characterized by a prevalence of high level peaks if it is assumed that the cycle package shown lasts one hour and is applied every hour, expressing the load level in 2000 h service as percentage of maximum load occurring in that time period it turns out a typical flat load spectrum shown in Fig. 12.9. In a flat spectrum the number of high loads is high and that of low loads is low. For example, referring to Fig. 12.9, high loads with a peak value exceeding 90% of the maximum occur about

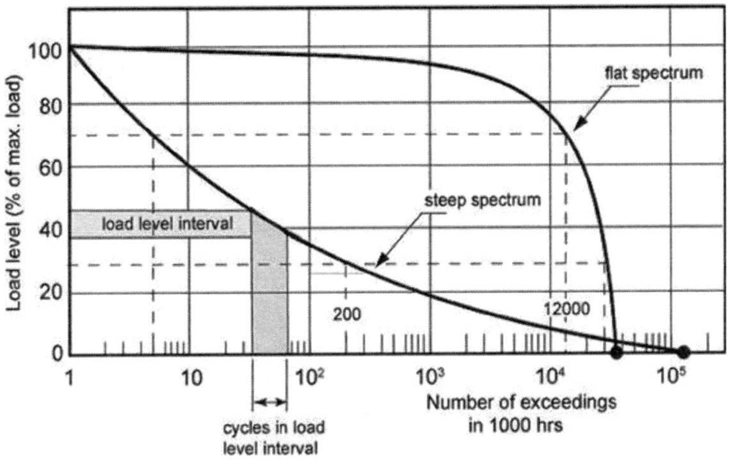

Fig. 12.9 Different types of load spectra (modified from [2])

12,000 times while those below 30% are 40,000–30,000 = 10,000 which is 25% of all cycles.

On the contrary, if lower amplitude loads prevail over the larger ones that are relatively few, then a steep load spectrum is obtained, as shown in Fig. 12.9. In this case, high loads with a peak value exceeding 90% of the maximum occur only 5 times whereas there are 40,000–200 = 39,800 cycles below 20% of the maximum which is 99.5% of the total.

12.2.2 Three-Point Cycle Counting

Symmetric load spectra represent a rather particular case in the time histories panorama. Usually, these are irregular, as in the example of Fig. 12.12. one method to count cycles in an irregular load spectrum is that of three-point cycle counting. The three-point cycle counting rule uses three consecutive points in a load-time history to determine whether or not a cycle is formed. Its rational is to separate low amplitude from high amplitude cycle, combining the residuals. Figure 12.10 shows the rules that identify the possible closed cycles in a time history where stress is the load parameter. The three consecutive stress points (S_1, S_2, S_3) define the two consecutive ranges as $\Delta S_1 = |S_1 - S_2|$, $\Delta S_2 = |S_2 - S_3|$. If $\Delta S_1 \leq \Delta S_2$, one cycle from S_1 to S_2 is extracted, and if $\Delta S_1 > \Delta S_2$, no cycle is counted. The final result will be two cycles counted, one extracted of amplitude ΔS_1 and the residual. In the most generic case of many cycles, the three-point cycle counting method requires that the stress time history be rearranged so that it starts with either the highest peak or the lowest valley, whichever is greater in absolute magnitude. Then, the cycle identification rule is applied to check every three consecutive points. When a cycle is formed it is

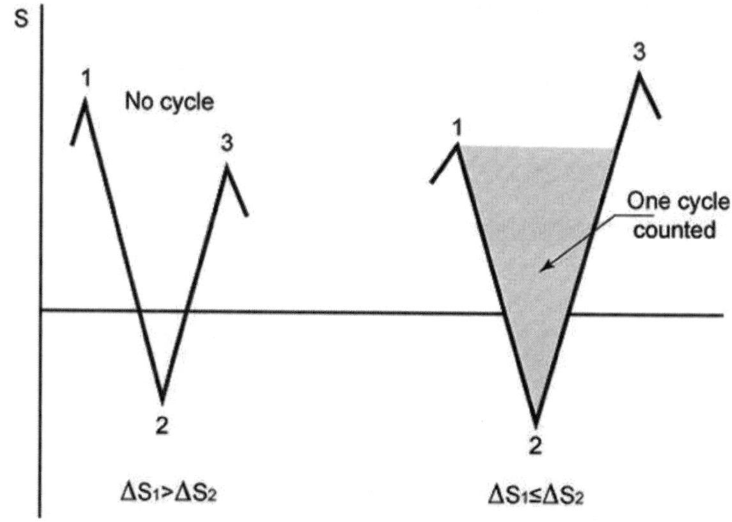

Fig. 12.10 Rules of three-point cycle counting

discarded and the remaining points are connected to each other. This procedure is repeated from the beginning until the remaining data are exhausted.

12.2.3 Four-Point Cycle Counting

Similar to the three-point cycle counting method, the four-point cycle counting rule uses four consecutive points to extract a cycle. Figure 12.11 illustrates the principles for two possible cycles counted in a most generic time history. The four consecutive stress points (S_1, S_2, S_3, S_4) define the inner $\Delta S_2 = |S_2 - S_3|$ and the outer stress range $\Delta S_1 = |S_1 - S_4|$, Fig. 12.11a.

If the inner stress range is less than or equal to the outer stress range $\Delta S_2 \le \Delta S_1$ and the points comprising the inner stress range are bounded by (between) the points of the outer stress range, the inner cycle from S_2 to S_3 is extracted, the two inner points discarded, and the two outer points S_1 and S_4 connected to each other to form a main cycle. Otherwise, no cycle is counted. The same procedure is adopted for the next four consecutive stress points (S_2, S_3, S_4, S_5) until no data remain. Unlike the three-point cycle counting method, this technique does not guarantee that all the data points will form closed cycles. The remaining data points that cannot constitute a cycle are called the residue. This technique has the particularity to be very easy to use in conjunction with as-recorded data acquisition and data reduction, because it does not require rearrangement of the load-time history.

An example of four-point cycle counting is given in Fig. 12.12. Start from point A to select the first four points A, B, C and D to apply the procedure previously

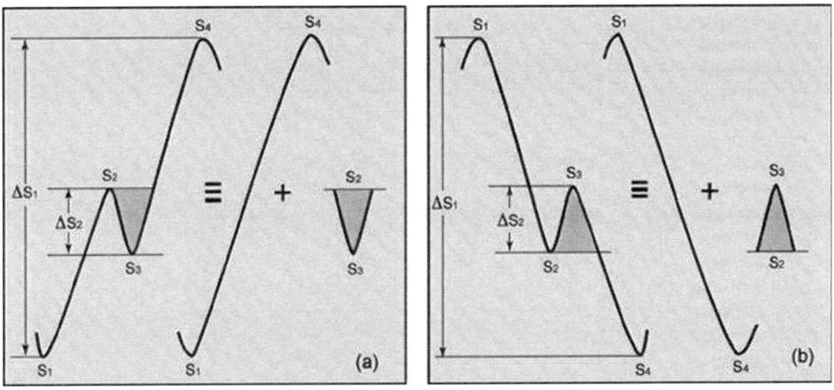

Fig. 12.11 Intermediate load reversal ΔS_2 that can be separated using the four-point cycle counting rule. Two cycles of stress range ΔS_1 and ΔS_2, respectively, can be counted

described and separate the internal cycle BCB′ since the inner stress range $\Delta S_2 = |S_B - S_C|$ and the outer stress range $\Delta S_1 = |S_A - S_D|$ are such that $\Delta S_2 \leq \Delta S_1$. Once points B and C are eliminated continue the four by four consecutive points approach till the last cycle and In other words, the peak values of the intermediate small load reversal should be inside the range of the two peak values of the larger range. A possible drawback of the method is in the residue counting. It is questionable, in fact, whether count of residue cycles, as those in Fig. 12.12c, is still meaningful from a fatigue damage accumulation stand point, since in general they may be separated by many small amplitude cycles. It, then, depends on the material memory, which in fact is not our memory, whether and how they actually accumulate.

12.2.4 Rainflow Cycle Counting

The method of cycle counting most widely used to-day is that proposed by Matsuishi and Endo in 1968 [3] known as *rainflow counting*. Dowling [4] considered this method as the most accurate to foresee experimental results. Actually, since its introduction rainflow counting has become a generic term that describes any cycle counting method which attempts to identify and separate cycles in a complex time history. In this respect, both the three-point cycle counting and the four-point cycle counting can be considered variants of the rainflow counting method. The term *rain flow* comes from the analogy of a drop falling from a pagoda roof. This is why it is also called pagoda roof method. This falling is schematized in Fig. 12.13 for a generic time history. The spectrum is rearranged so as to start and terminate at the highest peak stress or strain value to eliminate counting of half cycles, as shown in Fig. 12.13b.

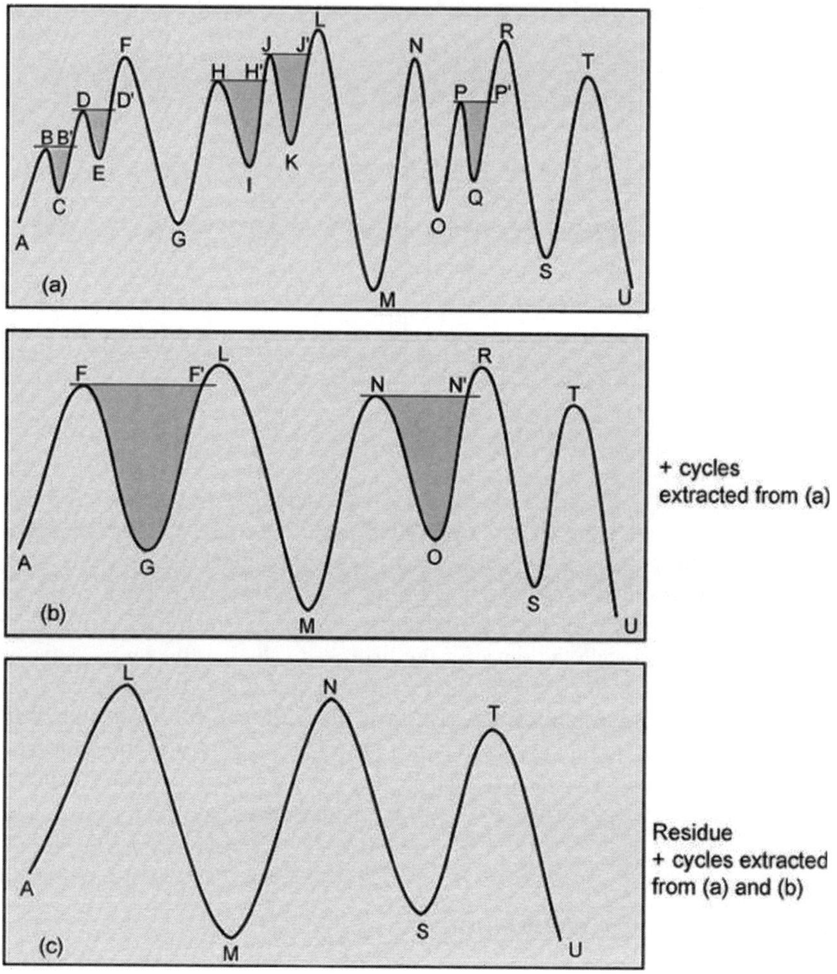

Fig. 12.12 Four-point cycle counting method

The time axis is then rotated by 90° and the loads time history becomes a series of peaks and valleys resembling a pagoda roof. It can be agreed that the tops of the roofs are on the right of the axis, valleys are on the left or vice versa. One may start from the highest peak (maximum) or from a valley (minimum) and continue accordingly always on the same side of the pagoda. If the fall starts from a peak it continues to flow till it arrests for one of the following reasons.

(a) it meets an opposing peak larger than that from which it came,
(b) it encounters a previous rain flow.

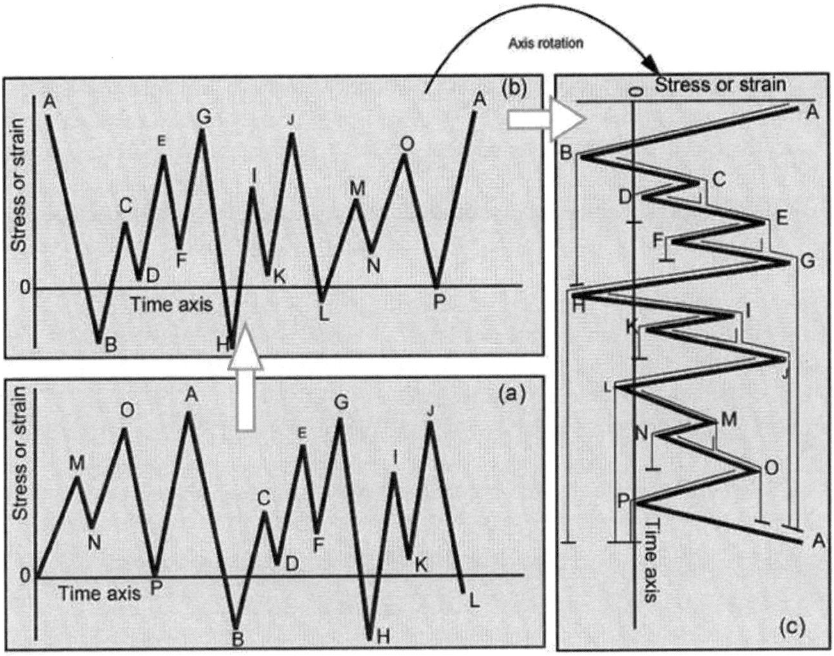

Fig. 12.13 Analogy between the rain falling on the roofs of a pagoda and the rainflow cycle counting method

The drop can fall on another roof and to continue to slip according to rules (a) and (b). If the fall begins from a valley it continues till it arrests for one of the following reasons.

(c) the drop meets a valley deeper than that of departure
(d) the fall crosses the path of a drop coming from a preceding valley.

The drop can fall on another roof and continue according to rules (c) and (d). The above mentioned rules are schematized in Fig. 12.13. As it can be seen, the original time history, Fig. 12.13a, has been rearranged as to start from the largest peak, Fig. 12.13b. All peaks preceding peak A have been reintroduces at the end of the spectrum. Then, the time axis has been rotated by 90° and the load history becomes a series of roofs and valleys with peaks on the left and valley on the right. The rain fall may start from peaks or valleys as well. The result shall be the same.

12.2.5 Hysteresis Loop Counting

Another method for the cycle counting is to look at the hysteresis loops. This has been already seen in sample problem N. 3 of Sect. 6.3 and is proposed in Fig. 12.14

coupled to a rainflow counting. The procedure is that previously indicated and shown in Fig. 12.13.

The initial strain history is that of Fig. 12.14a. Cycles have been rearranged to start from the highest positive peak C, Fig. 12.14b. Peaks A and B have been moved to the end of the spectrum. Then the new spectrum has been rotated, Fig. 12.14c, to let the rain drops flow over the roofs. Note that from valley B the rain stops since it crosses the path of a drop coming from a preceding valley. With the rainflow procedure four cycles can be counted that are: CH, DE, FG and AB. These four cycles generate four hysteresis loops that can be seen in Fig. 12.14c and separated in Fig. 12.15. These hysteresis loops can be analyzed individually. Each of them introduces a damage D that is proportional to the cycle ratio $n_i/N_i = 1/N_i$. The total damage is the sum of individual damages (see Sect. 12.3). If the spectrum of Fig. 12.14 is repeated m times, then damages cumulate and the total one becomes

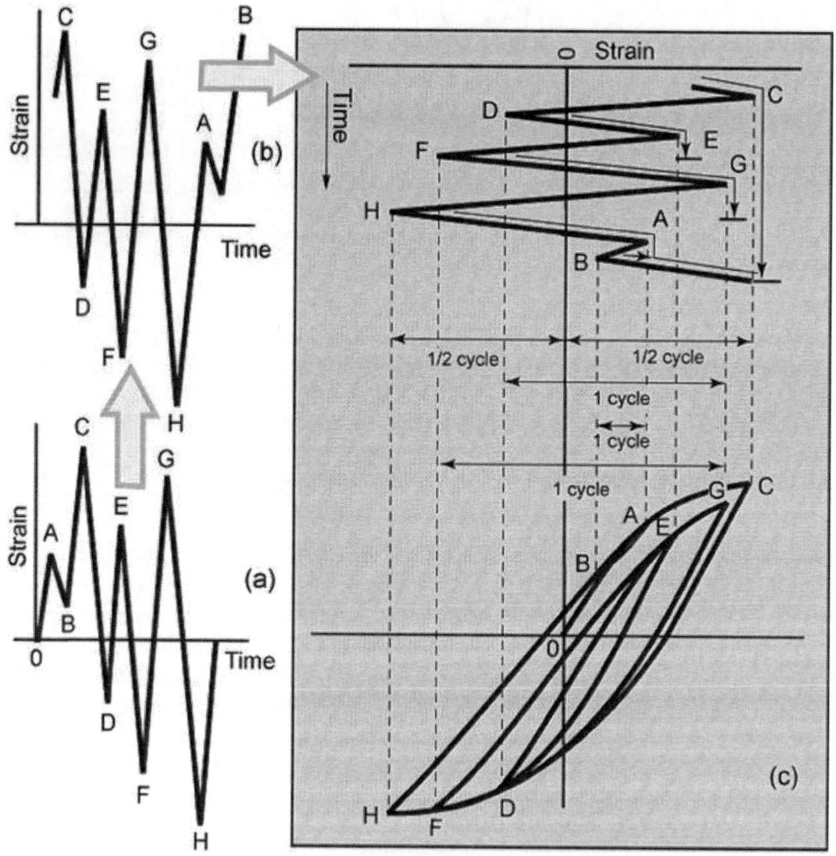

Fig. 12.14 Relationship between rainflow counting and hysteresis loops

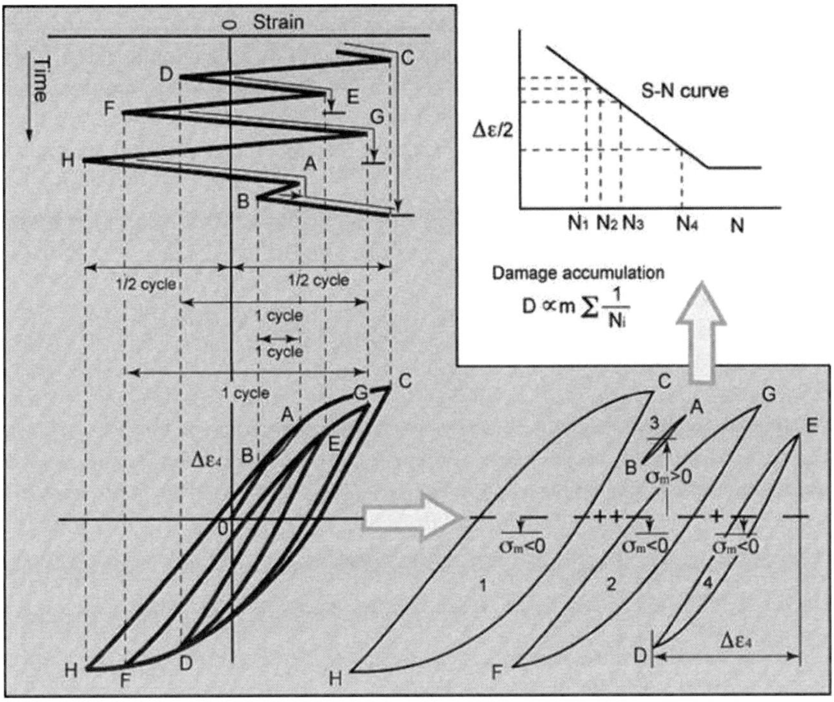

Fig. 12.15 Load spectrum decomposed in four individual cycles

$$D_{total} \propto m \sum \frac{1}{N_i} \tag{12.3}$$

Note that cycle 3 is the only one to have a positive mean stress. In all other cycles mean stress is negative and then ineffective.

12.3 Damage Progression and Accumulation

The use of standard *S–N* fatigue curve or Wöhler curve is limited to constant stress amplitude applications. In this particular case each individual cycle introduces a damage *D* in the material which we know is unrecoverable unless some heat or mechanical treatment, such as annealing or surface polishing, are performed to remove it. This damage is said to *progress* with the cycling and eventually leads to failure. If fatigue failure does not occur with the application of the first load spectrum and a second spectrum of different but constant stress amplitude is applied, then new damage is introduced in the material that is said to *accumulate* with the previous one. In general, progression is a term that refers and characterizes damage introduced by cycles of same stress amplitude while accumulation identifies damage generated by cycles or blocks of cycles of different stress amplitude. These simple definitions of damage progression and accumulation may become complex to recognize and use when constant amplitude spectra are replaced by a variable amplitude spectra in which stress amplitude is continuously changing. In this more general case it is said that damage introduced by each individual cycle accumulates with that introduced by previous cycling. The issue that will be addressed in this section is extremely important and concerns precisely the damage progression during the repeated application of cycles of a constant load spectrum and its accumulation when spectra of different load amplitudes are consecutively applied. It will be also shown how these concepts of damage progression and accumulation can be implemented in the case of variable load spectra. The starting point resides in the concept of fatigue damage D and its definition. However, it will be shown that from an operative point of view it is indeed not necessary to precisely define or measure damage, but it is sufficient to introduce and refer to a concept of relative damage which, being relative, needs not be defined in absolute terms.

12.3.1 Miner's Rule

As to the damage progression during the application of a train or history of loads of equal amplitude σ_i or S_i the first hypothesis that comes to mind is that damage is related to the ratio between the number of applied cycles n_i and the allowable cycles N_i at that stress or strain amplitude, known as *cycle ratio*

$$D_i = \left(\frac{n_i}{N_i}\right)^{m_i} \tag{12.4}$$

in which the exponent m can assume any value depending on the stress amplitude. When m equal one the damage progression is said to be linear. It owes to the Swedish engineer Pålmgren [5] the first hypothesis, already done in 1924, of the linear progression of damage. Studying the fatigue resistance of ball bearings, he arrived to the conclusion that damage D introduced by each individual cycle progressed linearly with load repetition. He also noticed that damages D_i produced by load spectra of different amplitude accumulated in a linear fashion, which actually means that the exponent m in Eq. (12.4) is equal one for any stress amplitude. Few years later, in 1937 Langer [6], working for Westinghouse in the electric power generation field, independently proposed a similar linear rule for pressure vessel and piping components made of steel. In 1945 Miner [7] in the United States, working for Douglas Aircraft built on Langer's work and applied the linear damage rule to tension–tension axial fatigue data for aircraft skin material (aluminum alloy 24S-T ALCLAD). Miner demonstrated excellent agreement between the predictions from the linear damage rule and his experimental results formulating what is to-day known as Miner's rule or Palmgren–Miner rule of linear damage progression and accumulation. The rule states that the fraction of life used by a series of n_i cycles of equal amplitude S_i is proportional to the correspondent cycle ratio, as per Eq. (12.4) with $m = 1$. This fraction of life is then a damage fraction D_i in the sense that it introduces a fraction n_i/N_i of the critical damage D_{cr} that produces failure by fatigue. The total damage produced by all load spectra of different amplitude accumulates also linearly so that it is given by

$$D_{total} = \sum D_i = \sum D_{cr} \frac{n_i}{N_i}. \tag{12.5}$$

Note that it is not done, at this time, any hypothesis about the time history of load spectra application that can be arbitrary. Fatigue failure occurs when this summation reaches the value of the critical damage D_{cr} for the particular material. Consider, for example, a block of three-step spectra of constant amplitude stress $S_1 > S_2 > S_3$. Values of the respective cycle ratios n_i/N_i, with $i = 1,2,3$, can be deducted from the *S–N* diagram of the material, as shown in Fig. 12.16. Fatigue failure occurs when, as said, the total damage introduced by the three load spectra reaches the value of the critical damage

$$D_{cr} = D_{cr}\left(\frac{n_1}{N_1}\right) + D_{cr}\left(\frac{n_2}{N_2}\right) + D_{cr}\left(\frac{n_3}{N_3}\right) \tag{12.6}$$

or

$$1 = \frac{n_1}{N_1} + \frac{n_2}{N_2} + \frac{n_3}{N_3} \tag{12.7}$$

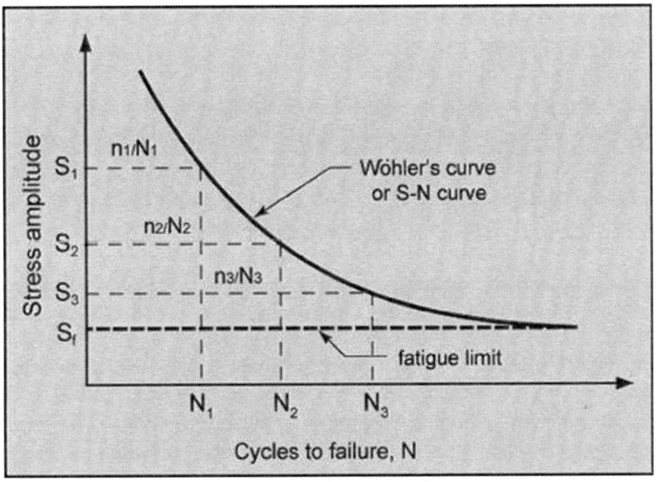

Fig. 12.16 A block of three-step high-low sequence $S_1 > S_2 > S_3$

In general, under a system of a given number of load spectra each of constant amplitude S_i failure will occur by fatigue when it is

$$1 = \frac{n_1}{N_1} + \frac{n_2}{N_2} + \frac{n_3}{N_3} + \frac{n_4}{N_4} + \cdots = \sum \frac{n_i}{N_i}. \tag{12.8}$$

Note that deriving Eq. (12.8) no hypothesis has been done about the value of the critical damage D_{cr} or how to define and measure this critical damage. It was enough to say that each load spectrum introduces a relative damage D_i which is a fraction of the critical damage D_{cr}. It must be also noted that at equal number of applied cycles n_i the damage fraction produced by higher stress amplitudes is larger than that caused by low amplitude stresses since the corresponding number of allowable cycles N_i is lower and this reduction follows a power low because of the exponential nature of the Wöheler curve. Below the fatigue limit damage is null since the allowable number of cycles N is infinity or unlimited high. This may not seem consistent with the definition of fatigue limit as the highest stress possible that cannot propagate damage; even dough damage may have been generated in some grains. The congruence is reintroduced by specifying that from the point of view of life ratio, null damage is damage that cannot be propagated. Graphically in a D/D_{cr}-n/N diagram linear damage progression and accumulation formulated by Eq. (12.8) is a 45° straight line, as shown in Fig. 12.17. If, for example, the stress amplitude S_1 is applied n_1 times over a total of N_1 allowable cycles producing a relative damage D_1/D_{cr}, then according to Eq. (12.8) there will be

$$n_2 = N_2\left(1 - \frac{n_1}{N_1}\right) \tag{12.9}$$

cycles of a second stress amplitude S_2 spectrum before reaching the critical damage D_{cr} and, therefore, fatigue failure. If stress cycles n_2 are less that allowed by Eq. (12.9) then there will be room for a third spectrum of n_3 cycles of amplitude S_3 given by

$$n_3 = N_3\left[1 - \left(\frac{n_1}{N_1} + \frac{n_2}{N_2}\right)\right] \tag{12.10}$$

as schematized in Fig. 12.17.

12.3.2 Nonlinear Damage Progression and Accumulation

The Palmgren–Miner hypothesis of linear damage progression and accumulation has a major shortcoming in that it assumes that equal cycle ratios produce equal damage independently of the stress amplitude or order of application, which is not true at all. As to the stress amplitude, its effect on damage progression can be inferred from French's curve of Fig. 1.59 relative to mild steel that can be considered as the

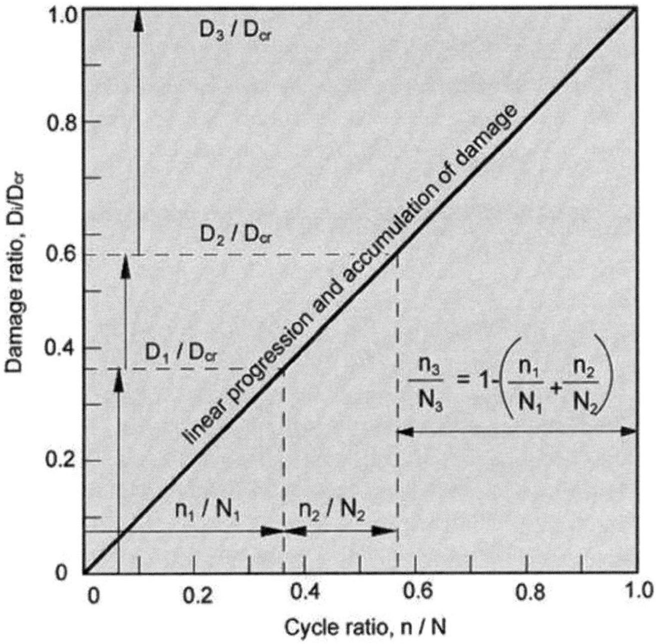

Fig. 12.17 Linear damage progression and accumulation according to the Palmgren–Miner rule room left for other

non-propagating limit of submicroscopic damage beyond which MSC (mechanically short cracks, see Sect. 1.5) start to appear and grow. These micro-cracks that form in persistent slip bands can be assumed the first microscopic sign of damage. Now, it turns out that to introduce this initial damage it takes about 30% of the total life N if the stress amplitude is close to the fatigue limit and about 5% only if it is equal to 220 MPa. At higher stress amplitude the same damage would be introduced at just 0.5% of the total life of the material or even before after just few cycles. What this actually means is that there cannot be just a single line of damage for all stress amplitudes, as shown in Fig. 12.17, but each of them will have its own damage progression curve.

The same conclusions can be drawn looking at Fig. 1.63 which is general for steels. At very high stress range $\Delta\sigma$ initiation takes place immediately in the material while at very low stress range close to the fatigue limit it will take an enormous fraction of life. Also damage accumulation cannot be linear. A linear dependence means that independently of the order of application of two or more load spectra the final damage is the same. But this also is clearly not true. If low stress amplitude cycles are applied first and their number is such to result on the left of French's curve they would not even initiate damage so that they would be totally ineffective. But if high amplitude cycles are given first, then damage would certainly be introduced in the material so that the following low amplitude cycles, that before were ineffective, now have a damage on which to act and make it progress.

Low amplitude cycles may not generate damage, but they can propagate damage already introduced by other causes. If damage is already present in the material even cycles below the fatigue limit of the material could be effective, as it will be described in Sect. 12.3.3. In practice, going back to the example of the three-step high-low sequence $S_1 > S_2 > S_3$ of Fig. 12.16, fatigue failure will happen before the cycle ratio summation (12.8) will reach unity, as schematized in Fig. 12.18a. But if the sequence is inverted as in three-step low–high sequence of Fig. 12.18b, then fatigue failure may occur much later and the cycle ratio summation (12.8) exceeds unity. It is still possible that Miner's rule apply and this happens when cycles of different amplitudes are applied in a random sequence, as schematized in Fig. 12.18c. Studies performed by different researches, including Miner himself, applying blocks of cycles of different amplitude in different order [8–14] have demonstrated that effectively high-low sequence of loads application results in a value of summation (12.8) lower than one that may go as low as 0.18 while in the opposite low–high sequence of application can reach even 22.

Significant, to this purpose, is the research performed by Jacoby [15] and reported by Ahrensdorf [16], on steel, titanium and aluminum alloy specimens loaded by blocks of cycles of different amplitude in any possible sequence. The distribution of results in terms of Miner summation (12.8) computed for 384 possible cases is shown in Fig. 12.19. The theoretical value of one predicted by Miner's rule was obtained in 30% of cases. Most of data falls between 0.3 and 3. It is argued that Miner's rule is valid only in cases where the load sequence is not based on blocks of cycles of equal amplitude, but on randomly applied individual cycles. In these cases, Miner's rule approximates unity the more equally distributed are cycles of different amplitude, as schematized in Fig. 12.18c.

Therefore, the idea of a single progression and accumulation line common to all stress amplitudes must be abandoned. We shall necessarily direct towards a power low dependence of the type

$$\frac{D}{D_{cr}} \propto \left(\frac{n_i}{N_i}\right)^{m_i} \tag{12.11}$$

where the exponent m_i depends on the stress amplitude S_i. The relative curves of damage progression are schematized in Fig. 12.20. The high cycle fatigue domain is characterized by values of the m exponent larger than one ($m > 1$). As the stress amplitude increases m decreases toward unity, i.e., toward the 45° Miner straight line. Above the Miner line it opens the domain of low cycle fatigue.

The exponent m is lower than one and decreases as the stress amplitude increases. At equal cycles ratio n_i/N_i higher amplitude loads introduce a higher damage, as schematized in Fig. 12.20 for a cycle ratio equal to 0.5. Many researchers [9, 13, 14, 17–19] have addressed the problem of damage progression and accumulation on the base of a power low dependence of the type expressed by Eq. (12.11). Consider now two blocks of constant, but different, stress amplitudes $S_1 > S_2$ and the relative

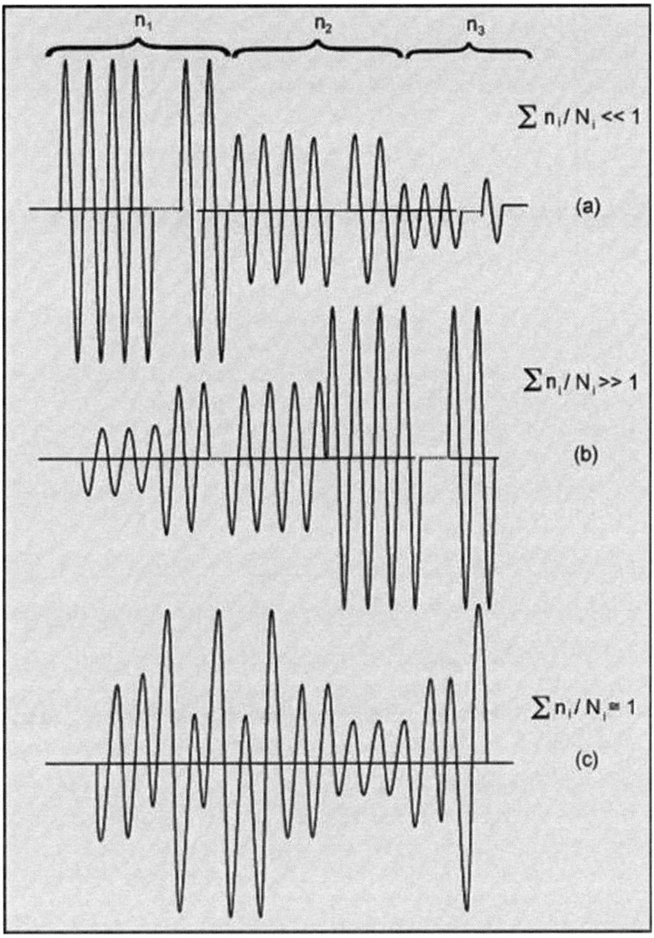

Fig. 12.18 Cumulative damage produced by load spectra of different stress amplitude depends on the order of application. **a** Miner's rule may be lower than unity in high-low sequence or higher than unity if the order is the opposite, case (**b**), or may be equal to one for a random sequence as in (**c**)

damage curves 1 and 2, as shown in Fig. 12.21. Imagine applying block 1 of stress amplitude S_1 first, Fig. 12.18a. It consists of n_1 overload cycles and produces a relative high damage D_1/D_{cr} correspondent to point A on curve 1 in Fig. 12.21. Subsequent application of block 2 of lower stress amplitude S_2 will find this damage already present in the material so it will start to progress and accumulate damage from this level of damage that on curve 2 is point A′. The visible consequence is that now there will be a sudden jump from A to A′ with the cycle fraction between A and A′ that needs not be spent any more to reach point A′. The new damage accumulation can be devastating leading to fatigue failure before summation (12.8) reaches unity.

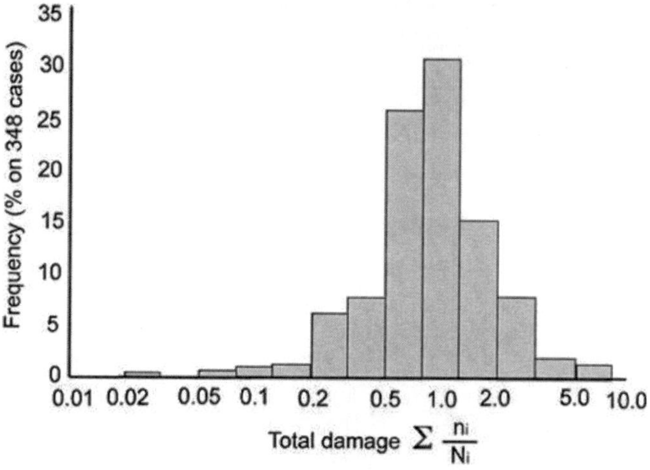

Fig. 12.19 Experimental results of Miner summation obtained by applying blocks of cycles of different amplitude in 348 different combinations [16]

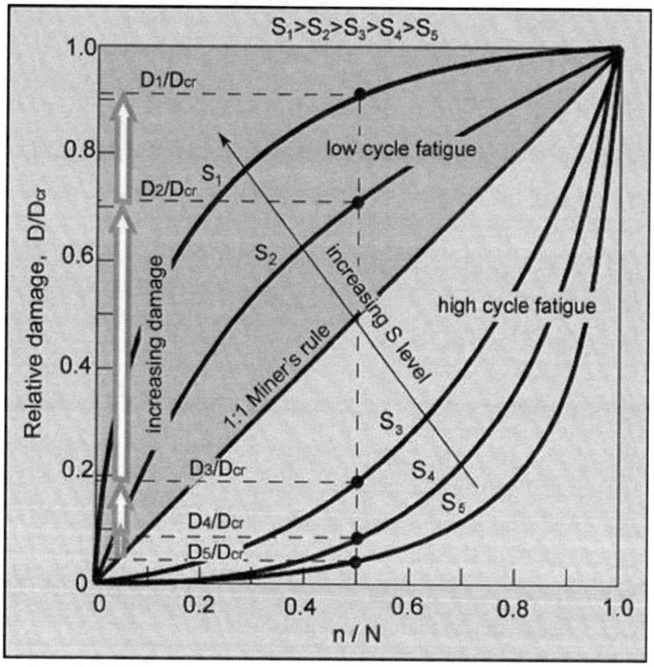

Fig. 12.20 Power low dependence of damage progression and accumulation for different load amplitudes

In fact, the combined action of blocks 1 and 2 will progress along path is 0AA′B of Fig. 12.21 and the cumulative damage till fatigue failure is given by

$$\sum \frac{n}{N} = \frac{n_1}{N_1} + \left(1 - \frac{n_2}{N_2}\right) \tag{12.12}$$

Since relative damage is expressed as

$$\frac{D_1}{D_{cr}} = \left(\frac{n_1}{N_1}\right)^{m_1} = \left(\frac{n_2}{N_2}\right)^{m_2} = \frac{D_2}{D_{cr}} \tag{12.13}$$

Equation (12.12) becomes

$$\sum \frac{n}{N} = \frac{n_1}{N_1} - \left(\frac{n_1}{N_1}\right)^{m_1/m_2} + 1 < 1. \tag{12.14}$$

Equation (12.14) summation will always be lower than unity since is $m_1 < m_2$. On the contrary, if low amplitude block 2 is applied first, Fig. 12.18b, its n_2 cycles that before lead to failure after block 1, now will introduce damage represented by

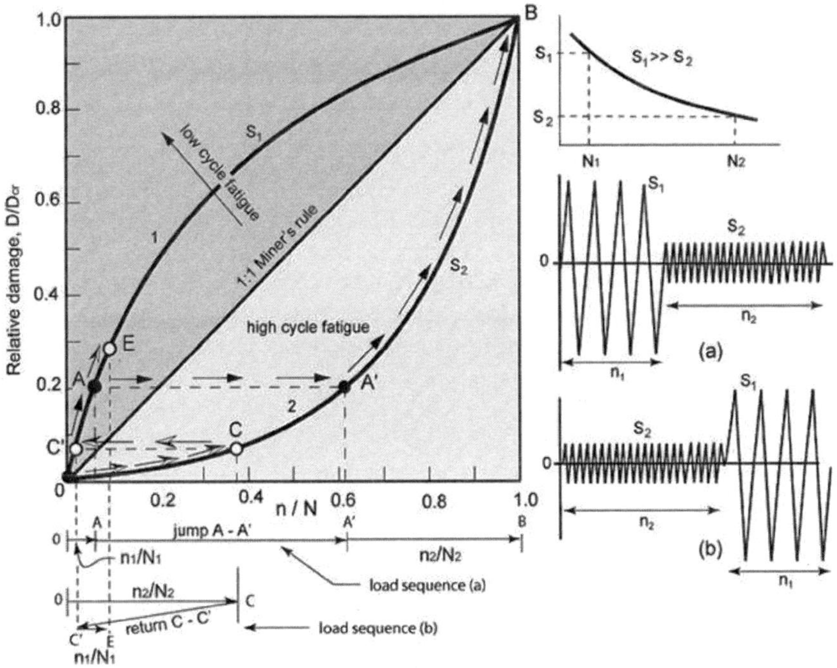

Fig. 12.21 Two blocks of different amplitude applied in two different orders **a** and **b** produce different cumulative damage

C on curve 2. This damage corresponds to point C′ on curve 1 so that when block 1 is applied there will be a return back from C to C′ *losing*, this time, cycles that before were gained. The new cumulative damage path is, then, 0CC′E arresting in E without failure. The new damage cumulates as

$$\sum \frac{n}{N} = \frac{n_2}{N_2} - \left(\frac{n_2}{N_2}\right)^{m_2/m_1} + 1 > 1. \tag{12.15}$$

The summation is always greater that one. In the more general case of many load spectra or individual cycles of different amplitudes the procedure to assess damage progression and accumulation is basically the same. Figure 12.22 schematizes the most generic situation where different spectra of different stress amplitude are applied to the work piece till failure occurs. It is a continuous passage from one damage curve relative to a particular stress amplitude to another one and vice versa. If the simple two-step load test is replaced by a multi-block load sequence, the procedure to follow would be basically the same, as schematically shown in Fig. 12.22. The key issue in this kind of approach is the knowledge of exponents m_i relative to each individual curve of damage progression for the material of interest. The knowledge of these m_i exponents is matter of experimental work and represents a *knowhow* of the designer. There is, in fact, no theoretical formulation for the m values. Corten and Dolan [17] suggested an approach to predict the life N in a two-step block sequence as

$$N = \frac{N_1}{\alpha + (S_2/S_1)^d (1 - \alpha)} \tag{12.16}$$

where α defines the two-block length, $n_1 = n\alpha$ and $n_2 = n(1\text{-}\alpha)$, and d is a parameter to be determined experimentally from two-level tests that was thought to have the same value also in a multi-level sequence. The advantage of the Corten and Dolan approach over that based on the m_i exponents is, then, that the parameter d would be an invariant and a two-step test procedure could be sufficient to determine its value. Unfortunately, it was found that the value of d obtained from a two-level test could not always be used to predict the life in more complicated block tests and could also be non-conservative in some circumstances.

12.3.3 Cumulative Damage of Load Spectra with Stress Amplitude Below the Fatigue Limit

It has been shown in Sect. 1.6 that stress amplitudes below the fatigue limit can also lead to fatigue failure if the damage they would not be able to propagate to a critical size living it dormant in a persistent slip band had already reached that critical size. In that section this already existing critical defect was assumed to be a surface inclusion, larger than a MSC, or a pre-existing macro defect (see Fig. 1.69 or

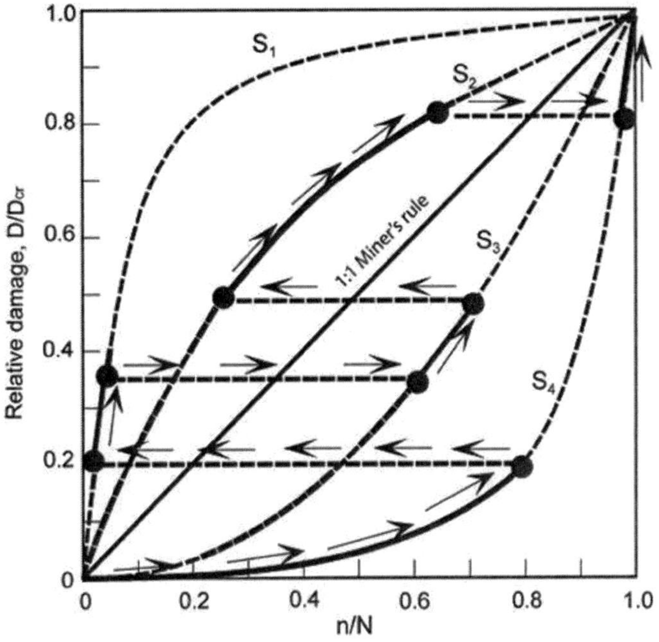

Fig. 12.22 Damage progression and accumulation in a four-block load sequence

Fig. 1.76). In other words the damage on which the stress amplitude below the fatigue limit would be effective was not originated by slip, but introduced by other means. In the light of nonlinear damage accumulation it is possible to assert that even slip process damage can progress and accumulate by stress amplitude below the fatigue limit if the material is subjected to variable loading conditions. Consider the case examined in Fig. 12.21 of two-block cycling. If the stress amplitude S_2 of block 2 were below the fatigue limit of the material according to Miner's rule it would not have any effect at all independently of the order of application of the two blocks since its contribution to summation (12.8) would always be null. In other words, the Miner-rule ignore the fatigue damage contribution of stress amplitudes below the fatigue limit simply because for them is $N = \infty$ and $n/N = 0$.

Conversely, in a nonlinear progression and damage accumulation regime these below-the-fatigue-limit stress amplitudes would indeed have an effect if they were preceded by a load spectrum capable of generating and growing a slip damage to a size that they can now afford to propagate. This is because after the first block of cycles S_2 the actual fatigue limit of the material decreased. This can be understood recalling the kitagawa–Takahashi diagram of Fig. 1.54 showing how the fatigue limit of a mild steel decreases with increasing defect size. Beyond a certain damage size, which we know is in the range of dozens of microns, the fatigue limit starts to diminish. Therefore, cycles that before were under the fatigue limit now are above the fatigue limit and contribute to damage. The actual question is where and how to find the corresponding damage curve on a diagram like that of Fig. 12.21 or Fig. 12.22. An expedient could be to refer to a damage curve whose stress amplitude is equal to the actual plus the decrease of the fatigue limit in a kitagawa–Takahashi diagram corresponding to a damage size equal to that produced by the first loading block. This size may be taken in a first approximation as the critical length that can be tolerated in the material under the stress range of block 1 multiplied by the relative damage introduced by block 1.

12.3.4 Lemaitre and Plumtree Nonlinear Cumulative Damage

Lemaitre and Plumtree [20] using a formulation based on damage mechanics proposed a damage progression low of the type

$$\frac{D}{D_{cr}} = 1 - \left(1 - \frac{n}{N}\right)^{1/m} \tag{12.17}$$

similar to that expressed by Eq. (12.4). However, Eq. (12.17) ascribes less weight, i.e., less damage accumulation to low n/N values. This is shown in Fig. 12.23 for three different values of the m exponent equal to 1.2, 2.5 and 20, respectively. It can be seen the different damage progression between prediction made by Eq. (12.17) (solid lines) and that of Eq. (12.8) (dotted lines) below the Miner-line. Above the Miner-line in low cycle fatigue predictions made by Eq. (12.17) overestimate damage progression with respect to that of Eq. (12.8).

12.3.5 Marin Cumulative Damage Approach

Imagine to have a group of damage curves, as shown in Fig. 12.24. select a constant damage level, for instance $D/D_{cr} = 50\%$. On the six curves this damage level corresponds to six points each of which intercepts a cycle ratio n_i/N_i on the abscissa axis. Report each of these values in a new S–N diagram obtaining a constant damage curve, as shown in Fig. 12.25. same procedure repeated for several values of damage leads to the construction of a family of curves of finite-life at constant damage. The base curve obtained for $D = 0$ is the base Wöhler curve for $R = -1$. As expected, the fatigue S–N curves moves continuously down as damage progresses. Marin [21] defined this damage as

$$\left(\frac{n_1}{N_1}\right) + \left(\frac{n_2}{N_1}\right)\left(\frac{S_2}{S_1}\right)^y + \left(\frac{n_3}{N_1}\right)\left(\frac{S_3}{S_1}\right)^y + \cdots + \left(\frac{n_i}{N_1}\right)\left(\frac{S_i}{S_1}\right)^y = 1 \tag{12.18}$$

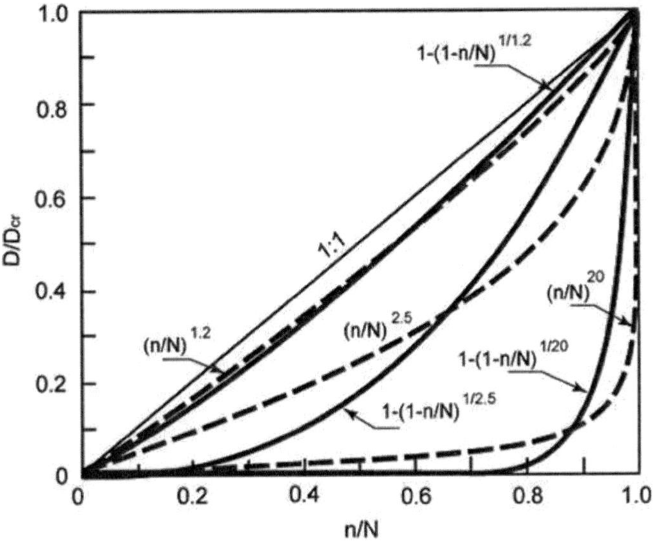

Fig. 12.23 Damage accumulation according to Lemaitre and Plumtree for several values of the exponent *m* (solid lines) compared with the cumulative damage curve approach (dotted lines)

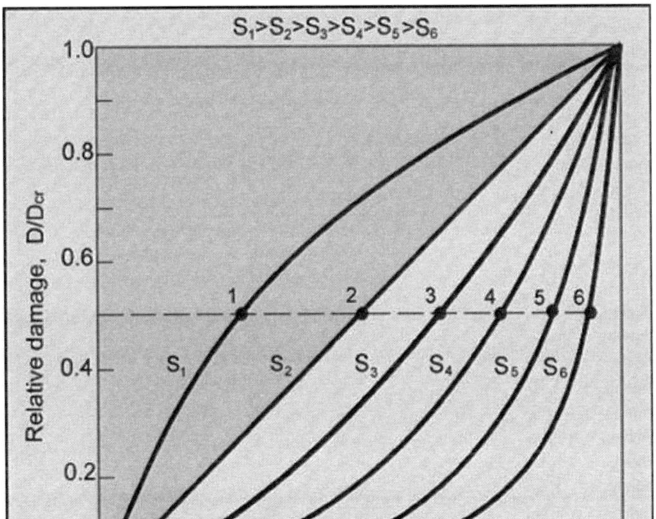

Fig. 12.24 Family of damage curves for different stress amplitudes

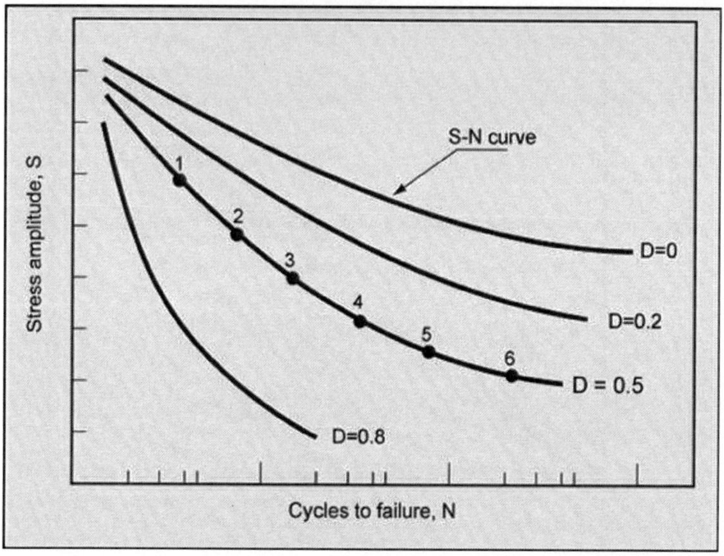

Fig. 12.25 Constant damage fatigue finite life *S–N* curves

where S_1 represents the highest stress level. It is interesting to note how Eq. (12.18) is exactly the expression obtained by Corten and Dolan [17] with their cumulative damage theory. For the steel they were working with they obtained a y exponent variable from 6.2 to 6.9 with an average of 6.57 and for 7075-T6 aluminum alloy they found the exponent was 6. Recalling the expression of the Basquin line (5.22) it is possible to write

$$S^x \cdot N = k \tag{12.19}$$

with $x = -1/b$ and $k = \left(\sigma_f'\right)^{-1/b}$ where b is the exponent of the Basquin line and σ_f' the fatigue strength coefficient that is equal to the true stress at failure. From Eqs. (12.18) and (12.19) it stems out

$$\left(\frac{n_1}{N_1}\right) + \left(\frac{n_2}{N_2}\right)\left(\frac{S_2}{S_1}\right)^q + \left(\frac{n_3}{N_3}\right)\left(\frac{S_3}{S_1}\right)^q + \cdots + \left(\frac{n_i}{N_i}\right)\left(\frac{S_i}{S_1}\right)^q = 1 \tag{12.20}$$

in which it is

$$q = y - x = y + \frac{1}{b} \tag{12.21}$$

Equation (12.21) allows to foreseen fatigue strength under conditions of constant stress spectra combined in any fashion if the values of the y and b exponent were known.

12.3.6 Henry Cumulative Damage Approach

If, as it has been seen in the previous section, fatigue life *S–N* curve moves down as damage accumulates, then it is possible to express this damage as the ratio of fatigue limit $S_{f,o}$ of the *virgin* material to the reduced one

$$D = \frac{S_{f,o} - S_f}{S_{f,o}}. \tag{12.22}$$

Equation (12.22) can be solved for S_f yielding

$$S_f = \frac{S\left(1 - \frac{n}{N}\right)}{\left(\frac{S - S_{f,o}}{S_{f,o}}\right) + \left(1 - \frac{n}{N}\right)} \tag{12.23}$$

with S representing the applied stress amplitude. Equation (12.23) [22] provides the value of the fatigue limit according to Henry after n cycles of S amplitude have been applied and for which the fatigue life is N. If after a spectrum of n_1 cycles of amplitude S_1 a second sequence is applied with n_2 cycles at a stress amplitude S_2, Eq. (12.23) is adjourned posing $S = S_2$ and $S_{f,o} = S_{f,1}$ previously obtained with $S = S_1$.

12.3.7 Linear Damage Accumulation

The Palmgren–Miner linear accumulation has been shown to be not realistic in most cases. However, there is a case where damage accumulation can be linear even if individual progression at constant stress amplitude is not linear, but develops according to a power low given by an equation of the type (12.4) in which the m exponent is positive, but not one. This nonlinear progression and linear accumulation behavior is, in fact, typical of Stage II of fatigue (see Sect. 1.5.2 and Fig. 1.64) when a macro-crack is already present in the material and propagates at each load cycle on a plane normal to external load. If the crack is present, it is possible to define a relative damage that is proportional to the crack ratio

$$\frac{D}{D_{cr}} = \frac{a}{a_{cr}} = \frac{1}{a_{cr}}\left[a_o + (a_{cr} - a_o)\left(\frac{n}{N}\right)^{\alpha}\right] \tag{12.24}$$

in which a is the current crack length, and a_{cr} the length of the critical crack that produces sudden fracture Eq. (12.24) is based on fracture mechanics fatigue crack growth and fitted with a large amount of test data for loading with two-step life/stress levels, where n is the number of loading cycles applied to achieve a crack length a from an initial crack length a_o. N represents the number of cycles applied to achieve the crack length a_{cr} at final fracture. The exponent α is empirically determined and has the following form:

$$\alpha = \frac{2}{3} N^{0.4}. \tag{12.25}$$

This relative damage definition implies that fatigue failure occurs when D/D_{cr} is equal to unity, i.e., $a = a_{cr}$. Equation (12.23) is graphically shown in Fig. 12.26. Damage accumulation in this case is still linear because it does not depend on the applied stress amplitude $\sigma_{a,i}$. Equation (12.23) is based on Paris' postulate (see section) according to which if a is the current crack length, the corresponding crack growth rate per cycle da/dN is given by

$$\frac{da}{dN} = C \cdot \Delta K^n \tag{12.26}$$

with C and n material constants. ΔK is the so called stress intensity factor given by

$$\Delta K = \Delta\sigma\sqrt{\pi a} \cdot f(a) \tag{12.27}$$

where $\Delta\sigma$ is the stress range and $f(a)$ a non-dimensional factor that depends only on the geometry of the system. By integrating Eq. (12.27) with respect to dN the number of allowable cycles N_1 is obtained, i.e., the number of cycles that produce fatigue failure under the stress range S_1 by growing the crack from the initial length

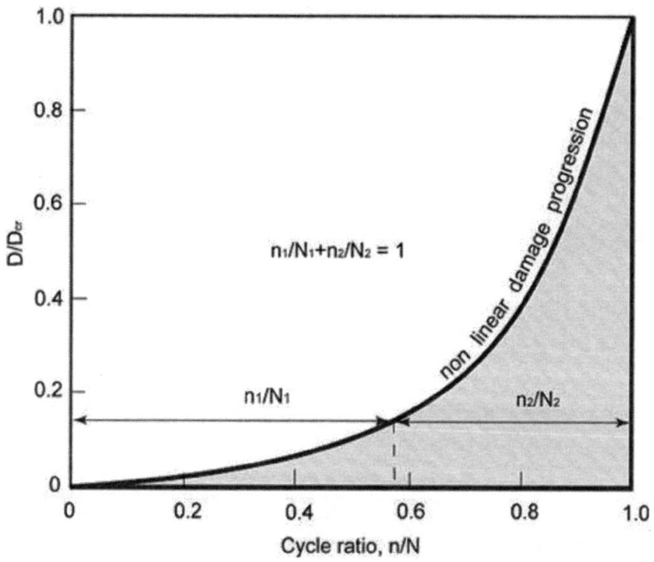

Fig. 12.26 Nonlinear damage progression with linear damage accumulation characteristic of Stage II of fatigue

a_o to the final length a_{cr}

$$N_1 = \frac{1}{2^n C \cdot S_1^n \pi^{n/2} f(a)^n \left(\frac{n}{2} - 1\right)} \left[\frac{1}{a_o^{(n/2-1)}} - \frac{1}{a_{cr}^{(n/2-1)}} \right] \tag{12.28}$$

that can be written as

$$N_1 = \frac{C_1}{S_1^n} \left(\frac{1}{a_o^m} - \frac{1}{a_{cr}^m} \right). \tag{12.29}$$

The same equation will held for a stress S_2

$$N_2 = \frac{C_1}{S_2^n} \left(\frac{1}{a_o^m} - \frac{1}{a_{cr}^m} \right). \tag{12.30}$$

If it is assumed that cycles of stress level S_1 are applied in a number $n_1 < N_1$ such that the initial crack grows from a_o to $a_1 < a_{cr}$ without producing failure

$$S_1^n n_1 = C_1 \left(\frac{1}{a_o^m} - \frac{1}{a_1^m} \right) \tag{12.31}$$

then n_2 cycles of the second stress level S_2 can be applied to produce failure by growing the crack from a_1 to a_{cr} so that it can be written

$$S_2^n n_2 = C_1 \left(\frac{1}{a_1^m} - \frac{1}{a_{cr}^m} \right). \tag{12.32}$$

By substituting S_1 and S_2 in Eqs. (12.31) and (12.32) with the equivalent given by Eqs. (12.30) and (12.31) it yields

$$\begin{aligned} C_1 \left(\frac{1}{a_o^m} - \frac{1}{a_1^m} \right) &= \frac{n_1}{N_1} C_1 \left(\frac{1}{a_o^m} - \frac{1}{a_{cr}^m} \right) \\ C_1 \left(\frac{1}{a_1^m} - \frac{1}{a_{cr}^m} \right) &= \frac{n_2}{N_2} C_1 \left(\frac{1}{a_o^m} - \frac{1}{a_{cr}^m} \right) \end{aligned} \tag{12.33}$$

and by summing, it is

$$1 = \frac{n_1}{N_1} + \frac{n_2}{N_2} \tag{12.34}$$

that is precisely the Palmgren–Miner rule of damage accumulation (12.8). Therefore, in Stage II of fatigue damage not only accumulates linearly, but follows a progression curve that is the same for all stress levels since exponent α in Eq. (12.24) is constant and given by Eq. (12.15). Reassuming, it can be said that under variable amplitude loadings as Stage I of fatigue concludes with the macro-crack formation after a

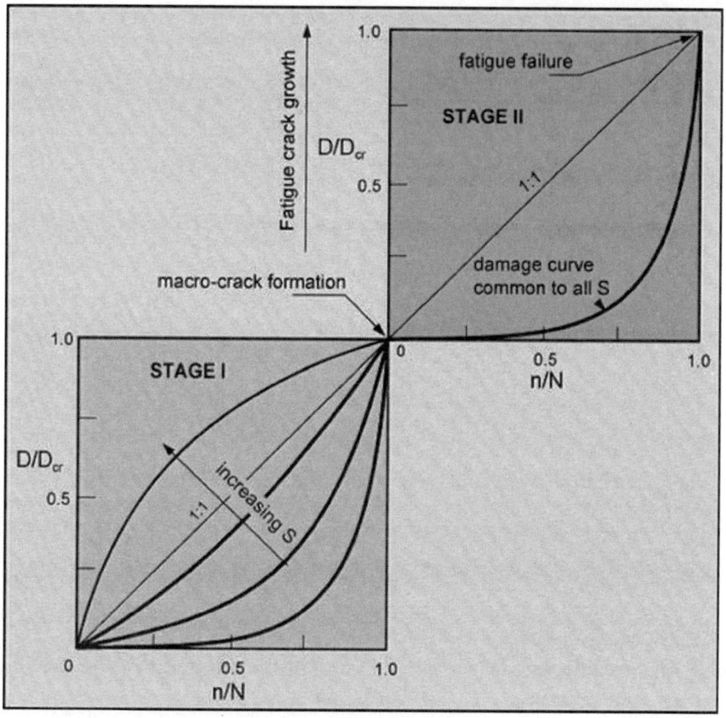

Fig. 12.27 Damage evolution in Stage I and II of fatigue. In Stage I damage progression and accumulation is not linear and depends on stress amplitude. In Stage II damage progression is not linear, but accumulation is linear

continuous process of damage progression and accumulation that is not linear at all and depends on the stress amplitude, Stage II initiates where the macro-crack is grown by each loading cycles following a damage progression curve that is not linear, but unique for all load levels and yet the accumulation of damage is linear so that Miner's rule apply. This is schematically shown in Fig. 12.27.

12.3.8 Double Linear Damage Rule

Based on the observation that fatigue is at least a two-phase process, Stage I and II (see Sect. 1.5.2 and Fig. 1.64), the models for the damage curves can be assumed to be bilinear. In Stage I damage develops as micro-pores in persistent slip bands and terminates with the macro-crack formation. We know that Stage I of damage nucleation and macro-crack formation take a life fraction that is as much longer as the stress amplitude is lower (see Figs. 1.18 and 1.63). Stage II is a crack growth phase that brings the specimen or work piece or component to final failure. Figure 12.28 schematizes

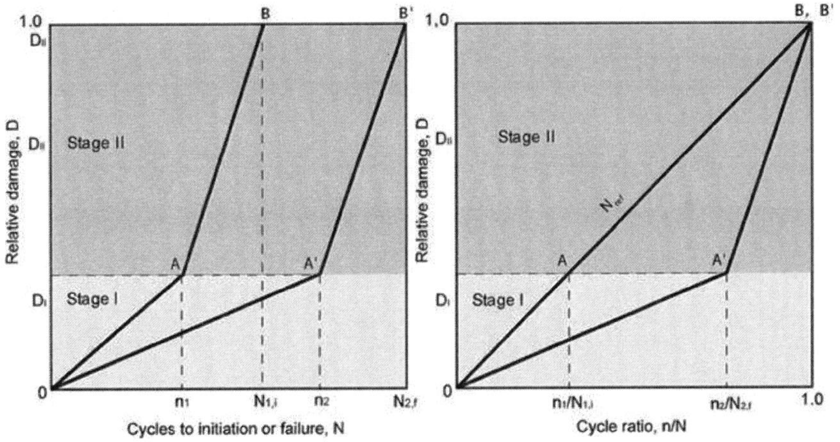

Fig. 12.28 Double linear damage progression and accumulation

this approach. The bilinear model represents an equivalent damage model between the Miner's rule and the damage curve accumulation rule examined in Sect. 12.3.3. In this new model, in fact, damage progression and accumulation is assumed to be linear, as in the Miner's rule and the nonlinear nature of damage is taken into consideration by the double line with two distinct slopes. The double line is characterized by a *knee* that is located at the separation between Stage I and II of fatigue. This bi-linear schematization of damage was first proposed by Grover [23]. However, he did not proposed any method to determine the knee.

Manson and Halford [24, 25] derived the required criteria to determine the coordinates of the knee point, i.e., the intersection between the two straight lines of the bilinear curves. It is suggested that the straight line connecting (0, 0) and (1, 1) be the reference damage line for the lowest life level. Figure 12.28 indicates that the total damage can be decomposed into Stage I damage (D_I) and Stage II damage (D_{II}). The Stage I linear damage accumulation rule states that prior to reaching the damage D_I, the cycle ratios can be summed linearly and are independent of the loading sequence (i.e., from OA to OA′ or from OA′ to OA). When the sum of the cycle ratios reaches unity, Stage I damage is completed. From this moment that consists in the macro-crack formation, the Stage II linear damage progression and accumulation rule applies. Regardless of the loading sequence, the damage accumulation depends only on the total sum of cycle fractions at each stress level. Based on the considerable amount of test data developed for two-step loading on many materials, Manson and Halford discovered that the knee point between Phase I and Phase II damage in a two-block sequence depends on the ratio of the cycles to macro-crack initiation $N_{1,i}$ under stress amplitude 1 to the cycles $N_{2,f}$ of final failure under stress level 2 instead of the physical significance of macro-crack initiation and propagation. This finding is illustrated in Fig. 12.29 for several steel and titanium alloys.

The knee has coordinates

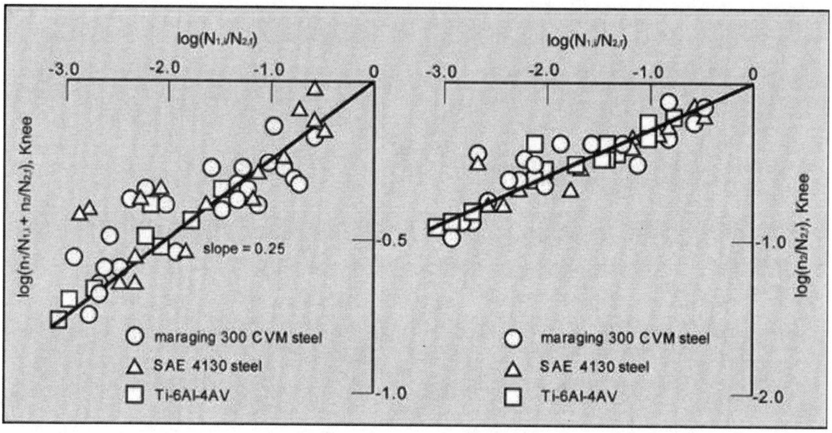

Fig. 12.29 Coordinates of the knee in the double linear damage model

$$\left[\frac{n_1}{N_{1,i}}\right]_{knee} = 0.35 \cdot \left(\frac{N_{1,i}}{N_{2,f}}\right)^{0.25}$$
$$\left[\frac{n_2}{N_{2,f}}\right]_{knee} = 0.65 \cdot \left(\frac{N_{1,i}}{N_{2,f}}\right)^{0.25}. \tag{12.35}$$

It is important to note that the knee coordinates are independent of the specific material. Hence, the knee points would be the same for all materials. Their location is dependent only on maximum and minimum lives. The drawback in the Manson and Halford rule is that linear damage accumulation is possible only in Stage II of fatigue, as shown in Sect. 12.3.7 and the damage progression is not linear, as shown in Fig. 12.26, dough damage progress curves reduces to one single curve for all stress levels. Probably, they though to overcome the nonlinear progression and accumulation process by distinguishing between the two Stages of fatigue. Concluding, the actual damage progression and accumulation process remains that illustrated in Fig. 12.26.

12.3.9 Damage Progression and Accumulation in Low Cycle Fatigue

In low cycle fatigue the issue of damage accumulation can be addressed through an elegant treatment based on damage mechanics introduced by Kachanov [26] and Lemaitre e Chaboche [27] on the base of continuum mechanics. Consider a specimen subjected to fatigue loadings. After a certain number of applied cycles suppose to extract a generic small volume of material that can be considered as representative of the entire volume. The volume is a damaged solid since it contains a certain degree of fatigue damage. The amount of such damage depends on the stress amplitude and number of cycles applied. However, no matter what the damage may be it is always represented by isolated or coalesced micro-pores that progress so as to form eventually elongated micro-voids randomly spaced inside the volume. These micro-pores and micro-voids will continue to increase in size and number during the fatigue cycling increasing the void fraction in the material. Therefore, if a generic section ABCD of the volume normal to the applied load is considered its effective resisting area A_{eff} is decreased with respect to its original value in the *virgin* no-damaged material. This is schematized in Fig. 12.30 in which the material volume is affected by extensive fatigue void formation. The ABCD surface has been extracted and compared to the original one (dark) of area A_o. The effective area will be given by the difference between the initial A_o and the total void cross area

$$A_{eff} = A_o - \sum \Delta A_i. \tag{12.36}$$

With such an hypothesis, Lemaitre and Chaboche observed that damage could be defined as

$$\frac{D}{D_{cr}} = 1 - \frac{A_{eff}}{A_o}. \tag{12.37}$$

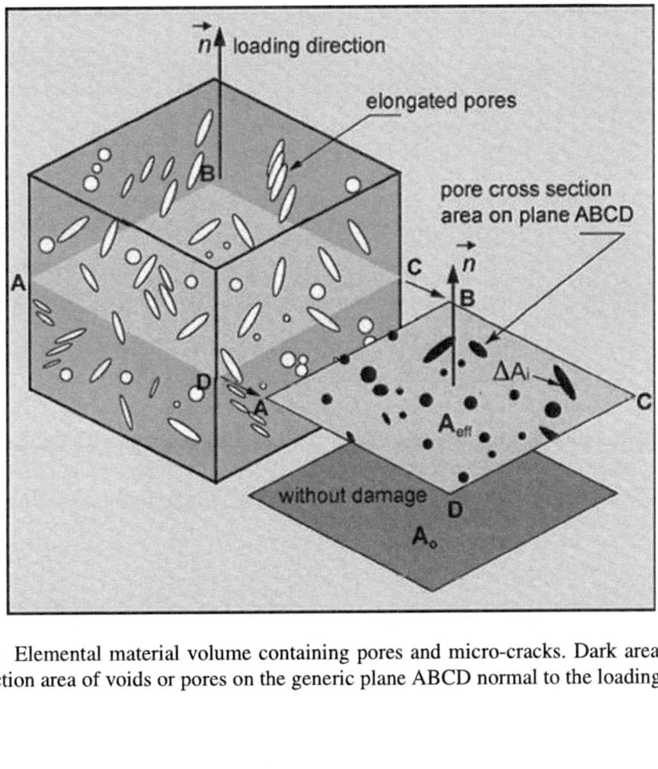

Fig. 12.30 Elemental material volume containing pores and micro-cracks. Dark areas represent the net section area of voids or pores on the generic plane ABCD normal to the loading direction

Equation (12.37) dough interesting cannot be a valid tool to evaluate damage since the value A_{eff} is clearly unknown nor can be directly measured. However, it can be measured indirectly if it is assumed that the progressive reduction of the resisting area results in a parallel increase of the compliance of the small volume that will be seen through a progressive reduction of the Young's modulus E' of the material with respect to the initial one E, then it can be written

$$\frac{D}{D_{cr}} = 1 - \frac{E'}{E} \tag{12.38}$$

This time the progressive reduction of the Yong's modulus can be continuously measured at each loading cycle by the slope of the relative hysteresis loop. An example is shown in Fig. 12.31 [28] for an aluminum alloy type AA6066-T6 of 469 MPa yield strength and 492 MPa ultimate at ambient temperature with initial Young's modulus of 74 GPa. Figure 12.31 shows the various hysteresis loops at selected number N of cycles obtained under strain controlled conditions that in the specific case was $\Delta\varepsilon_t = 0.01$. It can be seen the progressive slope reduction with increasing number of cycles N that documents the continuous increase of the compliance of the specimen due to the reduction of the resisting net section area following the damage progression. Note how the damage effect is measurable through the continuous reduction of the Young's modulus only in traction. In compression, damage that certainly develops since in Stage I of fatigue it is due to shearing stresses has no effect on the Young's modulus because void formation does not change the compliance. It is clear that such procedure can be applied only in low cycle fatigue where damage through large void fraction formation can result in measurable Young's modulus variations. In high cycle fatigue where the overall material behavior is elastic Young's modulus does not change.

12.4 Analysis of Variable Amplitude Spectra

In Sect. 12.2 it has been shown how to separate and count cycles of equal amplitude in a complex variable amplitude (VA) spectrum. In Sect. 12.3 rules have been given to analyze these individual or blocks of cycles of equal amplitude and combine their effects. There is another method to analyze the effect of a VA spectrum that consists in calculating how many times the block of cycles can be repeated before producing fatigue failure. In other words, the VA spectrum is not decomposed in its stress amplitude content, but it is rather applied as a whole to the specimen or work piece or component. This spectrum can be representative, as it has been shown in Sect. 12.1, of an entire commercial airplane *mission*, Fig. 12.1, or be the operational daily time history of a pressure vessel or a typical of a standard automotive unit trek, see Figs. 6.15 and 12.2, etc. To design an automotive multileaf spring or shock absorber it can be considered one of these time histories that refer, for instance, to a typical one kilometer or ten kilometer bumped and disconnected trek. It may also

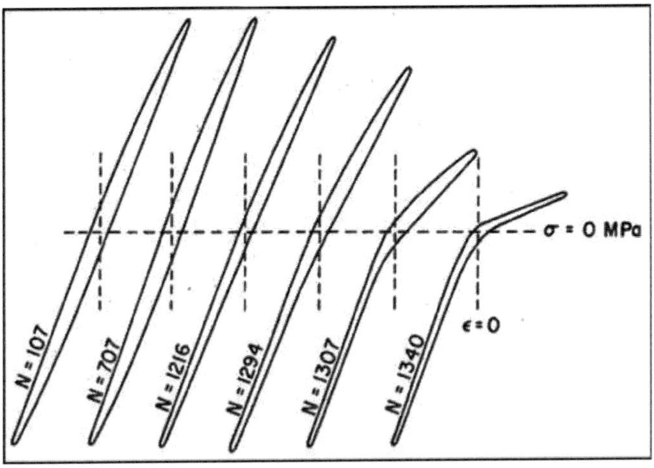

Fig. 12.31 Hysteresis loops at 1% cyclic deformation in a fatigue specimen of AA6066-T6 aluminum alloy [28]

be an umbrella spectrum taken by the designer as a deterministic worst case of real working conditions. Such spectrum acquires the meaning of a reference spectrum for fatigue design. The basic hypothesis behind this kind of approach is that the work piece will see during its life only the same reference spectrum applied as many time as possible.

The target of the designer is precisely to assess this *possible* number of reference spectrum repetitions or *periods*. Figure 12.32 presents a series of typical reference VA blocks [29] used in aeronautical field. It must be understood, however, that blocks having the same composition, in terms of stress amplitudes and number of cycles at each amplitude, may have completely different effect on the work piece depending on the distribution of cycles within the block and the mean stress they can generate. The useful feature of the approach is that the wing of the plane or the automotive spring or the pressure vessel can be instrumented with strain gauges and accelerometers that provide directly the block or unit package of cycles to be used in designing the component. Unfortunately, the designer is not given to know the actual time history of load application and it has been shown in Sect. 6.1.2 and Fig. 12.19 that, since damage progression and accumulation is not a linear process, the time distribution of the same type and number of cycles within a reference spectrum can make the difference. Probably, the first to adopt a procedure based on the reference block or unit block of cycles to fatigue design was Gassner, in Germany in 1939 [30].

His so-called block-program fatigue test was that shown in Fig. 12.33 consisting of low–high low (LoHiLo) sequence of CA cycles. The sequence was repeated till fatigue failure occurred.

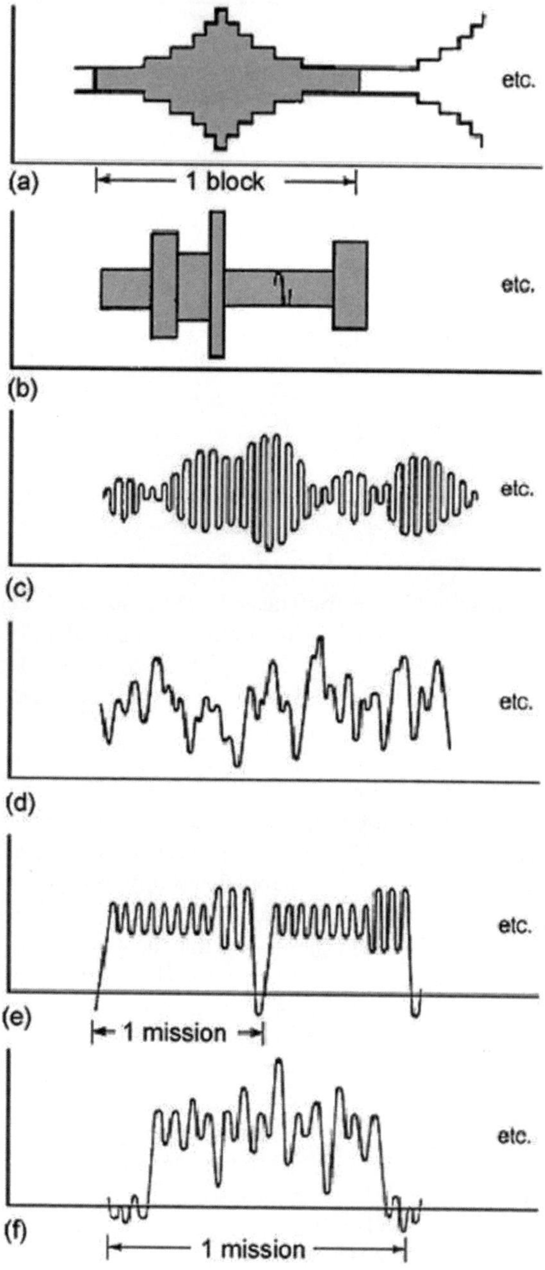

Fig. 12.32 Reference spectrum type **a** programmed sequence, **b** random sequence, **c** narrow-band sequence, **d** broad-band sequence, **e** mission simulation and **e** complex flight mission simulation [29]

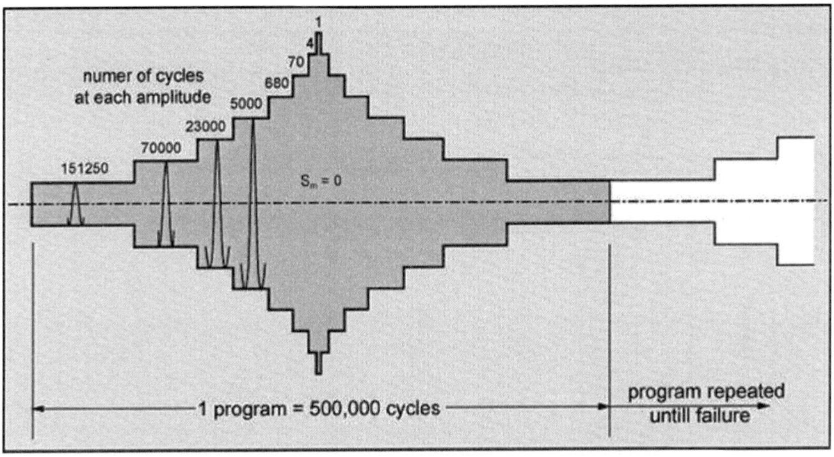

Fig. 12.33 Block-program fatigue test introduced by Gassner in Germany [30] without failing

12.4.1 Block-Program Procedure

Imagine that the design test-block consists of j packages each of h_i cycles of constant stress amplitude σ_I or S_i for a total of N cycles

$$\sum_{i=1}^{j} h_i = N \tag{12.39}$$

The design target will be that of assessing the number m of blocks that can be sustained by the component.

$$m\sum_{i=1}^{j} h_i = \sum_{i=1}^{j} n_i. \tag{12.40}$$

Figure 12.34 schematizes on the Wöhler's curve of the material the reference unit block and the m repeated blocks (in the example of Fig. 12.34 m = 3). The Miner's rule would indicate

$$\sum_{i=1}^{j} \frac{n_i}{N_i} = m\sum_{i=1}^{j} \frac{h_i}{N_i} = 1 \tag{12.41}$$

with N_i being the allowable number of cycles at the i-stress amplitude σ_i. Recalling Eq. (12.5.34) it is

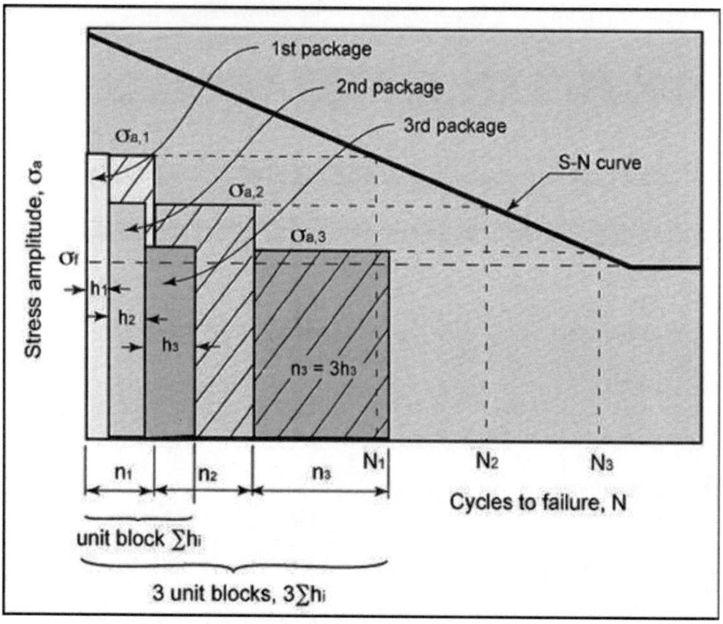

Fig. 12.34 Unit block of three packages at constant stress amplitude used in a Miner-calculation

$$m\sum_{i=1}^{j}\frac{h_i}{N_i} = m\sum_{i=1}^{j}\frac{h_i}{N_f\left(\frac{\sigma_{a,i}}{\sigma_f}\right)^{1/b}} = 1 \tag{12.42}$$

where N_f is the number of cycles at the fatigue limit σ_f. Equation (12.42) yields the value of m

$$m = N_f\sum_{i=1}^{j}\frac{1}{h_i}\left(\frac{\sigma_{a,i}}{\sigma_f}\right)^{1/b}. \tag{12.43}$$

Alternatively, recalling the Basquin Eq. (5.22)

$$N_i = \frac{1}{2}\left(\frac{\varepsilon_{a,i}E}{\sigma_f'}\right)^{1/b} = \frac{1}{2}\left(\frac{\sigma_{a,i}}{\sigma_f'}\right)^{1/b} \tag{12.44}$$

or in case the i-stress amplitude is given with a mean stress $\sigma_{m,i}$, recalling Eq. (5.48)

$$N_i = \frac{1}{2}\left[\frac{\sigma_{a,i}}{\sigma_f'}\left(\frac{\sigma_u}{\sigma_u - \sigma_{m,i}}\right)\right]^{1/b} \tag{12.45}$$

However, the use of a Miner-calculation as per Eq. (12.43) is always introducing the question as to whether or not the summation may be different than unity. Consider, to this purpose, the study of Schijve e Jacobs [31] on aluminum alloys shown in Fig. 12.35 that resembles what has been already presented in Fig. 12.18. The low–high (LoHi) sequence $S_a = 172.5$ MPa-$S_a = 345$ MPa yields a value of the Miner-summation higher than one whereas the HiLo sequence brings that summation below one down to 0.63. An analogous study by Naumann et al. on 2024-T3 and 7075-T6 aluminum notched specimens in which the loading sequence was varied has yielded surprising results presented in Fig. 12.36. It may come as somewhat of a surprise and apparently contradicting what has been said about Miner's rule shortcoming, this time for both alloys the best results are obtained by the HiLo sequence and, in particular, with a positive mean stress acting. The reason may be found in the stress concentration factor relatively high, $k_t = 4$. Under such circumstances during the loading phase local stress at notch root can overcome yielding generating a plastic zone ahead of the notch tip. During unloading a state of residual compressive stresses can be generated that protect the notch tip during subsequent traction, as shown in Fig. 5.2. Bat it may be also another reason that explains this behavior. This is connected to the hardening effect exerted by a notch, of which it has been said in Sect. 7.12. Both 2024-T3 and 7075-T6 aluminum alloys present a cyclic strain hardening behavior, as shown in Fig. 1.27 and listed in Table 6.1. This actually means that the notch root material, where local stress can exceed yielding, is forced to harden under cyclic loads over yielding. The hardened material withstands fatigue much better than base metal that remains in the elastic field.

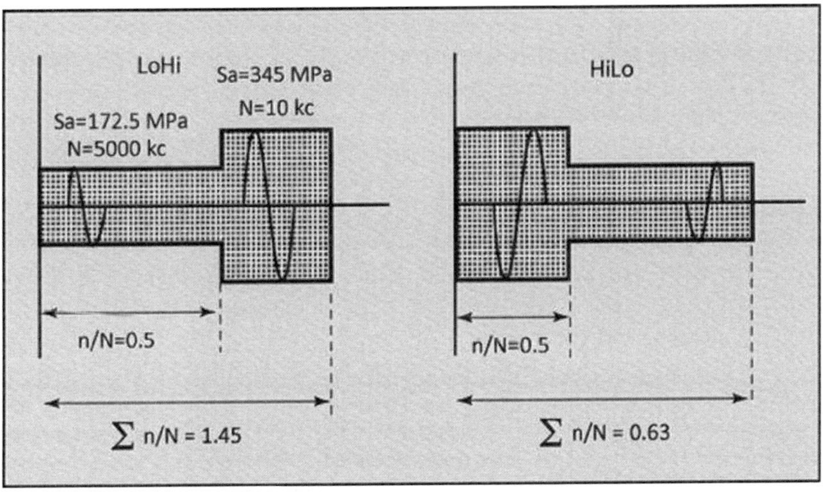

Fig. 12.35 Low–High and High-Low load sequence effects on cumulative damage in terms of Miner-summation [32]

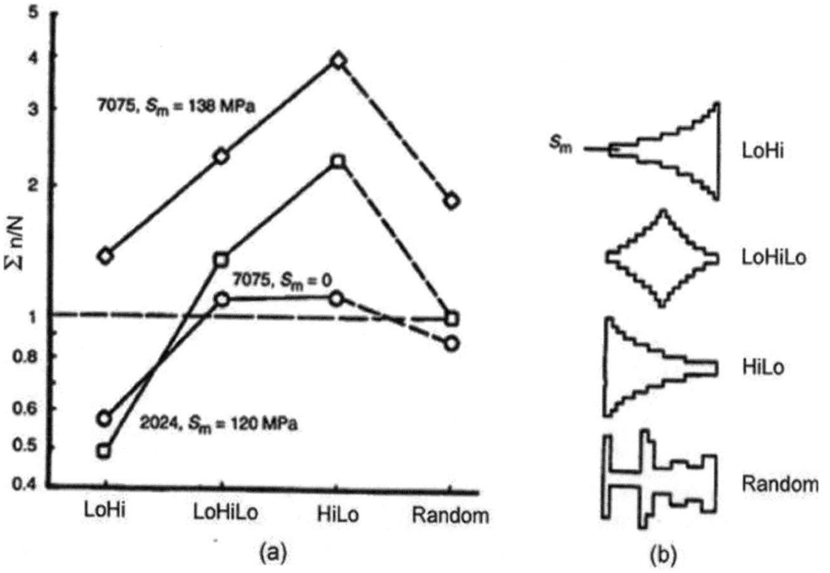

Fig. 12.36 Results of block-program fatigue tests of NASA on notched specimens of aluminum alloys [32]

12.4.2 Fatigue Damage Under Narrow-Band Random Processes

A narrow-band process has been defined as a spectrum whose irregularity factor γ given by Eq. (12.2) is equal unity. This actually means one zero upcrossing per peak. More generally, a narrow-band is a spectrum whose irregularity factor γ is close to unity or, even better, the spectral width parameter λ, Eq. (12.2a) goes to zero. The frequency domain (see Sect. 12.5) provides a better definition in that a narrow-band is a spectrum in which most if not all of the spectral power density (SPD) is concentrated in just two or few closely spaced frequencies. The limiting case is that of a single frequency that refers to a constant amplitude spectrum. A typical example of narrow-band spectrum is shown in Fig. 12.42 with its two component of constant amplitude obtained through the Fourier analysis at closely-spaced frequencies. At variance, in a broad-band spectrum the SPD is distributed over a wide field of frequencies or on frequencies far from each other. The starting point is the knowledge of the loads or stress time history, as schematized in Fig. 12.37b. This time history can be computed at any point of the structure knowing the load time history or can be obtained experimentally in term of strain history instrumenting the particular component under service conditions with strain gauges or accelerometers. Next, the stress or strain time history is converted into an equivalent PSD by using the Fourier transform (see Sect. 12.5.1), as shown in Fig. 12.37c. The narrow-band variable amplitude spectrum is decomposed into a sequence of just few blocks of constant amplitudes, Fig. 12.37d. at this point it is possible to use the block program procedure described in Sect. 12.4.1. Damage D is, then, defined according to Eq. (12.11) with $m = 1$ (linear accumulation, Miner's rule) as

$$D = \sum_{i=1}^{s} \frac{n_i}{N_i} \tag{12.46}$$

where n_i is the total number of cycles in the ith block of constant-stress amplitude $\sigma_{a,i}$, N_i is the number of cycles to failure under $\sigma_{a,i}$, and s is the total number of blocks. Recalling the Basquin Eq. (5.22) valid in the linear domain, the number of cycles to failure N_i can be written as

$$N_i = \left(\frac{\sigma_{a,i}}{\sigma'_f}\right)^{1/b} \tag{12.47}$$

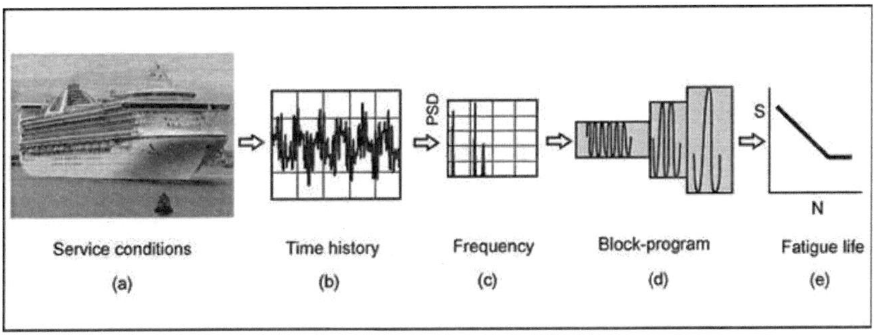

Fig. 12.37 General procedure for life analysis of narrow-band random process

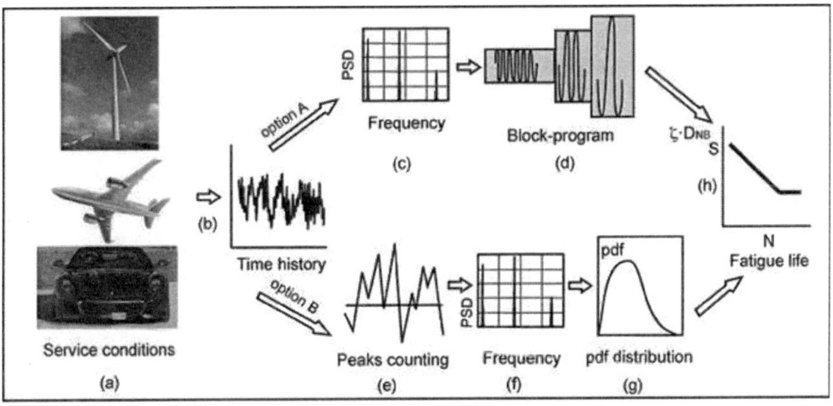

Fig. 12.38 General procedure for fatigue life analysis of a wide-band time history

where σ'_f is the fatigue strength coefficient and b is the fatigue strength exponent. The probability (frequency) f_i associated to a stress amplitude $\sigma_a = \sigma_{a,i}$ is given by

$$f_i = \frac{n_i}{\sum_{i=1}^{s} n_i} \tag{12.48}$$

thus, f_i is the *pdf* of the random variable σ_a. The total fatigue damage can be written as

$$D = \sum_{i=1}^{s} \frac{n_i}{N_i} = \sum_{i=1}^{s} \frac{f_i \sum_{i=1}^{s} n_i}{N_i} \tag{12.49}$$

By using the Basquin-based expression (12.47), the fatigue damage is then

$$D = \sum_{i=1}^{s} n_i \sum_{i=1}^{s} f_i \left(\frac{\sigma_{a,i}}{\sigma'_f} \right)^{-1/b} = \frac{\sum_{i=1}^{s} n_i}{\left(\sigma'_f\right)^k} \sum_{i=1}^{s} f_i \sigma_{a,i}^k \tag{12.50}$$

with $k = -1/\mathrm{b}$. In Eq. (12.50) the total count of cycles Σn_i is equal to the rate of zero upcrossing multiplying the total time period

$$\sum_{i=1}^{s} n_i = E[0] \cdot T \tag{12.51}$$

Equation (12.49) is based on a *pdf* expression derived from the cycles ratio given by Eq. (12.47). There is another expression for the *pdf* based on the PSD. For the narrow-band process this expression is

$$p(\sigma) = \frac{\sigma}{4M_o} e^{-\frac{\sigma^2}{8M_o}} \tag{12.52}$$

where M_o is the 0-order moment of the PSD (see Sect. 12.5.2). The expected value of σ_a^k is

$$P\left(\sigma_a^k\right) = \sum_{i=1}^{s} f_i \sigma_i^k. \tag{12.53}$$

Assuming as cumulative probability function the Weibull expression $P(\sigma_a)$ given by Eq. (5.45), then the expected value of σ_a^k will be

$$P\left(\sigma_a^k\right) = \sigma_o^k \Gamma\left(\frac{k}{m} + 1\right) \tag{12.54}$$

where σ_o and m are the scale factor and the Weibull exponent, respectively and Γ the Gamma function (4.18) or (4.19). The expected total fatigue damage is

$$D = \frac{\sum_{i=1}^{s} n_i}{\sigma_f'} \sigma_o^k \cdot \Gamma\left(\frac{k}{m} + 1\right) \tag{12.55}$$

In deriving Eq. (12.50) or (12.55) it has been assumed that all stress cycles have a mean value equal to zero. If a mean stress exists it must be taken into consideration. According to Miner's rule (see Sect. 12.3.1) failure occurs when $D = 1$. However, a shortcoming of Miner's rule is that damage does not progress and accumulate in linear fashion (see Sect. 12.3.2).

12.4.3 Fatigue Damage Under Wide-Band Random Processes

The fatigue analysis of a wide-band random process requires a more complex approach. This time the narrow-band method of analysis cannot be used without, at least corrections. There are at least two fundamental reasons for this. The first reason is that, even dough also for a wide-band spectrum it is possible to identify and separate the fundamental components of the time history through the Fourier analysis, the resulting blocks can be so many, even almost infinite, to complicate the analysis too-much. But even overcoming the very cumbersome procedures of very many blocks of constant stress there is a problem that cannot be overcome. This is connected to the mean stress that in a wide-band time history can be continuously changing and some time in a rather unpredictable fashion. Consider, for instance, the two frequencies time history of Fig. 12.42b. It can be considered a wide-band spectrum in that the frequencies of the two superimposed components are very far from each other, as shown in Fig. 12.43b. Using the λ parameter of Eq. (12.2a) in a time period T there about six upcrossing and 24 peaks so that the irregularity factor $\gamma = E[0]/E[P] = 6/26 = 0.23$ and $\lambda \sim 1$. It can be easily seen how the mean stress associated to the low amplitude high frequency cycles is continuing changing from zero to a maximum at the peak value of the low frequency high stress cycle. Therefore, some different solution must apply. The first option, proposed by Wirsching and Light [33] in 1980 and schematized in Fig. 12.38 as option A, can be adopted when the frequency components are few and is based on a modified narrow-band approach where the frequency content, PSD, is first found, Fig. 12.38c and d. By using the narrow-band approach as a starting point, a model for predicting fatigue damage under stationary wide-band Gaussian stress process has been developed leading to the general expression for damage D_{WB} over a time interval t

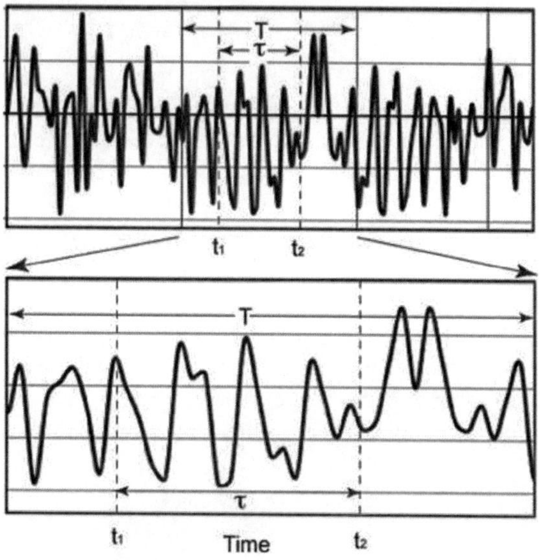

Fig. 12.39 Schematic of sampling a random variable at two given instants t_1 and t_2. **a** time history, **b** detail of time history in the interval T

$$D_{WB} = \zeta_W \cdot D_{NB} \tag{12.56}$$

where D_{NB} is the fatigue damage under a narrow-band random process seen in Sect. 12.4.2 and ζ_W is the rainflow correction factor. ζ_W is an empirical factor derived from extensive Monte Carlo simulations that include a variety of spectral density functions. It is expressed as follows

$$\zeta_W = a_S + (1 - a_S)(1 - \lambda)^{c_W} \tag{12.57}$$

where

$$a_S = 0.926 - 0.033\ k c_W = 1.587k - 2.323 \tag{12.58}$$

Note that k is the slope of the *S–N* curve defined in Eq. (5.22), and λ is the spectral width parameter defined in Eq. (12.2a). There is another expression for the correction factor ζ_O to be used with the narrow-band damage D_{NB} due to Oritz and Chen [34] that is

$$D = \zeta_O \cdot D_{NB}$$
$$\zeta_O = \frac{1}{\gamma}\sqrt{\frac{M_o M_m}{M_o M_{m+2}}} \ and\ m = \frac{2.0}{k} \tag{12.59}$$

Instead of using the damage correction factor to shift from the narrow-band approach to the wide-band method, Dirlik [35] developed in 1985 an based on extensive Monte Carlo simulations of the stress amplitudes, option B in Fig. 12.312.

Dirlik's solutions were successfully verified by Bishop [36] in theory. Dirlik's damage model for a time period of T is as follows

$$D_{WB} = E[P] \cdot \frac{T}{A} \cdot \int_0^{\infty} \sigma_a^k p(\sigma_a) d\sigma_a \tag{12.60}$$

where *E[P]* is the number of peaks per second in the sample period T and A derives from the Basquin Eq. (5.22)

$$A = \left(\sigma'_f\right)^K \tag{12.61}$$

with σ'_f being the fatigue strength coefficient of the material. The Dirlik empirical closed-form expression for the *pdf* is given in Sect. 12.5.2. In Eq. (12.61) σ'_f is the fatigue strength coefficient and $k = -1/b$ with b being the fatigue strength exponent. It is opinion of Bishop and Sherratt [37] and Bishop [38] that the Dirlik solution gives the better results as compared with the corresponding time domain results. The frequency domain method for predicting fatigue damage from PSD assumes that the *pdf* of peaks is equal to the *pdf* of stress amplitudes. The problem with this solution is that positive troughs and negative peaks are ignored and all positive peaks are matched with corresponding troughs of similar magnitude regardless of whether they actually form stress cycles. The method, therefore, may overestimate the probability of large stress ranges and the corresponding damage may result to be conservative.

12.5 Fourier Analysis. The Frequency Domain

Fatigue analysis so far described has always been performed in the time domain, in which all input loading and output stress or strain response are time-based signals. In many instances it may be convenient to transfer time-based signals to the frequency domain. The rational is that it is always possible to form any function as a summation of a series of sine and cosine terms of increasing frequency. In other words, any time varying data can be transformed into a different domain called the frequency domain or frequency space. For a time history of the type shown in Fig. 12.43b or c it is relatively easy to separate the two sinusoidal components and analyze the effects, but in real life more general time histories are faced for which it is impossible to separate the fundamental components or harmonicas, as in the case shown in Fig. 12.43d. The idea of decomposing a function of time into a summation of a series of sine and cosine terms of increasing frequency was first proposed by the French mathematician Joseph Fourier in the nineteenth century, but never used since, in general, it does not exist. For two centuries the so called Fourier transform has been confined to the heuristic world of theoretical knowledge until the introduction of digital computers that has made it possible to solve by a digital approach what has been impossible from a classical mathematical analysis point of view: habemus mathematicarum optimus peritus! Then, if $X(t)$ is a periodic function of time with a period T, $X(t)$ can be expressed by the sum of an infinite trigonometric series in the following form:

$$X(t) = A_o + \sum_{k=1}^{\infty} A_k \cos\frac{2\pi kt}{T} + B_k \sin\frac{2\pi kt}{T} \tag{12.62}$$

where

$$\begin{aligned} A_o &= \frac{1}{T}\int_{-T/2}^{T/2} X(t)dt \\ A_k &= \frac{2}{T}\int_{-T/2}^{T/2} X(t)\cos\frac{2\pi kt}{T}dt \\ B_k &= \frac{2}{T}\int_{-T/2}^{T/2} X(t)\sin\frac{2\pi kt}{T}dt. \end{aligned} \tag{12.63}$$

Recalling the Euler's formula

$$e^{i2\pi kt} = \cos(2\pi kt) + \sin(2\pi kt) \tag{12.64}$$

the Fourier series can be expressed in terms of basic harmonicas by using complex coefficients as

$$X(t) = \sum_{k=1}^{\infty} C_k e^{i2\pi kt/T} \tag{12.65}$$

where the complex coefficients C_k are given by

$$C_k = \frac{1}{T}\int_{-T/2}^{T/2} X(t)e^{i2\pi kt/T}dt \tag{12.66}$$

This passage from sine and cosine to complex exponentials gives both the amplitude of the wave present in the function and the phase of the wave. The Fourier transform can be considered as the limit of the Fourier series of $X(t)$ as T approaches infinity. This can be illustrated as follows by rewriting Eq. (12.62) with infinite T.

$$X(t) = \lim_{T\to\infty} \sum_{k=-\infty}^{\infty} \left(\frac{1}{T} \int_{-T/2}^{T/2} X(t) e^{-i2\pi kt/T} dt \right) e^{i2\pi kt/T} \tag{12.67}$$

If the frequency of the kth harmonic, ω_k, in radians per second, is

$$\omega_k = \frac{2\pi k}{T} \tag{12.68}$$

and the spacing between adjacent periodic functions, $\Delta\omega$, is

$$\Delta\omega = \frac{2\pi}{T} \tag{12.69}$$

Equation (12.67) becomes

$$X(t) = \lim_{T\to\infty} \sum_{k=-\infty}^{\infty} \left(\frac{\Delta\omega}{2\pi} \int_{-T/2}^{T/2} X(t) e^{-ik\Delta\omega t} dt \right) e^{ik\Delta\omega} \tag{12.70}$$

As T goes to infinity, the frequency spacing, $\Delta\omega$, becomes infinitesimally small, denoted by dω, and the sum becomes an integral. As a result, Eq. (12.56) can be expressed by the well-known Fourier transform pair $X(t)$ and $f(\omega)$.

$$\begin{aligned} X(\omega) &= \frac{1}{2\pi} \int_{-\infty}^{\infty} X(t) e^{-i\omega t} dt \\ X(t) &= \int_{-\infty}^{\infty} X(\omega) e^{i\omega t} d\omega \end{aligned} \tag{12.71}$$

The function $X(\omega)$ is the forward Fourier transform of $X(t)$, and $X(t)$ is the inverse Fourier transform of $X(\omega)$.

12.5.1 Power Spectral Density

As said, the Fourier transform of a stationary random process $X(t)$ usually does not exist. However, the Fourier transform of the autocorrelation function $R(\tau)$ always exists. To understand what is an autocorrelation function consider a generic random spectrum like that shown in Fig. 12.39 and sample the value of the variable $X(t)$ at two different times t_1 and $t_2 = t_1 + \tau$. For digital data the autocorrelation function $R(\tau)$ is a function of the time interval $\tau = | t_2 - t_1|$ defined as

$$R(\tau) = \lim_{N\to\infty} \frac{1}{N} \sum_{t=1}^{N} X(t)X(t+\tau) \tag{12.72}$$

Then, the autocorrelation function $R(\tau)$ of a random process is the mean value of the product $X(t_1)\cdot X(t_2)$ that can be approximated by the average value of the same product which can be obtained by sampling the random variable σ at times t_1 and t_2.

If sampling is taken at $r\Delta t$ intervals (suppose Δt is 0.01 s then $r\Delta t = 0, 0.01, 0.02, 0.03..$) and, for instance, $X(t) = \sigma(t)$ then

$$R(r\Delta t) = \lim_{N\to\infty} \frac{1}{N-r} \sum_{t=1}^{N} X(t)X(t+r\Delta t) \equiv \frac{1}{N-r} \sum_{n=1}^{N-r} \sigma_n \sigma_{n+1} \tag{12.73}$$

with $r = 0,1,2,3,\ldots m$ where $m < n$. For instance, the autocorrelation function $R(\tau)$ of the sine function is a sine function having half the amplitude of the base function that continues indefinitely. This is shown in Fig. 12.40. If sampling is taken for a finite length of the signal (finite number of data points) the autocorrelation function $R(\tau)$ decays in time as shown in Fig. 12.41. The power spectral density (PSD) of a normalized (zero mean value) random process $X(t)$ is given by the forward Fourier transform of the autocorrelation function $R(\tau)$

$$S(\omega) = \frac{1}{2\pi} \int_{-\infty}^{\infty} R(\tau)e^{-i\omega\tau} d\tau \tag{12.74}$$

while the inverse Fourier transform is

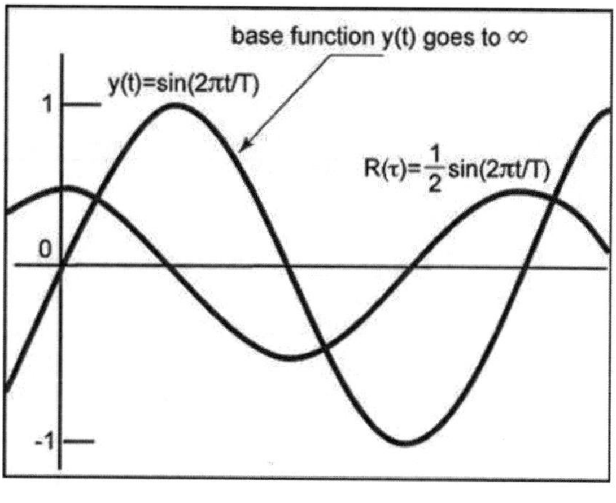

Fig. 12.40 Sine base function and its autocorrelation function $R(\tau)$

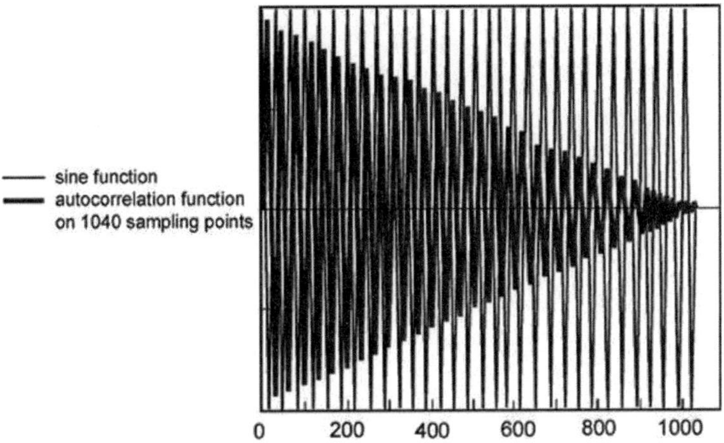

Fig. 12.41 Autocorrelation function calculated on an array of 1040 data points

Fig. 12.42 One-sided power spectral density function

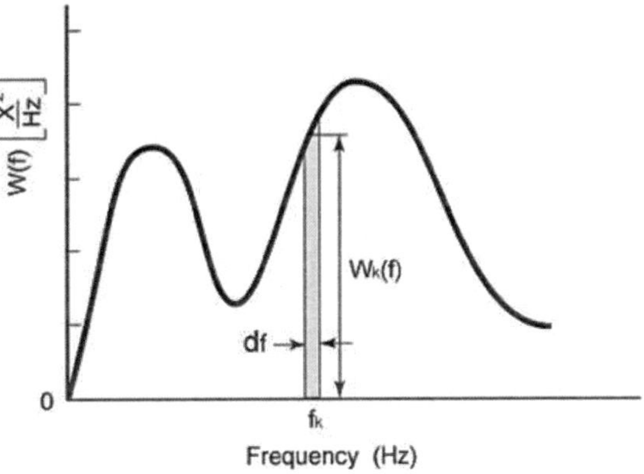

$$R(\tau) = \int_{-\infty}^{\infty} S(\omega) e^{i\omega\tau} d\omega \tag{12.75}$$

When the time interval is $\tau = 0$, Eq. (12.73) reduces to

$$R(0) = \int_{-\infty}^{\infty} S(\omega) d\omega = \Delta^2 \tag{12.76}$$

which means that the area under the curve of a PSD is the variance, see Eq. (4. 6), and its square root is the root mean square (RMS) of a normalized stationary random process (mean value equal zero). The idea of a negative frequency has been

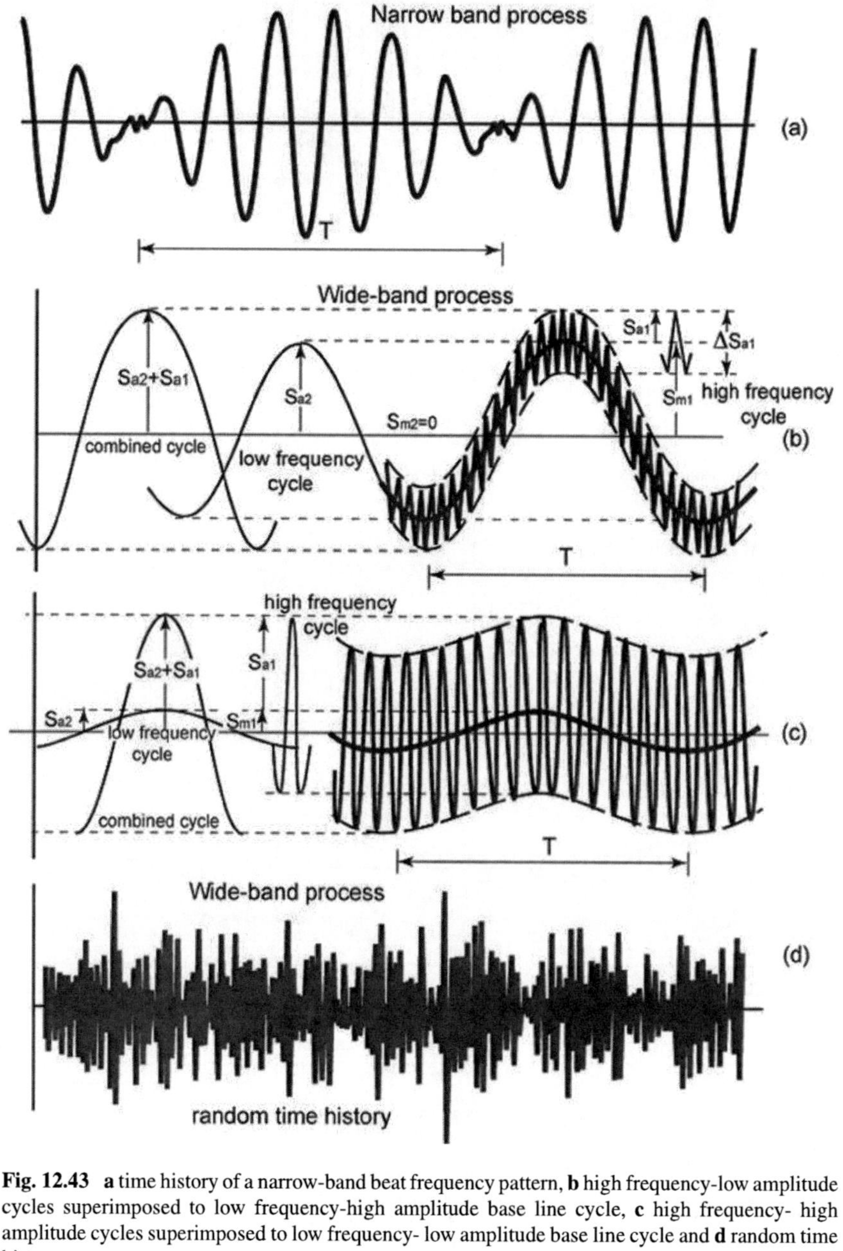

Fig. 12.43 **a** time history of a narrow-band beat frequency pattern, **b** high frequency-low amplitude cycles superimposed to low frequency-high amplitude base line cycle, **c** high frequency- high amplitude cycles superimposed to low frequency- low amplitude base line cycle and **d** random time history

introduced for mathematical completeness. However, it has no physical meaning. It is common practice to consider the frequency from zero to infinity and to have the frequency expressed in hertz (cycle/second) rather than radians/second. Therefore, the two-sided spectral density, $S(\omega)$, can be transformed into an equivalent one-sided spectral density $W(f)$ as follows

$$\Delta^2 = \int_o^\infty W(f)d\omega \tag{12.77}$$

where

$$W(f) = 4\pi S(\omega);\ f = \frac{\omega}{2\pi} \tag{12.78}$$

Note that here f is a frequency not a function. The typical units of a PSD are acceleration (g^2/Hz) or (($m/s^2)^2$/Hz) versus frequency (Hz). It can also be given in strain^2/Hz or Mpa^2/Hz, depending on the gauge used for the time history measurement. As previously noted, Eq. (12.74), the amplitude (area) of each frequency is actually RMS^2/Hz, where RMS is the root-mean-square. The RMS value of a normalized signal is equal to the standard deviation, assuming a zero mean. A pure sinusoidal function has the following relationship

$$peak = \sqrt{2} \cdot RMS \tag{12.79}$$

This actually means that, once the time history has been transferred to the frequency domain and PSD assessed, the value of a peak, for instance the stress amplitude S_a, can be calculated by multiplying the area under the peak curve by 1.414. $\sqrt{2}$ is also called crest factor.

12.5.2 Moments of the Power Spectral Density

From the characteristics of the PSD, nth moments M_j can be obtained. The relevant spectral moments M_j are easily computed from a one sided power spectral density, $W(f)$, Eq. (12.78) and Fig. 12.42, using the following expression

$$M_j = \int_o^\infty f^j W(f)df = \sum_{i=1}^{k} f_i^j W(f)\Delta f \tag{12.80}$$

Summation in Eq. (12.80) can be used in case where the PSD is not a continuous function but concentrates in more or less frequencies, as shown in Fig. 12.44. The determination of the moments M_j is fundamental in the fatigue analysis since

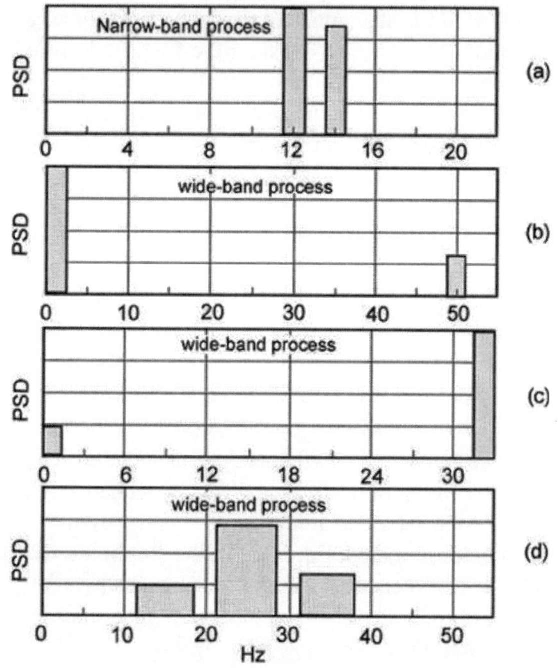

Fig. 12.44 PSD of the four time history of Fig. 12.42

it makes it possible to calculate an empirical closed-form expression for the *pdf* probability density function of a random spectrum variable $X(t)$ [35], the number of zero upcrossing *E[0]* and peak *E[P]* per second in the sample period T and the irregularity factor γ given by Eq. (12.61). In theory, all the possible moments could be calculated, however, in practice, M_o, M_1, M_2 and M_4 are sufficient to calculate all of the information for the fatigue analysis. This information is

- the probability density function of a random spectrum variable $X(t)$

$$p(\sigma_a) = \frac{\frac{D_1}{Q} e^{-\frac{Z}{Q}\sigma_a} + \frac{D_2 Z}{R^2} e^{-\frac{1}{2}\left(\frac{Z\sigma_a}{R^2}\right)^2} + D_3 Z e^{-\frac{1}{2}(Z\sigma_a)^2}}{2\sqrt{M_o}} \tag{12.81}$$

with

$$D_1 = \frac{2\left(X_k - \gamma^2\right)}{1 + \gamma^2}; \; R = \frac{\gamma - X_k - D_1^2}{1 - \gamma - D_1 + D_1^2}; \; D_2 = \frac{1 - \gamma - D_1 + D_1^2}{1 - R}$$
$$D_3 = 1 - D_1 - D_2; \; Q = \frac{1.25 \cdot (\gamma - D_3 - D_2 \cdot R)}{D_1} \tag{12.82}$$
$$Z = \frac{1}{2\sqrt{M_o}}; \; X_k = \frac{M_1}{M_o}\sqrt{\frac{M_2}{M_4}}$$

- the number of zero upcrossing *E[0]* per second in the sample period T

$$E[0] = \sqrt{\frac{M_2}{M_o}} \tag{12.83}$$

- the number of peak *E[P]* per second in the sample period T

$$E[P] = \sqrt{\frac{M_4}{M_2}} \tag{12.84}$$

- The irregularity factor γ and the λ function

$$\gamma = \frac{M_2}{\sqrt{M_o M_4}}; \lambda = \sqrt{1-\gamma^2} = \sqrt{1 - \frac{M_2^2}{M_o M_4}} \tag{12.85}$$

12.5.3 Cycles Superposition

The case spectra made of superposition of cycles is the most frequent in real life and can be analyzed by the Fourier transform, as seen in Sect. 12.5. Some key examples from the simplest to the most complex superposition are offered in Fig. 12.43. The time history (a) of Fig. 12.43 is the simplest case made of two sine waves of slightly different amplitude and closely-spaced frequencies. Combined the two waves show the characteristic beat frequency pattern. It is a narrow-band process because in the time period T it can be counted 5 zero upcrossing and 5 peaks with an irregularity factor $\gamma = 1$, see Eq. (12.2). Case (b) of Fig. 12.43 refers to a high frequency load cycle of amplitude S_{al} and angular frequency $\omega_1 t$ superimposed to a low frequency cycle of amplitude S_{a2} and angular frequency $\omega_2 t$, with $S_{al} < S_{a2}$ and $\omega_2 < < \omega_1$ or, case (c) of Fig. 12.43, with $S_{al} > > S_{a2}$. It can be written

$$S = S_m + S_1 \sin \omega_1 t + S_2 \sin \omega_2 t \tag{12.86}$$

It is also possible to imagine that stress amplitude S_{al} of case (b) is very small, even below the fatigue limit of the material. It may have an effect anyhow. This is because of cumulative damage. Moreover, small amplitude cycles have also the effect to increase, dough by little, the amplitude of the high cycle spectrum, as shown in Fig. 12.43b, from S_{a2} to $S_{al} + S_{a2}$. The resulting overload can more easily breach the material providing a micro-crack to low amplitude cycles that now can exert a cumulative effect that, otherwise, would have not be seen (see Sect. 12.3.3). Small amplitude cycle, then, can result in a devastating effect because of the combined action of high cycle amplitude increase and high frequency that discharges an enormous number of cycles in relatively short period of time. Moreover, the effect of the main low frequency cycle is also that to provide a mean stress S_{m1} to low amplitude cycles, as shown in Fig. 12.43b.

Almost half the cycles, in fact, are applied at high mean stress that reaches a maximum when the wave $S_{a2}\cdot sin\omega_2 t$ also reaches its maximum value. This fact sharply increases the potential of low amplitude cycles (see sample problem N. 2 in Sect. 5.6) to fatigue failure. However, this mean stress that can be particularly high when combined to high peaks low frequency cycles, is not constant since it reduces periodically to become also negative and then regrows. This *waving* mean stress may introduce same real difficulty in the assessment of fatigue damage. Case (b) may seem similar to case (a), but it is not since it is, actually, a wide-band process. In the time frame *T*, in fact, it is possible to count six zero upcrossing and 26 peaks so that the irregularity factor is $\gamma = E[0]/E[P] = 6/26 = 0.23$ close to zero and the λ parameter of Eq. (12.2a) is ~ 1 characteristic of the narrow-band process. The same arguments can be used for case (c) of Fig. 12.43. This time the high amplitude cycles have the higher frequency that means many cycles of high amplitude, and some of them are also applied with a positive mean stress.

Things become complicated when one or both spectra do not show a constant angular frequency as in Fig. 12.43 that represents a most general random process. The use of the Fourier transform can present the three spectra under a different point of view transferring the four functions from the time domain to the frequency domain. The PSD of case (a) time history reveals two closely-spaced peaks at 12 Hz and 14 Hz where the entire power concentrates, Fig. 12.44a. This is characteristic of a narrow-band process. At variance, the PSD of case (b) also reveals two peaks, bat at frequencies of 1 Hz and 50 Hz very far from each other. This is characteristic of a wide-band process. The same PSD is seen for case (c) with the inversion of power distribution. Finally, case (d) yields a PSD completely different made of three wide peaks at 15, 25 and 35 Hz. This also is a wide-band spectrum.

12.5.4 Method of the Equivalent Spectrum

Equation (12.78) provides a rather effective way to define an equivalent spectrum to be used in the fatigue analysis under random spectra. Assume to have a time history whose PSD has yielded the frequency content and assume that those frequencies are *j*. Through Eq. (12.79) it is possible to evaluate the RMS of each component wave and define an $\mathrm{RMS_{spectr}}$ of the equivalent spectrum

$$RMS_{spectr} = \sqrt{\frac{1}{j}\sum_{i=1}^{j}(RMS_i)^2} \tag{12.87}$$

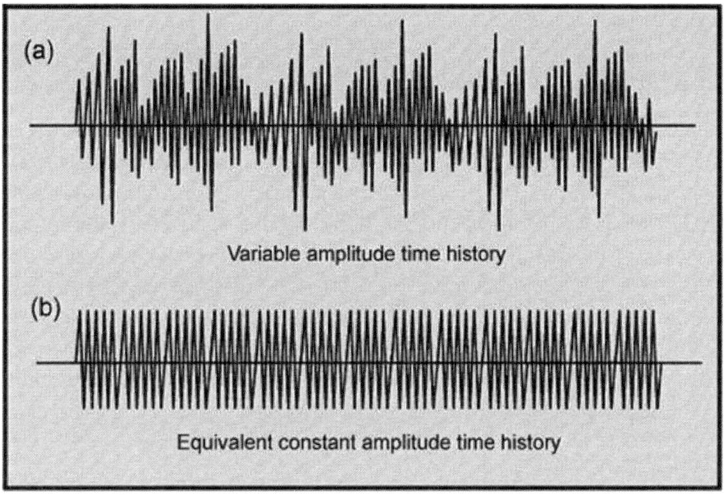

Fig. 12.45 **a** random spectrum and **b** equivalent constant stress amplitude spectrum

The equivalence is schematized in Fig. 12.45. This actually means that N cycles of constant amplitude S_a given by Eq. (12.87) produce the same damage of the real N cycles of variable amplitude. The RMS method is largely used in seismic engineering in the evaluation of the behavior of systems having anchor points on different structures with different response spectra not necessarily in phase among each other. It has also been proved effective in the evaluation of fatigue crack growth rate in Stage II of fatigue. But we know that Stage II of fatigue is characterized by linear damage accumulation (see Sect. 12.3.7). It may be also the case for random spectra since it has been shown in Sect.12.3.2, Fig. 12.18, that Miner's rule could be effective for random spectra.

12.6 Impact Fatigue

We complete this treatment of variable amplitude loads by addressing the rather complicate issue of impact fatigue. The impact fatigue loadings usually observed can be divided into two types according to their characteristics: the regularly repeated impact loads and the randomly repeated impact loads, which are frequently superimposed with ordinary fatigue loads. This section deals with fatigue caused by regularly repeated impact loads. In general, an impact load is defined as a load that is applied to the structure in a time frame lower than half the period corresponding to the natural frequency of the structure or, which is the same, a load applied with a frequency at least double than the natural frequency. These loads, also called impulsive loads, can be grouped into three categories.

(1) rapidly varying loads;
(2) dynamic loads;
(3) sudden loads.

To the first category belong all rapidly moving loads such as those produced on suspensions by a car running at high speed on a paving and actually include all cyclic loads even dough they are not precisely impacting, according to the above definition. This kind of loads has been the subject of all previous sections. Traditional fatigue, in fact, can be studied using frequencies as high as 120 Hz, which means one cycle in 8×10^{-3} s or a peak reached in 2×10^{-3} s. Loads generated by explosions or internal combustion in a cylinder of an internal-combustion engine belong to the second category. To the last category real impact-loads belong as those produced by sudden strikes, like drop forge, hammering, knocking, etc. It is well known that forces and deformations resulting from impact loads are generally greater than those produced by a static force of equal value. The ratio of dynamic to static response is called impact factor. By *static* force it is meant a load gradually varying and by static deflection a displacement that takes on the load gradually. The actual value of the impact factor depends on the loading-structure geometry and compliance. The higher the compliance of the system the lower the impact factor. For the simplest case of a stiff structure without any dumper or spring support the impact factor is equal two. Much remains to be learned about the impact fatigue owing to the difficulty of impact-fatigue testing and the large variety of conditions that can lead to impact fatigue. Just think to the effect of the coupling specimen-testing machine that results in a particular shock wave and successive reflections which produce energy re-entering into action. Moreover, it must be taken into consideration the material response to impact loads that depends on the loading rate or, better, on the strain rate $\dot{\varepsilon} = d\varepsilon/dt$. Normal traction tests are run at a stain rate of the order of 10^{-4} to 10^{-3} s^{-1}. In a fatigue test at 100 Hz the strain rate can be as high as 4–10 s^{-1}. Impact loads can result in strain rates by far higher than that. Figure 12.46 shows a shock wave generated in a cylindrical steel specimen by hammering [39]. The load rising time is 0.5 ms that compared to high frequency fatigue is at least ten times shorter. Therefore, the strain rate will be ten times larger. At such a strain rate material response may change dramatically. Yielding depends on dislocations movement. Edge dislocations act like a wedge in the crystal lattice with the upper side under compression and the lower, at the base, in traction. Compressive stresses tend to push interstitial solute atoms out of the upper side into the lower one where tensile stresses dilate the lattice and attract them. This interstitial solute atoms form what is known as the Cottrell's atmosphere. When a dislocation moves its atmosphere tends to move with it. But dislocation movement depends on the velocity and intensity of the applied forces while the Cottrell's atmosphere formed at its base moves according to atoms diffusion laws that depend on temperature and not on external forces. This movement obedience to different laws causes a conflicting behavior between the dislocation and its atmosphere that worsens as the strain rates increases (high loading rate) and the temperature decreases.

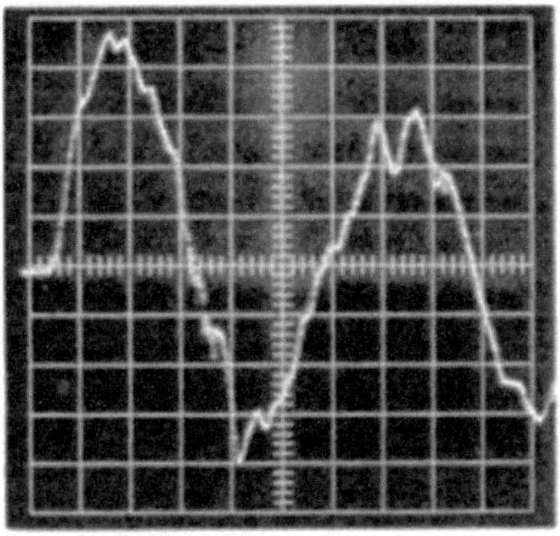

Fig. 12.46 Impacting load wave [39]

The Cottrell's atmosphere cannot follow its dislocation and exerts a drag force that can be so high as to block the dislocation, depending on the applied strain rate and the temperature. The resulting effect is the increase of the yield strength of the material that is prevented from flowing with the same rate as the applied force. This phenomenon of strain rate hardening is particularly seen on ductile materials having BCC lattice (ferritic steels) and almost absent in FCC materials, like aluminum and its alloys, austenitic stainless steels and nickel alloys in which the Cottrell's atmosphere is particularly diluted. The yield strength can double or increase even more. Also the ultimate strength increases, dough at a lower rate so eventually the two merge.

This is shown in Fig. 12.47 [40] for mild steel. As it can be seen, mild steel, ductile and low strength material at low strain rates, becomes strong like high strength steel and, therefore, it is expected that also its fatigue strength improves. Effectively, Iguchi et al. [39] observed an improvement of fatigue strength of cylindrical specimens of AISI 1020 mild steel, 240 MPa yield strength and 412 MPa ultimate with 32% deformation and 68% of area reduction at failure, when subjected to impulsive loads shown in Fig. 12.46. The results of their experiments are shown in Fig. 12.48 [39]. As it can be seen, impact fatigue curves run over the fatigue *S–N* curve obtained by *traditional* fatigue obtained at a cyclic frequency of 30 Hz, notwithstanding that all impact curves are obtained at higher R ratios. In particular, for $R = -0.17$ (impact fatigue) and $R = -1.0$ (conventional fatigue), it is shown the experimental data relative to damage nucleation (N_s data) with persistent slip band formation (see Sect. 1.5 and Fig. 1.54) and macro-crack formation (N_c data) (see Fig. 1.55). N_f denotes the cycles to failure. Interesting enough, the real gain is in the low cycle fatigue region. The reason is schematized in Fig. 12.49 that shows the *static* and dynamic (impact) force–deformation, F-ε, response of the material. When the applied static force F_1 acts in the elastic domain it produces a deformation ε_1, corresponding to point A. The same force F_1 applied dynamically doubles its effect because of the impact factor, becoming equal to $2F_1$.

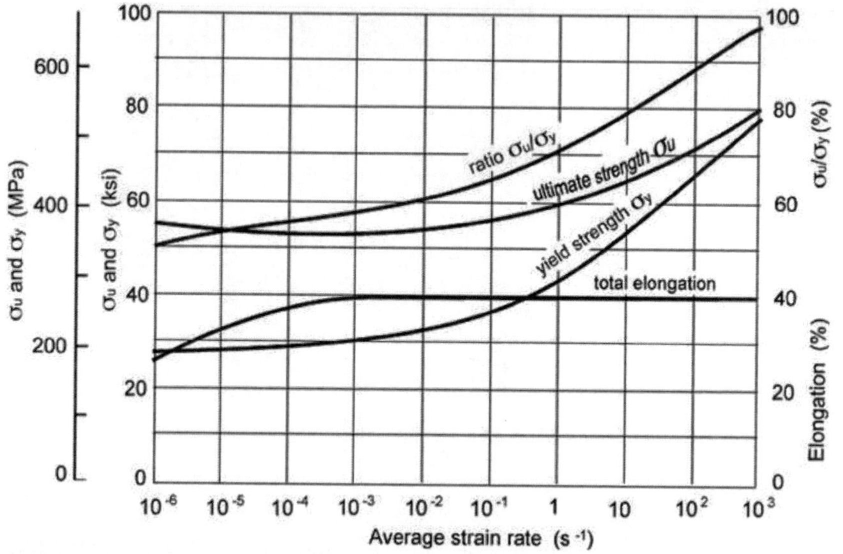

Fig. 12.47 Effect of strain rate on tensile properties of mild steel at room temperature (modified from [40])

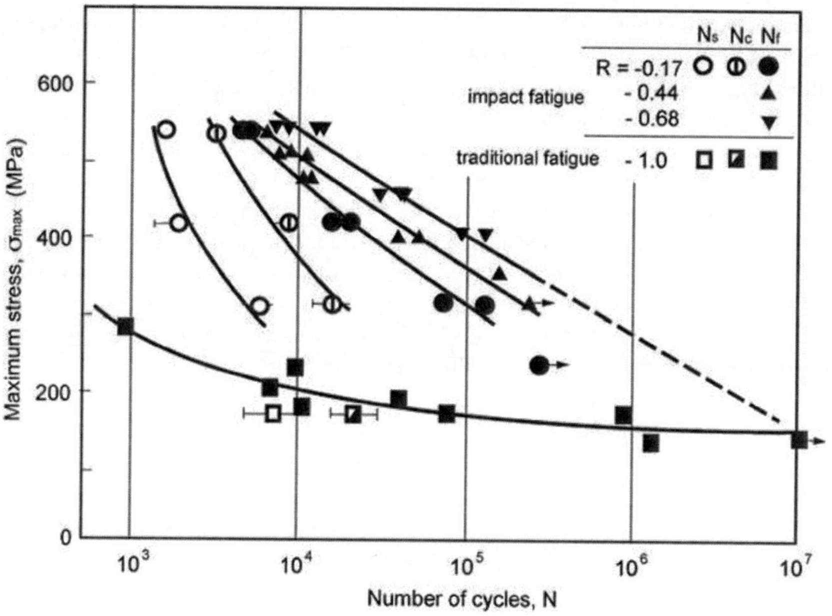

Fig. 12.48 Comparison between traditional fatigue and impulsive fatigue on smooth specimens of ductile AISI 1020 steel subjected to cycles of Fig. 12.46 (modified from [39])

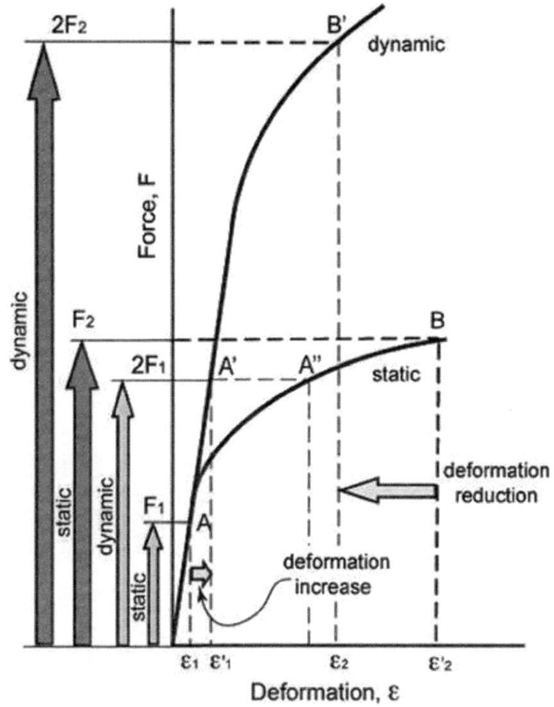

Fig. 12.49 Impact loads increase deformations in the elastic domain, but reduce them in the plastic one

But also the material response increases so that the working point will be A′ and the corresponding deformation ε_1', lower than that what would be obtained by applying a force $2F_1$ statically (point A″). Nevertheless, the elastic deformation increases from ε_1 to ε_1' producing more fatigue damage. But when the applied static force increases to F_2 causing the material to enter into the plastic domain and reach a plastic deformation ε_2, point B of Fig. 12.49, its dynamic application with an impact factor two, $2F_2$, will find a much more resisting material, point B′. This time the corresponding deformation ε_2' will be lower than ε_2 and the fatigue damage will reduce. This actually means that the beneficial effect, if any, of impact fatigue observed on mild steel will be obtained only in the low cycle region of fatigue.

In effect, plastic deformations produced by impulsive loads can be even 10 times lower than the corresponding deformation obtained under a static load of equal value. The damage mechanism of impact fatigue is the same as in conventional fatigue, at least up to a certain strain rate. Persistent slip bands are formed within which damage nucleates. Figure 12.50 is an interesting example of persistent slip band generated in a titanium alloy by impulsive loading in which voids nucleate, Fig. 12.50a, and coalesce forming a micro-crack, Fig. 12.50b [41]. This is similar to Fig. 1.43 already seen for an aluminum alloy under static stress. Recently, Tanaka et al. [42] working with smooth specimens of high ductility, high toughness and low strength Sn-40Pb alloy obtained results shown in Fig. 12.51. Also in this case it is possible to observe a clear improvement of fatigue strength under impulsive cyclic loads, with respect to conventional fatigue applied at 10 Hz. As said, this beneficial effect, if any, is observed only on soft and ductile materials and up to a certain load speed (strain rate).

If high strength materials are involved, then it should be expected a remarkable reduction of the fatigue response. In fact, the embrittling effect of the dynamic load prevails over the strain rate hardening effect because in these alloys there is very little room left to any further hardening. The rapidly applied load can jeopardize any possible dislocation movement that will not allow any slip band formation, but results in a diffused micro-crack formation of the type shown in Fig. 12.52 [41]. This actually means that the micro-void nucleation, coalescence and micro-crack formation phases are bypassed because micro-cracks are immediately introduced in the material.

The brittle tendency of hard materials favors the micro-fracturing of the crystal eliminating Stage I of fatigue. This different behavior between soft and ductile materials and hard and brittle materials is now well documented [42–46]. Figure 12.53 [47] presents results obtained by Okabe et al. in terms of fatigue limit versus the

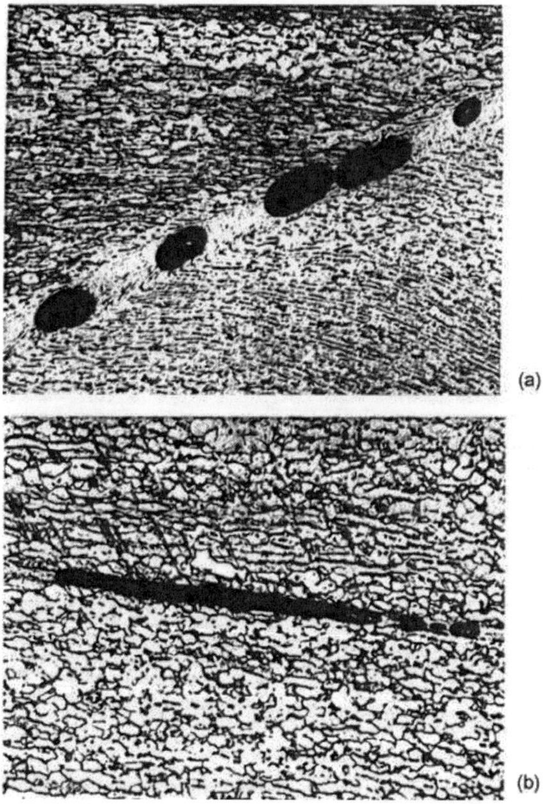

Fig. 12.50 **a** voids formation and **b** coalescence in an adiabatic shear band in titanium alloy under impulsive loads [41]

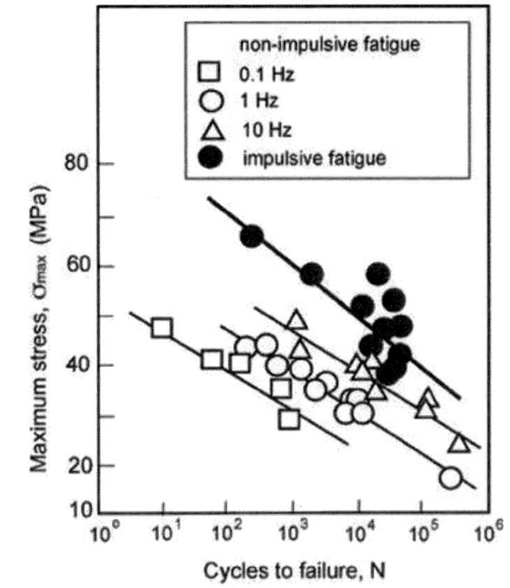

Fig. 12.51 Improvement of fatigue response of an Sn-40Pb alloy under impulsive loadings [42]

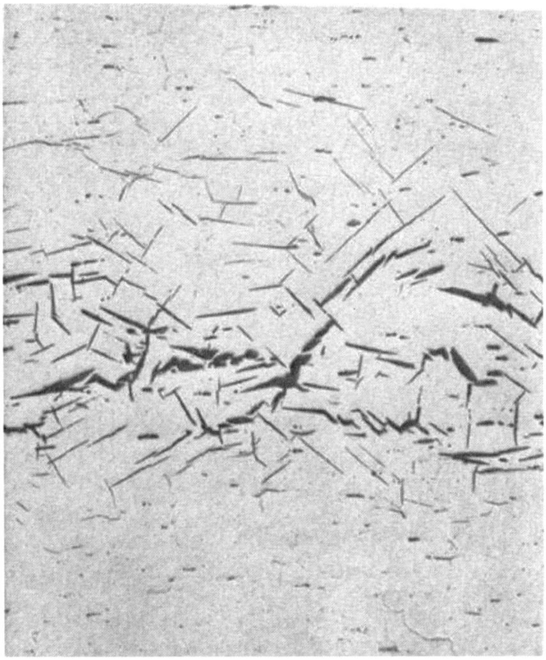

Fig. 12.52 Micro-cracks formation following an impact loading by detonation [41]

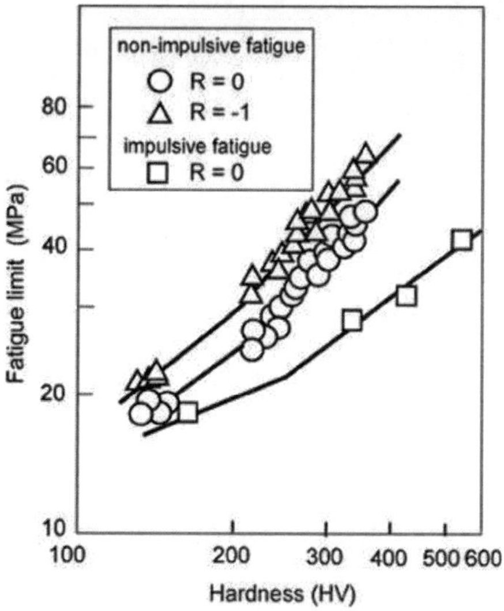

Fig. 12.53 Comparison between ordinary fatigue and impact fatigue in an S45C alloy as function of its hardness [47]

hardness of a S45C alloy. Observe how the impact fatigue becomes more and more devastating with respect to conventional fatigue as the hardness increases. However, in many cases it is not possible at all to evidence a real fatigue limit.

12.7 Sample Problems

12.7.1 Sampling a Random Time History

The probability density function (PDF) of a time history $X(t)$ can be obtained by calculating its statistical properties. Figure 12.54 shows a sample time history for a random process, $x(t)$, during the time interval T, where $X(t)$ exits between the values of x and $x + dx$ for a total time of $t_{tot} = (dt_1 + dt_2 + dt_3 + dt_4 + dt_5)$. The probability P that $x \leq X(t) \leq x + dx$ is therefore.

$$P = \frac{dx_1 + dx_2 + dx_3 + dx_4 + dx_5}{T}$$

.

If the duration T is long enough, the *pdf* (x) is given by.

$$p(x) = \frac{dx_1 + dx_2 + dx_3 + dx_4 + dx_5}{T} = \frac{\sum_{i=1}^{\infty} dt_i}{T}$$

.

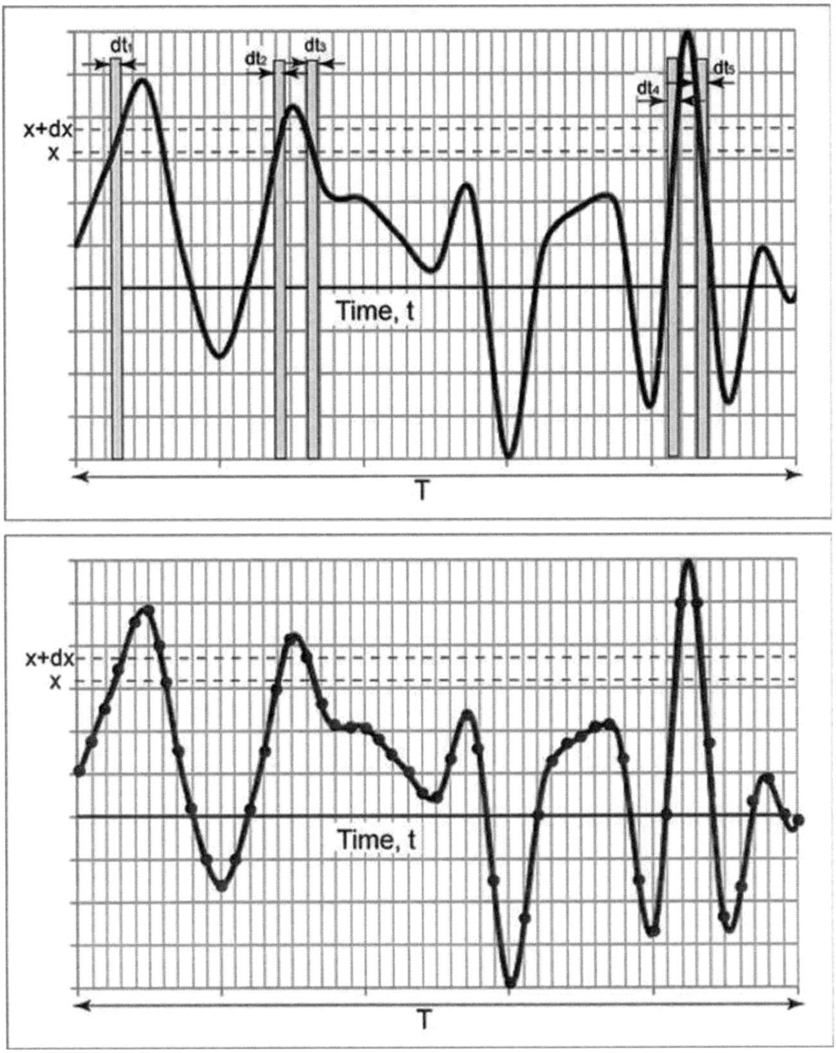

Fig. 12.54 Sampling a time history to infer the relative *pdf*. **a** continuous, **b** digitalized

The expression of $p(x)$ is correct if the time duration T goes to infinity that is to say if the sample time history continues indefinitely. Measurement of the time segments dt_i and their subsequent summation, Σ, may be really very cumbersome. Alternatively, the *pdf* can be determined by the fraction of the total number of samples in the band between x and $x + dx$. This can be done by digitizing the time history at a certain sampling rate in the time interval T as shown in Fig. 12.54b. The *pdf* is then given by

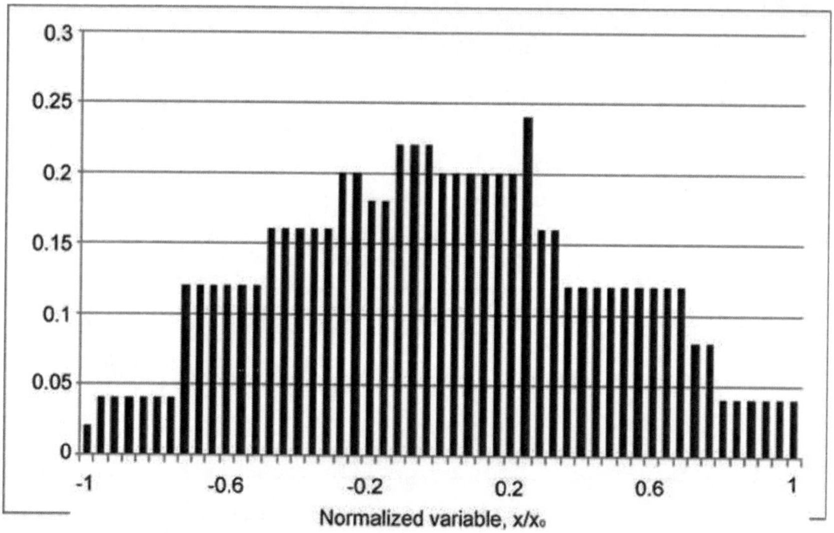

Fig. 12.55 Probability density function of the random variable $x(t)$ of Fig. 12.54 obtained by sampling the relative time history

$$p(x) = \frac{\#sample_{band}}{\#sample_{total}} = \frac{3}{50}$$

where the band is x and $x + dx$, as shown in Fig. 12.54b. the result is shown in Fig. 12.55.

Once the *pdf*, $p(x)$, of the random variable x is known also the mean value of the random process μ and the variance Δ^2 are known

$$\begin{aligned} \mu &= \int_{-\infty}^{\infty} xp(x)dx \cong \frac{1}{T}\sum_{o}^{T} X(t)dt \\ \Delta^2 &= \int_{-\infty}^{\infty} (x-\mu)^2 p(x)dx \cong \frac{1}{T}\sum_{o}^{T} (X(t)-\mu)^2 dt \end{aligned} \tag{12.88}$$

12.7.2 Fatigue Damage-Block-Package Method

The gears of a track are made of the same steel considered in sample problem 4 of Sect. 7.9. The vehicle has been tested on a one kilometer long test trek and the corresponding unit block-program for the fifth-gear derived. It is made of four packages of torque values T_i listed in Table 12.1. The corresponding stress amplitudes on the root of tooth are also shown in Table 12.1. Assess the life of teeth.

The steel is a MoCr type 20MoCr4 cemented and shot peened with a yield strength $\sigma_y = 819$ MPa and ultimate $\sigma_u = 1800$ MPa. Since the loading cycle is in released tension ($R = 0$) the mean stress effect must be considered on fatigue strength by modifying (reducing) the fatigue limit according to Eq. (12.5.48)

$$\sigma'_{f,m} = \sigma'_f\left(1 - \frac{\sigma_m}{\sigma_r}\right) = 819 \cdot \left(1 - \frac{410}{1800}\right) = 632 \text{ MPa}$$

The new *S–N* curve is the full line of Fig. 12.48 whereas the dotted line represents the *S–N* curve under fully reversed condition ($R = -1$). It can be seen how the two lines merge at 10^3 cycles since mean stress is losing its effect in low cycle fatigue. The slope $1/b$ of the line is equal to $-1/0.129 = -7.75$. Since there will be stress amplitudes below the fatigue limit that can have an effect because of the variable amplitude nature of loads (see Sect. 12.3.3) the Haibach correction must be applied, see Sect. 5.2.2, Eq. (12.5.35). The Haibach line is

$$N = N_f\left(\frac{\sigma_a}{\sigma'_f}\right)^{(2k-1)} = 10^6 \cdot \left(\frac{\sigma_a}{632}\right)^{-16.5} = 1.6 \times 10^{52}(\sigma_a)^{-16.5}$$

The Haibach line is shown in Fig. 12.56. Table 12.1 list the corresponding life cycles N_i below the fatigue limit and the ratios h_i/N_i. Equation (12.43) yields the number m of allowable blocks

$$m = \frac{1}{\sum_{i=1}^{j} \frac{h_i}{N_i}} = \frac{1}{5,3 \times 10^{-6} + 4,65 \times 10^{-8} + 1.35 \times 10^{-9} + 1.9 \times 10^{-14}} = 186.991$$

The gear, then, can withstand about 187,000 km. Note that the first package of cycles alone can be tolerated by the gear for 188,000 cycles which actually means the remaining three packages reduce the life of just 1000 km. The gear life is determined by the amplitude of the first package.

Table 12.1 Unit block-program with torque values, cycles h_i, stress amplitude σ_a and allowable cycles *Ni*

Package	Torque (Nm)	h_i	σ_a (MPa)	N_i	h_i/N_i
1	3300	4	550	7.5×10^6	5.3×10^{-6}
2	2500	20	416	$4.3\ 10^8$	4.65×10^{-8}
3	1650	230	275	1.7×10^{11}	1.35×10^{-9}
4	800	60	140	3.1×10^{15}	1.9×10^{-14}

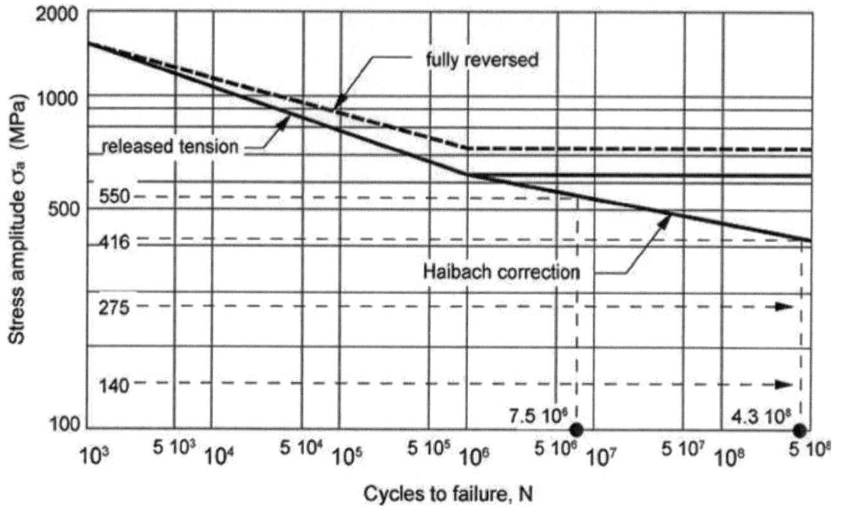

Fig. 12.56 *S–N* curve under completely reversed (dotted line, $R = -1$) and released tension (full line, $R = 0$) conditions and Haibach correction for the gear material

12.7.3 Block-Program-Volume Effect

A cantilever beam is 300 mm long, with a rectangular section of 100 × 10 mm. The beam is made of carbon steel already examined in sample problem 5 of Sect. 4.7.5 and is loaded at the edge by three different load histories A, B and C with $R = 0$ as follows.

(A) $F_A = 1650$ N *for* 20,000 cycles
(B) $F_B = 1350$ N *for* 80,000 cycles
(C) $F_C = 1100$ N *for* 1,000,000 cycles

The three load histories constitute the unit block-program. Knowing that 1000 work pieces will be built, calculate the number m of blocks that can be tolerated by the beam without fatigue failure with 99.9% of probability of survival $P_s = 1$- P, P being the cumulative probability function. The three moments at the fix end of the cantilever beam corresponding to the three loads are.

(A) $M_A = 495$ kNxmm *for* 20,000 cycles
(B) $M_B = 405$ kNxmm *for* 80,000 cycles
(C) $M_C = 330$ kNxmm *for* 1,000,000 cycles

The stress state corresponding to the three moments are

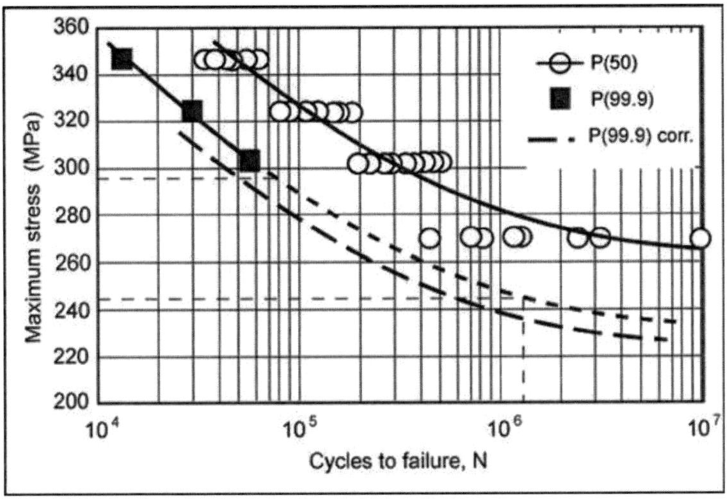

Fig. 12.57 Reference *P(50) S–N* curve, *P(99.9)S–N* curve and *P(99.9)S–N* curve corrected for volume effect

$$\sigma_A = \frac{M_A \cdot y}{I} = 6 \times \frac{M_A}{b \cdot h^2} = 6 \times \frac{495.000}{100 \times 10^2} = 297 \text{ MPa}$$

$$\sigma_B = 6 \times \frac{405.000}{100 \times 10^2} = 243 \text{ MPa}$$

$$\sigma_C = 6 \times \frac{330.000}{100 \times 10^2} = 198 \text{ MPa}$$

Reference is made to Fig. 4.27 that follows reported as Fig. 12.57. The allowable cycles for each load history are those of curve $P_s(99.9)$.

Then, it follows

(A) $N_A = 80{,}000$ cycles
(B) $N_B = 1{,}200{,}000$ cycles
(C) $N_C \cong \infty$ cycles

Note that it has not been considered the Haibach correction as in the previous problem, but it should. The number of allowable blocks is

$$m = \frac{1}{\sum_{i=1}^{3} \frac{h_i}{N_i}} = \frac{1}{\frac{20{,}000}{80{,}000} + \frac{80{,}000}{1{,}200{,}000} + \frac{1{,}200{,}000}{\infty}} = 3.15$$

The beam, then, could tolerate just three unit blocks. However, it must be understood that the *S–N* curve 99,9% probability of survival has been obtained using four series of 10 specimens having 50 mm^2 section and 50 mm long, respectively, to which it correspond an effective process volume of

$$V_{eff} \cong 0.5 \cdot 10 \cdot \pi \cdot D \cdot l = 0.5 \cdot 10 \cdot \pi \cdot 4 \cdot 50 = 3140 \text{ mm}^3$$

Note that in the assessment of the effective process volume it has been considered only a surface layer 0.5 mm thick because it is only that surface layer that takes place in the high cycle fatigue process. The beam under consideration has a much larger process volume. The bending stress reaches its maximum at the fix end on the farthest fibers from the neutral axis. The 90% of the maximum stress criterion is adopted (see Sect. 7.5) to evaluate the process volume. Stresses maintain equal or larger than 90% of the maximum value within a distance d from the surface equal to

$$d = 0.1 \cdot s/2 = 0.05 \cdot \text{s} = 0.5\text{mm}$$

with s equal to the width of the beam. The distance l from the fix end at which stress is equal to 90% of the maximum is

$$l = 0.1 \cdot L = 0.1 \cdot 300 = 30\text{mm}.$$

therefore, the effective volume $V_{(90\%)}$ is

$$V_{(90\%)} = 100 \cdot d \cdot l = 100 \cdot 0.5 \cdot 30/2 = 1500 \text{ mm}^3$$

Since it is expected that 1000 work piece will be built the total volume to be considered is $V_{total(90\%)} = 1{,}500{,}000 \text{ mm}^3$. The volume effect will be assessed using Eq. (12.4.48) by writing

$$\frac{\sigma_{f,90\%}}{\sigma_{V_{eff}}} = \left(\frac{1.5 \times 10^6}{3140}\right)^{-1/m} \approx (477.70)^{-1/m}$$

For the steel considered the Weibull exponent $m = 36$ so that the value of the new fatigue limit and S–N curve that take into account the volume effect shall be lowered by a factor $477{,}7^{-1/36} = 0.87$. This is shown in Fig. 12.57. The new allowable number of blocks is

$$m = \frac{1}{\sum_{i=1}^{3} \frac{h_i}{N_i}} = \frac{1}{\frac{20.000}{50.000} + \frac{80.000}{600.000} + \frac{1.000.000}{\infty}} = 1.87$$

New results indicate that because of the volume effect we shall expect that the allowable number of blocks would be less than two. Since the probability of survival is 99.9% it must be expected that in about two blocks one beam out of 1000 will fail.

12.7.4 Block-Program-Volume Effect-Henry Approach

Reference is made to the previous sample problem 3. Calculate the minimum section of the beam to withstand with 99.9% probability of survival at least 10 unit reference blocks using Miner's rule and Marin approach. Let's proceed by trial and error. Increase the section thickness from 10 to 11 mm. the stress state will be

$$\sigma_A = \frac{M_A \cdot y}{I} = 6 \times \frac{M_A}{b \cdot h^2} = 6 \times \frac{495.000}{100 \times 11^2} = 245\ \text{MPa}$$

$$\sigma_B = 6 \times \frac{405.000}{100 \times 11^2} = 201\ \text{MPa}$$

$$\sigma_C = 6 \times \frac{330.000}{100 \times 11^2} = 164\ \text{MPa}$$

The new fatigue lives will be

(A) $N_{245} = 400{,}000$ cycles
(B) $N_{201} = \infty$ cycles
(C) $N_{164} = \infty$ cycles.

Therefore

$$m = \frac{1}{\sum_{i=1}^{3} \frac{h_i}{N_i}} = \frac{1}{\frac{20.000}{400.000} + \frac{80.000}{\infty} + \frac{1.000.000}{\infty}} = 20$$

Just increasing the thickness by one millimeter has resulted in a 10 folds increase in fatigue life. Let's use, now, the Henry approach described by Eq. (12.23). Apply the 245 Mpa stress amplitude package first. The new reduced fatigue life is

$$S_{f,A} = \frac{245 \times \left(1 - \frac{20 \times 20{,}000}{400{,}000}\right)}{\left(\frac{245-220}{220}\right) + \left(1 - \frac{20 \times 20{,}000}{400{,}000}\right)} = 0\ \text{MPa}$$

The first package of cycles will reduce the fatigue limit to zero, which actually means that there is no room for any other package of cycles. According to Henry, then, the beam cannot withstand 20 unit blocks. Let' try to reduce the number of blocks to 10, then

$$S_{f,A} = \frac{245 \cdot \left(1 - \frac{10 \times 20{,}000}{400{,}000}\right)}{\left(\frac{245-220}{220}\right) + \left(1 - \frac{10 \times 20{,}000}{400{,}000}\right)} = 199.6\ \text{MPa}$$

$$S_{f,B} = \frac{199.6 \times \left(1 - \frac{10 \times 80{,}000}{2{,}000{,}000}\right)}{\left(\frac{201-199.6}{199.6}\right) + \left(1 - \frac{10 \times 80{,}000}{2{,}000{,}000}\right)} = 197\ \text{MPa}$$

This time the beam will withstand the fatigue cycles. Increasing the number of blocks to 13 it yield

$$S_{f,A} = \frac{245 \cdot \left(1 - \frac{13\times 20{,}000}{400{,}000}\right)}{\left(\frac{245-220}{220}\right) + \left(1 - \frac{13\times 20{,}000}{400{,}000}\right)} = 185\ \text{MPa}$$

$$S_{f,B} = \frac{201 \cdot \left(1 - \frac{13\times 80{,}000}{1{,}004{,}000}\right)}{\left(\frac{201-185}{185}\right) + \left(1 - \frac{13\times 80{,}000}{1{,}040{,}000}\right)} = 0\ \text{MPa}$$

which actually means that the second package will saturate the fatigue life and no further room is left to the third package. Therefore, according to Henry the number of allowable blocks will be 12 and not 20 as predicted by Miner's rule. This finding should not come as a surprise. Miner's rule, in fact, does not consider the order of application of loads that, in the case examined, was HiLo.

Consider, to this purpose, the inverse order of application of cycle packages: LoHi and use the Henry approach. The 164 Mpa package will be applied first and will not introduce any damage because the stress amplitude is below the fatigue limit. The same with the 201 Mpa cycles. When the 245 Mpa package will be finally applied it will not find any damage at all in the material. If its 20,000 cycles are applied 20 times it will make a total of 400,000 cycles at 245 Mpa. But at 245 Mpa the life N is equal to about 400,000 cycles or something more with the result that 20 blocks of LoHi order of application will be tolerated, as predicted by Miner's rule.

12.7.5 Wide-Band Time History Fatigue Analysis

A component is made of carbon steel considered in sample problems 1 and 2 of Sect. 5.3. The fatigue strength coefficient σ'_f and the strain hardening exponent n' are 1061 MPa and − 0.165, respectively. The stress time history at a critical location is shown in Fig. 12.58a.

The corresponding PSD is that of Fig. 12.58b. The PSD has three main frequencies of 1, 10 and 30 Hz, respectively, corresponding to 25,000, 20,000 and 9,000 Mpa^2/Hz, respectively. Determine the fatigue life of the component. The three frequencies that compose the time history are quite far from each other, 1, 10 and 30 Hz, respectively. The value of the irregularity factor γ and of the λ parameter (see Sect. 12.2.1) is assessed by Eq. (12.84) knowing the three moments M_o, M_2 and M_4.

$M_o = 54{,}000$; $M_2 = 10{,}125{,}000$; $M_4 = 7{,}490{,}025{,}000$

$$\gamma = \frac{M_2}{\sqrt{M_o M_4}} = 0.50;\ \lambda = \sqrt{1-\gamma^2} = \sqrt{1 - \frac{M_2^2}{M_o M_4}} = 0.87$$

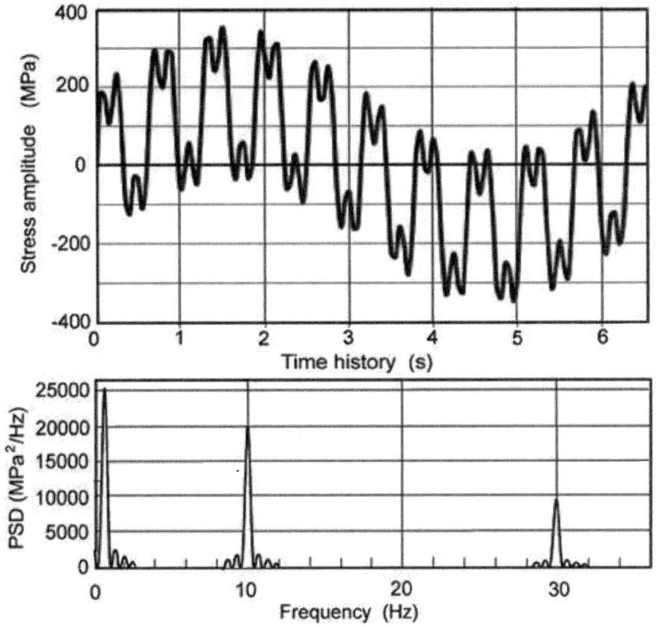

Fig. 12.58 **a** time history of the spectrum affecting the component and **b** PSD of the spectrum

confirming that the time history is closer to a wide-band process. The square root of the area under each curve in the PSD diagram is the RMS of the stress amplitude times the crest factor, see Eq. (12.79). Therefore, it is

$$\sigma_{a,1} = \sqrt{0.5 \cdot 2 \cdot 25,000} \cdot \sqrt{2} = 158 \text{ MPa}$$
$$\sigma_{a,2} = \sqrt{0.5 \cdot 2 \cdot 20,000} \cdot \sqrt{2} = 141.4 \text{ MPa}$$
$$\sigma_{a,3} = \sqrt{0.5 \cdot 2 \cdot 9,000} \cdot \sqrt{2} = 95 \text{ MPa}$$

Knowing the stress amplitudes the lives N_1, N_2 and N_3 can be inferred from the Basquin Eq. (5.22)

$$N_1 = \left(\frac{\sigma_{a,1}}{\sigma'_f}\right)^{1/b} = \left(\frac{158}{1061}\right)^{-1/0.165} = 1,394,135 cycles$$

$$N_2 = \left(\frac{\sigma_{a,2}}{\sigma'_f}\right)^{1/b} = \left(\frac{141.4}{1061}\right)^{-1/0.165} = 4,789,020,481 cycles$$

$$N_3 = \left(\frac{\sigma_{a,3}}{\sigma'_f}\right)^{1/b} = \left(\frac{95}{1061}\right)^{-1/0.165} = 396,357,043,568$$

Table 12.2 Values of moments M_i, probabilities $f(\sigma i)$ and correction factor ζo

M_o	M_2	M_4	M_m	M_{m+2}	$f(\sigma_1)$	$f(\sigma_2)$	$f(\sigma_3)$	ζo
54,000	10,125,000	7.49E + 09	71,975.66	18,040,241	0.000776	0.000686	0.000448	1.717

Table 12.3 Probabilities $f(\sigma i)$ associated with stress amplitude σi

$f(\sigma_1)$	$f(\sigma_2)$	$f(\sigma_3)$
0.000874	0.000939	0.001147

In a 1-s time interval, the sine waves at 1, 10 and 20 Hz represent 1 cycle, 10 cycles and 20 cycles, respectively. Then, using the Miner's rule the total linear damage calculation for this time interval is

$$D = \frac{n_1}{N_1} + \frac{n_2}{N_2} + \frac{n_3}{N_2} = \frac{1}{1,394,135} + \frac{10}{4,789,020,481} + \frac{20}{396,357,043,568} = 2.88 \times 10^{-9}$$

that corresponds to 11 years (347,090,924 = 1/2.88·10^{-9} s). Now we can calculate the component life using option A approach of Fig. 12.38 with the Oritz and Chen correction factor ζ_O given by Eq. (12.59). First we need know the probabilities $f(\sigma_i)$ of the process and the moment M_i. All calculated data necessary to assess damage are reported in Table 12.2.

Damage, given by Eq. (12.59), is

$$D = \zeta_O \sum_{i=1}^{s} n_i \sum_{i=1}^{s} f_i \left(\frac{\sigma_{a,i}}{\sigma'_f} \right)^{-1/b} = 2.81 \times 10^{-9}$$

which is exactly the value obtained with Miner's solution. Let's now use option B approach of Fig. 12.38 with the Dirlik solution given by Eq. (12.60). the probabilities related to the thee stress amplitude are listed in Table 12.3.

Therefore, the damage according to Dirlik is

$$D = 0.91 \times 10^{-9}$$

which correspond to 34 years life, three times the life calculated with both Miner's rule and Oritz approach.

References

1. Petrone, N., Saraceni, M., Crugnola F.: Sviluppo di un simulatore di strada per prove complesse su forcelle motociclistiche. Associazione Italiana Analisi delle Sollecitazioni, XXXIV Convegno Nazionale, Milano, pp. 14–17 settembre (2005)
2. Schijve, J.: Fatigue of Structures and Materials. Kluwer Academic Publishers (2004)
3. Matsuishi, M., Endo, T.: Fatigue of metals subjected to various stress. In: Preceedings of JSME, Fukuoka, Japan, 19612.
4. Dowling, N.E.: Fatigue at notches and the local strain and fracture mechanics approaches. Am Soc Testi Materials ASTM **677** (1979)
5. Palmgren, A.: The fatigue life of roll-bearings (in German) Z. Ver. Deut. Ing. **68**, 339–341 (1924)
6. Langer, B.F.: Fatigue failure from stress cycles of varying amplitude. J Appl Mech 59:A160–A162 (1937). Table 2.4 Four-Level Step-Stress Fatigue Test Data
7. Miner, M.A.: Cumulative damage in fatigue. J. Appl. Mech. **12**(A), 159 (1945)
8. Dolan, T.J., Richart, F.E., Work, C.E.: Influence of fluctuations in stress amplitude on the fatigue of metals. ASTM Proc. **49** (1949)
9. Marco, S.M., Starkey, W.L.: ASME Trans. **76**, 627 (1954)
10. Corten, H.T., Sinclair, G.M., Dolan, T.J.: ASTM Proc. **54**, 737 (1954)
11. Plantema, F.J.: Colloquium on Fatigue, p. 218. Springer-Verlag, Berlin (1956)
12. Miner, M.A.: Estimation of fatigue life with particular emphasis on cumulative damage. In: Sines, G., Waisman, J.L. (eds.) Metal Fatigue, Chap. 12, McGraw-Hill Book C., New York (1959)
13. Richart, F.E., Newmark, N.M.: Proceedings ASTM **48**, 767, 19412.
14. Newmark, N.M.: Fatigue and Brittle Fracture, John Wiley, New York, p. 197 (1950)
15. Ahrensdorf, K.: "Fatigue Design Practice", Specialist Meeting on design against Fatigue, Conference Proceedings AGARD –CP-141, pp. 2–5 (1973)
16. Jacoby, G.: Vergleich der Lebensdauer aus Betriebsfestigkeits, Einzelflug und Digital Programmierten Random-Versuchen sowie nach der Linearen Shadens-Accumulationshypothese", VDI-Z Fortschritts-Berichte: Labensdaueranalyse bei Unregelmaßig Schwankender Beanspruchung (random Load Fatigue) (1969)
17. Corten, H.T., Dolan, T.J.: Cumulative fatigue damage. In: Proceedings of International Conference on Fatigue of Metals, Institution of Mechanical Engineers, ASME and IME, p. 235 (1956)
18. Spitzer, R., Corten, H.T.: Proc. ASTM **61**, 719 (1961)
19. Freudenthal, A.M., Symposium on Acoustic Fatigue, ASTM STP-284, p. 26, 1960.
20. Lemaitre, J., Plumtree, A.: J. Eng. Mater. Tech. **101**, 284–292 (1979)
21. Marin, J.: Mechanical Behavior Of Engineering Materials. Prentice-Hall, Englewood Cliffs, N.J. (1962)
22. Henry, D.L.: Theory of fatigue damage accumulation in steel. ASME Trans. **77**, 913 (1955)
23. Grover, H.J.: An observation concerning the cycle ratio in cumulative damage. ASTM STP **274**, 120–124 (1960)
24. Manson, S.S., Halford, G.R.: Practical implementation of the double linear damage rule and damage curve approach for testing cumulative fatigue damage. Int. J. Fract. **17**(2), 169–192 (1981)
25. Manson, S.S., Halford, G.R.: Re-examination of Cumulative Fatigue Damage Analysis-An Engineering Perspective. Eng. Fract. Mech. **25**(5/6), 359–371 (1986)
26. Kachanov, L.M.: On the time to failure under creep condition. Int. J. Fracture **97**(1–4), 11–18 (1999)
27. Lemaitre, J., Chaboche, J.L.: Mechanics of Solid Materials. Cambridge University Press (1990)
28. O'Connor, B.P.D., Plumtree, A.: Fatigue crack propagation behavior and damage accumulation relationships in an aluminum alloy. In: Fracture Mechanics: Nineteenth Symposium, ASTM STP 969, pp. 787–799 (1988).

29. Schijve, J.: The accumulation of fatigue damage in aircraft materials and structures. In: Symposium on Random Load fatigue, Conference Proceedings AGARD-CP-118, pp. 3–84 (1972)
30. Gassner, E.: Strength experiments under cyclic loading in aircraft structures. Luftwissen 6:61–64 (in German) (1939)
31. Schijve, J., Jacobs, F.A.: Research on cumulative damage in fatigue of riveted aluminum alloy joints. Nat Aeronautical Research Ins., Amsterdam, Report 1999, January 1956.
32. Naumann, E.C., Hardrath H.R., Guthrie, E.C.: Axial load fatigue tests of 2024-T3 and 7075-T6 aluminum alloy sheet specimens under constant and variable-amplitude-loads, Report TN D-212, NASA (1959)
33. Wirsching, P.H., Light, M.C.: Fatigue under wide band random stresses. ASCE J Struct Div **106**, 1593–1607 (1980)
34. Oritz, K., Chen, N.K.: Fatigue damage prediction for stationary wide-band stresses. In: Presented at the 5th International Conference on the Applications of Statistics and Probability in Civil Engineering, Vancouver, Canada, (1987)
35. Dirlik, T.: Application of Computers in Fatigue Analysis. Ph.D. Thesis, Warwick University (1985)
36. Bishop, N.W.M.: The Use of Frequency Domain Parameters to Predict Structural Fatigue", Ph.D. Thesis, Warwick University (1988)
37. Bishop, N.W.M., Sherratt, F.: Fatigue life prediction from power spectral density data. Part 2: Recent Development". Environ Eng 2(1 and 2), 5–10 (1989)
38. Bishop, N.W.M.: Spectral methods for estimating the integrity of structural components subjected to random loading. In: Carpinteri, A. (Ed.), Handbook of Fatigue Crack Propagation in Metallic Structures, vol. 2, Elsevier, Dordrecht, pp. 1685–1720 (1994)
39. Iguchi, H., Tanaka, K., Taira, S.: Failure Mechanisms in Impact Fatigue of Metals. Fatigue of Engineering Materials and Structures **2**, 165–176 (1979)
40. Manjoine, M.J.: Influence of rate of strain and temperature on yield strength of mild steel. J. of Appl. Mech. **66**, A211–A218 (1944)
41. Meyers, M.A.: Dynamic Behavior of Materials. John Wiley & Sons, New York, pp. 449–450 (1994)
42. Tanaka, T., Nakayama, H.: J. Jap. Soc. Mater. Sci. **23**(252), 678 (1974)
43. Okabe, N., Yano, T., Mori, T.: Trans. Jap Soc Mech Eng. **449A**, 41 (1984)
44. Tanaka, T., Nakayama, H.: J. Jap. Soc. Mater. Sci. **34**(387), 1483 (1985)
45. Nose, H., Sakamoto, I., Nakayama, H.: Mater. Syst. **6**, 181 (1987)
46. Tanaka, T., Nakayama, H., Mori, T.: Int. J. JSME, Ser. 1 **31**(4), 760 (1988).
47. Okabe, N., et al., Fatigue '84, Vol. 2, (U.K.: Engineering Advance Advisory Services ltd), p. 1181 (1984)

Chapter 13
Multiaxial Fatigue

13.1 Introduction

So far fatigue has been treated assuming that simple uniaxial loads were acting. The same *S–N* fatigue curves were obtained under uniaxial conditions, generally a push-and-pull tension or a fully reversed bending. Unfortunately, uniaxial loads are not that common as it would be expected. In most real cases stresses acting on a work piece or on a part of it are not uniaxial even dough the external load is uniaxial. It suffices to consider a component with a hole or a discontinuity where localized triaxial stress state develops. Besides the localized triaxiality, multiaxial loading represents the most general working condition for real components subjected to stresses acting simultaneously in different directions or, which is equivalent, to combined bending and torsion. The latter case is quite common. A torsion arm of a car rear axle or a helical spring or a gear shaft or a crankshaft are typical examples. The most general multiaxial stress state is composed of the three tension stresses σ_x, σ_y e σ_z acting on the three directions x, y and z plus the six shear components τ_{ij}, two per each plane. However, it has been shown in Sects. 1.4 and 1.5 that, at least in high cycle fatigue, damage nucleation always occurs on the free surface of the material in the form of localized slips on persistent slip bands that eventually lead to the macro-crack formation. Generally, the free surface is not interested by any external force so that, in fact, the most common stress state associated with fatigue is the biaxial one and the nine stress components reduce to just three: σ_x, σ_y and τ_{xy}. There are some cases, however, in which the free surface is affected by a stress state, like a pressure vessel, where besides the longitudinal stress σ_l and the hoop stress σ_c there is a third radial component of stress σ_r acting on the inner surface and equal to the pressure. Speaking of multiaxial fatigue it must be distinguished between proportional loads or in-Phase (IP), in which the various stress or strain components change simultaneously of the same quantity so that the principal axes maintain a fixed orientation and non-proportional loadings (NP) in which the stress components are not in phase and the principal axes rotate in time. In terms of fatigue

P. P. Milella, *Fatigue and Corrosion in Metals*,
https://doi.org/10.1007/978-3-031-51350-3_13

damage, proportional loading is often considered as equivalent to uniaxial loading on the basis of effective stress and strain. But there are important differences in the formation of small fatigue cracks among different stress states, even for proportional loading. In uniaxial straining there is an infinite number of octahedral and maximum shear planes making equal angles with the axis of tension–compression loading. In contrast, pure torsion may be identified with a single set of octahedral or maximum shear planes. A typical case of rotating torsion and bending out-of-phase resulting in a non-proportional stress state is offered by the crankshaft of any internal combustion engine. An example of proportional loads is given in Fig. 13.1 that shows the two components of longitudinal and circumferential strain measured by two strain gauges at 90° between each other on the external free surface of a pressure piping subjected to pulsating pressure variations on a constant service pressure. An example of non-proportional loads is given in Fig. 13.2 for a cylindrical axle subjected to 75° out-of-phase alternating traction F and torque T. Stresses σ_x and τ_{xy} are normalized to one. Note in Fig. 13.2b that the principal directions are continuously changing with the time history of Fig. 13.2a. In A (pure shearing, $\sigma_x = 0$) principal planes are at 45° with respect to x and y axes and rotate continuously. Figure 13.3 presents the Mohr's circle of the four stress states A, B, C and D of Fig. 13.2a and the corresponding principal planes. It must be kept in mind that the continuous rotation of the principal directions results in material challenging along all possible directions. Therefore, the anisotropy always possible in a material plays a fundamental role in the fatigue strength assessment since the material will fail along a direction that minimizes the ratio between the fatigue strength and some cyclic equivalent stress. But there is also another effect that may take place as a consequence of the rotation of principal directions. This is the interaction among the many active slip planes that may lead to additional hardening of the material. The first attempts to investigate the effects of multiaxial fatigue dates back to the end of XIX century when Lanza [1] published results of tests concerning combined bending and torsion loadings.

In the early decades of last century researchers like Manson [2], Haigh [3], Nishiara and Kavamoto [4] and Gough [5] and Gough et al. [6] proposed empirical relations based on experimental data. However, a real systematic study of multiaxial fatigue did not started before the early 1970s because of costs and difficulties connected with experimental procedures. This has resulted in a relatively paucity of experimental data. But the importance of multiaxial fatigue has been lately recognized and the subject given the proper attention both from theoretical and experimental point of view, as well. Often, because of difficulties connected with the analysis of cyclic loads not-in-phase it is assumed that they are always in-phase. This assumption is still under questioning since it may be conservative in some cases and non-conservative in other, as will be described in Sect. 13.8.

In the following sections the state-of-the-art of fatigue correlations for such combined stress states will be briefly reviewed. Fatigue crack "*initiation*" parameters are discussed, ranging from simple equivalent stress and strain concepts to triaxiality factor (FT) based approach to the more recent critical plane theories. Also differences between low-cycle fatigue (LCF) and high-cycle fatigue (HCF) behaviors are discussed. Finally, several other features of multiaxial fatigue are considered,

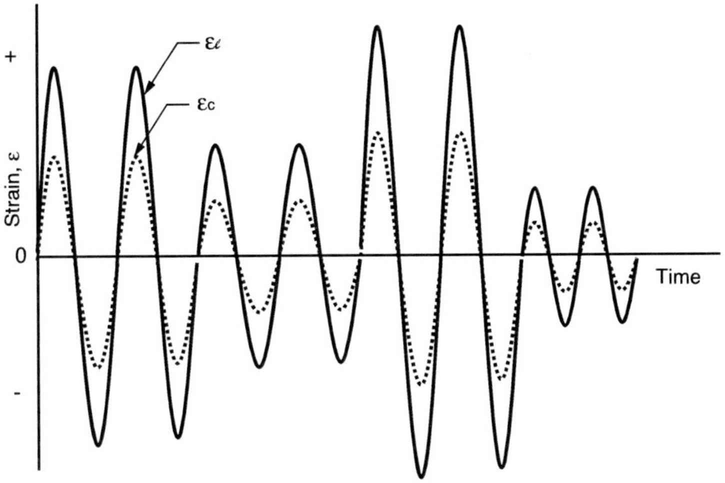

Fig. 13.1 Proportional loads acting in a pressure pipe subjected to pressure excursions

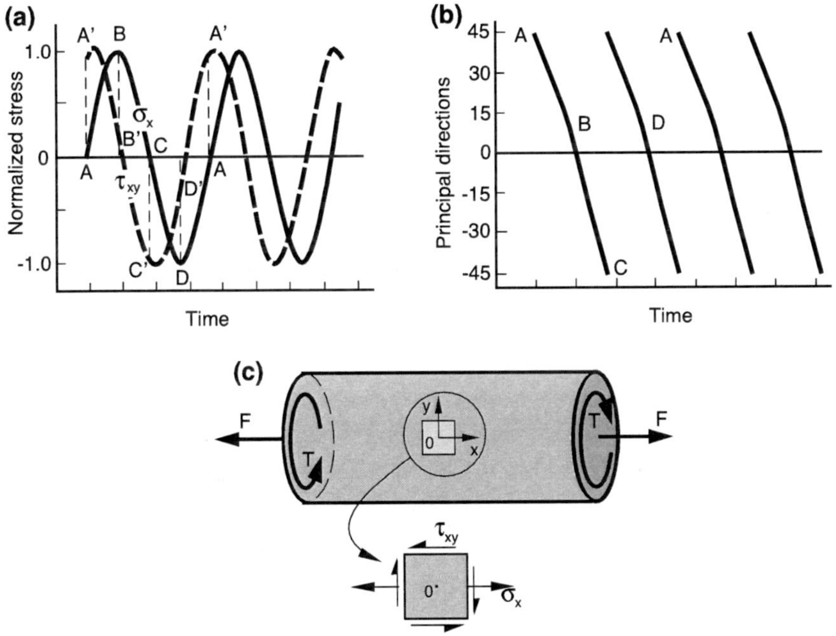

Fig. 13.2 Non-proportional loads in an axle subjected to cyclic traction F and torque T out-of-phase that **a** generate out-of-phase stress components. **b** Principal directions are continuously. **c** stress state generated by force F and torsion T

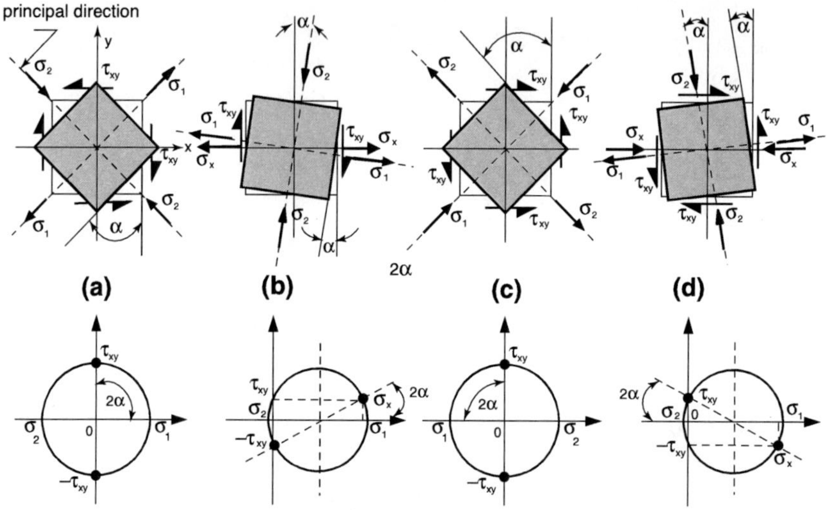

Fig. 13.3 Principal directions and Mohr's circles for the four reference positions **a**, **b**, **c** and **d** of Fig. 13.2

including mean stress effects, sequences of stress/strain amplitude or stress state, non-proportional loading and HCF limits. It must be clear that by the term *initiation* it is intended to refer to the first formation of a macro-crack, generally in HCF on the external surface, that crack is no longer than 300 to 400 μm (see Sect. 1.5). Cracks of larger sizes will be treated in Chap. 10 by means of Fracture Mechanics. So far no consensus has yet been reached on the best approach to predict fatigue failure, in particular under multiaxial out-of-phase stresses.

13.2 Failure Theories

The various approaches can be divided into five groups based on empirical equivalent stress, stress triaxiality, stress invariants, average stress and critical plane stress, respectively. The initial theories proposed to predict fatigue failure under combined loading belonged to the first group and were basically an extension of the existing failure theories for static multiaxial state of stress such as the maximum normal stress theory or the maximum shear stress or Tresca theory or the von Mises theory. The idea, in fact, was to enter the uniaxial fatigue *S–N* curve with an equivalent stress amplitude, $\sigma_{a,e}$, borrowed from one of those theories. The aim of the following three sections is at discussing precisely these three theories and their application to cyclic stresses and fatigue failure. Note that the most general stress condition in fatigue analysis is represented by a biaxial stress state since fatigue is acting on the free surface of the material, at least in HCF, where the external component of stress is null ($\sigma_z = 0$). The case of in phase loads (IP) is apparently simpler, in particular when no mean stress is acting, and we will start with that.

13.2.1 Maximum Normal Stress

This theory states that failure occurs when one of the three principal stresses, σ_1, σ_2 and σ_3 reaches a permissible strength. Normally, for ductile materials failure theories are rather yield theories so that the permissible strength is, in fact, the yield strength σ_y. For a biaxial stress state the theory is represented by the square domain ABCD of Fig. 13.4a intercepting on the σ_1–σ_2 axes the yield strength in tension σ_y and compression $-\sigma_y$ assumed to be equal (except for the sign). In the case where the yield stress $-\sigma_y$ in compression is not equal but greater than that in traction σ_y, as for *uneven* materials like cast iron, the maximum theory domain becomes that represented by the rectangle AB′C′D′A. Note that, at variance with classical mechanics, in Fig. 13.4 the principal stress σ_1 is assumed to act in the y direction which is, in general, the direction of the specimen axis. This theory is regarded to be valid for real brittle materials whose yield strength σ_y is practically coincident with the ultimate strength σ_u (strain at failure negligible, almost zero).

A modification of the maximum normal stress theory for *uneven* materials where the yield strength in compression is higher than that in traction is the Coulomb-Mohr or internal friction theory in which the failure domain is constructed by connecting

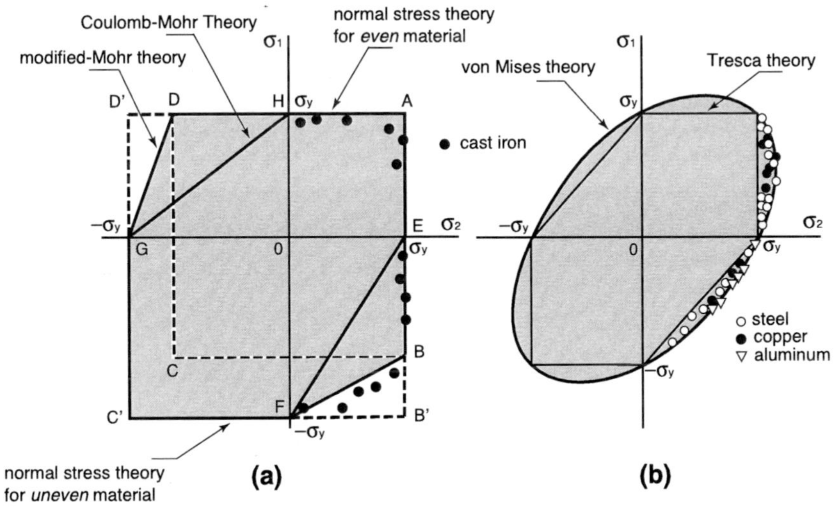

Fig. 13.4 Failure theories in a two-dimensional space and comparison with experimental results; **a** brittle materials having equal or different yield strength in traction and compression, **b** ductile materials

the opposite corners of quadrant I and III. The result is the safety envelop shown in Fig. 13.4a by the hexagon AEFC′GHA. A further variant is the modified Mohr's theory that predicts that a material will fail if a stress state is on the envelope that is tangent to the three Mohr's circles corresponding to uniaxial ultimate stress in tension, uniaxial ultimate stress in compression and pure shear, respectively. The corresponding safety domain is that of Fig. 13.4a by the hexagon ABFC′GDA. This last seems to well predict the behavior of cast iron that effectively has uneven properties in traction and compression. The equivalent stress amplitude $\sigma_{a,e}$ derived from the maximum normal stress criterion knowing the three stress components $\sigma_{a,x}$, $\sigma_{a,y}$ and $\tau_{a,xy}$, is given by

$$\sigma_{a,e} = \frac{\sigma_{a,x} + \sigma_{a,y}}{2} + \frac{1}{2}\sqrt{\left(\sigma_{a,x} - \sigma_{a,y}\right)^2 + 4\tau_{a,xy}^2}. \tag{13.1}$$

For a combination of cyclic traction and cyclic torsion Eq. (13.1) becomes

$$\sigma_{a,e} = \frac{1}{2}\left(\sigma + \sqrt{\sigma^2 + 4\tau_{xy}^2}\right) \tag{13.2}$$

which in terms of principal stresses σ_1 and σ_2 becomes

$$\sigma_{a,e} = \sigma_{1,\max} \quad \text{or} \quad \sigma_{a,e} = \sigma_{2,\max} \tag{13.3}$$

whichever is larger between σ_1 and σ_2. If a mean stress $\sigma_{1,m}$ or $\sigma_{2,m}$ is acting combining the maximum stress theory with the modified Goodman relations (5.62) fatigue failure will occur if

$$\begin{array}{ll} \sigma_{1,\max} - 2\sigma_{1,m} - \sigma_y \geq o & \text{for } -\sigma_y \leq \sigma_{1,m} \leq \left(\sigma_N - \sigma_y\right) \\ \sigma_{1,\max} - \sigma_{1,m} - \sigma_N \geq o & \text{for } \left(\sigma_N - \sigma_y\right) \leq \sigma_{1,m} \leq 0 \\ \sigma_{1,\max} - \sigma_{1,m}(1-r) - \sigma_N \geq o & \text{for } 0 \leq \sigma_{1,m} \leq \left(\frac{\sigma_y - \sigma_N}{1-r}\right) \\ \sigma_{1,\max} - \sigma_y \geq 0 & \text{for } \left(\frac{\sigma_y - \sigma_N}{1-r}\right) \leq \sigma_{1,m} \leq \sigma_y \end{array} \tag{13.4}$$

or if

$$\begin{array}{ll} \sigma_{2,\max} - 2\sigma_{2,m} - \sigma_y \geq o & \text{for } -\sigma_y \leq \sigma_{2,m} \leq \left(\sigma_N - \sigma_y\right) \\ \sigma_{2,\max} - \sigma_{2,m} - \sigma_N \geq o & \text{for } \left(\sigma_N - \sigma_y\right) \leq \sigma_{2,m} \leq 0 \\ \sigma_{2,\max} - \sigma_{2,m}(1-r) - \sigma_N \geq o & \text{for } 0 \leq \sigma_{2,m} \leq \left(\frac{\sigma_y - \sigma_N}{1-r}\right) \\ \sigma_{2,\max} - \sigma_y \geq 0 & \text{for } \left(\frac{\sigma_y - \sigma_N}{1-r}\right) \leq \sigma_{2,m} \leq \sigma_y \end{array} \tag{13.5}$$

with σ_N being the allowable stress amplitude at a given life N or the same fatigue limit σ_f and r the ratio between σ_N and the ultimate strength σ_u, $r = \sigma_N/\sigma_u$. Fatigue design is governed by the most critical among the eight relations (13.4) and (13.5).

13.2.2 Maximum Shear Stress Theory

This theory was first proposed by Tresca in 1865 and experimentally verified by Guest in 1900. It states that in a multiaxial stress state failure occurs when the maximum shear stress exceeds the maximum shear stress at failure in a monotonic tensile traction test. In a tensile test it is

$$\tau_{\max} = \frac{\sigma_y}{2} \tag{13.6}$$

hence for a generic biaxial stress state failure (yielding) occurs when

$$\tau_{\max} = \frac{\sigma_1 - \sigma_2}{2} = \frac{\sigma_y}{2} \tag{13.7}$$

The theory leads to the hexagonal safety domain shown in Fig. 13.4b. It is worth noting that for ductile materials the theory may be conservative, as shown in Fig. 13.4b for steel, copper and aluminum. Combining this theory with the modified Goodman relations (5.62) fatigue failure will occur if

$$\begin{aligned}
&2\tau_{\max} - 4\tau_m - \sigma_y \geq 0 && \text{for } -\sigma_y \leq 2\tau_m \leq (\sigma_N - \sigma_y)\\
&2\tau_{\max} - 2\tau_m - \sigma_N \geq 0 && \text{for } (\sigma_N - \sigma_y) \leq 2\tau_m \leq 0\\
&2\tau_{\max} - 2\tau_m(1-r) - \sigma_N \geq 0 && \text{for } 0 \leq 2\tau_m \leq \left(\tfrac{\sigma_y - \sigma_N}{1-r}\right)\\
&2\tau_{\max} - \sigma_y \geq 0 && \text{for } \left(\tfrac{\sigma_y - \sigma_N}{1-r}\right) \leq 2\tau_m \leq \sigma_y
\end{aligned} \tag{13.8}$$

with σ_N being the allowable stress amplitude at a given life N or the same fatigue limit σ_f and r the usual ratio between σ_N and the ultimate strength σ_u, $r = \sigma_N/\sigma_u$. Fatigue design is governed by the most critical among the four relations (13.8). The equivalent stress amplitude $\sigma_{a,e}$ derived from the maximum shear stress criterion knowing the three stress components σ_x, σ_y and τ_{xy}, is given by

$$\sigma_{a,e} = \sqrt{(\sigma_x - \sigma_y)^2 + 4\tau_{xy}^2} \tag{13.9}$$

that for a combination of cyclic traction and cyclic torsion become

$$\sigma_{a,e} = \sqrt{\sigma^2 + 4\tau^2} \tag{13.10}$$

An important drawback of the Tresca criterion is that it does not consider the effect of the normal stress on the damage progression (see Fig. 13.11). In fact, a component under uniaxial traction σ and another under pure torsion $\tau = \sigma/2$ have the same equivalent stress σ_e but while the former loading condition results in the existence of a stress, σ_n, normal to the plane of $\tau_{\max}$ opening the micro-crack as it forms, the latter condition has no normal stress and the micro-crack remains closed, as

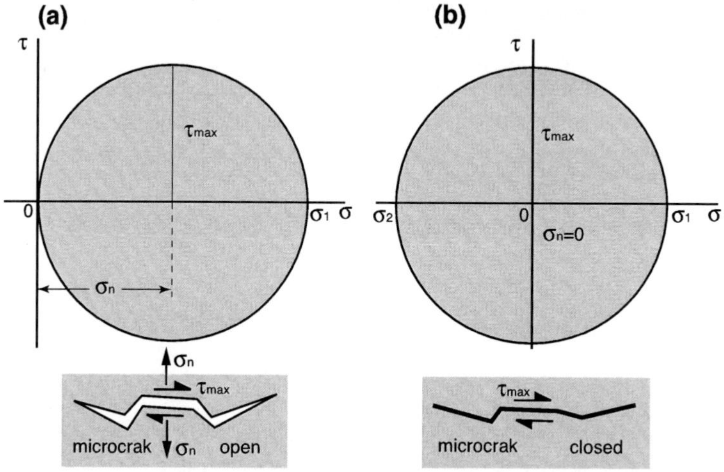

Fig. 13.5 Mohr circle for two stress states **a** uniaxial traction and **b** pure torsion having the same Tresca equivalent stress σ_e and $\tau_{\max}$ but differing in the presence of a stress σ_n normal to $\tau_{\max}$

it can be easily seen from the respective Mohr circles shown in Fig. 13.5. This closure introduces friction and mechanical interlocking effects that retard the micro-crack growth.

13.2.3 Distorsion Strain Energy Theory

Although formulated by Maxwell in 1865 and later in 1904 by Huber who wrote a paper in Polish, this criterion is generally attributed to von Mises and commonly referred to as the Maxwell-Huber-Henky-von Mises theory. The distortion energy theory states that in a multiaxial stress state failure (yielding) occurs when the distortion strain energy in the material equals or exceeds the distortion strain energy of a specimen of the same material at yield in a tensile test. For a general triaxial case (3D-case) the energy u stored in the material as strain energy per unit volume is given by

$$u = \frac{1}{2}(\sigma_1\varepsilon_1 + \sigma_2\varepsilon_2 + \sigma_3\varepsilon_3). \tag{13.11}$$

Applying Hooke's law it becomes

$$u = \frac{1}{2E}\left[\sigma_1^2 + \sigma_2^2 + \sigma_3^2 - 2\nu(\sigma_1\sigma_2 + \sigma_2\sigma_3 + \sigma_1\sigma_3)\right]. \tag{13.12}$$

In this energy there are two components: the hydrostatic or mean stress σ_m and the deviatoric one σ_d given by

$$\sigma_m = \frac{\sigma_1 + \sigma_2 + \sigma_3}{3}$$
$$\sigma_{1,d} = \sigma_1 - \sigma_m;\ \sigma_{2,d} = \sigma_2 - \sigma_m;\ \sigma_{3,d} = \sigma_3 - \sigma_m \tag{13.13}$$

respectively. In an isotropic material the hydrostatic component changes the volume of the material without modifying the shape. At variance, the deviatoric component modifies the shape without changing the volume (Poisson's coefficient ν becomes 0.5). The energy due to the hydrostatic stress σ_m is

$$\begin{aligned} u_m &= \frac{1}{2E}\left[\sigma_m^2 + \sigma_m^2 + \sigma_m^2 - 2\nu(\sigma_m\sigma_m + \sigma_m\sigma_m + \sigma_m\sigma_m)\right] \\ &= \frac{1}{2E}\left[3\sigma_m^2(1-2\nu)\right] \\ &= \frac{1-2\nu}{6E}\left(\sigma_1^2 + \sigma_2^2 + \sigma_3^2 + 2\sigma_1\sigma_2 + 2\sigma_2\sigma_3 + 2\sigma_1\sigma_3\right) \end{aligned} \tag{13.14}$$

Huber noted that the hydrostatic component did not produce failure while Henky deduced that if the hydrostatic component was not responsible for yielding then it should have been the other component and offered a physical interpretation of the criterion suggesting that yielding begins when the elastic energy of distortion reaches a critical value. Therefore, by subtracting the hydrostatic component (13.14) from the total strain energy (13.12) the distortion component u_d is obtained

$$\begin{aligned} u_d = u - u_m &= \frac{1+\nu}{3E}\left(\sigma_1^2 + \sigma_2^2 + \sigma_3^2 - \sigma_1\sigma_2 - \sigma_2\sigma_3 - \sigma_1\sigma_3\right) \\ &= \frac{1+\nu}{6E}\left[(\sigma_1 - \sigma_2)^2 + (\sigma_2 - \sigma_3)^2 + (\sigma_1 - \sigma_3)^2\right] \end{aligned} \tag{13.15}$$

and comparing this distortion energy to the distortion energy u_{ten} stored in a specimen under uniaxial tensile stress at yielding, obtained from Eq. (13.15) by substituting $\sigma_I = \sigma_y$ and $\sigma_2 = \sigma_3 = 0$, it yields

$$u_{ten} = \frac{1+\nu}{3E}\sigma_y^2 = u_d = \frac{1+\nu}{6E}\left[(\sigma_1 - \sigma_2)^2 + (\sigma_2 - \sigma_3)^2 + (\sigma_1 - \sigma_3)^2\right]. \tag{13.16}$$

From Eq. (13.16) it can be derived that failure (yielding) will occur when

$$\sigma_y = \frac{\sqrt{2}}{2}\sqrt{\left[(\sigma_1 - \sigma_2)^2 + (\sigma_2 - \sigma_3)^2 + (\sigma_1 - \sigma_3)^2\right]} = \sigma_{eq}. \tag{13.17}$$

Equation (13.17) defines the equivalent stress σ_{eq} according to the von Mises theory. Under plain stress conditions (biaxial) Eq. (13.17) becomes

$$\sigma_{eq} = \sigma_y = \sqrt{\sigma_1^2 + \sigma_2^2 - \sigma_1 \sigma_2} \tag{13.18}$$

or using the three component of stress σ_x, σ_y and τ_{xy}

$$\sigma_{eq} = \sqrt{\sigma_x^2 - \sigma_x \sigma_y + \sigma_y^2 + 3\tau_{xy}^2}. \tag{13.19}$$

The introduction of the von Mises equivalent stress to fatigue, $\sigma_{a,e}$, requires the substitution of the principal stresses σ_1, σ_2 and σ_3 with the correspondent stress amplitudes $\Delta\sigma_{1a}/2 = \sigma_{1a}$, $\Delta\sigma_{2a}/2 = \sigma_{2a}$ and $\Delta\sigma_{3a}/2 = \sigma_{3a}$. For a combination of cyclic traction and cyclic torsion Eq. (13.19) becomes

$$\begin{aligned} \sigma_{a,e} &= \frac{\Delta\sigma_{a,e}}{2} = \sqrt{\sigma_a^2 + 3\tau_a^2} \\ \varepsilon_{a,e} &= \frac{\Delta\varepsilon_{a,e}}{2} = \frac{1}{\sqrt{2}(1+\overline{\nu})}\sqrt{2\varepsilon_a(1+\overline{\nu})^2 + \frac{3}{2}\gamma_a^2} \end{aligned} \tag{13.20}$$

with

$$\overline{\nu} = \frac{\nu_e \Delta\varepsilon_{a,e}^{el} + \nu_p \Delta\varepsilon_{a,e}^{pl}}{\Delta\varepsilon_{a,e}} \tag{13.20a}$$

where $\varepsilon_{a,e}^{el}$, $\varepsilon_{a,e}^{pl}$, and $\varepsilon_{a,e}$ are the equivalent elastic, plastic and total strains, ν_e and ν_p are the elastic and plastic Poisson's ratios equal to about 0.3 and 0.5, respectively. From Eq. (13.15) of distortion energy we can also derive the maximum distortion energy, the mean distortion energy and the minimum distortion energy for fatigue loads

$$\begin{aligned} u_{d,\max} &= \left(\frac{1+\nu}{3E}\right)\left[\frac{(\sigma_{1,\max} - \sigma_{2,\max})^2}{2} + \frac{(\sigma_{2,\max} - \sigma_{3,\max})^2}{2} + \frac{(\sigma_{1,\max} - \sigma_{3,\max})^2}{2}\right] \\ u_{d,\min} &= \left(\frac{1+\nu}{3E}\right)\left[\frac{(\sigma_{1,\min} - \sigma_{2,\min})^2}{2} + \frac{(\sigma_{2,\min} - \sigma_{3,\min})^2}{2} + \frac{(\sigma_{1,\min} - \sigma_{3,\min})^2}{2}\right] \\ u_{d,m} &= \left(\frac{1+\nu}{3E}\right)\left[\frac{(\sigma_{1,m} - \sigma_{2,m})^2}{2} + \frac{(\sigma_{2,m} - \sigma_{3,m})^2}{2} + \frac{(\sigma_{1,m} - \sigma_{3,m})^2}{2}\right] \end{aligned} \tag{13.21}$$

Note that

$$u_{d,m} \neq \frac{u_{d,\max} + u_{d,\min}}{2}. \tag{13.22}$$

In a uniaxial stress state, as in the specimen under simple traction, Eq. (13.21) yield

$$
\begin{aligned}
u_{d,\max} &= \frac{1+\nu}{3E}\sigma_{\max}^2 \\
u_{d,\min} &= \frac{1+\nu}{3E}\sigma_{\min}^2 \\
u_{d,m} &= \frac{1+\nu}{3E}\sigma_m^2
\end{aligned}
\tag{13.23}
$$

that solved for $\sigma_{\max}$, $\sigma_{\min}$ and σ_m yield

$$
\begin{aligned}
\sigma_{\max} &= \sqrt{\frac{3E}{1+\nu}}\sqrt{u_{d,\max}} \\
\sigma_{\min} &= \sqrt{\frac{3E}{1+\nu}}\sqrt{u_{d,\min}} \\
\sigma_m &= \sqrt{\frac{3E}{1+\nu}}\sqrt{u_{d,m}}
\end{aligned}
\tag{13.24}
$$

so that combining Eq. (13.24) with the modified Goodman relations (5.62) fatigue failure will occur if

$$
\begin{aligned}
&\sqrt{\frac{3E}{1+\nu}}\left(\sqrt{u_{d,\max}} - 2\sqrt{u_{d,m}}\right) - \sigma_y \geq 0 \\
&\sqrt{\frac{3E}{1+\nu}}\left(\sqrt{u_{d,\max}} - 2\sqrt{u_{d,m}}\right) - \sigma_N \geq 0 \\
&\sqrt{\frac{3E}{1+\nu}}\left[\sqrt{u_{d,\max}} - 2\sqrt{u_{d,m}}\left(1-\sqrt{r}\right)\right] - \sigma_N \geq 0 \\
&\sqrt{\frac{3E}{1+\nu}}\sqrt{u_{d,m}} - \sigma_y \geq 0
\end{aligned}
\tag{13.25}
$$

with σ_N being the allowable stress amplitude at a given life N or the same fatigue limit σ_f and r the usual ratio between σ_N and the ultimate strength σ_u, $r = \sigma_N/\sigma_u$. Fatigue design is governed by the most critical among the four relations (13.25) realizing that some of them could not have, in certain cases, any physical meaning. Distortion energy is always positive or null. Therefore, it cannot be possible to operate with the same distinction done with the two previous criteria. A clear limit of the von Mises criterion is that the equivalent stress given by Eqs. (13.19) or (13.20) is a scalar and is always positive. Its lack of direction and sign makes it suitable only for proportional or near-proportional multiaxial loadings impeding, by fact, a distinction between positive and negative stresses. Another drawback of the vonMises equivalent stress is that it does not consider the effect of the hydrostatic stress σ_m, Eq. (13.13) that may have an effect on multiaxial fatigue. To overcome this limitation Sines [21] has proposed a fatigue failure criterion given by Eq. (13.38) based on the von Mises equivalent stress and the hydrostatic stress.

13.3 Failure Theories Experimental Evidence

The three failure theories presented in Sect. 13.2 lead to the definition of three equivalent stresses $\sigma_{a,e}$ given by Eqs. (13.2), (13.10) and (13.20), respectively. The simplest and straightforward verification of the most suitable equivalent stress $\sigma_{a,e}$ in fatigue design can be accomplished by comparing fatigue limits obtained under fully reversed bending, $S_{f,bend}$ and cyclic torsion, $S_{f,tors}$, acting separately. On the base of Eq. (13.2) the ratio between the two fatigue limits, $S_{f,tors}/S_{f,bend}$, will be equal to unity if the criterion of maximum principal stress holds, will be equal to 0.5 if the Tresca criterion of Eq. (13.10) prevails and equal to 0.577 in the case of the von Mises theory according to Eq. (13.20). Dated studies on the issue performed on carbon steels, alloyed steels and nonferrous alloys [7–14] have evidenced the better correlation of the von Mises criterion with the experimental results, but also brought to the finding that the fracture planes are always principal planes. Table 13.1 presents a synoptic picture for various alloys of the ratio between the fatigue limit obtained in fully reversed torsion, $S_{f,tors}$, and that under rotating bending, $S_{f,bend}$.

It is interesting to note how the behavior of cast iron, which is a rather brittle material, is closer to the prediction of the maximum stress theory than to von Mises criterion (see also Fig. 13.4a). However, later studies, also dated, made by Gough [5, 6] on fatigue strength of steel specimens applying several combinations of alternating bending, alternating torsion and five different combinations of alternating in-phase bending stresses of amplitude σ_a and torsional shear stresses τ_a, using 7.6 mm diameter plain specimens, showed that none of the usual theories of failure was really effective in predicting all the results. Figure 13.6 presents a diagram of experimental results in a plane $\sigma_a - \tau_a$ normalized to $S_{f,bend}$ obtained by Gough [5].

Nevertheless, Gough et al. [6] found that for wrought steels having a tensile strength ranging from 400 to 1850 MPa the experimental results plotted in a $\sigma_a - \tau_a$ plane were following the trend shown in Fig. 13.7. They also found that experimental points could be represented by the equation

$$\frac{\sigma_a^2}{S_{f,bend}^2} + \frac{\tau_a^2}{S_{f,tors}^2} = 1 \tag{13.26}$$

Table 13.1 Ratio of fatigue limit $S_{f,tors}$ to $S_{f,bend}$

Material	Ref	$S_{f,tors}/S_{f,bend}$
Carbon steel	[7, 11, 14]	0.55
Alloyed steel	[7, 11, 14]	0.58–0.62
Wrought alloys	[8, 12, 14]	0.56–0.57
Average	0.56–0.58	
Cast iron	[11, 13, 14]	0.81–0.95
Nonferrous alloys	[7, 14]	0.52
Duralumin	[13, 14]	0.64

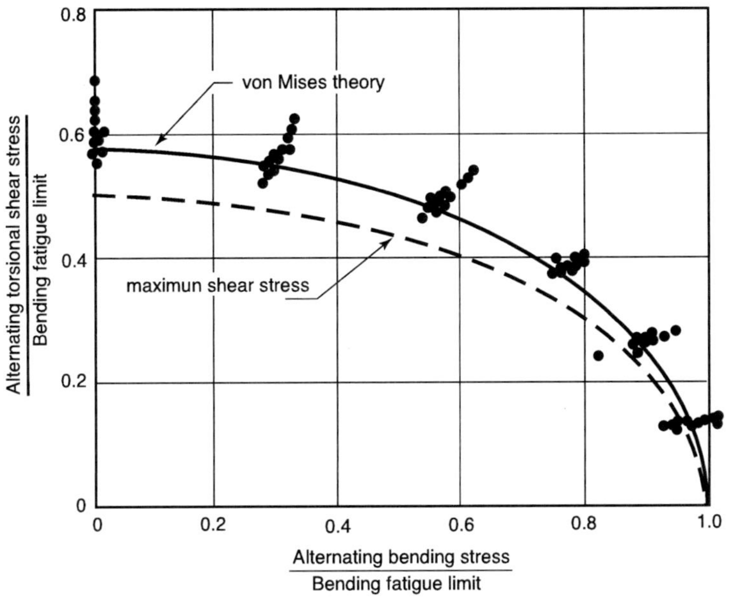

Fig. 13.6 Correlation of failure theories with experimental data for steels under combined reversed-bending and reversed-torsion stresses (modified from [5])

that they called the ellipse quadrant. A question arises as to whether Eq. (13.26) has some relation with any of the previously mentioned failure theories and equivalent

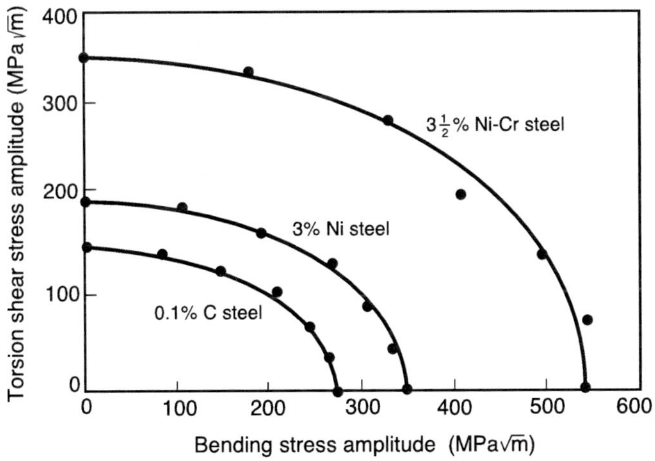

Fig. 13.7 Fatigue limits of three steels under combined reversed-bending and reversed-torsion (modified from [6])

stresses. As matter of fact, few passages can answer the question. In fact it is

$$\frac{\sigma_a^2}{S_{f,bend}^2} + \frac{\tau_a^2}{S_{f,tors}^2} = 1 \rightarrow \sigma_a^2 + \frac{S_{f,bend}^2}{S_{f,tors}^2}\tau_a^2 = S_{f,bend}^2$$

$$S_{f,bend} = \sqrt{\sigma_a^2 + \frac{S_{f,bend}^2}{S_{f,tors}^2}\tau_a^2} = \sqrt{\sigma_a^2 + \left(\frac{1}{0.58}\right)^2 \tau_a^2}$$

$$\approx \sqrt{\sigma_a^2 + 3\tau_a^2}. \tag{13.27}$$

The right side of Eq. (13.27) is the von Mises equivalent stress (13.20). For $S_{f,bend}/S_{f,tors} = 0.5$ it would represent the Tresca equivalent criterion of Eq. (13.10). Looking at Table 13.1 this could be the case of nonferrous alloys. In apparent disagreement are results obtained by Bundy and Marin [15] on thin pipes of 4½% Cu-aluminum alloy subjected to in-phase pulsating oil pressure and cyclic axial load that showed how none of the previously mentioned failure criteria was really correlating the experimental data. However, they found that the alloy was not isotropic with respect to fatigue properties. Fatigue strength in the circumferential direction was about 60% of that in the longitudinal direction. By taking this anisotropy into consideration the von Mises criterion proved to be effective in correlating experimental data.

This could be the corrected lecture key: in those materials in which fatigue failure initiates by slip in surface grains, fatigue strength under combined loads always starts on principal planes whose orientation with respect to the work piece depends on loads combination so that a possible anisotropy plays a fundamental role. The effect of anisotropy has been treated in Sect. 4.4 where Fig. 4.11 lets clearly envisage that fatigue properties of that particular low-alloyed carbon steel cannot be the same in longitudinal and transverse direction, as it really is.

But this is also the case of high strength steels affected by inclusions and wrought alloys or forgings. About the equivalent stress issue, interesting are the results obtained by Savaidis e Seeger [16] on Fe 460 type carbon steel, 500 MPa yield and 643 MPa ultimate and type Al 5083 aluminum alloy, 169 MPa yield and 340 ultimate, shown in Figs. 13.8 and 13.9, respectively. Three types of specimens were used. The first was a cylindrical unnotched specimen used as reference. The other two were either a pipe under pulsing pressure that produced an in-phase biaxial stress state or an hourglass type cylindrical specimen under cyclic uniaxial traction. In this last specimen a multiaxial stress state was generated in the mid-section by the non-uniform geometry. The equivalent stress used to compare the results was the von Mises one. The number of cycles reported is that N_i to macro-crack initiation and not to failure. The agreement between uniaxial and multiaxial specimen results can be defined excellent in both cases. However, it must be kept in mind that the von Mises equivalent stress is based on energy considerations and, ass such, is always positive. Therefore, when the applied stress state becomes negative as during stress reversal, the equivalent stress remains positive and the equivalent stress amplitude

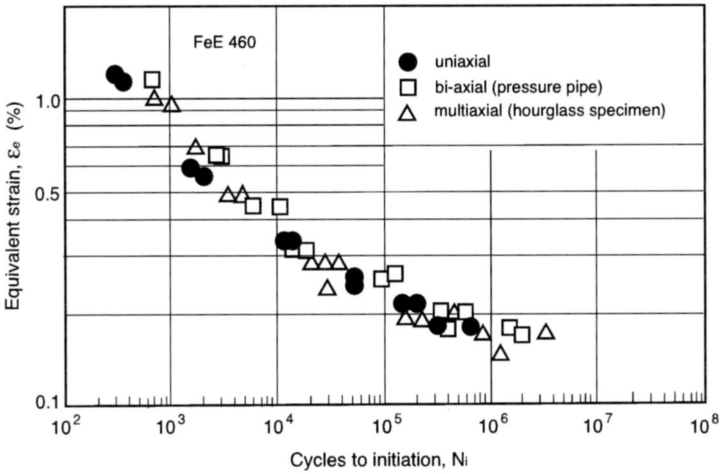

Fig. 13.8 Comparison among experimental results obtained by Savaidis and Seeger [16] on different specimens of FeE 460 steel loaded in uniaxial and multiaxial in-phase conditions

cannot be inferredit may be given a conventional negative sign and the total $\Delta\sigma_e$ calculated accordingly but it remains an arbitrary decision. The above-mentioned work of Gough et al. [6] suggested authors to distinguish between ductile and brittle materials. As to the former the ellipse quadrant Eq. (13.26) seems to be adequate while for the latter materials they proposed an ellipse arc equation (Fig. 13.10)

$$\left(\frac{\sigma_a}{S_{f,fles}}\right)^2\left(\frac{S_{f,fles}}{S_{f,tors}}-1\right)+\left(\frac{\sigma_a}{S_{f,fles}}\right)\left(2-\frac{S_{f,fles}}{S_{f,tors}}\right)+\frac{\tau_a^2}{S_{f,tors}^2}=1. \quad (13.28)$$

Experiments made always by Gough et al. [6] on cast iron and wrought high carbon steel have shown that the ellipse quadrant Eq. (13.26) could not well represents the results that, on the contrary, could be described in terms of Eq. (13.28), as shown in Fig. 13.14. An operative definition of ductile and brittle materials can be done on the base of the ratio between the fatigue limit in traction and that in torsion. For ductile materials it is

$$\frac{S_{f,fles}}{S_{f,tors}} \geq \sqrt{3} \quad (13.29)$$

While for brittle materials it is

$$1.2 < \frac{S_{f,fles}}{S_{f,tors}} < \sqrt{3} \quad (13.30)$$

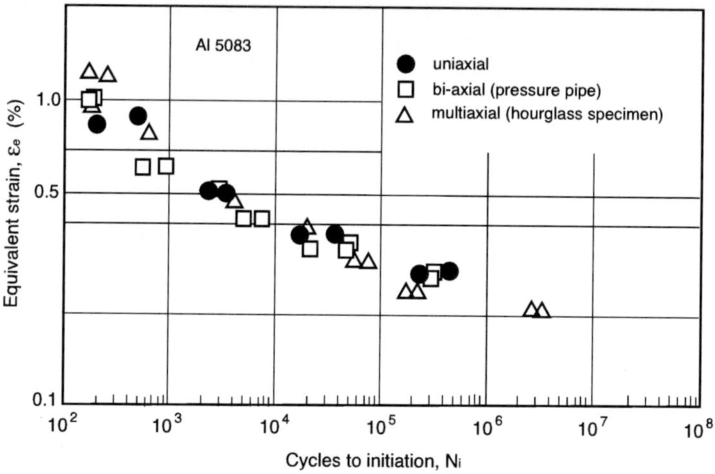

Fig. 13.9 Comparison among experimental results obtained by Savaidis and Seeger [16] on different specimens of Al 5083 aluminum alloy loaded in uniaxial and multiaxial in-phase conditions

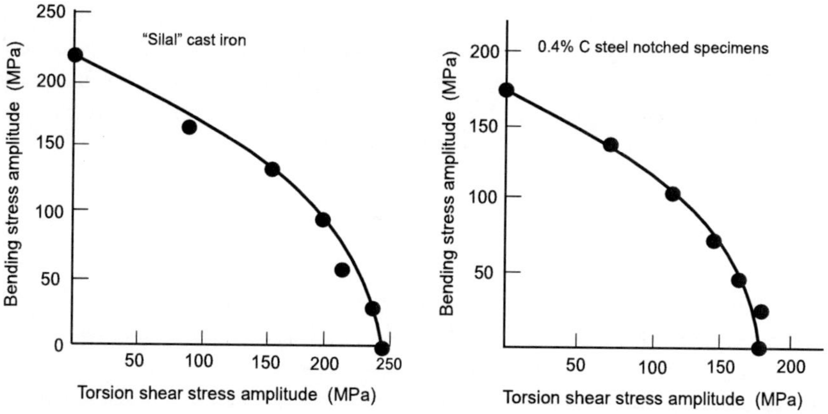

Fig. 13.10 Fatigue behavior of cast iron and 0.4% steel specimens subjected to combined fully reversed bending and torsion (modified from [6])

An equation similar to (13.28) has been derived by Papadopulos et al. [17] for brittle materials

$$\left(\frac{\sigma_{a,fles}}{S_{f,fles}}\right)^2\left(\frac{2}{\sqrt{3}}\frac{S_{f,fles}}{S_{f,tors}}-1\right)+\left(\frac{\sigma_{a,fles}}{S_{f,fles}}\right)\left(2-\frac{2}{\sqrt{3}}\frac{S_{f,fles}}{S_{f,tors}}\right)+\frac{\tau_{a,tors}^2}{S_{f,tors}^2}=1. \tag{13.31}$$

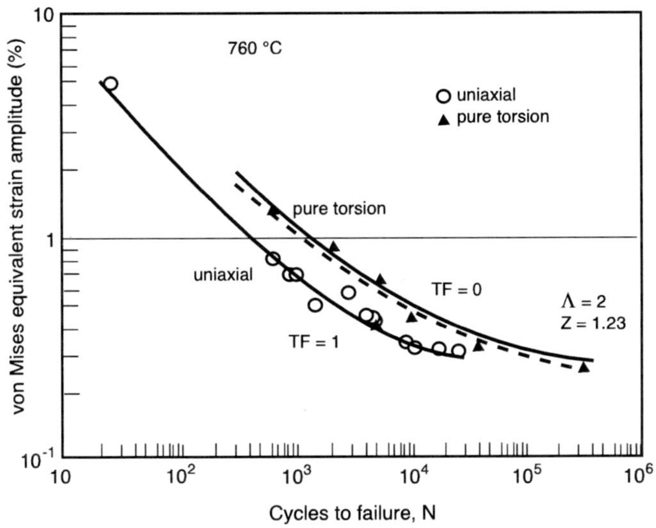

Fig. 13.11 Correlation of equivalent strain amplitude vs. cycles to failure *N* for axial and torsional fatigue of Haines 188 cobalt-base alloy at 760 °C (modified from [19])

At variance with Eqs. (13.28), (13.31) is judged to be equally valid also for out-of-phase loads. It is common opinion that as far as brittle materials are concerned the equivalent stress based on the maximum principal stress criterion is effective.

13.4 Correlation Based on Triaxiality Factor

It has been shown in Sect. 8.6.1 how a multiaxial stress state can modify the ductility of a material. To this purpose the triaxiality factor (TF) was introduced with Eq. (8.41) as the ratio of the hydrostatic stress to the von Mises equivalent stress. It is quite logical to think that the TF should have an effect on fatigue strength since it may harden the material or softening it, depending on the respective signs of the stress components. Figure 8.24 shows how biaxiality can reduce the applied plastic strain. Manson and Halford [18], Zamrik et al. [19] have used the TF to correlate fatigue crack initiation to the combined stress state for a wide range of HCF and LCF conditions. Experimental results have been obtained by Zamrik et al. [19] on Haynes 188 cobalt-base alloy at 760 °C, as shown in Fig. 13.11. The correlation that they used was

$$\varepsilon_a = \varepsilon_e + \varepsilon_p = Z^{(1-TF)} \frac{\sigma'_f}{E} N^b + \Lambda^{(1-TF)} \varepsilon'_f N^c \tag{13.32}$$

with

$$Z = \frac{3}{2(1+\nu)}\frac{\tau'_f}{G}\frac{E}{\sigma'_f}\frac{\varepsilon'_f}{\gamma'_f} \tag{13.33}$$

where E and G are the Young's modulus and shear modulus, respectively, and τ'_f and γ'_f are the coefficients in a pure torsion strain-life relation analogous to the fatigue strength coefficient σ'_f and to the fatigue ductility coefficient ε'_f of Eq. (6.10).

Λ is a ductility parameter ~ 2. Note how the new *S–N* curve for pure torsion is shifted towards the right with respect to the uniaxial *S–N* curve. The use of the triaxiality factor offers somewhat improved correlation of uniaxial and torsional fatigue data.

13.5 Stress Invariant Criteria

Stress invariant criteria are based on the value of the hydrostatic stress σ_H and the second invariant of the stress deviator J_2

$$\begin{aligned} \sigma_H &= \frac{\sigma_1+\sigma_2+\sigma_3}{3} \\ J_2 &= \frac{1}{6}\left[(\sigma_1-\sigma_2)^2+(\sigma_1-\sigma_3)^2+(\sigma_2-\sigma_3)^2\right] \end{aligned} \tag{13.34}$$

Marin [20] suggested a criterion based on the values of the amplitude $J_{2,a}$ and mean stress $J_{2,m}$ of the second invariant of the stress deviator J_2. According to Marin, fatigue failure would not occur when

$$\left(\frac{\sqrt{3J_{2,a}}}{\sigma_f}\right)^{\lambda} + \left(\kappa\frac{\sqrt{J_{2,m}}}{\sigma_u}\right)^{\mu} \le 1 \tag{13.35}$$

where σ_f is the fatigue limit of the material under fully reversed bending and σ_u its ultimate strength. He suggested to use $\kappa = 1$ and $\mu = \lambda = 2$. Application of Eq. (13.35) to cyclic torsion with a combined constant shear stress, $\tau = \tau_a \cdot (2\pi t/P) + \tau_m$, leads to the following inequality

$$\left(\frac{\sqrt{3}\tau_a}{\sigma_f}\right)^2 + \left(\frac{\tau_m}{\sigma_u}\right)^2 \le 1 \tag{13.36}$$

which indicates a dependence of fatigue strength on the static shear stress τ_m that experience excludes. Applying Eq. (13.35) to the case of cyclic bending with combined static mean stress, $\sigma = \sigma_a \cdot (2\pi t/P) + \sigma_m$, it yields

$$\left(\frac{\sigma_a}{\sigma_f}\right)^2 + \left(\frac{\sigma_m}{\sqrt{3}\sigma_u}\right)^2 \le 1 \tag{13.37}$$

that fails to distinguish between a static tensile stress σ_m and a compressive one $-\sigma_m$ because of the power 2. Probably, the most popular criterion for HCF is that of Sines [21] expressed by

$$\sqrt{J_{2,a}} + \kappa\sigma_{H,m} \leq \lambda \tag{13.38}$$

with $\sigma_{H,m}$ being the mean hydrostatic stress. The parameters κ and λ can be inferred from a fully reversed torsion test $\left(\sqrt{J}_{2,a} = \tau_f, \sigma_{H,m} = 0\right)$ and a pulsating bending test with $R = \sigma_{\min}/\sigma_{\max} = 0$ ($\sqrt{J}_{2,a} = \sigma_{fo}/\sqrt{3}, \sigma_{H,m} = \sigma_{fo}/3$ with σ_{fo} being the fatigue limit in pulsating bending)

$$\kappa = \frac{3\tau_f}{\sigma_{fo}}; \lambda = \tau_f \tag{13.39}$$

This criterion respects the finding that a constant torsion stress does not affect fatigue strength while a constant normal stress does. The application to a fully reversed bending leads to

$$\frac{\tau_f}{\sigma_f} \leq \frac{1}{\sqrt{3}} \tag{13.40}$$

Although Eq. (13.40) is verified for most materials, the ratio of the fatigue limits in torsion and bending τ_f/σ_f normally varies from 0.5 to 1.0, with this last value for brittle materials. To infer the value of the fatigue limit in pulsating bending with $R = 0$, σ_{fo}, it can be used the Goodman relation (5.48) so that Eq. (13.38) becomes

$$\sqrt{J_{2,a}} + \left(\frac{\sqrt{3}\sigma_f}{\sigma_u}\right)\sigma_{H,m} \leq \lambda \tag{13.41}$$

The criterion was successively modified by Sines and Ohgi [22] to include an additional term to take care of nonlinear effects due to possible higher mean stresses

$$\sqrt{J_{2,a}} + \nu \cdot J_{2,m}^{\mu} + \kappa\sigma_{H,m} \leq \lambda \tag{13.42}$$

Similar to the Sines criterion (13.38) is that formulated by Crossland [23] who considered the maximum hydrostatic stress $\sigma_{H,\max}$ rather than the medium one $\sigma_{H,m}$

$$\sqrt{J_{2,a}} + \kappa\sigma_{H,\max} \leq \lambda. \tag{13.43}$$

The parameters κ and λ that can be inferred from a fully reversed torsion test $\left(\sqrt{J}_{2,a} = \tau_f, \sigma_{H,\max} = 0\right)$ and a fully reversed bending test $\left(\sqrt{J}_{2,a} = \sigma_f/\sqrt{3}, \sigma_{H,\max} = \sigma_f/3\right)$

$$\kappa = \frac{3\tau_f}{\sigma_f} - \sqrt{3}; \lambda = \tau_f. \tag{13.44}$$

The criterion reflects the independence of the fatigue limit of the mean static torsion stress but not of the static normal stress. Finally, Kakuno and Kawada [24] suggested separating the effects of the hydrostatic stress amplitude and of the mean value of the hydrostatic stress

$$\sqrt{J_{2,a}} + \kappa\sigma_{H,a} + \lambda\sigma_{H,m} \leq \mu \tag{13.45}$$

with

$$\kappa = \frac{3\tau_f}{\sigma_f} - \sqrt{3}; \lambda = \frac{3\tau_f}{\sigma_{fo}} - \sqrt{3}; \mu = \tau_f \tag{13.46}$$

13.6 Average Stress Theories

The average stress approach is based on the averages of normal and shear stresses acting on a generic material plane within an elementary volume. This quantity is treated as an equivalent stress and correlated to the fatigue damage. Papadopoulos et al. [25] proposed a fatigue criterion based on the average values of the normal stress N and of the shear stress Ta within an elementary volume V

$$\sqrt{\langle T_a^2 \rangle} + \kappa \left(\max_t \langle N \rangle \right) \leq \mu \tag{13.47}$$

where $\langle\ \rangle$ indicates the mean value of the argument, $\max_t \langle N \rangle$ is the maximum value that the spatial mean normal stress reaches during the loading cycle, and κ and μ are material constants to be derived from the fully reversed endurance limits, σ_f and τ_f, for bending and torsion, respectively. The formula is similar to Crossland's criterion except by the second invariant of the stress deviator, which is replaced by the average stress quantity. This model is limited to materials in which the relation σ_f/τ_f is between 0.577 and 0.8. Such criterion leads to conclusion that non-proportional loading has no effect on material behavior in multiaxial fatigue, which is strongly opposed since in conflict with experimental observations made by You and Lee [26].

13.7 Critical Plane Theory

As we know fatigue cracks initiate in persistent slip bands of just few surface grains that are most favorably oriented with respect to the external loads to be subjected to the maximum resolved shear stress (see Sect. 1.4). Grain orientation is not known and, therefore, it is not known a priori which grain will slip easier under the action of cyclic loads and generate fatigue damage. A practical solution is to replace the maximum resolved shear stress with the maximum shear stress and identify the most severely loaded planes in the material. Along these planes in some surface grain slips occur and damage will progress till just one of those few slips will propagate through the grains giving life to the macro-crack. This propagation process proceeds through the creation of irregular surfaces that remain in close contact to each other and this would difficult the micro-crack embryo to growth due to mechanical interlocking and friction effects. Under these circumstances any stress and strain acting normal to the damage plane would open the micro-crack and release the friction allowing the micro-crack to grow easier, as schematized in Fig. 13.12. From this point of view any stress and strain acting normal to the most severely loaded planes in the material would govern the fatigue process with tensile and compressive mean stresses having, respectively, a detrimental or beneficial effect on fatigue life. These observations led to the proposal of several theories known as the critical plane methods. The operative procedure of these theories assumes that during the load cycle, stresses and strains are assessed on several planes and applying some criterion, the more severely loaded plane, or critical pane, where the fatigue damage is expected to nucleate, is then identified.

Since damage is associated to slip processes, the simplest criterion is to defined the critical plane as the plane on which the shear stress amplitude τ_a achieves the maximum value τ_{max}. The orientation of this critical plane is described by the spherical coordinates, ϕ_c and θ_c, of its unit normal vector n_c, as schematized in Fig. 13.13 for a generic plane. Experimental fatigue curves were obtained by Kaufman and Topper [27] evidencing the effect of static stress normal to the maximum shear

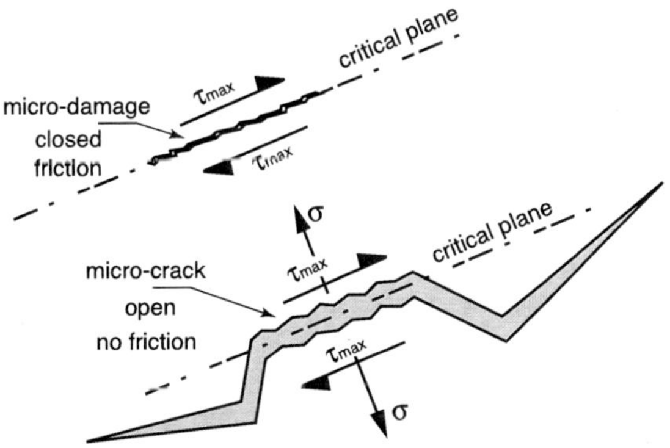

Fig. 13.12 A normal stress on the critical plane enables the micro-crack to open overcoming friction interlocking effects

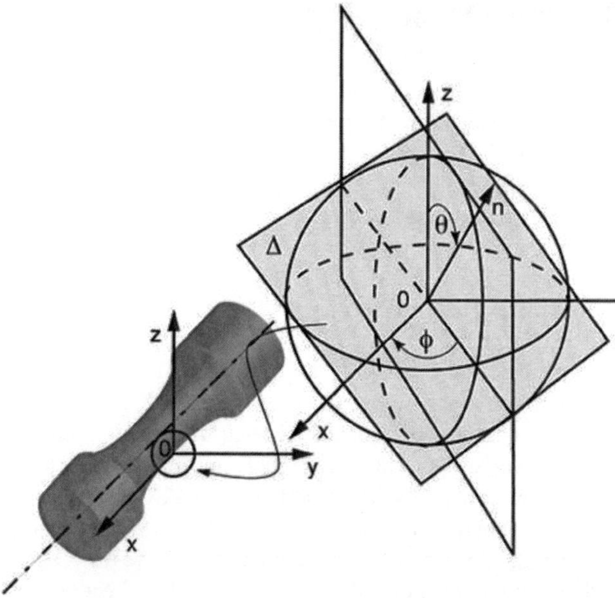

Fig. 13.13 Reference axes on point 0 on the specimen surface and definition of the spherical coordinates ϕ and θ that define the orientation of a generic plane Δ with respect to the reference x–y plane

plane for hard (BHN 456) and soft (BHN 203) SAE 1040 steel. Results, shown in Fig. 13.14 for the hard steel [27], indicate that the fatigue strength actually decreases with increasing static mean stress normal to the planes of maximum shear stress and compressive static stresses have no effect.

Findley [28], among the first to use the critical plane concept, determined this critical plane maximizing the linear combination of the shear stress amplitude and maximum normal stress acting on that plane

$$(\phi_c, \theta_c) = \max_{(\phi,\theta)}[\tau_a(\varphi, \theta) + \kappa\sigma_{\max}(\varphi, \theta)]. \tag{13.48}$$

Once the critical plane is determined the criterion can be written as

$$\tau_a(\varphi_c, \theta_c) + \kappa\sigma_{\max}(\varphi_c, \theta_c) \leq \lambda \tag{13.49}$$

with κ and λ material constants to be determined. The application of the Findley criterion to cyclic torsion with static mean shear stress, $\tau_{\max} = \tau_a \cdot (2\pi t/P) + \tau_m$, leads to the relation

$$\tau_a \leq \sqrt{\frac{\kappa^2\tau_m^2}{\left(1+\kappa^2\right)^2} - \left(\kappa^2\tau_m - \lambda^2\right)} - \frac{\kappa\tau_m}{\left(1+\kappa^2\right)} \tag{13.50}$$

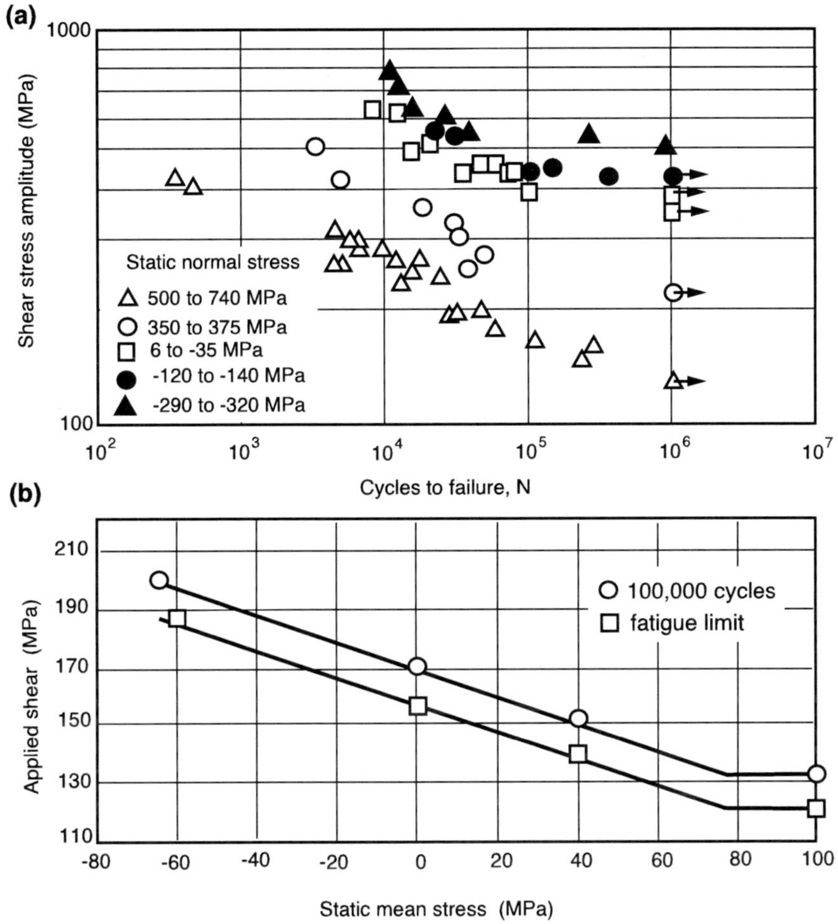

Fig. 13.14 **a** fatigue *S*–*N* curves for various static mean stress normal to the planes of maximum alternating shear stress; **b** relationship between alternating shear stress on the maximum shear planes and static mean stress for SAE 1045 steel (BHN 456) (modified from [26])

that, erroneously, anticipates a dependence of the fatigue limit on the static mean shear stress that does not exist. Under cyclic uniaxial traction with cyclic stress amplitude between σ_{max} and σ_{min} with $R = \sigma_{min}/\sigma_{max}$ it can be shown that Findley's criterion can be written as

$$0.5\sigma_a\left[\sqrt{1+\left(\frac{2\kappa}{1-R}\right)^2}+\frac{2\kappa}{1-R}\right]=\lambda \tag{13.51}$$

which quantifys the mean stress effect. Therefore, it is possible to estimate the fatigue limit σ_f (*R*) relative to any ratio *R* from the value of κ knowing the fatigue limit σ_f (*R* = *0*) as

$$\frac{\sigma_{f,R}}{\sigma_{f,R=0}} = \frac{\sigma_{a,R}}{\sigma_{a,R=0}} = \frac{\sqrt{1+\kappa^2}+\kappa}{\sqrt{1+\left(\frac{2\kappa}{1-R}\right)^2}+\frac{2\kappa}{1-R}} \tag{13.52}$$

This dependence is graphically shown in Fig. 13.15 [29]. Matake [30] simplified the Findley hypothesis (13.48) assuming as critical plane that on which the shear stress amplitude reached the maximum value

$$(\phi_c, \theta_c) = \max_{(\phi,\theta)}[\tau_a(\varphi, \theta)]. \tag{13.53}$$

The Matake criterion can be written as the Findley criterion given by Eq. (13.49). The two parameters κ and λ can be determined, then, applying a fully reversed torsion load, $\tau_a(\phi_c\theta_c) = \tau_f$ without any static stress, $\sigma_{\max}(\phi_c\theta_c) = 0$, and a fully reversed bending load, $\sigma_a(\phi_c\theta_c) = \sigma_f/2$ and $\sigma_{\max}(\phi_c\theta_c) = \sigma_f/2$ that yield

$$\kappa = \frac{2\tau_f}{\sigma_f} - 1;\ \lambda = \tau_f \tag{13.54}$$

The criterion respects the independence of the fatigue limit of the static shear stress and considers a linear dependence on the static normal stress. Brown and Miller [31] introduced the concept of Case A and Case B cracks. Case A cracks propagate along the component surface while Case B cracks propagate into the surface, as schematically shown in Fig. 13.16. The Brown and Miller model consider the mean

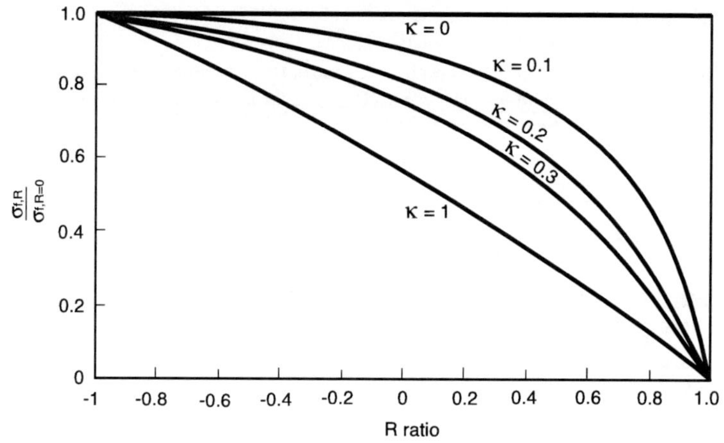

Fig. 13.15 Fatigue limit $\sigma_{f,R}$ as a function of the ratio *R*, according to Findley (modified from [29])

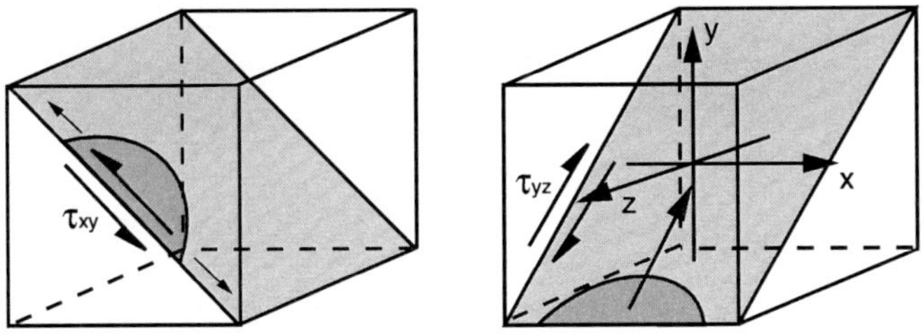

Fig. 13.16 Case A: crack propagates along the surface, Case B crack propagates into the surface

stress effects combining the maximum range of shear strain $\Delta\gamma_{\max}$ to the range of strain $\Delta\varepsilon_n$ normal to the plane of maximum shear strain to assess the fatigue life N using the Manson-Coffin Eq. (6.10) in which is taken into consideration the mean normal stress σ_n perpendicular to the plane of maximum shear strain

$$\frac{\Delta\gamma_{\max}}{2} + \kappa_{BM}\Delta\varepsilon_n = \beta_1\frac{\sigma'_f - 2\sigma_n}{E}(N)^b + \beta_2\varepsilon'_f(N)^c \tag{13.55}$$

where κ_{BM} is a fitting parameter ($\kappa_{BN} \cong 0.3$ for ductile metals at or near the fatigue limit). This equation was adapted from Morrow to fit uniaxial test data where the mean stress σ_m is equal to $2{\cdot}\sigma_n$ since σ_n is acting perpendicularly to the plane of $\Delta\gamma_{\max}$ it is worth ½ σ_m. the values of β_1 and β_2 can be obtained assuming uniaxial traction, as shown in Fig. 13.17

$$\left.\begin{array}{l}\Delta\gamma_{\max} = (1+\nu)\Delta\varepsilon \\ \Delta\varepsilon_n = (1-\nu)\Delta\varepsilon/2\end{array}\right\} \Rightarrow \frac{\Delta\gamma_{\max}}{2} + \kappa_{BM}\Delta\varepsilon_n = \frac{\Delta\varepsilon}{2}[(1+\nu) + \kappa_{BM}(1-\nu)] \tag{13.56}$$

Coefficients β_1 and β_2 can be obtained from eqs. (13.55) and (13.56)

$$\beta_1 = 1 + \nu + (1-\nu)\kappa_{\mathrm{BM}}; \beta_2 = 1.5 + 0.5\kappa_{\mathrm{BM}}. \tag{13.57}$$

The Brown and Miller model has a major drawback in that two micro-cracks having the same $\Delta\gamma_{\max}$ and $\Delta\varepsilon_n$ may have different fatigue lives if one is opened in traction and the other is closed in compression due to the mean load effect. Therefore, Fatemi and Socie [32] suggested replacing $\Delta\varepsilon_n/2$ with the maximum stress $\Delta\sigma_{n,\max}/2$ normal to the plane of maximum shear strain

$$\gamma_{\max}\left(1 + \kappa_{FS}\frac{\sigma_{n,\max}}{\sigma_y}\right) = \text{constant} \tag{13.58}$$

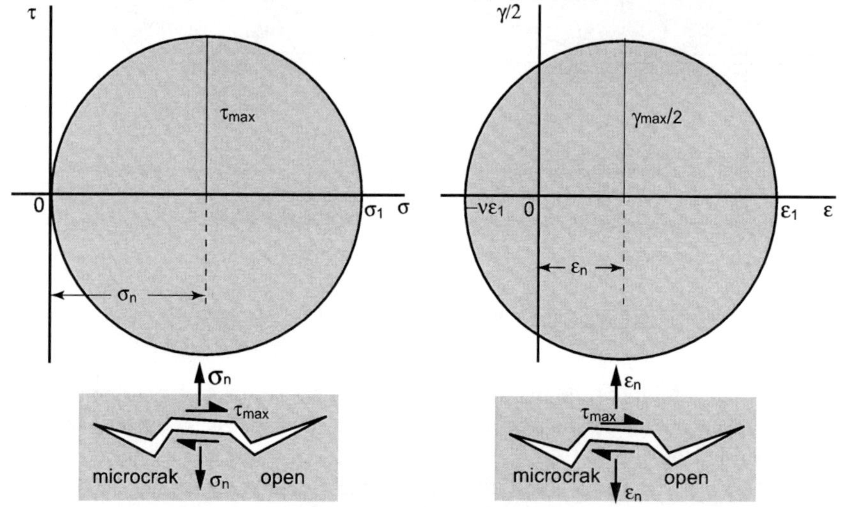

Fig. 13.17 Mohr circles for stress and strain in uniaxial traction

The maximum normal stress $\sigma_{n,\max}$ has been normalized to the yield strength σ_y to maintain the dimensionless feature of strain. Combining Eq. (13.58) with the Manson and Coffin Eq. (6.10) it yields

$$\gamma_{\max}\left(1+\kappa_{FS}\frac{\sigma_{n,\max}}{\sigma_y}\right)=(1+\nu_e)\frac{\sigma'_f}{E}(N)^b+\frac{\kappa_{FS}}{2}(1+\nu_e)\frac{\sigma'^2_f}{E\sigma_y}(N)^b$$
$$+\left(1+\nu_p\right)\varepsilon'_f(N)^c+\frac{\kappa_{FS}}{2}\left(1+\nu_p\right)\frac{\varepsilon'_f\sigma'_f}{\sigma_y}(N)^{b+c} \quad (13.59)$$

where $\nu_e = 0.3$ and $\nu_p = 0.5$ are the Poisson's coefficients in the elastic and plastic domain, respectively. The Smith–Watson–Topper [33] damage model (see Sect. 6.4.2) can be also used in the multiaxial fatigue substituting in Eq. (6.22) the maximum stress σ_{max} with the maximum stress $\sigma_{n,\max}$ normal to the plane of maximum shear strain

$$\frac{\Delta\varepsilon_1}{2}\sigma_{n,\max}=\frac{\sigma'^2_f}{E}(N)^{2b}+\sigma'_f\varepsilon'_f(N)^{b+c} \quad (13.60)$$

where $\Delta\varepsilon_1$ is the range of the maximum principal strain and $\sigma_{n,\max}$ the peak stress in the direction normal to ε_1. Figure 13.18 summarizes the three strain-based models with the parameters used [29].

Using the Brown and Miller concept of Case a and case B, McDiarmid [34, 35] developed a criterion based on a critical plane defined as that plane on which the shear stress amplitude reaches the maximum value, precisely as the Matake hypothesis, but at variance with the Matake and Findley criterion (13.49) it was differentiating

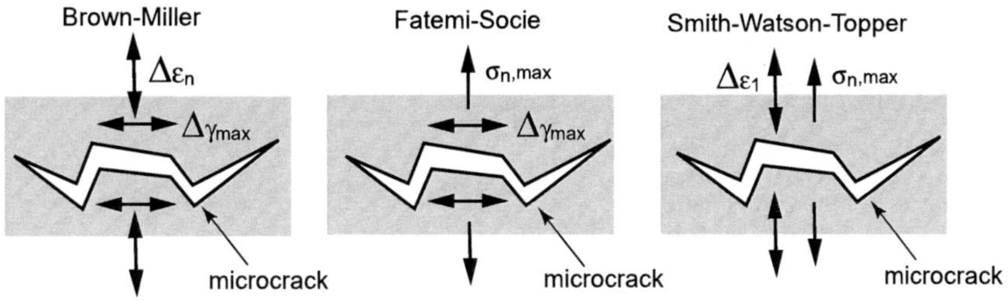

Fig. 13.18 Strain-based models with the parameters used (modified from [29])

between case A and case B

$$\tau_a(\varphi_c, \theta_c) + \frac{\tau_{f,A,B}}{2\sigma_u}\sigma_{n,\max}(\varphi_c, \theta_c) \leq \tau_{f,A,B} \tag{13.61}$$

where the values $\tau_{f,A}$ and $\tau_{f,B}$ are used depending on whether the critical plane, identified by Eq. (13.49), indicates as potential propagation path that of Case A or Case B, respectively. In the case of pure torsion Case A applies and, therefore, $\tau_{f,A} = \tau_f$. Since Case A is also valid for combined bending and torsion the McDiarmid criterion can be written as

$$\tau_a(\varphi_c, \theta_c) + \frac{\tau_f}{2\sigma_u}\sigma_{n,\max}(\varphi_c, \theta_c) \leq \tau_f. \tag{13.62}$$

Also this criterion is respecting the fatigue limit independence of the static shear stress and the dependence on the static normal stress.

13.8 Out-of-Phase Cyclic Loads

In this section we shall try to shed some light on the intriguing world of the out-of-phase (OP) or non-proportional (NP) cyclic loads. It has been said that in the past one of the most common method to overcome the rock of the OP loads was to treat them as if they were in-phase (IP). It was also thought that such approach would be even conservative. As matter of fact, such a loophole has count in the past as many supporters as disdainful opinions. Today opposers to this practice gained a rather large consensus since OP loads have been found more damaging than the equivalent IP loading in the low cycle fatigue regime or Region I (see Fig. 1.5), in particular [36–39]. Figure 13.19 presents the results obtained on three different steels, namely Hard steel with $\sigma_u = 704$ MPa and $\tau_f = 196$ MPa [40], mild steel with $\sigma_u = 519$ MPa [41] and $\tau_f = 137$ MPa and 42CrMo4 steel with $\sigma_u = 1025$ MPa and $\tau_f = 260$ MPa [42], using the McDiarmid criterion (13.62). The OP loadings yield non-conservative results with few exceptions that lie within a $\pm 5\%$ band. Actually, it is indeed difficult

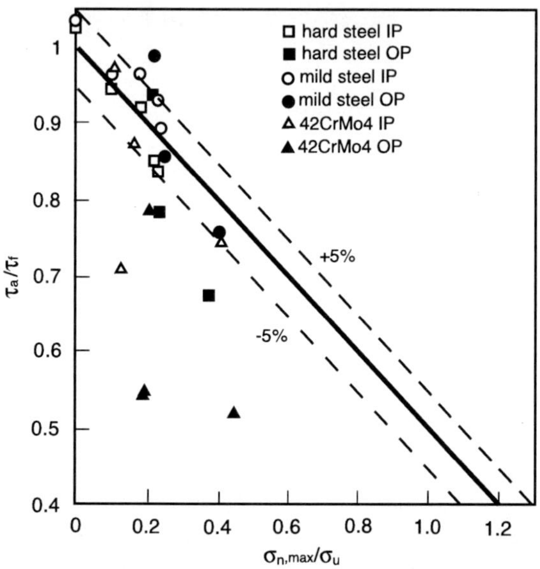

Fig. 13.19 Results obtained applying the McDiarmid criterion (data from [40, 41])

to formulate a one and for all proposition on this issue since the result of OP tests depend on many different factor among which it must be included:

- load path;
- non proportional hardening;
- phase difference;
- frequency difference.

The following sections will address each of the aforementioned factors.

13.8.1 Load Path

Figure 13.20 schematizes some possible and typical load paths. Figure 13.20a shows an IP cycle in which the generic material point is forced to move from A to B and vice versa keeping always the same γ-ε ratio. Apparently, the same cycle in a 90° out-of-phase condition is shown in Fig. 13.20b. This time the same generic material point moves along the circumference of diameter $AB = a = b$. The γ-ε ratio varies as it moves. When the shear deformation reaches the maximum value $\gamma_{\max,OP}$ the normal strain ε is zero and vice versa. When the material point reaches points A or B it is subjected to the same combination γ-ε as in the IP cycle. But this time the maximum shear strain $\gamma_{\max,OP}$ and the maximum normal strain $\varepsilon_{\max,OP}$ are much larger than $\gamma_{\max,IP}$ and $\varepsilon_{\max,IP}$ of the corresponding IP cycle. If we want to maintain the same maximum values than the OP cycle must be that of the inner circle having

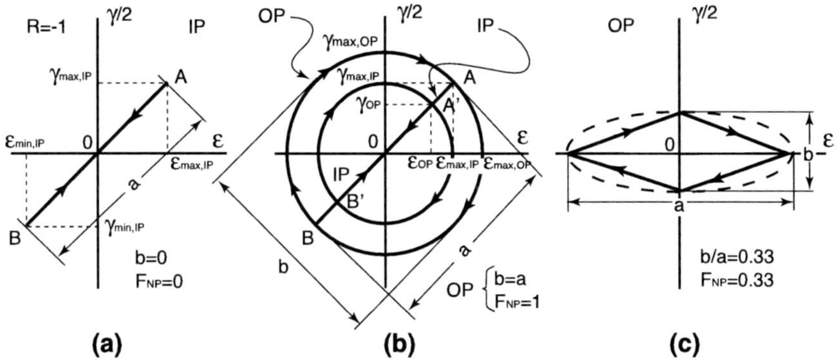

Fig. 13.20 Example of IP and OP load paths and illustration of NP factor

the smaller diameter A′B′. But in this case the maximum values γ_{OP} and ε_{OP} of the OP cycle will be much smaller than those of the IP cycle. In any case the two IP and OP cycles cannot be compared because the latter cycle will yield either a non-conservative or a conservative result with respect to the IP cycle, depending on the circle chosen.

This is why opinions about the procedure of considering OP loadings as if they were IP are so different. Finally, Fig. 13.20c shows an OP cycles so called diamond shaped. The IP or OP cycles can be classified on the base of the so called biaxial strain ratio $\Delta\gamma/\Delta\varepsilon$ or NP factor F_{NP}. One such method to evaluate the factor of non-proportionality, F_{NP}, was proposed by Kanazawa et al. [42] as a factor of ellipticity of the strain path in the $\gamma/2$-ε plot, as shown in Fig. 13.20, or the ratio of the minor to major axis of the circumscribed ellipse (Socie and Marquis [43]). The two OP cycles of Fig. 13.20b and c have $F_{NP} = 1$ and $F_{NP} = 0.33$, respectively.

13.8.2 Non-Proportional Hardening

Some materials under OP or non-proportional (NP) cyclic loading can harden in addition to the cyclic hardening phenomenon that has been discussed in Sect. 1.3. This phenomenon of additional hardening, called NP hardening, that does not take place under IP or proportional loading, depends on the load history through the factor F_{NP}, defined in Sect. 13.8.1, and a constant α that modifies the plastic term in the Ramberg–Osgood type Eq. (6.3) as

$$\frac{\Delta\varepsilon_{p,e}}{2} = \left[\frac{\Delta\sigma_e}{2 \cdot K'(1 + \alpha F_{NP})}\right]^{1/n'} \tag{13.63}$$

where E is the Young's modulus, K' the plastic modulus or proportional (or uniaxial) cyclic strength coefficient and n' the proportional (uniaxial) cyclic strain hardening exponent and α is the material dependent non-proportional hardening coefficient. The subscript e in Eq. (13.63) reminds us that the quantities are equivalent plastic strain and stress, respectively, since we are dealing with multiaxial stress states. Note that since it is $0 \leq \alpha \leq 1$ the NP hardening can increase the plastic modulus K' by a factor up to 2. The largest NP hardening occurs with 90° OP loading circle of Fig. 13.20b or in the diamond shaped loading path of Fig. 13.20c in which it is $F_{NP} = 1$. Itoh et al. [44] and Kida et al. [45] proposed a formulation for the factor of non-proportionality computed from the strain path that allows evaluating the stress response directly from the strain history. They considered the non-proportionality parameter as the angle change of principal strain direction and strain path length after changing direction on a $\gamma/\sqrt{3}$ plot as

$$F_{NP} = \frac{\pi}{2T\varepsilon_{1,\max}} \int_0^T [|\sin\xi(t)|\varepsilon_1(t)]dt \tag{13.64}$$

where $\varepsilon_1(t)$ is the principal strain at time t, $\varepsilon_{1,\max}$ is the maximum absolute value of $\varepsilon_1(t)$, $\xi(t)$ is the angle between $\varepsilon_1(t)$ and $\varepsilon_{1,max}$ and T is the time length of the cycle. These quantities are defined in Fig. 13.21 [46]. The strain path length is defined as a straight line connecting two points with the utmost distance on the strain path. Austenitic stainless steels are particularly sensitive to the NP hardening.

Type 316 austenitic stainless steel has $\alpha \cong 1$, type 1045 medium carbon steel has $\alpha \cong 0.3$, Al 7075 and Al 6061 alloys have $\alpha \cong 0$ and $\alpha \cong 0.44$, respectively. This phenomenon is characteristic of low stacking fault energy materials and well-spaced dislocations where the slip bands generated by proportional loadings are always planar (see Sect. 1.4 and Fig. 1.44). In these materials, due to the rotation of principal axes and maximum shear stress planes, NP loadings activate cross slip bands giving rise to additional hardening ($\alpha \gg 0$) that cannot be seen under IP

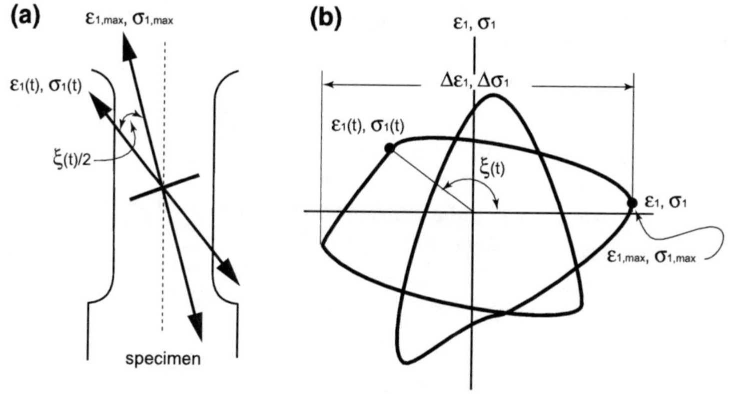

Fig. 13.21 **a** Definition of $\xi(t)$ and **b** polar representation of $\Delta\varepsilon_1$ [46]

conditions. In type 316 austenitic stainless steel the stacking fault energy is of the order of 23 mJ/m^2 while in aluminum it reaches 160–250 mJ/m^2. Also titanium and nickel have very high stacking-fault energy, this last of the order of 260–300 mJ/m^2. Figure 13.22 [47] presents the equivalent stress σ_e-equivalent strain ε_e response of 1050 QT steel and 304L stainless steel under both IP and OP cyclic loading. As to 304L SS it is also shown the monotonic stress–strain curve. Type 1050 QT steel has high stacking fault energy and $\alpha = 0$ while low carbon 304L stainless steel has low stacking fault energy and $\alpha = 0.55$.

It can be seen how type 1050 QT steel does not present any NP hardening in the 90° out-of-phase conditions with respect to the IP test conditions while type 304L stainless steel shows a remarkable NP hardening. It is clear that this additional hardening to the normal cyclic hardening presented by the stainless steel, as Fig. 13.22 indicates, must have an influence on the fatigue strength.

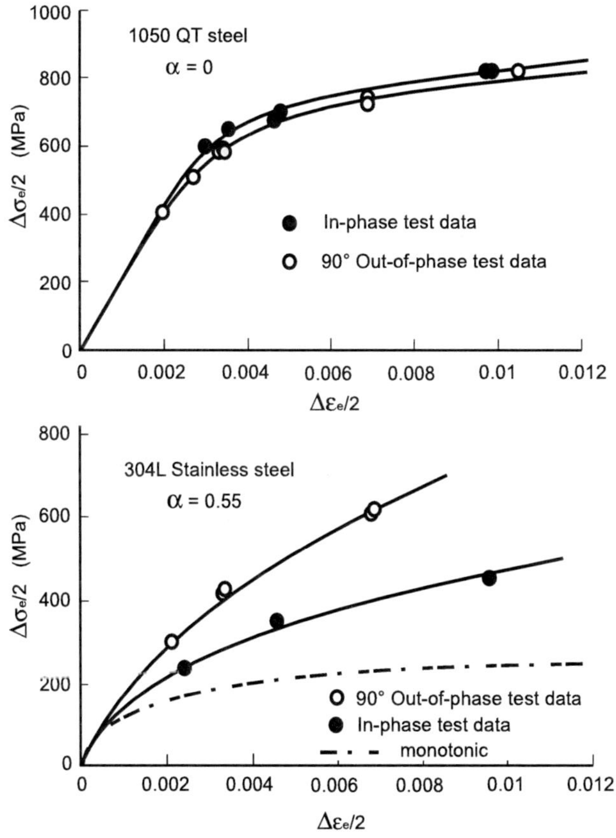

Fig. 13.22 Equivalent stress–strain response of 1050 QT and 304L SS under non NP hardening and NP hardening conditions (modified from [47])

Actually, it is rather difficult to evidence the contribution of NP hardening to the overall fatigue response just comparing the two responses under IP and OP conditions since in OP loading there might also be a contribution due to some effect other than NP hardening. This can be seen in Fig. 13.23 [47]. OP loading always results in shorter fatigue life as compared to IP loading, but this effect is much more pronounced in the 1050 QT steel that is not sensitive to NP hardening with respect to 304L stainless steel.

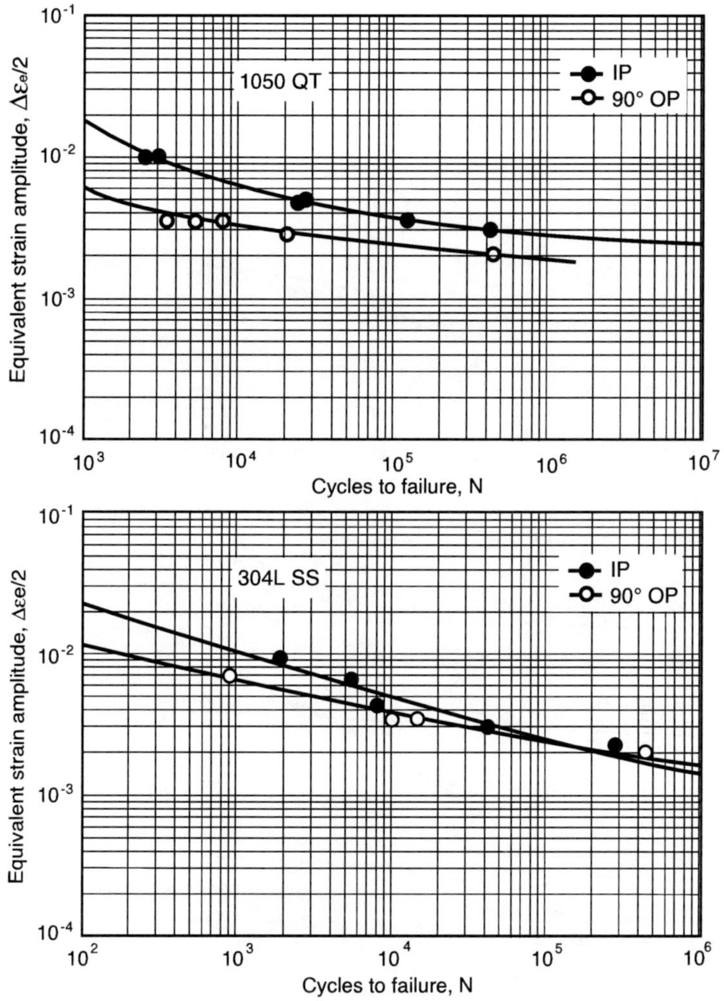

Fig. 13.23 Comparison of IP and 90° OP equivalent strain amplitude versus cycles to failure for 1050 QT steel and 304L SS (modified from [47])

Furthermore, the NP hardening observed in the 304L stainless steel increases the fatigue life in the high cycle fatigue region ($N \gg 100{,}000$ cycles) because it retards slip and damage formation while reducing the fatigue strength in the low cycle fatigue region (see Sect. 1.3 and Fig. 1.17).This actually means that NP hardening reduces fatigue life only under strain controlled situations that increase the applied stress. Conversely, under stress controlled conditions this additional hardening continuously reduces the applied strain increasing the fatigue life. The higher sensitivity of 1050 QT steel to OP conditions may be due to inclusions and to the time-varying direction of principal axes that scan the metallurgical conditions of the material in all possible directions. Consider that this NP hardening is not taken into consideration by the Basquin-Manson-Coffin Eq. (6.10). This implies that traditional ε-N equations developed to model uniaxial fatigue can lead to non-conservative predictions if used in multiaxial NP loading under strain controlled conditions and be conservative under stress controlled conditions.

It must also be said that as we enter the high cycle fatigue regime (elastic stresses) the NP hardening effect reduces till disappearing. Materials for which cyclic hardening was observed (i.e. 1050 N steel and 304L stainless steel), exhibited some non-proportional cyclic hardening as well, whereas no non-proportional cyclic hardening was observed for a cyclic softening material (i.e. 1050 QT steel). Therefore, Shamsaei concluded that, quoted, it appears that materials with cyclic hardening exhibit additional hardening due to the non-proportionality of loading. Considering the fact that both cyclic hardening and non-proportional cyclic hardening phenomena can be qualitatively related to the staking-fault-energy, there should be also a qualitative relation between these two hardening phenomena.

13.8.3 Constitutive Equation and NP Hardening Models

As matter of fact, the α parameter in not that easy to obtain. Several empirical methods to predict the value of the non-proportional hardening coefficient α are available. Among these, the Borodii [48] method is based on the $\Delta\varepsilon_r/N$ fatigue curve where $\Delta\varepsilon_r$ is the reduced strain range defined as

$$\Delta\varepsilon_r = (1 + \alpha F_{NP})(1 + k \sin\varphi)\Delta\varepsilon \tag{13.65}$$

where k is a material property related to proportional deformation and φ a rotation angle of an arbitrary cycle path with the principal axis. Borodii pointed out that in a log scale there is a linear relationship between the reduced strain range and the number N of cycles to crack initiation, as in the uniaxial ε/N curve. Therefore, it suffices to run three tests, as shown in Fig. 13.24, two of which in a uniaxial deformation state in the low-cycle regime at two different $\Delta\varepsilon_1$ and $\Delta\varepsilon_2$ strain ranges and one at any non-proportional biaxial deformation with a strain ranges $\Delta\varepsilon_1$ or $\Delta\varepsilon_2$ and measure the relative life $\overline{N}_2$ to derive the α coefficient as

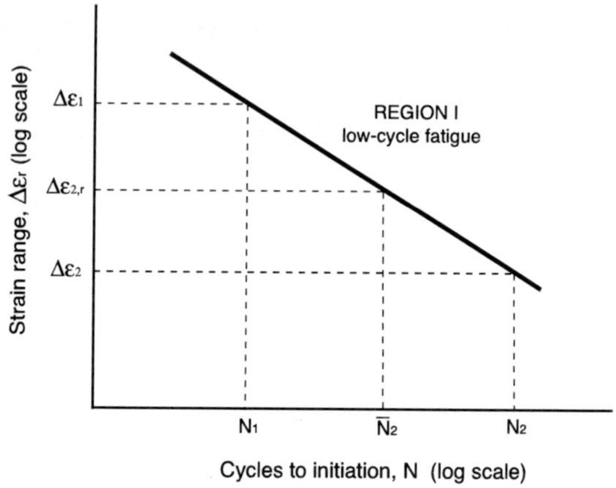

Fig. 13.24 $\Delta\varepsilon_r$-N diagram to determine material constant α and k according to Borodii

$$\alpha = \frac{1}{F_{NP}}\left[\left(\frac{\Delta\varepsilon_1}{\Delta\varepsilon_2}\right)^{\log\frac{N_2}{\overline{N}_2}\Big/\log\frac{N_2}{N_1}} - 1\right]. \tag{13.66}$$

If the slope c of the $\Delta\varepsilon$-N line in the uniaxial deformation is known the determination of α becomes simpler

$$\alpha = \frac{1}{F_{NP}}\left[\left(\frac{\overline{N}_2}{N_2}\right)^{c} - 1\right]. \tag{13.67}$$

As to the k parameter it can be assessed as

$$k = \frac{1}{\varphi}\left[\left(\frac{\Delta\varepsilon_1}{\Delta\varepsilon_2}\right)^{\log\frac{N_2}{\overline{N}_2}\Big/\log\frac{N_2}{N_1}} - 1\right]$$

$$k = \frac{1}{\varphi}\left[\left(\frac{\overline{N}_2}{N_2}\right)^{c} - 1\right]. \tag{13.68}$$

With such method Borodii [48] calculated the values reported in the following Table 13.2 for several metals and alloys. In a more recent study Borodii and Shukaev [49] proposed an approximate relation for the non-proportional cyclic hardening coefficient. This relation is based on monotonic strength properties of the material and is expressed as

$$\log|\alpha| = 0.705\left(\frac{\sigma_u}{\sigma_y} - 1\right) - 1.22 \tag{13.69}$$

Table 13.2 α and k factors calculated by Borodii [48]

Material	α	k
304 SS	0.44	0.2
6061 Al	0.4–0.44	0.0
Aluminum	0.0	0.0
Copper	0.9	0.0
310 steel	0.8–0.84	0.55–0.8

where σ_u and σ_y are the ultimate strength and the yield strength of the material, respectively. This relation is limited to the cyclic strain amplitudes from $0.02\% < \Delta\varepsilon_p/2 < 1\%$. However, significant scatter of data was observed by Shamsaei [50] when relating the non-proportional cyclic hardening to only monotonic properties. This is due to the fact that non-proportional cyclic hardening also depends on the material slip system induced under cyclic loading.

Therefore, non-proportional cyclic hardening coefficient cannot be predicted on the base of just monotonic properties. Shamsaei-Fatemi [51] proposed a simple model to assess the non-proportional cyclic hardening coefficient, α, based on the monotonic and cyclic curves of the material, as shown in Fig. 13.25. From the two curves it is inferred a cyclic hardening coefficient h based on the ratio of the monotonic and cyclic stress values corresponding to a given strain ε and from this the NP hardening coefficient α as

$$\begin{aligned} h &= 1 - \frac{\sigma_M}{\sigma_C} \\ \alpha &= 1.6h^2 + 0.6h \\ &= 1.6\left(\frac{\sigma_M}{\sigma_C}\right)^2 - 3.8\left(\frac{\sigma_M}{\sigma_C}\right) + 2.2. \end{aligned} \tag{13.70}$$

It is obvious that such definition relates the NP cyclic hardening coefficient α to the strain level selected for the computation. As matter of fact if the α coefficient is assessed in the strain range over 0.6% where both cyclic and monotonic curves start to bend over and flatten, this dependency becomes less and less pronounced. Comparison of experimental and predicted non-proportional cyclic hardening coefficient based on Eq. (13.70) for 1050 steel at different hardness levels, 304L stainless steel, and several other materials from the literature including 1045 normalized (N) steel, pure titanium, titanium alloy BT9, 16MnR steel, 321 stainless steel, 800H stainless steel at room temperature (RT) and 800°C and 316L stainless steel is presented in Fig. 13.26 [51]. The range of data used in this figure is $0.05\% < \mathrm{D}\varepsilon_p/2 < 1.2\%$.

Equation (13.70) can be rewritten in terms of uniaxial monotonic and cyclic deformation properties as well as of the level of plastic strain in Eq. (6.3), as

$$\alpha = 1.6 \cdot \left(\frac{K}{K'}\right)^2 \left(\frac{\Delta\varepsilon}{2}\right)^{2(n-n')} - 3.8 \cdot \left(\frac{K}{K'}\right)\left(\frac{\Delta\varepsilon}{2}\right)^{(n-n')} + 2.2. \tag{13.71}$$

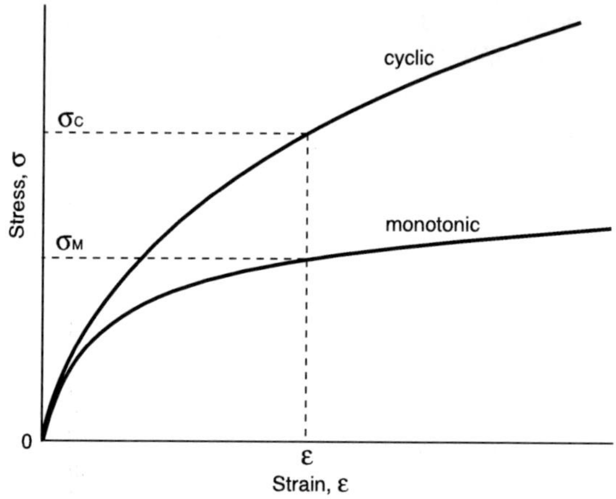

Fig. 13.25 Monotonic and cyclic curves and corresponding stresses at a given strain

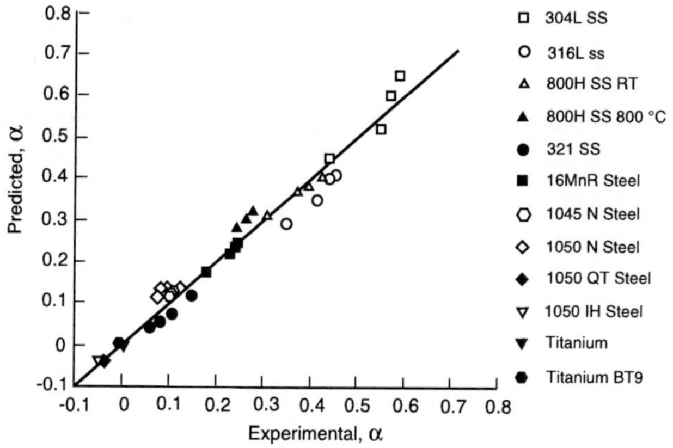

Fig. 13.26 Comparison of experimental and predicted NP cyclic hardening coefficient (modified from [51])

13.8.4 *Effect of Phase Difference*

A phase shift between the alternating shear stress and the alternating normal stress may result in a slight increase in fatigue strength. Liu and Zenner [52] reported a series of tests made by several researchers [53–57] on different alloys where the fatigue limit remained almost unchanged up to a phase shift of 90° afterwards the phase shift $\delta_{\sigma\tau}$ started to have an effect, but that effect was not the same neither for all alloys nor within the same material since it could range from plus 20% to minus 15%, as shown in Fig. 13.27.

Therefore, Liu and Zenner concluded that it was not possible to prove the existence of a unique material dependence. For the case of two cyclic normal stresses, only

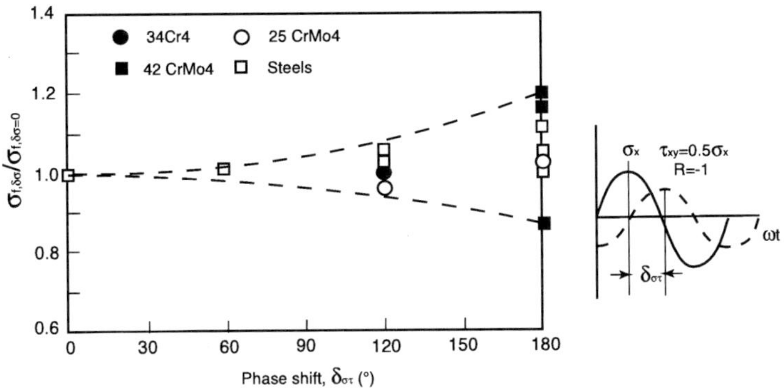

Fig. 13.27 Effect of phase shift between cyclic normal stress σ_x and cyclic shear stress τ_{xy} (modified from [52]. Data from [53–56])

test results with pulsating loads are available, because of experimental difficulties. The effect of a phase shift between two pulsating normal stresses is small within the range from 0° to 60° and results in a significant decrease in the fatigue limit within the range from 90° to 180°, Fig. 13.28. A slight increase in strength is observed in the vicinity of 60°, but this is just a single point that can be well within the normal data scatter. A phase shift $\delta_\sigma = 180°$ between two normal stresses of equal magnitude corresponds to the case of torsional loading. The minimum fatigue limit is obtained right at 180° phase shift and is rather massive (about 30%).

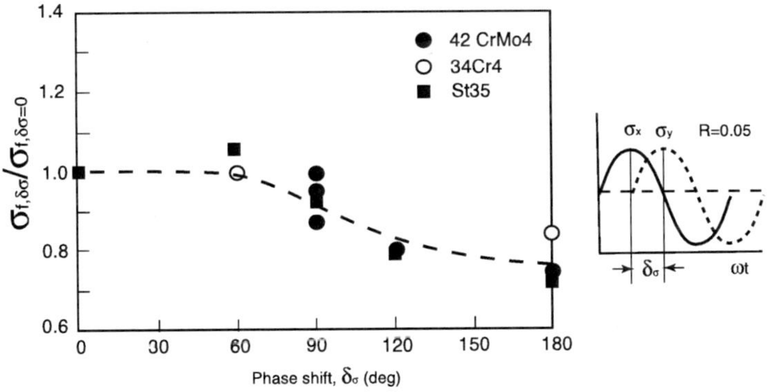

Fig. 13.28 Effect of phase shift between two cyclic normal stresses σ_x and σ_y on fatigue limit (modified from [52]. Data from [53, 57, 58])

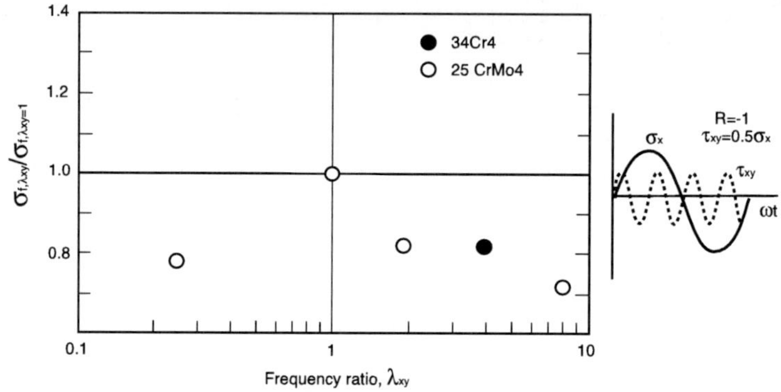

Fig. 13.29 Effect of frequency difference between cyclic normal stress and cyclic shear stress (modified from [52]. Data from [53,55])

13.8.5 Effect of Frequency

A word shall be said on frequency effect. Under uniaxial loading the frequency of the cyclic loads has no effect at all unless corrosion starts to assist fatigue damage. Any environment is potentially aggressive even air itself, if humid. But to fully evidence corrosion assisted fatigue it is necessary to decrease the frequency of fatigue cycles down or below one cycle per second because corrosion needs time to exert its devastating effect combined to fatigue (see Section .). At variance with the case of uniaxial loading, in multiaxial loading the frequency difference between the stress components plays an important role. However, considerably less attention has been paid to the influence of differences in frequency of the stress components on the fatigue strength with respect to the phase shift experimental study. The effect of the frequency ratio $\boldsymbol{\lambda}_{xy}$ between a shear stress τ_{xy} and a normal stress σ_x is shown in Fig. 13.29 [52]. Test results indicate, at least for the steels considered, that the effect of frequency on the fatigue limit is null when the frequency ratio $\boldsymbol{\lambda}_{xy}$ is equal one and decreases as this ratio gets either lower or larger than one. For the case considered in the study with $\tau_{xy} = 0.5\,\sigma_x$ and $R = -1$ the fatigue strength was decreasing by 20% or 30% depending on whether the frequency ratio $\boldsymbol{\lambda}_{xy}$ was lower or higher than 1.

13.9 Out-of-Phase Experimental Evidence

The first assessment of fatigue theories to predict the OP behavior of materials will be done checking the von Mises criterion, already analyzed in Sect. 13.3 for the IP situations. In particular, Eq. (13.20) will be used combined with the Basquin-Manson-Coffin Eq. (8.10) to yield the equivalent strain versus life N

$$\varepsilon_{a,e} = \frac{\Delta\varepsilon_{a,e}}{2} = \frac{1}{\sqrt{2(1+\overline{\nu})}}\sqrt{2\varepsilon_a(1+\overline{\nu})^2 + \frac{3}{2}\gamma_a^2} = \frac{\sigma'_f}{E}(N)^b + \varepsilon'_f(N)^c \quad (13.72)$$

where σ'_f is the fatigue strength coefficient, b the fatigue strength exponent, ε'_f the fatigue ductility coefficient, c the fatigue ductility exponent and E is the average of the first cycle modulus of elasticity. Actually, σ'_f, ε'_f, b and c should be substituted with the corresponding equivalent von Mises quantities. Figure 13.30 shows the correlation of IP and OP test data for 1050 steel in three different hardening condition as well as 304L stainless steel with predictions based on Eq. (13.72). As it can be seen the difference between OP and IP data is higher for 1050 QT steel with no non-proportional hardening than for 304L stainless steel with 55% non-proportional hardening. Significant differences in fatigue lives observed for IP and OP loadings at the same equivalent strain amplitudes suggest that the von Mises criterion underestimates the fatigue damage under non-proportional loadings.

Moreover, crack orientation angle cannot be predicted by von Mises criterion, the distortion energy being a scalar quantity. This is not consistent with the experimental observations indicating the fact that all cracks are formed along a preferred direction, as it can be seen in Fig. 13.31 for IP and OP loading of 1050 N steel, 1050 QT steel, and 304L stainless steel, respectively. Under IP loading conditions the planes of maximum shear stress are at 20° and 110° with while under OP conditions the

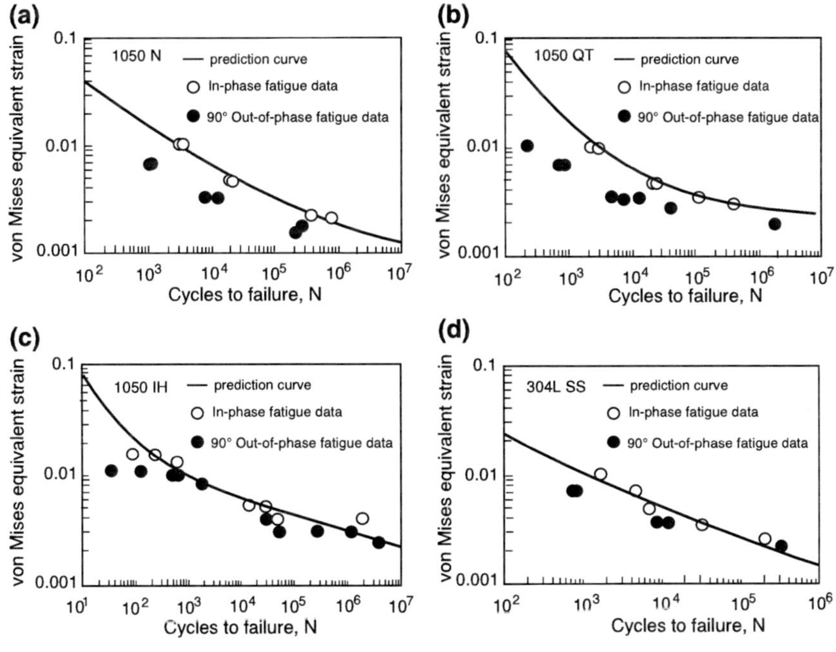

Fig. 13.30 Correlation of IP and OP test data for 1050 steel in three different hardening condition and 304L stainless steel with predictions based on von Mises equivalent strain (modified from [50])

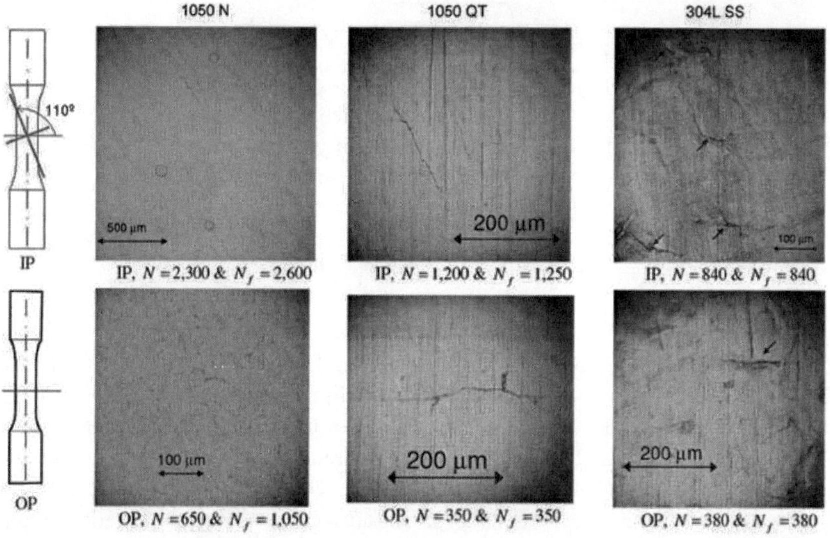

Fig. 13.31 Cracks observed under IP and OP loading in 1050 N and 1050QT steels and in 304L SS [50]

maximum shear plane is normal to the specimen axis. It can be seen that effectively cracks form along those directions.

This finding suggests that a critical plane theory should well predict the fatigue behavior of those four steels. To this end, the Fatemi–Socie criterion will be used. As previously done with the von Mises equivalent strain, Eq. (13.58) will be written coupling with the Basquin-Manson-Coffin Eq. (6.10) to yield

$$\gamma_{\max}\left(1+\kappa_{FS}\frac{\sigma_{n,\max}}{\sigma_y}\right)=\frac{\tau'_f}{G}(N)^{b_o}+\gamma'_f(N)^{c_o} \tag{13.73}$$

k is a material constant and can be found by fitting fatigue data from uniaxial tests to fatigue data from pure torsion tests, τ'_f is the shear fatigue strength coefficient, b_o is the shear fatigue strength exponent, γ'_f is the shear fatigue ductility coefficient, c_o is the shear fatigue ductility exponent, and G is the cyclic shear modulus. The maximum normal stress on the maximum shear strain plane, $\sigma_{n,\max}$, takes into account constitutive behavior including additional hardening due to the non-proportionality of loading. Since shear strain-life fatigue properties (i.e. τ'_f, γ'_f, b_o and c_o) were not available, they were found from the equivalent fatigue properties obtained from in-phase tests (σ'_f, ε'_f, b and c) as $\tau'_f=\sigma'_f/\sqrt{3}$ and $\gamma'_f=\varepsilon'_f\sqrt{3}$. Results are shown in Fig. 13.32. Compared with those presented in Fig. 13.30 a better correlation is achieved. A last word should be spent about the critical plane approach.

As said, the criterion is based on the assumption that damage initiation is due to surface slips. This is important to remember because if the initiation process were

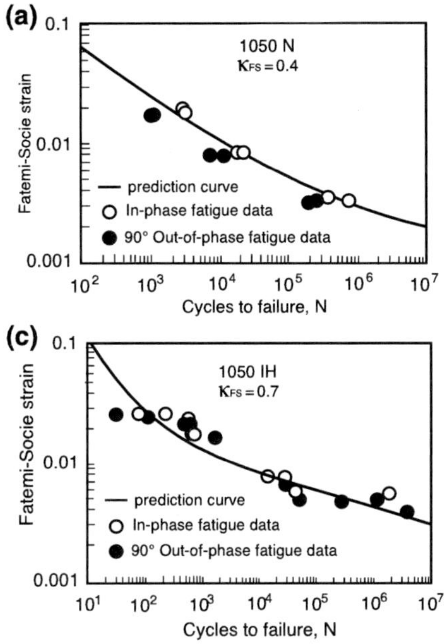

Fig. 13.32 Correlation of IP and OP test data for 1050 steel in three different hardening condition and 304L stainless steel with predictions based on the Fatemi–Socie criterion (modified from [50])

different the criterion may not be appropriate. This is the case of nodular cast iron in which defects of even one or more millimeters are already present. These defects are not represented by degenerated graphite, but are due to macro-shrinkage cavities as that presented in Fig. 3.23.

If a defect is already present it is not necessary to invoke the slip initiation mechanism and the critical plane with the maximum shear stress is not working any more. Now it is expected that the maximum principal stress is the controlling parameter.

Neugebauer [59] working with nodular cast iron type GGG40 and GTS45 with τ_{xy}/σ_x ratio equal to 1.0 and 0.575 and phase differences of 0°, 45° e 90°, found that in OP loadings the fracture plane, with the exception of one case, was always a principal plane of maximum stress as shown in Table 13.3. Moreover, Milella analyzing Neugebauer experimental results assuming in-phase stresses obtained that both cast irons type GGG40 and GTS45 were behaving according to the quadrant of ellipse criterion, with the exception of test at $\tau_{xy}/\sigma_x = 0.575$ and phase angle of 90°, as shown in Fig. 13.33.

Table 13.3 Fracture planes and principal planes in ductile cast iron [59]

τ_{xy}/σ_x	Phase angle (deg)	Fracture plane GGG40 (deg) GTS45		Principal plane (deg)
1.0	0	26	25	32
1.0	45	39	–	32.2
1.0	90	33	24	35.5
0.575	0	20.3	24.5	24.5
0.575	45	23.2	–	22.8
0.575	90	18	21.0	0

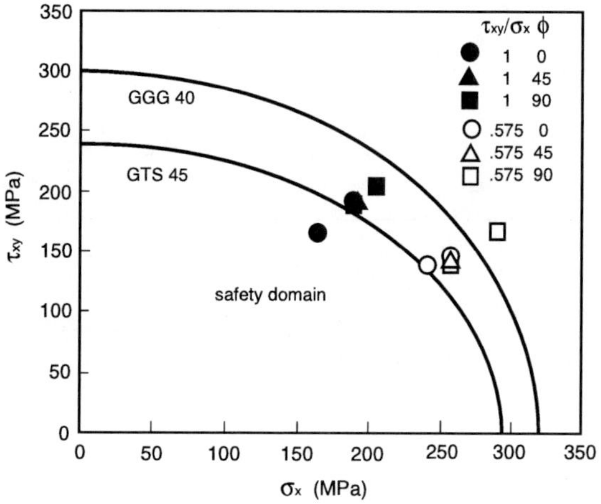

Fig. 13.33 IP and OP experimental results for globular cast iron analyzed with the quadrant of ellipse method. OP data have been considered as if they were in-phase. Data from [59]

References

1. Lanza, G.: Strength of shafting subjected to both twisting and bending. Trans. ASME **8**, 121–196 (1886)
2. Mason, W., (ed).: Alternating stress experiments. IMechE (1917)
3. Haigh, B.P.: The thermodynamic theory of mechanical fatigue and hysteresis in metals. Rep. British Association for the Advancement of Science, pp. 358–368 (1923)
4. Nishiara, T., Kawamoto, M.: The Strength of Metals under Combined Alternating Bending and Torsion, Memoirs, College of Engineering, vol. 10. Kyoto Imperial University, Japan (1941)
5. Gough, H.J.: Engineering steels under combined cyclic and static stresses. J. Appl. Mech. **72**, 113–125 (1950)
6. Gough, H.J., Pollard, H.V., Clenshaw, W.J.: Some experiments on the resistance of metals to fatigue under combined stresses, Aero Research Council, RSM, 2522, Part 1. H.M.S.O, London (1951)
7. Radaj, D., Sonsino, C.M.: Fatigue Assessment of Welded Joints by Local Approaches. Abington Publishing, Cambridge (1998)
8. Gough, H.J.: The Fatigue of Metals. Scott, Greenwood and Son, London (1924)
9. Moore, H.F., Kommers, J.B.: The Fatigue of Metals. McGraw-Hill, New York (1927)
10. Cazaud, R.: Fatigue of Metals. Chapmann and Hall, London (1953)
11. Mann, J.Y.: Proc. ASTM **60**, 602 (1956)

12. Snow, A.L., Langer, B.F.: Low-cycle fatigue of large diameter bolts. J. Eng. Ind., Series. B **89**(1), 53 (1967)
13. Forrest, P.G., Tate, A.E.L.: The influence of grain size on the fatigue behavior of 70/30 brass. J. Inst. Met. 93, 438 (1964–1965)
14. Frost, N.E., Marsh, K.J., Pook, L.P.: Metal Fatigue. Clarendon Press, Oxford (1974)
15. Bundy, R.W., Marin, J.: Proc. Am. Soc. Test. Mater. **52**, 755 (1954)
16. Savaidis, G., Seeger, T.: Consideration of multiaxiality in fatigue life prediction using the closure concept. Fatigue Fract. Eng. Mater. Struct.Fract. Eng. Mater. Struct. **20**(7), 985–1004 (1997)
17. Papadopulos, I.V., Davoli, P., Gorla, C., Filippini, M., Bernasconi, A.: A comparative study of multiaxial high-cycle fatigue criteria for metals. Int. J. Fatigue **19**, 219–235 (1996)
18. Manson, S.S., Halford, G.R.: Multiaxial low-cycle fatigue of type 304 stainless steel. ASME J. Eng. Mater. Techn. 283–285 (1977)
19. Zamrik, S.Y., Mirdamadi, M., Davis, D.C.: A proposed model for biaxial fatigue analysis using the triaxiality factor concept, advances in multiaxial fatigue. ASTM, STP **1191**, 85–106 (1993)
20. Marin, J.: Proceedings of International Conference on Fatigue of Metals, Institution of Mechanical Engineering, pp. 184–194 (London, 1956)
21. Sines, G.: Behavior of metals under complex static and alternating stresses. In: Sines, G., Waisman, J.L. (eds.) Metal Fatigue, pp. 145–169. McGraw-Hill, New York (1959)
22. Sines, G., Ohgi, G.: Transactions ASTM. J. Eng. Mater. Technol. **103**, 82 (1891)
23. Crossland, B.: Proceedings of International Conference on Fatigue of Metals, Institution of Mechanical Engineering, pp. 138–149 (London, 1956)
24. Kakuno, H., Kawada, Y.: Fatigue Eng. Mater. Struct. **2**, 229 (1979)
25. Papadopoulos, I.V.: A new criterion of fatigue strength for out-of-phase bending and torsion of hard metals. Int. J. Fatigue **16**, 377–384 (1994)
26. You, B.R., Lee, S.B.: A critical review on multiaxial fatigue assessments of metals. Int. J. Fatigue **18**, 235–244 (1996)
27. Kaufman, R.P., Topper, T.: The influence of static mean stresses applied normal to the maximum shear planes in multiaxial fatigue. In Carpinteri, A., de Freitas, M., Spagnoli, A. (eds.), Biaxial/ Multiaxial Fatigue and Fracture, 31, p. 133 (Elsivier, ESIS Publication, 2003)
28. Findley, W.N.: A theory for the effect of mean stress on fatigue of metals under combined torsion and axial load or bending. Trans. ASME, J. Eng. Ind. **B81** (1959)
29. Meggiolaro, M.A., de Castro, J.T.P., de Olivera Miranda, A.C.: Evaluation of multiaxial stress-strain models and fatigue life prediction methods under proportional loadings. In da Mattos, C., Alves, M. (eds.), Mechanics of Solids in Brazil, Brazilian Society of Mechanical Sciences and Engineering (2009)
30. Matake, T.: An explanation on fatigue limit under combined stress. Bull. JSME **20**, 257–263 (1977)
31. Brown, M.W., Miller, K.J., (eds).: A Theory for Fatigue Failure under Multiaxial Stress Strain Condition, Vol. 187, pp. 745–755. (Institution of Mechanical Engineering, 1973)
32. Fatemi, A., Socie, D.F.: A critical plane approach to multiaxial fatigue damage including out-phase loading. Fatigue Fract. Engng. Mater. Struct. **11**, 149–165 (1988)
33. Smith, R.N., Watson, P., Topper, T.H.: A stress-strain parameter for the fatigue of metals. J. Mater. **5**(4), 767–778 (1970)
34. McDiarmid, D.L.: A general criterion for high cycle multiaxial fatigue failure. Fatigue Fract. Eng. Mater. Struct.Fract. Eng. Mater. Struct. **14**, 429–454 (1991)
35. McDiarmid, D.L.: A General Criterion for High Cycle Multiaxial Fatigue Failure. Fatigue Fract. Eng. Mater. Struct.Fract. Eng. Mater. Struct. **17**, 1475 (1994)
36. Fatemi, A., Stephens, R.I.: Biaxial fatigue of 1045 steel under In-phase and 90° out-of-phase loading. SAE AEXX: Multiaxial Fatigue (1987)
37. Zamrik, S.Y., Frishmuth, R.E.: The effect of out-of-phase biaxial strain cycling on low cycle fatigue. J. Exp. Mech. **13**, 204–208 (1973)
38. Kanazawa, K., Miller, K.J., Brown, M.V.: Low cycle fatigue under out-of-phase loading conditions. J. Eng. Mater. Tech. **99**, 222–228 (1977)

39. Socie, D.F.: Multiaxial fatigue damage models. J. Eng. Mater. Tech. **109**, 293–298 (1987)
40. Carpinteri, A., Spagnoli, A.: Multiaxial high-cycle fatigue criterion for hard metals. Int. J. Fatigue **23**, 135–145 (2001)
41. Papadopoulos, I.V., Davoli, P., Gorla, C., Fillippini, M., Bernasconi, A.: A comparative study of multiaxial high-cycle fatigue criteria for metals. Int. J. Fatigue **19**(3), 219–235 (1997)
42. Kanazawa, K., Miller, K.J., Brown, M.W.: Cyclic deformation of 1% Cr-Mo-V steel under out-of-phase loads. In Fatigue of Engineering Materials and Structures, Vol. 2, pp. 217–228 (1979)
43. Socie, D.F., Marquis, G.B.: Multiaxial Fatigue. (Society of Automotive Engineers, Inc., 2000)
44. Itoh, T., Sakane, M., Ohnami, M., Socie, D.F.: Non-proportional low cycle fatigue criterion for type 304 stainless steel. ASME J. Eng. Mater. Technol. **117**, 285–292 (1995)
45. Kida, S., Itoh, T., Sakane, M., Ohnami, M., Socie, D.F.: Dislocation structure and non-proportional hardening of type 304 stainless steel. Fatigue Fract. Eng. Mater. Struct.Fract. Eng. Mater. Struct. **20**, 1375–1386 (1997)
46. Itoh, T., Kameoka, M., Obataya, Y.: A new model for describing a stable cyclic stress-strain relationship under non-proportional loading based on activation state of slip systems. Fatigue Fract. Eng. Mater. Struct.Fract. Eng. Mater. Struct. **27**, 957–966 (2004)
47. Shamsaei, N.: Fatigue life prediction under general multiaxial loading employing simple material properties. In: Advisor Ali Fatemi, SAE Fatigue Design and Evaluation committee Meeting, Chrysler Head. Technical Center (19 Oct, 2010)
48. Borodii, M.V.: Obtaining a low-cycle fatigue strain criterion. Strength Mater. 33(3) (2001)
49. Borodii, M.V., Shukaev, S.M.: Additional cyclic strain hardening and its relation to material structure, mechanical characteristics, and life time. Int. J. Fatigue **29**, 1184–1191 (2007)
50. Shamsaei, N.: A Dissertation Entitled Multiaxial Fatigue and Deformation Including Non-proportional Hardening and Variable Amplitude Loading Effects, Submitted to the Graduate Faculty as Partial Fulfillment of the Requirements for the Doctor of Philosophy Degree in Engineering. (The University of Toledo, Aug, 2010)
51. Shamsaei, N., Fatemi, A.: Effect of hardness on multiaxial fatigue behavior and some simple approximations for steels. J. Fatigue Fract. Eng. Mater. Struct. **32**, 631–646 (2009)
52. Liu, J., Zenner, H.: Fatigue limit of ductile metals under multiaxial loading. In: Carpinteri, A., de Freitas, M., Spagnoli, A. (eds.), Biaxial/Multiaxial Fatigue and Fracture, pp. 147–164. (ESIS Publication, 2003)
53. Heidenreich, R., Zenner, H., Richter, I.: Dauerschwingfestigkeit bei Mehrachsiger Beanspruchung. Forschungshefte FKM, Heft 105 (1983)
54. Lempp, W.: Festigkeitsverhalten von Stahlen bei Mehrachsiger Dauerschwingbeanspruchung durch Norrnalspannungen mit Uberlagerten Phasengleichen und Phasenverschobenen Schubspannungen. Diss. University, Stuttgart (1977)
55. Mielke, S.: Festigkeitsverhalten Metallischer Werkstoffe unter Zweiachsig Schwingender Beanspmchung mit Verschiedenen Spannungszeitverlaufen. (Diss. RWTHAachen, 1980)
56. Nishihara, T., Kawamoto, M.: The strength of metal under combined alternating bending and torsion with phase difference, memoirs of the college of engineering, Kyoto Imperial University **11**, 85 (1945)
57. Issler, L.: Festigkeitsverhalten Metallischer Werkstoffe bei Mehrachsiger Phasenverschobener Schwingbeanspruchung. Diss. University, Stuttgart (1973)
58. Bhongbhibhat, T.: Festigkeitsverhalten von Stiihlen unter Mehrachsiger Phasenverschobener Schwingbeanspruchung mit Unterschiedlichen Schwingungsformen und Frequenzen. Diss. University, Stuttgart (1986)
59. Neugebauer, J.: Fatigue strength of cast iron materials under multiaxial stresses of different frequencies. Report FB-175, Fraunhofer Institute für Betribsfestigkeit, Darmstadt (1986)

Chapter 14
Fracture Mechanics Approach to Fatigue Crack Propagation

14.1 History and Overview

So far it has been analyzed the phenomenology that precedes the macro-crack formation and studied all those factors, both mechanical and metallurgical, that have a role in the macro-crack formation that closes Stage I of fatigue (see Sect. 2.7 and Fig. 2.47). The next phase, Stage II, is that in which the macro-crack is directly opened by stresses normal to the crack plane and grows at each applied cycle. Therefore, Stage I is the initiation phase of the fatigue process while Stage II is the growth phase (see Fig. 2.47). Generally, the life or cycles N reported in conventional S–N curves are the cycles to failure and not those N_i to crack initiation (phase 3 in Fig. 1.19). But the failure considered is that of the test specimen that may not necessarily be that of the real structure. This is due to the fact that in a small specimen when a macro-crack if formed there is very little room left to its growth and macro-crack initiation practically coincides or is very close to failure, depending on the stress amplitude. But in real structures there could be room left to further growth before failure occurs. Therefore, design S–N curves should actually be S–N_i curves terminating at the initiation phase that requires the same number of cycles N_i for both the specimen and the real structure, apart from the process volume that in large structure can anticipate initiation because of the larger number and size of inclusions and defects (see Sect. 4.5.1). Also the lack of valid knowledge to treat the crack growth process may have contributed to neglect, in the past, the difference between initiation and final failure. It was known that for a given geometry containing a crack the fatigue crack growth rate (FCGR) or fatigue crack propagation (FCP) was related to the stress excursion and to same sort of crack length function. This had led, in the past, to the proposition of many empirical and semi-empirical formulas based on a rather poor number of experimental data. Such formulas, however, had no theoretical basis and could be applied at most within the restricted boundary of their derivation. Often, they were also in open conflict among each other. It was in 1960 that Paris and Erdogan [1] and later Paris himself, Gomez and Anderson [2],

P. P. Milella, *Fatigue and Corrosion in Metals*,
https://doi.org/10.1007/978-3-031-51350-3_14

found that fracture mechanics could be effectively applied to the study of fatigue crack propagation providing, for the first time, an analytical tool to predict, cycle by cycle, the fatigue growth of a macroscopic crack as a function of its dimension, system geometry and applied load. It was then formulated what would have been improperly called the Paris law. Its introduction shed great enthusiasm among the scientific community that even thought to have solved the fatigue question once and for all. Fracture mechanics, developed to shed some light over the dark mystery of brittle fracture of Liberty ships during World War II, was successfully entering the field of fatigue. Unfortunately, the so called Paris law revealed its actual nature of simple empirical equation. The equation, in fact, could explain neither the threshold phenomenon nor the fatigue crack propagation of small cracks or long cracks under variable amplitude loads. Many uncertainties still remain as, for instance, the impact of metallurgical structure and barriers that were not considered in the Paris equation or the fatigue crack growth retardation by overloads and large scale plasticity. However, the Paris law, though declassed to the rank of common postulate, represented, quite right, the turning point in the fatigue study introducing a new and very powerful tool such as the Irwin stress intensity factor K that a decade before had given birth to fracture mechanics. It is to-day a fact, fully accepted within the scientific community, that the FCGR da/dN is related to the stress intensity factor excursion ΔK by a power law relation known as the Paris-Erdogan equation. Subsequence developments have modified that simple, fundamental equation trying to take into consideration all possible effects that can change the FCP such as the mean stress, crack size (small cracks) and, lately, the crack closure effect introduced by Elder has given a plausible explanation to the apparent crack growth retardation observed in practice, with respect to the theoretical predictions based on the stress intensity factor excursion ΔK, by introducing an effective ΔK_{eff} factor, lower than the theoretical ΔK, due to a wake of residual deformations left behind the crack tip as it propagates. The Elber model has gained wide acceptance in current practice. Stage II of fatigue and FCGR predictions and measurements is the subject of the following sections.

14.2 Fatigue Crack Growth

If experiments are run with specimens containing a crack of length a by applying a cyclic load of excursion ΔF it will be seen that the crack grows at each load step by an amount Δa that, albeit very small, has a finite value that can be measured. If such a growth is recorded at each load cycle as a function of the load excursion ΔF and the number of the applied cycles N, a diagram like that shown in Fig. 14.1 will be obtained [3]. In the specific case of Fig. 14.1 the specimens used were of the type WOL (wedge opening load) one inch thick (see Appendix A). The material was a NiMoV steel at ambient temperature. Note that the tangent at each curve at any point yields the correspondent FCGR da/dN.

Curves like those of Fig. 14.1 suggested researches that the FCGR da/dN should be a function of the crack size a and the applied stress σ

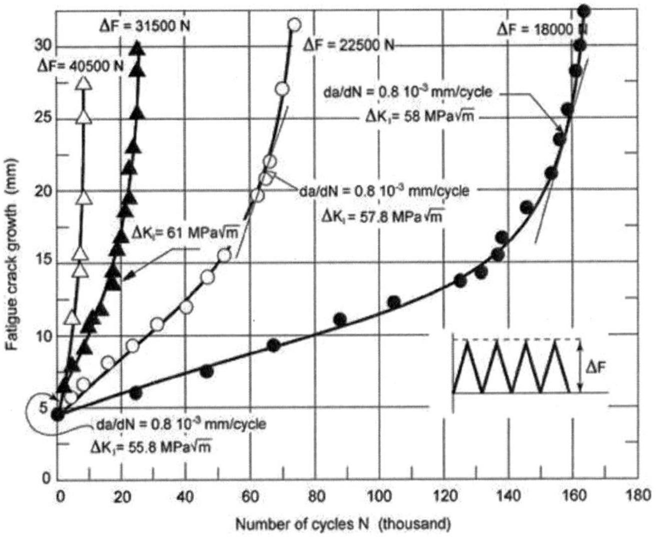

Fig. 14.1 Measurements of crack length vs. elapsed cycles observed at different load excursions ΔF on a 1 T-WOL specimen of NiMoV at ambient temperature (modified from [3])

$$\frac{da}{dN} \propto f(a, \sigma). \tag{14.1}$$

However, one of the first knots to loosen if not the first one was the stress entering Eq. (14.1). It was, in fact, necessary to choose which stress to consider among the external stress σ, the average stress σ_m on the section containing the crack and the local stress σ_l ahead of the crack tip, but in this latter case the distance from the crack tip at which to calculate the local stress remained unknown. Generally, the attention of researchers was concentrated on either the external stress σ or the mean stress σ_m on the cracked section, both of simple determination at least on regular sections as those of the specimens used. The functional dependency more considered, always of empirical nature, was of the type [4, 5]

$$\frac{da}{dN} \propto \sigma^m a^n. \tag{14.2}$$

Liu [5] found that m was equal 2 and n equal 1, which is rather close to what would have been found by applying fracture mechanics to fatigue crack propagation, as it will be seen. Weibull [6], instead, argued a dependency of FCGR on mean stress σ_m on the cracked section. The choice of considering the local crack tip stress σ_l seemed to be the more realistic of them all since it was focusing the attention right on the point where crack was growing, the so called crack tip process zone. However, at variance with the other two stresses, the local stress was not unique, but characterized by a steep gradient typical of any sharp discontinuity so that the question arose as to what point to consider calculating the stress. Furthermore, before linear elastic fracture mechanics (LEFM) was introduced, i.e., before the 50 s, it was not even so easy to quantify the stresses ahead of the crack tip, point by point besides that the theoretical stress at the tip was infinitely large. It was G. Irwin [7] that, among the first, provided the theoretical analytical expressions of the entire stress field for an infinite plate containing a central crack of length $2a$ under a remote bi-axial stress state σ

$$\begin{aligned} \sigma_x &= \frac{K_I}{\sqrt{2\pi r}} \cdot \cos\frac{\theta}{2}\left(1 - \sin\frac{\theta}{2}\sin\frac{3\theta}{2}\right) \\ \sigma_y &= \frac{K_I}{\sqrt{2\pi r}} \cdot \cos\frac{\theta}{2}\left(1 + \sin\frac{\theta}{2}\sin\frac{3\theta}{2}\right) \\ \tau_{xy} &= \frac{K_I}{\sqrt{2\pi r}} \cdot \sin\frac{\theta}{2}\cos\frac{\theta}{2}\cos\frac{3\theta}{2} \end{aligned} \tag{14.3}$$

or with a single equation

$$\sigma_{ij} = \frac{K_I}{\sqrt{2\pi r}} \cdot f_{ij}(\theta). \tag{14.4}$$

The reference system for Eqs. (14.3) and (14.4) is the polar one shown in Fig. 14.2. Similar expressions exist for the shear stresses. The function $f_{ij}(\theta)$ is an non-dimensional factor whose value depends only on the anomaly θ. Note that the stress field given by Eqs. (14.3) exists only in a near vicinity of the crack tip where the terms of higher order of the ratio a/r can be neglected. In practice, at few millimeters from the crack tip this stress field already does not exists any longer. Equations (14.3) and (14.4) indicate that the crack tip stress field σ_{ij} has a singularity of the type $1/\sqrt{r}$, r being the distance from the tip, whose intensity is given by the term K_I that Irwin called stress intensity factor. In the infinite plate considered it is simply equal to

$$K_I = \sigma\sqrt{\pi a}. \tag{14.5}$$

The subscript I means that we are examining the Mode I of aperture of a crack, among the three possible shown in Fig. 14.3. The first, or Mode I, is certainly the most severe since it requires less energy to produce the same effects among the three (fracture of the work-piece). Equation (14.3) indicates that the crack tip stress field is always self-similar, independent of geometry. Only the intensity varies, depending on the applied stress σ and the crack length a. However, the Irwin solution is based on the theory of elasticity that assumes an ideal linear-elastic behavior of the material. At variance, real materials always present a certain amount of plasticity that, although small in the brittle ones, is high enough to cut the theoretically infinite peak stress at $r = 0$ to the level of the yield strength σ_y or higher in strain hardening materials.

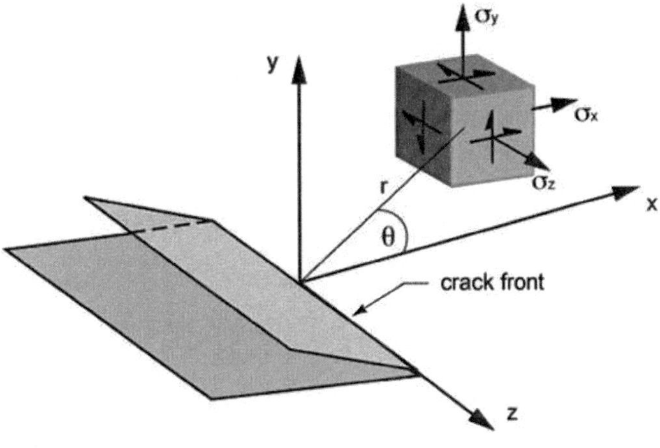

Fig. 14.2 Polar coordinates to describe the stress field ahead of a crack

Therefore, a crack tip plastic zone develops and the elastic stress field moves ahead of the plastic enclave.

If this apical plastic zone is small, in practice below one millimeter radius, it can be completely neglected and the elastic stress field continues to be described by Eq. (14.4) and keeps its $1/\sqrt{r}$ dependency. These are the so called small scale yielding or SSY conditions. From the first of Eq. (14.3) it follows that the radius r_p of the plastic zone is

$$r_p = \frac{1}{2\pi}\left(\frac{K_I}{\sigma_y}\right)^2 \tag{14.6}$$

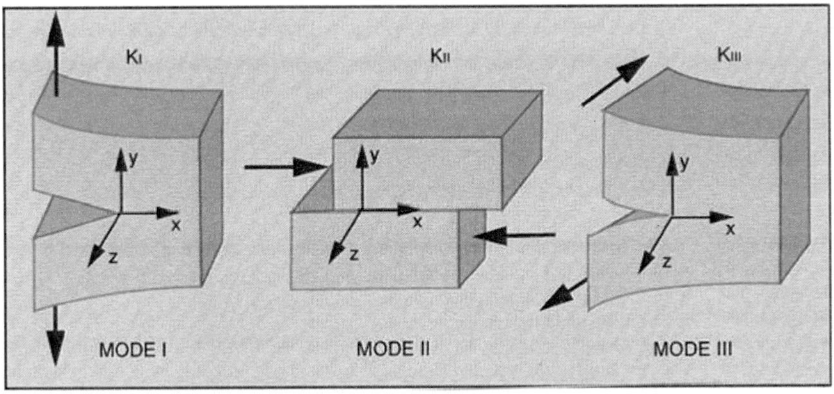

Fig. 14.3 Schematic view of the three possible modes of aperture of a crack

where σ_y is the yield stress of the material. Equation (14.6) provides a first approximation of the plastic zone size assumed to be circular and called the Irwin plastic zone whose real shape is given by two lobes emanating from the crack tip at 60° with the crack plane. Equation (14.6) indicates that the higher the yield strength of the material the smaller the plastic zone and the better SSY conditions are achieved. Moreover, if plain strain conditions are maintained ahead of the crack tip, which is typical of cracks in thick members, than the yield strength σ_y grows by triaxiality (see Fig. 8.23) to 2.5 times the monotonic yield strength farther favoring the existence of SSY conditions. Under the conditions of validity of LEFM, in practice wright ahead of the crack tip and in case of SSY, the elastic stress field is not influenced at all by the small crack tip plastic zone and continues to be given by Eq. (14.4). For $\theta = 0$, i.e., on the crack plane, the non-dimensional factor $f_{ij}(\theta)$ is exactly equal one. The stress field is schematically shown in Fig. 14.4 in which the crack tip plastic zone is intentionally enlarged just for illustration purpose to show that, notwithstanding it continues to enlarge with increasing stress level, in as long as the SSY conditions are maintained the crack tip elastic stress field remains always self-similar with a singularity of the type $1/\sqrt{r}$. Equation (14.3) refer, as it has been said, to an infinite plate under remote biaxial loading. For any other geometry and loading conditions, the stress intensity factor K_I takes the general form of the type (see Appendix A)

$$K_I = \sigma\sqrt{\pi a} \cdot f_1(a/w)$$
$$K_I = \frac{F\sqrt{a}}{sw} \cdot f_2(a/w) \tag{14.7}$$

depending on whether the system containing the crack of semi length a is subjected to an external stress σ or force F. In Eq. (14.7) the specimen width w enters the corrective factor $f(a/w)$ just to mean that this non-dimensional factor, that is not the angular factor $f(\theta)$ of Eq. (14.3), depends on the geometry of the system. In the infinite plate with central crack $2a$ it is worth one. In the International System unit the stress intensity factor K is measured in MPa$\sqrt{\text{m}}$. In the English unit it is measured in ksi$\sqrt{\text{in}}$, with 1 MPa$\sqrt{\text{m}}$ = 0.910048 ksi$\sqrt{\text{in}}$.

14.3 The Paris Postulate

Knowing the crack tip stress field we can try to relate the FCGR *da/dN* to the ΔK excursion. The first check to do to demonstrate the applicability of fracture mechanics and ΔK to fatigue is to check that to equal FCGR *da/dN* correspond equal ΔK. This can be done using the experimental diagram of Fig. 14.1 selecting a point on a curve, for instance point A relative to the force excursion $\Delta F = 18{,}000$ N and measuring the value of the tangent to that point. The tangent provides the value of the FCGR that in A is $da/dN = 0.8 \cdot 10^{-3}$ mm/cycle. Reading the corresponding value Δa we know the total crack length a and, therefore, the value of ΔK by using the second

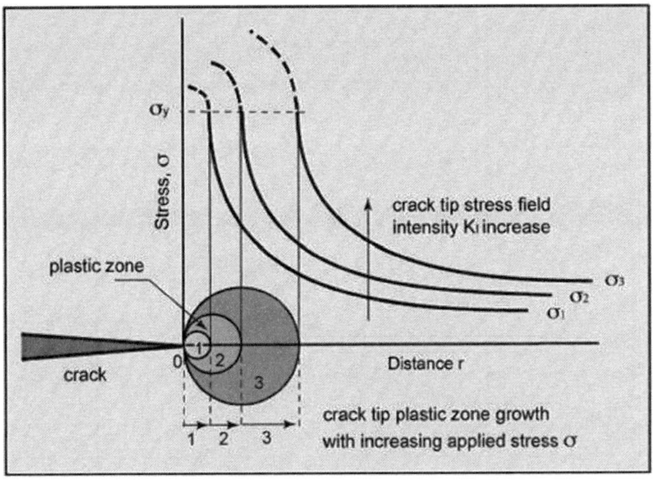

Fig. 14.4 Schematic view of the stress field ahead of the crack tip plastic zone

of Eq. (14.7). The value of $f_2(a/w)$ for the WOL geometry considered is given in Appendix A. By doing so it turns out that it is $\Delta K = 58$ MPa$\sqrt{}$m. Next step is to find point B on the second curve, $\Delta F = 22{,}500$ N, where the FCGR is equal to $0.8 \cdot 10^{-3}$ mm/cycle. This is shown in Fig. 14.1. by using the same procedure it yields $\Delta K = 57.8$ MPa$\sqrt{}$m. Indeed, the same value previously found. The same result would be obtained working with the other two curves corresponding to $\Delta F = 31{,}500$ N and 40,500 N, respectively, as indicated in Fig. 14.1. This actually means that equal applied ΔK result in equal FCGR *da/dN* and vice versa. Therefore, ΔK is the right key to calculate the FCGR in a given structure, note ΔK and not just σ. It is the entire stress field ahead of the crack tip and not just a single value of stress, whatever it may be σ, σ_m or $\sigma_{\max}$ to determine the FCGR. The first demonstration of what has just been said, i.e., that the FCGR should depend on the applied ΔK, was given by Paris [8] and Swanson et al. [9] through experiments that would become famous. The former used two different specimens, a panel with a central crack of length $2a$ under remote traction CCP(T) (Central Crack Panel in Traction, see Appendix A) and a panel of the same geometry, but loaded in a different way (wedge loaded), as schematically shown in Fig. 14.5. Both panels were made of 7075-T6 aluminum alloy. The ΔK expressions for the two panels are

$$Panel\ under\ remote\ traction \rightarrow \begin{array}{l} K_I = \sigma\sqrt{\pi a} \\ \Delta K_I = \Delta\sigma\sqrt{\pi a} \end{array} \tag{14.8}$$

$$Wedge\ loaded\ panel \rightarrow \begin{array}{l} K_I = \frac{F}{\sqrt{\pi a}} \\ \Delta K_I = \frac{\Delta F}{\sqrt{\pi a}} \end{array} \tag{14.9}$$

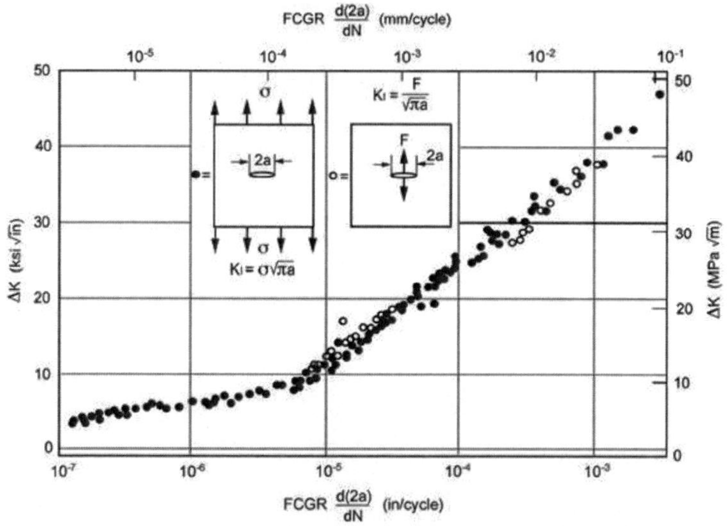

Fig. 14.5 Comparison of FCGR data generated in 7075-T6 aluminum alloy (modified from [8])

respectively. Note the two panels were chosen to differentiate between the effects of the applied mean stress σ_m, the nominal stress σ and ΔK. In the first panel, in fact, as the crack grows both ΔK and the mean stress on the section containing the crack increase. In the second, instead, ΔK decreases (see Eq. 14.9) while σ_m increases. It is clear that if σ_m were the governing parameter than the FCGR should increase with increasing crack length a and ΔK would fail in predicting the growth. The results obtained by Paris with the two panels are shown in Fig. 14.5 [8]. They confirmed that the FCGR *da/dN* was obeying to ΔK and not to σ_m that had a completely opposite trend in the two panels. The second interesting experiment was done by Swanson, et al. [9]. They made use of type 7079-T6 aluminum alloy sheets with a central crack remotely loaded in traction. The load was changed in such a mode to maintain constant either ΔK_I or the average stress σ_m on the section as the crack grew. When ΔK_I was kept constant also the rate *da/dN* was constant. At variance, when it was the average stress σ_m to be constant the FCGR was first constant, but then it sharply decreased with increasing crack length a. These results were indeed confirming that the FCGR *da/dN* was governed by the ΔK. The Paris finding that the FCGR was linear in a log–log scale *da/dN* versus ΔK, as shown in Fig. 14.5, indicated that the FCGR was described by a power law relationship

$$\frac{da}{dN} = C \cdot \Delta K^n \tag{14.10}$$

in which C and n are materials characteristics. Paris and Erdogan were apparently the first to discover the power law relationship (14.10) that became widely known as the Paris law or Paris-Erdogan equation. They proposed an exponent of four. Subsequent studies, still continuing, have shown that the exponent n is not necessarily four, but ranges from two to seven, depending on the material. Note that the results obtained by Liu [5], Eq. (14.2), with $m = 2$ and $n = 1$ correspond to a Paris law with $n = 2$ that is, as we shall see, very close to the experimental results obtained on most steels. Knowing the values of the constant C and n it is possible to infer the number of cycles needed to grow a crack from an initial size a_o to a final one a_f by just integrating Eq. (14.10)

$$N = C \cdot \int_{a_o}^{a_f} \frac{da}{\Delta K^n}. \tag{14.11}$$

Vice versa, the growth of a crack after N cycles will be given by

$$a = \cdot \int_{0}^{N} \Delta K_I^n dN. \tag{14.12}$$

Paris findings had an immediate echo in the fatigue and fracture mechanics world and were followed by large experimental campaigns with almost any material. Effectively, these campaigns showed the validity of the Paris-Erdogan Eq. (14.10), but they also revealed its restricted conditions of validity where the fatigue cycles have constant amplitude and there is no mean stress acting. Both the empirical nature of Eq. (14.10) and the particular field of validity induce researchers to express in terms of Paris postulate rather than Paris law.

14.4 Empirical FCP Equations

The typical trend *da/dN* versus ΔK for metals is schematized in Fig. 14.6 in a log–log scale. It has a ∫ shape known as sigmoidal curve. Three different zones can be distinguished. The central part of the curve, Region II, is that of the linear growth governed by the Paris law (14.10). In this zone the FCGR *da/dN* is largely independent of the microstructure and of the inclusions present in the material, in as long as corrosive phenomena do not happen. The first zone, Region I, is the so called threshold zone. Its characteristic is that below a certain threshold\u00a0value, indicated as ΔK_{th}, there is no apparent growth. It is the equivalent of the fatigue limit in smooth specimens. This is the zone of macro-cracks just formed by conventional fatigue, Stage II of fatigue\u00a0(see Fig. 2.47). They are no longer than 300 μm to 500 μm (see Sect. 2.3) so that the corresponding ΔK_I is usually very small, smaller than 5–6 MPa$\sqrt{m}$. In general, this is the zone of small cracks that have a particularly steep FCGR gradient (see Sect. 15.2), but it is also the zone of long cracks subjected to low stress state that results in a very low applied ΔK_I. Being the zone of very small cracks (fatigue macro-cracks), it should not come as a surprise that, at variance with Region II, the FCGR *da/dN* is strongly dependent on the microstructure and mechanical characteristics. Inclusions, microscopic defects and microstructural barriers, in general, have, in fact, a size of the same order of the macro-cracks and, therefore, interference must be expected. Also the *R* ratio has a strong influence on the FCGR (see Sect. 11.4). Finally, the third region, Region III, is that of low cycle fatigue where stress excursion is close or above the yield strength of the material. Now the FCGR *da/dN* is particularly high increasing more and more as the stress intensity factor excursion ΔK_I approaches the toughness of the material K_{Ic} at which fracture occurs at ¼ cycle, Stage III of fatigue (see Fig. 2.47). Also in this zone as in Region I microstructure and material properties play an important role. A number of researchers have developed equations that describe all or part of the sigmoidal FCGR *da/dN*-ΔK relationship. Many are empirical, dough based on physical considerations. Forman [10] proposed the following relationship for Regions II and III

$$\frac{da}{dN} = \frac{A \cdot \Delta K^{m-1}}{\frac{K_c}{K_{\max}} - 1}. \tag{14.13}$$

Thus the FCGR becomes infinite as K_{max} approaches the toughness of the material K_c, which actually means final fracture, Stage III of fatigue. Note that in the Foreman Eq. (14.13) A and m are material characteristic whose value is different from C and n of the Paris law (14.10). Weertman [11] proposed an alternative equation for Regions II and III

$$\frac{da}{dN} = \frac{B \cdot \Delta K^4}{K_c^2 - K_{\max}^2}. \tag{14.14}$$

The ΔK exponent in the Weertman Eq. (14.14) may be variable and not necessarily four. Klesnil and Lucas [12] proposed an equation for Regions I and II modifying Eq. (14.10) to account for the threshold ΔK_{th}

$$\frac{da}{dN} = D\left(\Delta K^p - \Delta K_{th}^p\right). \tag{14.15}$$

McEvily [13] developed another equation that can be fit to the entire sigmoidal curve

$$\frac{da}{dN} = H\left(\frac{\Delta K - \Delta K_{th}}{K_c - K_{\max}}\right)^q. \tag{14.16}$$

Equations (14.13) to (14.16) are based on LEFM and, therefore, assume SSY conditions. Indeed, it is quite difficult to believe that in Region III the elastic behavior is still maintained and SSY conditions are respected. Dowling and Begley [14] overcame this shortcoming by using the *J*-Integral approach to fatigue crack propagation

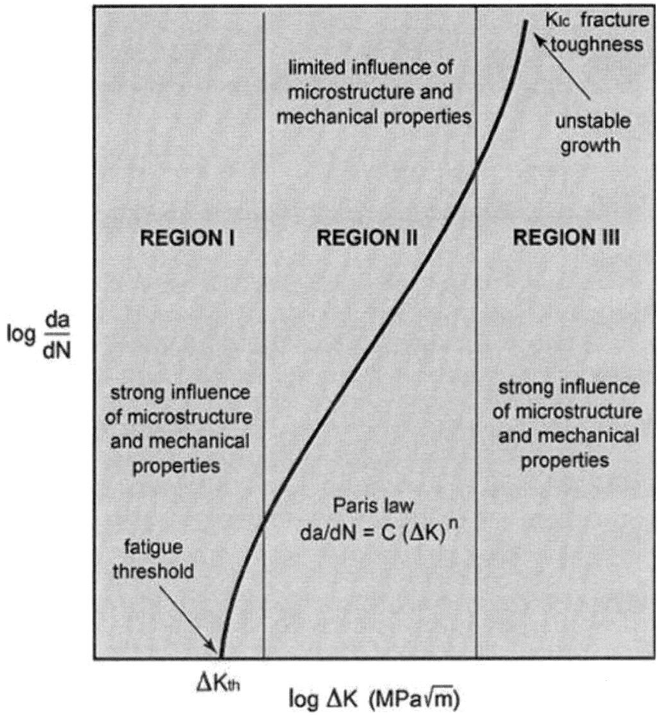

Fig. 14.6 Typical sigmoidal FCGR *da/dN* in metals

under large scale yielding conditions (LSY) where K is no longer valid. They fit the FCGR data to a power law expression in ΔJ

$$\frac{da}{dN} = C \cdot \Delta J^m. \tag{14.17}$$

An interesting equation that describes the entire sigmoidal curve was developed by Newman [15]

$$\frac{da}{dN} = C_1 \Delta K_{eff}^n \frac{1 - \left(\frac{\Delta K_{op}}{\Delta K_{eff}}\right)^2}{1 - \left(\frac{K_{\max}}{K_{c,cicl}}\right)^2} \tag{14.18}$$

in which it is

$$\Delta K_o = C_2 \left(1 - C_3 \frac{K_{op}}{K_{\max}}\right)$$
$$\Delta K_{eff} = \Delta K - \Delta K_{op} \tag{14.19}$$

where $K_{c,cicl}$ indicates the toughness of the material determined under cyclic loads, K_{op} is the so called *K opening* that derives from the crack closure model that will be discussed in Sect. 11.4 and C1, C2 and C3 are material constant that for the 2219-T851 aluminum alloy were found to be

$$C_1 = 2.486 \cdot 10^{-10};\ n = 3.115$$
$$C_2 = 2.97;\ C_3 = 0.8;\ K_{c,cicl} = 77\ \text{MPa}\sqrt{\text{m}}. \tag{14.20}$$

The importance of Eq. (14.18) resides in the fact that the toughness of the material enters into an equation predicting the FCGR. It is not precisely the first time that it happens because precisely in the same year Milella proposed an equation to predict the FCGR in which the toughness of the material played a fundamental role. Milella equation will be discussed in Sect. 15.1.

14.5 Experimental FCGR Measurements

Paris observations and the publication of the Paris-Erdogan Eq. (14.10) vitalized extensive experimental campaigns, in practice, with any metallic material. The specimens used for the measurement of Δa and of the relative ΔK were typical fracture mechanics specimens (see Appendix A). Figure 14.7 provides a schematic synoptic picture. A first classification of these specimens can be done on the basis of the applied ΔK as

1. specimens with applied K_I that increases with the ratio *a/w*;
2. constant K_I specimens;
3. specimens with applied K_I that decreases with the ratio *a/w*.

where a is the crack length and w the specimen width. Historically, the first to specimens be used were DEC and CCP specimens in traction. Today these specimens have been discarded because of their limited capability to guarantee SSY conditions. The preferred specimens are the C(T) or compact and the TPB for their high capability to sustain plane stress conditions. Also the RNB is excellent, from the point of view of SSY conditions, perhaps the best. Wedge opening specimens in which the applied ΔK_I decreases with the growth of the crack are particularly effective in the study of stress corrosion (see Chapter 11). The analytical expressions for the calculation of ΔK_I associated to each geometry are given in Appendix A. Normally, a sharp V shaped notch is introduced by machining. A sharp notch cannot be considered a crack, therefore after its introduction the specimen is fatigue pre-cracked at a ΔK_I lower than the minimum ΔK_I expected to be applied during the testing. This is done to avoid the formation during the pre-cracking of a plastic zone at the crack tip larger than that corresponding to the ΔK_I applied for the testing. In fact, if the plastic zone is buried into a larger one preceding the testing the fatigue crack growth would be retarded, as explained in Sect. 12. During the testing the length of the crack changes

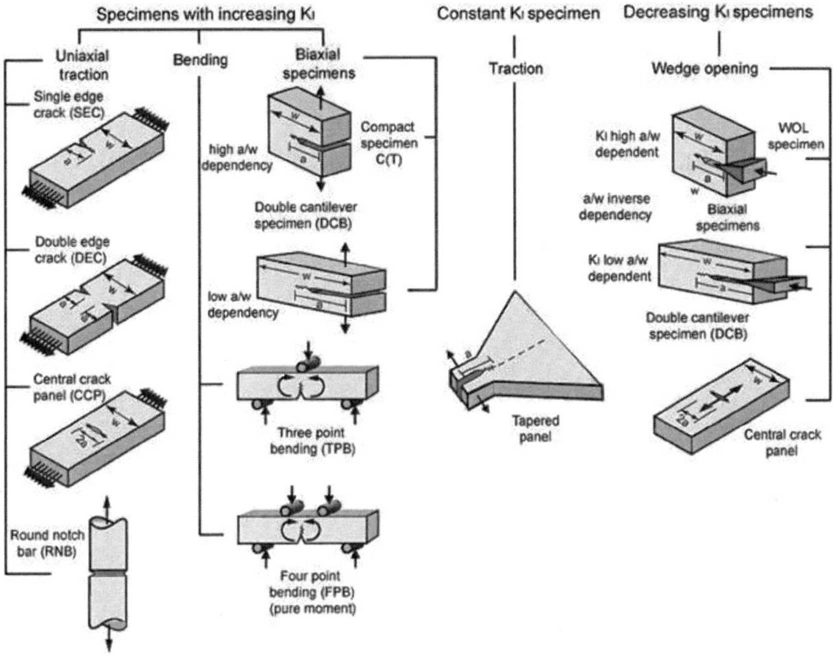

Fig. 14.7 Synoptic picture of fracture mechanics specimens used in the FCGR study

and must be monitored. One of the first and most common technique to develop length *a* versus cycles *N* data consisted in visual measurements of crack growth made with a microscope at relatively low power magnification (~20x). For visual examination it is advisable to put reference marks on the specimen surface, such as tiny Vickers indentations or a grid printed by a photo emulsion technique, to avoid continuous recalibrating during the test.

The visual examination procedure can have an accuracy of 0.1 mm, but it has also a drawback in that it is not possible to machine V grooves on the sides along the crack path to avoid possible crack growth out of the initial plane and to introduce triaxiality also on the surface of the specimen to reduce the thickness. Another drawback of the visual examination is that the crack can grow preferentially in the center where triaxiality is maintained rather than on the surface. This is shown in Fig. 14.8 [16] for a compact C(T) specimen. It can be seen the nail shaped or tunneling growth of the crack that may lead to an undervaluation of the crack size. A procedure, complex and sophisticated, was introduced by Westinghouse Nuclear [17] in his testing laboratories that during the ′60 s and ′70 s, in particular, were among the most active worldwide in the field of fracture mechanics.

The method was based on an ultrasonic monitoring of the growing crack, schematically shown in Fig. 14.9 with a WOL compact specimen. The procedure was used to permit either semi-automatic or completely automatic monitoring of the crack growth

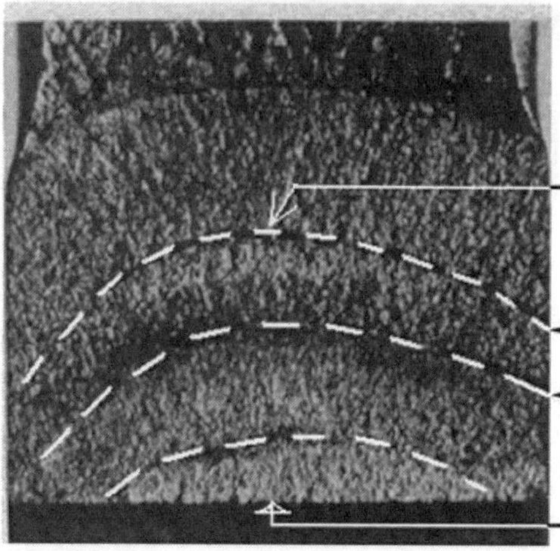

Fig. 14.8 Fatigue crack growth by tunneling (modified from [16])

behavior and had a sensitivity of about 0.25 mm. Today, direct visual measurements of the crack size have been substituted by indirect readings based on the specimen compliance. This is measured by a clip gauge mounted on the specimen at the crack mouth that provides the continuous reading of the so called COD (crack opening displacement) or, better, CMOD (crack mouth opening displacement) during crack propagation, as shown in Fig. 14.14. As the K_I also COD is a function of the crack length a (see Eq. A.2.8 in Appendix A). Therefore, at each applied cycle N, the current crack length a is deduced from the COD reading, as schematized in Fig. 14.10c, together with the applied ΔK_I. From the tangent to the Δa versus N curves it is also known the value of the FCGR *da/dN* and the diagram of *da/dN* versus ΔK_I finally obtained, as schematized in Fig. 14.10d.

A technique that has received a particular attention in the last decades and is widely used today is that based on the measurement of the electric resistance of the specimen under fatigue testing, also known as direct current potential drop o DCPD [18–23]. The specimen is traversed by an electric direct current of about 10 A while the electric potential is measured between two points on either sides of the crack front. As the crack grows, the electric resistance between the two points increases and a consequent potential drop is measured. By calibrating the system, the current crack length is inferred. The method is already tested by more than 20 years and already reached an appreciable experimental maturity. Its sensitivity is particularly high, of the order of 0.0015 mm. However, the DCPD technique seems to be unreliable under rapidly varying loads. In fact, it appears that the presence of a rapidly varying stress field generates magneto-static effects that interfere with the direct current field modifying the potential drop. This has led to the development of a new technique based on high frequency alternated current known as *alternated current potential drop* or ACPD [17, 23]. Figure 14.11 is a schematic of the ACPD test rig with a TPB

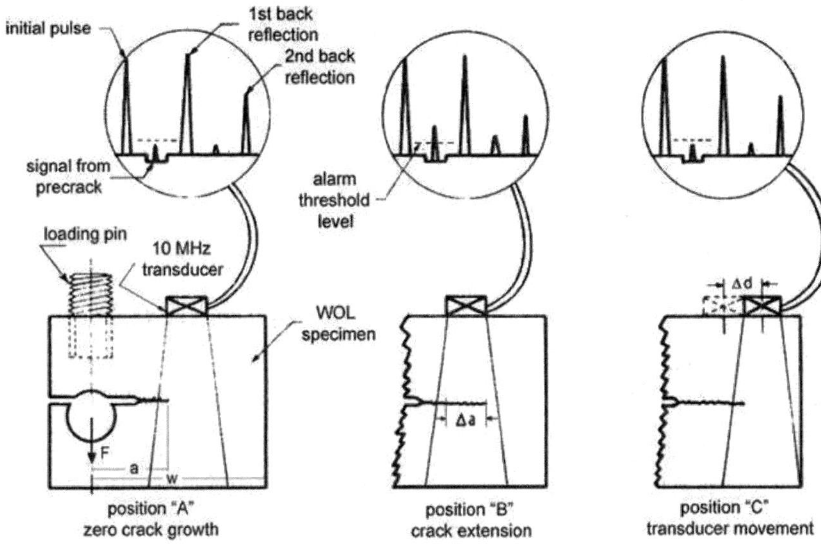

Fig. 14.9 Schematic representation of the Westinghouse ultrasonic technique to measure crack growth [17]

specimen (three point bending). As it can be seen, the potential reading points are placed on either side of the crack ahead of the initial tip. Figure 14.12 [24] shows a comparison of the ACPD technique measurements and the COD derived crack growth readings of Fig. 14.14.

The specimens used were TPB of low alloy carbon steel type A 533B of 445 MPa yielding and 597 MPa ultimate used in the nuclear industry for the construction of pressure vessels. The star in Fig. 14.12 indicates the crack final length seen at naked eye. Whatever the measuring technique may be, the result of constant amplitude load testing is always of the type already shown in Fig. 14.1 and the procedure to infer the FCGR *da/dN* is that of Fig. 14.14. Figure 14.13 present the result obtained using the technique described with WOL specimens of NiMoV steel for forgings, 585 MPa yielding [3]. The load was varying from the minimum value to the maximum with a frequency of 1800 cycles per minute with the ratio $R = F_{min}/F_{max}$ of 0.1.

Effectively, in accordance with the Paris postulate, a power law equation can be fit to the experimental FCGR data

$$\frac{da}{dN} = 2.9 \cdot 10^{-8} (\Delta K_I)^{2.5}. \tag{14.21}$$

Another example of FCGR measurements obtained with 1″ WOL specimens is given in Fig. 14.14 for a high strength steel type 4340, with 910 MPa yielding and 1400 MPa ultimate strength [25]. The equation interpolating the experimental data is

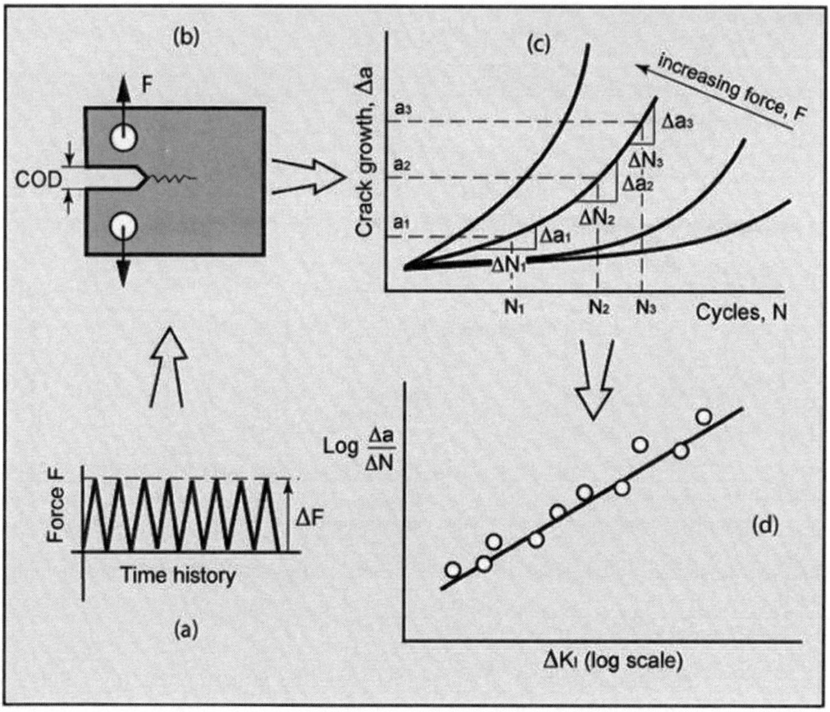

Fig. 14.10 Schematic of FCGR $\Delta a/\Delta N$ measurements from COD readings

$$\frac{da}{dN} = 3.7 \cdot 10^{-8}(\Delta K_I)^{2.5}. \tag{14.22}$$

Interesting enough, the strength of the material seems not influence the FCGR that has the same exponent 2.5 of the NiMoV previously seen.

14.6 Striations and Fatigue Crack Growth

We have already said in Sects. 2.4.1 and 2.4.3 that the fatigue crack growth leaves some characteristic microscopic features on the fracture surface of the metal called striations. The mechanics of their formation has been studied and said that each striation marks a single cycle, at least beyond a minimum. FCGR *da/dN* (see Fig. 2.29). Striation, then, are a kind of fingerprints left by the crack as it grows by fatigue. Consequently, the Paris-Erdogan Eq. (14.10) must necessarily provide an analytical relationship between striation spacing and the ΔK_I, which actually means that SEM observations and instrumental readings, of whatever technique, of crack advance per cycle must yield same result. The matter has been subject of discussion and studies

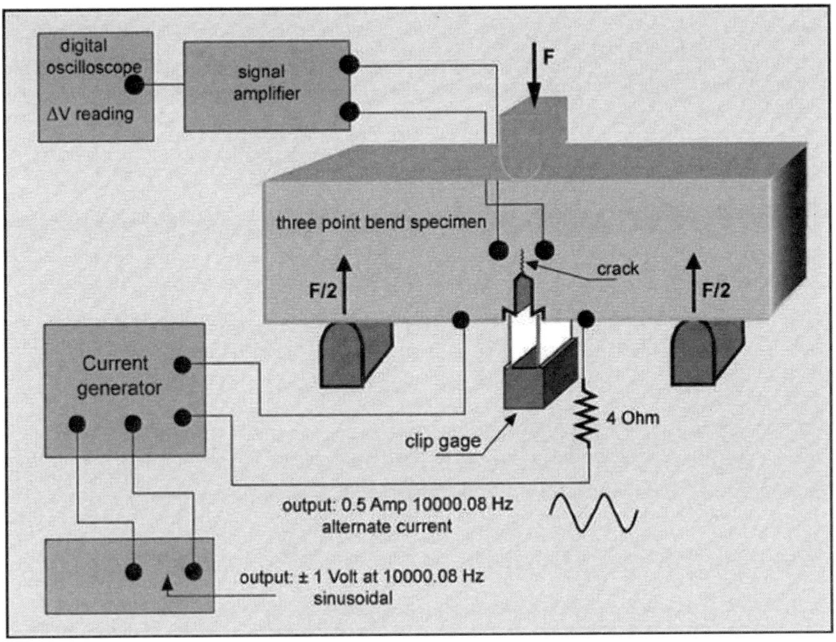

Fig. 14.11 Schematic of the ACPD technique used in rapidly varying loads fatigue tests

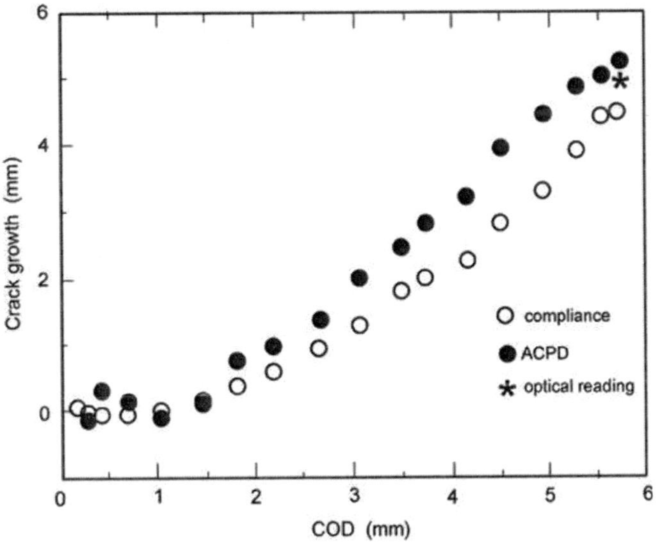

Fig. 14.12 Comparison between fatigue crack growths obtained by compliance and ACPD techniques (modified from [24])

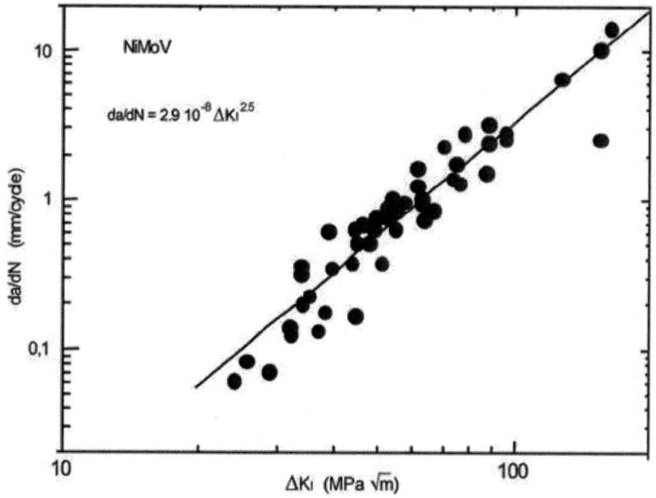

Fig. 14.13 Experimental FCGR *da/dN* obtained with specimens of NiMoV steel for forgings (modified from [25])

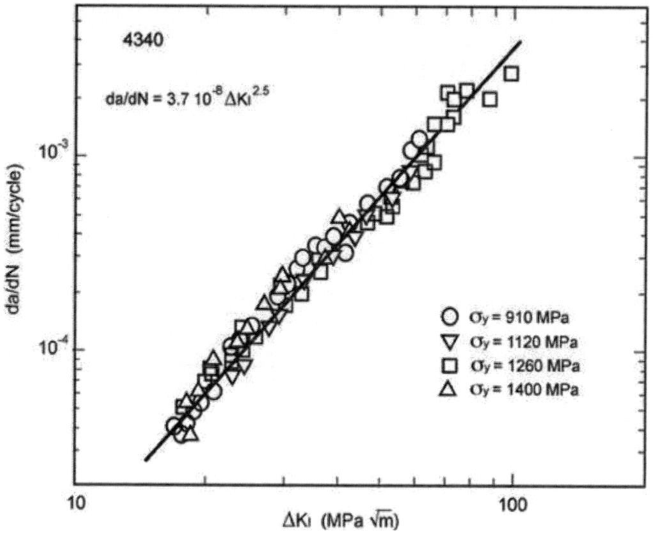

Fig. 14.14 Experimental FCGR *da/dN* obtained with specimens of 4340 steel hardened to four different yield strengths (modified from [25])

ever since the introduction of fracture mechanics to fatigue [26–36]. Few examples will be shown starting with results obtained by Hertzberg and Paris [26] on an aluminum alloy type 2024-T3 shown in Fig. 14.15. Open circles refer to instrumental readings of crack propagation per cycle while full circles are SEM observation of striations. As it can be seen the agreement is rather good. It is convenient to recall that so far no striation smaller than 0.1–0.2 μm has ever been observed, as indicated in Fig. 14.15 and already said. This is not a lower limit of SEM resolution, but is due to the fact that below 0.1 μm of growth each cycle seems not be capable of marking a visible sign on the metal surface. It takes hundreds or thousands of cycles before a single striation of this size is formed. Other examples are shown in Fig. 14.16 [32] and Fig. 14.17 [33] that refer to inconel X-750 and type 2017-T4 aluminum alloy, respectively.

Also in this case a very good correlation is observed that cannot go below 0.1 μm of growth for the reason just explained. In the case of the aluminum alloy, 270 MPa yielding and 435 MPa ultimate strength, the study has been done at different R ratios. It must be observed the significant increase of FCGR da/dN moving from $R = -1$ to high mean stress cycling with $R = 0.8$. The R ratio effect will be discussed in Sect. 15.4. It suffices to change from symmetrical fatigue ($R = -1$) to pulsating fatigue with $R = 0.1$ to increase the rate by a factor ten. Bates and Clark [27], among the first to make a systematic study of the subject, found an empirical relationship between the striation spacing and the stress intensity factor excursion normalized to the Young's modulus of the material $\Delta K_I/E$

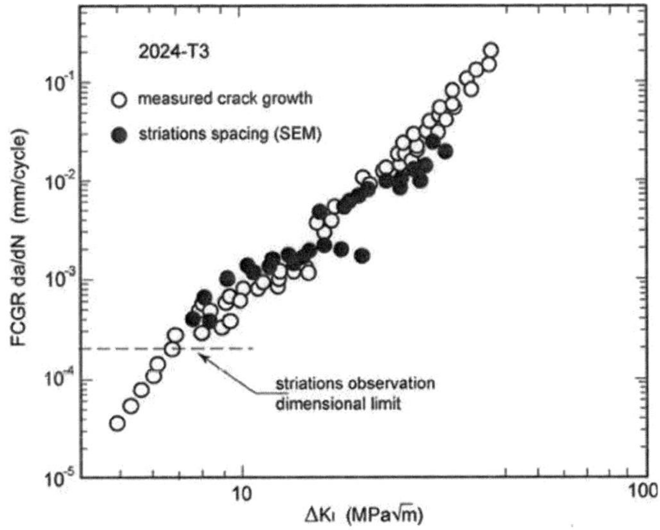

Fig. 14.15 Comparison between measured crack growths and striation spacing observed on 2024-T3 aluminum alloy (modified from [26])

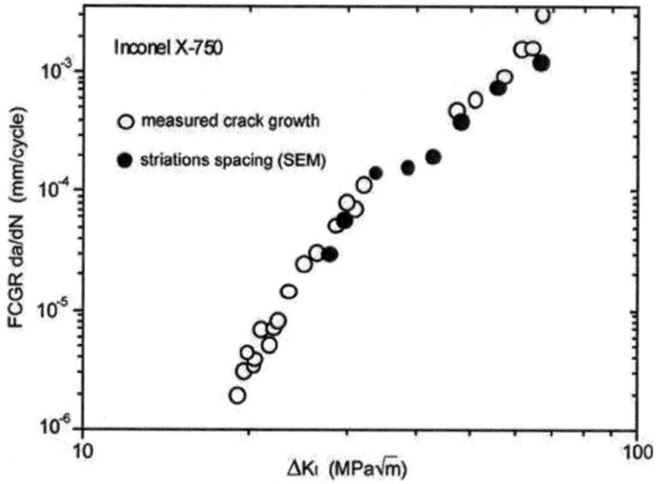

Fig. 14.16 Comparison between measured crack growths and striation spacing observed on inconel X-750 alloy (modified from [32])

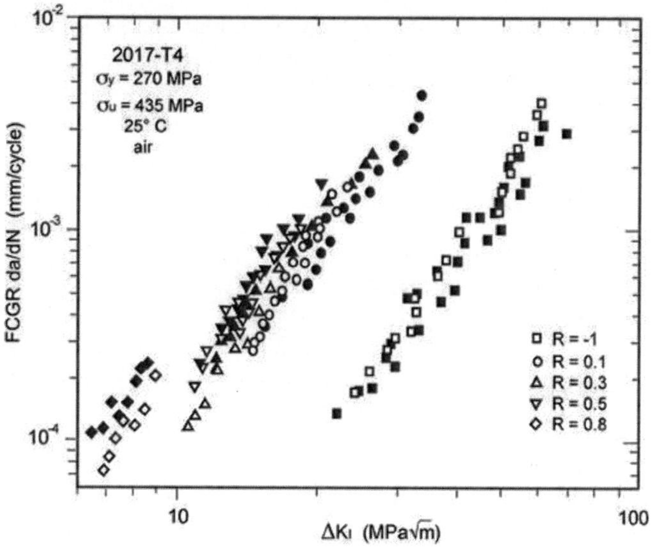

Fig. 14.17 Comparison between measured crack growths and striation spacing observed on 2017-T4 aluminum alloy at different R ratios (modified from [33])

$$Striation\ spacing(\text{mm}) = 6 \cdot 10^3 \cdot \left(\frac{\Delta K_I}{E}\right)^2. \tag{14.23}$$

Figure 14.18 [27] presents the empirical basis of the Bates and Clark finding. The exponent of two that appears in Eq. (14.23) induces to believe that striation spacing is somehow related to the Irwin plastic zone size (14.6) or to the crack tip opening displacement or CTOD that, as the plastic zone radius, is proportional to the square of the stress intensity factor K_I

$$\begin{aligned} CTOD &= \psi \cdot \frac{K_I^2}{\sigma_y E} \\ &= \frac{\psi}{\varepsilon_e} \cdot \left(\frac{K_I}{E}\right)^2 \end{aligned} \tag{14.24}$$

where ε_e is the elastic strain at yielding and ψ a non-dimensional factor. This dependency on the square of K_I is in accordance with all theoretical models of striation formation [16, 34, 35]. However, the exponent n of the Paris-Erdogan Eq. (14.10) varies from 2 to even 5, as in cast iron. Values of n higher than 2 cannot be explained in terms of crack tip plastic zone or CTOD. Of particular interest is the comparison of the Bates and Clark empirical Eq. (14.23) with FCGR experimental results obtained later on different materials, from martensitic steels to ferritic and austenitic, from aluminum alloys to titanium alloys, from magnesium to molybdenum etc. This is shown in Fig. 14.19.

The upper and lower bound lines have the ΔK_I exponent equal to 3.5, much higher than 2, but inside the band individual slopes can be distinguished having exponent

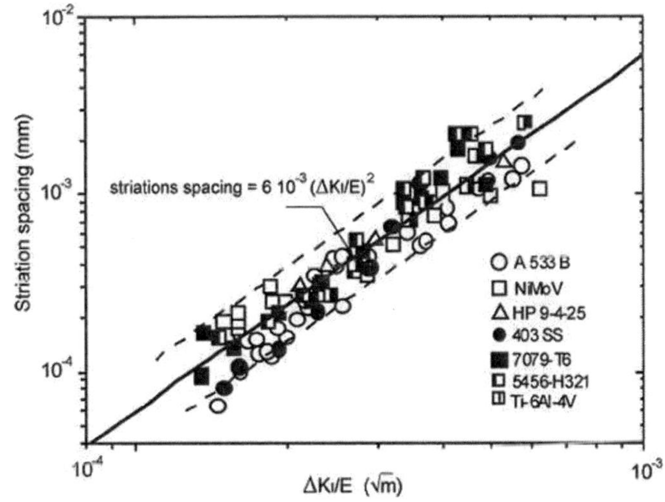

Fig. 14.18 Data base used to derive Eq. (14.20) (modified from [26])

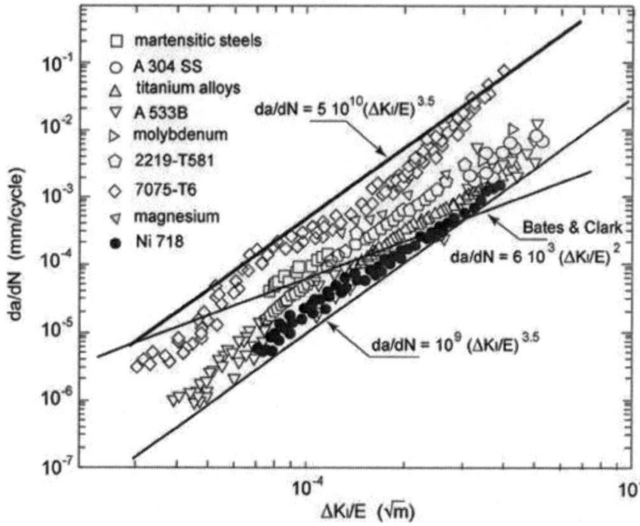

Fig. 14.19 Comparison between Bates and Clark empirical equation and FCGR experimental results

from 2.5 to 4. As it can be seen, the Bates and Clark empirical Eq. (14.23) tends, in general, to overestimate the experimental growth measurements at the lower $\Delta K_I/E$ values while it underestimate the growth at higher values. Effectively, this tendency has been verified by many researchers [26–31, 34]. This is because the Bates and Clark empirical equation is averaged on a large variety of materials therefore its exponent of two must be regarded as a mean value and not a real one. Often, two types of striations appear at the SEM analysis; one has a morphology so called large or multiple that collects many single and smaller striation of the second type, the fine one.

It can be seen in Fig. 14.20 [35] for a C-Mn steel of 340 MPa yield and 570 MPa ultimate strength subjected to a load excursion resulting in $\Delta K_I = 35$ MPa$\sqrt{\text{m}}$. The larger striation spacing is in good agreement with an estimate of half the maximum crack tip opening displacement (CTOD) while smaller striations are in good correlation with the macroscopic crack growth rate shown in Fig. 14.21 [35]. Striations do not appear in all regions of fatigue. They are typical of Region II of Fig. 14.6. The variation of the fracture surface appearance with ΔK is is shown in Fig. 14.22 for Inconel X-750 [32]. The FCGR, *da*/*dN*, plotted versus ΔK shows that at ΔK higher than 30 MPa$\sqrt{\text{m}}$ the fatigue fracture surface exhibits well-defined striations that continue to appear also over 60 MPa$\sqrt{\text{m}}$, but this time they are coupled with rather poorly-defined microvoids that anticipate the beginning of Region III where sudden failure occurs by overload (dimple rupture). At progressively lower values of ΔK a combination of fatigue fissures (associated with small secondary cracks) and striations can be observed.

Fig. 14.20 Formation of multiple striations containing finer striations in C-Mn steel [35]

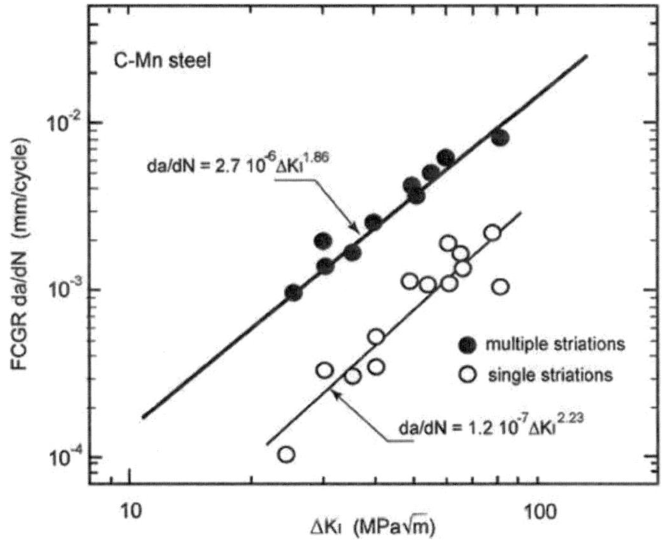

Fig. 14.21 Spacing of multiple and single striations in C-Mn steel [35]

Finally, below 30 MPa$\sqrt{}$m (Region I), striations are no longer visible, but a highly faceted fracture surface appears due to crystallographic fracture along intense slip bands. Precisely the same general appearance can be seen on stainless steel type 305 as shown in Fig. 14.23 [36].

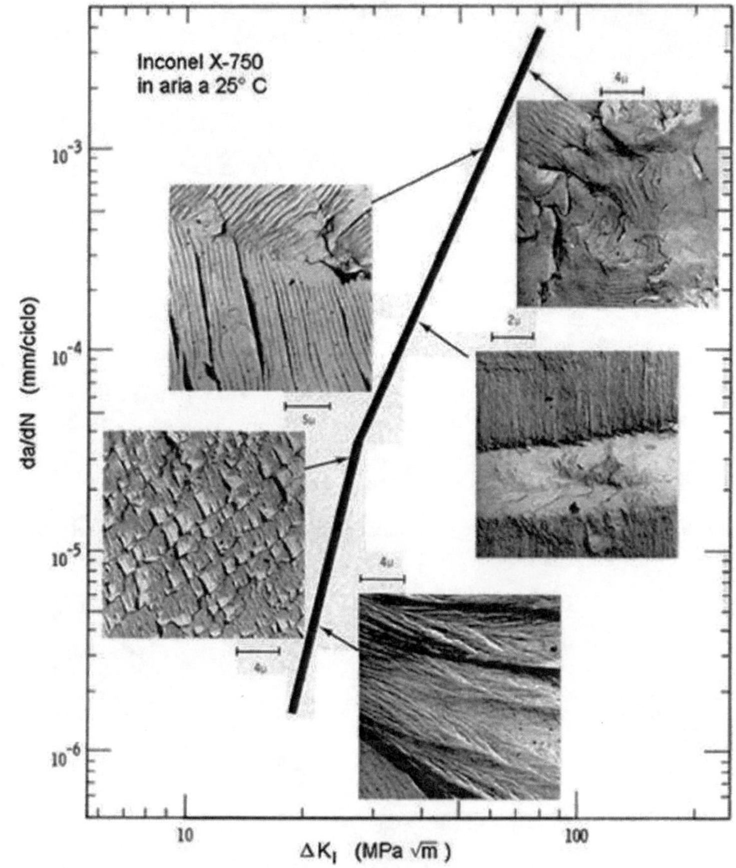

Fig. 14.22 Fatigue fracture appearance in Inconel X-750 [32]. Striations are well-defined in Region II of fatigue

14.6.1 MODE II and MODE III Crack Growth

Striations are a characteristic feature of Mode I of aperture of a crack, shown in Fig. 14.3 and do not appear in the other two modes, Mode II and Mode III. This can be used as another element of diagnosis: if striations appear on the fracture surface it means that the crack has been opened by stresses σ normal to the fracture surface and not by the shear stresses τ either parallel to the crack plane or transverse to the crack plane. Figure 14.24 [37] is a SEM observation of Mode II fatigue failure of a railway steel specimen with 937 MPa ultimate strength. The crack propagated from the tip of a 60° V-shape chevron notch visible in Fig. 14.24a. Figure 14.24b is a higher magnification view of the area in the small rectangle (a) in Fig. 14.24a. There are no striations. The array of almost parallel lines is not striations, being parallel

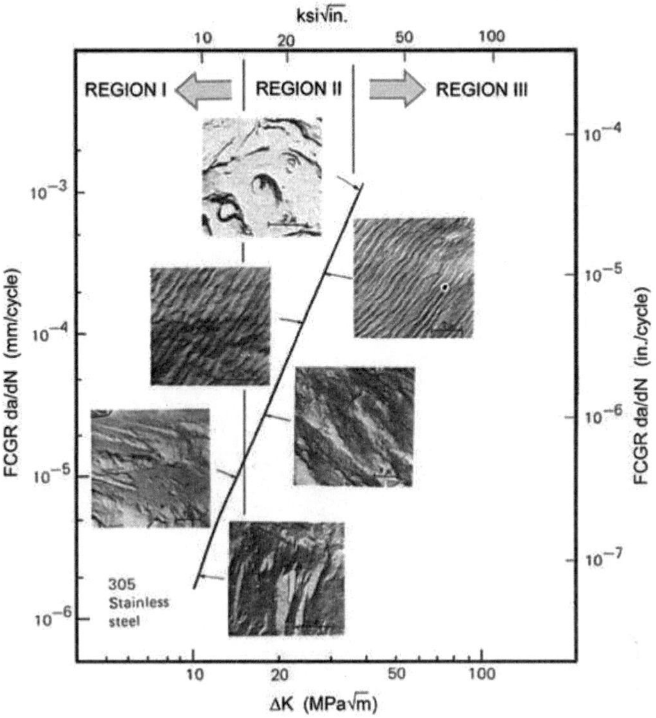

Fig. 14.23 Fatigue fracture appearance in type 305 stainless steel. Striations are well-defined only in Region II of fatigue (modified from [36])

to the direction of crack propagation. Figure 14.25a [38] is the macrograph of the fracture surface of a specimen with a circumferential crack failed by torsional fatigue (Mode III) with a superimposed small static Mode I of just $K_I = 1$ MPa$\sqrt{\text{m}}$. The material is AISI 1018 steel 260 MPa yield and 412 MPa ultimate strength. The static Mode I reduced the crack faces interaction and the crack closure effect so that the driving Mode III term was only slightly reduced.

The ΔK_{III} excursion was rather small near the fatigue threshold. The flat surface typical of Mode I fatigue has been replaced by a so called factory roof type or saw-tooth fracture. The same features are visible in Fig. 14.25b [39] that refers to a 3.5% NiCrMoV steel, 682 MPa yield and 832 MPa ultimate strength, fatigued in mixed mode I + III.

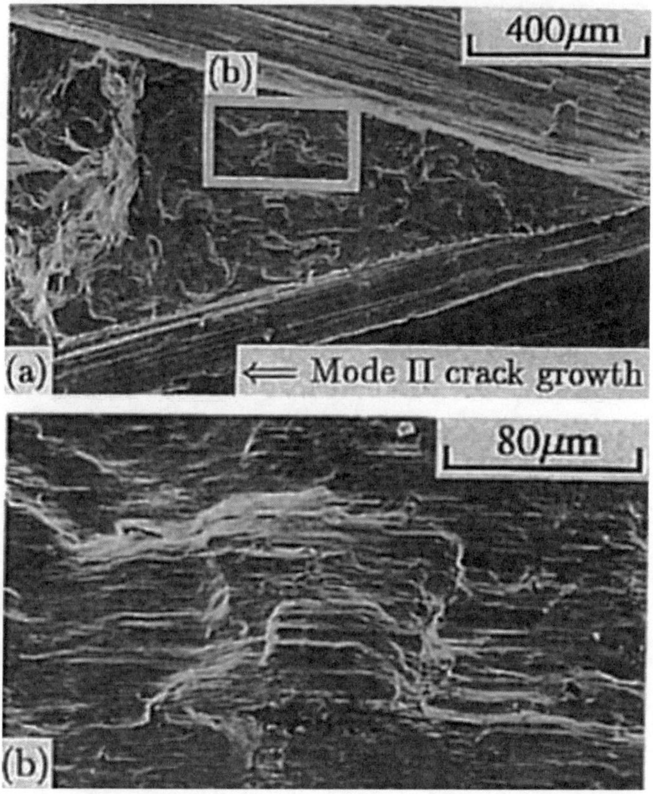

Fig. 14.24 SEM examination of the fracture surface of a railway steel specimen fatigued under Mode II of crack opening. (b) detail of the small area within the rectangle marked (b) [37]

14.7 FCGR in Metals

14.7.1 Martensitic Steels

It has been shown in Sect. 3.4 how metallurgical variables, such as structure, grain size and inclusions, in particular, can affect the fatigue strength of materials in Stage I of fatigue (see Fig. 1.64). The same effects can be seen in Stage II of fatigue where the macro-crack propagates, depending on what region of fatigue we are (see Fig. 14.6). In Region II, where the Paris-Erdogan equation applies, there is not that much influence. Consider Fig. 14.14 that refers to a family of martensitic steels type 4340. The FCGR, da/dN, seems not to be affected by the yield strength of the material, dough it is varying from 910 to 1400 MPa. In all cases the R ratio was almost the same, varying from 0 to a maximum of 0.1. In effect, studies performed on martensitic steels with yield strength ranging from 550 to 2070 MPa would indicate

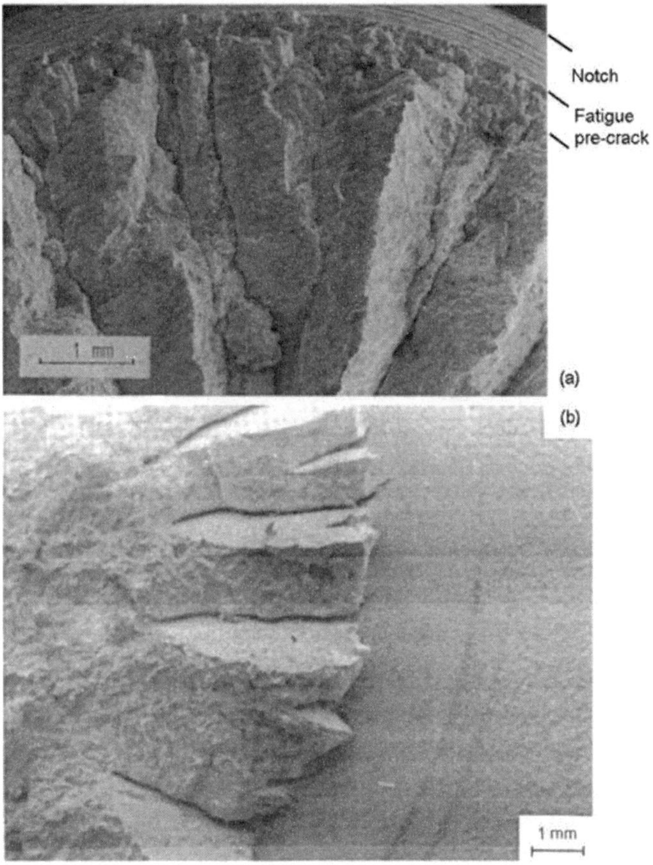

Fig. 14.25 Fractography of the fracture surface of a specimen with circumferential notch that failed by Mode III torsional fatigue **a** AISI 1018 steel [38], **b** 5.5 NiCrMoV steel [39]

a rather complete absence of any mechanical properties or metallurgical influence on the FCGR. Figure 14.26 is a useful compendium of experimental results obtained on different martensitic steel such as: 12Ni-5Cr-3Mo, 10Ni-8Co-1Mo, HY 80 and HY130 [40, 41], 5Ni of 609 MPa yielding and 756 MPa ultimate strength and 9Ni of 749 MPa yielding and 814 MPa ultimate strength [33]. All data falls within a narrow scatter band that is the same of 4340 steel already seen in Fig. 14.14. All FCGR, *da/dN*, share the same exponent $n = 2.5$ and are fitted by a line of equation

$$\frac{da}{dN} = 4 \cdot 10^{-8} \cdot (\Delta K_I)^{2.5} (\text{mm}/cycle) \tag{14.25}$$

in which *da/dN* is in mm and ΔK_I in MPa$\sqrt{}$m. Also the experimental data obtained by Matocha et al. [42] confirms this finding.

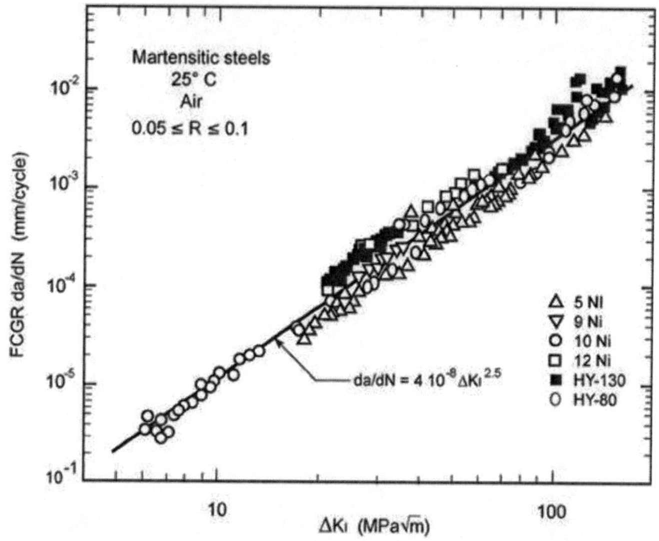

Fig. 14.26 FCGR data, *da/dN*, for martensitic steels at RT in air (test data are reported in [33, 40, 41])

14.7.2 Carbon Steels

Carbon steels present the same characteristic on independence of FCGR, *da/dN*, of the metallurgical factors in Region II of growth, but show a different general trend. This can be seen in Fig. 14.27 for type A 533 Gr B and A 508-2 steels, used in the construction of nuclear pressure vessels.

The 280 experimental data come from different authors [43–48] and were obtained in air at RT with $0.1 \leq R \leq 0.2$. The FCGR, *da/dN*, in Region II can be fitted to the line of equation

$$\frac{da}{dN} = 1.26 \cdot 10^{-9} \cdot (\Delta K_I)^{3.38} (\mathrm{mm}/cycle) \tag{14.26}$$

in which, as usual, *da/dN* is in mm and ΔK_I in MPa$\sqrt{}$m. The value of the constant C is much lower than that of martensitic steels, see Eq. (14.22), but the exponent n has grown to 3.5 from 2.5. Region I and Region III are distinguishable. For this steel the toughness at room temperature is in the range of 120–150 MPa$\sqrt{}$m and this variation can be seen by the spread of data in Region III where the metallurgical variables do affect the FCGR of the material.

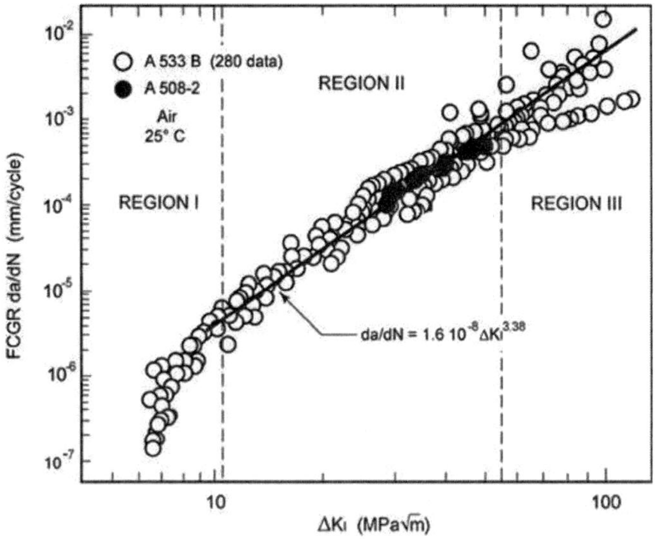

Fig. 14.27 FCGR data, *da/dN*, for carbon steel type A 533 B and A 508-2 at RT in air (test data are taken from [43–48])

14.7.3 Stainless Steels

Stainless steels can be divided into two main families: austenitic and ferritic. The different structure, γ for the *austenite* and α for the *ferrite*, seems to have an appreciable effect on FCGR properties of steels. This can be seen in Fig. 14.30 that collects 383 data obtained at RT by different researchers [29, 30, 49–53] on four austenitic stainless steels of the series 300, type 304, 316, 321 and 350, the last two with columbium-tantalum and titanium, respectively for high temperatures use (430–820 °C), an austenitic-ferritic (14% δ ferrite) steel type 351 cast [54, 55], a ferritic stainless steels, type 18Cr-Nb [56] and experimental data relative to 403 type martensitic stainless steel [57] (Fig. 14.28).

The yield strength of the austenitic steels varied from 210 to 350 MPa and the ultimate strength from 520 to 685. The experimental data can be fitted to the line of equation

$$\begin{matrix} austenitic\ steels\ \frac{da}{dN} = 6.2 \cdot 10^{-9}(\Delta K_I)^3 \\ ferritic\ steel\ \frac{da}{dN} = 3.1 \cdot 10^{-10}(\Delta K_I)^{3.3} \end{matrix} \ (\mathrm{mm}/cycle). \tag{14.27}$$

Note that for the ferritic stainless steel the exponent n equals that of ferritic steels. The test frequency for the austenitic stainless steels was varying from 6 to 600 cycles per minute while for the ferritic stainless steel it was equal to 15 Hz and has no effect on the test results. In fact, it has been already recognized that the test frequency must drop below 0.5 cycles per minute to give enough time to corrosion to take place. Also

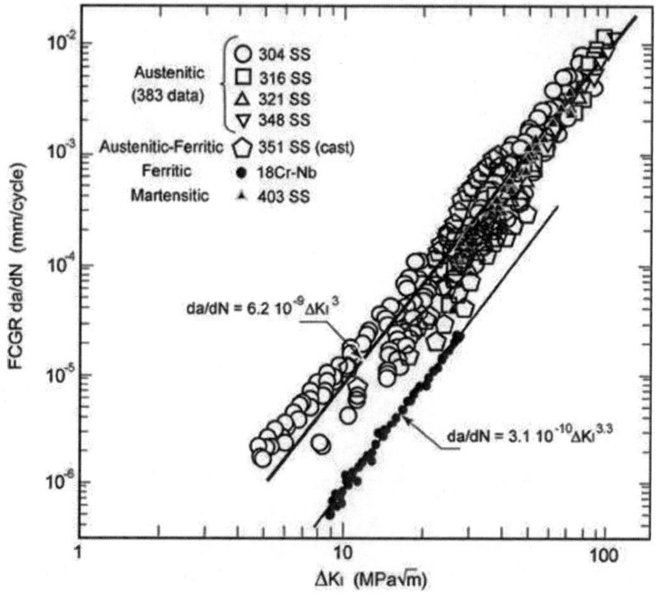

Fig. 14.28 FCGR data, *da/dN*, for stainless steels at RT in air (data are extracted from [29, 30, 49–57]). For comparison data for ferritic and martensitic steel are also shown

air can be potentially aggressive if it is not dry, but contains humidity in a certain percentage.

14.7.4 Aluminum Alloys

Aluminum alloys have the same FCC structure of austenitic stainless steels. The FCGR, *da/dN*, of different aluminum alloys is presented in Fig. 14.29. The figure is based on data obtained by different researches [8, 28, 31, 58–62]. Experimental results seem to be bounded by alloys 7075-T6 and 5456-H321 having yield strength of 500 MPa and 259 MPa, respectively, whose FCGR data, *da/dN*, are fitted to the lines having equations

$$\begin{matrix} \frac{da}{dN} = 5.0 \cdot 10^{-7} \cdot (\Delta K_I)^{3.3} \\ \frac{da}{dN} = 2.1 \cdot 10^{-8} \cdot (\Delta K_I)^{3.3} \end{matrix} (\text{mm}/cycle) \tag{14.28}$$

respectively.

At variance with carbon steels or martensitic steels or and, in particular, stainless steels the wide family of aluminum alloys seems not to merge in Region II of the FCGR diagram where the Paris-Erdogan equation apply. Between the two alloys,

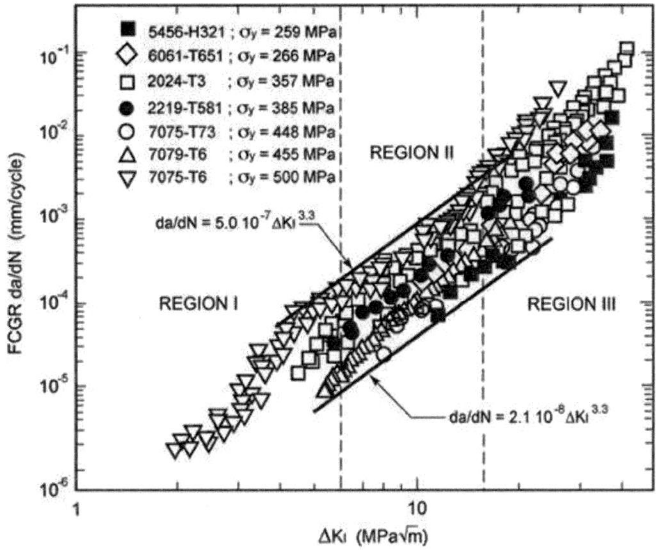

Fig. 14.29 FCGR data, *da/dN*, for different aluminum alloys at RT in air (data are extracted from [8, 28, 31, 58–62])

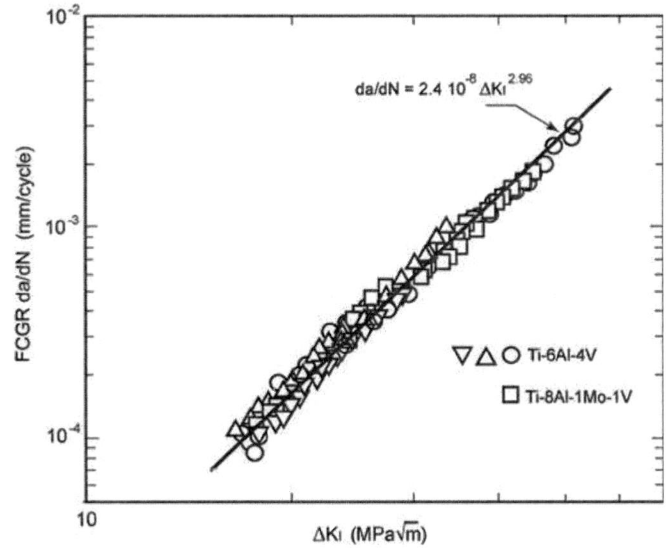

Fig. 14.30 FCGR data, *da/dN*, for two titanium alloys at RT in air (test data are extracted from [62–65])

7075-T6 and 5456-H321, that bound all the others aluminum alloys behavior there is almost a factor of ten in the FCGR. Note that at equal ΔK_I, the FCGR, *da/dN*, in aluminum alloys is much higher than in steels, sometimes also by a factor of one hundred. This should not come unexpected since the toughness of aluminum alloys is much lower than that of steels. Therefore, at equal ΔK_I, aluminum alloys are closer to Region III and to the critical toughness $\Delta K_{Ic,}$ i.e., to instability, than steels. This can also explain the reason of such large a variation in the FCGR among the different aluminum alloys since also the toughness is much different.

14.7.5 Titanium Alloys

Titanium has a hexagonal structure. Its alloys can have either α or β structure or a mixed one $\alpha + \beta$. The FCGR, *da/dN*, of two high strength $\alpha + \beta$ titanium alloys, with strong α prevalence, namely Ti-6Al-4 V of 740 MPa yield strength and 818 MPa ultimate strength [62, 63] and Ti-8Al-1Mo-1 V of 794 yield strength and 894 MPa ultimate strength [63], respectively, are presented in Fig. 14.30. As for martensitic steels and aluminum alloys there is almost no difference between the two alloys in Region II of the FCGR, *da/dN*, diagram that can be represented by the equation

$$\frac{da}{dN} = 2.4 \cdot 10^{-8} \cdot (\Delta K_I)^{2.96} (\mathrm{mm}/cycle). \tag{14.29}$$

14.7.6 Nickel Alloys

Nickel and its alloys have a FCC structure as aluminum and stainless steel. Figure 14.31 presents the FCGR, *da/dN*, of three different nickel alloys known as inconel 600, 718 and X750, respectively. The frequency of load application varied from 1 to 600 cycles per minute. The experimental data shown in Fig. 14.31 were obtained by several researchers [32, 66, 67].

As for the other alloys, the testing frequency has no effect on the FCGR of these nickel alloys. The exponent of the Paris-Erdogan equation varies from 3.23 to 3.94 as in carbon steels or aluminum alloys, but the value of the constant C reaches a value among the lowest of them all

$$\frac{da}{dN} = 10^{-9} \cdot (\Delta K_I)^{3.23} \, (\mathrm{mm}/cycle). \tag{14.30}$$

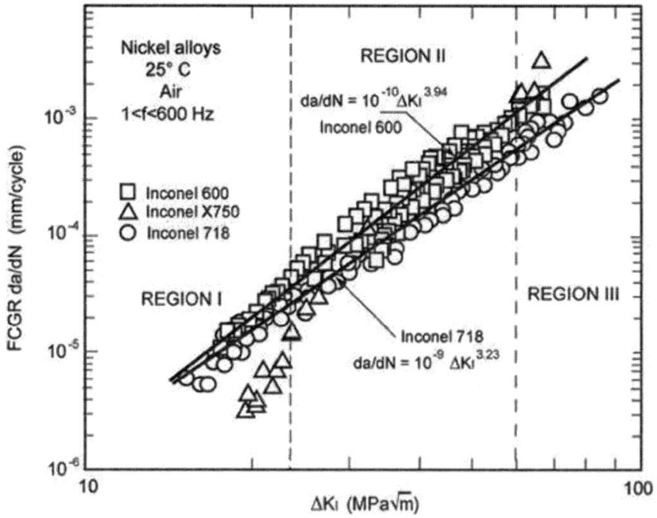

Fig. 14.31 FCGR data, *da/dN*, for inconel at RT in air (test data are extracted from [32, 66, 67])

14.8 Effect of Toughness on FCGR

A synoptic picture of all data considered is presented in Fig. 14.32. Also shown in Fig. 14.32 is the FCGR, *da/dN*, of nodular cast iron (Milella) and the scatter band of ferrite-pearlite steels type A36, A302 B and A 537 A [64]. The *n* exponent varies from a minimum of 2.5 for martensitic steels to a maximum of 5 for cast iron. As it can be seen, light metals and alloys, as magnesium and aluminum alloys, are placed on the upper portion of the diagram. The analysis of the experimental data obtained in air at RT and for $0 \leq R \leq 0.2$ clearly indicates that in Region II of the fatigue diagram *da/dN* versus ΔK_I the FCGR of alloys of the same metal having the same metallurgical structure are almost the same. Chemistry and microstructure have a negligible effect. Another finding is that the value of the exponent *n* is not the same for all alloys. The aluminum alloys seem to violate the general behavior of being independent of the metallurgy and yield strength, in particular, since in Region II there is a factor of ten between the maximum and the minimum FCGR of different alloys, as seen in Fig. 14.29. The reason for this apparent exception must be found in the fact that among the different aluminum alloys there is a considerable variation of the fracture toughness that plays a fundamental role in the FCGR of materials. A direct comparison between different alloys, in fact, cannot be done without considering the individual toughness. It has been already said that the Young's modulus cannot be the reading key of the wide spread field of striation spacing, as shown in Fig. 14.19. Even normalizing the striation spacing to the Young's modulus of the material, a factor of 15 still remains between the maximum and the minimum rate. Striation

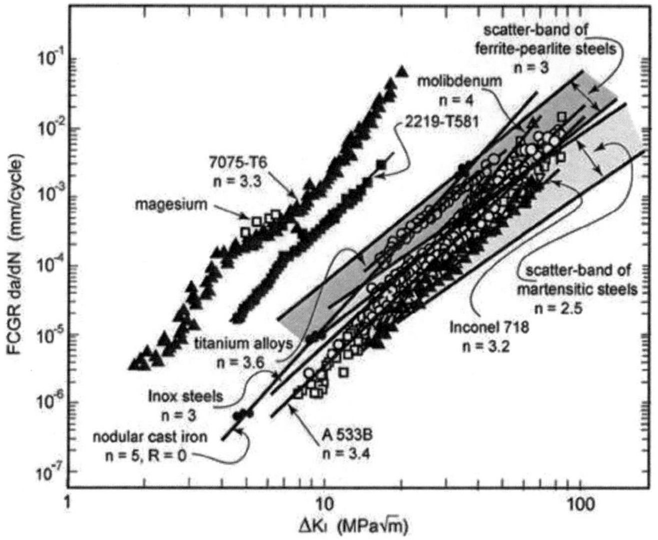

Fig. 14.32 FCGR of all data presented in previous figures (test data are reported in [8, 31, 34, 41–61])

spacing is a measurement of FCGR therefore the Young's modulus cannot be the reading key also for the different FCGR presented by different materials.

This is clearly shown in Fig. 14.19 where all metal alloys are bounded by the two lines of equation

$$\frac{da}{dN} = 5 \cdot 10^{10} \cdot \left(\frac{\Delta K_I}{E}\right)^{3.5}$$
$$\frac{da}{dN} = 10^{9} \cdot \left(\frac{\Delta K_I}{E}\right)^{3.5}. \tag{14.31}$$

A difference of two decades exists between the fastest and slowest growing rate. Figure 14.32 shows that at $\Delta K_I = 30$ MPa$\sqrt{}$m, for instance, cracks in an aluminum alloy type 7075-T6 can growth at a rate even higher than 0.3 mm/cycle, while in a martensitic steel this rate collapses to $5 \cdot 10^{-5}$ mm/cycle. A factor of 10^4 separates the FCGR of these alloys at that stress intensity factor excursion ΔK_I. But in a 4340 steel 30 MPa$\sqrt{}$m are quite far from its critical toughness K_{Ic} that can reach also 100 MPa$\sqrt{}$m, while in 7075-T6 it is very close to sudden fracture that occurs at just $K_{Ic} = 35$ MPa$\sqrt{}$m. When a metal is fatigued at a ΔK_I close to its K_{Ic} the FCGR jumps to stars. Sudden fracture can even be expected. This is the reason why light metals and alloys are placed on the upper part of the FCGR diagram: they have the smaller toughness of them all. Therefore, the comparison between different materials must be done at equal ratio $\Delta K_I/K_{Ic}$. This is the reading key and this is shown in Fig. 14.33 for all materials of Fig. 14.32. The data collapsing in Region II of fatigue

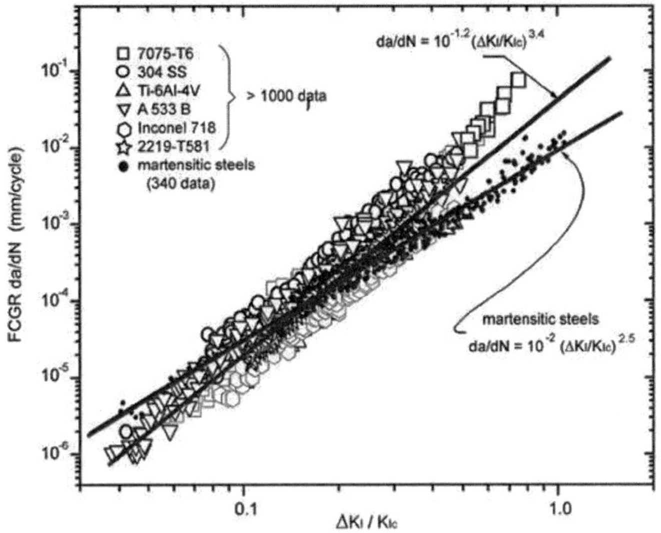

Fig. 14.33 FCGR, *da/dN*, versus normalized $\Delta K_I/K_{Ic}$, for materials presented in Fig. 14.32

crack growth is impressive. The spread in FCGR data between aluminum and Inconel that was affected by a factor of about 10^4 now reduces to a factor of 14. A common trend given by

$$\frac{da}{dN} = 0,063 \cdot \left(\frac{\Delta K}{K_c}\right)^{3.4} \text{mm}/ciclo \tag{14.32}$$

can be drawn. It is important to recall that the data base of Fig. 14.33 is composed by over 1000 experimental points. It is a rather impressive amount of experimental data that provides a granitic solidity to the assertion of the role of fracture toughness. However, it must be kept in mind that the toughness K_{Ic} considered to normalize ΔK_I may not be the actual seen by the material under fatigue testing.

This is because the size of the fatigue specimens used to measure the FCGR can be smaller than that required by the fracture toughness testing to maintain plain strain conditions and yield the minimum value K_{Ic}. This results in a crack that grows in a material with higher apparent toughness. But also the cycling itself can have an impact on the toughness seen by the growing crack. If the material cyclically softens, as in hard steels, the apparent toughness increases. The opposite may happen with hard alloys that have the tendency to soften.

Therefore the difference in FCGR between hard and soft alloys further reduces when compared in terms of $\Delta K_I/K_o$, where K_o indicates the apparent toughness or cyclic toughness and not the minimum toughness K_{Ic}. This may be the reason why martensitic steels in Fig. 14.32 seem to follow a different trend; the have the tendency to cyclically soften and this is particularly true at higher ΔK_I so that experimental

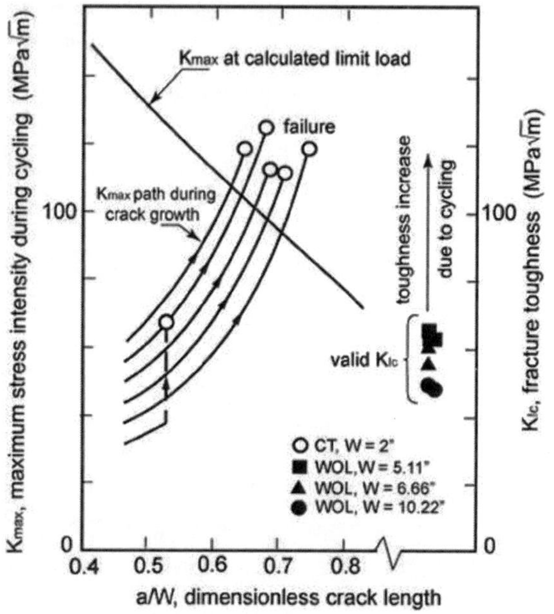

Fig. 14.34 Unexpected failures at $K_{\max}$ almost double than K_{Ic} following fatigue crack growth in A 470 steel (modified from [68])

points from $\Delta K_I/K_c$ in Fig. 14.32 should move backwards and realign with other data. Significant to sustain this point of view are the experimental unpublished data obtained by Begley and reported by Dowling [68] on high strength (646 MPa yield strength) low toughness steels (45–65 MPa$\sqrt{m}$), type A 470 (3,5NiCrMo¼V), for rotor forgings in which it was evidenced the capability of this material to be cycled at ΔK_I higher than the critical toughness K_{Ic}, without fracturing. Begley was using CT specimens 2″ thick cycled at constant amplitude. With the exception of one case, in all other (14.5) cases the crack would grow by fatigue and failure would occur at ΔK_I almost double than the K_{Ic} of the material, as shown in Fig. 14.34 [68].

The only reason that can explain such a behavior is the large cyclic softening of the alloy that would allow the toughness to almost double.

14.9 Effect of Temperature on FCGR

It has been shown in Sect. 3.8 how temperature exerts a rather important influence on fatigue. An increase in the temperature of materials leads to a systematic reduction of fatigue life. At variance, the effect of temperature on the FCGR of metals seems not be that systematic and simple, but it is rather complex and often controversial. Figure 14.35 [48] presents the FCGR, *da/dN*, of a family of stainless steels, type 304, 316, 321 and 348 tested at 593 °C compared to results obtained at room temperature, RT. Two characteristic features must be caught. The first is that, within the range

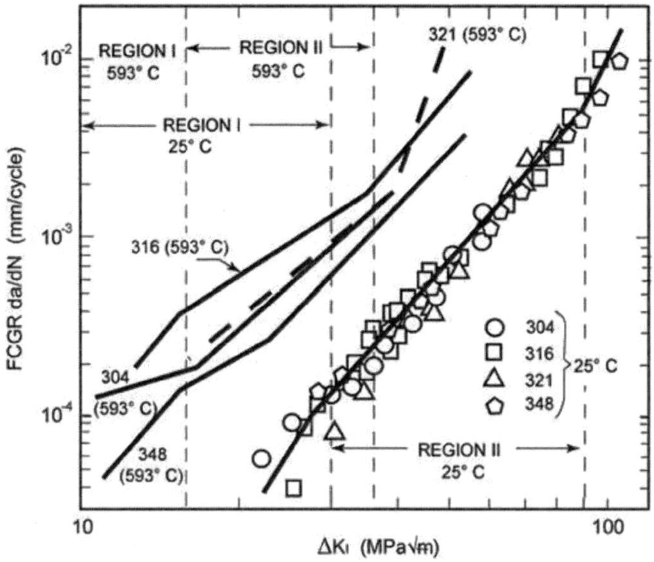

Fig. 14.35 Effect of temperature on the FCGR, *da/dN,* of stainless steels (modified from [48])

considered, temperature indeed increases the FCGR by a factor from six to ten. Secondly, fatigue response that at RT was almost the same for all types of steel now differentiates appreciably. The picture further complicates moving to carbon steel type A 533 B, already seen in Fig. 14.27. Figure 14.36 shows the experimental data obtained at several temperatures, from − 18 °C to 300 °C [48]. Now it is not possible at all to identify a trend common to all temperatures. At 120 °C the FCGR, *da/dN*, reaches its minimum whilst the maximum is at 25 °C. Between these two temperatures all other FCGR lay, but in a rather random order not certainly sequential.

It is definitely necessary a reading key and this key is precisely the fracture toughness K_{Ic} already discussed in Sect. 14.8 and used in Fig. 14.33 to compact all FCGR curves. To this purpose, Fig. 14.37 presents the same results of Fig. 14.36 [64], but in a different fashion. Figure 14.37a [48], in fact, is a summary of the effect of temperature on the fatigue crack length versus elapsed cycles behavior observed for the same steel with 3″ WOL specimens. Note that the number of cycles required to extend the initial crack length (0.3 inch) a given amount increases in the following order of temperatures: 288, 25, − 18, 205 and 120 °C. This method of presenting the number of cycles required to extend the initial crack from 0.3 inch to 2.2 inches allows relating the fatigue properties to the fracture toughness of the material. This is shown, by comparison, in Fig. 14.37b.

The increase in the number of cycles required to extend by fatigue the initial crack a given amount, i.e., decrease in crack growth rate, can be attributed to the parallel increase of the fracture toughness K_{Ic}. In fact, the maximum fracture toughness of the material is reached at 100–120 °C while at ambient temperature it has the same

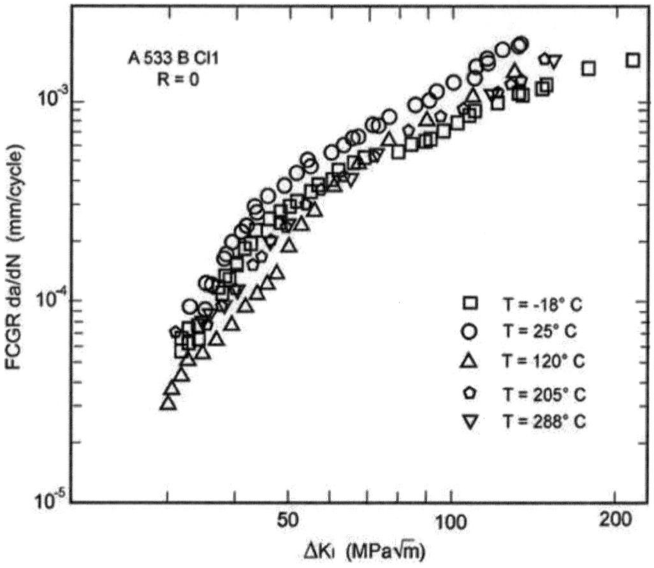

Fig. 14.36 Effect of temperature on FCGR, *da/dN*, of carbon steel Type A 533 B Cl1 (modified from [48])

value that at 288 °C. This actually means that at equal applied ΔK_I the ratio ΔK_I / K_{Ic} is minimum at 100–120 °C and is higher at both 25 °C and 288 °C.

Therefore, at 100 ÷ 120 °C we shall expect the minimum FCGR, while both at 25 °C and 288 °C the FCGR is almost the same and minimum, at least in the interval of temperatures considered. The reason why below 0 °C there is a sudden decrease of the FCGR that at − 18 °C reaches a lower value may be due to the influence of cyclic strain softening already seen in Fig. 14.34 since this steel at low temperature reaches the same high yield strength of A 470 steel. A confirmation of the importance of the ratio $\Delta K_I / K_{Ic}$ or of the ratio toughness/FCGR is offered by the behavior of the submerged arc weld material used to weld A 533-B plates. This is shown in Fig. 14.38. This time, the FCGR, *da/dN*, is almost continuously growing from RT to 288 °C. The same procedure used to evaluate the cycles needed to grow an initial crack of 0.3 inch to a final length of 2.2 inches is shown in Fig. 14.39a [48] together with the fracture toughness, Fig. 14.39b. The fatigue crack growth observed for the weld-metal follows the K_{Ic} behavior. Another interesting example is offered in Figs. 14.40 and 14.41 by the temperature effect on the FCGR of 1TC(T) specimens of martensitic steels 5Ni and 9Ni having the yield strength equal to 609 MPa and 749 MPa, respectively, and the ultimate strength equal to 756 MPa and 814 MPa, respectively.

The temperature range is 25 °C to − 195 °C. In the 5Ni alloy, Fig. 14.40 [32], the FCGR increases going from RT to − 195 °C. Also in this case if we look at the fracture toughness K_{Ic} of the material we find that it is higher at 25 °C and reduces

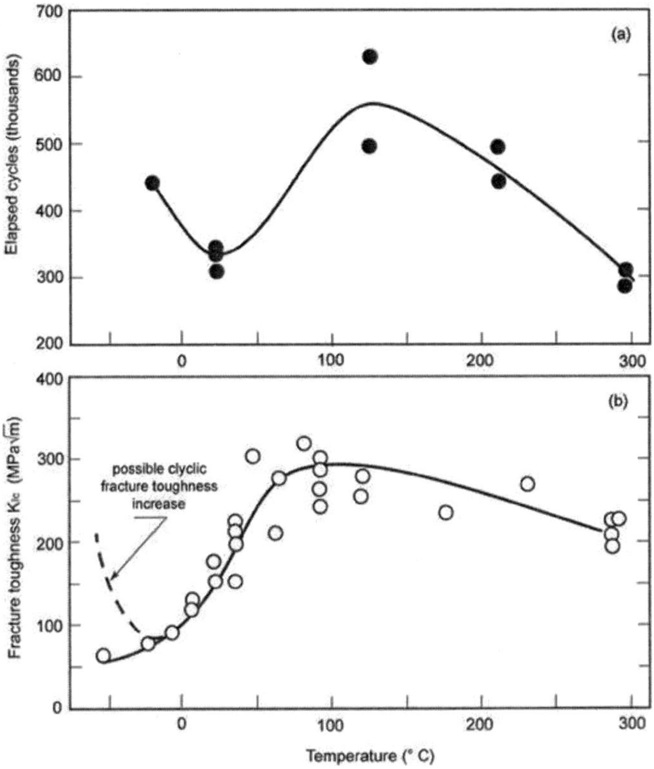

Fig. 14.37 **a** influence of temperature on the fatigue crack growth behavior for A533-B steel (modified from [48]) and **b** influence of temperature on the toughness of A533-B steel

progressively to 69–86 MPa$\sqrt{m}$ at − 195 °C. But if we move from the 5Ni alloy to the 9Ni steel we will not note any difference, in practice, in the FCGR reducing the temperature from RT to − 195 °C. This is shown in Fig. 14.41 [32]. The FCGR is practically the same at any temperature and equal to that of the 5Ni alloy already seen in Fig. 14.40. Well, it is interesting to note that in this alloy the toughness, in practice, is not changing that much at those testing temperatures keeping a value around 109–143 MPa$\sqrt{m}$ that represent the toughness of 5Ni alloy at room temperature. The same temperature effect is visible for nickel alloys. Figure 14.42 [32] shows the FCGR, *da/dN*, for type X750 and Inconel 600 alloys at several temperatures varying from − 269 °C to 649 °C for the X750 and from RT to 538 °C for the Inconel 600, respectively. Now, in the X750 alloy the FCGR increases by a factor of about 15 from − 269 to 649 °C.

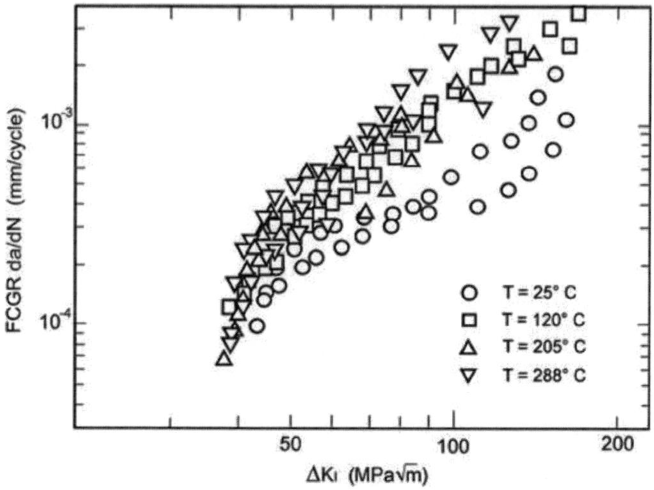

Fig. 14.38 Summary of the effect of temperature on the FCGR of A533-B steel weld-metal (modified from [48])

14.10 Sample Problems

14.10.1 Fatigue Crack Growth-Cylindrical Geometry, Continuous Crack

A cylindrical steel specimen type 4340, having 900 MPa yield strength and 1200 MPa ultimate strength, 5.0 mm of diameter D and 20 mm length L, has been subjected to fatigue with stress amplitude equal to 500 MPa. After a certain number of cycles, a circumferential crack appears having a depth $a_o = 0.3$ mm. It must be assessed the number of cycles N_f at which failure occurs and the ratio N_f/N_{iniz} between the cycles to failure and those N_{iniz} required to initiate the crack a_o. With the first appearance of a circumferential crack, the specimen is equivalent to a cylindrical round-notch bar as that shown in Fig. 14.43. The corresponding stress intensity factor K_I is given by the relationship (see Appendix A)

$$K_I = \sigma_{net}\sqrt{\pi d/D} \cdot f(d/D) \tag{14.33}$$

where σ_{net} is the stress acting on the crack plane

$$\sigma_{net} = \sigma\left[\frac{D}{D-2a}\right]^2 = \sigma\left(\frac{D}{d}\right)^2 \tag{14.34}$$

and *f(d/D)* given by

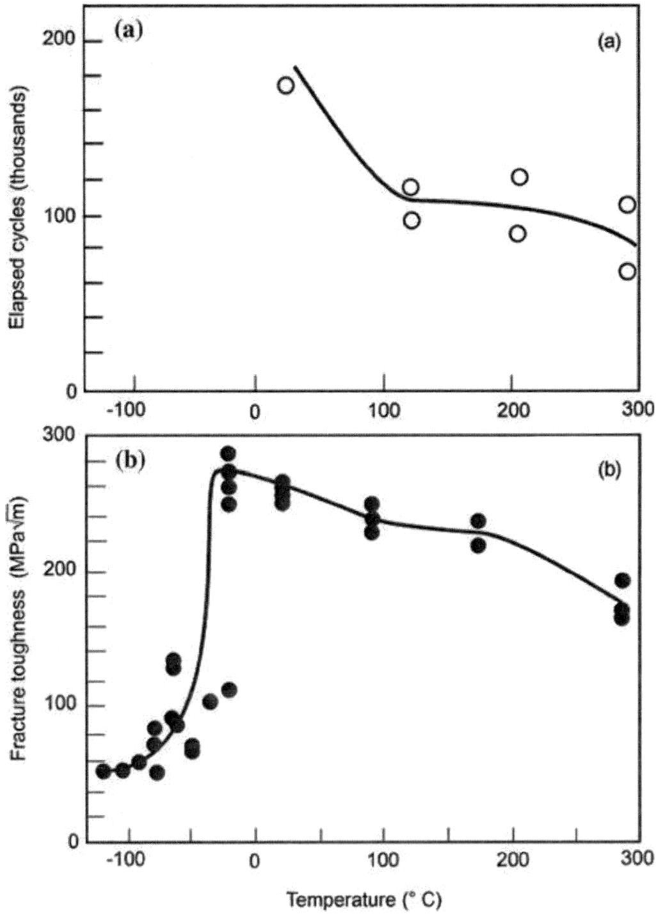

Fig. 14.39 **a** influence of temperature on the fatigue crack growth behavior for A533-B steel base weld-metal (modified from [48]) and **b** influence of temperature on the toughness of the weld metal

$$f\left(\frac{d}{D}\right)=\frac{1}{2}\left[1+\frac{1}{2}\frac{d}{D}+\frac{3}{8}\left(\frac{d}{D}\right)^{2}-0.363\left(\frac{d}{D}\right)^{3}+0.731\left(\frac{d}{D}\right)^{4}\right]$$
$$\left[1+0.1\left(\frac{2a}{D}\right)^{1/2}\left(1-\frac{2a}{D}\right)^{2}\right]. \tag{14.35}$$

Under the action of cyclic loads, the crack propagates according to the Paris-Erdogan Eq. (14.10). By integrating the Paris equation it will be possible to infer the number of cycles N_f necessary to make the crack grow from the initial size $a_o =$ 0.3 mm to the final one a_f that causes complete fracture

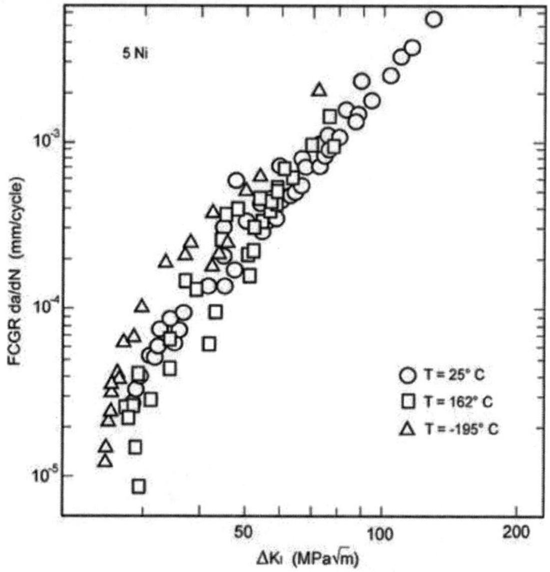

Fig. 14.40 FCGR, *da/dN*, of martensitic steel type 5Ni at various temperatures (modified from [32])

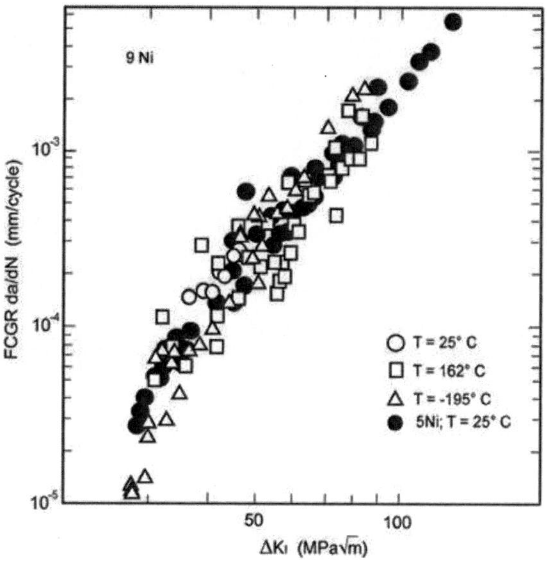

Fig. 14.41 FCGR, *da/dN*, of martensitic steel type 9Ni at various temperatures (modified from [32]). Also shown for comparison is the 5Ni FCGR of Fig. 14.40

$$N = C \int_{a_o}^{a_f} \frac{da}{\Delta K_I^n}. \tag{14.36}$$

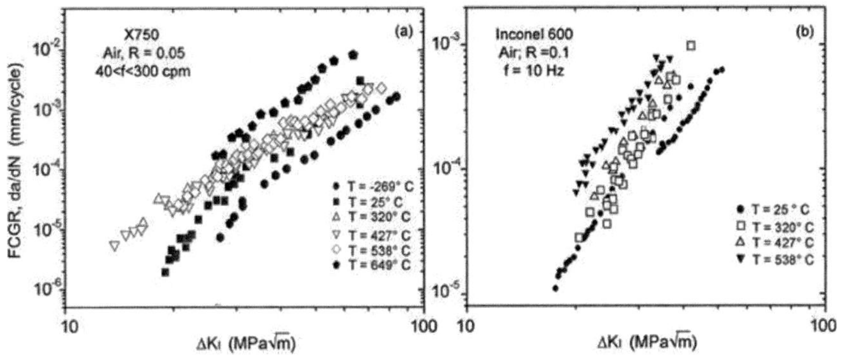

Fig. 14.42 FCGR, *da/dN*, for nickel alloys **a** type X750 and **b** type Inconel 600 at various temperatures (modified from [32])

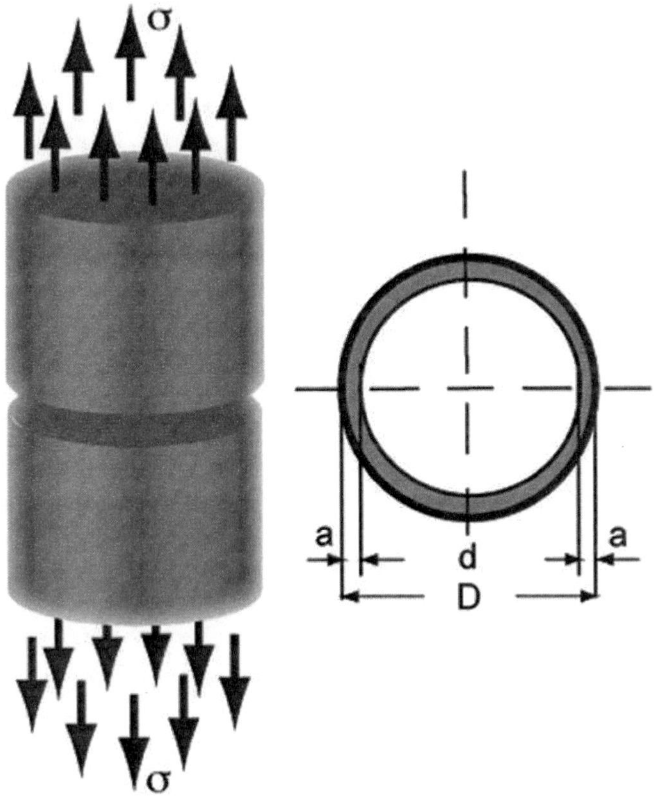

Fig. 14.43 Round-notched bar

The final length a_f is that producing fracture of the residual net section and can be assessed by equating the stress σ_{net} to the brittle or ductile fracture failure stress. Since the specimen is operating at ambient temperature which is over the nihil ductily temperature (NDT) of the material, it can be said that it will fail by plastic instability and not by brittle cleavage. Therefore, it can be assumed that the plastic collapse stress is the flow stress of the material σ_{flow} assumed to be equal to the average between the yield strength and the ultimate strength

$$\sigma_{flow} = \frac{\sigma_y + \sigma_u}{2} = \frac{900 + 1200}{2} = 1050 \text{ MPa}. \tag{14.37}$$

Hence, the section will collapse when the final section diameter d_f and the final crack depth a_f will be

$$\begin{aligned} \sigma\left(\frac{D}{d_f}\right) &= \sigma_{flow} \rightarrow d_f = D\sqrt{\frac{\sigma}{\sigma_{flow}}} \\ a_f &= \frac{D - d_f}{2} = \frac{D}{2}\left(1 - \sqrt{\frac{\sigma}{\sigma_{flow}}}\right) \\ &= \frac{5}{2}\left(1 - \sqrt{\frac{500}{1050}}\right) = 0.77 \text{ mm}. \end{aligned}$$

Now it is possible to solve the integral (14.36) and it will be done, for simplicity, by a finite differences integration procedure. To that end, the total crack growth $\Delta a = a_f - a_o$ is divided into a convenient number of parts such as, for example, 30 so that in each single elementary interval $da = \Delta a/30$ the net section stress σ_{net} and the corresponding stress intensity factor ΔK_I are considered constant

$$a_f = a_o + \Delta a;\ \Delta a = 0.77 - 0.3 \text{ mm};da = \frac{\Delta a}{30} = 0.0156 \text{ mm}$$

so that

$$N = \sum_{i=1}^{30} dN_i = \sum \frac{1}{C}\frac{da_i}{K_I^n}. \tag{14.38}$$

Constant C and n for the 4340 steel are now needed. It can be used the curve of Fig. 14.14 with $C = 3.7 \cdot 10^{-8}$ mm/cycle; $n = 2.5$.

Now summation given by Eq. (11.35) can be solved step-by-step. Results are shown in Table 14.1. It can be seen that when the crack has reached a depth, for example, of $a = 051$ mm the diameter is reduced to $d = 3.99$ mm, the function $f(d/D)$ is equal to 0.90, the stress $\sigma_{net} = 786$ MPa, $K_I = 25$ MPa$\sqrt{}$m, the FCGR $da/dN = 1.12 \cdot 10^{-4}$ mm/cycle, the crack growth is $\Delta a = 0.205$ mm for which 2,688 cycles were necessary, whilst to grow from 0.189 mm to 0.205 mm 142 cycle, only, were needed. The final result is that 4,285 cycles of 500 MPa amplitude will be sufficient to grow the initial crack from $a_o = 0.3$ mm to $a_f = 0.77$ mm at which plastic collapse occurs. But how many cycles were elapsed to produce the initial crack of 0.3 mm? This question can be answered looking at the Wöhler curve of the material shown in Fig. 14.43. Under a stress amplitude of 500 MPa the specimen fails after 50,000 cycles. Of these 50,000 cycles, 4285 need to grow the crack to failure so that the remaining N_{iniz} = 50,000–4,285 = 45,715 cycles have initiated the crack to a_o = 0.3 mm. The ratio N_f/N_{iniz} is

$$\frac{N_f}{N_{iniz}} = \frac{4,285}{45,715} = 0.09$$

This actually means that 9% only of the total life is spent to grow the crack to failure and 91% to initiate the crack. Figure 14.45 shows how the crack grows during the cycling.

If the stress amplitude is reduced to 400 MPa, using the same procedure it will be found the results listed in Table 14.2.

This time, the cycles required to produce final fracture have grown to 8548, but also the allowable cycle N_f obtained from the Wöhler diagram of Fig. 14.44 increased to 770,000. Therefore the ratio N_f/N_{iniz} becomes

$$\frac{N_f}{N_{iniz}} = \frac{8548}{770,000} = 0.01.$$

Finally, if the stress amplitude is further reduced to 300 MPa, the cycles required to grow the crack to failure become equal to 19,086, as listed in Table 14.3, but it will take an infinite number of cycles to fail the specimen without a crack, since the stress amplitude is below the fatigue limit. Figure 14.45 shows how the cracks grow during the cycling at each stress amplitude considered.

14.10.2 Fatigue Crack Growth-Cylindrical Geometry, Semi-elliptical Crack

The same specimen of the previous example is showing, after a certain cycling, a nail type semi-elliptical surface crack rather than a continuous circumferential crack. Its initial depth is $a_o = 0.3$ mm with a length $2c_o = 0.6$ mm, as shown in Fig. 14.46. Calculate the life of the specimen after the first appearance of the crack. In this case the stress intensity factor K_I varies along the crack front. It reaches the maximum at the deepest point A where it is given by (see Appendix A)

$$K_I = \sigma\sqrt{\pi \cdot s} \cdot f(a/c)$$
$$s = r \cdot arcsen\left(\frac{c}{r}\right) \qquad (14.39)$$

Table 14.1 Step by step solution of Eq. (14.38) for stress amplitude of 500 MPa

a (mm)	d (mm)	$f(d/D)$	σ_{net} (MPa)	ΔK_I (MPam$^{1/2}$)	da/dN (mm/cycle)	$\Sigma\Delta a$ (mm)	ΔN cycles	N cycles
0.30	4.40	0.99	646	18.34	0.000000	0.0000	0	0
0.32	4.37	0.98	655	18.88	0.000053	0.0158	297	297
0.33	4.34	0.97	665	19.42	0.000057	0.0317	276	573
0.35	4.31	0.97	674	19.95	0.000061	0.0475	257	831
0.36	4.27	0.96	684	20.48	0.000066	0.0633	241	1.071
0.38	4.24	0.95	695	21.01	0.000070	0.0791	225	1.296
0.39	4.21	0.94	705	21.53	0.000075	0.0950	211	1.508
0.41	4.18	0.94	716	22.05	0.000080	0.1108	199	1.707
0.43	4.15	0.93	727	22.57	0.000084	0.1266	187	1.894
0.44	4.12	0.92	738	23.09	0.000090	0.1425	177	2.071
0.46	4.08	0.92	750	23.60	0.000095	0.1583	167	2.238
0.47	4.05	0.91	761	24.12	0.000100	0.1741	158	2.396
0.49	4.02	0.91	773	24.64	0.000106	0.1899	150	2.546
0.51	3.99	0.90	786	25.16	0.000112	0.2058	142	2.688
0.52	3.96	0.89	798	25.68	0.000117	0.2216	135	2.822
0.54	3.93	0.89	811	26.21	0.000124	0.2374	128	2.950
0.55	3.89	0.88	825	26.73	0.000130	0.2532	122	3.072
0.57	3.86	0.88	838	27.26	0.000137	0.2691	116	3.188
0.58	3.83	0.87	852	27.80	0.000144	0.2849	110	3.298
0.60	3.80	0.86	866	28.34	0.000151	0.3007	105	3.403
0.62	3.77	0.86	881	28.88	0.000158	0.3166	100	3.503
0.63	3.74	0.85	896	29.43	0.000166	0.3324	95	3.599
0.65	3.70	0.85	911	29.98	0.000174	0.3482	91	3.690
0.66	3.67	0.84	927	30.54	0.000182	0.3640	87	3.776
0.68	3.64	0.84	943	31.11	0.000191	0.3799	83	3.859
0.70	3.61	0.83	960	31.68	0.000200	0.3957	79	3.939
0.71	3.58	0.83	977	32.26	0.000209	0.4115	76	4.014
0.73	3.55	0.82	995	32.85	0.000219	0.4274	72	4.087
0.74	3.51	0.82	1.013	33.45	0.000229	0.4432	69	4.156
0.76	3.48	0.81	1.031	34.06	0.000239	0.4590	66	4.222
0.77	3.45	0.81	1.050	34.67	0.000250	0.4748	63	4.285

where 2 s is the length of the arc corresponding to the cord 2c and r the radius of the specimen. The function $f(a/c)$ depends on the ratio a/r and for $a = c$ being constant it is

Table 14.2 Step by step solution of Eq. (14.38) for stress amplitude of 400 MPa

a (mm)	d (mm)	$f(d/D)$	σ_{lnet} (MPa)	ΔK_I (MPam$^{1/2}$)	da/dN (mm/ciclo)	$\Sigma\Delta a$ (mm)	ΔN cycles	N cycles
0.30	4.40	0.99	517	15	0.000000	0.0000	0.00	0.00
0.32	4.36	0.98	527	15	0.000031	0.0219	718	718
0.34	4.31	0.97	538	16	0.000034	0.0438	649	1.367
0.37	4.27	0.96	549	16	0.000037	0.0657	590	1.958
0.39	4.22	0.95	560	17	0.000041	0.0876	539	2.497
0.41	4.18	0.94	572	18	0.000044	0.1095	495	2.992
0.43	4.14	0.93	584	18	0.000048	0.1314	455	3.447
0.45	4.09	0.92	597	19	0.000052	0.1533	420	3.867
0.48	4.05	0.91	610	19	0.000056	0.1752	389	4.255
0.50	4.01	0.90	623	20	0.000061	0.1971	360	4.616
0.52	3.96	0.89	637	20	0.000065	0.2190	335	4.951
0.54	3.92	0.89	651	21	0.000070	0.2409	312	5.263
0.56	3.87	0.88	666	22	0.000075	0.2628	291	5.554
0.58	3.83	0.87	681	22	0.000081	0.2847	272	5.825
0.61	3.79	0.86	697	23	0.000086	0.3066	254	6.079
0.63	3.74	0.85	714	23	0.000092	0.3285	238	6.317
0.65	3.70	0.85	731	24	0.000098	0.3504	223	6.540
0.67	3.66	0.84	748	25	0.000105	0.3723	209	6.748
0.69	3.61	0.83	767	25	0.000112	0.3942	196	6.944
0.72	3.57	0.82	786	26	0.000119	0.4161	184	7.128
0.74	3.52	0.82	805	27	0.000127	0.4380	173	7.301
0.76	3.48	0.81	826	27	0.000135	0.4599	162	7.463
0.78	3.44	0.80	847	28	0.000144	0.4818	152	7.615
0.80	3.39	0.80	869	29	0.000153	0.5037	143	7.758
0.83	3.35	0.79	892	29	0.000163	0.5256	135	7.893
0.85	3.31	0.78	915	30	0.000173	0.5475	127	8.020
0.87	3.26	0.78	940	31	0.000184	0.5694	119	8.139
0.89	3.22	0.77	966	32	0.000196	0.5913	112	8.251
0.91	3.17	0.77	993	32	0.000208	0.6132	105	8.356
0.94	3.13	0.76	1.021	33	0.000221	0.6351	99	8.455
0.96	3.09	0.75	1.050	34	0.000236	0.6570	93	8.548

$$f\left(\frac{a}{c}\right) = 0.7433\left(\frac{a}{r}\right)^2 - 0.2194\left(\frac{a}{r}\right) + 0.6757 \tag{14.40}$$

As in the previous example, to integrate the Paris equation it is necessary to calculate the final depth and shape a_f of the crack. To this purpose, the most likely hypothesis is that the crack grows keeping always the same shape, i.e., the same

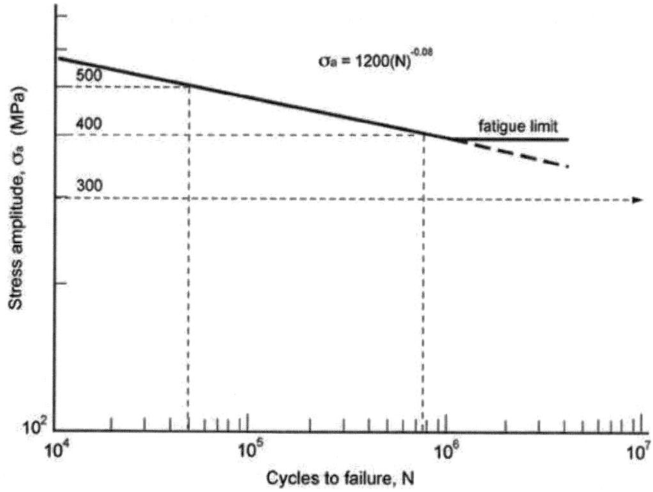

Fig. 14.44 Wöhler diagram for 4340 steel

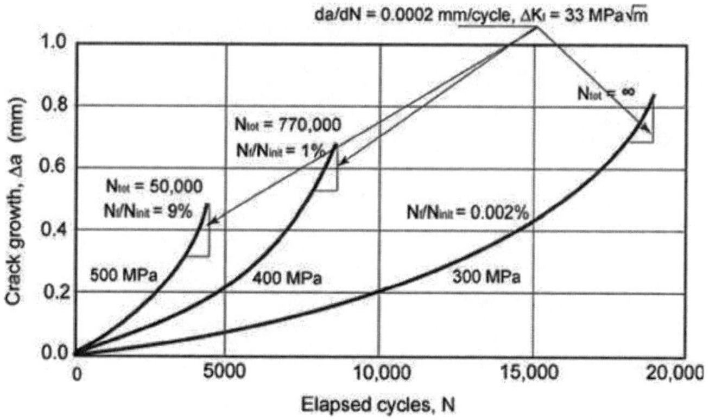

Fig. 14.45 Fatigue crack growth during cycling at different stress amplitude

aspect ratio $a/c = 1$. In this case, a simple calculation yields a final depth $a_f =$ 2.12 mm under a stress amplitude $\sigma_a = 300$ MPa. As in the previous example, the Paris-Erdogan equation will be integrated by finite differences method. The overall growth of $\Delta a = 2.12–0.3 = 1.82$ mm is divided into 30 equal parts $da = 1.82/30$ $= 0.06$ mm in each of which the corresponding ΔK_I will be considered constant. Results are shown in Table 14.4. The number of cycles N_f necessary to fail the specimen (growth from 0.3 mm to 2.12 mm) increases from 4285 of the previous example to 123,863. If the stress amplitude is increased to 400 MPa it is obtained the results listed in Table 14.5. The cycles needed to grow the crack to failure decreases to 55,929. Figure 14.47 provides a comparison with the results of the previous example.

Table 14.3 Step by step solution of Eq. (14.38) for stress amplitude of 300 MPa

a (mm)	d (mm)	$f(d/D)$	σ_{net} (MPa)	ΔK_I (MPam$^{1/2}$)	da/dN (mm/ciclo)	$\Sigma\Delta a$ (mm)	ΔN cicli	N cicli
0.30	4.40	0.99	387	11	0.000000	0.0000	0.00	0.00
0.33	4.34	0.97	398	12	0.000015	0.0288	1.938	1.938
0.36	4.28	0.96	408	12	0.000017	0.0576	1.700	3.638
0.39	4.23	0.95	420	13	0.000019	0.0864	1.504	5.142
0.42	4.17	0.94	431	13	0.000021	0.1152	1.341	6.483
0.44	4.11	0.92	444	14	0.000024	0.1439	1.203	7.686
0.47	4.05	0.91	456	14	0.000027	0.1727	1.084	8.770
0.50	4.00	0.90	469	15	0.000029	0.2015	981	9.751
0.53	3.94	0.89	483	16	0.000032	0.2303	891	14.642
0.56	3.88	0.88	498	16	0.000035	0.2591	812	11.453
0.59	3.82	0.87	513	17	0.000039	0.2879	741	12.195
0.62	3.77	0.86	529	17	0.000042	0.3167	679	12.874
0.65	3.71	0.85	545	18	0.000046	0.3455	622	13.496
0.67	3.65	0.84	563	19	0.000050	0.3743	571	14.067
0.70	3.59	0.83	581	19	0.000055	0.4031	525	14.593
0.73	3.54	0.82	600	20	0.000060	0.4318	484	15.076
0.76	3.48	0.81	620	20	0.000065	0.4606	445	15.521
0.79	3.42	0.80	641	21	0.000070	0.4894	410	15.932
0.82	3.36	0.79	663	22	0.000076	0.5182	378	16.310
0.85	3.31	0.78	686	23	0.000083	0.5470	349	16.659
0.88	3.25	0.78	711	23	0.000089	0.5758	322	16.981
0.90	3.19	0.77	737	24	0.000097	0.6046	297	17.277
0.93	3.13	0.76	764	25	0.000105	0.6334	274	17.551
0.96	3.08	0.75	793	26	0.000114	0.6622	252	17.803
0.99	3.02	0.74	823	27	0.000124	0.6910	232	18.035
1.02	2.96	0.74	856	27	0.000135	0.7197	214	18.249
1.05	2.90	0.73	890	28	0.000146	0.7485	197	18.446
1.08	2.85	0.72	926	29	0.000159	0.7773	181	18.627
1.11	2.79	0.72	965	30	0.000173	0.8061	166	18.793
1.13	2.73	0.71	1.006	31	0.000189	0.8349	153	18.946
1.16	2.67	0.70	1.050	33	0.000206	0.8637	140	19.086

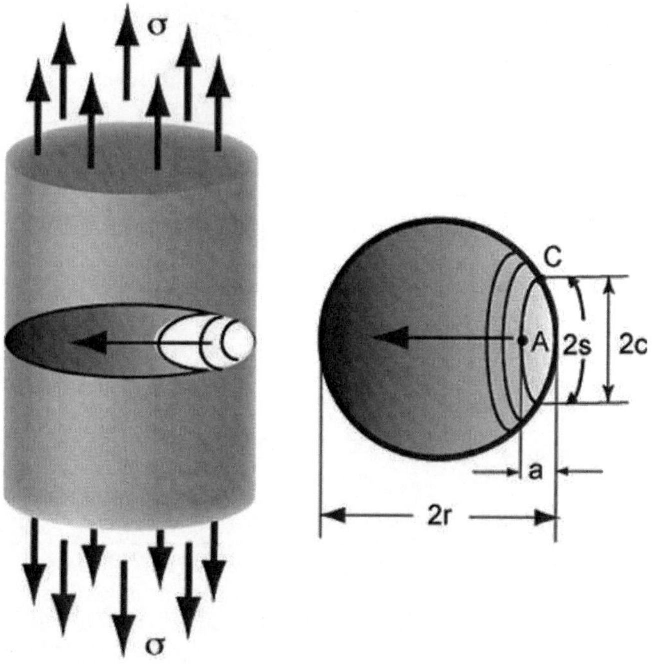

Fig. 14.46 Round bar with a semi-elliptical surface crack

14.10.3 Fatigue Crack Growth-Cylindrical Geometry, Semielliptical Crack, Rotating Bending

Always the same material as in the previous examples, but this time the work-piece is a shaft of 150 mm diameter, subjected to rotating bending. It must be assessed the life under a bending stress amplitude of 300 MPa with an initial crack $a_o = 0.3$ mm depth, $2c_o = 0.6$ mm. The K_I expression remains the same given by Eq. (14.39), but the this time the function *f(a/c)* becomes

$$f\left(\frac{a}{c}\right) = 0.3342\left(\frac{a}{r}\right)^2 - 0.4723\left(\frac{a}{r}\right) + 0.6556 \tag{14.41}$$

since the shaft is working under rotating bending. Proceeding as in the previous examples, calculation yield a final crack length to fracture $a_f = 126.5$ mm. the integration of the Paris-Erdogan equation between the limits $a_o = 0{,}3$ mm and $a_f = 126{,}5$ mm yields a number of cycles $N = 540{,}000$ by far higher than the 123,000 found before for the 5 mm specimen in cyclic traction.

Table 14.4 Step by step calculation of the growth of semi-elliptical surface rack under stress amplitude of 300 MPa

a (mm)	*a/r*	*area A* a/c = 1	σ_{net} (MPa)	*s* (mm)	*f(a/c)* a/c = 1	ΔK (MPa$\sqrt{m}$)	*dN* cycles	ΣN cycles	$\Sigma \Delta a$ (mm)	*da/dN* cycles
0	0.000	1.96E−05	300	0.00E+00	0.676	0	0	0	0	0.E+00
0.3	0.120	1.93E−05	304	3.01E−04	0.660	6	17.925	17.925	0.06	1.E−04
0.360578	0.144	1.92E−05	306	3.62E−04	0.660	7	14.254	32.179	0.12	1.E−04
0.421156	0.168	1.91E−05	309	4.23E−04	0.660	7	11.706	43.885	0.18	2.E−04
0.481733	0.193	1.89E−05	312	4.85E−04	0.661	8	9.833	53.718	0.24	2.E−04
0.542311	0.217	1.87E−05	315	5.47E−04	0.663	8	8.396	62.114	0.30	2.E−04
0.602889	0.241	1.85E−05	319	6.09E−04	0.666	9	7.257	69.371	0.36	3.E−04
0.663467	0.265	1.82E−05	323	6.72E−04	0.670	9	6.330	75.701	0.42	3.E−04
0.724044	0.290	1.80E−05	327	7.35E−04	0.675	10	5.561	81.262	0.48	3.E−04
0.784622	0.314	1.77E−05	333	7.98E−04	0.680	10	4.911	86.174	0.55	4.E−04
0.8452	0.338	1.74E−05	339	8.62E−04	0.686	11	4.356	90.529	0.61	4.E−04
0.905778	0.362	1.70E−05	345	9.27E−04	0.694	11	3.875	94.405	0.67	5.E−04
0.966355	0.387	1.67E−05	353	9.92E−04	0.702	12	3.456	97.861	0.73	5.E−04
1.026933	0.411	1.63E−05	361	1.06E−03	0.711	12	3.088	100.950	0.79	6.E−04
1.087511	0.435	1.59E−05	370	1.13E−03	0.721	13	2.763	103.713	0.85	7.E−04
1.148089	0.459	1.55E−05	380	1.19E−03	0.732	13	2.475	106.188	0.91	7.E−04
1.208667	0.483	1.50E−05	392	1.26E−03	0.743	14	2.218	108.406	0.97	8.E−04
1.269244	0.508	1.46E−05	404	1.33E−03	0.756	15	1.989	114.395	1.03	9.E−04
1.329822	0.532	1.41E−05	418	1.40E−03	0.769	15	1.784	112.179	1.09	1.E−03
1.3904	0.556	1.36E−05	434	1.47E−03	0.784	16	1.600	113.780	1.15	1.E−03
1.450978	0.580	1.30E−05	452	1.55E−03	0.799	17	1.435	115.215	1.21	1.E−03

(continued)

Table 14.4 (continued)

a (mm)	a/r	*area A* a/c = 1	σ_{net} (MPa)	s (mm)	$f(a/c)$ a/c = 1	ΔK (MPa$\sqrt{}$m)	dN cycles	ΣN cycles	$\Sigma \Delta a$ (mm)	da/dN cycles
1.511555	0.605	1.25E−05	473	1.62E−03	0.815	17	1.287	116.502	1.27	1.E−03
1.572133	0.629	1.19E−05	496	1.70E−03	0.832	18	1.154	117.656	1.33	2.E−03
1.632711	0.653	1.13E−05	523	1.78E−03	0.849	19	1.034	118.690	1.39	2.E−03
1.693289	0.677	1.06E−05	554	1.86E−03	0.868	20	926	119.616	1.45	2.E−03
1.753866	0.702	9.97E−06	591	1.94E−03	0.888	21	829	120.445	1.51	2.E−03
1.814444	0.726	9.29E−06	634	2.03E−03	0.908	22	742	121.187	1.58	2.E−03
1.875022	0.750	8.59E−06	686	2.12E−03	0.929	23	663	121.851	1.64	3.E−03
1.9356	0.774	7.86E−06	749	2.21E−03	0.951	24	593	122.444	1.70	3.E−03
1.996177	0.798	7.11E−06	828	2.31E−03	0.974	25	529	122.972	1.76	3.E−03
2.056755	0.823	6.34E−06	928	2.42E−03	0.998	26	471	123.444	1.82	4.E−03
2.117333	0.847	5.55E−06	1061	2.53E−03	1.023	27	419	123.863	1.88	4.E−03

Table 14.5 Step by step calculation of the growth of semi-elliptical surface rack under stress amplitude of 400 MPa

a (mm)	a/r	A a/c = 1	σ_{net} (MPa)	$f(a/c)$ a/c = 1	ΔK (MPa$\sqrt{}$m)	dN cycles	ΣN cycles	da/dN cycles	s (mm)	$\Sigma \Delta a$ (mm)
0	0.000	1.96E−05	400	0.676	0	0	0	0.E+00	0.00E+00	0
0.3	0.120	1.93E−05	406	0.660	8	7.063	7.063	2.E−04	3.01E−04	0.049
0.349	0.140	1.92E−05	408	0.660	9	5.851	12.914	3.E−04	3.50E−04	0.098
0.398	0.159	1.91E−05	410	0.660	9	4.958	17.872	3.E−04	4.00E−04	0.147
0.447	0.179	1.90E−05	413	0.660	10	4.272	22.144	3.E−04	4.49E−04	0.196
0.496	0.198	1.89E−05	416	0.661	10	3.728	25.872	4.E−04	4.99E−04	0.245
0.545	0.218	1.87E−05	420	0.663	11	3.286	29.159	4.E−04	5.49E−04	0.294
0.594	0.238	1.85E−05	424	0.666	12	2.920	32.078	5.E−04	6.00E−04	0.343
0.643	0.257	1.83E−05	428	0.668	12	2.610	34.688	6.E−04	6.50E−04	0.392
0.692	0.277	1.81E−05	433	0.672	13	2.345	37.033	6.E−04	7.01E−04	0.441
0.741	0.296	1.79E−05	439	0.676	13	2.115	39.148	7.E−04	7.52E−04	0.49
0.79	0.316	1.77E−05	444	0.681	14	1.914	41.063	8.E−04	8.04E−04	0.539
0.839	0.336	1.74E−05	451	0.686	14	1.737	42.800	8.E−04	8.56E−04	0.588
0.888	0.355	1.71E−05	458	0.692	15	1.580	44.380	9.E−04	9.08E−04	0.637
0.937	0.375	1.69E−05	465	0.698	15	1.439	45.819	1.E−03	9.60E−04	0.686
0.986	0.394	1.66E−05	474	0.705	16	1.313	47.132	1.E−03	1.01E−03	0.735
1.035	0.414	1.63E−05	483	0.712	16	1.199	48.331	1.E−03	1.07E−03	0.784
1.084	0.434	1.59E−05	493	0.720	17	1.096	49.427	1.E−03	1.12E−03	0.833
1.133	0.453	1.56E−05	503	0.729	18	1.002	50.429	1.E−03	1.18E−03	0.882
1.182	0.473	1.52E−05	515	0.738	18	917	51.346	2.E−03	1.23E−03	0.931
1.231	0.492	1.49E−05	528	0.748	19	840	52.186	2.E−03	1.29E−03	0.98

(continued)

Table 14.5 (continued)

a (mm)	*a/r*	*A* a/c = 1	σ_{net} (MPa)	*f(a/c)* a/c = 1	ΔK (MPa$\sqrt{m}$)	*dN* cycles	ΣN cycles	*da/dN* cycles	*s* (mm)	$\Sigma \Delta a$ (mm)
1.28	0.512	1.45E−05	542	0.758	20	769	52.955	2.E−03	1.34E−03	1.029
1.329	0.532	1.41E−05	558	0.769	20	704	53.659	2.E−03	1.40E−03	1.078
1.378	0.551	1.37E−05	575	0.781	21	645	54.303	2.E−03	1.46E−03	1.127
1.427	0.571	1.32E−05	593	0.793	22	590	54.894	2.E−03	1.52E−03	1.176
1.476	0.590	1.28E−05	614	0.805	23	541	55.434	3.E−03	1.58E−03	1.225
1.525	0.610	1.23E−05	637	0.818	23	495	55.929	3.E−03	1.64E−03	1.274
1.574	0.630	1.18E−05	663	0.832	24	453	56.382	3.E−03	1.70E−03	1.323
1.623	0.649	1.14E−05	691	0.847	25	415	56.797	4.E−03	1.77E−03	1.372
1.97	0.788	7.44E−06	1055	0.964	33	219	57.016	7.E−03	2.27E−03	1.421

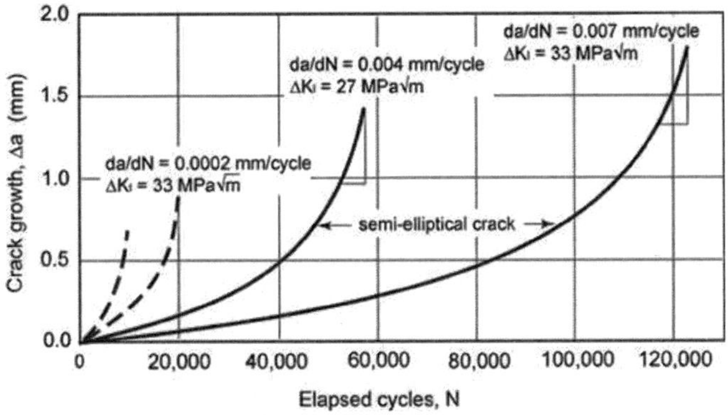

Fig. 14.47 Comparison of results obtained in example N. 1 and 2

14.10.4 Fatigue Crack Growth-Surface Crack in Plane Geometry

A column of low-alloyed carbon steel, type Fe520, 400 MPa yield strength and 550 MPa ultimate strength, 400 mm wide and 150 mm thick, has been found affected by a semi-circular surface crack, as shown in Fig. 14.48. An accurate examination revealed that its dimensions are approximately $a_o = 0.5$ mm and $c_o = a_o$. The column is subjected to fatigue pulsating from zero to $\sigma_{\max} = 150$ MPa, as schematized in Fig. 14.47. Calculate the residual life. As usual, the assessment is based on the integration of the Paris-Erdogan Eq. (14.11) between the initial crack size a_o that is known and the final a_f that must be assessed. For the geometry considered, the stress intensity factor K_I is given by

$$K_I = \lambda\sigma\sqrt{\frac{\pi a}{Q}}f(\theta)$$

$$\lambda = \left[1.13 - 0,09\left(\frac{a}{c}\right)\right]\left[1 + 0.1(1 - \sin\theta)^2\right];\ Q = 1 + 1.464\left(\frac{a}{c}\right)^{1.65}$$

$$f(\theta) = \left[\sin^2\theta + \left(\frac{a}{c}\right)^2\cos^2\theta\right]^{1/4}. \tag{14.42}$$

This time, there are to shape factors *f*(θ) and λ that do not depend on the ratio *a*/2*c*, but also on the angle θ relative to the point on the crack surface where the applied K_I is considered. The examination of the two factors reveals that for $a/c = 1$ the *f*(θ) factor is practically independent of θ and equal one. Therefore, the crack is likely to grow in both directions, as schematized in Fig. 14.47c rather than in a preferential direction, as schematized in Fig. 14.47d, keeping the same aspect ratio $a/c = 1$. The assessment of the final crack depth a_f that produce fracture is based on the calculation of the final area A_f on which the flow stress σ_{flow} is reached

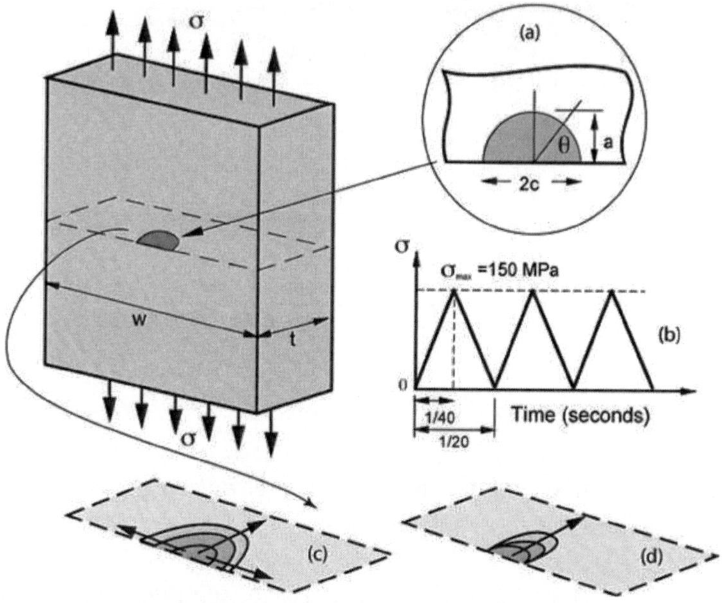

Fig. 14.48 Geometry of the column, load history and crack growing modes

$$\sigma_{flow} = \frac{\sigma_s + \sigma_r}{2} = \frac{400 + 550}{2} = 475 \text{ MPa} \tag{14.43}$$

$$\frac{A_o}{A_f}\sigma = \sigma_{flow}$$

$$A_f = \frac{150}{475} \cdot 600 = 189.5 \text{ cm}^2$$

$$A_f = A_o - \frac{1}{2}\pi \cdot r_f^2 \rightarrow r_f = a_f = \sqrt{\frac{2(A_o - A_f)}{\pi}} = 16.2 \text{ cm.}$$

Therefore, the crack can grow up to a depth of 1620 mm over a thickness of only 150 mm, which actually means that it can became a trough-wall crack before producing fracture. However, since the working ambient may not be heated for some time because a possible black-out, it must be assumed that the lowest temperature can be also equal to 0 °C. Since the material is a ferritic steel it has a brittle-to-ductile transition temperature NDT. it will be necessary, then, to perform also a brittle fracture analysis. When the applied K_I, given by Eq. (14.41), becomes equal to the toughness K_{Ic} of the material fracture occurs. It is necessary to assess the material toughness considering that the load is not applied statically, but dynamically. Looking at the load history of Fig. 14.47 it is possible to infer that the load rises from zero to the maximum in 1/40 s. During this time interval the stress rises from 0 to 150 MPa with a rate

$$\dot{\sigma} = \frac{\Delta\sigma}{\Delta t} = 40 \cdot 150 = 6.000 \text{ MPa/s}$$

so that it can be assumed that the applied $\dot{K}_I$ rate will be much higher, because of the stress concentration. Figure 14.49 [69] shows the variation of toughness of a carbon steel as a function of the applied. Results shown in Fig. 14.49 have been obtained using CT specimens of different size, from 1 to 8 inches thick. At 0 °C the toughness can be estimated of about 80 MPa$\sqrt{\text{m}}$. With $a/c = 1$ the function λ that appears in Eq. (14.41) is practically zero, while it is $Q = 2.46$. Therefore, the applied K_I is

$$K_I = \sigma\sqrt{\frac{\pi a}{2.46}} = 1.13 \cdot \sigma\sqrt{a} \tag{14.44}$$

Equating (14.44) to 80 MPa$\sqrt{\text{m}}$ and solving for a yields

$$a = \left(\frac{80}{1.13 \cdot 150}\right)^2 = 0.22 \text{ m}$$

This means that the crack can become a through-wall crack without failure occurs. Therefore, the thickness $t = 150$ mm will be assumed as final crack depth a_f. now, Eq. (14.11) can be integrated

$$N = \frac{1}{C\left(\Delta\sigma \cdot \sqrt{\pi}\right)^n} \int_{a_o}^{a_f} a^{-2} da$$

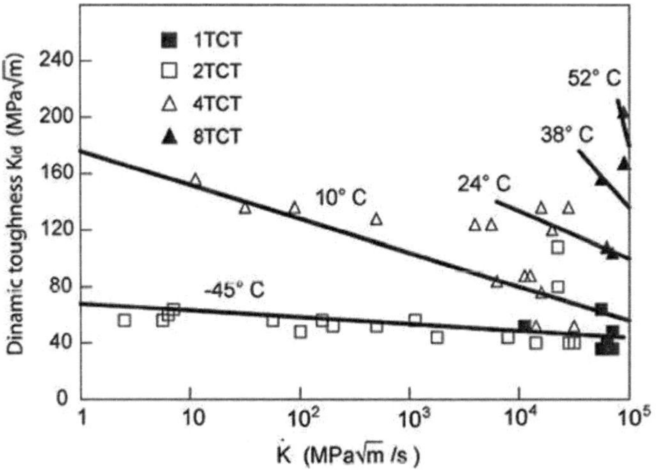

Fig. 14.49 Toughness variation as a function of temperature and $\dot{K}_I$ [69]

$$= \frac{a_o^{(1-n/2)} - a_f^{(1-n/2)}}{C\left(\frac{n}{2} - 1\right)\left(\Delta\sigma \cdot \sqrt{\pi}\right)^n} \tag{14.45}$$

and the values of constant C and n are those characteristic of a carbon steel $C = 1.26 \cdot 10^{-9}$ mm/cycle; $n = 3.38$.

Therefore it will be $N = 218{,}871$ cycles.

14.10.5 Fatigue Crack Growth in Cast Iron

Two flat specimens of nodular cast iron having a Sect. 14.3 × 35 mm and 30 cm long, have been subjected to fatigue in traction pulsating from zero ($R = 0$) with 81 MPa stress amplitude. One has developed a fatigue crack that after 320,000 cycles had reached 75% of the section thickness, the other had not initiated any crack after 90,000,000 cycles. Explain the reason of that discrepancy and find the FCGR *da/dN* of the crack. The stress amplitude equal to 81 MPa is below the fatigue limit of the material, therefore the behavior of the second specimen is precisely as expected. The reason of the fatigue cracking observed in the first specimen is due to metallurgical variability that in ductile cast iron can be extremely high, coupled to a process volume of the specimen also unusually high (see Sects. 3.5 and 4.5.1). As matter of fact, such large volume specimens were chosen precisely to study the metallurgical variability of the material. In the first specimen it was found a large shrinkage cavity, about one millimeter long, that had become a crack after relatively few cycles. The FCGR, *da/dN*, can be found integrating the Paris-Erdogan equation, but this time knowing the number of elapsed cycles and the final length of the crack the integration will yield the value of the constant C of the Paris-Erdogan equation

$$C = \frac{1}{N} \int_{a_o}^{a_f} \frac{da}{\Delta K_I^n}. \tag{14.46}$$

Equation (14.46) contains also the unknown variable n. however, it is known that the exponent n for ductile cast iron is equal to 5. For the geometry of the flat specimen containing a surface edge crack the expression of the stress intensity factor is

$$\begin{aligned} K_I &= \sigma\sqrt{\pi a} \cdot f(\alpha) \\ f(\alpha) &= 1.12 - 0.231\alpha + 10.55\alpha^2 - 21.72\alpha^3 + 30.39\alpha^4 \\ \alpha &= \frac{a}{w} \end{aligned} \tag{14.47}$$

in which a is the crack depth and w the specimen width. The polynomial expression of $f(\alpha)$ given by Eq. (14.47) transform the integral of Eq. (14.46) into five simple

Table 14.6 Nondimensional $f(a)$ functions and relative stress intensity factors, K_I, for several crack depths

a (m)	*a/w*	*f(a)*	K_I
5.00E−04	1.43E−02	1.12E+00	4.45E+00
1.00E−03	2.86E−02	1.12E+00	6.31E+00
2.00E−03	5.71E−02	1.14E+00	8.25E+00
5.00E−03	1.43E−01	1.25E+00	1.27E+01
1.00E−02	2.86E−01	1.61E+00	2.31E+01
1.50E−02	4.29E−01	2.27E+00	4.00E+01
2.00E−02	5.71E−01	3.62E+00	7.35E+01
2.20E−02	6.29E−01	4.49E+00	9.57E+01
2.50E−02	7.14E−01	6.33E+00	1.44E+02
3.00E−02	8.57E−01	1.14E+01	2.83E+02

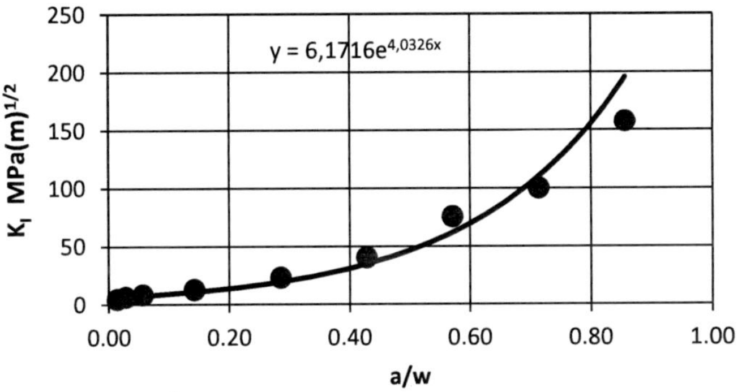

Fig. 14.50 Trend of the stress intensity factor K_I versus *a/w* for the specime geometry

integrals, but it is rather useful to follow another approach. To this purpose, it is convenient to calculate the value of the stress intensity factor K_I for some different crack lengths, as shown in Table 14.6. Results are shown in graphic fashion in Fig. 14.50 and are fitted to the curve of equation

$$K_I = 6.17e^{4a/w}. \tag{14.48}$$

Using Eq. (14.48) in Eq. (14.46) and integrating it yields

$$C = \frac{w}{A^n \cdot B \cdot mN}\left[e^{-Bn\alpha_o} - e^{-Bn\alpha_f}\right] = 4.14 \cdot 10^{-13}$$

being $A = 6.17$, $B = 4$ and $\alpha = a/w$. The so found constant C provides the value of *da/dN* in m/cycle.

References

1. Paris, P.C., Erdogan, F.: A critical analysis of crack propagation lows. J. Basic Eng. **85**, 528–534 (1960)
2. Paris, P.C., Gomez, M.P., Anderson, W.P.: A rational analytical theory of fatigue. Trend Eng. **13**, 9–14 (1961)
3. Clark, W.G. Jr.: How cracks grow in structural steels. In: Metal Progress, pp. 81–86 (1970)
4. Frost, N.E.: J. Mech. Eng. Sci. **1**(2), 151 (1959)
5. Liu, H.W.: Allied Mater. Res. **3**(4), 229 (1964)
6. Weibull, W.: Acta Metall. **11**(2), 725 (1963)
7. Irwin, G.: Analysis of stresses and strains near the end of a crack traversing a plate. J. Appl. Mech. **24**, 361–364 (1957)
8. Paris, P.C.: The fracture mechanics approach to fatigue. In: Proceedings of the 10th Sagamore conference, p. 107, Syracuse Press (1965)
9. Swanson, S.R., Cicci, F., Hoppe, W.: Crack propagation in clad 7079-T6 aluminum alloy sheet under constant and random amplitude fatigue loading. ASTM-STP 415 (1868)
10. Forman, R.G., Kearney, V.E., Engle, R.M.: Numerical analysis of crack propagation in cyclic loaded structures. J. Basic Eng. **89**, 459–464 (1967)
11. Weertman, J.: Rate of growth of fatigue cracks calculated from the theory of infinitesimal dislocations distributed on a plane. Int. J. Fract Mech **2**, 460–467 (1966)
12. Klesnil, M., Lukas, P.: Influence of strength and stress history on growth and stabilization of fatigue cracks. Eng. Fract. Mech.Fract. Mech. **4**, 77–92 (1972)
13. McEvily, A.J.: On closure in fatigue crack growth. ASTM STP, Am. Soc. Test. Mater. **982**, 35–43 (1988)
14. Dowling, N.E., Begley, J.A.: Fatigue crack growth during gross plasticity and the J integral. ASTM STP, Am. Soc. Test. Mater. **590**, 82–103 (1976)
15. Newman, J.C., Jr.: American society for testing and materials. ASTM STP, Meth. Models Predicting Fatigue Crack Growth Random Loading **748**, 55–84, (1981)
16. McClintok, F.A.: Discussion of the influence of metallurgical structure on the mechanisms of FCP by C. Laird. ASTM STP, Am. Soc. Test. Mater. **425**, 170–174 (1967)
17. Clark, W.G. Jr., Ceschini, L.J.: An Ultrasonic Crack Growth Monitor. Westinghouse Scientific Paper 68-7D7-BFPWR-P1 (1968)
18. Joice, J.A., Schneider, C.S.: Application of Alternating Current Potential Difference to Crack Length Measurement During Rapid Loading, US-NRC, NUREG/CR-4699 (1986)
19. Steigerwald, E.A., Hanna, G.L.: Proc. Am. Soc. Test. Mater. **62**, 885 (1962)
20. Wei, R.P., Talda, P.M., Li, C.-Y.: Fatigue crack propagation. ASTM STP, Am. Soc. Test. Mater. **415**, 460 (1967)
21. Ritchie, R.O., Garret, G.G., Knott, J.F.: Int. J. Fract.Fract. **7**, 462 (1971)
22. Schwalbe, K.H., Hellman, D.: The application of the electric potential method to crack length measurement using Johnson's formula. JTEVA **9**(3), 218.221 (1981)
23. Vassilaros, M.G., Hackett, E.M.: ASTM STP, Am. Soc. Test. Mater. **833**, 535–552 (1984)
24. Wei, R.P., Brazill, R.L.: An assessment of A–C and D–C potential systems for monitoring fatigue crack growth. ASTM STP, Am. Soc. Test. Mater. **738**, 103–119 (1981)
25. Imhof, E.J., Barsom, J.M.: Fatigue and corrosion fatigue crack growth of 4340 steel at various yield strength. ASTM STP, Am. Soc. Test. Mater. **536** (1973)
26. Hertzberg, R.W., Paris, P.C.: Application of electron fractography and fracture mechanics to fatigue crack propagation. In: Proceedings of 1st International Conference on Fracture, Sendai, Japan, vol. 1, pp. 459–478 (1965)
27. Bates, R.C., Clark, W.G., Moon, D.M.: Correlation of fractographic features with fracture mechanics data. ASTM STP, Am. Soc. Test. Mater. **453**, 192–214 (1969)
28. Kershaw, J., Liu, H.W.: Electron fractography and fatigue crack propagation in 7075–T6 aluminum sheet. Int. J. Fract.Mech.Fract.Mech. **7**, 269–276 (1971)
29. Bathias, C., Pelloux, R.M.: Fatigue crack propagation in martensitic and austenitic steels. Metall. Trans. **4**, 1265–1273 (1973)

30. Pickard, A.C, Ritchie, R.O., Knott, J.F.: Fatigue crack propagation in type 316 stainless steel weldment. Metall. Technol. 253–263 (1975)
31. Yokobory, T., Sato, K.: The effect of frequency on fatigue crack propagation rate and striation spacing in 2024–T3 aluminium alloy and SM-50 steel. Eng. Fract. Mech.Fract. Mech. **8**, 81–88 (1976)
32. Mills, W.J., James, L.A.: Effect of temperature on the fatigue crack propagation behaviour of inconel X-750. Fatigue Eng. Mater. Struct. **3**(2), 159–175 (1980)
33. Paris, P.C.,
34. Pelloux, R.M.N.: Mechanism of formation of ductile fatigue striations. Trans. ASM **62**, 281–285 (1969)
35. McMinn, A.: Fractographic analysis in the understanding of corrosion fatigue mechanisms. In: Proceeds. International Atomic Energy Agency, Specialists Meeting on Subcritical Crack Growth, Nureg/CP-0044, vol. 2, pp. 3–26, Freiburg, Germany (1981)
36. Hertzberg, R.W., Mills, W.J.: Am. Soc. Test. Mater. ASTM STP **600**, 220 (1976)
37. Murakami, Y., Hamada, S.: A new method for the measurement of MODE II fatigue threshold stress intensity factor range $\Delta K_{\tau th}$. Fatigue Fract. Eng. Mater. Struct.Fract. Eng. Mater. Struct. **20**(6), 863–870 (1997)
38. Tschegg, E.K., Stanzl, S.E., Mayer, H.R., Czegley, M.: Crack face interaction and near-threshold fatigue cracl growth. Fatigue Fract. Eng. Mater. Struct.Fract. Eng. Mater. Struct. **16**(1), 71–82 (1992)
39. Yates, J.R., Miller, K.J.: Mixed mode (I+III) fatigue threshold in a forging steel. Fatigue Fract. Eng. Mater. Struct.Fract. Eng. Mater. Struct. **12**(3), 259–270 (1989)
40. Barsom, J.M.: J. Eng. Ind., Trans., ASME, Series B **93**(4) (1971)
41. Barsom, J.M., Imhof, E.J., Rolfe, S.T.: J. Eng. Fracture Mech. **2**(4) (1971)
42. Matocha, K., Wozniak, J., Siegl, J.: The Effect of Strain Aging on the Propagation of Fatigue Cracks in NiMoV (AISI 1018) Low Alloy Steel
43. Paris, P.C., Bucci, R.J., Wessel, E.T., Clark, G.W., Mager, T.R.: Extensive study of low crack growth rates in A533 and A508 steels. In: Proceedings of the 1971 National Symposium on Fracture Mechanics, Part I, ASTM STP vol. 513, pp. 141–176 (1972)
44. Tsuji, H., Yokoyama, N., Nakajima, H., Kondo, T.: Statistical analysis of variability/ reproducibility of environmentally assisted cyclic crack growth rate data relative to ΔK modes. In: Proceedings of the 3rd International Atomic Energy Agency Specialists' Meeting on Subcritical Crack Growth, NUREG/CR–0112, vol. 1, pp. 231–251 (1990)
45. Cullen, W.H., Loss, F.J., Watson, H.E.: Design Operation and Inspection Relevant Factors of Fatigue Crack growth rates for Pressure Vessel and Piping Steels, Nureg/CP-0044, IAEA, Specialists Meeting on Subcritical Crack Growth, vol. 1, p. 127, (1981)
46. Kondo, T., Kuniya, J., Takaku, H., Tokimasa, K., Arii, M., Kurihara, M.: Recent study on cyclic crack growth of reactor pressure boundary materials in high temperature water environment in Japan. In: Proceedings 2nd IAEA Specialists Meeting on Subcritical Crack Growth, NUREG/ CP-0067, vol. 1, pp. 219–249 (1985)
47. Mager, T.R., Landes, J.D., McLuoghlin, V., Moon, D.M.: The Effect of Low Frequencies on the Fatigue Crack Growth Characteristics of A533B Class 1 Plate in an Environment of High Temperature Primary Grade Nuclear Reactor Water, HSST Report No. 35 (1973)
48. Clark, W.G., Jr.: Fatigue Crack Growth Characteristic of ASTM A533 Grade B, Class 1 Steel Base Plate, Weld Metal and Heat Affected Zone, Westinghouse Research Report 69-7E7-BFPWR-R2, October, 1969. Also Basic Fracture Mechanics for Nuclear Applications, Westinghouse Presentation by Mager, T., Chirigos, J., McGowan, J. to ENEA-ENEL, Rome, Italy, (6–8 March, 1978)
49. Shahinian, P., Watson, H.E., Smith, H.H.: Fatigue crack growth in selected alloy for reactor applications. J. Mater. **7**(4), 1972
50. James, L.A.: Fatigue Crack Propagation Behavior of Pressure Vessel Steels, Bulletin N. 194, Welding Research Council, N.Y. (1974)
51. James, L.A.: Fatigue crack propagation in austenitic stainless steels. At. Energy Rev. **14**, 34–86 (1976)

52. Banford, W.: Fatigue crack growth of stainless steel piping in PWR environment, ASME paper 77-PVP-34. J. Press. Vessel Technol. (1979)
53. Bernard, J.L., Slama, G, Amzallag, C., Rabbe, P.: Influence of PWR Environment on Fatigue Crack Growth Beaviour of Stainless Steels, Time and Load Dependent Degradation of Pressure Boundary Materials, International Atomic Ener. Agency, IWG-RRPC-79/2, Innsbruck, Austria, (1978)
54. James, L.A.: Fatigue Crack Propagation in a Cast Stainless Steel in a Pressurized Water Reactor Environment, ASME Paper 77-PVP 34 (1977)
55. Joyce, J.A., Hackett, E.M., Roe, C.: Effect of Cyclic Loading on the Deformation and Elastic-Plastic Fracture Behavior of a Cast Stainless Steel, David Taylor research Center, DTRC/SME-91-11 (1991)
56. Makhlouf, K., Jones, J.W.: Effect of temperature and frequency on fatigue crack growth in 18%Cr ferritic stainless steel. Int. J. Fatigue **15**, 163–171 (1993)
57. Clark, W.G. Jr., Bates, R.C.: Microscopic Aspects of Fracture Toughness, Westinghouse Scientific Paper 69-1E7-RDAFC-P1 (4 Dec, 1969)
58. Ambriz, R.R., Mesmacque, G., Benhamena, A., Ruiz, A., Amrouche, A., Lopez, V.H.: Fatigue Crack Growth Behaviour in 6061-T6 Aluminum Alloy Welds Obtained by MIEA
59. Miller, M.S., Gallagher, J.P.: An analysis of several fatigue crack growth rate (FCGR) description. ASTM STP **738**, 205–251 (1981)
60. Marukamai, Y., Shiraishi, N., Furukawa, K.: Estimation of service loading from the width and heigth of fatigue striations of 2017–T4 Al alloy. Fatigue Fract. Engng. Mater. Struct. **14**(9), 897–906 (1991)
61. Crooker, W.T.: Crack Propagation in Aluminum Alloys under High Amplitude Cyclic Loads, Naval Research Laboratory, Report No. 7286, (12 July, 1971)
62. Clark, Jr., W.G.: How Fatigue Crack Initiation and Growth Properties Affect Material Selection and Design Criteria, Metals Engineering Quarterly, pp 16–22 (August 1974)
63. Yoder, G.R., Cooley, L.A., Crooker, T.W.: Procedure for precision measurement of fatigue crack growth rate using crack-opening displacement techniques. ASTM STP **738**, 85–102 (1981)
64. Rolfe, S.T, Barsom, J.M.: Fracture and Fatigue Control in Structures. Application of Fracture Mechanics, Prentice-Hall, Inc. (1977)
65. Hopkinns, S.W., Rau, C.A., Jr.: Prediction of structural crack growth behavior under fatigue loading. ASTM STP **738**, 255–270 (1981)
66. James, L.A., Mills, W.J.: An evaluation of the round compact specimen for fatigue crack growth rate testing. ASTM STP **738**, 70–82 (1981)
67. Amzallag, C, Baudry, G., Bernard, J.L.: Effect of PWR Environment on the Fatigue Crack Growth of Different Stainless Steels and Inconel Type Alloy, Nureg/CP-0044, IAEA, Specialists' Meeting on Subcritical Crack Growth, vol. 1, p. 283 (1981)
68. Dowling, N.E.: Fatigue Crack Growth Rate Testing at High Stress Intensity, Wesinghouse Scientific Paper 76-IE7-MSLRA-P1 (9 March, 1976)
69. Shabbits, W.O.: Dynamic Fracture Toughness Properties of Heavy Section ASTM A 533 Grade B, Class 1 steel Plate, ORNL Heavy Section Technology Program, Technical Report N. 16, December 1970, Westinghouse WCAP-7561

Chapter 15
Crack Tip Plastic Zone Effect on Fatigue Crack Propagation

15.1 The Fatigue Mechanism

It has been shown in the previous section that the fatigue crack propagation in metallic materials follows a trend that can be described by the Paris-Erdogan Eq. (10.10). However, the Paris postulate is valid only under particular circumstances as constant load cycling in Region II of fatigue, long cracks and stress ratio *R* equal zero. If these circumstances are not met the Paris-Erdogan equation fails to represent the FCGR of the material. The fundamental point to understand for any further step is that the crack propagation is controlled by what is actually happening ahead of the crack tip in the plastic zone. Therefore, before even asking whether the Paris postulate is valid or not, it is necessary to address the issue of the crack tip plastic zone. The equilibrium of a crack under a stress state was first treated by the Englishman Griffith in 1920 [1]. He considered a fixed ends infinite plate of ideally elastic material carrying a central transverse crack of initial length $2a_o$. The plate, of unit thickness, is loaded at its ends by a stress σ, as schematized in Fig. 15.1.

If the crack extends over a length *da* a certain amount of elastic energy $\mathcal{G} = dU/da$ is released. Since the plate is fixed at its ends, so that no kinetic energy can be introduced and its behavior is ideally elastic, this released elastic energy will be converted into the only possible available energy that is the surface energy of the two new surfaces created $2\Gamma da$, Γ being the surface energy of the metal per unit area *A*. Under the stress state σ the crack will become unstable and propagate without arresting if the elastic energy release rate$\mathcal{G}$,nalso called *driving* force, will equal the surface energy absorption rate $d(2\Gamma da)/dA$ that represents the resistance *R* of the material to crack propagation. Therefore, the instability or critical condition is

$$\mathcal{G}_c \leq R \tag{15.1}$$

where the suffix *c* denotes criticality. Griffith evaluated the energy release rate $\mathcal{G}$ of the geometry of Fig. 15.1 as $\pi\sigma^2 a/E$, with *E* being the Young's modulus, therefore

P. P. Milella, *Fatigue and Corrosion in Metals*,
https://doi.org/10.1007/978-3-031-51350-3_15

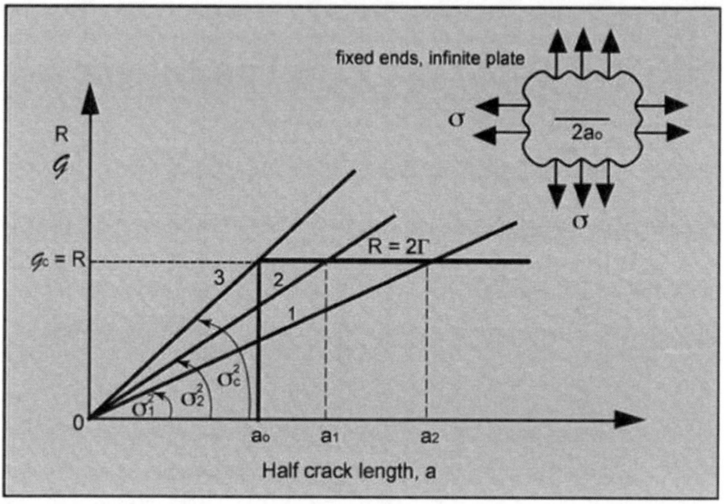

Fig. 15.1 Schematic instability condition for a crack of length $2a$ in an ideally elastic material

Eq. (15.1) becomes

$$\frac{\pi\sigma_c^2 a}{E} = 2 \cdot \Gamma. \tag{15.2}$$

In a $\mathcal{G}$-R versus crack length a space, the driving force is a straight line with angular coefficient given by σ^2 and R is a constant equal to 2Γ independent of the crack length, as schematized in Fig. 15.1. It is an *in–out* situation in that a crack under a stress σ either becomes unstable when $\sigma = \sigma_c$ and propagates indefinitely, line 3 in Fig. 15.1, or stays there as it is, line 1 or 2. Griffith demonstrated the validity of his theory making experiments on glass that he considered a material most likely similar to an ideally brittle (linear elastic) material. Real materials, though brittle can be, always present a certain degree of plasticity that make it possible to generate a plastic zone ahead of the crack tip. Therefore, in real materials plastic energy released or stored as plastic deformation on the crack surface and, in particular, in the plastic zone takes the dominant role in the equilibrium of the crack under stress. Plastic energy, in fact, even in very brittle materials can be thousands of times larger than the surface energy that can be neglected completely. Accordingly, in real materials the driving force $\mathcal{G}$ and the resistance R are no longer linear and constant, respectively, as in ideally brittle materials, but depend on the crack length a since the size of the plastic zone depends on the crack length, see Eqs. (10.6) and (10.5). Now, they follow a power law relationship, as schematized in Fig. 15.2. The consequence is that now the critical condition becomes the tangency between $\mathcal{G}$ and R, as shown in Fig. 15.2. But there is another extremely important consequence from the fatigue stand point. The new situation is no longer the *in–out* situation of the ideally elastic case of Fig. 15.1. in real materials, even though a stress state and crack length combination doesn't

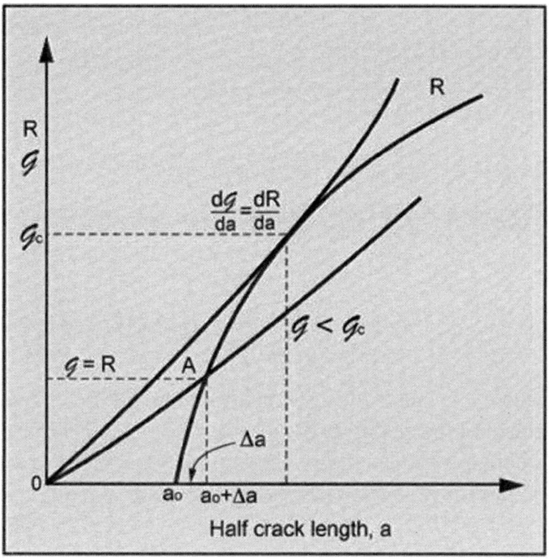

Fig. 15.2 Schematic of the driving force $\mathcal{G}$ and resistance R in real elastic–plastic materials

result in a critical state, that is to say $\mathcal{G} < \mathcal{G}_c$, there is enough energy into the system to drive the crack up to point A, as shown in Fig. 15.2, where the equilibrium between $\mathcal{G}$ and R is reached. The process is called stable tearing because the crack tip tears a little bit without breaking through catastrophically. Upon unloading, a new crack is left into the system whose length is $a_o + \Delta a$, as shown in Fig. 15.2. Actually, this Δa is not a fatigue crack growth since it is much larger than a single cycle growth or striation, but it is the precursor of that growth.

When the system is reloaded it would be expected that the equilibrium between the driving force $\mathcal{G}$ and the resistance R already reached in the previous cycle still holds and nothing would happen. But experience indicates the crack would grow further by fatigue. The author [2, 3] supposed that if this happens it is because during reloading the plastic zone ahead of the crack tip can no longer absorb energy with the same rate as before so that the crack has to grow further to widen the initial plastic zone into new virgin material and regain the energy lost. This is fatigue. He also made the hypothesis that the loss of energy release rate was due to a shake-down effect (see Sect. 7.4) that in the reloading prevents the material in the plastic zone from absorbing the same amount of energy as before. An amount of energy proportional to the elastic energy, about 10 to 30%, is lost. Based on these assumptions, Milella [2, 3] derived an equation to assess the fatigue crack growth per cycle

$$\Delta a = \frac{61.5 \cdot 10^{-3}\varphi}{\alpha^2}\frac{\Delta K^{2(n+1)}}{\sigma_y}\left[\frac{\alpha\sigma_y^2 - \sigma^2}{2K_{Ic}^2\left(\alpha\sigma_y^2 - \sigma_c^2\right) - \Delta K^2\left(\alpha\sigma_y^2 - \sigma^2\right)}\right]^n \qquad (15.3)$$

where σ_y is the yield strength, n the exponent of the power low representing the R curve of the material

$$R = C \cdot \Delta a^n \tag{15.4}$$

with C characteristic of the material, K_{Ic} is the toughness of the material, σ_c the stress at fracture that results in K_{Ic}, α is related to the plastic constraint factor pcf by the relationship

$$\alpha = 2(pcf)^2 = 2\left(\frac{\sigma_e}{\sigma_y}\right)^2 \tag{15.5}$$

where σ_e is the equivalent yield strength of the material under multiaxial stress state ahead of the crack tip, that can be as high as $2.5 \cdot \sigma_y$ (G. Irwin suggested to adopt the value of 6 for α), and φ a material factor related to the stretched-zone size (SZ)

$$SZ = 10^{-3}\varphi\left(\frac{K_{Ic}}{\sigma_y}\right)^2. \tag{15.6}$$

Fractography shows that between fatigue striations and final dimple or cleavage fracture by overload there is a featureless zone known as stretched-zone. This zone must necessarily be the fractographic evidence of the maximum fatigue crack growth $\Delta a_{\max}$ that can be possible in the material before fracture occurs considering the final overload as 1/4 cycle. Figure 15.3 shows a stretched-zone in aluminum 7178-T651 alloy [4]. Table 15.1 presents the values of the φ factor found by Milella [3] for several alloys.

It interesting to note how Eq. (15.3) contains the fracture toughness of the material that was found to have an important role on the FCGR of materials (see Sect. 14.8).

15.2 Fatigue Threshold

In the study of fatigue through the *S–N* curves or Wöhler curves it was identified a Region III, see Fig. 1.5, of the fatigue limit where the material is no longer subjected to fatigue fracture. This is because the fatigue limit σ_f represents the highest stress amplitude that cannot propagate an existing micro defect possibly existing in the material. Below σ_f, in fact, persistent slip bands (see Sect. 1.4.1) are either not produced or, if they are, the associated microscopic defect cannot be driven by the applied stress through adjacent grains, remaining confined. Something analogous happens also as far as fatigue crack propagation is concerned. Below a certain value of the stress intensity factor, called threshold factor or K_{th}, Fig. 10.6, cracks remain dormant precisely as the micro cracks were dormant below the fatigue limit σ_f (see Sect. 1.5.1). Measurements of ΔK_{th} were not that easy since it is in the range of

Fig. 15.3 Fractograph of stretched-zone observed in 7178-T651 aluminum alloy [4]

Table 15.1 φ factors for the evaluation of the stretched-zone

Alloy	φ	Alloy	φ
A 316	0.26	A 533 B	0.5
20MnMoNi55	0.43	HT 80	0.59
A 304	0.46	4340	0.56
A 572	0.44	A 469	0.60
5083–0	0.46	2017-T3	0.52
2024	0.49	7075-T6	0.88
Ti-6Al-4V	1.58		

10^{-8} mm/cycle and to produce a crack growth of just one millimeter at that ΔK requires 10^8 cycles and times extremely long with the usual cycling frequencies. To-day the introduction of resonant-type machines, see Fig. 3.3, working at 100–150 Hz has made things simpler reducing that time down to just few days. Results so far obtained judge the fatigue threshold to be located between 4 and 10 MPa$\sqrt{m}$ for most materials. At these low levels of ΔK_I there is a sharp break in the FCGR, *da/dN*, versus ΔK_I relationship. The existence of the fatigue threshold produces the characteristic sigmoidal shape of the FCGR, *da/dN*, curve shown in Fig. 10.6. In the vicinity of ΔK_{th} cracks grow with a much lower rate but much higher dependency on the ΔK_I variation than in Region II of the Paris-Erdogan equation. However, fatigue threshold is not a unique characteristic of the materials, but depends on many factors. Among these, certainly the most important are the R-ratio defined as the ratio between the minimum, $K_{\min}$ and maximum, $K_{\max}$, applied stress intensity factor

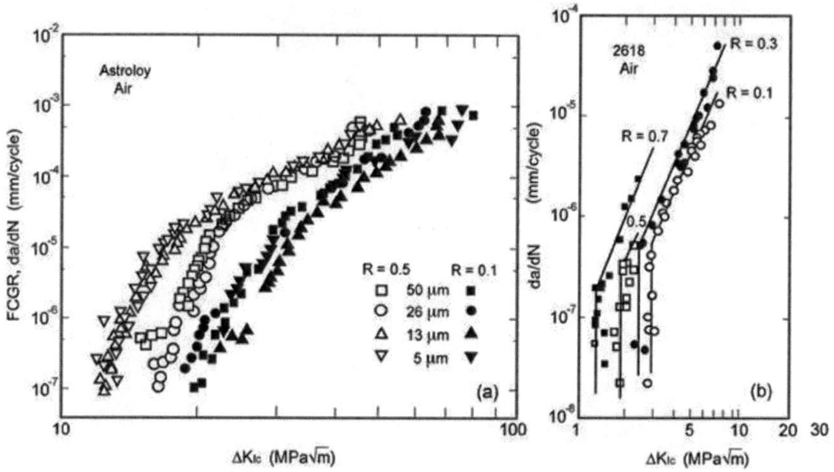

Fig. 15.4 **a** FCGR in Astroloy as function of *R* and GS (data assembled from [5]) and **b** in 2618 aluminum alloy [6]

$$\Delta K = K_{\max} - K_{\min}$$
$$R = \frac{K_{\min}}{K_{\max}} \tag{15.7}$$

and the metallurgical effects. Figure 15.4 is an interesting example of $\Delta K_{I,th}$ variation with *R*. In particular, Fig. 15.4a [5] refers to a low-carbon Ni-base Astroloy and shows how the $\Delta K_{I,th}$ variation as function of both the *R*-ratio and the grain size (GS), at ambient temperature and in dry air.

Samples with different grain size were obtained by altering the hot isostatic pressing and solution treatment procedures. The yield strength was varying from a minimum of 910 MPa for the 50 μm GS material to a maximum of 1110 MPa for the 5 μm GS material and the ultimate strength from 1350 MPa for the 50 μm GS material to a maximum of 1550 MPa for the 5 μm GS material. Figure 15.4b [6] shows the $\Delta K_{I,th}$ variation for an aluminum alloy type 2618 as a function of *R* at ambient temperature and in dry air. Some characteristic features must be caught. First, the *R*-ratio has a large influence on the threshold value $\Delta K_{I,th}$. In the Astroloy, the $\Delta K_{I,th}$ almost reduces to a half going from $R = 0.1$ to $R = 0.5$. Furthermore, the metallurgical effect of GS that is not apprized at $R = 0.1$, becomes well evident at $R = 0.5$. The response of the 5 and 13 μm GS alloys separates completely from that of the 26 and 50 μm GS alloys. Hence, the metallurgical effect gets out at higher *R*-ratios. But there is one more feature that must be stressed; even though the factor *R* and the metallurgy influence the threshold value, the different FCGR curves seem to merge in Region II of the *da/dn* versus ΔK_I diagram.

The same metallurgical effect, in terms of different ultimate strength, can be seen in the titanium alloy type Ti-6Al-4V, as shown in Fig. 15.5 [7].

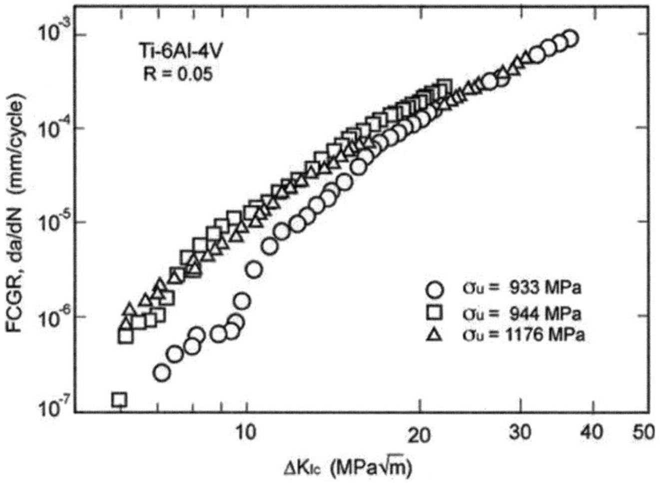

Fig. 15.5 Separation of FCGR curves with strength in the threshold region for Ti-6Al-4 V alloy [7]

An overall view of the impact of the *R*-ratio is offered in Fig. 15.6 that show the effect of the *R*-ratio on the FCGR, *da/dN*, behavior observed at very low ΔK_I levels for low-alloyed carbon steel type A 533 B Cl1 and A 508 Cl2 used in the construction of nuclear pressure vessels [8]. The various curves are given at constant rate $\Delta a/\Delta N$ in the vicinity of the threshold and at the threshold $\Delta K_{I,th}$. Note that as the *R*-ratio increases from 0.1 to 0.8, the threshold level decreases passing from about 6–7 MPa$\sqrt{m}$ to just 3 MPa$\sqrt{m}$. It is also interesting to note that the *R*-ratio effect reduces as the FCGR, *da/dN*, increases. Figure 15.7 presents another very interesting point that is the effect of temperature on the low FCGR behavior of the same steels of Fig. 15.6 at an *R*-ratio of 0.1. Note that the threshold level appears to reach a minimum at about 100–120 °C. Recalling Fig. 10.37, in that interval of temperatures the toughness K_{Ic} of the material reaches its maximum, then it drops at both lower and higher temperatures. This actually means that despite the apparent concave trend, the ratio $\Delta K_{I,th}/K_{Ic}$ remains the same at all temperatures.

This further confirms what has been said in Sect. 10.8 about the role of toughness on FCGR of materials, i.e., that the effective value of a ΔK_I excursion on the FCGR, *da/dN*, behavior of a material should be assessed by indexing ΔK_I to the relative K_{Ic}. Tanaka [9], studying a large number of Fe-alloys with yield strength varying from 255 to 1590 MPa, found that the threshold stress intensity factor $\Delta K_{I,th}$ varied with both the *R*-ratio and the yield strength of the material. The results of his study are shown in Fig. 15.8 [9]. The dependence of $\Delta K_{I,th}$ on the *R*-ratio appears to be linear for all steels considered and all curves converge theoretically towards zero. In practice, saturation occurs at about 2 to 4 MPa$\sqrt{m}$ for *R*-ratio higher than 0.7–0.8. This saturation effect was already observed by Barsom e Rolfe [10] on another series of steels that, anyhow, included also stainless steel, type 18/8, and mild steel

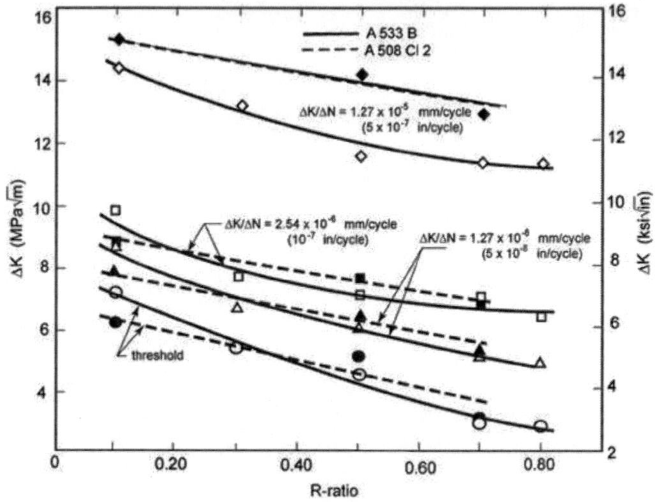

Fig. 15.6 ΔK_I range variations versus *R*-ratio for constant crack propagation rates of A 533 and A 508 steels in air at room temperature (modified from [8])

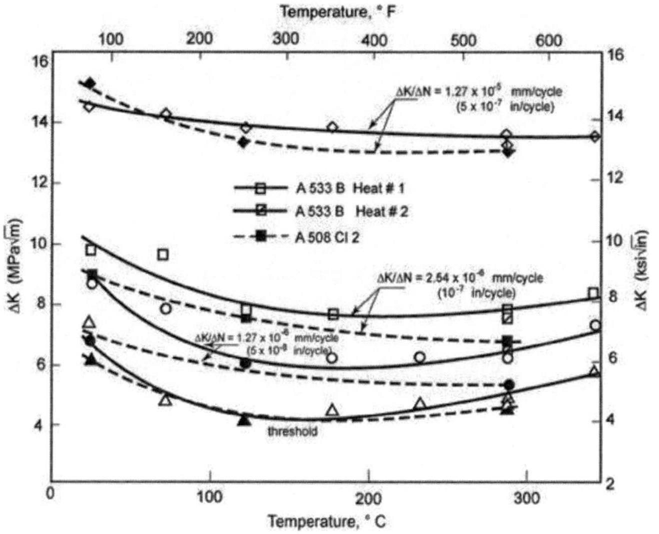

Fig. 15.7 ΔK_I range variations versus temperature for constant crack propagation rates of A 533 and A 508 steels (modified from [8])

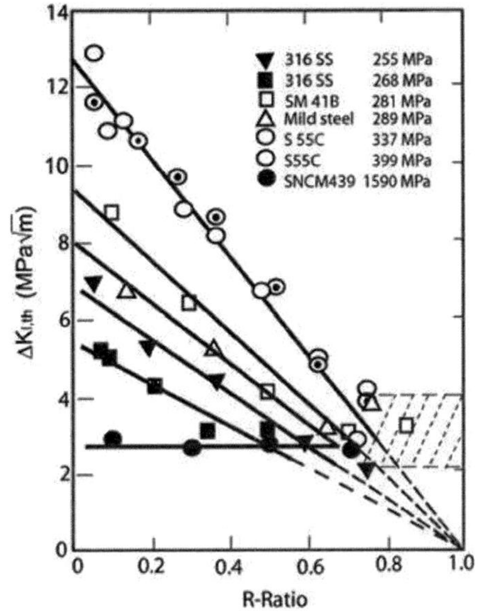

Fig. 15.8 *R*-ratio effect on the $\Delta K_{I,th}$ threshold for 7 different steels [9]

considered by Tanaka. Characteristic, in Fig. 15.8, is the behavior of high strength steel SNCM439 of 1590 yield strength, whose response appears to be independent of the *R*-ratio and fixes the lower bound of $\Delta K_{I,th}$. This dependency of the threshold stress intensity factor on the *R*-ratio induce to believe that $\Delta K_{I,th}$ is composed of two components, one is constant and is an intrinsic characteristic of the material $\Delta K_{th(in)}$ the other, instead, is dependent on external factor among which the *R*-ratio is the most effective, $\Delta K_{th,R}$

$$\Delta K_{th} = \Delta K_{th,in} + \Delta K_{th,R}. \tag{15.8}$$

Klesnil e Lukas [11] proposed an empirical relationship between and the *R*-ratio analogous to that proposed by Walker between ΔK and R (see Sect. 15.4)

$$\Delta K_{th} = \Delta K_{th,o} \cdot (1 - R)^s \tag{15.9}$$

in which $\Delta K_{th,o}$ is the threshold value that corresponds to $R = 0$ and s a fitting exponent.

15.3 Plastic Zone and *R*-Ratio Effect on Fatigue Threshold

The *R*-ratio effect on the threshold stress intensity factor can be explained by the residual stress system set up ahead of the crack tip during the unloading phase. What is actually happening is schematized in Fig. 15.9. During the load application phase a plastic zone develops at the crack tip, Fig. 15.9a, which allows the tip to open and blunt. Beyond this plastic enclave, assumed as a first approximation to be circular in shape (G. Irwin assumption), there is the elastic high amplitude stress field that reach the yield strength and go higher in strain hardening materials. On the crack flanks the so called plastic wake of residual deformations remains as memory of previous deformation cycles. During unloading, crack closes, but because of the shake-down effect (see Sect. 7.4) most of the material ahead of the plastic zone will go into residual compression and will be balanced by a system of tensile residual stresses, Fig. 15.9b.

The compressive residual stress state certainly exerts a beneficial effect on the fatigue behavior since in the following reloading it will be necessary to overcome it before the crack opens and restarts to grow. This actually means that it will not be the entire load excursion $\Delta\sigma$ to cause fatigue, but rather only the part of it that will open the crack, $\Delta\sigma_{op}$. This will reduce the effective applied ΔK. Let's follow what happens in the entire plastic zone ahead of the crack tip. This is schematized in Fig. 10.10. Under the load history (a), point A, right at the tip of the crack, during unloading will undergo the largest hysteresis loop between A and A′ on the stress–strain curves relative to $\Delta K = K_{I,\max} - K_{I,\min}$ with $K_{I,\min} = 0$. On the border of the plastic zone opposite to point A, the corresponding point E moves on the elastic

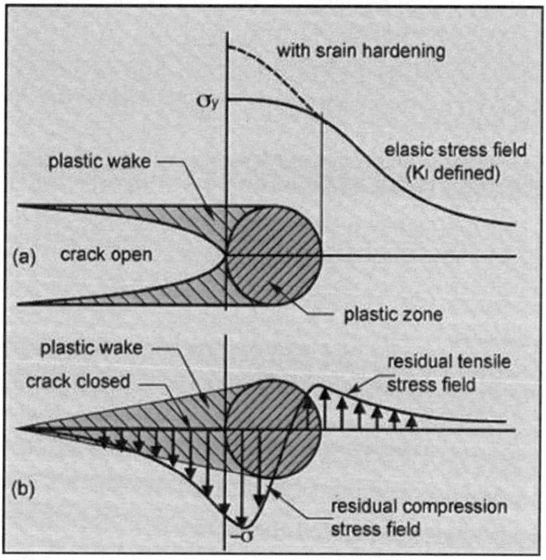

Fig. 15.9 Schematic of the crack tip plastic zone and residual stress state

straight line 0E. Between those two extreme points behavior there is any other point of the plastic zone that, upon unloading, enters into e residual compressive stress state, as B′, C′ and D′, for example. As said before, the consequence is that upon reloading the effective ΔK_{eff} that will challenge the crack will not be given by $K_{I,\max} - K_{I,\min}$, but will be $\Delta K_{eff} = K_{I,\max} - K_{I,op} < \Delta K$. Note that if the $K_{I,min}$ increases, load history (b) of Fig. 15.10, which means that the *R*-ratio increases, the residual compressive stress state reduces, points A″, B″, C″ and D″, so that also the difference between ΔK and ΔK_{eff} reduces up to a point when $K_{I,\min}$ reaches $K_{I,op}$, , as shown by the time history (c) in Fig. 15.10, when the entire ΔK will be effective, $\Delta K = \Delta K_{eff}$, because the residual compressive stress state disappears completely, points A‴, B‴, C‴ and D‴. Now, as the load increases the crack opens immediately. If this simple schematization of the crack behavior can explain why a minimum $K_{I,\min}$ below $K_{I,op}$ cannot be effective, it cannot account for any reduction of fatigue strength when *R*-ratio further increases and it becomes $K_{I,\min} < K_{I,op}$. To fully understand this point it is necessary to go back to the fatigue mechanism explained in Sect. 15.1. The crack grows, it was said, because at any reloading the material in the plastic zone ahead of the crack tip is no longer able to absorb energy with the same rate as in the previous cycle. This forces the crack to grow further and regain the losses thorough the extension of the plastic zone into new virgin material. As clearly explained in [3] this capability to absorb new additional energy to compensate the loss of energy absorption rate decreases more and more as the $K_{I,\min}$ increases because the area under the curve of Fig. 15.10 reduces more and more, so that the crack has to grow at increasing rates with increasing *R*-ratios.

15.4 *R*-Ratio Effect on the FCCG

It has been shown in the previous section that the progressive elevation of the $K_{I,\min}$ value and, consequently, of the ratio $R = K_{\min}/K_{\max}$ leads to a parallel reduction of the threshold stress intensity factor ΔK_{th}. The R-ratio effect persists, though in a less fashion, in Region II of the FCGR, see Fig. 10.6, where the Paris-Erdogan Eq. (10.10) holds and is further emphasized in Region III. As to Region II, Fig. 15.11 shows this effect in a titanium alloy type Ti-6Al-4V [12]. The *R*-ratio is varying from a rather high 0.7, representative of a very high mean stress $\sigma_m > 0$, down to negative values representative of a stress cycle with negative mean stress, $\sigma_m < 0$. The FCGR, *da/dN*, varies by a stunning factor of one hundred moving from $R = -5.0$ to $R = 0.7$ to which, by the way, it corresponds a very low threshold value of just 4 MPa$\sqrt{}$m. Walker [13] proposed an empirical relationship to correct the Paris-Erdogan equation to take into account the *R*-ratio effect

$$\frac{da}{dN} = C \cdot \Delta K_{eff}^{n} \tag{15.10}$$

where ΔK_{eff} is given by

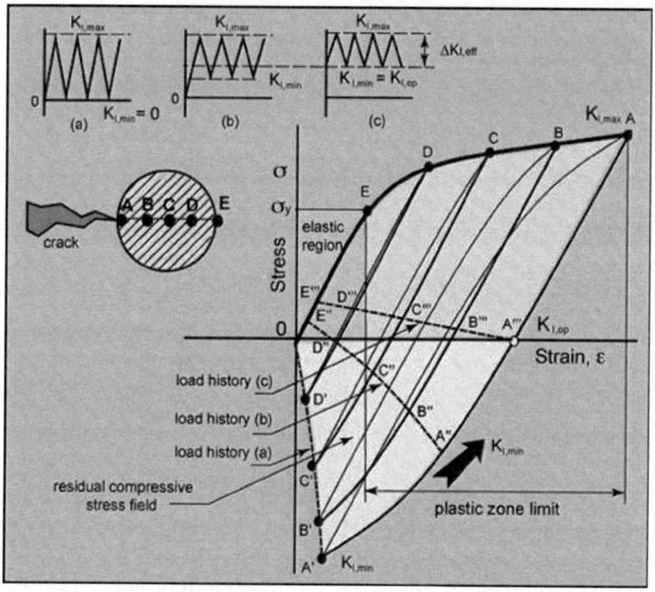

Fig. 15.10 Increasing the $K_{I,\min}$ to $K_{I,op}$ reduces the residual compressive stress field to zero

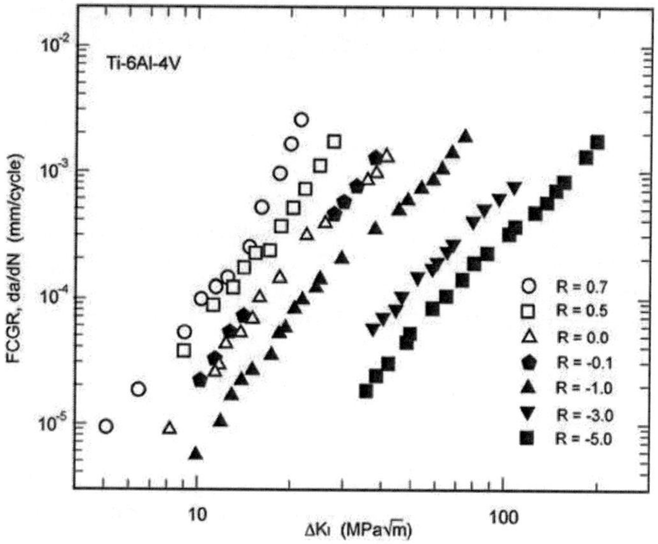

Fig. 15.11 FCGR, *da/dN*, versus ΔK_I diagram for a titanium alloy type Ti-6Al-4 V, as function of the *R*-ratio [12]

$$\Delta K_{eff} = K_{\max}(1 - R)^m \tag{15.11}$$

in which m is a material characteristic to be experimentally determined. Since it is

$$K_{\max} = \frac{K_{\min}}{R} \tag{15.12}$$

it also is

$$\Delta K = K_{\max} - K_{\min} = K_{\min}\left(\frac{1 - R}{R}\right)$$

$$K_{\max} = \frac{\Delta K}{(1 - R)} \tag{15.13}$$

and Eq. (15.11) can be written as

$$\Delta K_{eff} = \Delta K \cdot (1 - R)^p \tag{15.14}$$

with $p = m - 1$. Therefore, Eq. (15.10) becomes

$$\frac{da}{dN} = C \cdot \Delta K^n (1 - R)^s \tag{15.15}$$

with $s = n\ (m - 1)$. The experimental points of Fig. 15.11 have been retraced as function of ΔK_{eff} given by Eq. (15.15), with $p = -0.4$. Results are shown in Fig. 15.12. It can be seen the strong compacting action that has led, in practice, to the superposition of all experimental data obtained for R varying from 0.7 to −1.0. Only data relative to $R > -1$ remain out of the common trend even though they compacted in a separate group. This deficiency of the Walker ΔK_{eff} to represent data for negative R-ratios appears to be a characteristic common to all corrective equations. However it has been shown how the compression cycle doesn't really contribute to the crack opening and, therefore, to the crack growth. It can be convenient, then, to revisit the experimental data of Fig. 15.11 in terms of $K_{I,\max}$ rather than of the entire ΔK_I. This can be seen in the diagram of Fig. 15.13 that appears to upturn results of Fig. 15.12. This time, the compacting action applies only to negative R-ratios or $R = 0$, at most, while data relative to positive R-ratios seem to spread and deviate, resulting in lower growth under higher ΔK_I.

Another interesting example of FCGR dependency on R-ratio is shown in Fig. 15.14 [8] for type 7075-T6 clad aluminum alloy fatigued in dry air at ambient temperature at 3400 cpm. Introducing the Walker correction given by Eq. (15.15) with $p = -0.5$, the experimental data of Fig. 15.14 rearrange as shown in Fig. 15.15.

As for the titanium alloy, also for the aluminum clad there is a remarkable competing. Crooker et al. [14, 15] studying the fatigue behavior of a martensitic alloy type HY-80 of 965 MPa yield strength (see Fig. 10.26) in the range $-2 \le R \le$

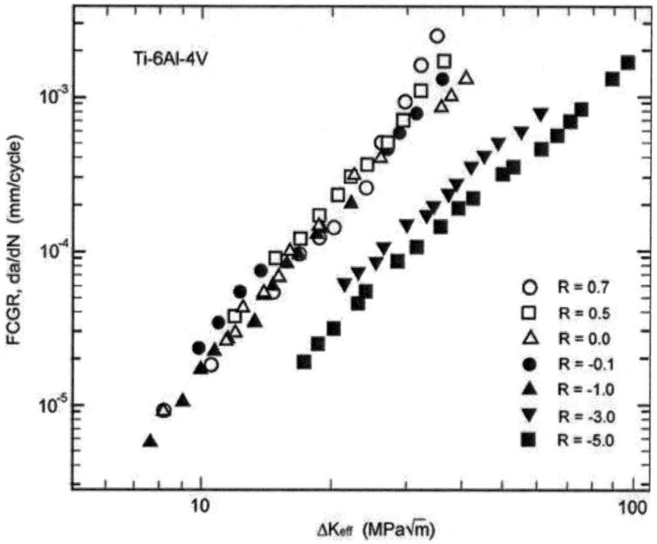

Fig. 15.12 FCGR, *da/dN*, versus ΔK_{eff} diagram for a titanium alloy type Ti-6Al-4 V, as function of the *R*-ratio [12]

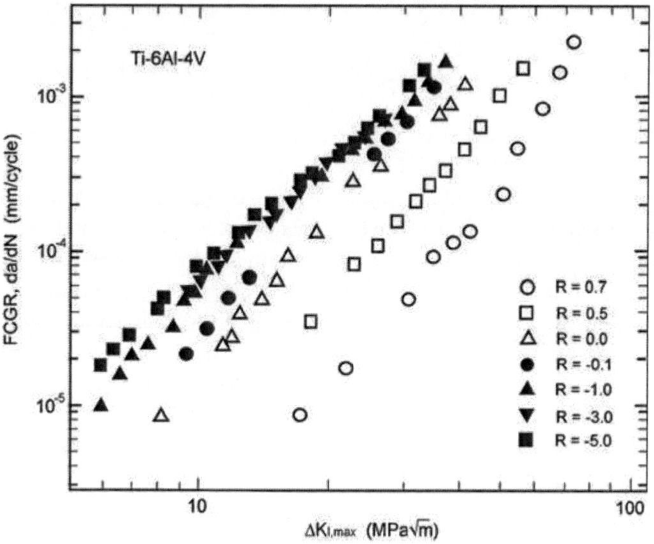

Fig. 15.13 FCGR, *da/dN*, versus $\Delta K_{I,max}$ diagram for a titanium alloy type Ti-6Al-4 V, as function of the *R*-ratio (experimental data from [12])

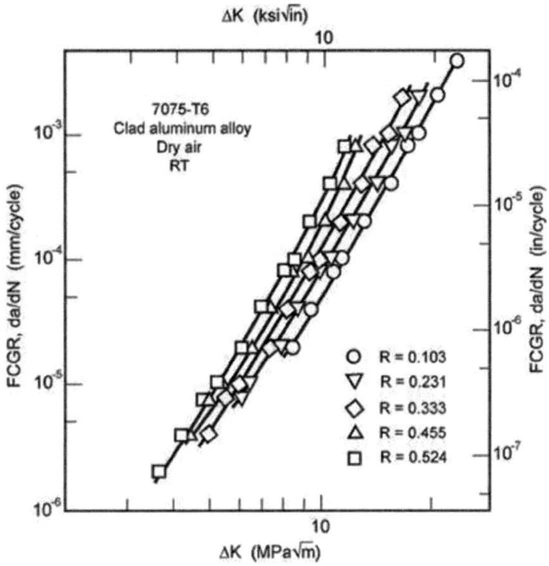

Fig. 15.14 FCGR, *da/dN*, versus ΔK_I diagram for 7075-T6 clad aluminum alloy, as function of the *R*-ratio [8]

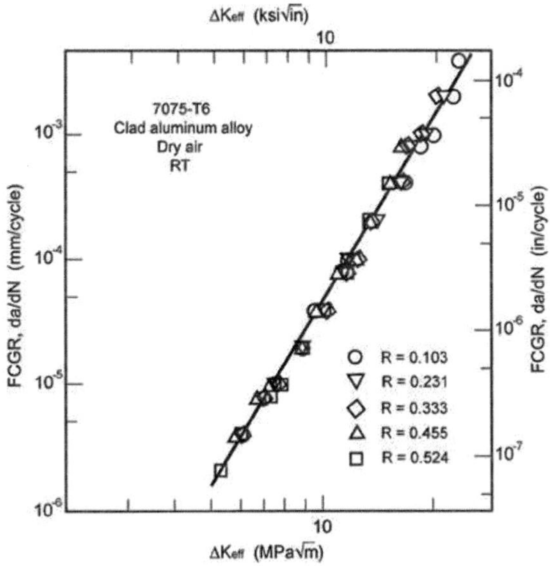

Fig. 15.15 FCGR, *da/dN*, versus ΔK_{eff} diagram for 7075-T6 clad aluminum alloy, as function of the *R*-ratio [8]

0.75 found that the Walker Eq. (15.14) was well representing the FCGR using $p = -0.5$.

However, as the *R*-ratio increases, the ΔK needed to produce the unstable crack growth decreases, which actually means that the *da*/*dN* curves in Region III of the FCGR diagram, see Fig. 10.6, diverge as it happened in Region I. this can be seen

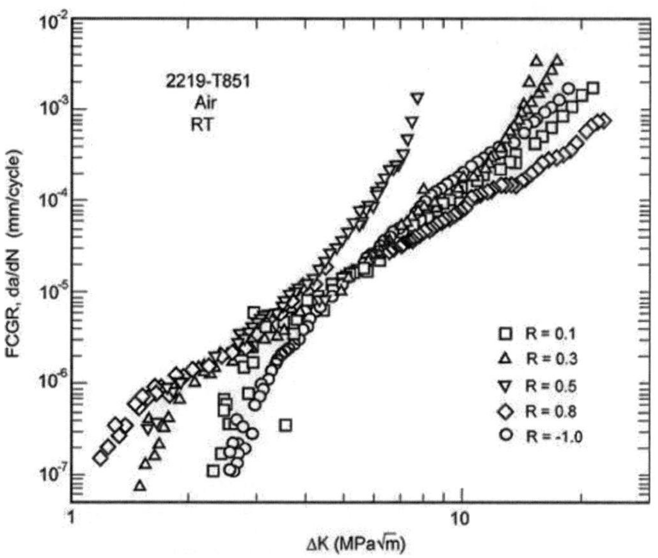

Fig. 15.16 FCGR, *da/dN*, versus ΔK_I diagram for 2219-T851 aluminum alloy, as function of the *R*-ratio [16]

in Fig. 15.16 [16] relative to an aluminum alloy type 2219-T851. This raveling in Region I and III, respectively, of the *da/dN* curves that in Region II get closer is schematized in Fig. 15.17.

15.5 Crack Closure

It has been shown, in Sect. 15.3, how the *R*-ratio effect can be explained in terms of ΔK_{eff} that represents that part of the entire stress intensity factor excursion ΔK active in producing the growth of the crack. This ΔK_{eff} is, in practice, the component that effectively opens the crack. Another approach to define ΔK_{eff}, that has gained wide acceptance, it that proposed by Elber in 1970 [17], referred to as crack closure. Working with some types of fracture mechanics specimens, Elber observed that during unloading the crack was already closed before the tension stress became zero. In other words, during unloading the compliance of the specimens carrying a crack, C_{cr}, was equal that of the loading phase, predicted by theoretical formulas, but only up to a certain load level, generally high, below which it decreased reaching the value that competed to the non-cracked specimens. This is schematically shown in Fig. 15.18a; if a specimen containing a crack is loaded up to B on its, F-δ, load–deflection curve and then unloaded, it starts moving on the line BC parallel to the initial load line that represents the stiffness $k_{cr} = F/\delta_{cr}$ of the cracked specimen ($k_{cr} = 1/C_{cr}$), but upon reaching point C instead of continuing on the same course CD

Fig. 15.17 Schematic of the continuous $K_{I,th}$ reduction and instability advance due to increasing R-ratio

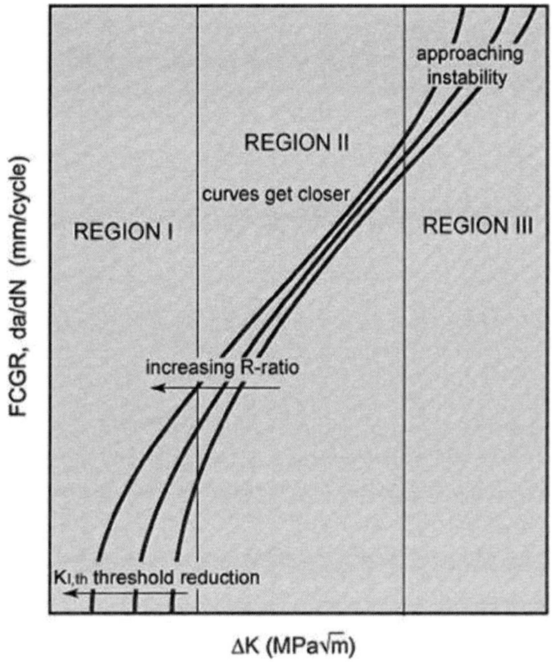

it deviates along line CE that represents the stiffness of the non-cracked specimen $k = F/\delta$. Elber ascribed such deviation from the initial course to the rough and uneven nature of the crack surfaces that during unloading did not match perfectly, but get in contact well before the load went to zero. Therefore, the two crack surfaces were able to exert a reaction F_{op} opposing the complete closure of the crack or, from a different point of view, the crack closed before the load went to zero. In terms of ΔK this effect of crack closure results in the specimen being not cycled between $K_{\min}$ and $K_{\max}$ because it will never return in the state of $K_{\min}$, but it will remain in a rather higher K_{op} that corresponds to the force F_{op} acting at the moment when the crack surfaces get in contact, as schematized in Fig. 15.18b. Therefore, the effective ΔK_{eff} challenging the crack will no longer be equal to $K_{\max} - K_{\min}$, but to

$$\Delta K_{eff} = K_{\max} - K_{op}. \tag{15.16}$$

This, in practice, is worth a reduction of the RCCF, da/dN. As such, Elber proposed a modified Paris-Erdogan equation

$$\frac{da}{dN} = C \cdot \Delta K_{eff}^{n}. \tag{15.17}$$

It must be underlined that the K_{op} is not a material characteristic, since it is the results of a crack closure produced by the roughness of the crack surfaces. As matter of

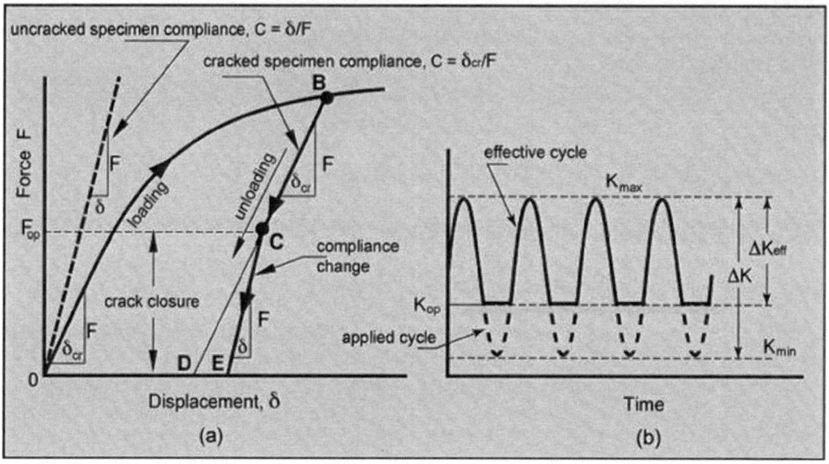

Fig. 15.18 **a** Schematic of the loading–unloading behavior of a cracked specimen that shows crack closure and **b** effective cycle

fact, besides the material characteristics the crack roughness depends also on the load level or $K_{\max}$; to different $K_{\max}$ is likely that different crack roughness correspond. An interesting example is provided by the experimental observation of the surface roughness, shown in Fig. 15.19 [7], produced by fatiguing a titanium alloy, type Ti-6Al-4V, thermally aged to three different ultimate strength already considered in Sect. 10.7.5. Indeed, there is a material dependency of the surface roughness that, ceteris paribus, results more pronounced in the lower strength alloy that allows larger plastic deformation, but it is worth noting that at higher FCGR associated to higher ΔK, the surface roughness is lower. In the three cases of Fig. 15.19 the R-ratio was kept constant and equal to 0.05, therefore the largest roughness is associated to the lowest K_{max}. Elber also verified the dependency of the K_{eff} on the R-ratio. He introduced an non dimensional parameter U as ratio between ΔK_{eff} and ΔK

$$U \equiv \frac{\Delta K_{eff}}{\Delta K} = \frac{K_{\max} - K_{op}}{K_{\max} - K_{\min}} \tag{15.18}$$

so that Eq. (15.18) becomes

$$\frac{da}{dN} = C \cdot (U \cdot \Delta K)^n. \tag{15.19}$$

Working with an aluminum alloy type 2023-T3, Elber found that the parameter U was not constant, but could be linked to the R-ratio by a relationship of the type

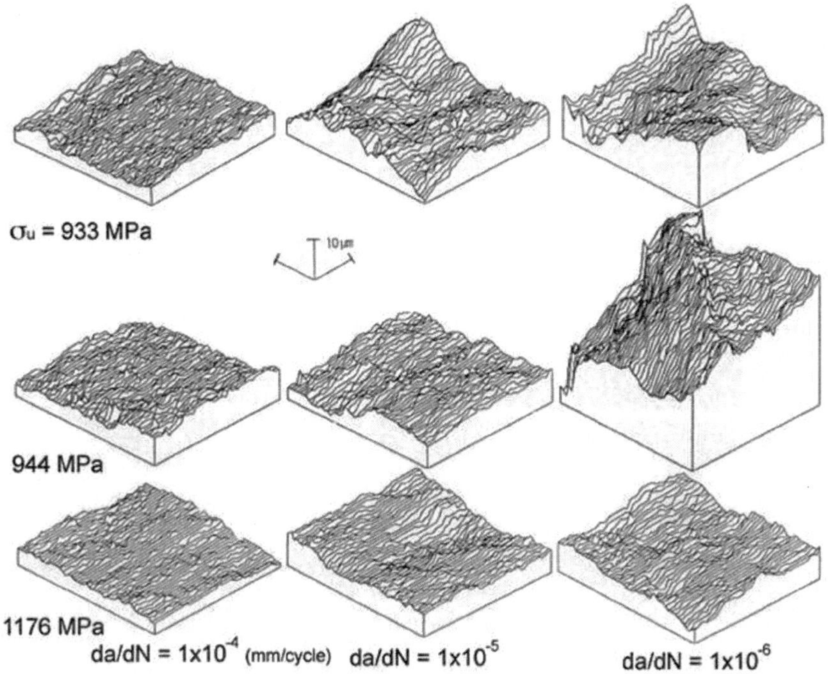

Fig. 15.19 Experimental measurements of the surface roughness in a Ti-6Al-4 V alloy heat hardened to three different strength and subjected to different ΔK that result in different FCGR, *da/dN* [7]

$$U = 0.5 + 0.4 \cdot R. \tag{15.20}$$

Other researchers have confirmed the results obtained by Elber [19–22]. An interesting example regarding the Elber ΔK_{eff} is provided in Fig. 15.20 that reports the experimental data obtained by Zhang et al. [18] on aluminum alloy type 7475-T7351. It can be observed the same *R*-ratio effect already seen in Fig. 15.11 for the titanium alloy and Fig. 15.16 for the aluminum alloy type 2219-T851. Zhang et al. Found that the parameter U was effectively depending on the *R*-ratio through an expression of the type

$$U = \frac{\Delta K_{eff}}{\Delta K} = 0.618 + 0.365 \cdot R + 0.139 \cdot R^2. \tag{15.21}$$

Reanalyzing the data of Fig. 15.20 as function of the ΔK_{eff} given by Eq. (15.22) it is obtained the diagram of Fig. 15.21. The result offers interesting considerations. First, the compacting action operated by the Elber ΔK_{eff} exists, but not for the data obtained at high negative *R*-ratio that stand apart. This failure of the ΔK_{eff} corrective action was evidenced also for the Walker ΔK_{eff} in Sect. 15.4, see Fig. 15.12. Secondly,

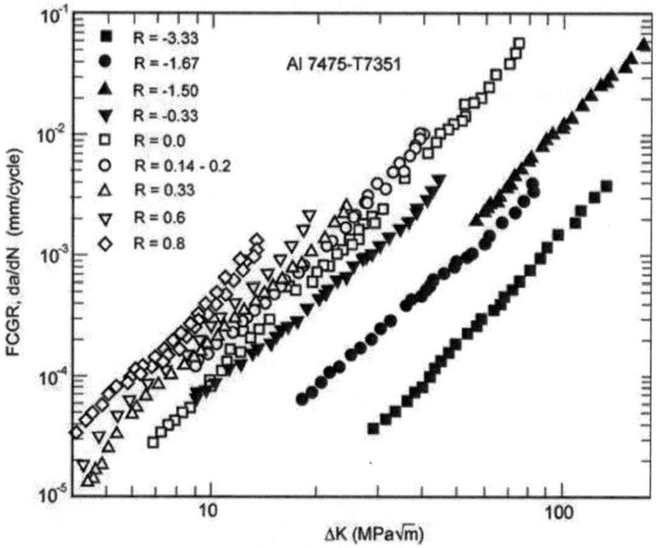

Fig. 15.20 FCGR, *da/dN*, versus ΔK observed by Zhang et al. [18] for Al 7475-T7351 aluminum alloy as a function of the *R*-ratio

since it is $\Delta K = K_{max}(1 - R)$, for $R > 0$ at equal ΔK the corresponding $K_{\max}$ will increase with increasing *R*. therefore, if in terms of ΔK_{eff} the experimental data come closer it means that K_{op} is a function of K_{max} and increases as the K_{max} decreases, according to the reasoning about the surface roughness. This actually means that for high *R*-ratio, let's say $R > 0.7$, the corresponding K_{op} is so small that, in practice, the ΔK_{eff} coincides with ΔK. An analogous effect has already been observed for the threshold stress intensity factor ΔK_{th} that for $R > 0.7$ practically saturated, as shown in Fig. 15.8.

An experimental verification of this circumstance was given by Newman [23, 24] working with an aluminum alloy, type 2219-T851. Note that this coincidence between ΔK_{eff} and ΔK for $R \geq 0.7$ can be used for the experimental measurement of K_{op}. In fact, working at high *R*-ratio ($R \geq 0.7$) it is possible to infer the values of constants *C* and *n* in Eq. (15.18) simply assuming that ΔK is equal to ΔK_{eff}. Once these values are known, interpolating the experimental results obtained at lower *R*-ratio it will be possible to infer the applied ΔK_{eff} and, therefore, the K_{op} value via Eq. (15.17). As to the first consideration about the failure of ΔK_{eff} to compact the FCGR, *da/dN*, in the negative *R* field, the data of Fig. 15.20 can be reanalyzed as function of the applied $K_{I,max}$, as already done in Sect. 15.4 and shown in Fig. 15.13 for the titanium alloy. Results are presented in Fig. 15.22.

Again, the new diagram presents a superposition of experimental data obtained at $R < 0$ or positive but close to zero, but those relative to $R > 0$ remain apart. About the surface roughness, it must be said that it certainly plays an important role, but it is not the only factor affecting the crack closure. Suresh e Ritchie [25] have identified

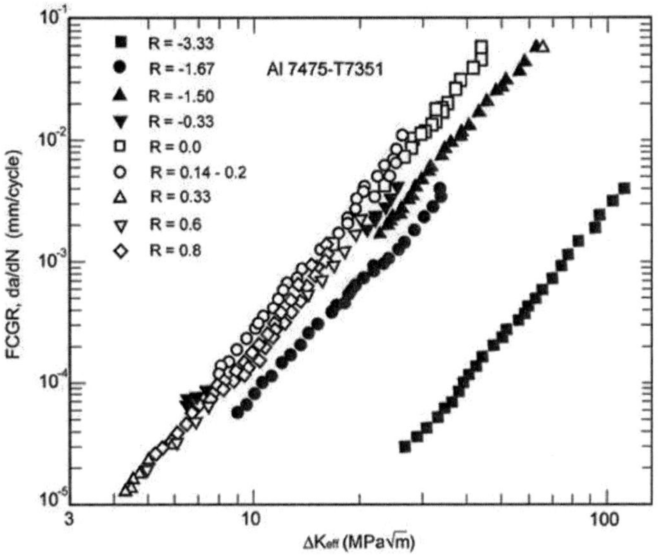

Fig. 15.21 FCGR, *da/dN*, of Fig. 15.20 reanalyzed in terms of Elber ΔK_{eff}

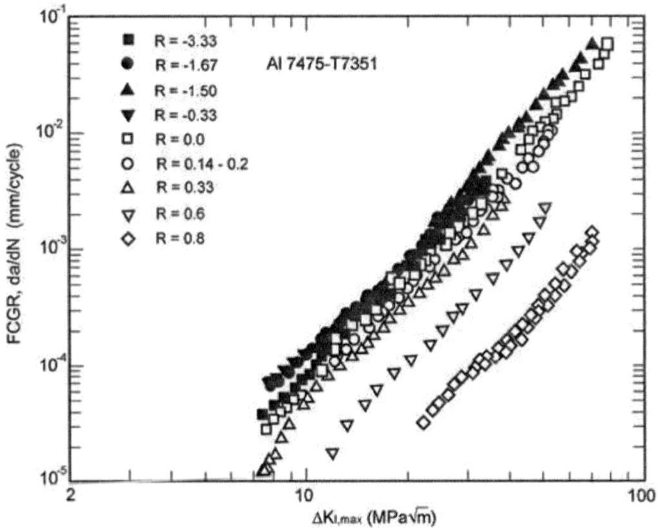

Fig. 15.22 Reanalysis of data of Fig. 15.20 by Zhang et al. [18] as function of the $K_{I,max}$

five different causes affecting crack closure that are schematized in Fig. 15.23. The first, Fig. 1.23a, is the surface roughness already analyzed aggravated by a though minimum Mode II aperture that can be activated on a microscopic scale and inhibits the perfect merging of crack surfaces at unloading.

Another important crack closure effect is due to the wake of residual local deformations left behind the crack tip as it propagates, Fig. 15.23b. Inside the monotonic plastic zone formed at maximum load, there is the cyclic plastic zone in

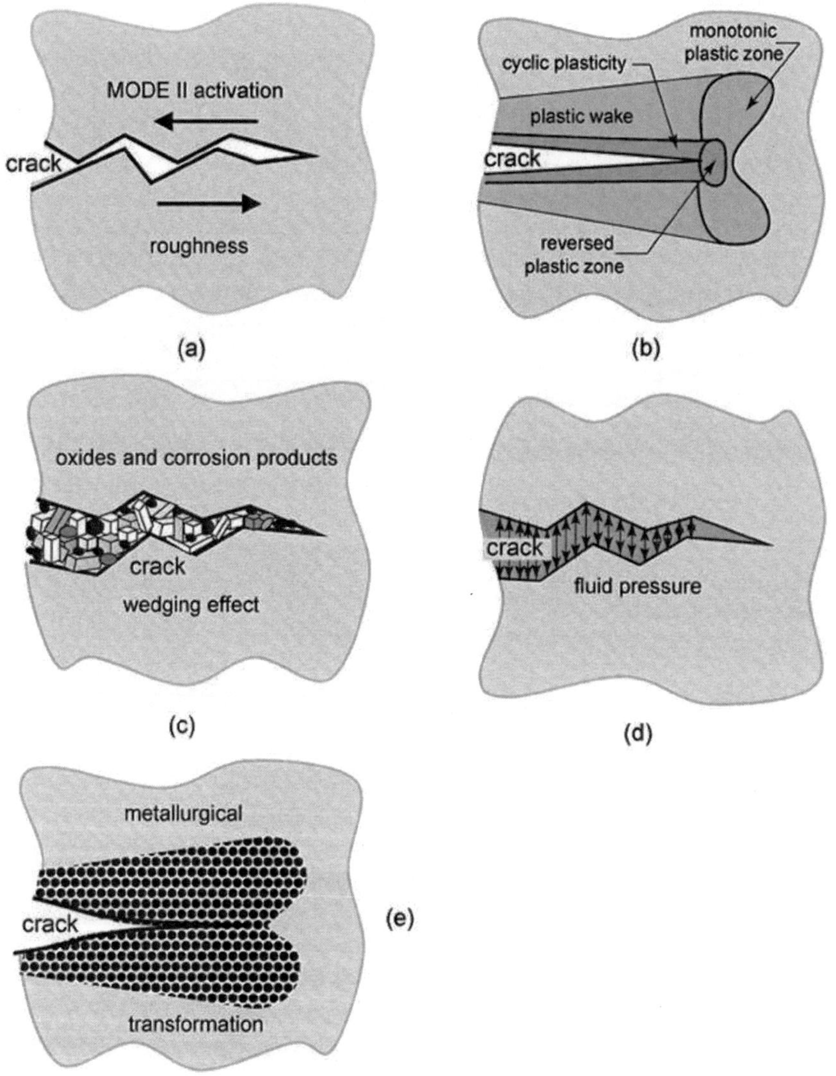

Fig. 15.23 Possible causes for crack closure identified by Suresh e Ritchie [25]

which reversed plasticity occurs. According to the shake-down model explained in Sect. 5.4.2, reversed plasticity requires that the stress exceeds twice the yield stress. This was also pointed out by Rice [26]. Based on Eq. (10.6), this implies that the reversed plastic zone should be ¼ of the monotonic plastic zone. Also this zone leaves a wake along the growth of the crack, but much smaller than the monotonic plastic wakes. Because of the difference in the process volumes of the small plastic wake on the rim of the crack and the large monotonic plastic wake this latter prevails. This phenomenon is referred to as plastic induced crack closure. Moreover, if the cycling is given in a potentially aggressive environment, even humid air, the resulting oxide formation will act as a wedge inside the crack anticipating its closure, Fig. 15.23c.

Also, a possible internal pressure exerted by a fluid can affect the crack closure, as schematized in Fig. 15.23d. Finally, Fig. 15.23e, a possible metallurgical transformation operated by the plastic deformation of the crack tip, which is equivalent to a cold-working effect, can result in a residual opening or squeezing of the crack tip, depending whether the new phase has a larger or smaller structure. For instance, the transformation of metastable austenite into martensite is not resulting into any significant cell size modification. At variance, the transformation of ferrite or perlite into martensite happens with volume enlargement of the elemental cell structure. It is possible to conclude that if from the one end the crack closure effect can explain the R-ratio effect on the FCGR, da/dN, the empirical dependency of the ΔK_{eff} on R makes its use not dissimilar from any other empirical relationship.

15.6 Overload Retardation

It has been long observed by many researchers [27–35] that any single overload imposed during a constant amplitude cycling results in a retardation of the crack growth in the following cycles. Figure 15.24 [33] is a significant example of this phenomenon known as overload retardation. A series of specimens of type 2024-T3 aluminum alloy have been subjected to different load histories indicated as A, B and C in Fig. 15.24. If the fatigue crack growth is measured versus the elapsed number of applied cycles N for each load history, the corresponding curves A, B and C of Fig. 15.24 are obtained. For the constant load history A the crack growth has a regular trend, but if the constant amplitude load cycling is regularly interrupted by a single overload (load history B) it will be observed a period of reduced crack growth rate after each overload. It will take many cycles before the FCGR returns to be the previous one.

This delay depends also on the type of overload. In the sequence B of Fig. 15.24 the overload was fully reversed, as the constant load cycling. But if the overload is pulsating, as in load history C, the delaying effect will last longer, as shown in Fig. 15.24. In all cases, however, the crack will return to grow at the same rate as before the overload application. The retardation effect can be explained by considering the residual compressive stress state set up ahead of the crack tip, already discussed in Sect. 15.3 and shown in Fig. 15.9. The overload, in fact, generates a plastic zone that is

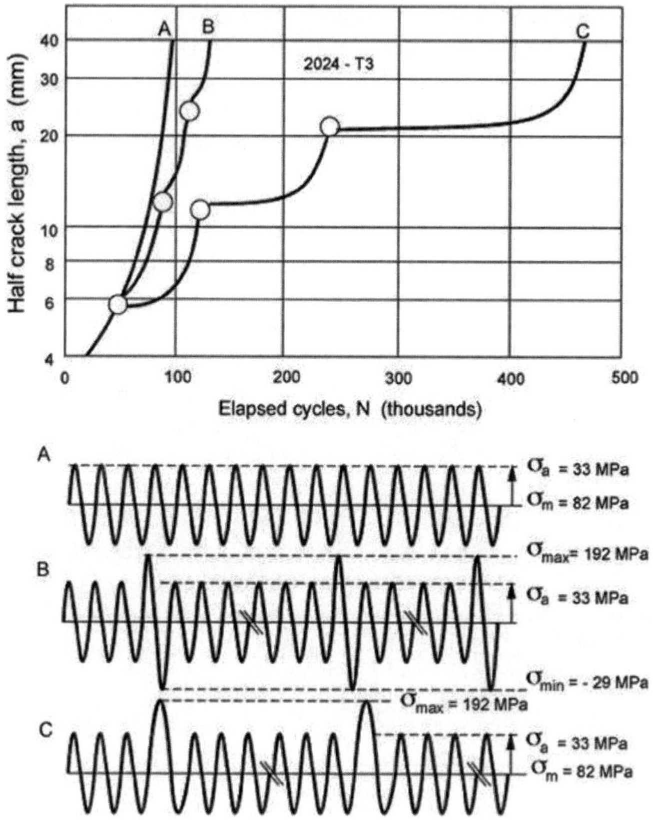

Fig. 15.24 Crack growth dependency on the load history in 2024-T3 aluminum alloy [33]. Load histories A and B result in growth retardation

larger than that associated to the smaller constant amplitude cycling. Therefore, when the smaller amplitude loads are applied they have to overcome the high compressive residual stresses left ahead of the crack tip.

With reference to Fig. 15.10 let's assume that the overload has produced the hysteresis loop AA′ while the lower amplitude cycles result in the DD′ one. The consequence is that, starting from A′, the effective ΔK_{eff} associated to cycle DD′ is enormously reduced. In the limit, if the maximum ΔK is smaller than the ΔK_{op} associated to the overload the crack will not grow anymore. Wheeler [31, 32] has ascribed this retardation effect precisely to the residual stress state and it evolution during the cycling subsequent to the overload. He supposed that the retardation effect could persist as long as the new and smaller plastic zone remained confined within the larger one, as schematized in Fig. 15.25. Figure 15.25a refers to the first cycle after overload; the plastic zone produced by normal cycles is smaller than the plastic enclave created by the overload. In (b) the plastic zone of the running crack is still

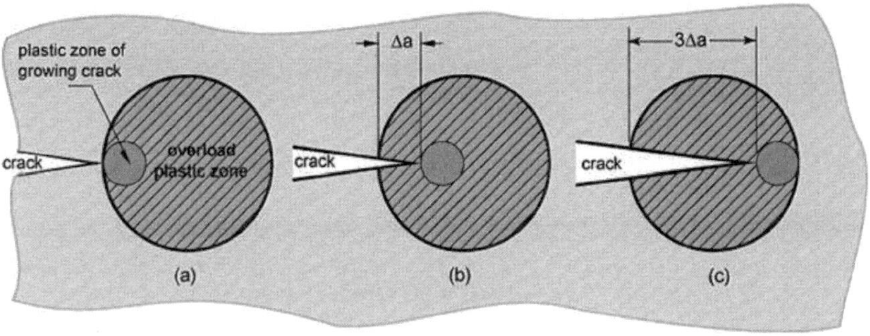

Fig. 15.25 Wheeler retardation model

within the overload plastic enclave; retardation persists. Finally, Fig. 15.25c, the plastic zone has reached the border of the larger overload plastic enclave; retardation is about to come to an end. Recalling that the crack tip plastic zone radius r_p is given by

$$r_p = \frac{1}{\alpha\pi}\left(\frac{K_{\max}}{\sigma_y}\right)^2 \tag{15.22}$$

with σ_y being the yield strength of the material and α equal to 2 for plain stress conditions and 6 for plain strain conditions, it is likely to assume that the retardation effect will last up to the moment when the crack growth Δa will equal the overload plastic zone size $2{\cdot}r_{p,ovrld}$

$$\Delta a \cong 2r_{p,ovrld}. \tag{15.23}$$

Figure 15.26 [34] presents the crack growth measurements done for a prismatic specimen type SEN (see Appendix A) 100 × 20 × 10 mm of 9Cr1Mo steel, after a series of applied cycles with K_a equal to 12 MPa$\sqrt{}$m and $R = 0.1$. The overload was equal to three times the normal cyclic load and resulted in an applied $3{\cdot}K_a =$ 36 MPa$\sqrt{}$m. It is interesting to see how from the moment of overload application the crack has grown of about 1.8 mm before returning to grow with the same rate as before. The alloy considered has yield strength of about 450 MPa. Considering that the growth measurement were done on the surface where triaxiality is practically null and the plain stress conditions prevail, Eq. (15.23) yields the value $r_{p,ovrld}$ ~ 1 mm and, therefore, the plastic zone size is about 2 mm wide, very close to the 1.8 mm crack growth. Wheeler introduced an empirical retardation factor Φ_R

$$\Phi_R - \left(\frac{\Delta a + r_p}{r_{p,ovrld}}\right)^{\gamma} \tag{15.24}$$

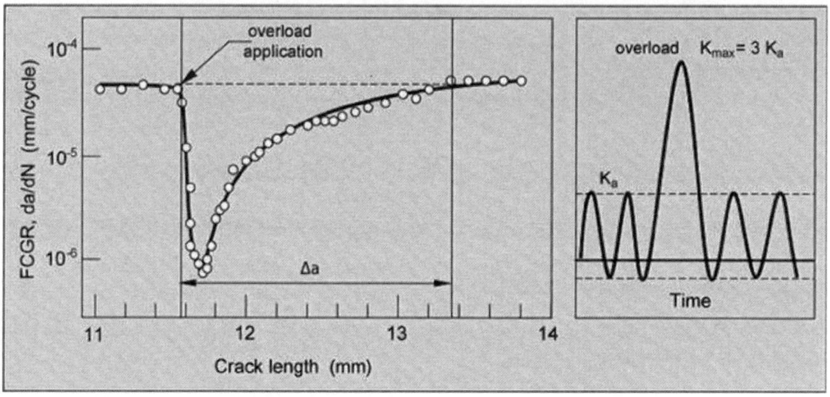

Fig. 15.26 FCGR, *da/dN*, before and after the imposition of an overload [34]

in which γ is a material constant that can be inferred from the interpolation of experimental data. The Φ_R factor corrects the FCGR, *da/dN*, according to the relationship

$$\left(\frac{da}{dN}\right)_R = \Phi_R \frac{da}{dN} \tag{15.25}$$

from which it can be assessed the number of cycles N_R the crack needs to extend through the plastic zone generated by the overload

$$N_R = \frac{1}{C} \int_{a_o}^{a_f} \frac{da}{\Phi_R \cdot \Delta K_{eff}^n} \tag{15.26}$$

with a_o being the initial crack length at the moment of overload imposition and a_f the final one after the growth through the overload plastic zone $r_{p,ovrld}$ when the retardation effect vanishes

$$a_f = a_o + r_{p,ovrld} - r_p \tag{15.27}$$

where r_p is the radius of the crack tip plastic zone produced by the constant load cycling. Note that in Eq. (15.26) the Paris-Erdogan equation takes into consideration the possibly existing *R*-ratio effect through the use of the ΔK_{eff}. Actually, things can be more complicated than so far supposed. In fact, von Euw et al. [30] observed that retardation can be, in turn, delayed, as schematically shown in Fig. 15.27.

This delayed retardation doesn't occur immediately after the overload imposition. The crack does not follow path AC of Fig. 15.27, but rather an initial phase AB in which the FCGR is still high, dough lower than before overload, followed by the real retardation BC.

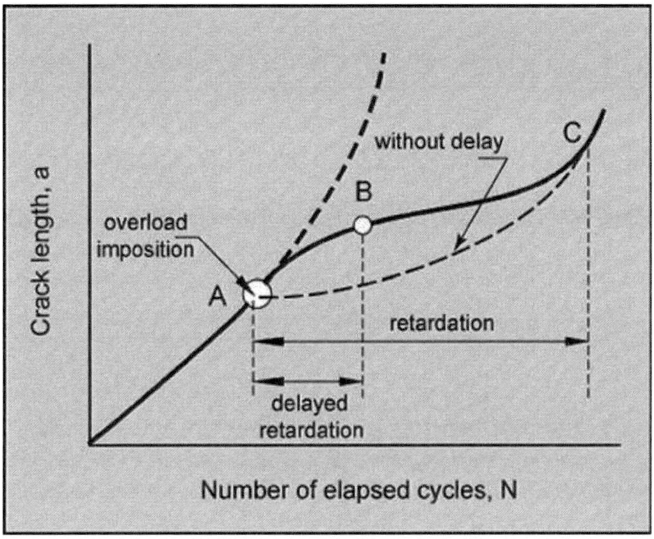

Fig. 15.27 Delayed retardation after overload imposition

Elber [35] and Bernard et al. [37] ascribed the overload retardation effect to the system of residual stresses created in the plastic zone invoking a crack closure effect, such that illustrated in Fig. 15.23b. Also the sequence of different load histories application can have an impact on the retardation effect. If, for example, two load histories are applied to the system, each having a constant amplitude equal to σ_{aI} and σ_{a2}, respectively, with $\sigma_{aI} \gg \sigma_{a2}$, the sequence of application $\sigma_{aI} \rightarrow \sigma_{a2}$ will give rise to a fatigue crack growth retardation that, on the contrary, will not be seen at all in the opposite sequence $\sigma_{a2} \rightarrow \sigma_{aI}$. This is schematized in Fig. 15.28 [36] in which AA′ and BB′ are the crack growth curves under the two loading sequences, respectively. more recently, Kim et al. [38] working with specimens of AISI 4340 steel found that the fatigue crack after retardation bifurcated.

Bifurcation leads to a reduction of the stress intensity factor K_I that has to distribute upon two crack tips, not necessarily in equal part. Therefore, bifurcation produces a FCGR reduction that persist up to the moment when one of the two crack tips, the one having the higher ΔK_I that advances faster, takes the lead and eventually obscures the other rebuilding the single crack ΔK_I. However, this circumstance seems not be confirmed by other investigators so that bifurcation must be associated to that single experience. Temperature can strongly reduce the overload retardation effect.

To this end, referring to the previously mentioned experience of Cotterill and Knott [34] shown in Fig. 15.26, while at room temperature the fatigue crack growth Δa preceding the overload imposition was restored after a crack growth approximately equal to the overload plastic zone width at 525 °C this recovery happened before, after about 10 to 20% of the plastic zone size. This recovery must be attributed to the overload residual stress relaxation operated by temperature that eventually, cancel

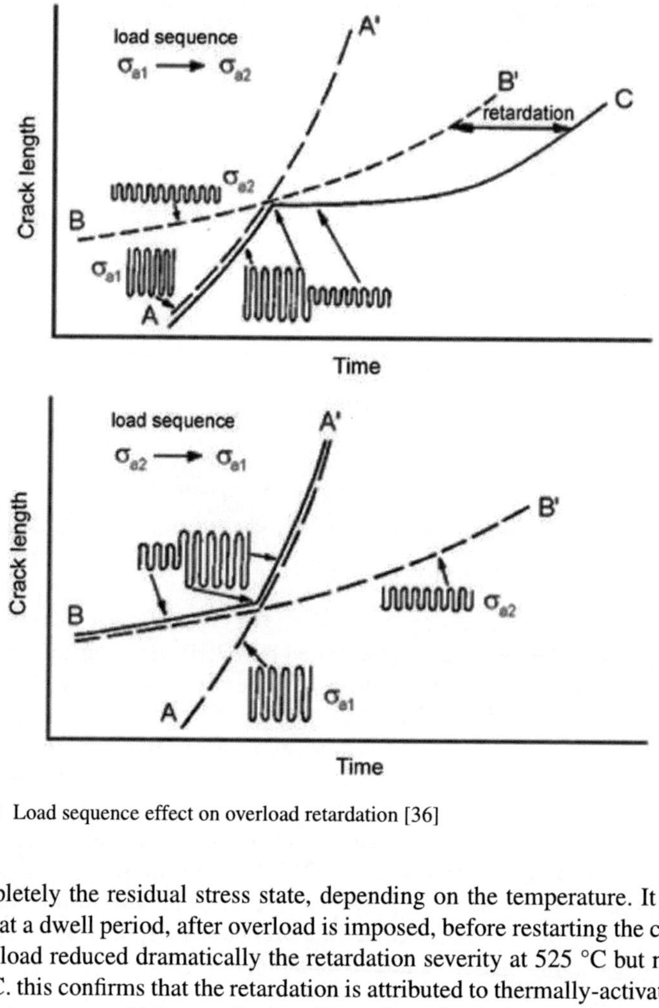

Fig. 15.28 Load sequence effect on overload retardation [36]

out completely the residual stress state, depending on the temperature. It is worth noting that a dwell period, after overload is imposed, before restarting the cycling at constant load reduced dramatically the retardation severity at 525 °C but not at RT or 225 °C. this confirms that the retardation is attributed to thermally-activated time-dependent process, which cancels the effect of overload due to plasticity-induced closure.

15.7 Growth of Short Cracks

The definition of micro-crack and macro-crack has already been introduced in Sect. 1. 5. Micro-cracks, also called MSC or mechanically small cracks, are defects generated during Stage I of fatigue, see Sect. 1.5.2 and Fig. 1.64, within a grain or few grains and have a dimension varying from just 3 μm to about 300 μm. We shall consider MSC as small cracks, just increasing the size a little bit up to 1 mm. Long cracks go beyond 1 mm length. Such a distinction is not just heuristic or related to the validity of continuum mechanics or fracture mechanics. Point is that during the last decades it has been found that small or short cracks grow earlier (smaller ΔK_{th}) and at faster rate than long cracks [39–44]. Worth mentioning are results shown in Fig. 15.29 obtained by Pineau et al. [44] on a ductile or nodular cast iron 3.8% C, 3.21 Si, 0.27 Mn, 0.04 Mg, 0.01 P, with yield strength equal to 270 MPa, ultimate of 450 MPa, HV30 hardness equal to 185 and fatigue limit at 10^7 cycles of 205 MPa. The technique to obtain small cracks is that of using notched specimens fatigue pre-cracked that are machined to remove a surface layer up to leave a crack of wanted size, in any case smaller than one millimeter. The diagram of Fig. 15.29 [44] shows the comparison of FCGR, *da/dN*, obtained with specimens carrying long cracks (10 mm) and specimens with short cracks of 240, 330 and 550 μm, respectively. Some features are worth noting.

First, it can be seen that in Region I of fatigue the threshold stress intensity factor ΔK_{th} depends on the crack size; about 5 MPa$\sqrt{m}$ for the 240 μm crack, about 6 MPa$\sqrt{m}$ for the 330 μm crack and about 7 MPa$\sqrt{m}$ for the 550 μm crack, versus the 8.2 MPa$\sqrt{m}$ of long cracks. Second, the FCGR, *da/dN*, of short cracks is Region I of fatigue depends on the crack length and is higher for smaller cracks. Finally, data coming from all types of cracks merge in Region II of fatigue, as already seen for the *R*-ratio effect. However, in the case of short cracks this findings may have other

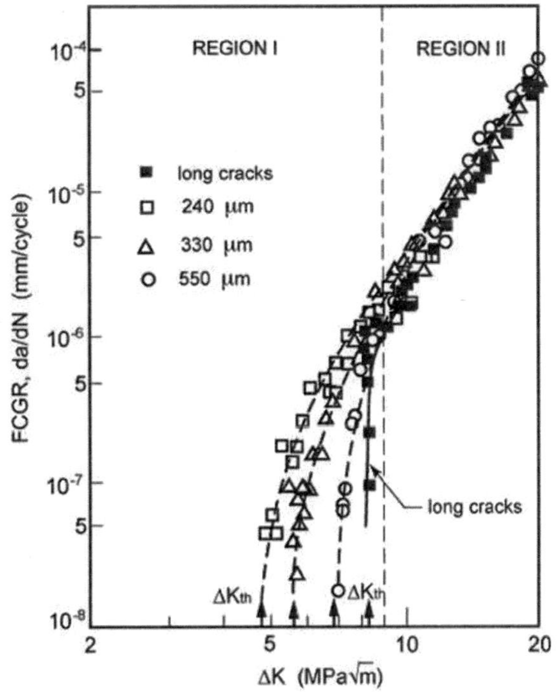

Fig. 15.29 FCGR, *da/dN*, obtained by Pineau et al. [44] for specimens of ductile cast iron carrying long cracks (10 mm) and short cracks

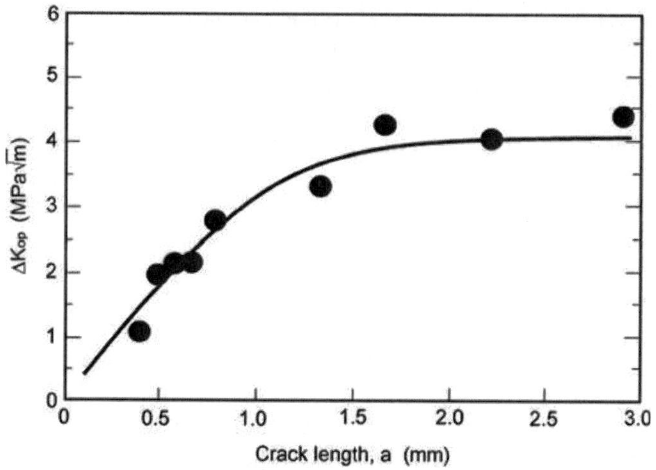

Fig. 15.30 Measurements of ΔK_{op} as function of crack size [44]

reasons. One of the conditions of validity of linear elastic fracture mechanics (LEFM) and K_I is that the crack tip plastic zone size be small, negligible with respect to all other dimensions. Now, since the plastic zone radius given by Eq. (15.23) already at $\Delta K_I = 10$ MPa$\sqrt{}$m is of the order of 300 $\div$ 400 μm, i.e., equal or larger than the same crack size one may argue that LEFM is not valid any longer.

On the contrary, data coming from long cracks are valid in terms of ΔK_I, since the crack size is at least an order of magnitude larger than the crack tip plastic zone size. Moreover, in Region I a crack closure effect can be responsible of the larger FCGR, dough with the limitation already evidenced about the applicability of LEFM. Measurements of ΔK_{op} for the various crack sizes have yielded the results of Fig. 15.30 [44].

It can be seen that below one millimeter of crack length, the ΔK_{op} reduces progressively, whilst for cracks longer than one millimeter it remains constant at about 4 MPa$\sqrt{}$m. On the base of Eq. (15.17) this finding results in a parallel increase of ΔK_{eff} for short cracks that could explain the larger FCGR in Region I of fatigue. Reanalyzing the experimental data of Fig. 15.29 with respect to ΔK_{eff} it is obtained the diagram of Fig. 15.31 [44]. The discrepancy among the various FCGR reduced very much.

However, this behavior considered characteristic of small cracks is not always confirmed by experimental data. Several researchers [45–48] reports episodes of complete independency of FCGR of the crack length. An example of such a behavior is offered in Fig. 15.32 for quenched and tempered 4340 steel, 1187 MPa yield strength, whose FCGR of small cracks is practically the same of long cracks. Small cracks were penny shaped with depth a from 15 to 200 μm. Fatigue loads were pulsating with $R = 0.1$. A comprehensive review of existing FCGR data for short

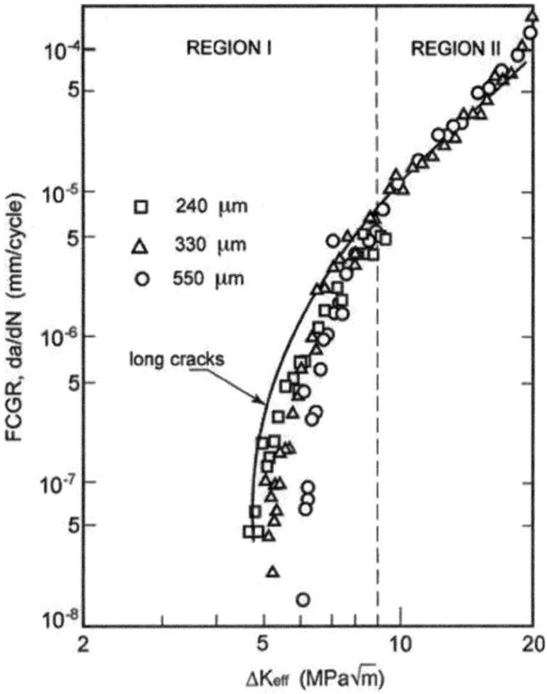

Fig. 15.31 Reanalysis of data of Fig. 15.29 with respect to ΔK_{eff} [44]

and long cracks has been done by Lankford [49] in order to understand differences in the fatigue behavior of short and long cracks.

Examining 11 different alloys for which LEFM was applicable to the FCGR measurement, Lankford tried to establish a relationship between the short crack capability to grow faster than long cracks and a metallurgical characteristic M that could be the grain size or, in the case of martensitic steels, the martensite lamellae. He found that the discrepancy between small and long cracks held when the radius r_p of the crack tip plastic zone was smaller than this metallurgical characteristic M that is to say when the plastic zone was well within the grain. Table 15.2 provides a synoptic picture of Lankford results [49]. It is shown the minimum crack size, $a_{\min}$, of small cracks and the FCGR with respect to long cracks; *yes* means equal behavior, whilst *no* indicates different behavior. Characteristic, to this end, are results obtained for aluminum alloys, practically identical, DTD 5050 and 7075-T651 [39–48] shown in Fig. 15.33 [49]. Some small cracks arrested completely within the grain (curve A), others, after some growth delay with a minimum about the average grain size, restarted to grow (curves B, C and D). Others, finally, grew with a rate higher than that of long cracks up to merge into Region II of the da/dN versus ΔK_I diagram (curve E). The different behavior was ascribed to metallurgical barriers, as the grain boundaries, on cracks too short, as already discussed in Sect. 1.5.

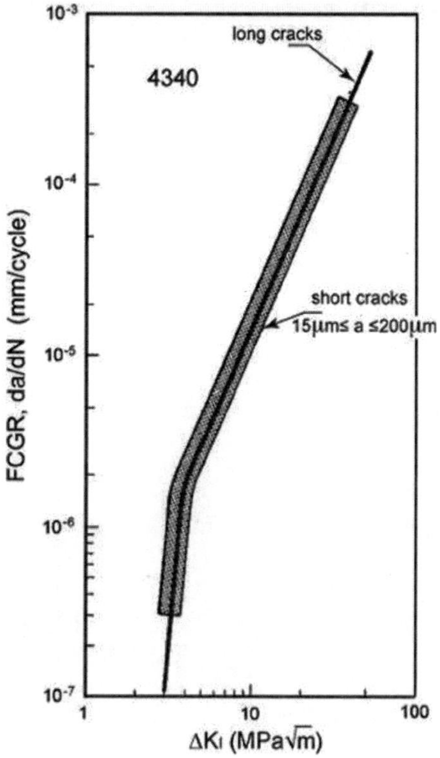

Fig. 15.32 FCGR, *da/dN*, observed for small and long cracks in 4340 steel [45]

Table 15.2 Materials, RCCF and mechanical-metallurgical parameters [49]

Material	*RCCF*	R	σ_y (MPa)	$a_{\min}$ (μm)	M	r_p (μm)
4130 (quenched & tempered)	Yes	0.1	1325	50	~ 1	3
4340 (quenched & tempered)	Yes	0.1	1187	15	~ 1	1.2
Steel (quenched & tempered)	Yes	− 1	760	20	~ 1	1.2
Ni alloy IN 100	Yes	0.05	1100	60	7	14
IMI 685 Ti	No	0	865	250	5000	12
Ti-6Al-4 V	No	0.2	865	20	4.7 ÷ 11.7	2.2
Ti-8Al	No	0.1 ÷ 0.6	865–780	10	20 ÷ 100	2.6
7075-T651	No	0.05	508	15	20 ÷ 50	12
DTD 5050 Al	No	− 0.14	508	15	20 ÷ 50	12
Steel 3%Si	No	− 1	431/365	20	130 ÷ 600	3.6
Al-Bronze	No	0.1	345	60	100	18

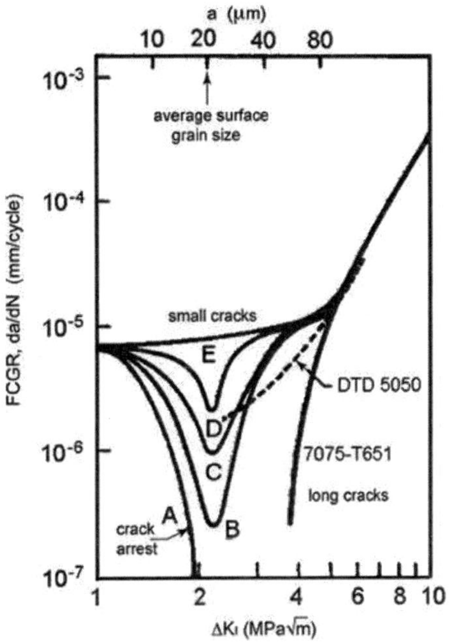

Fig. 15.33 FCGR of small and long cracks in aluminum alloys type DTD 5050 and 7075-T651 [49]

15.8 Variable-Amplitude Load Fluctuation

It has been considered, in Sect. 15.6, the case of single or multiple overloads imposed to constant load cycling. We shall now examine the more general case of a spectrum of variable-amplitude load fluctuation. Many attempts of addressing the issue and many models have been proposed with varying degrees of success [50–53]. A rather interesting approach has been proposed by Barsom [50] and is referred to as the root-mean-square (rms) model.

The model is based on the usual approach of defining a single stress intensity factor ΔK_{eff} to be used in a Paris-Erdogan expression that results in the same FCGR as that produced by the actual load time history

$$\frac{da}{dN} = A\left(\Delta K_{eff}\right)^n. \tag{15.28}$$

The Barsom approach is based on the root-mean-square

$$\Delta K_{eff} = \Delta K_{rms} = \sqrt{\frac{\sum_{i=1}^{k} \Delta K_I^2}{n}} \tag{15.29}$$

where ΔK_I is the stress intensity factor excursion relative to each peak of the loading sequence. The method is similar to that successfully used in seismic engineering when it must be assessed the stress response of a complex system, such as a long and branched piping, anchored to many different points of different buildings or structures having different individual floor response spectra not-in-phase in case of a seismic event. The same method has been inspiring the approach suggested in Sect. 8.5.4 referred to as equivalent spectrum. In his original work Barson considered two different loading spectra as reference spectra. These to spectra are shown in Fig. 15.34 [50]. They are characterized by a non-symmetric probability density function (*pdf*, see Chap. 4) with a modal value or peak value P_{mod} (see Fig. 4.3) different from the mean one. The first spectrum, Fig. 15.34a, is characterized by a ratio of the standard deviation Δ to peak value, Δ/P_{mod}, equal 1. The second, Fig. 15.34b, by $\Delta/P_{\text{mod}} = 0.5$. Specimens of WOL type (see Appendix A) were repeatedly subjected to load time history (a) and (b), respectively, and the crack length was measured as function of the elapsed number of cycles N. results are shown in Fig. 15.35 [50]. The stress excursion $\Delta\sigma$ indicated in Fig. 15.35 is that relative to the peak or modal value σ_{mod} of each spectrum.

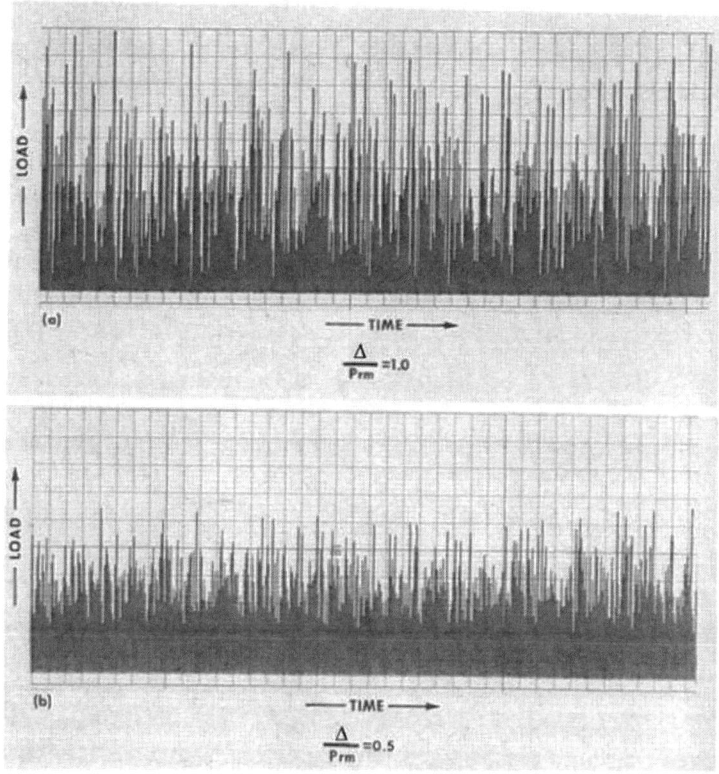

Fig. 15.34 Two variable-amplitude random-sequence load fluctuation investigated by Barsom [50]

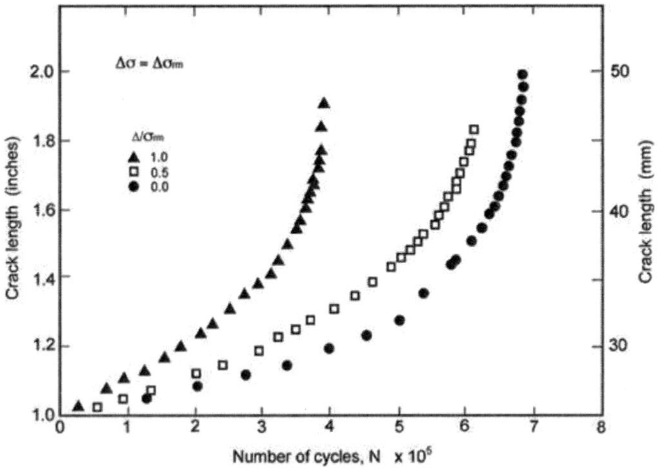

Fig. 15.35 Fatigue crack growth under spectra loads of Fig. 15.34 [50]

Figure 15.35 also reports the results obtained with a constant load spectrum, $\sigma_a = \sigma_{mod}$, for which the standard deviation is zero. As it can be seen, the fastest FCGR is that pertaining to the load spectrum (a) of Fig. 15.34 whilst the lowest FCGR competes to the constant load spectrum. The tangent at any point of the curves provides the FCGR, *da/dN*, relative to the selected crack length. To this length a ΔK_{mod} can be associated through the calibration function $f(a/w)$ that for the specimen considered is known. The result is a FCGR versus ΔK for all three curves of Fig. 15.35, as shown in Fig. 15.36a [50]. If instead of the ΔK_{mod}, the ΔK_{rms} is used the three different curves of Fig. 15.36a merge into almost a single one, as shown in Fig. 15.36b [50]. Barsom completed his experimental campaign using different load spectra as those shown in Fig. 15.37 [50]. Spectra were always including also the constant load time history as reference. Results in terms of FCGR versus ΔK_{rms} are shown in Fig. 15.38 [50]. The give credit to the method proposed.

15.9 Sample Problems

15.9.1 FCGR in Helicopter Blade

During a scheduled inspection, the blade of a helicopter rotor has been found affected by a tiny surface crack on the side put in traction by the wing lift, as schematized in Fig. 15.39. Its dimensions are $a_o = 0.3$ mm depth and $c_o = 1.4 \cdot a = 0.42$ mm length. The thickness of the metal, an aluminum alloy type 6061-T651, is $t = 4.5$ mm. The alloy has a yield strength of 266 MPa. The blade has been instrumented with strain gauges that found a tension stress variable from 26 to 75 MPa. Knowing that

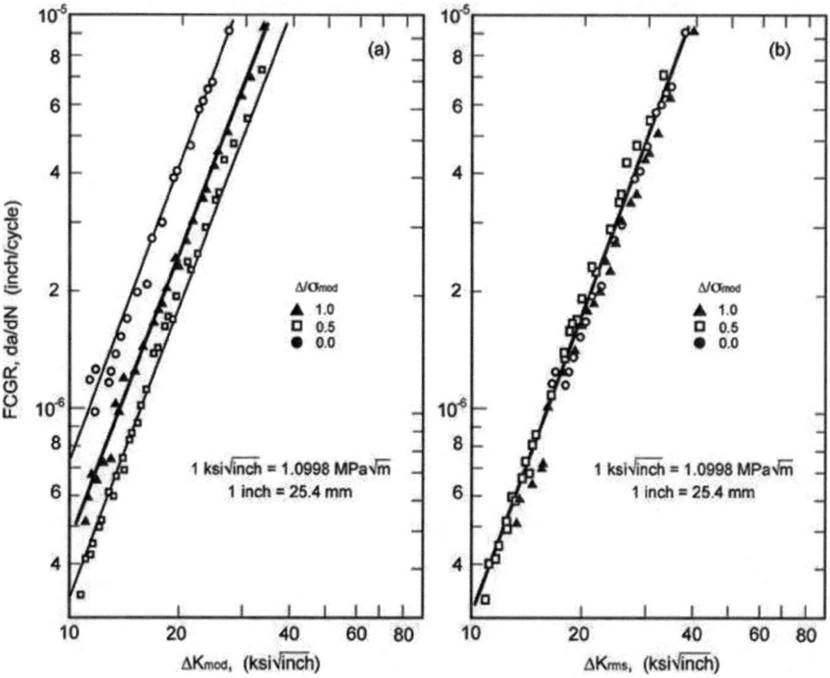

Fig. 15.36 **a** FCGR, *da/dN*, as a function of the modal stress intensity factor ΔK_{mod} and **b** of the ΔK_{rms} [50]

the rotor runs at 200 rpm calculate how many life hours are left before the crack becomes a through-wall-crack, assuming that the crack grows maintaining always the same aspect ratio 1:1.4.

The blade section can be approximated by a long and thin plate. For this latter geometry the expression of the stress intensity factor K_I is (see Appendix A).

$$K_I = \sigma\sqrt{\frac{\pi a}{Q}} \cdot F\left(\frac{a}{c}; \frac{a}{s}; \frac{a}{w}; \theta\right) \tag{15.30}$$

The two non-dimensional functions F and Q, whose expression are given in Appendix A, depend on the crack size a as shown in the diagram of Fig. 15.40. this means the explicit integration of the Paris-Erdogan equation is rather difficult. It will be used the finite difference integration approach.

Since the stress is varying from a minimum of 26 MPa to a maximum of 75 MPa, the stress amplitude $\Delta\sigma = 75–26 = 49$ MPa is applied with an R-ratio of 0.35. its effect will be affecting the FCGR, *da/dN*. To consider this effect, the Walker correction will be used, see Eq. (15.11) and the ΔK replaced by the ΔK_{eff}

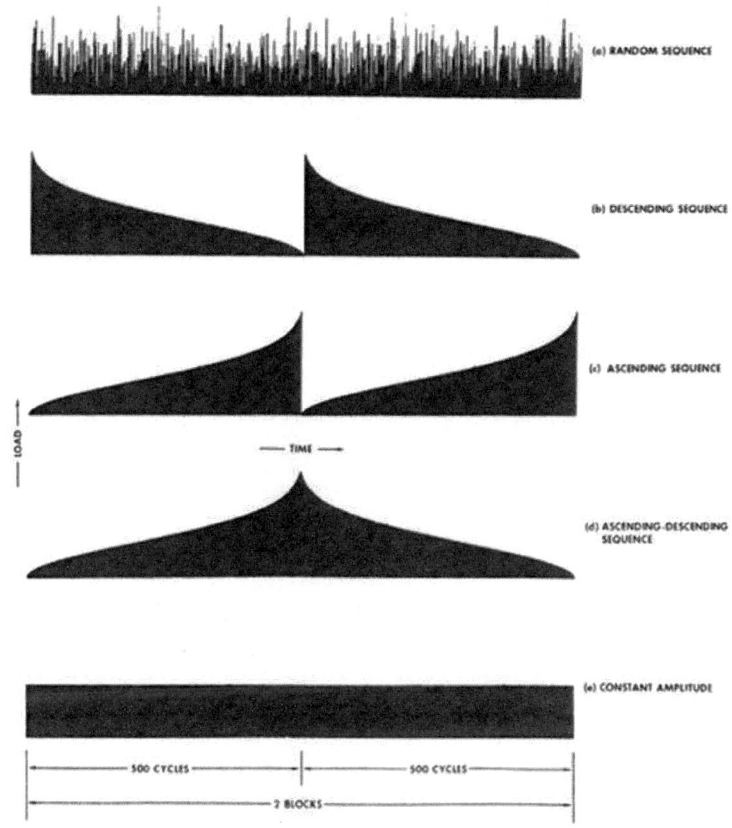

Fig. 15.37 Various random sequence used by Barsom [50]

$$\frac{da}{dN} = C \cdot \Delta K_{eff}^{n}$$
$$\Delta K_{eff} = \Delta K(1 - R)^{p} \tag{15.31}$$

For the aluminum alloy used for the blade, the constant C, n and p are $C = 7.7 \cdot 10^{-8}$, $n = 3.45$ and $p = -0.5$

The integration of the Paris-Erdogan Eq. (10.10) by the finite differenced method (see example 1 of Sect. 10.10) is done dividing the possible crack growth interval $\Delta a = 4.5\text{–}0.3 = 4.2$ mm into 30 equal parts in each of which the ΔK_{eff} and the F and Q factors can be considered constant. Results are shown in Table 15.3. To become a through-wall-crack there will be necessary $1.7 \cdot 10^{7}$ cycles. Considering that the rotor angular velocity is 200 rpm, it will take

$$t = \frac{1.7 \cdot 10^{7}}{200 \cdot 60} = 1416.67 \text{ h}.$$

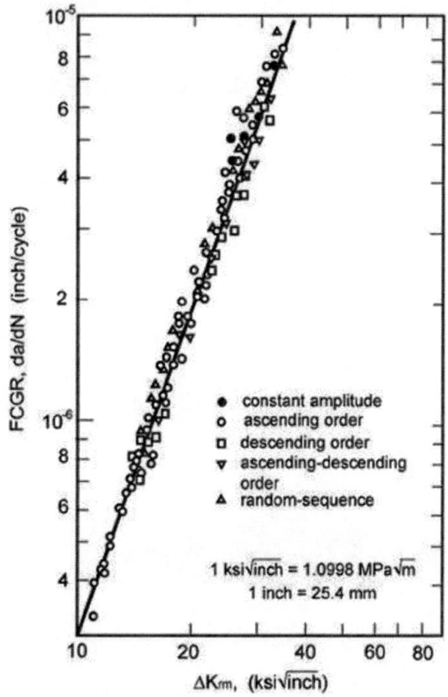

Fig. 15.38 Summary of FCGR, *da/dN*, under the load fluctuations sequences of Fig. 15.37 [50]

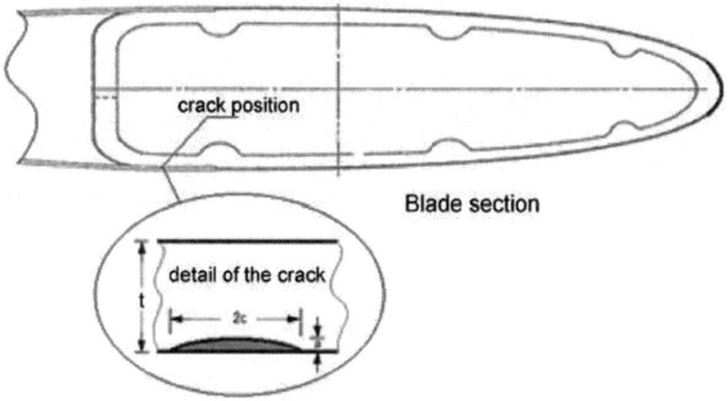

Fig. 15.39 Blade section and crack detail

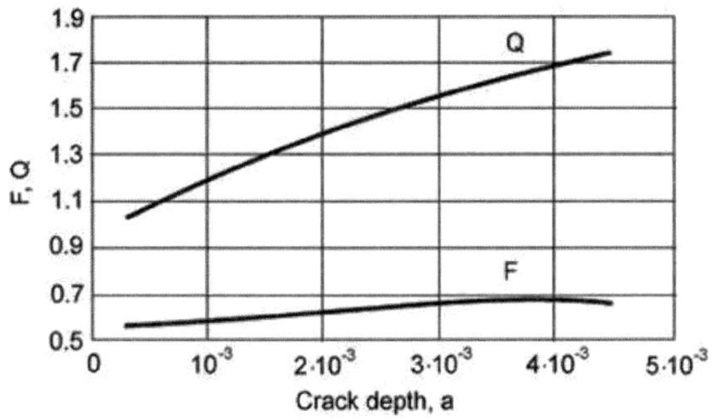

Fig. 15.40 F and Q functions versus crack depth for $c/a = 1.4$

Table 15.3 Results of finite difference integration of the Paris-Erdogan equation

a (m)	c (m)	F	ΔK	ΔK_{eff}	ΔN	$\Sigma \Delta N$
3.00E−04	4.20E−04	0.533818	0.58	0.71	6.25E+06	6.25E+06
4.50E−04	6.30E−04	0.535014	0.71	0.88	3.07E+06	9.32E+06
6.00E−04	8.40E−04	0.536676	0.82	1.02	1.84E+06	1.12E+07
7.50E−04	1.05E−03	0.538793	0.92	1.14	1.23E+06	1.24E+07
9.00E−04	1.26E−03	0.541349	1.01	1.26	8.85E+05	1.33E+07
1.05E−03	1.47E−03	0.544326	1.10	1.36	6.65E+05	1.39E+07
1.20E−03	1.68E−03	0.547702	1.18	1.47	5.16E+05	1.45E+07
1.35E−03	1.89E−03	0.551451	1.26	1.57	4.11E+05	1.49E+07
1.50E−03	2.10E−03	0.555544	1.34	1.66	3.33E+05	1.52E+07
1.65E−03	2.31E−03	0.559948	1.42	1.76	2.75E+05	1.55E+07
1.80E−03	2.52E−03	0.564628	1.49	1.85	2.30E+05	1.57E+07
1.95E−03	2.73E−03	0.569545	1.57	1.94	1.94E+05	1.59E+07
2.10E−03	2.94E−03	0.574656	1.64	2.04	1.65E+05	1.61E+07
2.25E−03	3.15E−03	0.579914	1.71	2.13	1.42E+05	1.62E+07
2.40E−03	3.36E−03	0.585269	1.79	2.22	1.23E+05	1.63E+07
2.55E−03	3.57E−03	0.59067	1.86	2.31	1.07E+05	1.64E+07
2.70E−03	3.78E−03	0.596058	1.93	2.39	9.42E+04	1.65E+07
2.85E−03	3.99E−03	0.601375	2.00	2.48	8.32E+04	1.66E+07
3.00E−03	4.20E−03	0.606556	2.07	2.57	7.38E+04	1.67E+07
3.15E−03	4.41E−03	0.611535	2.14	2.65	6.60E+04	1.68E+07
3.30E−03	4.62E−03	0.616242	2.21	2.74	5.92E+04	1.68E+07
3.45E−03	4.83E−03	0.620601	2.27	2.82	5.35E+04	1.69E+07
3.60E−03	5.04E−03	0.624537	2.34	2.90	4.86E+04	1.69E+07

(continued)

Table 15.3 (continued)

a (m)	*c (m)*	*F*	ΔK	ΔK_{eff}	ΔN	$\Sigma\Delta N$
3.75E−03	5.25E−03	0.627967	2.40	2.97	4.45E+04	1.70E+07
3.90E−03	5.46E−03	0.630808	2.45	3.04	4.09E+04	1.70E+07
4.05E−03	5.67E−03	0.632972	2.51	3.11	3.78E+04	1.70E+07
4.20E−03	5.88E−03	0.634366	2.56	3.18	3.53E+04	1.71E+07
4.35E−03	6.09E−03	0.634896	2.61	3.24	3.31E+04	1.71E+07
4.50E−03	6.30E−03	0.634463	2.65	3.29	3.13E+04	1.71E+07

References

1. Griffith, A.A.: The phenomena of rupture and flow in solids. Roy. Soc. Trans., (Lond.) Ser. A **221**, 163–198 (1920)
2. Milella, P.P.: A Fatigue Crack Growth Theory Based upon Energy Considerations, IAEA Specialists Meeting on Subcritical Crack Growth, Freiburg, Federal Republic of Germany, pp. 484–508 (1981)
3. Milella, P.P.: A Fatigue Crack Growth Theory Based upon Energy Considerations. Further Development on Small Crack Behavior and R Ratio Effect, Fatigue and Fracture Mechanics, Twenty-Ninth Volume, ASTM-STP 1332 (1999)
4. Wessel, E.T., Clarck, W.G., Jr.: Fracture Prevention Procedure for Heavy Section components, ASM Conference on Fracture Control (1970)
5. Vecchio, R.S., Crompton, J.S., Hertzberg, R.W.: The influence of specimen geometry on near threshold fatigue crack growth. Fatigue Fract. Engng. Mater. Struct. **10**(4), 333–342 (1987)
6. Amzallag, C., Rabbe, P., Bathias, C., Benoit, D., Truchon, M.: Influence of variou parameters on the determination of the fatigue crack arrest threshold. In: Hudac, S.J., Jr., Bucci, R.J. (eds.) ASTM STP 738, American Society for Testing and Materials, pp. 29–44 (1981)
7. Ogawa, T., Tokaji, K., Ohya, K.: The effect of microstructure and fracture surface roughness on fatigue crack propagation in a Ti-6Al-4V alloy. Fatigue Fract. Engng. Mater. Struct. **16**(9), 973–982 (1993)
8. Basic Fracture Mechanics for Nuclear Applications, Westinghouse course on fracture mechanics to ENEA-ENEL, Rome, Westinghouse data from W. G., Clark, Jr., Italy (6–8 March 1978)
9. Tanaka, K.: Mechanics and micromechanics of fatigue crack propagation. Am. Soc. Test. Mater., ASTM STP **1020**, 151–183 (1989)
10. Rolfe, S.T., Barsom, J.M.: Fracture and fatigue control in structure. Prentice-Hall, Englewood Cliff, N.J. (1977)
11. Klesnil, M., Lukas, P.: Effects of stress cycle asymmetry on fatigue crack growth. Mater. Sci. Eng. **9**, 231–240, 1072.
12. Hopkinns, S.W., Rau, C.A., Jr.: Prediction of structural crack growth behavior under fatigue loading. ASTM STP **738**, 255–270 (1981)
13. Walker, E.K.: The effect of stress ratio during crack propagation and fatigue for 2024–T3 and 7075–T6 aluminum, ASTM STP 462. American Society for Testing and Materials, Philadelphia (1970)
14. Crooker, T.W.: Effect of Tension-Compression Cycling on Fatigue Crack Growth in High-Strength Alloy, Naval Research Laboratory Report 7220 (Washington, D.C., 1971)
15. Crooker, T.W., Krauser, D.J.: The influence of stress ratio and stress level on fatigue crack growth rates in 140-Ksi HY steel. Naval Research Laboratory Report, Washington, D.C. (1972)
16. Miller, M.S., Gallagher, J.P.: An analysis of several fatigue crack growth rate (FCGR) description. ASTM STP **738**, 205–251 (1981)

17. Elber, W.: Fatigue crack closure under cyclic tension. Eng. Fract. Mech.Fract. Mech. **2**, 37–45 (1970)
18. Zhang, S., Marissen, R., Shulte, K., Trautmann, K.K., Nowak, H., Schijve, J.: Crack propagation studies on Al 7475 on the basis of constant amplitude and selective variable amplitude loading histories. Fatigue Fract. Eng. Mater. Struct.Fract. Eng. Mater. Struct. **10**(4), 315–332 (1987)
19. Gomez, M.P., Ernst, H., Vazquez, J.: On the validity of Erber, s results on fatigue crack closure for 2024–T3 aluminum. Int. J. Fract.Fract. **12**, 178–180 (1976)
20. Clerivet, A., Bathias, C.: Study of crack tip opening under cyclic loading taking into account the environment and R ratio. Eng. Fract. Mech.Fract. Mech. **12**, 599–611 (1979)
21. Schijve, J.: Some formulas for the crack opening stress level. Eng. Fract. Mech.Fract. Mech. **14**, 461–465 (1981)
22. Castro, D.E., Marci, G., Munz, D.: A generalized concept of a fatigue threshold. Fatigue Fract. Eng. Mater. Struct.Fract. Eng. Mater. Struct. **10**(4), 305–314 (1987)
23. Newman, J.C., Jr.: American Society for Testing and Materials, ASTM STP 748, Methods and Models for Predicting Fatigue Crack Growth under Random Loading, pp. 55–84 (1981)
24. Newman, J.C., Jr.: Prediction of fatigue crack growth under variable-amplitude using a closure model. Am. Soc. Test. Mater., ASTM STP **761**, 255–277 (1982)
25. Suresh, S., Ritchie, R.O.: Propagation of short fatigue cracks. International Metallurgical Reviews, Vol. 29, pp. 455–476
26. Rice, J.R.: The mechanism of crack tip deformation and extension by fatigue. Fatigue Crack Propag., ASTM STP **415**, 247–309 (1967)
27. Hardrath, H.F., McEvily, A.T.: Engineering aspects of fatigue crack propagation, Proceeding of Crack Propagation Symposium, Vol. 1 (Cranfield, England, Oct 1961)
28. Schijve, J.: Significance of fatigue cracks in micro-range and macro-range. Fatigue Crack Propagation, ASTM STP 415 (Philadelfia, 1967)
29. McMillan, J.C., Pelloux, R.M.N.: Fatigue crack propagation under program and random loads. Fatigue Crack Propagation, ASTM STP 415 (Philadelfia, 1967)
30. Von Euw, E.F.J., Hertzberg, R.W., Roberts, R.: Delay Effects in Fatigue Crack Propagation, ASTM STP 513, pp. 230–259 (Philadelfia, 1972)
31. Wheeler, O.E.: Spectrum Loading and Crack Growth, General Dynamic Report FZM-5602, Fort Worth (30 June, 1970)
32. Wheeler, O.E.: Spectrum loading and crack growth. J. Basic Eng. **44**, 181–186 (1972)
33. Schijve, J.: Fatigue crack propagation in light alloy sheet material and structures. Advances in Aeronautical Science, Vol. 3, p. 387 (Pergamon Press, 1961)
34. Cotterill, P.J., Knott, J.F.: Overload retardation of fatigue crack growth in 9%Cr 1%Mo steel at elevated temperatures. Fatigue Fract. Eng. Mater. Struct.Fract. Eng. Mater. Struct. **16**(1), 53–70 (1993)
35. Elber, W.: The significance of fatigue crack closure. ASTM STP **486**, 230–242 (1971)
36. Schijve, J.: The accumulation of fatigue damage in aircraft materials and structures, AGARD Conference No. 118, Symposium on Random Load Fatigue, pp. 3–82 (1972)
37. Bernard, P.J., Lindley, T.C., Richard, C.E.: Mechanisms of overload retardation during fatigue crack propagation. ASTM STP **595**, 78–97 (1976)
38. Kim, S., Tai, W.: Retardation and arrest of fatigue crack growth in AISI 4340 Steel by introducing rest periods and overload. Fatigue Fract. Eng. Mater. Struct.Fract. Eng. Mater. Struct. **15**(6), 519–530 (1992)
39. Pearson, S.: Initiation of fatigue cracks in commercial aluminum alloys and the subsequent propagation of very short cracks. Eng. Fract. Mech.Fract. Mech. **7**, 235 (1975)
40. Kitagawa, H., Takahashi, S.: Applicability of fracture mechanics to very small cracks or the cracks in the very early stage. Proceedings of the 2nd International Conference Mechanical Behavior of Materials, p. 627 (Boston, Ma., 1976)
41. Lankford, J.: Initiation and early growth of fatigue cracks in high strength steel. Eng. Fract. Mech.Fract. Mech. **9**, 617–624 (1977)
42. Taylor, D., Knott, J.F.: Fatigue crack propagation behaviour of short cracks; the effect of microstructure. Fatigue Eng. Mater. Struct. **4**, 147–155 (1981)

43. Brown, C.W., Hicks, M.A.: A study of short fatigue crack growth behavior in titanium alloy IMI 685. Fatigue Eng. Mater. Struct. **6**, 46–67 (1983)
44. Clement, P., Angeli, J.P., Pineau, A.: Short crack behavior in nodular cast iron. Fatigue Eng. Mater. Struct. **7**(4), 251–265 (1984)
45. Lankford, J.: The influence of microstructure on the growth of small fatigue cracks. Fatigue Fract. Eng. Mater. Struct.Fract. Eng. Mater. Struct. **8**(2), 161–175 (1985)
46. Yokobori, T., Kuribayashi, H., Kawagishi, M., Takeuchi, N.: Study of Initiation and Propagation of Fatigue Cracks in Tempered Martensitic High Strength Steel by Plastic-replication Method and Scanning Microscope, Reports of the Research Institute for Strength and Fracture of the Materials, Tohoku University, Sendai, Japan, Vol. 7, pp. 1–23 (1971)
47. Lankford, J.: Initiation and early growth of fatigue cracks in high strength steels. Eng. Fract. Mech.Fract. Mech. **9**, 617–624 (1977)
48. Lankford, J., Cook, T.S., Sheldon, G.P.: Fatigue microcracks growth in nickel base superalloy. Int. J. Fract.Fract. **17**, 143–155 (1981)
49. Lankford, J.: The effect of the environment on the growth of small fatigue cracks. Fatigue Eng. Mater. Struct. **6**, 15–31 (1983)
50. Barsom, J.M.: Fatigue crack growth under variable amplitude loading in ASTM A514 grade B steel. Am. Soc. Test. Mater., ASTM STP **536**, 147–167 (1973)
51. Wei, R.P., Shih, T.T.: Delay in fatigue crack growth. Int. J. Fracture **10**(1) (1974)
52. Wheeler, O.E.: Spectrum loading and crack growth, Genaral Dynamics Report FZM-5602 (1970)
53. Willemborg, J., Engle, R.M., Wood, H.A.: A Crack growth Retardation Model Using an Effected Stress Concept, Technical Memorandum 71-1-FBR, Air Force Flight Dynamics Laboratory (1971)

Chapter 16
Fatigue in Welds

16.1 Introduction

More than half a century is passed from World War II when welding heavily entered into engineering practice and in the naval construction technique, in particular, often with catastrophic results. As in the development of any technological field it took its toll of failures and casualties before welding would become a discipline of its own and institutes, research centers, universities and industries would develop enough knowledge to provide good practices and standards. Today, welds are used in practically any field of metal constructions; ships, bridges, cranes, off shore structures, pressure vessels, piping, metal containers, buildings etc. Generally speaking, it is possible to assert that a well-executed weld has both fatigue and toughness properties superior to those of the base metal. Problem is that welds can still be done with low attention and poor metallurgical and technological knowledge coupled with the firm belief to have good practice so that they become the primary cause of steel cracking. Welds, in fact, are the most probable location of defects generation and high residual stresses. Another problem is that most of these defects are also the most serious of them all; cracks that are rather difficult to be detected. Welding technology and inspection practices, in fact, have long been separated by cultural barriers and this has been another serious problem. As matter of fact, inspections were originally focused on material acceptance criteria and standards and were aiming at detecting, in particular, pores and inclusions more than cracks. The nuclear field, in particular, would act as *trait d'union* between inspection technique and design requirements, i.e., crack detection. In the nuclear field, in fact, safety is a mandatory issue and it has long been struggling for reliable cracks detection techniques recognizing that welds in piping and pressure vessels, in particular, could be a dangerous problem. Moreover, austenitic stainless steel pipe welds could develop intergranular corrosion cracks that conventional non-destructive examinations were not capable to detect because cracks were intergranular and austenitic steels rather opaque to ultrasonics. In 1980 in-service inspections of the reactor pressure vessel of the Trino nuclear power plant, a PWR of

P. P. Milella, *Fatigue and Corrosion in Metals*,
https://doi.org/10.1007/978-3-031-51350-3_16

the first generation, were repeated in Italy. The pressure vessel, in fact, had already been inspected in 1978 according to the ASME Section XI procedures without any indication. But calculations made by DISP (the Italian Nuclear Safety and Health Protection Directorate) based on fracture mechanics revealed that even a tiny under cladding crack 3–5 mm long could become critical in case of small LOCA (loss of coolant water accident) because of the rather low toughness steel used (the pressure vessel was manufactured in 1960). The Italian Nuclear Directorate did not judge the ASME Section XI procedures of that time to be capable to detect such small size and surface crack. So a new ultrasonics technique was chosen based on 70° angle probes and 20% DAC for calibration procedures. The new inspections gave indications of about 5 mm. The Trino nuclear power plant was never allowed to re-enter operation. A decade of round robin nondestructive ultrasonics testing of steel plates containing flaws of almost any type and size initiated in 1976 under a cooperative international program called PISC (Plate Inspection Steering Committee), sponsored by OECD-NEA, with USA, Japan, UK, Germany, Italy and France participation led to a substantial improvement of calibration procedure and ultrasonics techniques. Today, both nondestructive crack detection techniques and weld technology has reached a sound maturity. Yet, welds remain the most challenging factor also because of the process volume effect already discussed in Sect. 6.6.1. Welds, in fact, are often done in hundreds of meters, sometimes even in hundreds or thousands of kilometers in open air where local environmental conditions, such as wind or humidity or chloride ions, can affect the welding quality. In these cases the probability that a crack is introduced somewhere must be given for granted. It is not the intention of this section to deeply enter into the welding science. Its ambitions is just that of giving some fundamentals of welding, outlining the major issues of welding impact on fatigue and crack propagation and warning about the major problems related to welds.

16.2 Weld Defects

As previously said, a well-done welding has both fatigue and toughness properties superior to those of the base metal. Problem is that welds can still be done with low attention and poor metallurgical knowledge so that they become the primary cause of steel cracking. Welds, in fact, are the most probable location of defects generation. Figure 16.1 schematizes some of the most characteristic defects found in (a) a butt-join weld and (b) fillet weld, respectively, such as center cracking, longitudinal crack in the heat affected zone (HAZ), transverse cracking, incomplete fusion or penetration, undercut or overlap, excessive porosity etc., and (c) under a cladding deposition. Figure 16.2 shows how some of the defects schematized in Fig. 16.1 can appear in real life in piping of carbon steel of 10″ diameter. Figure 16.2a shows an undercut combined with a lack of side fusion. The detail of Fig. 16.2b shows that the tip of the lack of side fusion of Fig. 16.2a is, in fact, a real crack being very sharp and having broken in a half several grains. Figure 16.2c, d show undercut and lack of penetration, respectively, that are sharp notches, while Fig. 16.2e presents an inclusion that during weld solidification has developed cracks at both sides. It is no longer a forgiving inclusion, but a real crack. Note that it took a destructive examination to recognize the crack-like nature of the inclusion. Figure 16.3 offers an example of transverse crack in a butt weld joining two 2 inches thick mild steel plates.

Transverse cracking are usually associated with hydrogen embrittlement that appears as intergranular separation (see Section). The detail of Fig. 16.3b is an enlargement of the crack tip encircled in Fig. 16.3a that shows precisely the intergranular nature of the transverse crack due to hydrogen embrittlement. It is interesting to note that the crack originated in the weld is entered in the associated HAZ and then deep into the base metal where it could not be driven by residual stresses since they were demolished by the growth. To sustain the crack growth outside the weld-HAZ area was fatigue. Of all possible defects, cracks are by far the most dangerous. Whereas there are acceptable limits for slag inclusions and porosity, cracks are never acceptable. However, we must be very cautious with inclusions because, as shown in Figs. 16.2e and 16.5, they can often develop cracks and must be considered as real cracks. Depending on whether the filler metal is harder than the base metal, cracks will affect the weld or the base metal in the HAZ. Transverse cracks are generally due to a filler metal harder than the base metal. The change of the electrode for another softer than base metal resolves the problem in most cases. Cracks can be grouped in two main categories; hot cracking and cold cracking. Hot cracks are those that occur at high temperatures in the weld during solidification while cold cracks form either in the weld or in the HAZ after the weld metal has cooled to room temperature and may be hydrogen related. Centerline cracking, see Fig. 16.1a, is characterized by the separation in the center of the weld bead. If the weld is a multiple pass weld this type of crack may not be right in the center. It is caused by one of these three phenomena; segregation induced cracking, bead shape induced cracking or surface profile induced cracking. Unfortunately the three phenomena result in the same type of cracking and cannot be distinguished from each other. Segregation induced cracking originates when low melting point constituents containing zinc, copper, phosphorous, sulfur and tin, in particular, separate during weld solidification.

Low melting point constituents segregate to the center of the molten metal because they are the last to solidify and cause the weld to separate and crack in the center region. Note that these low melting point constituents usually come from the base metal at the interface with the weld pool. High heat input welds favor the migration from base metal to molten pool. When sulfur is high in the iron ore (< 0.02%) to overcome possible harmful effect on the steel, manganese is added to form manganese sulfide (MnS) which is a high melting point constituent (~ 1590 °C). In this manner when MnS is formed it does not segregate. The second type of centerline cracking, known as bead shape induced cracking, is associated with deep penetration welds, such as SAW (submerged arc weld) and CO_2 shielded FCAW (flux core arc welding), that are more deep than wide. In these cases, solidification of dendrites proceeds from both sides toward the center of weld where they merge, but do not gain fusion across the joint. The last process, the surface profile induced cracking, is associated to concave weld profiles where shrinkage stresses place the surface of the weld metal

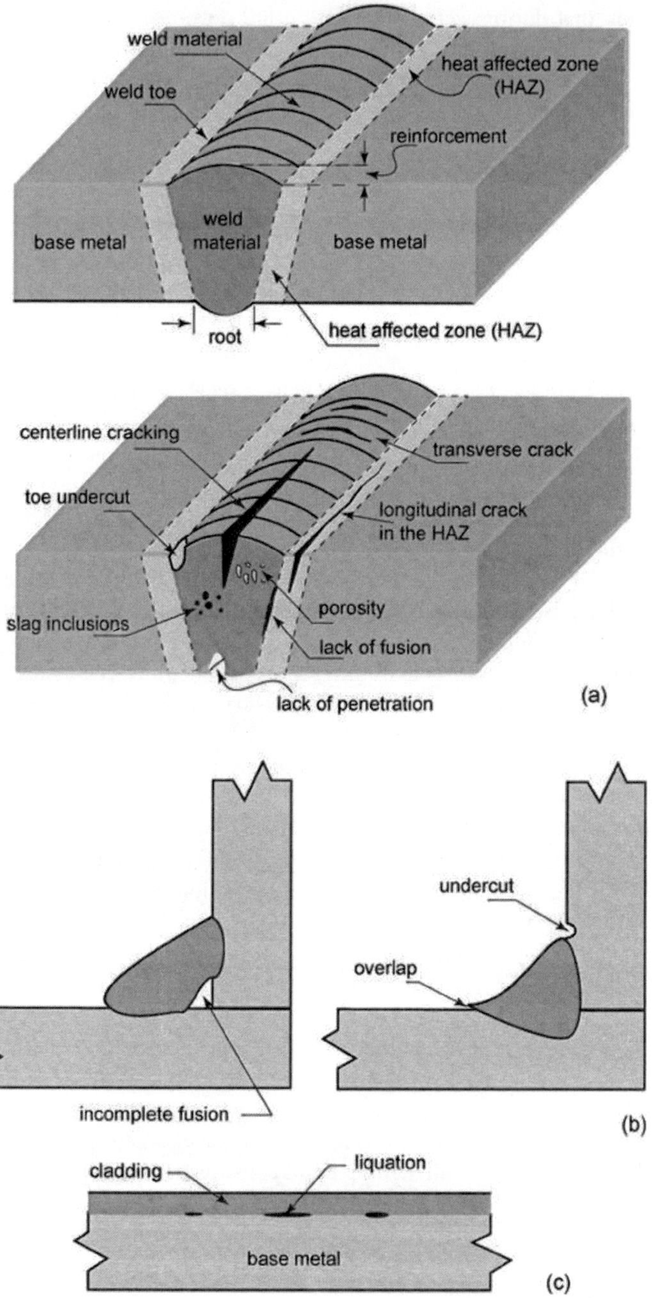

Fig. 16.1 Some typical weld defects in **a** butt welds and **b** fillet welds and **c** under cladding cracks

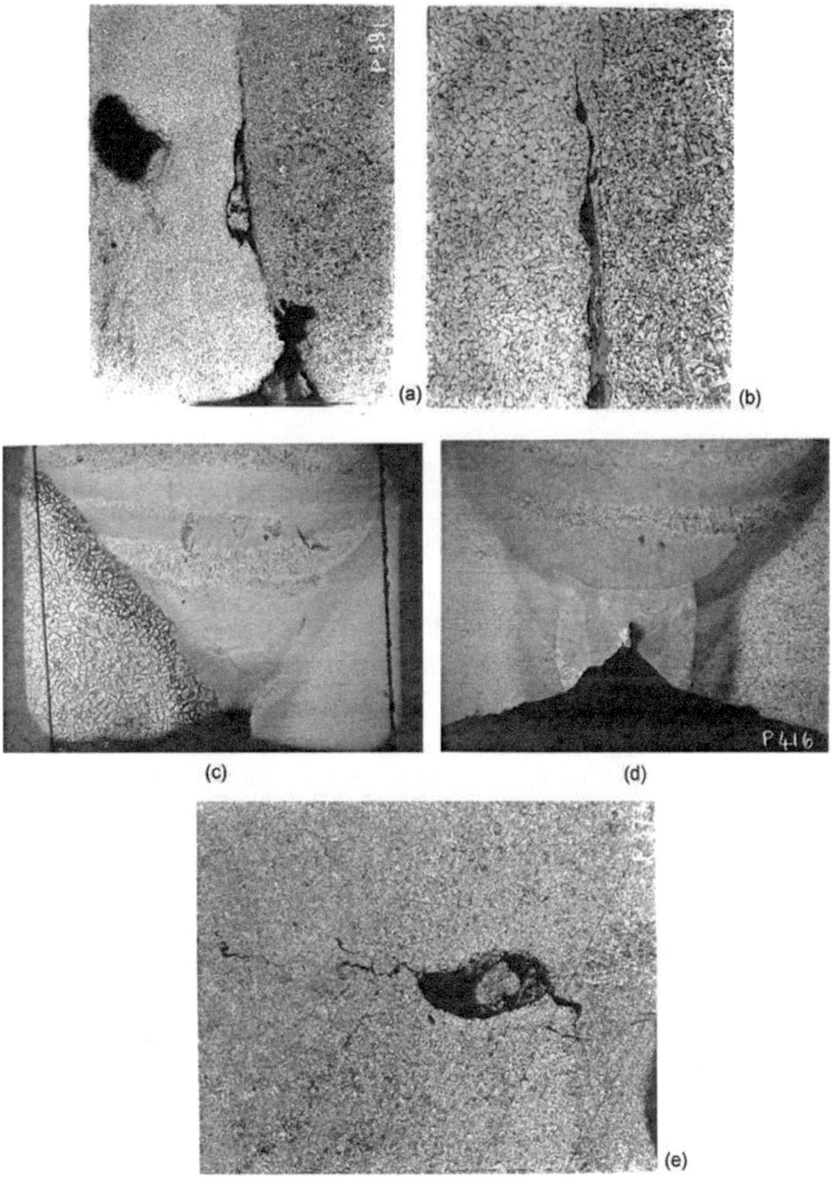

Fig. 16.2 Defects in a 10″ carbon pipe. **a** lack of side fusion, **b** detail of **a**–**d** undercut and lack of penetration, respectively and **e** inclusion that has developed cracks at both sides

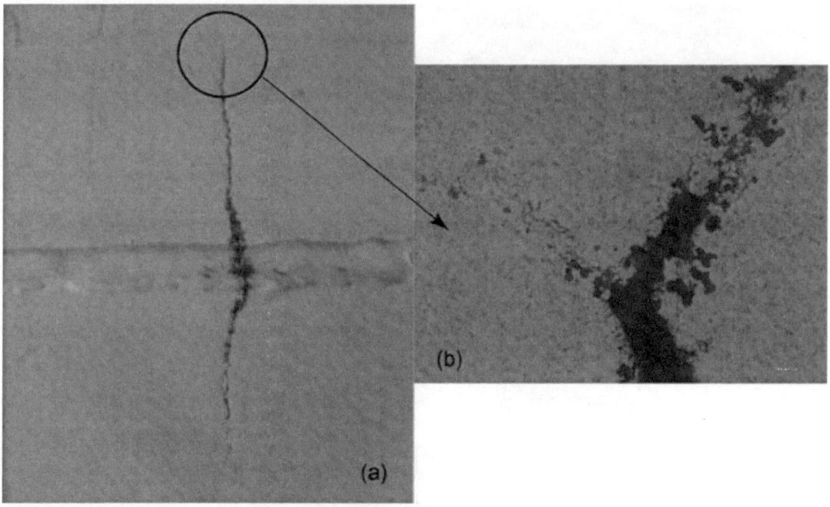

Fig. 16.3 **a** Transverse crack in a butt weld, **b** detail of the crack tip in the circle of **a**

into high tension, as schematized in Fig. 16.4. At variance, the surface of convex welds remains in a less stressed state.

Another type of hot cracking is the so called liquation, Fig. 16.1c. When carbon steel is cladded by a stainless steel layer to be protect from corrosion, it may happen that some low melting point inclusions present at the interface base metal-cladding melt and detach from the steel matrix. An internal cavity is formed. During cooling the different thermal expansion coefficient between austenitic stainless steel and

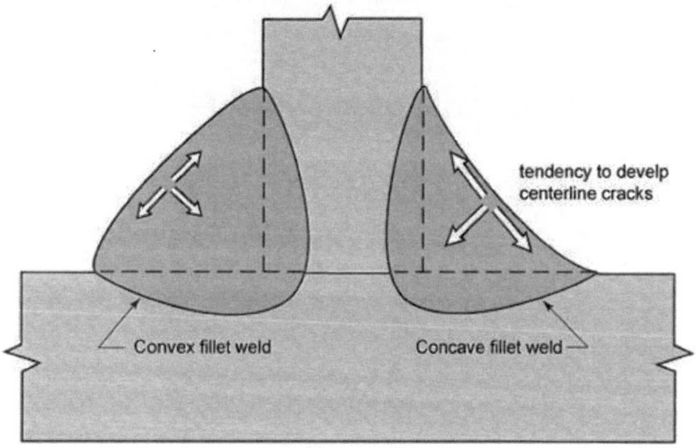

Fig. 16.4 Convex and concave fillet welds

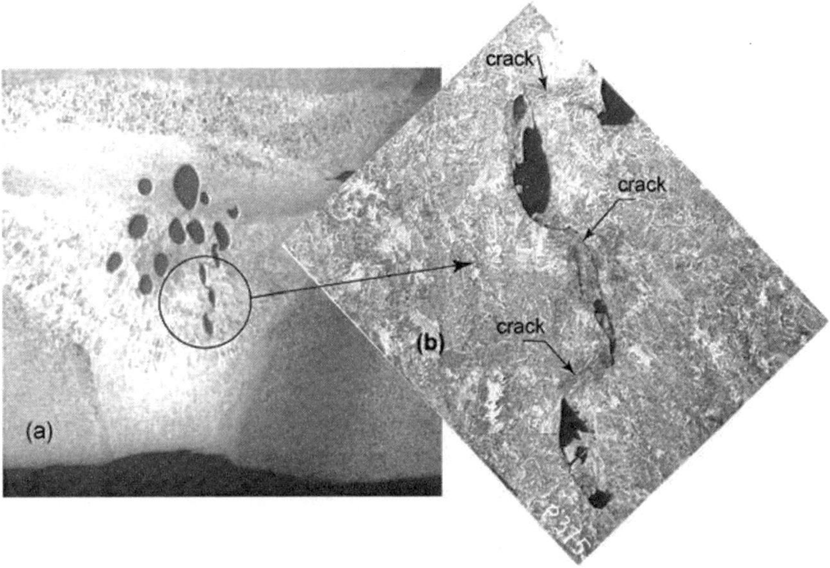

Fig. 16.5 **a** Porosities and slag inclusions in a butt weld and **b** detail of the area in the circle

carbon steel put the interface heavily in traction. The resulting tensile stresses can reach yielding and can break the liquation cavity introducing a crack.

Liquation is favored when cladding is deposited by strip electrodes of high heat input and the base metal is rich in S and P inclusions. Cold cracking may form in the weld or in the HAZ. HAZ cracking, Fig. 16.1a, is characterized by separation that occurs along the weld side in the base material. This type of longitudinal crack develops after the steel has cooled below approximately 200 °C.

As with transverse cracking, Fig. 16.3, this type of crack is usually associated with hydrogen and, therefore, is also called hydrogen assisted cracking. Hydrogen can enter the weld pool from a variety of sources. Air humidity, moisture or organic compounds or grease left on the steel to be welded are the primary sources of hydrogen. But it can also enter from the electrodes or from their shielding material. To limit hydrogen incoming from electrodes, the weld electrodes must be properly stored and maintained in field furnaces during welding because hydrogen is absorbed by steel at low temperatures and diffuses (escapes) at high temperatures over 80–100 °C.

It must also remember that HAZ is a poor metallurgical properties zone, as it will be described in the next section. In both cases, transverse and HAZ cracking, preheat may alleviate cracking. It will reduce residual stresses and favor hydrogen diffusion out of the metal. The weld toe, see Fig. 16.1a, is a point of stress concentration in a low toughness zone, the HAZ. It is, therefore, one of the most probable crack initiation points, especially when the weld reinforcement, see Fig. 16.1a, is too much or in case of toe undercut. It is also the point preferred by corrosion to start. Porosity and slug inclusions are generally three dimensional defects with large end radii. However, sometimes they can be too many and coalesce into a single crack as shown in Fig. 16.5 that refers to a circumferential butt weld in a 10″ carbon steel pipe. The detail of Fig. 16.5b clearly shows that the porosities and slag inclusions in the circle of Fig. 16.5a have joined by cracks extending from the extremities.

16.3 Heat Affected Zone (HAZ)

From a metallurgical point of view weld metal has a particular structure so called dendritic (from Greek dendron = tree) because it forms during the solidification of the weld pool in form of trees growing and extending in some preferred directions. This process initiates around germs of solidification, usually foreign particles, and proceeds from the weld pool to the fusion line of the parent metal. As the weld pool solidifies dendrites form that may change from equiaxed to columnar, cellular and planar as schematically shown in Fig. 16.6a. Figure 16.6b shows the same change as it appears in 1100 aluminum welded with 4047 filler metal [1]. Equiaxed and columnar dendrites form where the ratio between the temperature gradient G and the crystal growth rate R, G/R, is low. The growth rate R is low along the fusion line while the temperature gradient G is higher. The opposite happens at the centerline. Therefore, weld microstructure varies noticeably from the edge to the centerline of the weld.

As said in the previous section, heat affected zone or HAZ plays an important role on the soundness and robustness of a weld. It develops at the interface between weld and base material. This zone, at variance with the weld pool, does not reach the melting temperature yet can be heated at temperatures over the critical point that for steel is at about 723 °C, to undergo a series of processes among which,

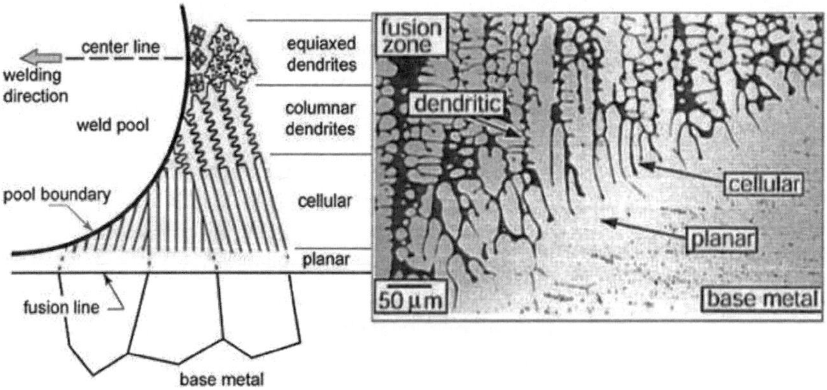

Fig. 16.6 **a** Schematic of dendrites forming during solidification of the weld pool and **b** as it appears in type 1100 aluminum alloy weld [1]

depending on the cooling rate and carbon content, also non-tempered martensite formation is included. Within the HAZ some characteristic metallurgical features can be distinguished, as schematized in Fig. 16.7. The first, near the weld fusion line, is a coarse-grained zone in which temperature has forced grains to grow. It is normally a low fatigue strength low toughness zone because the large grain size makes it easier the slip bands formation and, hence, introduces the fatigue damage earlier. Moreover, it reduces the ratio between the overall grain boundary surface and foreign phases and particles present in the material with the result of increasing their concentration on grain boundaries. This can sharply reduce the toughness of the material. Then a fine-grained recrystallization area appears of higher material properties followed by a partial austenitization and, finally, an annealed structure where the base metal has been heated over 600 °C, but it has never exceeded the critical transformation temperature. Beyond this zone there is the base metal that has not been affected by the welding process.

The actual metallurgical characteristics of the HAZ depend on the temperature reached by the base metal and the time of permanence at that temperature, i.e., the cooling rate, on the thermal input and the metal composition and thickness. The longer the time of permanence at the maximum temperature the larger the grain coarsening and the annealing effect. Figure 16.8 shows the HAZ of a butt weld on low alloyed-carbon steel. It is possible to note the dendritic structure of the weld (radial growth) and the HAZ showing a grain coarsening followed by a fine-grained structure. The two structures can be seen also in Fig. 16.3c. A post weld heat treatment can be very effective in restoring the pre-welding mechanical properties of the steel.

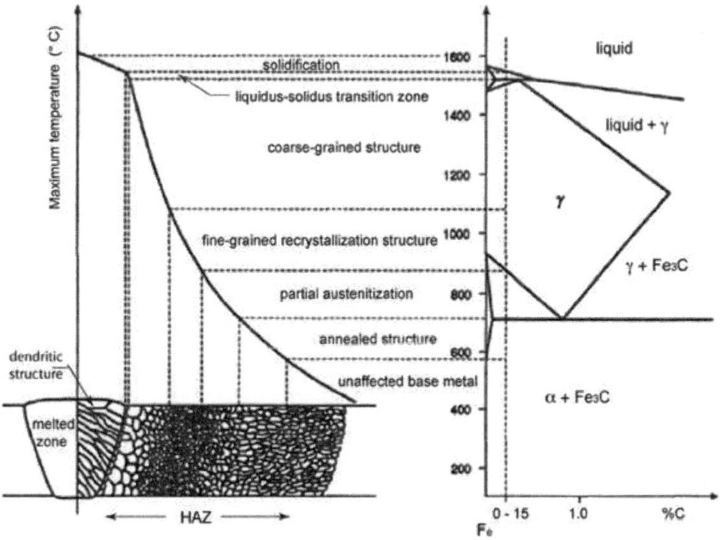

Fig. 16.7 HAZ and metallurgical transformations (modified from [2])

Fig. 16.8 Dendritic structure of a weld on carbon steel followed by grain coarsening and grain refinement

16.4 Residual Stresses

The cooling and solidification of the weld pool is accompanied by a parallel shrinkage of the weld metal itself that is prevented by the adjacent base metal that doesn't want to follow the weld pass shrinkage.

The result is the build-up of an internal stress state known as welding residual stress state. The intensity of such a stress state depends on the volume of the melted pool and of the two sides of base metal to be welded. To better understand the phenomenon it is convenient to look at the weld metal as to the molten mass that wants to shrink during cooling and solidification and to the base metal as to the resisting volume that wants to remain where it is exerting a drag force on the weld metal. Therefore, at the end of solidification and cooling, the weld metal will remain in a tensile stress state and the adjacent base metal in a compression state. since residual stresses are self-limiting internal stresses the overall tension must be balanced by an equivalent compression, i.e., their algebraic sum must be zero. The values of the resulting maximum tensile and compression stress will depend on the ratio of those two volumes; the forcing volume of the weld metal and the resisting one of the base metal.

Two considerations emerge from this simple schematization. First, that the larger the thickness of the sides to be welded the higher the residual stress state. Second, to check and qualify the chosen weld process it is necessary to make use a full scale process test-piece. If this is too complex or expensive and small test-pieces are used instead, it must be understood that to approximate the real situation it is at least necessary to block the two pieces of base metal in order to prevent them from moving and follow the weld metal contraction during solidification. This is too often forgotten and when small sample welds are tested they behave much better than the actual full scale weld because residual stresses have been strongly reduced with the entire negative consequences they may lead to, including the sort of cracks that have been presented in Sect. 16.2.

The most credited distribution of residual stresses in a butt weld is schematized in Fig. 16.9 [3, 4]. Longitudinal residual stresses, σ_l, are generally higher than transverse stresses, σ_t, unless the work-piece has an external constraint than can in part or fully prevent the weld shrinkage. In the case of complete weld shrinkage prevention the resulting residual stresses, σ_l and σ_t, may reach the yielding strength of the weld material. As it can be seen in Fig. 16.9, longitudinal stresses, σ_l, reach the maximum at the center of the weld, where the shrinkage of the molten pool during solidification is maximum. These tensile stresses must be balanced by compressive stresses in the base material. According to Saint Venant principle residual stresses must extinguish within a distance two–three times the characteristic dimension of their cause, which is the weld size. Figure 16.10 presents an experimentally determined distribution of longitudinal residual stresses in a butt weld [5] and two possible interpolating curves proposed by Masubuchi and Martin [6]

16.5 Warm Prestressing

Postweld heat treatment (PWHT) is normally used to reduce, when not eliminate almost completely, welding residual stresses so that it is also called stress relief thermal treatment. But PWHT does more than simply that. It can also restore the metallurgical and mechanical properties that base metal has lost as consequence of the severe thermal insult suffered during welding and minimize the potential for hydrogen induced cracking. As consequence of a PWHT the steel becomes softer and more ductile through the precipitation of iron carbide. It must be understood that all these recovery benefits require different temperatures and holding times. To reduce the potential for hydrogen induced cracking it is adequate to maintain the metal at 230–290 °C for one hour per inch (25.4 mm) of thickness. About residual stresses, experience indicates that at a working temperature of about 500 °C they disappear after something as 20,000 h (more than two years), which is too long a period to wait.

$$
\begin{aligned}
&curve\ a\ \ \sigma_l = \sigma_o e^{-0.5(x/c)^2}\left[1-(x/c)^2\right] \\
&curve\ b\ \ \sigma_l = \sigma_o \frac{1-(x/c)^2}{1+(x/c)^4}
\end{aligned}
\tag{16.1}
$$

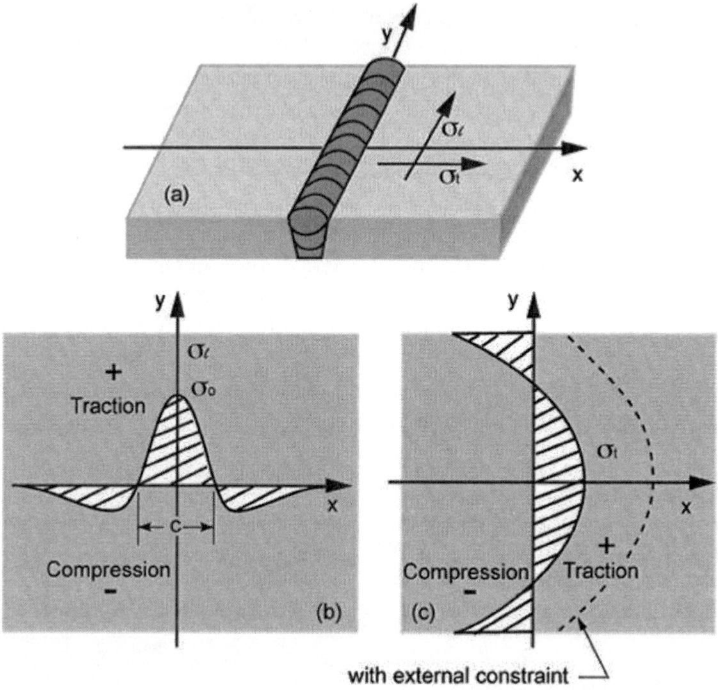

Fig. 16.9 Typical distribution of **b** longitudinal σ_l and **c** transverse σ_t residual stresses in **a** a butt weld (modified from [3])

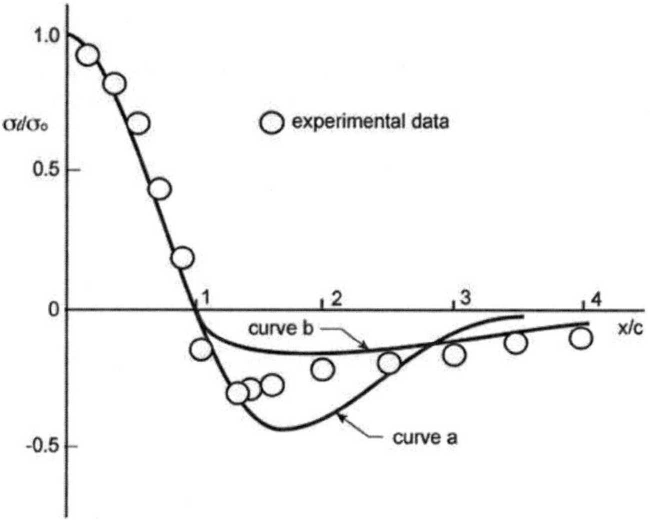

Fig. 16.10 Experimentally determined residual stress distribution in a butt weld (modified from [5])

in which σ_o is the maximum stress attained on the weld centerline and c the width of the tension zone, as shown in Fig. 16.9b. Therefore, the relief residual stresses it is necessary to go higher that 500 °C to at least 600 °C with PWHT unless the pieces to be welded have been preheated so that to reduce the level of residual stresses. Finally, to restore the metallurgical properties of the HAZ it is necessary for steels to go heat beyond 730 °C at controlled heating rate and then cool-down under controlled cooling rate.

Unfortunately, PWHT cannot be done on thin structures that would deform permanently nor on very wide structures that cannot be put in a furnace. In this latter case, on-site PWHT are performed through the use of oxy-acetylene flames on the surface of the welded areas. But this is not an easy job and a real controlled procedure cannot be guaranteed especially when the structure is exposed to strong marine winds. PWHT cannot be performed also on austenitic stainless steels because austenite sensitizes to stress corrosion cracking when maintained over 600 °C (see Sect. 18.2). As a valid alternative to PWHT it can be used a mechanical stress relief procedure. The technique consists of applying a mechanical traction to the welded piece. The effect is that described in Sect. 3.7.1. It is called prestressing since the stress is applied before the work-piece enters into service. Figure 16.11, taken from Kihara et al. [7], clearly shows the effectiveness of the method in reducing the residual stress state introduced in mild steel by welding. Any applied stress is effective (Fig. 16.12).

The amount of the reduction actually depends on the value of the applied stress, so that, with a tensile stress of only 50 MPa the measured reduction will be modest, curve A-50 of Fig. 16.11. But traction of 250 MPa will reduce the peak stress of about 310 MPa down to 60 MPa, as shown by curve A-250. The entire residual stress field has been sharply demolished. It is desirable to use the highest prestressing load practical without causing gross distortion or brittle failure. It goes without saying that the technique must be applied with caution. If a crack already exists in the weld the application of a rather high traction may lead to a catastrophic fracture especially because the HAZ has very poor toughness properties. Prestressing, then, must be applied at a temperature at which we are sure that the material will not undergo a brittle fracture, i.e., at least 20 °C over its transition temperature NDT (nihil ductility temperature), depending on the thickness of the work-piece. this kind of prestressing is called warm prestressing (WPS). Carbon steel and, in general all materials having a BCC structure have a brittle to ductile transition temperature NDT. At variance, FCC metals and alloys, like austenitic stainless steel, nickel and its alloys, aluminum and its alloys etc., have not a transition temperature. The practice of prestressing to relief residual stresses was introduced at the beginning of last century. According to Nichols, who conducted an extensive survey on the subject [8], probably the earliest reference to this effect is by Liljeblad in the Swedish ASEA Journal of 1937 [9].

Liljeblad has reported that ASEA have used it for fabricated beams, steam boilers and welded frames. It is, however, only in recent years that more specific and quantitative use of controlled WP has been proposed, with claims for residual stress relief, fatigue strength improvements and brittle fracture prevention. A detailed study of the effect was made in Japan by Kihara et al. [7, 10]. Their results obtained on tensile tests of notched specimens, welded specimens of 18 mm thick mild steel plate, are

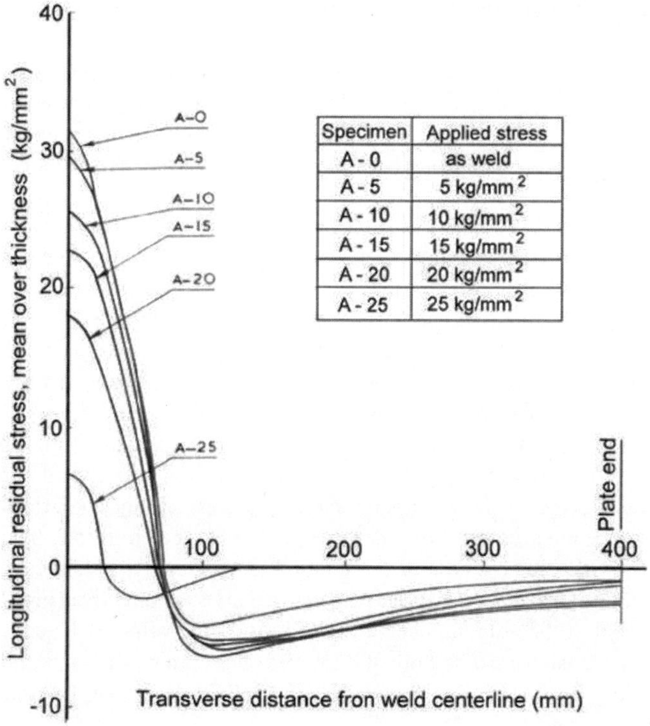

Fig. 16.11 Distribution of weld residual longitudinal stresses in as welded and mechanically stress-relieved condition (reported by [8] from [7])

shown in Fig. 16.16. In these tests, specimens were tested at – 25 and – 50 °C after they were loaded to various preloading levels at 20 °C, i.e., 20 °C over the transition temperature NDT of the material that was equal to 0 °C.

The specimens previously loaded at a temperature above the transition temperature only failed in subsequent low temperature tests when the applied stress was nearly equal or somewhat higher than the preloading stress. For example, a specimen preloaded at 230 MPa (level M 20) at 20 °C upon unloading, cooled to – 25 °C and then reloaded at that temperature failed at 252 MPa, a little bit over the prestressing load. Note that this specimen failed at about the same stress of a stress-relieved one and this indicates that residual stresses were removed almost completely by WPS. The same specimen tested at – 25 °C without WPS would fail at just 25 MPa. The higher the prestressing the higher was the failure stress in low-temperature tests until it approached the stress for general yielding. This stress was equal to that being needed to produce general yielding in other specimens which had been thermally stress-relieved. Since those early studies much has been done worldwide [11–15]. It has been also recognized that the use of periodic WPS would not only relieve residual stresses, but even increase the fracture stress with respect to the WPS level.

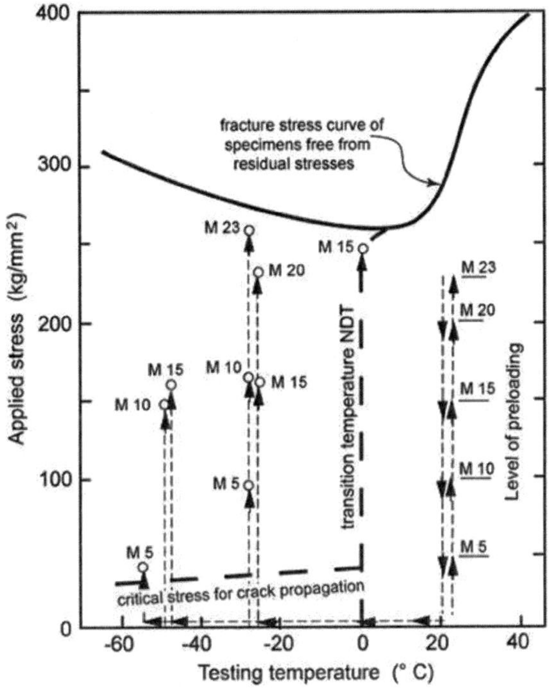

Fig. 16.12 Summary results of notch tensile tests on mechanically-stress-relieved specimens (modified from [10])

The mechanism by which WPS can effectively relieve residual stresses and improve the fracture toughness of the material even below the NDT of the material is recognized to be due to the plastic zone created ahead of the notch or crack possibly existing in the material. This plastic enclave introduced above the NDT of the material, at which it would not fail by brittle fracture, is then kept in the material and when it is reloaded at low temperature this *frozen* plastic zone would prevent fracture as if the actual temperature seen by the material would be that at which the plastic enclave was originally built-up. It has been recognized that the load history between WPS and fracture testing plays a role on the effectiveness of WPS. This is schematically shown in Fig. 16.13 [13]. The first two loading paths were investigated by Harrison and Farnehough [11]. In load path LCF the specimen is loaded (L), cooled (C) and then fractured (F) (never unloaded). In load path LCURF the specimen is loaded (L), cooled (C), unloaded (U) and then reloaded (R) to fracture (F). All specimens were prestressed at room temperature T_{WPS} and fractured at T_{test} = − 196 °C. In all cases, specimens sustained a load at fracture that they would never reach at − 196 °C, temperature at which they behaved fully ductile. Load path LU-CURF was investigated by McGowan [14] and was simulating a thermal shock on a pressure vessel with sudden depressurization and then re-pressurization. F. Loss et al. studied the last path LUCRF [16].

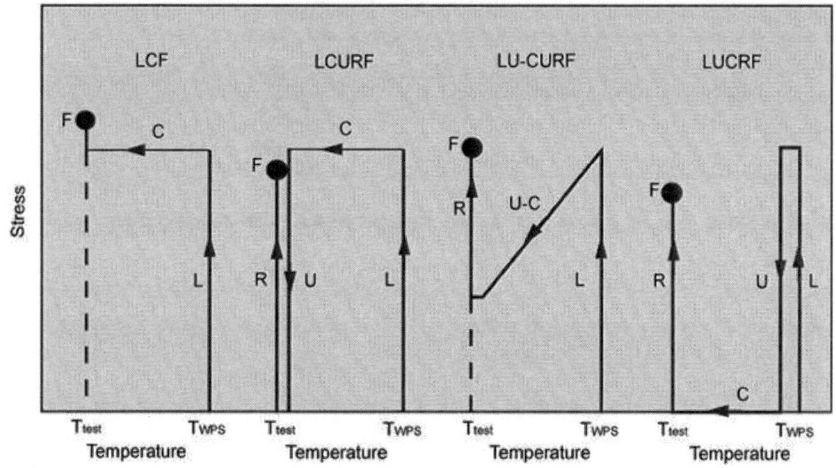

Fig. 16.13 Schematic illustration of loading sequences WPS effect (loading paths from several investigators [11, 14, 16])

16.6 S–N Curves of Welded Joints

It is rather difficult to provide S–N curves for welds, not just because there are many types of welds, but also because there are many different weld geometries and types (butt weld, fillet weld etc.). Generally speaking, a drastic decrease of the fatigue strength must be expected because of residual stresses and poor metallurgical quality of the HAZ. Figure 16.14 [17] provides a picture of fatigue strength of a fillet weld on steel BS 5360 Grade B, compared with those of a plate with and without an internal hole.

The plate with the hole suffers a relatively modest reduction in fatigue strength while the high cycle fatigue strength of the welded joint is reduced to about 10% of that of the base material. It is interesting to note how the fatigue limit reduction is accompanied by a shift of the toe of the *S–N* curve towards higher lives by a factor of ten, about 2×10^7 against about 2×10^6 of the base material. This is because the sharp reduction of the fatigue strength to very low stress amplitudes allows the metal to resist at higher number of cycles. This is also the sign that the weld was not containing cracks or large inclusions otherwise also the number of cycles at the toe of the *S–N* curve would reduce with respect to the plate without weld. Interesting is to see, Fig. 16.15 [17], how the strength of the material is affecting the fatigue strength at 10^6 cycles.

It can be seen that the fatigue strength of steel plate with or without the central hole increases with increasing tensile strength, while that of the welded joints is independent of the tensile strength. This fact should be taken into consideration in selecting the steel for welded structures. Looking at the *R* ratio effect, Fig. 16.16 [18] reports the results obtained by Maddox on fatigue strength of as welded steel

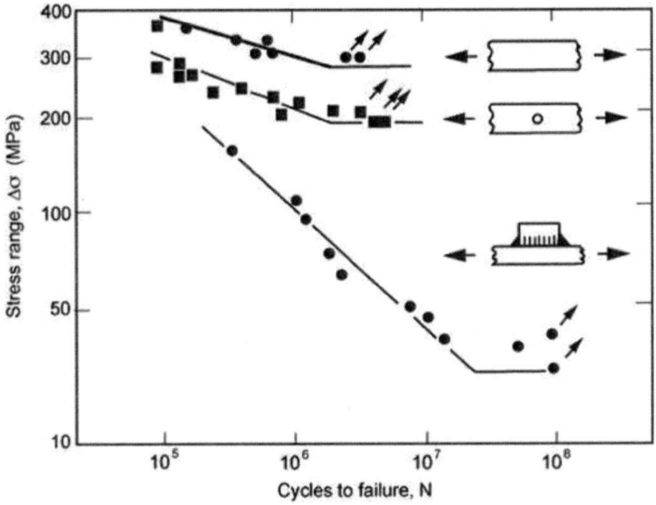

Fig. 16.14 Comparison of fatigue *S–N* curves of a plate with a fillet weld and a plate with or without a central hole (modified from [17])

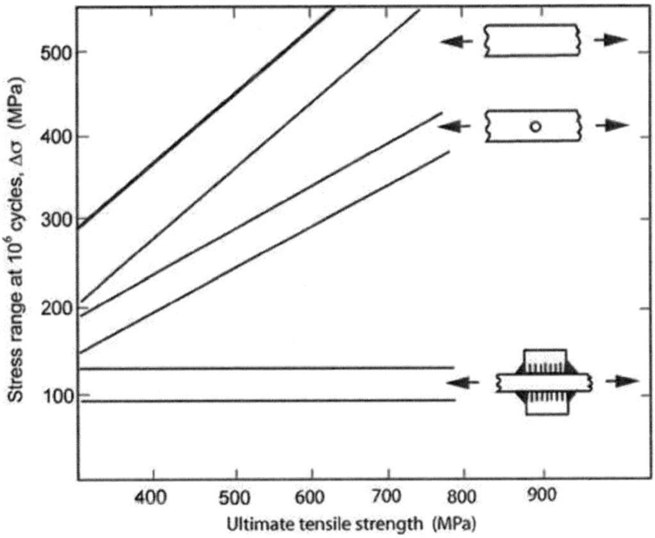

Fig. 16.15 Effect of tensile strength on fatigue strength of steel plates in different conditions (modified from [17])

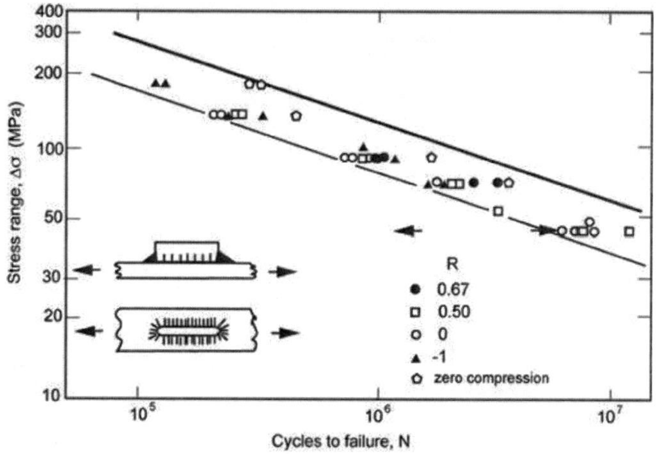

Fig. 16.16 Effect of applied stress ratio on fatigue strength of as-welded steel joints (modified from [18])

joints. As it can be seen, even if there is a compression load, the presence of tensile residual stresses will result in tensile cyclic stress being experienced at the weld toe. For this reason, the fatigue strength of joint in the as-welded condition is the same in tension and compression. The British Standard Institution [19] groups welds into Class A, B, C, D, E, F, F2 and G.

These classes and the relative fatigue design curves are discussed in [20, 21]. Figure 16.17 shows the large variation in fatigue *S–N* curves existing among the different classes presented on the side. Class G welds cover a large variety of poor welds. The one shown in Fig. 16.17 is just one of those types.

16.7 Weld Fatigue Improvement Techniques

Weld fatigue improvement technique are basically based on one or more of the following mechanisms

1. reduction of the notch stress concentration factor by improving the shape of the weld;
2. removal of pre-existing crack-like defects;
3. residual stresses relief.

As to point 1., the toe of the weld bead is the most critical point as crack starter because of its concentration factor that can be quite high in case of bad welds with large reinforcement or sharp corners.

With reference to the geometries and symbols of Fig. 16.18.

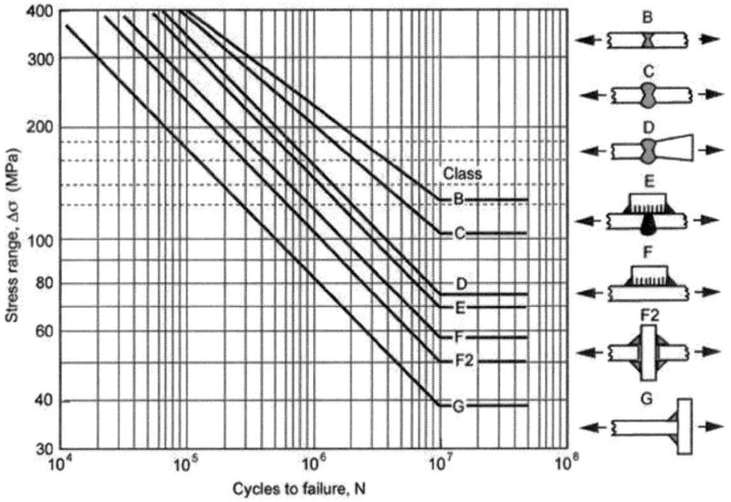

Fig. 16.17 Fatigue *S–N* curves for various class of steel joints (modified from [20, 21])

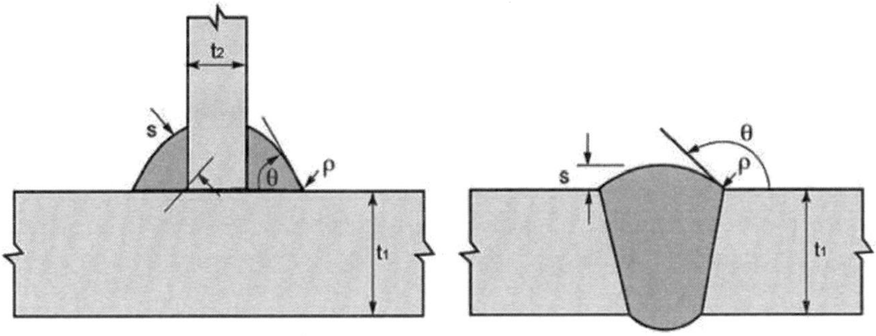

Fig. 16.18 Weld geometries and symbols

Niu and Glinka [22] proposed a stress concentration factor k_w for the notch-like toe given by

$$k_w = 1 + 0.5121 \cdot \theta^{0.572} \cdot \left(\frac{t_1}{\rho}\right)^{0.469} \tag{16.2}$$

where θ is in radians. Typical weld toe radii for manual welds are in the range of 0.1–0.2 mm. Figure 16.19 [23] shows that different welding processes results in different weld profile and this can dramatically affect the fatigue strength of the same weld metal.

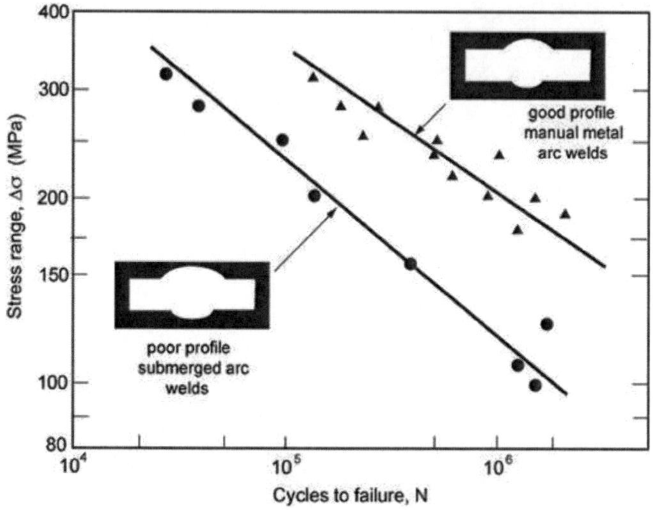

Fig. 16.19 Influence of type and weld profile on fatigue strength of welds (modified from [23])

16.7.1 Weld Grinding

To ameliorate the weld toe profile and to remove possible surface or near surface defects burr grinding can be used. Burr grinding is carried out using high speed pneumatic, hydraulic or electric grinder driving rotary burrs at a rotational speed of between 15,000 and 40,000 rpm. In full profile burr grinding the complete weld face is machined to remove surface defects, to blend the weld metal with the base plate and to reshape the weld toe reducing the stress concentration factor, as schematized in Fig. 16.19. The benefit of a burr grinding is estimated to improve the fatigue limit at 2×10^6 cycles from 50 to 200%, depending on the type of joint. A disc grinder can also be used to remove slag inclusions and undercut defects modifying the weld shape. The process is less time consuming and, thus, less costly. However an inexperienced operator may grind too much material and leave grinding marks that behave as cracks initiators. Therefore, disc grinding can result less effective than burr grinding and produce wider scattered results. A disc grinding is estimated to improve the fatigue limit at 2×10^6 cycles from 20 to 50%, depending on the type of joint. Figure 16.20 presents the results obtained by Shimokava et al. [24] from gusset weld joints specimens of 800 MPa steel, that can be classified as Class F2, in three different conditions; as welded, weld grinding finish and smooth design weld. It can be possible to see the improvement gained with the fillet smooth design with respect to grinder finish and, in particular, to the as welded specimen. Experimental results are compared with the BS Class B, F2 and G of Fig. 16.17. It must be remembered that grinding is not allowed on austenitic stainless steel since it may sensitize the steel to intergranular stress corrosion cracking and transform metastable austenite into brittle martensite (see Sect. 18.2).

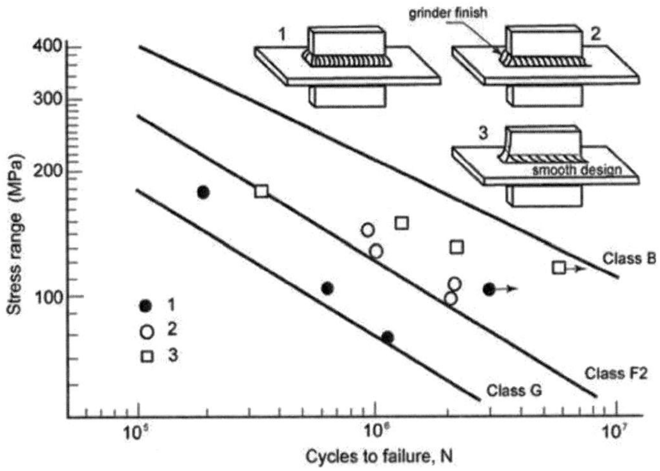

Fig. 16.20 Fatigue test results for gusset weld joints (modified from [24])

16.7.2 Water Jet Eroding

The water jet eroding technique uses a jet of high pressure water that contains abrasive particles. The jet is positioned at 45° to the weld axis. The abrasive particles erode the weld face material removing the weld toe area including possible undercut and surface defects and reducing the weld toe angle to provide a smoother transition, as with the grinding techniques. The rate of erosion is from 20 to 45 m/h (65–150 ft/h) as compared to 0.5–2.5 m/h (1.5–8 ft/h) for grinding. However, this fast rate of erosion requires special operator training and control since there can be a risk that just a little hesitation or modest slowdown of the rather fast moving water jet nozzle can be removing too much material.

16.7.3 Weld Toe Remelting Technique (TIG Dressing)

The remelting or weld toe dressing process is carried out using Tungsten Inert Gas (TIG) or plasma welding equipment. TIG Dressing has been traditionally used to describe weld toe remelting technique. Current terminology for TIG welding is Gas Tungsten Arc Welding (GTAW). Using this technique the weld toe is remelted to a shallow depth that improves weld toe shape, removes slag inclusions and toe undercut and increases the hardness of the HAZ. In high carbon C-Mn steel, the weld toe remelting process produces excessive hardness in the HAZ. Because of this, a second TIG pass procedure has been introduced that temper the weld toe region hardened by the first TIG run and results in a better transition between the weld and the base plate. The effectiveness of the TIG Dressing procedure is very dependent on the operator

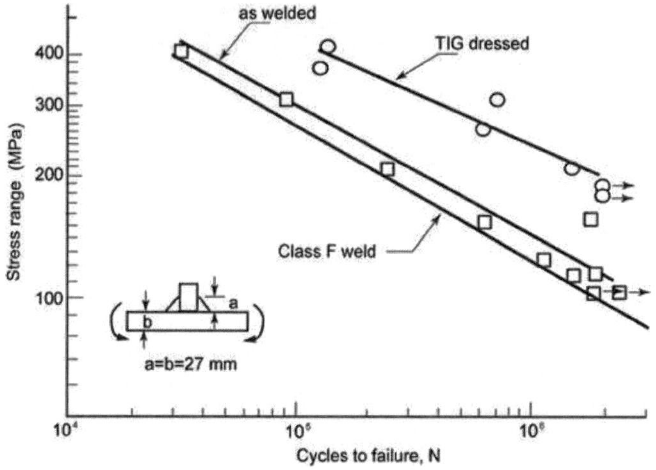

Fig. 16.21 Effect of TIG Dressing on the *S–N* curve of a medium strength steel (modified from [25])

skill. Typical results obtained from TIG Dressing on FG 47 CT medium strength steel, at load ratio R = 0.1, are shown in Fig. 16.21 [25], compared to BS Class F welds. Optimum conditions for TIG Dressing have been proposed by Kado et al. [26].

16.8 Fatigue Crack Growth in Welds

The problem of measuring the FCGR in welds and HAZ is that by cutting a small specimen out of the welded area residual stresses may go lost. These stresses, in fact, are self-balancing where the balance is due to the base material. Therefore, there must be enough base material in the specimen to be sure that residual stresses are still there in full amount. Figure 16.22 [27] schematizes the CT specimen cutting positions with respect to weld, HAZ and base material to make FCGR measurements from a butt weld. Figure 16.23 presents the results presented by Clark [28] for A 533 Gr B Cl1 carbon steel plate 15 cm thick containing a butt weld. After welding, the plate was given a PWHT at 620 °C for 40 h, followed by furnace cooling. Therefore, it must be assumed that the HAZ properties were restored and residual stresses removed almost entirely. In fact, it can be seen how the FCGR in base metal and HAZ are much the same. The weld metal, instead, behaves much better. FCGR obtained with CT specimens 2 inches (2T-CT) and 4 inches (4T-CT) thick, is by far lower than that of both the base metal and the HAZ, respectively, beyond 40 MPa$\sqrt{m}$. It is very interesting to note how in a case (full heptagons) the crack initially running in the HAZ at about 40 MPa$\sqrt{m}$ has changed its course entering the toe of the weld and following the FCGR of the weld metal.

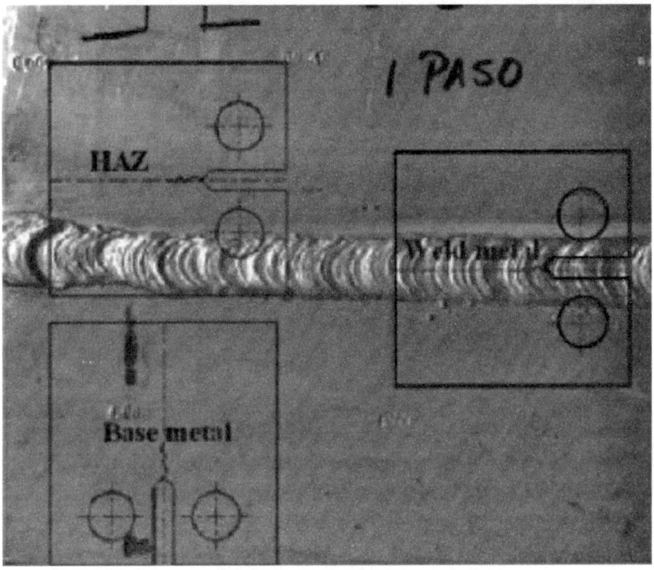

Fig. 16.22 Schematic of CT specimens position to measure FCGR in weld, HAZ and base metal [27]

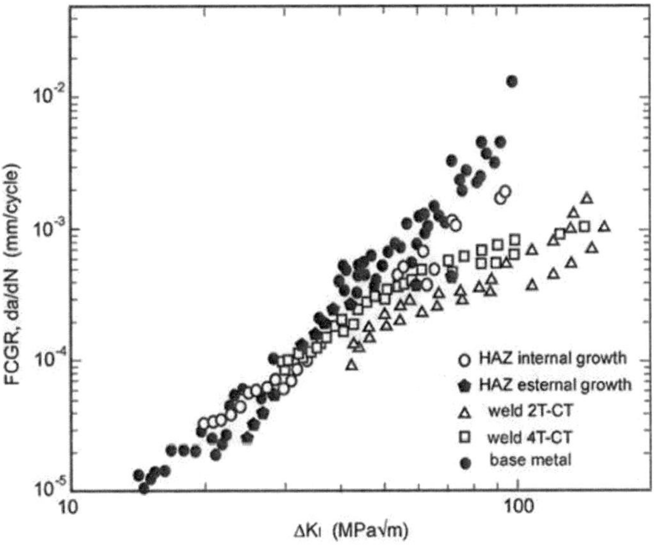

Fig. 16.23 FCGR, *da/dN,* for A 533 GrB, Cl1 carbon steel, HAZ and weld material (data from [28])

An interesting example of FCGR, *da/dN,* is presented in Fig. 16.24 [29] for a high strength carbon steel type F 514F, 745 MPa (108 ksi) yield strength and 789 MPa ultimate strength of 814 MPa (118 ksi) and two weld metals having yield strength of 680 and 820 MPa, respectively, and ultimate strength of 745 MPa (108 ksi) and 883 MPa (128 ksi), respectively, in as welded and PWHT conditions. The experimental data were obtained with CT specimens 3″ × 3″ × 5/8″. At variance with the previous case, now in the as welded conditions the FCGR, *da/dN,* in the weld metal is much lower than that of the base metal. Upon a PWHT at about 600 °C for one hour the weld metal worsened presenting a behavior similar to that of the base plate. To the best of my knowledge this is the only case where PWHT worsen the FCGR properties of a weld material.

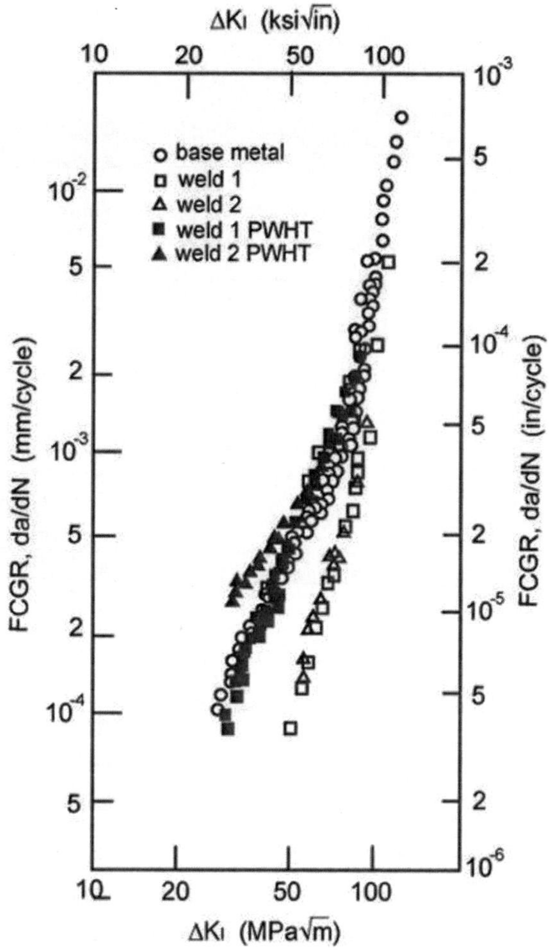

Fig. 16.24 FCGR, *da/dN,* for A514F steel and two welds, with and without PWHT (data taken from [29])

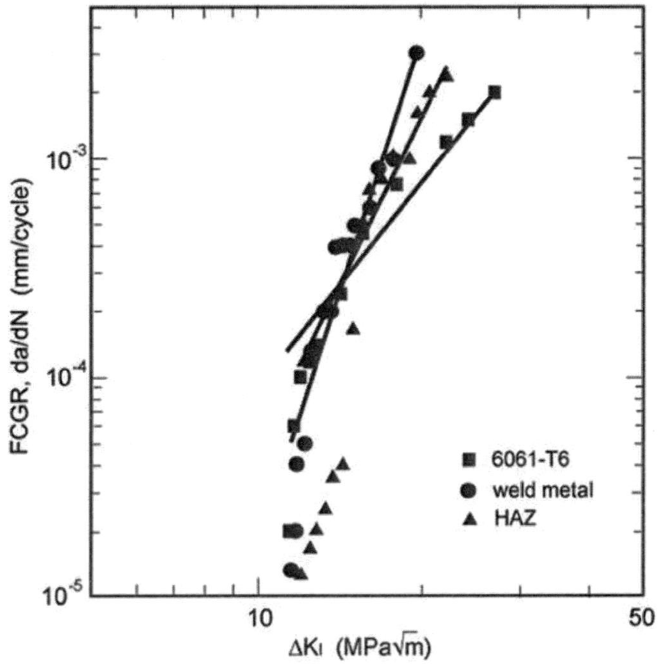

Fig. 16.25 FCGR, *da/dN,* for 6061-T6 aluminum alloy, weld metal and HAZ [27]

The reason is not known, but it is an experimental result that must be reported. Finally, Fig. 16.25 [27] presents the FCGR, *da/dN*, measured on aluminum alloy, type 6061-T6, specimens welded by modified indirect electric arc (MIEA). The tensile properties of the base material, HAZ and weld metal are shown in Fig. 16.26 [27]. It can be seen how the strength properties of the HAZ are poor compared to that of the weld metal and, in particular, of the base material. Nevertheless, the higher FCGR, *da/dN*, is that of the weld metal followed by the HAZ and base material. The FCGR, *da/dN*, obeys to a Paris law relationship

$$\begin{aligned} &6061-T6\ \tfrac{da}{dN} = 5.88 \times 10^{-8} \Delta K_I^{3.17}\ \text{mm/cycle} \\ &weld\ \tfrac{da}{dN} = 3.22 \times 10^{-13} \Delta K_I^{7.72}\ \text{mm/cycle} \\ &HAZ\ \tfrac{da}{dN} = 4.07 \times 10^{-10} \Delta K_I^{5.06}\ \text{mm/cycle} \end{aligned} \tag{16.3}$$

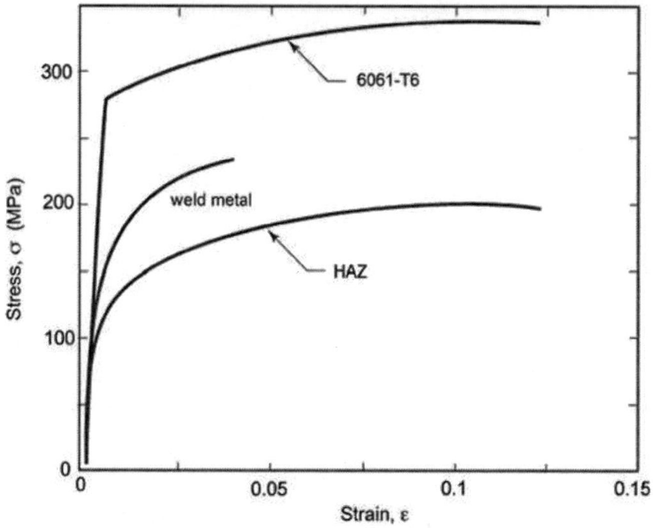

Fig. 16.26 Stress–strain characteristics of 6061-T6 base material, HAZ and weld metal [27]

References

1. Solidification and Phase Transformations in Welding. Suranaree University of Technology, Thailand (2007)
2. La Vecchia, G.M.: Il Giunto Saldato: Microstruttura e Principali Difetti (in Italian). University of Brescia, Department of Mechanical Engineering, Bresia, Italy
3. Masubuchi, K.: Control of Distortion and Shrinkage in Welding. In: Welding Research Council Bulletin No. 149 (April 1970)
4. UKAEA: An Assessment of the Integrity of PWR Pressure Vessels. 2nd Report by Dr. W. Marshall Study Group (March 1982)
5. Kanazawa, T., Oba, H., Susei, S.: The effect of welding residual stress upon brittle fracture propagation, Report II. Jpn Soc. Naval Arch. **110**, 359–368 (1961)
6. Masubuchi, K., Martin, D.C.: Investigation of Residual Stresses in Steel Weldments. Battelle Memorial Inst., Columbus, Ohio, Defense Technical Information Center (1998)
7. Kihara, H., Masubuchi, K., Kusuda, T., Iida, K.: IIW-X-219 (1959)
8. Nichols, R.W.: The use of overstressing technique to reduce the risk of subsequent brittle fracture. In: A Review Prepared for Commission X, vol. IIW, pp. X-409–67E (1967)
9. Liljeblad, R.: ASEA J. 23 (1937)
10. Kihara, H., Masubuchi, T., Iida, K., Oba, H.: Report to the IIW Commission vol. X, p. X-218 (1959)
11. Harrison, T.C., Farnehough, G.D.: The influence of warm prestressing on the brittle fracture of structures containing sharp defects. Trans. Am. Soc. Mech. Eng. **94**(Series D), 373–386 (1972)
12. Loss, F.J., Gray, R.A. Jr., Hawtorne, J.R.: Significance of Warm Prestressing to Crack Initiation During Thermal Shock. Naval Research Laboratory (September, 1977)
13. Loss, F.J., Gray, R.A. Jr., Hawtorne, J.R.: Investigation of Warm Prestressing for the Case of Small ΔT During a reactor LOCA. Naval Research Laboratory (March, 1978)
14. McGowan, J.J.: An Assessment of the Beneficial Effects of Warm Prestressing on the Fracture Properties of Nuclear Reactor Vessels Under Severe Thermal Shock. Westinghouse Nuclear Energy Systems, WCAP-9178 (March, 1978)

15. Stonesifer, R.B., Rybicki, E.F., McCabe, D.E.: Warm Prestress Modeling: Comparison of Models and Experimental Results. Material Engineering Associates, Inc., NUREG/CR-5208, MEA-2305 (1989)
16. Loss, F.J.: Structural Integrity of Water reactor Pressure Boundary Components. Naval Research Laboratory, NRL/NUREG-3512 (May, 1977)
17. Maddox, S.J.: An Introduction to Fatigue of Welded Joints, Improving the Fatigue Strength of Welded Joints. TWI (1983)
18. Maddox, S.J.: Fatigue assessment of weld structures. IIW Weld. World **32**, 33–42 (1993)
19. Code of Practice of Fatigue Design and Assessment of Steel Structures, BS 7608: 1993. British Standard Institution, London (1993)
20. Gurney, T.R.: Fatigue of Welded Structures, 2nd edn. Cambridge University Press, UK (1979)
21. Maddox, S.J.: Fatigue Strength of Welded Structures, 2nd edn. Abington Publishing, Cambridge (1991)
22. Niu, X., Glinka, G.: The weld profile effect on stress intensity factor in weldments. Int. J. Fract.Fract. **35**, 3–20 (1987)
23. Gordon, J.R.: Improving the fatigue life of welded structures. In: Fatigue Improvement Seminar. American Bureau of Shipping (April, 1993)
24. Shimokawa, H., Takena, K., Itoh, F., Miki, K.: Fatigue strength of large gasset joints of 800 MPa class steel. JSCE **1** (1985)
25. Haangsen, P.J.: Improving the strength of welded joints. In: Almar-Naess, A. (ed.) Fatigue Handbook. Tapir, Trondheim (1985)
26. Kado, S., et al.: Influence of the Conditions in TIG Dressing on the Fatigue strength in Welded High Tensile Strength Steels. IIW Document XIII-771-75 (1975)
27. Ambriz, R.R., Mesmacque, G., Benhamena, A., Ruiz, A., Amrouche, A., López, V.H.: Fatigue Crack Growth Behavior oin 6061-T6 Aluminum Alloy Welds Obtained by MIEA (2010)
28. Clarck, W.G., Ir.: Fatigue Crack Growth Characteristics of ASTM A 533 Grade B, Class 1 Steel Base Plate, Weld Metal and Heat-Affected-Zone. Westinghouse Research Report 69-7E7-BFPWR-R2 (1969)
29. Parry, M., Nordberg, H., Herzberg, R.W.: Fatigue Crack Propagation in A514 Base Plate and Welded Joints

Chapter 17
Corrosion

17.1 History and Overview

Very few events, if any, have such a devastating effect from the point of view of structural, economical and human casualties as corrosion. Virtually any metallic material is subjected to corrosion even in a clean air environment. Structures are continuously attacked and sometime even completely demolished by corrosion. The damage caused in the world to metallic structures by corrosion can be assessed in terms of billions of Euro every year. Corrosion is certainly a very complex and many-sided phenomenon hardly tied to a single parameter theory. A first, generic classification divides corrosion into *generalized* or uniform and *localized* corrosion. Both are fundamentally electrochemical processes, but while the first affects almost uniformly the entire surface exposed to the corrosive agent, the second attacks the metal locally and selectively. Generalized corrosion is very common in carbon steel where it results in the formation of the so called rust which advances very slowly so that it is effectively dangerous only in very thin materials or very long terms. Localized corrosion, instead, initiates locally, unexpected and once started it proceeds very rapidly along intergranular or transgranular paths to go through the thickness of a work piece in matters of minutes or days, at most. Initiation may last years, but growth is fast. Localized corrosion is a subtle and continuous process in which stresses play a role, which is referred to as stress corrosion or SCC (stress corrosion cracking) or fatigue. Materials, such as stainless steel, aluminum alloys, high strength steels and titanium alloys, previously considered to be corrosion resistant were found to be very susceptible to SCC. On the contrary, iron that is known to be susceptible to generalized corrosion may be more resistant to localized corrosion. The old metallurgists saying that iron is a gentleman perhaps is true. In this chapter only localized corrosion will be considered. The documented history of corrosion dates back more than a century when cavalry used to keep brass cartridges under saddles [1]. Cartridges were so exposed to ammonia vapors coming from horse urine. The environment revealed particularly corrosive also because coupled to a residual stress state coming

P. P. Milella, *Fatigue and Corrosion in Metals*,
https://doi.org/10.1007/978-3-031-51350-3_17

from cold-drawn manufacturing. Probably, the most scenic event was the failure of the Quebec Bridge on Saint Lawrence River in Canada happened in 1916. During the second attempt to raise into position the central span of the bridge, it fell into the river killing 13 workers. Figure 17.1 is a picture of the disaster [2]. The cause was found to be stress corrosion of the hoisting devices. The history of engineering counts thousands of corrosion failures of steel boilers exposed to caustic solutions, mild steels in nitrates, stainless steels or aluminum alloys in chloride environments (sea water) and martensitic steels in air and space applications. It is the combination of a specific material and a specific environment that determines the actual severity of SCC. However, many materials that are susceptible to SCC in one specific environment are susceptible to SCC in all environments. Very important questions are often without answer, such as: when and why corrosion occurs and how long will it take to initiate corrosion? In this chapter localized corrosion will be considered mainly from its characteristic electrochemical aspects. Without this basic information the treatment of SCC and fatigue assisted corrosion cannot be fully understood.

17.2 Physics-Chemical Aspects of Corrosion

From a general point of view besides the alloy composition and metallurgical structure, the most important physics-chemical parameters that control SCC are fundamentally five:

- electrochemical potential E,
- pH,
- concentration and type of chemical species present in the potentially aggressive environment,
- applied stress,
- temperature.

Their influence on SCC is very complex and can be generally expressed in terms of penetration depth versus time through an equation of the type [3]

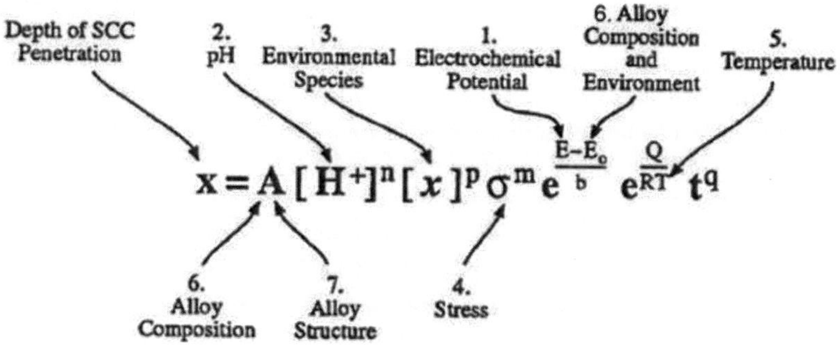

$$x = A\,[H^{+}]^{n}\,[x]^{p}\,\sigma^{m}\,e^{\frac{E-E_{o}}{b}}\,e^{\frac{Q}{RT}}\,t^{q} \qquad (17.1)$$

Fig. 17.1 Quebec Bridge accident on Saint Lawrence River in September 1916 [2]

in which the pH is expressed as hydrogen ions concentration H^+, Q and R are the activation energy and the perfect gas universal constant, respectively and n, m, p, q and r characteristic exponents. Certainly, the electrochemical potential E and the pH of the solution are, by far, the most important parameters in the SCC occurrence. Note that the term *solution* must be intended in the most general meaning. From the SCC point of view also water vapor (humidity) in the environment is a solution to which the material is exposed. Humid air is potentially aggressive. The functional link between the two parameters, E and pH, is expressed through the so called Pourbaix diagram [4] that indicates the thermodynamics equilibriums relative to all possible reactions between metal and solution and is normally given at ambient temperature, the so called normal condition. To understand the Pourbaix diagram it is necessary to recall some fundamental notions of inorganic chemistry. Corrosive processes involve both chemical and. Electrochemical reactions differ from chemical equilibria in that they involve the passage of free electric charges, or electrons, in the so called redox reactions. Chemical reactions are described by the Van't Hoff equation

$$\Delta G = -RT \ln K \tag{17.2}$$

in which ΔG is the variation of the Gibbs free energy relative to the passage from the first chemical species to the second, R is the ideal gas constant equal to 8.3144 J/(mol·K) or 1.987 cal/(mol·K), T the absolute temperature and K the equilibrium constant of the specific reaction. The equilibrium constant of a reaction involving two (or more) species A and B that react to form two new species C and D is a number that represents the ratio between the product of the concentrations or moles per liter of the two final species to that of the initial species, each to the power of the stoichiometric factor c, d, a and b of the chemical reaction

$$aA + bB \rightarrow cC + dD$$
$$K = \frac{[C]^c[D]^d}{[A]^a[B]^b} \tag{17.3}$$

where [A], [B], [C] and [D] are the concentrations of the various species. For example, in the reaction

$$H_2 + I_2 = 2HI. \tag{17.4}$$

The equilibrium constant K is given by

$$K = \frac{[HI]^2}{[H_2][I_2]}. \tag{17.5}$$

If the concentrations of the participating species are:

$[H_2] = 0.000862$ mol/liter
$[I_2] = 0.00263$ mol/liter
$[HI] = 0.0102$ mol/liter.

The equilibrium constant K will be

$$K = \frac{[HI]^2}{[H_2][I_2]} = \frac{(0.0102)^2}{(0.000862)(0.00263)} = 45.9. \tag{17.6}$$

The electrochemical reactions or redox reactions are defined by the Nernst equation

$$\Delta G = -RT \ln K + RT \ln \frac{[A]^a[H^+]^d}{[B]^b[H_2O]^c} \tag{17.7}$$

that refers to the generic reaction

$$bB + cH_20 \rightarrow aA + dH^+ + ne^- \tag{17.8}$$

in which ne^- free electrons are involved and a, d, c e d are the concentration of the species A, B, H_2O and H, respectively. The free energy variation (reduction) provides the work W that a thermodynamic system can perform during the transformation $-\Delta G = W$. In our case, the work performed is an electrical work and is given by the product of an electric charge times the electric field E of the redox reaction

$$\Delta G = n \cdot F \cdot E \tag{17.9}$$

where F is the Faraday constant that represents the electrical unit used in electrochemistry equivalent to 96,480 C. Equation (17.9) states that if n faraday moves between two electrodes under an electrical field E the work performed is precisely $-\Delta G$. Equation (17.7) can then be written as

$$E = -\frac{RT}{nF}\ln K + \frac{RT}{nF}\ln\frac{[A]^a[\mathrm{H}^+]^d}{[B]^b[\mathrm{H_2O}]^c}. \tag{17.10}$$

In Eq. (17.10) the term $RT/nF \cdot lnK$ is a constant for any given temperature and will be indicated as E_o, so that at ambient temperature (K = 298) taking the water concentration equal one, Eq. (17.10) becomes

$$\begin{aligned} E &= E_o + \frac{RT}{nF}\ln\frac{[A]^a[\mathrm{H}^+]^d}{[B]^b[\mathrm{H_2O}]^c} \\ &= E_o + \frac{0.0591}{n}\log\frac{[A]^a}{[B]^b} - \frac{0.0591 \cdot d}{n}\mathrm{pH} \end{aligned} \tag{17.11}$$

in which the natural logarithm has been converted into a decimal logarithm (ln = 23,026 log) and pH = $-\log[\mathrm{H}^+]$. The E_o values for the reaction are obtainable from the standard free energy change ΔG_o that are tabulated thermodynamic data. Note that that the argument of the second logarithm of Eq. (17.10) has the same expression of the equilibrium constant K of reaction (17.8), but the value of the concentrations [A], [B], [H_2O] and [H^+] are not those at equilibrium so that the two term at the right side of Eq. (17.10) have not the same value and their arithmetic sum is not zero besides that, as the reaction progresses, equilibrium conditions are reached. When this actually happens, the potential E becomes effectively equal to zero, $E = 0$, and the system can no longer make any work. Also note that in an *E-pH* diagram Eq. (17.11) represents a straight line so that the Pourbaix diagram is constructed by tracing the equilibrium lines of the various possible reactions. Figure 17.2 shows the Pourbaix diagram for low-alloyed carbon steel in water at ambient temperature and one atmosphere of pressure. In the diagram the only solid species considered are iron, Fe_3O_3, $Fe(OH)_2$ and $Fe(OH)_3$ [3]. It is obvious that other solid species exist, but are not shown because it is just an example of what a Pourbaix diagram is. Generally speaking, the Pourbaix diagram is not significantly altered by the value of pressure since thermodynamic properties are relatively insensitive to pressure, as compared to temperature. In Fig. 17.2b it is also shown the normal level of concentration of soluble iron ions in water that is equal to 10^{-6} g-atom per liter, often assumed as representative of complete absence of corrosion. In the Pourbaix diagram only three types of straight lines appear that can be horizontal, vertical or having any slope. In particular

(a) vertical lines refer to reactions that involve a solid substance, a dissolved substance and hydrogen ions water without free electrons and are, therefore, independent of the potential E (in Eq. (17.11) it is $n = 0$ and the slope is ∞);
(b) horizontal lines refer to reactions that involve a solid substance, a dissolved substance and free electrons in water, but without hydrogen ions and are independent of the pH (in Eq. (17.11) it is $d = 0$ and the slope is 0);
(c) any slope lines refer to reactions that involve a solid substance, a dissolved substance, free electrons and hydrogen ions in water and have a slope equal to $-0.0591\ d/n$.

The diagram of Fig. 17.2 identifies the equilibrium regions among four substances that are Fe, $Fe(OH)_2$, $Fe(OH)_3$ and Fe_2O_3, respectively, plus three in solution as ionic species that are Fe^{++}, Fe^{+++} and $HFeO_2{}^-$, respectively. Areas in which iron is present as ion and not in a solid species are those in which the metal is said to hydrate, therefore it corrodes.

There are others that are not shown as, for example, Fe_3O_4 or $Fe(OH)_2{}^+$. Solid thick lines limit the solid substances. Solid thin lines limit, instead, equilibrium fields between solid and ionic species. The area delimitated by the two lines (a) and (b) is that of thermodynamic stability of water with its oxygen, H_2O/O_2 and with its hydrogen H_2O/H_2, respectively. Above line (a) water is unstable with respect to oxygen evolution, therefore it releases oxygen. Below line (b) water is unstable with respect to hydrogen evolution, therefore it releases hydrogen. Corrosion is possible in the areas where iron is present as ionic species Fe^{++} or Fe^{+++}. In the other areas iron is stable and combined as a solid substance. The horizontal line (c), for example, refers to the equilibrium reaction to which a solid substance, Fe_2O_3, one in solution, Fe^{+++}, plus hydrogen ions H^+ in water participate, but without free electrons. The corresponding equilibrium reaction is

$$2Fe^{+++} + 3H_2O = Fe_2O_3 + 6H^+. \tag{17.12}$$

Assuming a concentration of H_2O and Fe_2O_3 equal one, the constant of equilibrium K of reaction (17.12) can be written as

$$\begin{aligned} K &= \frac{[H^+]^6}{[Fe^{+++}]^2} \\ \log K &= 6\log[H^+] - 2\log[Fe^{+++}] \\ &= -6pH - 2\log[Fe^{+++}]. \end{aligned} \tag{17.13}$$

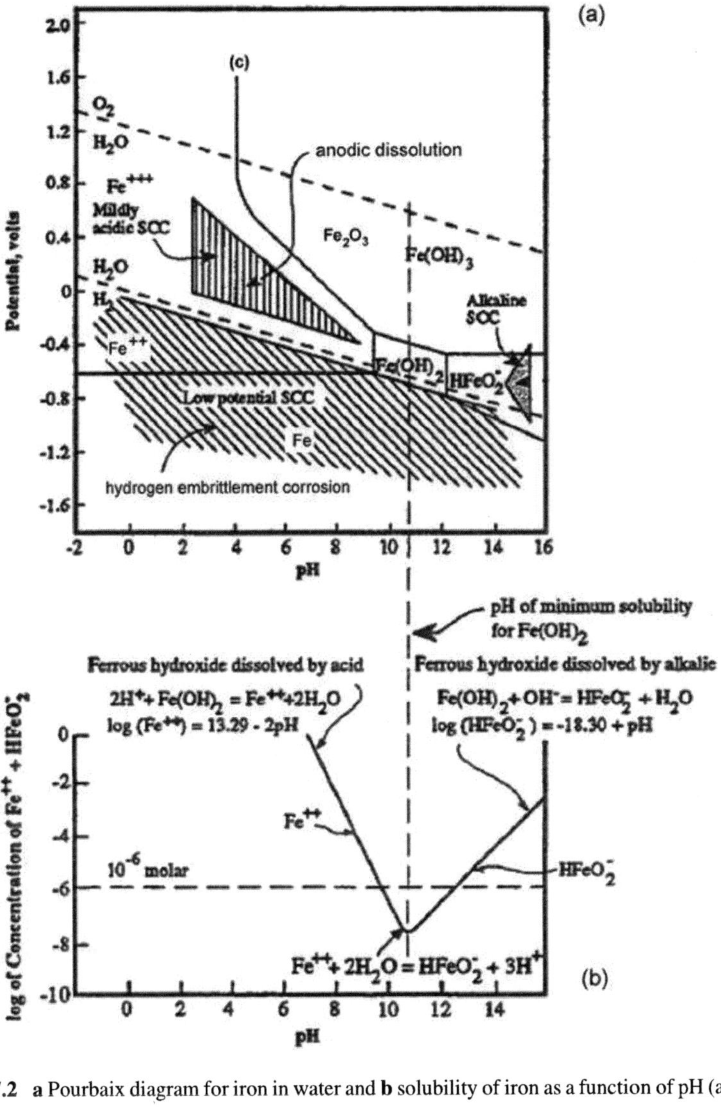

Fig. 17.2 **a** Pourbaix diagram for iron in water and **b** solubility of iron as a function of pH (adapted from [3])

By using the Van't Hoff Eq. (17.2) and tabulated thermodynamic data, it is possible to assess log (K)

$$\Delta G = -RT \ln K = -1.987 \cdot 298 \cdot 2.303 \log K$$
$$\log K = \frac{-1970}{-1.987 \cdot 298 \cdot 2.303} = 1.43 \tag{17.14}$$

Substituting in Eq. (17.13) the value of log (K) just found the Pourbaix generalized equation is obtained

$$\log(Fe^{+++}) = -0.0776 - 3\ pH. \qquad (17.15)$$

At equilibrium concentration of iron ions Fe^{+++} that is equal to 10^{-6} mol, a pH = 1.76 corresponds, which is indicated in Fig. 17.2 by the vertical line (c). In the area between lines (a) and (b), i.e., the one of stability of iron in solution as ions Fe^{+++}, there is the hatched triangle where mild acid SCC is possible as anodic dissolution (see Sect. 17.7). In this area metallic ions hydrate entering into solution as Fe^{+++}. In a symmetrical position with respect to the equilibrium line H_2O/H_2 it opens the area of possible anodic dissolution SCC, but this time iron enters into solution as Fe^{++} ions. In these zones corrosion is possible because the layer of oxide that forms on the metal surface is not stable, therefor is not passive-protecting, but results particularly soluble in acidic environment, as shown in Fig. 17.2b. Below this area it opens the area of the so called low potential SCC in which iron is stable, but the presence of free hydrogen ions may lead to hydrogen embrittlement stress corrosion cracking (see Sect. 14.4). There is, then, a small area where alkaline SCC is possible. This time the oxide that forms is $HFeO_2^-$ which is soluble, but only in a limited field of potential/pH combination, as shown in Fig. 17.2a and b. The two corrosions, the acidic and the alkaline environment SCC, are separated by a zone where iron hydroxide $Fe(OH)_2$ forms that is stable, showing a minimum of solubility right at pH10, as shown in the diagram of Fig. 17.2b. The diagram also shows that if corrosion from dissolution of iron as Fe^{+++} is possible, for example at a potential of 0.4 V and pH 4, the simple pH increase above 6 stop the corrosion because it enters the area of stable ferric hydroxide $Fe(OH)_3$ or Fe_2O_3 that passivizes the metal surface. It must be kept in mind that the Pourbaix diagram, even dough provides a very powerful tool for predicting SCC of steel in a water environment, is actually referring to a general water chemistry whose potential E and pH is under control, but this environment that may no longer exists in some restricted areas of the metal such as interstitial sites or crevices. In these tiny de-aerated zone local chemistry can be very different and may give rise to the so called crevice corrosion (Sect. 17.7).

17.3 Electrochemical Aspects of Corrosion

Ulrick R. Evans, the British scientist considered to be the *father of Corrosion Science* said that "corrosion is largely an electrochemical phenomenon, which may be defined as destruction by electrochemical or chemical agencies". The study of corrosion, therefore, cannot be done disregarding completely the electrochemical process at its base. In the electrochemical interpretation of corrosion, the progressive dissolution of the crystal structure in contact with a liquid, that is called electrolyte, is due to a continuous passage into solution of metallic ions, or ions that are hydrated, that feed an anodic current. For instance, if a zinc bar is immersed into water, some ions are hydrated and will leave the metal surface entering the solution as Zn^{++}

$$Zn \rightarrow Zn^{++} + 2e^{-} \qquad (17.16)$$

As a consequence, the metal acquires a negative charge because two negative free electrons e^{-} remain for each hydrated ion of Zn^{++}, while the solution immediately in contact with the zinc bar charges positively since enriches of Zn^{++} ions. Therefore, it forms a double electrical layer at the metal-water interface, i.e., a potential difference ΔV between metal (negative) and solution (positive). This double layer structure must be considered as the result of a dynamic process in that metal ions enter continuously the solution and abandon the metal while others move from the solution to the metal surface attracted by the negative charges, neutralizing, so that the potential difference ΔV is kept constant. This process can be seen as an elemental stage of corrosion. The tendency to abandon the metal as ionic species Me^{n+} depends on the potential difference ΔV that establishes between the metal and the solution. The higher ΔV, the higher is the predisposition to corrosion. However, at equilibrium, the metal that enters the solution as ionic species is extraordinary small (some billionth of gram) because the metal that becomes more and more negative, sends ions into solution with more and more difficulty. Coming back to the Nernst Eq. (17.7) in which the decrease of free energy provides the work that a system can do during a transformation, at the equilibrium the system is no longer capable of performing any work and, hence, is $\Delta G = 0$. Therefore, the spontaneous corrosion of a metal is practically null. To sustain corrosion, then, it is necessary metallic ions Me^{n+} entered into solution, through what is referred to as anodic reaction or oxidation reaction, and free electrons remained into metal are continuously removed. This is possible only if in the metal ions Me^{++} that leave the metal react with another ionic species of opposite sign existing nearby in the solution returning in a neutral state. The process is referred to as *metal reduction* because it reduces or completely nulls their valence. In the meantime, free electrons remained in the metal are captured oxidizing some other chemical species. It owes to these reduction–oxidation reactions or redox reactions that corrosion can proceed. More precisely, in electrochemistry it is defined.

(a) oxidation reaction any reaction, such as (17.16), also called anodic reaction, that produces metal ions (cations) that enter into solution. The area of the surface metal where the reaction occurs is called anode. The oxidation reaction releases electrons;
(b) reduction reaction any reaction, also called cathodic, by which a metal or non-metal or a species as hydrogen ions or oxygen molecules experience a reduction of their valence. The area of the surface metal where the reaction occurs is called cathode. The cathodic reaction absorbs electrons.

Corrosion, then, is the consequence of a couple of redox reactions. In particular, what is actually occurring in water can be expressed as

(1) anodic reaction: $Me \rightarrow Me^{n+} + ne^{-}$ anodic dissolution
(Me oxidation)

(2) water hydrolysis: $H_2O \leftrightarrow OH^- + H^+$
(3) cathodic reaction: $2H^+ + 2e^- \rightarrow H_2$
(hydrogen reduction)
(4) metal cations reduction: $Me^{n+} + nOH^- \rightarrow Me(OH)_n$
(5) prosecution of anodic dissolution of point (1): $Me \rightarrow Me^{n+} + ne^-$ (17.17)

Cathodic reaction (1) represents what has been called anodic dissolution of metal. Metal, in other words, dissolves by entering the solution or hydrating and this is precisely corrosion. As long as reaction (1) is sustained corrosion will continue. It is clear that to sustain reaction (1) metal ions must be removed as they hydrate otherwise an electrical barrier creates that arrests the process (potential difference ΔV between metal (negative) and solution (positive) becomes zero). In carbon steels, reaction (4) is one of the possible reactions that is doing this job. It leads to the formation of $Fe(OH)_2$ that is one of the constituents of the so called *rust*. Rust precipitates on the surface of the metal covering it, but without adhering so that the surface is not passivized and the reaction can continue till the complete transformation of metal into rust. Atomic hydrogen reduction produced by the hydrolysis of water, reaction (2), is the cathodic reaction (3) that contributes to the subtraction of free electrons produced by reaction (1), feeding the anodic reaction. However, the availability of hydrogen produced by the hydrolysis of water and that of OH^- ions is bound to the concentration of H^+ and OH^- ions in equilibrium with H_2O. By consequence, the rate by which hydrogen ions H^+ can subtract free electrons through reaction (3) or that by which OH^- hydride can reduce metal cations, which is necessarily the same, is linked precisely to that of water dissociation in ions H^+ and OH^- that under normal condition, pH ~ 7, is extremely modest. It is precisely the cathodic reaction that determines the rate of corrosion progress. This is why acids are so effective in the corrosion process. Even a light acidity that lower the pH from 7 to 5, for example, increases by a factor of 100 the concentration of hydrogen ions in the solution. Therefore, in neutral or basic conditions the metal dissolution is possible only if another cathodic reaction takes place. For example, another important reduction reaction is that provided by the oxygen in the solution

(6) cathodic reaction: $O_2 + 2H_2O + 4e^- \rightarrow 4OH^-$
(election suppression by oxygen reduction)
and ions OH^- production to feed reaction (4) (17.18)

or, in case of low oxygen

$$2H_2O + 2e^- \rightarrow H_2 + 2OH^-. \quad (17.19)$$

Cathodic reaction (6) of oxygen reduction is very important since it is this reaction that determines the effective corrosion rate at steady state. The higher the rate of elimination of anodic electrons, the higher the rate of anodic dissolution. If the cathodic reaction (6) is contained or completely blocked because there is no oxygen to be reduced the anodic dissolution is also blocked or, at least, is strongly reduced since it will be linked only to the hydrolysis of water. However, if from the one end the abundance of oxygen in the solution favors the continuation of the corrosion process, it may also stop it in steels and alloys that do not corrode by rust because it allows passivation, as it will be discussed later. Moreover, if there is oxygen availability in an acidic solution it will be possible also to feed another cathodic reaction of hydrogen reduction

$$4H^+ + O_2 + 4e^- \rightarrow 2H_2O \tag{17.20}$$

Table 17.1 is a synoptic picture of cathodic reaction that can take place depending on the pH of the solution. At low pH the solution is rich in hydrogen and it is hydrogen to be reduced. At medium or high pH it is oxygen that must be reduced.

Figure 17.3 schematizes reactions (17.17) and (17.18) for iron in water. Iron hydrates into Fe^{2+} ions and free electrons e^- while oxygen O_2 is reduced by electrons and reacts with water forming OH^- ions, according to Eq. (17.18). OH^- ions, in turn, allow iron reduction that precipitates as $Fe(OH)_2$ or $Fe(OH)_3$ (rust). The iron hydroxide $Fe(OH)_2$ that forms from reaction (4) in Eq. (17.17), in fact, precipitates because it is insoluble in water solution or is further oxidized into ferric hydroxide $Fe(OH)_3$ that gives rust that characteristic reddish color, through the reaction

$$2Fe(OH)_2 + H_2O + 1/2\ O_2 \rightarrow 2Fe(OH)_3. \tag{17.21}$$

Not all metals are equally subjected to corrosion. Better or lesser tendency of metals to hydrate and corrode can be assessed from their oxidation or reduction potential, also called redox potential. In general, the higher the oxidation potential the lesser the tendency to corrode. However, some metals corrode less than other metal with higher redox potential.

For example, chromium (−0.74 V), zinc (−0.76 V), titanium (−0.89 V), aluminum (−1.71 V) etc. withstand corrosion much better than iron (−0.42 V). This is due to the fact that the surface of these metals coats with an insoluble very thin layer, just a veil, of hard-bitten oxide not reactive at all that, at variance with rust, passivizes the surface blocking the prosecution of corrosion. Table 17.2 provides a synoptic picture of the standard potentials, the so called electrode potential, relative

Table 17.1 Possible cathodic reactions that feed metal corrosion

		Oxygen content in the solution	
		Low	High
pH	Low	$2H^+ + 2e^- \rightarrow H_2$	$4H^+ + O_2 + 4e^- \rightarrow 2H_2O$
	High	$2H_2O + 2e^- \rightarrow H_2 + 2OH^-$	$2H_2O + O_2 + 4e^- \rightarrow 4OH^-$

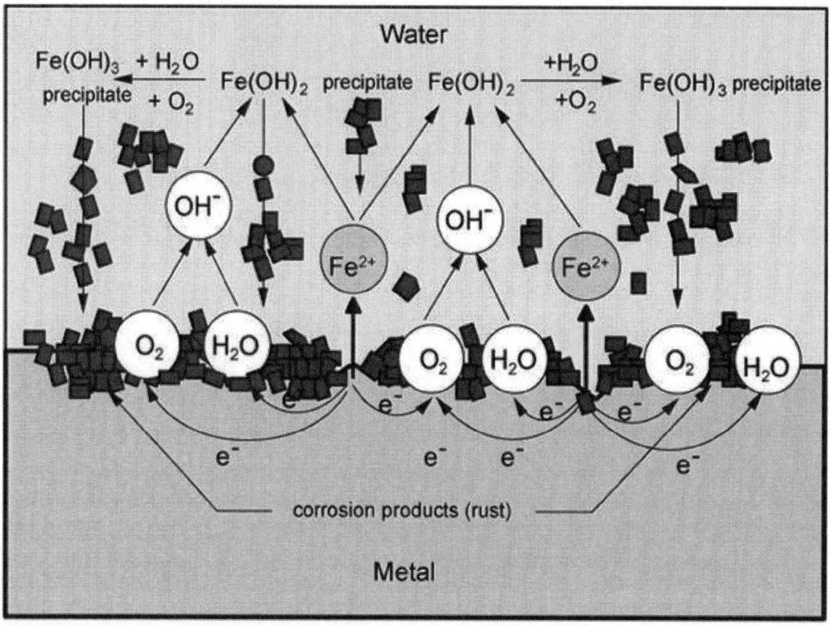

Fig. 17.3 Schematic view of the iron corrosion with rust production

to oxidation reactions of various metals. The standard electrode potential, abbreviated as E°, is given in volts and is the measure of the potential of any individual metal electrode which is with solute at an effective concentration of 1 mol/dm^3 at 1 atm of pressure. These potentials are referred to a hydrogen electrode whose reference potential is assumed equal to zero. This is because it is not possible to measure experimentally the value of the difference of potential ΔV between an electrode and its solution as, for example, in the case of zinc reaction (17.16), because any device used for making the measurement must be inserted in the circuit with two electrodes of which one is put in contact with the metal electrode of interest and the other with the solution.

Now, this second electrode creates necessarily another interface metal-solution and the potential difference provided by the system is that between the two metals, without any possibility to infer the absolute value of each of them. This is why it is necessary to introduce a reference electrode, which any other potential can be referred to. To the reference electrode, usually hydrogen, it is given a zero value of potential. Metals more reactive of hydrogen have negative potentials and act as anodes with respect to the former. These metals oxidize to form ions, reaction (1) in Eq. (17.17), whilst hydrogen ions are reduced by reaction (3). Metals less reactive than hydrogen have positive potential and are cathodic with respect to the former. In this latter case it takes metallic ions to be reduced to metal atoms, while hydrogen gas is oxidized to hydrogen ions.

Table 17.2 Standard electrode potentials at 25 °C (data from [5–7])

	Oxidation reaction (corrosion)	Electrode potential (volt)
Noble ↑ More cathodic (less tendency to corrode) ↑	$Au \rightarrow Au^+ + e^-$	+1.700
	$Au \rightarrow Au^{3+} + 3e^-$	+1.498
	$2H_2O \rightarrow O_2 + 4H^+ + 4e^-$	+1.229
	$Pt \rightarrow Pt^{2+} + 2e^-$	+1.200
	$2Hg \rightarrow Hg^{2+} + 2e^-$	+0.799
	$Ag \rightarrow Ag^+ + e$	+0.799
	$Fe^{2+} \rightarrow Fe^{3+} + e^-$	+0.788
	$4(OH)^- \rightarrow O_2 + 2H_2O + 4e^-$	+0.771
	$Cu \rightarrow Cu^{2+} + 2e^-$	+0.401
	$Sn^{2+} \rightarrow Sn^{4+} + 2e^-$	+0.337
↑	$H_2 \rightarrow 2H^+ + 2e^-$	0.000
↓ More anodic (higher tendency to corrode) ↓ Active	$Pb \rightarrow Pb^{2+} + 2e^-$	–0.126
	$Sn \rightarrow Sn^{2+} + 2e^-$	–0.136
	$Ni \rightarrow Ni^{2+} + 2e^-$	–0.250
	$Co \rightarrow Co^{2+} + 2e^-$	–0.277
	$Cd \rightarrow Cd^{2+} + 2e^-$	–0.403
	$Fe \rightarrow Fe^{2+} + 2e^-$	–0.440
	$Cr \rightarrow Cr^{3+} + 3e^-$	–0.744
	$Zn \rightarrow Zn^{2+} + 2e^-$	–0.763
	$Pb \rightarrow Pb^{2+} + 2e^-$	-1.260
	$Al \rightarrow Al^{3+} + 3e^-$	–1.662
	$Mg \rightarrow Mg^{2+} + 2e^-$	–2.363
	$Na \rightarrow Na^+ + e^-$	–2.714
	$Ca \rightarrow Ca^{2+} + 2e^-$	-2.760
	$K \rightarrow K^+ + e^-$	-2.920
	$Li \rightarrow Li^+ + e^-$	-3.040

17.4 Galvanic Cell

It comes as logical consequence of what has been said in the previous section and in Table 17.2 that if the different metals are put in the same solution they originate a bimetallic galvanic cell or voltaic pile from the names of the Italian philosopher, physicist and physician Luigi Galvani who first discover it in 1780 and the Italian philosopher and physicist Alessandro Volta who further investigate it in 1800 and invented the first electric battery. The cell is fed by the difference of potential between the two metals having different electrochemical potential following each individual anodic reaction. Figure 17.4 is a schematic of a galvanic cell. The two electrodes are iron and magnesium, respectively, electrically connected by a metal circuit. Inserted in the circuit there is a voltmeter. Table 17.2 shows that magnesium is more electronegative than iron. It will be the magnesium electrode, then, to corrode. On its surface the anodic reaction (1) of Eq. (17.17) causes the metal to hydrate (oxidation process), i.e., to dissolve into the solution while free electrons left in the metal migrate towards the cathode of iron where they recombine with the iron ions passed into solution (reduction process) allowing them to precipitate and return back on

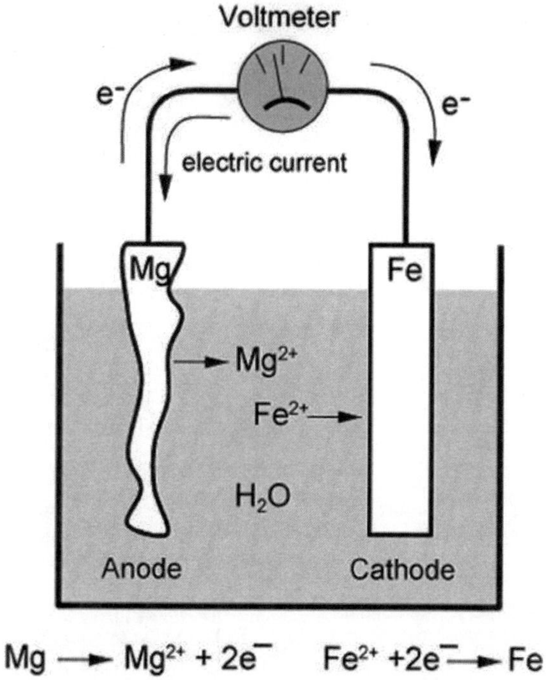

Fig. 17.4 Galvanic chemical cell with electrodes of iron and magnesium. The magnesium behaves as anode and corrode

the cathode as metallic iron. Therefore, the iron electrode will undergo a process of continuous iron hydration and iron ions recombination without any corrosion. The rection will continue till the complete corrosion of magnesium. The voltmeter will monitor this passage of electric charges (electric current) from cathode to anode by showing a difference of potential ΔV. In real life, as already said about magnesium, aluminum and zinc, in particular, the reaction may stop before the complete corrosion of magnesium.

This may happen when the magnesium ions entered into solution react with oxygen in the water forming an oxide that covers the electrode that becomes watertight and passives. To continue the reaction and the corrosion, the solution must be sufficiently acidic to dissolve this tight layer of oxide.

Note the galvanic cell can be obtained also using two different alloys of the same metal, but having different structure. Notorious is the coupling of ferritic steel with austenitic steel. Ferritic steel is electronegative with respect to the nobler austenitic steel. Figure 17.5 shows a galvanic corrosion test between two electrodes of ferritic steel and austenitic 304L steel, respectively. The two electrodes are in a 3.5% NaCl solution and are electrically connected. After 20 days the electrodes presented as shown in Fig. 17.4b and c. The austenitic stainless steel electrode, Fig. 17.4c, didn't show any corrosion sign, besides the light about which it will be said in the following. On the contrary, the ferritic electrode, Fig. 17.4b, is completely covered by rust.

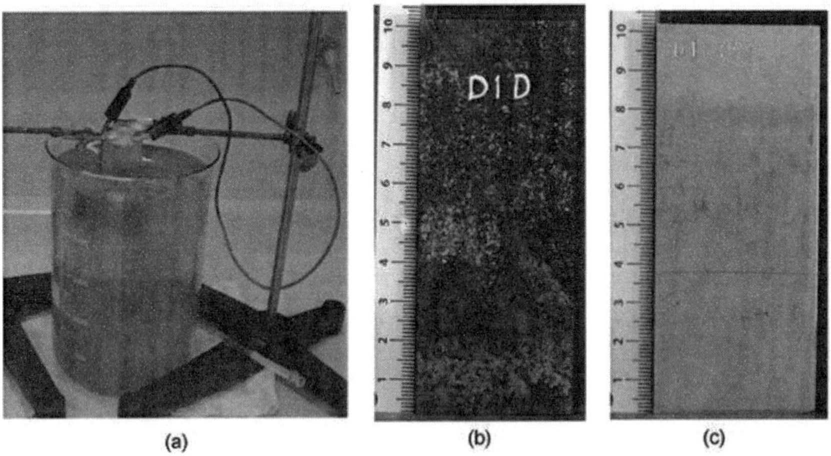

Fig. 17.5 **a** Galvanic cell with **b** anode of carbon steel and **c** cathode of austenitic stainless steel

The weight loss was of about 1.0 g/cm^2 against 0.01 of the other; a factor of 100 difference. On the other hand, a galvanic cell can be composed by coupling also two pieces of the same metal, but having different stories. This is the case of an iron cold worked electrode coupled to an annealed electrode. The former electrode becomes the anode to the latter that behaves as a cathode because of the larger dislocation density and residual stresses introduced by cold working. This, for instance, is the case of a cold-rolled container not annealed that can be joined or simply put in contact with the same metal, but annealed or the case of cold-rolled austenitic stainless steel work-piece on an austenitic stainless steel base. Austenitic stainless steel, in fact, cannot be annealed otherwise sensitization would occur.

Since many metals and alloys form passivizing surface films, as pointed out before about zinc, in a galvanic cell they may not have the behavior expected from the standard electrode potentials tabulated in Table 3.2. Therefore, for practical applications to know the effective anodic–cathodic behavior, the so called galvanic series have been developed, each referring to a specific environment. Table 17.3 lists galvanic series for same alloys exposed to running sea water. If the sea water is still potentials may change. This is the case, in particular, of stainless steel that in still sea water can become active as low-alloyed steels. It must be pointed out the need to develop such series for each particular working environment.

17.4.1 Cathodic Protection

Cathodic protection is a technique used to control the corrosion of a metal or alloy surface by making it the cathode of an electrochemical or galvanic cell. Cathodic protection was first described by Sir Humphry Davy in a series of papers presented

Table 17.3 Galvanic series of metals and alloys in sea water

↑ Noble $E^{\circ} > 0$ ↑ $E^{\circ} < 0$ ↓	Graphite Platinum (Pt) Titanium Ni–Cr-Mo alloy
Active ↓	Stainless steel Hastelloy C Monel Inconel Nickel (Ni) Cu-Ni alloys Aluminum bronze Copper (Cu) Tin (Sn) Yellow and red brass Aluminum brass α-brass Cast iron Low-alloyed steels Low carbon steels Aluminum alloys Zinc (Zn) Magnesium (Mg)

to the Royal Society in 1824 [8] and developed by his pupil Michael Faraday [9]. Common application are; steel water or oil and gas pipeline, storage tanks, pier piles, ships and boats, offshore oil wells and platforms.

Cathodic protection can be applied in two different way:

(a) using a DC power source that makes electrons flow on the area to be protected evoing the metal oxidation, as schematized in Fig. 17.6;
(b) using a sacrificial anode of a metal more active (electronegative) than the one to be protected electrically connected to this last, as schematized in Fig. 17.7.

Magnesium is one of the most suitable metals to be used since it appears to be one of the most active metals, see Table 17.3. At variance with air, a marine environment is enough corrosive to destroy the oxide film that otherwise would passive magnesium avoiding its use as sacrificial anode. Soil may be enough acidic to allow cathodic protection to be used. Typically, the lower the resistivity of a soil, the higher will be its corrosivity. Generally, soil resistivity decreases with increasing water content and ionic species concentration. Sandy soils, for example, have high resistivity and appear to be the least corrosive while clay soils are on the opposite end of the corrosive spectrum since they are excellent in retaining water. Therefore, before using any cathodic protection the electrical resistance of the soil must be assessed by appropriate methods. Zinc too is very efficient as sacrificial anode. Cathodic protection is really effective in preventing the pitting formation (see Sect. 15.2) that represents the initial stage of corrosion cracks.

However, it must be kept in mind that cathodic protection may not work if the system to protect contains deep crevices whose local electric potential, at the apex,

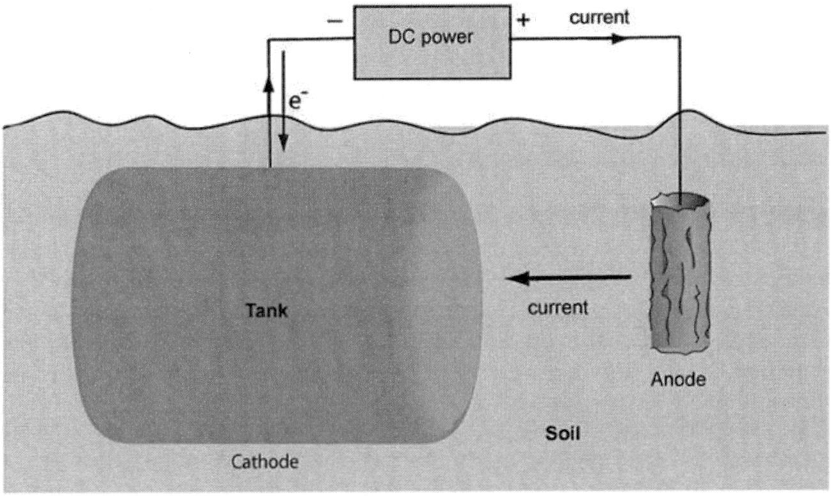

Fig. 17.6 Cathodic protection applied by using a DC power source

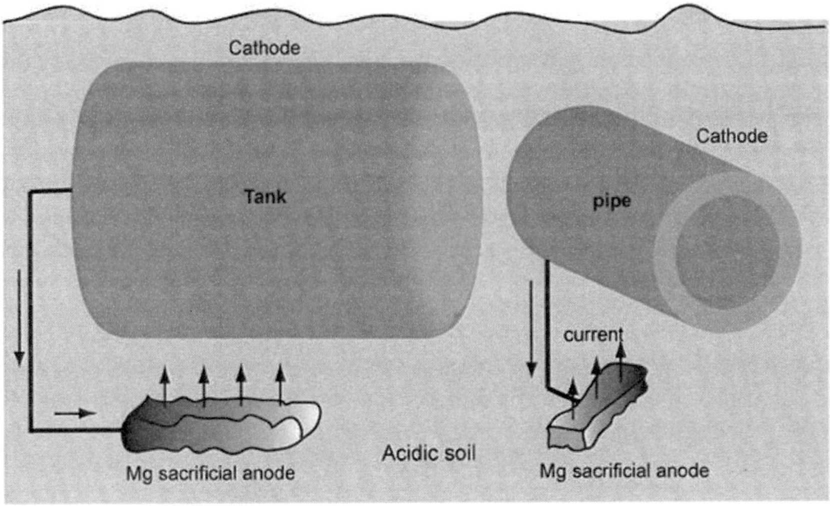

Fig. 17.7 Cathodic protection applied by using a sacrificial anode

is not affected by any type of protection, either DC or sacrificial anode. This actually means that once started, anodic protection cannot be interrupted. If a pitting event or crack initiate during the interruption, the restarting of the cathodic protection may not be effective any more on that pitting or crack.

17.5 Corrosion and Polarization Reactions

Assume to immerse a zinc anode in a hydrochloric acid solution. The anodic reaction is that already seen in Eq. (17.16)

$$Zn \rightarrow Zn^{2+} + 2e^{-} \qquad (17.22)$$

The rate at which Zn is dissolved as Zn^{2+} and, therefore, corrodes depends on the cathodic reaction (3) in Eq. (17.17) by which hydrogen is reduced taking electrons available from reaction (17.22). Note that in the case considered, hydrogen is coming in large quantities from hydrochloric acid. It must be observed that the working point of zinc in equilibrium with its ions, i.e., in the absence of corrosion, is that corresponding to its standard electrode potential $E^o = -0.763$ V, as shown in Table 17.2 to which it corresponds a current density $i_{o,Zn^{2+}/Zn}$, also called exchange current. When measured, this exchange current would yield the value of 10^{-7} A/cm^2. In a diagram electrode potential- exchange current, known as Evans diagram, to the above anodic reaction corresponds a point of coordinates (E_o; $i_{o,Zn^{2+}/Zn}$) indicated in Fig. 17.8 [10] by the letter A. Also the hydrogen reduction has its working point shown in Fig. 17.8 by the letter B. Its standard electrode potential is zero by definition, $E^o{}_{H2/H+} = 0$ V and the exchange current is equal to 10^{-10} A/cm^2. As the zinc corrosion starts in the acidic solution, the cathodic reaction is fed increasing the anodic current and its potential. Also the cathodic current increases, while the potential of the cathodic zone decreases so that the two trends will merge at the equilibrium point C of Fig. 17.8.

At this point the dissolution rate of Zn is equal to that of hydrogen ions production. The corresponding current density i_{corr} is called corrosion current of zinc in hydrochloric acid. Therefore, on the contact surface zinc-acidic solution microscopic galvanic cells are formed of the type shown in Fig. 17.3 in which the electrons made available by zinc that oxidizes reduce the acid determining a new equilibrium point in which anode and cathode have the same potential E_{corr}. The variation of the electrode potentials with respect to their equilibrium values that leads to the common value E_{corr} is called *polarization*. The thermodynamic *driving force* that sustains corrosion is given by the difference

$$\Delta E_{term} = E^o_{H_2/H^+} - E^o_{Zn/Zn^{2+}} = E^o_c - E^o_a \qquad (17.23)$$

between the cathodic equilibrium potential $E^o{}_c$ and the anodic one $E^o{}_a$. Figure 17.9 schematizes the same anodic dissolution process driven two distinct cathodic reactions of potentials E'_c and E''_c, respectively, with the equilibrium currents i'_{corr} and i''_{corr}. Me is the metal and O the oxidizing species. In general, the higher the difference between the thermodynamic potential E_{term}, the higher the intensity of anodic corrosion current i_{corr} will be, as shown in Fig. 17.9a and, hence, corrosion itself. The value of $\Delta E_{\mathrm{therm}}$ is not the only parameter controlling the corrosion rate. Figure 17.9b shows the same situation as in Fig. 17.9(a) except the two cathodic reactions possess

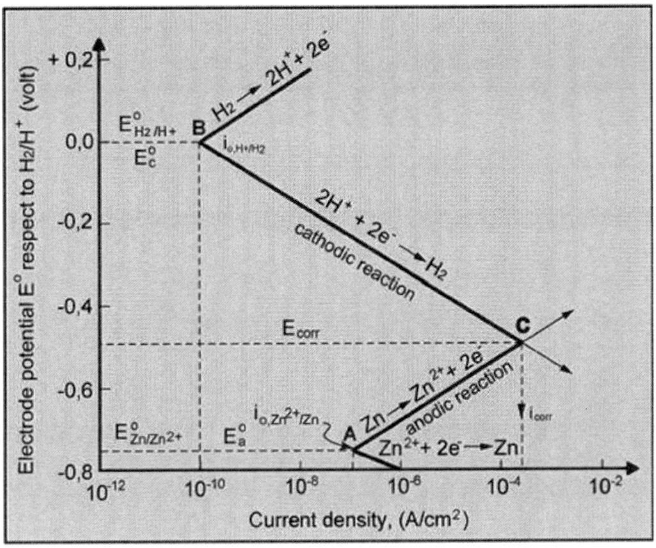

Fig. 17.8 Polarization diagram of a zinc anode in acidic solution (adapted from [10])

very different polarization characteristics. Despite the fact that $E''_{o,c} > E''_{o,c}$, the activation driving force $\Delta E''_{\text{therm}}$ is less than $\Delta E'_{\text{therm}}$ therefore, the corrosion couple with the largest thermodynamic driving force produces the lowest corrosion current. This situation often occurs for the corrosion of a metal in acid compared to its corrosion in dissolved oxygen. Even though the thermodynamic driving force is greater for corrosion in dissolved oxygen, corrosion often proceeds more quickly in acid. This is due to the slowness of the kinetics of oxygen reduction and can be appreciated by comparing the kinetic characteristics for the two processes on iron. Figure 17.10 schematizes the electrochemical corrosion of zinc in hydrochloric acid. It is clear that the corrosion process develops through successive stages. Hydrogen ions H^+, in fact, must migrate towards the metal surface where they meet free electron e^- and are reduced as hydrogen atoms. These atoms combine to form hydrogen molecules which, in turn, combine to form hydrogen gas bubbles. The slowest of these stages is actually controlling the cathodic reaction. It is obvious that any factor affecting the kinetics of hydrogen diffusion to the surface has a dramatic influence on corrosion. Therefore, temperature and the turbulent motion of the electrolyte by increasing the ions mobility can accelerate corrosion. Also the ratio of the cathodic surface to the anodic one is very important. A large cathode with respect to the anode increases the cathodic reaction that controls the corrosion rate.

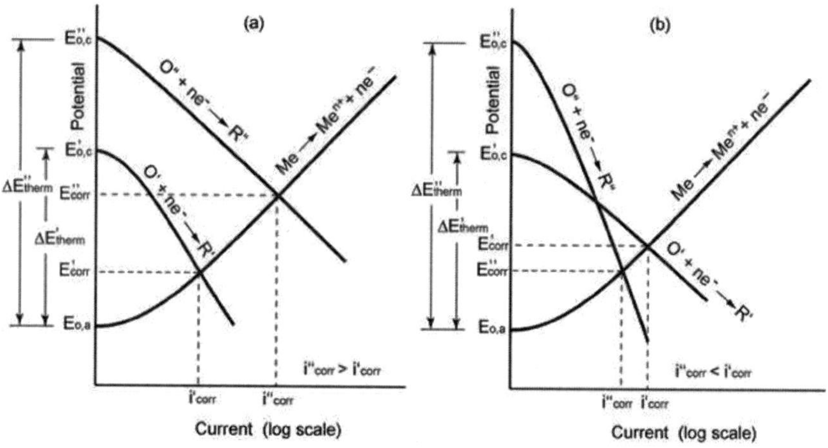

Fig. 17.9 Polarization diagram of anodic dissolution of a metal in two cathodic reactions with different ΔE_{therm}; **a** to higher ΔE_{therm} it corresponds higher anodic current, **b** higher ΔE_{therm} with lower anodic current

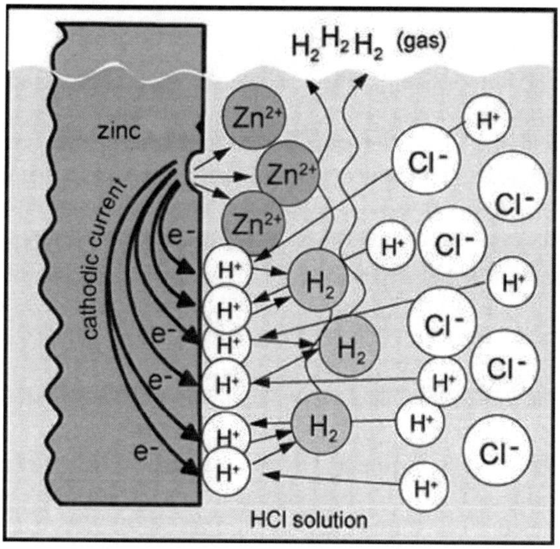

Fig. 17.10 Schematization of electrochemical dissolution of zinc in hydrochloric acid

17.6 Passivation

Imagine that the metal or alloy is exposed to a corrosive oxidizing environment as air can be or water. Its surface covers with a film of corrosion products (oxides) more or less tight. When this layer is so tight adherent to the surface protecting it from the prosecution of corrosion the metal is passivized. Many metals and alloy have this particular property that decreases their chemical activity becoming corrosion resistant in environments more or less oxidizing. Stainless steels, high strength steels, nickel, titanium, aluminum, copper, zinc and their alloys have this particular capability. Passivation can be evidenced through a polarization curve that characterizes both the reactivity and the passivity field of the metal in the particular environment, as shown in Fig. 17.11 [11]. In A the metal is at its equilibrium potential $E_{M^+/M}$ to which it corresponds an exchange current density $i_{oM^+/M}$. If its potential increases also the metal reactivity increases, i.e., the dissolution rate and, therefore, the current. The current increase is exponential and appears linear in a log scale. When the potential becomes E_{corr} (polarization potential) the current density i_a reaches a maximum to which it corresponds the highest corrosion rate. The formation of a passivation oxide film is observed to start near the peak B of the polarization curve AB and the potential of the peak is referred to as the critical primary passivation potential, E_{pp},. Beyond E_{pp} the anodic dissolution current decreases rapidly to reach a passivation value, i_{pass}, that can be even three or four orders of magnitude lower than the maximum i_a. It is clear that corrosion anodic current i_a must be balanced by a cathodic current of the same magnitude, therefore

$$i_a = -i_c = i_{corr} \tag{17.24}$$

This can happen at the common corrosion potential E_{corr} that must be

$$E_{o,H/H^+} < E_{corr} < E_{o,M/M^+} \rightarrow E_{o,c} < E_{corr} < E_{o,a} \tag{17.25}$$

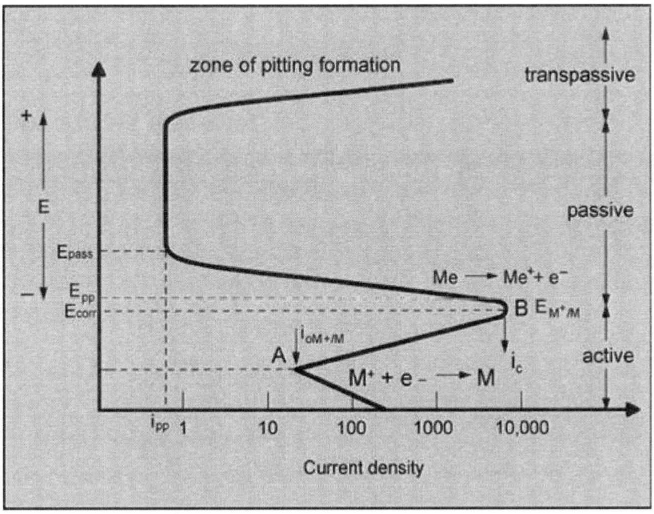

Fig. 17.11 Polarization curve showing passivation (modified from [11])

Let's consider, for example, the cathodic reaction of Fig. 17.12 [12]. Its equilibrium potential $E_{o,c1}$ is lower than E_{pp}. Since, according to Eq. (17.24), when the corrosion potential $E_{corr,1}$ is between $E_{o,c1}$ and $E_{o,a}$ a corrosion couple is always created, it happens that the condition $E_{corr} > E_{pp}$ is never reached and a corrosive event takes place, though at low power (low $i_{corr,1}$). The same happens when reaction (2) in which the potential remains in the active region of the polarization diagram, but if reaction (3) is considered it happens that the condition $E_{corr} > E_{pp}$ is respected and the metal passivizes. In the men time the corrosion current precipitates to its minimum passivation value i_{pp}. The increase in potential through the passive region is accompanied by a progressive thickening of the oxide film formed on the metal surface such that the electric field within the oxide, and therefore the dissolution current, remain constant. This current is controlled by the rate of dissolution of the film (a chemical, as opposed to an electrochemical, process) and is potential independent. The current is just sufficient to replace the dissolving film. Any disruption of the passive film is a dangerous situation, and film breakdown at localized points leads to the initiation of such localized corrosion processes as pitting and cracking (see Sect. 15.2). These processes are characterized by very high local rates of metal dissolution and can lead to very rapid penetration of metal.

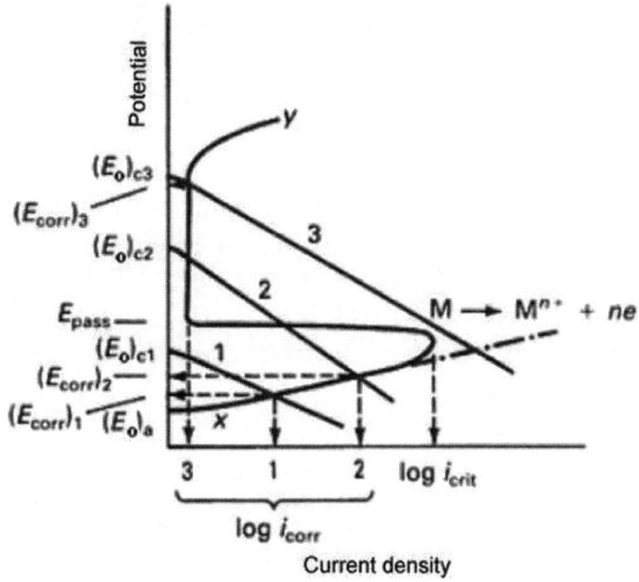

Fig. 17.12 Polarization curve for a metal/metal ion system that undergoes an active to passive transition [12]

17.7 Crevice Corrosion

A particular anodic dissolution process is that associated with crevices and in general with deep and narrow contact regions of stagnant solution under a stress state. These regions may have a local chemistry much different from that interesting the work piece in which they are present. This favors the development of an accelerated and localized corrosion process known as crevice corrosion. Crevice corrosion may be expected in clamped interfaces, lap joints, gasket joints, rolled joints or under bolt and rivet heads or any deep and narrow structural discontinuity. It has been said how oxygen is important in the corrosive process of the metal, see Eq. (17.18) and Table 17.1, but in the crevice corrosion it is also its concentration gradient between the innermost area of the crevice and the mouth that plays a fundamental role. The deeper it goes into the stagnant crevice region the more the oxygen depletion is encountered, compared to the mouth where the oxygen concentration depends on its solubility at the working temperature. The solution can be water or simply humidity condensation. This difference in the oxygen concentration is responsible for the formation of a galvanic cell, see Sect. 17.4. The cell is activated by a redox reaction of oxygen reduction/metal oxidation, described by the theory of differential aeration introduced by the British scientist Evans in 1924 [13]. This theory is based on the observation that the oxidizing power of oxygen increases with increasing oxygen pressure, i.e., concentration. Therefore, if two areas of a wet metal have different oxygen concentrations, the oxygen rich zone becomes positive (cathodic) compared to the other. The point is that the crevice sides can be passivized by the tight oxide film that covers them. Therefore, the working hypothesis is that in the deepest point of the crevice that may be similar to a sharp notch, if not a real crack, the plastic deformation caused by the stress state may break the passive film that protects the metal. This protective film is made, in general, of oxides being extremely brittle. Its rupture, therefore, can happen under very small deformations of the order of 10^{-3} microstrains. This causes the tip of the crevice to be exposed to the potentially aggressive environment, as schematized in Fig. 17.13.

The process is referred to as the film rupture/anodic dissolution mechanism because it is activated by local slip, as shown in Fig. 17.13. It is precisely there, at the tip on the bare metal, that the anodic dissolution takes place, reaction (1) of Eq. (17.17). This causes the crevice to become a crack. At the mouth of the crevice crack oxygen is reduced by the cathodic reaction (6) Eq. (17.18). The crevice corrosion progression is based on a delicate thermodynamic equilibrium between crack tip propagation, the dissolution rate on the two adjacent crack sides and the atomic mobility of oxygen inside the crevice. This is shown schematically in Fig. 17.14.

Once the passive film is broken, the crack tip must propagate faster (V_{ct}) than the dissolution rate on the adjacent crack sides (V_{cs}), $V_{ct} >> V_{cs}$, as shown in Fig. 17.14b. In fact, if the dissolution rate of the crack sides is too high because there is too much oxygen penetration and the passive oxides are not though-tight to the surfaces, then $V_{cs} >> V_{ct}$, Fig. 17.14a and the crevice becomes larger and larger, the tip blunts. The crevice is no longer a crack which reduces the tip strain and the tip passivation

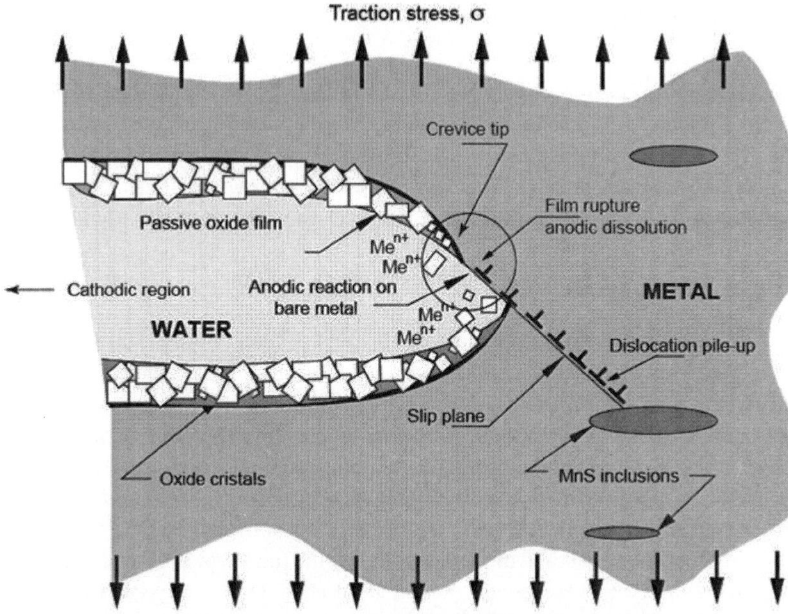

Fig. 17.13 Schematic of the passive film rupture with activation of the anodic reaction

rebuilds. More and more oxygen can enter to feed the process and to increase the tip oxidation process, stopping the tip dissolution. Under these conditions the reaction stops and the cell switches off. The crack speed must never go below the oxygen diffusion rate which depends on oxygen pressure and temperature. Therefore, there will be a temperature range in which crevice corrosion can happen.

The lower limit marks the temperature below which the atomic mobility is too low to move ions inside the crevice and the pile switches off. The upper limit, instead, identifies the temperature at which the few oxygen available in the water is so mobile to quickly enter the crevice and switches off the pile. Figure 17.14 also schematizes the galvanic cell that creates between the tip and the mouth of the crevice. At the tip, once the passive film is broken, the metal hydrates entering the solution as ion, Me^{n+}. The free electrons made available move to the crevice mouth as cathodic current where they make possible the reduction of oxygen that produces OH^- ions that enter the crevice and feed the anodic reaction; the galvanic cell is closed. Situation depicted in Fig. 17.14a usually happens in iron or low-alloyed carbon steels that are not resistant to generalized corrosion.

In these materials the corrosion of the lateral sides of the crevice is faster than the possible crack tip growth, $V_{cs} > > V_{ct}$, Fig. 17.14a, because the oxide layers are not hard-tight and do not passivize; corrosion can continue. At variance, in high-strength steels, stainless steels, aluminum and titanium alloys that are corrosion resistant, the surface oxide is thermodynamically stable and hard-tight. Once it forms a very thin

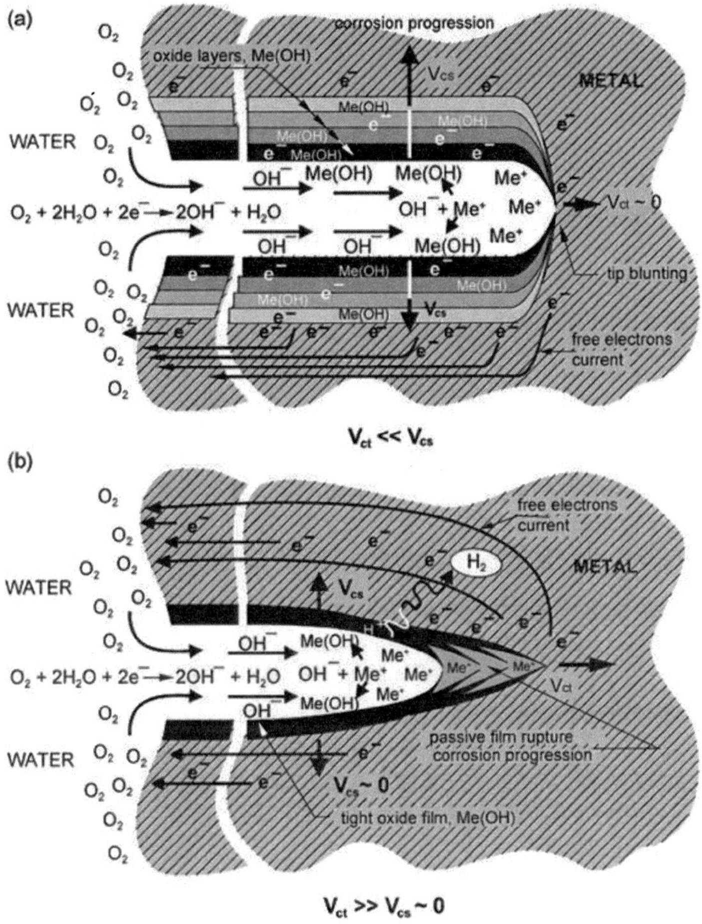

Fig. 17.14 Schematic of the macro cell that originates between the mouth and the tip of a crevice. In **a** the lateral sides dissolve faster than the tip. The opposite happens in **b** allowing corrosion to drive the crack

film it passivizes the metal. In these alloys if the crevice tip passive film is broken by tension stresses, Fig. 17.13, the crack tip remains sharp and $V_{ct} >> V_{cs}$. It is clear, from what has been said, that the stress corrosion crack growth is not uniform, but happens through consecutive elemental acts of acceleration-deceleration. In fact, the process starts with a strong initial acceleration, once the tip passive film is broken and the tip dissolution process takes place. Then, the few available oxygen will start to oxidize the bared tip trying to passivize it. During this phase the metal dissolution slows down and the crack decelerates. But, then, the passive layer not completely restored is broken again and the process will continue. It is also clear that if the stress state is not high enough the passive film is not broken. The crevice acuity is also important. Without a sharp notch it may be difficult to have such a high stress state. This oxygen effect is well documented in Fig. 17.15 [14] for a high-strength steel type H-11, 1610 MPa yield strength, tested in humid nitrogen environment. It is interesting to note how each oxygen injection blocks corrosion arresting the crack growth. Upon oxygen removal, corrosion starts again making the crack grow with the same rate as before.

Interesting enough is the diagram of Fig. 17.16 obtained using experimental data obtained by Indig [15, 16] and McDonald et al. [17] for carbon steel that shows the variation of the corrosion potential with the concentration of oxygen in water solution at various temperatures. It can be seen that as the O_2 concentration increases also the corrosion potential increases entering to positive field for 10 ppm of O_2, noble potentials, allowing corrosion only at very low pH (acidic environment), as it can be seen from the Pourbaix diagram of Fig. 17.2. It is recalled that O_2concentration in water at ambient temperature and 1 atm is ~ 8 ppm.

Figure 17.17 shows the susceptibility of carbon steel type A 508 Cl.3 in high purity demineralized water and low oxygen concentration ($0_2 < 0.2$ ppm) as a function of temperature. The corrosion potential refers to the SHE (standard hydrogen electrode) [18]. Note that the minimum corrosion potential, below which there is no more susceptibility to corrosion, is obtained at a temperature between 200 and 250 °C. For higher temperatures and, above all, lower temperatures the potential slips towards noble values that make corrosion more difficult to happen. In the temperature interval between 200 and 250 °C any electrochemical potential over – 200 mV(SHE) is, in practice, capable to feed a stress corrosion mechanism.

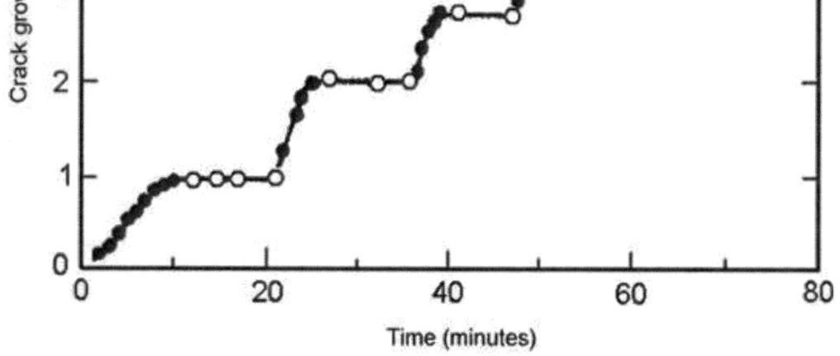

Fig. 17.15 Oxygen effect on the corrosion crack growth rate for steel type H-11 in humid nitrogen [14]

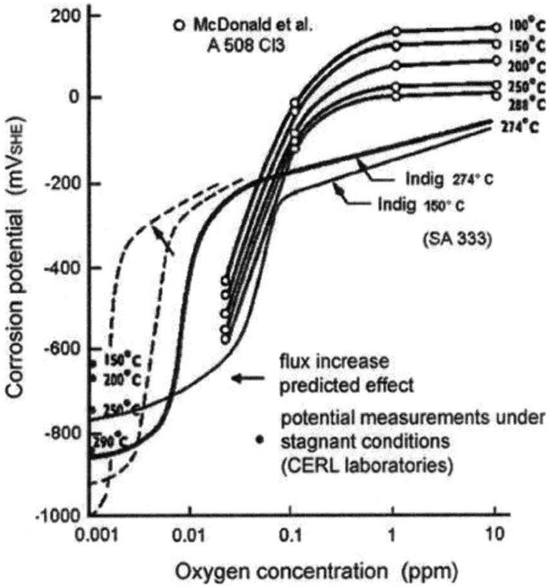

Fig. 17.16 Corrosion potential as a function of oxygen concentration in water and temperature for carbon steel [15]

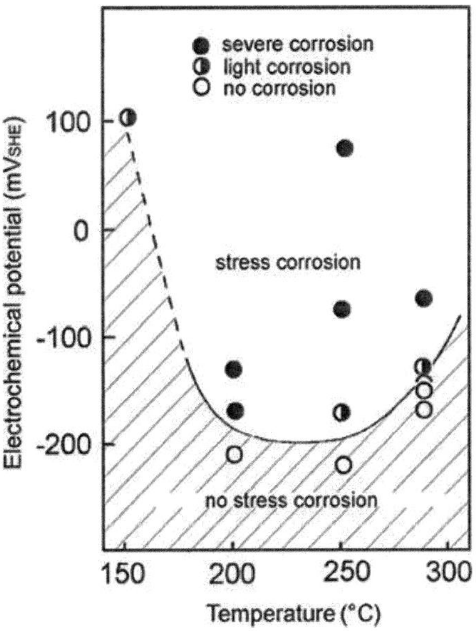

Fig. 17.17 Critical electrochemical potential versus temperature in A 508 Cl1 steel [18]

17.8 Corrosion Crack Growth Models

It has been shown, in the previous section, the anodic dissolution/passive film rupture model from a phenomenological point of view. In this section it will be discussed the same model from an analytical point of view. In the anodic dissolution model, two hypotheses must be done to make corrosion advance into metal. The first is that the crack sides or flanks are protected against generalized corrosion, i.e., that $V_{ct} > > V_{cs}$, as shown in Fig. 17.14b. If this hypothesis is not respected there will be a continuous crack mouth opening and crack tip blunting that will switches off the pile and the crack stops. It will be seen, in Chap. 15, that the continuous passive film rupture can be facilitated by a superposed fatigue cycling that helps localized corrosion even when the applied constant stress and the couple environment/material are oriented towards the condition $V_{ct} << V_{cs}$, see Fig. 17.14a. In these conditions, in fact, it can happen that the total stress plus fatigue crack growth rate is higher than the crack sides dissolution reestablishing the condition $V_{ct} > > V_{cs}$. The second condition, very important, is that strain accumulates on the crack tip to break the passive oxide film. As the film forms, the metal surface reduces its exposure to the potentially aggressive environment. Therefore, the oxidation current, i, on the free surface of the crack tip shows a decay of the type [19].

$$i = i_o \left(\frac{t}{t_o} \right)^{-n} \tag{17.26}$$

in which t_o represents the incubation time of the crack tip oxidation process and n is a decay exponent that depends on the alloy, the corrosive environment, the oxygen concentration and other possible anions as S^- or Cl^- and the flux rate. Equation (17.26) is plotted in Fig. 17.18 for a carbon steel type A 533 in pure water at 288° C [19], however, the trend is general and valid for any other alloy. After the incubation time t_o, the oxidation process starts again its reconstruction action of the oxide film that had been broken by stresses. Once the film is restored, corrosion stops and the oxidation current goes to zero. The crack tip rupture of the passive film happens with a periodicity characterized by a time t_f, as schematized in Fig. 17.18. the area under the curve of the current i, between a rupture event and the next one represents the oxidation density charge Q. Invoking the Faraday low, it can be possible to write that the equivalent weight p of metal entered into solution by anodic dissolution is linked to the oxidation density charge Q that crosses the solution in t_f seconds by the relationship.

$$p = \frac{M}{z\rho F} \cdot Q \tag{17.27}$$

where M is the atomic weight of metal, ρ its density, F the Faraday constant equal to 96,494 C and z the number free electrons involved in the oxidation process of the metal (valence). Considering that the quantity p of metal interested by the process of anodic dissolution determines the corrosion mean velocity growth at the crack tip *(ct)* $\overline{V}_{ct} = da/dt$, with a being the crack length, it will be possible to infer $\overline{V}_{ct}$ dividing Eq. (17.27) by the characteristic time t_f

$$\overline{V}_{ct} = \frac{da}{dt} = \frac{M}{z\rho F} \cdot \frac{Q}{t_f} \tag{17.28}$$

Ford and Emig [20] proposed the use of the strain rate at the crack tip, $\dot{\varepsilon}_{ct}$, , as fundamental parameter to describe the corrosion crack growth (CCG) process. By posing $\dot{\varepsilon}_{ct} = \varepsilon_f / t_f$, with ε_f the deformation at failure of the passive film at the crack tip, it follows that the characteristic time is $t_f = \varepsilon_f / \dot{\varepsilon}_{ct}$. Considering that the oxidation density charge Q can be obtained by integrating Eqs. (17.26), (17.28) becomes

$$\overline{V}_{ct} = \frac{M}{z\rho F} \cdot \frac{Q}{\varepsilon_f} \dot{\varepsilon}_{ct} = \frac{M}{z\rho F} \cdot \frac{i_o t_o^n}{(1-n)} \left(\frac{\dot{\varepsilon}_{ct}}{\varepsilon_f} \right)^n = f(n) (\dot{\varepsilon}_{ct})^n. \tag{17.29}$$

Note that for incubation times much long and, above all, for high strain rate $\dot{\varepsilon}_{ct}$, Eq. (17.28) can be approximated by

$$\overline{V}_{ct} = \frac{M}{z\rho F} \cdot i_o = \text{constant}. \tag{17.30}$$

Equation (17.30) express the circumstance in which under applied $\dot{\varepsilon}_{ct}$ much high, to which high CCG corresponds, the material at the crack tip has no time to passivizes and, therefore, it remains always uncovered and continuously exposed to the environment with an oxidation current always equal to i_o. Equation (17.19), therefore,

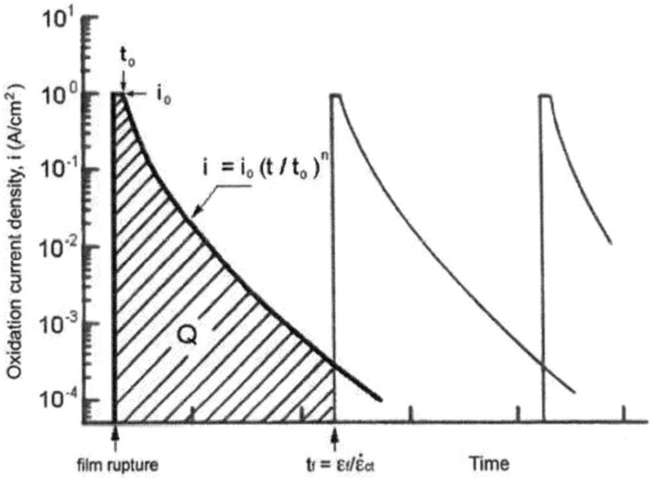

Fig. 17.18 Crack tip oxidation current density versus time after the failure of the passive film [19]

represents the upper limit of CCG by anodic dissolution for that particular pairing metal/environment that is independent of the applied strain rate.

As matter of fact, parallel measurements of the corrosion crack growth rates (CCGR) and mean oxidation current densities have evidenced the connection existing between these two variables, as shown in Fig. 17.19 [21]. The kinetics of the crack tip oxidation processes, therefore the oxidation density current Q with time, can be experimentally determined by a variety of techniques that, dough not simple to accomplished, are based on the progressive removal, either mechanical or chemical, of the oxide layer and subsequent measurement of the electric charge that passes during the transient. The mechanical methods of surface removal of oxide thermodynamically stable go from grinding [22] to stretching or scratching [23–29]. An example of oxidation density current measurement for an A 533 B carbon steel obtained with the stretching method is given in Fig. 17.20 [29]. The specimen was a sub-standard traction NACE specimen (24.4 mm length and 2.54 mm diameter) stretched with strain rate of $2 \cdot 10^{-3}\ s^{-1}$ in borated solution at 288° C to maintain a pH7 to simulate deareate conditions. Specimens were isolated fron the traction cell and put under potentiometric control at – 720 mV SHE.

Cell current was measured during the loading cycle and its increase associated to an oxidation current by passive film rupture. Note that $2 \cdot 10^{-3}\ s^{-1}$ do not represent the strain rate of the specimen, but that applied to the crack tip which is much higher since it is augmented by the stress intensity factor. Also note how the oxidation current follows the stress–strain trend with sudden increases at yielding. The current depends, besides on the applied electrochemical potential ECP, also on the strain rate $\dot{\varepsilon}$.

It increases as $\dot{\varepsilon}$ increases, as shown in Fig. 17.21 [28]. Figure 17.21 also shows a comparison of the results obtained by the stretching metod (CER test, see Sect. 15.5.2) with those measured by potential pulsing method. This latter method in a chemical

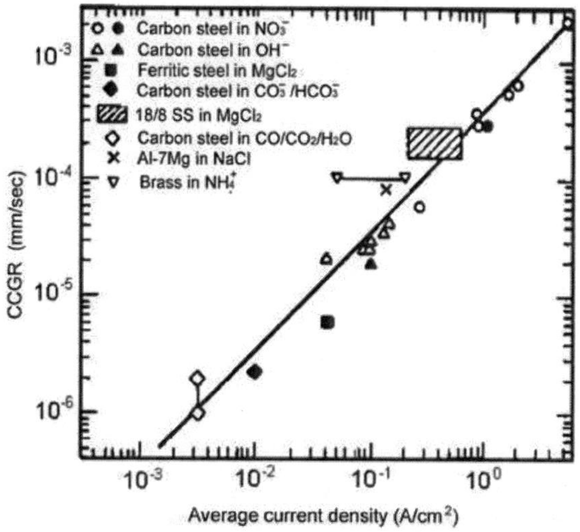

Fig. 17.19 CCGR versus average oxidation current density [21]

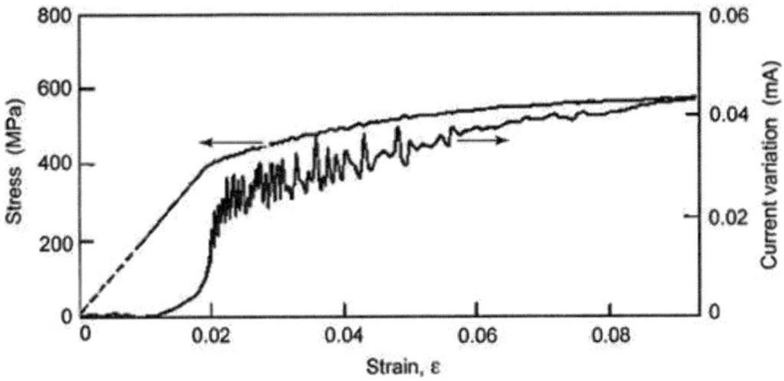

Fig. 17.20 Correlation between change in oxidation rate and stress upon straining at $2{\cdot}10^{-3}$ s^{-1} in borated electrolyte containing 1.5 ppm sulfate [29]

one of cathodic reduction of oxide followed by a rapid pulsation at the required potential. This technique has been used by Ford [29] to measure the anodic density current for a carbon steel type A 533 B in a water environment pulsating the ECP potential from − 1520 mV SHE to − 820 or + 180 mV SHE with pH varying from 7 to 10.3. Figure 17.22 [29] shows an example of such measurements in a water solution of Na_2MoO_4 0.01 M at 288° C varying the ECP from − 1520 mV SHE to − 620 mV SHE with various *S* concentrations.

It must be caught the strong influence of sulfur that increases the current density by an order of magnitude and, hence, the corrosion rate. It can be also noted the decreasing trend of the oxidation current with *n* equal to about 0.5 and oxide formation incubation time t_o equal to about 0.1 s. the dissolution model based on the periodicity of crack tip passive film rupture, Eq. (17.29), is schematically shown in Fig. 17.23 [29]. It must be said that during the crack tip anodic dissolution, hydrogen may enter the metal attracted by the high triaxiality stress state, see Sect. 14.4.3. When this

Fig. 17.21 Comparison of strain rate dependence of the oxidation kinetics during constant straining [28]

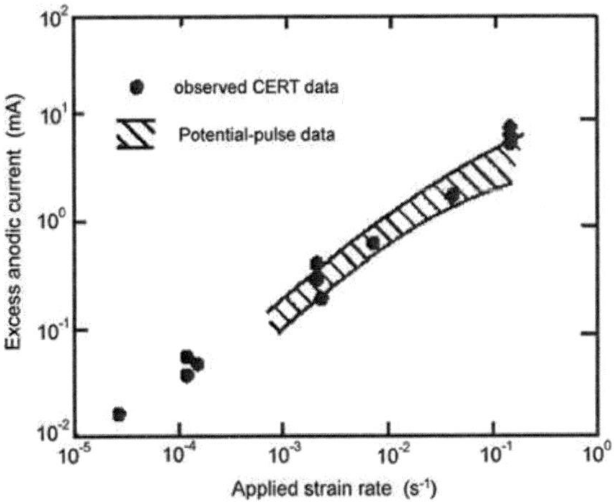

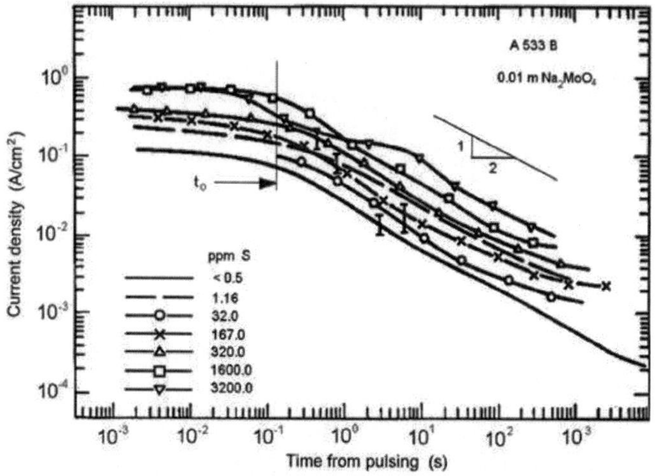

Fig. 17.22 Oxidation current density versus time transient following potential pulsing [29]

happens, a contribution of hydrogen embrittlement penetration by cleavage, da^*, in addition to the dissolution penetration process must be taken into consideration. The crack tip average penetration rate, then, becomes

$$\overline{V}_{pc} = \left(\frac{M}{z\rho F} \cdot Q + a^*\right)\frac{1}{t_f} = \frac{M}{z\rho F} \cdot \frac{i_o t_o^n}{(1-n)}\left(\frac{\dot{\varepsilon}_{ct}}{\varepsilon_f}\right)^n + a^* \frac{\dot{\varepsilon}_{ct}}{\varepsilon_f}. \tag{17.31}$$

The extent of this additional environmental-assisted cleavage a^* is of the order of 0.1–1 μm, depending on the metal-environment combination. This further growth is not continuous, but takes place by sudden flicks, dough decelerated. The model is schematized in Fig. 17.24 [31]. Figure 17.25 shows an example of corrosion crack growth according to the model of film-rupture/anodic-dissolution shown in Fig. 17.23 for carbon steel in de-oxygenated water obtained at 288 °C for different S contents and flow rate conditions.

The solution characteristics are: $M = 55.8$ gm mol., $z = 2$, $\rho = 7.8$ gr/cm^3, $F =$ 96,500 C/gm-equivalent, $n = 0.5$, $\varepsilon_r = 0.001$, $t_o = 0.1$ s and two different values of i_o equal to 1 A/cm^2 and 0.1 A/cm^2, respectively, with two sulfur content < 0.5 ppm and $1.6 \cdot 10^3$ ppm, respectively. These limits are assumed to span the condition of a critical activity of sulfide anions at the crack tip, and are characteristic of the crack tip sulfur content in *low* and *high*-sulfur steels. The growth limits are faradaic imposed by the anodic dissolution and are given by Eq. (17.30). Beyond that growth rate corrosion alone cannot go. Figure 17.25 reports also the CCGR in case of hydrogen absorption embrittlement.

The growth acceleration is impressive. As said, the lower limit is given by the condition $V_{ct} = V_{cs}$ when generalized corrosion of the crack sides prevails. As the crack advances, also the stress field ahead of the crack tip advances and increases

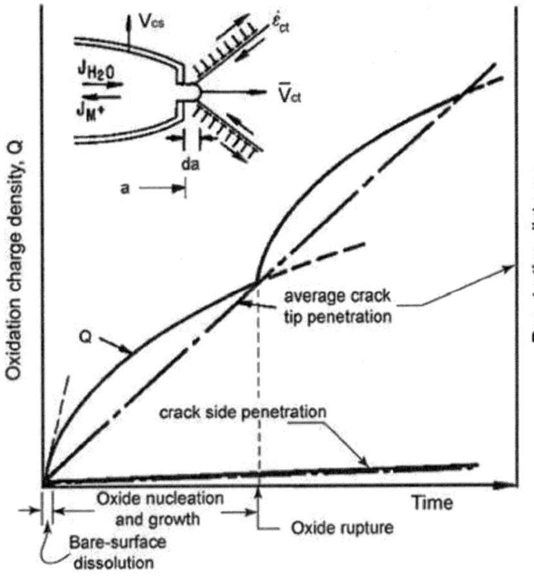

Fig. 17.23 Schematic of crack penetration with time in the film-rupture/ anodic-dissolution model [29]

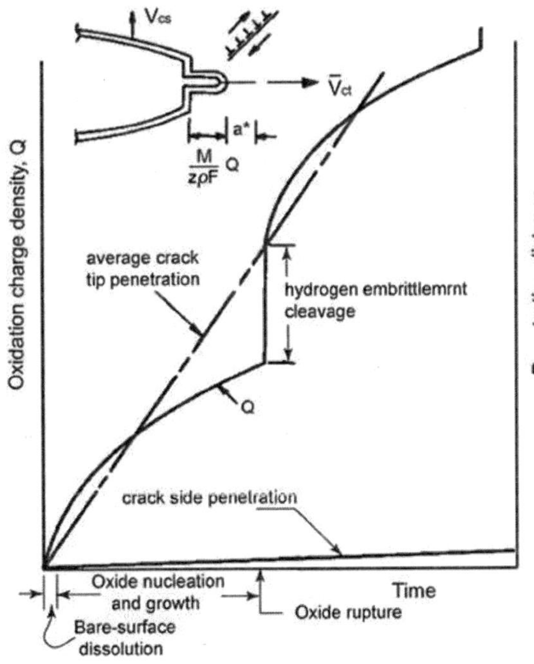

Fig. 17.24 Schematic of crack penetration with time in the film-rupture/ anodic-dissolution model accelerated by hydrogen absorption embrittlement [30]

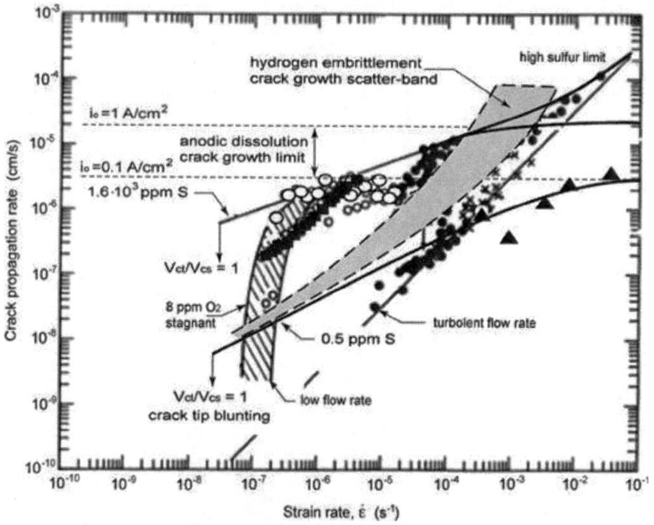

Fig. 17.25 Theoretical crack propagation rate/crack tip strain rate relationships for low alloy steel in stagnant deaerated water at 288 °C and experimental results (data from [32])

its amplitude because of the longer crack. This stress field movement and growth may lead to a new dislocation source activation with a consequent crack advance L^*. This hypothesis has led several authors [33–37] to believe that the prerequisite for the continuous process of film-rupture/anodic dissolution must be found just in this characteristic length L^* as minimum crack step advance. It is evident that L^* is linked to the faradaic oxidation charge Q^* by the relationship

$$L^* = \overline{V}_{ct} \cdot t^* = \frac{M}{z\rho F} \cdot Q^*. \tag{17.32}$$

With respect to the formulation given by Eq. (17.28) in which a characteristic time was considered t_f, now a constant charge density $Q^*$is fixed and the time t_f depends on Q that, in turn, varies with environmental conditions. As a proof of this hypothesis, it has been observed [35] that L^* is of the order of 0.5–1 μm that represents just the typical diameter of dislocation cells. Once Q^* is fixed, Ford [36] introduces an apparently contradictory concepts stating that now Q^* is also function of the stress field ahead of the crack tip whose amplitude is defined by the stress intensity factor $K_I = \sigma\sqrt{\pi a}{\cdot}f(a)$, according to a relationship of the type

$$Q^* = A \cdot K_I^{-1} \tag{17.33}$$

where A is a constant that depends on the crack length a and on the dislocation source spacing. This position, dough contradictory as said, has great consequences because it will suffice to use a WOL specimen (see Appendix A) to measure even one single crack penetration velocity V_{ct} in correspondence of the applied K_I to obtain the value of the constant A. in fact, it can be written

$$V_{ct} = 3.65 \cdot 10^{-5} \frac{Q*}{t}. \tag{17.34}$$

Combined with (17.33) it yields

$$A = \frac{V_{ct} \cdot t \cdot K_I}{3.65 \cdot 10^{-5}}. \tag{17.35}$$

The last unknown, the time t, can be assessed, as first approximation, as the ratio of strain at failure of the oxide layer, equal to ~ 10^{-3}, to the strain rate $\dot{\varepsilon}$. For example, if $\dot{\varepsilon} = 2.1 \times 10^{-6}\ s^{-1}$, the time t will be equal to 476 s. knowing the value of the constant A, it is possible to obtain the entire course of V_{ct} as a function of K_I simply scaling the growth value obtained to that applied K_I, using the $Q*$ value obtained with Eq. (17.33) for any other K_I. An example is shown in Fig. 17.26 for a stainless steel type 304 in water 1.5 ppm O_2 at 95 °C [37]. The ECP of solution was equal to + 193 mV SHE. The experimental point used for the evaluation of the A constant is that relative to $K_I = 27$ MPa$\sqrt{m}$, to which it corresponds a crack propagation rate of about $4.7 \cdot 10^{-8}$ cm/s. Equation (17.35) yields

$$A = 1.6 \times 10^2. \tag{17.36}$$

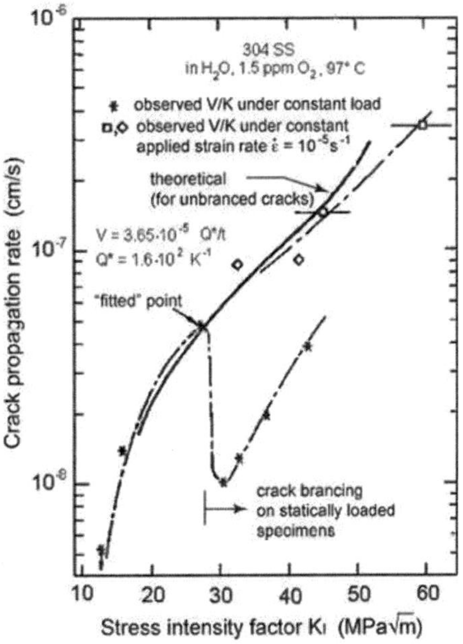

Fig. 17.26 Comparison of the experimental CCGR with the theoretical one for a stainless steel type 304 in water with 1.5 ppm di O_2 at 98° C [37]

Experimental data are adequately described by the theory in as long as a phenomenon of bifurcation does not occur. In this latter case, the K_I is reduced because it has to distribute over the two new crack tips created. This actually means that any crack bifurcation can drastically reduce corrosion.

References

1. Brown, B.F. (ed.): Stress Corrosion Cracking in High-Strength Steels and in Titanium and Aluminum Alloys. NRL, US Govt. Printing Office (1972)
2. Barker, H.: Quebec bridge suspended span falls into river in hoisting. Eng. News **76**, 524–528 (1916)
3. Staehle, R.W.: Bases for predicting the earliest penetrations due to SCC for the alloy 600 on the secondary side of steam generators. US NRC NUREG/CR-6737, p. 15 (2001)
4. Pourbaix, M.: Atlas of Electrochemical Equilibria in Aqueous Solutions. NACE, Huston, Texas, and Centre Belge d'Etude de la Corrosion (CEBELCOR), Brussels (1974)
5. Jones, D.A.: Principles and Prevention of Corrosion, pp. 39–70. Macmillan Publishing Company, New York (1992)
6. Carter, G.F.: Principles of Physical and Chemical Metallurgy, pp. 316–317. American Society for Metals, Metals Park, Ohio (1979)
7. Smith, W.F.: Foundations of Materials Science and Engineering, 3rd edn. The McGraw-Hill Co. Inc. (2004)
8. Davy, H.: Transactions of the Royal Society, 114, 151, 242 and 328 (1824)
9. Ashworth, V.: Corrosion, vol. 2, 2nd edn (1994)
10. Fontana, M.G., Greene, N.D.: Corrosion Engineering, 2nd edn, p. 314. McGraw-Hill (1978)
11. Fontana, M.G., Greene, N.D.: Corrosion Engineering, 2nd edn, p. 321. McGraw-Hill (1978)
12. Shoesmith, D.W.: Kinetics of aqueous corrosion. In: Corrosion, ASM Handbook, vol. 13 (1987)
13. Evans, U.R.: J. Inst .Metall. **30**, 265, 267 (1923); Corrosion of Metals, Arnold Pb, p. 91 (1926)
14. Johnson, H.H., Paris, P.C.: Subcritical flaw growth. Eng. Fract. Mech. **1**, 3–45 (1968)
15. Indig, M.E.: EPRI CAC Workshop Presentation. Palo Alto, USA (1979)
16. Indig, M.E., Weber, J.E., Weinstain, D.: Environmental aspects of carbon steel stress corrosion in high purity water. Rev. Coating Corrosion **5**, 173 (1982)
17. McDonald, D., Smalowska, Z., Pednekar, S.P., Mizumo, I., Choi, H.: EPRI Progress Report Project, t 115-5 (1980)
18. Shoji, T.: The critical cracking potential of reactor pressure vessel steels in PWR. In: Proceedings of the 3rd IAEA Specialist's Meeting on Subcritical Crack Growth, vol. 2, pp. 109–116. US NRC NUREG/CP–0112, Moscow (1990)
19. Ford, F.P., Andersen, P.L.: Stress corrosion cracking of low-alloy pressure vessel steels in 288° C water. In: Proceedings of the Third IAEA Specialists' Meeting on Subcritical Crack Growth, vol. 2, pp. 37–56. NUREG/CP–0112 (1990)
20. Ford, F.P., Emig, P.W.: Prediction of the maximum corrosion fatigue crack propagation rate in the low-alloy steels/deoxygenated water system at 288 °C. European/Corrosion Federation Meeting, Munich (September 1984)
21. Parkins, R.N.: Br. Corrosion J. **14**, 5 (1979)
22. Ambrose, J.R., Kruger, J.: Corrosion **28**, 30 (1973)
23. Tomashou, N.D., Verastinina, L.P.: Electrochemica Acta **15**, 501 (1976)
24. Lees, D.J., Hoar, T.P.: European Federation of Corrosion Electrochemical Methods for Stress Corrosion Cracking. Firmy, France (September 1978)
25. Hoar, T.P., Jones, R.V.: Corros. Sci. **13**, 725 (1973)
26. Hoar, T.P., Ford, F.P.: J. Electrochem. Soc. **120**, 1013 (1973)
27. Diegle, R.B., Vermilyea, D.A.: J. Electrochem. Soc. **122**, 180 (1975)

28. Combrade, P.: Prediction of environmental crack growth on reactor pressure vessels. EPRI Contract RP 2006-8. Final Report on Task 310-Anodic Dissolution (February 1985)
29. Ford, F.P.: Electrochemical reaction rates on bare surface and their use in a crack prediction model for the low alloy steels/water system. In: Proceedings of the Second IAEA Specialists' Meeting on Subcritical Crack Growth, vol. 2, pp. 231–268. NUREG/CP–0067 (1985)
30. Ford, F.P.: A Mechanism of Environmentally-Controlled Crack-Growth of Structural Steels in High-Temperature Water. Report N. 81CRD125, GE Co., Schenectady, New York (1981)
31. Ford, F.P.: Overview of collaborative research into the mechanism of environmentally controlled cracking in the low alloy pressure vessel steel/water system. In: Proceedings of the Second IAEA Specialists' Meeting on Subcritical Crack Growth, vol. 2, pp. 3–71. NUREG/CP–0067 (1985)
32. PMDA PIRT REPORT-Appendix B.1 (2005)
33. Newman, J.F.: On the Mechanism of Stress Corrosion of a 3%CrMo Steel in Sodium Hydroxide in the Potential Range −900 to −650 mV Hg/HgO. Central Electricity Research Lab, Report RD/L/N78/75 (1975)
34. Newman, J.F.: The application of a film rupture mechanism to stress-corrosion of a 3% CrMo steel in sodium hydroxide in the potential range −900 to −650 mV Hg/HgO. In: Proceedings of Conference Mechanisms of Environmental Sensitive Cracking in Materials, p. 19. University of Surrey, April 4–7 (1977)
35. Scully, J.: Corros. Sci. **15**, 207 (1975)
36. Vermilyea D.A.: Stress Corrosion Cracking and Hydrogen Embrittlement of Iron-Based Ally, p. 208. Firminy, France (1973) (Strahle, R.W., Hochmann, J., McCright, R.D., Slater, J.E. (Eds.) NACE Huston Publication, 1977)
37. Ford, F.P.: A mechanism of environmentally-controlled crack-growth of structural steels in high-temperature water. In: Proceedings of the IAEA Specialists' Meeting on Subcritical Crack Growth, vol. 2, pp. 249–294. NUREG/CP–0044 (1981)

Chapter 18
Hydrogen Embrittlement and Sensitization Cracking

18.1 Introduction

Even dough sensitization and hydrogen embrittlement failures can be classified as stress corrosion cracking, they are so distinctive and important a form of corrosion that deserve to be treated in a section of their own. Therefore, this chapter will address just these two most devastating corrosive events that can jeopardize the integrity of metal structures: hydrogen embrittlement or cracking and stress corrosion of sensitized materials. Even dough they are completely different phenomena, they both initiate with the same event that is the slip dissolution process discussed in Sect. 13.7. As matter of fact, the starting event is always the destruction, at least locally, of the passive and tight film that protects high strength steels, stainless steels, nickel alloys and aluminum alloys, in particular. Precisely those alloys developed to resist general corrosion are very susceptible to localized corrosion. A typical example is offered by austenitic stainless steel. This alloy is known for its excellent resistance to generalized corrosion. Brearly in England and Strauss and Maurer in Germany were the first to exploit the corrosion resistance afforded to iron by chromium. Brearly developed the 11–14% chromium ferritic stainless steel and probably coined the term "stainless steel". At the same time, Strauss and Maurer developed the austenitic Fe–Cr-Ni alloys. The stress is one of the active ingredients in corrosion of these alloys; the other two are hydrogen in one case (hydrogen cracking) and, in the case of sensitization cracking, the precipitation of some phase at grain boundaries, which actually means: temperature. This latter form of corrosion is always intergranular and is a subtle form of microscopic galvanic cells that create between base metal and precipitates.

P. P. Milella, *Fatigue and Corrosion in Metals*,
https://doi.org/10.1007/978-3-031-51350-3_18

18.2 Sensitization

Sensitization is defined as the state of a metal in which the precipitation of some compound or phase at the grain boundaries makes it susceptible to stress corrosion cracking. Generally, but not necessarily, the precipitation is the consequence of a thermal treatment. A typical example is offered by austenitic stainless steel that becomes susceptible to intergranular stress corrosion cracking when subjected to heat treatments in the temperatures range of 500° to 850 °C. In this temperature field carbon precipitates as complex chromium carbide $Cr_{23}C_6$ or iron carbide $Fe_{23}C_6$. Welding is one of the most common thermal process that sensitizes austenitic stainless steel. During welding, in fact, the base metal adjacent to the fusion zone undergoes precisely those temperatures needed to sensitize the metal (see Fig. 12.7). The morphology and distribution of the precipitate is strongly time and temperature dependent. The grain boundary precipitate can be classified into three types: separate geometric particles, dendrites and sheets of connected geometric particles. Figure 18.1 [1] presents the relationship between temperature and grain boundary precipitates in austenitic stainless steel type 304.

In the low field of sensitizing temperatures, complex carbides precipitate creating a kind of continuous net along all the grain boundary structure, as schematized in Fig. 18.2, case A [1]. Once corrosion starts, it will follow the continuous path penetrating the full thickness without arresting. On the contrary, at higher sensitizing

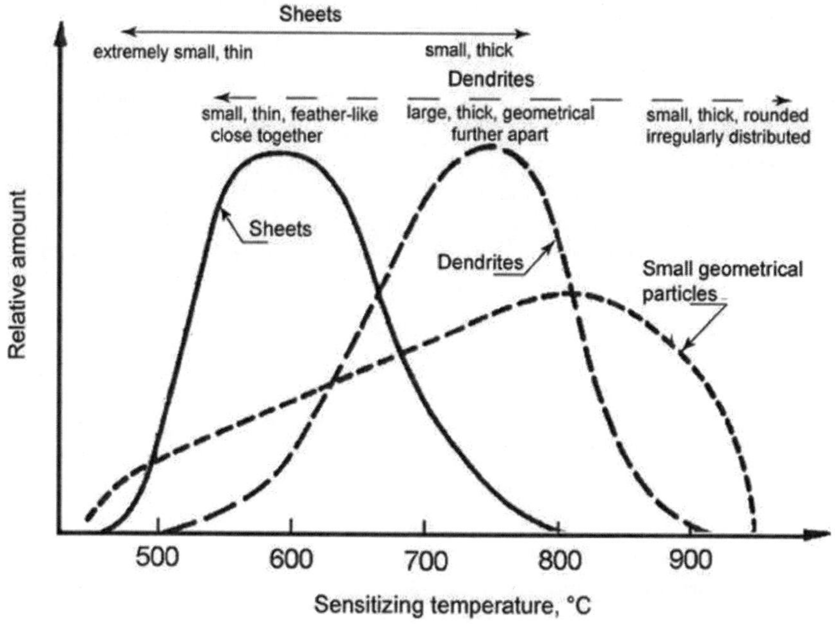

Fig. 18.1 Temperature dependency of morphology of grain-boundary carbides [1]

temperature precipitates are by far more abundant, but they are isolated dendrite particles or beads, as schematized in Fig. 18.2, case B [1]. These dendrites become at higher temperature, further and further separated and, therefore, it will become more and more difficult for the corrosion to penetrate the material. An example of sensitization produced by welding in austenitic stainless steel, type 304, is shown in Fig. 18.3. It is possible to note the complex carbides precipitated to the grain boundaries in a continuous fashion. Another method to determine whether or not the metal is sensitized is based on the so called potentio-dynamic polarization method [2].

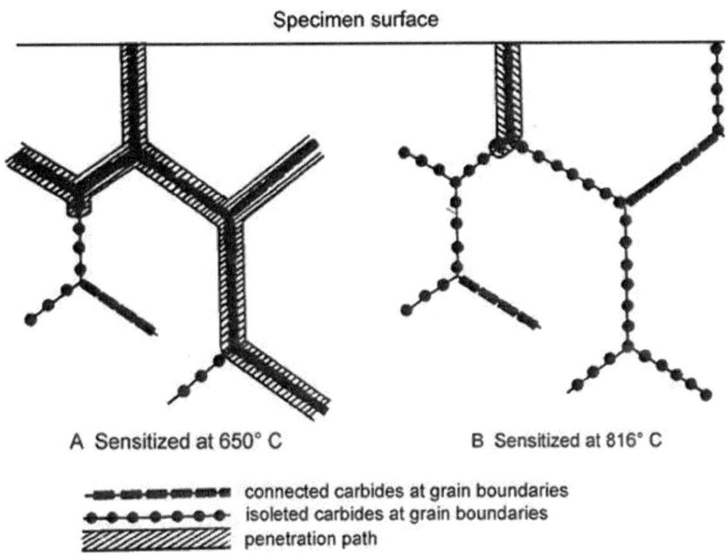

Fig. 18.2 Mechanism of intergranular corrosion in sensitized stainless steel [1]

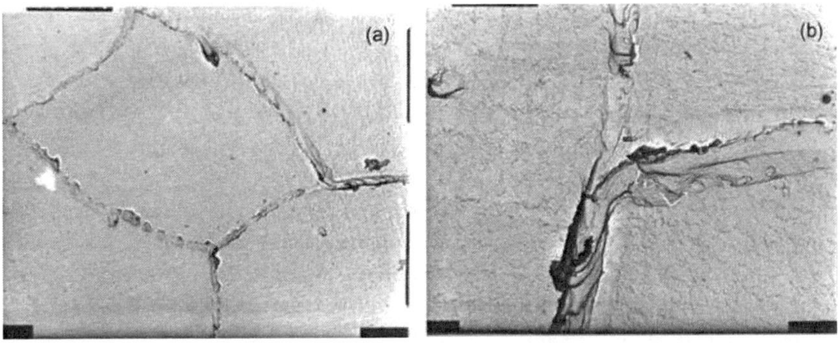

Fig. 18.3 SEM on replica of welding sensitization of 304 stainless steel. **a** Thin, feather-like carbaides joined together, **b** Detail of A

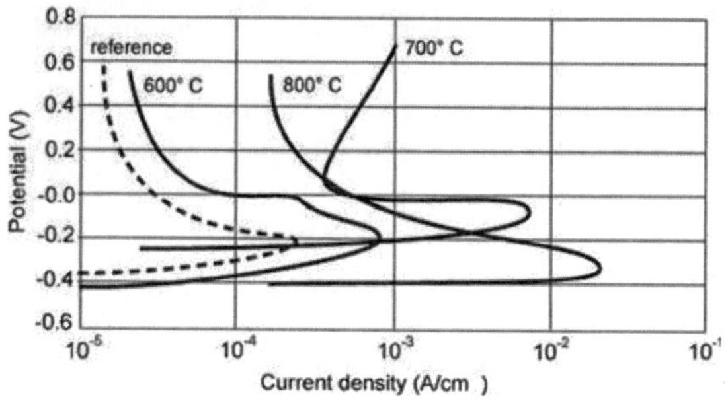

Fig. 18.4 Potentio-dynamic response for 304L stainless steel specimens sensitized for 1/2 hour at 600°, 700° and 800 °C, respectively. Also shown is the reference specimen

The specimen to be examined is mirror-polished and put into an acidic solution. The solution is made oxygen free by flowing nitrogen. Between the specimen and a platinum electrode a difference of electric potential is applied whose value increases over and over. When the specimen becomes more electronegative than the electrode the metal starts co corrode by anodic dissolution generating a current that reaches a maximum i_c at the critical passivation potential E_{pp} (see Fig. 13.11). If sensitization does not occur, after this maximum the current density decreases indicating that the material is passivizing by the formation of an oxide protective film. On the contrary, if sensitization takes place a second peak appears and/or the passivation disappears with the current density reaching values some order of magnitude higher than that observed in non-sensitized metal. An example of potentio-dynamic polarization is shown in Fig. 18.4 for an austenitic stainless steel type 304L previously subjected to isothermal treatments at 600°, 700° and 800 °C for 1/2 hour, respectively. It is also shown the reference curve relative to the same material non-heat-treated.

Note that the critical current i_c of the heat-treated specimens increases with respect to that of the reference specimen, with the maximum obtained at 700° C, which is two order of magnitude higher, but at 800° C passivation occurs at the critical passivation potential E_{pp} of –0.3 V. The current peaks of anodic dissolution range from ~ $2 \cdot 10^{-4}$ A/cm^2 at E_{pp} of about – 0.2 V for the reference specimen, to about $8.8 \cdot 10^{-4}$ A/cm^2 for the specimen at 600 °C and to $7.0 \cdot 10^{-3}$ A/cm^2 at E_{pp} of –0.1 V for that at 700 °C. At 800 °C the peak current increases to $2.0 \cdot 10^{-2}$ A/cm^2 at E_{pp} of –0.3 V. Strauss also noted that complex carbides precipitated at grain boundaries were fed, in particular, by chromium so that the metal adjacent the grain boundaries would be chromium depleted. Basing on this observation, Strauss et al. [3] proposed a mechanism for the intergranular stress corrosion cracking based on the impoverishment of chromium in the grain boundaries. The chromium impoverishment theory was claimed to be proven by Chevenard et al. [4]. In effect, the chromium depleted zone surrounding grain boundaries exists, even dough extremely narrow, as schematized in Fig. 18.5. The grain boundary may be chromium enriched up to 70% while in the adjacent areas, from 300 to about 5.000 Å, chromium drops below the critical limit of 12% that makes the material susceptible of severe attack.

Recent studies [1, 5–9] have definitely demonstrated that chromium impoverishment theory is not precisely the real cause of corrosion that must be found in a

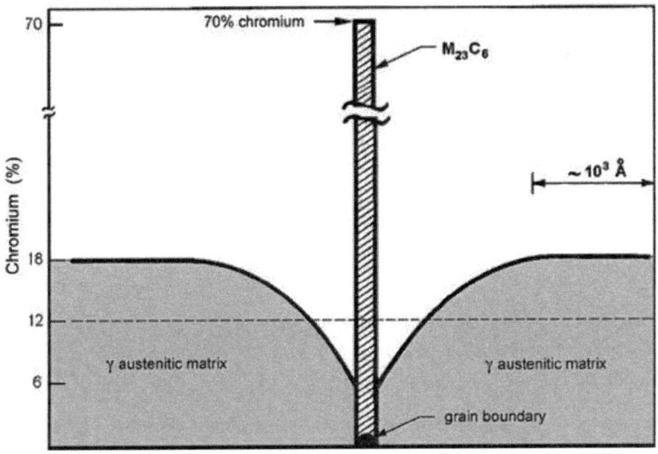

Fig. 18.5 Typical chromium concentration profile in the surroundings of a grain boundary in sensitized 304 stainless steel

galvanic anodic dissolution. Complex chromium or iron carbides have, in fact, corrosion potential much different with respect to the surrounding metal matrix in which they precipitate, hence the presence of any electrolyte, even humidity condensation, can trigger a series of microscopic galvanic cell reactions.

Since chromium carbides are noble with respect to the austenitic matrix, it is the less noble metal matrix that corrodes by anodic dissolution. It is, then, clear why and how a morphology of sheet-like continuous carbides, such as the one described in Fig. 18.2a and shown in Fig. 18.3, can be extremely dangerous because it provides a continuous path to corrosion that penetrates the full thickness without arresting. Also time–temperature relationship and carbon content play a major role in sensitization. Figure 18.6 [10] provides, in the temperature range of 400–900 °C, the approximate time needed to produce sensitization in an austenitic stainless steel, type 304, for different carbon contents. Carbides precipitate in the areas to the right of the various carbon content curves. Results of Fig. 18.6 are based on tests in the so called Strauss solution, copper–copper sulfate-16% sulfuric acid.

Note that the exposition to Strauss solution or Strauss test, while can establish whether or not a metal is sensitized, is not so effective as potentio/dynamic or metallographic observation to establish the actual *strength* of sensitization. This is because the Strauss test, by etching the metal exposed to the solution, provides a result that depends on the total amount of precipitates and not on the type, continuous or disconnected carbides, which is the real indicator of sensitization. This can be seen in Fig. 18.7 relative to six specimens isothermally sensitized at six different temperatures, which are: A at 482 °C, B at 566 °C, C at 649 °C, D at 732 °C and E at 816 °C, F at 899 °C, respectively, and then exposed for 45 h to the Strauss solution. The severity of etching increases with increasing amount of precipitations and is most severe for the samples E, in Fig. 18.7, sensitized at about 816 °C. But

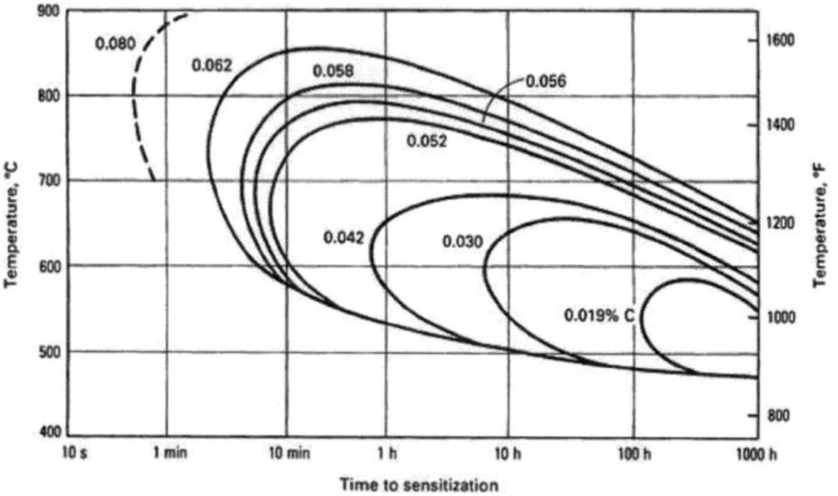

Fig. 18.6 Time–temperature-carbon content sensitization curves for type 304 stainless steel in a mixture of $CuSO_4$ and H_2SO_4 containing free copper (Strauss solution) [10]

this maximum etching of grain boundaries differs from heat treatments that result in the most severe sensitization that occurs, instead, from 565 to 650 °C, samples B and C of Fig. 18.7. From Fig. 18.6 it is possible to note how carbon content is essential in the sensitization process. The higher the carbon content, the sooner sensitization occurs at any temperature at which it can actually appear. For instance, with C $\geq$ 0.08% it suffices some minutes to sensitize steel at 750 °C, but if the carbon content reduces to 0.03% there will not be sensitization at all. Note how, the lower the carbon content, the lower the temperature at which sensitization occurs and the longer the time needed.

To produce sensitization in austenitic steel with less than 0.02% carbon it is necessary to heat it below 600 °C for more than 100 h, which actually means that with such steel any welding process will not be able to produce sensitization. This has led, in the ′70 s, to introduce in the nuclear field austenitic stainless steel with C $\leq$ 0.03% to avoid sensitization. These austenitic stainless steels were referred to as N steels from Nuclear grade. To recover the strength lost by a so heavy carbon reduction, these steels were added titanium or niobium or vanadium. Today, austenitic stainless steels with C $\leq$ 0.03% are known as L grade steels from Low carbon. Another method of avoiding sensitization is to use stabilized steels. Such stainless steels contain titanium and/or niobium. These elements have an affinity for carbon and form carbides readily; this allows the chromium to remain in solution even for extremely long exposures to temperatures in the sensitizing range. The presence of a possible stress state can contribute to accelerate corrosion in sensitized steels. The phenomenon is known as stress corrosion cracking and is particularly severe in an environment rich of chlorides, such as sea water. However, curves like those of Fig. 18.6 must always be verified for the steel used. For example, 304L austenitic

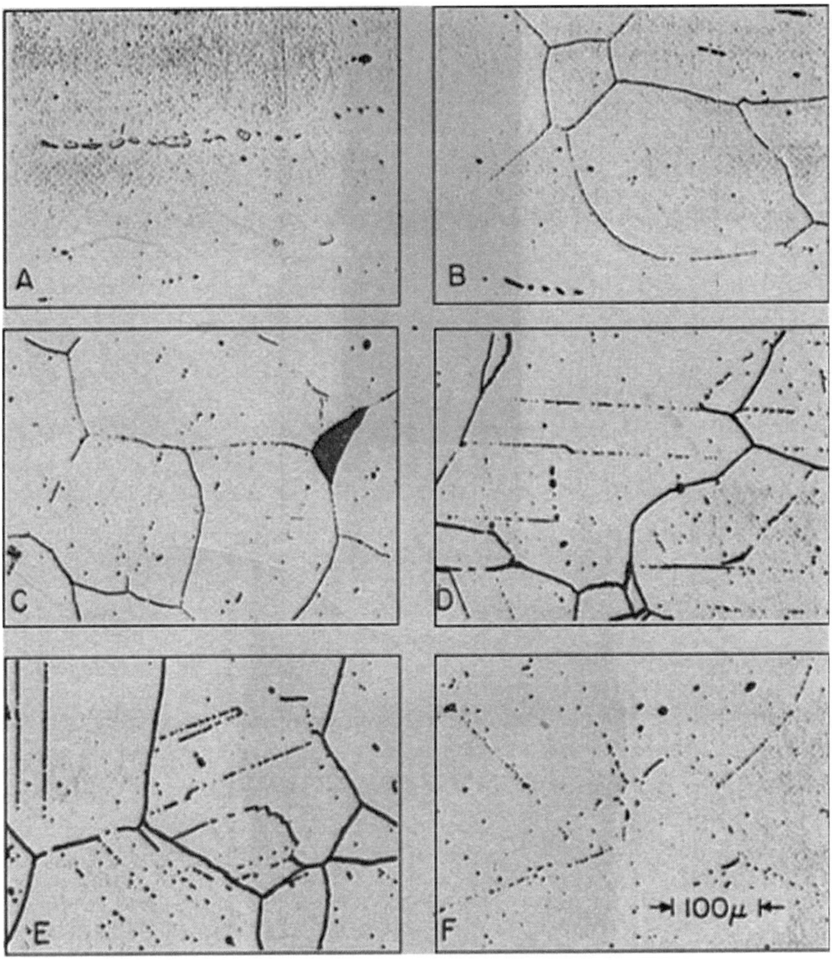

Fig. 18.7 Etching by Strauss solution of specimens sensitized at different temperatures [1]

stainless steel that from Fig. 18.6 should not be susceptible to sensitization below 700 °C or for less than 1 h exposure to any temperature was found to sensitize at 800 °C in 1/2 hour. This is shown in Fig. 18.8 that refers to two samples of 304L of the same heat.

Sample A did not sensitize at all, but sample B did, which is also an indication of metallurgical variability. Even these extra low steels, then, precipitates carbides under certain conditions of time and temperature. Although such steels are definitely less susceptible to intergranular corrosion than the regular type 304 or 316, it is apparent that the complete immunity to intergranular corrosion of unstabilized austenitic stainless steels can be achieved only when the carbon content is less than

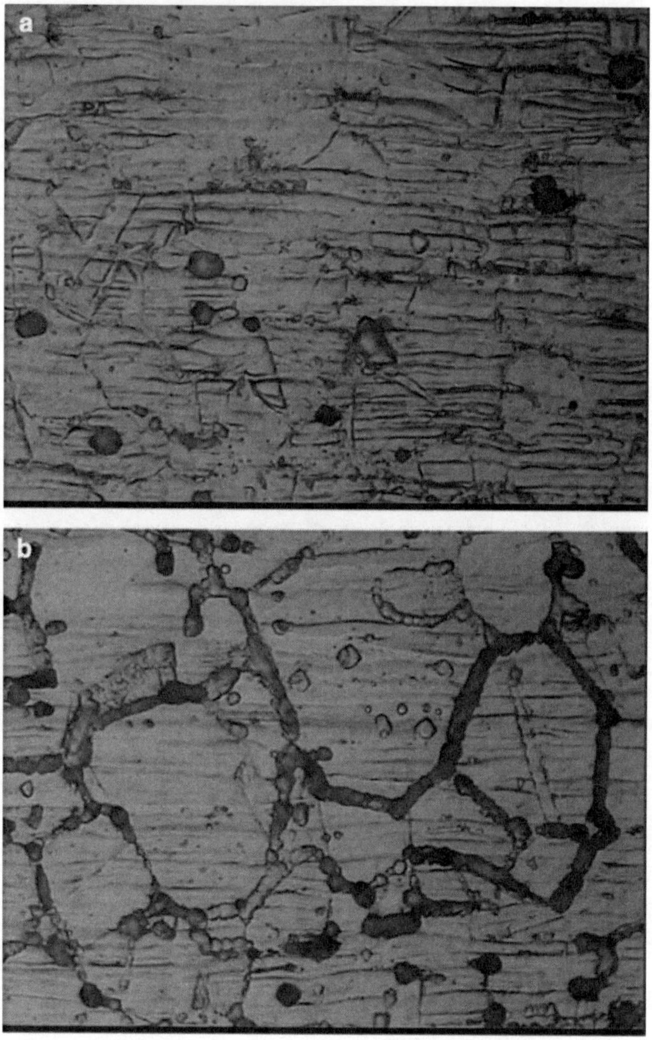

Fig. 18.8 Two samples of the same heat of 304L isothermally treated at 800° C for 1/2 hr. Sample **a** did not sensitized, sample **b** did

the limit of solid solubility. It is now generally believed that the limit of solid solubility of carbon is less than the value of 0.02% usually assumed. Therefore, the only real alternative to stabilization to eliminate sensitization of austenitic stainless steels is the annealing at high temperature, between 1070 to 1090 °C. The purpose of annealing is to solubilized complex chromium carbides so that carbon precipitated at grain boundary reenters into solution and sensitization is recovered. Figure 18.9 [11] show the variation of growth rate and type of corrosion in austenitic stainless steel,

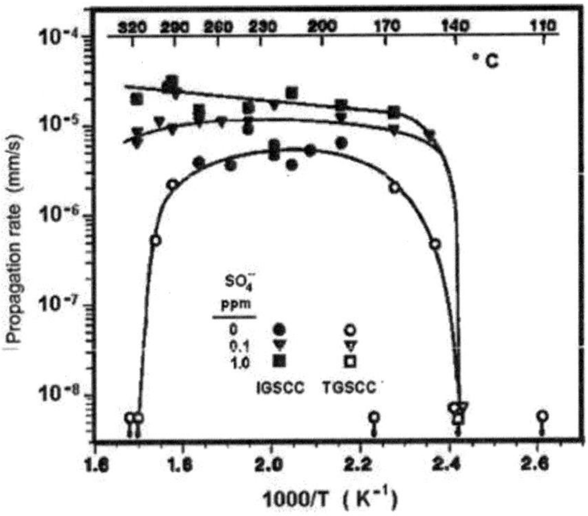

Fig. 18.9 Temperature effect on the IGSCC propagation rate in austenitic stainless steel type 304 lightly sensitized [11]

type 304, specimens in water containing 0.2 ppm of dissolved oxygen and various sulfites impurities introduced as H_2SO_4, equal to 0, 0.1 and 1.0 ppm, respectively.

The specimens were subjected to CERT test (see Sect. 15.5.2) with 10^{-6} s^{-1}strain rate, up to failure. Steel was in a light-sensitized condition. The maximum effect can be seen between 170° C and 250° C and results in an IGSCC (microscopic grain-boundary galvanic cell) while outside that temperature range fracture is transgranular (TGSCC) and crack growth rates fall below by a factor of 10^3, indicating absence of corrosion.

At 320° C as at 140 °C corrosion is practically absent. Sulfides represent impurities of water chemistry that can increase corrosion rate sustaining IGSCC, also beyond temperature at which it disappears in clear water.

18.2.1 Sensitization of Nickel and Aluminum Alloys

Nickel and its alloys are extremely susceptible to temperature sensitization by trace impurities. As in stainless steel, also in the nickel alloy type 800, Cr21Ni33TiAl, the presence of chromium over 20 wt%. and carbon (0.05–0.07 wt%) results in the same type of complex carbide precipitation at grain boundaries. Also the 600 series and in particular the 690 alloy with chromium in the range of 20-31 wt%, is susceptible to sensitization [12]. These alloys used for steam generators tubing experienced IGSCC in hot leg that is operating around 320 °C dome 30 to 40 °C hotter than the cold leg. It was clear, then, that temperature had a significant influence on IGSCC, indicating a strongly thermally activated process. The apparent activation energy from fitting the temperature dependence of the Arrhenius equation is rather high (~180 kJ/mole) so that a typical temperature difference of 30 °C could easily account for a factor of four to five increase in the time to onset of detectable cracking. Alloys X750 and

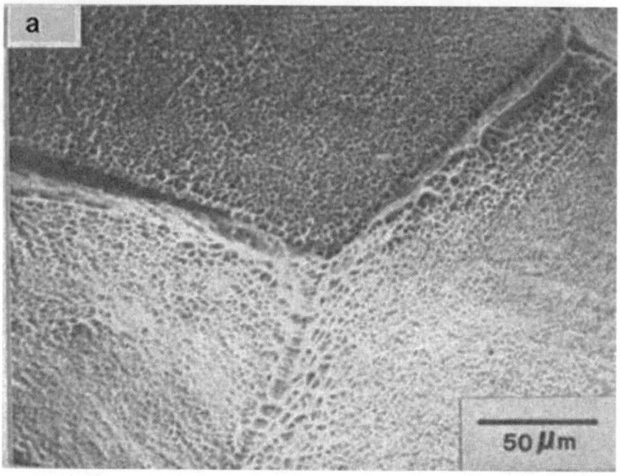

Fig. 18.10 Sulfur segregation at grain boundaries in nickel 200 alloy [13]

718 are hardened to precipitate phase γ' $Ni_3(Ti, Al)$ and/or γ''. They have strength from 790 to even 1380 MPa. The series X750 has experienced many more cases of IGSCC than 718 at temperatures of steam generators.

However, sulfur, a common impurity in commercial ore, is the real destroyer. Its effect is so severe that it was not until about 1870 that the first malleable nickel was produced by the addition of Mn and Mg. these elements react with S and produce innocuous precipitates. Nickel sensitization is caused precisely by sulfur precipitation at grain boundary. An example of sulfur effect in nickel 200 alloy isothermally treated at 600 °C for 24 h is visible in Fig. 18.10 [13]. Note that sulfur segregation at grain boundaries occurred despite the extremely low S concentration that was just a mere 0.005 wt%. This kind of sensitization has long been raging over jet engine turbine blades till the introduction of single crystal nimonic alloys. Without grain boundaries there cannot be sulfur precipitation. Also aluminum alloys are susceptible to thermal instability. In the 5XXX series aluminum alloys a magnesium presence just over 3 wt% can cause precipitation of β phase (Al_3Mg_2) along grain boundaries. This process can happen also at relatively low temperatures (~70 °C), but it needs a very long period of time (10–20 years). Figure 18.11 [14] shows the sensitization in AA5083, Al-4.5 Mg-1.0Mn alloy, after 150 h at 150 °C. A continuous film of β precipitates is formed.

18.2.2 Cold Work Sensitization

It has long been recognized worldwide that also a cold work process particularly severe can contribute to corrosion in austenitic stainless steels by sensitizing the material [15–19]. Grinding must be completely rejected, in particular when high speed without lubricant is used. The sensitization process is due to two simultaneous actions. First, any heavy cold-working may leave a system of traction residual stresses

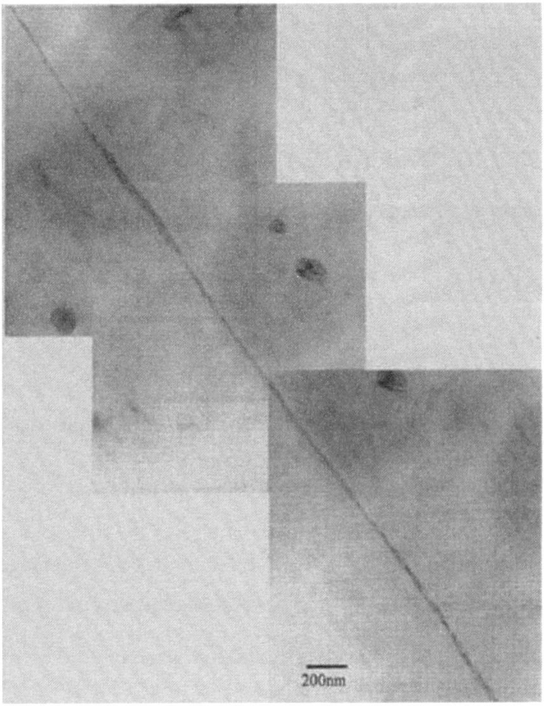

Fig. 18.11 Sensitization in AA5083, Al-4.5 Mg-1.0Mn alloy, after 150 h at 150 °C [14]

(see Sect. 4.5 and Fig. 4.15) that favor stress corrosion. Secondly, plastic deformations induced by cold working may transform part of the metastable austenite into brittle martensite. Martensite is inclined to stress corrosion. Micro-fissuring can easily occur in brittle areas of martensite on the surface of the specimen exposed to the potentially aggressive environment.

Stress corrosion may start from this surface scratches and grow even faster that in austenitic stainless steel. This austenite–martensite transformation can trigger a sensitization process favoring the precipitation of complex carbides at grain boundary. Figure 18.12 presents the surface of austenitic stainless steel heavily cold-worked by grinding. It can be seen the slip bands (plastic deformations) in surface grains in which austenite transformed into martensite.

18.3 Morphology of IGSCC

The intergranular nature of corrosion in sensitized alloys is clearly reveled by SEM or optical microscopy. Figure 18.13 [20] shows the typical appearance of IGSCC in weld sensitized type 304 stainless steel failed after 695 h of permanence in demineralized water containing 24 ppm of O_2 at 288 °C, under a stress of 276 MPa. Single grain

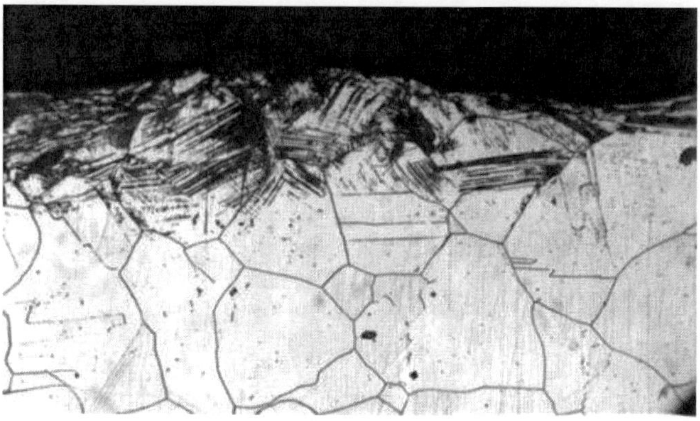

Fig. 18.12 Cold work sensitization and martensite formation in surface grains of austenitic stainless steel, type 304, after grinding

facets can be seen denouncing the grain boundary separation process of intergranular stress corrosion cracking. A characteristic of intergranular stress corrosion cracking is also the bifurcation and branching.

An example of stress corrosion branching is shown in Fig. 18.14 for a Ni–Cr steel after heat treatment [21]. It can be clearly seen the intergranular nature of branching that grows along grain boundaries in a continuous path destroying the metal tissue that separates along grains. An interesting example of cracking morphologies is presented in Fig. 18.15 [22] for a specimen of 304 stainless steel containing a pre-cracked notch. The specimen was cut from a block containing a butt weld. The round

Fig. 18.13 SEM analysis of IGSCC in stainless steel type 304 weld sensitized [20]

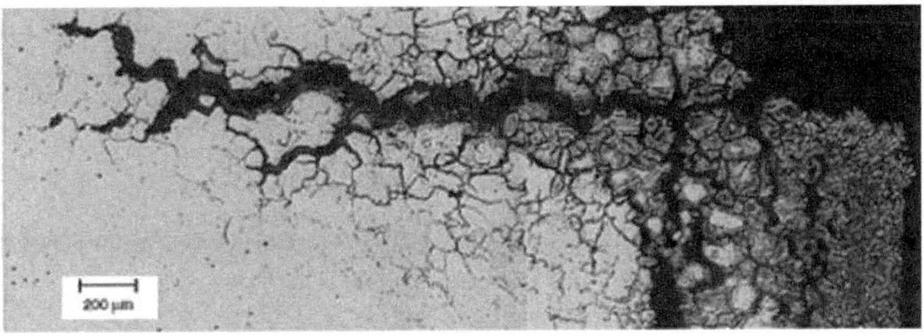

Fig. 18.14 Branching by IGSCC in a heat treated Ni–Cr steel [21]

notch tip is visible with its pre-crack right before the weld. The steel was sensitized at 620 °C for 2 h and fatigue cycled in demineralized water at 288 °C. The characteristic of this picture is in the fact that two cracks originated from the pre-crack. One is the growth of the original pre-crack towards the weld. The second originated on the pre-crack side and grew in the heat affected zone. Interesting enough, the fatigue crack towards the weld is transgranular (TG) because it is *fed* by fatigue and breaks the grains apart, while the crack moving in the HAZ is intergranular because it is an IGSCC not sustained by fatigue but, rather, by residual stresses.

This difference in morphology between fatigue cracks and stress corrosion cracks is of general type. It can be recognized in Fig. 18.16 [23] that shows a 1 T-CT specimen of stainless steel type 304 sensitized at 620 °C for 24 h and put in clear water at 289 °C with 8 ppm dissolved oxygen. The specimen was fatigued at 2 cycles/hr between 29 and 33 MPa$\sqrt{m}$ up to a total crack growth of 1.8 mm after which the load had been maintained constant at 33 MPa$\sqrt{m}$ for a further 0.5 mm crack growth and, eventually, broken open by overload. These three phases are indicated on the right of Fig. 18.16a. The respective appearance at 10 × enlargement of fatigue and stress corrosion cracking of the areas enclosed in the small rectangles are shown in the pictures on left. It is interesting to note how the fatigue crack growth was transgranular (TG), Fig. 18.16a, while under static load the crack growth was intergranular (IG), Fig. 18.16b, by stress corrosion (IGSCC).

18.4 Hydrogen Embrittlement

We shall now enter the universe of hydrogen embrittlement. Few pages or even books would not be enough to fully describe it, so that the following treatment of the subject will be a universe in a nutshell. The term hydrogen embrittlement does not describe a single or specific process of degradation. It is rather generic and describes a group of phenomena, all involving hydrogen as the principal cause of degradation of materials. Hydrogen embrittlement is a very complicated process that can affect virtually any metal or alloy and is nearly always unpredictable. Even

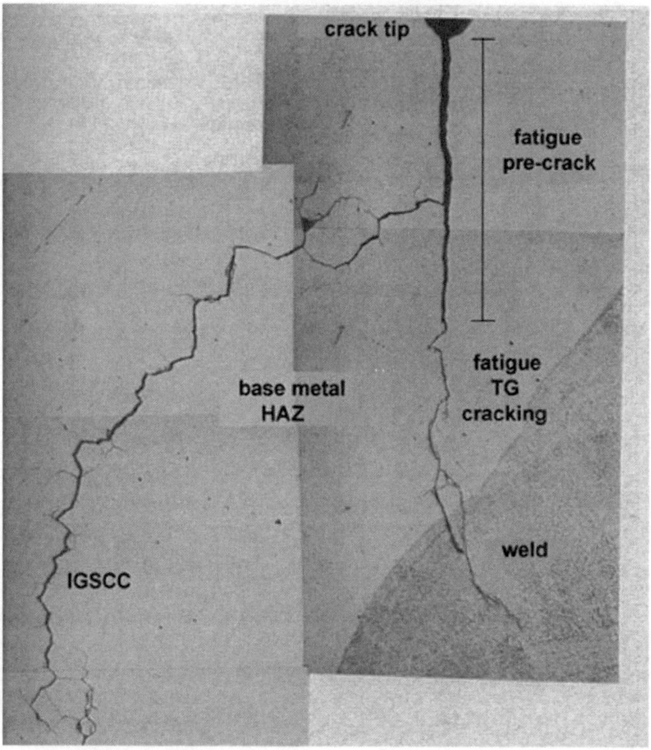

Fig. 18.15 TG fatigue crack growth and IGSCC originated from a notch pre-crack in a specimen of type 304 SS containing a weld [22]

dough hydrogen is always coming from the outside environment, it can be already present in the bulk of the material during the application of loads, which is referred to as internal hydrogen embrittlement (IHE), or it can enter the material as result of the exposure to hydrogen containing environment, which is referred to as external hydrogen embrittlement (EHE).

18.4.1 Internal Hydrogen Embrittlement (IHE)

Extremely large hydrogen concentrations can build up as consequence of chemical or electrochemical pre-operational processing like cleaning and pickling cycles, electroplating, phosphating solutions, cathodic cleaning solutions and stripping. When new components fail by hydrogen embrittlement, usually after short times in service, this is mainly because the manufacturing procedures for finishing processes have not been followed correctly or were not correct. Notorious is the internal hydrogen

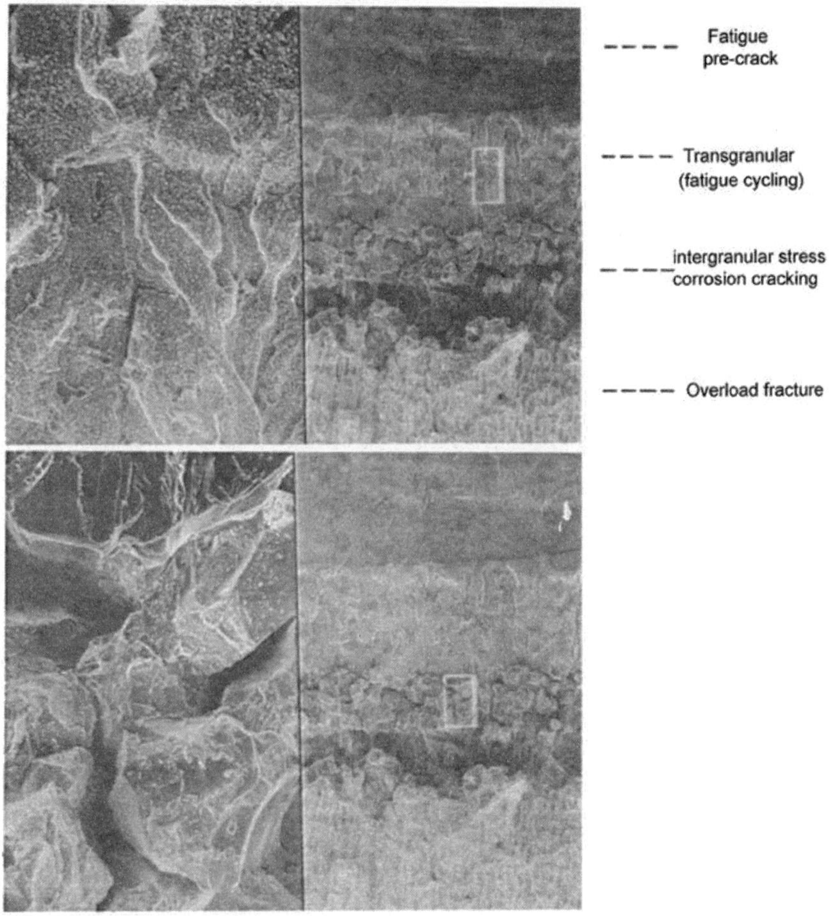

Fig. 18.16 SEM examination of a CT specimen fracture surface of type 304 SS showing both **a** TG fatigue and **b** IG stress corrosion crack growth [23]

embrittlement of screws protected by hot bluing or tarnishing process not correctly applied.

Figure 18.17 [24] presents the SEM fractography of the fracture surface of a bolt from S99 low alloy steel heat-treated to a tensile strength of about 1380 MPa, bright-cadmium electroplated cracked by IHE. The typical intergranular stress corrosion cracking by hydrogen demolition appears. Once entered in the bulk metal, hydrogen starts its demolishing action unless an appropriate thermal treatment is carried out just to eliminate itThe process is also termed *baking-out* and involves both diffusion within the metal and outgassing. However, it must be kept in mind that baking-out of internal hydrogen must be done within a time limit from its diffusion in the metal, usually as soon as possible but no more than four hours, otherwise cracking may have already occurred. Time-temperatures needed to remove hydrogen depend on the strength of the alloy. Figure 18.18 presents the classic work of Troiano [25] for delayed failure of sharp notched specimens of high strength AISI 4340 steel after cadmium plating and baking for various times. As it can be seen, the plateau failure stress level increases as the baking time increases indicating that the amount of retained hydrogen has been continuously reduced with time.

As to the failure time (delayed failure), it is rather short, less than 0.1 h, for 0.5 h baking at 149° C and this indicates that some hydrogen is still inside the metal.

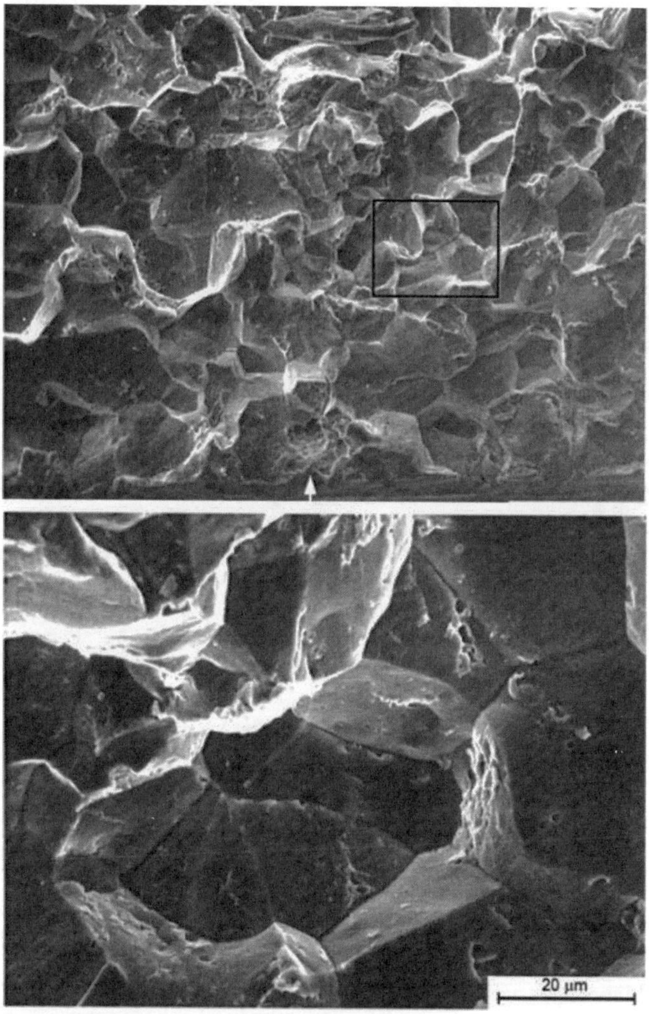

Fig. 18.17 Example of IGSCC of a bolt failed by IHE. The framed area is shown in more detail in the lower picture [24]

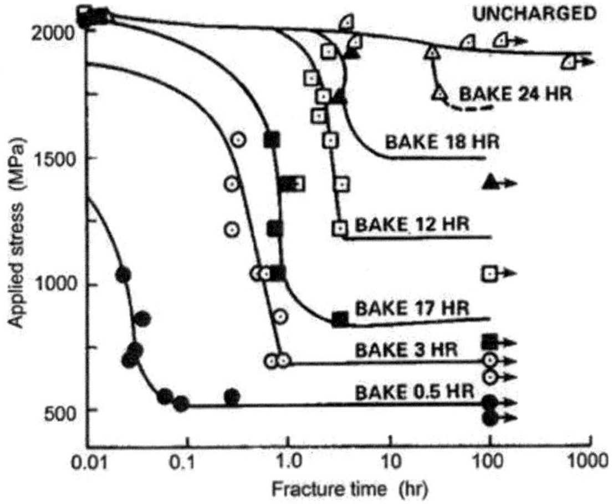

Fig. 18.18 Delayed failure curves for cadmium-plated 4340 steel after baking at 149 °C for different times [25]

Interesting enough, even after 24 h baking at 149° C there still is a small amount of hydrogen retained in the matrix that after about 50–60 h results in specimen failure at a stress slightly lower than that needed to produce failure in an uncharged specimen. Dough in small concentration, hydrogen can be very dangerous as shown in Fig. 18.19 [26] that presents the influence of gross internal hydrogen concentration on the stress intensity factor threshold K_{ISCC} (see Sect. 15.3) to initiate IGSCC in three high strength steels, 18 Ni maraging steel (2000 MPa or 290 ksi), 18 Ni maraging steel (1862 MPa or 270 ksi) and 300 M steel, respectively. Just few ppm of retained hydrogen are enough to make the K_{ISCC} precipitate.

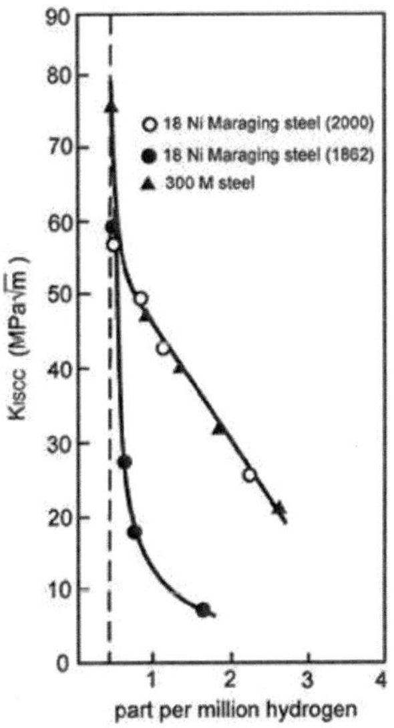

Fig. 18.19 Effect of hydrogen concentration on the K_{ISCC} for three high strength steels [26]

Moreover, the actual time needed to bake hydrogen out depends on the thickness of the work piece. This time increases with the square of the section so that to outgassing hydrogen from some millimeter thick section may take few hours but with a thick section it can mean hundreds of hours at any reasonable temperature. With austenitic stainless steels things get more complicated because from the one hand solubility of hydrogen in FCC (γ austenite) is higher than in BCC α-iron (ferrite), from the other hand its diffusivity is some order of magnitude lower. For example, at 100 °C and 1 atm of external hydrogen gas, the concentration at equilibrium in austenite is 1000 times higher than in ferrite, but its diffusivity is about 10^6 times lower. However, hydrogen damage in austenite is not a real problem unless some of the austenite transformed into martensite by cold working, which is very sensitive to hydrogen embrittlement.

18.4.2 External Hydrogen Embrittlement (EHE)

Hydrogen can enter the work piece from the working environment as atomic hydrogen. Molecular hydrogen cannot enter and diffuse in metals. At ambient temperature and normal pressure, hydrogen can be present as molecular hydrogen gas. However, it has been found that molecular hydrogen can chemically react with iron becoming atomic hydrogen easily absorbed at low temperatures (chemisorption). Chemisorption of dry hydrogen is very rapid, practically instantaneous, and this can explain the nearly absence of an incubation time. However, when oxygen is mixed to dry hydrogen or dry hydrogen sulfide, H_2S, hydrogen embrittlement is blocked and can start again only after the complete removal of oxygen. This suggest the hypothesis that metals reacts preferentially with oxygen rather than with hydrogen. This concept is consistent with the well-known affinity of iron with oxygen forming oxides that provide a barrier to hydrogen entrance into the metal. When oxygen is removed, hydrogen can reduce oxide to metal iron, in case of iron or steel, eliminating the barrier. This is shown in Fig. 18.20 [27] for a high strength steel type H-11, 1610 MPa yield strength. Dry hydrogen is very aggressive, but as oxygen is admitted in the environment, even in few amount as low as 0.6%, the embrittlement stops immediately. It is only upon O_2 complete removal that the hydrogen corrosion can start again, with the same rate as before.

18.4.3 Film Rupture/Anodic Dissolution Model

Virtually any metal or alloy is subjected to the hydrogen attack. However, high strength steels and special alloys are the most susceptible to hydrogen corrosion. Besides heat treatment sensitization to corrosion, see Sect. 18.2, it is precisely hydrogen embrittlement the major mechanism of corrosion attack under static stress or stress corrosion cracking (SCC). The starting mechanism is always that of film

Fig. 18.20 Oxygen effect on the corrosion rate in a high strength steel type H-11 [27]

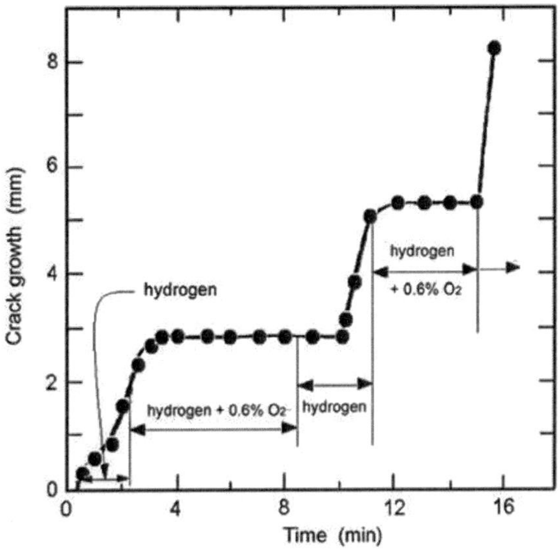

rupture/anodic dissolution model already discussed in Sect. 17.7, see Fig. 17.13, but with a different mechanism of crack propagation. This new mechanism is schematized in Fig. 18.21. Once the protective film is broken by slip dissolution, hydrogen can be absorbed through the bare metal surface at the crack tip.

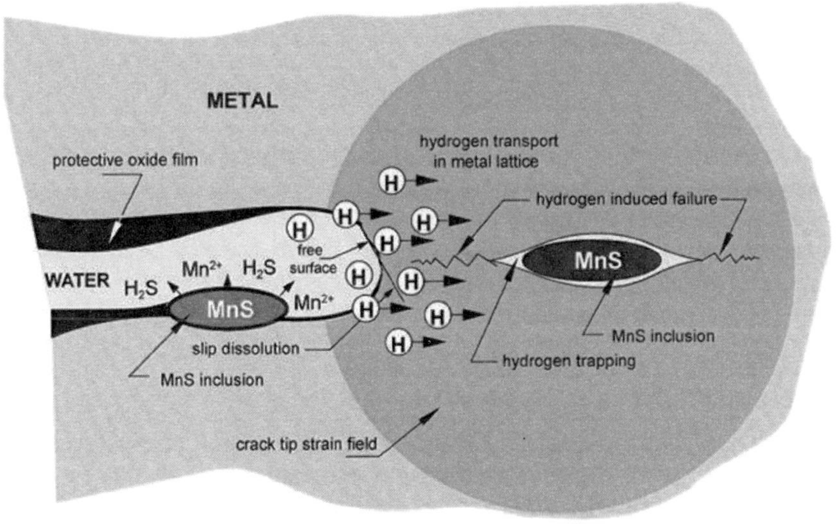

Fig. 18.21 Schematic of the creation of a free surface by film rupture/anodic dissolution process through which hydrogen may enter the bare metal (modified from [28])

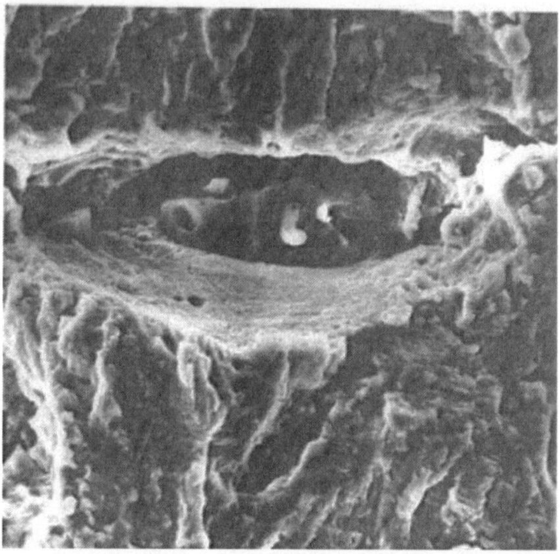

Fig. 18.22 MnS inclusion in a carbon steel dissolved in water at 288 °C [29]

In water environment, the local potential is extremely important since it determines whether or not hydrogen production is thermodynamically feasible. Figure 13.2a indicates that in carbon steels at ambient temperature, hydrogen can separate from water below – 0.6 V to give rise to the so called low potential SCC. In gaseous system, it is the hydrogen partial pressure that plays a similar role. Often, inclusions, such as manganese sulfide (MnS) can give rise to hydrogen. In fact, if crack propagation finds one of those inclusions and open it to the water environment, as schematized in Fig. 18.21 [28], the MnS inclusion can dissolve into water especially at higher temperatures. Figure 18.22 [29] shows a manganese sulfide inclusion in carbon steel type A 508 Cl3 dissolved in water at 288 °C.

The same type of inclusion has been fragmented on the surface of a 304 stainless steel plate following a cold work, as shown in Fig. 18.23 [30]. Fragmentation increases the surface exposed to the water environment thus accelerating dissolution. The dissolution of MnS inclusions produces H_2S and HS^- in the crack tip region. These species are good hydrogen donors and enhance hydrogen absorption on the bare metal surface. They also change completely the local chemistry creating an acidic environment. Regions with MnS inclusions are very suitable zone for the formation of corrosion pits due to the creation of localized anodic regions. Usually, the brittle, cleavage-like cracking starts from MnS inclusions and spreads like a fan, as shown in Fig. 18.39. Moreover, it is known, by now, that hydrogen sulfide acts as catalyzer in the hydrogen absorption process in that it can be considered a kind of poison that delay or prevent the recombination of atomic hydrogen in molecular hydrogen. It is also now clear that hydrogen absorption is enhanced by the existence of a high stress state and, in particular, of a triaxial or hydrostatic stress state [31]. It has been observed, in fact, that hydrogen embrittlement is not a problem when the metal prior of being loaded is exposed to a hydrogen rich environment. It seems that

Fig. 18.23 Fragments of a MnS inclusion on the surface of 304 stainless steel [30]

without a high stress state hydrogen does not enter the metal. However, still doubts persist about the stress state. For instance, if an hydrostatic stress state is necessary, then hydrogen embrittlement should not be observed in Mode III of crack aperture, see Fig. 10.3, i.e., under a pure shear stress that deforms the volume without changing the volume or without generating any hydrostatic stress state. Studies performed on high strength steels and titanium alloys indicate, effectively, complete absence of hydrogen cracking [32–34]. However, the same studies on brass in ammonia solution [33] have shown same results independently of the loading conditions, whether hydrostatic or pure shear.

As to high strength aluminum alloys in aqueous solution of Cl^-/CrO_3^-, it has been observed [33] hydrogen embrittlement in Mode I that reduced in Mode III conditions. Figure 18.24 [35] shows how the loading conditions can influence the response of a cylindrical specimen of 7075 aluminum alloy carrying a circumferential notch in 3.5% NaCl solution.

Fracture times in Mode I (traction, solid lines) are from one to two order of magnitude lower than those competing to Mode III (torsion, dashed lines), depending on the initial conditions of the alloy that were perfect aging (P), over-aged (OA) or under-aged (UA). Under traction, stresses needed to produce failure are by far lower and below the yield strength. A similar behavior is observed in 304 stainless steel in boiling solution of $MgCl_2$. The threshold stress intensity factor for IGSCC in Mode I was half that in Mode III and once triggered, stress corrosion by hydrogen embrittlement proceeded by an order of magnitude faster than in Mode III [36]. Figure 18.25 [37] presents the influence of the loading mode on the time to rupture of a stainless steel type 304 lightly sensitized under traction or torsion at 288° C in water with 7 ppm O_2 and 0.1 ppm sulfide as H_2SO_4. Specimens of 11 mm diameter had a circumferential notch, 60° V shaped and 2.6 mm depth, with a root radius of

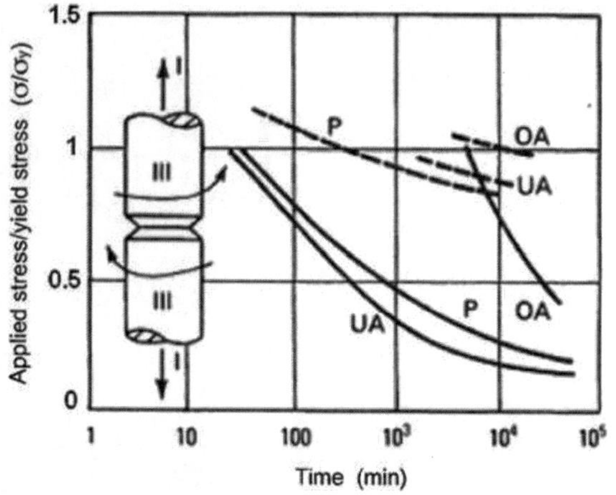

Fig. 18.24 Susceptibility of of 7075 aluminum alloy to loading conditions [35]

0.1 mm, with the exception of some that were pre-cracked and loaded in Mode III. The stress is normalized to the collapse stress σ_m of the notched section.

Yield strength in traction was equal to 340 MPa and 202 MPa in torsion, respectively. In Mode III no failure occurred even after 1000 h under a stress level close to that of plastic collapse of the pre-cracked specimens. On the contrary, in Mode I loading conditions intergranular fracture always occurred, as shown in Fig. 18.26 [37], with a percentage varying from 60 to 75 of the entire surface, while in Mode III no sign of corrosion was observable. The influence of the hydrostatic stress, dough still argument of discussion, is widely recognized as fundamental to cause hydrogen absorption into high strength alloys, in particular. This triaxiality influence on the hydrogen effect in high strength steels is well documented in Fig. 18.27 [38] that

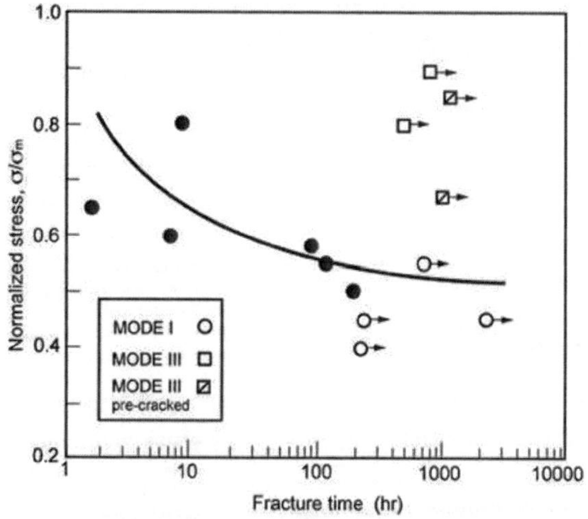

Fig. 18.25 Normalized stress versus time to fracture for notched specimens of lightly sensitized 304 stainless steel in water at 288 °C with ~ 7 ppm di O_2 and 0.1 ppm sulfide [37]

reports the failure stress versus time to fracture for high strength steel specimens having different notch tip radius. The degree of triaxiality existing ahead of the notch depends on the radius of curvature of its tip, increasing as the notch becomes sharper and sharper. It can be seen how in blunted notch specimens (5–6 mm radius) waiting times are enormously extended, even under very high stress state, while in those with sharp notch (0.025 mm radius) the waiting time reduces to just 6 min already at a stress of 550 MPa.

Fig. 18.26 IGSCC by hydrogen embrittlement in mode I loading [37]

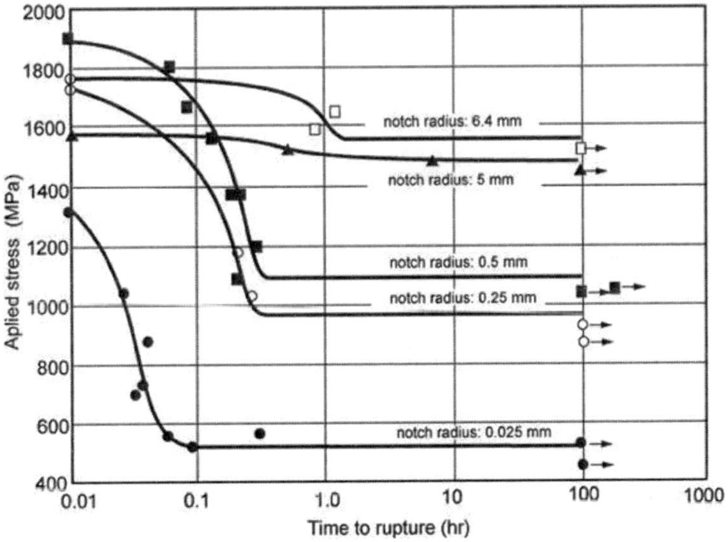

Fig. 18.27 Influence of notch root radius (triaxiality) on the time to failure by hydrogen embrittlement in high strength steel specimens [38]

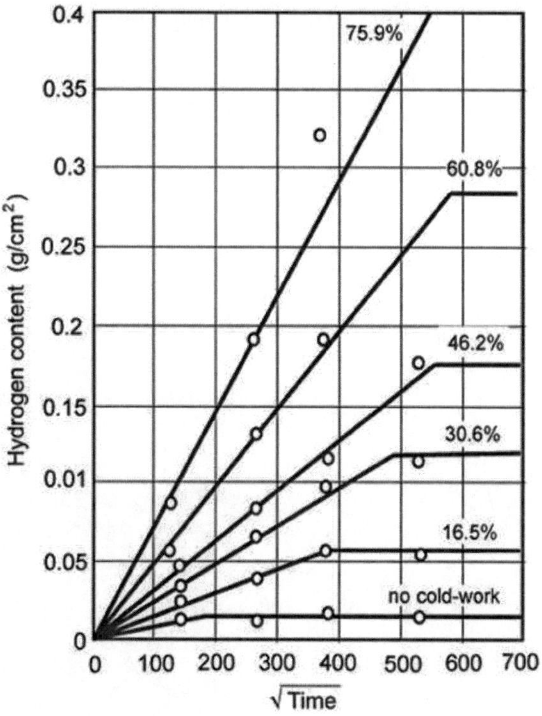

Fig. 18.28 Cold work effect on the absorbed hydrogen concentration [39]

Dough small, a 0.025 mm notch radius is rather far from being a crack whose root radius is no larger than 0.005 mm and results in the maximum triaxiality. Therefore, a crack is the most dangerous initiator of SCC. Furthermore, the high strain field ahead of the notch or crack tip favors the mobility and diffusion of hydrogen. Figure 18.28 [39] is, in fact, showing how plastic strains influence the time needed to absorbed hydrogen to reach a given concentration.

18.5 Mechanisms of Hydrogen Embrittlement

We shall now describe how hydrogen can effectively attack and jeopardize the integrity of metals and alloys. Dough this aggression always initiates with the protective film rupture, hydrogen embrittlement at variance with what has been long thought, is not just a single process of degradation or a specific change of the mechanical behavior of engineering alloys. For years it has been generically referring to as hydrogen embrittlement to describe a series of phenomena very dissimilar from each other that can be brought back to one of these five major mechanisms, which are

1. Atomic decohesion [40–43];
2. Dislocation interaction [44–48];
3. Cathodic hydrogen absorption [49, 50];
4. Brittle phases formation and/or rupture of existing phases [51–55];
5. Hydrogen blistering [56–58].

18.5.1 Atomic Decohesion Embrittlement

The hypothesis that hydrogen could influence the strength of the crystal lattice in which diffused was formulated for the first time by Petch and Stables [59] and reformulated by Barnett and Troiano in 1957 [60], dough with different aspects. These latter supposed that hydrogen could interfere with crystal lattice weakening locally the atomic bonds. This atomic decohesion, dough recognized, has never been adequately formulated in its specific mechanisms of action. It is based on the thermodynamic possibility of hydrogen supersaturation in areas of high hydrostatic stresses. For this reason it was considered to affect only hard, high-very high strength materials, since it is difficult to create a high triaxiality stress state ahead of a crack tip in ductile materials. In ductile metals, in fact, there is enough ductility to blunt the crack or notch tip that loses its acuity making the stress field amplitude drop down. However, recently this phenomenon has been observed also in low-alloyed steels that were considered immune. Hydrogen accumulated at the notch tip reduces atomic cohesion forces making metal fracture easier. In the original Petch and Stables idea, the embrittlement mechanism, also called embrittlement by absorption, was due to a decrease of the surface energy of the metal following the hydrogen absorption. However, this model leaves some doubts. In fact, recalling the Griffith theory [61], in an infinite, ideally brittle body containing a central flaw of length $2a$ under remote loading the critical stress σ_c is given by

$$\sigma_c = \sqrt{\frac{2E\Gamma}{\pi a}} \tag{18.1}$$

with Γ being the surface energy density of the material. A decrease of Γ results in a reduction of σ_c. However, this may be true for ideally brittle materials in which the only energies that can be retained in the matrix and are the surface energy that appears in Eq. (18.1) and the elastic deformation energy. They are irreversible form of energy that can turn into each other. But in real materials, dough brittle they can be, the contribution of Γ is completely negligible with respect to the plastic deformation energy E_p that always precedes fracture that is not reversible at all. This actually means that to invoke a contribution of absorbed hydrogen to the dechoesion embrittlement process it must necessarily come from a decrease of the plastic deformation energy E_p that substitutes the term Γ in Eq. (18.1). But if this is true, it must be associated to the failure mechanism of dislocation interaction, which it will be said next. Several authors have also been pointing out that such a mechanism should result in a propagation of corrosion of the same order of magnitude of brittle fracture, which is of the order some thousands of meters per second, while stress corrosion cracking is by far slower, of millimeters per second at most.

However, it must also be considered that while in brittle fracture it is the entire metal matrix that fractures at once, in the case of hydrogen embrittlement it is only few areas locally embrittled to fracture in a matrix that is still ductile. Therefore, each elementary act of growth must be followed by a further absorption process that will take some time, dough very short, with the result that it will be each elementary act of growth to occur with the dynamic of brittle fracture and not the entire propagation that is slowed down by waiting times. Hydrogen absorption interests the notch or crack tip high strain zone where it acts locally on atomic bonds or, in particular, in the high triaxiality stress field, as schematized if Fig. 18.29a and b. in case (a) it is the crack to propagate, in case (b) it is the micro-fracture ahead of the tip to propagate towards the main crack. In both cases, the process is discontinuous, joggling. It is thought that the process interests just some atomic distance, hence the resulting fracture is of the order of tens of Angstrom, i.e., of about 10^{-3} μm, afterwards it must wait a new hydrogen charging. Figure 18.30 [62] presents an interesting example of phosphorus and sulfur concentration effect on the energy (resilience) absorbed by high-strength steel specimens. The resilience drops dramatically as the P and S content increases.

The dependence of resilience on S and P content is not a new fact, but the authors claim that there was a strong contribution to the loss of ductility by hydrogen absorption embrittlement due to the presence of these two elements that inhibit hydrogen recombination as H_2 and act as catalytic centers.

18.5.2 Dislocation Interaction Embrittlement

The second mechanism of hydrogen embrittlement, probably the most important, is that of local hardening by dislocation locking. Among all interstitial atoms that can be present in the metal matrix, as carbon, nitrogen, oxygen, etc., hydrogen is the smallest. The arrangement of these foreign atoms in a BCC or FCC crystal is rather difficult because of the extremely limited interatomic spaces, so that they have to deform the crystal lattice. These atoms, then, tend to reach a situation of lower potential energy that is offered by dislocation. Dislocations, in fact, act as a wedge in the metal matrix putting the upper zone, over the dislocation line, in compression and the lower one in traction. Foreign atoms are pushed into this enlarged traction zone. Therefore, they gather below the dislocation line forming what is referred to as the Cottrell atmosphere. Plastic slips are activated by dislocation movements which, in turn, must displace and carry on their Cottrell atmosphere of foreign atoms. The respective motions can be very different since they obey to different principles. While, in fact, dislocation movements respond to actions directly exerted by stresses, therefore, to a mechanical action, the movements of the atoms in the Cottrell atmosphere are governed by atomic diffusivity laws that depend on temperature. Depending on temperature, then, this may cause a delay of the Cottrell atmosphere motion with respect to that of their dislocation that causes a reaction to the dislocation movement. It is, then, clear that this locking action exerted by the Cottrell atmosphere on its dislocation is more or less effective depending on the temperature and atoms concentration.

Hydrogen atoms, because of their minimal dimension among all interstitial atoms, can overpopulate the Cottrell atmosphere and block dislocations. As temperature decreases, the locking action increases and this is seen from the outside as temperature dependence of yield strength on temperature. Metals having concentrated Cottrell atmospheres, like ferrite and in general BCC materials, have a yield strength strongly dependent on temperature, while metals having a dilute Cottrell atmosphere do not, like austenite and FCC metals and alloys in general. These latter materials, in fact, do not show a temperature dependent component of the yield strength. For this reason austenitic stainless steel are less susceptible to hydrogen embrittlement caused by dislocation locking. But also martensitic high-strength steels do not seem to be that susceptible to dislocation blocking by hydrogen absorption. This is because their hardness is already obtained by a mechanism of dislocation locking operated by precipitates that cannot be further worsened by hydrogen absorption. These steels, instead, are very susceptible to hydrogen embrittlement, but caused by another mechanism: that of cathodic hydrogen absorption. Significant, from the point of view of dislocation locking, is Fig. 18.31 [63] that presents the response of a flat specimen of a high strength cast steel, type H50, subjected to a traction test at various temperatures.

While in vacuo or in air the material respond with a ductility (reduction of area) that increases with increasing temperature, in a hydrogen atmosphere it embrittles the more the higher the pressure. Note how ductility reduces the most at those temperature at which the hydrogen absorption is maximum (see also Fig. 15.40), i.e., low temperature, while it rises back to that shown in air at higher temperatures at which hydrogen presents a high diffusivity and is baked-out and outgassed (see Fig. 18.18). It is clear that the morphology will be typically transgranular and characteristic of low ductility steels. Traction tests were run at normal strain rate, about 1.1×10^{-3} s^{-1}. Note the difference with austenitic stainless steels of Fig. 18.9 with a maximum propagation rate that occurs between 180 °C and 250 °C.

18.5.3 Cathodic Hydrogen Absorption

Also for this mechanism of corrosion there still is not a precise description, for the very same reason already explained that is the fact that traditionally just the simple hydrogen presence was, by itself, sufficient to justify SCC. Fundamentally, the process is still based on cathodic hydrogen absorption, that is hydrogen that activates reaction 3) of Eq. (13.17) with low oxygen. Hydrogen, this time, cannot combine to form molecular hydrogen because of the presence of impurities such as sulfur, phosphorus, arsenic, antimony, tin or bismuth, in particular, that act as poisons inhibiting the recombination ot hydrogen atoms into molecules. The mechanism hypnotized by Latanision et al. [49] is schematized in Fig. 18.32. tin and antimony precipitated on grain boundaries of nickel favor absorption of hydrogen that does not recombine in molecular hydrogen because of their presence. Hence, a preferential path for corrosion is created because the grain boundaries become brittle. In effect, it is not precisely a really new mechanism, but rather one in which impurities enable the hydrogen incoming along grain boundaries where they have precipitated. However, a critical review [50] of combined effects of impurities-hydrogen embrittlement revealed that impurities would equally behave in the presence of either cathodic or hydrogen gas and would not really favor hydrogen absorption.

The effect would rather oppose recombination of atomic hydrogen in molecular hydrogen or dissociate molecular hydrogen in atomic one. A phenomenon like this is observed in high-strength steels type 4130 and 4340 in an atmosphere of hydrogen sulfide [64, 65]. An example is offered in Fig. 18.33 for martensitic steel 4340 in hydrogen sulfide [64]. The stress corrosion crack growth rate (SCCGR) is that corresponding to steady state conditions that do not depend, any more, on the applied K_I (Region II of Fig. 19.11). it can be observed the same temperature dependent behavior already seen in austenitic stainless steel of Fig. 18.9 and the same intergranular nature of the corrosive aggression. Probably, the SCCCR fall down above a certain temperature is due to the hydrogen diffusion and escaping from the metal matrix. In effect, at those temperatures fracture return to be transgranular. The H_2S gas pressure shifts the temperature at which the metal becomes practically immune to hydrogen embrittlement towards some higher values.

A similar effect is seen in maraging steels, as shown in Fig. 18.34 [66, 67] for a steel type 18Ni(250) of 1650 MPa yield strength. Also in this case it can be seen the negative effect of hydrogen gas pressure and the that of temperature. Beyond 290–300 K (15–27 °C) the fracture becomes transgranular because hydrogen leaves the metal by diffusing out. While in (b) fracture occurs at a stress intensity factor of just 28 MPa$\sqrt{m}$ and is intergranular, in (a) it is transgranular and requires a stress intensity factor of 56 MPa$\sqrt{m}$.

18.5.4 Brittle Phases Formation and/or Rupture of Existing Phases

The mechanism of hydrogen embrittlement through the formation of brittle phases (hydrides) is still thought to be questionable by some researchers since stable hydrides have never been found in metals with hydrogen cracks. Fujita [68] proposed that hydrogen can accumulate in iron in form of hydride lamellae, favored by the existence of high triaxial stress states. Hydrides are very brittle, but also very difficult to observe since they are not stable. Micro-cracks produced by the fracture of these phases join between themselves and with the main crack, so that the resulting growth is of jerking nature. The presence of MnS inclusions makes the process of internal micro-fracturing even easier because, especially when they debond from metal matrix, hydrogen fill the micro-cavity exerting its action right there where there is a local stress concentration favored by the long and thin shape of such inclusions. Today, it is certain that this mechanism can occur also in medium-strength steel under cyclic stress, in particular [69, 70]. Severe embrittlement is caused by hydrogen absorption and hydride formation in titanium, zirconium and vanadium metals and alloys.

18.5.5 Hydrogen Blistering

The last possible mechanism of hydrogen rupture of the crystal matrix is due to the hydrogen gas formation. This rupture interests, in particular, medium–low strength steels. When hydrogen diffuses into the metal matrix it can eventually find internal micro-cavities, often associated with MnS or other inclusions dedonded from the matrix, or pores where it enters forming molecular hydrogen. Very high pressure may build up at such locations due to continued hydrogen absorption and recombination into hydrogen gas. It is not known precisely at what pressure this mechanism will end, but it is thought that the equilibrium constant of this process, at which it terminates, is at extremely high values of pressure, in excess of even 100,000 bar.

Therefore, when it starts, the process must eventually lead to the rupture of these cavities by internal over pressurization. The phenomenon is known as hydrogen blistering because it forms bubbles like blisters in steels, aluminum alloys and titanium alloys. Since it is a diffusion related process, hydrogen blistering needs a suitable temperature that allows atomic transport and is favored by the existence of temperature traps where the metal is locally cooler. It is there the hydrogen can preferentially accumulate inside existing micro-cavities. An example of hydrogen blistering is offered in Fig. 18.35 [71]. The figure refers to two flat steel specimens of carbon steel of (a) 636 MPa of ultimate strength, 521 MPa yield strength and area reduction equal to 69.5% and (b) 677 MPa of ultimate strength, 542 MPa yield strength and area reduction equal to 57%, respectively. Specimens were subjected to cathodic hydrogen charging for six months in a solution 0.004 wt% of H_2SO_4 in distilled water at pH 3.2 with a current density of 0.1 mA/cm^2. In Fig. 18.35a and b it is possible to see the blisters appeared at the surface, while in (c) and (d) the same blisters appear in a cross section of the specimens. At higher temperatures, usually over 200 °C, in carbon steels the presence of iron carbide (cementite) can also favor the methane formation through the reaction

$$FeC + 2H_2 \Leftrightarrow 3Fe + CH_4 \quad (gas) \tag{18.2}$$

Also in this case, the methane gas pressure continuously grows till final fracture. Moreover, the carbon impoverishment, because of the methane gas formation, weakens the metal.

18.6 Morphology of Hydrogen Embrittlement

To recognize a hydrogen embrittlement failure is not that simple. First because the failure mechanism is not unique, as discussed in Sect. 18.4, secondly because the detection of hydrogen in metals and the measurement of its concentration is difficult and costly, when not impossible. Its detection, in fact, on the fracture surfaces is practically impossible, given the extreme volatility of the element that evaporates immediately after the exposure of the surface to the air. On the other hand, measurements of its concentration somewhere inside the metal, dough possible, provides really very few information on what could have been the concentration elsewhere, in particular in the fractured area. The morphology remains as the most important element of diagnosis. However, at variance with what commonly thought, the aspect of hydrogen embrittlement failure is not necessarily intergranular, dough it is the most common one. At the optical microscope or SEM analysis the appearance of the hydrogen induced fracture depends on the type of damage caused by a possible exposure to hydrogen, as said in the previous sections. Generally, fracture produced by internal hydrogen, see Sect. 18.4.1, as consequence of chemical or electrochemical pre-operational processing like cleaning and pickling cycles, electroplating, phosphating solutions, cathodic cleaning solutions and stripping appears intergranular, as shown in Fig. 18.17. In these cases there could be not just one but a series of evidences leading to the internal hydrogen cause of failure. This is shown, for example, in Fig. 18.36 that refers to a bolt of high-strength steel type 34CrMo4 of 1250 MPa yield strength that failed in service after few weeks. Figure 18.36a is a macrograph of the cross section of the bolt that failed. It is possible to recognize the origin of failure in the form of a tiny nail shaped crack, indicated by the arrow, on a completely flat surface apparently without any sign of plastic deformation (brittle).

If attention were focused on the initiation nail it will appear the intergranular nature of failure, as it can be possible to see in Fig. 18.36b in which intergranular cracks appear. The rest of the surface is interested by transgranular cleavage characteristic of sudden fracture. This may be the first evidence of a hydrogen embrittlement that initiated the crack that triggered the fracture. If, then, the attention is moved to the shank of the bolt, it will appear a network of intergranular microscopic cracks, as visible in Fig. 18.36c, that affect just few microns of material. The same examination made on bolts that never entered into service, but were coming from the same lot of the bolts that failed, reveled the same network of surface microscopic intergranular cracks. This set of evidences convinced that the failure was due to a not well done hot bluing. The operation had introduced hydrogen in the metal that could not be completely removed by the following baking procedure because the oxide layer that protected the external surface acted as a barrier to hydrogen outgassing or, probably, the baking of the bolts was not done immediately after the bluing operation so that hydrogen was given enough time to act and form the network of micro-cracks. In these cases, as in the case of electroplating with cadmium that acts as a barrier, it is necessary to bake the work pieces as soon as possible and at temperatures higher than the usual 150–200 °C and for longer times.

Another evidence of hydrogen embrittlement in ductile carbon steels stems out of the presence of brittle cracks. This time it is the external hydrogen that is acting. In the case shown in Fig. 18.37 the cracks emanated from sharp MnS inclusions [69] in a carbon steel type A 302 Gr B in pure water at 288 °C. This phenomenon is rather usual in medium-to-low strength carbon steels in which hydrogen may be trapped in the surroundings of inclusions, the most common of which is MnS, to form molecular hydrogen or methane if the temperature is over 200 °C. If the inclusion is sharp the buildup of internal pressure cracks the metal ahead of it.

It is significant the long crack apparently not associated to any inclusion. The fracture has a stepped aspect, typical of this type of attack already seen in Fig. 18.35d. But when the hydrogen or methane gas accumulates along grain boundaries then fracture turns from transgranular into intergranular, as it can be seen in Fig. 18.38 [72] that refers to a carbon steel pipe of A 106 severely damaged by hydrogen attack. Fracture occurred after 15 months of service in hydrogen gas at 35.4 MPa and 320 °C. Another example of hydrogen-MnS inclusion fracture can be seen in Fig. 18.39 [72]. This time it is a carbon steel type A 533 B Cl 1 specimen that failed in high purity water at 288 °C. It is possible to identify the three initiation sites indicated by the letters A, B and C, respectively. The three initiation sites are MnS inclusions. these sites are localized immediately below the crack front introduced in the CT specimen by fatigue. The elevated triaxial stress state at the crack tip has favored the hydrogen absorption, see Fig. 18.29. Note the fan shaped fracture emanating from the initiation sites, flat and without any apparent sign of plastic deformation. The SEM analysis revealed the transgranular nature of the fracture, dough caused by hydrogen embrittlement. An example of hydrogen induced stress corrosion cracking is visible in Fig. 18.40a [73] for a 18Ni-Cr–Mo steel specimen type WOL having 1765 MPa yield strength.

This time the morphology is that of a branched mainly intergranular crack typical stress corrosion cracking in high strength steels. Fully intergranular is, instead, the one observed in a 12Ni-5Cr-3Mo steel of 1262 MPa yield strength, shown in Fig. 18.39b. In this latter case fracture started from a weld following a interdendritic path and arriving in the base material. the same morphological aspects appear at the SEM analysis. Figure 18.39c shows the characteristic brittle fracture appearance, while Fig. 18.39d presents the characteristic features of the intergranular fracture in 12Cr-Mo-V steel that separates along grain boundaries [73]. Reassuming, it is possible to say that, in general, brittle transgranular fracture characterizes medium-to-low strength carbon steel in which hydrogen embrittlement is mainly due to blistering by methane gas accumulation. When this accumulation is in grain boundaries the fracture appear as intergranular. In high strength steels hydrogen causes preferentially an intergranular stress corrosion attack.

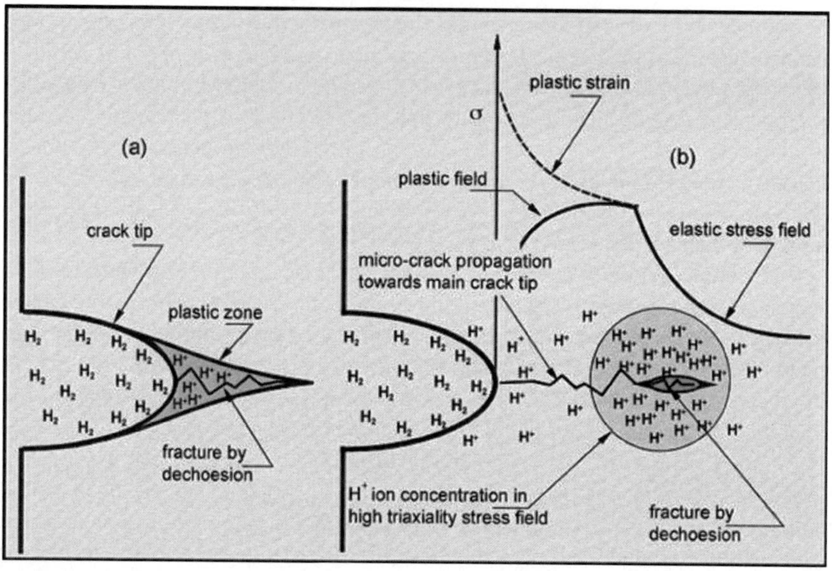

Fig. 18.29 Schematic of the dechoesion mechanism by hydrogen absorption, **a** cracking of the crach tip and **b** cracking ahead of the tip

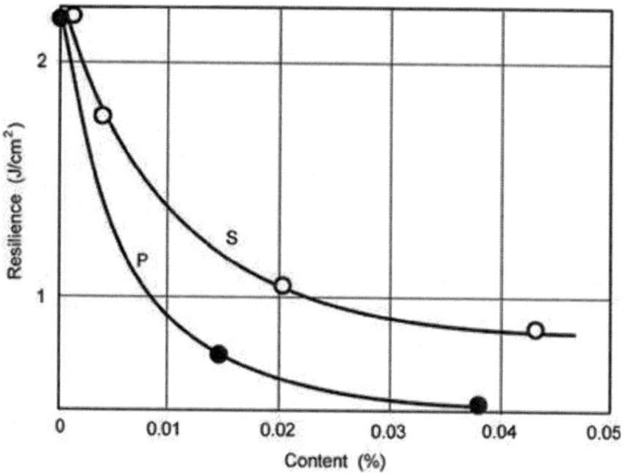

Fig. 18.30 High-strength steel hardening due to hydrogen embrittlement enhanced by Sulfur (S) and Phosphorus (P) concentration [62]

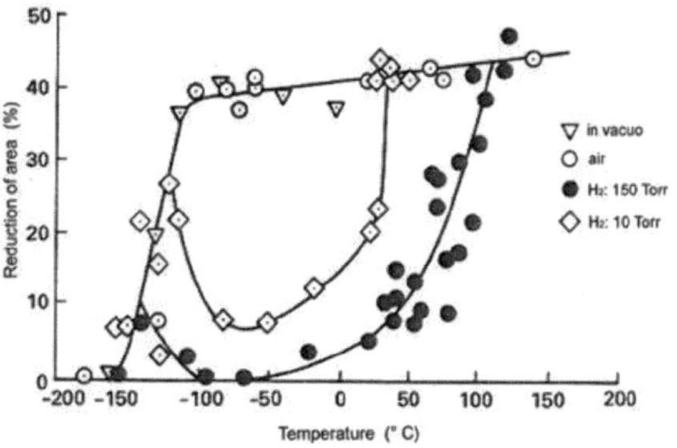

Fig. 18.31 Reduction of area in a smooth bar of high-strength steel type H50 versus tempreature for several environments [63]

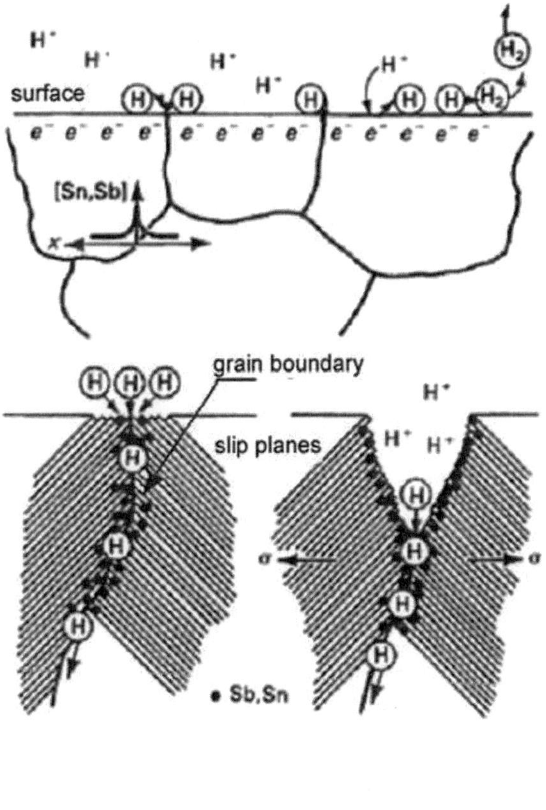

Fig. 18.32 Schematic showing how antimony and tin favor the intergranular absorption of cathodic hydrogen [49]

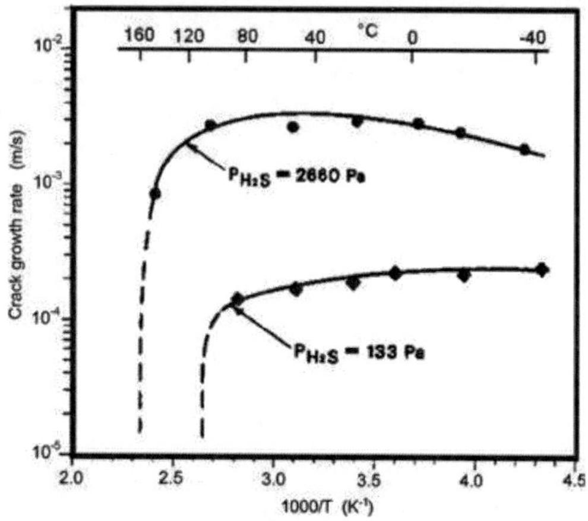

Fig. 18.33 Temperature dependance of SCCCR in 4340 steel in hydrogen sulfide [64]

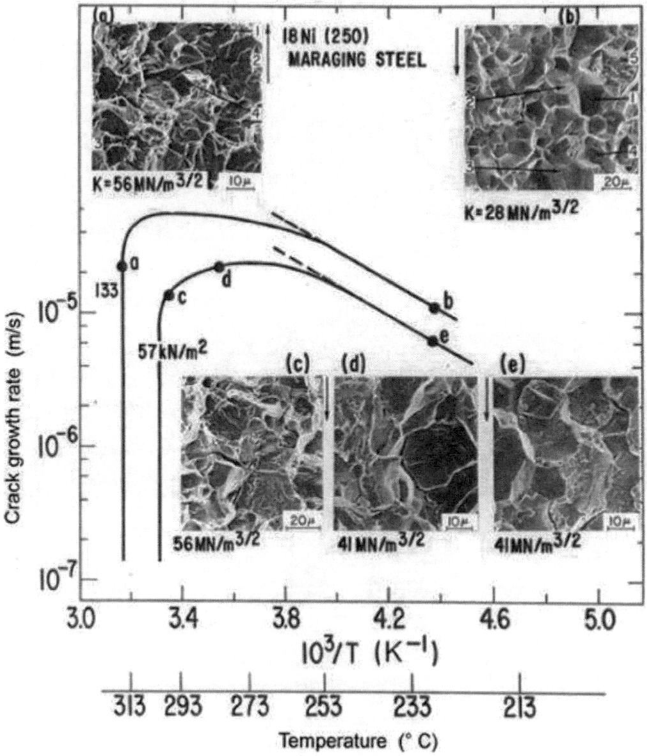

Fig. 18.34 Dependency of SCCCR of a high-strength steel type 18Ni(250) on temperature for two different hydrogen gas pressures [66, 67]. Note the passage from intergranular to transgranular fracture with increasing temperature

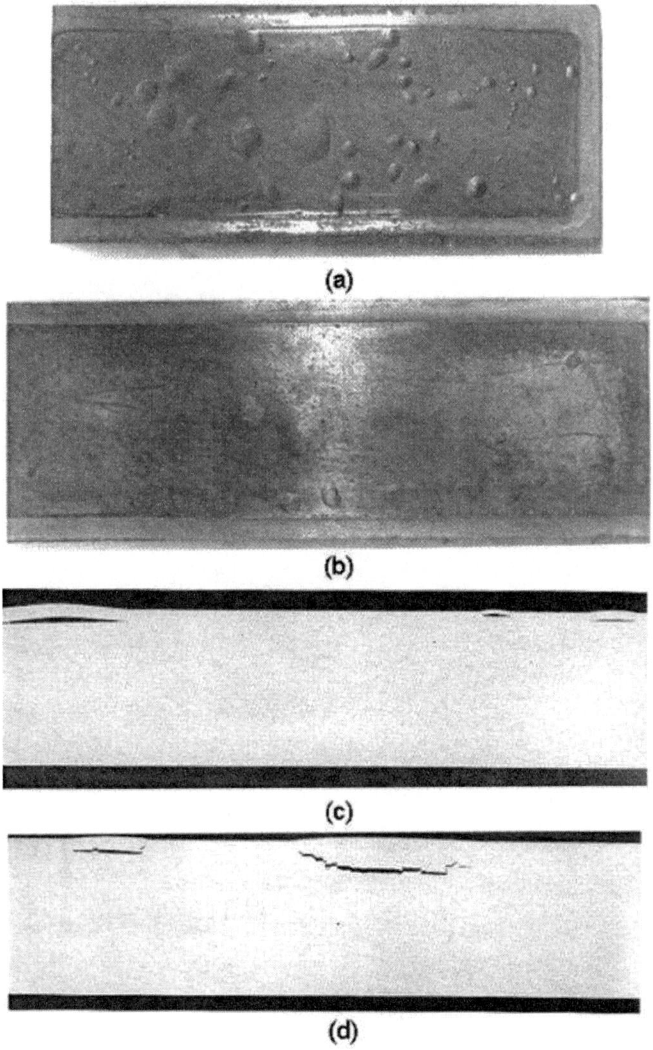

Fig. 18.35 Appearance of the external surface of two flat specimens a and b subjected to hydrogen charging and blistering [71]. c and d show a cross-section view of the same specimens [71]

Fig. 18.36 High-strength bolt that failed by hydrogen embrittlement after few weeks of service

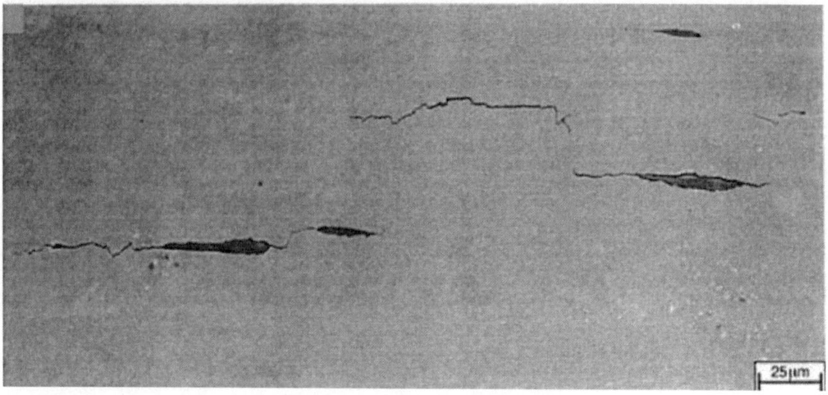

Fig. 18.37 Cracks formation ahead of MnS inclusions by hydrogen embrittlement in carbon steel at 288 °C in water [69]

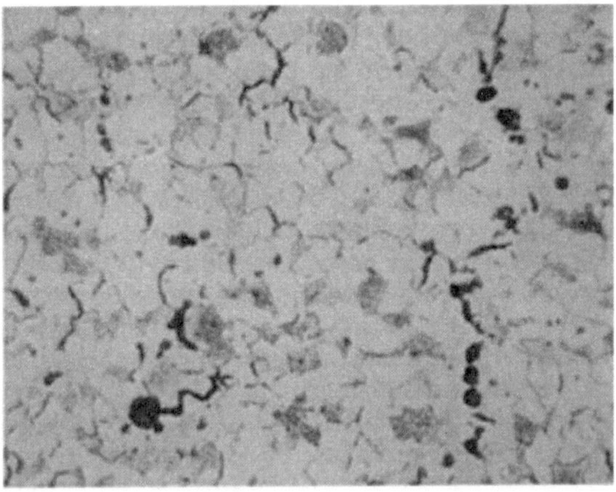

Fig. 18.38 Micrograph of the intergranular attack in a carbon steel pipe type A 106 severely damaged by hydrogen embrittlement [72]

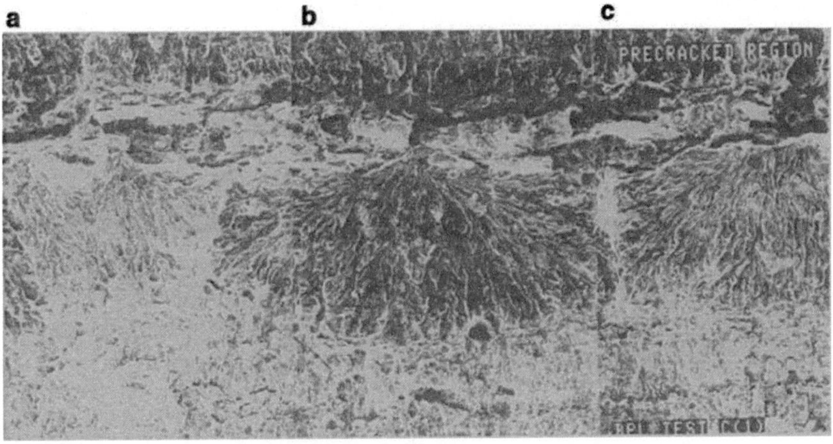

Fig. 18.39 Hydrogen induced fan shaped fracture emanating from three initiation sites **a**, **b** and **c** identified as MnS inclusions in carbon steel type A 533 [72]

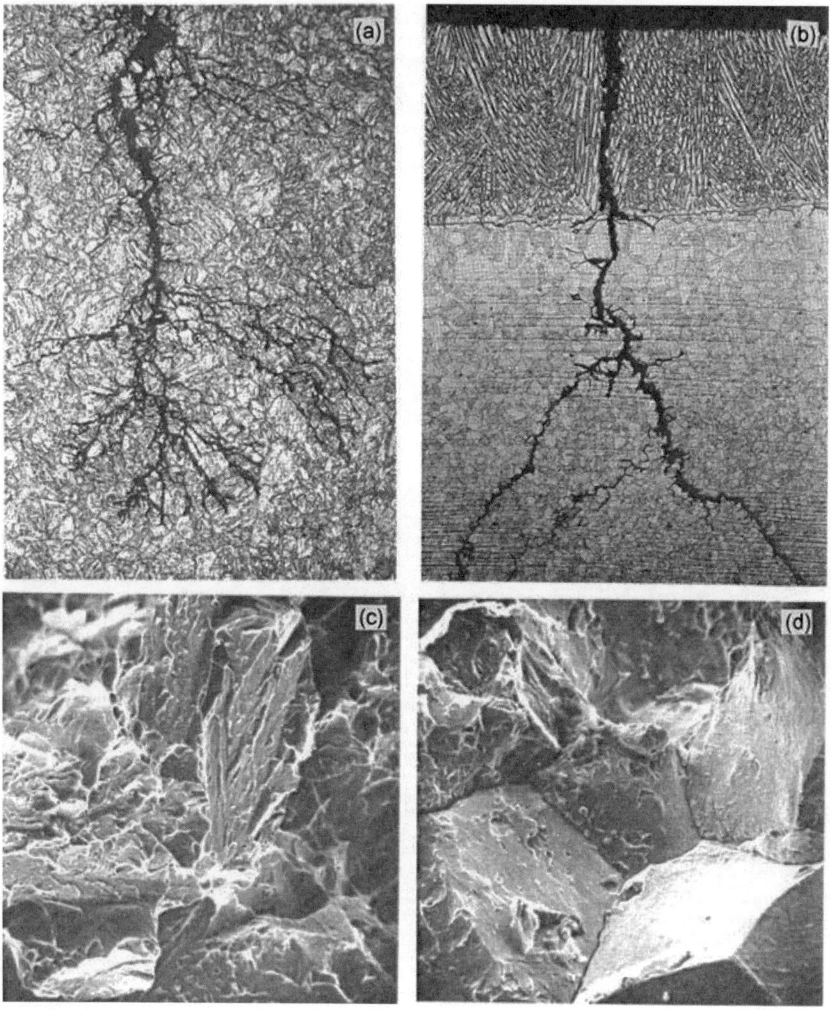

Fig. 18.40 Morphology of stress corrosion cracking **a** trans-intergranular with branching, **b** intergranular with branching. SEM analysis: **c** brittle fracture, **d** intergranular [73]

References

1. Stickler, R., Vinckier, A.: Morphology of grain-boundary carbides and its influence on intergranular corrosion of 304 stainless steel. Trans. Amer. Soc. Metals **54**, 362–380 (1961)
2. ASTM G-59.: Standard Method for Conducting Potentio-Dynamic Polarization Resistance Measurements (1997)
3. Strauss, B., Schottky, H., Innueber, J.: Die Carbidausseheidung beim Gluehen von Nichtrostenden, Unmagnetischen Cr-Ni Stahl. Z. Anorg. Allg. Chem. **188**, 309–324 (1930)

4. Chevenard, P., Portevin, A.: Etude Expérimentale de l'Héterogénéité de Métaux et Alliages. Rev. Mét. **33**, 96–113 (1936)
5. Conserva, M. Leoni, M., Met. Trans. A, Physical Metallurgy and Mate. Sci. 189 (1975)
6. Osozawa, K., Hengell, H,J.: Corrosion Sci. 389 (1966)
7. Keddam, M., Mattos, O.R., Takenouti, H.: Electrochimica Acta 1147 (1986)
8. Dobbelaar, J.A.L., Herman, E.C.M., De Wit, J.H.W.: Corrosion Sci. 765 (1992)
9. Szclarska-Smialovska, Z.: Pitting and Crevice Corrosion. NACE Press, Huston, Texas (2005)
10. Davidson, R.M., DeBold, T., Johnson, M.J.: In: Metals Handbook, vol. 13, 9th edn, pp. 551. ASM International (1987)
11. Soppet, W.K., Kassner, T.F.: Evaluation of environmental corrective actions. In: Environmentally Assisted Cracking il Light Water Reactors, NUREG/CR-4287, ANL-85–33, Annual Report, June (1985)
12. Staehle, R.W.: Bases for predicting the earliest penetrations due to SCC for the alloy 600 on the secondary side of steam generators. US NRC NUREG/CR-6737, pp. 15. (2001)
13. Chaung, H.E., Lumsden, J.B.: Grain boundary segregation in nickel. EPRI Report FCC **7704**, 9 (1977)
14. Searles, J.L., Gouma, P.I., Buchheit, R.G.: Stress corrosion cracking of sensitized AA5083 (Al-4.5Mg-1.0Mn). Metallurgical and Mater. Trans A **32A**, 2859–2867 (2001)
15. Dillamore, I.L., Roberts, W.T., Wilson, D.V.: In: The Mechanical Properties of Stainless Steel with Particular Reference to Crystallographic Anisotropy. Iron and Steel Institute, vol. 117, pp. 37–49. Publication N., London (1969)
16. Griffiths, A.J., Wright, J.C.: Mechanical properties of austenite and metastable stainless steel sheet and their relationship with press forming behaviour, Ibidem
17. Tedmon, C.S., Vermilyea, D.A., Broeker, D.E.: technical note-effect of cold work on intergranular corrosion of sensitized stainless steel. Corrosion, NACE, March, vol. 27 (1971)
18. Indig, M.E., Vermilyea, D.A.: Corrosion of sensitized stainless steel in hot aqueous solution under natural and electrochemical control. Corrosion, NACE, February, vol. 31 (1975)
19. Vermilyea, D.A., Indig, M.E.: Corrosion and electrochemical studies in aqueous solution at 289 °C. In: Proceedings of 5th International Congress on Metallic Corrosion, NACE (2020)
20. Clarke, W.L.: Studies on AISI type 304 stainless steel piping weldments for use in BWR application. EPRI NP-944, Final Report, p. 3–60, December (1978)
21. Lai, G.Y.: In: High Temperature Corrosion of Engineering Alloys, ASM International (1990)
22. Jewett, C.W.: The growth and stability of stress corrosion cracks in large-diameter BWR piping. EPRI NP-2472, vol. 2, Final Report, July, pp. G-25. (1982)
23. Caine, T.A., Bensch, M.M.: Role of the loading mode on crack growth rates in sensitized 304 stainless steel. In: The Growth and Stability of Stress Corrosion Cracks in Large Diameter Piping, Vol. 2: Appendixes, EPRI NP-2472, Vol.2, Final Report, July, pp. H-86. (1982)
24. Cox, A.F., Pollok, W.J.: Investigation of the cracking in HS748 propeller blade retaining bolts. In: ARL File BM2/03/17, Material div., Ref. M89/83/AFC/WJP, DSTO, Melbourne, Australia (1984)
25. Troiano, A.R.: Delayed failure of high strength steel. Corrosion **15**, 207t–212t (1959)
26. Dautovic, P.D., Floreen, S.: Metallurgical Trans. **4**, 2627–2630 (1973)
27. Johnson, H.H., Paris, P.C.: Subcritical flaw growth. Eng. Fract. Mech. **1**, 3–45 (1968)
28. Bulloch, J.H., Buchanan, L.W.: Fatigue crack growth behavior of A 533-B steel in simulated PWR water. Corros. Sci. **24**(8), 661–674 (1984)
29. Atkinson, J.D., Forrest, J.E.: The role of MnS inclusion in the development of environmentally assisted cracking of nuclear reactor pressure vessel steels. In: Proceedings of the 2nd IAEA Specialist's Meeting on Subcritical Crack Growth, Moscow, US NRC NUREG/CP—0067, vol. 2, pp. 177. (1985)
30. Studies on AISI type 304 Stainless Steel Piping Weldments for Use in BWR Application. EPRI NP-944, Final Report, December, pp. 2–92. (1978)
31. Scott. P.M., Traswell, A.E.: The influence of water chemistry on fatigue crack propagation in LWR pressure vessel steels. In: Proceedings of the IAEA Specialists' Meeting on Subcritical Crack Growth, NUREG/CP—0044, vol. 2, pp. 91–126. (1981)

32. Green, J.A.S., Hayden, H.W.: In: Hydrogen in Metals. ASM, Metals Park, OH, pp. 235. (1974)
33. Green, J.A.S., Hayden, H.W., Montague, W.G.: In: Effect of Hydrogen on Behavior of Materials. TMS-AIME, New York, pp. 200. (1976)
34. Thompson, A.W., Bernstein, I.M.: In: Fontana, M.G., Staehle, R.W. (eds) Advances in Corrosion Sciences and Technology, vol. 7, pp. 53. Plenum Press (1980)
35. Riecke, R.M., Athens, A., Smith, I.O.: Mater. Sci. Technol. **2**, 1066 (1986)
36. John, C.F.: St., Scripta Metallurgica, vol. 9, pp. 141. (1974)
37. Ruther, W.E., Kassner, T.F., Nichols, F.A.: Mechanistic studies in environmentally assisted cracking in light water reactors. NUREG/CR-4287, ANL-85–33, Annual Report, June, pp. 121. (1985)
38. Johnson, H.H., et al.: Hydrogen crack initiation and delayed failure in steel. Trans. AIME **212**, 526–536 (1958)
39. Darken, L.S. et al.: Behavior of hydrogen in steel during and after immersion in acid. In: Hydrogen Damage, ASM, pp. 60–75. (1977)
40. Germer, L.H., Macrae, A.U.: Adsorption of hydrogen on a (110) nickel surface. J. Appl. Phys. **37**, 1382 (1962)
41. Van der Sluys, W.A.: Mechanisms of environmental induced subcritical flaw growth in AISI 4340 steel. Eng. Fract. Mech. **1**, 447–462 (1968)
42. Oriani, R.A., Josephic, P.H.: Equilibrium aspects of hydrogen-induced cracking of steels. Acta Metall. **22**(9), 1065–1074 (1974)
43. Oriani, R.A.: In: Stress-Corrosion Cracking and Hydrogen Embrittlement in Iron-Base Alloys. pp. 351. Houston, NACE Pubblication (1977)
44. Bastien, P., Azou, P.: In: Proceeding World Metallurgical Congress, 1st American Society of Metals, pp. 535–552. (1951)
45. Beachem, C.D.: Metallurgical Trans. **3**, 437–451 (1972)
46. Tien, J.K., et al.: Metallurgical Trans. **7A**, 821–829 (1976)
47. Lynch, S.P.: Metal Forum **2**, 189–200 (1979)
48. Hirth, J.P.: Metallurgical Trans. **11A**, 861–890 (1980)
49. Latanision, R.M., Opperhauser, H.: Metall. Trans. In: Scientific and Technical Book Service, vol. 5, pp. 483. (1974)
50. Jones, R.H.: A review of combined impurity segregation-hydrogen embrittlement processes. In: Latanision, R.M., Fischer, T.E. (eds) Advances in the Mechanics and Physics of Surfaces, Scientific and Technical Book Service (1986)
51. Raj, R., Varadan, V.K.: Mechanism of environment sensitive cracking of materials. University of Surrey, U.K., The Metals Society, Editors P.R. Swann, F.P. Ford and A.R.C. Westwood, pp. 426. (1977)
52. Scully, J.C., Powell, D.T.: Corros. Sci. **10**, 719 (1970)
53. McCoy, R.A., Gerberich, W.W.: Metallurgical Trans. **4**, 539 (1973)
54. McCoy, R.A.: In: Bernstein, I.M., Thompson, A.W. (eds) Hydrogen in Metals. pp. 169. ASM Publication (1977)
55. Bernstein, I.M., Thompson, A.W.: The role of metallurgical variables in hydrogen assisted environmental fracture. Rockwell Science Center Report, SCPP-75-63
56. Zapffe, C.A., Sims, C.E.: Hydrogen embrittlement, internal stress and defects in steel. Trans. AIME **145**, 225–259 (1941)
57. Tetelman, A.J., Robertson, W.D.: Transactons AIME. vol. 224. pp. 775. (1962)
58. Tetelman, A.J.: In: Fracture of Solids. pp. 671. J. Wiley, New York (1963)
59. Petch, N.O., Stables, P.: Nature. London **169**, 842–843 (1952)
60. Barnett, W.J., Troiano, A.R.: J. Metals **209**, 486–494 (1957)
61. Griffith, A.A.: The phenomena of rupture and flow in solids. Trans. Royal Soc. London **221** (1920)
62. Zubko, A.M., Malkin, V.I., Medvedev, É.A., Pokidyshev, V.V., Khokhlov, S.F., Shnol, E.M.: Effect of phosphorus and sulfur on hydrogen absorption in high-strength steel and its susceptibility to corrosion cracking. Metal Sci and Heat Treatm **15**(12), 1071–1073 (1973)
63. Bowker, P., Hardie, D.: Metal Sci. **9**, 432 (1975)

64. Lu, M., Pao, P.S., Weir, T.W., Simmons, G.W., Wei, R.P.: Rate controlling processes for crack growth in hidrogen sulfide for an AISI 4340 steel. Met. Trans. **12**, 805–811 (1981)
65. Gao, M., Lu, M., Wei, R.P.: Crack path and hydrogen assisted crack growth response in AISI 4340 steel. Met. Trans. **15**, 735–746 (1984)
66. Gangloff, R.P., Wei, R.P.: Gaseous hydrogen embrittlement of high strength steels. Met. Trans. **8**, 1043–1053 (1977)
67. Gangloff, R.P., Wei, R.P.: In: Fractography in Failure Analysis, ASTM STP 645, ASTM International, West Conshohocken, PA, pp. 87–106. (1978)
68. Fujita, F.E.: Theory of hydrogen induced delayed fracture of steels. In: Azou, P. (ed) 2nd International Congress on Hydrogen in Metals, Paris, Paper 2B10 (1977)
69. Chopra, O.K., Shack, W.J.: Effect of LWR coolant environments on fatigue design curves of carbon and low-alloy steels. NRC NUREG/CR-6583 (1997)
70. Hanninen, H., Torronen, K., Kemppainen, M., Solonen, S.: Corrosion Sci. **23**, 663 (1983)
71. Groeneveld, T.P., Fessler, R.R.: Hydrogen stress cracking overview and controls. In: 6th Symposium on Line Pipe Research, Pipeline Research Committee of American Gas Ass, Huston, TX, pp. Y-17–Y-18. (1979)
72. Atkinson, J.D., Bulloch, J.H., Forrest, J.E.: A fractographic study of fatigue cracks produced in A533B pressure vessel steel exposed to simulated PWR primary water environment. In: Proceedings of the 2nd IAEA Specialist's Meeting on Subcritical Crack Growth, Moscow, US NRC NUREG/CP−0067, vol. 2, pp. 290. (1985)
73. Phelps, E.H.: Microscopic identification of stress-corrosion cracking in steels with high yield strength. In: Specialists Meeting on Stress Corrosion Testing Methods, AGARD Conference, AGARD-CP-98, pp. 24–10. (1972)

Chapter 19
Fracture Mechanics Approach to Stress Corrosion

19.1 Introduction

Stress corrosion cracking (SCC) is used as a generic term to describe any behavior in which a combination of a static load and an aggressive environment results in the generation and propagation of a crack by corrosion. Stress corrosion has been treated from the electrochemical point of view in Chap. 17 and with a chemical-metallurgical approach in Chap. 18 for the important cases of hydrogen embrittlement and sensitization embrittlement of metals. We shall now study SCC from a mechanical point of view showing how fracture mechanics, in particular, originally developed to treat brittle fracture of World War II ships and later used in the treatment of fatigue crack propagation (see Chap. 10) can be also applied to treat the mysterious world of stress corrosion. It has been said in Sect. 17 that the study of corrosion had evidenced the fundamental role played by the combination of a specific material with a specific environment. Austenitic stainless steels, for example, are very sensible to chlorides, high strength martensitic steels to sulfides, etc. This concept of material-environment combination dominating the stress corrosion behavior is interpreted and well evidenced in Fig. 19.1 with the three circles each referring to a main variable that is: environment, material and stresses. SCC is possible only in the area common to three circles in which the proper conditions for SCC to occur are met. Engineering target as far as SCC is concerned is precisely that of identifying that common area and deriving a functional link among those three variables. However, many materials which are susceptible to SCC in a specific environment are susceptible to SCC in all environments.

So far, notwithstanding the large body of knowledge accumulated in the study of SCC, it still remains somewhat questionable whether and when it will happen. Probably, this is also because this knowledge has not been developed in a coherent fashion and all the tails of the mosaic sometimes do not fit together very well. The great deal of variables involved and its multidiscipline nature that includes chemistry (electrochemistry, chemical kinetics and analytical chemistry), metallurgy (mechanical and

P. P. Milella, *Fatigue and Corrosion in Metals*,
https://doi.org/10.1007/978-3-031-51350-3_19

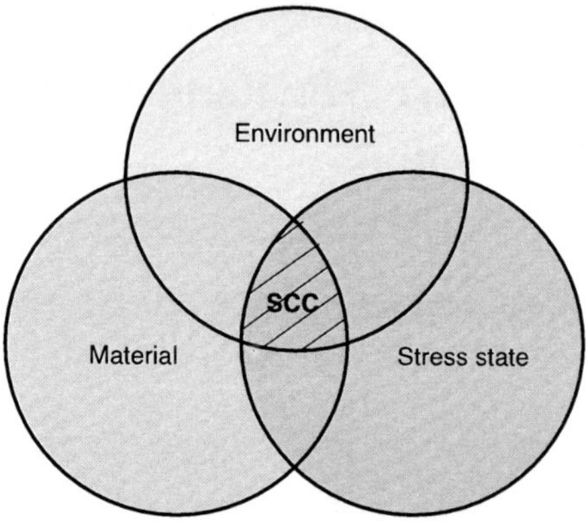

Fig. 19.1 SCC requires a specific environment-material-stress combination to initiate and is possible only in the domain common to the three circles

physical), material science and mechanics (both classical and fracture mechanics) have led, in past, to a variety of approaches to the problem that was not favoring the solution. The multidiscipline nature of SCC have resulted in multilingual technical terminology used among experts having different education with the result that identical phenomena have been described differently and different phenomena could not be distinguished. In this enormous entanglement of data some light has been finally shed by fracture mechanics application. It has been discovered, in fact, that notwithstanding its electrochemical nature, SCC is well described by linear elastic fracture mechanics (LEFM) that can predict the conditions for it to initiate in a given material and environment. It is a result of great importance that opens new and very interesting perspectives to SCC study.

To LEFM application to SCC it has been arrived almost casually. First studies of SCC were, in fact, done on smooth specimens subjected to a traction stress in the given environment. Normally, these specimens required long-to-very long waiting times before SCC would occur and the location of occurrence was apparently casual. This had led to the introduction of a notch in the specimen to provide a privileged initiation point and reduce the waiting times. One of such specimens is shown in Fig. 19.2. The notch tip root radius R was varying from 0.05 to 1.27 mm with very different results on initiation times. A typical results obtained with notched specimens is shown in Fig. 19.3 [1] that refers to a high-strength steel, type 4340, tested at room temperature and at $-$ 18 °C in hydrogenated environment.

These tests have been referred to as static fatigue tests. It can be seen that waiting times at initiation and subsequent fracture are within one hour and depend on temperature. Furthermore, SCC always occurs at the notch tip. Without notches waiting times would swell even to months. Moreover, the use of notched specimens led to another very important result. Materials that were considered immune to corrosion when used in smooth conditions reveled very sensitive to SCC when carrying a sharp notch. This was the case of austenitic stainless steel, aluminum and titanium alloys in sea water. Sharp notches were entering the field of SCC and so did LEFM.

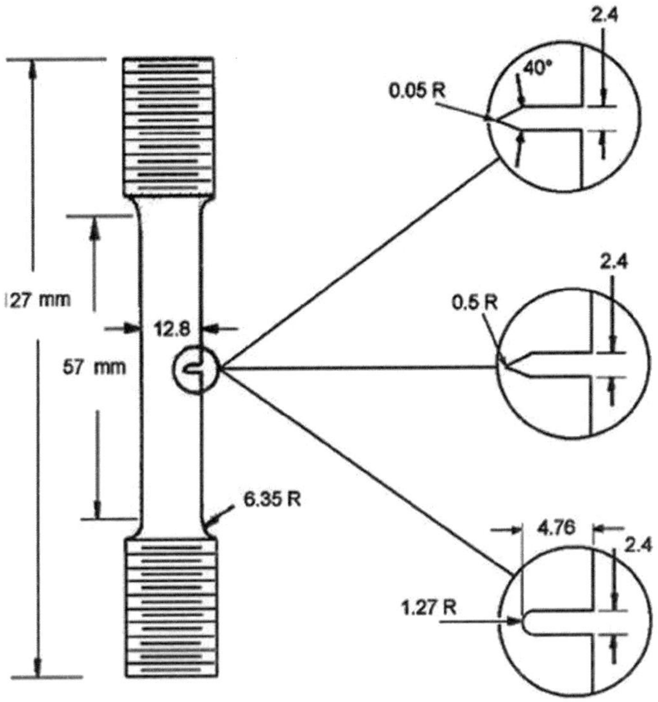

Fig. 19.2 Notched bar for SCC study

19.2 Pitting

It has been said in the previous paragraph, that materials that were considered practically immune to SCC were found to be sensitive to it when affected by a crack. The existence of a crack is, therefore, fundamental for SCC to develop. Local chemistry that exists at the tip of a notch may be completely different from the general chemistry of the environment in which the material is operating and this can trigger a SCC that was never supposed to start.

The effect of such a notch or crevice on corrosion has been studied in Sect. 17.7. It was shown how the passivizing film rupture at notch tip would lead to a slip dissolution process and, therefore, to corrosion. This initial notch or crack may not exist in the material, yet it can easily form as well. Fatigue may be the cause of crack initiation, but also a static stress can do it. In these cases we have what is referred to as pitting.

The mechanism by which a pit is formed is always that of the film rupture/slip dissolution process, already mentioned. The passivizing oxide film that protects the metal surface, at least in corrosion resistant alloys where it is tight and stable, breaks somewhere. This can be the consequence of fatigue or a localized change of corrosion potential (local chemistry) that enters into the transpassive zone where pitting occurs,

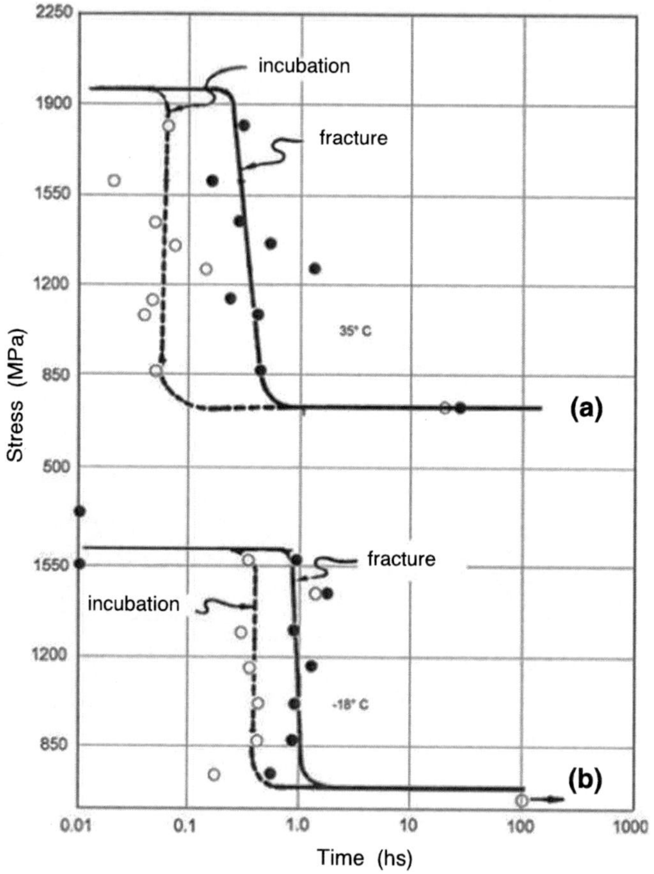

Fig. 19.3 Static fatigue curves or SCC obtained in a high strength steel type 4340 in hydrogenated environment making use of notched specimens [1]

see Fig. 17.11. Pitting, in fact, is a form of extremely localized corrosion that leads to the creation of a small hole in the metal. The presence of anions like chlorides Cl^- (sea water) or sulfides S^-, from manganese sulfide MnS inclusions in particular, aggravates the conditions for formation and growth of pits. On the contrary, the existence of anions from weak acids inhibits pitting by increasing the pitting potential. Pitting potential of iron in the presence of borate is more than 100 times higher than in its absence. The same is for aluminum when soluble salts of acetate are present. The driving power for pitting corrosion is due to a small area of bare-metal that becomes anodic in a surrounding area that behaves as a cathode (see Sect. 17.3).

A local galvanic cell is formed (see Sect. 17.4) and metal starts to dissolve into the potentially corrosive environment that can be even humid air. Temperature is also a critical factor for pitting corrosion, because many materials will not pit at

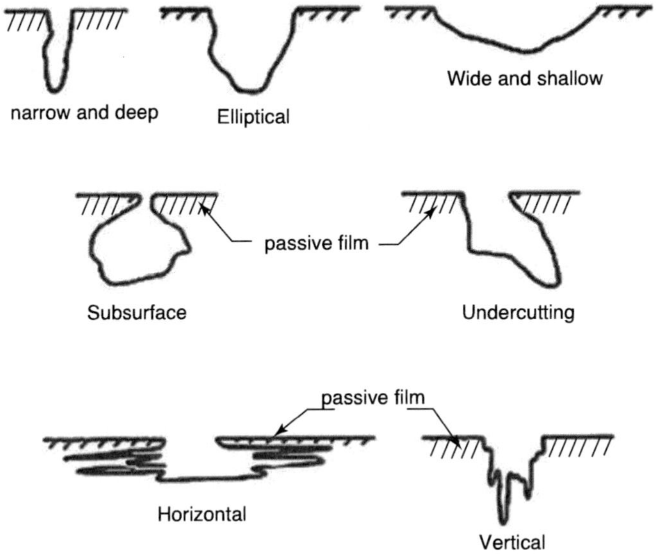

Fig. 19.4 Possible shapes and appearance of corrosion pits [3]

a temperature below a certain value. Also the surface conditions are important. In general, a rough surface finish is more susceptible to pitting and exhibits a lower pitting potential. For example, the pitting potential of type 302 stainless steel with a 120-grit finish was shown to be approximately 150 mV lower than that for the same material with a 1200-grit finish over a range of chloride concentrations [2]. Rougher surface, in fact, have more occluded sites, which can sustain the conditions required for active metal dissolution. For stainless steel, any heat treatment or grinding are detrimental to pitting (see Sect. 18.2 and Fig. 18.12). A table of standard types of pits is presented in Fig. 19.4 [3]. Once the passive film breaks down and a pit forms it is very possible that the pit will grow and become a crack.

Figure 19.5 [4] presents, in the small frame on the bottom left, the surface of a 2024-T3 aluminum alloy exposed to a 0.5M NaCl solution (sea water) at 25 °C. The surface is covered by pits. The cross section in the larger picture shows two corrosion pits. One is very severe, shallow and about 0.3 mm deep which makes it be a real crack (see Sect. 2.3). The second is of subsurface type, see Fig. 19.4. This alloy has a yield strength of 355 MPa and an ultimate strength of 480 MPa with elongation at failure of 17%. Figure 19.6 shows a corrosion, nail shaped crack that triggered a fast brittle fracture in a high strength 54SiCrV6 steel used for suspension springs, 1650 MPa yield strength and 2000 MPa ultimate. The crack originated from a corrosion pit. The SEM examination of the crack surface, Fig. 19.6b, indicated that hydrogen embrittlement was the cause of corrosion (see Sect. 18.5). High strength steels are very susceptible to pitting as they are to stress corrosion. It is fundamental

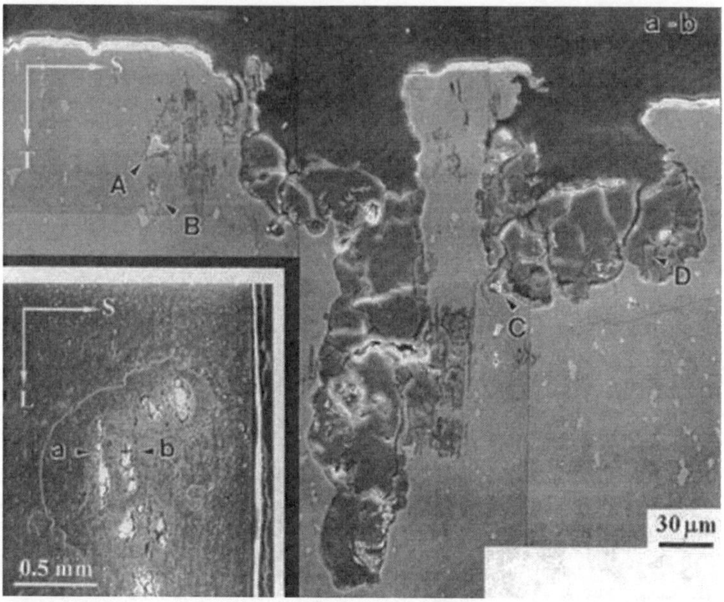

Fig. 19.5 Cross section of two pits in a type 2024-T3 aluminum alloy. The first is narrow and deep, the second is of subsurface type [4]

to stress the importance of avoiding pitting in high strength and special alloys, in particular.

Once a pit is formed it can be very difficult to avoid corrosion even dough a passivation system is enforced (see Sects. 17.4 and 17.6). Notorious are cases of off-shore structures found heavily affected by cracks. These structures had been operating for decades under a cathodic protection system without any problem. The simple temporarily and accidentally switch off of the protective system had resulted in a massive pitting. The reintroduction of the cathodic protection could not prevent those pits from growing to become long cracks.

19.3 Threshold Stress Intensity Factor for Corrosion

One of the first popular test method, entered by now into the history of corrosion, is shown in Fig. 19.7 [5]. The method uses a cantilever beam specimen which is precracked and loaded with a dead weight loading apparatus. The system may have an environmental chamber in which a potentially aggressive solution is acting on the crack. The progression of corrosion is monitored by the dial gauge through the beam compliance increase due to crack growth. The apparatus is simple and a little bit primitive, but it worked. It was clear that, by introducing a crack, the

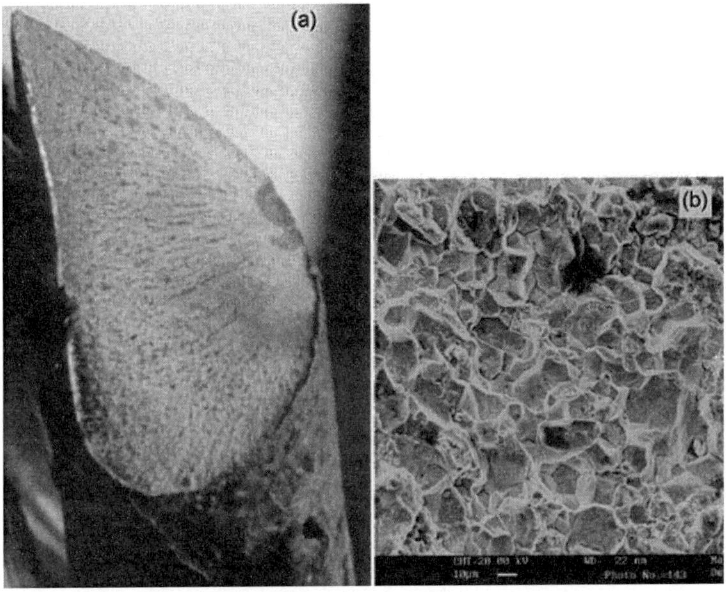

Fig. 19.6 **a** Hydrogen embrittlement corrosion crack initiated from a pit in high strength steel for suspension spring and **b** SEM intergranular appearance of the corrosion area

study of intergranular stress corrosion cracking (IGSCC) would suggest a fracture mechanics application introducing, in particular, the stress intensity factor K that is its fundamental concept in a fashion similar to what had be done by Paris in the field of fatigue crack propagation (see Chap. 14). Also in this kind of corrosion test the K_I is recorded, which is a function of the beam-crack geometry and applied weight, waiting for the moment of crack propagation. A characteristic of such a loading apparatus is that of being a load-controlled system. Therefore, if the crack initiates to grow by corrosion, then it will not stop leading to the complete failure of the beam.

An example of results obtained in sea water for a maraging steel type 18Ni, of 1750 MPa yield strength and 2100 MPa ultimate strength, is shown in Fig. 19.7 [6, 7]. Two things can be noted. First, by increasing the strength of the steel and, in general, of the alloy the susceptibility to SCC also increases. Secondly, by decreasing the applied K_I the time needed to initiate a SCC event, also referred to as incubation time, increases according to a power law and eventually becomes infinite, meaning that no SCC occurs, when the applied K_I becomes lower than a threshold value labeled K_{ISCC}. Below K_{ISCC} no SCC can ever occur. This value was originally considered a material property, which was dependent on both material and environment. The discovery of a K_{ISCC} was an extremely important event in the study of SCC. Today it still remains the most important parameter in the application of a SCC technology to engineering problems and represents that longed functional link between the three circles of Fig. 19.1. However, at the time of its introduction LEFM was not a so

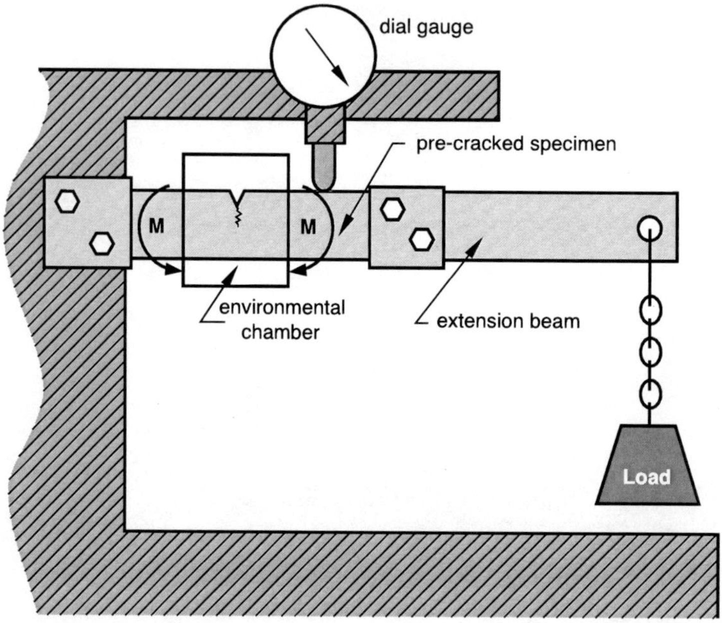

Fig. 19.7 Schematic illustration of the cantilever beam corrosion test (modified from [5])

widely used science, in particular in the electrochemical and corrosion field where the stress was considered instead, as shown by Eq. (13.1) (Figure 19.8).

Therefore, the K_{ISCC} concept was initially resisted by a large part of researchers. That was why an experiment was conceived to solve once and for ever the question of whether or not the stress intensity factor was the most proper parameter in determining the initiation of a SCC event. Smith et al. [8], in fact, in 1968 demonstrated the dependency of the SCC phenomenon on the stress intensity factor K_I rather than on stress. The experiment, similar to that used by Paris et al. to derive the Paris-Erdogan Eq. (14.10), was making use of a wedge loaded panel of Ti-8Al-1Mo-1V titanium alloy of thickness t with a central crack of length $2a$, shown in Fig. 19.9a, tested in sea water. The wedge loaded panel has the property that K_I decreases with increasing crack length under a constant applied load. In fact, for a constant load F the applied K_I is given by the expression

$$K_I = \frac{F}{t}\sqrt{\frac{2}{w}} \cdot f\left(\frac{2a}{w}\right) \tag{19.1}$$

in which the nondimensional function $f(2a/w)$ depends on the geometry of the panel. The value of this function decreases as the crack length increases, as shown in Fig. 19.9b. On the contrary, the net stress σ_{net} acting on the cracked section increases as the crack advances and varies according to the expression

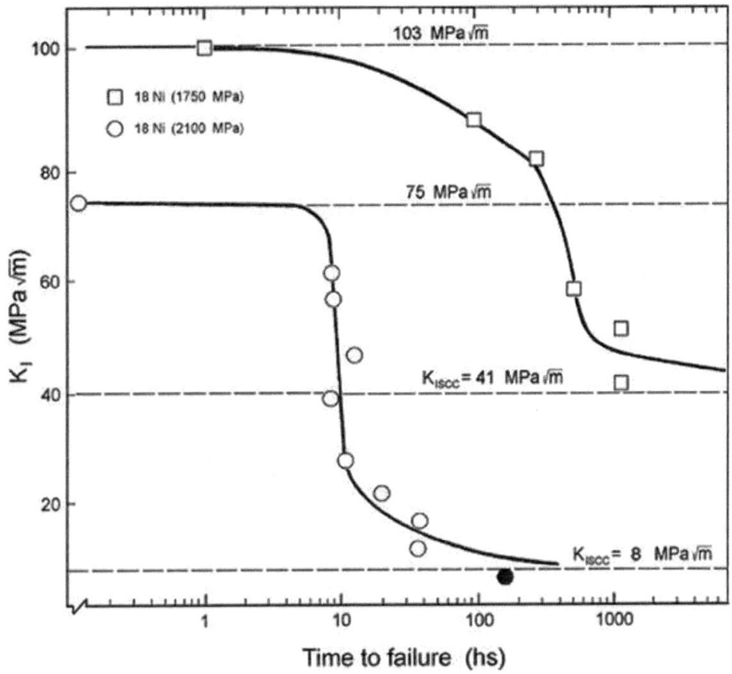

Fig. 19.8 SCC in a maraging steel, type 18Ni, in sea water (data from [6, 7])

$$\sigma_{net} = \frac{F}{(w - 2a) \cdot t}. \tag{19.2}$$

Its trend is also shown, for comparison, in Fig. 19.9b. The two reference quantities K_I and σ_{net} have opposing trends. The results of this experiment are shown in Fig. 19.10 [8]. The panel was initially loaded to a K_I value above the alleged K_{ISCC}. The crack would immediately grow. Crack growth rate was rapid at the beginning, but as the crack advanced and the applied K_I decreased the growth rate would slow down and eventually the crack would arrest as the K_{ISCC} was approached. This procedure could be repeated with increasing loads, but the response was identical, with crack arrest occurring as K_I approached K_{ISCC}. In terms of net section stress σ_{net} the result was precisely the opposite since the crack would grow faster at lower net section stress and then it would arrest right when the net section stress would reach its maximum, which was against any logic.

The experiment demonstrated beyond any reasonable doubt that it was the stress intensity factor and not the stress the controlling parameter. This dependence of the stress corrosion crack growth rate (SCCGR) has been found for all alloys susceptible to stress corrosion and can be, therefore, considered as general. It was also found that the SCCGR, *da/dt*, has a very particular trend. This is shown schematically in Fig. 19.11 by a curve that has three different regions. The first or Region 1 is labeled

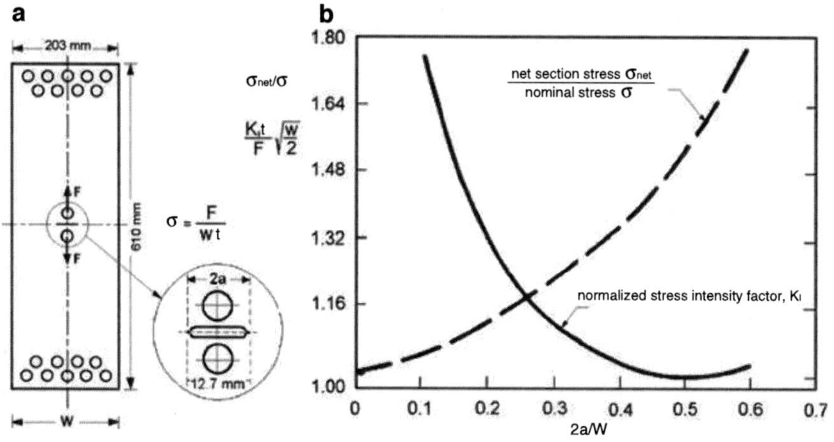

Fig. 19.9 **a** Central crack panel wedge loaded with a constant load and **b** trend of K_I and net section stress (modified from [8])

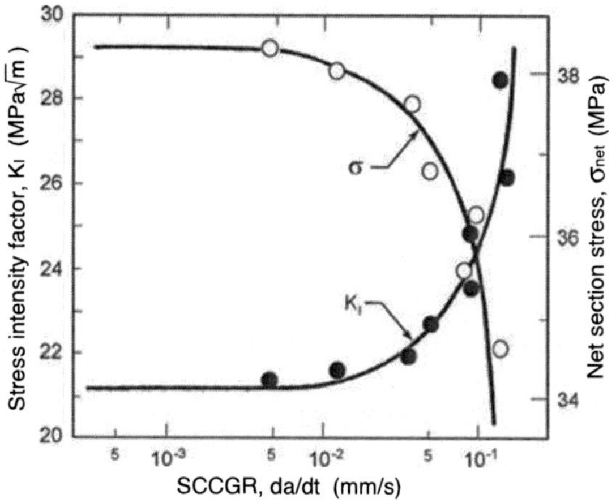

Fig. 19.10 Results for Ti-8Al-1Mo-1 V wedge loaded panel tested in sea water showing the effect of K_I σ on the crack growth rate [8]

the K_I dependent region. It shows that SCCGR, *da/dt*, does not occur until a K_{ISCC} value is reached. Once reached, the crack starts to grow fast as a function of the applied K_I. The second region or Region 2 is labeled the K_I independent region or constant SCCGR, *da/dt*, region. This region represents a kind of unexpected surprise since the applied K_I has no effect at all. This actually means that in this region the corrosion mechanism is only time-dependent. Therefore, the SCCGR is simply limited by the environment. Finally, the third region or Region 3 appears to be again K_I dependent. However, in this region the sudden increase of the SCCGR, *da/dt*, is attributed to the fact that the specimen is approaching the K_{Ic} of the material, i.e., final failure. The existence of the Region 1 is by now fully demonstrated ad universally

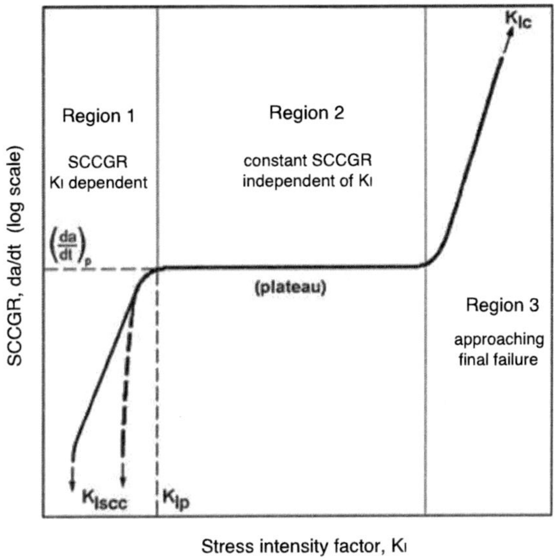

Fig. 19.11 Schematic of SCCGR, *da/dt*, versus K_I showing the three regions of SCCGR

accepted as demonstrated by Johnson and Willner on H-11 steel in sea water [9], Steigerwald and Benjamin [10] and Peterson [11] on 4340 steel in distilled water and sea water.

19.4 Incubation Time

In defining K_{ISCC} as a material property from a constant displacement test the question arises as to whether or not the test has been given enough time to actually measure the K_{ISCC}. The effect of test time in determining K_{ISCC} is not just a heuristic question. Sometimes long incubation times are encountered before any crack extension occurs. Figure 19.8 shows the delayed SCC observed in two maraging steels, type 18Mi, in sea water. Landes [12] also reports tests done on 1240 MPa (180 ksi) yield steel in sea water at room temperature in which the apparent K_{ISCC} was depending on the test time. His results are summarized in the following table.

Time (hr)	Apparent K_{ISCC} (MPa$\sqrt{m}$)
100	187
1000	126
10,000	27

Such a dramatic effect of the waiting time could make the measurement of K_{ISCC} questionable. As matter of fact, a test that presents no apparent crack growth may do so because the applied K_I is below the K_{ISCC} or because the test time was too short compared with the incubation time. This time may depend on material, environment, temperature and, of course, on the applied K_I. It may also depend on the test procedure. For instance, to bolt load in air a WOL specimen, of the type shown in Fig. 19.19, of high strength material and then place it in a hydrogen environment may require high waiting time before the crack grows, unless it would be loaded at very high initial K_I, while a specimen of the same material bolt loaded directly in hydrogen could not show any incubation time. Consider that it may not be that simple to load a specimen directly in hydrogen. But it may not be that simple either to overload the specimen in order to reduce or even extinguish the incubation period. This is because the specimen is containing a crack and under an applied K_I too much in excess with respect to the K_{ISCC} that is to be measured it may require a too large specimen to maintain plain strain conditions.

It may also happen that in a decreasing K_I test, a too large initial K_I would make the crack grow sub-critically and then arrest unloading the specimen before the test even started. The problem of the incubation time was observed by Wei et al. [13] that referred of analogous observations made by Li et al. [14] and Barsom and is schematized in Fig. 19.12. Following an initial transient growth, the crack appears to stop growing or it can enter an incubation period before accelerating again to some steady state rate of growth, Region 2 of Fig. 19.11, depending on the applied K_I. If it is $K_I > K_{ISCC}$, the incubation time will last forever. During the transient the initial slope of the curve provides a *da/dt* rate which is very high, bur only apparent. The real crack growth rate will come later. Figure 19.13 reports the data obtained by Landes and Wei [15] on 4340 in distilled water. The specimens used were of the type tapered with constant *K*, see Fig. 10.7. Therefore, the SCCGR, *da/dt*, did not change with the crack length. However, the SCCGR, *da/dt*, was varying with time up to the steady state growth rate. Note that it is not the occurrence of the two regions 1 and 2 of Fig. 19.11 that appears in Fig. 19.13 since the applied K_I is constant, but the growing transient.

Fig. 19.12 Schematic of the incubation period on loading for 4340 steel [13]

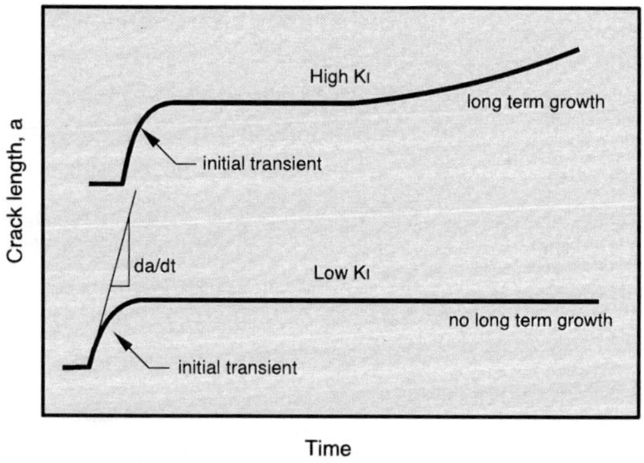

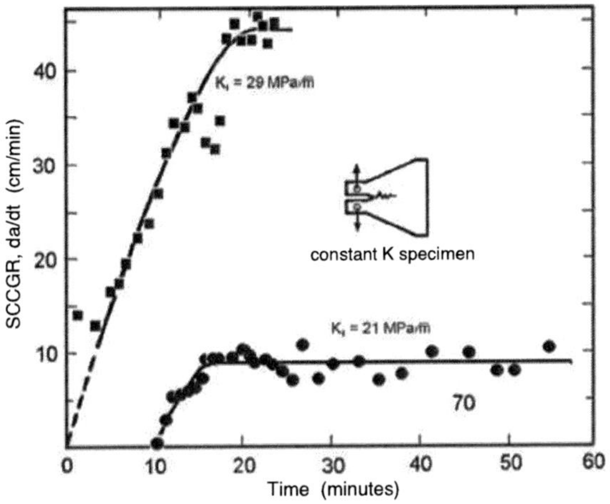

Fig. 19.13 RCCC, *da/dt*, obtained with constant *K* specimens of 4340 in distilled water [15]

Using the same material and the same environment, but a K_I decreasing specimen the results of Fig. 19.14 [15] would be obtained. It is well visible the kinetics of the crack growth that reaches the steady state, which is the same for any applied K_I, after a waiting time that depends on the initial value of $K_{I,i}$. Wei et al. [13] mention non published results obtained by Novak on WOL specimens of 1260 MPa yield strength steel in sea water at ambient temperature, presented in Fig. 19.15 [13]. The influence of K_I on the incubation time is evident. At $K_I = 176$ MPa$\sqrt{}$m the incubation time is about 300 h that become 1400 h (~ 58 days) if the applied K_I reduces to 147 MPa$\sqrt{}$m and grows to even 3500 h (~146 days) under an applied K_I of 88 MPa$\sqrt{}$m. it can be thought that the total waiting time preceding the final fracture is composed of the incubation time t_{inc} and a time t_{scc} of steady stress corrosion crack growth

$$t_f = t_{inc} + t_{scc} \tag{19.3}$$

This is schematically shown in Fig. 19.16. For applied K_I values higher than K_{Iscc} the corrosive process initiates. The higher the applied K_I, the shorter both the incubations time t_{inc} and the growth time t_{scc}.

19.4.1 SCC and Fracture Toughness

The upper limit assumed in Fig. 19.16 for the applied K_I is, obviously, that of the static toughness of the material K_{Ic}. Once this value is reached brittle fracture is triggered. The lower bound of K_I is K_{Iscc} so that it is $K_{Iscc} < K_{Ic}$. When this is the case, K_{Iscc} replaces K_{Ic} as the controlling factor for the safety assessment of a structure since SCC is so fast that time to failure cannot be considered in a structural analysis. The ratio of K_{Iscc}/K_{Ic} is, therefore, a measure of *toughness reduction* caused by SCC. In

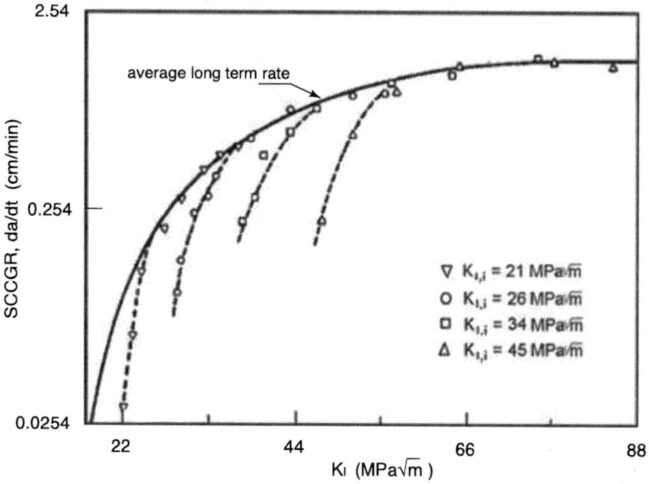

Fig. 19.14 RCCC, *da/dt*, for a 4340 steel in distilled water at ambient temperature for different [15]

Fig. 19.15 Influence of K_I on the incubation time in a 1260 MPa yield strength steel in sea water (data coming from Novak)

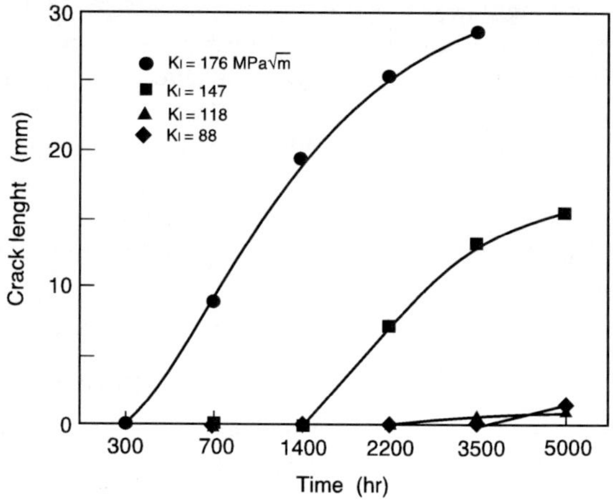

reality it is not precisely a toughness reduction since K_{Ic} remains the same, but rather a substitution of K_{Ic} with K_{Iscc}.

Both K_{Ic} and K_{Iscc} depend on the strength of the material, decreasing as the strength increases but, in general, at room temperature or lower, the K_{Iscc} is smaller than K_{Ic}, at least in medium–high strength materials. This is shown in Fig. 19.17 [16] for a family of 4340 steels in flowing sea water. The K_{Ic}, of course, is not affected by the environment, but K_{Iscc} is. Interesting enough, the K_{Iscc} approaches the toughness K_{Ic} as the material strength decreases.

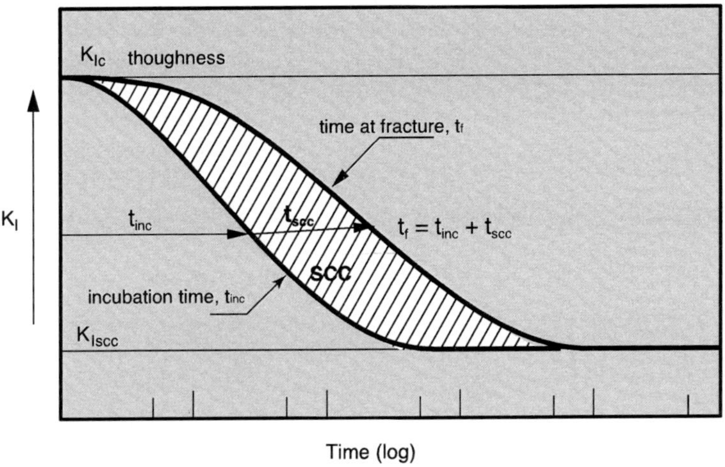

Fig. 19.16 Schematic of dependency of incubation time and fracture time under static loads

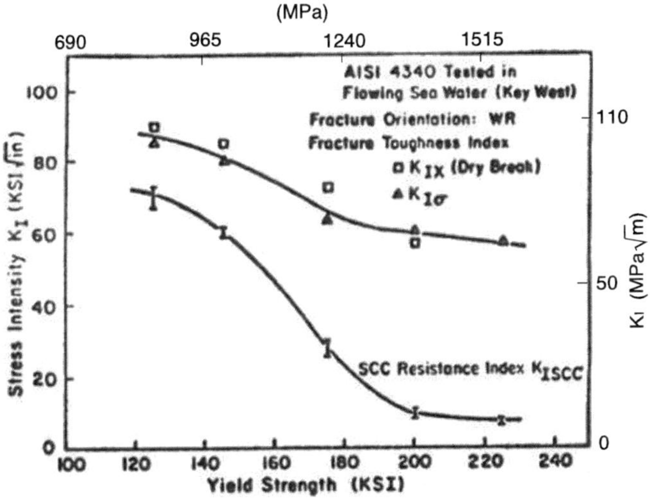

Fig. 19.17 Threshold K_{Iscc} and fracture toughness K_{Ic} versus yield strength for 4340 in sea water (modified from [16])

While for a 1500 MPa yield strength the K_{Iscc} is about 10 MPa$\sqrt{m}$ against 70 MPa$\sqrt{m}$ of the toughness K_{Ic} at 800 MPa of yield strength this difference sharply reduces and could be completely canceled by scatter due to metallurgical variation (see Sect. 19.4 sample problem N. 6). This is because the lower the strength, the lower the susceptibility to SCC of steels. This makes the K_{Iscc} curve grow faster than

K_{Ic} with reducing strength. Moreover, both factors, K_{Ic} and K_{Iscc}, depend on temperature. The former because of metallurgical factors, the latter because of hydrogen absorption kinetics. Figure 19.18 shows how K_{Iscc} and K_{Ic} vary with temperature in a high strength steel type 4340. They both increase with increasing temperature but, again, with a different rate. Up to about 90 °C it is stress corrosion by hydrogen embrittlement that occurs first, which actually means that effectively it is $K_{ISCC} < K_{Ic}$. But beyond that temperature material toughness K_{Ic} may become lower than K_{ISCC} because hydrogen absorption makes way to hydrogen diffusion and, therefore, hydrogen exit from the metal becomes more probable than its entrance.

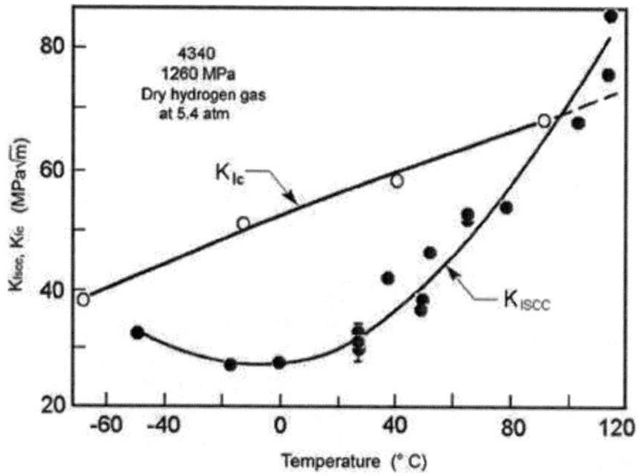

Fig. 19.18 Threshold K_{Iscc} and fracture toughness K_{Ic} versus temperature for 4340 in dry hydrogen at 5.4 atm

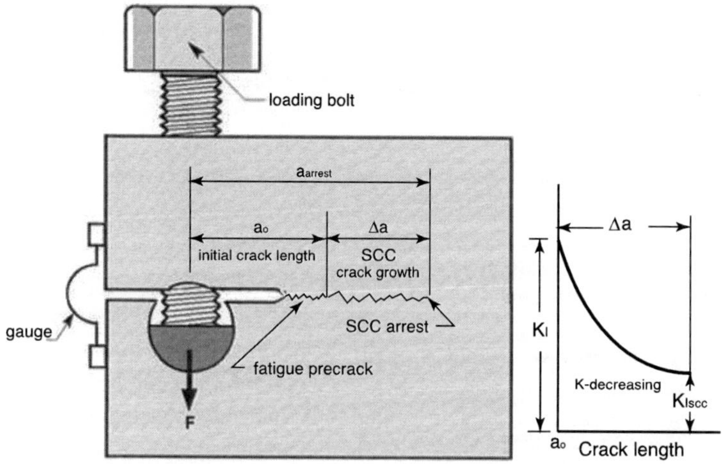

Fig. 19.19 WOL specimen used for K_{Iscc} measurements

19.5 Measurement of SCC Threshold K_{Iscc}

One of the simplest and direct method to measure the threshold K_{Iscc} is certainly using a fracture mechanics specimen characterized by a stress intensity factor K decreasing with increasing crack length a. Any wedge loaded specimen loaded under displacement controlled conditions holds this characteristic. A typical and most used specimen for measurements is the precracked WOL (wedge opening load) shown in Fig. 19.19 in which the crack is bolt loaded at an initial K_I level in excess to the K_{Iscc} that is thought to be measured.

The initial K_I is deduced from a mouth opening gauge reading, since the specimen is calibrated (see Appendix A). Once loaded, the specimen is placed in the environment under study. Caution must be done to what has been said in Sect. 5. 4 about the enormous difference in the incubation time that may arise whether the specimen is first loaded and then exposed to the environment or vice versa. After the incubation time the crack would start to grow in a decreasing applied stress intensity factor, K_I, field so that, eventually, it would arrest at the K_{Iscc} value of the material, as schematized in Fig. 19.19. Note that a real arrest of SCC could not happen because corrosion would continue asymptotically towards the lower limit K_{Iscc} that would be reached after an infinite time. Therefore, it is conventionally referred to as threshold K_{Iscc} the value at which the SCCGR, *da/dt*, went down to the minimum level of measurement, which is generally equal to 10^{-10} m/s that is to say a growth lower than 0.2 mm in 30 days. Any K_I-decreasing specimen of the type shown in Fig. 14.7 can be used to this purpose. Other types of specimens, with or without an initial crack, are shown in Fig. 19.20 and described by ASTM G 39 Standard [17] for specific use.

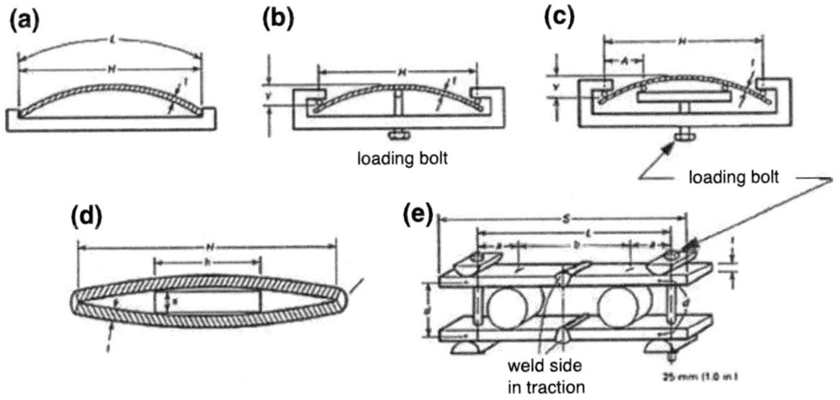

Fig. 19.20 Displacement controlled specimens used for SCC K_{Iscc} measurements [17]

19.5.1 K_{Iscc} as a Material Characteristic

It has been said that it is the combination of the material and the environment which determines the actual severity of SCC. However, many materials which are susceptible to SCC in one of those severe environment are susceptible to SCC in all of them. Moreover, many experimental results show that K_{Iscc} is strictly a material parameter and does not depend on the environment. This is shown in Fig. 19.21 [18] that refers to a 835 M 30 steel loaded under *K*-decreasing conditions in three different environments that are: hydrogen sulfide, hydrogen and sea water. Interesting enough, independently of the environment the K_{Iscc} seems to have always the same value of about 12–15 MPa$\sqrt{m}$. What is actually changing is the incubation time that is almost null for hydrogen sulfide and rather long for the sea water (about 8 h). The same data plotted as a function of the applied K_I show that the environment has certainly a dramatic effect on the SCCGR, *da/dt*, which goes from about 10^{-3} mm/s in sea water to about $2{\cdot}10^{-2}$ mm/s in hydrogen and up to about 10 mm/s in hydrogen sulfide, as shown in Fig. 19.22, but again it has no influence on the K_{Iscc} value [18]. Two more significant examples are shown in Figs. 19.23 and 19.24 for a 4340 steel of 1260 MPa yield strength [19] and for a 18Mn-5Cr steel [20], respectively. This is actually indicating that the environment does have an effect on the SCCGR, *da/dt*, in Region 2 of SCCGR, see Fig. 19.11, but not on the threshold K_{Iscc}. Also activation energy would confirm this finding.

Since SCC is a thermally activated phenomenon, measurements of activation energies from crack groth rate studies. It was found that for high strength steel, even

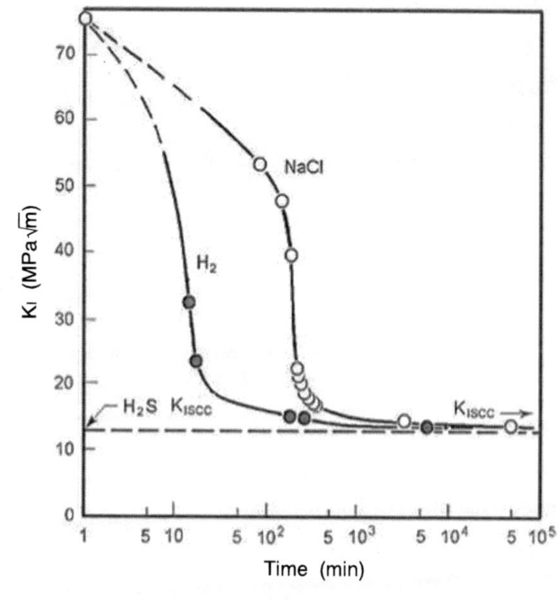

Fig. 19.21 Time to failure of 835 M 30 steel in three different environments [18]

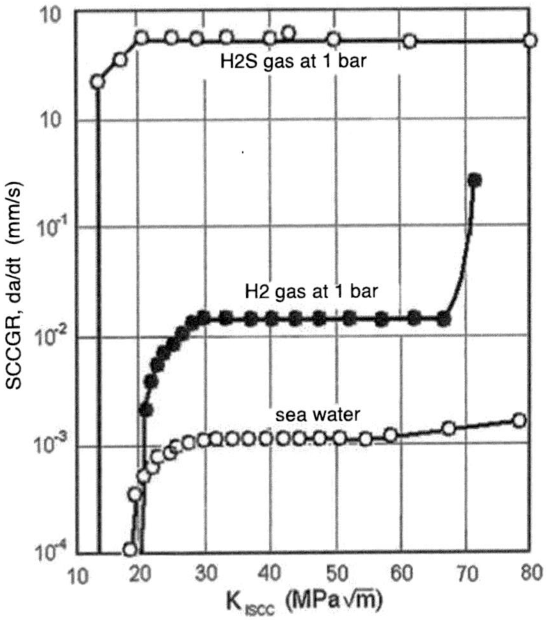

Fig. 19.22 SC crack growth rate for type 835 M 30 steel in three different environments [18]

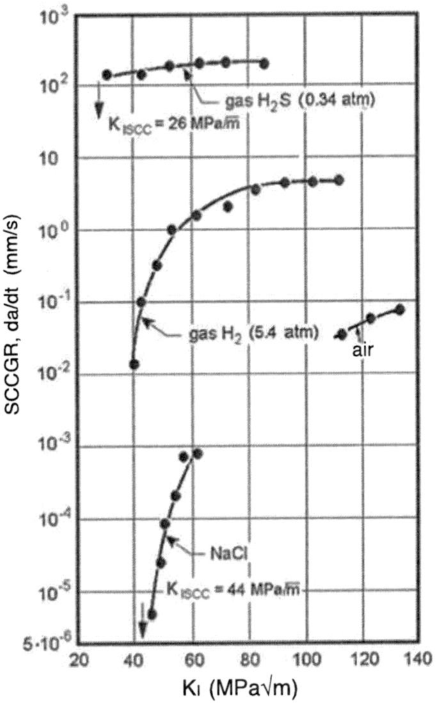

Fig. 19.23 SC crack growth rate for type 4340 steel of 1260 MPa 835 M 30 steel in four different environments [19]

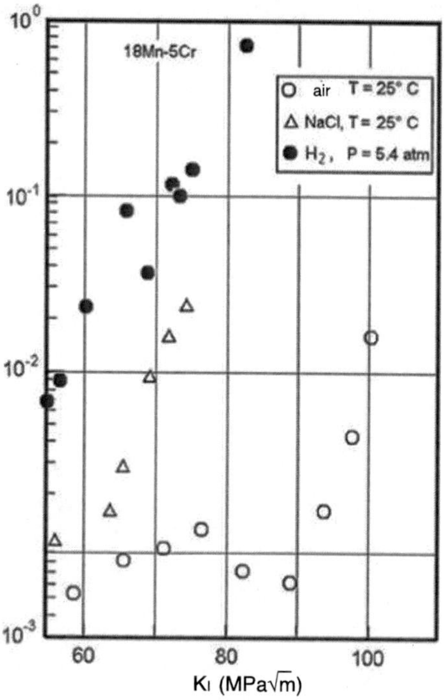

Fig. 19.24 SC crack growth rate for type 18Mn-5Cr steel in three different environments [20]

dough the threshold stress intensity factor K_{Iscc} for SCC is similar in both water and hydrogen, the activation energy differs in the two environments.

This supports the assumption that K_{Iscc} is controlled by the material and not by the environment. The finding can be used to shorten as much as possible when not even eliminate at all the incubation time by using the *most severe environment* just to measure the threshold stress intensity K_{Iscc} for SCC. In general, for high strength steel the worst environment possible id hydrogen sulfide H_2S. However, still some reasonable doubts exist on the K_{Iscc} being a material property independent of the environment.

An example of this is shown in Fig. 19.25 for 4340 steel in a hydrogen gas environment [21]. The SCCGR, *da/dt*, increases as the hydrogen gas pressure increases while the threshold K_{Iscc} remains the same at about 20–23 MPa$\sqrt{m}$, the same seen in Fig. 19.23, for pressures ranging from 11 to 77 kPa, confirming that indeed it is a material characteristic, but lowering the hydrogen pressure to 2.5 kPa K_{Iscc} sudden jumps to 36 MPa$\sqrt{m}$, which is also consistent with data of Fig. 19.23.

The same effect can be seen on aluminum alloys. Figure 19.26 [22] refers to 7079-T651 aluminum alloy in several environments, from dry air (0.1% relative humidity) to sea water (3.5% NaCl). Both in the marine and industrial environment the threshold K_{Iscc} is practically the same, ranging from 3 to 5 MPa$\sqrt{m}$, but in the laboratory atmosphere (47% relative humidity) it goes up to about 12 MPa$\sqrt{m}$. In Fig. 19.26 it is also indicated the toughness K_{Ic} of the material that is much higher,

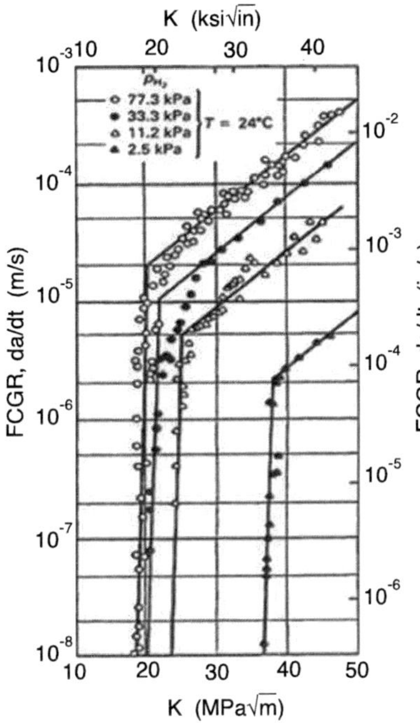

Fig. 19.25 Effect of hydrogen gas pressure on the SCCGR threshold K_{Iscc} for 4340 steel [21]

about 22 MPa$\sqrt{}$m, than the K_{Iscc}. Therefore, at room temperature stress corrosion will always occur first. No corrosion at all was ever registered in dry air even after 3 years. A synoptic picture about SCC behavior for aluminum alloys is provided in Fig. 19.27. The environment was 0.6 M NaCl + 0.02M $Na_2Cr_3O_7$ + 0,07M $NaC_2H_3O_2$ + $HC_2H_3O_2$ to pH 4. As said, a problem that arises in testing aluminum is the wedging of corrosion product. Tests run in a decreasing K field may reach the point where corrosion products have built up on the crack surfaces and wedge the crack open. This effect does not allow K to decrease according to the reached crack length and provide misleading results. In fact, because of this wedging effect the crack could continue to grow or even accelerate. Under these conditions the threshold will never be found. This is shown in Fig. 19.26 by the dashed horizontal lines.

19.5.2 Constant Extension Rate Test (CERT)

The two major drawbacks to the K_{Iscc} measurement through the use of K-decreasing specimens are, as said in Sect. 19.5, the long waiting times due to incubation and the subcritical crack growth that may occur immediately during the same application of an initial K_I too much in excess with respect to the K_{Iscc} that must be measured,

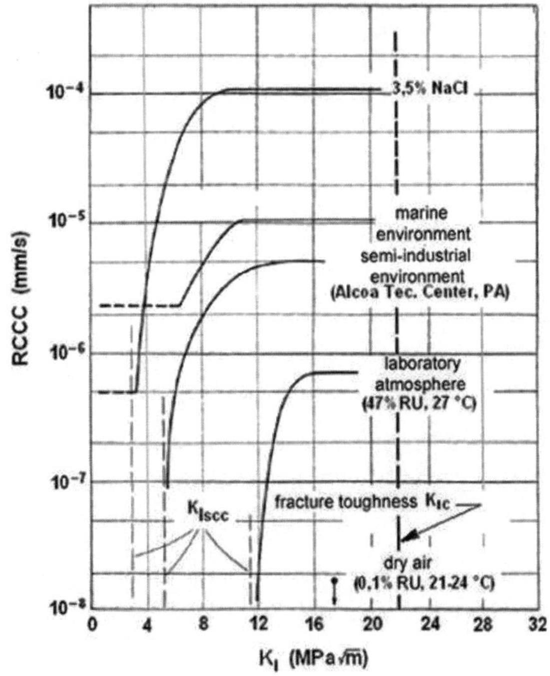

Fig. 19.26 Effect environment on the SCCGR threshold K_{Iscc} for 7079-T651 aluminum alloy [22]

in order to reduce the waiting times. It was necessary, then, to develop a new test that could overcome these obstacles. The discovery that the K_{Iscc} could be a material property not depending on the specific environment, mentioned in Sect. 5.1, dough still questionable, was getting rid of the long waiting times by using the worst possible environment. Still the second drawback remained. This led to the development, in the nuclear field in particular, of a new SCC test method in which a notched specimen was subjected to a low rising test under the worst environmental conditions, which is generally the hydrogen sulfide.

When the K_{Iscc} is reached and the crack starts to propagate the load displacement record will show a sudden deviation from linearity. To clearly separate and non-confusing this SCC non linearity from that occurring at yielding, the test is first run with a specimen of the same material in a non-aggressive environment such as, for example, dry air. This is schematically shown in Fig. 19.28. The dashed line refers to the load–displacement diagram first obtained in dry air. If a second specimen is tested in the environment under study, when the SCC event initiates the point of deviation from linearity occurs before, as shown in Fig. 19.28. The K_{Iscc} value is then taken at this point of deviation from linearity. An example of rising load test is shown in Fig. 19.29 [19] for a type 4340 steel of 1240 MPa (180 ksi) yield strength in three different environments: air, hydrogen gas and hydrogen sulfide. In air there is no SCC while in the other two environments SCC occurs. Note that in H_2S at 5 psig pressure the deviation from linearity is sharp and the load remains flat while in

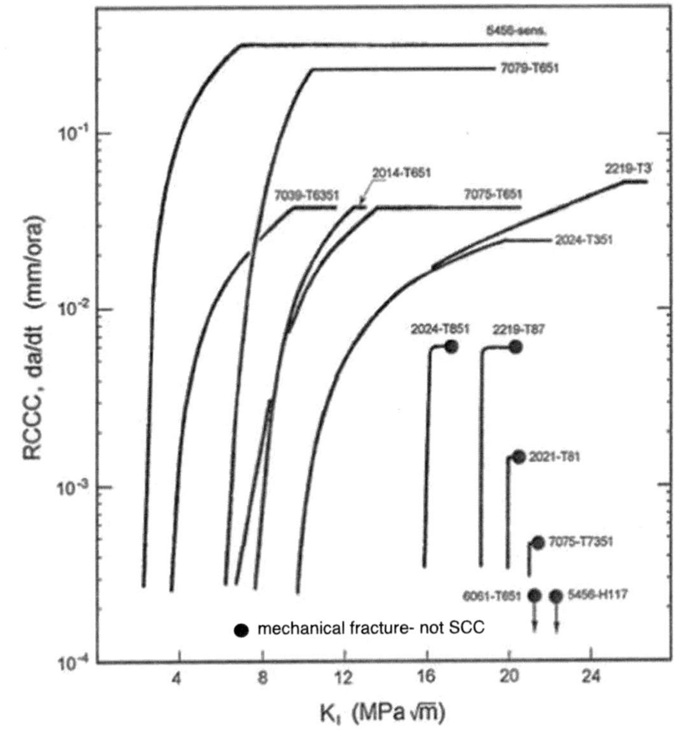

Fig. 19.27 SCCGR, *da/dt*, for several aluminum alloy in sea water (data from Westinghouse Nuclear, Pittsburgh)

H_2 gas at 80 psig pressure it is smooth, but in both cases the point of deviation is the same and results in a threshold K_{Iscc} value of ~ 5.5 MPa$\sqrt{}$m (5 ksi$\sqrt{}$in).

Also note that the test was run at slow rate of $\dot{K} = 0.1$ ksi$\sqrt{}$in/min. The influence of loading rate on the apparent K_{Iscc} is shown in Fig. 19.30 [19] for two 4340 steels having yield strength equal to 1260 MPa (180 ksi) and 1142 MPa (160 ksi), respectively, in hydrogen and hydrogen sulfide environment at different pressures. While in H_2S environment there is no influence of the loading rate on the apparent K_{Iscc}, in a hydrogen gas the slower growth rate gives a loading rate effect.

Also H_2S pressure has no apparent effect on the K_{Iscc} that remains between 28 to 37 MPa$\sqrt{}$m for the 1260 yield steel. A simple decrease of the yield strength from 1260 down to 1142 MPa increases the apparent K_{Iscc} to about 52 MPa$\sqrt{}$m. The reason for the rate effect must be found in the time SCC needs to occur, especially in the less severe environments. The proper testing rate actually depends on the type of corrosion. It has been shown, in Chap. 18, that there are two basic modes of corrosion: sensitization cracking and hydrogen embrittlement, this latter being a particular form of the more general passive film rupture/anodic dissolution process described in Sect. 17.7 (see also Fig. 17.17). If the corrosion mechanism is that of

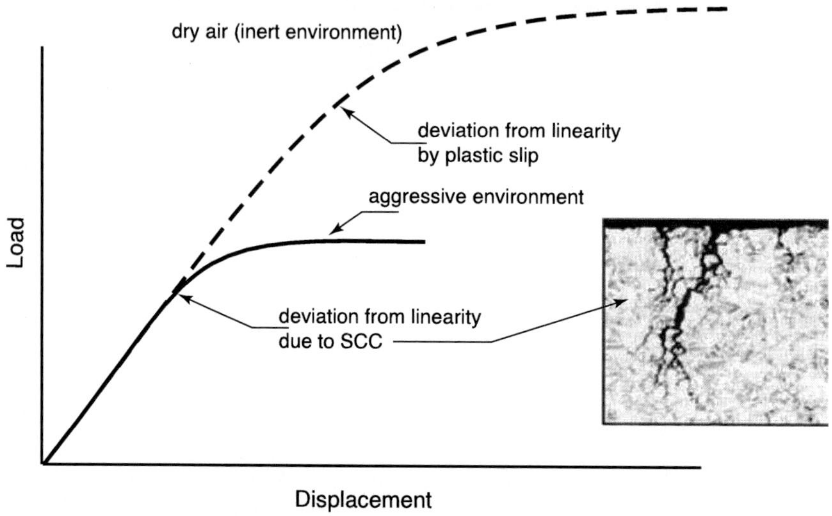

Fig. 19.28 Schematic of a rising load testing

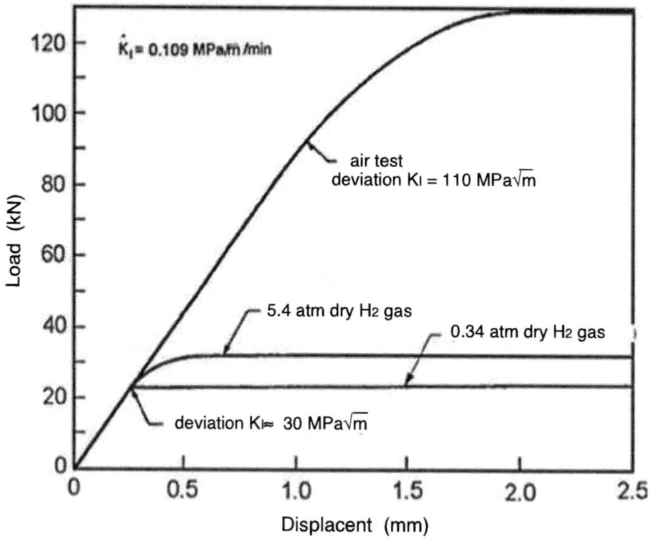

Fig. 19.29 Effect of the environment on the load–displacement diagram in a rising load test (modified from [19])

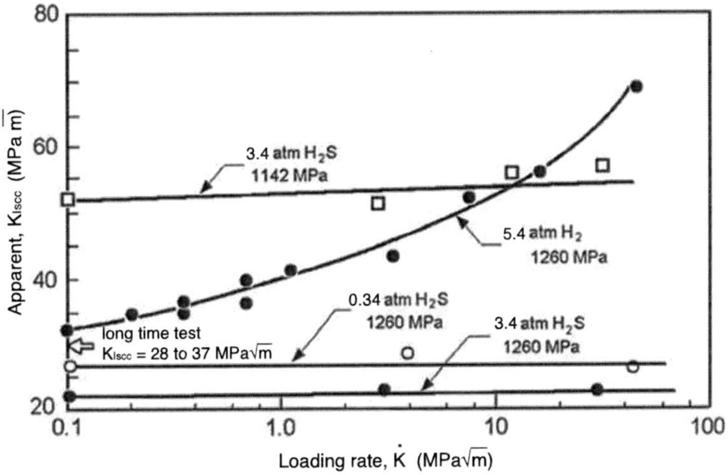

Fig. 19.30 Effect of loading rate on apparent K_{Iscc} (modified from [19])

sensitization of the material or is due to the dissolution of the metal at the crack tip, then it needs some time to occur. Therefore, the strain rate at which the specimen is subjected must be very low. The test can also use specimens without any notch. In this case the strain rate $\dot{\varepsilon} = d\varepsilon/dt$ should be as low as 10^{-6} s^{-1} or 10^{-7} s^{-1} against the 10^{-3} s^{-1} or 10^{-4} s^{-1} of a regular traction test, i.e., about thousand times lower. The unnotched specimen is particularly useful for sensitization studies. An example is shown in Fig. 19.31 [22] for austenitic stainless steel type 304 loaded in traction with a constant strain rate of $2.2 \cdot 10^{-6}$ s^{-1} in air and in NaCl 1M and 4M, respectively. The response of the material in a solution NaCl 1M is showing a sharp reduction of ductility that is an indication that the material has been sensitized to SCC. In the NaCl 4M solution, the material dissolution starts before the yielding is reached and is so rapid to break the specimens in few percent of deformation.

Since the specimen is loaded in traction at constant extension rate, the test is also referred to as CERT (constant extension rate test). The CERT procedure with an unnotched specimen cannot yield the value of the K_{Iscc}, but it is just an indication of the susceptibility of the material to SCC. However, a strain rate $\dot{\varepsilon}$ too slow, generally lower that 10^{-7} s^{-1}, could not trigger SCC anymore because the material can be restoring the passive film that protects its surface with a rate higher than that by which the strain can destroy it, unless the environment, acid or basic, is so aggressive to destroy itself the protective film. Therefore, there will be a strain rate range within which SCC occurs and this event will be identified in the CER test through a reduction of ductility with respect to a traction test in dry air. This is schematized in Fig. 19.32a. Point B marks the loss of ductility due to SCC. At the same time, a stress corrosion crack will appear whose growth rate, *da*/*dt*, has a jump precisely in the same range of strain rates in which the loss of ductility occurs and reach a maximum exactly when the latter reaches the minimum value, as shown by point B in Fig. 19.32b.

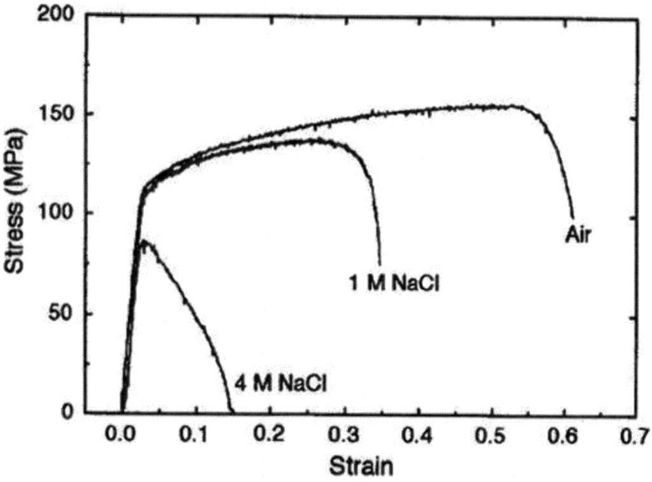

Fig. 19.31 CER test on austenitic stainless steel type 304 in sensitized conditions [22]

In the strain range field marked by the letter A the load application is too slow therefore the passivation kinetics can rebuild the protective film that broken by the mechanical stretching. In C it is too fast for the corrosive event to start. It is clear that the aggressiveness of the environment plays a role in the definition of the critical $\dot{\varepsilon}$ range. If the corrosion mechanism is hydrogen embrittlement the range of active strain rates is that schematized in Fig. 19.33 [23] by curve B.

The loss of ductility, which is the sign of corrosive aggression, in no longer confined to a particular range of $\dot{\varepsilon}$ values, but initiates with a threshold value below which it continues progressively without any apparent passivation effect. This is because the hydrogen absorption continues in particular when impurities of S, P, As, An, Sn and Sb are present that prevent the recombination of atomic hydrogen

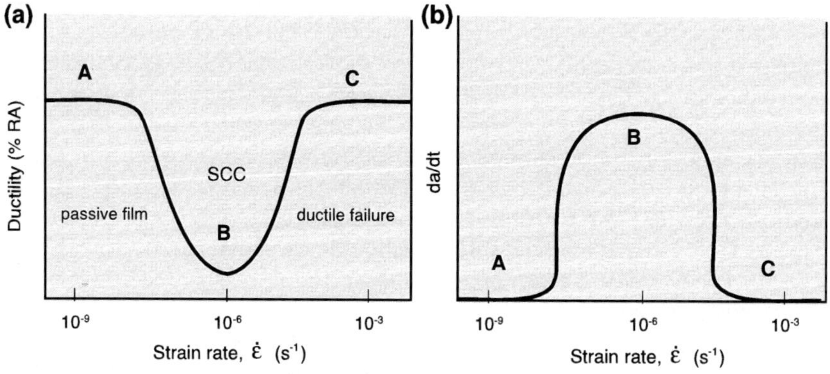

Fig. 19.32 Schematic of the range in which SCC develops during CER test

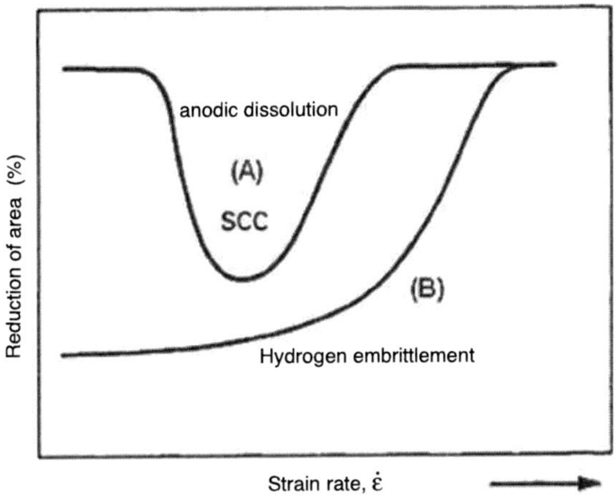

Fig. 19.33 Schematic showing the effect of strain rate on anodic dissolution SCC and hydrogen-induced cracking [23]

into molecular hydrogen. The choice of the most appropriate $\dot{\varepsilon}$ value to run a CER test is fundamental in inferring the real K_{Iscc} of the material, in particular when the environment is not so aggressive or the material is not so sensitive to SCC as medium–low strength carbon steels. Perhaps, this choice and the very long incubation times may be the true reason why in these low strength steels it is yet to demonstrate that K_{Iscc} is a material characteristic independent of the environment. At variance, in β-titanium alloys no dependency of K_{Iscc} the on the strain rate is observed at all.

19.5.3 Effect of Electric Potential on CERT Results

It has been shown in Sect. 13.6 that the maximum anodic current. i.e., the maximum metal dissolution in the solution occurs at the polarization potential E_{corr}. This potential for a sensitized 304L stainless steel is shown in Fig. 18.4 to be between – 3.0 to – 1.0 V, depending on the degree of sensitization. This effect will be seen in a CER test as a variation of ductility being maximum et and practically null in the passive region of electric potential. This is shown in Fig. 19.34 for a sensitized stainless steel type 304 exposed at 50 °C to a 4M NaCl solution at various electric potentials [24]. The applied strain rate was $2.2{\cdot}10^{-6}\ s^{-1}$. In air there is no SCC and the metal present a large deformation, about 60%, at failure. At variance, in the test solution at the polarization potential of – 110 mV the corrosion is very severe, as the drastic reduction of deformation at fracture would indicate. SCC reduces as the potential grows. At – 310 mV, SCC is almost completely prevented and the deformation at failure approaches that observed in air.

This provides a strong indication about the possible control of SCC through the applied potential. The same effect could be obtained by increasing the electric potential in the positive range of values. But it must be cautious in doing that without a

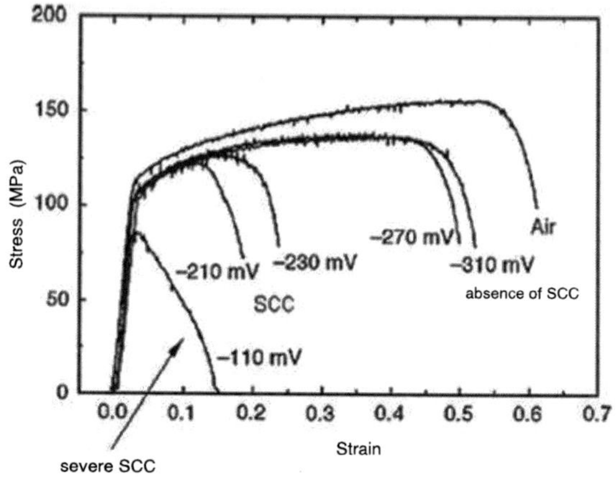

Fig. 19.34 Effect of applied corrosion potential on the CER test results for a sensitized austenitic stainless steel type 304 [24]

real knowledge of polarization potentials for the particular steel in the sensitizing conditions because it could be possible to enter the pitting formation potential (see Fig. 17.11). The same effect can be seen in sensitized aluminum alloy type AA5083 in a 3.5% NaCl solution. This alloy undergoes severe heat treatment sensitization in which the β-phase (Al_3Mg_2) at grain boundaries (see Fig. 18.11). CERT data for this alloy are shown in Fig. 19.35 [25]. At the open circuit potential (OCP) of the β-phase, in practice the electric potential of precipitates, a devastating metal dissolution is observed. But by applying an electric potential of − 0.98 V the ductility of the metal is completely recovered since in the 3.5% NaCl solution it shows the same load–displacement behavior that is has in air. This actually means complete SCC occurrence prevention.

From a practical point of view it may be rather difficult to perform a CER test in the worst environmental condition since it requires an environmental chamber in which the load cell is inserted, as schematically shown in Fig. 19.36 for a CER test in hydrogen gas [19]. The gas pressure and temperature are continuously monitored. The conduction of the test can be even worse if the environment is a solution since also the chemistry of the solution must be kept under control. It is not an easy and economical job. If the environment is sea water, to which austenitic stainless steels, titanium and aluminum alloys are very sensible, then it can be used a simple solution like that used by the author and shown in Fig. 19.37. It consists of a Perspex hose in which a precracked cylindrical or flat specimen is inserted. The hose is simply sealed by a rubber cup plugged at base carrying a hole through which the specimen is tightly inserted. To improve the aggressiveness of the solution its pH can be slightly decreased to 6 or even 5 by adding the proper amount of HCl or H_2SO_4. Never use weak organic acids, like acetic acid, since they may have a counter effect acting as buffer solution. If the test solution is chosen too acid and there is some fear that the specimen can fail breaking the hose and letting the acid flow on the test rig corroding some important part of the loading cell, then it can be used a cloth wetted with the acid solution and wrapped in tight contact all around the specimen.

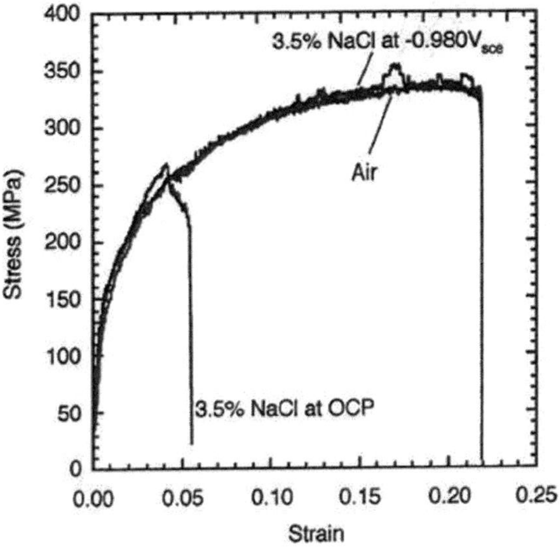

Fig. 19.35 CERT data showing ductility recovery at − 0.98 V polarization during testing of type AA5083 aluminum alloy [25]

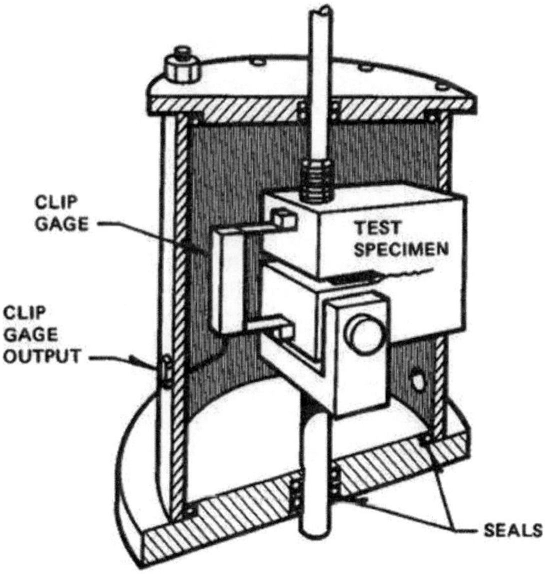

Fig. 19.36 Schematic of the CER test in an environmental chamber [19]

19.6 Factors that Affect SCC

Besides the material properties and corrosion potential, temperature and hydrogen pressure are the most important parameters that have a definite effect on SCC. Figure 19.38 [26] shows the loss of strength in a CER test for high strength steels

Fig. 19.37 Precracked notched specimen (three notches) inserted into a Perspex hose sealed by a rubber cup

as a function of temperature and strain rate. Specimens were precracked. The environment was hydrogen gas. It is particularly interesting to note how the loss of strength in hydrogen has a minimum around 0 °C while in the environment without hydrogen is practically constant. The loss of strength stops at a certain temperature that increases with increasing strain rate up to approaching the hydrogen free behavior. Note that the same temperature dependence was found for the threshold stress intensity factor K_{Iscc} in 4340 steel, as shown in Fig. 19.18. Again, the same temperature dependence observed in martensitic steels is also characterizing 18Ni high strength maraging steel, as shown in Fig. 19.39 [27]. Since SCC in these steels is caused by hydrogen embrittlement, it is evident that the same temperature dependence must be observed for hydrogen absorption. In effect, studies conducted by Chornet and Coughlin on hydrogen absorption in iron have evidenced precisely this dependency on temperature, as shown in Fig. 19.40 [28].

The maximum absorption in iron is obtained at low temperatures, while increasing the temperature hydrogen mobility and diffusion in iron prevails. This favors the elimination of hydrogen that cannot remain in iron. At 120 °C the baking-out of hydrogen from iron is almost complete, at the lower hydrogen pressure. For higher hydrogen pressures the temperature to avoid hydrogen absorption increases. As to

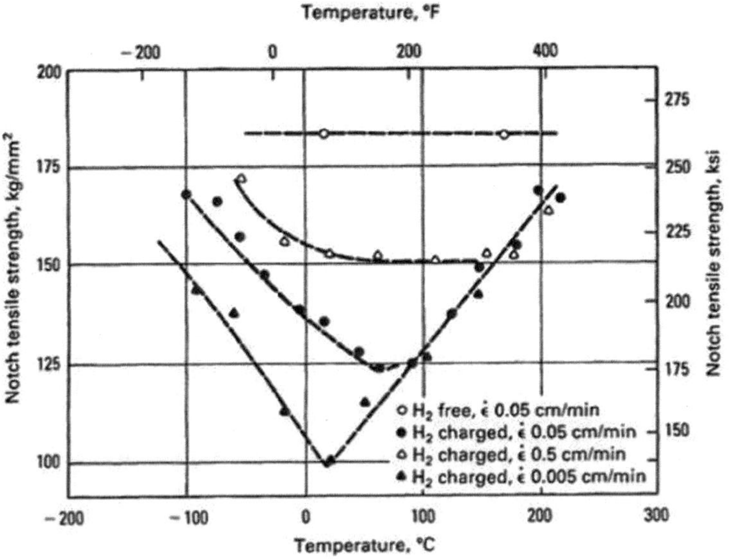

Fig. 19.38 Notch tensile strength of high-strength steel plotted against testing temperature for three strain rates (crosshead speeds) [26]

Fig. 19.39 Variation of K_{Iscc} with temperature in high strength steels martensitic and maraging steels [27]

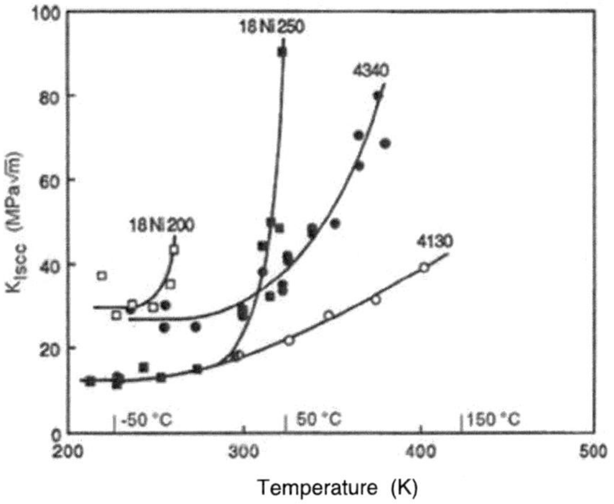

this latter factor, the hydrogen pressure, Fig. 19.41 [12] shows its effect on 4340 steel. Figure 19.41 also presents this effect of strength on SCC. Increasing the strength, in fact, also the susceptibility of the material to SCC increases. This can be seen in Fig. 19.42 for NiCrMoV rotor steel [19] subjected to CER tests at ~ 90 N/min (20 lbs/min) of loading rate equivalent to ~ 1 MPa$\sqrt{m}$/min, in a hydrogen sulfide environment and in Fig. 19.43 for 18Ni maraging steel in hydrogen gas at 2 atm of

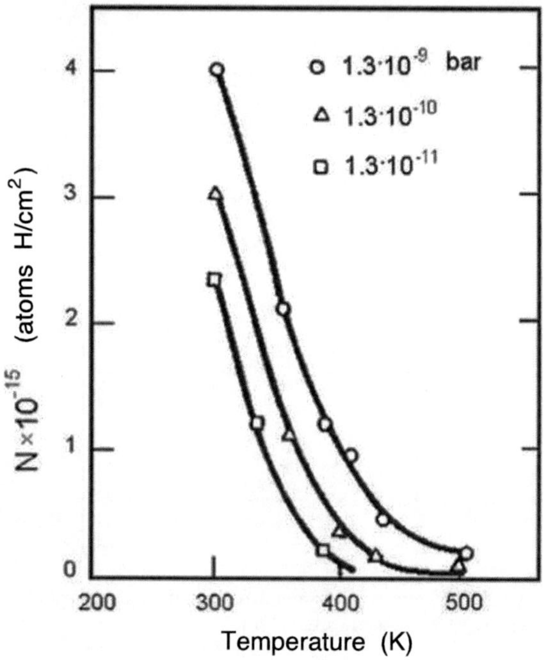

Fig. 19.40 Temperature and pressure dependency of hydrogen absorption in iron [28]

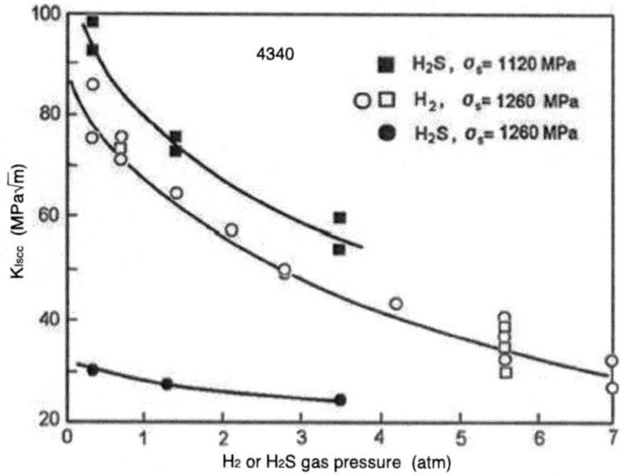

Fig. 19.41 Effect of hydrogen gas pressure on the apparent threshold K_{Iscc} for 4340 steel (modified from [12])

pressure [29, 30]. The same effect can be seen in terms of grain size for two 4340 steel [31] (Fig. 19.44).

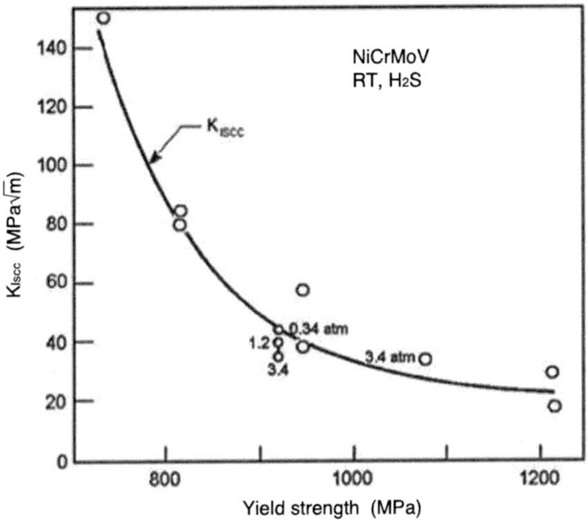

Fig. 19.42 K_{ISCC} dependency on the yield strength for NiCrMoV rotor steel in hydrogen sulfide [19]

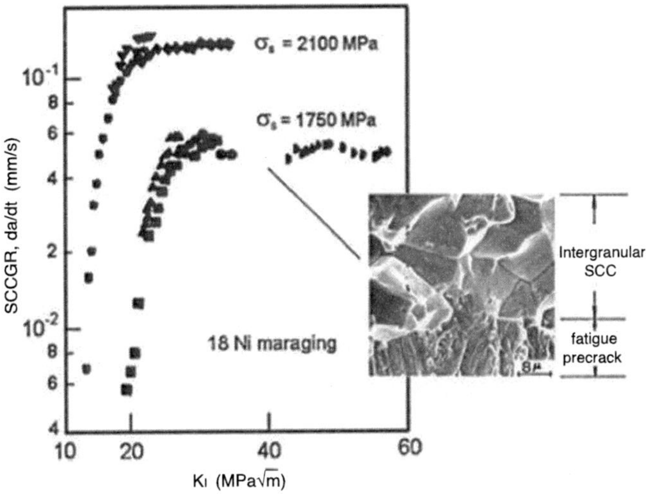

Fig. 19.43 SCCGR, *da/dt*, for 18 Ni maraging steel in hydrogen gas (composite diagram from [29, 30])

Fig. 19.44 Refinement of grain size improves resistance to hydrogen cracking as measured by the time to failure of two 4340 steels [31]

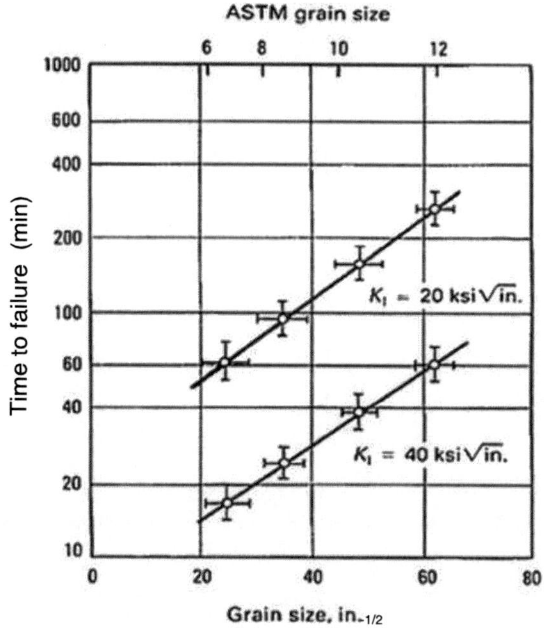

References

1. Steigerwald, E.A.: Delayed failure of high-strength steel in liquid environment. Proc. ASTM **60**, 750–760 (1960)
2. Laycock, N.J., Newmann, R.C.: Corros. Sci.. Sci. **39**, 1771 (1997)
3. Compton, K.G., Mendizza, A., Bradley, W.W.: Atmospheric galvanic couple corrosion. Corrosion **II**, 383 (1955)
4. Chen, G.S., Liao, C.M., Wan, K.C., Gao, M., Wei, R.P.: Pitting corrosion and fatigue crack nucleation. Am. Soc. Test. Mater., ASTM STP **1298**, 18–33 (1997)
5. Weiderhorn, S.: Moisture assisted crack growth in ceramics. Inten. J. Fracture Mecganics **4**(2), 171 (1968)
6. Novak, S.R., Rolfe, S.T.: Comparison of fracture mechanics and nominal stress analysis in stress corrosion cracking. Corrosion **26**(4), 121–130 (1970)
7. Carter, C.S.: The effect of silicon on stress corrosion resistance of low-alloy high-strength steels. Corrosion **25**, 423–431 (1969)
8. Smith, H.R., Piper, D.E., Downey, F.K.: A study of stress corrosion cracking by wedge-force loading. Eng. Fract. Mech.Fract. Mech. **1**, 123–128 (1968)
9. Johnson, H.H., Willner, A.M.: Moisture and stable crack growth in high strength steels. Appl. Mater. Res. **4**, 34 (1965)
10. Steigerwald, E.A., Benjamin, W.D.: Stress Corrosion Cracking Mechanisms in Martensitic High Strength Steels, 3rd Quarter Progress Report, Contr. No. AF 33(615)-3651, Air Force Materials Laboratory (1967)
11. Peterson, M.H., Brown, B.F., Newbegin, R.L., Groover, R.E.: Stress corrosion cracking of high strength steels and titanium alloys in chloride solution at ambient temperature. Corrosion **23**, 142 (1967)
12. Landes, J.D.: Stress Corrosion Crack Growth, Lecture on the Use of Fracture Mechanics at Westinghouse, Pittsburgh, PA (1974)

13. Wei, R.P, Novak, S.R., Williams, D.P.: Some Important Considerations in the Development of Stress Corrosion Cracking Test Methods, AGARD Conference Proceedings No. 98, pp. 5–1, 8 (1972)
14. Che-yu Li, P., Talda, M., Wei, R.P.: Int. J. Fracture Mech. **3**, 29 (1967)
15. Landes, J.D., Wei, R.P.: The kinetics of subcritical crack Groth and deformation in a high strength steel. Journal of Engineering Materials and Technology, ASME Series H, Vol. 1, pp. 2/9 (Jan, 1973)
16. Johnson, H.H., Paris, P.C.: Sub-critical flaw growth. Eng. Fract. Mech.Fract. Mech. **1**, 3–45 (1968)
17. ASTM, G.: 39, Metal Corrosion, Erosion, and Wear Standards, Annual Book of ASTM, American Society for Testing and Materials, Vol. 03.02, Section 3
18. McIntyre, P.: The relationship between stress corrosion cracking and sub critical flaw growth in hydrogen and hydrogen sulphide gases. In: Stress Corrosion Cracking and Hydrogen Embrittlement of Iron Base Alloys, National Association of Corrosion Engineers, p. 788 (1977)
19. Clark, W.G. Jr., Landes, J.D.: The Effect of Hydrogen Gas Environment on the Crack Initiation, Growth and Fracture Properties of 180 ksi Yield Strength Type 4340 Steel, Westinghouse Research Report 73-7E7-ETIWA-R1, (5 April, 1973)
20. Clark, W.G. Jr., An Evaluation of the Crack Growth and Fracture Properties of 18Mn-5Cr Steel in Generator Environment, Westinghouse Research Report 73-1E7-MAGRR-R1, (3 May, 1973)
21. Nelson, H.G., Williams, D.P.: Quantitative observations of hydrogen induced slow crack growth in a low alloy steel. In: Stress Corrosion Cracking and Hydrogen Embrittlement of Iron Base Alloys, National Association of Corrosion Engineers, p. 390 (1977)
22. Kwon, H.S., Cho, E.A., Yeom, K.A.: Predeiction of stress corrosion cracking susceptibility of stainless steels based on repassivation kinetics. Corrosion, NACE Int. **56**(1), 40 (2000)
23. Kim, C.D., Wilde, B.E.: A review of constant strain-rate stress corrosion cracking test. In: Ugianski, G.M., Payer, J.H. (Eds.), Stress Corrosion Cracking—The Slow Strain Rate Technique, ASTM STP 665, pp. 97–112. (American Society for Testing and Materials, 1979)
24. Kwon, H.S., Cho, E.A., Yeom, K.A.: Prediction of stress corrosion cracking susceptibility of stainless steels based on repassivation kinetics. Corrosion **56**(1), 37 (2000)
25. Searles, J.L., Gouma, P.I., Buchheit, R.G.: Stress corrosion cracking of sensitized AA5083. Metall. Mater. Trans. A **32A**, 2865 (2001)
26. Graville, B.A., Baker, R.G., Watkinson, F.: Br. Weld. J. **14**, 337 (1967)
27. Moody, N.R., Robinson, S.L., Garrison, W.M., Jr.: Hydrogen effects on the properties and fracture modes of iron-based alloys. Res. Mech. **30**, 143–206 (1990)
28. Chornet, E., Coughlin, R.: Chemisorption of hydrogen in iron. J. Catal.Catal. **27**, 246–265 (1972)
29. Gangloff, R.P., Wei, R.P.: Gaseous hydrogen embrittlement of high strength steels. Met. Trans. 8°, 1043–1053 (1977)
30. Gangloff, R.P., Wei, R.P.: In fractography in failure analysis, ASTM STP 645, pp. 87–106. ASTM International, West Conshohocken, PA (1978)
31. Procter, R.P.M., Paxton, H.W.: Trans. ASM **62**, 989 (1969)

Chapter 20
Corrosion Fatigue

20.1 Introduction

It is rather difficult to establish a simple yet comprehensive picture of the combined effects of fatigue and environment. It is the author's view that to fully understand corrosion fatigue it is necessary to proceed through fundamental steps starting with concepts that may seem elementary. As first, it must be said that any environment is potentially aggressive, even dry air. To recognize this simple fact any fatigue *S–N* or *da/dN* crack propagation behavior must be first compared with what can be considered the absolute non-corrosive environment, which is vacuum. Absolute vacuum, in fact, is *par excellence* the absence of any environment and, as such, cannot be aggressive at all. It is already proved that dry air, apparently non-aggressive, has indeed an influence on fatigue of metals and this effect is due to oxygen in particular. To unveil this effect it is necessary to make a comparison with fatigue in vacuo. For example, Wadsworth and Hutchings [1] already in the '50s found that for each given strain amplitude the ratio of fatigue strength in air to that in vacuo was about 1–40 for copper, 1–5 for aluminum and 1–1 for gold. Gold seems not be affected by the environment. However this fall of fatigue strength in air was not accompanied by a parallel decrease of the number of cycles at which persistent slip bands first appeared in air with respect to vacuum (see Sect. 1.4). Wadsworth and Hutchings, then, came to the conclusion that air was playing a fundamental role in fatigue propagation of the first macro-crack or MSC (see Sect. 1.5) but not on micro-crack initiation. Furthermore, it was also found [2] that the rotating bending fatigue limit σ_f at 10^7 cycles for 4340 steel, for 4% Al titanium alloy, for 60/40 brass and for 6% Al magnesium alloys was higher in air with 3% relative humidity (dry air) that in air with more than 85% RH, dough within a mere 5%. This apparent incongruence can be explained through another fundamental point. The testing frequency was equal to 150 Hz, too high to evidence any environmental effect. Corrosion needs some time to act and cannot be seen if fatigue cycles are applied with a frequency higher than few cycles per minute. The third fundamental point is that the stress corrosion effect

P. P. Milella, *Fatigue and Corrosion in Metals*,
https://doi.org/10.1007/978-3-031-51350-3_20

must be separated from the fatigue one to study true corrosion fatigue. It has been shown, in the previous chapter, that a threshold stress intensity factor K_{Iscc} exists below which no SCC occurs. Therefore, in the study of fatigue crack propagation the applied stress intensity excursion $\Delta K = K_{max} - K_{min}$ must have the K_{max} below the SCC threshold, $K_{max} < K_{Iscc}$. If this is the case, fatigue will be referred to as true corrosion fatigue (TCF). This chapter will start with a description of variations of *S–N* fatigue curves moving from dry air to vacuum or argon gas, which is an inert environment almost equivalent to vacuum. Once this difference is digested, then dry air will become our reference environment.

20.2 In Vacuo Fatigue

Figure 20.1 [3] presents a comparison between fatigue strength *S–N* curves for aluminum alloy type 6060-T6 in dry air and humid air, for $R = -0.5$. It is significant that at equal amplitudes there is a finite life reduction of about 5 times. Another example of life reduction in air with respect to vacuum at 7×10^{-6} mm Hg, is provided in Fig. 20.2 for high purity iron [4]. Again, at equal stress amplitude there is a reduction of the finite life by a factor of 10, while the strain ε_f at the fatigue limit of 10^7 cycles reduces of about 15% from ~ 1.3×10^{-3} to ~ 1.1×10^{-3}. If water (humidity) and oxygen in air are the aggressive elements, then avoiding the contact with air would improve the fatigue strength almost as in vacuum. This can be done by using mineral, non-aggressive oil. In effect, already in 1935 it was found that by protecting steel wires with grease would improve their fatigue fife by a 5% [5]. However, in vacuum the fatigue life would have been even higher so that grease, dough beneficial, was itself aggressive. An example of mineral oil type SAE 30 protection is offered in Fig. 20.3 for mild steel in rotating fatigue in air [6].

Note that the cleaning of the specimen with benzene before the testing does not yield any beneficial effect. Humid air, has been said, is not as much detrimental in persistent slip bands formation as in the propagation of the first macro-crack that forms by fatigue in smooth specimens. This actually means that it is in Stage II of fatigue (see Sect. 2.7 and Fig. 2.47) that its effect must be better evidenced through the FCGR, *da/dN*, versus ΔK_I (Paris-Erdogan equation). Figure 20.4 [7] shows the FCGR, *da/dN*, found by Amzallag et al. for type 2618-T651 aluminum alloy, 402 MPa yield strength and 477 MPa ultimate strength, in air and in vacuum, respectively. it is possible to note how the FCGR, *da/dN*, follows the Paris-Erdogan power low (10.10) and how the environmental effect results in a variation of both the C constant and the n exponent. Also the threshold $K_{I,th}$ seems to decrease. Interesting enough, the new slope in air is lower than in vacuum. Therefore, it must be expected that the two curves merge into a common *da/dN* versus ΔK_I curve.

This is a general aspect of fatigue in potentially aggressive environment. This can be seen also in Fig. 20.5 [8] for a Ti-6Al-4V titanium alloy tested in air and in vacuum at ambient temperature and 300 °C, respectively. All curves merge in Region II of fatigue (see Fig. 14.6). The threshold stress intensity factor $K_{I,th}$ is decreased in air

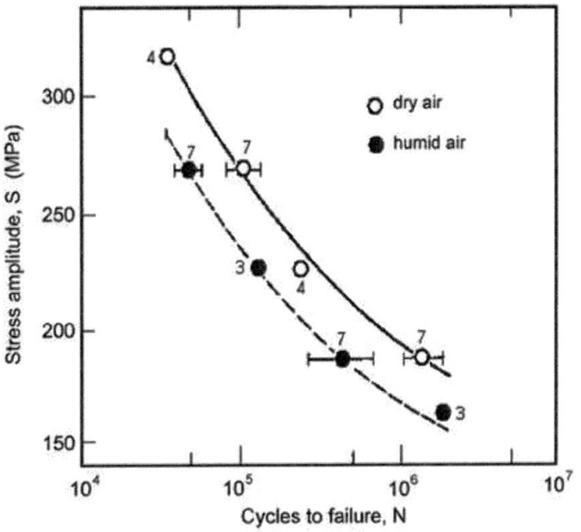

Fig. 20.1 *S–N* diagrams for 6060-T6 aluminum alloy in dry air and humid air (modified from [3])

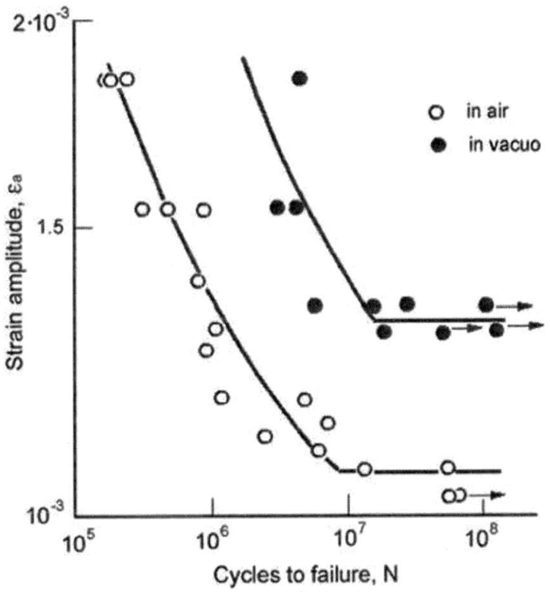

Fig. 20.2 Comparison of *S–N* diagrams for 0.5%C iron in air and in vacuo [4]

with respect to vacuum. In vacuum there is no difference between RT and 300 °C, as expected. The new interesting feature that stems out of Fig. 20.5 is the plateau in the FCGR, *da/dN*, at 300 °C. This is clearly indicating that a stress corrosion crack propagation event occurred that made the growth rate independent of the applied ΔK_I (Region II of Fig. 19.11).

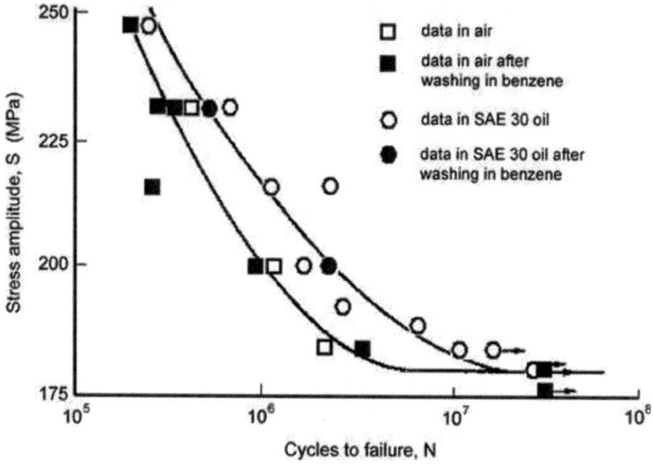

Fig. 20.3 Passivizing effect offered by SAE 30 mineral oil for a mild steel (modified from [6])

Fig. 20.4 Effect of air on the FCGR, *da/dN*, of type 2618-T651 aluminum alloy (modified from [7])

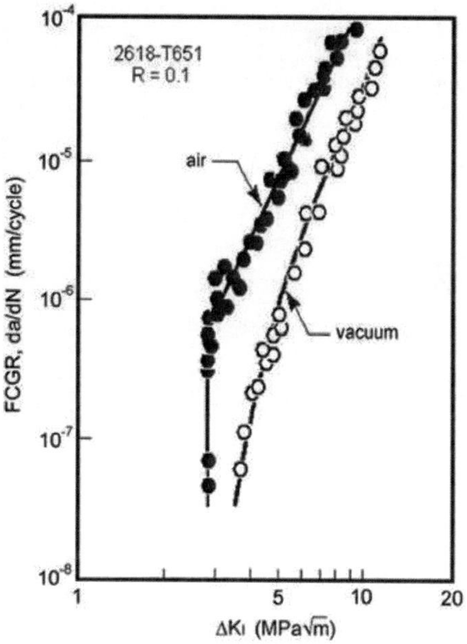

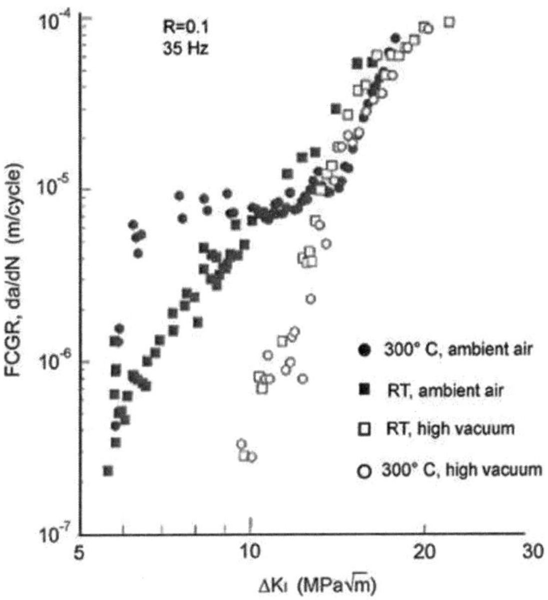

Fig. 20.5 FCGR, *da/dN*, versus ΔK_I for a titanium alloy, type Ti-6Al-4V, in air and vacuum at RT and 300 °C, respectively [8]

20.3 Environmental Fatigue

As an alternative to vacuum, also noble, inert gases like argon can be used as a reference environment. A comparison between FCGR, *da/dN*, in air and in argon gas is offered in Fig. 20.6 for a low-alloyed carbon steel type A 508 Cl 3 fatigued at a frequency of 1 Hz [9]. Almost no difference can be observed.

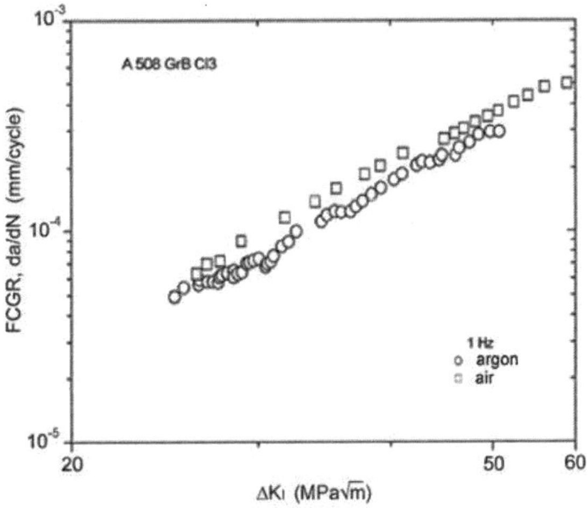

Fig. 20.6 Comparison of FCGR, *da/dN*, in air and argon gas for A 508 Cl. 3 steel [9]

From a practical point of view it is more convenient to use air and dry air, in particular, as the reference environment to compare and judge any environmental effect, especially when this latter is particularly aggressive. As an example, results obtained by Imhof and Barsom for two martensitic steels type 4340, heat treated to two different yield strengths of 910 MPa (130 ksi) and 1260 MPa (180 ksi), respectively, in NaCl solution simulating sea water are shown in Fig. 20.7 [10]. The FCGR in air is consistent with that already seen in Fig. 10.14. While in air there is no difference between the two steels, in the NaCl solution the fatigue data separate becoming larger by a factor of four for the higher strength steel. Both steels show a remarkable environmental effect with respect to the air. The values of the respective K_{Iscc} are also indicated. Note that for the 1260 MPa (180 ksi) steel fatigue crack propagation data have all been obtained below the corresponding K_{Iscc}, therefore they are true corrosion fatigue data, with no influence of stress corrosion (static fatigue). The same holds for the lower strength steel. Also note that the fatigue frequency was very low and equal to just 6 cpm. At higher frequencies now sign of corrosion fatigue would have been seen so that the data obtained in the sea water environment would be overlapping those obtained in air.

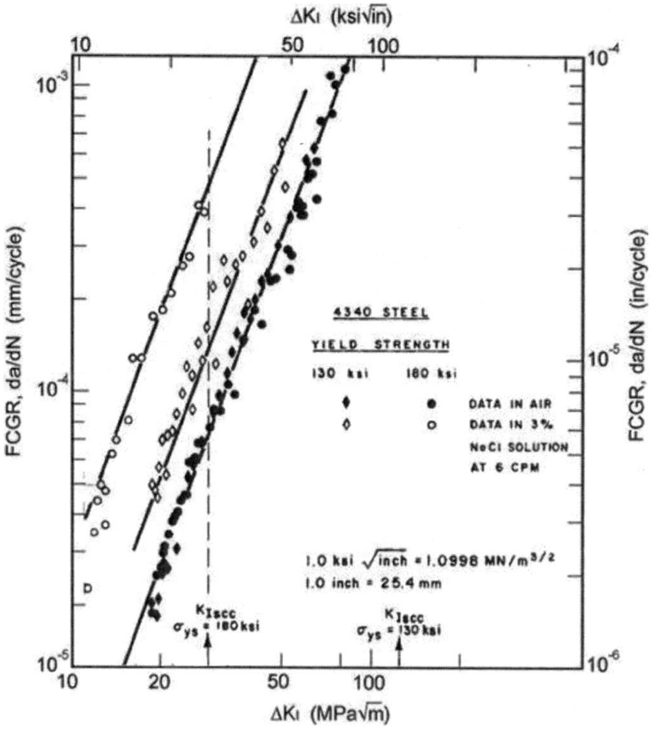

Fig. 20.7 Effect of sea water environment on the FCGR, *da/dN*, for 4340 steel (modified from [10])

The same environmental effect can be seen in Fig. 20.8 for type 7075-T6 aluminum alloy sheets in dry air and in a 3.5% NaCl solution, respectively [11]. The data shown refer to two different loading conditions, namely remote traction and wedge loading. A factor of ten characterizes the FCGR, *da/dN*, in sea water with respect to the air environment, while the loading conditions have no effect at all. Aluminum alloys are very sensible to the sea water environment. As for 4340 type steel, FCGR curves for the aluminum alloy in sea water seem to run parallel to that in dry air, which was not the case seen in Figs. 20.4 and 20.5. Also high strength steels type 12Ni5Cr3Mo, HY 130 and 10NiCrMoCo present the same characteristic environmental effect in sea water, as shown in Fig. 20.9 [12]. Interesting enough, the three steels have practically the same behavior in air, while in the aggressive environment the FCGR, *da/dN*, they separate depending on the alloy. The testing was done at ΔK_I below the K_{Iscc} of the material.

The environmental effect with a variation of the slope of the FCGR, *da/dN*, is found again for austenitic stainless steel type 13Cr-8Ni-2Mo PH of 1230 MPa yield strength tested in simulated sea water (3.5% NaCl), as shown in Fig. 20.10 [13]. Again, the testing was done at ΔK_I below the K_{Iscc} of the material. Austenitic stainless steels are very sensitive to chlorides.

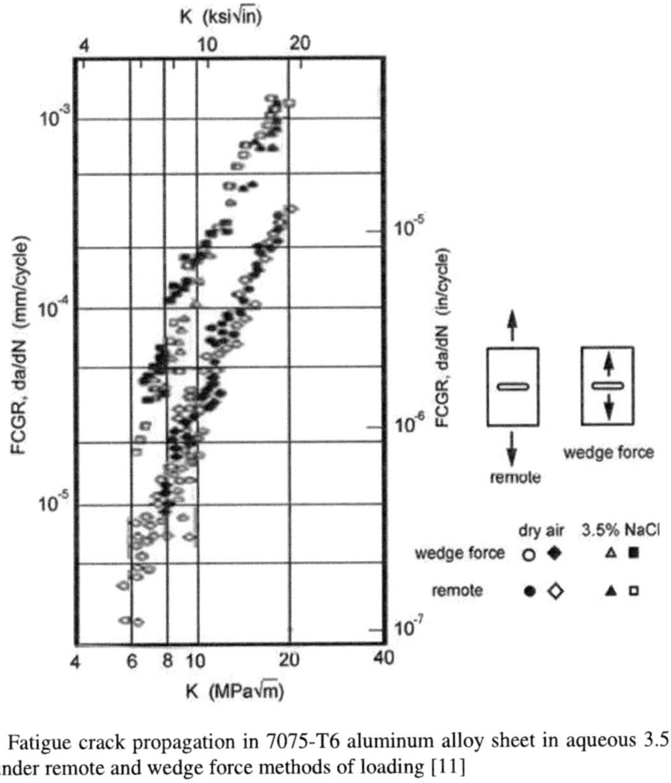

Fig. 20.8 Fatigue crack propagation in 7075-T6 aluminum alloy sheet in aqueous 3.5% sodium chloride under remote and wedge force methods of loading [11]

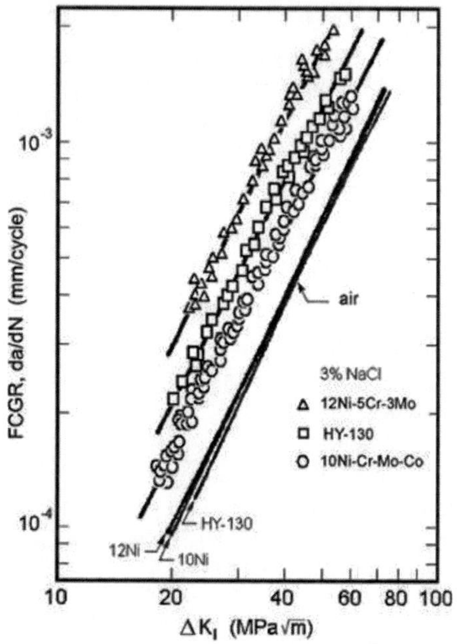

Fig. 20.9 FCGR, *da/dN*, for three different steels in air and sea water environment [12]

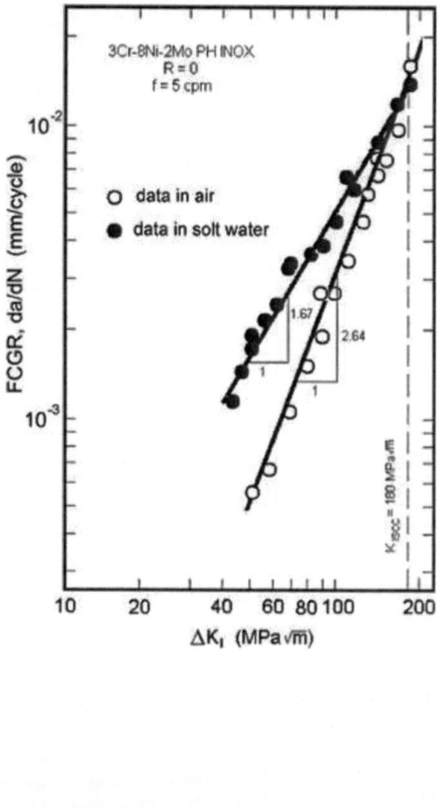

Fig. 20.10 FCGR, *da/dN*, for austenitic stainless steel type 13Cr-8Ni-2Mo PH in air and sea water [13]

20.3.1 Frequency Effect

It is clear that any environmental effect, if it ever exists, must be depending on time. It has been said, in fact, that corrosion, whether by metal dissolution or hydrogen embrittlement, depends on the time of exposure to the potentially aggressive environment. In fatigue the time variable is resident in the frequency of the cyclic loads. Therefore, it must be expected a FCGR, *da/dN*, dependency on the frequency and, in particular, a larger effect at lower frequencies. In effect, in early 1972 Kondo [14] reported data indicating that FCGR for A 533 GrB Cl1 low alloyed steel in high temperature pure water environment were significantly increased at low cyclic frequencies. The drastic decrease of the frequency of the load application results, first, in the appearance of the environmental effect and then in a worsening of the effect itself up to a point where the waiting time before the maximum load is reached is so long that the bare metal exposed to the environment can passivize by restoring the protective oxide film. This is shown in Fig. 20.11 [15] for high strength steel, type 12Ni-5Cr-3Mo of 1330 MPa yield strength, tested by Barsom and Rolfe in air and in a 3% NaCl solution. In air the frequency effect is null, but in the marine environment the progressive reduction from 600 to 6 cpm of the loading frequency results in a systematic increase of the environmental effect up to a factor of three on the FCGR, *da/dN*. At 600 cpm the exposure time to the potentially aggressive environment is limited to 0.1 s while at 6 cpm it grows to 10 s, a one hundred times higher. The slope of the curves remains the same. Another example is presented in Fig. 20.12 [16] for 4340 steel tested in dehumidified argon and in water vapor at a pressure of 585 Pa (4.4 tor) and $R = 0.1$.

The frequency reduction from 20 to 10 Hz seems not have any significant effect, but by further reducing the frequency the environmental effect comes out growing continuously. At 0.1 Hz the FCGR, *da/dN*, is about 20 times higher than in dry argon. Also in the water vapor environment fatigue curves remain parallel to that obtained in argon (inert environment). They can be fitted to a power law curve of the Paris-Erdogan type

$$\frac{da}{dN} = C(t) \cdot (\Delta K)^n \tag{20.1}$$

in which the slope n does not vary with frequency. The C factor is no longer a constant, but depends on the test frequency and, therefore, on time t, $C = C(t)$.

The slope n is precisely that found for high strength steels shown in Fig. 10.26; $n = 2.5$ and in argon it is $C = 4 \times 10^{-8}$ as for dry air. Particularly interesting are the results obtained for A 508 GrB Cl3 steel in argon, air and demineralized water at room temperature and different testing frequencies. Test data are shown in Fig. 20.13 [9].

At the test frequency of 10 Hz fatigue data are practically the same, almost independent of the test environment with a slight increase for those obtained in water, as it can be recognized in Fig. 20.13a. But when the testing frequency is lowered to 1.0 Hz then differences among the three environments start to appear, in particular

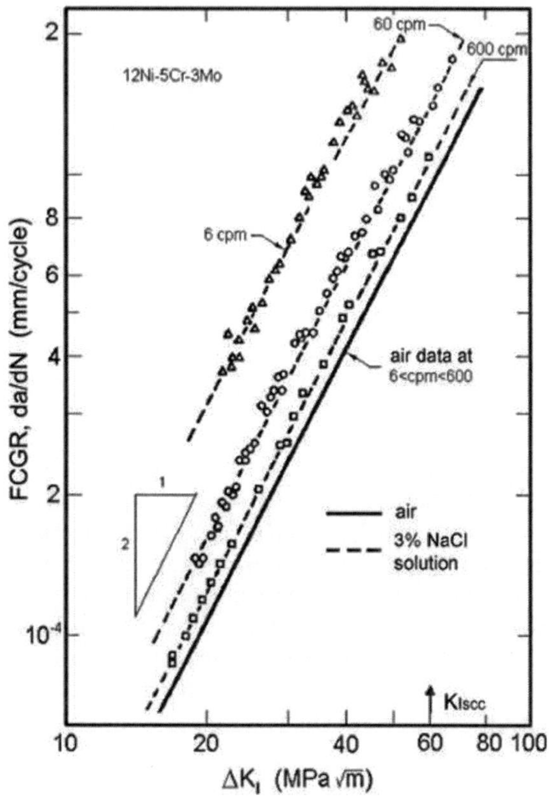

Fig. 20.11 FCGR, *da/dN*, for type 12Ni-5Cr-3Mo maraging steel in 3% NaCl solution at various test frequencies (modified from [15])

for the water environment in which the FCGR, *da/dN*, increases by a factor of about 2.5 with respect to inert argon. But the real outcome appears at 0.1 Hz (6 cpm) of testing frequency. At this very low frequency, in fact, the growth characteristics are completely changed in that not only there is an astonishingly higher FCGR, but it becomes also independent of the applied ΔK_{I}, as shown in Fig. 20.14 [9]. This finding that the growth rate is controlled by time rather than ΔK is explicitly indicating that the crack growth is sustained by a stress corrosion mechanism (see Sect. 19.3). The material, in fact, is entering Region II of the stress corrosion diagram of Fig. 19.11. The tests were run with an *R* ratio equal to 0.2. Therefore, with a $\Delta K_I = K_{I,max} - K_{I,min}$ equal, for example, to 25 MPa at which SCC is already present, the maximum stress intensity factor is $K_{I,max} = 31$ MPa$\sqrt{}$m that should be close to the K_{Iscc} of the material. In terms of crack growth increment per time increment, *da/dt*, versus K_{Imax} which is obtained by multiplying the FCGR, *da/dN*, by the frequency *f*, it is obtained $da/dt = 4 \times 10^{-4} \cdot 0.1 = 4 \times 10^{-5}$ mm/s.

The same behavior was observed by Mager et al. [17] always on A 533 Cl3 steel in demineralized water at 288 °C simulating the environment of either a PWR (pressure water reactor) or BWR (boiling water reactor) nuclear reactor. Mager et al.

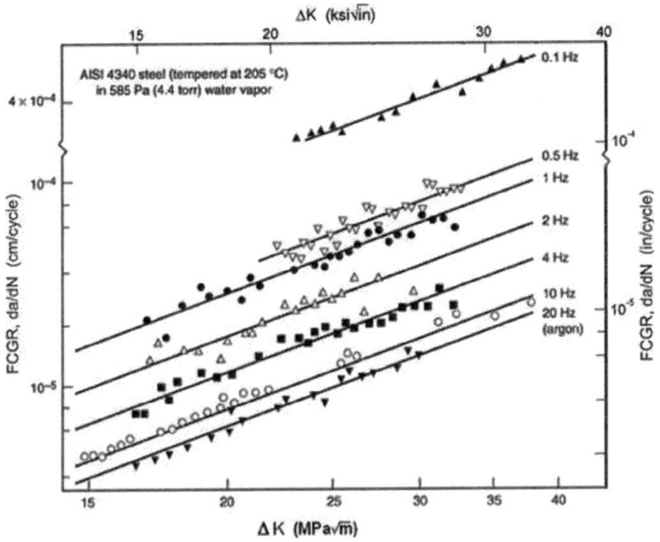

Fig. 20.12 FCGR, *da/dN*, for type 4340 martensitic steel tested in argon and water vapor (modified from [15])

data replotted in terms of *da/dt* are shown in Fig. 20.15. They clearly indicate the appearance of a SCC mechanism whose rate is precisely 4×10^{-5} mm/s. From Fig. 20.15 diagram it is possible to infer the K_{Iscc} of the material that is between 35 and 45 MPa$\sqrt{}$m. What is interesting to stress out is that tests under static stresses did not reveal any sign of SCC and, as such, were not able to measure the K_{Iscc} of the material. Therefore, for this type of low-alloyed, medium strength steel the incubation time (see Sect. 19.4) is so long that SCC cannot be seen under static loadings, but fatigue can shorten that time dramatically. In that specific case specimen had been subjected to static loads for two weeks.

This actually means that it is fatigue to assist SCC rather than being assisted by corrosion. But it can also explain the common and probably wrong conviction that medium–low strength carbon steels are not subjected to SCC. Another interesting feature that stems out of the diagrams of Fig. 20.15 is that at low ΔK_I or K_{Imax} the FCGR in water is much larger than in air, but the opposite happens at high ΔK_I or K_{Imax}. This is due to the characteristic of SCC of branching (see Fig. 18.14 and Sect. 18.5). When bifurcation or branching appears, generally at 2–4 times the K_I at which the plateau of Region II initiates, the crack tip stress intensity factor halves or cuts in three or four with the consequence to dampen the K_I at each of the many crack tips the more the higher the branching. It may happen, then, that the reduction brings the applied ΔK_I below the K_{Iscc} but not below the threshold $\Delta K_{I,th}$ for corrosion fatigue. This allows fatigue to continue and regain rate over the SCC at higher ΔK_I. The same frequency effect can be observed on titanium alloys. Figure 20.16 [18] reports data obtained by Dawson and Pelloux for Ti-6Al-4V titanium alloy. The

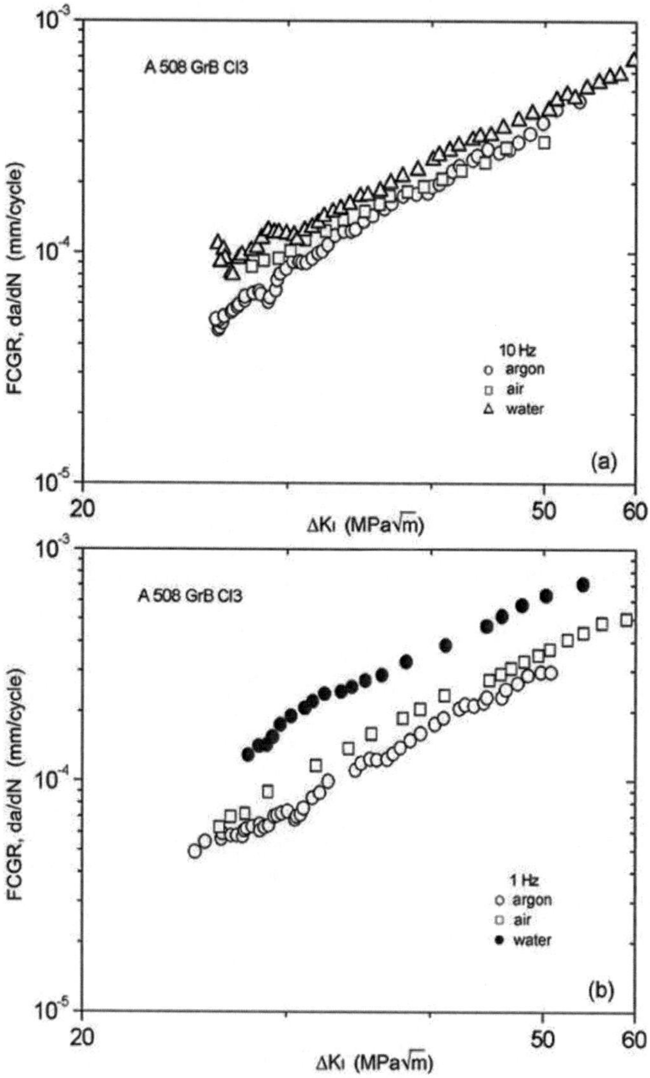

Fig. 20.13 **a** Test data obtained at a frequency of 10 Hz in argon, air and water, **b** test data obtained at 1.0 Hz in argon, air and water (modified from [9])

particularity of this research stands in the fact that the applied ΔK_I was imposed with a K_{Imax} equal to the K_{Iscc} of the material.

Since the load was cyclic and not constant, it could be argued that the SCC mechanism would not be triggered because not enough time was given to this mechanism to occur. This time, actually, depends on the frequency. As matter of fact, in ambient air at 10 Hz of load frequency the usual sigmoidal curve of the FCGR appears with

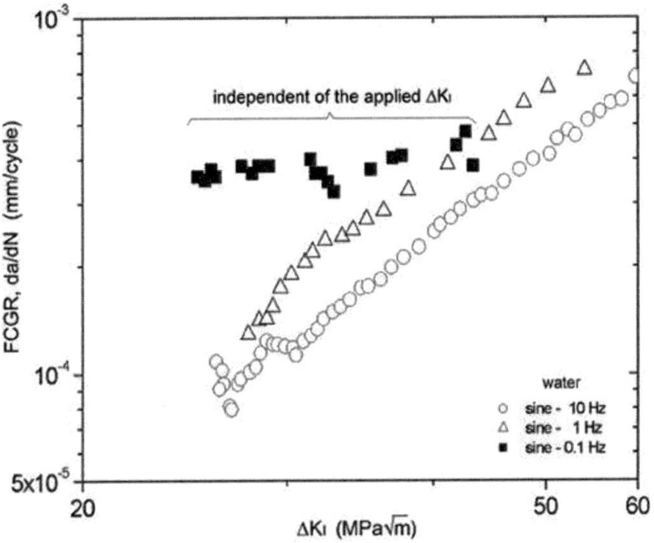

Fig. 20.14 Appearance of the Region II plateau for type A 508 Cl3 steel in demineralized water at RT and 0.1 Hz load frequency (modified from [9])

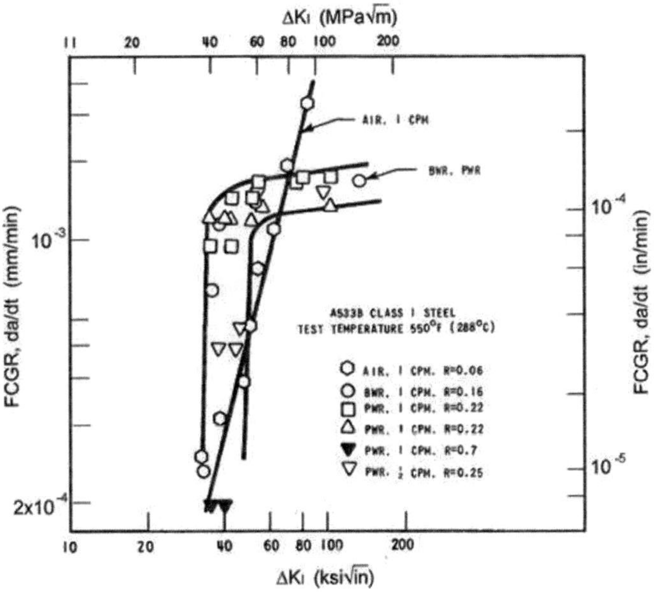

Fig. 20.15 FCGR, *da/dN*, for A 533 GrB Cl3 steel in either PWR or BWR water environment showing the appearance of SCC at 1 cpm [14]

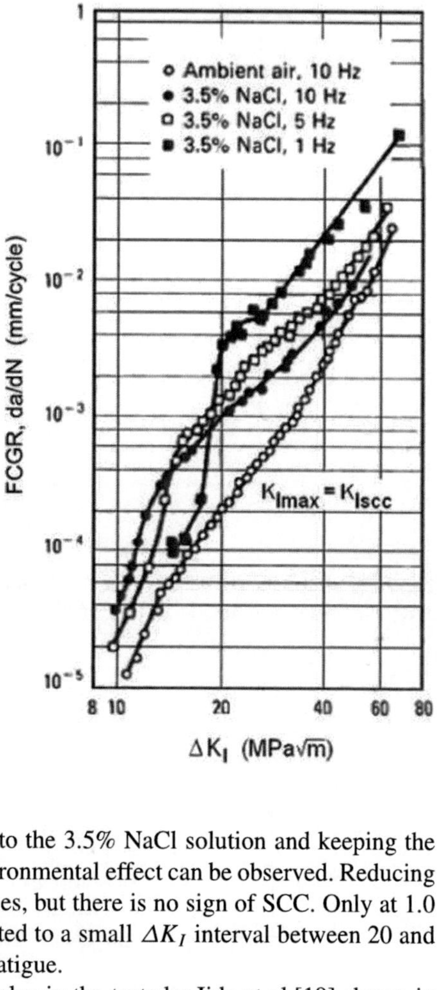

Fig. 20.16 Effect of frequency on the FCGR of Ti-6Al-4V titanium alloy [18]

no apparent sign of corrosion. Moving to the 3.5% NaCl solution and keeping the 10 Hz frequency a certain degree of environmental effect can be observed. Reducing the frequency to 5 Hz this effect increases, but there is no sign of SCC. Only at 1.0 Hz SCC it suddenly appears, but is limited to a small ΔK_I interval between 20 and 30 MPa$\sqrt{}$m to shift again to corrosion fatigue.

This time related effect can be seen also in the tests by Iida et al. [19] shown in Fig. 20.17 for a carbon steel type STS42, similar to ASTM A 333–6, in demineralized water containing 8 ppm of dissolved oxygen O_2 at 250 °C. Strain rates $\dot{\varepsilon}$ ranged from 0.4 to 0.0001%/s in the loading phase and were equal to 0.4%/s in the unloading one. The R ratio was equal to − 1. The N_{25} value of the abscissa scale refers to the number of cycles relative to 25% of the life of the specimen, in practice close to the macro-crack formation ($a \geq 300$ μm, see Sect. 2.3).

At strain rates of about 4%/s there is almost no influence of the environment, but at 0.0001 Hz fatigue life reduces by about two orders of magnitude. This finding recalls the CER test for stress corrosion (see Sect. 15.5.2). However, this frequency or time dependent effect is observed up to a lower limit of the load frequency. Studies by Van Der Sluys [20] and Cullen [21] on martensitic steels in demineralized water at 288 °C with $R = 0.2$ revealed that, in effect, the FCGR, *da/dN,* decreases with decreasing

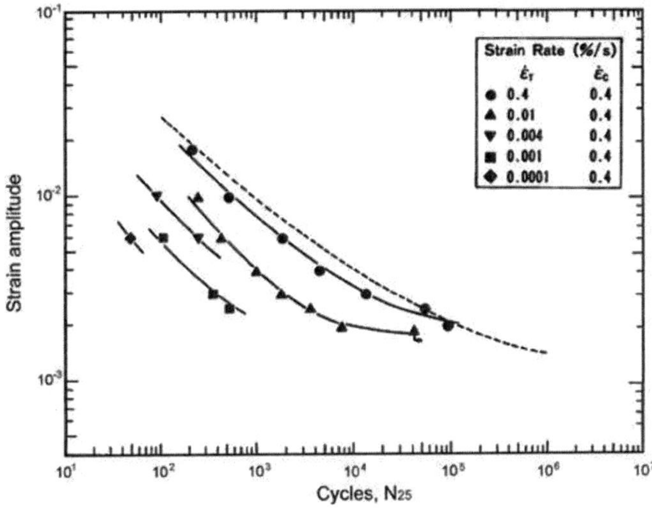

Fig. 20.17 *S-N*$_{25}$ curves obtained for carbon steel in air and water at 250 °C versus applied strain rate $\dot{\varepsilon}$ [19]

frequency up to a limit value of about 0.017 Hz (1 cpm). Below that value it does not grow any further, but stabilizes or even decreases. This is because by lowering too much the frequency enough time is given to the metal to passivize blocking the environmental effect. This minimum frequency at which the maximum FCGR is obtained is referred to as saturation frequency. Experimental practice indicates that the effect of reducing the frequency of the cyclic load is felt almost immediately by the material through a correspondent increase of the FCGR, *da*/dN, while a frequency increase requires same time to fully activate the correspondent FCGR decrease.

This is shown in Fig. 20.18 for 4340 steel in water vapors at 4.4 torr [22]. The frequency decrease from 10 to 1 Hz and then to 0.1 Hz is followed by an immediate response of the material that increases its FCGR while the frequency reduction requires a certain time to be fully followed. The frequency effect can be resumed as schematized in Fig. 20.19 for the two possible conditions: (a) the FCGR, *da/dN* versus ΔK environmental curves run parallel to each other and to the correspondent non-environmental fatigue curve (see Figs. 20.11, 20.12, 20.13 and 20.22), (b) they have a slope different from that pertaining to fatigue in dry air or argon (see Figs. 20.5 and 20.10). In both cases the FCGR, *da/dN*, increases with decreasing frequency f of the cyclic load so that at high frequency (f > 10 Hz) any environmental effect disappears and the response curves become equal to that obtained in the inert environment, which is independent of the test frequency. There is a saturation effect at a frequency of about 1 cpm below which the growth rate does not increases any more or it can even decreases, probably because re-passivation of the metal surface occurs due to long waiting time to reach the maximum stress. Frequency f is also affecting the threshold stress intensity factor ΔK_{th} that becomes lower and lower as f

increases. In case (a) the passage to Region III of fatigue (see Fig. 14.6) is anticipated by the frequency decrease. In the case schematized in Fig. 20.18b all curves converge towards that obtained in the inert or quasi-inert environment.

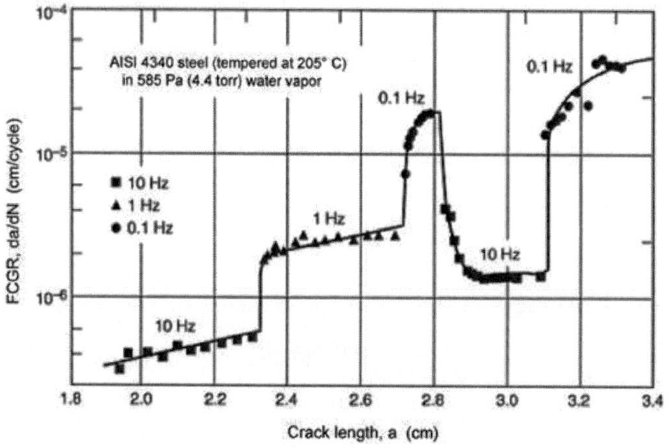

Fig. 20.18 FCGR response from changes in cyclic load frequency [22]

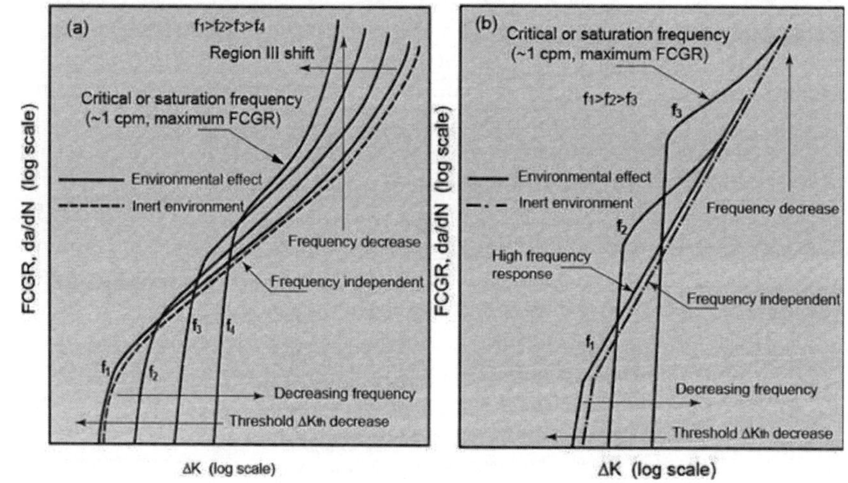

Fig. 20.19 Schematic of the frequency effect on FCGR, **a** the environment does not change the slope of the fatigue curves and **b** it does change the slope

20.3.2 Cyclic Wave Form Effect

If time has an effect on corrosion fatigue the question arises as to whether or not also the cyclic wave form mat have an effect. Figure 20.20 presents various forms of cyclic-stress fluctuations used to check this effect. They are: (a) sinusoidal load, (b) triangular load, (c) square load, (d) positive-sawtooth load and (e) negative-sawtooth load, respectively.

The load profile is characterized by the rising time to the maximum load and the time of permanence at that load. This latter is maximum for the square load which has the lowest possible rising time. On the contrary, the positive-sawtooth load has the maximum rising time and the minimum possible time of permanence to the maximum load. An interesting result of the waveforms effect in presented in Fig. 20.21 [23]. The data shown were obtained by Barsom for type 12Ni-5Cr-3Mo maraging steel in air and 3% NaCl solution at 6 cpm. Note that in air the scatter due to different waveforms is constrained within a 20%, but in the aggressive environment the load profile plays an important role. Square load and negative-sawtooth load give results that lay in the upper part of the air data scatter band.

On the contrary, sinusoidal load, positive-sawtooth load and triangular load accelerates the FCGR by a factor of three. It must be deducted that the rising time is more important than the time of permanence at the maximum load. This was already evidenced in Sect. 19.5.2 introducing the CER test in which the load rising time is extremely important. The unloading phase has no effect at all. The fatigue crack growth develops entirely during the loading phase. This finding may suggest

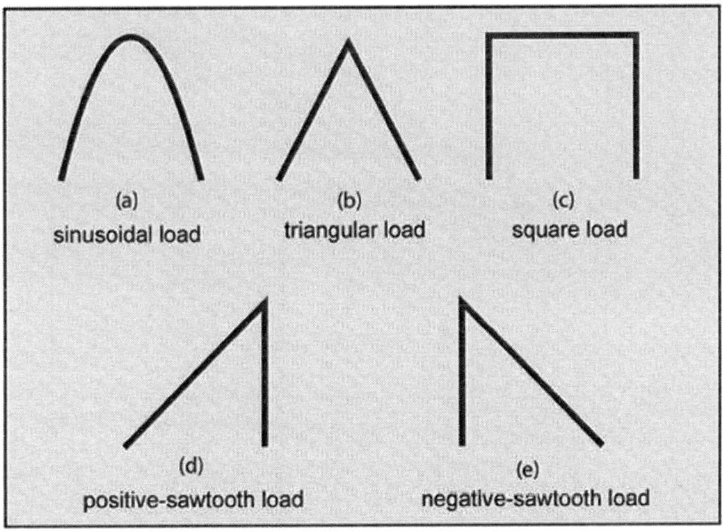

Fig. 20.20 Schematic of possible forms of cyclic-stress fluctuations used in corrosion fatigue study

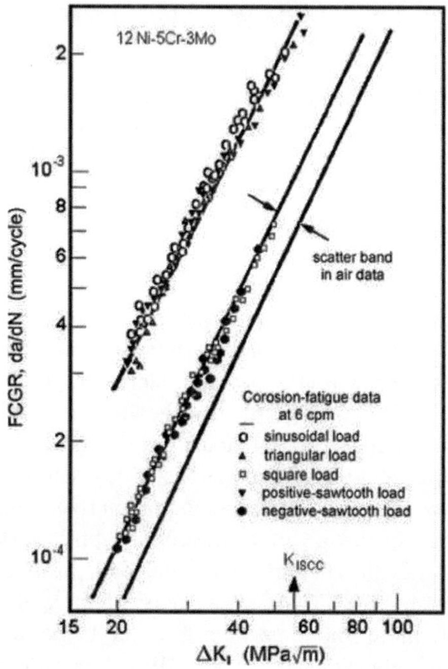

Fig. 20.21 Corrosion FCGR, *da/dN*, for type 12Ni-5Cr-3Mo maraging steel under various cyclic-stress fluctuations in air and sea water (modified from [23])

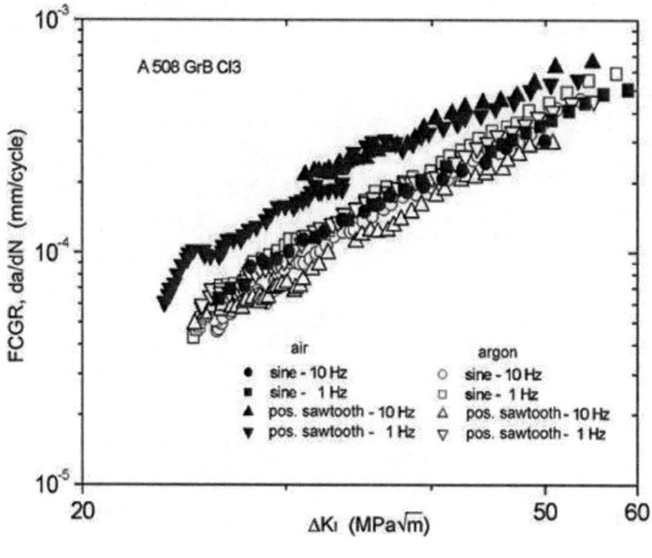

Fig. 20.22 Comparison between sinusoidal and positive-sawtooth loadings for carbon steel type A 508 Cl3 [9]

repeating the tests of Fig. 20.6 for carbon steel type A 508 Cl_3 using both the positive-sawtooth form and the sinusoidal one as types of loading. Results can be seen in Fig. 20.22 [9]. It is worth noting that the most effective loading form is the positive-sawtooth one. While in argon there is no difference between both the load form and the frequency, in the humid air environment the form of the cyclic stress makes the difference.

20.3.3 Effect of R Ratio

Also the *R* ratio plays an important role in the FCGR, *da/dN*, in environmental fatigue as it does in conventional fatigue or non-environmental fatigue (see Sect. 15.4). This effect is shown in Fig. 20.23 for Ti-8Al-1Mo-1V alloy tested in argon and 3.5% NaCl solution [24]. But the *R*-ratio can have also another effect. In fact, by increasing the *R*-ratio SCC appears that otherwise would not occur. This can be seen in Fig. 20.24 for type A 533 GrB Cl1 steel and welds for nuclear reactors applications tested in demineralized water at 288 °C [25]. The testing frequency was 1 cpm and the load form sinusoidal. While the weld metal does not seem to experience any SCC event, the base metal is sensible to SCC that appears already at $R = 0.2$ (see also Fig. 20.14) but becomes particularly heavy for $R = 0.7$. At the same time, the threshold stress intensity factor decreases from about 18 $MPa\sqrt{m}$ to about 10 $MPa\sqrt{m}$.

20.3.4 Effect of Inclusions and Chemical Species

The importance of chemical species and in particular sulfur has been already discussed in Sects. 17.8 and 18.4.3. Sulfur, in particular, comes from manganese sulfide, MnS, inclusions rather common in carbon steels. Their dissolution in a water environment (see Figs. 18.21, 18.22 and 18.23) creates a local acidic environment at the crack tip with the release of hydrogen ions highly dangerous. Today, the sulfur content in a carbon steel is recognized as the principal factor in the FCGR acceleration [23, 26–29]. A sulfur content of the order of 0.025–0.03% is considered to be too high from the FCGR point of view, in particular at low frequency of load application (see Sect. 20.3.1).

The resulting FCGR, *da/dN*, can speed up by even two orders of magnitude with respect to the inert environment. This clearly emerged in the ′70s when it was used to test specimens of high quality steels. It was rather a mistake that once recognized led to the testing of specimens of materials still under the AISI or SAE standards, but with the sulfur in the upper range of values allowed by the standards. An example of the influence of the sulfur content is offered in Fig. 20.25 that presents data obtained by Kondo et al. [29] for A 533 GrB Cl1 steel in demineralized water at 288 °C with ~ 8 ppm of dissolved oxygen, typical of a BWR, at 1 cpm (critical frequency) and $R = 0.2$. The steels tested had two distinct sulfur contents equal to 0.004% (low S) and

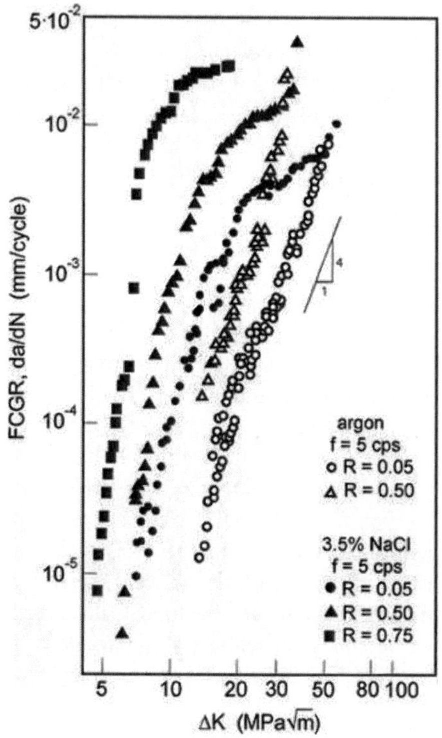

Fig. 20.23 *R*-ratio effect on the FCGR of Ti-8Al-1Mo-1V alloy [24]

0.014% (medium S), respectively. Experimental data for the low S steel seem to be consistent with those obtained in air, meaning that such low sulfur cannot trigger an environmental effect, but for the medium sulfur steel the FCGR are almost a factor of 5 higher.

They are aligned with the ASME Sec. XI design fatigue curve [30] which is now based right on medium–high sulfur content. It is now a fact that the accelerating effect exerted by sulfur is not only due to that contained in the steel as inclusion but also to that possibly existing in the solution. One ppm of H_2SO_4 is enough to introduce a serious environmental effect [31–33]. Sulfur, in its various forms as SO_4^-, to which steels are very sensitive, seems not be jeopardizing aluminum and titanium alloys that are, instead, very sensible to chlorides, brome and iodine species, Cl^-, Br^- and I^-. At variance, magnesium alloys are sensible to all the elements above mentioned.

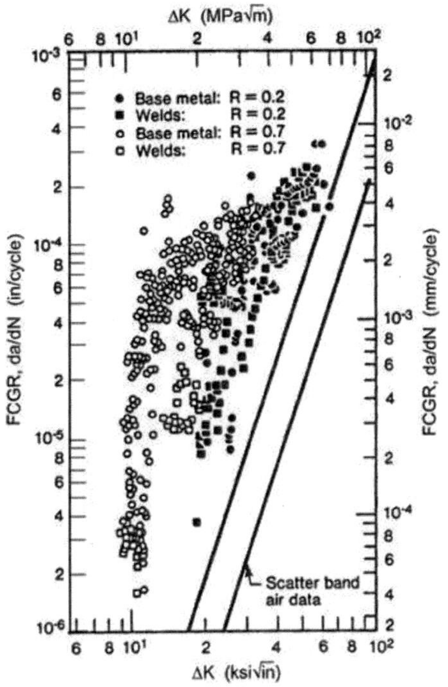

Fig. 20.24 Effect of *R*-ratio on A 533 GrB Cl1 and A 508 GrB Cl2 base metal and welds in demineralized water at 288 °C [25]

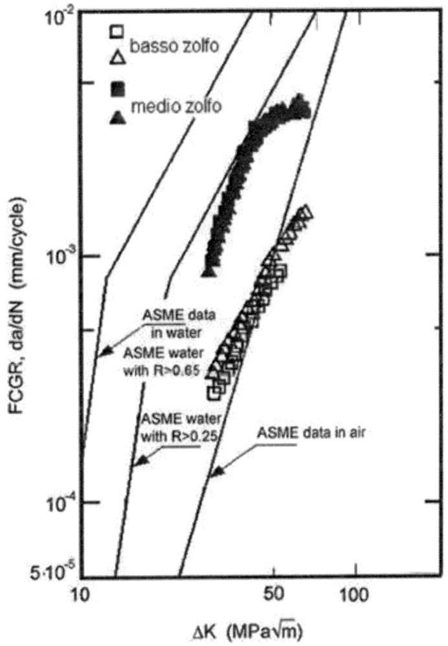

Fig. 20.25 FCGR, *da/dN*, obtained Kondo et al. for A 533 B steel with two sulfur content in demineralized water at 288 °C [29]

20.4 Morphology of Corrosion and Fatigue

Basically, there are 4 modes through which corrosion appears on materials surface, as shown schematically in Fig. 20.26. Generalized corrosion acts on the entire surface exposed to the environment (rust for example) and is not treated in this book. Pitting and SCC have been considered in Chaps. 14 and 15. SCC can be either intergranular or transgranular. Fatigue is always transgranular unless it is coupled to SCC which remains the main cause of crack growth and fatigue is only providing a stress initiation mechanism of passive film rupture (see Sect. 14.4.3). As such, true corrosion fatigue must display the typical striations of fatigue (see Sect. 2.4).

This time, striations can be ductile or brittle (see Sect. 2.4.2 and Fig. 2.37) depending on whether they are caused by true corrosion fatigue or assisted by hydrogen embrittlement. After the removal of the corrosion layer that covers the surface of the metal that failed by fatigue, which is not that simple and sometimes almost impossible, these characteristic features appear at the microscope. An example of ductile striations is offered in Fig. 20.27 [34] for a carbon steel type A 533 GrB Cl1 at 288 °C. The applied were equal to 42 MPa$\sqrt{m}$ and 48 MPa$\sqrt{m}$, respectively. The black arrow indicates the direction of fatigue propagation. The main difference with respect to traditional striations can be seen on the crests and borders in general that are smoothed out by corrosion and appear blunted and not sharp.

If a SCC and, in particular, a hydrogen embrittlement occurs, brittle striation appear whose morphological features is fan-shaped appearance on crystallographic planes with concentric circles intersected by normal radii. Figure 20.28 presents a case of brittle striations originated from a MnS inclusion in A 533 GrB Cl1 steel, medium sulfur content, in demineralized water at 288 °C [35].

The sunburst pattern and the cracked MnS inclusion are well visible. Brittle striation can be observed also on austenitic stainless steel hydrogen embrittled. This

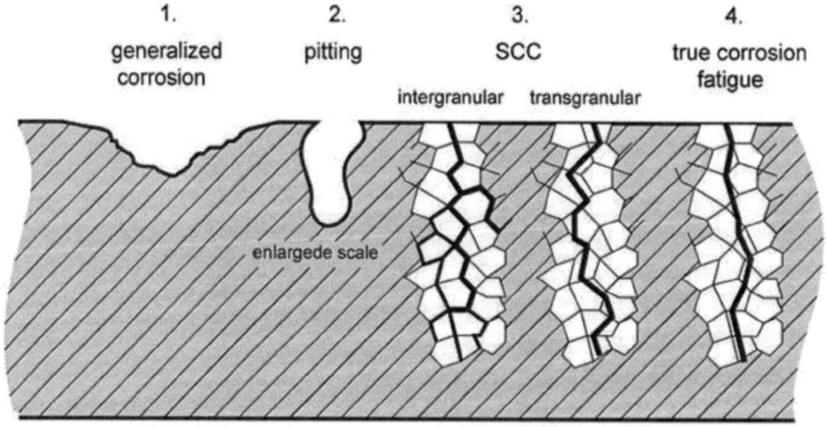

Fig. 20.26 Schematic of the four modes of environmental aggression

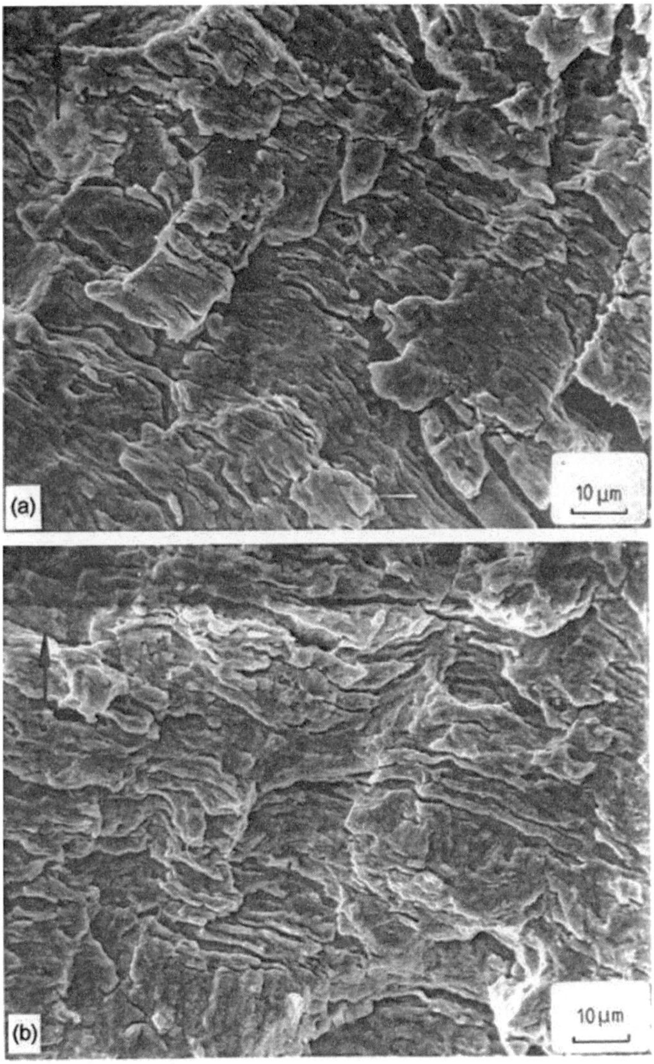

Fig. 20.27 Ductile fatigue striations observed on A 533 GrB Cl1 steel in water at 288 °C [34]

can be seen in Fig. 20.29 for 304 and 304NG (304 L) stainless steels in different testing environment [36]. In air, Fig. 20.29a, at 288 °C the normal fatigue striations appear. In water at 288 °C with high dissolved oxygen (DO) concentration these last leave the place to typical brittle striations, Fig. 20.29b that can be also seen in Fig. 20.29c in low oxygen demineralized water at 288 °C, but with presence of some hydrogen coming from hydrazine typical of PWR nuclear reactors. Note the different appearance of striations that in air run along metal grains and in water develops on

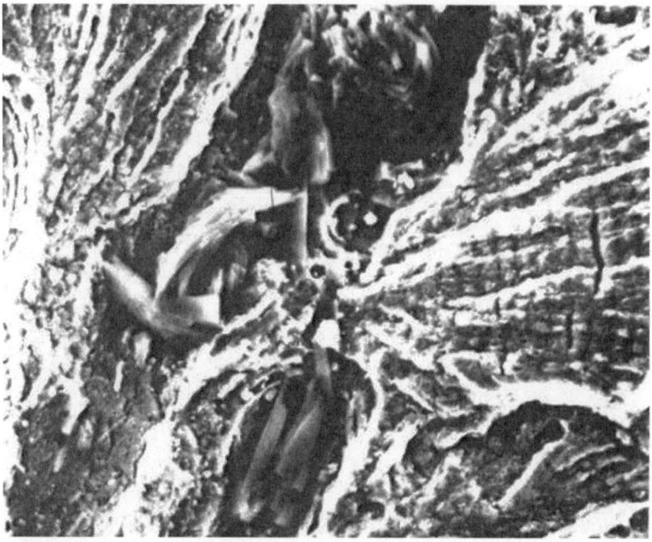

Fig. 20.28 Brittle striations propagating from a MnS inclusion [35]

crystallographic flat planes with the appearance of radii normal to striation circles, pointing towards the origin of fatigue. The diagram of Fig. 20.30 presents the correlation found by Atkinson et al. [33] between the percentage of brittle striations and the ratio, $(da/dN)_{env}/(da/dN)_{air}$, of the FCGR in the aggressive environment to that in air, for A 533 GrB Cl1 steel in demineralized water at 288 °C. Note how the percentage of brittle striations increases with the acceleration in the FCGR, *da/dN*, due to the hydrogen effect. When hydrogen embrittlement is acting through an intergranular separation in the presence of impurities such as Z, P, An, Bi, Sn and As, in particular, (see Fig. 18.32) striations may not appear leaving the place to the characteristic morphology of Fig. 18.40d.

20.5 Environmental Fatigue and SCC

In the previous section it has been examined the case in which the maximum value K_{Imax} of the applied stress intensity factor ΔK was always lower than the SCC threshold K_{Iscc} characteristic of the material (see Sect. 19.5.1). It is important to realize that under this circumstance the mechanism of crack growth is just that of fatigue, dough increased by environmental effects that for this reason will be referred to as *true corrosion fatigue* (TCF). It may be questioned the kind of mechanism that can activate corrosion below the threshold K_{Iscc}. In effect, the mechanism is always the same anodic dissolution already discussed in Sect. 17.7. The only difference is

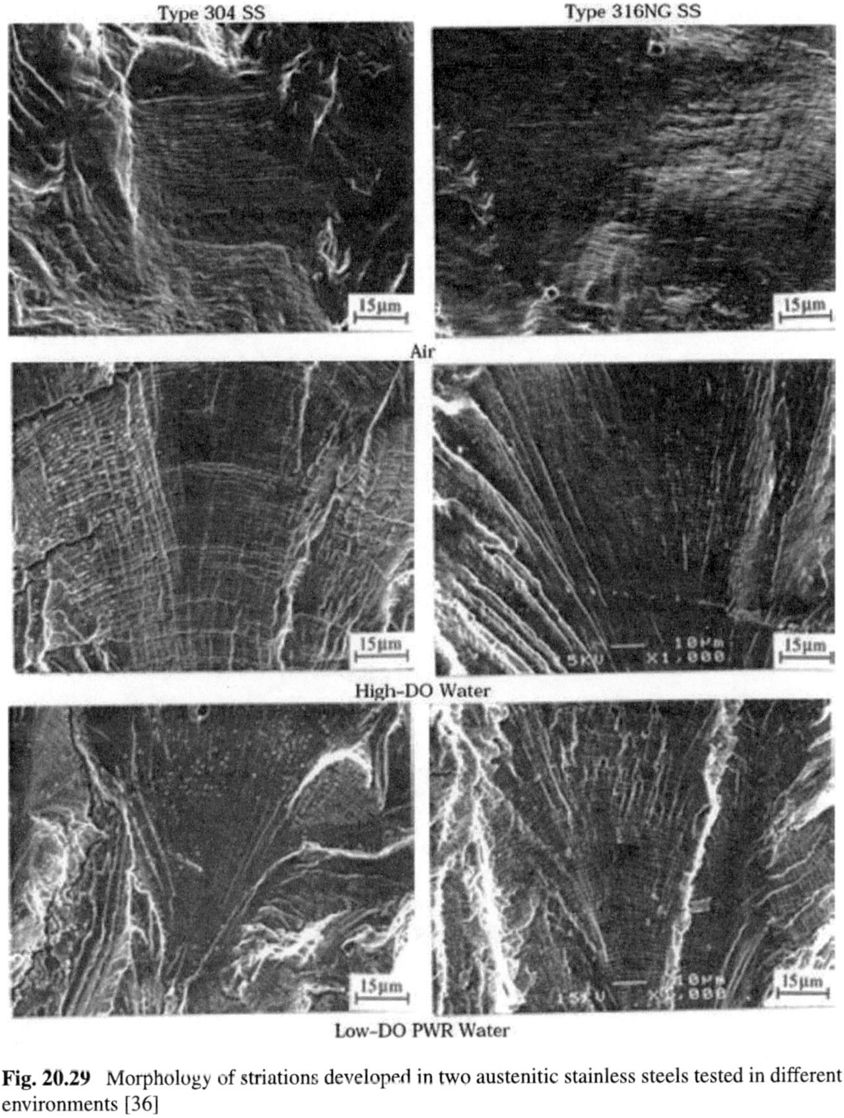

Fig. 20.29 Morphology of striations developed in two austenitic stainless steels tested in different environments [36]

that now it takes fatigue cycles to break the passivizing film and activate the anodic dissolution of the crack tip bare metal.

This is done at each cycle and, therefore, the material is not given the chance to re-passivize permanently and this is explaining the increased crack growth rate of corrosion fatigue with respect to the inert environment. However, it has been also shown that the reduction of the frequency f of the cyclic load or the increase of the R

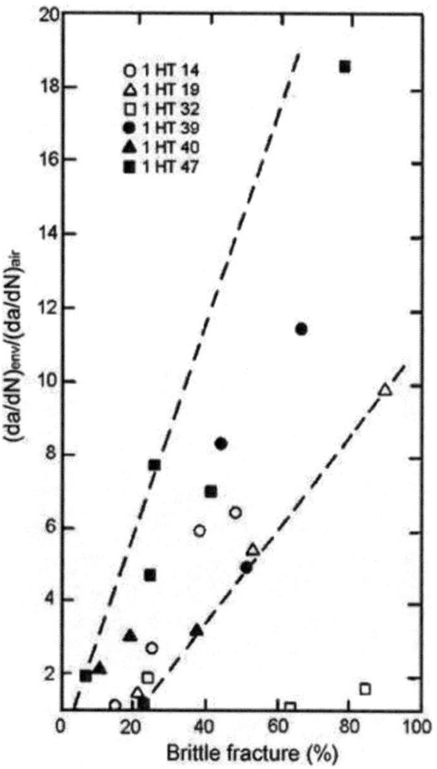

Fig. 20.30 Correlation between brittle striations percentage with the ratio between FCGR in the aggressive environment and in air [33]

ratio can trigger a SCC mechanism that, otherwise, would have not been experienced by materials less prone to SCC as carbon steels, in particular.

It is also clear that if in the fatigue cycle the applied K_{Imax} exceeds the K_{Iscc} of the material then fatigue can activate the SCC. If this is the case then SCC becomes the main crack growth mechanism while fatigue is just borrowing the stress needed to activate the process. There are, therefore, a combination of possibilities that are schematizes by the three circles of the Venn diagram shown inFig. 20.31 [37]. The type of stress, whether static of cyclic, and the *R*-ratio or frequency combining one another determine the kind of corrosion attack that shall be expected. The crosshatched region of the three circles diagram allows any kind of aggression possible. On this ground, three distinct fatigue phenomena can be distinguished, as described by Austen [38] and revisited in Fig. 20.32. The first case (A), $K_{Imax} < K_{Iscc}$, is that already seen of TCF. The environmental effect generally determines a decrease of the threshold stress $\Delta K_{I,th}$ to initiate fatigue and results in a FCGR, *da/dN*, increase with respect to the inert environment. Corrosion accelerates the crack growth, but does not modify the functional dependency of *da/dN* on ΔK that remains that of the Paris-Erdogan power low

$$\frac{da}{dN} = C \cdot \Delta K_I^n. \tag{20.2}$$

The fracture morphology is transgranular with ductile striations of the type seen in Fig. 20.27. The second case (B) of Fig. 20.32 is that in which the K_{Iscc} of the material is lower than the threshold $\Delta K_{I,th}$ for fatigue therefore SCC will appear

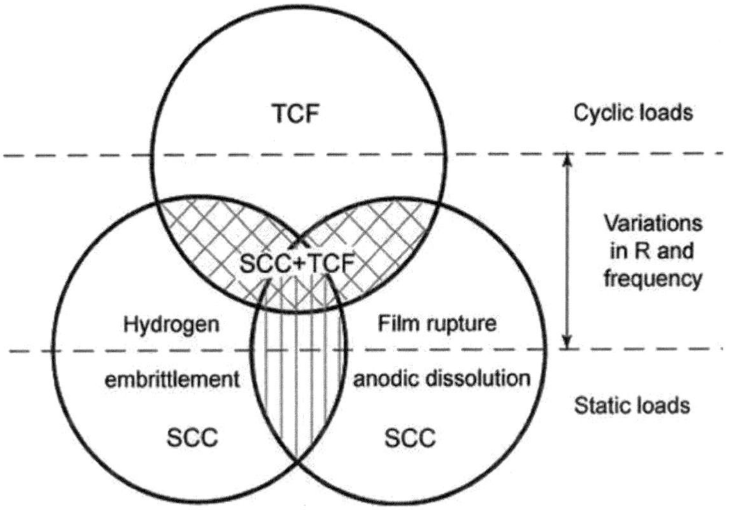

Fig. 20.31 Modified Venn diagram of interdependence among fatigue, film rupture-anodic dissolution and hydrogen embrittlement (modified from [37])

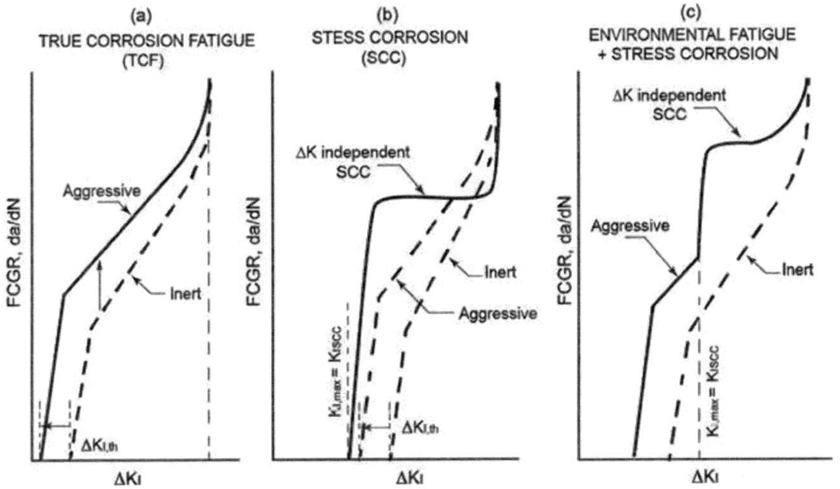

Fig. 20.32 Fundamental types of corrosion-fatigue (modified from [38])

first, unless the frequency f of load cycles is too large ($f \gg 1$ cpm). The case is that interesting the crosshatched zone of Fig. 20.31. Now, as discussed in Sect. 15.3, the FCGR, *da/dN*. The fracture morphology is that already discussed in Sects. 14.3 and 14.5. However, at higher ΔK_I values SCC can disappear and TCF reappears.

This is because the FCGR, *da/dN*, prevails over the SCC *da/dt*, as shown in Fig. 20.32b. An interesting example of this crack growth behavior is shown in Fig. 20.33 [39] that reports the experimental data obtained by Speidel et al. for 7079-T651 aluminum alloy in dry argon and several aqueous halide solutions at the free corrosion potential and a load frequency of 4 Hz, with $R = 0$. In dry argon the mechanism of crack growth is that typical of fatigue. The passage from dry argon to distilled water gives rise to TCF. The FCGR, *da/dN*, increases by a factor of 5 because of the environmental effect. Note that the K_{Iscc} is very low, between 2 and 3 MPa$\sqrt{m}$. If SCC has not appeared it is because in distilled water the frequency of 4 Hz is still too high to trigger this corrosion mechanism and the *R*-ratio too low (see Sects. 20.3.1 and 20.3.3).

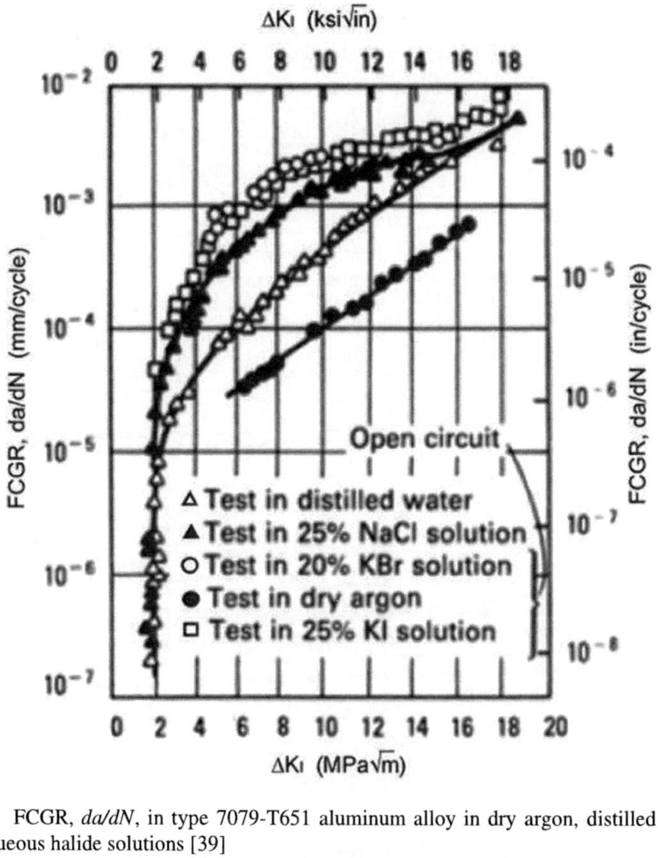

Fig. 20.33 FCGR, *da/dN*, in type 7079-T651 aluminum alloy in dry argon, distilled water and several aqueous halide solutions [39]

However, when the environment becomes sea water then things change. In sea water, in fact, the 4 Hz frequency is sufficient to activate SCC and the typical plateau appears where the FCGR, *da/dN*, is no longer depending on the applied ΔK_I. In both 3.8 M KBr and 5.0 M KI potassium iodide the appearance of SCC is even more evident. However, at higher ΔK_I values of about 17 MPa$\sqrt{\text{m}}$, SCC disappears and TCF returns to take the lead. This passage is more evident in Fig. 20.34 relative to a magnesium alloy type ZK 60A-T5 tested in the same aqueous halide solutions of the previous aluminum alloy plus a 1.4 M Na_2SO_4 solution [39]. The test frequency was always equal to 4 Hz and $R = 0$. Also in this case SCC does not appear in distilled water probably because the load frequency of 4 Hz is too high and the *R*-ratio too low for this environment to initiate a SCC process. But once the environment becomes more aggressive then SCC appears, but it terminates at about 12–14 MPa$\sqrt{\text{m}}$ switching back to TCF because this last growth rate is higher than that produced by SCC. It must be kept in mind that when a SCC process is activated in a material the FCGR, *da/dN* loses its meaning and must be abandoned. Actually, to continue to measure crack extension in terms of *da/dN* may become even confusing or misleading. Astonishingly high growth rate can be measured that are only fictitious and not real because the crack growth rate does no longer depend on the number of cycles *N* but, rather, on time. Therefore, it is necessary to shift from *da/dN* to *da/dt* using the change of variable $t = N/f$, where t is the time and f is the frequency of the cyclic load. Therefore it can be written

$$\frac{da}{dt} = \frac{da}{dN} \cdot f \tag{20.3}$$

Since *da/dt* is constant and independent of the applied ΔK_I it can be possible to sustain that also the product *da/dN* $\cdot f$ will be constant. By moving from *da/dN* to *da/dt* using Eq. (20.3) the apparent acceleration of the FCGR, *da/dN*, with decreasing frequency, apparently observed for any material, vanishes. Gallagher and Sinclair [40] were the first to realize that the FCGR, *da/dN*, had to be converted into a *da/dt* to obtain a constant rate of growth. The example of Fig. 20.35 [41] can clarify this reasoning. The experimental data were found by Barsom for high strength maraging steel in sea water (3.5% NaCl solution). At 64 Hz the steel tested in sea water shows a remarkable reduction of the fatigue threshold with respect to vacuum and a parallel increase of the FCGR, *da/dN*. This is a typical case of TCF, case (A) of Fig. 20.32. However, as soon as the test frequency is reduced from 64 to 8 Hz, the FCGR, *da/dN*, increases and a first, timid sign of SCC appears.

For this high strength steel 8 Hz are sufficiently low to initiate SCC. By further reducing the frequency to 4 Hz, SCC becomes more significant and even more evident at 1 Hz and 0.25 Hz. Note that in all cases the K_{Iscc} is always the same and equal to about 18 MPa$\sqrt{\text{m}}$. The lowering of the frequency seems to increase the apparent FCGR, *da/dN*, but converting this crack growth rate into a *da/dt* value by using Eq. (20.2) it turns out that the growth rate is always the same for all frequencies and equal to about 4.5×10^{-4} mm/s. This would not happen in TCF. For example, reanalyzing the experimental data of Fig. 20.12, relative to 4340 steel tested in water

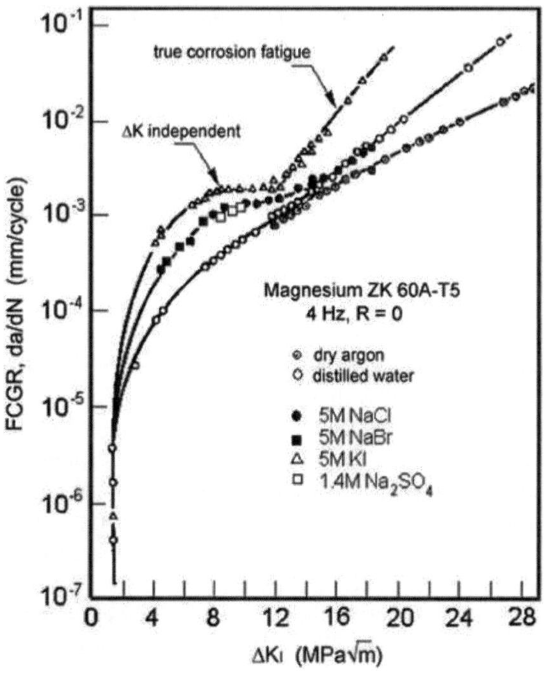

Fig. 20.34 FCGR, *da/dN*, in type ZK 60 A-T5 magnesium alloy in dry argon, distilled water, several aqueous halide solutions and 1.4 M Na_2SO_4 [39]

vapor at 4.4 torr of pressure and different frequencies, in terms of *da/dt* versus applied ΔK the diagram of Fig. 20.36 is obtained. At variance with the case of Fig. 20.35, now the crack growth rate *da/dt* is not constant at all ΔK and frequencies because the crack is growing by TCF and not by SCC. Yet, it is possible to recognize that by lowering the frequency to 0.1 Hz this FCGR dependency on the applied ΔK progressively reduces and this is a sign that the material is approaching a SCC event. It is almost certain that at 1 cpm this dependency would disappear at all because SCC would initiate. The same effect can be seen in Fig. 20.37 for a titanium alloy type Ti-6Al-4V at ambient temperature in a 3.5% NaCl solution [22].

The maximum effect is actually reached at a combination of applied ΔK and frequency f of the cyclic load. At low ΔK the maximum FCGR, *da/dN*, is very modest and obtained at higher frequencies while increasing the applied ΔK the maximum FCGR requires lower frequencies to be reached. But in terms of *da/dt* the maximum growth rate is always the same and equal to about 0.001 mm/s. The same effect can be seen in Fig. 20.38 [42] for a carbon steel type X-65 for pipes in air and salt water. In salt water at the frequency of 10 Hz there is a very small plateau in which the FCGR, *da/dN*, is independent of the applied ΔK. Afterall the crack growth rate resembles that in air. But by reducing the frequency the occurrence of a large plateau becomes more and more evident. But, again, the tremendous increase in the FCGR is only apparent because by using Eq. (20.2) it turns out that the rate *da/dt* is always the same and equal to about 10^{-4} mm/s. The last case (C) of Fig. 20.31 is that

Fig. 20.35 SCC activated by cyclic loadings in 12Ni-5Cr-3Mo high strength steel in sea water [41]

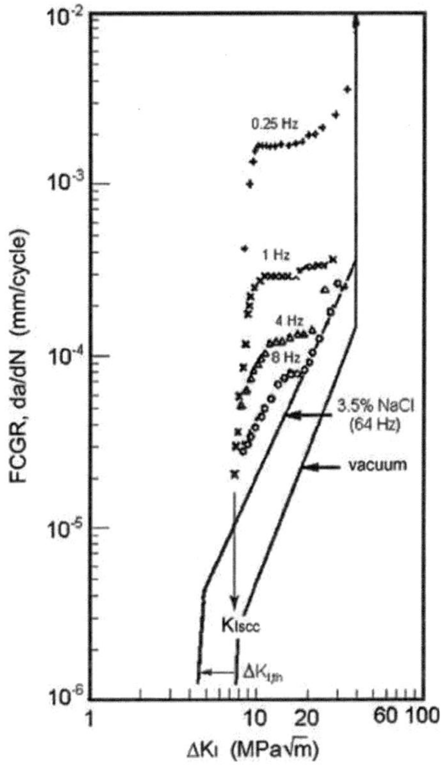

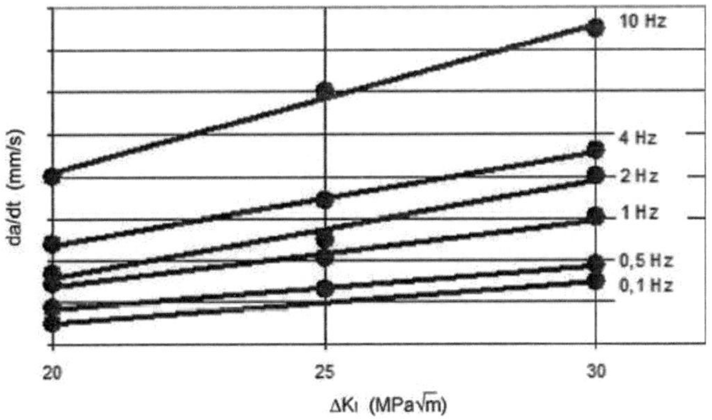

Fig. 20.36 Crack growth rates *da/dt* versus applied ΔK_I for 4340 steel in water vapor at different testing frequencies

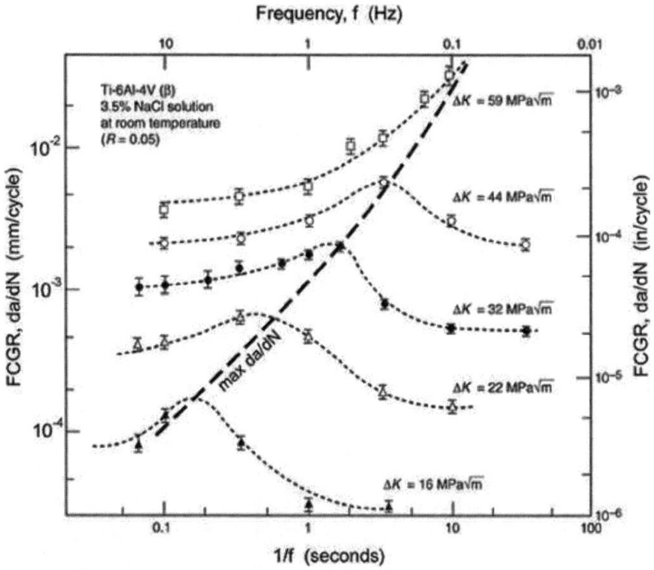

Fig. 20.37 Influence of frequency on FCGR in Ti-6Al-4V alloy exposed to 3.5% NaCl solution at room temperature and $R = 0.1$ [22]

in which the corrosion fatigue threshold $\Delta K_{I,th}$ is lower than that for SCC K_{Iscc}. In this case TCF starts moving the crack up to the point when the applied ΔK_I becomes larger than the threshold K_{Iscc} for SCC.

20.6 Models of Corrosion-Fatigue Crack Growth

Starting from experimental observation, Wei and Landes [43] made the hypothesis of a separate action of fatigue (F) and corrosion (C) formulating a model of superposition of the effects in which the crack growth rate was equal to the sum of individual effects without any synergic effect

$$\left.\frac{da}{dN}\right|_{FC} = \left.\frac{da}{dN}\right|_{F} + \left.\frac{da}{dN}\right|_{C}. \tag{20.4}$$

In Eq. (20.4) it is

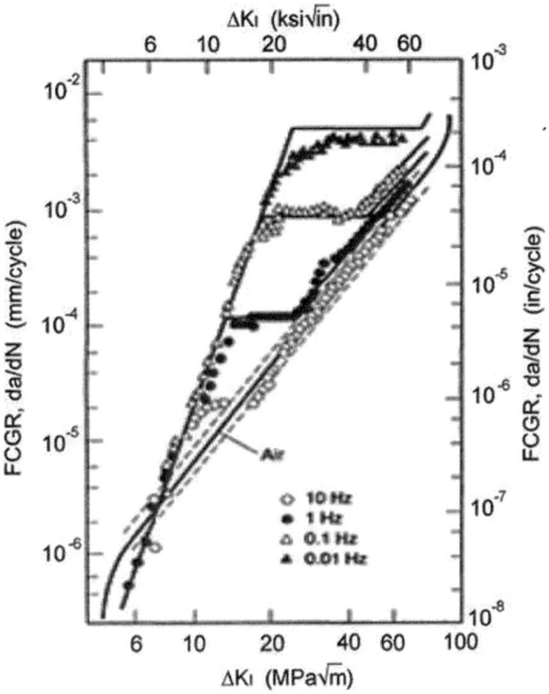

Fig. 20.38 Effect of frequency on the FCGR of carbon steel type X-65 for pipes [42]

$$\left.\frac{da}{dN}\right|_F = C \cdot \Delta K^n$$
$$\left.\frac{da}{dN}\right|_C = \int_T \left(\frac{da}{dt}\right)_{SCC} dt \tag{20.5}$$

in which T is the duration of that part of the cycle in which the K_{Iscc} is overcome. Note that the first of Eq. (20.5) is the Paris-Erdogan equation for mechanical fatigue. Wei and Landes suggested a graphic method to derive the growth by corrosion-fatigue, as schematized in Fig. 20.39 [43]. Of each load cycle, assumed to have a sinusoidal form in Fig. 20.39a, only the part where the relative K is larger than K_{Iscc} is considered, i.e., the arc BAB. At each point of this part of sinusoid a K value is inferred to which a value of the rate *da/dt* is associated on the SCC growth curve of the material indicated in Fig. 20.39b, obviously obtained experimentally. In this manner it is possible to build the bell-shaped curve of Fig. 20.39c that will be used to integrate the second of Eq. (20.5), which provides the contribution of SCC to fatigue. The final result is shown in Fig. 20.39d. However, the model holds some drawbacks that are:

(a) it is applicable as first approximation to material having high yield strength (σ_y > 1400 MPa) that are very sensible to SCC whose crack growth rate is at least two orders of magnitude larger than $(da/dN)_F$ in inert environment;
(b) in its original formulation it does not explicitly include the dependency on frequency and *R*-ratio. However, this dependency can be expressed through the ΔK_{eff}

$$\Delta K_{eff} = K_{\max}(1 - R) \tag{20.6}$$

(c) it does not take into consideration the synergic effects that arise for critical values of the frequency in the transition zone between the SCC dominant condition, typical of low frequencies, and that of dominant mechanical fatigue, typical of high frequencies.

Note that restraint a) on the stress corrosion crack growth rate that must overrun the FCGR in inert environment, $(da/dN)_F$, by at least two orders of magnitude implies that mechanical fatigue is practically negligible and irrelevant on the total growth rate. Therefore, it take SCC to drive the crack and fatigue is just providing the stress level necessary to initiate SCC by a protective film rupture mechanism already discussed in Sect. 17.7. Equation (20.6) allows taking into consideration the *R*-ratio effect on the fatigue and stress corrosion mechanism as schematized in Fig. 20.40.

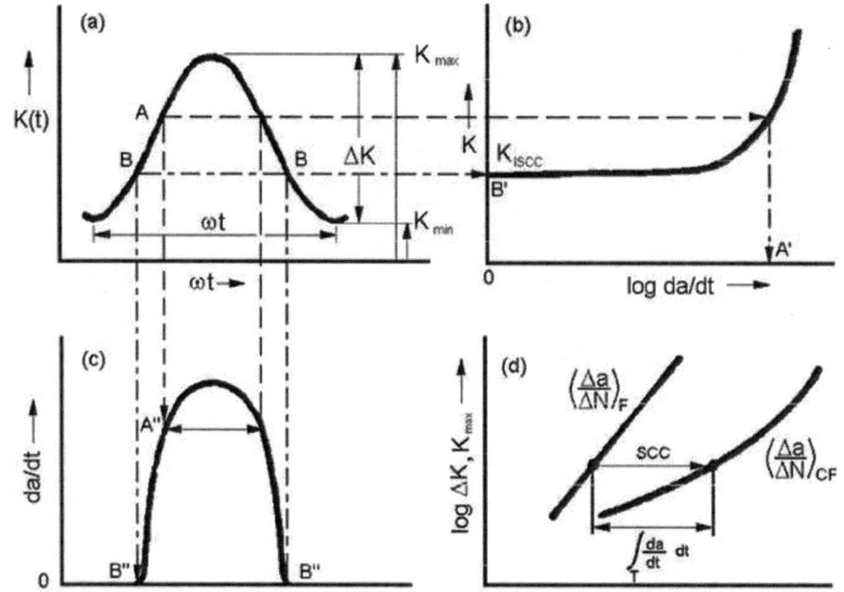

Fig. 20.39 Schematic of the Wei and Landes model for evaluating the crack growth by corrosion-fatigue [43]

The finding that the fatigue behavior of a material in a given environment depends on both the R-ratio and the frequency f, as schematized in Figs. 20.19 and 20.40, has suggested Walker and May et al. [44, 45] to formulate a model that is referred to as the process-competition model. In this model the crack growth rate for corrosion and fatigue (CT) is given by the larger of the two possible growth

$$\left(\frac{da}{dN}\right)_{CF} = \max\begin{cases} \left(\frac{da}{dN}\right)_{TCF} \\ \left(\frac{da}{dN}\right)_{SCC} \cdot \frac{1}{f} \end{cases} \cdot \qquad (20.7)$$

In their model the R-ratio is considered to act only on SCC, Fig. 20.32b, and not on TCF, Fig. 20.32a. this actually means that increasing R determines the passage from TCF to SCC. In the process-competition model the condition for the passage from TCF to SCC is represented by the tangency of the TCF line to the SCC curve. Recalling the plateau of Fig. 19.11, Eqs. (20.1), (20.2) and (20.6) can be written as

$$\begin{aligned} \left(\frac{da}{dN}\right)_P &= C \cdot \Delta K_{I,P}^n \\ \left(\frac{da}{dt}\right)_P \cdot \frac{1}{f} &= C \cdot \left[\Delta K_{I,P}(1-R)\right]^n \end{aligned} \qquad (20.8)$$

where the subscript P stands for *plateau*. The critical frequency f^* and R^*-ratio that determines the passage from TCF to TCF + SCC can be inferred as

Fig. 20.40 Schematic of the R-ratio effect on fatigue and corrosion at constant frequency

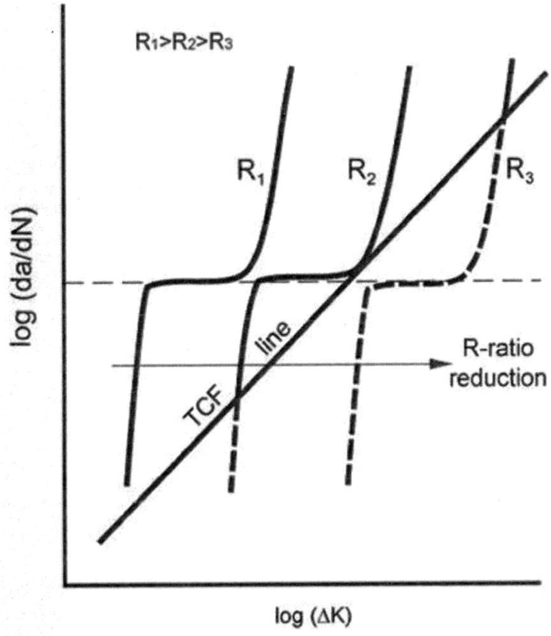

$$f^* = \left(\frac{da}{dt}\right)_P \cdot \frac{1}{(1-R)^n \cdot C \cdot K_{I,P}^n}$$
$$R^* = 1 - \left\{\left[\left(\frac{da}{dt}\right)_P \cdot \frac{1}{f} \cdot \frac{1}{C}\right]^{\frac{1}{n}} \cdot \frac{1}{K_{I,P}}\right\}. \tag{20.9}$$

Note that the term $C{\cdot}(K_{I,P})^n$ that appears on the right of the first of Eqs. (20.9) is the FCGR by TCF while $(da/dt)_P$ is SCC growth rate. At the tangency the two values are equal, therefore, it will take the term $(1-R)^n$ to establish how much the frequency f has to decrease with respect to the value that allowed the passage from TCF to SCC. This last is not so simple to be determined, as the same $R*$ value. As first, the da/dN versus ΔK curve must be experimentally assessed with its possible shift as a function of the frequency f as shown, for example, in Figs. 20.11, 20.12 and schematized in Fig. 20.19. In their model Walker and May do not consider the effect of the R-ratio on the FCGR. For this reason the experimental determination of the RCCF, da/dN, must be done at high frequency to avoid the interference of the F-effect. Normally, in the technical literature the R and f effects are treated separately, but this is not fully correct. Therefore, the question is complicated by the possible synergic effect between the two variables R and f, as Van Der Sluys et al. [44] pointed it out testing with specimens at constant ΔK. secondly, it must be considered that in a loading cycle the stress and, therefore, the applied K are not constant while the model considers that SCC acts under a constant driving force given by the maximum K_{Imax} value of the cycle. This last factor could be overrun using the approach proposed by Wei and Landes shown in Fig. 20.39. Imagining loading the material with a square wave, then six possible situations emerge where K_{Imin} and K_{Imax} can be found with respect to the K_{Iscc} and $K_{I,P}$ of the material, as schematically shown in Fig. 20.41 [45].

In the first case, $K_{Imax} < K_{Iscc}$, TCF occurs if the frequency is sufficiently low. In the second case, a SCC will be activated by K_{Imax} that exceeds K_{Iscc}. In the third case, the FCGR by corrosion fatigue is given by the second of Eqs. (20.7) that is not yet true SCC since the plateau value $K_{I,P}$ has not been overcome. It is possible to make the hypothesis that, as for mechanical fatigue, also for the growth between K_{Iscc} and $K_{I,P}$ the Paris-Erdogan relationship holds

$$\frac{da}{dt} = B \cdot K_I^m. \tag{20.10}$$

In this case the corrosion-fatigue crack growth rate will be

$$\left(\frac{da}{dN}\right)_{CF} = \frac{B}{f} \cdot \left(\int_o^{1/2} K_{\max}^m \cdot dN + \int_{1/2}^{1} K_{\min}^m \cdot dN\right)$$
$$\left(\frac{da}{dN}\right)_{CF} = \frac{B}{2f} \cdot \left(K_{\max}^m + K_{\min}^m\right). \tag{20.11}$$

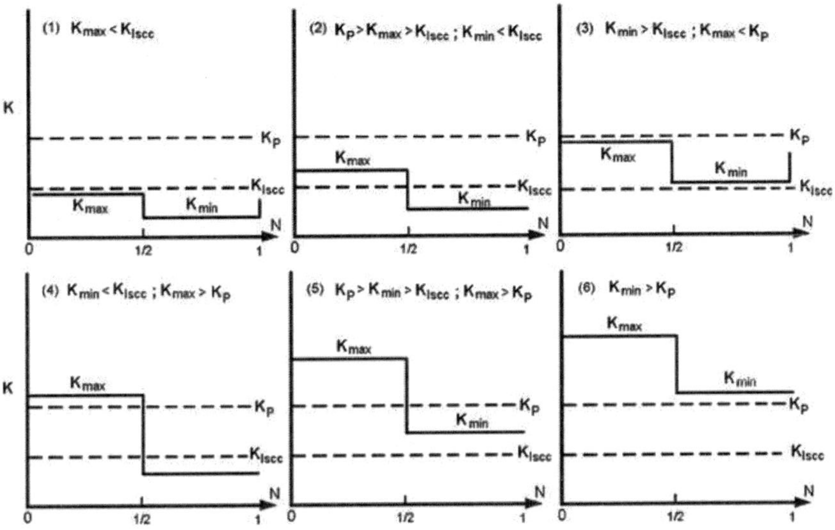

Fig. 20.41 Schematic of the six possible situations characterizing fatigue and corrosion [45]

Applying the procedure to the six cases of Fig. 20.41 and the five cyclic wave forms of Fig. 20.20, namely sinusoidal, triangular, square, positive-sawtooth and negative-sawtooth, $6 \times 5 = 30$ expression analogous to Eq. (20.11) will be obtained.

References

1. Wadsworth, N.J., Hutchings, J.: Phil. Mag. **3**, 1154 (1958)
2. Shives, T.R., Bennet, J.A.: NASA Report N. CR-267 (1965)
3. Bennet, J.A.: Fatigue—An Interdisciplinary Approach, p. 209. Siracuse University Press (1964)
4. Wadsworth, N.J.: Phil. Mag. **6**, 397 (1961)
5. Goodacre, R.: Engineering **139**, 457 (1935)
6. Frost, N.E.: Appl. Mater. Res. **3**, 131 (1964)
7. Amzallag, C., Rabbe, P., Bathias, C., Benoit, D., Trushon, M.: Influence of various parameters on the determination of the fatigue crack arrest threshold. Am. Soc. Test. Mater. ASTM STP **738**, 29–44 (1981)
8. Sarrazin-Baudoux, C., Lesterlin, S., Petit, J.: Atmospheric influence on fatigue crack propagation in titanium alloys at elevated temperature. ASTM STP **1297**, 117–130 (1997)
9. Achilles, R.D., Bulloch, J.H.: The influence of frequency, waveform and environment on the fatigue crack growth behaviour of SA 508 RPV Steel. In: Proceedings of the Second IAEA Specialists' Meeting on Subcritical Crack Growth, NUREG/CP-0067, vol. 1, pp. 379–422 (1986)
10. Imohf, E.J., Barsom, J.M.: Fatigue and corrosion-fatigue crack growth of 4340 steel at various yield strengths. In: Progress in Flaw Growth and Fracture Toughness Testing, ASTM STP 536 (1973)
11. Sprowls, D.O.: Evaluation of corrosion fatigue. In: ASM Metals Handbook, Corrosion, vol. 13 (1992)

12. Barsom, J.M., Sovak, J.F., Imohf, E.: Corrosion fatigue crack propagation below K_{ISCC} in four high-yield strength steels. In: Laboratory Report 89.021-024(3), U.S. Steel Co. (1970)
13. Crooker, T.W., Lange, E.A.: Corrosion fatigue crack propagation for some new high strength structural steels. J. Basic Eng. Trans. ASME 91 (1969)
14. Kondo, T., et al.: Fatigue crack propagation behavior of ASTM A 533 GrB Cl1 plate in an environment of high temperature primary grade nuclear reactor water. In: HSST 6th Annual Information Meeting, Paper N. 6 (April, 1972)
15. Barsom, J.M., Rolfe, S.T.: Fracture and Fatigue Control in Structures. Application of Fracture Mechanics, p. 320. Prentice-Hall, Inc. (1977)
16. Pao, P.S., Wei, W., Wei, R.P.: Environment-Sensitive Fracture of Engineering Materials, p. 565. TMS-Time (1979)
17. Mager, T.R., Landes, J.D., McLoughlin, V., Moon, D.M.: The Effect of Low Frequencies on the Fatigue Crack Growth Characteristic of A533 Grade B Class I Plate in an Environment of High-Temperature Primary Grade Nuclear Reactor Water, Oak Ridge, HSST Report No. 35, December, 1973
18. Dawson, D.B., Pelloux, R.M.: Corrosion fatigue crack growth of titanium alloys in aqueous environments. Metall. Trans. A **5A**, 723–731 (1974)
19. Iida, K., Kobayashi, H., Higuchi, M.: Predictive method of low cycle fatigue life of carbon and low alloy steels in high temperature water environments. In: Proceedings of the Second IAEA Specialists' Meeting on Subcritical Crack Growth, NUREG/CP-0067, vol. 2, pp. 385–409 (1986)
20. Van Der Sluys, W.A., Emanuelson, R.: Overview of data trends in cyclic crack growth results in LWR environments. In: Proceedings of the Second IAEA Specialists' Meeting on Subcritical Crack Growth, NUREG/CP-0067, vol. 1, pp. 199–218 (1986)
21. Cullen, W.H.: Fatigue crack growth rates in pressure vessel and piping steels in LWR environment. In: USNRC Report NUREG/CR-4724 (1987)
22. Pao, P.S., Wei, W., Wei, R.P.: Environment-sensitive fracture of engineering materials. Tms-Time, p. 565 (1979)
23. Chiou, S., Wei, R.P.: Corrosion-fatigue cracking response of beta annealed Ti-6Al-4V alloy in 3.5% NaCl solution. In: NADC-83126-60, U.S. Naval Air Development Center (1984)
24. Bucci, R.J.: Environment Enhanced fatigue and Stress Corrosion Cracking of a Titanium Alloy Plus a Simple Model for the Assessment of Environmental Influence on Fatigue Behavior. PhD Dissertation, Lehigh University, Bethlehem, PA (1970)
25. Banford, W.H., Moon, D.M., Ceschini, L.J.: Crack growth rate testing in reactor pressure vessel steels. In: Proceedings Fifth Water Reactor Safety Information Meeting, Gaithersburg, MD (November, 1977)
26. Amzallag, C., Bernard, J.L., Slama, G.: Effect of loading and metallurgical parameters on the fatigue crack growth rates of pressure vessels steels in PWR environment. In: International Conference on Environmental Degradation of Materials in Nuclear Power System—Power Reactors, Myrtle Beach, South Carolina, pp. 15–17, August 22–25 (1983)
27. Atkinson, J.D., Bulloch, J.: The Effect of Bulk Sulfur Content on da/dN. In: Minutes of IGCGR Meeting, Akron, Ohio (October, 1984)
28. Banford, W.H., Jacko, R.J., Wilson, I.L.W., Ceschini, L.J.: The Effect of Material and Environmental Variables on Corrosion Fatigue Crack Growth in Pressure Vessel Steels. NRC NUREG/CP-0058, vol. 4 (1984)
29. Kondo, T., Kuniya, J., Takaku, H., Arii, M., Kurihara, M.: Recent study on cyclic crack growth of reactor pressure boundary materials in high temperature water environment in Japan. In: Proceedings of the Second IAEA Specialists' Meeting on Subcritical Crack Growth, NUREG/CP-0067, vol. 1, pp. 219–249 (1986)
30. ASME XI, Appendix A, American Society of Mechanical Engineers (1986)
31. Tice, D.R., Atkinson, J.D., Scott, P.M.: Proceedings of the 2nd IAEA Specialist's Meeting on Subcritical Crack Growth, Sendai, Japan, US NRC NUREG/CP-0067, vol. 1, pp. 251–282 (1986)

32. Scott, P.M., Truswell, A.R.: Proceedings of the IAEA Specialists' Meeting on Subcritical Crack Growth, Freiburg, Germany, US NRC NUREG/CP-0044, vol. 1, p. 376 (1981)
33. Atkinson, J.D., Bulloch, J.H., Forrest, J.E.: A Fractographic Study of Fatigue Cracks Produced in A533B Pressure Vessel Steel Exposed to Simulated PWR Primary Water Environment. In: Proceedings of the 2nd IAEA Specialist's Meeting on Subcritical Crack Growth, Sendai, Japan, US NRC NUREG/CP-0067, vol. 1, p. 290 (1986)
34. Torronen, K., Hänninen, H., Cullen, W.H., Jr.: Mechanisms of environmental assisted cyclic crack growth of nuclear reactor pressure vessel steels. In: Proceedings of the IAEA Specialists' Meeting on Subcritical Crack Growth, NUREG/CP-0044, vol. 2, p. 42, Freiburg, Germany (1981)
35. Amzallag, C., Bernard, G.L., Slama, G.: French studies of cyclic crack growth behaviour of RPV steels in PWR environment. In: Proceedings of the 2nd IAEA Specialists' Meeting on Subcritical Crack Growth, NUREG/CP-0067, vol. 1, p. 305 (1985)
36. Torronen, K., Kemppainen, M., Hanninen, H.: Fractographic Evaluation of Specimens of A 533 B Pressure Vessel Steels, Final Report of EPRI, Contract RP 1325-7, Report NP 3483 (1984)
37. Fontana, M.G.: Corrosion Engineering, 3rd edn. (1986)
38. Austen, I.M., Walker, E.F., May, M.J.: Factors affecting the rate of growth of cracks by corrosion-fatigue. In: Technical Report STF-6210/KE/18/802 (1976)
39. Speidel, M.O., Blackburn, M.J., Beck, T.R., Feeney, J.A.: Corrosion fatigue and stress corrosion crack growth in high strength aluminum alloys, magnesium alloys and titanium alloys exposed to aqueous solutions. In: Corrosion Fatigue, International Corrosion Conference Series, vol. NACE-2 (1972)
40. Gallagher, J.P., Sinclair, G.M.: Environmentally assisted fatigue crack growth rates in SAE 4340 steel. J. Basic Eng. Trans. ASME **21**, 508 (1969)
41. Barsom, J.M.: Corrosion fatigue crack propagation below K_{ISCC}. Eng. Fract. Mech.Fract. Mech. **3**(1), 15–18 (1971)
42. Vosikovsky, O.: Fatigue-crack growth in an X-65 line-pipe steel at low cyclic frequencies in aqueous environments. Trans. ASME Ser. H **97**, 298–305 (1975)
43. Wei, R.P., Landes, J.D.: Correlation between sustained load and fatigue crack growth in high strength steels. Mater. Res. Standard ASTM **9**, 25 (1969)
44. Austen, I.M., Walker, E.F., May, M.J.: Factors affecting the rate of growth of cracks by corrosion-fatigue. Br. Steel Ref. 6210, Ke 18/802 (1978)
45. Walker, E.F., May, M.J., Irvine, K.J.: Mechanisms and control of crack growth in steels. In: Technical Report STF/85/76, 6210/KE/8/802 and STF/771, 6210/KE/8/802 (1976)
46. Van Der Sluys, W.A., Emanuelson, R.: Cyclic crack growth of reactor pressure vessel steels in light water reactor environment. Eng. Mater. Technol. **108**, 1–26 (1985)

Appendix A
Linear Elastic Fracture Mechanics. Compendium of Stress Intensity Factors Solutions

A.1 Introduction

Linear elastic fracture mechanics (LEFM) studies the behavior of materials, work pieces and structures in which cracks are present. Actually, the term fracture refers and identifies those failures caused by the presence of a crack. A crack is defined as an extremely sharp structural discontinuity characterized by a root radius no larger than 0.005 mm. Structural discontinuity having root radii larger than 0.005 are classified as notches. In the presence of a notch LEFM is no longer valid. However, LEFM can still be applied supposing that, conservatively, sooner or later at the tip of a notch a crack will initiate by fatigue or corrosion, in particular. The fundamental result of LEFM is that ahead of a crack an elastic stress field exists that is always self-similar (see Fig. 14.4). Its analytical expression for an infinite body containing a through wall crack of length *2a* under remote loading, known as Griffith crack, schematized in Fig. A.1 is given by the G. Irwin expression (see Eq. 14.4)

$$
\begin{aligned}
\sigma_{ij} &= \frac{K_I}{\sqrt{2\pi r}} f_{ij}(\theta) \\
K_I &= \sigma\sqrt{\pi a}
\end{aligned}
\qquad \text{(A.1)}
$$

in which $f(\theta)$ is a non-dimensional factor that depends on the angle θ, r is the distance from the crack tip and a the semi-crack length, as schematized in Fig. A.2. The elastic stress field presents a singularity of the type $1/\sqrt{r}$. Its amplitude is given by the Irwin stress intensity factor K_I.

The subscript I indicates that the stress intensity factor K_I refers to the first of the three fundamentals mode of aperture of a crack, schematized in Fig. A.3. Any other mode can be considered as a combination of two or more fundamental modes.

For any real case in which the geometry is not infinite, loaded under whatever conditions and the crack is not central, the expression of the stress intensity factor K_I is always given by the second of eqs. (A.1) with the addition of a multiplying factor $f(a)$

P. P. Milella, *Fatigue and Corrosion in Metals*,
https://doi.org/10.1007/978-3-031-51350-3

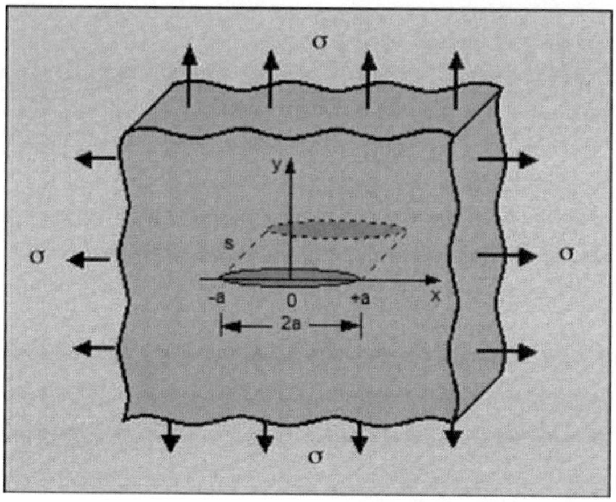

Fig. A.1 Central through-wall crack of length $2a$ in an infinite body remote loaded

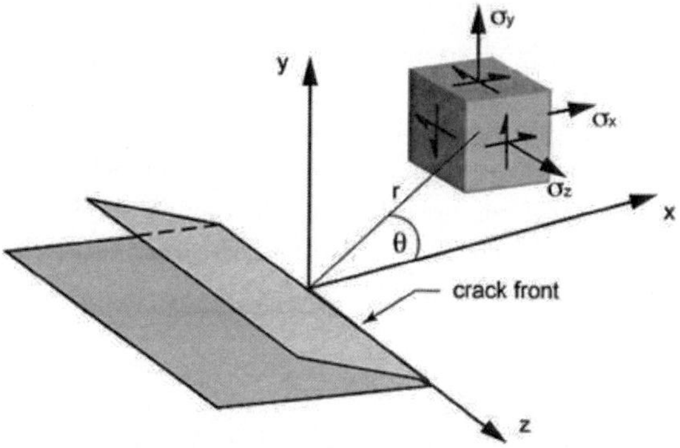

Fig. A.2 Polar coordinates ahead of a crack tip

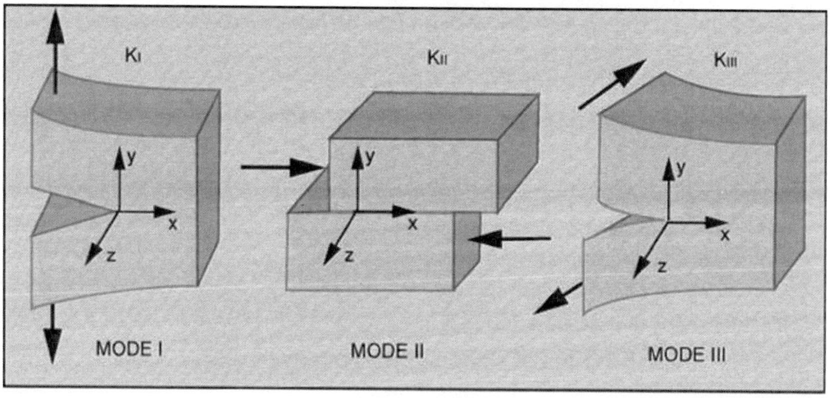

Fig. A.3 Schematic of the three modes of aperture of a crack

$$K_I = \sigma\sqrt{\pi a}\cdot f(a). \tag{A.2}$$

The non-dimensional function *f(a)* depends uniquely on the geometry of the system and length *a* of the crack. One of the main objective of LEFM is the assessment of the function *f(a)* relative to the particular geometry under study. Once the *f(a)* function is known, the relative K_I can be inferred via Eq. (2). On K_I depends either the FCGR, *da/dN*, through the Paris-Erdogan power law Eq. (14.10), or the occurrence of SCC when the applied K_I becomes equal to the threshold stress intensity factor K_{Iscc}. Also brittle fracture occurs when the applied K_I reaches the critical value of the toughness of the material K_{Ic}. Several solutions relative to simple, yet important geometries will be given in the next Sections.

A.2 Fracture Mechanics Specimens with Increasing K_I

A.2.1 Plain Uniaxial Specimens

Same simple geometries will be considered first. These are plane geometries belonging to the category of K_I-increasing specimens with increasing crack length (see Fig. 14.7). The specimens considered in this section where the first to be used in fracture mechanics applications and are shown in Fig. A.4.

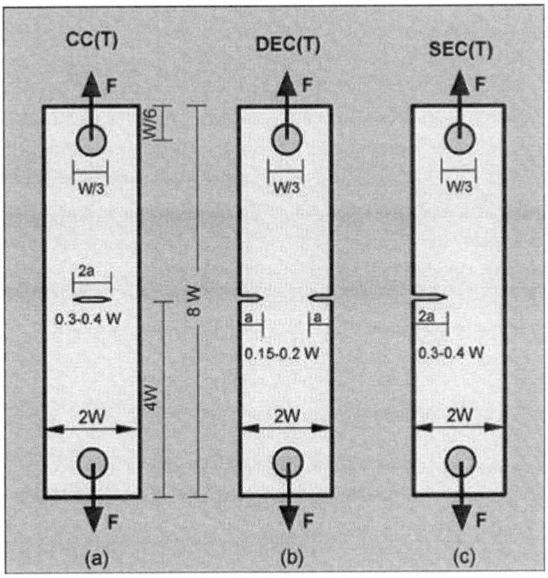

Fig. A.4 Geometries of standard fracture mechanics plane specimens under remote force *F*

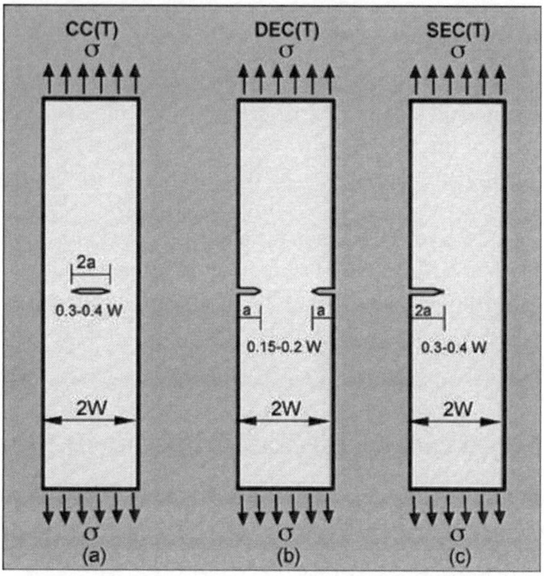

Fig. A.5 Geometries of standard fracture mechanics plane specimens under remote stress σ

For all the geometries considered the general expression of K_I is

$$K_I = \frac{F}{s\sqrt{w}} \cdot h\left(\frac{a}{w}\right)$$
$$K_I = \sigma\sqrt{\pi a} \cdot f\left(\frac{a}{w}\right). \tag{A.3}$$

They have side or central notches fatigued to develop a crack at the tip and are subjected to a uniaxial stress state. The first geometry, Fig. A.4a, is the central crack panel CC(T) (the T stands for *traction*). The second, Fig. A.4b, is the double edge crack panel or DEC(T). The third one, Fig. A.4c, is the single edge crack panel or SEC(T). For each geometry the relative *f(a)* or *h(a)* function are given, depending on whether the K_I expression is given in terms of force F, Fig. A.4, or stress σ, Fig. A.5 acting on the extremities of the specimen. Figures A.4 and A.5 also indicates the standard dimensions of the specimens.

The *f(a)* and *h(a)* functions, that in the specific geometries considered are *f(a/w)* and *h(a/w)* functions, are:

CC(T) Panel [1, 2]

$$h\left(\frac{a}{w}\right) = \sqrt{\frac{\pi a}{4w}\sec\frac{\pi a}{2w}} \cdot \left[1 - 0.025 \cdot \left(\frac{a}{w}\right)^2 + 0.06 \cdot \left(\frac{a}{w}\right)^4\right]$$
$$f\left(\frac{a}{w}\right) = \sqrt{\sec\frac{\pi a}{2w}} \cdot \left[1 - 0.25 \cdot \left(\frac{a}{w}\right)^2 + 0.06 \cdot \left(\frac{a}{w}\right)^4\right]. \tag{A.4}$$

The functions given by Eq. (A.4) are shown in Fig. A.6

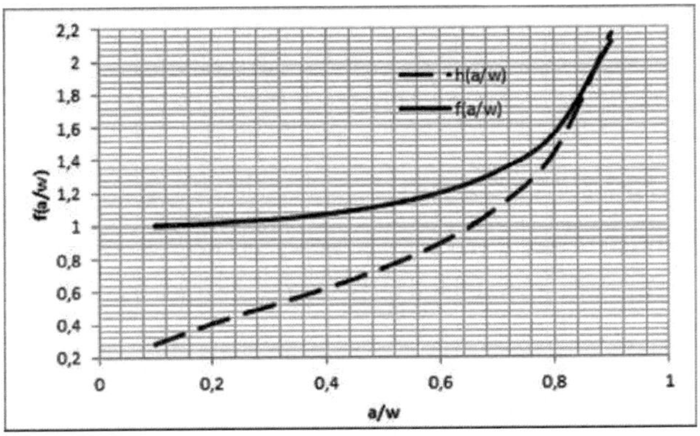

Fig. A.6 Diagram of the $f(a/w)$ and $h(a/w)$ functions for a CC(T) panel

DEC(T) Panel [1, 2]

$$h\left(\frac{a}{w}\right) = \sqrt{\frac{\frac{\pi a}{2w}}{1-\frac{a}{w}}}\left[1.122 - 0.561\left(\frac{a}{w}\right) - 0.205\left(\frac{a}{w}\right)^2 + 0.471\left(\frac{a}{w}\right)^3 + 0.190\left(\frac{a}{w}\right)^4\right]$$

$$f\left(\frac{a}{w}\right) = \frac{1.122 - 0.561 \cdot (a/w) - 0.205 \cdot (a/w)^2 + 0.471 \cdot (a/w)^3 - 0.190 \cdot (a/w)^4}{\sqrt{1 - a/w}} \tag{A.5}$$

Graphically they are shown in Fig. A.7.

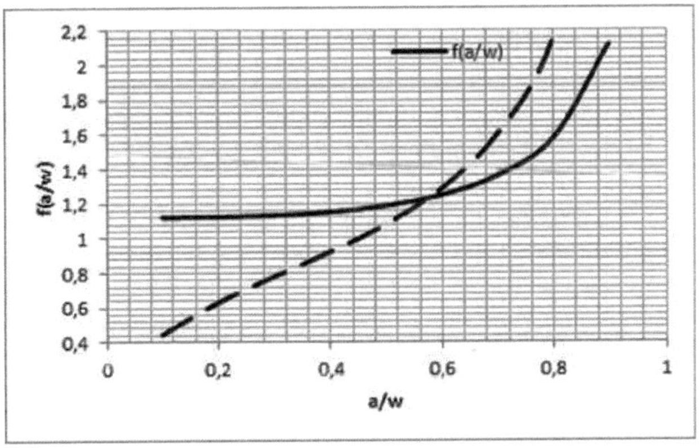

Fig. 7 Diagram of the $f(a/w)$ and $h(a/w)$ functions for a DEC(T) panel

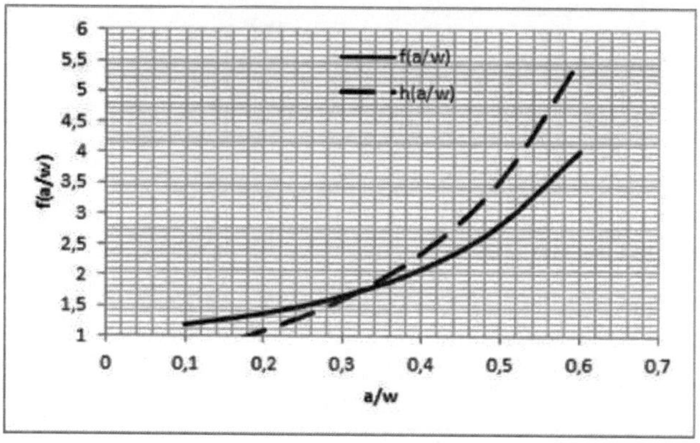

Fig. A.8 Diagram of the $f(a/w)$ and $h(a/w)$ functions for a SEC(T) panel

SEC(T) Panel [1, 3]

$$h\left(\frac{a}{w}\right) = \frac{\sqrt{2\tan\frac{\pi a}{2w}}}{\cos\frac{\pi a}{2w}}\left[0.752 + 2.02\left(\frac{a}{w}\right) + 0.37\left(1 - \sin\frac{\pi a}{2w}\right)^3\right]$$

$$f\left(\frac{a}{w}\right) = 1.12 - 0.231\cdot\left(\frac{a}{w}\right) + 10.55\cdot\left(\frac{a}{w}\right)^2 - 21.72\cdot\left(\frac{a}{w}\right)^3 + 30.39\cdot\left(\frac{a}{w}\right)^4. \quad \text{(A.6)}$$

Graphically the two functions are shown in Fig. A.8.

For this last geometries the expression are valid only for $a < 0.6w$ because for larger a value a non-negligible bending component must be taken into consideration.

A.2.2 *Biaxial Specimens*

Soon after the introduction of CC(T) and DEC(T) panels the attention of fracture mechanics researchers focused on a particular specimen that could introduce a certain degree of biaxial stresses. This was due to the fact that in the early '60s, in particular, concerns were arising about the triaxial or plain strain state on the toughness of materials. A material that under uniaxial stress state exhibited large ductility and, therefore, had an apparent high toughness could become rather brittle under a plain strain state condition. Concerns were arising from the high pressure industries and, in particular, from the nuclear industry that built heavy section steel pressure vessels that were subjected to high pressure and biaxial stress state. It was an engineer of the nuclear industry, Manjoine [4] that at Westinghouse Nuclear introduced the historical WOL (wedge opening load) type X specimen shown in Fig. A.9a.

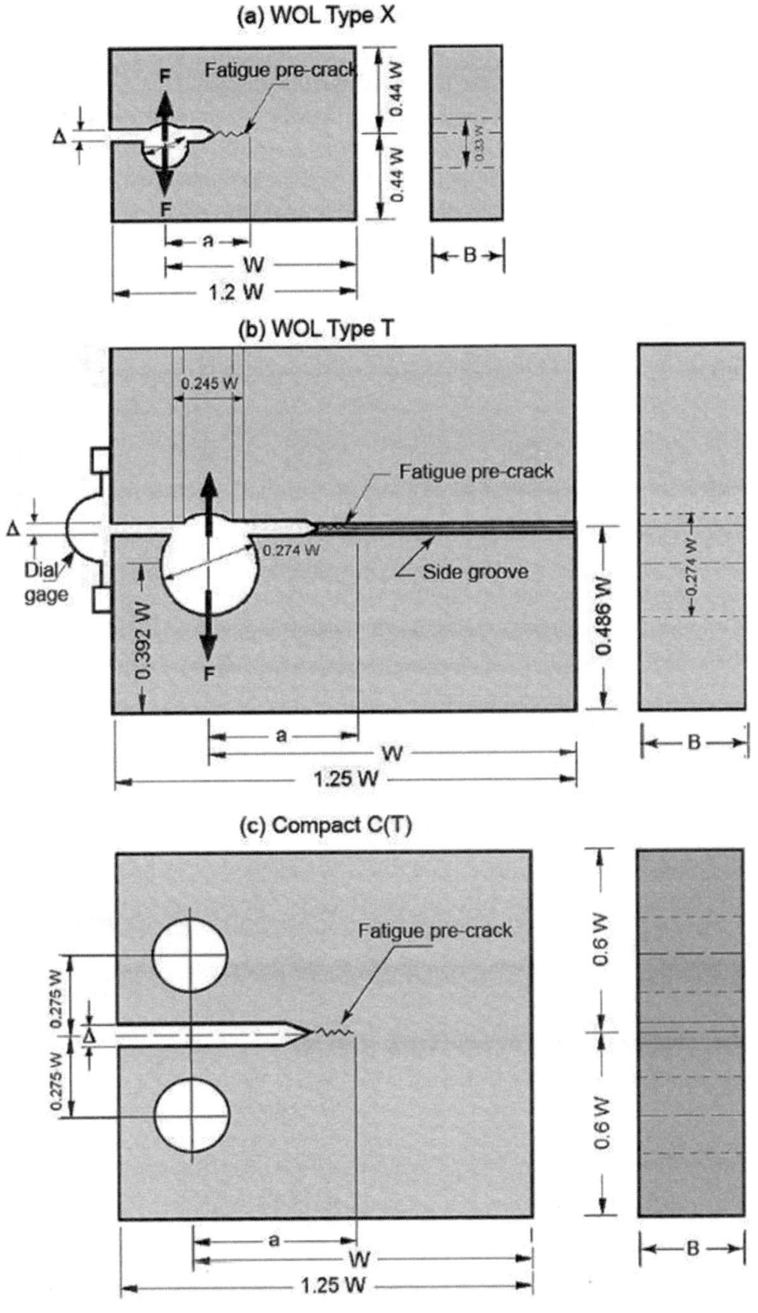

Fig. A.9 Fracture mechanics specimens type WOL-X, WOL-T and compact C(T)

The reduced dimensions of these specimens were due to the limited space available in a nuclear power reactor in which they were introduced close to core for neutron embrittlement surveillance programs. Type X specimen is carrying a V shaped central notch extending at least to 40% of the total length W fatigued to develop a sharp crack ($\rho \leq 0.005$ mm). The specimen is loaded by a pin-clevis system pulling the lower side of the hole while the upper face of the specimen is screw-fixed to the loading cell. The stress state on the crack plane is equivalent to the biaxial stress state existing in the wall of a pressure vessel. As shown in Fig. A.10, on the A-A section of area S_1 is acting a stress whose maximum value σ_1 at the crack tip is given by the moment $M_1 = F{\cdot}a$ plus traction F

$$\sigma_1 = \frac{Fa}{W_1} + \frac{F}{S_1}. \tag{A.7}$$

On the B-B section of area S_2 the maximum bending stress σ_2

$$\sigma_2 = \frac{Fa}{W_2} \tag{A.8}$$

with W_1 and W_2 being the strength moduli of sections S_1 and S_2, respectively. Westinghouse gave a significant impulse to the development of this type introducing the WOL type T specimen of Fig. A.10b [5], which was larger and thicker so to overcome the excessive bending experienced by type X specimen.

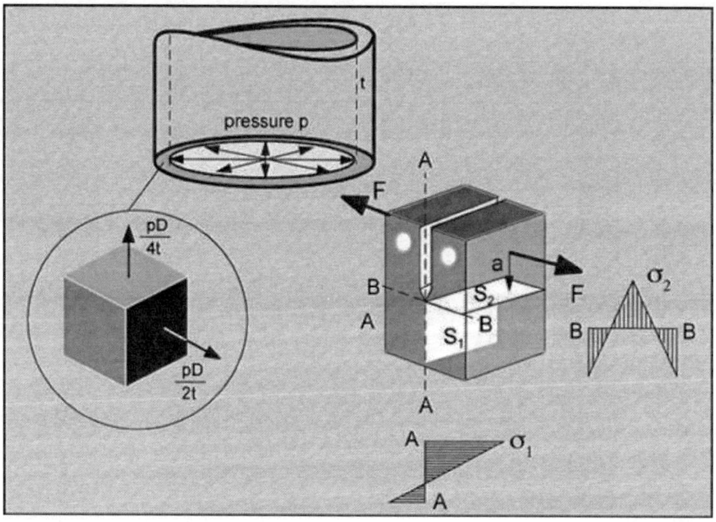

Fig. A.10 Analogy between the biaxial stress state in a pressure vessel wall and that existing ahead of the crack tip in a WOL-X or WOL-T or compact C(T)

WOL Type T specimen is wedge bolt-loaded as shown in Fig. 19.19. Type X specimen is no longer used. The development process continued till the introduction of the third type of specimen worldwide known as the Compact C(T) specimen of Fig. A.9c. Today the C(T) type specimen is the most used one in fracture mechanics applications. However, WOL Type T specimen is still used in SCC applications for measuring the stress intensity threshold K_{Iscc} (see Sect. 15.5). The *h(a/w)* and *V(a/w)* functions for the calculation of the corresponding K and Δ for types WOL-T and C(T) specimens are listed in the following [6]. The symbols refer to Fig. A.9b and c for types WOL-T and C(T) specimens, respectively. Δ represents the crack mouth opening displacement (CMOD), i.e., the displacement measured at the notch opening on the specimen surface, as shown in Fig. A.9.

Compact Specimen C(T)

$$K_I = \frac{F}{s\sqrt{w}} h\left(\frac{a}{w}\right)$$

$$h\left(\frac{a}{w}\right) = \frac{\left(2+\frac{a}{w}\right)\left[0.886 + 4.64\left(\frac{a}{w}\right) - 13.32\left(\frac{a}{w}\right)^2 + 14.72\left(\frac{a}{w}\right)^3 - 5.6\left(\frac{a}{w}\right)^4\right]}{\left[1-\left(\frac{a}{w}\right)\right]^{3/2}}$$

$$\Delta = \frac{F}{E's} V\left(\frac{a}{w}\right)$$

$$V\left(\frac{a}{w}\right) = \left[1 + \frac{0.25}{\left(\frac{a}{w}\right)}\right]\left[\frac{1+\left(\frac{a}{w}\right)}{1-\left(\frac{a}{w}\right)}\right]^2 \left[1.6137 + 12.678\left(\frac{a}{w}\right) - 14.231\left(\frac{a}{w}\right)^2 - 16.61\left(\frac{a}{w}\right)^3 + 35.05\left(\frac{a}{w}\right)^4 - 14.494\left(\frac{a}{w}\right)^5\right]. \quad \text{(A.9)}$$

Functions *h(a/w)* and *V(a/w)* are graphically shown in Figs. A.11 and A.12, respectively.

WOL Type T Specimen (Figs. A.13 and A.14).

$$K_I = \frac{F}{s\sqrt{a}} h\left(\frac{a}{w}\right)$$

$$h\left(\frac{a}{w}\right) = \left(\frac{a}{w}\right)\left[30.96 - 195.8\left(\frac{a}{w}\right) + 730.6\left(\frac{a}{w}\right)^2 - 1186.3\left(\frac{a}{w}\right)^3 + 754.6\left(\frac{a}{w}\right)^4\right]$$

$$\Delta = \frac{F}{E's} V\left(\frac{a}{w}\right)$$

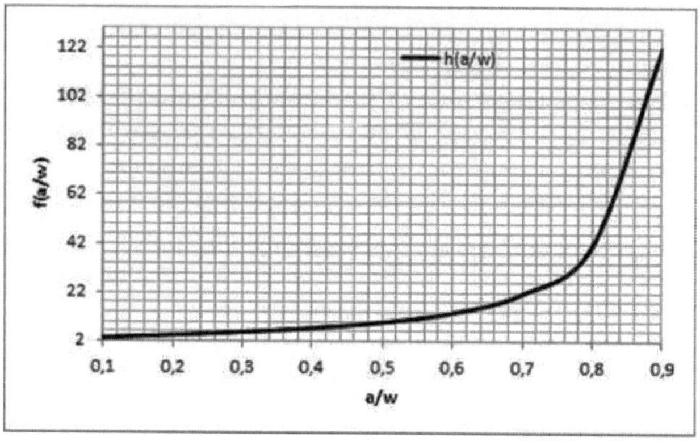

Fig. A.11 Diagram of the function *h(a/w)* for compact C(T) specimen

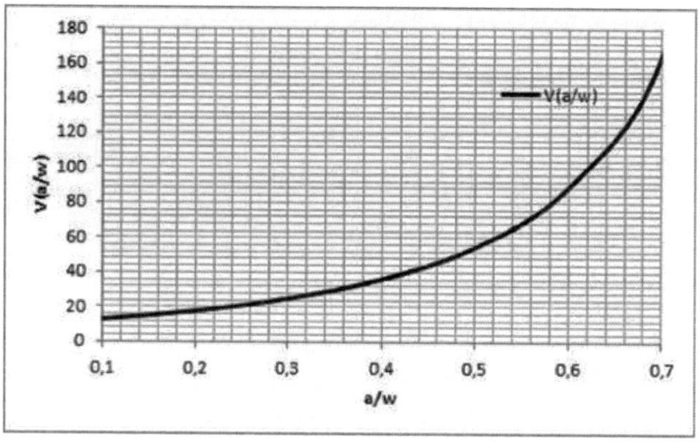

Fig. A.12 Diagram of the function *V(a/w)* for compact C(T) specimen

$$V\left(\frac{a}{w}\right) = \exp\left[4.495 - 16.13\left(\frac{a}{w}\right) + 63.838\left(\frac{a}{w}\right)^2 - 89.125\left(\frac{a}{w}\right)^3 + 46.815\left(\frac{a}{w}\right)^4\right] \quad (A.10)$$

A.2.3 *High Triaxiality Specimens*

A particular citation deserve cylindrical specimens containing a circumferential crack and those in three point bending since they have a very high triaxiality, even with smaller thickness. These are the round notch bar in traction or RNB(T) of Fig. A.15 and the three point bending TP(B) or single edge crack in bending, SE(B) of Fig. A.17,

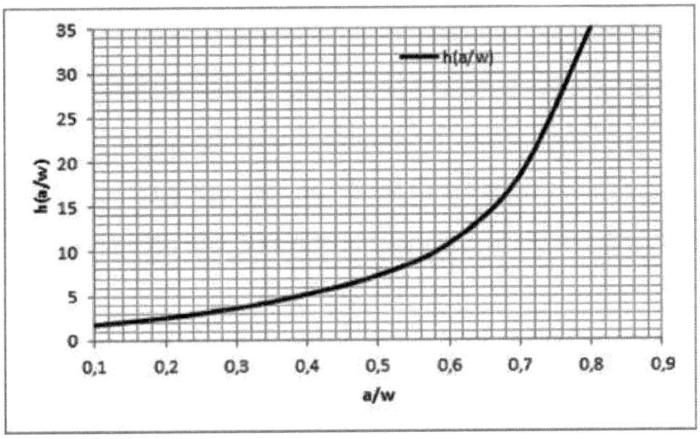

Fig. A.13 Diagram of the function *h(a/w)* for WOL type T specimen

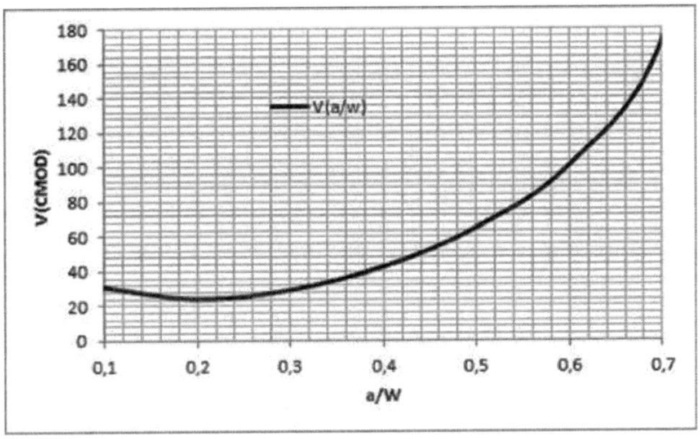

Fig. A.14 Diagram of the function *V(a/w)* for WOL type T specimen

respectively. Their calibration function *h(a/w)* are given in the following Figs. A.16 and A.18, respectively [7].

TP(B) or SE(B) Specimen

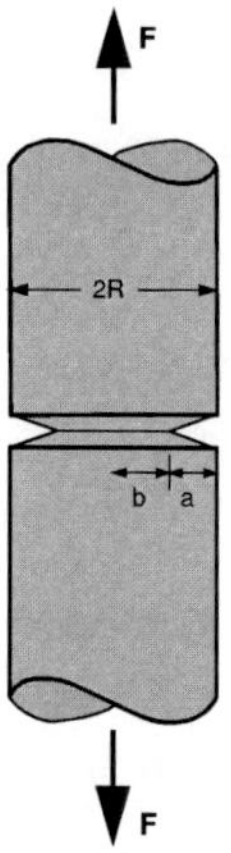

RNB(T) SPECIMEN

$$K_I = \frac{F}{\pi \cdot b^2}\sqrt{\pi \cdot b} \cdot h\left(\frac{b}{R}\right)$$

$$h\left(\frac{b}{R}\right) = \frac{1}{2}\left[1 + \frac{1}{2}\left(\frac{b}{R}\right) + \frac{3}{8}\left(\frac{b}{R}\right)^2 - 0.363\left(\frac{b}{R}\right)^3 + 0.731\left(\frac{b}{R}\right)^4\right]\sqrt{1 - \left(\frac{b}{R}\right)} \quad \text{(A.11)}$$

Fig. A.15 Cylindrical specimen with circumferential crack RNB(T)

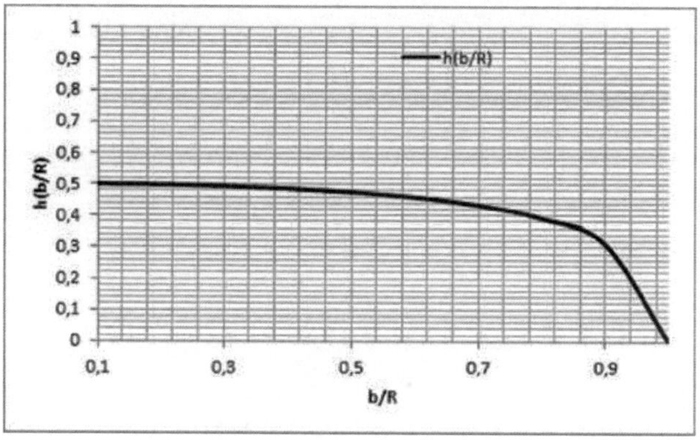

Fig. A.16 Diagram of the function *h(a/w)* for RNB(T) specimen

Fig. A.17 Schematic of SE(B) specimen

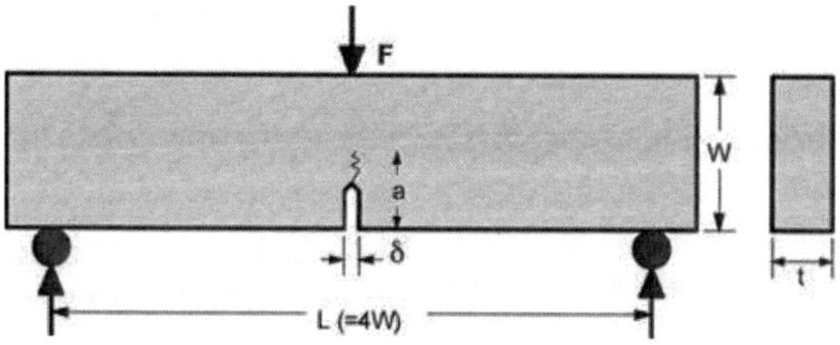

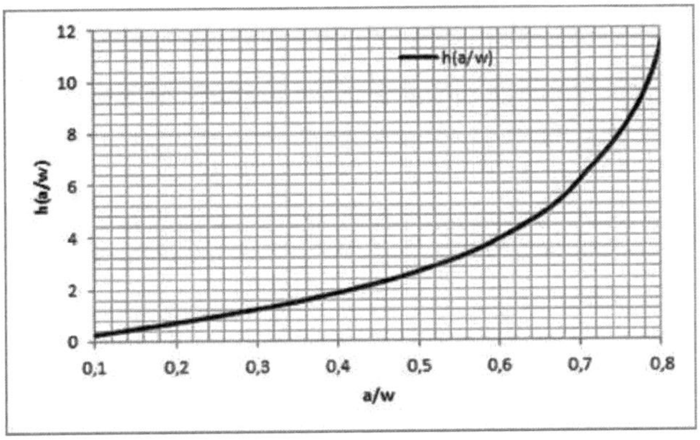

Fig. A.18 Diagram of the function $h(a/w)$ for SE(B) specimen

$$K_I = \frac{FL}{sw^{3/2}} h\left(\frac{a}{w}\right)$$

$$h\left(\frac{a}{w}\right) = 3 \cdot \left(\frac{a}{w}\right)^{1/2} \frac{1.99 - \left(\frac{a}{w}\right)\left(1 - \frac{a}{w}\right)\left[2.15 - 3.93 \cdot \frac{a}{w} + 2.7 \cdot \left(\frac{a}{w}\right)^2\right]}{2 \cdot \left[1 + 2 \cdot \left(\frac{a}{w}\right)\right]\left(1 - \frac{a}{w}\right)^{3/2}} \quad \text{(A.12)}$$

$$V = F \cdot \left(\frac{L}{E' s w}\right) q\left(\frac{a}{w}\right)$$

$$q\left(\frac{a}{w}\right) = 6 \cdot \left(\frac{a}{w}\right)\left[0.76 - 2.28 \cdot \left(\frac{a}{w}\right) + 3.87 \cdot \left(\frac{a}{w}\right)^2 - 2.04 \cdot \left(\frac{a}{w}\right)^3 + \frac{0.66}{\left(1 - \frac{a}{w}\right)^2}\right] \quad \text{(A.13)}$$

A.3 Three-Dimensional Geometries. Surface and Internal Cracks

Semi-Elliptical Surface Crack with a ≤ c [8] (Fig. A.19).

Semi-Elliptical Surface Crack with a > c [8] (Fig. A.20).

Elliptical Central Crack [8] (Fig. A.21).

1/4 Ellipse Corner Crack with a ≤ c [8] (Fig. A.22).

1/4 Ellipse Corner Crack with a > c [8] (Fig. A.23).

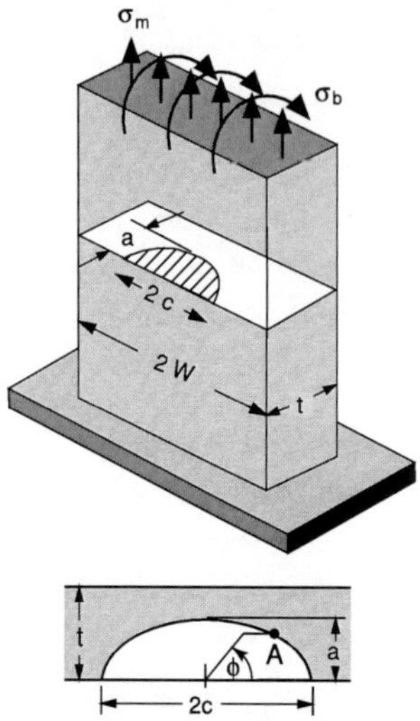

σ_m = membrane stress
σ_b = bending stress

$$\sigma_b = \frac{Ms}{2I}$$

$$I = \frac{wb^3}{6}$$

$$K_I = (\sigma_m + H\sigma_b)\sqrt{\frac{\pi a}{Q}} f\left(\frac{a}{s}, \frac{a}{c}, \frac{c}{w}, \Phi\right)$$

$$Q = 1 + 1.464\left(\frac{a}{c}\right)^{1.65}$$

$$f = \left[M_1 + M_2\left(\frac{a}{t}\right)^2 + M_3\left(\frac{a}{t}\right)^4\right] g_\Phi g_w h$$

$$M_1 = 1.13 - 0.09 \cdot \left(\frac{a}{c}\right)$$

$$M_2 = -0.54 + \frac{0.89}{0.2 + \left(\frac{a}{c}\right)}$$

$$M_3 = 0.5 - \frac{1}{0.65 + \left(\frac{a}{c}\right)} + 14 \cdot \left(1.0 - \frac{a}{c}\right)^{24}$$

$$g_\Phi = \left[\left(\frac{a}{c}\right)^2 \cos^2\Phi + \sin^2\Phi\right]^{1/4}$$

$$g_w = \left[\sec\left(\frac{\pi c}{2w}\sqrt{\frac{a}{t}}\right)\right]^{1/2}$$

$$h = 1 + \left[0.1 + 0.35 \cdot \left(\frac{a}{t}\right)^2\right](1 - \sin\Phi)^2$$

$$H = H_1 + (H_2 - H_1)(\sin\Phi)^p$$

$$p = 0.2 + \frac{a}{c} + 0.6 \cdot \left(\frac{a}{t}\right)$$

$$H_1 = 1 - 0.34 \cdot \left(\frac{a}{t}\right) - 0.11 \cdot \left(\frac{a}{c}\right)\left(\frac{a}{t}\right)$$

$$H_2 = 1 - F_1\left(\frac{a}{t}\right) + F_2\left(\frac{a}{t}\right)^2$$

$$F_1 = -1.22 - 0.12 \cdot \left(\frac{a}{c}\right)$$

$$F_2 = 0.55 - 1.05 \cdot \left(\frac{a}{c}\right)^{0.75} + 0.47 \cdot \left(\frac{a}{c}\right)^{1.5}$$

Fig. A.19 Semi-elliptical surface crack with $a \leq c$ [8]

A.4 Cylindrical Geometries Under Pressure

Through-Wall Crack [9] (Fig. A.24).

Longitudinal Internal Surface [9] (Fig. A.25).

Finite Surface Internal Longitudinal Crack [9] (Fig. A.26).

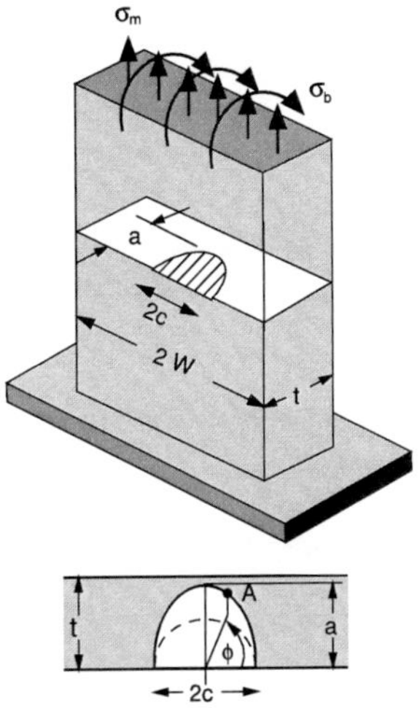

σ_m = membrane stress
σ_b = bending stress

$$\sigma_b = \frac{Ms}{2I}$$

$$I = \frac{ws^3}{6}$$

$$K_I = (\sigma_m + H\sigma_b)\sqrt{\frac{\pi a}{Q}} f\left(\frac{a}{s}, \frac{a}{c}, \frac{c}{w}, \phi\right)$$

$$Q = 1 + 1.464\left(\frac{c}{a}\right)^{1.65}$$

$$f = \left[M_1 + M_2\left(\frac{a}{t}\right)^2 + M_3\left(\frac{a}{t}\right)^4\right] g_\phi g_w h$$

$$M_1 = \sqrt{\frac{c}{a}}\left[1.0 + 0.04 \cdot \left(\frac{c}{a}\right)\right]$$

$$M_2 = 0.2 \cdot \left(\frac{c}{a}\right)^4$$

$$M_3 = -0.11 \cdot \left(\frac{c}{a}\right)^4$$

$$g_\Phi = \left[\left(\frac{c}{a}\right)^2 \sin^2\phi + \cos^2\phi\right]^{1/4}$$

$$g_w = \left[\sec\left(\frac{\pi c}{2w}\sqrt{\frac{a}{t}}\right)\right]^{1/2}$$

$$h = 1 + \left[0.1 + 0.35 \cdot \left(\frac{c}{a}\right)\left(\frac{a}{t}\right)^2\right](1 - \sin\phi)^2$$

$$H = H_1 + (H_2 - H_1)(\sin\phi)^p$$

$$p = 0.2 + \frac{c}{a} + 0.6 \cdot \left(\frac{a}{t}\right)$$

$$H_1 = 1 + F_{11} \cdot \left(\frac{a}{t}\right) + F_{12} \cdot \left(\frac{a}{t}\right)^2$$

$$H_2 = 1 + F_{21}\left(\frac{a}{t}\right) + F_{22}\left(\frac{a}{t}\right)^2$$

$$F_{11} = -0.04 - 0.41 \cdot \left(\frac{c}{a}\right)$$

$$F_{12} = 0.55 - 1.93 \cdot \left(\frac{c}{a}\right)^{0.75} + 1.38 \cdot \left(\frac{c}{a}\right)^{1.5}$$

$$F_{21} = -2.11 - 0.77 \cdot \left(\frac{c}{a}\right)$$

$$F_{22} = 0.55 - 0.72 \cdot \left(\frac{c}{a}\right)^{0.75} + 0.14 \cdot \left(\frac{c}{a}\right)^{1.5}$$

Fig. A.20 Semi-elliptical surface crack with $a > c$ [8]

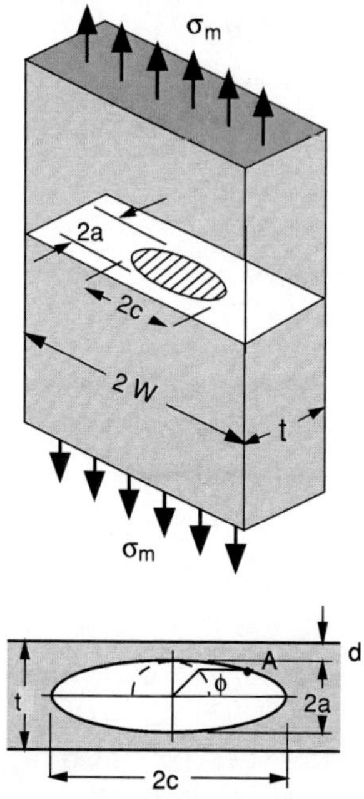

σ_m = membrane stress

$$K_I = \sigma_m \sqrt{\frac{\pi a}{Q}} f\left(\frac{a}{s}, \frac{a}{c}, \frac{c}{w}, \Phi\right)$$

$$\lambda = \frac{a}{a+d}$$

$$f = \left(M_1 + M_2\lambda^2 + M_3\lambda^4\right) g_\Phi g_w h$$

$$M_2 = \frac{0.05}{0.11 + \left(\frac{a}{c}\right)^{1.5}}$$

$$M_3 = \frac{0.29}{0.23 + \left(\frac{a}{c}\right)^{1.5}}$$

$$g_w = \left[\sec\left(\frac{\pi c}{2w}\sqrt{\frac{2a}{t}}\right)\right]^{1/2}$$

$$h = 1 - \frac{\lambda^4\sqrt{2.6 - 2\lambda}}{1 + 4\lambda}|\cos\Phi|$$

per $a/c \le 1$:

$$Q = 1 + 1.464\left(\frac{a}{c}\right)^{1.65}$$

$$g_\Phi = \left[\left(\frac{a}{c}\right)^2 \cos^2\Phi + \sin^2\Phi\right]^{1/4}$$

$$M_1 = 1$$

per $a/c > 1$:

$$Q = 1 + 1.464\left(\frac{c}{a}\right)^{1.65}$$

$$g_\Phi = \left[\left(\frac{c}{a}\right)^2 \sin^2\Phi + \cos^2\Phi\right]^{1/4}$$

$$M_1 = \sqrt{\frac{c}{a}}$$

Fig. A.21 Elliptical center crack [8]

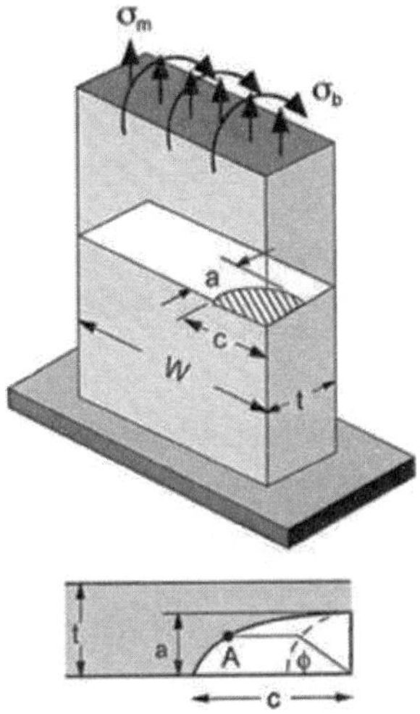

σ_m = membrane stress
σ_b = bending stress

$$\sigma_b = \frac{Ms}{2I}$$

$$I = \frac{ws^3}{6}$$

$$K_I = (\sigma_m + H\sigma_b)\sqrt{\frac{\pi a}{Q}} f\left(\frac{a}{s}, \frac{a}{c}, \frac{c}{w}, \phi\right)$$

$$Q = 1 + 1.464\left(\frac{a}{c}\right)^{1.65}$$

$$f = \left[M_1 + M_2\left(\frac{a}{t}\right)^2 + M_3\left(\frac{a}{t}\right)^4\right] h_1 h_2 g_\phi g_w$$

$$M_1 = 1.08 - 0.03 \cdot \left(\frac{a}{c}\right)^{1.65}$$

$$M_2 = \frac{-0.44 + 1.06}{0.3 + \left(\frac{a}{c}\right)}$$

$$M_3 = -0.5 - 0.25 \cdot \left(\frac{a}{c}\right) + 14.8 \cdot \left(1 - \frac{a}{c}\right)^{15}$$

$$g_\phi = \left[\left(\frac{a}{c}\right)^2 \cos^2\phi + \sin^2\phi\right]^{1/4}$$

$$g_w = \left[\sec\left(\frac{\pi c}{2w}\sqrt{\frac{a}{t}}\right)\right]^{1/2}$$

$$h_1 = 1 + \left[0.08 + 0.4 \cdot \left(\frac{a}{t}\right)^2\right](1 - \sin\phi)^3$$

$$h_2 = 1 + \left[0.08 + 0.15 \cdot \left(\frac{a}{t}\right)^2\right](1 - \cos\phi)^3$$

$$H = H_1 + (H_2 - H_1)(\sin\phi)^p$$

$$p = 0.2 + \frac{a}{c} + 0.6 \cdot \left(\frac{a}{t}\right)$$

$$H_1 = 1 - 0.34 \cdot \left(\frac{a}{t}\right) - 0.11 \cdot \left(\frac{a}{c}\right)\left(\frac{a}{t}\right)$$

$$H_2 = 1 - F_1\left(\frac{a}{t}\right) + F_2\left(\frac{a}{t}\right)^2$$

$$F_1 = -1.22 - 0.12 \cdot \left(\frac{a}{c}\right)$$

$$F_2 = 0.55 - 1.05 \cdot \left(\frac{a}{c}\right)^{0.75} + 0.47 \cdot \left(\frac{a}{c}\right)^{1.5}$$

Fig. A.22 1/4 elliptical corner crack with $a < c$

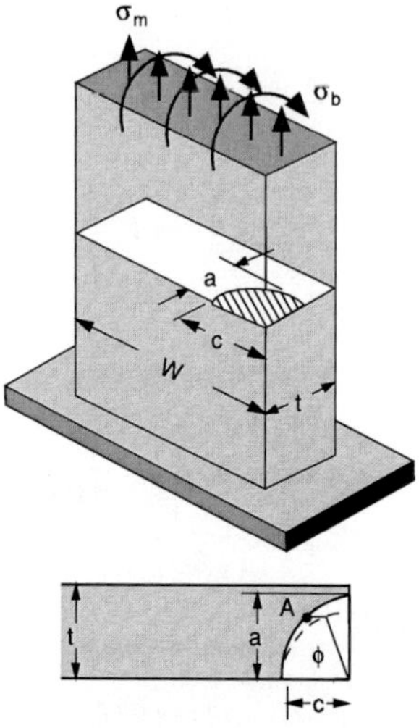

σ_m = membrane stress
σ_b = bending stress

$$\sigma_b = \frac{Ms}{2I}$$

$$I = \frac{ws^3}{6}$$

$$K_I = (\sigma_m + H\sigma_b)\sqrt{\frac{\pi a}{Q}} f\left(\frac{a}{s}, \frac{a}{c}, \frac{c}{w}, \phi\right)$$

$$Q = 1 + 1.464\left(\frac{c}{a}\right)^{1.65}$$

$$f = \left[M_1 + M_2\left(\frac{a}{t}\right)^2 + M_3\left(\frac{a}{t}\right)^4\right] g_\phi g_w h_1 h_2$$

$$M_1 = \sqrt{\frac{c}{a}}\left[1.08 + 0.03 \cdot \left(\frac{c}{a}\right)\right]$$

$$M_2 = 0.375 \cdot \left(\frac{c}{a}\right)^2 \qquad M_3 = -0.25 \cdot \left(\frac{c}{a}\right)^2$$

$$g_\Phi = \left[\left(\frac{c}{a}\right)^2 \sin^2\phi + \cos^2\phi\right]^{1/4}$$

$$g_w = \left[\sec\left(\frac{\pi c}{2w}\sqrt{\frac{a}{t}}\right)\right]^{1/2}$$

$$h_1 = 1 + \left[0.08 + 0.4 \cdot \left(\frac{c}{t}\right)^2\right](1 - \sin\phi)^3$$

$$h_2 = 1 + \left[0.08 + 0.15 \cdot \left(\frac{c}{t}\right)^2\right](1 - \cos\phi)^3$$

$$H = H_1 + (H_2 - H_1)(\sin\phi)^p$$

$$p = 0.2 + \frac{c}{a} + 0.6 \cdot \left(\frac{a}{t}\right)$$

$$H_1 = 1 + F_{11} \cdot \left(\frac{a}{t}\right) + F_{12} \cdot \left(\frac{a}{t}\right)^2$$

$$H_2 = 1 + F_{21}\left(\frac{a}{t}\right) + F_{22}\left(\frac{a}{t}\right)^2$$

$$F_{11} = -0.04 - 0.41 \cdot \left(\frac{c}{a}\right)$$

$$F_{12} = 0.55 - 1.93 \cdot \left(\frac{c}{a}\right)^{0.75} + 1.38 \cdot \left(\frac{c}{a}\right)^{1.5}$$

$$F_{21} = -2.11 - 0.77 \cdot \left(\frac{c}{a}\right)$$

$$F_{22} = 0.64 - 0.72 \cdot \left(\frac{c}{a}\right)^{0.75} + 0.14 \cdot \left(\frac{c}{a}\right)^{1.5}$$

Fig. A.23 1/4 elliptical corner crack with $a > c$

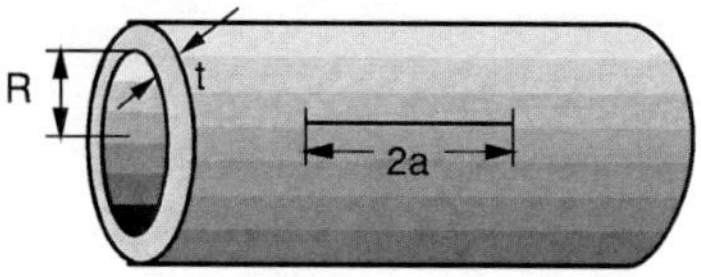

Valid for $R \gg s$

$$K_I = \sigma_c\sqrt{\pi a}\sqrt{1 + 0.52\chi + 1.29\chi^2 - 0.074\chi^3}$$

$$\sigma_c = \frac{pR}{s}$$

$$\chi = \frac{a}{\sqrt{Rs}}$$

Fig. A.24 Through-wall crack [9]

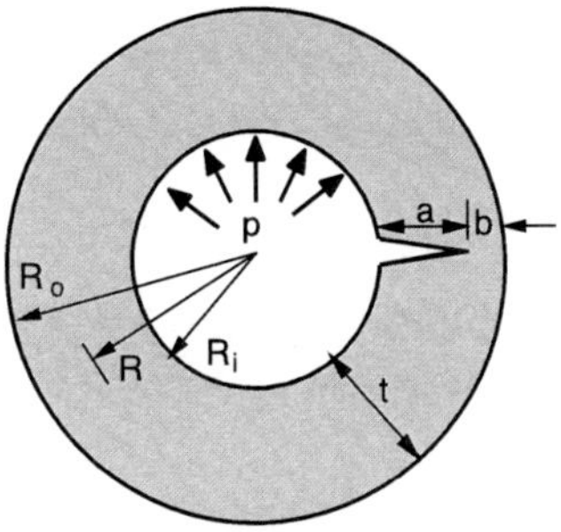

Valid for $5 \leq R/s \leq 20$ e $a/s \leq 0.75$

$$K_I = \frac{2pR_o^2}{R_o^2 - R_i^2}\sqrt{\pi a} \cdot f\left(\frac{a}{s}, \frac{R_i}{s}\right)$$

$$f = 1.1 + A\left[4.951 \cdot \left(\frac{a}{s}\right)^2 + 1.092 \cdot \left(\frac{a}{s}\right)^4\right]$$

$$A = \left(0.125 \cdot \frac{R_i}{s} - 0.25\right)^{0.25} per \quad 5 \leq \frac{R_i}{s} \leq 10$$

$$A = \left(0.2 \cdot \frac{R_i}{s} - 1.0\right)^{0.25}$$

Fig. A.25 Surface longitudinal crack [9]

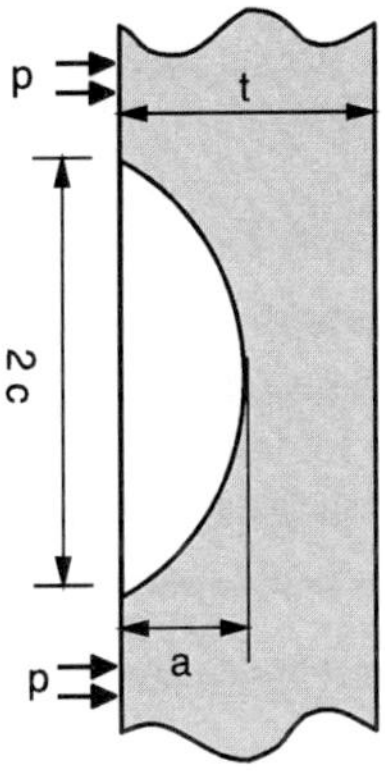

Valid for $5 \leq R/s \leq 20$ e $c/a \leq 0.80$ (il K_I refers to the innermostpoint)

$$K_I = \frac{pR}{s}\sqrt{\frac{\pi a}{Q}} f\left(\frac{a}{s}, \frac{a}{2c}, \frac{R}{s}\right)$$

$$f = 1.12 + 0.053 \cdot \xi + 0.0055 \cdot \xi^2$$

$$+\left(1 + 0.02 \cdot \xi + 0.0191 \cdot \xi^2\right)\frac{\left(20 - \frac{R}{s}\right)^2}{1400}$$

$$\xi = \frac{2c}{s} \quad Q = 1 + 1.464\left(\frac{a}{c}\right)^{1.65}$$

Fig. A.26 Surface finite longitudinal crack

References

1. Tada, H., Paris, P.C., Irwin, G.R.: The Stress Analysis of Cracks Handbook, 2nd ed. St. Louis (1985)
2. Tada, H., Paris, P.C., Irwin, G.R.: The Stress Analysis of Cracks Handbook. Del Research Co., Hellerton, PA (1973)
3. Brown, W.F., Srawley, J.E.: Plain Strain Fracture Toughness Testing of High Strength Metallic Materials. American Society for Testing and Materials, ASTM STP-410 (1967)
4. Manjoine, M.J.: Biaxial brittle fracture tests. J. Basic Eng., Trans. ASME 293–298 (June 1965)
5. Wilson, W.K.: Optimization of WOL brittle fracture test specimen. Westinghouse Research Report 66-B40-BTLFR-R1 (4 Jan, 1966)
6. E 399-90.: Annual Book of ASTM Standards, Section 3, Metal Test Methods and Analytical Procedures, pp. 506–536 (1990)
7. Benthem, J.P., Koiter, W.T.: Asymptotic Approximation to CracK Problems, Method of Analysis and Solutionsof Crack Problems, Edited by Sih, G.C., p. 131 (1973)
8. Newman, J.C., Raju, I.S.: Stress Intensity Factors Equations for Cracks in Three-Dimensional Finite Bodies Subjected to Tension and Bending Loads, NASA Technical Memorandum 85793, NASA Langley Research Center, Hampton, VA (April 1984)
9. Zahoor, A.: Closed form expression for fracture mechanics analysis of cracked pipes. J. Press. Vessel. Technol. **107**, 203–205 (1985)

Author Index

P. P. Milella, *Fatigue and Corrosion in Metals*,
https://doi.org/10.1007/978-3-031-51350-3

Subject Index

P. P. Milella, *Fatigue and Corrosion in Metals*,
https://doi.org/10.1007/978-3-031-51350-3